FLORIDA STATE
UNIVERSITY LIBRARIES

MAR 29 1995

TALLAHASSEE, FLORIDA

HANDBOOKS IN ECONOMICS

11

Series Editors

**KENNETH J. ARROW
MICHAEL D. INTRILIGATOR**

ELSEVIER
AMSTERDAM · LAUSANNE · NEW YORK · OXFORD · SHANNON · TOKYO

HANDBOOK OF GAME THEORY
with Economic Applications
VOLUME II

HANDBOOK OF GAME THEORY
with Economic Applications

VOLUME II

Edited by

ROBERT J. AUMANN
The Hebrew University of Jerusalem

and

SERGIU HART
The Hebrew University of Jerusalem

1994

ELSEVIER
AMSTERDAM · LAUSANNE · NEW YORK · OXFORD · SHANNON · TOKYO

ELSEVIER SCIENCE B.V.
Sara Burgerhartstraat 25
P.O. Box 211, 1000 AE Amsterdam, Netherlands

Library of Congress Cataloging-in-Publication Data
(Revised for vol. 2)

Handbook of game theory with economic applications.

(Handbooks in economics ; 11-   )
Includes bibliographical references and index.
1. Game theory. 2. Economics, Mathematical.
I. Aumann, Robert J. II. Hart, Sergiu. III. Series:
Handbooks in economics: bk. 11.
HB144.H36 1992   519.3   91-38429
ISBN 0-444-89427-6   (v. 2: alk. paper)

ISBN for this volume: 0 444 89427 6

©ELSEVIER SCIENCE B.V., 1994. All rights reserved

All rights reserved. No part of this publication may be reproduced, stored in a retrieval system, or transmitted, in any form or by any means, electronic, mechanical, photocopying, recording or otherwise, without the prior written permission of the publisher, Elsevier Science B.V., Copyright and Permissions Department, P.O. Box 521, 1000 AM Amsterdam, Netherlands.

Special regulations for readers in the USA – This publication has been registered with the Copyright Clearance Center Inc. (CCC), Salem, Massachusetts. Information can be obtained from the CCC about conditions under which photocopies of parts of this publication may be made in the USA. All other copyright questions including photocopying outside of the USA, should be referred to the copyright owner, Elsevier Science B.V., unless otherwise specified.

No responsibility is assumed by the publisher for any injury and/or damage to persons or property as a matter of products liability, negligence or otherwise, or from any use or operation of any methods, products instructions or ideas contained in the material herein.

This book is printed on acid-free paper.

Printed in The Netherlands.

INTRODUCTION TO THE SERIES

The aim of the *Handbooks in Economics* series is to produce Handbooks for various branches of economics, each of which is a definitive source, reference, and teaching supplement for use by professional researchers and advanced graduate students. Each Handbook provides self-contained surveys of the current state of a branch of economics in the form of chapters prepared by leading specialists on various aspects of this branch of economics. These surveys summarize not only received results but also newer developments, from recent journal articles and discussion papers. Some original material is also included, but the main goal is to provide comprehensive and accessible surveys. The Handbooks are intended to provide not only useful reference volumes for professional collections but also possible supplementary readings for advanced courses for graduate students in economics.

KENNETH J. ARROW and MICHAEL D. INTRILIGATOR

CONTENTS OF THE HANDBOOK*

VOLUME I

Preface

Chapter 1
The Game of Chess
HERBERT A. SIMON and JONATHAN SCHAEFFER

Chapter 2
Games in Extensive and Strategic Forms
SERGIU HART

Chapter 3
Games with Perfect Information
JAN MYCIELSKI

Chapter 4
Repeated Games with Complete Information
SYLVAIN SORIN

Chapter 5
Repeated Games of Incomplete Information: Zero-Sum
SHMUEL ZAMIR

Chapter 6
Repeated Games of Incomplete Information: Non-Zero-Sum
FRANÇOISE FORGES

Chapter 7
Noncooperative Models of Bargaining
KEN BINMORE, MARTIN J. OSBORNE and ARIEL RUBINSTEIN

*Detailed contents of this volume (Volume II of the Handbook) may be found on p. xix

Chapter 8
Strategic Analysis of Auctions
ROBERT WILSON

Chapter 9
Location
JEAN J. GABSZEWICZ and JACQUES-FRANÇOIS THISSE

Chapter 10
Strategic Models of Entry Deterrence
ROBERT WILSON

Chapter 11
Patent Licensing
MORTON I. KAMIEN

Chapter 12
The Core and Balancedness
YAKAR KANNAI

Chapter 13
Axiomatizations of the Core
BEZALEL PELEG

Chapter 14
The Core in Perfectly Competitive Economies
ROBERT M. ANDERSON

Chapter 15
The Core in Imperfectly Competitive Economies
JEAN J. GABSZEWICZ and BENYAMIN SHITOVITZ

Chapter 16
Two-Sided Matching
ALVIN E. ROTH and MARILDA SOTOMAYOR

Chapter 17
Von Neumann–Morgenstern Stable Sets
WILLIAM F. LUCAS

Chapter 18
The Bargaining Set, Kernel, and Nucleolus
MICHAEL MASCHLER

Contents of the Handbook

Chapter 19
Game and Decision Theoretic Models in Ethics
JOHN C. HARSANYI

VOLUME II

Preface

Chapter 20
Zero-Sum Two-Person Games
T.E.S. RAGHAVAN

Chapter 21
Game Theory and Statistics
GIDEON SCHWARZ

Chapter 22
Differential Games
AVNER FRIEDMAN

Chapter 23
Differential Games – Economic Applications
SIMONE CLEMHOUT and HENRY Y. WAN Jr.

Chapter 24
Communication, Correlated Equilibria and Incentive Compatibility
ROGER B. MYERSON

Chapter 25
Signalling
DAVID M. KREPS and JOEL SOBEL

Chapter 26
Moral Hazard
PRAJIT K. DUTTA and ROY RADNER

Chapter 27
Search
JOHN McMILLAN and MICHAEL ROTHSCHILD

Chapter 28
Game Theory and Evolutionary Biology
PETER HAMMERSTEIN and REINHARD SELTEN

Chapter 29
Game Theory Models of Peace and War
BARRY O'NEILL

Chapter 30
Voting Procedures
STEVEN J. BRAMS

Chapter 31
Social Choice
HERVÉ MOULIN

Chapter 32
Power and Stability in Politics
PHILIP D. STRAFFIN Jr.

Chapter 33
Game Theory and Public Economics
MORDECAI KURZ

Chapter 34
Cost Allocation
H.P. YOUNG

Chapter 35
Cooperative Models of Bargaining
WILLIAM THOMSON

Chapter 36
Games in Coalitional Form
ROBERT J. WEBER

Chapter 37
Coalition Structures
JOSEPH GREENBERG

Chapter 38
Game-Theoretic Aspects of Computing
NATHAN LINIAL

Contents of the Handbook

Chapter 39
Utility and Subjective Probability
PETER C. FISHBURN

Chapter 40
Common Knowledge
JOHN GEANAKOPLOS

CHAPTERS PLANNED FOR VOLUME III

Games of incomplete information
Two player non-zero-sum games
Conceptual foundations of strategic equilibrium
Strategic equilibrium
Stochastic games
Noncooperative games with many players
Bargaining with incomplete information
Oligopoly
Implementation
Inspection games
The Shapley value
Variations on the Shapley value
Values of large games
Values of non-transferable utility games
Values of perfectly competitive economies
Values in other economic applications
History of game theory
Macroeconomics
Experimentation
Psychology
Law

LIST OF CHAPTERS PLANNED FOR ALL THE VOLUMES[1]

Non-Cooperative

The game of chess (I, 1)
Games in extensive and strategic forms (I, 2)
Games of perfect information (I, 3)
Games of incomplete information
Two player zero-sum games (II, 20)
Two player non-zero-sum games
Statistics (II, 21)
Differential games (II, 22)
Economic applications of differential games (II, 23)
Conceptual foundations of strategic equilibrium
Strategic equilibrium
Communication, correlated equilibria, and incentive compatibility (II, 24)
Stochastic games
Repeated games of complete information (I, 4)
Repeated games of incomplete information: the zero-sum case (I, 5)
Repeated games of incomplete information: the non-zero-sum case (I, 6)
Noncooperative games with many players
Noncooperative models of bargaining (I, 7)
Bargaining with incomplete information
Auctions (I, 8)
Location (I, 9)
Entry and exit (I, 10)
Patent licensing (I, 11)
Signalling (II, 25)
Moral hazard (II, 26)
Search (II, 27)
Oligopoly
Implementation
Inspection games
Evolutionary biology (II, 28)
Peace and war (II, 29)
Voting procedures (II, 30)
Social choice (II, 31)

---

[1] "m, n" means "volume m, chapter n".

Cooperative

Cooperative models of bargaining (II, 35)
Games in coalitional form (II, 36)
The core and balancedness (I, 12)
Axiomatizations of the core (I, 13)
The core in perfectly competitive economies (I, 14)
The core in imperfectly competitive economies (I, 15)
Two-sided matching (I, 16)
Von Neumann–Morgenstern stable sets (I, 17)
The bargaining set, kernel and nucleolus (I, 18)
The Shapley value
Variations on the Shapley value
Values of large games
Values of non-transferable utility games
Values of perfectly competitive economies
Values in other economic applications
Coalition structures (II, 37)
Power and stability in politics (II, 32)
Public economics (II, 33)
Cost allocation (II, 34)

General

History of game theory
Computer science (II, 38)
Utility and subjective probability (II, 39)
Common knowledge (II, 40)
Macroeconomics
Experimentation
Psychology
Law
Ethics (I, 19)

PREFACE

This is the second of three volumes planned for the Handbook of Game Theory with Economic Applications. For an introduction to the entire *Handbook*, please see the preface to the first volume. Here we confine ourselves to providing an overview of the organization of this volume. As before, the space devoted in the preface to the various chapters is no indication of their relative importance.

We follow the rough division into "noncooperative," "cooperative," and "general" adopted in the first volume. Chapters 20 through 31 are mainly noncooperative; 32 through 37, cooperative; 38 through 40, general. This division should not be taken too seriously; chapters may well contain aspects of both approaches. Indeed, we hope that the *Handbook* will help demonstrate that noncooperative and cooperative game theory are two sides of the same coin, which complement each other well.

In game theory's early period, from the twenties through the early fifties, two-person zero-sum games – where one player's gain is the other's loss – were the main object of study; many thought that that is all there is to the discipline. Those days are long gone; but two-person zero-sum games still play a fundamental role, and it is fitting that we start the second volume with them. Chapter 20 is devoted to the basic minimax theory and its ramifications, with an appendix on duels. Chapter 21 covers the branch of statistical decision theory that deals with conservative or "worst case" analysis, i.e., in which nature is treated as a malevolent player who seeks to minimize the payoff to the decision maker. This is one of the earliest applications of two-person zero-sum games. The next two chapters treat differential games, i.e., games played over continuous time, like pursuit games. The mathematical theory – which both benefitted from and contributed to control theory and differential equations – is in Chapter 22; economic applications, zero-sum as well as non-zero-sum, in Chapter 23.

This brings us to the non-zero-sum world. Here, there is room for correlation and communication between the players. Chapter 24 treats correlated equilibria, communication equilibria and "mechanisms" in general; these concepts are extensions of the classical Nash strategic equilibrium.

"Actions speak louder than words": in addition to the explicit communication considered in the previous chapter, players may also communicate implicitly, by means of their actions – as in Spence's seminal 1972 work, in which education serves as a signal of ability. Signalling is the subject of Chapter 25, the first of three chapters in this volume on economic applications of noncooperative game theory (in addition to Chapters 8–11 on such applications in the first volume).

To serve as signals, actions must be observed by the other players. Chapter 26 discusses the opposite case, in which actions cannot be directly monitored. The principal–agent problem concerns relationships between owner and manager, patient and physician, insurer and insured, and so on. Characteristic of these relationships is that the principal must devise incentives to motivate the agent to act in the principal's interests. A particular aspect of this is the moral hazard problem, of which a typical instance is insurance: Insuring one's home for more than its value would create an incentive to burn it down.

Chapter 27 concerns game theoretic aspects of search models in economics. Not classical zero-sum search such as destroyer–submarine, but non-zero-sum problems like shopping and marketing, job hunting and recruiting, and so on.

We come next to one of the most promising newer applications of game theory: biology. Rather surprisingly, the inequalities defining a Nash strategic equilibrium are, when properly reinterpreted, identical to those that characterize an equilibrium between populations (or within a population), in the sense of evolutionary ecology. This field, initiated in 1972 by John Maynard Smith, has spawned a large literature, which is surveyed in Chapter 28.

The following five chapters deal with applications to political science and related topics; these chapters bridge the noncooperative and the cooperative parts of the volume. Chapter 29 surveys models of international conflict – an area to which game theory was applied already in the fifties, at the height of the cold war. Chapter 30 presents a game theoretic analysis of voting systems, such as proportional representation, plurality voting, approval voting, ranking methods, single transferable vote, and others. Much of the analysis deals with strategic considerations of voters facing the various systems. These questions may also be studied in the more general framework of "social choice" – group decision problems. Social choice constitutes a large and much studied area; Chapter 31 deals with its game theoretic aspects, i.e., how to devise schemes that implement certain outcomes when the participants act strategically in their own best interest, and whether this is at all possible. This concludes the noncooperative part of the volume.

The cooperative part starts with three chapters on applications, to political science and economics. All use the concepts of core and value. Chapter 32 deals with measures of power and notions of stability in various political applications. Chapter 33 is devoted to a subject with both economic and political content – public economics; this concerns taxation, provision of public goods, and so on.

In most applications of game theory, the "players" are either human individuals or collectives of humans, like companies, unions, political parties or nations. To some extent this is so even in the applications to statistics (Chapter 21) and computer science (Chapter 38); though there the "players" are not necessarily human, we ascribe to them human motives, which in one sense or another correspond to the goals of the architect or programmer. There is, however, another kind of application, where the mathematical formalism of game theory – the "equations defining the game" – is interpreted in a way that is quite different from standard. One such application is to evolutionary biology (Chapter 28). Another, surveyed in Chapter 34, is to the problem of allocating joint costs. For example, airport landing fees, overhead costs billed by universities, phone charges within an organization, and so on. Here, the players are individual aircraft landings; activities like specific research projects or student-hours taught in a specific course; single minutes of long-distance phone calls. The worth of a "coalition" of such activities is defined as the hypothetical cost of carrying out the activities in that coalition only.

Three chapters on theory end the cooperative part of this volume. The first is on bargaining problems, which were studied from the noncooperative viewpoint in Chapter 7 (Volume 1). Chapter 35 presents the axiomatic approach to these problems: solutions are sought that satisfy certain desirable properties. The axiomatic method studies various solution concepts from the viewpoint of their properties rather than their definitions, and helps us to compare them. Those that appear again and again in different setups, like the classic 1951 bargaining solution of Nash, gain credibility.

In bargaining problems, only the individual players and the "grand coalition" – that of all players – matter. In general cooperative games, also the intermediate coalitions play a role. Such "coalitional games" are presented and classified in Chapter 36. An important question arising in this connection is which coalitions actually form. Chapter 37 surveys some of the approaches to this problem, which have both cooperative and noncooperative aspects.

We have already said that game theory has significant ties to various disciplines, some of which may, at first glance, seem quite unrelated. In the case of computer science, the influence goes in both directions. For example, to model bounded rationality, game theory has borrowed from the theory of finite automata, Turing machines, and so on (some discussion of this is included in Chapter 4 of Volume 1). Chapter 38 surveys the other direction – game theoretic ideas in computer science. Two examples are the application of iterated knowledge in the study of distributed systems, and the use of power indices – from cooperative game theory – to estimate the vulnerability of a system to "crashes" because of possible failures by relatively small sets of parts.

The last two chapters in this volume pertain to the foundations of game theory. Chapter 39 deals with how the players evaluate the possible outcomes; the subjective measurement of utilities and probabilities is at the basis of decision theory, both interactive and one-person. Finally, information drastically affects the way in which games are played. A coherent model of knowledge, knowledge about others' knowledge, and so on, has become essential equipment in game theory. The theory of these levels of knowledge, culminating in "common knowledge," is the subject of Chapter 40.

ROBERT J. AUMANN and SERGIU HART

CONTENTS OF VOLUME II

Introduction to the Series	v
Contents of the Handbook	vii
Preface	xv

Chapter 20
Zero-Sum Two-Person Games 735
T.E.S. RAGHAVAN

References	757
Appendix, by T. Radzik and T.E.S. Raghavan. Duels	761

Chapter 21
Game Theory and Statistics 769
GIDEON SCHWARZ

1. Introduction	770
2. Statistical inference as a game	771
3. Payoff, loss and risk	772
4. The Bayes approach	773
5. The minimax approach	774
6. Decision theory as a touch-stone	776
Appendix. A lemma and two examples	777
References	779

Chapter 22
Differential Games 781
AVNER FRIEDMAN

0. Introduction	782
1. Basic definitions	782
2. Existence of value and saddle point	787
3. Solving the Hamilton–Jacobi equation	791
4. Other payoffs	793
5. N-person differential games	796
References	798

Chapter 23
Differential Games – Economic Applications 801
SIMONE CLEMHOUT and HENRY Y. WAN Jr.

1. Introduction 802
2. Technical preliminaries 803
3. Example for non-cooperation. I. Exhaustible common-property resource 804
4. Example for non-cooperation. II. Renewable common-property resource 812
5. Methodology reconsidered 815
6. "Tractable" differential games 815
7. The linear-quadratic differential game 817
8. Some other differential game models 818
9. Concluding remarks 820
References 823

Chapter 24
Communication, Correlated Equilibria and Incentive Compatibility 827
ROGER B. MYERSON

1. Correlated equilibria of strategic-form games 828
2. Incentive-compatible mechanisms for Bayesian games 835
3. Sender–receiver games 840
4. Communication in multistage games 845
References 847

Chapter 25
Signalling 849
DAVID M. KREPS and JOEL SOBEL

1. Introduction 850
2. Signalling games – the canonical game and market signalling 851
3. Nash equilibrium 852
4. Single-crossing 854
5. Refinements and the Pareto-efficient separating equilibrium 856
6. Screening 861
7. Costless signalling and neologisms 863
8. Concluding remarks 865
References 866

Chapter 26
Moral Hazard 869
PRAJIT K. DUTTA and ROY RADNER

1. Introduction 870
2. The principal–agent model 871

		2.1. The static model	871
		2.2. The dynamic model	873
	3.	Analyses of the static principal–agent model	874
	4.	Analyses of the dynamic principal–agent model	878
		4.1. Second-best contracts	879
		4.2. Simple contracts	881
	5.	Games of imperfect monitoring	886
		5.1. Partnership model	887
		5.2. Repeated games with imperfect monitoring	892
	6.	Additional bibliographical notes	897
	References		900

Chapter 27
Search   905
JOHN McMILLAN and MICHAEL ROTHSCHILD

1.	Search theory	906
2.	Stopping rules	907
3.	Price dispersion	912
4.	Search and bargaining	916
5.	Conclusion	921
References		922

Chapter 28
Game Theory and Evolutionary Biology   929
PETER HAMMERSTEIN and REINHARD SELTEN

1.	Introduction	931
2.	Conceptual background	932
	2.1. Evolutionary stability	932
	2.2. The Darwinian view of natural selection	933
	2.3. Payoffs	934
	2.4. Game theory and population genetics	935
	2.5. Players	936
	2.6. Symmetry	936
3.	Symmetric two-person games	937
	3.1. Definitions and notation	937
	3.2. The Hawk–Dove game	937
	3.3. Evolutionary stability	938
	3.4. Properties of evolutionarily stable strategies	940
4.	Playing the field	942
5.	Dynamic foundations	948
	5.1. Replicator dynamics	948
	5.2. Disequilibrium results	951
	5.3. A look at population genetics	952

6.	Asymmetric conflicts	962
7.	Extensive two-person games	965
	7.1. Extensive games	965
	7.2. Symmetric extensive games	966
	7.3. Evolutionary stability	968
	7.4. Image confrontation and detachment	969
	7.5. Decomposition	970
8.	Biological applications	971
	8.1. Basic questions about animal contest behavior	972
	8.2. Asymmetric animal contests	974
	8.3. War of attrition, assessment, and signalling	978
	8.4. The evolution of cooperation	980
	8.5. The great variety of biological games	983
References		987

Chapter 29
Game Theory Models of Peace and War   995
BARRY O'NEILL

1.	International relations game theory	996
	1.1. Game analyses of specific international situations	996
	1.2. The debate on realism and international cooperation	999
	1.3. International negotiations	1003
	1.4. Models of armsbuilding	1004
	1.5. Deterrence and signalling resolve	1006
	1.6. The myth that game theory shaped nuclear deterrence strategy	1010
	1.7. First-strike stability and the outbreak of war	1013
	1.8. Escalation	1014
	1.9. Alliances	1016
	1.10. Arms control verification	1017
2.	Military game theory	1018
	2.1. Analyses of specific military situations	1019
	2.2. Early theoretical emphases	1021
	2.3. Missile attack and defence	1023
	2.4. Tactical air war models	1024
	2.5. Lanchester models with fire control	1025
	2.6. Search and ambush	1026
	2.7. Pursuit games	1026
	2.8. Measures of effectiveness of weapons	1027
	2.9. Command, control and communication	1027
	2.10. Sequential models of a nuclear war	1028
	2.11. Broad military doctrine	1029
3.	Comparisons and concluding remarks	1029
References		1031

Contents of Volume II

Chapter 30
Voting Procedures — 1055
STEVEN J. BRAMS

1. Introduction — 1056
2. Elections and democracy — 1057
3. Sincerity, strategyproofness, and efficacy — 1058
   - 3.1. Voter preferences and dominance — 1059
   - 3.2. Dominance between strategies — 1061
   - 3.3. Admissible strategies — 1063
   - 3.4. Sincere voting and strategyproofness — 1067
   - 3.5. Efficacy — 1069
4. Condorcet properties — 1069
   - 4.1. Dichotomous preferences — 1070
   - 4.2. Runoff systems — 1072
   - 4.3. Condorcet possibility and guarantee theorems — 1073
5. Preferential voting and proportional representation — 1075
   - 5.1. The Hare system of single transferable vote (STV) — 1075
   - 5.2. The Borda count — 1078
   - 5.3. Cumulative voting — 1080
   - 5.4. Additional-member systems — 1081
6. Conclusions — 1084

References — 1085

Chapter 31
Social Choice — 1091
HERVÉ MOULIN

0. Introduction — 1092
1. Aggregation of preferences — 1094
   - 1.1. Voting rules and social welfare preorderings — 1094
   - 1.2. Independence of irrelevant alternatives and Arrow's theorem — 1095
   - 1.3. Social welfare preorderings on the single-peaked domain — 1096
   - 1.4. Acyclic social welfare — 1098
2. Strategyproof voting — 1101
   - 2.1. Voting rules and game forms — 1101
   - 2.2. Restricted domain of preferences: equivalence of strategyproofness and of IIA — 1103
   - 2.3. Other restricted domains — 1104
3. Sophisticated voting — 1105
   - 3.1. An example — 1105
   - 3.2. Dominance-solvable game forms — 1106
   - 3.3. Voting by successive vetos — 1109

4.	Voting by self-enforcing agreements	1110
	4.1. Majority versus minority	1110
	4.2. Strong equilibrium in voting by successive veto	1113
	4.3. Strong monotonicity and implementation	1116
	4.4. Effectivity functions	1118
5.	Probabilistic voting	1121
References		1123

Chapter 32
Power and Stability in Politics 1127
PHILIP D. STRAFFIN Jr.

1.	The Shapley–Shubik and Banzhaf power indices	1128
2.	Structural applications of the power indices	1130
3.	Comparison of the power indices	1133
4.	Dynamic applications of the power indices	1138
5.	A spatial index of voting power	1140
6.	Spatial models of voting outcomes	1144
Bibliography		1150

Chapter 33
Game Theory and Public Economics 1153
MORDECAI KURZ

1.	Introduction	1154
2.	Allocation of pure public goods	1155
	2.1. Pure public goods: Lindahl equilibrium and the core	1156
	2.2. Decentralization in public goods economies: are public goods large or small?	1161
	2.3. The core of the second best tax game	1165
	2.4. Shapley value of public goods games	1168
3.	Externalities, coalition structures and local public goods	1174
	3.1. On the core of games with externalities and public goods	1174
	3.2. Equilibria and cores of local public goods economies	1177
	3.3. Second best tax game revisited	1182
4.	Power and redistribution	1183
5.	Some final reflections	1189
References		1190

Chapter 34
Cost Allocation 1193
H.P. YOUNG

1.	Introduction	1194
2.	An illustrative example	1195
3.	The cooperative game model	1197

4.	The Tennessee Valley Authority	1198
5.	Equitable core solutions	1203
6.	A Swedish municipal cost-sharing problem	1206
7.	Monotonicity	1209
8.	Decomposition into cost elements	1211
9.	The Shapley value	1213
10.	Weighted Shapley values	1215
11.	Cost allocation in the firm	1217
12.	Cost allocation with variable output	1219
13.	Ramsey prices	1219
14.	Aumann–Shapley prices	1220
15.	Adjustment of supply and demand	1223
16.	Monotonicity of Aumann–Shapley prices	1223
17.	Equity and competitive entry	1224
18.	Incentives	1225
19.	Conclusion	1228
Bibliography		1230

Chapter 35
Cooperative Models of Bargaining 1237
WILLIAM THOMSON

1.	Introduction	1238
2.	Domains. Solutions	1240
3.	The main characterizations	1245
	3.1. The Nash solution	1245
	3.2. The Kalai–Smorodinsky solution	1249
	3.3. The Egalitarian solution	1251
4.	Other properties. The role of the feasible set	1254
	4.1. Midpoint domination	1254
	4.2. Invariance	1254
	4.3. Independence and monotonicity	1256
	4.4. Uncertain feasible set	1258
5.	Other properties. The role of the disagreement point	1260
	5.1. Disagreement point monotonicity	1261
	5.2. Uncertain disagreement point	1262
	5.3. Risk-sensitivity	1264
6.	Variable number of agents	1265
	6.1. Consistency and the Nash solution	1266
	6.2. Population monotonicity and the Kalai–Smorodinsky solution	1268
	6.3. Population monotonicity and the Egalitarian solution	1270
	6.4. Other implications of consistency and population monotonicity	1271
	6.5. Opportunities and guarantees	1271
	6.6. Replication and juxtaposition	1273

7. Applications to economics	1274
8. Strategic considerations	1275
Bibliography	1277

Chapter 36
Games in Coalitional Form   1285
ROBERT J. WEBER

1. Introduction	1286
2. Basic definitions	1286
3. Side payments and transferable utility	1287
4. Strategic equivalence of games	1288
5. Properties of games	1290
6. Balanced games and market games	1293
7. Imputations and domination	1294
8. Covers of games, and domination equivalence	1295
9. Extensions of games	1295
10. Contractions, restrictions, reductions, and expansions	1297
11. Cooperative games without transferable utility	1298
12. Two notions of "effectiveness"	1299
13. Partition games	1301
14. Games with infinitely many players	1301
15. Summary	1302
Bibliography	1302

Chapter 37
Coalition Structures   1305
JOSEPH GREENBERG

1. Introduction	1306
2. A general framework	1307
3. Why do coalitions form?	1308
4. Feasible outcomes for a fixed coalition structure	1311
5. Individually stable coalition structures	1313
6. Coalition structure core	1316
7. Political equilibrium	1319
8. Extensions of the Shapley value	1322
9. Abstract systems	1324
10. Experimental work	1327
11. Noncooperative models of coalition formation	1329
References	1332

Contents of Volume II

Chapter 38
Game-Theoretic Aspects of Computing — 1339
NATHAN LINIAL

1. Introduction — 1340
   1.1. Distributed processing and fault tolerance — 1340
   1.2. What about the other three main issues? — 1342
   1.3. Acknowledgments — 1343
   1.4. Notations — 1344
2. Byzantine Agreement — 1344
   2.1. Deterministic algorithms for Byzantine Agreement — 1346
   2.2. Randomization in Byzantine Agreements — 1350
3. Fault-tolerant computation under secure communication — 1352
   3.1. Background in computational complexity — 1357
   3.2. Tools of modern cryptography — 1360
   3.3. Protocols for secure collective computation — 1368
4. Fault-tolerant computation – The general case — 1372
   4.1. Influence in simple games — 1373
   4.2. Symmetric simple games — 1379
   4.3. General perfect-information coin-flipping games — 1381
   4.4. Quo vadis? — 1384
5. More points of interest — 1385
   5.1. Efficient computation of game-theoretic parameters — 1385
   5.2. Games and logic in computer science — 1386
   5.3. The complexity of specific games — 1387
   5.4. Game-theoretic consideration in computational complexity — 1388
   5.5. Random number generation as games — 1388
   5.6. Amortized cost and the quality of on-line decisions — 1390
References — 1391

Chapter 39
Utility and Subjective Probability — 1397
PETER C. FISHBURN

1. Introduction — 1398
2. Historical sketch — 1400
3. Ordinal utility and comparable differences — 1401
4. Expected utility and linear generalizations — 1407
5. Nonlinear utility — 1411
6. Subjective probability — 1413
7. Expected utility and subjective probability — 1416
8. Generalizations and alternatives — 1420
References — 1423

Chapter 40
Common Knowledge — 1437
JOHN GEANAKOPLOS

1. Introduction — 1438
2. Puzzles about reasoning based on the reasoning of others — 1439
3. Interactive epistemology — 1441
4. The puzzles reconsidered — 1444
5. Characterizing common knowledge of events and actions — 1450
6. Common knowledge of actions negates asymmetric information about events — 1453
7. A dynamic state space — 1455
8. Generalizations of agreeing to disagree — 1458
9. Bayesian games — 1461
10. Speculation — 1465
11. Market trade and speculation — 1467
12. Dynamic Bayesian games — 1469
13. Infinite state spaces and knowledge about knowledge to level N — 1476
14. Approximate common knowledge — 1480
15. Hierarchies of belief: Is common knowledge of the partitions tautological? — 1484
16. Bounded rationality: Irrationality at some level — 1488
17. Bounded rationality: Mistakes in information processing — 1490
References — 1495

Index — 1497

Chapter 20

ZERO-SUM TWO-PERSON GAMES

T.E.S. RAGHAVAN*

University of Illinois at Chicago

Contents

References	757
Appendix, by T. Radzik and T.E.S. Raghavan. Duels	761

*The author would like to thank Ms Evangelista Fe, Swaminathan Sankaran and the anonymous referees for the many detailed comments that improved the presentation of this article.

Handbook of Game Theory, Volume 2, Edited by R.J. Aumann and S. Hart
© *Elsevier Science B.V., 1994. All rights reserved*

Many parlor games like the game of "Le Her" or "Morra" [Dresher (1963)], involve just two players and finitely many moves for the players. When the number of actions available to a player is finite, we could in theory, reduce the problem to a game with exactly one move for each player, a move where actions for the players are various master plans for the entire game chosen from among the finitely many possible master plans. These master plans are called *pure strategies*. The original game is often analyzed using such a reduced form called games in normal form (also called strategic form). Certain information about the original game can be lost in the process [Kuhn (1953), Aumann and Maschler (1972)]. The reduction however helps to focus our understanding of the strategic behavior of intelligent players keen on achieving certain guaranteed goals against all odds.

Given a two-person game with just one move for each player, the players independently and simultaneously select one among finitely many actions resulting in a payoff for each player. If i, j are their independent choices in a play, then the game is defined by a pair of real matrices $A = (a_{ij})$, $B = (b_{ij})$ where a_{ij} is the payoff to player I and b_{ij} is the *payoff* to player II. The game is called zero sum if $a_{ij} + b_{ij} \equiv 0$. Thus in zero-sum games, what one player gains, the opponent loses. In such games A suffices to determine the payoff.

We can use the following example [Dresher (1963)] to illustrate what we have said so far.

Example. From a deck of three cards numbered 1, 2, 3 player I picks a card at will. Player II tries to guess the card. After each guess player I signals either High or Low or Correct, depending on the guess of the opponent. The game is over as soon as the card is correctly guessed by player II. Player II pays player I an amount equal to the number of trials he made.

There are three pure strategies for player I. They are:

α: Choose 1,
β: Choose 2,
γ: Choose 3.

For player II the following summarizes the possible pure strategies, excluding obviously "bad ones".

(a) Guess 1 at first. If the opponent says Low, guess 2 in the next round. If the opponent still says Low, guess 3 in the next round.
(b) Guess 1 at first. If the opponent says Low, guess 3 in the next round. If the opponent says High, guess 2 in the next round.
(c) Guess 2 at first. If the opponent says Low, guess 3; if the opponent says High, guess 1.
(d) Guess 3 at first. If the opponent says High, guess 1 in the next round. If the opponent says Low, guess 2 in the next round.

(e) Guess 3 at first. If the opponent says High, guess 2 in the next round. If the opponent still says High, guess 1 in the next round.

Thus the payoff matrix is given by

$$A = \begin{array}{c} \\ \alpha \\ \beta \\ \gamma \end{array} \begin{pmatrix} \text{(a)} & \text{(b)} & \text{(c)} & \text{(d)} & \text{(e)} \\ 1 & 1 & 2 & 2 & 3 \\ 2 & 3 & 1 & 3 & 2 \\ 3 & 2 & 2 & 1 & 1 \end{pmatrix}.$$

A pair of pure strategies (i^*, j^*) for a payoff matrix $A = (a_{ij})$ is called a *saddle point* if $a_{ij^*} \leqslant a_{i^*j^*} \leqslant a_{i^*j}$ $\forall i, \forall j$. Thus $a_{i^*j^*}$ is the guaranteed gain for player I against any "j" player II chooses. It is also the maximum loss to player II against any "i" player I chooses. In general such a saddle point may not exist. When it does we have

$$a_{i^*j^*} = \max_i \min_j a_{ij} = \min_j \max_i a_{ij}.$$

The above game has no saddle point. In fact $\max_i \min_j a_{ij} = 1$ and $\min_j \max_i a_{ij} = 2$. When players face intelligent opponents, the saddle point strategies are the best course of actions for both players.

When a game has no saddle point, without violating the rules of the game, it might be possible for players to enlarge the available set of masterplans and look for a saddle point in the enlarged game. For example a master plan for a player in this new game could be based on the outcome of a suitably chosen random device from an available set of random devices, where the possible outcomes are themselves masterplans of the original game. If each player selects his masterplan based on the outcome of his independently chosen random device we enter into a new game called *a mixed extension* of the game. Here the set of available random devices are called mixed strategies for the original game and they will be the pure strategies for the mixed extension game. The expected payoff $K(x, y)$ is the outcome for the mixed extension game when player I uses the random device x and player II uses the random device y.

In our example above player I could rely on a random device $x^* = (x_1^*, x_2^*, x_3^*) = (\frac{2}{5}, \frac{1}{5}, \frac{2}{5})$ where say x_1^* is the probability of choosing pure strategy α in the original game. Player I can guarantee an expected gain of $\frac{9}{5}$ no matter which pure strategy player II uses. Similarly, the random device $y^* = (y_a^*, y_b^*, y_c^*, y_d^*, y_e^*) = (0, \frac{1}{5}, \frac{3}{5}, \frac{1}{5}, 0)$ where say $y_c^* = \frac{3}{5}$ is the probability of selecting the pure strategy c by the random device y^*. Player II by using y^* can guarantee an expected loss no more than $\frac{9}{5}$ whatever player I does. Thus, if x, y are any arbitrary probability vectors for players I and II, then the expected payoff $K(x, y)$ to player I satisfies

$$K(x, y^*) \leqslant K(x^*, y^*) \leqslant K(x^*, y) \quad \forall x, \text{ and } \forall y.$$

Thus for the mixed extension game, (x^*, y^*) is a saddle point. Here x^*, y^* are called *optimal mixed strategies* for players I and II respectively and $\frac{9}{5}$ is called the *value* of the game.

While Borel [von Neumann (1953)] suspected that games may not in general possess value even in the mixed extension, von Neumann (1928) laid the foundations of Game Theory by proving the decisive *minimax* theorem. This celebrated theorem remained dormant for a while. It was the classic monograph of [von Neumann and Morgenstern (1944)] that arrested the attention of mathematicians and social scientists and triggered research activity in Game theory.

Of the many existence proofs for the minimax theorem [von Neumann (1928), Kakutani (1941), Ville (1938), Loomis (1946), Nash (1950), Owen (1970) and so on] Kakutani's proof and Nash's proof are amenable for extensions to more general situations. Nash's proof is based on the Brouwer's fixed point theorem. Brouwer's theorem asserts that any continuous self map ϕ of a compact convex set X in Euclidean n-space R^n admits a fixed point. That is $\phi(x) = x$ for some $x \in X$. We will now prove the minimax theorem using the Brouwer's fixed point theorem.

Minimax theorem. (von Neumann). *Let $A = (a_{ij})$ be any $m \times n$ real matrix. Then these exists a pair of probability vectors $x^* = (x_1^*, x_2^*, \ldots, x_m^*)$ and $y^* = (y_1^*, y_2^*, \ldots, y_n^*)$ such that for a unique constant v*

$$\sum_i a_{ij} x_i^* \geq v, \quad j = 1, 2, \ldots, n, \tag{1}$$

$$\sum_j a_{ij} y_j^* \leq v, \quad i = 1, 2, \ldots, m. \tag{2}$$

Equivalently if $K(x, y) = \sum_i \sum_j a_{ij} x_i y_j$ then (x^*, y^*) is a saddle point for $K(x, y)$. That is

$$\min_y \max_x K(x, y) = \max_x \min_y K(x, y),$$

where min *and* max *are taken respectively over the set of all probability vectors x for player I and probability vectors y for player* II.

Proof. Given any probability vector x for I and y for II if $p_i = p_i(x, y) = \sum_j a_{ij} y_j - \sum_i \sum_j a_{ij} x_i y_j$, $q_j = q_j(x, y) = \sum_i \sum_j a_{ij} x_i y_j - \sum_i a_{ij} x_i$, we are looking for a pair (x, y) with $p_i \leq 0$ and $q_j \leq 0$ for all i, j. Let $\varphi:(x, y) \to (\xi, \eta)$ be a map where ξ is a probability m-vector and η is a probability n-vector defined by coordinates

$$\xi_i = \frac{x_i + \max(p_i, 0)}{1 + \sum_1^m \max(p_k, 0)}, \quad \eta_j = \frac{y_j + \max(q_j, 0)}{1 + \sum_1^n \max(q_k, 0)}; \quad i = 1, 2, \ldots, m,$$

$$j = 1, 2, \ldots, n.$$

It can be easily checked that φ is a continuous self-map of the set of probability vector pairs. By the Brouwer's fixed point theorem, there exists an (x^*, y^*) such that $\varphi(x^*, y^*) = (x^*, y^*)$.

Thus $x_i^* \sum_k \max(p_k, 0) = \max(p_i, 0)$ for all i. For some $i, x_i^* > 0, p_i \leq 0$. Thus $\max(p_k, 0) = 0$ for all k. A similar argument shows that $\max(q_l, 0) = 0$ for all l. These are precisely the equalities we wanted for $v = \sum_i \sum_j a_{ij} x_i^* y_j^*$. The second assertion that (x^*, y^*) is a saddle point for the mixed extension $K(x, y)$ easily follows from multiplying respectively the inequalities (1) and (2) by y_j and x_i and taking the sum over j and i respectively. □

The existence theorem does not tell us how to compute a pair of optimal strategies. An algebraic proof was given by Weyl (1950). Thus if the entries of the payoff are rational, by the algebraic proof one can show that the value is rational and a pair of optimal strategies have rational entries. Brown and Von Neumann (1950) suggested a solution by differential equations that roughly mimics the above tatonnement process $(x, y) \to (\xi, \eta)$. It turned out that the minimax theorem can be proved via linear programming in a constructive way leading to an efficient computational algorithm a la the simplex method [Dantzig (1951)]. Interestingly the minimax theorem can also be used to prove a version of the duality theorem of linear programming.

Equivalence of the minimax theorem and the duality theorem

Given an $m \times n$ real matrix $A = (a_{ij})$ and given column m-vector b and column n-vector c we have the following two problems called dual linear programs in standard form.

Primal: $\max c \cdot x$
subject to $Ax \leq b$, $x \geq 0$,

Dual: $\min b \cdot y$
subject to $A^T y \geq c$, $y \geq 0$.

Any x satisfying the constraints of the primal is called a *feasible solution to the primal*. Feasible solutions to the dual are similarly defined. Here $c \cdot x$ and $b \cdot y$ are called the *objective functions* for the primal and the dual.

The following is a version of the fundamental theorem of linear programming due to von Neumann [Dantzig (1951)].

Duality Theorem. *If the primal and the dual problems have at least one feasible solution, then the two problems have optimal solutions; further, at any optimal solution, the value of the two objective functions coincide.*

Indeed the duality theorem and the minimax theorem are equivalent. We will see that either one implies the other.

Duality theorem ⇒ minimax theorem

Given a payoff matrix $A = (a_{ij})$, by adding a constant c to all the entries, we can assume $a_{ij} > 0$ for all i, j and $v > 0$. Our inequalities (1) and (2) can be rewritten as

$$\sum_i a_{ij}\left(\frac{x_i}{v}\right) \geq 1, \quad \text{for all } j,$$

$$\sum_j a_{ij}\left(\frac{y_j}{v}\right) \leq 1, \quad \text{for all } i,$$

where say, $p = (p_1, \ldots, p_m)$, $q = (q_1, \ldots, q_n)$ with

$$p_i = \frac{x_i}{v} \geq 0, \quad \sum_i p_i = \frac{1}{v}, \tag{3}$$

$$q_j = \frac{y_j}{v} \geq 0, \quad \sum_j q_j = \frac{1}{v}. \tag{4}$$

Given a matrix $A = a_{ij}$ with all entries positive consider the dual linear programs

Problem A. $\min \sum_i p_i$ subject to $\sum_i a_{ij} p_i \geq 1$ for all j and $p_i \geq 0$ for all i.

Problem B. $\max \sum_j q_j$ subject to $\sum_j a_{ij} q_j \leq 1$ for all i and $q_j \geq 0$ for all j.

Here $q = 0$ is feasible for Problem B. For large N the vector $p = (N, \ldots, N)$ is feasible for Problem A. By the duality theorem we have an optimal solution p^*, q^* to the two problems. Further $\sum_i p_i^* = \sum_j q_j^*$ at any optimal pairs. By normalizing p^* and q^* we have optimal x^* and y^* satisfying (1), (2) for the payoff A with value $v = 1/\sum_j q_j^*$. Thus the minimax theorem is equivalent to solving the above dual linear programs.

Before reducing the dual linear programs to a single game problem we need the following theorems.

Theorem on skew symmetric payoffs. *Let $A = -A^T$ be a payoff matrix. Then the value of the game is zero and both players have the same set of optimal strategies.*

Proof. Let if possible $v < 0$. Let y be optimal for player II. Thus $Ay = -A^Ty < 0$ and hence $A^Ty > 0$. This contradicts $v < 0$. Similarly we can show that $v > 0$ is not possible. Thus $v = 0$ and optimal strategies of one player are also optimal for the other player. □

Equalizer theorem. *Let $A = (a_{ij})$ be a payoff with value v. Let the expected payoff to player I when he uses any optimal strategy x and when player II uses a fixed*

column be v. Then player II has an optimal strategy which chooses this column with positive probability.

Proof. We can as well assume $v = 0$ and the particular column is the nth column. It is enough to show that the system of inequalities

$$\sum_j a_{ij} u_j \leq 0$$

$$u_n = 1$$

has a solution $u \geq 0$. We could normalize this and get the desired strategy. This is equivalent to finding a solution $u, w \geq 0$ to the matrix equation

$$\begin{bmatrix} A & I \\ e_n & 0 \end{bmatrix} \begin{bmatrix} u \\ w \end{bmatrix} = \mathbf{p} = \begin{bmatrix} 0 \\ 1 \end{bmatrix}.$$

(Here e_n is nth unit vector). It is enough to show that the point $\mathbf{p}$ is in the closed cone K generated by the columns of the matrix

$$B = \begin{bmatrix} A & I \\ e_n & 0 \end{bmatrix}.$$

Suppose not. Then by the strong separation theorem [Parthasarathy and Raghavan (1971)] we can find a strictly separating hyperplane $(\mathbf{f}, \alpha)$ between the closed cone K and the point $\mathbf{p}$. This shows that the vector $\mathbf{f} \geq 0$ and the scalar $\alpha < 0$. Further $\mathbf{f} \cdot a_j \geq 0$ for all columns a_j and $\mathbf{f} \cdot a_n + \alpha \geq 0$. Normalizing $\mathbf{f}$ we get an optimal strategy for player I which gives positive expectation when player II selects column n. This contradicts the assumption about column n. □

Minimax theorem ⇒ Duality theorem

Consider the following $m + n + 1 \times m + n + 1$ skew symmetric payoff matrix

$$B = \begin{bmatrix} 0 & A & -b \\ -A^T & 0 & c \\ b^T & -c^T & 0 \end{bmatrix}.$$

Since B is a skew symmetric payoff the value of B is 0. Any optimal mixed strategy (η^*, ξ^*, θ) for player II is also optimal for player I. If $\theta > 0$, then the vectors $y^* = (1/\theta) \cdot \eta^*$, $x^* = (1/\theta) \cdot \xi^*$ will be feasible for the two linear programming problems. Further when player II uses the optimal strategy and I uses the last row, the expected income to player I is $b \cdot y^* - c \cdot x^* \leq 0$. Since the inequality $b \cdot y^* - c \cdot x^* \geq 0$ is always true for any feasible solutions of the two linear programs, the two objective functions have the same value and it is easy to see that the problems have y^*, x^* as their optimal solutions. We need to show that for some optimal (η^*, ξ^*, θ),

$\theta > 0$. Suppose not. Then by the equalizer theorem player II pays a *penalty* if he chooses column n with positive probability, namely $-b \cdot \eta^* + c \cdot \xi^* > 0$. Thus either $b \cdot \eta^* < 0$ or $c \cdot \xi^* > 0$. Say $c \cdot \xi^* > 0$. For any feasible $\bar{x}$ for the primal and for the feasible y° of the dual we have $c \cdot \bar{x} - b \cdot \bar{y} \leq 0$. For large N, $\bar{x} = x^\circ + N\xi^*$ is feasible for the primal and $c \cdot (x^\circ + N\xi^*) - b \cdot y^\circ > 0$ a contradiction. A similar argument can be given when $b \cdot \eta^* < 0$.

Extreme optimal strategies. In general optimal strategies are not unique. Since the optimal strategies for a player are determined by linear inequalities the set of optimal strategies for each player is a closed bounded convex set. Further the sets have only *finitely many extreme points* and one can effectively enumerate them by the following characterization due to Shapley and Snow (1950).

Theorem. *Let A be an $m \times n$ matrix game with $v \neq 0$. Optimal mixed strategies x^* for player I and y^* for player II are extreme points of the convex set of optimal strategies for the two players if and only if there is a square submatrix $B = (a_{ij})_{i \in I, j \in J}$, such that*

   i. *B nonsingular.*
   ii. *$\sum_{i \in I} a_{ij} x_i^* = v \quad j \in J \subset \{1, 2, \ldots, n\}$.*
   iii. *$\sum_{j \in J} a_{ij} y_j^* = v \quad i \in I \subset \{1, 2, \ldots, m\}$.*
   iv. *$x_i^* = 0$ if $i \notin I$.*
   v. *$y_j^* = 0$ if $j \notin J$.*

Proof. (Necessary.) After renumbering the columns and the rows, we can assume for an extreme optimal pair (x^*, y^*), $x_i^* > 0$ $i = 1, 2, \ldots, p$; $y_j^* > 0$ $j = 1, 2, \ldots, q$. If row i is actively used (i.e. $x_i^* > 0$,) then $\sum_{j=1}^n a_{ij} y_j^* = v$. Thus we can assume $\tilde{A} = (a_{ij})_{i \in \bar{I}, j \in \bar{J}}$ such that

$$\sum_{i=1}^p a_i^j x_i^* \begin{cases} = v, & j \in \bar{J} \supset \{1, 2, \ldots, q\}, \\ > v, & \text{for } j \notin \bar{J}, \end{cases}$$

$$\sum_{j=1}^q a_{ij} y_j^* \begin{cases} = v, & i \in \bar{I} \supset \{1, 2, \ldots, p\}, \\ < v, & \text{for } i \notin \bar{I}. \end{cases}$$

We claim that the p rows and the q columns of $\tilde{A} = (a_{ij})_{i \in \bar{I}, j \in \bar{J}}$ are independent. Suppose not. For some $(\pi_1, \ldots, \pi_p) \neq 0$

$$\sum_{i=1}^p a_{ij} \pi_i = 0, \quad j \in \bar{J},$$

$$\sum_{i=1}^p a_{ij} x_i^* = v, \quad j \in \bar{J}.$$

Thus $\sum_{i=1}^p \pi_i \sum_{j=1}^q a_{ij} y_j^* = v \sum_{i=1}^p \pi_i = 0$. As $v \neq 0$, $\sum_{i=1}^p \pi_i = 0$. Since $\sum_i a_{ij} x_i^* > v$ for

$j \notin \bar{J}$ we can find $\varepsilon > 0$ sufficiently small such that $x_i' = x_i^* - \varepsilon \pi_i \geq 0$, $x_i'' = x_i^* + \varepsilon \pi_i \geq 0$, $x_i' = x_i'' = \pi_i \equiv 0$, $i > p$ and

$$\sum_{i=1}^{m} a_{ij}x_i' = \sum_{i=1}^{m} a_{ij}x_i^* - \varepsilon \sum_{i=1}^{m} a_{ij}\pi_i \geq v, \quad \text{for all } j,$$

$$\sum_{i=1}^{m} a_{ij}x_i'' = \sum_{i=1}^{m} a_{ij}x_i^* + \varepsilon \sum_{i=1}^{m} a_{ij}\pi_i \geq v, \quad \text{for all } j.$$

Thus x', x'' are optimal and $x^* = (x' + x'')/2$ is not extreme optimal, a contradiction. Similarly the first q columns of $\tilde{A}$ are independent. Hence a nonsingular submatrix B containing the first p rows and the first q columns of $\tilde{A} = (a_{ij})_{i \in \bar{I}, j \in \bar{J}}$ and satisfying conditions (i)–(iv) exists.

Conversely, given such a matrix B satisfying (i)–(iv), the strategy x^* is extreme optimal, otherwise $x' \neq x''$, $x^* = (x' + x'')/2$ and we have $\sum_i a_{ij}x_i' \geq v$, $\sum_i a_{ij}x_i'' \geq v$ for $j \in J$ with $\sum_i a_{ij}x_i^* = v$ for $j \in J$. Thus $\sum_{i \in I} a_{ij}x_i' = \sum_{i \in I} a_{ij}x_i'' = v$ for $j = J$ and the matrix B is singular. □

Since there are only finitely many square submatrices to a payoff matrix there could be only finitely many extreme points and as solutions of linear equations, they are in the same ordered subfield as the data field. The problem of efficiently loacting an extreme optimal strategy can be handled by solving the linear programming problem mentioned above. Among various algorithms to solve a linear programming problem, the simplex algorithm is practically the most efficient. Linear inequalities were first investigated by Fourier (1890) and remained dormant for more than half a century. Linear modeling of problems in industrial production and planning necessitated active research and the pioneering contributions of Kantorovich (1939), Koopmans (1951) and Dantzig (1951) brought them to the frontiers of modern applied mathematics.

Simplex algorithm. Consider the canonical linear programming problem

$$\max \sum_j b_j y_j$$

subject to

$$\sum_{j=1}^{n} a_{ij} y_j = d_i, \quad i = 1, 2, \ldots, m,$$

$$y_j \geq 0, \quad j = 1, 2, \ldots, n.$$

Any solution $y = (y_1, \ldots, y_n)$ to the above system of inequalities is called a feasible solution. We could also write the system as

$$y_1 C^1 + y_2 C^2 + \cdots + y_n C^n = d,$$

$$y_1, y_2, \ldots, y_n \geq 0,$$

where $C^1, C^2, \ldots, C^n$ are the columns of the matrix A and d is the column vector with coordinates $d_i, i = 1, 2, \ldots, m$. It is not hard to check that any extreme point $y = (y_1, y_2, \ldots, y_n)$ of the convex polyhedra of feasible solutions can be identified with a set of linearly independent columns $C^{i_1}, C^{i_2}, \ldots, C^{i_k}$ such that the y_j are zero for coordinates other than $i_1, i_2, \ldots, i_k$. By slightly perturbing the entries we could even assume that every extreme point of feasible solutions has exactly m coordinates positive.

We call two extreme points adjacent if the line segment joining them is a one dimensional face of the feasible set. Algebraically, two adjacent extreme points can be identified with two bases which differ by exactly one basis vector. The new basis is chosen by bringing a column from outside into the current basis which in turn determines the removal of an appropriate column from the current basis. An iteration consists of searching for an improved value of the objective function at an adjacent extreme point. The algorithm terminates if no improvements in the value of the objective function with adjacent extreme points is possible.

Fictitious play. An intuitively easy to implement algorithm to find the approximate value uses the notion of fictitious play [Brown (1951)].

Assume that a given matrix game, $A = (a_{ij})_{m \times n}$ has been played for t rounds with $(i_1, j_1), (i_2, j_2), \ldots, (i_t, j_t)$ as the actual choices of rows and columns by the two players. Let rows $1, 2, \ldots, m$ appear $k_1, k_2, \ldots, k_m$ times in the t rounds. For player II one way of learning from player I's actions is to pretend that the proportions $(k_1/t, \ldots, k_m/t)$ are the true mixed strategy choices of player I. With such a belief, the best choice for player II in round $t + 1$ is to choose any column j_{t+1} which minimizes the fictitious expected payoff $(1/t) \sum_{i=1}^{m} k_i a_{ij}$.

Suppose in the above data, columns $1, 2, \ldots, n$ appear $l_1, l_2, \ldots, l_n$ times in the first t rounds. Player I can also pretend that the true strategy of player II is $(l_1/t, l_2/t, \ldots, l_n/t)$. With such a belief the best choice for player I in round $t + 1$ is to choose any row i_{t+1} which maximizes the fictitious expected income $1/t \sum_{j=1}^{n} l_j a_{ij}$. The remarkable fact is that this naive procedure can be used to approximate the value of the game. We have the following

Theorem. *Let (x^t, y^t) be the strategies $(k_1/t, \ldots, k_m/t), (l_1/t, \ldots, l_n/t)$ $t = 1, 2, \ldots$ where (x^1, y^1) is arbitrary and (x^t, y^t) for $t \geq 2$ is determined by the above fictitious play. Then*

$$v = \lim_{t \to \infty} \min_j \frac{1}{t} \sum_{i=1}^{m} k_i a_{ij} = \lim_{t \to \infty} \max_i \frac{1}{t} \sum_{j=1}^{n} l_j a_{ij}.$$

The above procedure is only of theoretical interest. It is impractical and the convergence to the value is known to be very slow. Even though $v(t) = \min_j (1/t) \sum_{i=1}^{m} k_i a_{ij} \to v$ as $t \to \infty$, the mixed strategies $\xi(t) = (k_1/t, k_2/t, \ldots, k_m/t)$ and $\eta(t) = (l_1/t, \ldots, l_n/t)$ may not converge.

A proof can be given [Robinson (1951)] by showing that for any skew symmetric payoff A,

$$\lim_t \min_j \sum_i k_i a_{ij} = 0.$$

In general the sequence of strategies $\{(x^t, y^t)\}$ oscillates around optimal strategies.

Completely mixed games. A mixed strategy x for player I is *called completely mixed* iff $x > 0$ (i.e. all rows are essentially used). Suppose x, y are completely mixed optimal strategies for the two players. The inequalities $Ay \leqslant v \cdot 1$ are actually equalities, for otherwise $v = (x, Ay) < (x, v \cdot 1) = v$, a contradiction. In case $v = 0$, the matrix A is singular. We call a matrix game A *completely mixed* if every optimal strategy is completely mixef for both players.

Theorem. *If a matrix game A is completely mixed, then A is a square matrix and the optimal strategies are unique.*

Proof. Without loss of generality $v \neq 0$. In case $y' \neq y''$ are two extreme optimal strategies for player II, then $Ay' = v \cdot 1, Ay'' = v \cdot 1$ and $A(y' - y'') = 0$. Thus rank $A < n$. Since the extreme $y' > 0$, by the Shapley – Snow theorem, A has an $n \times n$ submatrix which is nonsingular. This contradicts rank $A < n$. Thus rank $A = n$ and the extreme optimal strategy is unique for player II. A similar argument applies for player I and shows that rank $A = m$, and the extreme optimal strategy is unique for player I. □

We have a formula to compute the value v for completely mixed games, and it is given by solving $Ay = v \cdot 1$. The unique solution y is optimal for player II. Since y is a probability vector $y = vA^{-1} \cdot 1$ gives $v = \det A/(\sum_i \sum_j A_{ij})$ where $\det A$ is the determinant of A and A_{ij} are the cofactors of A.

In case the payoff is a square matrix it can be shown that when one player has on optimal strategy which is not completely mixed then his opponent also possesses an optimal strategy that is not completely mixed [Kaplansky (1945)].

For Z-matrices (square matrices with off-diagonal entries nonpositive) if the value is positive, then the maximizer cannot omit any row (this results in a submatrix with a nonpositive column which the minimizer will choose even in the original game). Thus the game is completely mixed. One can infer many properties of such matrices by noting the game is completely mixed. It is easy to check that since $v > 0$ the matrix is non-singular and its inverse is nonnegative: For completely mixed games $A = (a_{ij})$ with value zero, the cofactor matrix (A_{ij}) has all $A_{ij} > 0$ or all $A_{ij} < 0$. This can be utilized to show that any Z-matrix with positive value has all principal minors positive [Raghavan (1978), (1979)].

The reduction of matrix games to linear programming is possible even when the strategy spaces are restricted to certain polyhedral subsets. This is useful in solving for stationary optimal strategies in some dynamic games [Raghavan and Filar (1991)].

Polyhedral constraints and matrix games. Consider a matrix game A where players I(II) can use only mixed strategies constrained to lie in polyhedra $X(Y)$. Say $X = \{x : B^T x \leq c, x \geq 0\}$ where $X \subset$ set of mixed strategies for player I. Let

$$Y = \{y : Ey \geq f, \ y \geq 0\}.$$

We know that $\max_{x \in X} \min_{y \in Y} x^T A y = \min_{y \in Y} \max_{x \in X} x^T A y$. The linear program $\min_{y \in Y} x^T A y$ has a dual

$$\max_{z \in T} f \cdot z,$$

where $T = \{z : E^T z \leq A^T x, z \geq 0\}$. Thus player I's problem is

$$\max_{(z, x) \in K} f \cdot z,$$

where $K = \{(z, x) : E^T z \leq x, B^T x \leq c, z \geq 0, x \geq 0\}$. This is easily solved by the simplex algorithm.

Dimension relations. Given a matrix game $A = (a_{ij})_{m \times n}$ let X, Y be the convex sets of optimal strategies for players I and II. Let the value $v = 0$. Let $J_1 = \{j : y_j > 0$ for some $y \in Y\}$. Let $J_2 = \{j : \sum_i a_{ij} x_i = 0$ for any $x \in X\}$. It is easy to show that $J_1 \subset J_2$. From the equalizer theorem we proved earlier that $J_2 \subset J_1$. Thus $J_1 = J_2 = J$, say. Similarly we have $I_1 = I_2 = I$ for player I. Since the rows outside I_2 and the columns outside J_2 are never used in optimal plays, we can as well restrict to the game $\tilde{A}$ with rows in I_2 and columns in J_2 and with value 0. Let $\tilde{X}, \tilde{Y}$ be the vector spaces generated by X, Y, respectively. The following theorem independently due to [Bohnenblust, Karlin and Shapley (1950)] and [Gale and Sherman (1950)] characterizes the intrinsic dimension relation between optimal strategy sets and essential strategies of the two players.

Dimension theorem. $|I| - \dim \tilde{X} = |J| - \dim \tilde{Y}$.

Proof. Consider the submatrix $\tilde{A} = (a_{ij})_{i \in I, j \in J}$ For any $y \in Y$ let $\bar{y}$ be the restriction of y to the coordinates $j \in J$. We have $\tilde{A} y = 0$. Let $\tilde{A} \pi = 0$ for some π in $R^{|J|}$. Since we can always find an optimal strategy y^* with $y_j^* > 0$ for all $j \in J$, we have $z = y^* - \varepsilon \pi \geq 0$ for small ε and $\tilde{A} z = 0$. Clearly $z \in \tilde{Y}$. Further since the linear span of y^* and z yield π the vector space $\{u : \tilde{A} u = 0\}$ coincides with $\tilde{Y}$. Thus dim $\tilde{Y} = |J| - \text{rank } \tilde{A}$. A Similar argument shows that dim $\tilde{X} = |I| - \text{rank } \tilde{A}$. Hence rank $\tilde{A} = |I| - \dim \tilde{X} = |J| - \dim \tilde{Y}$. □

Semi-infinite games and intersection theorems

Since a matrix payoff $A = (a_{ij})$ can be thought of as a function on $I \times J$, where $i \in I = \{1, 2, \ldots, m\}$ and $j \in J = \{1, 2, \ldots, n\}$, a straightforward extension is to prove the existence of value when I or J is not finite. If exactly one of the two sets I or J is assumed infinite, the games are called *semi-infinite games* [Tijs (1974)]. Among them the so-called S-games of Blackwell and Girshick (1954) are relevant in statistical decision theory when one assumes the set of states to be finite of nature.

Let Y be an arbitrary set of pure strategies for player II. Let $I = \{1, 2, \ldots, m\}$ be the set of pure strategies for player I. The bounded kernel $K(i, y)$ is the payoff to player I when "i" is player I's choice and $y \in Y$ is the choice of player II. Let $S = \{(s_1, s_2, \ldots, s_m) : s_i = K(i, y), i = 1, 2, \ldots, m; y \in Y\}$. The game can also be played as follows. Player II selects an $s \in S$. Simultaneously player I selects a coordinate i. The outcome is the payoff s_i to player I by player II.

Theorem. *If S is any bounded set then the S-game has a value and player I has an optimal mixed strategy. If con S (convex hull of S) is closed, player II has an optimal mixed strategy which is a mixture of at most m pure strategies. If S is closed convex, then player II has an optimal pure strategy.*

Proof. Let $t^* \in T = \text{con } \bar{S}$ be such that $\min_{t \in T} \max_i t_i = \max_i t_i^* = v$. Let $(\xi, x) = c$ be a separating hyperplane between T and the open box $G = \{x : \max_i x_i < v\}$. For any $\varepsilon > 0$ and for any $i, t_i^* - \varepsilon < v$. Thus $\xi_i \geq 0$ and we can as well assume that ξ is a mixed strategy for player I. By the Caratheodory theorem [Parthasarathy and Raghavan (1971)] the boundary point t^* is a convex combination of at most m points of $\bar{S}$. It is easy to check that $c = v$ and ξ is optimal for player I. When S is closed the convex combination used in representing t^* is optimal for player II; otherwise t^* is approximated by t in S which is in the ε neighborhood of t^*. □

The sharper assertions are possible because the set S is a subset of $\mathbf{R}^m$. Many intersection theorems are direct consequences of this theorem [Raghavan (1973)]. We will prove Berge's intersection theorem and Helly's theorem which are needed in the sequel. We will also state a geometric theorem on spheres that follows from the above theorem.

Berge's intersection theorem. *Let $S_1, S_2, \ldots, S_k$ be compact convex sets in $\mathbf{R}^m$. Let $\bigcap_{i \neq j} S_i \neq \phi$ for $j = 1, 2, \ldots, m$. If $S = \bigcup_{i=1}^k S_i$ is convex then $\bigcap_{i=1}^k S_i \neq \phi$.*

Proof. Let players I and II play the S-game where I chooses one of the indices $i = 1, 2, \ldots, k$ and II chooses an $x \in S$. Let the payoff be $f_i(x) = $ distance between x and S_i. The functions f_i are continuous convex and nonnegative. For any optimal

$\mu_1 I_{x_1} + \mu_2 I_{x_2} + \cdots + \mu_p I_{x_p}$ of player II, we have

$$v \geq \sum_j \mu_j f_i(x_j) \geq f_i\left(\sum_j \mu_j x_j\right).$$

Thus player II has an optimal pure strategy $x^\circ = \sum_j \mu_j x_j$. If an optimal strategy $\lambda = (\lambda_1, \ldots, \lambda_k)$ of player I skips an index say 1, then player II can always choose any $y \in \bigcap_{i \neq 1} S_i$ and then $0 = \sum_i \lambda_i f_i(y) \geq v$. Thus $v = 0$ and $x^\circ \in \bigcap_{i=1}^k S_i$. If $\lambda_i > 0 \forall i$, then $f_i(x^\circ) \equiv v$. Since $x^\circ \in S_i$, for some i we have $v = 0$ and $x^\circ \in \bigcap_{i=1}^k S_i$. □

Helly's theorem. *Let $S_1, S_2, \ldots, S_k$ be convex sets in $\mathbf{R}^m$. Let $\bigcap_{i \in I} S_i \neq \phi$ if $1 \leq |I| \leq m + 1$. Then $\bigcap_{i=1}^k S_i \neq \phi$.*

Proof. By induction we can assume that for any $|I| = r \geq (m+1)$, $\bigcap_{i \in I} S_i \neq \phi$ and prove it for $|I| = r + 1$. Say, $I = \{1, 2, \ldots, r+1\}$. Let $a_j \in S_i$ if $i \neq j$. Let $C = \text{con}\{a_1, \ldots, a_{(r+1)}\}$. Since $r > n$ by the Caratheodary theorem $C = \bigcup_{i \in I} C_i$ where $C_i = \text{con}\{a_1, \ldots, a_{i-1}, a_{i+1}, a_{r+1}\}$. (Here we define $a_0 = a_{r+1}$ and $a_{r+2} = a_1$). Further $C_i \subset S_i$. By Berge's theorem $\bigcap_{i \in I} C_i \neq \phi$. □

The following geometric theorem also follows from the above arguments.

Theorem. *Let $S_1, S_2, \ldots, S_m$ be compact convex sets in a Hilbert space. Let $\bigcap_{i \neq j} S_i \neq \phi$ for $j = 1, 2, \ldots, m$, but $\bigcap_{i=1}^m S_i = \phi$. Then there exists a unique $v > 0$ and a point x_0 such that the closed sphere $S(x_0, v)$ with center x_0 and radius v has nonnull intersection with each set S_i while spheres with center x_0 and with radius $< v$ are disjoint with at least one S_i. In fact no sphere of radius $< v$ around any other point in the space has nonempty intersection with all the sets S_i.*

When both pure strategy spaces are infinite, the existence of value in mixed strategies fails to hold even for very simple games. For example, if $X = Y =$ the set of positive integers and $K(x, y) = \text{sgn}(x - y)$, then no mixed strategy $p = (p_1, p_2, \ldots)$ on X can hedge against all possible y in guaranteeing an expected income other then the worst income -1. In a sense if p is revealed, player II can select a sufficiently large number y such that the chance that a number larger than y is chosen according to the mixed strategy p is negligible. Thus

$$\sup_p \inf_y K^*(p, y) = -1 \quad \text{where } K^*(p, y) = \sum_x p(x) K(x, y).$$

A similar argument with the obvious definition of $K^*(p, q)$ shows that

$$\sup_p \inf_q K^*(p, q) = -1 < \inf_q \sup_q K^*(p, q) = 1.$$

The failure stems partly from the noncompactness of the space P of probability measures on X.

Fixed point theorems for set valued maps

With the intention of simplifying the original proof of von Neumann, Kakutani (1941) extended the classical Brouwer's theorem to set valued maps and derived the minimax theorem as an easy corollary. Over the years this extension of Kakutani and its generalization [Glicksberg (1952)] to more general spaces have found many applications in the mathematical economics literature. While Brouwer's theorem is marginally easier to prove, and is at the center of differential and algebraic topology, the set valued maps that are more natural objects in many applications are somewhat alien to mainstream topologists.

Definition. For any set Y let 2^Y denote a collection of nonempty subsets of Y. Any function $\phi: X \to 2^Y$ is called a *correspondence* from X to Y. When X, Y are topological spaces, the correspondence ϕ is *upper hemicontinuous* at x iff given an open set G in Y and given $G \supset \phi(x)$, there exists a neighborhood $N \ni x$, such that the set $\phi(N) = \bigcup_{y \in N} \phi(y) \subset G$. The correspondence ϕ is *upper hemicontinuous on X* iff it is upper hemicontinuous at all $x \in X$.

Kakutani's fixed point theorem. *Let X be compact convex in $\mathbf{R}^n$. Let 2^X be the collection of nonempty compact convex subsets of X. Let $\phi: X \to 2^X$ be an upper hemi continuous correspondence from X to X. Then $x \in \phi(x)$ for some x.*

In order to prove minimax theorems in greater generality, Kakutani's theorem was further extended to arbitrary locally convex topological vector spaces. These are real vector spaces with a Hausdorff topology admitting convex bases, where vector operations of addition and scalar multiplication are continuous. The following theorem generalizes Kakutani's theorem to locally convex topological vector spaces [Fan (1952), Glicksberg (1952)].

Fan–Glicksberg fixed point theorem. *Let X be compact convex in a real locally convex topological vector space E. Let Y be the collection of nonempty compact convex subsets of X. Let ϕ be an upper hemicontinuous correspondence from X to Y. Then $x \in \phi(x)$ for some x.*

The following minimax theorems of [Ville (1938) and Ky Fan (1952)] are easy corollaries of the above fixed point theorem.

Theorem. (Ville). *Let X, Y be compact metric spaces. Let $K(x, y)$ be continuous on $X \times Y$. Then*

$$\min_{\nu} \max_{\mu} \int\int K(x, y) \, d\mu(x) \, d\nu(y) = \max_{\mu} \min_{\nu} \int\int K(x, y) \, d\nu(x) \, d\mu(y)$$

where μ, ν may range over all probability measures on X, Y, respectively.

Ky Fan's minimax theorem. *Let X, Y be compact convex subsets of locally convex topological vector spaces. Let $K: X \times Y \to \mathbf{R}$ be continuous. For every $\bar{x} \in X$, $\bar{y} \in Y$, let $K(\bar{x}, y): Y \to \mathbf{R}$ be a convex function and $K(x, \bar{y}): X \to \mathbf{R}$ be a concave function. Then*

$$\max_{x \in X} \min_{y \in Y} K(x, y) = \min_{y \in Y} \max_{x \in X} K(x, y).$$

Proof. Given any $\bar{x} \in X$, $\bar{y} \in Y$ let

$$\Lambda(\bar{x}) = \{v : v \in Y, \min_{y \in Y} K(\bar{x}, y) = K(\bar{x}, v)\}$$

and

$$\Gamma(\bar{y}) = \{u : u \in X, \min_{x \in X} K(x, \bar{y}) = K(u, \bar{y})\}.$$

The sets $\Lambda(\bar{x})$ and $\Gamma(\bar{y})$ are compact convex and the function $\phi: (x, y) \to \Gamma(\bar{y}) \times \Lambda(\bar{x})$ is an upper hemicontinuous map from $X \times Y$ to all nonempty compact convex subsets of $X \times Y$. Applying Fan–Glicksberg fixed point theorem we have an $(x^\circ, y^\circ) \in \phi(x^\circ, y^\circ)$. This shows that (x°, y°) is a saddle point for $K(x, y)$. □

In Ky Fan's minimax theorem, the condition that the function K is jointly continuous on $X \times Y$ is somewhat stringent. For example let $X = Y =$ the unit sphere S of the Hilbert space l_2 endowed with the weak topology. Let $K(x, y)$ be simply the inner product $\langle x, y \rangle$. Then the point to set maps $y \to \Lambda(y)$, $x \to \Gamma(x)$ are not upper hemicontinuous. [Nikaido (1954)]. Hence Ky Fan's theorem is inapplicable to this case. Yet $(0, 0)$ is a saddle point!

However in Ville's theorem, the function $K^*(\mu, \nu) = \iint K(x, y) \, d\mu(x) \, d\nu(y)$ is *bilinear* on $P \times Q$ where $P(Q)$ are the set of probability measures on $X(Y)$. Further K^* is *jointly continuous* on $P \times Q$ where P, Q are compact convex metrizable spaces viewed as subsets of $C^*(X)(C^*(Y))$ in their weak topologies. Ville's theorem now follows from Fan's theorem.

Definition. Let X be a topological space. A function $f: X \to R$ is called upper semicontinuous iff $\{x : f(x) < c\}$ is open for each c. If g is upper semicontinuous, $-g$ is called lower semicontinuous.

In terms of mixed extensions of K on $X \times Y$, the following is a strengthening of Ville's theorem.

Theorem. *Let X, Y be compact Hausdorff. Let $K: X \times Y \to \mathbf{R}$ be such that $K(x, \cdot)$ and $K(\cdot, y)$ are upper semicontinuous and bounded above for each $x \in X$, $y \in Y$. Then*

$$\inf_{\nu \in Q} \sup_{\mu \in P} K^*(\mu, \nu) = \sup_{\mu \in P} \inf_{\nu \in Q} K^*(\mu, \nu),$$

where $P(Q)$ are the set of probability measures with finite support on $X(Y)$.

Proof. Let X^*, Y^* denote regular Borel probability measures on X and Y respectively. When X or Y is finite the assertion follows from Blackwell's assertion for S-games. Let $\tilde{Y}$ be any finite subset of Y. For $\varepsilon > 0$, the set of ε-optimal strategies μ of the maximizer in the mixed extension K^* on $X^* \times \tilde{Y}^*$ is weakly compact. This decreasing net of ε-optimals have a nonempty intersection with

$$\inf_{v \in Q} \max_{\mu \in X^*} K^*(\mu, v) = \max_{\mu \in X^*} \inf_{v \in Q} K^*(\mu, v). \tag{7}$$

For Y metrizable, K^* is well-defined on $X^* \times Y^*$. However by (7)

$$\inf_{v \in Q} \max_{\mu \in X^*} K^*(\mu, v) \geqslant \inf_{v \in Q} \sup_{\mu \in P} K^*(\mu, v)$$

$$\max_{\mu \in X^*} \inf_{v \in Q} K^*(\mu, v) \leqslant \sup_{\mu \in P} \inf_{v \in Q} K^*(\mu, v).$$

Thus the assertion follows when Y is metrizable. The general case can be handled as follows. Associate a family G of continuous functions ϕ on Y with $K(x, y) \geqslant \phi(y)$ for some x. We can assume G to be countable with

$$\inf_{v \in Q} \sup_{\phi \in G} \int \phi(y) \, dv(y) = \inf_{v \in Q} \int \sup_{\mu \in P} K(\mu, v).$$

Essentially G can be used to view Y as a metrizable case. □

Other extensions to non-Hausdorff spaces are also possible [Mertens (1986)].

One can effectively use the intersection theorems on convex sets to prove more general minimax theorems.

General minimax theorems are concerned with the following problem: Given two arbitrary sets X, Y and a real function $K: X \times Y \to \mathbf{R}$, under what conditions on K, X, Y can one assert

$$\sup_{x \in X} \inf_{y \in Y} K(x, y) = \inf_{y \in Y} \sup_{x \in X} K(x, y).$$

A standard technique in proving general minimax theorems is to approximate the problem by the minimax theorem for matrix games. Such a reduction is often possible with some form of compactness of the space X or Y and a suitable continuity and convexity or quasi-convexity on the function K.

Definition. Let X be a convex subset of a topological vector space. Let $f: X \to \mathbf{R}$. For convex functions, $\{x: f(x) < c\}$ is convex for each c. Generalizing convex functions, a function $f: X \to \mathbf{R}$ is called *quasi-convex* if for each real c, $\{x: f(x) < c\}$ is convex. A function g is *quasi-concave* if $-g$ is quasi-convex.

Theorem. [Sion (1958)]. *Let X, Y be convex subsets of linear topological spaces with X compact. Let $K: X \times Y \to \mathbf{R}$ be upper semicontinuous in x (for each fixed y) and lower semicontinuous in y (for each x). Let $K(x, y)$ be quasi-concave in x and* *quasi-convex in y*. Then

$$\sup_{x \in X} \inf_{y \in Y} K(x, y) = \inf_{x \in Y} \sup_{x \in X} K(x, y).$$

Proof. Case (i). Both X, Y are compact and convex: Let if possible $\sup_x \inf_y K(x,y) < c < \inf_y \sup_x K(x,y)$. Let $A_x = \{y : K(x,y) > c\}$ and $B_y = \{x : K(x,y) < c\}$. Therefore we have finite subsets $A \subset X$, $B \subset Y$ such that for each $y \in Y$ and hence for each $y \in \text{Con } B$, there is an $x \in A$ with $K(x,y) > c$ and for each $x \in X$ and hence for each $x \in \text{Con } A$, there is a $y \in B$, with $K(x,y) < c$. Without loss of generality let A, B be with *minimal cardinality* satisfying the above conditions. We claim that there exists an $x_0 \in \text{Con } A$ such that $K(x_0, y) < c$ for all $y \in B$ and hence for all $y \in \text{Con } B$ [by quasi-convexity of $K(x_0, y)$]. Suppose not. Then if $B = \{y_0, y_1, \ldots, y_n\}$, for any $x \in X$ there exists a $y_j \in B$ such that $K(x, y_j) \geq c$. Let $C_i = \{x : K(x, y_i) \geq c\}$. The minimality of B means that $\bigcap_{i=0, i \neq j}^{n} C_i \neq \phi$. Further $\bigcap_{i=0}^{n} C_i = \phi$. By Helly's theorem the dimension of Con B is n. Thus it is an n-simplex. If G_i is the complement of C_i, then the G_i are open and since every open set G_i is an F_σ set the G_i's contain closed sets H_i that satisfy the conditions of Kuratowsky–Knaster–Mazurkievicz theorem [Parthasarathy and Raghavan (1971)]. Thus we have $\bigcap_{i=0}^{n} H_i \neq \phi$, and that $\bigcap_{i=0}^{n} G_i \neq \phi$. That is, for some $x_0 \in \text{Con } A$, $K(x_0, y) < c$ for all y. Similarly there is a $y_0 \in \text{Con } B$, such that $K(x, y_0) > c$ for all $x \in \text{Con } A$. Hence $c < K(x_0, y_0) < c$, a contradiction.

Case (ii). Let X be compact convex: Let $\sup_x \inf_y K < c < \inf_y \sup_x K$. There exists $B \subset Y$, B finite, such that for any $x \in X$, there is a $y \in B$ with $K(x,y) < c$. The contradiction can be established for K on $X \times \text{Con } B \subset X \times Y$. □

Often, using the payoff, a topology can be defined on the pure strategy sets, whose properties guarantee a saddle point. Wald first initiated this approach [Wald (1950)].

Given arbitrary sets X, Y and given a bounded payoff K on $X \times Y$, we can topologize the spaces X, Y with topologies $\mathcal{T}_X, \mathcal{T}_Y$ where a base for $\mathcal{T}_X$ consists of sets of the type

$$S(x_0, \varepsilon) = [x : K(x,y) - K(x_0, y) < \varepsilon \text{ for all } y\}, \quad \varepsilon > 0, \ x_0 \in X.$$

Definition. The space X is *conditionally compact* in the topology $\mathcal{T}_X$ iff for any given $\varepsilon > 0$, there exists a finite set $\{x_1, x_2, \ldots, x_{n(\varepsilon)}\}$ such that $\bigcup_{i=1}^{n(\varepsilon)} S(x_i, \varepsilon) = X$.

The following is a sample theorem in the spirit of Wald.

Theorem. *Let $K : X \times Y \to \mathbf{R}$. Let X be conditionally compact in the topology $\mathcal{T}_X$. For any $\delta > 0$ and finite sets $A \subset X$, $B \subset Y$ let there exist $\bar{x} \in X, \bar{y} \in Y$ such that $K(x, \bar{y}) \leq K(\bar{x}, y) + \delta$ for all $x \in A$, $y \in B$. Then $\sup_X \inf_Y K(x,y) = \inf_Y \sup_X K(x,y)$.*

Ch. 20: Zero-Sum Two-Person Games

Proof. Since X is conditionally compact in the topology $\mathscr{T}$, for $\varepsilon > 0$

$$\inf_Y \sup_X K \leqslant \inf_Y \max_A K + \varepsilon$$

for a finite set $A \subset X$. If B is any finite subset of Y, then by assumption $\inf_Y \max_A K \leqslant \sup_X \min_B K$. Thus

$$\inf_Y \sup_X K \leqslant \inf_{\mathscr{B}} \sup_X \min_B K + \varepsilon,$$

where $\mathscr{B}$ is the collection of finite subsets of Y.

Since X is $\mathscr{T}_X$ conditionally compact, the right side of the above inequality is finite or $-\infty$. We are through if $v = \inf_{\mathscr{B}} \sup_X \min_B K \leqslant \sup_X \inf_Y K + 2\varepsilon$. For otherwise if $\sup_X \inf_Y K < v - 2\varepsilon$, we have

$$\inf_Y K(x, y) < v - 2\varepsilon, \quad \text{for all } x$$

and thus for any finite set A

$$\inf_Y k(x_i, y) < v - 2\varepsilon, \quad \text{for all } x_i \in A.$$

That is for each x_i, and $\varepsilon > 0$, there exists a y_i such that $K(x_i, y_i) \leqslant K(x_i, y_i) + \varepsilon \leqslant v - \varepsilon$ for each i. Thus

$$\sup_X \min_B K(x, y_i) \leqslant v - \varepsilon, \quad \text{where } B = \{y_i, i \in I \text{ with } |I| \text{ finite}\}.$$

That is $v \leqslant v - \varepsilon$, a contradiction. □

Continuous payoffs. While the above general minimax theorems guarantee ε-optimals with finite steps, in mixed extensions, certain subclasses of games admit optimal mixed strategies *with finite steps*. These subclasses depend on the nature of the cones generated by $K(x, \cdot)$ and $K(\cdot, y)$. Let X, Y be compact metric. Let $K: X \times Y \to \mathbf{R}$ be continuous. Since K is bounded, we can assume $K > 0$. Let $C(X)$, $C(Y)$ be the Branch space of continuous functions on X, Y, respectively. The functions $h_\alpha(x) = K(x, \alpha)$ generate a cone whose closure we denote by C. Let $E = C - C$. Let P be the cone of nonnegative functions in $C(X)$. Any *positive linear operator* $A: E \to C(X)$ maps C into P.

Theorem. *Let C have non-null interior in $\bar{E}$ or let A be an isometry and the image cone $A(C)$ have non-null relative interior in its closed linear span. Then player* II *(the minimizer) has an optimal strategy with finite spectrum. If C has nonnull interior, both players have finite step optimals. Further, in this case $K(x, y)$ is separable. That is $K(x, y) = \sum_i \sum_j a_{ij} r_i(x) s_j(y)$.*

Proof. For any Borel probability measure v on Y the map $\tau: v \to K^*(v)$ is a continuous map of Y^* into $C(X)$. Further, $B = \tau(Y^*)$ is compact in $C(X)$. In fact when A is an isometry, $S = \text{con}(A(B), 0)$ is compact and the range cone $A(C) = \bigcup_{n=1}^{\infty} nS$. By the Baire Category Theorem the compact set S has nonnull interior and hence *finite dimensional*. The isometry A preserves extreme points. The function $K(\cdot, \alpha)$ is mapped into $M(\cdot, \alpha)$ under the isometry and any finite step optimal v_0 for $M(x, y)$ is optimal for K. Separability of K follows from the finite dimensional nature of C when C has interior. $\square$

Extreme optimals. Extreme optimal strategies are hard to characterize for infinite games except for some special cases. Let X, Y be compact metric with a continuous payoff $K(x, y)$. Then the following theorem extends Shapley–Snow theorem to certain infinite games.

Theorem. *Let $v \neq 0$ and let v_0 be optimal for player II with spectrum $\sigma(v_0) = Y$. Then an optimal strategy μ with spectrum $\sigma(\mu) = X_0$ is extreme iff $\{h_\alpha(x): K(x, \alpha), \alpha \in Y\}$ generates a dense linear manifold in $L_1(X_0, \mathscr{B}, \mu)$, where $\mathscr{B}$ is the class of Borel sets on X_0.*

The proof mimics Shapley–Snow arguments for finite games with the helpful hint that $L^\infty = L_1^*$.

Needless to say, the structure of optimal strategies can be quite complicated even for C^∞ payoffs K on the unit square. An example will settle what we want to convey [Glicksberg and Gross (1953)].

Example. Let μ_n, v_n be the nth moments of any arbitrary probability measures $\mu \neq v$ with infinite spectrum. Then the C^∞ kernel $K(x, y) = \sum_{n=0}^{\infty} (1/2^n)(x^n - \mu_n)(y^n - v_n)$ has μ, v as the unique optimal strategies.

Proof. The kernel K is analytic and for any optimal λ for player I we have $\sum_{n=0}^{\infty} (1/2^n)(\lambda_n - \mu_n)(y^n - v_n) \equiv 0$. Thus $\lambda_n = \mu_n$ for all $n \Rightarrow \lambda = \mu$. Similarly v is the unique optimal for the minimizer. $\square$

More generally, given compact convex sets S, T of probability measures on the unit interval, one would like to know when they would be the precise set of optimal mixed strategies for a continuous game on the unit square. The following theorem is a partial answer to this question [Chin, Parthasarathy and Raghavan (1976)].

Theorem. *Let S, T be compact convex sets of probability measures on the unit interval with only finitely many extreme points given by $\{\mu^1, \mu^2, \ldots, \mu^p\}$ and $\{v^1, v^2, \ldots v^q\}$ respectively. Let the spectrum of at least one $\mu \in S$, $v \in T$ be the entire unit interval. Further for any $\varepsilon > 0$, let*

$$\max_{1 \leq j \leq q} \sup_{i \neq j} \frac{\mu^i(E_j)}{\mu^j(E_j)} < \varepsilon, \quad \max_{1 \leq j \leq q} \sup_{i \neq j} \frac{\mu^i(H_j)}{\mu^j(H_j)} < \varepsilon.$$

for some $\mu^r(E_r) > 0$, $v^s(H_s) > 0$, $r = 1, \ldots, p$, $s = 1, \ldots, q$. *Then there exists a continuous payoff $K(x, y)$ on the unit square with S and T as the precise set of optimal strategies for the two players.*

Proof. We will indicate the proof for the case $q = 1$. The general case is similar. One can find a set of indices $i_1, i_2, \ldots, i_{k-1}$, such that the matrix

$$\begin{bmatrix} 1 & \mu_{i_1}^1 & \mu_{i_2}^1 & \cdots & \mu_{i_{k-1}}^1 \\ 1 & \mu_{i_1}^2 & \mu_{i_2}^2 & \cdots & \mu_{i_{k-1}}^2 \\ 1 & \cdots & \cdots & \cdots & \cdots \\ 1 & \mu_{i_1}^k & \mu_{i_2}^k & \cdots & \mu_{i_{k-1}}^k \end{bmatrix}$$

is nonsingular where $\mu_n^j = \int_i^1 x^n d\mu^j$. Using this it can be shown that

$$K(x, y) = \sum_n \frac{[a_n + b_n x^{i_1} + \cdots h_n x^{i_{k-1}} - x^n][y^n - v_n^1]}{2^n \cdot M_n}$$

is a payoff precisely with optimal strategy sets S, T for some suitable constants M_n. □

Many infinite games of practical interest are often solved by intuitive guess works and ad hoc techniques that are special to the problem. For polynomial or separable games with payoff

$$K(x, y) = \sum_i \sum_j a_{ij} r_i(x) s_j(y),$$

$$K^*(\mu, v) = \sum_i \sum_j a_{ij} \left[\int r_i(x) \, d\mu(x) \right] \left[\int s_j(y) \, dv(y) \right] = \sum_i \sum_j a_{ij} u_i v_j,$$

where $u = (u_1, \ldots, u_m)$ and $v = (v_1, \ldots, v_n)$ are elements of the finite dimensional convex compact sets U, V which are the images of X^* and Y^* under the maps $\mu \to (\int r_1 \, d\mu, \int r_2 \, d\mu, \ldots, \int r_m \, d\mu)$; $v \to (\int s_1 \, dv, \ldots, \int s_n \, dv)$. Optimal μ^0, v^0 induce optimal points u^0, v^0 and the problem is reduced to looking for optimal u^0, v^0. The optimal u^0, v^0 are convex combinations of at most $\min(m, n)$ extreme points of U and V. Thus finite step optimals exist and can be further refined by knowing the dimensions of the sets U and V [Karlin (1959)]. Besides separable payoffs, certain other subclasses of continuous payoffs on the unit square admit finite step optimals. Notable among them are the convex and generalized convex payoffs and Polya-type payoffs.

Convex payoffs. Let X, Y be compact subsets of $\mathbf{R}^m$ and $\mathbf{R}^n$ respectively. Further, let Y be convex. A continuous payoff $K(x, y)$ is called *convex* if $K(\alpha, \cdot)$ is a convex function of y for each $\alpha \in x$. The following theorem of [Bohnenblust, Karlin and Shapley (1950)] is central to the study of such games.

Theorem. *Let $\{\phi_\alpha\}$ be a family of continuous convex functions on Y. If $\sup_\alpha \phi_\alpha(y) > 0$ for all $y \in Y$, then for some probability vector $\lambda = (\lambda_1, \lambda_2, \ldots \lambda_{(n+1)})$, $\sum_{i=1}^{n+1} \lambda_i \phi_{\alpha_i}(y) > 0$ for all y, for some $n+1$ indices $\alpha_1, \alpha_2, \ldots, \alpha_{n+1}$.*

Proof. The sets $K_\alpha = \{y : \phi_\alpha(y) \leq 0\}$ are compact convex and by assumption $\bigcap_\alpha K_\alpha = \phi$. By Helly's intersection theorem

$$\bigcap_{i=1}^{n+1} K_{\alpha_i} = \phi. \tag{10}$$

That is $\phi_{\alpha_i}(y) > 0$ for some $1 \leq i \leq n+1$ for each y. For any mixed strategy $p = (p_1, \ldots, p_{n+1})$ on $\{\alpha_i, i = 1, \ldots, n+1\}$ the kernel

$$\bar{K}(p, y) = \sum_{i=1}^{n+1} p_i \phi_{\alpha_i}(y) \tag{11}$$

admits a saddle point with value $\bar{v}$ by Fan's minimax theorem. Let (λ, y°) be such a saddle point. By (10) the value $\bar{v} > 0$. Thus

$$\sum_{i=1}^{n+1} \lambda_i \phi_{\alpha_i}(y) > 0, \quad \text{for all } y \tag{12}$$

□

As a consequence we have the following

Theorem. *For a continuous convex payoff $K(x, y)$ on $X \times Y$ as defined above, the minimizer has a pure optimal strategy. The maximizer has an optimal strategy using at most $n + 1$ points in X.*

Proof. For any probability measure μ on X let $K^*(\mu, y) = \int_X K(x, y) d\mu(x)$. By Ky Fan's minimax theorem K^* has a saddle point (μ°, y°) with value v. Given $\varepsilon > 0$, $\max_\alpha K(\alpha, y) - v + \varepsilon > 0$. From the above theorem $\sum_{i=1}^{n+1} \lambda_i^\varepsilon K(\alpha_i^\varepsilon, y) - v + \varepsilon > 0$ for all y. Since X is compact, by an elementary limiting argument an optimal λ with at most $n + 1$ steps guarantee the value. Here y° is an optimal pure strategy for the minimizer. □

Weaker forms of convexity of payoffs still guarantee finite step optimals for games on the unit square $0 \leq x, y \leq 1$. The following theorem of Glicksberg (1953) is a sample (Karlin proved this theorem first for some special cases).

Theorem. *For a continuous payoff K on the unit square let $(\partial^l / \partial y^l) K(x, y) \geq 0$ for all x, for some power l. Then player II has an optimal strategy with at most $\frac{1}{2} l$ steps; $(0, 1$ are counted as $\frac{1}{2}$ steps$)$. Player I has an optimal strategy with at most l steps.*

Proof. We can assume $v = 0$ and $(\partial^l / \partial y^l) K(x, y) > 0$. If I_y is the degenerate measure at y, we have $K^*(\mu, I_y) \geq 0$ for all optimal μ of the maximizer. By assumption $K^*(\mu, y) = 0$ has at most l roots counting multiplicities. Since interior roots have *even* multiplicities the spectrum of optimal strategies of the minimizer lie in this set. Hence the assertion for player II. We can construct a polynomial $p(y) \geq 0$ of degree $\leq l-1$ with $K^*(\mu, y) - p(y)$ having exactly l roots. Let $y_1, y_2, \ldots, y_l$ be those roots. We can find for each $\bar{x}, K^*(\mu, y) - p(y) = K(\bar{x}, y) - p(\bar{x}, y)$, with the same roots. Next we can show $K(x, y) - p(x, y)$ has the same sign for all x, y. Indeed $p(x, y)$ is a polynomial game with value 0 whose optimal strategy for the maximizer serves as an optimal strategy for the original game $K(x, y)$. □

Bell-shaped kernels. Let $K : \mathbf{R}^2 \to \mathbf{R}$. The kernel K is called *regular bell shaped* if $K(x, y) = \varphi(x - y)$ where φ satisfies (i) φ is continuous on R; (ii) for $x_1 < x_2 < \cdots < x_n$; $y_1 < y_2 < \cdots < y_n$, $\det \| \varphi(x_i - y_j) \|$ is nonnegative; (iii) For each $x_1 < \cdots < x_n$ we can find $y_1 < y_2 < \cdots < y_n$ such that $\det \| \varphi(x_i - y_j) \| > 0$; (iv) $\int_R \varphi(u) \, du < \infty$.

Theorem. *Let K be bell shaped on the unit square and let φ be analytic. Then the value v is positive and both players have optimal strategies with finitely many steps.*

Proof. Let v be optimal for the minimizer. If the spectrum $\sigma(v)$ is an infinite set then

$$\int_0^1 K(x-y) \, dv(y) \equiv v$$

by the analyticity of the left-hand side. But for any φ satisfying (i), (ii), (iii) and (iv) $\int_0^1 K(x-y) \, dv(y) \to 0$ as $x \to \infty$. This contradicts $v > 0$. □

Further refinements are possible giving bounds for the number of steps in such finite optimal strategies [Karlin (1959)].

References

Aumann, R.J. and M. Maschler. (1972) 'Some thoughts on the minimax principle' *Management Science*, **18**: 54–63.
Blackwell, D and G.A. Girshick. (1954) *Theory of Games and Statistical Decisions*. New York: John Wiley.
Bohnenblust, H., S. Karlin and L.S. Shapley. (1950) 'Solutions of discrete two person games'. In *Contributions to the theory of games*. Vol. I, eds. Kuhn, H.W. and A.W.Tucker. Annals of Mathematics Studies, **24**. Princeton, NJ: Princeton University Press, pp. 51–72.
Bohnenblust, H., S. Karlin and L.S. Shapley. (1950) 'Games with continuous convex payoff'. In *Contributions to the thoery of games*. Vol I, eds. Kuhn, H.W. and A.W. Tucker. Annals of Mathematics Studies, **24**. Princeton, NJ: Princeton University Press, pp. 181–192.
Brown, George W. (1951) 'Interative solutions of games by fictitious play'. In *Activity Analysis of Production and Allocation*, Cowles commission Monograph 13. Ed. T.C. Koopmans. New York: John Wiley. London: Chapman Hall, pp. 374–376.
Brown, George W. and J. von Neumann (1950) 'Solutions of games by differential equations'. In *Contributions to the theory of games*. Vol. I, eds. Kuhn, H.W. and A.W. Tucker. Annals of Mathematics Studies, **24**. Princeton, NJ: Princeton University Press, pp. 73–79.

Chin, H., T. Parthasarathy and T.E.S. Raghavan. (1976) 'Optimal strategy sets for continuous two person games', *Sankhya*, Ser. A **38**: 92–98.
Dantzig, George B. (1951) 'A proof of the equivalence of the programming problem and the game problem'. In *Activity analysis of production and allocation*, Cowles Commission Monograph 13. Ed. T.C. Koopmans. New York: John Wiley. London: Chapman and Hall, pp. 333–335.
Dresher, M. (1962) *Games of strategy*. Englewood Cliffs, NJ: Prentice Hall.
Fan, Ky. (1952) 'Fixed point and minimax theorems in locally convex topological linear spaces', *Proceedings of the National Academy of Sciences, U.S.A.*, **38**: 121–126.
Flourier, J.B. (1890) 'Second Extrait'. In: *Oeuvres*. Ed., G. Darboux, Paris: Gauthiers Villars, pp. 325–328 [(English translation D.A. Kohler (1973)].
Gale, D. and S. Sherman. (1950) 'Solutions of finite two person games'. In *Contributions to the theory of games*. Vol. I, eds. Kuhn, H.W. and A.W. Tucker. Annals of Mathematics Studies, **24**. Princeton, NJ: Princeton University Press, pp. 37–49.
Glicksberg, I. (1952) 'A further generalization of Kakutani's fixed point theorem', *Proceedings of the American Mathematical Society*, **3**: 170–174.
Glicksberg, I., and O. Gross. (1953) 'Notes on games over the square'. In *Contributions to the theory of games*. Vol. II, eds. Kuhn, H.W. and A.W. Tucker. Annals of Mathematics Studies, **28** Princeton, NJ: Princeton University Press, pp. 173–184.
Glicksberg, I. (1953) 'A derivative test for finite solutions of games', *Proceedings of the American Mathematical Society*, **4**: 595–597.
Kakutani, S. (1941) 'A generalization of Brouwer's fixed point theorem', *Duke Mathematical Journal*, **8**: 457–459.
Kantarovich, L.V. (1939) 'Mathematical methods in the organization and production'. [English translation: *Management science* (1960), **6**: 366–422.]
Kaplansky, I. (1945) 'A contribution to von Neumann's theory of games', *Annals of Mathematics*, **46**: 474–479.
Karlin, S. (1959) *Mathematical Methods and Theory in Games, Programming and Economics*. Volumes I and II. New York: Addison Wesley.
Koopmans, T.C. (1951) 'Analysis of production as an efficient combination of activities'. In *Activity analysis of production and allocation*, Cowles Commission Monograph 13. Ed. T.C. Koopmans. New York: John Wiley. London: Chapman and Hall, pp. 33–97.
Kohler, D.A. (1973) 'Translation of a report by Fourier on his work on linear inequalities', *Opsearch*, **10**: 38–42.
Kuhn, H.W. (1953) 'Extensive games and the problem of information'. In *Contributions to the theory of games*. Vol. II, eds. Kuhn, H.W. and A.W. Tucker. Annals of Mathematics Studies, **28** Princeton, NJ: Princeton University Press, pp. 193–216.
Loomis, I.H. (1946) 'On a theorem of von Neumann', *Proceedings of the National Academy of Sciences, U.S.A*, **32**: 213–215.
Mertens, J.W. (1986) 'The minimax theorem for U.S.C.–L.S.C. payoff functions'. *International Journal of Game Theory*, **15**: 237–265.
Nash, J.F. (1950) 'Equilibrium points in n-person games', *Proceedings of the National Academy of Sciences, U.S.A.*, **36**: 48–49.
Nikaido, H. (1954) 'On von Neumann's minimax theorem', *Pacific Journal of Mathematics*, **4**: 65–72.
Owen, G. (1967) 'An elementary proof of the minimax theorem', *Management Science*, **13**: 765.
Parthasarathy, T and T.E.S. Raghavan (1971) *Some Topic in Two Person Games*. New York: American Elsevier.
Raghavan, T.E.S. (1973) 'Some geometric consequences of a game theoretic results', *Journal of Mathematical Analysis and Applications*, **43**: 26–30.
Raghavan, T.E.S. (1978) 'Completely mixed games and M-matrices', *Journal of Linear Algebra and Applications*, **21**: 35–45.
Raghavan, T.E.S. (1979) 'Some remarks on matrix games and nonnegative matrices', *SIAM Journal of Applied Mathematics*, **36**(1): 83–85.
Raghavan, T.E.S. and J.A. Filar (1991) 'Algorithms for stochastic games-A survey', *Zeitschrift für Operations Research*, **35**: 437–472.
Robinson, J. (1951) 'An iterative method of solving a game', *Annals of Mathematics*, **54**: 296–301.
Shapley, L.S. and R.N. Snow (1950) 'Basic Solutions of Discrete Games'. In *Contributions to the theory of games*. Vol. I, eds. Kuhn, H.W. and A.W. Tucker. Annals of Mathematics Studies, **24**. Princeton, NJ: Princeton University Press, pp. 27–35.

Sion, M. (1958) 'On general minimax theorem', *Pacific Journal of Mathematics*, **8**: 171–176.
Ville, J. (1938) 'Note sur la théorie géneralé des jeux oú intervient l'habilité des joueurs'. In *Traité du calcul des probabilités et de ses applicatioris*. Ed. by Emile Borel. Paris: Gauthier Villars, pp. 105–113.
von Neumann, J. (1928) 'Zur Theorie der Gesellschaftspiele', *Mathematische Annalen*, **100**: 295–320.
von Neumann, J. and O. Morgenstern. (1944) *Theory of Games and Economic Behavior*. Princeton, NJ: Princeton University Press.
von Neumann, J. (1953) 'Communication on the Borel notes', *Econometrica*, **21**: 124–127.
Wald, A. (1950) *Statistical Decision Functions*. New York: John Wiley.
Weyl, H. (1950) 'Elementary proof of a minimax theorem due to von Neumann'. In *Contributions to the theory of games*. Vol. I, eds. Kuhn, H.W. and A.W. Tucker. Annals of Mathematics Studies, **24**. Peinceton, NJ: Princeton University Press, pp. 19–25.

Appendix to Chapter 20. DUELS

T. Radzik*

Wroclaw Technical Institute

and

T.E.S. Raghavan

University of Illinois at Chicago

*The authors would like to thank Evangelista Fe, Swaminathan Sankaran and the anonymous referees for the many detailed comments that improved the presentation of this article.

Handbook of Game Theory, Volume 2, Edited by R.J. Aumann and S. Hart
© *Elsevier Science B.V., 1994. All rights reserved*

In many models of two person competition, timings of the actions is crucial. Often they are modelled as games on the unit square. Invariably the payoffs have discontinuities caused by delayed actions that simultaneously increase the player's accuracy and risks from the opponent.

Example. (One shot noisy duel). With just one arrow in their hands, players I and II walk at uniform speed towards a target, D feet away. If they use their arrows at the tth feet, then with probability $P(t)$, player I can hit the target. It is $Q(t)$ for player II. These *accuracy functions* $P(t), Q(t)$ are strictly increasing with $P(0) = Q(0) = 0$, $P(D) = Q(D) = 1$. The one who hits the target before the opponent, wins \$1 from the opponent. In all other cases there is no payment. If a player uses his arrow before the opponent, the outcome of this trial is immediately known to the opponent.

We can reduce this to a payoff on the unit square as follows: Let $\xi = \dfrac{x}{D}$ and $\eta = \dfrac{y}{D}$ where the players have planned to hit the target from the xth and yth feet. Let $p(\xi) = P(x/D)$, $q(\eta) = Q(y/D)$. Then let

$$K(\xi, \eta) = \begin{cases} p(\xi) - [1 - p(\xi)], & \xi < \eta \\ p(\xi) - q(\xi), & \xi = \eta \\ -q(\eta) + [1 - q(\eta)], & \xi > \eta. \end{cases}$$

Here $K(\xi, \eta)$ is the expected income to player I when the players use the strategy ξ, η in $0 \leq \xi, \eta \leq 1$. Even though player II may plan to use the arrow at a position $\eta > \xi$, knowing that player I has failed in his attempt, player II will postpone his action until the end and definitely hit the target.

Let $p(\xi^*) + q(\xi^*) = 1$. Since p, q are strictly increasing ξ^* is unique. One can easily check that ξ^* is an optimal pure strategy for both players and $v = p(\xi^*) - q(\xi^*)$.

Example. (One shot silent duel). The duel is the same as above; however the players have just one small invisible pebble in their hands to hit the target with. The difference is that any preemptive failure by a player is unknown to the opponent until the end of the game.

The payoff that reflects this secrecy can be defined by

$$K(\xi, \eta) = \begin{cases} p(\xi) - [1 - p(\xi)]q(\eta), & \xi < \eta \\ p(\xi) - q(\xi), & \xi = \eta \\ -q(\eta) + [1 - q(\eta)]p(\xi), & \xi > \eta. \end{cases}$$

For example when $P(\xi) = \xi$, $Q(\xi) = \xi$, this game has no saddle point. In this case

Appendix to Chapter 20. Duels

the value if it exists must be zero. One can directly verify that the density

$$f(t) = \begin{cases} 0 & 0 \leq t < \frac{1}{3} \\ \frac{1}{4}t^{-3} & \frac{1}{3} \leq t \leq 1. \end{cases}$$

is optimal for both players with value zero.

Example. (One shot silent vs. noisy duel). Here player I has just an invisible pebble and player II has an arrow. The rest are the same as in the first example.
One can verify that for $P(\xi) = Q(\xi) = \xi$,

$$K(\xi, \eta) = \begin{cases} (1)\xi - (1 - \xi)\eta, & \xi < \eta \\ 0, & \xi = \eta \\ (-1)\eta + (1 - \eta)(1), & \xi > \eta. \end{cases}$$

is a suitable payoff. Let $a = \sqrt{6} - 2$. Then the game has value $v = 1 - 2a = 5 - 2\sqrt{6}$. An optimal strategy for player I is given by the density

$$f(\xi) = \begin{cases} 0, & \text{for } 0 \leq \xi < a \\ \sqrt{2a}(\xi^2 + 2\xi - 1)^{-3/2}, & \text{for } a \leq \xi \leq 1. \end{cases}$$

An optimal strategy for player II is given by

$$g(\eta) = \begin{cases} 0, & \text{for } 0 \leq \xi < a \\ 2\sqrt{2}\frac{a}{2+a}(\eta^2 + 2\eta - 1)^{-3/2} & \text{for } \sqrt{6} - 2 \leq \eta \leq 1 \end{cases}$$

with an additional mass of $a/(2 + a)$ at $\eta = 1$.

One shot duels can be further complicated by additional assumptions or initial and final accuracies. For a wide class of such one shot duels the optimal strategies have jumps and densities mixed. If a player's efforts are futile for a while, say up to position "a", his opponent gains nothing by using his resources before reaching position "a". Thus the density parts have the same initial points for their spectrum. The payoffs for many such duels with exactly one pebble or arrow for each player has the form

$$K(\xi, \eta) = \begin{cases} L(\xi, \eta), & \xi > \eta \\ \phi(\xi), & \xi = \eta \\ M(\xi, \eta), & \xi > \eta. \end{cases}$$

Here L, M are defined on $0 \leq \xi \leq \eta \leq 1$ and $0 \leq \eta \leq \xi \leq 1$ respectively. Further, L, M are strictly increasing as functions of ξ and decreasing as functions of η except that the continuous partial derivative $(\partial L/\partial \xi) = (\partial L/\partial \eta) = 0$ at $(\xi, \eta) = (1, 1)$. Further, in nontrivial case $\phi(0)$ lies between $L(0, 0)$ and $M(0, 0)$ and $\phi(1)$ lies between $L(1, 1)$ and $M(1, 1)$.

Assuming the existence of optimal strategies of the type $\mu = (\alpha I_0, f_{ab}, \beta I_1)$, $v = (rI_0, g_{cd}, \delta I_1)$ with I_t, representing the Dirac measure at t, one can verify that

$a = c, b = d = 1$. When player II has an optimal strategy with a spectrum containing an interval the equalizer property for any optimal distribution function of player I yields

$$\int_0^1 K(\xi, \eta) \, dF(\xi) \equiv v \quad \text{on the interval } [a \, 1].$$

Differentiating left-hand side with respect to η and dividing by the positive entry $L(\eta, \eta) - M(\eta, \eta)$ yields for any $a < \eta < 1$

$$f(t) - \int_a^1 T(\xi, t) f(\xi) \, d\xi = \alpha p_0(t) + \beta p_1(t),$$

where

$$T(\xi, t) = \begin{cases} \left| \dfrac{-L_\eta(\xi, t)}{L(t, t) - M(t, t)} \right|, & a \leq \xi < t \leq 1 \\ \left| \dfrac{-M_\eta(\xi, t)}{L(t, t) - M(t, t)} \right|, & a \leq t < \xi \leq 1, \end{cases}$$

$$p_0(t) = \frac{-L_\eta(0, t)}{L(t, t) - M(t, t)}, \quad p_1(t) = \frac{-M_\eta(1, t)}{L(t, t) - M(t, t)}.$$

Thus $T(\xi, \eta)$ is a positive kernel and we can identify the above integral equation with a compact operator T_a on $C[a, 1]$ satisfying

$$(I - T_a) f = \alpha p_0 + \beta p_1.$$

The elegant theory developed by Krein and Rutman (1950) on positive operators which extends the well-known Perron–Frobenius theorem on nonnegative matrices [Gantmacher (1959)] to positive operators on Banach spaces is useful for identifying optimal densities. [Shiffman (1953)]. Further, the spectral radius $\lambda(a)$ of T_a has the following remarkable properties [Karlin (1959)]. The spectral radius $\lambda(a)$ is an eigenvalue of T_a with a unique positive eigenfunction f_a (modulo scalar multiples). Further $\lambda(a)$ is a strictly decreasing continuous function with $\lambda(a) \downarrow 0$ as $a \uparrow 1$. Also $f_a \to f_{a^*}$ uniformly in $[a^* + \delta, 1]$ for any δ. Thus $(I - T_a)$ has a bounded inverse if $\lambda(a) < 1$. Further when $L(1, 1) > M(1, 1)$ and $L(\xi, \xi) = M(\xi, \xi)$ for some ξ in $[0, 1)$, one can show that $\lambda(a) = 1$ for some $a > \xi$, based on the strict positivity of f_a. The optimal strategies for player I are of the type $(\alpha I_0, f_{[a1]}, \beta I_1)$ where I_0, I_1 are Dirac measures at $0, 1$ respectively. Optimal strategies fall into 14 different cases depending on the signs of $L(1, 1) - M(1, 1), \lambda(a) - 1$ and $\mu(a) - 1$ where $\mu(a)$ is the spectral radius of a similar operator U_a corresponding to player II [Karlin (1959)].

Noisy duels with asymmetric strengths. The simplest noisy duels with asymmetric initial strength occurs when say player I has one arrow and player II has two arrows in the beginning. Even assuming $D = 1$, $p(x) = Q(x) = x$ it is not apparent as to what players should do in this case. Intuitively player I would like player II to waste an arrow. Player II would like to take a risk in sacrificing his first arrow by using it earlier than $x = \frac{1}{2}$. (This is what they will do if both of them had just 1 arrow). The main question is: How early should player II gamble with his first arrow and what precautions player I should take if II is coming closer to the target without using either arrow.

For each $\Delta > 0$, small, the following strategy is Δ-optimal for player I. Wait till $x = \frac{1}{3}$. Choose a random point in $(\frac{1}{3}, \frac{1}{3} + \Delta)$. Use the only arrow at s if the opponent has not used his arrow before s; otherwise act as though they both have one arrow each and the initial accuracy is s for both. Player II has the following optimal strategy. Use the first arrow at $x = \frac{1}{3}$. If it fails, act as though both start with initial accuracy $\frac{1}{3}$ and with 1 arrow each. The value is $-\frac{1}{3}$. The following is an argument for the case when player II also selects to use his first arrow at a position $s \in [\frac{1}{3}, \frac{1}{3} + s]$.

Let $s = \frac{1}{3} + \lambda \Delta$ where $0 \leqslant \lambda \leqslant 1$. With probability λ the randomly chosen point of player I preempts s. When he does, the expected accuracy is $\frac{1}{3} + \frac{1}{2} \lambda \Delta$ and the expected inaccuracy is $\frac{2}{3} - \frac{1}{2} \lambda \Delta$. Thus player I has expectation $(\frac{1}{3} + \frac{1}{2} \lambda \Delta(1) + (\frac{2}{3} - \frac{1}{2} \lambda \Delta(-1))$. With probability $1 - \lambda$, player II's point s will be preempting player I's choice. Here the expectation is $(\frac{1}{3} + \lambda \cdot \Delta)(-1) + (1 - \frac{1}{3} - \lambda \cdot \Delta)(0)$. The zero in the second term is due to the symmetric game reached when II loses his excess arrow. Thus the expectation for player I is

$$\lambda \left[\left(\frac{1}{3} + \frac{\lambda \Delta}{2} \right) + \left(1 - \frac{1}{3} - \frac{\lambda \Delta}{2} \right)(-1) \right] + (1 - \lambda) \left[-\frac{1}{3} - \lambda \Delta \right].$$

For Δ small, this is close to $-\frac{1}{3}$. A similar argument can be built for all other cases. Since the above strategy for player I guarantees an expected payoff $-\frac{1}{3} + o(\Delta)$, against all strategies of the opponent it is $o(\Delta)$-optimal.

More generally say, player I has m arrows and player II has n arrows in the beginning. Even assuming $D = 1$, $P(x) = Q(x) = x$ it is not apparent as to what players should do in this case. Let us denote this duel by Γ_{mn}. The structure of optimal or ε-optimal strategies for the players in Γ_{mn} with $m, n > 1$, is much more complex than in duel $\Gamma_{1,1}$. The duel Γ_{mn} is solved in [Blackwell and Girschick (1954)]. They show that $v_{mn} = (m - n)/(m + n)$, is the value of duel Γ_{mn}. The generalization of this game to the case of arbitrary nondecreasing functions P and Q with $P(0) = Q(0) = 0$ and $P(1) = Q(1) = 1$, has been solved in Fox and Kimeldorf (1969). And although the method applied there is somewhat complicated, the structure of optimal strategies is identical to the one described in Blackwell and Girschick (1954).

Noisy duels with asymmetric strengths and asymmetric rewards. Noisy duels with asymmetric strengths in all their generality ($P(0), Q(0) \geqslant 0$ and $P(1), Q(1) \leqslant 1$) have been recently in [Radzik (1991)]. This is based on a class of constrained matrix games where strategies are restricted to some subset of admissible pairs.

Let players I and II, initially with m and n arrows approach a target of unit distance with uniform speed. Their accuracies from position t are $p_i(t)$ $i = 1, 2$. The functions $p_i(t)$ are nondecreasing and continuous with $0 \leqslant p_i(0) \leqslant p_i(1) \leqslant 1$ $i = 1, 2$. The final rewards to player I after the game is over is α if I wins, β if II wins, γ if both win, and δ if neither wins. Let $G_{m,n,t}$ be the subgame played with m, n arrows starting from position t.

The following theorem [Radzik (1991)], characterizes value and ε-optimal strategies (in general there are no optimal strategies). Let $t_* = \max\{t : p_1(t)p_2(t) = 0\}$.

Theorem. *There exist a set (t_{ij}) of numbers from $[t_*, 1]$ and a unique set $\{v_{ij}\}$, $i, j = 0, 1, 2, \ldots$ of numbers such that*

$$t_{i0} = t_{0j} = 1 \quad \text{and} \quad v_{00} = \gamma, \quad i, j \geqslant 1$$

$$v_{ij} = \begin{cases} \alpha p_1(t_{ij}) + [1 - p_1(t_{ij})]v_{i-1,j}, & i \geqslant 1, j \geqslant 0 \\ \beta p_2(t_{ij}) + [1 - p_2(t_{ij})]v_{i,j-1}, & i \geqslant 0, j \geqslant 1. \end{cases}$$

Further for all $i, j \geqslant 1$

$$t_* \leqslant t_{ij} < \min(t_{i-1,j}, t_{i,j-1}) \quad \text{if } t_* < \min(t_{i-1,j}, t_{i,j-1})$$

$$t_* = t_{ij} \quad \text{if } t^* = \min(t_{i-1,j}, t_{i,t-1}).$$

In case $i, j \geqslant 1$ $t^* < \min(t_{i-1,j}, t_{i,j-1})$ and $p_2(t_*) = 0$, then

$$t_* = t_{ij} \quad \text{iff} \begin{cases} i = 1, j = 2, p_2(1) = 1 \text{ or } p_1(t^*) = 1 \\ p_1(t_{i,j-1}) = p_1(t^*). \end{cases}$$

A similar assertion holds for $p_2(t)$ if $p_1(t^*) = 0$. Further for any $\bar\varepsilon > 0$, small, players I and II have $\bar\varepsilon$-optimal strategies $\{\xi_{ij}\}, \{\eta_{ij}\}$ $i, j = 0, 1, 2, \ldots$ of the following type. For all $t_{ij} > t_*, i, j = 1, 2, \ldots$ choose $\tau_{ij} > 0$ satisfying

$$0 < t_{ij} + \tau_{ij} < \min(t_{i-1,j}, t_{i,j-1})$$

and $p_1(t_{ij} + \tau_{ij}) < p_1(t_{ij}) + \bar\varepsilon$ and $p_2(t_{ij} + \tau_{ij}) < p_2(t_{ij}) + \bar\varepsilon$. Now define

(1) *for all $i \geqslant 1, \xi_{i0}$ reserve all the arrows till $t = 1$.*
(2) *for all $i \geqslant 1, j \geqslant 1$, if $t_{ij} = t^*$. Then exhaust s arrows at time t_* where $s = \max\{r : t_{i-r,j} = t_*\}$ and then use $\xi_{i-s,j}$ in the interval $(t^* 1]$. In case $t_{ij} > t_*$ choose a point T in $(t_{ij}, t_{ij} + \tau_{ij})$ at random. In case player II has exhausted s of his arrows before T, then use ξ_{ij-s} in $(T 1]$ else use an arrow at time T and then follow $\xi_{i-1,j}$ in $(T 1]$. Thus the game $G_{i,j,o}$ has value v_{ij} and for $\bar\varepsilon > 0$ small ξ_{ij}, η_{ij} are $\bar\varepsilon$ optimal.*

Proof. The proof is an extension of the inductive proof of Blackwell and is too detailed for inclusion. The intuitive idea behind optimal strategies is clear. Namely if the value is v_{ij} when I holds i arrows while II holds j arrows, then one of the players preempts with success or failure at a suitable time t_{ij}. If t_{ij} is the time for I to preempt then his winnings are α with chance $p_1(t_{ij})$, otherwise they play a new game in $(t_{ij} 1]$ with $i-1, j$ arrows with chance $1 - p_1(t_{ij})$. □

The randomization in a small interval is effectively used by the player with fewer arrows to encourage the opponent to deplete his excess resourse at lower levels of accuracy as in our example of noisy duels with exactly 1 and 2 arrows.

Silent duels with asymmetric strengths. Such games generalize one shot silent duel discussed earlier, where the numbers of invisible pebbles (m, n) and accuracy functions $P(x)$ and $Q(x)$ are arbitrary. Let G_{mn} denote the duel of this kind, with m pebbles for player I and n pebbles for player II. The solution for the general case of $G_{mn}(P(0) = Q(0) = 0$ and $P(1) = Q(1) = 1)$ can be found in Restrepo (1957). The optimal strategy F_1 for player I has the following structure: For some constants $0 < a_1 < a_2 < \cdots < a_m < 1$ and $0 \leqslant \alpha \leqslant 1$, he should use randomly his first pebble in the interval (a_1, a_2) according to some density $f_1(x)$, the second in (a_2, a_3) according to $f_2(x)$, and so on. The last pebble should be shot randomly with probability $1 - \alpha$ according to an $f_m(x)$ in $(a_m, 1)$, and with probability α at $x = 1$. The optimal strategy F_2 for player II has analogous structure described by constants $0 < b_1 < \cdots < b_n < 1$ and $0 \leqslant \beta \leqslant 1$, and some densities $g_j(x), j = 1, 2, \ldots n$. There exist densities and constants as mentioned above such that $a_1 = b_1$ and $\alpha\beta = 0$, and the strategies F_1 and F_2 are optimal for the players in G_{mn}. Also see Cegielski (1986). Silent duels G_{mn} under $P(0), Q(0) \geqslant 0$ and $P(1), Q(1) \leqslant 1$, are yet to be solved.

Silent–noisy discrete duels. Let SN_{mn} be the duel, where player I is in possession of m invisible pebbles and player II has n arrows. In general, that class of duels is not solved. We know the solution for such games only in some special cases. (for $SN_{1,2}$ with $P(x) = Q(x) = x$) and (for $SN_{m,1}$), see Kurisu (1983) and Styszyski (1974).

Non-discrete duels. The next essential class of games very closely related to the classical duels are the games with the following structure: There are two players I and II possessing some amounts M_1 and M_2 of "resources", respectively, to be arbitrarily distributed by them in the interval $[0, 1]$. By distributing part of his resources, a player has a chance to "overcome" the opponent. Here accuracy function $P(x)$ for player I is defined as the probability that he overcomes the opponent, provided that player I uses one unit of his resources at $x, 0 \leqslant x \leqslant 1$. In the same way, accuracy function $Q(x)$ for player II is defined. Strategies for player $i, i = 1, 2$, are measures μ_i on $[0, 1]$ satisfying $\mu_i[0, 1] \leqslant M_i$. The rest of the assumptions on the model are similar to those in classical discrete duels analyzed earlier. It is shown in Radzik (1988b) that there is a very close relationship between the payoff functions in non-discrete and discrete duels. Solutions for some subclasses of non-discrete duels can be found in Karlin (1959). A rigorous formulation and solution for non-discrete duels [under $P(0) = Q(0) = 0$ and $P(1) = Q(1) = 1$] is given in Lang and Kimeldorf (1976). This duel is solved in Radzik (1988a) in all its generality, when $P(0), Q(0) \geqslant 0$ and $P(1), Q(1) \leqslant 1$. It is shown that the value exists for this game, and optimal strategies for the players are measures with possible jumps at $x = 0$ and $x = 1$, and a continuous part with the support of the form $[a, 1]$. The complete characterization of optimal strategies given in Radzik (1988a) is similar to game of timing of class II as found in Karlin (1959).

Duels of mixed type. Suppose player I is in possession of m indivisible resources while player II has an amount M of divisible resource that can be used as in a non-discrete duel. The complete characterization of solutions for such duels under $m = 1$ is solved in Radzik (1988). For the case $m > 1$, only the optimal strategy for player II is known Radzik (1989).

References

Blackwell, D. and M.A. Girshick (1954) *Theory of Games and Statistical Decisions*. New York: John Wiley.
Cegielski, A. (1986) 'Tactical problems involving uncertain actions', *Journal of Optimization Theory and Applications*, **49**(1): 81–105.
Fox, M. and G.S. Kimeldorf (1969) 'Noisy duels', *SIAM Journal of Applied Mathematics*, **17**: 353–361.
Gantmacher, F. (1959) *Applications of the Theory of Matrices*, Vol. 2. New York: Interscience.
Karlin, S. (1959) *Mathematical Methods and Theory in Games, Programming, and Economics*. Reading, Massachusetts: Vol. 2. Addison-Wesley.
Krein, M.G. and M.A. Rutmann (1950) 'Linear Operators Leaving invariant a cone in a Banach space', *American Mathematical Society Translations*, Series 2, **26**: 1–128.
Kurisu, T. (1983) 'Two noisy versus one silent duel with equal accuracy functions', *Journal of Optimization Theory and Applications*, **39**: 215–235.
Lang, J.P. and G.S. Kimeldorf (1976) 'Silent duels with nondiscrete firing', *SIAM Journal on Applied Mathematics*, **31**: 99–100.
Restrepo, R. (1957) 'Tactical problems involving several actions'. In *Contributions to the theory of games*. Vol. III, eds. Dresher, M., A.W. Tucker and P. Wolfe, Annals of Mathematics Studies, **39**: pp. 313–335.
Radzik, T. (1991) 'General noisy duels', *Mathematica Japonica*, **36**(5): 827–857.
Radzik, T. (1988a) 'Games of timing related to distribution of resources', *Journal of Optimization Theory and Applications*, **58**(3): 443–471.
Radzik, T. (1988b) 'Games of timing with resources of mixed type', *Journal of Optimization Theory and Applications*, **58**(3): 473–500.
Radzik, T. (1989) 'Silent mixed duels' *Optimization*, **20**(4): 533–556.
Shiffman, M. (1953) 'Games of timing'. In *Contributions to the theory of games*. Vol. II, eds. H.W. Kuhn, and A.W. Tucker. Annals of Mathematics Studies, **28**, pp. 97–124.
Styszyński, A. (1974) 'An n-silent-vs.-noisy duel with arbitrary accuracy function', *Zas-tosowania Matematyki*, **14**(2): 205–225.

Chapter 21

GAME THEORY AND STATISTICS

GIDEON SCHWARZ

The Hebrew University of Jerusalem

Contents

1. Introduction	770
2. Statistical inference as a game	771
3. Payoff, loss and risk	772
4. The Bayes approach	773
5. The minimax approach	774
6. Decision theory as a touch-stone	776
Appendix. A lemma and two examples	777
References	779

Handbook of Game Theory, Volume 2, Edited by R.J. Aumann and S. Hart
© *Elsevier Science B.V., 1994. All rights reserved*

1. Introduction

Game theory, in particular the theory of two-person zero-sum games, has played a multiple role in statistics. Its principal role has been to provide a unifying framework for the various branches of statistical inference. When Abraham Wald began to develop sequential analysis during the second world war, the need for such a framework became apparent. By regarding the general statistical inference problem as a game, played between "Nature" and "the Statistician", Wald (1949) carried over the various concepts of game theory to statistics. After his death, Blackwell and Girshick (1954) elaborated on his work considerably.

It was first and foremost the concept of "strategy" that became meaningful in statistics, and with it came the fruitful device of transforming a game from the "extensive form" to "normal form". In this form the different concepts of optimality in statistics, and the relations between them, become much clearer.

Statistics, regarded from the game-theoretic point of view, became known as "decision theory". While unifying statistical inference, decision theory also proved useful as a tool for weeding out procedures and approaches that have taken hold in statistics without good reason.

On a less fundamental level, game theory has contributed to statistical inference the minimax criterion. While the role of this criterion in two-person zero-sum games is central, its application in statistics is problematic. Its justification in game theory is based on the direct opposition of interests between the players, as expressed by the zero-sum assumption. To assume that there is such an opposition in a statistical game – that the Statistician's loss is Nature's gain – is hardly reasonable. Such an assumption entails the rather pessimistic view that Nature is our adversary. Attempts to justify the minimax criterion in Statistics on other grounds are somewhat problematic as well.

Together with the minimax criterion, randomized, or mixed, strategies also appeared in decision theory. The degree of importance of randomization in statistics differs according to which player is randomizing. Mixed strategies for Nature are a priori distributions. In the "Bayes approach", these are assumed to represent the Statistician's states of knowledge prior to seeing the data, rather than Nature's way of playing the game. Therefore they are often assumed to be known to the Statistician before he makes his move, unlike the situation in the typical game-theoretic set-up. Mixed strategies for the Statistician, on the other hand, are, strictly speaking, superfluous from the Bayesian point of view, while according to the minimax criterion, it may be advantageous for the Statistician to randomize, and it is certainly reasonable to grant him this option.

A lemma of decision-theoretic origin, and two of its applications to statistical inference problems have been relegated to the appendix.

2. Statistical inference as a game

The list of ingredients of a generic statistical inference problem begins with the *sample space*. In fixed-sample-size problems, this space is the n-dimensional Euclidean space. The coordinates of a vector in sample space are the *observations*. Next, there is a family of (at least two) distributions on it, called *sampling distributions*. In most problems they are joint distributions of n-independent and identically distributed random variables. The distributions are indexed by the elements of a set called the *parameter space*. On the basis of observing a point of the sample space, that was drawn from an unknown sampling distribution, one makes inferences about the sampling distribution from which the point was drawn. This inference takes on different forms according to the type of problem. The most common problems in statistics are *estimation*, where one makes a guess of the value of a *parameter*, that is, a real-valued function, say $t(\theta)$ on the parameter space, and *hypothesis testing*, where the statistician is called upon to guess whether the index θ of the sampling distribution belongs to a given subset of the parameter space, called the *null-hypothesis*.

As a game in extensive form, the problem appears as follows. The players are *Nature* and *the Statistician*. There are three moves. The first move consists of Nature choosing a point θ in the parameter space. The second move is a chance move: a point X in the sample space is drawn from the distribution whose index was chosen by Nature. The third move is the Statistician's. He is told which point X came out in the drawing, and then chooses one element from a set A of possible responses to the situation, called *actions*. In the case of estimation, the actions are the values of the possible guesses of the parameter. In hypothesis testing the actions are "yes" and "no", also called "acceptance" (of the null-hypothesis) and "rejection".

For the conversion of the game to normal form, strategies have to be defined, and this is easily done. Nature, who is given no information prior to its move, has the parameter space for its set of (pure) strategies; for the Statistician a strategy is a (measurable) function on the sample space into the set of actions. The strategies of the Statistician are also called *decision procedures*, or *decision functions*.

Sequential analysis problems appear in the game theory framework as multiple-move games. The first move is, just like in the fixed-sample-size case, the choice of a point in parameter space by Nature. The second move is again a chance move, but here the sample space from which chance draws a point may be infinite-dimensional. There follows a sequence of moves by the Statistician. prior to his nth move, he knows only the n first coordinates of the point X drawn from sample space. He then chooses either "Stop" or "Continue". The first time he chooses the first option, he proceeds to choose an action, and the game ends. A (pure) strategy for the Statistician consists here of a sequence of functions, where the nth function maps n-space (measurably) into the set $A \cup \{c\}$, that contains the actions and one further element, "continue". There is some redundancy in this definition of a strategy, since a sequence of n observations will occur only if all its initial segments have been assigned the value "continue" by the Statistician's strategy, and for other sequences the strategy need not be defined. An alternative way of defining strategies for the Statistician separates the decision when to stop from the decision which action to takes after stopping. Here a strategy is a pair, consisting of a stopping time T, and a mapping from the set of sequences of the (variable) length T into the set of actions. This way of defining strategies avoids the redundancy mentioned above. Also, it goes well with the fact that it is often possible to find the optimal "terminal decision function" independently of the optimal stopping time.

3. Payoff, loss and risk

The definition of the statistical game is not complete unless the payoff is defined. For a fixed-sample-size problem in its extensive form, what is needed is a function that determines what one player pays the other one, depending on the parameter point chosen by Nature and on the action chosen by the Statistician. Usually the Statistician is considered the payer, or in game-theoretic terminology, the *minimizer*, and the amount he pays is determined by the *loss function L*. Ideally, its values should be expressed in utility units (there are no special problems regarding the use of utility thoery in statistics).

In hypothesis testing the loss is usually set to equal 0 when the Statistician chooses the "right" action, which is "accept" if Nature has chosen a point in the null-hypothesis, and "reject" otherwise. The loss for a wrong decision is nonnegative. Often it takes on only two positive values, one on the null-hypothesis, and one off it, their relative magnitude reflecting the "relative seriousness of the mistake". In estimation problems, the loss often, but not always, is a function of the *estimation error*, that is, of the absolute difference between the parameter value chosen by Nature, and the action chosen by the Statistician. For mathematical convenience, the square of the estimation error is often used as a loss function.

Sequential analysis is useful only when there is a reason not to sample almost for ever. In decision theory such a reason is provided by adding to the loss the cost of observation. To make the addition meaningful, one must assume that the cost is expressed in the same (utility) units as the loss. In the simplest cases the cost depends only on the number of observations, and is proportional to it. In principle it could depend also on the parameter, and on the values of the observations.

As in all games that involve chance moves, the payment at the end of the statistical game is a random variable; the payoff function for the normal form of the game is, for each parameter point and each decision procedure, the expectation of this random variable. For every decision procedure d, the payoff is a (utility valued) function $R_d(\theta)$ on Θ, called the *risk function* of d.

4. The Bayes approach

In game theory there is the concept of a strategy being *best* against a given, possibly mixed, strategy of his opponent. In a statistical game, a mixed strategy of Nature is an a priori distribution F. In the *Bayes approach* to statistics it is assumed to be known to the Statistician, and in this case choosing the strategy that is best against the a priori distribution F is the only reasonable way for the Statistician to act. In a certain sense the Bayes approach makes the statistical problem lose its game-theoretic character. Since the a priori distribution is known, it can be considered a feature of the rules of the game. Nature is no longer called upon to make its move, and the game turns into a "one-person game", or a problem in probability theory. The parameter becomes a random variable with (known) distribution F, and the sampling distributions become conditional distributions given θ.

The Bayes fixed-sample-size problem can be regarded as a variational problem, since the unknown, the optimal strategy, is a function of X, and what one wants to minimize, the expected payoff, is a functional of that function. However, by using the fact that the expectation of a conditional expectation is the (unconditional) expectation, the search for the *Bayes strategy*, that is best against F, becomes an ordinary minimum problem. Indeed, the expectation of the risk under the a priori distribution is

$$E(L(\theta, a(X))) = E[E(L(\theta, a(X))|X)]$$

and the Bayes strategy is one that, for each value of X, chooses the action $a(X)$ that minimizes the conditional expectation on the right, that is called the *a posteriori* expectation of the loss.

Defined in this way, the Bayes strategy is always "pure", not randomized. Admitting randomized strategies cannot reduce the expected loss any further. If the Bayes strategy is not unique, the Statistician may randomize between different Bayes strategies, and have the expected loss attain the same minimum; a randomization that assigns positive probability to strategies that are not Bayes against the given a priori distribution will lead to a higher expected loss. So, strictly speaking, in the Bayes approach, randomization is superfluous at best. Statisticians who believe that one's uncertainty about the parameter point can always be expressed as an a priori distribution, are called "Bayesians". Apparently strict Bayesians can dispense with randomization. This conclusion becomes difficult to accept, when applied to "statistical design problems", where the sampling method is for the Statistician to choose: what about the hallowed practice of random sampling? Indeed, an eloquent spokesman for the Bayesians, Leonard J. Savage, said (1962, p. 34) that an ideal Bayesian "would not need randomization at all. He would simply choose the specific layout that promised to tell him the most". Yet, during the following discussion (p. 89), Savage added, "The arguments against randomized analysis would not apply if the data were too extensive or complex to analyze thoroughly... It seems to me that, whether one is a Bayesian or not, there is still a good deal to clarify about randomization". This seems to be still true thirty years later.

From the Bayesian's point of view, the solution to any statistics problem is the Bayes strategy. For the so-called Nonbayesians, the class of Bayes strategies is still of prime importance, since the strategies that fulfill other criteria of optimality are typically found among this class (or the wider class that includes also the extended Bayes strategies described in the following section). Wald and other declared Nonbayesians often use the Bayes approach without calling it by name. In their attempt to prove the optimality of Wald's sequential probability ratio test (SPRT), for example, Wald and Wolfowitz (1948) found it necessary to cover themselves by adding a disclaimer: "Our introduction of the a priori distribution is a purely technical device for achieving the proof which has no bearing on statistical methodology, and the reader will verify that this is so."

In fact, Wald and Wolfowitz did use the Bayes approach, and then attempted to strengthen the optimality result, and prove a somewhat stronger optimality property of the SPRT, thereby moving away from the Bayesian formulation. The stronger form of optimality can be described as follows. The performance of a sequential test of simple hypotheses is given by four numbers: the probabilities of error and the expected sample-sizes under either of the two hypotheses. A test is easily seen to be optimal in the Bayes sense, if no other test has a lower value for one of the four numbers, without increasing one of the others. The stronger result is that one cannot reduce one of them without increasing *two* of the others. There was a small lacuna in the proof, and subsequently Arrow, Blackwell and Girshick (1949) completed the proof, using the Bayes approach openly.

5. The minimax approach

In game theory there is also the concept of a *minimax* strategy. For the minimizing player, this is a strategy that attains the minimax payoff, that is, minimizes the maximum of the payoff over all strategies of the receiving player. Here, unlike in the Bayes approach, randomization may be advantageous: a lower minimax, say m, may be attained by the minimizing player if the game is *extended* to include randomized strategies. Similarly, the maximizing player may attain a higher *maximin*, say M. The fundamental theorem of game theory states that $M = m$. Equivalently, the extended game has a *saddle-point*, or a *value*, and each player has a *good* strategy. The theorem was established first for finite games, and then generalized to many classes of infinite games as well. In statistics, a good strategy for Nature is called a *least favorable* a priori distribution, and the minimax strategy of the Statistician will be Bayes against it. In problems with infinite parameter space, there may be no proper least favorable a priori distribution, that is, no probability measure against which the minimax strategy is Bayes. In such cases the minimax strategy may still be formally Bayes against an infinite a priori measure, that fails to be a probability measure only by assigning infinite measure to the parameter space, and to some of its subsets. Such a strategy is called an *extended Bayes* strategy.

Wald recommends using a strategy that is minimax in the extended game, whenever in a statistics problem no a priori distribution is known. The justification for the minimax criterion cannot just be taken over from game theory, because the situation in statistics is essentially different. Since in statistics, one isn't really playing against an adversary whose gain is one's loss, the zero-sum character of the game does not actually obtain. The fundamental theorem is not as important in statistics, as in games played against an true adversary. The game-theoretic justification of minimax as a criterion of optimality does not hold water in statistics. The role of the criterion must be reassessed.

Clearly, if the Statistician knows Nature's mixed strategy, the a priori distribution F, the Bayes strategy against it is his only rational choice. So the minimax criterion could be applicable at most when he has no knowledge of F. But between knowing F with certainty, and not knowing it at all, the are many states of partial knowledge, such as guesses with various degrees of certainty. If we accept the idea that these states can also be described by (so-called "subjective") a priori distributions, the use of the minimax criterion in the case of no prior knowledge constitutes a discontinuity in the Statistician's behavior.

The justification of the minimax principle in statistics as the formalization of caution, protecting us as much as possible against "the worst that may happen to us", also appears to have a flaw. The minimax principle considers the worst Nature may have done to us, when it played its move and chose its strategy. But the second move, played by chance, is treated differently. Does not consistency require that we also seek protection against the worst chance can do? Is chance not to be regarded as part of the adversary, same as Nature? Answering my own question, I would aver that, under the law of large numbers, chance is barred from bringing us "bad luck" in the long run. So, for a statistician who is pessimistic but not superstitious, minimax may be a consistent principle and he may be led to include randomization among his options.

Sometimes, however, randomization is superfluous also from the minimax point of view. As shown by Dvoretzky, Wald and Wolfowitz (1951), this holds, among other cases, when the loss function is convex with respect to a convex structure on the action space. In this case it is better to take an action "between" two actions than to chose one of them at random. Consider the following simple examples from everyday life. If you don't remember whether the bus station that is closest to your goal is the fifth or the seventh, and you loss is the square of the distance you walk, getting off at the sixth station is better than flipping a coin to choose between the fifth and the seventh. But if you are not sure whether the last digit of the phone number of the party you want is 5 or 7, flipping a coin is much better than dialing a 6. The story does not end here, however: If 5 is the least bit more likely than 7, you should dial a five, and not randomize; and even if they seem exactly equally likely, flipping a coin is not worse, but also not better than dialing a 5. Only if the true digit was chosen by an adversary, who can predict our strategy, but not the coin's behavior, does randomization become advantageous.

Randomization can enter a game in two different ways. In the normal form, a player may choose a mixed strategy, that is, a distribution on the set of his available pure strategies. In the extensive form of the game, randomization can also be introduced at various stages of the game by choosing a distribution on the set of available moves. These two ways of randomizing are equivalent, by Kuhn's (1953) theorem, extended to uncountably many strategies by Aumann (1964), if the player has "perfect recall", meaning that whatever he knows at any stage of the game, he still knows at all later stages (including his own past moves). Perfect recall holds trivially for a player who has only one move, as in fixed-sample-size statistical games. In sequential games it mostly holds by assumption. For statistical games, the equivalence of the two ways of randomizing has also been established by Wald and Wolfowitz (1951).

6. Decision theory as a touch-stone

If the formulation of statistics as a game is taken seriously, it can provide a test for statistical concepts and theories. Concepts that do not emerge from the decision-theoretic reformulation as meaningful, and whose use cannot be justified in the new framework, are probably not worth keeping in the statisticians' arsenal.

In decision theory it is clear that the evaluation of the performance of a decision function should be made on the basis of its risk function alone. The Bayes approach, where an average of the risk function is minimized, and the minimax approach, where its maximum (or supremum) is minimized, both follow this rule. In practice, some statisticians use criteria that violate it. A popular one in the case of estimation is *unbiasedness*, which holds when, for all points in parameter space, the expectation of the estimated value equals the value of the parameter that is being estimated. Such "optimality criteria" as "the estimator of uniformly smallest risk *among the unbiased estimators*" are lost in the game-theoretic formulation of statistics, and I say, good riddance.

For the statistician who does not want to commit himself to a particular loss function, and therefore cannot calculate the risk function, there is a way to test whether a particular decision procedure involves irrelevancies, without reference to loss or risk. For expository convenience, consider the case where the sampling distributions all have a density (with respect to some fixed underlying measure). The density, or any function resulting by multiplying the density by a function of X alone, is called a *likelihood function*. Such a function contains all the information that is needed about X in order to evaluate the risk function, for any loss function. The behavior of the statistician should, clearly, depend on X only through the likelihood function. *This likelihood principle* is helpful in demasking assorted traditional practices that have no real justification.

An illuminating example of the import of this principle is given by Savage (1962, pp. 15–20). For Bernoulli trials, with unknown probability p of success, the likelihood function $p^S(1-p)^F$ clearly depends on the observations only through the number S of successes and the number F of failures. The order in which these outcomes where observed does not enter. Indeed, most statisticians will agree, that the order is irrelevant. In sequential analysis, however, S successes and F failures may have been observed by an observer who had decided he would sample $S+F$ times, or else by one who had decided to continue sampling "sequentially", until S successes are observed. The "classical" approach to sequential analysis makes a distinction between these two occurrences, in violation of the likelihood principle.

A practice that obviously *does* follow the likelihood principle, is *maximum likelihood estimation*, where the parameter point that maximizes the likelihood function is used to estimate the "true" parameter point. Actually the maximum likelihood estimate has the stronger justification of being very close to the Bayes strategy for the estimation problem, when the a priori distribution is uniform, and the loss function is zero for very small estimation error, and 1 otherwise.

Appendix. A lemma and two examples

A strategy whose risk function is a constant, say C, is called an *equalizer*. The following lemma establishes a relation between Bayes and minimax strategies. It provides a useful tool for finding one of the latter:

An equalizer d that is Bayes (against some a priori distribution F) is a minimax strategy.

Proof. If d were not minimax, the risk function of some strategy d' would have a maximum lower than C. Then the average of $R_{d'}(\theta)$ under F would also be less than C, but the average of $R_d(\theta)$ is C. Hence d would not be Bayes against F.

First we apply the lemma to what may be the simplest estimation problem: The statistician has observed n Bernoulli trials, and seen that S trials succeeded, and $n-S$ failed. He is asked to estimate the probability p of success. The loss is the square of the estimation error. A "common sense" estimator, that also happens to be the maximum likelihood estimator, is the observed proportion $Q = S/n$ of successes. We want to find a minimax estimator. In the spirit of the lemma, one will search first among the Bayes estimators. But there is a vast variety of a priori distributions F, and hence, of different Bayes estimators. One way of narrowing down the search, is to consider the *conjugate priors*. These are the members of the family of distributions that can appear in the probem as a posteriori distributions, when F is a uniform distribution (an alternative description is, the family of distributions whose likelihood function results from that of the sampling distributions by interchanging the parameter and the observation). For Bernoulli trials, the conjugate priors are the beta distributions $B(\alpha, \beta)$. Using the approach outlined in section 4, one can easily see that Bayes estimates for squared-error loss are the a posteriori expectations of the parameter. When F ranges over all beta distributions, the general form of the Bayes estimator, for the problem on hand, turns out to be a weighted average $wQ + (1-w)B$, where B is the a priori expectation of the unknown p, and w is a weight in the open unit interval. Symmetry of the problem leads one to search among symmetric priors, for which $B = \frac{1}{2}$. The next step is to find a w in the unit interval, for which the risk function of the estimator $wQ + (1-w)^{1/2}$ is constant. For general w, it is $[(1-w)^2 - w^2/n]p(1-p) + \frac{1}{4}(1-w)^2$. The risk function will be constant when the coefficient of $p(1-p)$ vanishes. This yields the unique positive solution $w = \sqrt{n}/(\sqrt{n}+1)$. The estimator $(\sqrt{n}Q + \frac{1}{2})/(\sqrt{n}+1)$ is therefore a minimax solution to the problem. It is also the Bayes strategy against the least favorable a priori distribution, which is easily seen to be $B(\frac{1}{2}\sqrt{n}, \frac{1}{2}\sqrt{n})$.

The second application of the lemma concerns a prediction problem. The standard prediction problem is actually a problem in probability, rather than statistics, since it does not involve any unknown distribution. One knows that joint distribution of two random variables X and Y, and wants to predict Y when X is given. under squared-error loss, the solution is the conditional expectation $E(Y|X)$. However, it is common statistical practice to use the linear regression instead. This is sometimes justified by the simplicity of linear functions, and sometimes by reference to the fact that if the joing distribution were normal, the conditional expectation would be linear, and thus coincide with the linear regression.

The use of linear regression is given a different justification by Schwarz (1987). Assume that nothing is known about the joint distribution of X and Y except its moments of orders 1 and 2. The problem has now become a statistical game, where Nature chooses a joint distribution F among those of given expectations, variances and covariance. A pair (X, Y) is drawn from F. The Statistician is told the value of X, and responds by guessing the value of Y. His strategy is a prediction function f. He loses the amount $E((f(X) - Y)^2)$. Note that in this game F is a pure strategy of nature, and due to the linearity of the risk in F, Nature can dispense with randomizing. The given moments suffice for specifying the regression line L, which is therefore available as a strategy. The given moments also determine the value of $E((L(X) - Y)^2)$, which is "the unexplained part of the variance of Y". It is also the risk of L, so L is an equalizer. Clearly, L is also Bayes against the (bivariate) normal distribution with the given moments. It is therefore a minimax solution. The normal distribution is a (pure) optimal strategy for Nature.

The result is extended by Schwarz (1991) to the case when the predicting variable is a vector, whose components are functions of X. In particular, polynomial regression of order n is optimal, in the minimax sense, if what is known about the joint distribution of X and Y is the vector of expectations and the covariance matrix of the vector $(X, X^2, \ldots, X^n, Y)$.

References

Arrow, K., D. Blackwell and A. Girshick (1949) 'Bayes and minimax solutions of sequential decision problems', *Econometrica*, **17**.
Aumann, R.J. (1964) 'Mixed and behavior strategies in infinite extensive games', in: *Advances in game theory*. Princeton: Princeton University Press, pp. 627–650.
Blackwell, D. and A. Girshick (1954) *Theory of games and statistical decisions*. New York: Wiley.
Dvoretzky, A., A. Wald and J. Wolfowitz (1951) 'Elimination of randomization in certain statistical decision procedures and zero-sum two-person games', *Annals of Mathematical Statistics*, **22**: 1–21.
Kuhn, H.W. (1953) 'Extensive games and the problem of information', *Annals of Mathematical Study*, **28**: 193–216.
Savage, L.J. et al. (1962) *The foundations of statistical inference*. London: Methuen.
Schwarz, G. (1987) 'A minimax property of linear regression', *J. A. S. A. 82/397, Theory and Methods*, 220.
Schwarz, G. (1991) 'An optimality property of polynomial regression', *Statistics and Probability Letters*, **12**: 335–336.
Wald, A. (1949) 'Statistical decision functions', *Annals of Mathematical Statistics*, **20**: 165–205.
Wald, A. and Wolfowitz, J. (1948) 'Optimum character of the sequential probability ratio test', *Annals of Mathematical Statistics*, **19**: 326–339.
Wald, A. and Wolfowitz, J. (1951) 'Two methods of randomization in statistics and the theory of games', *Annals of Mathematics (2)*, **53**: 581–586.

Chapter 22

DIFFERENTIAL GAMES

AVNER FRIEDMAN*

University of Minnesota

Contents

0.	Introduction	782
1.	Basic definitions	782
2.	Existence of value and saddle point	787
3.	Solving the Hamilton–Jacobi equation	791
4.	Other payoffs	793
5.	N-person differential games	796
	References	798

*This work is partially supported by National Science Foundation Grants DMS-8420896 and DMS-8501397.

Handbook of Game Theory, Volume 2, Edited by R.J. Aumann and S. Hart
© *Elsevier Science B.V., 1994. All rights reserved*

0. Introduction

In control theory a certain evolutionary process (typically given by a time-dependent system of differential equations) depends on a control variable. A certain cost is associated with the control variable and with the corresponding evolutionary process. The goal is to choose the "best" control, that is, the control for which the cost is minimized

In differential games we are given a similar evolutionary process, but it now depends on two (or more) controls. Each "player" is in charge of one of the controls. Each player wishes to minimize its own cost (the cost functions of the different players may be related). However, at each time t, a player must decide on its control without knowing what the other (usually opposing) players intend to do. In other words, the strategy of each player at time t depends only on whatever information the player was able to gain by observing the evolving process up to time t only.

Some basic questions immediately arise:

(i) Since several players participate in the same process, how can we formulate the precise goal of each individual player. This question is easily resolved only when there are just two players, opposing each other ("two-person zero-sum game").

(ii) How to define mathematically the concepts of "strategy" and of "saddle point" of strategies. The fact that the games progress in a continuous manner (rather than step-by-step in a discrete manner) is a cause of great difficulty.

(iii) Do saddle points exist and, if so, how to characterize them and how to compute them.

These questions have been studied mostly for the case of two-person zero-sum differential games. In this survey we define the basic concepts of two-person zero-sum differential game and then state the main existence theorems of value of saddle point of strategies. We also outline a computational procedure and give some examples. N-person differential games are briefly mentioned; these are used in economic models.

1. Basic definitions

Let $0 \leq t_0 < T_0 < \infty$. We denote by $\mathbf{R}^m$ the m-dimensional Euclidean space with variable points $x = (x_1, \ldots, x_m)$. Consider a system of m-differential equations

$$\frac{dx}{dt} = f(t, x, y, z), \quad t_0 < t < T_0, \tag{1}$$

with initial condition
$$x(t_0) = x_0, \tag{2}$$
where $f = (f_1, \ldots, f_m)$ and y, z are prescribed functions of t, say $y = y(t)$, $z = z(t)$ with values in given sets Y and Z respectively; Y is a compact subset in $\mathbf{R}^p$ and Z is a compact subset of $\mathbf{R}^q$. We assume

(A) $f(t, x, y, z)$ is continuous in $(t, x, y, z) \in [0, T_0] \times \mathbf{R}^m \times Y \times Z$, $x f(t, x, y, z) \leq k(t)$ $(1 + |x|^2)$, and, for any $R > 0$,
$$|f(t, x, y, z) - f(t, \bar{x}, y, z)| \leq k_R(t)|x - \bar{x}|$$
if $0 \leq t \leq T_0$, $y \in Y$, $z \in Z$, $|x| \leq R$, $|\bar{x}| \leq R$, where
$$\int_{t_0}^{T_0} k(t)\, dt < \infty, \quad \int_{t_0}^{T_0} k_R(t)\, dt < \infty.$$

We refer to the variables y, z as the *players* y, z and call Y and Z the control sets of the players y and z, respectively. A measurable function $y(t)$ [or $z(t)$] with values in Y (or Z) is called a *control function*, or an *admissible function*. In view of (A), for any control functions $y = y(t)$, $z = z(t)$ there exists a unique solution $x(t)$ of (1), (2) (called *trajectory*): clearly
$$x(t) = x_0 + \int_{t_0}^{t} f(s, x(s), y(s), z(s))\, ds.$$

We introduce a functional P, called *payoff*,
$$P(y, z) = g(x(T)) + \int_{t_0}^{T} h(t, x(t), y(t), z(t))\, dt \tag{3}$$

for some fixed $T \in (t_0, T_0]$, where

(B) $g(x)$ and $h(t, x, y, z)$ are continuous functions in $[0, T_0] \times \mathbf{R}^m \times Y \times Z$.

The objective of y is to maximize the payoff and the objective of z is to minimize it. However, the information available to each player is just that of the past; thus their strategies would depend only on the past performance, i.e., on observing $y(s)$, $z(s)$ for $t_0 < s < t$.

The simplest way of defining a strategy is as follows. Set
$$M(t) = \{\text{all control functions for } y \text{ on } t \leq s \leq T\},$$
$$N(t) = \{\text{all control functions for } z \text{ on } t \leq s \leq T\}.$$

Definition 1.1 [Elliott and Kalton (1992)]. A *strategy* α for y (beginning at time t_0) is a mapping
$$\alpha: N(t_0) \to M(t_0)$$

such that if $z, \hat{z} \in N(t_0)$ then

$z(\tau) = \hat{z}(\tau)$ a.e. in $t_0 \leq \tau \leq s$

implies $\alpha(z)(\tau) = \alpha(\hat{z})(\tau)$ a.e. in $t_0 \leq \tau \leq s$. (4)

Similarly a mapping

$\beta: M(t_0) \to N(t_0)$

is a strategy for z (beginning at time t_0) if for any $y, \hat{y} \in M(t_0)$,

$y(\tau) = \hat{y}(\tau)$ a.e. in $t_0 \leq \tau \leq s$

implies $\beta(y)(\tau) = \beta(\hat{y})(\tau)$ a.e. in $t_0 \leq \tau \leq s$. (5)

Denote by $\Gamma(t_0)$ the set of all strategies for y and by $\Delta(t_0)$ the set of all strategies for z.

Definition 1.2 [Elliott and Kalton (1972)]. The lower value $U^- = U^-(t_0, x_0)$ is defined by

$$U^-(t_0, x_0) = \sup_{y \in M(t_0)} \inf_{\beta \in \Delta(t_0)} P(y, \beta(y)), \tag{6}$$

and the upper value $U^+ = U^+(t_0, x_0)$ is defined by

$$U^+(t_0, x_0) = \inf_{z \in N(t_0)} \sup_{\alpha \in \Gamma(t_0)} P(\alpha(z), z). \tag{7}$$

By the *differential game* associated with (1)–(3) we mean the scheme whereby each player chooses a strategy with the purpose of maximizing the payoff for y and minimizing it for z. In general, $U^+(t_0, x_0) \geq U^-(x_0, t_0)$; if the equality holds then the common value

$$U(t_0, x_0) = U^+(t_0, x_0) = U^-(t_0, x_0)$$

is called the *value* of the game, and it seems the appropriate payoff when each player does its best.

In order to actually construct the optimal strategies we shall introduce another approach to the definition of strategies and value [Friedman (1971)].

Let n be any positive integer and let $\delta = (T - t_0)/n$. We divide the interval (t_0, T) into n intervals I_j of length δ:

$I_j = \{t; t_{j-1} < t \leq t_j\}$ where $t_j = t_0 + j\delta$, $1 \leq j \leq n$.

Denote by Y_j (Z_j) the set of control functions for y (z) on I_j. Let $\Gamma^{\delta,j}$ be any mapping from $Z_1 \times Y_1 \times Z_2 \times Y_2 \times \cdots \times Z_{j-1} \times Y_{j-1} \times Z_j$ into Y_j. The vector

$$\Gamma^\delta = (\Gamma^{\delta,1}, \ldots, \Gamma^{\delta,n})$$

is called an *upper δ-strategy* for y. Similarly we define an upper δ-strategy

$\Delta^\delta = (\Delta^{\delta,1},\ldots,\Delta^{\delta,n})$ for z, where $\Delta^{\delta,j}$ is a mapping from $Y_1 \times Z_1 \times \cdots \times Y_{j-1} \times Z_{j-1} \times Y_j$ into Z_j.

Next let $\Gamma_{\delta,j}$ ($1 \leq j \leq n$) be any mapping from $Y_1 \times Z_1 \times \cdots \times Y_{j-1} \times Z_{j-1}$ into Y_j and denote by $\Gamma_{\delta,1}$ any function in Y_1. The vector

$$\Gamma_\delta = (\Gamma_{\delta,1},\ldots,\Gamma_{\delta,n})$$

is called a *lower δ-strategy* for y; similarly we define a lower δ-strategy $\Delta_\delta = (\Delta_{\delta,1},\ldots,\Delta_{\delta,n})$ for z.

Given a pair $(\Delta_\delta, \Gamma^\delta)$, we can uniquely determine control functions $y^\delta(t)$, $z_\delta(t)$ with components y_j, z_j on I_j by $z_1 = \Delta_{\delta,1}$, $y_1 = \Gamma^{\delta,1}(z_1)$ and

$$z_j = \Delta_{\delta,j}(z_1, y_1, z_2, y_2,\ldots, z_{j-1}, y_{j-1}),$$
$$y_j = \Gamma^{\delta,j}(z_1, y_1, z_2, y_2,\ldots, z_{j-1}, y_{j-1}, z_j),$$

for $2 \leq j \leq n$. We call (y^δ, z_δ) the *outcome* of $(\Delta_\delta, \Gamma^\delta)$, and set

$$P[\Delta_\delta, \Gamma^\delta] \equiv P[\Delta_{\delta,1}, \Gamma^{\delta,1},\ldots,\Delta_{\delta,n}, \Gamma^{\delta,n}] = P(y^\delta, z_\delta). \tag{8}$$

The above scheme is called *upper δ-game*; here z commits itself first in the first δ-interval, then y acts; next z commits itself in the second δ-interval, followed by y, etc.

Definition 1.3. The *number*

$$V^\delta = \inf_{\Delta_{\delta,1}} \sup_{\Gamma^{\delta,1}} \cdots \inf_{\Delta_{\delta,n}} \sup_{\Gamma^{\delta,n}} P[\Delta_{\delta,1}, \Gamma^{\delta,1},\ldots,\Delta_{\delta,n}, \Gamma^{\delta,n}] \tag{9}$$

is called the *upper δ-value*.

Similarly we define a *lower δ-game*; here y chooses a lower δ-strategy Γ_δ and z chooses an upper δ-strategy Δ^δ, with outcome (y_δ, z^δ), and the corresponding payoff is

$$P[\Gamma_\delta, \Delta^\delta] \equiv P[\Gamma_{\delta,1}, \Delta^{\delta,1},\ldots,\Gamma_{\delta,n}, \Delta^{\delta,n}] = P(y_\delta, z^\delta). \tag{10}$$

The number

$$V_\delta = \sup_{\Gamma_{\delta,1}} \inf_{\Delta^{\delta,1}} \cdots \sup_{\Gamma_{\delta,n}} \inf_{\Delta^{\delta,n}} P[\Gamma_{\delta,1}, \Delta^{\delta,1},\ldots,\Gamma_{\delta,n}, \Delta^{\delta,n}] \tag{11}$$

is called the *lower δ-value*.

One can prove Friedman (1971) that the order of "inf" and "sup" in (9) is irrelevant; in fact,

$$V^\delta = \inf_{\Delta_\delta} \sup_{\Gamma^\delta} P[\Delta_\delta, \Gamma^\delta] = \sup_{\Gamma^\delta} \inf_{\Delta_\delta} P[\Delta_\delta, \Gamma^\delta]; \tag{12}$$

further,

$$V^\delta = \inf_{z_1} \sup_{y_1} \cdots \inf_{z_n} \sup_{y_n} P(z_1, y_1,\ldots,z_n, y_n), \tag{13}$$

where

$$P(y, z) = P(z_1, y_1, \ldots, z_n, y_n),$$

if y_j (z_j) is the component of y (z) in I_j.

We also have: $V^\delta \geqslant V_\delta$.

Theorem 1.1 [Friedman (1971)]. *The limits*

$$V^+ \equiv V^+(t_0, x_0) = \lim_{\delta \to 0} V^\delta,$$

$$V^- \equiv V^-(t_0, x_0) = \lim_{\delta \to 0} V_\delta,$$

exist.

Definition 1.4. V^+ and V^- are called the *upper and lower values* of the game.

As we shall see later on, $V^+ = U^+$ and $V^- = U^-$.

In general $V^+(t_0, x_0) \geqslant V^-(t_0, x_0)$; if equality holds then we denote the common value by $V(t_0, x_0)$ and call it the *value* of the game.

Definition 1.5. A sequence $\Gamma = \{\Gamma_\delta\}$ is called a *strategy* for y and a sequence $\Delta = \{\Delta_\delta\}$ is called a *strategy* for z.

Take any pair of strategies (Γ, Δ). At the δ-level there is a unique outcome (y_δ, z_δ); the corresponding trajectory is denoted by $x_\delta(t)$. Suppose, for a subsequence,

$$y_\delta \to \bar{y}, \quad z_\delta \to \bar{z} \quad \text{weakly in } L^1, \text{ and } x_\delta \to \bar{x} \text{ uniformly,} \tag{14}$$

and $\bar{x}$ is the trajectory corresponding to $\bar{y}, \bar{z}$. This we call $(\bar{y}, \bar{z})$ an *outcome* of the strategies (Δ, Γ). The set of all numbers $P(\bar{y}, \bar{z})[(\bar{y}, \bar{z})$ an outcome of $(\Delta, \Gamma)]$ is called the *payoff set* and is denoted by $P[\Delta, \Gamma]$.

Notation. Let A, B be two nonempty sets of real numbers. If $a \leqslant b$ for any $a \in A$, $b \in B$ then we write $A \leqslant B$ or $B \geqslant A$.

Definition 1.6. A pair of strategies (Δ^*, Γ^*) is called a *saddle point* if $P[\Delta^*, \Gamma^*]$ is nonempty and

$$P[\Delta^*, \Gamma] \leqslant P[\Delta^*, \Gamma^*] = \{V\} \leqslant P[\Delta, \Gamma^*] \tag{15}$$

for any strategies Γ for y and Δ for z.

Remark 1.1. Without loss of generality one may assume that $h(t, x, y, z) \equiv 0$ (otherwise introduce a new variable x_0 with $dx_0/dt = h$). In this case, since the cost function depends only on the trajectory, we can define more general concepts of payoff sets: (i) $P_0[\Delta, \Gamma]$ as the set of all uniform limits $\bar{x} = \lim x_\delta$ for subsequences $\delta \to 0$, $\bar{x}$ a trajectory, and (ii) $P_*[\Delta, \Gamma]$ as the set of all uniform limits $\bar{x} = \lim x_\delta$ for subsequences $\delta \to 0$. The latter concept has been used extensively by Krasovskii and Subbotin (1974). We note however that the concept of payoff set $P[\Delta, \Gamma]$ which requires (14) is much more useful, since we are interested in actually computing the associated control functions $\bar{y}, \bar{z}$.

Example 1.1. Let $y^*(t, x)$ be a function defined on $[t_0, T] \times \mathbb{R}^m$ with values in Y, and assume that y^* and $\partial y^*/\partial x_j$ ($1 \leq j \leq n$) are continuous. Define a strategy $\Gamma^* = \{\Gamma_\delta^*\}$ by

$$\Gamma_\delta^* = \{\Gamma_{\delta,1}^*, \ldots, \Gamma_{\delta,n}^*\}$$
$$\Gamma_{\delta,j}^*(z_1, y_1, \ldots, z_{j-1}, y_{j-1})(t) = y^*(t, x(t_{j-1})),$$

where $x(t)$ is the trajectory corresponding to y_i, z_i ($1 \leq i \leq j-1$). Let Δ be any strategy for z. If $(\bar{y}, \bar{z}, \bar{x})$ is an outcome of (Δ, Γ^*), then clearly

$$\bar{y}(t) = y^*(t, \bar{x}(t)).$$

We call Γ^* a *continuously differentiable strategy* corresponding to $y^*(t, x)$.

Remark 1.2. The approach of δ-games does not depend, in fact, on partitioning the interval $[t_0, T]$ into subintervals of *equal* length [Friedman (1971)]; any sequence of partitions with mesh decreasing to zero yields the same numbers V^+, V^-. We also mention the paper of Elliott et al. (1974) which extends the concept of δ-games, allowing each player to get information on the other player in some part of each subinterval. This leads to intermediate values, between V^+ and V^-.

2. Existence of value and saddle point

It will be asserted below that the value functions $V^\pm(t, x)$ are Lipschitz continuous functions which satisfy a.e. a partial differential equation of the form

$$v_t + H(t, x, \nabla_x v) = 0, \quad \text{for } 0 < t < T, \ x \in \mathbb{R}^m,$$

with $v(x, T) = g(x)$, and H is a function determined by the data f, h. There are in general many solutions for the same H and g. If however we restrict the class of solutions, as will now be explained, then the value function is the unique solution in this class.

Assume that $H: [0, T] \times \mathbb{R}^m \times \mathbb{R}^m \to \mathbb{R}$ is continuous and $g: \mathbb{R}^m \to \mathbb{R}$ is bounded and continuous. A bounded, uniformly continuous function $u: [0, T] \times \mathbb{R}^m \to \mathbb{R}$ is called a *viscosity solution* of the Hamilton–Jacobi equation

$$u_t + H(t, x, Du) = 0 \quad \text{in } (0, T) \times \mathbb{R}^m,$$
$$u(T, x) = g(x) \quad \text{in } \mathbb{R}^m, \tag{16}$$

provided for each $\phi \in C^1((0, T) \times \mathbb{R}^m)$,

(a) if $u - \phi$ attains a local maximum at $(t_0, x_0) \in (0, T) \times \mathbf{R}^m$, then
$$\phi_t(t_0, x_0) + H(t_0, x_0, D\phi(t_0, x_0)) \geq 0,$$

and

(b) if $u - \phi$ attains a local minimum at $(t_0, x_0) \in (0, T) \times \mathbf{R}^m$, then
$$\phi_t(t_0, x_0) + H(t_0, x_0, D\phi(t_0, x_0)) \leq 0.$$

A continuously differentiable solution of (16) is clearly a viscosity solution.

It can be shown [Grandall et al. (1984)] that if a viscosity solution u is differentiable at a point (t_0, x_0), then
$$u_t(t_0, x_0) + H(t_0, x_0, Du(t_0, x_0)) = 0,$$

where $Du(t_0, x_0)$ is the differential of u at (t_0, x_0).

Further:

Theorem 2.1 [Grandall et al. (1984), Grandall and Lions (1983)] *If*
$$|H(t, x, p) - H(t, x, \bar{p})| \leq C|p - \bar{p}|,$$
$$|H(t, x, p) - H(\bar{t}, \bar{x}, p)| \leq C[|x - \bar{x}| + |t - \bar{t}|](1 + |p|), \tag{17}$$

for all $x, \bar{x}, p, \bar{p} \in \mathbf{R}^m$, $0 \leq t, \bar{t} \leq T$, *then there exists at most one viscosity solution of* (16).

This result has been extended to unbounded viscosity solution [Ishi (1984)]. Introduce now the *upper* and *lower Hamiltonians*
$$H^+(t, x, p) = \min_{z \in Z} \max_{y \in Y} \{f(t, x, y, z) \cdot p + h(t, x, y, z)\},$$
$$H^-(t, x, p) = \max_{y \in Y} \min_{z \in Z} \{f(t, x, y, z) \cdot p + h(t, x, y, z)\},$$

and the Isaacs' equations
$$V_t^+ + H^+(t, x, DV^+) = 0, \quad 0 \leq t \leq T, x \in \mathbf{R}^m,$$
$$V^+(T, x) = g(x), \quad x \in \mathbf{R}^m, \tag{18}$$

and
$$V_t^- + H^-(t, x, DV^-) = 0, \quad 0 \leq t \leq T, x \in \mathbf{R}^m,$$
$$V^-(T, x) = g(x), \quad x \in \mathbf{R}^m. \tag{19}$$

Theorem 2.2 (Barron et al. (1984)]. *The lower value V^- is the viscosity solution of (19) and the upper value V^+ is the viscosity solution of (18).*

The fact that V^- is a Lipschitz continuous function satisfying (19) a.e. was proved earlier by Friedman (1971). In this proof the following principle of optimality plays a fundamental role:

$$\sup_{\Gamma_\delta} \inf_{\Delta^\delta} \{V_\delta(t^0 + \varepsilon, x_\delta(t_0 + \varepsilon)) - V_\delta(t_0, x_0) + \int_{t_0}^{t_0 + \varepsilon} h(s, x_\delta(s), y_\delta(s), z^\delta(s)) \, ds\} = 0, \quad (20)$$

where $\varepsilon = k\delta$ for some positive integer k, $t_0 + \varepsilon \leqslant T$, Γ_δ and Δ^δ and the δ-strategies on the interval $t_0 \leqslant s \leqslant t_0 + \varepsilon$, $(y_\delta(\cdot), z^\delta(\cdot))$ is the outcome of $(\Gamma_\delta, \Delta^\delta)$, and $x_\delta(\cdot)$ is the corresponding trajectory.

The proof of Theorem 2.2 uses the optimality principle. The importance of Theorem 2.2 lies in the fact that viscosity solutions are uniquely determined by Theorem 2.1. [In general, there may be infinitely many Lipschitz continuous solutions of (19)]. Thus, in particular:

Corollary 2.3. *If for all $0 \leqslant t \leqslant T$, $p \in \mathbf{R}^m$,*

$$H^+(t, x, p) = H^-(t, x, p) \quad \text{(minimax condition)},$$

then $V^+ = V^-$ and thus the game has value V.

Similarly to Theorem 2.2 there holds:

Theorem 2.4 [Evans and Souganidis (1984)]. *The lower value U^- is the viscosity solution of (19) and the upper value U^+ is the viscosity solution of (18).*

Corollary 2.5 $U^+ = V^+$ *and* $U^- = V^-$.

For earlier proofs of Corollary 2.3 see [Friedman (1971, 1974)] and, for $U^+ = U^-$, Elliott and Kalton (1972); see also Fleming (1957, 1961), Roxin (1969) and Varaiya and Lin (1969).

When the minimax condition is not satisfied, one can define intermediate differential games for which in the nth partition of $[t_0, T]$ the players have "reaction times" σ/n and $(1 - \sigma)/n$ respectively ($0 \leqslant \sigma \leqslant 1$). The resulting value function V^σ is then a viscosity solution of a Hamilton–Jacobi equation

$$V_t^\sigma + H^\sigma(t, x, DV^\sigma) = 0,$$

where $H^\sigma = H^+$ if $\sigma = 1$ and $H^\sigma = H^-$ if $\sigma = 0$; for more details see Elliott et al. (1974). The method of Barron et al. (1984) can undoubtedly be used to simplify the proof by Elliott et al.

To prove the existence of a saddle point we assume:

$$f(t, x, y, z) = f^0(t, x) + F^1(t, x)y + F^2(t, x)z,$$
$$h(t, x, y, z) = h^0(t, x) + H^1(t, x)y + H^2(t, x)z, \quad (21)$$

when $F^1(t, x)$ is an $m \times p$ matrix, $F^2(t, x)$ is an $m \times q$ matrix, $H^1(t, x)$ is a $1 \times p$

matrix and $H^2(t,x)$ is a $1 \times q$ matrix. We further assume that

Y, Z are convex. (22)

Theorem 2.6 [Friedman (1971)]. *If (21), (22) hold then there exists a saddle point.*

The assumptions (21), (22) ensure that the payoff set $P[\Delta, \Gamma]$ is nonempty. If we weaken (21), requiring only that

for each $(t,x) \in [t_0, T_0] \times \mathbf{R}^m$ the set $f(t, x, Y, Z)$ is convex, (23)

and if the payoff has the form

$$P(y,z) = g(x(T)) + \int_{t_0}^T h(t, x(t)) \, dt,$$

then Theorem 2.6 is still valid provided $P[\Delta, \Gamma]$ is replaced by $P_0[\Delta, \Gamma]$ (see Ramark 1.1)

Berkovitz (1985, 1988) has introduced the concept of upper and lower values by

$$W^+ = \inf_\Delta \sup_\Gamma P_*[\Delta, \Gamma], \qquad W^- = \sup_\Gamma \inf_\Delta P_*[\Delta, \Gamma]$$

(cf. Remark 1.1) and showed (using a method of Lions and Souganidis (1985)) that these are viscosity solutions of (18), (19); thus $W^\pm = V^\pm = U^\pm$. If the minimax condition is satisfied then he shows the existence of a saddle point (Δ_*, Γ_*) in the sense that

$$P_*[\Delta_*, \Gamma] \leq P_*[\Delta_*, \Gamma_*] \leq P_*[\Delta, \Gamma^*]$$

for any Δ, Γ.

Example 2.1. (War of attrition and attack) [Friedman (1971, 1974), Isaacs (1965)]. The differential system is

$$\frac{dx_i}{dt} = a_i - \sum_{j=1}^{m_2} p_{ij} y_j \psi_j, \quad 1 \leq i \leq m_1,$$

$$\frac{dy_i}{dt} = b_i - \sum_{j=1}^{m_1} q_{ij} x_j \phi_j, \quad 1 \leq i \leq m_2,$$

with

$x_i(t_0) = \alpha_i, \quad y_i(t_0) = \beta_i$

and payoff

$$P(\psi, \phi) = \int_{t_0}^T \left(\sum_{i=1}^{m_2} r_i(1 - \psi_i) y_i - \sum_{i=1}^{m_1} s_i(1 - \phi_i) x_i \right) dt.$$

Ch. 22: Differential Games

The control sets are given by

$$0 \leq \phi_i \leq 1, \quad 0 \leq \psi_i \leq 1$$

and ψ is the maximizer; the constants $a_i, b_i, p_{ij}, q_{ij}, r_i, s_i$ are all positive. We also assume that α_i, β_i are positive and sufficiently large so that the trajectories remain in the domain $x_i > 0$ ($1 \leq i \leq m_1$), $y_i > 0$ ($1 \leq i \leq m_2$). Let $m = m_1 + m_2$.

Theorem 2.7 [Friedman (1971, 1974)]. *There exists a saddle point (Δ^*, Γ^*) such that for any Δ, Γ the outcome of (Δ^*, Γ) is $(\bar{\phi}, \psi)$ and the outcome of (Δ, Γ^*) is $(\phi, \bar{\psi})$; ψ depends on Γ, ϕ depends on Δ, and $\bar{\phi}, \bar{\psi}$ are independent of Δ, Γ and have the following form: there is a permutation J of the vector $(\bar{\psi}_1, \ldots, \bar{\psi}_{m_2}, \bar{\phi}_1, \ldots, \bar{\phi}_{m_1})$ into a vector $(\bar{\chi}_1, \ldots, \bar{\chi}_m)$ and times $\tau_m \leq \tau_{m-1} \leq \cdots \leq \tau_1 < T$ such that if $\tau_{k+1} \leq t_0 < \tau_k$ then*

$$\bar{\chi}_\lambda(t) = \begin{cases} 0, & \text{if } \tau_\lambda < t < T \\ 1, & \text{if } t_0 < t < \tau_\lambda \end{cases}$$

for $1 \leq \lambda \leq k$, and $\bar{\chi}_\lambda(t) = 0$ if $t_0 < t < T$, $k < \lambda < m$.

J can be computed by solving some transcendental equations.

The proof involves calculations of optimal δ-strategies. In the next section we shall develop another approach for computing a saddle point.

Theorem 2.7 has the following meaning: A saddle point strategy is such that each player applies each component of the control by exploiting the maximal available power ($\phi_j = 1$ or $\psi_j = 1$) for a certain time interval $0 \leq t \leq \sigma_j$; then he immediately switches off that component for the rest of the game. Such a type of behavior is quite common in the theory of optimal control.

3. Solving the Hamilton–Jacobi equation

In this section we assume that the minimax condition holds. Then the value $V(t, x)$, if continuously differentiable, can be computed in principle by solving the Hamilton–Jacobi equation. We wish also to find a continuously differentiable saddle point (if existing) which is independent of the initial conditions; we refer to it also as a *synthesis* of a saddle point. To do all this, we define the concept of *feedback control*:

Let $y^0(t, x, p)$, $z^0(t, x, p)$ be continuous functions from $[0, T] \times \mathbf{R}^m \times R^m$ into Y and Z respectively such that

$$\min_{z \in Z} \max_{y \in Y} \{f(t, x, y, z) \cdot p + h(t, x, y, z)\}$$

$$= \min_{z \in Z} \{f(t, x, y^0(t, x, p), z) \cdot p + h(t, x, y^0(t, x, p), z)\}$$

$$= \max_{y \in Y} \{f(t, x, y, z^0(t, x, p)) \cdot p + h(t, x, y, z^0(t, x, p))\}. \tag{24}$$

We call $y^0(t, x, p)$ and $z^0(t, x, p)$ *feedback controls* for y and z, respectively. The Hamilton–Jacobi equation then becomes

$$\begin{aligned}
&V_t + f(t, x, y^0(t, x, \nabla_x V), z^0(t, x, \nabla_x V)) \cdot \nabla_x V \\
&\quad + h(t, x, y^0(t, x, \nabla_x V), z^0(t, x, \nabla_x V)) = 0 \text{ in } (0, T) \times \mathbf{R}^m, \\
&V(T, x) = g(x) \text{ in } \mathbf{R}^m.
\end{aligned} \quad (25)$$

Theorem 3.1 [Friedman (1971)]. *Suppose V is a continuously differentiable solution of (25) and suppose*

$$\begin{aligned}
y^*(t, x) &= y^0(t, x, \nabla_x V(t, x)), \\
z^*(t, x) &= z^0(t, x, \nabla_x V(t, x)),
\end{aligned} \quad (26)$$

are continuous together with their first x-derivatives. Then (y^*, z^*) form a continuously differentiable saddle point.

In order to solve (25) we apply the Hamilton–Jacobi theory [Friedman (1971) ch. 4] which reduces the solution of (25) to the solution of a system of $2m$ ordinary differential equations for $x_1, \ldots, x_m, V_1, \ldots, V_m$:

$$\begin{aligned}
\frac{dx_k}{dt} &= f_k(t, x, y^0(t, x, \nabla_x V), z^0(t, x, \nabla_x V)), \quad 1 \leq k \leq m, \\
\frac{dV_k}{dt} &= -\sum_{i=1}^{m} V_i \frac{\partial f_i}{\partial x_k}(t, x, y^0(t, x, \nabla_x V), z^0(t, x, \nabla_x V)) \\
&\quad + \frac{\partial h}{\partial x_k}(t, x, y^0(t, x, \nabla_x V), z^0(t, x, \nabla_x V)), \quad 1 \leq k \leq m,
\end{aligned} \quad (27)$$

where $\nabla_x V$ stands for the vector $(V_1, \ldots, V_m)$. The terminal manifold is parametrized by $t = T, x_i = y_i$. The V_k are supposed to represent $\partial V / \partial x_i$. Hence

$$x_k = y_k, \quad V_k = \frac{\partial g(y)}{\partial y_k} \quad \text{on } t = T. \quad (28)$$

Suppose we have solved (27), (28) for $0 \leq t \leq T$; say,

$$x_i = x_i(y, t), \quad V_i = V_i(y, t). \quad (29)$$

If we then solve $y = y(x, t)$ from the first set of equations and substitute into V_i, we get $V_i = V_i(y(x, t), t) = \tilde{V}_i(x, t)$. The function

$$V(t, x) = \int_T^t \left(\sum_{i=1}^{m} \tilde{V}_i f_i(t, x, y^0(t, x, \nabla_x \tilde{V})) - H \right) dt + g(x) \quad (30)$$

is then the solution of (25).

Although the above scheme can be used to compute V and a synthesized saddle point, one encounters several difficulties:

(a) The feedback controls are often not differentiable in p.
(b) It is difficult to solve the y_j in terms of the x_j.
(c) The solution V is often not continuously differentiable; cf. Example 2.1.

These difficulties are not just technical; they are inherent and, in fact, a consequence of the phenomenon that optimal strategies $y^*(t,x)$, $z^*(t,x)$ are often piecewise continuous, exhibiting surfaces of discontinuity [called *singular surfaces*; see Isaacs (1965)].

The above procedure has been somewhat extended to include cases with singular surfaces; see Berkovitz (1967) and Friedman (1971).

Example 3.1. (The isotropic rocket game). The differential system is

$$\dot{x} = u, \quad \ddot{y} = v, \quad x(0) = x_0, \quad y(0) = y_0, \quad \dot{y}(0) = y_1,$$

where x, y, u, v are m-dimensional variables, the control sets are given by $|u| \leq \alpha$, $|v| \leq \beta$ and the payoff is

$$P(u,v) = |x(T) - y(T)|,$$

where u is the maximizer. Solving the Hamilton–Jacobi equation by the above procedure we derive the following formulas for the synthesis of a saddle point:

$$u^*(t,x,y,\dot{y}) = \alpha(x - y - (T-t)\dot{y})/|x - y - (T-t)\dot{y}|,$$
$$v^*(t,x,y,\dot{y}) = \beta(x - y - (T-t)\dot{y})/|x - y - (T-t)\dot{y}|. \tag{31}$$

Actually these formulas make no sense on the manifold $x - y - (T-t)\dot{y} = 0$. To overcome this difficulty one restricts the class of all strategies to a subclass, taking only such Γ_δ for which $P[\Delta_\delta, \Gamma_\delta] > V - \eta$ for some suitable $\eta > 0$. In that case it can be proved [Friedman (1971) Section 4.8] that (31) is valid and $x - y - (T-t)\dot{y} \neq 0$ for any outcome, provided $x_0 - y_0 - Ty_1 \neq 0$ and $\alpha t - \frac{1}{2}\beta t^2 < \eta_0$ if $0 \leq t \leq T$, $|x(T) - y(T)| > \eta_0$ for some $\eta_0 > 0$.

One may think of x as representing an attacking aircraft and of y as representing a rocket which tries to close in on the aircraft. Formula (31) for y, say, means that the rocket uses the full power (i.e. $|v^*| = 1$) at all times, but the direction of the control v^* is such that the acceleration $\ddot{y} = v^*$ is in the direction of the vector from $y(t) + (T-t)\dot{y}(t)$ to $x(t)$; this direction keeps changing with time.

4. Other payoffs

The results of Sections 1–3 can be extended to include games which do not terminate at a fixed time T, and games in which the phase set (where x varies) is not the entire space $\mathbf{R}^m$.

Let F be a closed set in the (t, x) space satisfying

$$F \supset (T_1, \infty) \times R^m \quad \text{for some } t_0 < T_1 < T_0. \tag{32}$$

We shall call F the *terminal set* and denote by $\tilde{t}$ the first time when $(t, x(t)) \in F$; $\tilde{t}$ is called the *capture time*. We replace the payoff $P(x, z)$ defined in (3) by

$$P(y, z) = g(\tilde{t}, x(\tilde{t})) + \int_{t_0}^{\tilde{t}} h(t, x(t), y(t), z(t)) \, dt \tag{33}$$

and proceed with the definitions of Section 1 without change.

We shall need the assumptions:

F is a closed domain with C^2 boundary, \hfill (34)

$$v_0 + \min_{z \in Z} \max_{y \in Y} \sum_{j=1}^{m} v_j f_j(t, x, y, z) < 0 \quad \text{for all } (t, x) \in \partial F,$$

where $(v_0, \ldots, v_m)$ is the exterior normal to F at $(t, x) \in \partial F$, \hfill (35)

$$v_0 + \min_{y \in Y} \max_{z \in Z} \sum_{j=1}^{n} v_i f_i(t, x, y, z) < 0, \quad \text{for all } (t, x) \in \partial F. \tag{36}$$

These conditions mean that each player is able to quickly terminate the game once the trajectory is near the terminal set.

Theorem 4.1 [Friedman (1971)]. *Assume that (34)–(36) hold. Then the assertions of Theorem 1.1 and Corollary 2.3 and Theorem 2.4 hold.*

The differential game associated with (33) is called a *game of survival*. The special case

$$P(y, z) = \int_{t_0}^{\tilde{t}} h(t, x(t), y(t), z(t)) \, dt, \quad h \geq 0 \tag{37}$$

is called a *pursuit–evasion* game; in this case Theorem 4.1 remains valid even if we drop the condition (36) [Elliott and Friedman (1975)]. Another proof of this result as well as of Theorem 4.1 were given more recently in Berkovitz (1986, 1988).

The most celebrated example of a pursuit-evasion game is the *homicidal chauffeur game* whereby a car tries to hit a pedestrian and the payoff is the time $\tilde{t}$. The techniques of Section 3 can be extended to games of survival and, in particular, one can obtain a solution for the homicidal chauffeur game for some initial data. Here the pedestrian y is represented by

$$\dot{x}_1 = w_1 \sin \theta, \quad \dot{x}_2 = w_1 \cos \theta, \quad \theta \in \mathbf{R}^1,$$

and the chauffeur z is represented by

$$\dot{x}_3 = w_2 \sin x_5, \quad \dot{x}_4 = w_2 \cos x_5, \quad \dot{x}_5 = \frac{w_2}{R} \phi, \quad -1 \leq \phi \leq 1,$$

with w_1, w_2, R positive constants, $w_2 > w_1$; the control variables are θ (for y) and ϕ for z, and the terminal set is given by

$$(x_3 - x_1)^2 + (x_4 - x_3)^2 \leq r^2,$$

for more details see Isaacs (1965) and Friedman (1971) p. 157.

Consider next the case when x is restricted to a closed domain X in $\mathbf{R}^m$; if one thinks of some of the x_i as representing some commodities, then the choice

$$X = \{x_k \geq 0, \ldots, x_m \geq 0\} \quad \text{for some } k \geq 1$$

seems very natural.

In order to define a differential game with $x(t) \in X$ we have to divide ∂X into two disjoint sets ∂X_y and ∂X_z such that if $x(t)$ exits X by crossing ∂X_y then y is "severely punished" by making $P(y, z) = -M$ with M very large (or $+\infty$); similarly z is punished by setting $P(y, z) = M$ if $x(t)$ exists X on ∂X_z.

For simplicity we shall take

$$X = \{x; x_m \geq 0\}, \quad \partial X_y = \partial X, \quad \partial X_z = \emptyset. \tag{38}$$

We need the following assumptions:

(P) (a) $Y = Y_1 \times Y_2$ where $Y_1 \subset R^{p_1}, Y_2 \subset R^{p_2}, p_1 + p_2 = p$.

(b) Write $y = (y', y'')$ where $y \in Y, y' \in Y_1, y'' \in Y_2$ and $x' = (x_1, \ldots, x_{m-1})$. Then, for any $t \in [t_0, T_0]$, $x' \in \mathbf{R}^{m-1}$, $y'' \in Y_2$,

$$\max_{y' \in Y_1} \min_{z \in Z} f_m(t, x', 0, y', y'', z) > 0.$$

(c) There is a positive number ε^* such that if $0 \leq x_m \leq \varepsilon^*$ then the variable y' does not appear in the functions $f_i(t, x', x_m, y, z)$ for $1 \leq i \leq m-1$, i.e., for any $y' \in Y_1, \bar{y}' \in Y_1, 0 \leq x_m \leq \varepsilon^*$,

$$f_i(t, x', x_m, y', y'', z) = f_i(t, x', x_m, \bar{y}', y'', z) \text{ if } 1 \leq i \leq m - 1.$$

(Q) (a) There exists a positive number ε_0 such that $\partial f_i(t, x, y, z)/\partial x_j (1 \leq i, j \leq n)$ exist and are continuous if $0 \leq x_m \leq \varepsilon_0$.

(b) For any control functions $y = y(t)$, $z = z(t)$ for which the corresponding trajectory satisfies $0 \leq x_m(t) \leq \varepsilon_0$ in some interval $(\bar{t}, \bar{\bar{t}}) \subset (t_0, T_0)$, set

$$b_{ij}(t) = \partial f_i/\partial x_j)(t, x'(t), 0, y(t), z(t)),$$

and let $(Z_1(t), \ldots, Z_m(t))$ be the solution of

$$dZ_i/dt = \sum_{j=1}^m b_{ij}(t) Z_j, \quad 1 \leq j \leq m, \quad \bar{t} \leq t \leq \bar{\bar{t}}$$

$$Z_m(\bar{t}) = 1, Z_i(\bar{t}) = 0, \quad \text{if } 1 \leq i \leq m - 1.$$

Then $Z_m(t) \geq \theta_0 > 0$ if $\bar{t} \leq t \leq \bar{\bar{t}}$ where θ_0 is independent of $y(t), z(t), \bar{t}, \bar{\bar{t}}$.

Condition P(b) implies that y cannot be forced into $x(t)$ exiting from X; thus y can attempt to navigate $x(t)$ nearby ∂X without worrying about getting out of X.

Theorem 4.2 [Friedman (1971)]. *If (38), (P) (Q) hold then the assertions of Theorem 1.1 and Corollary 2.3 hold.*

This result can be extended to general domains X [see Friedman (1971), Scalzo (1974)].

The proof involves an approximation of the original payoff by smooth payoffs for trajectories which are not restricted to X. Thus, in principle, the Hamilton–Jacobi theory can also be applied in order to compute the value and a saddle point.

So far the controls $y(t)$, $z(t)$ have been measurable functions with values in Y and Z respectively. Suppose we now assume that $y(t)$, $z(t)$ are to be Lipschitz continuous functions:

$$|y(t) - y(\bar{t})| \leq M|t - \bar{t}|,$$
$$|z(t) - z(\bar{t})| \leq L|t - \bar{t}|.$$

Then we obtain a differential game with dynamics

$$\frac{dx}{dt} = f(t, x, y, z), \quad \frac{dy}{dt} = u, \quad \frac{dz}{dt} = v$$

and control variables $u(t)$, $v(t)$ satisfying

$$|u(t)| \leq M, \quad |v(t)| \leq L.$$

Denote the value of the game by $V_{M,L}(x, t)$. It can be shown that as $M \to \infty$, $L \to \infty$, $V_{M,L} \to V^+$ where V^+ is the value of the game associated with (1)–(3) and with $Y = \{y: 0 \leq y_i \leq 1 \text{ for } i = 1, \ldots, p\}$, $Z = \{z: 0 \leq z_j \leq 1 \text{ for } j = 1, \ldots, q\}$, provided f, h are periodic with period 1 as functions of y_i and z_j; see Barron (1976) and Barron (1984).

5. N-person differential games

Consider a system

$$\frac{dx}{dt} = f(t, x, u_1, \ldots, u_N), \qquad (39)$$

$$x(t_0) = x_0, \qquad (40)$$

with f satisfying the condition (A) of Section 1 with (y, z) replaced by $(u_1, \ldots, u_N)$; the variables u_i are control variables, i.e., measurable functions with values in some given control sets U_i. We introduce cost functions

$$J_i(u_1, \ldots, u_N) = g_k(x(T)) + \int_{t_0}^{T} h_i(x, t, u_1, \ldots, u_N) \, dt, \qquad (41)$$

when g, h_i are continuous functions. For any choice $u_i(t)$ ($1 \leq i \leq N$) we compute the trajectory $x(t)$ from (39), (40) and then the costs J_i. We think of the player i (or u_i) as being in charge of the choice of $u_i(t)$, and its purpose is to minimize the cost J_i. If $N = 2$ and $J_1 + J_2 = 0$, then we are in the situation considered in Section 1, which is referred to as a *two-person zero-sum game*. In general we have an N-person game (which is *zero-sum* if $J_1 + J_2 + \cdots + J_N = 0$).

There have been several attempts to define a useful concept of a saddle point [see Friedman (1971)]; the most successful so far has been the concept of *Nash equilibrium*: $(\Gamma_*^{(1)}, \ldots, \Gamma_*^{(N)})$ is a Nash equilibrium point if

$$J_*[\Gamma_*^{(1)}, \ldots, \Gamma_*^{(k-1)}, \Gamma^{(k)}, \Gamma_*^{(k+1)}, \ldots, \Gamma_*^{(N)}] \geq J_k[\Gamma_*^{(1)}, \ldots, \Gamma_*^{(N)}], \quad k = 1, \ldots, N \tag{42}$$

for any strategies $\Gamma^{(k)}$ for u_k. Thus, if one player u_k unilaterally deviates from the strategy $\Gamma_*^{(k)}$ then its cost will increase.

If a Nash equilibrium exists then the vector $V = (V_1, \ldots, V_N)$ where $V_k = J_k[\Gamma_*^{(1)}, \ldots, \Gamma_*^{(N)}]$ is called an *equilibrium value*. One can find a synthesis $\phi_k(t, x)$ for $\Gamma_*^{(k)}$, as for the two-person zero-sum games, by resorting to Hamilton–Jacobi equations. Let

$$H_j(t, x, u_1, \ldots, u_N, p_i) = f(t, x, u_1, \ldots, u_N) \cdot p_i + h_i(t, x, u_1, \ldots, u_N),$$

$p = (p_1, \ldots, p_N)$, and suppose there exist continuously differentiable functions $u_i = u_i^0(t, x, p_i)$ such that

$$\min_{u_i \in U_i} H_i(t, x, u_1^0(t, x, p_1), \ldots, u_{i-1}^0(t, x, p_{i-1}), u_i, u_{i+1}^0(t, x, p_{i+1}), \ldots, u_N^0(t, x, p_N), p_i)$$

$$= H_i(t, x, u_1^0(t, x, p_1), \ldots, u_N^0(t, x, p_N), p_i). \tag{43}$$

Thus we call $u^0(t, x, p) = (u_1^0(t, x, p_1), \ldots, u_N^0(t, x, p_N))$ a *feedback control*.

Consider the Hamilton–Jacobi equations

$$\frac{\partial V_k}{\partial t} + H_k(t, x, u_1^0(t, x, \nabla_x V_1), \ldots, u_N^0(t, x, \nabla_x V_N), \nabla_x V_k) = 0, \quad 1 \leq k \leq N, \tag{44}$$

with the terminal conditions

$$V_k(T, x) = g_k(x), \quad 1 \leq k \leq N. \tag{45}$$

If this system has a continuous solution $(V_1, \ldots, V_N)$ with continuous second x-derivatives, then we obtain from it a continuously differentiable saddle point by [see Friedman (1971)]

$$u_j^*(t, x) = u_j^0(t, x, \nabla_x V_k(t, x)). \tag{46}$$

unfortunately there is no known general method for solving (44), (45) as in the case of a two-person zero-sum game.

In case f is linear in $x, u_1, \ldots, u_N$ and J_i is quadratic, namely,

$$J_i(u) = \langle x(T) - \xi, W_i[x(T) - \xi_i] \rangle$$
$$+ \int_{t_0}^{T} \langle z_i(t) - C_i(t)x(t), Q_i(t)[z_i(t) - C_i(t)x(t)] \rangle \, dt$$
$$+ \int_{t_0}^{T} \langle u_i(t), R_i(t)u_i(t) \rangle \, dt$$

($\langle \cdot, \cdot \rangle$ denotes the scalar product), then one speaks of *linear–quadratic game*. In this case, if $U_i \equiv \mathbf{R}^{q_i}$ and if $R_i(t)$ is a positive definite symmetric matrix and W_i, C_i, Q_i are positive semidefinite symmetric matrices, one can solve the Hamilton–Jacobi equations in the form

$$V_i(t, x) = \langle x, K_i(t)x \rangle + 2\langle g_i(t), x \rangle + \Phi_i(t),$$

the K_i satisfy a system of Ricatti equations for which a solution exists if T is small enough, the g_i satisfy linear differential equations and Φ_i is obtained by integrating a function depending on the K_i, g_i. If $f = Ax + \sum B_i u_i$ then (46) becomes

$$u_j^*(t, x) = -R_j^{-1}(t)B_j^*(t)[K_j(t)x + g_j(t)], \tag{47}$$

this solution is called a *closed-loop* equilibrium strategy.

An *open-loop* equilibrium point is a vector $(u_1^*(t), \ldots, u_N^*(t))$ such that

$$J_i(u_1^*, \ldots, u_{i-1}^*, u_i, u_{i+1}^*, \ldots, u_N^*) \geq J_i(u_1^*, \ldots, u_N^*), \quad \text{for } 1 \leq i \leq N,$$

such a vector, if existing, may depend in general on the initial conditions. For linear–quadratic games with the additional restrictions on R_i, W_i, C_i, Q_i as above, one can establish the existence of open-loop controls [Varaiya (1970), Scalzo (1974)] [see also Friedman (1971)]; if $U_i \equiv \mathbf{R}^{q_i}$ then there exist equilibrium values of open-loop equilibrium points which do not agree with the equilibrium value of closed-loop equilibrium strategies of (46) [see Friedman (1971) p. 322]; it is not known whether they are Pareto optimal.

References

Barron, E.N. (1976) 'Differential games with Lipschitz control functions and application to games with partial differential equations', *Transactions American Mathematical Society*, **219**: 39–76.

Barron, E.N., L.C. Evans and R. Jensen (1984) 'Viscosity solutions of Isaacs' equations and differential games with Lipschitz controls', *Journal of Differential Equations*, **53**: 213–233.

Berkovitz, L.D. (1967) 'Necessary conditions for optimal strategies in a class of differential games and control problems', *SIAM Journal on Control and Optimization*, **5**: 1–24.

Berkovitz, L.D. (1986) 'The existence of value and saddle point in games of fixed duration', *SIAM Journal on Control and Optimization*, **23**: (1985), 172–196; errata, **24**.

Berkovitz, L.D. (1986) 'Differential games of generalized pursuit and evasion', *SIAM Journal of Control and Optimization*, **24**: 361–373.

Berkovitz, L.D. (1988) 'Characterization of the values of Differential games', *Applied Mathematics and Optimization*, **17**: 177–183.
Berkovitz, L.D. (1988) 'Differential games of survival,' *Journal of Mathematical Analysis and Applications*, **129**: 493–504.
Crandall, M.G., L.C. Evans and P.L. Lions (1984) 'Some properties of viscosity solutions of Hamilton–Jacobi equations', *Transactions American Mathematical Society*, **282**: 487–502.
Crandall, M.G. and P.L. Lions (1983) 'Viscosity solutions of Hamilton–Jacobi equations', *Transactions American Mathematical Society*, **277**: 1–42.
Elliott, R.J. and A. Friedman (1975) 'A note on generalized pursuit–evasion games', *SIAM, Journal on Control and Optimization*, **13**: 105–109.
Elliott, R.J., A. Friedman and N.J. Kalton (1974) 'Alternate play in differential games', *Journal of Differential Equations*, **15**: 560–588.
Elliott, R.J. and N. Kalton (1972) 'The existence of value in differential games', *Memoires of the American Mathematical Society*, **126**.
Evans, L.C. and P.E. Souganidis (1984) 'Differential games and representation formulas for solutions of Hamilton–Jacobi equations' Indiana *University Mathematical Journal*, **33**: 773–797.
Fleming, W. (1957) *A note on differential games of prescribed duration, Contribution to the theory of Games*, Vol. III. Annals of Mathematic Studies, No. 39. New Jersey: Princeton University Press, pp. 407–416.
Fleming, W. (1961) 'The convergence problem for differential games', *Journal of Mathematical Analysis and Applications*, **3**: 102–116.
Friedman, A. (1971) *Differential games*. New York: Wiley.
Friedman, A. (1974) *Differential Games*. CBMS Regional Conference Series in Mathematics, No. 18, American Mathematical Society, Providence, RI.
Isaacs, R. (1965) *Differential games*. New York: Wiley.
Ishi, H. (1984) 'Uniqueness of unbounded viscosity solutions of Hamilton–Jacobi equations', *Indiana University Mathematical Journal*, **33**: 21–748.
Krasovskii, N.N. and A.I. Subbotin (1974) *Positional Differential Games*. Moscow: Nauka [Russian].
Lions, P.L. and P.E. Souganidis (1985) 'Differential games, optimal control and directional derivatives of viscosity solutions of Bellman's and Isaacs' equation', *SIAM Journal on Control and Optimization*, **23**: 566–583.
Roxin, E. (1969) 'The axiomatic approach in differential games', *Journal of Optimization Theory and Applications*, **3**: 153–163.
Scalzo, R.C. (1974) 'N-person linear quadratic differential games with constraints', *SIAM Journal on Control and Optimization*, **12**: 419–425.
Scalzo, R.C. (1974) 'Differential games with restricted phase coordinates', *SIAM Journal on Control and Optimization*, **12**: 426–435.
Varaiya, P. (1970) 'N-person non-zero sum differential games with linear dynamics', *SIAM Journal on Control and Optimization*, **8**: 441–449.
Varaiya, P. and J. Lin (1969) 'Existence of saddle points in differential games', *SIAM Journal on Control and Optimization*, **7**: 142–157.

Chapter 23

DIFFERENTIAL GAMES – ECONOMIC APPLICATIONS*

SIMONE CLEMHOUT and HENRY Y. WAN Jr.

Cornell University

Contents

1.	Introduction	802
2.	Technical preliminaries	803
3.	Example for non-cooperation. I. Exhaustible common-property resource	804
4.	Example for non-cooperation. II. Renewable common-property resource	812
5.	Methodology reconsidered	815
6.	"Tractable" differential games	815
7.	The linear-quadratic differential game	817
8.	Some other differential game models	818
9.	Concluding remarks	820
References		823

*We appreciate the very helpful comments and suggestions of Dr. James Case.

Handbook of Game Theory, Volume 2, Edited by R.J. Aumann and S. Hart
© *Elsevier Science B.V., 1994. All rights reserved*

1. Introduction

Two-person, zero-sum differential game theory was developed by Rufus Isaacs, with von Neumann's encouragement, in the early 1950s at the Rand Corporation. With few exceptions, its applications are military, like the problem of pursuit and evasion. But social sciences generally and economics in particular rarely deal with two players having diametrically opposed interests[1]. Indeed, it is because Morgenstern was able to persuade von Neumann on this matter that game theory itself exists now in the form we know it today.

The theory of Isaacs (1965) was thus generalized by Case (1969) (under the influence of Thrall) as well as Starr and Ho (1969) into a many-player theory. This employs a simultaneous system of Hamilton–Jacobi partial differential equations in its attempt to characterize the "most advantageous" strategies for the individual players.

Now, in the N-person *game theory*, what is "most advantageous" is usually defined in the sense of Nash, and not in dominant strategies, for which equilibrium does not exist in many economically interesting contexts. By the same token, the many-player *differential games* are more complex than their two-person zero-sum brethren, both computationally and conceptually. It is precisely in this perspective, that the recent findings in differential games are highly significant. It is established conclusively and intuitively that mutual expectations matter in the macro-economic theory of endogenous growth, through the existence of a large set of equilibria.

While exploitable internal structure have not yet been found for the partial differential equations, so that their integration cannot rely on the method of characteristics, for an ever-expanding collection of worked examples, the theory of differential games is effective for what it does best, in characterizing qualitatively the time-profile of equilibrium plays. When compared with what can be obtained in the much simpler situation of dynamic optimization, the theory has delivered quite detailed information for much more complex situations. If and when more progress is made in the qualitative theory of partial differential equations, more findings in differential games are to be expected. In summarizing this literature, we shall focus upon the post-1976 contributions, since the early works have been covered in Case (1979). Our definitions and symbols will be consistent with those of chapter 22.

We shall discuss some technical preliminaries in the next section. This is followed by two extended examples to illustrate the general model structure, showing the

---

[1] An exception is the problem of stabilizing an uncertain system. See Gutman and Leitmann (1976), Leitmann and Wan (1978) and Deissenberg (1988), where the problem is viewed as a 'game vs. Nature'. See also Basar and Bernhard (1991). An earlier discussion of the macro-economic issue is in Brito (1973).

qualitative insights which have been shed on both the general issue of multiplicity and the characteristics for equilibrium in various models. A survey of different types of differential games is then given, before some concluding remarks.

2. Technical preliminaries

Differential Game Theory is not, as it often appears to the novice, unrelated to the main body of game theory. In particular, it is not a branch either of systems engineering, or of optimal control theory, despite its frequent use of terminology and computational techniques borrowed from those disciplines[2]. On the contrary, differential games are simply games in extensive form whose players have more alternatives to choose from, and more frequent opportunities to choose, than originally anticipated by von Neumann. Typical differential games, indeed, involve a continuum of each.

Mastery of the subject requires the identification, in this expanded context, of von Neumann's three "elements of competition", namely strategies, moves, and information sets. When this is accomplished, the strategies will necessarily exhaust "all possible ways of playing the game". And because there are necessarily so many ways to exercise such a multiplicity of options, many-player differential games tend to possess impenetrably large and complex strategy spaces.

Accordingly, most students of the subject have come to agree with [Holt (1985)] that "... an explicitly dynamic analysis can be extremely difficult unless the class of feasible dynamic strategies is restricted". Though Holt's remarks were directed specifically toward dynamic models of oligopolistic competition, they are equally valid for other differential games as well. Indeed the entire literature offers no single example of a continuous differential game solved, in any of the usual senses, in wholly unrestricted strategy spaces.

The task of restricting strategy spaces is impeded by the fact that there exists no generally applicable notion of "strategic dominance" for many-player games. Consequently the required restriction is invariably performed during the modelling phase, rather than the analysis phase, of each individual investigation. One must always begin by associating, with any particular differential game, a "reduced strategy space" appropriate to that specific game. Certain generic responses to the need for strategic restriction do, however, exist.

The more drastic of these is the restriction to so-called "open-loop" strategies. It constitutes an assumption that the players can receive information from one source only, namely the clock. It may be shown that, in almost every situation, such an assumption is overly restrictive in that it excludes obviously advantageous modes of play.

---

[2]See Basar and Olsder (1982) and Mehlmann (1988) about how to solve differential games.

A less drastic restriction is to "closed-loop" strategies[3], whereby each player is assumed able to observe all variable quantities not directly under the control of some specific player (i.e. "state variables"), but unable to recall past observations. It may be shown that this is no restriction whatever for two-player zero-sum games, but for many-player differential games, past observations are needed to enforce cooperation with credible threat.

The fact that these two restrictions have become standard *does not* eliminate the need to justify the employment of either one of them for the analysis of any particular differential game.

Weaker restrictions are possible, because additional "state variables" may always be incorporated into the formulation of an individual differential game to permit limited recall, but the opposite extension of the theory, to situations wherein some state variables are not observable, appears beyond the reach of present methods even in the two-person zero-sum case.

The best developed branch of the many-player theory is that which concerns the subgame perfect Nash equilibria of non-cooperative games. Such equilibria constitute, by definition, N-tuples of "closed-loop" strategies. Results here apply to games in which the various subgames are indexed by points in a Euclidean space, and stages of play by continuous time.

Other solution concepts, both cooperative and otherwise, have on occasion been proposed for many-player differential games. We concentrate here on closed-loop Nash equilibria, because these seem still to constitute the main stream of the subject, and to best meet the requirements of economic theory.

3. Example for non-cooperation. I. Exhaustible common-property resource

For illustration, we now consider one of the simplest types of N-person, general-sum differential games, first studied by Levhari–Mirman (1980) in a discrete time version: resource exploitation with open access. We start with an even simpler subcase of this problem [Clemhout and Wan (1989)].
Let:

- $i = 1, 2, \ldots, N$ be the players.
- $X = \Re_+$ be the state space (phase set) for the non-negative stocks of exhaustible resources (say, petroleum reserves).
- $T = \Re$ be the real line (continuous time[4]).
- $C_i = \Re_+$ be the control space for the non-negative rates of exploitation, $h_i(x, t) = u_i(c_i) \exp(-r_i t)$ be the instantaneous payoff of player i, where $u_i \in$

---

[3]In Basar and Olsder (1982), such strategies are referred to as "feedback strategies". If strategies only depend upon the "state" but not "time", then one sometimes refer such strategies as Markovian.

[4]In economics, one usually assume that the player maximizes his payoff over an infinite horizon (rather than minimizing his cost over a finite horizon).

$C^2(\Re_{++}, \Re_+)$ is the utility function which is assumed to be strictly increasing and strictly concave, and $r_i > 0$ is the time preference rate for i.

Remarks. (1) "$X = \Re_+$" may be replaced by the less compact statement: "$X = [0, x_0]$, for whatever initial stock $x_0 > 0$", as a consequence of the exhaustible nature of x.

(2) "$C_i = \Re_+$" is a shorthand for some lengthier notion to be discussed later (see ASAP exploitation).

(3) The strict concavity of u_i implies that the choice of c_i by player i is always unique, as seen from (3.15), (3.16) below. It is this fact which enables any player to deduce the (unknown) concurrent moves of the other players at any instant from the state of the game. Thus we have a situation tantamount to a "perfect information" game so that there is neither the need for the other player(s) to use mixed strategies, nor the occasion to "refine" further the closed-loop Nash solution here (which is already subgame-perfect).

(4) When one player is indifferent over two distinct moves, the use of mixed strategy is in order, as the "cohato" move in the two-person, zero-sum differential game of "the Princess and the Monster". [The source is Isaacs (1965). See Ho and Olsder (1983) for a simple discussion.]

The system evolution follows the *state equation*:

$$x' = -\sum_i c_i, \quad c_i \in C_i. \quad (\text{' stands for the time derivative.}) \tag{3.1}$$

The payoff for any player is

$$J_i = \int_t^\infty u_i[c_i(\tau)] \exp(-r_i \tau) \, d\tau \tag{3.2}$$

and the admissible (closed-loop) strategy set for player i is $S_i = C^1(X, C_i)$, the class of C^1 functions from X to C_i, or $\Re_+$ to $\Re_+$. Thus, under strategy $s_i \in S_i$,

$$c_i(t) = s_i(x(t)), \quad \text{all } t\,[5].$$

Now set $S = S_1 \times S_2 \times \cdots \times S_N$, the Cartesian product of all players' admissible strategy sets, then, if an initial stock x is available at time t,

$$J(x, t; s) = (J_1(x, t; s), \ldots, J_N(x, t; s)), \quad x \in X, \quad t \in \Re, \quad s \in S, \tag{3.3}$$

is the vector of payoff functions.

We can now introduce the following:

Definition.

$$s^* = (s_1^*, \ldots, s_N^*) \in S, \tag{3.4}$$

---

[5] Since time only enters the problem through the discounting factor at a constant discount rate, the strategies for such an autonomous problem only depend on the state variable [cf. Kamien and Schwartz (1981)].

is a vector of *closed-loop Nash equilibrium strategies*, if for any player i, and at any (x, t), the following inequality holds:

$$J_i(x, t; s^*) \geq J_i(x, t; s_1^*, \ldots, s_{i-1}^*, s_i, s_{i+1}^*, \ldots, s_N^*). \tag{3.5a}$$

Denote,

$$s_{-i}^* = (s_1^*, \ldots, s_{i-1}^*, s_{i+1}^*, \ldots, s_N^*), \quad B_i(s_{-i}^*) = \{s_i^*; (3.5a) \text{ holds}\}$$

then one can also define a vector of *closed-loop Nash equilibrium strategies* as a vector s^*, with

$$\text{for all } (x, t), \quad s_i^* \in B_i(s_{-i}^*), \quad \text{all } i. \tag{3.5b}$$

Remark. B_i is the "best reply correspondence" for player i which we shall assume to be unique, unless it is explicitly stated otherwise.

Definition. $V_i(x, t) = J_i(x, t; s^*)$ is the *value function* of the game for player i.

The interest of economists is to decide (i) the structure of the optimal strategies (e.g., are exploitation rates proportional to the remaining stock?), (ii) the time path of the state variable, x (e.g., would the stock decline exponentially? if so, at what rate?), (iii) the distributive issues (i.e., which player receives what share of the output), (iv) the allocative issues (i.e., how does inefficiency arise from the open access of a scarce resource), (v) sensitivity of these configurations with respect to the number of players (e.g., would a large number of players improve or reduce efficiency?) and (vi) sensitivity of these configurations with respect to the characteristics of the players (i.e., their time preference, and the "curvature" of their utility indices). Furthermore, (vii) can a tax-cum-subsidy system be set up to eliminate inefficiency?

Simple as this problem appears, the answers to the above questions are only known under very special conditions. Several cases will be considered.

Case A. Players are identical, and their marginal utility is constant elastic

$$r_i = r, \quad cu_i''/u_i' = E \in [-1, -1 + (1/N)), \quad \text{all } i. \tag{3.6}$$

For concreteness, we assume that

$$u_i = c^{1+E}, \quad \text{for } E \in (-1, -1 + (1/N)), \quad \text{and } u_i = \ln c, \quad \text{for } E = -1.$$

One strategy equilibrium of all players is

$$s_i(x) = \{r/[1 - N(1 + E)]\}x, \quad \text{for } E \in [-1, -1 + (1/N)), \quad \text{all } i \tag{3.7}$$

and the value function is

$$V_i(x, t) = \{r/[1 - N(1 + E)]\}^E x^{1+E} \exp(-rt), \quad \text{for } E \in (-1, 1 + (1/N)), \quad \text{all } i \tag{3.8a}$$

or,
$$V_i(x,t) = [(\ln r + \ln x - N)/r]\exp(-rt), \quad \text{for } E = -1, \text{ all } i. \tag{3.8b}$$

These are readily verified from the Hamilton–Jacobi equation:
$$0 = \partial V_i/\partial t + \max_c \{u(c)\exp(-rt) - [(N-1)s(x) + c]\partial V_i/\partial x\}, \tag{3.9}$$

with $s(x) = s_i(x)$, by symmetry. Clearly,
$$x(t) = x(0)\exp\{-rt/[1 - N(1+E)]\}. \tag{3.10}$$

Moreover, simple computation shows that all these identical players would gain if they exploit the resource under coordinated management: extract the resource as if they were a single player and then share the output equally among each other. A corrective, non-linear tax-cum-subsidy schedule has been suggested for the logarithmic case in Clemhout and Wan (1985a). No tax proceeds need be actually collected.

One can easily confirm that the extraction rate, as a proportion to the remaining stock, is positively associated with N, the number of players, r, the time preference, and E, the negative of the index for "relative risk aversion". The larger is the value E, algebraically speaking, the more is the wastage due to competitive extraction. For the logarithmic case ($E = -1$), the entry of new players would not change the proportional intensity of exploitation, c_i/x. For $E > -1$, the entry of another player would actually speed up the exploitation activity of the incumbent players. The reason is obvious; what is not taken by oneself today may be taken by others. This does not necessarily stampede the older players to extract as fast as one can; to deny the access to other players, one must face perpetual deprivation for oneself. Yet, a "stampede" is not out of the question, should the utility function approach closer to linearity ($E = 0$); no admissible, continuous solution can exist anymore, for $E \geq -1 + (1/N)$. See discussion below. We leave the case of $E < -1$ (where $u = -c^{1+E}$) to the interested reader.

Case B. Two asymmetric players, constant elastic marginal utility. This was explored in Clemhout and Wan (1989). What concerns us here is the "distributive" issue of who takes what share. To simplify notations, the unit of time is so normalized that
$$r_1 + r_2 = 1,$$

also that, without losing generality, we may assume

$r_1 \leq \frac{1}{2} \leq r_2$. (Player 1 is no more impatient than player 2.)

E_1 and E_2 may take any constant, negative values. We claim that the equilibrium strategies for the two players are
$$s_1(x) = [-r_2 + E_2/(1 + E_1 + E_2)]x,$$
$$s_2(x) = [-r_1 + E_1/(1 + E_1 + E_2)]x, \tag{3.11}$$

with the "rate of decay" for the stock being

$$-x'/x = -1/(1 + E_1 + E_2), \tag{3.12}$$

provided that

$$(1 + E_1 + E_2) < 0, 0 < E_2 - r_2(1 + E_1 + E_2) < 1. \tag{3.13}$$

Thus, a relative difference index for exploitation may be defined as

$$[s_2(x) - s_1(x)]/x = (r_2 - r_1) + (E_1 - E_2)/(1 + E_1 + E_2), \tag{3.14}$$

with the implication that differences in time preferences can be outweighed by the differences in the curvature of the utility functions.

To verify that (3.11) is an equilibrium, we may write $s_i' = s_i(x)/x$, for $i = 1, 2$, and substitute $xs_j', j \neq i$, $i = 1, 2$, for the linear term, $(N-1)s(x)$, in (3.9) to solve the simultaneous Hamilton–Jacobi equations. Detailed computations are omitted here.

Note that here the best reply to a linear strategy happens to be another linear strategy. In fact, if we regard $\xi = \ln x$ rather than x as the state variable, and the terms $\gamma_i = c_i/x$ rather than c_i as the controls, then we obtain another "tractable" class of differential games, a topic which we shall discuss later.

A digression: ASAP (as soon as possible) exploitations. The statement $C_i = \Re_+$ is equivalent to $c_i \in [0, \infty]$, which allows for the possibility: $c_i = \infty$. This latter is a shorthand for the statement: "For any sequence of well-posed problems with ∞ replaced by some positive $B(q)$, $\{B(q)\}$ being an increasing *unbounded* sequence, then $c_i = B(q)$ for each q, such that $B(q) > B$, for some B." This avoids the alternative of imposing on c_i, the exact finite maximum B, as decided by technical or relativistic (!) considerations, case by real life case, with c_i equal to each specific value of B, in all such cases.

Technically, the space of continuous functions over a finite (or infinite) interval is not complete. The space of continuous functions satisfying a uniform Lipschitz condition is. For illustration, consider the (optimal control) problem for the owner of a finite resource stock, with a positive time preference ($r > 0$), but a linear utility ($E = 0$): the choice must be ASAP depletion, unless a maximum speed of adjustment is specified.

The above discussion is important for the game of exhaustible common property resources, where the ASAP depletion of $c_i = \infty$, all i, is *always an* equilibrium, and *sometimes the* only equilibrium, i.e., $E \geq -1 + (1/N)$ for Case A, in (3.6), and when the restrictions in (3.13) are not satisfied for Case B.

Case C. Generalization. The results of Cases A and B are suggestive, but not much more. Two problems remain. First, there is no reason why marginal utilities must be constant elastic for every player. Second, the equilibrium strategies are linear in the state variable; they are convex, not strictly convex. Facing other players' linear strategies, the "conditional" control problem of a game player barely satisfies the Mangasarian condition for existence. Thus, the solution may not survive an infinitesimal perturbation, which is not a satisfactory situation.

For our study, we continue to assume that $N = 2$ and introduce the Hamiltonian format, which has been proven valuable for both the control theory and the two-person, zero-sum differential games.

Define the "current value" Hamiltonian functions,

$$H_i = u_i(c_i) + p_i[-c_i - s_j(x)], \quad i,j = 1, 2, \quad i \neq j. \tag{3.15}$$

Consider the conditions

$$c_i = \operatorname{argmax} H_i, \quad i = 1, 2 \text{ (the maximum principle)} \tag{3.16}$$

and the *adjoint equations*,

$$p_i' = [r_i + ds_j(x)/dx]p_i, \quad i = 1, 2, \tag{3.17}$$

where $p_i = [\partial V_i(x)/\partial x] \exp(r_i t)$ is the current value adjoint variable for i.

These are precisely the necessary conditions for optimality from the viewpoint of player i, taking the strategy of player $u, s_j(\cdot)$, as given. Together with (i) the state equation in (3.1), (ii) the initial condition, $x(t_0) = x_0$, and (iii) the terminal conditions $s_i(0) = 0$, $i = 1, 2$, these will characterize the equilibrium strategies, the equilibrium evolution of x, etc. For instance, the solutions to Cases A and B satisfy this set of conditions. Yet, in general, one cannot *readily* deduce the equilibrium strategies of the two players from the above conditions and the information about $(u_i(\cdot), r_i)$, $i = 1, 2$. The difficulty is the presence of the "cross effect" terms $ds_j(x)/dx$ in (3.17). These are absent for optimal control models and identically zero for two-person, zero-sum differential games.

We illustrate below that sometimes, like in the present example, one can show that (a) a Nash equilibrium exists for a non-negligible subclass of all possible models, (b) computable relations can be derived associating the equilibrium strategies with the preferences of the players (i.e., their time preferences and their felicity indices), and finally, (c) some sensitivity analysis can be conducted.

For expository case, we shall start with the symmetric equilibrium,

$$s_i(x) = s(x), \quad i = 1, 2, \tag{3.18}$$

for the case of symmetric players, so that

$$u_i(\cdot) = u(\cdot); \quad r_i = r, \quad i = 1, 2. \tag{3.19}$$

Now, normalize $r = \frac{1}{2}$, and we have

$$x'(t) = -2s(x(t)) \tag{3.1'}$$

$$u'(s(x(t))) = p(t) \tag{3.16'}$$

$$p'(t)/p(t) = (\tfrac{1}{2}) + s'(x(t)). \tag{3.17'}$$

We next pose the "inverse differential game" problem: whether there exists any model with a symmetric equilibrium of the given particular form in (3.18). Under the assumption that $s(\cdot)$ is differentiable, (weakly) convex, and has the boundary condition: $s(0) = 0$ (zero exploitation at zero stock), we can equally well start with a given $s'(\cdot)$. This allows us to define

$$s(x) = \int_0^x s'(\xi) \, d\xi.$$

Differentiate (3.16') logarithmically and use (3.1') to eliminate x' and (3.17') to eliminate p and p', we have

$$\tfrac{1}{2} + s' = -2(su''/u')(s') = -2s'E, \tag{3.20}$$

where E is defined over $\Re_+$.

This implies the solution to the inverse game is,

$$E(s(x)) = -\tfrac{1}{2}[1 + (\tfrac{1}{2}s'(x))], \tag{3.20'}$$

with E as an "elasticity representation" of the felicity index, $u(\cdot)$, up to an increasing affine transform, since

$$u(s) = \int^s \exp\left(\int^\sigma [E(\lambda)/\lambda] \, d\lambda\right) d\sigma. \tag{3.21}$$

The two lower limits of integration supply the two degrees of freedom for the (non-essential) choice of the origin and the unit of measure.

The alternative approach is to start with $E(\cdot)$ or $u(\cdot)$, and transform (3.20) into a differential equation:

$$s' = -\tfrac{1}{2}[1 + 2E(s)], \tag{3.20''}$$

which is integrable, by variable separation, into the inverse of the equilibrium strategy:

$$x = s^{-1}(c)$$
$$= -2\left(c + 2 \int_0^c E(\gamma) \, d\gamma\right). \tag{3.22}$$

Noting that (3.22) is implied by the first-order necessary condition for an equilibrium but not in itself sufficient, one may impose the requirement that $s(\cdot)$ should be convex, or $s'' \geq 0$, so that one can invoke the Mangasarin condition.

In terms of $E(\cdot)$, the requirements now are

$$E(s) < \tfrac{1}{2}, E'(s) \geq 0 \; (\text{``decreasing relative risk aversion''}). \tag{3.23}$$

Thus, for point (a) above, as long as (3.23) is met, the equilibrium strategy can be solved by quadrature.

Moreover, for point (b), qualitatively, it is not hard to show that, (i) for a specific model, $E'(c) > 0$, for all c, then for the equilibrium $s(\cdot)$, the intensive measure of extraction, $s(x)/x$, is also systematically higher at higher x, and (ii) suppose that for two separate models, both functions $E^{(1)}(\cdot)$ and $E^{(2)}(\cdot)$ satisfy (3.23), and

$$E^{(1)}(c) \geq E^{(2)}(c), \quad \text{for all } c, \text{ and } E^{(1)}(c) > E^{(2)}(c), \quad \text{for some } c,$$

then one can deduce that the former implies a faster pace of exploitation:

$$s^{(1)}(x) \geq s^{(2)}(x), \quad \text{for all } x, \text{ and } s^{(1)}(x) > s^{(2)}(x), \quad \text{for some } x.$$

Turning to point (c), before we can conclude that the existence of closed-loop Nash equilibrium is "generic", and not dependent upon that the players have very special type of preferences (say, with constant elastic marginal utilities), we must relax the assumption of symmetry. This is sketched below.

Relax (3.19), and only require that $r_1 \in (0, \frac{1}{2}]$, $r_2 = 1 - r_1$, as in Case B. Consider again the inverse game problem, with given convex strategies, $s_1(\cdot)$, $s_i(0) = 0$, $i = 1, 2$. We can deduce as before the relationships

$$E_i(s_i) = -[s_i/(s_i + s_j)][(r_i + s_j')/s_i'], \quad i, j = 1, 2, \quad i \neq j. \tag{3.24}$$

On the other hand, if one starts from the $E_i(\cdot)$, to solve for the $s_i(\cdot)$, one may invoke the L'hôspital rule and the Brouwer's fixed point theorem to evaluate $s_i'(0)$, $i = 1, 2$. See Clemhout and Wan (1989) for details. Summing up, the existence of an equilibrium in such models is not by coincidence.

In general, the inverse problem has a one-dimensional indeterminacy. Given any function pair, (s_1, s_2), one can find a function pair, (E_1, E_2), for each value of $r_1 \in (0, \frac{1}{2}]$. They are observationally equivalent to each other.

The open-loop Nash equilibrium. In passing one can consider the case of "open-loop Nash equilibrium" which may be defined as follows.

Let the admissible (open-loop) strategy set for player i be $S_i = C^1(T, C_i)$, the class of C^1-functions from T to C_i or $\Re$ to $\Re_+$. Thus, under strategy $s_i \in S_i$,

$$c_i(t) = s_i(x(t - t_0), x_0), \quad \text{all } t, (x_0, t_0) \text{ being the initial condition.}$$

Now set $S = S_1 \times S_2 \times \cdots \times S_N$, the Cartesian product of all players' admissible strategy sets, then, if an initial stock x_0 is available at time t_0,

$$J(x_0, t_0; s) = (J_1(x_0, t_0; s), \ldots, J_N(x_0, t_0; s)), \quad x_0 \in X, \quad t_0 \in \Re, \quad s \in S, \tag{3.3'}$$

is the vector of payoff functions.

We can now introduce the following:

Definition.

$$s^* = (s_1^*, \ldots, s_N^*) \in S, \tag{3.4'}$$

is a vector of *open-loop Nash equilibrium strategies*, if for any player i, and at any (x, t), the following inequality holds:

$$J_i(x, t; s^*) \geq J_i(x, t; s_1^*, \ldots, s_{i-1}^*, s_i, s_{i+1}^*, \ldots, s_N^*). \tag{3.5'}$$

In the Hamiltonian format, $s_j(x)$ is replaced with $s_j(t-t_0)$. More important, for the adjoint equation in (3.17), the cross-effect term, $ds_j(x)/dx$, disappears. This makes the solution much easier to obtain, and appeals to some researchers on that ground.

It seems however, which equilibrium concept to use should be decided on the relative realism in a particular context, and nothing else. A player would use the open-loop strategy if he perceives that the exploitation rates of the other players depend only upon the calendar time, and not on how much the stock remains. In other words they act according to a behavior distinctly different from his own (In using the Hamiltonian format, he acts by the state)[6]. He would use the closed-loop strategy if he perceives that the other players decide "how much to extract" by "how much the resource is left": a behavior which is a mirror image of his own. Thus be expects that the less he consumes now means the more will be consumed by others later. This dampens his motive to conserve and forms the source of allocative inefficiency for dynamic common property problems.

4. Example for non-cooperation. II. Renewable common-property resource

We now consider the case of renewable resources where a "recruitment function" $f(\cdot)$ is added to the right-hand side of the *state* equation, (3.1), to yield:

$$x' = f(x) - \sum_i c_i, \quad c_i \in C_i. \tag{4.1}$$

$f(\cdot)$ is assumed to be non-negative valued, continuously differentiable and concave with $f(0) = 0$. It represents the growth potential of natural resources like marine life. f', the derivative of f, is the "biological rate of interest" which should be viewed as a deduction against the time preference rate in the conserve-or-consume decision of a player. This means, in the determination of the inter-relationship of "shadow utility values", the adjoint equation (3.17) should be so modified to reflect this fact:

$$p_i' = \{(r_i - f') + \sum_{j \neq i} [ds_j(x)/dx]\}p_i, \quad \text{all } i. \tag{4.2}$$

What is qualitatively novel for the renewable resources (vs. the exhaustible resources) is that instead of eventual extinction, there is the possibility of an asymptotical convergence to some strictly positive steady state. The first question is, whether more than one steady state can exist. The second question is, whether in the dynamic common property problem, some resources may be viable under coordinated management but not competitive exploration. both turn out to be true [see Clemhout and Wan (1990)].

---

[6]This may be cognitively inconsistent, but not necessarily unrealistic. A great majority of drivers believe that they have driving skills way above average.

The other questions concern the generalizability of the analysis to situations where (i) there are two or more types of renewable resources, (ii) there are random noises affecting the recruitment function, and (iii) there is some jump process for random cataclysms which would wipe out the resources in question. These were studied in Clemhout and Wan (1985a, c).

Under (i), it is found that (a) the different resource species may form prey–predator chains and a player may harvest a species heavily because of its predator role, and (b) neutrally stable non-linear cycles may persist in the long run. Under (ii), it is found that with logarithmic felicity, Gompertz growth, and Gausian noise, the equilibrium harvest strategies may be linear in form and the resource stock, in logarithmic scale, obeys an Orstein–Uhlenbeck process and any invariant distribution of stock levels will be log-normal. Under (iii), it is found that (a) although all species become extinct in the long run, players behave as if the resources last forever, but the deterministic future is more heavily discounted, and (b) under such a context, the model is invariant only over a change of *unit* for the felicity index, but not of the *origin*!

Differential games make two types of contributions: detailed information on various issues, as we have focused on so far, and qualitative insights about dynamic externalities, such as the "fish war" and endogenous growth, which is to be discussed below. This latter aspect arises from recent findings [Clemhout and Wan (1994a, b) and Shimemura (1991)] and its significance cannot be overstated.

The central finding is that with dynamic externalities, if there exists any one (closed-loop) equilibrium, then there exists a continuum of equilibria, including the former as a member. These equilibria may be Pareto-ranked and their associated equilibrium paths may never converge. The full implication of this finding is to be elaborated under three headings below.

(1) *Fact or artifact?* In "fish war", the possible co-existence of a finite set of equilibria is well known, but here there is a continuum of equilibria. This latter suggests that some economic mechanism is at work, more than the coincidental conjunction of particular function forms. In "fish-war", the equilibrium for a player calls for the equality between the discounted marginal rate of intertemporal substitution and the net marginal return of fishery to this player. The latter denotes the difference between the gross marginal return of fishery and the aggregate marginal rate of harvest, over all other players. In contrast to any known equilibrium, if all other players are now more conservation-minded, causing lower marginal rates of harvest, then the net marginal return for the first player would be higher, justifying in turn a more conservation-minded strategy of this player, as well. This shows the source of such multiple equilibria is an economic mechanism, and not the particular features of the differential game (among various dynamic games), such as time is assumed to be continuous, as well as that players are assumed to decide their actions by the current state and not the entire past history.

(2) *Context-specific or characteristics-related?* The economic reasoning mentioned above is useful in showing that such multiplicity need not be restricted to the fish

war. What is at issue are the characteristics of the economic problem: (a) Do players select strategies as "best replies" to the other players' strategies? (b) Do players interact through externalities rather than through a single and complete set of market prices which no player is perceived as capable to manipulate? (c) Does the externality involve the future, rather than only the present? For the non-convergence of alternative equilibria, one is concerned also with (d) Is the resource renewable rather than exhaustive? The "fish war" is affirmative to (a)–(d), and so is the problem of endogenous growth.

(3) *Nuisance or insight?* The presence of a continuum of equilibria may be viewed as a nuisance, if not "indeterminacy", since it concerns the findings of previous works in differential games (ours included) where, invariably, one equilibrium is singled out for study. Several points should be made: (a) since each equilibrium is subgame-perfect and varying continuously under perturbations, refinement principles are hard to find to narrow down the equilibrium class. Yet empirically based criteria might be obtained for selection. The literature on auction suggests that experimental subjects prefer simpler strategies such as linear rules. (b) Second-order conditions may be used to establish bounds for the graphs of equilibrium strategies. There might be conditions under which all alternative equilibria are close to each other in some metric, so that each solution approximates "well" all others. (c) There might be shared characteristics for all equilibria, and finally, (d) much of the interest of the profession is sensitivity analysis (comparative statics and dynamics), which remains applicable in the presence of multiplicity, in that one can study that in response to some perturbation, what happens to (i) each equilibrium, (ii) the worst (or best) case, and (iii) the "spread" among all equilibria.

But the presence of multiplicity is a promise, no less than a challenge, judging by the context of endogenous growth. Specifically, it reveals that (a) mutual expectations among representative agents matters, as "knowledge capital" is non-rivalrous, (thus, an economy with identical agents is not adequately representable as an economy with a single agent), (b) since initial conditions are no longer decisive, behavior norms and social habits may allow a relatively backward economy to overtake a relatively advanced one, even if both are similar in technology and preferences. Even more far-reaching results may be in order in the more conventional macro-economic studies where agents are likely to be heterogeneous.

Still far afield lies the prospect of identifying the class of economic problems where the generic existence of a continuum of equilibrium pevails, such as the overlapping-generation models on money as well as sunspots beside the studies of incomplete markets. The unifying theme seems to be the absence of a single, complete market price system, which all agents take as given. But only the future can tell.

5. Methodology reconsidered

The findings reported in the above two sections, coming from our own "highly stylized" research, are not the most "operational" for the study of common property resources. For example, the study by Reinganum and Stokey (1985) on staged cooperation between oil producers and by Clark (1976) on competing fishing fleets are models closer to issues in real life. Common property problems are not necessarily the most significant among the economic applications of differential games. Macro-economic applications and oligopolistic markets are at least as important. We do not claim that the closed-loop Nash equilibrium is the most appropriate in all contexts. More will be said below.

What we illustrated is that first, the problem of renewable common property resources forms a readily understood example for the likely existence of a non-denumerable class of Pareto-ranked equilibria. Second, many other problems, such as the theory of endogenous growth, have model structures highly similar to common property problems. Third, what is highlighted is the economically significant fact that the mutual expectations among individuals matter, and hence the differences in attitudes, habits and norms may be decisive in performance variations. Finally, for problems with the suitable *model structure* and given the equilibrium concept appropriate for the substantive issues, differential games may provide the most detailed information about the game. This is so under quite diverse hypotheses[7], without necessarily relying upon particular assumptions on either the function forms (e.g., logarithmic felicity) or the players (e.g., they are identical). This is quite contrary to the common perception among economists today.

Ours is a "constructive" approach, seeking to obtain the equilibrium strategies by solving the differential equations. An elegant alternative is found in Stokey (1986), where an "existential" approach is adopted via a contraction mapping argument.

The problem structure we dealt with above is not the only "tractable" case. The incisive analysis of Dockner et al. (1985) has dealt with other types of tractable games.

6. "Tractable" differential games

Two features of the models in Sections 4 and 5 make them easy to solve. First, there is only a single state variable so that there is neither the need to deal with partial differential equations, not much complication in solving for the dynamic path (more will be said on the last point, next section). Second, the model structure has two particularly favorable properties: (i) the felicity index only depends upon

---

[7]Whether the world is deterministic, or with a Gaussian noise, or subject to a jump process, with players behaving as if they are with a deterministic, though heavily discounted future.

the control c_i and (ii) c_i is additively separable from the other terms in the state equation. Together this means:

$$\partial(\text{argmax } H_i)/\partial x = 0. \tag{61}$$

With both the strict monotonicity of u_i, and the linearity of the state equation in c_i, property (6.1) allows the substitution of the control c_i for the adjoint variable, p_i. In control theory, this permits the characterization of the optimal policy by a fruitful means: a phase diagram in the two-dimensional c–x space. Now for differential games, even in the two-person case, a similar approach would lead to the rather daunting three-dimensional phase diagrams[8]. Instead, we opt for a straightforward analytic solution.

Whether similar properties can be found in other model structures deserve investigation. One obvious candidate is the type of models where the "advantage" enjoyed by one decision maker matches the "disadvantage" hindering the other, and this comparative edge (the state variable) may be shifted as a consequence of the mutually offsetting controls of the two players exercised at some opportunity cost. An on-going study of Richard Schuler on oligopolistic price adjustment pursues this line.

The other important class of models concern games with closed-loop Nash equilibrium strategies which are either dependent upon time only, or are constants over time. One of such types of games was introduced in the trilinear games of Clemhout and Wan (1974a), as an outgrowth of the oligopolistic studies of Clemhout et al. (1973). It was applied to the issue of barrier-to-entry by pre-emptive pricing in Clemhout and Wan (1974b)[9]. Subsequent research indicates that there is an entire class of games with state-independent closed-loop strategies. Within this genre, the studies on innovation and patent race by Reinganum (1981, 1982) have much impact on these issues. The contribution of Haurie and Leitmann (1984) indicates how can the literature of global asymptotic stability be invoked in differential games. The paper by Dockner et al. (1985) shows that a strengthening of the condition (6.1) yields the property of state-separation in the form of:

$$\partial(\text{argmax } H_i)/\partial x = 0 = \partial^2 H_i/\partial x^2. \tag{6.2}$$

This is the cause for the closed-loop strategies to be "state-independent". This paper also provides reference to the various applications of this technique by Feichtinger via a phase diagram approach. For aspiring users of the differential game, it is a manual all should consult. Fershtman (1987) tackles the same topic through the elegant approach based upon the concepts of equivalent class and degeneracy.

---

[8]Three-dimensional graphic analysis is of course feasible [cf. Guckenheimer and Holmes (1983)].

[9]In subsequent models, the entrant is inhibited in a two-stage game by signalling credibly either the intention to resist (by precommitment through investment), or the capability to win (through low price in a separating equilibrium). This paper explored the equally valid but less subtle tactics of precluding profitable entry by strengthening incumbency through the brand loyalty of an extended clientele.

7. The linear-quadratic differential game

The model of quadratic loss and linear dynamics has been studied in both control theory and economics [see, e.g., Bryson and Ho (1969), Holt et al. (1960)]. It was the principal example for Case (1969) when he pioneered the study of the N-person, non-zero sum differential games. The general form may be stated as follows:

The state equation takes the form of

$$x' = Ax + \sum_j B_j Y_j + w, \quad j = 1, \ldots, N, \tag{7.1}$$

where the form w may represent a constant or white noise.

Player i has the payoff

$$J_i = \int_0^T \left\{ d_i x + \sum_j c_{ij} y_j + \frac{1}{2} \left[x' Q_i x + \sum_j x' G_{ij} Y_j + \left(\sum_k y'_j R_{ijk} Y_k \right) \right] \right\} dt, \tag{7.2}$$

where x is the state vector and y_i is the control vector of player j, with all vectors and matrices having appropriate dimensions.

In almost all the early treatments of non-zero differential games, some economics-motivated linear-quadratic examples are ubiquitously present as illustration. Simaan and Takayama (1978) was perhaps one of the earliest among the article length micro-economic applications of this model.

The recent studies of stricky price duopolies by Fershtman and Kamien (1987, 1990) are perhaps among the most successful examples in distilling much economic insights out of an extremely simple game specification, where:

$N = 2$, $x \in \Re_+$ denotes price, y_i denotes sales of firm $i = 1, 2$.

$A = B_i = -e^{-rt}$, $w = e^{-rt}$, $d_i = 0, c_{ij} = -ce^{-rt}$, if $j = i$; $c_{ij} = 0$, if $j \neq i$.

$Q_i = 0$, $G_{ij} = 2e^{-rt}$, if $j = i$; $G_{ij} = 0$, $j \neq i$.

$R_{ijk} = -e^{-rt}$, if $i = j = k$; $R_{ijk} = 0$, otherwise.

Thus, we have

$$x' = [(1 - \sum y_i) - x], \tag{7.1'}$$

where $(1 - \sum y_i)$ is the "virtual" inverse demand function, giving the "target price" which the actual price x adjust to, at a speed which we have normalized to unity. Any doubling of the adjustment speed will be treated as halving the discount rate r. Also we have

$$J_i = \int_0^\infty e^{-rt} \{x - [c_i + (y_i/2)]\} y_i \, dt, \tag{7.2'}$$

where $[c_i + (y_i/2)]$ is the linearly rising unit cost and $\{x - [c_i + (y_i/2)]\}$ is the unit profit.

It is found that for the finite horizon model, the infinite horizon equilibrium path serves as a "turnpike" approximation, and for the infinite horizon model, output is zero at low prices and adjusts by a linear decision rule, as it is expected for such linear-quadratic models. The sensitivity analysis result confirms the intuition, that the equilibrium price is constant over time and the adjustment speed only depends upon how far the actual price differs from equilibrium. Furthermore, as noted above, the higher is the discount rate the slower is the adjustment. The GAS literature is invoked in determining the equilibrium path.

Tsutsui and Mino (1990) added a price ceiling to the Fershtman–Kamien model and found a continuum of solutions which reminds us of the "folk theorem". Their model has the unusual feature that both duopolists' payoff integrals terminate, once the price ceiling is reached. However, as shown above a continuum of solutions also arises in the models of fish war [see Clemhout and Wan (1994a) and the references therein] as well as endogenous growth [see Clemhout and Wan (1994b)].

At the first sight, linear-quadratic models seem to be attractive for both its apparent versatility (in dimensions) and ready computability (by Ricatti equations). However, a high-dimensional state space or control space ushers in a plethora of coefficients in the quadratic terms which defy easy economic interpretations. Nor is the numerical computability very helpful if one cannot easily determine the approximate values of these terms. Another caveat is that the linear-quadratic format is not universally appropriate for all contexts. Linear dynamics cannot portray the phenomenon of diminishing returns in certain models (e.g., fishery). Likewise, the quadratic objective function may be a dubious choice for the felicity index, since it violates the law of absolute risk aversion.

Linear-quadratic games played an extensive role in the macro-economic literature, mostly in discrete time and with open-loop as the solution concept. See a very brief remark below.

8. Some other differential game models

Space only allows us to touch upon the following topics very briefly. The first concerns a quasi-economic application of differential games, and the second and the third are points of contacts with discrete time games.

(1) Differential games of arms race. Starting from Simaan and Cruz (1975), the recent contributions on this subject are represented by Gillespie et al. (1977), Zinnes et al. (1978) as well as van der Ploeg and de Zeeuw (1989). As Isard (1988, p. 29) noted, the arrival of differential games marks the truly non-myopic modeling of the arms race. These also include useful references to the earlier literature. Some but not all of these are linear-quadratic models. Both closed-loop and open-loop equilibria concepts are explored. By simulation, these models aim at obtaining broad insights into the dynamic interactions between the two superpowers who

balance the non-economic interests in security against the economic costs of defense.

(2) Differential games of bargaining under strikes. Started by Leitmann and Liu (1974), and generalized in Clemhout et al. (1975, 1976), Chen and Leitmann (1980) and Clemhout and Wan (1988), this research has developed another "solvable" model, motivated by the phenomenon of bargaining under strike.

The model may be illustrated with the "pie" (e.g., Antarctic mineral reserves), under the joint control of two persons endowed with individual rationality but lacking group rationality. The mutually inconsistent demands of the two players are the two state variables, and the players' preference conform to what are now know in Rubinstein's axioms as "pie is desirable" and "time is valuable". Under Chen and Leitmann (1980) as well as Clemhout and Wan (1988), the admissible (concession) strategies include all those that are now known as "cumulative distribution function strategies" in Simon and Stinchcombe (1989). The information used may be either the current state or any history up to this date. Under the "Leitmann equilibrium", each player will give concessions to his bargaining partner at exactly such a pace so that the other player is indifferent between his initial demand or any of his subsequent scaled down demand. This characterizes the *least Pareto-effecient subgame-perfect equilibrium* among two individually rational agents. Inefficiency takes the form of a costly, delayed settlement. Worse results for either player would defy rationality, since that is dominated by immediate and unconditional acceptance of the opponent's initial demand.

Characterizing the (self-enforced), delayed, least efficient equilibrium is not an end in itself, especially since no proposal is made for its prevention. It may be viewed, however, as symmetrical to the research program initiated by Nash (1950) [as well as Rubinstein (1982), along the same line]; the characterizing of a self-enforced, instant, fully efficient equilibrium. Both may help the study of the real-life settlements of strikes which is neither fully, nor least efficient, and presumably with some but not the longest delays.

(3) Macro-economic applications. Starting from the linear, differential game model of Lancaster (1973) on capitalism, linear-quadratic game of policy non-cooperation of Pindyck (1977), and the Stackelberg differential game model on time-inconsistency by Kydland and Prescott (1977), there is now a sizeable collection of formal, dynamic models in macro-economics. (See Pohjola (1986) for a survey.) So far, most of these provide useful numerical examples to illustrate certain "plausible" positions, e.g., the inefficiency of uncoordinated efforts for stabilization. The full power of the dynamic game theory is dramatically revealed in Kydland and Prescott, op. cit.: It establishes, once and for all, that the "dominant player" in a "Stackelberg differential games" may gain more through the "announcement effect" by pre-committing itself in a strategy which is inconsistent with the principle of optimality. Thus, if credible policy statement can be made and carried out, a government may do better than using naively any variational principles for dynamic optimization.

To clarify the above discussions for a two-person game, note that by (3.3) and (3.5'), the payoff for 1 (the dominant player) by adopting the (most favorable[10]) closed-loop Nash equilibrium strategy at some specific (x, t) may be written as

$$J_1(x, t; s^*) = \max_{s_1} J_1(x, t; s_1, B_2(s_1)), \tag{8.1}$$

subject to

$$s_1 = B_1(B_2(s_1)). \tag{8.2}$$

A "time-consistent" policy for player 1 is identifiable as the solution to (8.1) and (8.2). A "time-inconsistent" optimal policy (or a "Stackelberg strategy" for player 1) is the maximizer for (8.1), *un*constrained by (8.2). It is usually better than the time-consistent policy.

Informally, the impact of dynamic game theory is pervasive, though still incomplete. In his Jahnsson Lecture, Lucas (1985) states that the main ideas of "rational expectations" school are consequences of dynamic game theory. The new wave has brought timely revitalization to the macro-economic debates. Yet it appears that if the logic of dynamic game theory is fully carried through, then positions of the "rational expectations" school in macro-economics may need major qualification as well. For example, as argued in Wan (1985) on "econometric policy evaluation", if a policy is to be evaluated, presumably the policy can be changed. If it is known that policies can be changed – hence also *re*changed[11] – then the public would not respond discontinuously to announced policy changes. If such (hedging) public response to past policy changes is captured by Keynesian econometrics, surely the latter is not flawed as it is critiqued.

In fact, much of the present day macro-economics rests upon the hypothesis that the *public act as one single player*, yet all economic policy debates suggest the contrary. As has been seen, even when all agents are identical, the collective behavior of a class of interacting representative agents gives rise to a continuum of Pareto-ranked Nash equilibria, in the context of endogenous growth, in sharp contrast with a dynamic Robinson-Crusoe economy. What agent heterogeneity will bring remains to be seen.

9. Concluding remarks

From the game-theoretic perspective, most works in the differential games literature are specific applications of *non-cooperative games* in *extensive form* with the *(subgame-perfect) Nash equilibrium* as its *solution concept*. From the very outset, differential games arose, in response to the study of military tactics. It has always

---

[10] If the equilibrium is not unique.
[11] In any event, no incumbent regime can precommit the succeeding government.

been more "operational" than "conceptual", compared against the norm in other branchs of the game theory literature. Its justification of existence is to describe "everything about something" (typical example: the "homicidal chauffeur"). In contrast, game theory usually derives its power by abstraction, seeking to establish "something about everything" (typical example: Zermelo's Theorem, see Chapter 3 in Volume I of this Handbook).

To characterize in detail the dynamics of the solution for the game, differential games are so formulated to take advantage of the *Hamilton–Jacobi theory*. Exploiting the duality relation between the "state variables" (i.e., elements of the *phase set*) and the *gradient* of the *value function*, this theory allows one to solve an optimal control problem through ordinary differential equations, and not the *Hamilton–Jacobi* partial differential equation. The assumptions (a) a continuous time, (b) *closed-loop strategies* which decide controls only by time and the current state (but not by history), and (c) the simultaneous moves by players are adopted to gain access to that tool-kit. Thus the differential game is sometimes viewed more as a branch of control theory rather than game theory, or else that game theory and control theory are the two "parents" to both differential games (in continuous time) and "multi-stage games" (in discrete time) in the view of Basar and Olsder (1982, p. 2).

Assumptions (a), (b) and (c) are not "costless". Like in the case of control theory, to provide detailed information about the solution, one is often forced either to stay with some particularly tractable model, for better for worse (e.g., the linear-quadratic game), or to limit the research to problems with only one or two state variables. Moreover, continuous time is inconvenient for some qualitative issues like asymmetric information, or Bayesian–Nash equilibrium. For example, continuous Bayesian updating invites much more analytical complications than periodic updating but offers few additional insight in return. Thus, such models contribute relatively less in the 1980s when economists are searching for the "right" solution concepts. In fact, many collateral branches of dynamic game models rose, distancing themselves from the *Hamilton–Jacobi theory* by relaxing the assumptions (c) [e.g., the "Stackelberg differential games" pioneered in Chen and Cruz (1972) and Simaan and Cruz (1973a and b)], (b) [e.g., the memory strategies studied by Basar (1976) for uniqueness, and Tolwinski (1982) for cooperation], as well as (a) [see the large literature on discrete time models surveyed in Basar and Olsder (1982)].

Further, the same Hamilton–Jacobi tool-kit has proved to be much less efficacious for N-person, general-sum differential games than for both the control theory and the two-person, zero-sum differential games. In the latter cases, the *Hamilton–Jacobi theory* makes it possible to bypass the *Hamilton–Jacobi* partial differential equation for backward induction. Thus, routine computation may be applied in lieu of insightful conjecture. The matter is more complex for N-person, general-sum differential games. Not only separate *value functions* must be set up, one for each player, the differential equations for the "dual variables" (i.e., the gradient of the value functions) are no longer solvable in a routine manner, in general. They are no longer ordinary differential equations but contain the (yet unknown) gradients of the closed-loop strategies. Progress is usually made by identifying more and more classes of "tractable" games and matching them with particular classes of solutions. Such classes may be either identified with particular function forms (e.g., the linear-quadratic game), or by features in the model structure (e.g., symmetric games). An analogy is the 19th century theory of differential equations, when success was won, case by case, type by type.

Of course, how successful can one divide-and-conquer depends upon what need does a theory serve. Differential equations in the 19th century catered to the needs of physical sciences, where the measured regularities transcend time and space. Devising a context-specific theory to fit the data from specific, significant problems makes much sense. More complicated is the life for researchers in differential games.

Much of the application of N-person, general-sum differential games is in economics, where observed regularities are rarely invariant as in natural sciences. Thus, expenditure patterns in America offer little insights upon the consumption habits in Papua-New Guinea. Out of those differential games which depend on specific function forms (e.g., the linear-quadratic game), one may construct useful theoretic examples, but not the basis for robust predictions. In contrast, in the case of optimal control, broad conclusions are often obtained by means of the "globally analytic" phase diagram, for those problems with a low-dimension state space.

On the other hand, from the viewpoint of economics, there are two distinct types of contributions that differential games can offer. First, as has been seen regarding the multiplicity of solutions, differential game can yield broad conceptual contributions which do not require the detailed solution(s) of a particular game. Second, above and beyond that, there remains an unsatisfied need that differential games may meet. For situations where a single decision maker faces an impersonal environment, the system dynamics can be studied fruitfully with optimal control models. There are analogous situations where the system dynamics is decided by the interactions of a few players. Here, differential games seem to be the natural tool. What economists wish to predict is not only the details about a single time profile, such as existence and stability of any long run configuration and the monotonicity and the speed of convergence toward that limit, but also the findings from the sensitivity analysis: how such a configuration responds toward parametric variations. From the patent race, industrial-wide learning, the "class" relationship in a capitalistic economy, the employment relationships in a firm or industry, to the oligopolistic markets, the explotation of a "common property" asset, and the political economy of the arms race, the list can go on and on.

To be sure, the formulation of such models should be governed by the conceptual considerations of the general theory of non-cooperative games. The closed-loop Nash equilibrium solution may or may not always be the best. But once a particular model – or a set of alternative models – is chosen, then differential games may be used to track down the stability properties and conduct sensitivity analysis for the time path. For the field of economics, the choice of the solution concept may be the first on the research agenda – some of the time – but it cannot be the last. Some tools like differential games are eventually needed to round out the details.

Today, the state of art in differential game is adequate for supplying insightful particular examples. To fulfill its role further in the division of labor suggested above, it is desirable for differential games to acquire "globally analytic" capabilities, and reduce the present dependence on specific functional forms. Toward this objective, several approaches toward this goal have barely started.

By and large, we have limited our discussions mainly to the closed-loop Nash equilibrium in continuous time models. Many of the contributions to the above mentioned research have also worked on the "cooperative differential games". Surveys by Ho and Olsder (1983) and Basar (1986) have commented on this literature. Papers by Tolwinski et al. (1986) and Kaitala (1986) are representatives of this line of work. The paper of Benhabib and Radner (1992) relate this branch of research to the repeated game literature. The "extreme strategy" in that model correspond to the fastest exploitation strategy in Clemhout and Wan (1989) and the ASAP depletion strategy in Section 3 above, for a "non-productive" resource (in the terminology of Benhabib and Radner). But this is treated elsewhere in this Handbook.

References

Basar, T. (1976) 'On the uniqueness of the Nash solution in linear-quadratic differential games', *International Journal of Game Theory*, **5**: 65–90.
Basar, T. and G.T. Olsder (1982) *Dynamic non-cooperative game theory*. New York: Academic Press.
Basar, T. and P. Bernhard (1991) 'H^∞-optimal control and related minimax design problems: A dynamic game approach', in: T. Basar and A. Haurie, eds., *Advances in Dynamic Games and Applications*. Boston: Birkhäuser.
Benhabib, J. and R. Radner (1992) 'Joint exploitation of a productive asset; a game theoretic approach', *Economic Theory*, **2**: 155–90.
Brito, D.L. (1973) 'Estimation prediction and economic control', *International Economic Review*, **14**: 3, 646–652.
Bryson, A.E. and Y.C. Ho (1969) *Applied optimal control*. Waltham, MA: Ginn and Co.
Case, J.H. (1969) 'Towards a theory of many player differential games', *SIAM Journal on Control*, **7**: 2, 179–197.
Case, J.H. (1979) *Economics and the competitive process*. New York: New York University Press.
Chen, C.I. and J. B. Cruz, Jr. (1972) 'Stackelberg Solution for two-person games with biased information patterns', *IEEE Transactions on automatic control*, **AC-17**: 791–797.
Chen, S.F.H. and G. Leitmann (1980) 'Labor-management bargaining models as a dynamic game', *Optimal Control, Applications and Methods*, **1**: 1, 11–26.
Clark, C.W. (1976) *Mathematical bioeconomics: The optimal management of renewable resources*. New York: Wiley.
Clemhout, S., G. Leitmann and H.Y. Wan, Jr. (1973) 'A differential game model of oligopoly', *Cybernetics*, **3**: 1, 24–39.
Clemhout, S. and H.Y. Wan, Jr. (1974a) 'A class of trilinear differential games', *Journal of Optimization Theory and Applications*, **14**: 419–424.
Clemhout, S. and H.Y. Wan, Jr. (1974b) 'Pricing dynamic: intertemporal oligopoly model and limit pricing', Working Paper No. 63, *Economic Department, Cornell University*, Ithaca, NY.

Clemhout, S., G. Leitmann and H.Y. Wan, Jr. (1975) 'A model of bargaining under strike: The differential game view', *Journal of Economic Theory*, **11**: 1, 55–67.
Clemhout, S., G. Leitmann and H.Y. Wan, Jr. (1976) 'Equilibrium patterns for bargaining under strike: A differential game model', in: Y.C. Ho and S.K. Mitter, eds. *Direction in Large-Scale Systems, Many-Person Optimization, and Decentralized Control*. New York: Plenum Press.
Clemhout, S. and H.Y. Wan, Jr. (1985a) 'Dynamic common property resources and environmental problems', *Journal of Optimization, Theory and Applications*, **46**: 4, 471–481.
Clemhout, S. and H.Y. Wan, Jr. (1985b) 'Cartelization conserves endangered species?', in: G. Feichtinger, ed., *Optimal Control Theory and Economic Analysis 2*. Amsterdam: North-Holland.
Clemhout, S. and H.Y. Wan, Jr. (1985c) 'Common-property exploitations under risks of resource extinctions', in: T. Basar, ed., *Dynamic Games and Applications in Economics*. New York: Springer-Verlag.
Clemhout, S. and H.Y. Wan, Jr. (1988) 'A general dynamic game of bargaining – the perfect information case', in: H.A. Eiselt and G. Pederzoli, eds., *Advances in Optimization and Control*. New York: Springer-Verlag.
Clemhout, S. and H.Y. Wan, Jr. (1989) 'On games of cake-eating', in: F. van der Ploeg and A. De Zeeuw, eds., *Dynamic policy Games in Economics*. Amsterdam: North-Holland, pp. 121–152.
Clemhout, S. and H.Y. Wan, Jr. (1990) 'Environmental problem as a common property resource game', *Proceedings of the 4th World Conference on Differential Games*, Helsinki, Finland.
Clemhout, S. and H.Y. Wan, Jr. (1994a) 'The Non-uniqueness of Markovian Strategy Equilibrium: The Case of Continuous Time Models for Non-renewable Resources', in: T. Basar and A. Haurie., *Advances in Dynamic Games and Applications*. Boston: Birkhäuser.
Clemhout, S. and H.Y. Wan, Jr. (1994b) 'On the Foundations of Endogeneous Growth', in: M. Breton and G. Zacoeur, eds., *Preprint volume, 6th International Symposium on Dynamic Games and Applications*.
Deissenberg, C. (1988) 'Long-run macro-econometric stabilization under bounded uncertainty', in: H.A. Eiselt and G. Pederzoli, eds., *Advances in Optimization and Control*. New York: Springer-Verlag.
Dockner, E., G. Feichtinger and S. Jorgenson (1985) 'Tractable classes of non-zero-sum, open-loop Nash differential games: theory and examples', *Journal of Optimization Theory and Applications*, **45**: 179–198.
Fershtman, C. (1987) 'Identification of classes of differential games for which the open look is a degenerate feedback Nash equilibrium', *Journal of Optimization Theory and Applications*, **55**: 2, 217–231.
Fershtman, C. and M.I. Kamien (1987) 'Dynamic duopolistic competition with sticky prices', *Econometrica*, **55**: 5, 1151–1164.
Fershtman, C. and M.I. Kamien (1990) 'Turnpike properties in a finite horizon differential game: dynamic doupoly game with sticky prices', *International Economic Review*, **31**: 49–60.
Gillespie, J.V. et al. (1977) 'Deterrence of second attack capability: an optimal control model and differential game', in: J.V. Gillespie and D.A. Zinnes, eds., *Mathematical Systems in International Relations Research*. New York: Praeger, pp. 367–85.
Guckenheimer, J. and P. Holmes (1983) *Nonlinear oscillations, dynamical systems and bifurcations of vector fields*. New York: Springer-Verlag.
Gutman, S. and G. Leitmann (1976) 'Stabilizing feedback control for dynamic systems with bounded uncertainty', *Proceedings of the IEEE Conference on Decision and Control*, Gainesville, Florida.
Haurie, A. and G. Leitmann (1984) 'On the global asymptotic stability of equilibrium solutions for open-loop differential games', *Large Scale Systems*, **6**: 107–122.
Ho, Y.C. and G.J. Olsder (1983) 'Differential games: concepts and applications', in: M. Shubik, ed., *Mathematics of Conflicts*. Amsterdam: North-Holland.
Holt, C., F. Modigliani, J.B. Muth and H. Simon (1960) *Planning production, inventories and work force*. Englewood Cliffs: Prentice Hall.
Holt, C.A. (1985) 'An experimental test of the consistent-conjectures hypothesis', *American Economic Review*, **75**: 314–25.
Isaacs, R. (1965) *Differential games*. New York: Wiley.
Isard, W. (1988) *Arms races, arms control and conflict analysis*. Cambridge: Cambridge University Press.
Kaitala, V. (1986) 'Game theory models of fisheries management – a survey', in: T. Basar, ed., *Dynamic Games and Applications in Economics*. New York: Springer-Verlag.
Kamien, M.I. and Schwartz, N.L. (1981) *Dynamic optimization*. New York: Elsevier, North-Holland.

Kydland, F.E. and E.C. Prescott (1977) 'Rules rather than discretion: The inconsistency of optimal plans', *Journal of Political Economy*, **85**: 473–493.

Lancaster, K. (1973) 'The dynamic inefficiency of capitalism', *Journal of Political Economy*, **81**: 1092–1109.

Leitmann, G. and P.T. Liu (1974) 'A differential game model of labor management negotiations during a strike', *Journal of Optimization Theory and Applications*, **13**: 4, 427–435.

Leitmann, G. and H.Y. Wan, Jr. (1978) 'A stabilization policy for an economy with some unknown characteristics', *Journal of the Franklin Institute*, 250–278.

Levhari, D. and L.J. Mirman (1980) 'The great fish war: an example using a dynamic Cournot-Nash solution', *The Bell Journal of Economics*, **11**: 322–334.

Lucas, R.E. Jr. (1987) *Model of business cycles*. Oxford: Basil Blackwell.

Nash, J.F. Jr. (1950) 'The bargaining problem', *Econometrica*, **18**: 155–62.

Mehlmann, A. (1988) *Applied differential games*. New York: Plenum Press.

Pindyck, R.S. (1977) 'Optimal planning for economic stabilization policies under decentralized control and conflicting objectives', *IEEE Transactions on Automatic Control*, **AC-22**: 517–530.

Van der Ploeg, F. and A.J. De Zeeuw (1989) 'Perfect equilibrium in a competitive model of arms accumulation', *International Economic Review*, forthcoming.

Pohjola, M. (1986) 'Applications of dynamic game theory to macroeconomics', in: T. Basar, ed., *Dynamic Games and Applications to Economics*. New York: Springer-Verlag.

Reinganum, J.F. (1981) 'Dynamic games of innovation', *Journal of Economic Theory*, **25**: 21–41.

Reinganum, J.F. (1982) 'A dynamic game of R and D: Patent protection and competitive behavior', *Econometrica*, **50**: 671–688.

Reinganum, J. and N.L. Stokey (1985) 'Oligopoly extraction of a common property natural resource: The importance of the period of commitment in dynamic games', *International Economic Review*, **26**: 161–173.

Rubinstein, A. (1982) 'Prefect equilibrium in a bargaining model', *Econometrica*, **50**: 97–109.

Shimemura, K. (1991) 'The feedback equilibriums of a differential game of capitalism', *Journal of Economic Dynamics and Control*, **15**: 317–338.

Simaan, M. and J.B. Cruz, Jr. (1973a) 'On the Stackelberg strategy in nonzero-sum games', *Journal of Optimization Theory and Applications*, **11**: 5, 533–555.

Simaan, M. and J.B. Cruz, Jr. (1973b) 'Additional aspects of the Stackelberg strategy in nonzero-sum games', *Journal of Optimization Theory and Applications*, **11**: 6, 613–626.

Simaan, M. and J.B., Cruz, Jr. (1975) 'Formulation of Richardson's model of arms race from a differential game viewpoint', *The Review of Economic Studies*, **42**: 1, 67–77.

Simaan, M. and T. Takayama (1978) 'Game theory applied to dynamic duopoly problems with production constraints', *Automatica*, **14**: 161–166.

Simon, L.K. and M.B. Stinchcombe (1989) 'Extensive form games in continuous time: pure strategies', *Econometrica*, **57**: 1171–1214.

Starr, A.W. and Y.C. Ho (1969) 'Non zero-sum differential games': *Journal of Optimization Theory and Applications*, **3**: 184–206.

Stokey, N.L. (1980) The dynamics of industry-wide learning, in: W.P. Heller et al., eds., *Essays in Honour of Kenneth J. Arrow.*, Cambridge: Cambridge University Press.

Tolwinski, B. (1982) 'A concept of cooperative equilibrium for dynamic games', *Automatica*, **18**: 431–447.

Tolwinski, B., A. Haurie and G. Leitman (1986) 'Cooperative equilibria in differential games', *Journal of Mathematical Analysis and Applications*, **119**: 182–202.

Tsutsui, S. and K. Mino (1990) 'Nonlinear strategies in dynamic duopolistic competition with sticky prices', *Journal Economic Theory*, **52**: 136–161.

Wan, H.Y. Jr. (1985) 'The new classical economics – a game-theoretic critique', in: G.R. Feiwel, ed., *Issues in Contemporary Macroeconomics and Distribution*. London: MacMillian, pp. 235–257.

Zinnes, et al. (1978) 'Arms and aid: A differential game analysis', in: W.L. Hollist, ed., *Exploring Competitive Arms Processes*. New York: Marcel Dekker, pp. 17–38.

Chapter 24

COMMUNICATION, CORRELATED EQUILIBRIA AND INCENTIVE COMPATIBILITY

ROGER B. MYERSON

Northwestern University

Contents

1.	Correlated equilibria of strategic-form games	828
2.	Incentive-compatible mechanisms for Bayesian games	835
3.	Sender–receiver games	840
4.	Communication in multistage games	845
References		847

1. Correlated equilibria of strategic-form games

It has been argued [at least since von Neumann and Morgenstern (1944)] that there is no loss of generality in assuming that, in a strategic-form game, all players choose their strategies simultaneously and independently. In principle, anything that a player can do to communicate and coordinate with other players could be described by moves in an extensive-form game, so that planning these communication moves would become part of his strategy choice itself.

Although this perspective may be fully general in principle, it is not necessarily the most fruitful way to think about all games. There are many situations where the possibilities for communication are so rich that to follow this modeling program rigorously would require us to consider enormously complicated games. For example, to model player 1's opportunity to say just one word to player 2, player 1 must have a move and player 2 must have an information state for every word in the dictionary! In such situations, it may be more useful to leave communication and coordination possibilities out of the explicit model. If we do so, then we must instead use solution concepts that express an assumption that players have implicit communication opportunities, in addition to the strategic options explicitly described in the game model. We consider here such solution concepts and show that they can indeed offer important analytical insights into many situations.

So let us say that a game is *with communication* if, in addition to the strategy options explicitly specified in the structure of the game, the players have a very wide range of implicit options to communicate with each other. We do not assume here that they have any ability to sign contracts; they can only talk. Aumann (1974) showed that the solutions for games with communication may have remarkable properties, even without contracts.

Consider the two-player game with payoffs as shown in table 1, where each player i must choose x_i or y_i (for $i = 1, 2$). Without communication, there are three equilibria of this game: (x_1, x_2) which gives the payoff allocation $(5, 1)$; (y_1, y_2) which gives the payoff allocation $(1, 5)$; and a randomized equilibrium which gives the expected payoff allocation $(2.5, 2.5)$. The best symmetric payoff allocation $(4, 4)$ cannot be achieved by the players without contracts, because (y_1, x_2) is not an equilibrium. However, even without binding contracts, communication may allow the players to achieve an expected payoff allocation that is better for both than

Table 1

	x_2	y_2
x_1	5,1	0,0
y_1	4,4	1,5

(2.5, 2.5). Specifically, the players may plan to toss a coin and choose (x_1, x_2) if heads occurs or (y_1, y_2) if tails occurs. Even though the coin has no binding force on the players, such a plan is *self-enforcing*, in the sense that neither player could gain by unilaterally deviating from this plan.

With the help of a *mediator* (that is, a person or machine that can help the players communicate and share information), there is a self-enforcing plan that generates the even better expected payoff allocation $(3\frac{1}{3}, 3\frac{1}{3})$. To be specific, suppose that a mediator randomly recommends strategies to the two players in such a way that each of the pairs (x_1, x_2), (y_1, y_2), and (y_1, x_2) may be recommended with probability $\frac{1}{3}$. Suppose also that each player learns only his own recommended strategy from the mediator. Then, even though the mediator's recommendation has no binding force, there is a Nash equilibrium of the transformed game with mediated communication in which both players plan always to obey the mediator's recommendations. If player 1 heard the recommendation "y_1" from the mediator, then he would think that player 2 may have been told to do x_2 or y_2 with equal probability, in which case his expected payoff from y_1 would be as good as from x_1 (2.5 from either strategy). If player 1 heard a recommendation "x_1" from the mediator then he would know that player 2 was told to do x_2, to which his best response is x_1. So player 1 would always be willing to obey the mediator if he expected player 2 to obey the mediator, and a similar argument applies to player 2. That is, the players can reach a self-enforcing understanding that each obey the mediator's recommendation when he plans to randomize in this way. Randomizing between (x_1, x_2), (y_1, y_2), and (y_1, x_2) with equal probability gives the expected payoff allocation

$$\tfrac{1}{3}(5, 1) + \tfrac{1}{3}(4, 4) + \tfrac{1}{3}(1, 5) = (3\tfrac{1}{3}, 3\tfrac{1}{3}).$$

Notice that the implementation of this correlated strategy $(\frac{1}{3}[x_1, x_2] + \frac{1}{3}[y_1, y_2] + \frac{1}{3}[y_1, x_2]$ without contracts required that each player get different partial information about the outcome of the mediator's randomization. If player 1 knew when player 2 was told to choose x_2, then player 1 would be unwilling to choose y_1 when it was also recommended to him. So this correlated strategy could not be implemented without some kind of mediation or noisy communication. With only direct unmediated communication in which all players observe anyone's statements or the outcomes of any randomization, the only self-enforcing plans that the players could implement without contracts would be randomizations among the Nash equilibria of the original game (without communication), like the correlated strategy $0.5[x_1, x_2] + 0.5[y_1, y_2]$ that we discussed above. However, Barany (1987) and Forges (1990) have shown that, in any strategic-form and Bayesian game with four or more players, a system of direct unmediated communication between pairs of players can simulate any centralized communication system with a mediator, provided that the communication between any pair of players is not directly observable by the other players. (One of the essential ideas behind this result is that, when there are four or more players, each pair of players can use two other players as parallel mediators. The messages that two mediating players carry can be suitably encoded so that neither of the mediating players can, by himself, gain by corrupting his message or learn anything useful from it.)

Consider now what the players could do if they had a bent coin for which player 1 thought that the probability of heads was 0.9 while player 2 thought that the probability of heads was 0.1, and these assessments were common knowledge. With this coin, it would be possible to give each player an expected payoff of 4.6 by the following self-enforcing plan: toss the coin, and then implement the (x_1, x_2) equilibrium if heads occurs, and implement the (y_1, y_2) equilibrium otherwise.

However, the players beliefs about this coin would be *inconsistent*, in the sense of Harsanyi (1967, 1968). That is because there is no way to define a prior probability distribution for the outcome of the coin toss and two other random variables such that player 1's beliefs are derived by Bayes's formula from the prior and his observation of one of the random variables, player 2's beliefs are derived by Bayes's formula from the prior and her observation of the other random variable, and it is common knowledge that they assign different probabilities to the event that the outcome of the coin toss will be heads. (See Aumann (1976) for a general proof of this fact).

The existence of such a coin, about which the players have inconsistent beliefs, would be very remarkable and extraordinary. With such a coin, the players could make bets with each other that would have arbitrarily large positive expected monetary value to both! Thus, as a pragmatic convention, let us insist that the existence of any random variables about which the players may have such inconsistent beliefs should be explicitly listed in the structure of the game, and should not be swept into the implicit meaning of the phrase "game with communication." (These random variables could be explicitly modelled either by a Bayesian game with beliefs that are not consistent with any common prior, or by a game in generalized extensive form, where a distinct subjective probability distribution for each player could be assigned to the set of alternatives at each chance node.) When we say that a particular game is played "with communication" we mean only that the players can communicate with each other and with outside mediators, and that players and mediators have implicit opportunities to observe random variables that have objective probability distributions about which everyone agrees.

In general, consider any finite strategic-form game $\Gamma = (N, (C_i)_{i \in N}, (u_i)_{i \in N})$, where N is the set of players, C_i is the set of pure strategies for player i, and $u_i: C \to \mathbb{R}$ is the utility payoff function for player i. We use here the notation

$$C = \underset{i \in N}{\times} C_i.$$

A mediator who was trying to help coordinate the players's actions would have (at least) to tell each player i which strategy in C_i was recommended for him. Assuming that the mediator can communicate separately and confidentially with each player, no player needs to be told the recommendations for any other players.

Without contracts, player i would then be free to choose any strategy in C_i after hearing the mediator's recommendation. So in the game with mediated communication, each player i would actually have an enlarged set of communication strategies that would include all mappings from C_i into C_i, each of which represents a possible rule for choosing an element of C_i to implement as a function of the mediator's recommendation in C_i.

Now, suppose that it is common knowledge that the mediator will determine his recommendations according to the probability distribution μ in $\Delta(C)$, so that $\mu(c)$ denotes the probability that any given pure strategy profile $c = (c_i)_{i \in N}$ would be recommended by the mediator. (For any finite set X, we let $\Delta(X)$ denote the set of probability distributions over X.) The expected payoff to player i under this correlated strategy μ, if everyone obeys the recommendations, is

$$U_i(\mu) = \sum_{c \in C} \mu(c) u_i(c).$$

Then it would be an equilibrium for all players to obey the mediator's recommendations iff

$$U_i(\mu) \geq \sum_{c \in C} \mu(c) u_i(c_{-i}, \delta_i(c_i)), \quad \forall i \in N, \quad \forall \delta_i : C_i \to C_i, \tag{1}$$

where $U_i(\mu)$ is as defined in (6.1). [Here $(c_{-i}, \delta_i(c))$ denotes the pure strategy profile that differs from c only in that the strategy for player i is changed to $\delta_i(c_i)$.] Following Aumann (1974, 1987), we say that μ is a *correlated equilibrium* of Γ iff $\mu \in \Delta(C)$ and μ satisfies condition (1). That is, a correlated equilibrium is any correlated strategy for the players in Γ that could be self-enforcingly implemented with the help of a mediator who can make nonbinding confidential recommendations to each player. The existence of correlated equilibria can be derived from the general existence of Nash equilibria for finite games, but elegant direct proofs of the existence of correlated equilibria have also been given by Hart and Scheidler (1989) and Nau and McCardle (1990).

It can be shown that condition (1) is equivalent to the following system of inequalities:

$$\sum_{c_{-i} \in C_{-i}} \mu(c) [u_i(c) - u_i(c_{-i}, e_i)] \geq 0, \forall i \in N, \quad \forall c_i \in C_i, \quad \forall e_i \in C_i. \tag{2}$$

[Here $C_{-i} = \times_{j \in N - i} C_j$, and $c = (c_{-i}, c_i)$.] To interpret this inequality, notice that, given any c_i, dividing the left-hand side by

$$\sum_{c_{-i} \in C_{-i}} \mu(c),$$

would make it equal to the difference between player i's conditionally expected payoff from obeying the mediator's recommendation and his conditionally expected payoff from using the action e_i, given that the mediator has recommended c_i. Thus, (2) asserts that no player i could expect to increase his expected payoff by using some disobedient action e_i after getting any recommendation c_i from the mediator. These inequalities (1) and (2) may be called *strategic incentive constraints*, because they represent the mathematical inequalities that a correlated strategy must satisfy to guarantee that all players could rationally obey the mediator's recommendations.

The set of correlated equilibria is a compact and convex set, for any finite game in strategic form. Furthermore, it can be characterized by a finite collection of linear inequalities, because a vector μ in $\mathbb{R}^N$ is a correlated equilibrium iff it satisfies the strategic incentive constraints (2) and the following *probability constraints*:

$$\sum_{e \in C} \mu(e) = 1 \text{ and } \mu(c) \geq 0, \quad \forall c \in C. \tag{3}$$

Thus, for example, if we want to find the correlated equilibrium that maximizes the sum of the players' expected payoffs in Γ, we have a problem of maximizing a linear objective $[\sum_{i \in N} U_i(\mu)]$ subject to linear constraints. This is a linear programming problem, which can be solved by any one of many widely-available computer programs. [See also the general conditions for optimality subject to incentive constraints developed by Myerson (1985).]

For the game in table 1, the correlated equilibrium μ that maximizes the expected sum of the players' payoffs is

$$\mu(x_1, x_2) = \mu(y_1, x_2) = \mu(y_1, y_2) = \tfrac{1}{3} \text{ and } \mu(x_1, y_2) = 0.$$

That is, $\mu = \tfrac{1}{3}[x_1, x_2] + \tfrac{1}{3}[y_1, y_2] + \tfrac{1}{3}[y_1, x_2]$ maximizes the sum of the players' expected payoffs $U_1(\mu) + U_2(\mu)$ subject to the strategic incentive constraints (2) and the probability constraints (3). So the strategic incentive constraints imply that the players' expected sum of payoffs cannot be higher than $3\tfrac{1}{3} + 3\tfrac{1}{3} = 6\tfrac{2}{3}$.

It may be natural to ask why we have been focusing attentions on mediated communication systems in which it is rational for all players to obey the mediator. The reason is that such communication systems can simulate any equilibrium of any game that can be generated from any given strategic-form game by adding any communication system. To see why, let us try to formalize a general framework for describing communication systems that might be added to a given strategic-form game Γ as above. Given a communication system, let R_i denote the set of all strategies that player i could use to determine the reports that he sends out, into the communication system, and let M_i denote the set of all messages that player i could receive from the communication system. For any $r = (r_i)_{i \in N}$ in $R = \times_{i \in N} R_i$ and any $m = (m_i)_{i \in N}$ in $M = \times_{i \in N} M_i$, let $v(m|r)$ denote the conditional probability that m would be the messages received by the various players if each player i were sending reports according to r_i. This function $v: R \to \Delta(M)$ is our basic mathematical characterization of the communication system. (If all communication is directly between players, without noise or mediation, then every player's message would be composed directly of other players' reports to him, and so $v(\cdot|r)$ would always put probability 1 on some vector m; but noisy communication or randomized mediation allows $0 < v(m|r) < 1$.)

Given such a communication system, the set of pure communication strategies that player i can use for determining the reports that he sends and the action in C_i that he ultimately implements (as a function of the messages that he receives) is

$$B_i = \{(r_i, \delta_i) \mid r_i \in R_i, \quad \delta_i : M_i \to C_i\}.$$

Player i's expected payoff depends on the communication strategies of all players according to the function $\bar{u}_i$, where

$$\bar{u}_i((r_j, \delta_j)_{j \in N}) = \sum_{m \in M} v(m \mid r) \, u_i((\delta_j(m_j))_{j \in N}).$$

Thus, the communication system $v : R \to \Delta(M)$ generates a *communication game* Γ_v, where

$$\Gamma_v = (N, (B_i)_{i \in N}, (\bar{u}_i)_{i \in N}).$$

This game Γ_v is the appropriate game in strategic form to describe the structure of decision-making and payoffs when the game Γ has been transformed by allowing the players to communicate through the communication system v before choosing their ultimate payoff-relevant actions. To characterize rational behavior by the players in the game with communication, we should look among the equilibria of Γ_v.

However, any equilibrium of Γ_v is equivalent to a correlated equilibrium of Γ as defined by the strategic incentive constraints (2). To see why, let $\sigma = (\sigma_i)_{i \in N}$ be any equilibrium in randomized strategies of this game Γ_v. Let μ be the correlated strategy in $\Delta(C)$ defined by

$$\mu(c) = \sum_{(r,\delta) \in B} \sum_{m \in \delta^{-1}(c)} \left(\prod_{i \in N} \sigma_i(r_i, \delta_i) \right) v(m \mid r), \quad \forall c \in C,$$

where we use the notation:

$$B = \underset{i \in N}{\times} B_i, \quad (r, \delta) = ((r_i, \delta_i)_{i \in N}),$$

$$\delta^{-1}(c) = \{m \in M \mid \delta_i(m_i) = c_i, \quad \forall i \in N\}.$$

That is, the probability of any outcome c in C under the correlated strategy μ is just the probability that the players would ultimately choose this outcome after participating in the communication system v, when early player determines his plan for sending reports and choosing actions according to σ. So μ effectively simulates the outcome that results from the equilibrium σ in the communication game Γ_v. Because μ is just simulating the outcomes from using strategies σ in Γ_v, if some player i could have gained by disobeying the mediator's recommendations under μ, when all other players are expected to obey, then he could have also gained by similarly disobeying the recommendations of his own strategy σ_i when applied against σ_{-i} in Γ_v. More precisely, if (1) were violated for some i and δ_i, then player i could gain by switching from σ_i to $\hat{\sigma}_i$ against σ_{-i} in Γ_v, where

$$\hat{\sigma}_i(r_i, \gamma_i) = \sum_{\zeta_i \in Z(\delta_i, \gamma_i)} \sigma_i(r_i, \zeta_i), \quad \forall (r_i, \gamma_i) \in B_i, \quad \text{and}$$

$$Z(\delta_i, \gamma_i) = \{\zeta_i \mid \delta_i(\zeta_i(m_i)) = \gamma_i(m_i), \quad \forall m_i \in M_i\}.$$

This conclusion would violate the assumption that σ is an equilibrium. So μ must satisfy the strategic incentive constraints (1), or else σ could not be an equilibrium of Γ_v.

Thus, any equilibrium of any communication game that can be generated from a strategic-form game Γ by adding a system for preplay communication must be equivalent to a correlated equilibrium satisfying the strategic incentive constraints (1) or (2). This fact is known as the *revelation principle* for strategic-form games. [See Myerson (1982).]

For any communication system v, there may be many equilibria of the communication game Γ_v, and these equilibria may be equivalent to different correlated equilibria. In particular, for any equilibrium $\bar{\sigma}$ of the original game Γ, there are equilibria of the communication game Γ_v in which every player i chooses a strategy in C_i according to $\bar{\sigma}_i$, independently of the reports that he sends or the messages that he receives. (One such equilibrium σ of Γ_v could be defined so that

if $\delta_i(m_i) = c_i$, $\forall m_i \in M_i$, then $\sigma_i(r_i, \delta_i) = \bar{\sigma}_i(c_i)/|R_i|$,

and

if $\exists \{m_i, \hat{m}_i\} \subseteq M_i$ such that $\delta_i(m_i) \neq \delta_i(\hat{m}_i)$ then $\sigma_i(r_i, \delta_i) = 0$.)

That is, adding a communication system does not eliminate any of the equilibria of the original game, because there are always equilibria of the communication game in which reports and messages are treated as having no meaning and hence are ignored by all players. Such equilibria of the communication game are called *babbling equilibria*.

The set of correlated equilibria of a strategic-form game Γ has a simple and tractable mathematical structure, because it is closed by convex and is characterized by a finite system of linear inequalities. On the other hand, the set of Nash equilibria of Γ, or of any specific communication game that can be generated from Γ, does not generally have any such simplicity of structure. So the set of correlated equilibria, which characterizes the union of the sets of equilibria of all communication games that can be generated from Γ, may be easier to analyze than the set of equilibria of any one of these games. This observation demonstrates the analytical power of the revelation principle. That is, the general conceptual approach of accounting for communication possibilities in the solution concept, rather than in the explicit game model, not only simplifies our game models but also generates solutions that are much easier to analyze.

To emphasize the fact that the set of correlated equilibria may be strictly larger

Table 2

	x_2	y_2	z_2
x_1	0.0	5.4	4.5
y_1	4.5	0.0	5.4
z_1	5.4	4.5	0.0

than the convex hull of the set of Nash equilibria, it may be helpful to consider the game in table 2, which was studied by Moulin and Vial (1978). This game has only one Nash equilibrium,

$$(\tfrac{1}{3}[x_1] + \tfrac{1}{3}[y_1] + \tfrac{1}{3}[z_1],\ \tfrac{1}{3}[x_2] + \tfrac{1}{3}[y_2] + \tfrac{1}{3}[z_2]),$$

which gives expected payoff allocation (3,3). However, there are many correlated equilibria, including

$$((\tfrac{1}{6}[(x_1,y_2)] + \tfrac{1}{6}[(x_1,z_2)] + \tfrac{1}{6}[(y_1,x_2)] + \tfrac{1}{6}[(y_1,z_2)] + \tfrac{1}{6}[(z_1,x_2)] + \tfrac{1}{6}[(z_1,y_2)]),$$

which gives expected payoff allocation (4.5, 4.5).

2. Incentive-compatible mechanisms for Bayesian games

The revelation principle for strategic-form games asserted that any equilibrium of any communication system can be simulated by a communication system in which the only communication is form a central mediator to the players, without any communication from the players to the mediator. The one-way nature of this communication should not be surprising, because the players have no private information to tell the mediator about, within the structure of the strategic-form game. More generally, however, players in Bayesian game [as defined by Harsanyi (1967, 1968)] may have private information about their types, and two-way communication would then allow the players' actions to depend on each others' types, as well as on extraneous random variables like coin tosses. Thus, in Bayesian games with communication, there may be a need for players to talk as well as to listen in mediated communication systems. [See Forges (1986).]

Let $\Gamma^b = (N,\ (C_i)_{i \in N},\ (T_i)_{i \in N},\ (p_i)_{i \in N},\ (u_i)_{i \in N})$, be a finite Bayesian game with incomplete information. Here N is the set of players, C_i is the set of possible actions for player i, and T_i is the set of possible types (or private information states) for player i. For any $t = (t_j)_{j \in N}$ in the set $T = \times_{i \in N} T_i$, $p_i(t_{-i}|t_i)$ denotes the probability that player i would assign to the event that $t_{-i} = (t_j)_{j \in N - i}$ is the profile of types for the players other than i if t_i were player i's type. For any t in T and any $c = (c_j)_{j \in N}$ in the set $C = \times_{j \in N} C_j$, $u_i(c,t)$ denotes the utility payoff that player i would get if c were the profile of actions chosen by the players and t were the profile of their actual types. Let us suppose now that Γ^b is a game with communication, so that the players have wide opportunities to communicate, after each player i learns his type in T_i but before he chooses his action in C_i.

Consider mediated communication systems of the following form: first, each player is asked to report his type confidentially to the mediator; then, after getting these reports, the mediator confidentially recommends an action to each player. The mediator's recommendations may depend on the players' reports in a deterministic or random fashion. For any c in C and any t and T, let $\mu(c|t)$ denote the conditional probability that the mediator would recommend to each player i that he should use action c_i, if each player j reported his type to be t_j. Obviously, these numbers $\mu(c|t)$ must satisfy the following *probability constraints*

$$\sum_{c \in C} \mu(c|t) = 1 \text{ and } \mu(d|t) \geq 0, \quad \forall d \in C, \quad \forall t \in T. \tag{4}$$

In general, any such function $\mu: T \to \Delta(C)$ may be called a *mediation plan* or *mechanism* for the game Γ^b with communication.

If every player reports his type honestly to the mediator and obeys the recommendations of the mediator, then the expected utility for type t_i of player i from the plan μ would be

$$U_i(\mu|t_i) = \sum_{t_{-i} \in T_{-i}} \sum_{c \in C} p_i(t_{-i}|t_i) \mu(c|t) u_i(c, t),$$

where $T_{-i} = \times_{j \in N - i} T_j$ and $t = (t_{-i}, t_i)$.

We must allow, however, that each player could lie about his type or disobey the mediator's recommendation. That is, we assume here that the players' types are not verifiable by the mediator, and the choice of an action in C_i can be controlled only by player i. Thus, a mediation plan μ induces a communication game Γ^b_μ in which each player must select his type report and his plan for choosing an action in C_i as a function of the mediator's recommendation. Formally Γ^b_μ is itself a Bayesian game, of the form

$$\Gamma^b_\mu = (N, (B_i)_{i \in N}, (T_i)_{i \in N}, (p_i)_{i \in N}, (\bar{u}_i)_{i \in N}),$$

where, for each player i,

$$B_i = \{(s_i, \delta_i) | s_i \in T_i, \quad \delta_i : C_i \to C_i\},$$

and $\bar{u}_i : (\times_{i \in N} B_j) \times T \to \mathbb{R}$ is defined by the equation

$$\bar{u}_i((s_j, \delta_j)_{j \in N}, t) = \sum_{c \in C} \mu(c|(s_j)_{j \in N}) u_i((\delta_j(c_j))_{j \in N}, t).$$

A strategy (s_i, δ_i) in B_i represents a plan for player i to report s_i to the mediator, and then to choose his action in C_i as a function of the mediator's recommendation according to δ_i, so that he would choose $\delta_i(c_i)$ if the mediator recommended c_i. The action that player i chooses cannot depend on the type-reports or recommended actions of any other player, because each player communicates with the mediator separately and confidentially.

Suppose, for example, that the true type of player i were t_i, but that he used the strategy (s_i, δ_i) in the communication game Γ_μ^b. If all other players were honest and obedient to the mediator, then i's expected utility payoff would be

$$U_i^*(\mu, \delta_i, s_i | t_i) = \sum_{t_{-i} \in T_{-i}} \sum_{c \in C} p_i(t_{-i} | t_i) \mu(c | t_{-i}, s_i) u_i((c_{-i}, \delta_i(c_i)), t).$$

[Here (t_{-i}, s_i) is the vector in T that differs from $t = (t_{-i}, t_i)$ only in that the i-component is s_i instead of t_i.]

Bayesian equilibrium [as defined in Harsanyi (1967–68)] is still an appropriate solution concept for a Bayesian game with communication, except that we must now consider the Bayesian equilbria of the induced communication game Γ_μ^b, rather than just the Bayesian equilibria of Γ. We say that a mediation plan μ is *incentive compatible* iff it is a Bayesian equilibrium for all players to report their types honestly and to obey the mediator's recommendations when he uses the mediation plan μ. Thus, μ is incentive compatible iff it satisfies the following general *incentive constraints*:

$$U_i(\mu | t_i) \geq U_i^*(\mu, \delta_i, s_i | t_i), \quad \forall i \in N, \quad \forall t_i \in T_i, \quad \forall s_i \in T_i, \quad \forall \delta_i : C_i \to C_i. \tag{5}$$

If the mediator uses an incentive-compatible mediation plan and each player communicates independently and confidentially with the mediator, then no player could gain by being the only one to lie to the mediator or disobey his recommendations. Conversely, we cannot expect rational and intelligent players all to participate honestly and obediently in a mediation plan unless it is incentive compatible.

In general, there may be many different Bayesian equilibria of a communication game Γ_μ^b, even if μ is incentive compatible. Furthermore, as in the preceding section, we could consider more general communication systems, in which the reports that player i can send and the messages that player i may receive are respectively in some arbitrary sets R_i and M_i, not necessarily T_i and C_i. However, given any general communication system and any Bayesian equilibrium of the induced communication game, there exists an equivalent incentive-compatible mediation plan, in which every type of every player gets the same expected utility as in the given Bayesian equilibrium of the induced communication game. In this sense, there is no loss of generality in assuming that the players communicate with each other through a mediator who first asks each player to reveal all of his private information and who then reveals to each player only the minimum information needed to guide his action, in such a way that no player has any incentive to lie or disobey. This result is the *revelation principle* for general Bayesian games.

The formal proof of the revelation principle for Bayesian games is almost the same as for strategic-form games. Given a general communication system $v: R \to \Delta(M)$ and communication strategy sets $(B_i)_{i \in N}$ as in Section 1 above, a Bayesian equilibrium of the induced communication game would then be a vector σ that specifies, for each i in N, each (r_i, δ_i) in B_i, and each t_i in T_i, a number $\sigma_i(r_i, \delta_i | t_i)$ that represents the probability that i would report r_i and choose his final action according to δ_i (as a function of the message that he receives) if his actual type were t_i. If σ is such a Bayesian equilibrium of the communication game Γ^b_v induced by the communication system v, then we can construct an equivalent incentive-compatible mediation plan μ by letting

$$\mu(c|t) = \sum_{(r,d)\in B} \sum_{m\in\delta^{-1}(c)} \left(\prod_{i\in N}\sigma_i(r_i,\delta_i|t_i)\right) v(m|r), \quad \forall c\in C, \quad \forall t\in T,$$

where $\delta^{-1}(c) = \{m\in M \mid \delta_i(m_i) = c_i, \quad \forall i\in N\}$.

This construction can be described more intuitively as follows. The mediator first asks each player (simultaneously and confidentially) to reveal his type. Next the mediator computes (or simulates) the reports that would have been sent by the players, with these revealed types, under the given equilibrium. Then he computes the recommendations or messages that would have been received by the players, as a function of these reports, under the given communication system or mechanism. Then he computes the actions that would have been chosen by the players, as a function of these messages and the revealed types in the given equilibrium. Finally, the mediator tells each player to do the action computed for him at the last step. Thus, the constructed mediation plan simulates the given equilibrium of the given communication system. To check that this constructed mediation plan is incentive compatible, notice that any type of any player who could gain by lying to the mediator or disobeying his recommendations under the constructed mediation plan (when everyone else is honest and obedient) could also gain by similarly lying to himself before implementing his equilibrium strategy or disobeying his own recommendations to himself after implementing his equilibrium strategy in the given communication game, which is impossible (by definition of a Bayesian equilibrium).

If each player's type set consists trivially of only one possible type, so that the Bayesian game is essentially equivalent to a strategic-form game, then an incentive-compatible mechanism is a correlated equilibrium. So incentive-compatible mechanisms are a generalization of correlated equilibria to the case of games with incomplete information. Thus, we may synonymously use the term *communication equilibrium* (or *generalized correlated equilibrium*) to refer to any incentive-compatible mediation plan of a Bayesian game.

Like the set of correlated equilibria, the set of incentive-compatible mediation plans is a closed convex set, characterized by a finite system of inequalities [(4) and (5)] that are linear in μ. On the other hand, it is generally a difficult problem to characterize the set of Bayesian equilibria of any given Bayesian game. Thus, by the revelation principle, it may be easier to characterize the set of all equilibria of all games that can be induced from Γ^b with communication than it is to compute the set of equilibria of Γ^b or of any one communication game induced from Γ^b.

For a simple, two-player example, suppose that $C_1 = \{x_1, y_1\}$, $C_2 = \{x_2, y_2\}$, $T_1 = \{1.0\}$ (so that player 1 has only one possible type and no private information),

Table 3

	$t_2 = 2.1$			$t_2 = 2.2$	
	x_2	y_2		x_2	y_2
x_1	1,2	0,1	x_1	1,3	0,4
y_1	0,4	1,3	y_1	0,1	1,2

$T_2 = \{2.1, 2.2\}$, $p_1(2.1|1.0) = 0.6$, $p_1(2.2|1.0) = 0.4$, and the utility payoffs (u_1, u_2) depend on the actions and player 2's type as in table 3.

In this game, y_2 is a strongly dominated action for type 2.1, and x_2 is a strongly dominated action for type 2.2, so 2.1 must choose x_2 and 2.2 must choose y_2 in a Bayesian equilibrium. Player 1 wants to get either (x_1, x_2) or (y_1, y_2) to be the outcome of the game, and he thinks that 2.1 is more likely than 2.2. Thus the unique Bayesian equilibrium of this game is

$$\sigma_1(\cdot|1.0) = [x_1], \quad \sigma_2(\cdot|2.1) = [x_2], \quad \sigma_2(\cdot|2.2) = [y_2].$$

This example illustrates the danger of analyzing each matrix separately, as if it were a game with complete information. If it were common knowledge that player 2's type was 2.1, then the players would be in the matrix on the left in table 3, in which the unique equilibirum is (x_1, x_2). If it were common knowledge that player 2's type was 2.2, then the players would be in the matrix on the right in table 3, in which the unique equilibrium is (y_1, y_2). Thus, if we looked only at the full-information Nash equilibria of these two matrices, then we might make the prediction "the outcome of the game will be (x_1, x_2) if 2's type is 2.1 and will be (y_1, y_2) if 2's type is 2.2."

This prediction would be absurd, however, for the actual Bayesian game in which player 1 does not initially know player 2's type. Notice first that this prediction ascribes two different actions to player 1, depending on 2's type (x_1 if 2.1, and y_1 if 2.2). So player 1 could not behave as predicted unless he got some information from player 2. That is, this prediction would be impossible to fulfill unless some kind of communication between the players is added to the structure of the game. Now notice that player 2 prefers (y_1, y_2) over (x_1, x_2) if her type is 2.1, and she prefers (x_1, x_2) over (y_1, y_2) if her type is 2.2. Thus, even if communication between the players were allowed, player 2 would not be willing to communicate the information that is necessary to fulfill this prediction, because it would always give her the outcome that she prefers less. She would prefer to manipulate her communications to get the outcomes (y_1, y_2) if 2.1 and (x_1, x_2) if 2.2.

Suppose that the two players can communicate, either directly or through some mediator, or via some tatonnement process, before they choose their actions in C_1 and C_2. In the induced communication game, could there ever be a Bayesian equilibrium giving the outcomes (x_1, x_2) if player 2 is type 2.1 and (y_1, y_2) if player 2 is type 2.2, as naive analysis of the two matrices in table 3 would suggest? The answer is No, by the revelation principle. If there were such a communication game, then there would be an incentive-compatible mediation plan achieving the same outcomes. But this would be the plan satisfying

$$\mu(x_1, x_2 | 1.0, 2.1) = 1, \quad \mu(y_1, y_2 | 1.0, 2.2) = 1.$$

which is not incentive compatible, because player 2 could gain by lying about her type. In fact, there is only one incentive-compatible mediation plan for this example, and it is $\bar{\mu}$, defined by

$$\bar{\mu}(x_1, x_2 | 1.0, 2.1) = 1, \quad \bar{\mu}(x_1, y_2 | 1.0, 2.2) = 1.$$

This is, this game has a unique communication equilibrium, which is equivalent to the unique Bayesian equilibrium of the game without communication.

Notice this analysis assumes that player 2 cannot choose her action and show it verifiably to player 1 before he chooses his action. She can say whatever she likes to player 1 about her intended action before they actually choose, but there is nothing to prevent her from choosing an action different from the one she promised if she has an incentive to do so.

In the insurance industry, the inability to get individuals to reveal unfavorable information about their chances of loss is known as adverse selection, and the inability to get fully insured individuals to exert efforts against their insured losses is known as moral hazard. This terminology can be naturally extended to more general game-theoretic models. The need to give players an incentive to report their information honestly may be called *adverse selection*. The need to give players an incentive to implement their recommended actions may be called *moral hazard*. In this sense, we may say that the incentive constraints (5) are a general mathematical characterization of the effect of adverse selection and moral hazard in Bayesian games.

3. Sender–receiver games

A sender–receiver game is a two-player Bayesian game with communication in which player 1 (the sender) has private information but no choice of actions, and player 2 (the receiver) has a choice of actions but no private information. Thus, sender–receiver games provide a particularly simple class of examples in which both moral hazard and adverse selection are involved. [See Crawford and Sobel (1982).]

A general *sender–receiver game* can be characterized by specifying (T_1, C_2, p, u_1, u_2), where T_1 is the set of player 1's possible types, C_2 is the set of player 2's possible actions, p is a probability distribution over T_1 that represents player 2's beliefs about player 1's type, and $u_1 : C_2 \times T_1 \to \mathbb{R}$ and $u_2 : C_2 \times T_1 \to \mathbb{R}$ are utility functions for player 1 and player 2 respectively. A sender–receiver game is finite iff T_1 and C_2 are both finite sets.

A *mediation plan* or *mechanism* for the sender–receiver game as above is any function $\mu: T_1 \to \Delta(C_2)$. If such a plan μ were implemented honestly and obediently by the players, the expected payoff to player 2 would be

$$U_2(\mu) = \sum_{t_1 \in T_1} \sum_{c_2 \in C_2} p(t_1) \mu(c_2|t_1) u_2(c_2, t_1)$$

and the conditionally expected payoff to player 1 if he knew that his type was t_1 would be

$$U_1(\mu|t_1) = \sum_{c_2 \in C_2} \mu(c_2|t_1) u_1(c_2, t_1).$$

The general incentive constraints (5) can be simplified in sender–receiver games. Because player 1 controls no actions, the incentive constraints on player 1 reduce to purely informational incentive constraints. On the other hand, because player 2 has no private information, the incentive constraints on player 2 reduce to purely strategic incentive constraints, as in (1) or (2). Thus, a mediation plan μ is *incentive compatible* for the sender–receiver game if any $\mu: T_1 \to \Delta(C_2)$ such that

$$\sum_{c_2 \in C_2} \mu(c_2|t_1) u_1(c_2, t_1) \geq \sum_{c_2 \in C_2} \mu(c_2|s_1) u_1(c_2, t_1), \forall t_1 \in T_1, \quad \forall s_1 \in T_1, \tag{6}$$

and

$$\sum_{t_1 \in T_1} p(t_1)[u_2(c_2, t_1) - u_2(e_2, t_1)] \mu(c_2|t_1) \geq 0, \forall c_2 \in C_2, \quad \forall e_2 \in C_2. \tag{7}$$

The informational incentive constraints (6) assert that player 1 should not expect to gain claiming that his type is s_1 when it is actually t_1, if he expects player 2 to obey the mediator's recommendations. The strategic incentive constraints (7) assert that player 2 should not expect to gain by choosing action e_2 when the mediator recommends c_2 to her, if she believes that player 1 was honest to the mediator.

For example, consider a sender–receiver game [due to Farrell (1993)] with $C_2 = \{x_2, y_2, z_2\}$ and $T_1 = \{1.a, 1.b\}$, $p(1.a) = 0.5 = p(1.b)$, and utility payoffs (u_1, u_2) that depend on player 1's type and player 2's action as in table 4.

Suppose first there is no mediation, but that player 1 can send player 2 any message drawn from some large alphabet or vocabulary, and that player 2 will be sure to observe player 1's message without any error or noise. Then, as Farrell (1993) has shown, in every equilibrium of the induced communication game, player

Table 4

	x_2	y_2	z_2
1.a	2,3	0,2	−1,0
1.b	1,0	2,2	0,3

2 will choose action y_2 for sure, after any message that player 1 might send with positive probability. To see why, notice that player 2 is indifferent between choosing x_2 and z_2 only if she assesses a probability of 1.a of exactly 0.5, but with this assessment she prefers y_2. Thus, there is no message that can generate beliefs that would make player 2 willing to randomize between x_2 and z_2. For each message that player 1 could send, depending on what player 2 would infer from receiving this message, player 2 might respond either by choosing x_2 for sure, by randomizing between x_2 and y_2, by choosing y_2 for sure, by randomizing between y_2 and z_2, or by choosing z_2 for sure. Notice that, when player 1's type is 1.a, he is not indifferent between any two different responses among these possibilities, because he strictly prefers x_2 over y_2 and y_2 over z_2. Thus, in an equilibrium of the induced communication game, if player 1 had at least two messages (call them "α" and "β") that are sent with positive probability and to which player 2 would respond differently, then type 1.a would be willing to send only one of these messages (say, "α"), and so the other message ("β") would be sent with positive probability only by type 1.b. But then, player 2's best response to this other message ("β") would be z_2, which is the worst outcome for type 1.b of player 1, so type 1.b would not send it with positive probability either. This contradiction implies that player 2 must use the same response to every message that player 1 sends with positive probability. Furthermore, this response must be y_2, because y_2 is player 2's unique best action given here beliefs before she receives any message. (This argument is specific to this example. However, Forges (1994) has shown more generally how to characterize the information that can be transmitted in sender–receiver games without noise.)

Thus, as long as the players are restricted to perfectly reliable noiseless communication channels, no substantive communication can occur between players 1 and 2 in any equilibrium of this game. However, substantive communication can occur when noisy communication channels are used. For example, suppose player 1 has a carrier pigeon that he could send to player 2, but, if sent, if would only arrive with probability $\frac{1}{2}$. Then there is an equilibrium of the induced communication game in which player 2 chooses x_2 if the pigeon arrives, player 2 chooses y_2 if the pigeon does not arrive, player 1 sends the pigeon if his type is 1.a, and player 1 does not send the pigeon if his type is 1.b. Because of the noise in the communication channel (the possibility of the pigeon getting lost), if player 2 got the message "no pigeon arrives," then she would assign a $\frac{1}{3}$ probability to the event that player 1's type was 1.a (and he sent a pigeon that got lost), and so she would be willing to choose y_2, which is better than x_2 for type 1.b of player 1. (See Forges (1985), for a seminal treatment of this result in related examples.)

Thus, using this noisy communication channel, there is an equilibrium in which player 2 and type 1.a of player 1 get better expected payoffs than they can get in equilibrium with direct noiseless communication. By analyzing the incentive constraints (6) and (7), we can find other mediation plans $\mu: T_1 \to \Delta(C_2)$ in which they both do even better. The informational incentive constraints (6) on player 1 are

$$2\mu(x_2|1.a) - \mu(z_2|1.a) \geq 2\mu(x_2|1.b) - \mu(z_2|1.b),$$

$$\mu(x_2|1.b) + 2\mu(y_2|1.b) \geq \mu(x_2|1.a) + 2\mu(y_2|1.a),$$

and the strategic incentive constraints (7) on player 2 are

$$0.5\mu(x_2|1.a) - \mu(x_2|1.b) \geq 0,$$

$$1.5\mu(x_2|1.a) - 1.5\mu(x_2|1.b) \geq 0,$$

$$-0.5\mu(y_2|1.a) + \mu(y_2|1.b) \geq 0,$$

$$\mu(y_2|1.a) - 0.5\mu(y_2|1.b) \geq 0,$$

$$-1.5\mu(z_2|1.a) + 1.5\mu(z_2|1.b) \geq 0,$$

$$-\mu(z_2|-1.a) + 0.5\mu(z_2|1.b) \geq 0.$$

(The last of these constraints, for example, asserts that player 2 should not expect to gain by choosing y_2 when z_2 is recommended.) To be a mediation plan, μ must also satisfy the probability constraints

$$\mu(x_2|1.a) + \mu(y_2|1.a) + \mu(z_2|1.a) = 1,$$

$$\mu(x_2|1.b) + \mu(y_2|1.b) + \mu(z_2|1.b) = 1,$$

and all $\mu(c_2|t_1) \geq 0$.

If, for example, we maximize the expected payoff to type $1.a$ of player 1

$$U_1(\mu|1.a) = 2\mu(x_2|1.a) - \mu(z_2|1.a)$$

subject to these constraints, then we get the mediation plan

$$\mu(x_2|1.a) = 0.8, \quad \mu(y_2|1.a) = 0.2, \quad \mu(z_2|1.a) = 0,$$

$$\mu(x_2|1.b) = 0.4, \quad \mu(y_2|1.b) = 0.4, \quad \mu(z_2|1.b) = 0.2.$$

Honest reporting by player 1 and obedient action by player 2 is an equilibrium when a noisy communication channel or mediator generates recommended-action messages for player 2 as a random function of the type-reports sent by player 1 according to this plan μ. Furthermore, no equilibrium of any communication game induced by any communication channel could give a higher expected payoff to type $1.a$ of player 1 than the expected payoff of $U_1(\mu|1.a) = 1.6$ that he gets from this plan.

On the other hand, the mechanism that maximizes player 2's expected payoff is

$$\mu(x_2|1.a) = \tfrac{2}{3}, \quad \mu(y_2|1.a) = \tfrac{1}{3}, \quad \mu(z_2|1.a) = 0,$$

$$\mu(x_2|1.b) = 0, \quad \mu(y_2|1.b) = \tfrac{2}{3}, \quad \mu(z_2|1.b) = \tfrac{1}{3}.$$

This gives expected payoffs

$$U_1(\mu|1.a) = 1.333, \quad U_1(\mu|1.b) = 1.333, \quad U_2(\mu) = 2.5.$$

Once we have a complete characterization of the set of all incentive-compatible mediation plans, the next natural question is: which mediation plans or mechanisms should we actually expect to be selected and used by the players? That is, if one or more of the players has the power choose among all incentive-compatible mechanisms, which mechanisms should we expect to observe?

To avoid questions of interpersonal equity in bargaining, which belong to cooperative game theory, let us here consider only cases where the power to select the mediator or design the communication mechanism belongs to just one of the players. To begin with suppose that player 2 can select the mediation plan. To be more specific, suppose that player 2 will first select a mediator and direct him to implement some incentive-compatible mediation plan, and then player 1 can either accept this mediator and communicate with 2 thereafter only through him, or 1 can reject this mediator and thereafter communicate with 2 only face-to-face.

It is natural to expect that player 2 will use her power to select a mediator who will implement the incentive-compatible mediation plan that is best for 2. This plan is worse than y_2 for 1 if his type is $1.b$, so one might think that, if 1's type is $1.b$, then he should reject player 2's proposed mediator and insist on communicating face-to-face. However, there is an equilibrium of this mediator-selection game in which player 1 always accepts always player 2's proposal, no matter what his type is. In this equilibrium, if 1 rejected 2's mediator, then 2 might reasonably infer that 1's type was $1.b$, in which case 2's rational choice would be z_2 instead of y_2, and z_2 is the worst possible outcome for both of 1's types.

Unfortunately, there is another sequential equilibrium of this mediator-selection game in which player 1 always rejects player 2's mediator, no matter what mediation plan she selects. In this equilibrium, player 2 infers nothing about 1 if he rejects the mediator and so does y_2, but if he accepted the mediator then she would infer (in this zero-probability event) that player 1 is type $1.b$ and so she would choose z_2.

Now consider the mediator-selection game in which the informed player 1 can select the mediator and choose the mediation plan that will be implemented, with the only restriction that player 1 must make the selection after he already knows his own type, and player 2 must know what mediation plan has been selected by player 1. For any incentive-compatible mediation plan μ, there is an equilibrium in which 1 chooses μ for sure, no matter what his type is, and they thereafter play honestly and obediently when μ is implemented. In this equilibrium, if any mediation plan other than μ were selected then 2 would infer from 1's surprising selection that his type was $1.b$ (she might think "only $1.b$ would deviate from μ"), and therefore she would choose z_2 no matter what the mediator might subsequently recommend. Thus, concepts like sequential equilibrium cannot determine the outcome of such a mediator-selection game beyond what we already knew from the revelation principle.

4. Communication in multistage games

Consider the following two-stage two-player game. At stage 1, player 1 must choose either a_1 or b_1. If he chooses a_1 then the game ends and the payoffs to players 1 and 2 are (3,3). If he chooses b_1 then there is a second stage of the game in which each player i must choose either x_i or y_i, and the payoffs to players 1 and 2 depend on their second-stage moves as follows:

Table 5

	x_2	y_2
x_1	7,1	0,0
y_1	0,0	1,7

The normal representation of this game in strategic form may be written as follows:

Table 6

	x_2	y_2
$a_1 x_1$	3,3	3,3
$a_1 y_1$	3,3	3,3
$b_1 x_1$	7,1	0,0
$b_1 y_1$	0,0	1,7

In this strategic-form game, the strategy $b_1 y_1$ is strongly dominated for player 1. So it may seem that any theory of rational behavior should imply that there is zero probability of player 1 choosing b_1 at the first stage and y_1 at the second stage.

However, this conclusion does not hold if we consider the original two-stage game as a game with communication. Consider, for example, the following medation plan. At stage 1, the mediator recommends that player 1 should choose b_1. Then, at stage 2, with probability $\frac{1}{2}$ the mediator will recommend the moves x_1 and x_2, and with probability $\frac{1}{2}$ the mediator will recommend the moves y_1 and y_2. In either case, neither player will be able to gain by unilaterally disobeying the mediator at stage 2. At stage 1, disobeying the mediator would give player 1 a payoff of 3, which is less than the expected payoff of $\frac{1}{2} \times 7 + \frac{1}{2} \times 1 = 4$ that he gets from obedience. Thus, this mediation plan is incentive compatible, and it can lead player 1 to choose b_1 and then y_1 with probability $\frac{1}{2}$.

The key to this mediation plan is that player 1 must not learn whether x_1 or y_1 will be recommended to him at stage 2 until after it is too late to go back and choose a_1. That is, when we study a multistage game with communication, we should take account of the possibility of communication at every stage of the game. For multistage games, the revelation principle asserts that any equilibrium of any communication game that can be induced by adding a communication structure is equivalent to some mediation plan of the following form: at the beginning of each stage, the players confidentially report their new information to the mediator; then the mediator determines the recommended actions for the players at this stage, as a function of all reports received at this and all earlier stages, by applying some randomly selected feedback rule; then the mediator confidentially tells each player the action that is recommended for him at this stage; and (assuming that all players know the probability distribution that the mediator used to select his feedback rule) it is an equilibrium of the induced communication game for all players to always report their information honestly and choose their actions obediently as the mediator recommends. The probability distributions over feedback rules that satisfy this last incentive-compatibility condition can be characterized by a collection of linear incentive constraints, which assert that no player can expect to gain by switching to any manipulative strategy of lying and disobedience, when all other players are expected to be honest and obedient. For a formal statement of these ideas, see Myerson (1986a).

The inadequacy of the strategic-form game in table 6 for analysis of the incentive-compatible mediation in the above example shows that we may have to make a conceptual choice between the revelation principle and the generality of the strategic form, when we think about multistage games. If we want to allow communication opportunities to remain implicit at the modeling stage of our analysis, then we get a solution concept which is mathematically simpler than Nash equilibrium (because, by the revelation principle, communication equilibria can be characterized by linear incentive constraints) but which cannot necessarily be analyzed via the normal strategic-form representation. If we want in general to study any multistage game via the strategic form, then all communication opportunities must be made an explicit part of the extensive game model before we construct the normal representation in strategic form.

Sequential rationality and termbling-hand refinements of equilibrium for games with communication have been considered by Myerson (1986a, b). In Myerson (1986a), *acceptable correlated equilibria* are defined as a natural analogue of Selten's (1975) trembling-hand perfect equilibria for strategic-form games. The main result of Myerson (1986b) is that these acceptable correlated equilibria satisfy a kind of *strategy-elimination* property, which may be stated as follows: for any strategic-form game with communication, there exists a set of *codominated* strategies such that the set of acceptable correlated equilibria of the given game is exactly the set of correlated equilibria of the game that remains after eliminating all the codominated strategies. A similar strategy-elimination characterization of sequentially rational correlated equilibria for multistage games with communication is derived in Myerson (1986a).

References

Aumann, R.J. (1974) 'Subjectivity and Correlation in Randomized Strategies', *Journal of Mathematical Economics*, **1**: 67–96.
Aumann, R.J. (1976) 'Agreeing to Disagree', *Annals of Statistics*, **4**: 1236–1239.
Aumann, R.J. (1987) 'Correlated Equilibria as an Expression of Bayesian Rationality', *Econometrica*, **55**: 1–18.
Barany, I. (1987) 'Fair Distribution Protocols or How the Players Replace Fortune', CORE Discussion Paper 8718, Université catholique de Louvain. *Mathematics of Operations Research*, **17**: 327–340.
Crawford, V. and J. Sobel (1982) 'Strategic Information Transmission', *Econometrica*, **50**: 579–594.
Grossman, S. and M. Perry (1986) 'Perfect Sequential Equilibrium', *Journal of Economic Theory*, **39**: 97–119.
Farrell, J. (1993) 'Meaning and Credibility in Cheap-Talk Games', *Games and Economic Behavior*, **5**: 514–531.
Forges, F. (1985) 'Correlated Equilibria in a Class of Repeated Games with Incomplete Information', *International Journal of Game Theory*, **14**: 129–150.
Forges, F. (1986) 'An Approach to Communication Equilibrium', *Econometrica*, **54**: 1375–1385.
Forges, F. (1994) 'Non-zero Sum Repeated Games and Information Transmission', in: N. Megiddo, ed., *Essays in Game Theory*. Berlin: Springer, pp. 65–95.
Forges, F. (1990) 'Universal Mechanisms'. *Econometrica*, **58**: 1341–1364.
Harsanyi, J.C. (1967–8) 'Games with Incomplete Information Played by 'Bayesian' Players', *Management Science*, **14**: 159–182, 320–334, 486–502.
Hart, S. and D. Schmeidler (1989) 'Existence of Correlated Equilibria'. *Mathematics of Operations Research*, **14**: 18–25.
Maskin, E. and J. Tirole (1990) 'The Principal-Agent Relationship with an Informed Principal: the Case of Private Values', *Econometrica*, **58**: 379–409.
Moulin, H. and J.-P. Viral (1978) 'Strategically Zero-Sum Games: The Class Whose Completely Mixed Equilibria Cannot be Improved Upon', *International Journal of Game Theory*, **7**: 201–221.
Myerson, R.B. (1982) 'Optimal Coordination Mechanisms in Generalized Principal-Agent Problems'. *Journal of Mathematical Economics*, **10**: 67–81.
Myerson, R.B. (1983) 'Mechanism Design by an Informed Principal', *Econometrica*, **51**: 1767–1797.
Myerson, R.B. (1985) 'Bayesian Equilibrium and Incentive Compatibility', In: L. Hurwicz, D. Schmeidler and H. Sonnenschein, eds., *Social Goals and Social Organization*. Cambridge: Cambridge University Press, pp. 229–259.
Myerson, R.B. (1986a) 'Multistage Games with Communication', *Econometrica*, **54**: 323–358.
Myerson, R.B. (1986b) 'Acceptable and Predominant Correlated Equilibria', *International Journal of Game Theory*, **15**: 133–154.
Myerson, R.B. (1989) 'Credible Negotiation Statements and Coherent Plans', *Journal of Economic Theory*, **48**: 264–303.
Nau, R.F., and K.F. McCardle (1990) 'Coherent Behavior in Noncooperative Games', *Journal of Economic Theory*, **50**: 424–444.
Neumann, J. von and O. Morgenstern (1944) *Theory of Games and Economic Behavior*. Princeton: Princeton University Press. 2nd edn., 1947.
Selten, R. (1975) 'Reexamination of the Perfectness Concept for Equilibrium Points in Extensive Games', *International Journal of Game Theory*, **4**: 25–55.

Chapter 25

SIGNALLING

DAVID M. KREPS

Stanford University and Tel Aviv University

JOEL SOBEL

University of California at San Diego

Contents

1. Introduction — 850
2. Signalling games – the canonical game and market signalling — 851
3. Nash equilibrium — 852
4. Single-crossing — 854
5. Refinements and the Pareto-efficient separating equilibrium — 856
6. Screening — 861
7. Costless signalling and neologisms — 863
8. Concluding remarks — 865
References — 866

1. Introduction

One of the most important applications of game theory to micro-economics has been in the domain of market signalling. The standard story is a simple one: Two parties wish to engage in exchange; the owner of a used car, say, wants to sell the car to a potential buyer. One party has information that the other lacks; e.g., the seller knows the quality of the particular car. (Think in terms of a situation where the quality of a car is outside of the control of the owner; quality depends on things done or undone at the factory when the car was assembled.) Under certain conditions, the first party wishes to convey that information to the second; e.g., if the car is in good condition, the seller wishes the buyer to learn this. But direct communication of the information is for some reason impossible, and the first party must engage in some activity that indicates to the second what the first knows; e.g., the owner of a good car will offer a limited warranty. The range of applications for this simple story is large: A worker wishes to signal his ability to a potential employer, and uses education as a signal [Spence (1974)]. An insuree who is relatively less risk prone signals this to an insurer by accepting a larger deductable or only partial insurance [Rothschild and Stiglitz (1976), Wilson (1977)]. A firm that is able to produce high-quality goods signals this by offering a warranty for the goods sold [Grossman (1981)]. A plaintiff with a strong case demands a relatively high payment in order to settle out of court [Reinganum and Wilde (1986), Sobel (1989)]. The purchaser of a good who does not value the good too highly indicates this by rejecting a high offer or by delaying his own counteroffer [Rubinstein (1985), Admati nd Perry (1987)]. A firm that is able to produce a good at a relatively low cost signals this ability to potential rivals by charging a low price for the good [Milgrom and Roberts (1982)]. A strong deer grows extra large antlers to show that it can survive with this handicap and to signal its fitness to potential mates [Zahavi (1975)].

The importance of signalling to the study of exchange is manifest, and so the original work on market signalling [Spence (1974), Rothschild and Stiglitz (1976), Wilson (1977)] received a great deal of attention. But the early work, which did not employ formal game theory, was inconclusive; different papers provided different and sometimes contradictory answers, and even within some of the papers, a welter of possible equilibria was advanced. While authors could (and often did) select among the many equilibria they found with informal, intuitive arguments, the various analyses in the literature were ad hoc. Formal game theoretic treatments of market signalling, which began to appear in the 1980s, added discipline to this study. It was found that the different and conflicting results in the early literature arose from (implicit) differences in the formulation of the situation as a game. And selections among equilibria that were previously based on intuitive arguments were rationalized with various refinements of Nash equilibrium, most especially refinements related to Kohlberg and Merten's notion of a stable equilibrium (cf. the Chapter on strategic stability in Volume III of this Handbook) and to out-of-equilibrium beliefs in the sense of a sequential equilibrium. This is not to say that game theory showed that some one of the early analyses was correct and the others were wrong. Rather, game theory has contributed a language with which those analyses can be compared and evaluated.

In this chapter, we survey some of the main developments in the theory of market signalling and its connection to noncooperative game theory. Our treatment is necessarily restricted in scope and in detail, and in some concluding remarks we point the reader toward the vast number of topics that we have omitted.

2. Signalling games – the canonical game and market signalling

Throughout this chapter, we work with the following canonical game. There are two players, called **S** (for sender) and **R** (for receiver). This is a game of incomplete information: **S** starts off knowing something that **R** does not know. We assume that **S** knows the value of some random variable t whose support is a given set T. The conventional language, used here, is to say that t is the *type* of **S**. The prior beliefs of **R** concerning t are given by a probability distribution ρ over T; these prior beliefs are common knowledge. Player **S**, knowing t, sends to **R** a signal s, drawn from some set S. (One could have the set of signals available to **S** depend on t.) Player **R** receives this signal, and then takes an action a drawn from a set A (which could depend on the signal s that is sent). This ends the game: The payoff to **S** is given by a function $u: T \times S \times A \to R$, and the payoff to **R** is given by $v: T \times S \times A \to R$.

This canonical game captures some of the essential features of the classic applications of market signalling. We think of **S** as the seller or buyer of some good: a used car, labor services, an insurance contract; **R** represents the other party to the transaction. **S** knows the quality of the car, or his own abilities, or his propensity towards risk; **R** is uncertain a priori, so t gives the value of quality or ability or risk propensity. **S** sends a signal r that might tell **R** something about t. And **R** responds – in market signalling, this response could be simply acceptance/rejection of the deal that **S** proposes, or it could be a bid (or ask) price at which **R** is willing to consummate the deal.

To take a concrete example, imagine that **S** is trying to sell a used car to **R**. The car is either a *lemon* or a *peach*; so that $T = \{\text{lemon}, \text{peach}\}$. Write t_0 for "lemon" and t_1 for "peach". **S** knows which it is; **R** is unsure, assessing probability $\rho(t_1)$ that the car is a peach. **S** cannot provide direct evidence as to the quality of the car, but **S** can offer a warranty – for simplicity, we assume that **S** can offer to cover all repair expenses for a length of time s that is one of: zero, one, two, three or four months. We wish to think of the market in automobiles as being competitive in the sense that there are many buyers; to accommodate this within the framework of our two-person game is nontrivial, however. Two ways typically used to do this are: (1) Imagine that there are (at least) two identical buyers, called R_1 and R_2. The game is then structured so that $S = \{0, 1, 2, 3, 4\}$ (the possible warranties that could be offered); S announces one of these warranty lengths. The R_1 and R_2 respond by simultaneously and independently naming a price $a_i \in [0, \infty)$ ($i = 1, 2$) that they are willing to pay for the car, with the car (with warranty) going to the high bidder at that bid. In the usual fashion of Bertrand competition or competitive bidding when bidders have precisely the same information, in equilibrium $a_1 = a_2$ and each is equal to the (subjective) valuation placed on the car (given the signal s) by the two buyers. If we wished to follow this general construction but keep precisely to our original two-player game formulation, we could artificially adopt a utility function for R that causes him to bid his subjective valuation for the car. For example, we could suppose that his utility function is minus the square of the difference between his bid and the value he places on the car. (2) Alternatively, we can imagine that $S = \{0, 1, 2, 3, 4\} \times [0, \infty)$. The interpretation is that S makes a take-it-or-leave-it offer of the form: a length of warranty (from $\{0, 1, 2, 3, 4\}$) and a price for the car (from $[0, \infty)$). And R either accepts or rejects this offer. Since S has all the bargaining power in this case, he will extract all the surplus. We hereafter call this the *take-it-or-leave-it* formulation of signalling.

Either of these two games captures market signalling institutional details in a manner consistent with our canonical game. But there are other forms of institutions that give a very different game-theoretic flavor to things. For example, we could imagine that R (or many identical Rs) offer a menu of contracts to S; that is, a set of pairs from $\{0, 1, 2, 3, 4\} \times [0, \infty)$, and then S chooses that contract that he likes the most. Models of this sort, where the uninformed party has the leading role in setting the terms of the contract, are often referred to as examples of market *screening* instead of *signalling*, to distinguish them from institutions where the informed party has the leading role. We will return later to the case of market screening.

3. Nash equilibrium

We describe a Nash equilibrium for the canonical signalling game in terms of behavior strategies for S and R. First take the case where the sets T, S and A are all finite. Then a behavior strategy for S is given by a function $\sigma: T \times S \to [0, 1]$, such that $\sum_s \sigma(t, s) = 1$ for each t. The interpretation is that $\sigma(t, s)$ is the probability that S sends message s if S is of type t. A behavior strategy for R is a function $\alpha: S \times A \to [0, 1]$ where $\sum_a \alpha(s, a) = 1$ for each s. The interpretation is that R takes action a with probability $\alpha(s, a)$ if signal s is received.

Proposition 1. Behavior strategies α for R and σ for S form a Nash equilibrium if and only if

$$\sigma(t,s) > 0 \text{ implies } \sum_a \alpha(s,a) u(t,s,a) = \max_{s' \in S} \left(\sum_a \alpha(s',a) u(t,s',a) \right), \quad (3.1)$$

and, for each s such that $\sum_t \sigma(t,s) \rho(t) > 0$,

$$\alpha(s,a) > 0 \text{ implies } \sum_t \mu(t;s) v(t,s,a) = \max_{a'} \sum_t \mu(t;s) v(t,s,a'), \quad (3.2a)$$

where we define

$$\mu(t;s) = \frac{\sigma(t,s)\rho(t)}{\sum_{t'} \sigma(t',s)\rho(t')} \text{ if } \sum_t \sigma(t,s)\rho(t) > 0. \quad (3.2b)$$

Condition (3.1) says that σ is a best response to α, while (3.2) says (in two steps) that α is a best response to σ; (3.2b) uses Bayes' rule and α to compute **R**'s posterior beliefs $\mu(\cdot;s)$ over T upon hearing signal s (if s is sent with positive probability), and (3.2a) then states that $\alpha(s,\cdot)$ is a conditional best response (given s). Note well that the use of Bayes' rule when applicable [in the fashion of (3.2b)] and (3.2a) is equivalent to α being a best response to σ in terms of the ex ante expected payoffs of **R** in the associated strategic-form game.

All this is for the case where the sets T, S and A are finite. In many applications, any or all of these sets (but especially S and A) are taken to be infinite. In that case the definition of a Nash equilibrium is a straightforward adaptation of what is given above: One might assume that the spaces are sufficiently nice (i.e., Borel) so that a version of regular conditional probability for t given s can be fixed (where the joint probabilities are given by ρ and σ), and then (3.2a) would use conditional expectations computed using that fixed version.

We characterize some equilibria as follows:

Definition. An equilibrium (σ, α) is called a *separating* equilibrium if each type t sends different signals; i.e., the set S can be partitioned into (disjoint) sets $\{S_t; t \in S\}$ such that $\sigma(t, S_t) = 1$. An equilibrium (σ, α) is called a *pooling* equilibrium if there is a single signal s^* that is sent by all types; i.e., $\sigma(t, s^*) = 1$ for all $t \in T$.

Note that we have not precluded the possibility that, in a separating equilibrium, one or more types of **S** would use mixed strategies. In most applications, the term is used for equilibria in which each type sends a single signal, which we might term a pure separating equilibrium. On the other hand, we have followed convention in using the unmodified expression *pooling equilibrium* for equilibria in which all types use the same pure behavior strategy. One can imagine definitions of pooling equilibria in which **S** uses a behaviorally mixed strategy (to some extent), but by virtue of Proposition 2 following, this possibility would not come up in standard formulations. Of course, these two categories do not exhaust all possibilities. We can have equilibria in which some types pool and some separate, in which all types pool with at least one other type but in more than one pool, or even in which some types randomize between signals that separate them from other types and signals that pool them with other types.

4. Single-crossing

In many of the early applications, a "single-crossing" property held. The sets T, S and A each were simply ordered, with $\geq$ used to denote the simple order in each case. (By a simple order, we mean a complete, transitive and antisymmetric binary relation. The single-crossing property has been generalized to cases where S is multi-dimensional; references will be provided later.) We will speak as if T, S and A are each subsets of the real line, and $\geq$ will be the usual "greater than or equal to" relationship. Also, we let $\mathscr{A}$ denote the set of probability distributions on A and, for each $s \in S$ and $T' \subseteq T$, we let $\mathscr{A}(s, T')$ be the set of mixed strategies that are best responses by **R** to s for some probability distribution with support T'. Finally, for $\alpha \in \mathscr{A}$, we write $u(t, s, \alpha)$ for $\sum_{a \in A} u(t, s, a)\alpha(a)$.

Definition. The data of the game are said to satisfy the *single-crossing property* if the following holds: If $t \in T$, $(s, \alpha) \in S \times \mathscr{A}$ and $(s', \alpha') \in S \times \mathscr{A}$ are such that $\alpha \in \mathscr{A}(s, T), \alpha' \in \mathscr{A}(s', T), s > s'$ and $u(t, s, \alpha) \geq u(t, s', \alpha')$, then for all $t' \in T$ such that $t' > t$, $u(t', s, \alpha) > u(t', s', \alpha')$.

In order to understand this property, it is helpful to make some further assumptions that hold in many examples. Suppose that S and A are compact subsets of the real line and that u is defined for all of $T \times \mathrm{co}(S) \times \mathrm{co}(A)$, where $\mathrm{co}(X)$ denotes the convex hull of X. Suppose that u is continuous in its second two arguments, strictly decreasing in the second argument and strictly increasing in the third. In terms of our example, think of *peach > lemon*, and then the monotonicity assumptions are that longer warranties are worse for the seller (if $s \in S$ is the length of the warranty) and a higher price is better for the seller (if $a \in A$ is the purchase price).

Then we can draw in the space $\mathrm{co}(S) \times \mathrm{co}(A)$ indifference curves for each type t. The monotonicity and continuity assumptions will guarantee that these indifference curves are continuous and increasing. And the single-crossing property will imply that if indifference curves for types t and t' cross, for $t' > t$, then they cross once, with the indifference curve of the higher type t' crossing from "above on the left" to "below on the right". The reader may wish to draw this picture. Of course, the single-crossing property says something more, because it is not necessarily restricted to pure strategy responses. But in many standard examples, either u is linear in the third argument [i.e., $u(t, s, a) = u'(t, s) + a$] or v is strictly concave in a and A is convex [so that $\mathscr{A}(s, T')$ consists of degenerate distributions for all s], in which case single-crossing indifference curves will imply the single-crossing property given above.

The single-crossing property can be used to begin to characterize the range of equilibrium outcomes.

Proposition 2. Suppose that the single-crossing property holds and, in a given equilibrium (σ, α), $\sigma(t, s) > 0$ and $\sigma(t', s') > 0$ for $t' > t$. Then $s' \geq s$.

That is, the supports of the signals sent by the various types are "increasing" in the type. This does not preclude pooling, but a pool must be an interval of types pooling on a single signal, with any type in the interior of the interval sending only the pooling signal (and with the largest type in the pool possibly sending signals larger than the pooling signal and the smallest type possibly sending signals smaller than the pooling signal).

While the single-crossing property holds in many examples in the literature, it does not hold universally. For example, this property may not hold in the entry deterrence model of Milgrom and Roberts (1982) [see Cho (1987)], nor it is natural in models of litigation [see Reinganum and Wilde (1986) and Sobel (1989)].

More importantly to what follows, this entire approach is not well suited to some variations on the standard models. Consider the example of selling a used car with the warranty as signal, but in the variation where **S** offers to **R** a complete set of terms (both duration of the warranty and the purchase price), and **R**'s response is either to accept or reject these terms. In this game form, **S** is not simply ordered, and so the single-crossing property as given above makes no sense at all.

Because we will work with this sort of variation of the signalling model in the next section, we adapt the single-crossing property to it. First we give the set-up.

Definition. A signalling game has the *basic take-it-or-leave-it setup* if: (a) T is finite and simply ordered; (b) $S = M \times (-\infty, \infty)$ for M an interval of the real line [we write $s = (m, d)$]; (c) $A = \{yes, no\}$; (d) $u(t, (m, d), a) = U(t, m) + d$ if $a = yes$ and $u(t, (m, d), a) = 0$ if $a = no$; and (e) $v(t, (m, d), a) = V(t, m) - d$ if $a = yes$ and $= 0$ if $a = no$; where U, W and V are continuous in m; V is strictly increasing in t and nondecreasing in m; and U is strictly decreasing in m.

The interpretation is that m is the message part of the signal s (such as the length of the warranty on a used car) and d gives the proposed "dollar price" for the exchange. Note that we assume proposed dollar prices are unconstrained. For expositional simplicity, we have assumed that the reservation values to S and to R if there is no deal are constants set equal to zero.

Definition. In a signalling game with the basic take-it-or-leave-it setup, the *single-crossing property* holds if $t' > t$, $m' > m$ and $U(t, m') + d' \geq U(t, m) + d$, then $U(t', m') + d' > U(t', m) + d$.

Proposition 2'. Fix a signalling game with the basic take-it-or-leave-it setup for which the *single-crossing property* holds. In any equilibrium in which type t proposes (m, d), which is accepted with probability p, and type t' proposes (m', d'), which is accepted with probability $p', t' > t$ and $p' \geq p$ imply $m' \geq m$.

5. Refinements and the Pareto-efficient separating equilibrium

Proposition 2 (and 2') begins to characterize the range of Nash equilibria possible in the standard applications, but still there are typically many Nash equilibria in a given signalling game. This multiplicity of equilibria arises in part because the response of R to messages that have zero prior probability under σ are not constrained by the Nash criterion; i.e., (3.2a) restricts $\alpha(s, \cdot)$ only for s such that $\sum_t \sigma(t, s) \rho(t) > 0$. However the response of **R** to so-called *out-of-equilibrium* messages can strongly color the equilibrium, since the value of sending out-of-equilibrium messages by **S** depends on the equilibrium response to those messages. That is, **S** may not send a particular message $s°$ because **R** threatens to blow up the world in response to this message; since (therefore) **S** will not send $s°$, the threatened response is not disallowed by (3.2a). This is clearly a problem coming under the general heading of "perfection", as discussed in the Chapters on "Strategic Equilibrium" and "Conceptual Foundations of Strategic Equilibrium" in volume III of this Handbook. And one naturally attacks this problem by refining the Nash criterion. The first step in the usual attack is to look for sequential equilibria. Hereafter, μ will denote a full set of beliefs for **R**; i.e., for each $s \in S$, $\mu(\cdot, s)$ is a probability distribution on T.

Proposition 3. Behavior strategies (σ, α) and beliefs μ for a signalling game constitute a sequential equilibrium if (3.1) holds, (3.2a) holds for every $s \in S$, and (3.2b) holds as a condition instead of a definition.

(The only thing that needs proving is that strategies and beliefs are consistent in the sense of sequential equilibrium, but in this very simple setting this is true automatically. The formal notion of a sequential equilibrium does not apply when T, S, or A is infinite; in such cases is it typical to use the conditions of Proposition 3 as a definition.)

Restricting attention to sequential equilibria does reduce the number of equilibria, in that threats to "blow up the world" are no longer credible. But multiple equilibria remain in interesting situations. For example, in applications where the types $t \in T$ and the responses $a \in A$ are simply ordered, u is increasing in the response a, and the responses by **R** at each signal s increase with stochastic increases in **R**'s assessment as to the type of **S**, we can construct many sequential equilibrium outcomes where **R** "threatens with beliefs" – for out-of-equilibrium messages s, **R** holds beliefs that put probability one on the message coming from the worst (smallest) type $t \in T$, and **R** takes the worst (smallest) action consistent with those beliefs and with the message. This, in general, will tend to keep **S** from sending those out-of-equilibrium messages.

Accordingly, much of the attention in refinements applied to signalling games has been along the lines: At a given equilibrium (outcome), certain out-of-equilibrium signals are "unlikely" to have come from types of low index. These out-of-equilibrium signals should therefore engender out-of-equilibrium beliefs that put relatively high probability on high index types. This then causes **R** to respond to those signals with relatively high index responses. And this, in many cases, will cause the equilibrium to fail.

There are a number of formalizations of this line of argument in the literature, and we will sketch only the simplest one here, which works well in the basic take-it-or-leave-it setup. (For more complete analysis, see Banks and Sobel (1987) and Cho and Kreps (1987). See also Farrell (1993) and Grossman and Perry (1987) for refinements that are similar in motivation.)

The criterion we use is the so-called *intuitive criterion*, which was introduced informally in Grossman (1981), used in other examples subsequently, and was then codified in Cho and Kreps (1987).

Definition. For a given signalling game, fix a sequential equilibrium (σ, α, μ). Let $u^*(t)$ be the equilibrium expected payoff to type t in this equilibrium. Define $B(s) = \{t \in T : u(t, s, \alpha) < u^*(t) \text{ for every } \alpha \in \mathscr{A}(s, T)\}$. The fixed sequential equilibrium *fails the intuitive criterion* if there exist $s \in S$ and $t^* \in T \setminus B(s)$ such that $u(t^*, s, \alpha) > u^*(t^*)$ for all $\alpha \in \mathscr{A}(s, T \setminus B(s))$.

The intuition runs as follows. If $t \in B(s)$, then, relative to following the equilibrium, type t has no incentive to send signal s; no matter what **R** makes of s, **R**'s response leaves t worse off than if t follows the equilibrium. Hence, if **R** receives the signal s, **R** should infer that this signal comes from a type of **S** drawn from $T \setminus B(s)$. But if **R** makes this inference and acts accordingly [taking some response from $\mathscr{A}(s, T \setminus B(s))$], type t^* is sure to do better than in the equilibrium. Hence, type t^* will defect, trusting **R** to reason as above.

This criterion (and others from the class of which it is representative) puts very great stress on the equilibrium outcome as a sort of status quo, against which defections are measured. The argument is that players can "have" their equilibrium values, and defections are to be thought of as a reasoned attempt to do better. This aspect of these criteria has been subject to much criticism: If a particular equilibrium is indeed suspect, then its values cannot be taken for granted. See, for example, Mailath et al. (1993). Nonetheless, this criterion can be justified on theoretical grounds.

Proposition 4. [Banks and Sobel (1987), Cho and Kreps (1987)]. If T, S and A are all finite, then for generically chosen payoffs any equilibrium that fails the intuitive criterion gives an outcome that is not strategically stable in the sense of Kohlberg and Mertens (1986).

This criterion is very strong in the case of basic take-it-or-leave-it games that satisfy the single-crossing property and that are otherwise well behaved. Please note the conflict with the hypothesis of Proposition 4. In the take-it-or-leave-it games, S is uncountably infinite. Hence, the theoretical justification for the intuition criterion provided by Proposition 4 only applies "in spirit" to the following results. Of course, the intuitive criterion itself does not rely on the finiteness of S.

The result that we are headed for is that, under further conditions to be specified, there is a single equilibrium outcome in the basic take-it-or-leave-it game that satisfies the intuitive criterion. This equilibrium outcome is pure separating, and can be loosely characterized as the first-best equilibrium for S subject to the separation constraints. This result is derived in three steps. First, it is shown that pooling is impossible except at the largest possible value of m. Then, more or less by assumption, pooling at that extreme value is ruled out. This implies that the equilibrium outcome is separating, and the unique separating equilibrium outcome that satisfies the intuitive criterion is characterized.

Proposition 5. Fix a basic take-it-or-leave-it game satisfying the single-crossing property. Then any equilibrium in which more than one type sends a given signal (m, d) with positive probability to which the response is *yes* with positive probability can satisfy the intuitive criterion only if m equals its highest possible value.

A sketch of the proof runs as follows: Suppose that, in an equilibrium, more than one type pooled at the signal (m, d) with m strictly less than its highest possible value, the response to which is *yes* with positive probability. Let t^* be the highest type in the pool. Then it is claimed that for any $\varepsilon > 0$, there is an unsent (m', d') with (a) $m < m' < m + \varepsilon$ and $d < d' < d + \varepsilon$, (b) $U(t^*, m') + d' > U(t^*, m) + d$, and (c), for all $t < t^*$, $U(t, m') + d' < U(t, m) + d$. (The demonstration of this is left to the reader. The single-crossing property is crucial to this result.)

Suppose that type t^* proposes (m', d'). By (c), $t \in B(s)$ for all $t < t^*$. Hence, **R** in the face of this deviation must hold beliefs that place probability one on types t^* or greater. Since V is strictly increasing in t, for ε sufficiently small, if **R** was willing to accept (m, d) with positive probability at the pool, with beliefs restricted to types t^* or greater, **S** must accept (m', d') with probability one. But then (b) ensures that t^* would deviate; i.e., the intuitive criterion fails.

Proposition 5 does not preclude the possibility that pooling occurs at the greatest possible level of m. To rule this possibility out, the next step is to make assumptions sufficient to ensure that this cannot happen. The simplest way is to assume that M is unbounded above, and we proceed for now on that basis. Hence by Proposition 5, attention can be restricted to equilibria where types either separate or do not trade with positive probability.

Proposition 6. Fix a basic take-it-or-leave-it game that satisfies the following three supplementary assumptions. (a) For all t, $\max_m U(t,m) + V(t_0, m) > 0$, where t_0 is the type with lowest index. (b) For all t, $U(t,m) + V(t,m)$ is strictly quasi-concave in m. (c) For each type t, if t' is the next-lowest type, then the problem $\max_m U(t,m) + V(t,m)$ subject to $U(t',m) + V(t,m) \leqslant u$ has a solution for every value of u. (If t is the lowest type, ignore the constraint.) Then there is a unique equilibrium outcome for sequential equilibria that satisfy the intuitive criterion and in which types either separate or do not trade.

We sketch the argument (and show how to construct this unique equilibrium outcome). Fix a sequential equilibrium that survives the intuitive criterion and that is separating (or where some types trade with probability zero). Let m_0 solve $\max_m U(t_0, m) + V(t_0, m)$, where t_0 is the lowest index type. [By supplementary assumption (b), this solution is unique.] By proposing $(m_0, V(t_0, m_0) - \varepsilon)$ for $\varepsilon > 0$, type t_0 can be sure of acceptance, because **R**'s beliefs can be no worse than that this proposal comes from t_0 with certainty. Hence, t_0 can be sure to get utility $u^*(t_0) := U(t_0, m_0) + V(t_0, m_0)$ in any sequential equilibrium. [Moreover, a similar argument using supplementary assumption (a) ensures that the no-trade outcome will not be part of the equilibrium.] Since the equilibrium is separating, this is also an upper bound on what t_0 can get; so we know that the equilibrium outcome for t_0 in the equilibrium gives $u^*(t_0)$ to t_0, and (hence) t_0, in this equilibrium, must propose precisely $(m_0, V(t_0, m_0))$, which is accepted with probability one. Let t_1 be the type of second-lowest index. By use of the intuitive criterion, it can be shown that this type can be certain of utility $\max_m \{ U(t_1, m) + V(t_1, m) : U(t_0, m) + V(t_1, m) \leqslant u^*(t_0) \}$, since by proposing the (unique) maximizing m and a payment a bit less than $V(t_1, m)$ for this m, acceptance is guaranteed; since t_0 receives strictly less than $u^*(t_0)$ whether or not **R** accepts the proposal, $t_0 \in B((m, V(t_1, m)^-)$. This is also an upper bound for t_1's payoff in any separating equilibrium in which t_0 gets $u^*(t_0)$, and so this gives the outcome for t_1 in the equilibrium. And so on. We can build up the equilibrium outcome by, for each type, maximizing its payoff, assuming it is separated and that it is constrained to send a signal that the next-lower type would not (given its equilibrium value, determined a stage earlier).

Remarks. (1) This particular equilibrium, the optimal equilibrium for **S** subject to a separation constraint, goes back in the literature to well before its "justification" by stability-related refinements. The notion that, in equilibrium, there will be overinvestment in signals for purposes of separation can be found in the very first work on market signalling, by Spence (1974) and by Rothschild and Stiglitz (1976). This precise equilibrium is singled out by Riley (1979) on other grounds (see Section 6 following).

(2) Supplementary assumption (a) in Proposition 6 can be paraphrased as "trade is beneficial", even at terms appropriate for the lowest index type. It is rather

strong and should (and can) be relaxed, however there is no general treatment of this written down yet.

(3) We avoided equilibria in which there is pooling at the highest value of m by fiat, by assuming that M is unbounded. But then we needed to assume that U and V are sufficiently well-behaved so that supplementary assumption (c) holds. Alternatively, we could assume that M is compact [so that (c) is no longer required] and either make assumptions sufficient to guarantee that there is no pooling at the "top" or deal with the possibility of such a pool. For a careful treatment of this, see Cho and Sobel (1990).

(4) This argument is finely tuned to the take-it-or-leave-it game form, since in this game form, any type of **S**, having the ability to propose an entire "deal", has very fine ability to distinguish himself from lower types. In the more standard game form, **S** proposes only m (this becomes the signal), and **R** (or more than one **R**) responds with a proposed price d (which becomes a). When there are more than two types of **S**, the intuitive criterion is generally not strong enough to imply full separation. In such cases, stronger restrictions than the intuitive criteria such as universal divinity [Banks and Sobel (1987)] are required. The basic idea is that for each signal s, the set $B(s)$ is enlarged to include any type t for which there is some other type t' with: If $u(t, s, \alpha) \geqslant u^*(t)$ for $\alpha \in \mathscr{A}(s, T)$, then $u(t', s, \alpha) > u^*(t')$. [This is not quite universal divinity, but instead is the slightly weaker D1 restriction of Cho and Kreps (1987).] Or, in words, $t \in B(s)$ if for any response to s that would cause t to defect, t' would defect. Using this enlarged $B(s)$ amounts to an assumption that **R**, faced with s, infers that there is no chance that s came from t; if type t of **S** were going to send s instead of sticking to the equilibrium, than type t' would certainly do so. Although less intuitive than the intuitive criterion, this is still an implication (in generic finite signalling games) of strategic stability. Together with the single-crossing property (and some technical conditions), it implies that Proposition 2 can be extended to out-of-equilibrium signals: If type t' sends signal s with positive probability, then for all $t < t'$ and $s' > s$, we must be able to support the equilibrium with beliefs that have $\mu(t, s') = 0$. With further monotonicity conditions, it then gives results similar to Propositions 5 and 6 above. See Cho and Sobel (1990) and Ramey (1988) for details.

(5) Perhaps the intuitively least appealing aspect of these results is that a very small probability of bad types can exert nonvanishing negative externality on good types. That is, if ninety-nine percent of all the cars in the world are peaches and only one percent are lemons, still the peach owners must offer large warranties to distinguish their cars from the lemons. You can see this mathematically by the fact that if t_0 is a lemon and t_1 a peach, then the peach owners must choose an m_1 satisfying $U(t_0, m_1) + V(t_1, m_1) \leqslant u^*(t_0)$. this constraint is independent of the number of peaches and lemons; it is there as long as there is a single lemon in the world. But if there are no lemons, then this constraint vanishes. Pooling equilibria, in which the lemon owners exert a negative externality on the peach owners that vanishes at the fraction of lemons goes to zero, seem intuitively more reasonable.

Mailath et al. (1993) require that the receiver interpret an out-of-equilibrium message as an attempt by some types of sender to shift to another, preferred equilibrium whenever such an interpretation is possible. This leads to what they call *undefeated equilibria*; in the current context, this selects a pooling outcome.

(6) The argument given depends on m being one-dimensional; more generally, the single-crossing condition developed in the previous section is formulated for the case of a one-dimensional signal. Engers (1987) generalizes the single-crossing condition to multi-dimensional signals, following earlier specific examples. Although Engers works with a screening model (see the next section), Cho and Sobel (1990) and Ramey (1988) use versions of single-crossing conditions for multi-dimensional signals to characterize equillibrium outcomes in the canonical signalling game.

Humans are not the only animals that communicate through indirect signals. The songs or plumage of birds, the antlers of deer, and the colors of fish may be viewed as signals designed to inform a potential mate about reproductive fitness. Zahavi (1975, 1977) argues that meaningful signals arise in biological settings when they can be reliably interpreted by their intended audience, and he informally proposes that a single-crossing property holds. Grafen (1990) provides a game theoretic treatment of Zahavi's ideas. Grafen's model is virtually identical with the basic Spence model with a continuum of types. He argues that the same pure separating equilibrium outcome given prominence in the economics literature is the unique evolutionarily stable outcome.

(7) The construction of the unique equilibrium outcome that survives the intuitive criterion made use of the assumption that T is a finite set. (The argument that no equilibrium with pooling would survive the intuitive criterion, on the other hand, did not require this assumption.) Many of the early examples, however, assumed that T is an interval, so it is useful to provide extensions in this direction. Mailath (1987) provides an analysis of separating equilibria when there is an interval of types, giving useful characterizations and an existence result. Ramey's (1988) analysis is posed in this sort of setting.

(8) It is also desirable to extend this sort of analysis to cases where types of the sender are drawn from a set which is multi-dimensional (i.e., only partially ordered). Not much has been done along these lines, although Kohlleppel (1983) and Quinzii and Rochet (1985) provide interesting analyses of specific cases.

6. Screening

In the canonical signalling game of Sections 2 and 3 and in the variations on this game described above, the party with private information takes the lead in deciding which signals will be sent. That is to say, the set of signals that can be sent is given exogenously, as part of the description of the game, and the uninformed party reacts to the signal that is chosen by the informed party.

In many applications to market signalling, it may seem more appropriate to model things with the uninformed party taking the lead. The simplest game form that one can imagine for this, specialized to the very concrete model of the take-it-or-leave-it setup, has a number (> 1) of identical buyers (the uninformed side to a transaction) simultaneously and independently proposing a set of contracts, each contract taking the form (m, d). (Assume T is finite and that buyers are restricted to proposing a finite set of pairs.) Sellers (or the single seller, the informed side) look over the set of contracts on offer and choose at most one – the deal is then consummated at the terms chosen. Since the choice of a contract is a "signal", this formulation falls under the general rubric of signalling, although in the literature this sort of formulation, where the uninformed take the lead, is referred to as *screening* [Stiglitz and Weiss (1990)].

For this game form, the only possible pure-strategy equilibrium outcome is the separating outcome sketched in the previous section. The argument is very similar to the argument given in previous section, except that one starts with the observation that, by the usual Bertrand argument, each contract that is taken with positive probability must just "break even", and then think of a uninformed party making the sorts of offers that we had the informed parties making in the argument to break any other pure-strategy equilibrium. But, as observed first by Rothschild and Stiglitz (1976) and as refined by Riley (1979), this separating outcome is (often) not an equilibrium outcome either; it is often possible for an uninformed party to propose a pooling a contract which attracts a number of types and which is profitable.

By general existence results [Dasgupta and Maskin (1986)], we know an equilibrium exists for this game (under certain regularity conditions that hold in many applications). Hence, the only equilibria are in mixed strategies. This lacks appeal in many of the economic contexts which this is meant to model. Hence early authors were moved to modify the equilibrium concept employed.

There are two basic modifications that have been studied. The first basic modification is Wilson's (1977) E2 equilibrium. Here, roughly put, an outcome is an equilibrium if there is no contract that can be added to the set of contracts that are offered that will be profitable *after all contracts which then sustain losses are removed from the set on offer*. Wilson shows that an E2 equilibrium always exists (under standard conditions), that pooling can be an E2 equilibrium, and that there are robust examples with multiple E2 equilibria.

The second is Riley's (1979) *reactive equilibrium*. Again roughly put, an outcome is a reactive equilibrium if no additional contract added to the set on offer would make a profit if other firms are given the opportunity to react and *add* still more offers. Engers and Fernandez (1987) study this equilibrium notion in substantial generality, showing that under standard conditions, the only reactive equilibrium is the separating equilibrium of Section 5.

Both Wilson and Riley advance notions of equilibrium that are meant to be reduced form solution concepts for an unspecified dynamic process of competition.

The linkages to noncooperative (Nash) equilibrium of completely well-specified game forms came later. Hellwig (1986) studies the game form where buyers make offers simultaneously, sellers choose contracts, but then buyers have the right to refuse any seller. This game admits many Nash (and even sequential) equilibria, but Hellwig announces the result that the only stable equilibrium outcome is the pooling contract which is most favorable to the highest type, at least for the case where there are only two types. Engers and Fernandez (1987), on the other hand, study a game form where buyers simultaneously name contracts; their contract choices are revealed, and then buyers are given the opportunity to add to the set of contracts offered; if any additions are made, buyers are given another opportunity to add contracts; and so on, until on some given round buyers all "pass" on the opportunity to add contracts, at which point sellers (or the single seller) chooses his most preferred contract. Any reactive equilibrium of the original game is a Nash equilibrium for this game, but there are many other Nash equilibria; and Engers and Fernandez do not explore whether any of the standard refinements will help pin things down.

A further variation on market screening concerns the case of a single uninformed party who offers the informed party a menu of contracts on a take-it-or-leave-it basis. That is, all the bargaining power is given to the uninformed party. The seminal reference to this sort of analysis is Stiglitz (1977), who shows that this reduces formally to a problem of a monopolist supplier who is able to offer nonlinear prices. He therefore conducts analysis analogous to that in the literature on optimal income taxation [Mirrlees (1971)] to characterize the monopolist's optimal set of contracts. Complications due to strategic interactions disappear and there is no problem guaranteeing the existence of equilibrium. In contrast to the outcomes obtained in signalling games or in competitive screening models, the monopolist's optimal set of contracts need not be fully separating (when there are at least three types of informed party).

7. Costless signalling and neologisms

We next specialize to the case where messages are costless; i.e., where $u(t, s, a) = u(t, a)$ and $v(t, s, a) = v(t, a)$. In this case, even though "talk is cheap", signalling may be of use, insofar as there is some commonality of interests between S and R.

Green and Stokey (1980) and Crawford and Sobel (1982) provide the first analyses of this case. Crawford and Sobel assume: (a) $T = [0, 1]$, and the prior distribution of t on T is absolutely continuous. (b) The space A is a connected interval from the real line, for every t, $u(t, \cdot)$ and $v(t, \cdot)$ are concave, and, denoting partials of u and v by subscripts in the usual fashion, $u_2(t, \cdot) = 0$ and $v_2(t, \cdot) = 0$ have solutions in A. (c) $u_{12} > 0$ and $v_{12} > 0$, so that the solutions of $u_2(t, \cdot) = 0$ and $v_2(t, \cdot) = 0$ as functions of t both increase with t. (d) Finally, if we write $a_u(t)$ for the solution of $u_2(t, \cdot) = 0$ and $a_v(t)$ for the solution of $v_2(t, \cdot) = 0$, then $a_u(t) \neq a_v(t)$ for all t. With these assumptions, they show that all equilibria are essentially of the following form: There is a finite partition of T into intervals, and **S**'s signal indicates which of the cells of the partition contains the true value of t, with **R** responding according to the posterior so generated. The strong monotonicity assumptions guarantee that the response taken "increases" with increases in the cell of the partition. Moreover, with a further monotonicity assumption, there are (essentially) finitely many of these equilibria; one for a partition which is trivial (no information is communicated), and one for partitions of each integer size up to some largest-sized partition. Further, the expected utility of both **S** and **R** increases as the number of cells in the partition increases. Finally, they give a sense in which one can say that preferences of **S** and **R** are more closely aligned, and they establish that the more closely aligned are these preferences, the greater the cardinality of the maximal partition.

All cheap-talk games have a no-communication or babbling equilibrium in which **S**'s signal is not informative, and **R** responds to all signals (on the equilibrium path) with the action that is optimal given a posterior that is equal to the prior. [To be precise, this is so if the game with no possibility of communication has an equilibrium; cf. Seidman (1992) and Van Damme (1987).] There may be other equilibria. In particular, if the preferences of **S** and **R** coincide, there exists a separating equilibrium. Yet the refinements discussed in Section 5, which depend on different signals being more costly for some types of **S** than for others, do not help at all to reduce the set of sequential equilibria. Nothing prevents **R** from interpreting an out-of-equilibrium signal s in exactly the same way as a signal s' that is sent with positive probability. When signalling is costless, this sort of response establishes that the signal s is not "bad" for any type that sends s'. Hence no sequential equilibrium can fail the intuitive criterion (or any more restrictive refinements derived from strategic stability). Put another way, in cheap-talk games signals have no natural meaning at all, so no interpretation of them can be ruled out.

It is possible to refine the equilibrium set for costless signalling games by requiring that **R** believe what **S** says if it is in **S**'s interest to speak the truth. Farrell (1993) allows **S** to invent a new signal, called a neologism, which is credible if and only if there exists a nonempty set J of T such that precisely the types in J prefer the neologism to candidate equilibrium payoffs when **R** responds optimally to the literal meaning of the new signal (that is, "$t \in J$"). Formally, take a sequential equilibrium (σ, α, μ), and let $u^*(t)$ be the equilibrium payoff of type t. $J \subseteq T$ can send a *credible neologism* if and only if $J = \{t : u(t, \alpha(J)) > u^*(t)\}$, where $\alpha(J)$ is **R**'s (assumed unique, for simplicity) optimal response to the prior distribution conditioned on $t \in J$. (When ρ is not nonatomic, things are a bit more complex than this, since some types may be indifferent between their equilibrium payoff and the payoff from the neologism.) If **R** interprets a credible neologism literally, then some types would send the neologism and destroy the candidate equilibrium. Accordingly, Farrell emphasizes sequential equilibria for which no subset of T can send a credible neologism. These *neologism proof* equilibria do not exist in all cheap-talk games, including the game analyzed in Crawford and Sobel (1982) for some specification of preferences.

Rabin (1990) defined credibility without reference to a candidate equilibrium. He assumed that a statement of the form "t is an element of J" is credible if (a) all types in J obtain their best possible payoff when **R** interprets the statement literally, and (b) **R**'s optimal response to "t is an element of J" does not change when he takes into account that certain types outside of J might also make the statement. Roughly, a sequential equilibrium is a credible message equilibrium if no type of **S** can increase his payoff by sending a credible message. He proves that credible message equilibria exist in all finite, costless signalling games, and that "no communication" need not be a credible message equilibrium in some cases.

Dynamic arguments may force cheap talk to take on meaning in certain situations. Wärneryd (1993) shows that in a subset of cheap-talk games in which the interests of the players coincide, only full communication is evolutionarily stable. Aumann and Hart (1990) and Forges (1990) show that allowing many rounds of communication can enlarge the set of equilibrium outcomes.

8. Concluding remarks

In this chapter, we have stayed close to topics concerned with applications to market signalling. Even this brief introduction to the basic concepts and results from this application has exhausted the allotment of space that was given. In a more complete treatment, it would be natural to discuss at least the following further ideas: (1) In the models discussed, the set of possible signals (or, more generally, contracts) is given exogenously. A natural question to ask is whether there might be some way to identify a broader or even universal class of possible contracts and to look for contracts that are optimal among all those that are possible. The literature on mechanism design and the revelation principle should be consulted; see Chapter 24. It is typical in this literature to give the leading role to the uninformed party, so that this literature is more similar to screening than to signalling, as these terms are used above. When mechanism design is undertaken by an informed party, the process of mechanism design by itself may be a signal. Work here is less well advanced; see Myerson (1985) and Crawford (1985) for pioneering efforts, and see Maskin and Tirole (1990, 1992) for recent work done more in the spirit of noncooperative game theory. (2) We have discussed cases where individuals wish to have information communicated, at least partially. One can easily think of situations in which one party would want to stop or garble information that would otherwise flow to a second party, or even where one party might wish to stop or garble the information that would otherwise flow to himself. For an introduction to this topic, see Tirole (1988). (3) Except for a bare mention of multistage cheaptalk, we have not touched at all on applications where information may be communicated in stages or where two or more parties have private information which they may wish to signal to one another. There are far too many interesting analyses of specific models to single any out for mention – this is a broad area in which unifying theory has not yet taken shape. (4) The act of signalling may itself be a signal. This can cut in (at least) two ways. When the signal is noisy, say, it consists of a physical examination which is subject to measurement error, then the willingness to take the exam may be a superior signal. But then it becomes "no signal" at all. Secondly, if the signal is costly and develops in stages, then once one party begins to send the signal, the other may propose a Pareto-superior arrangement where the signal is cut off in midstream. But this then can destroy the signal's separating characteristics. For some analyses of these points, see Hillas (1987).

All this is only for the case where the information being signalled is information about some exogenously specified type, and it does not come close to exhausting that topic. The general notion of signalling also applies to signalling past actions or, as a means of equilibrium selection, in signalling future intentions. We cannot begin even to recount the many developments along these lines, and so we end with a warning to the reader that the range of application of non-cooperative game theory in the general direction of signallng is both immense and rich.

References

Admati, A. and M. Perry (1987) 'Strategic delay in bargaining', *Review of Economic Studies*, **54**: 345–364.
Aumann, R. and S. Hart (1990) 'Cheap talk and incomplete information', private communication.
Banks, J.S., and J. Sobel (1987) 'Equilibrium selection in signaling games', *Econometrica*, **55**: 647–662.
Cho, I-k. (1987) 'Equilibrium analysis of entry deterrence: A re-examination', University of Chicago, mimeo.
Cho, I-k. and D.M. Kreps (1987) 'Signaling games and stable equilibria', *Quarterly Journal of Economics*, **102**: 179–221.
Cho, I-k. and J. Sobel (1990) 'Strategic stability and uniqueness in signaling games', *Journal of Economic Theory*, **50**: 381–413.
Crawford, V. (1985) 'Efficient and durable decision rules: A reformulation', *Econometrica*, **53**: 817–837.
Crawford, V. and J. Sobel (1982) 'Strategic information transmission', *Econometrica*, **50**: 1431–1451.
Dasgupta, P. and E. Maskin (1986) 'The existence of equilibrium in discontinuous economic games, I: Theory', *Review of Economic Studies*, **53**: 1–26.
Engers, M. (1987) 'Signaling with many signals', *Econometrica*, **55**: 663–674.
Engers, M. and L. Fernandez (1987) 'Market equilibrium with hidden knowledge and self-selection', *Econometrica*, **55**: 425–440.
Farrell, J. (1993) 'Meaning and credibility in cheap-talk games', *Games and Economic Behavior*, **5**: 514–531.
Forges, F. (1990) 'Equilibria with communication in a job market example', *Quarterly Journal of Economics*, **105**: 375–98.
Grafen, A. (1990) 'Biological signals as handicaps,' *Journal of Theoretical Biology*, **144(4)**: 517–46.
Green, J. and N. Stokey (1980) 'A two-person game of information transmission', Harvard University mimeo.
Grossman, S. (1981), 'The role of warranties and private disclosure about product quality', *Journal of Law and Economics*, **24**: 461–83.
Grossman, S. and M. Perry (1987) 'Perfect sequential equilibrium', *Journal of Economic Theory*, **39**: 97–119.

Hellwig, M. (1986) 'Some recent developments in the theory of competition in markets with adverse selection', University of Bonn, mimeo.
Hillas, J. (1987) 'Contributions to the Theory of Market Screening', Ph.D. Dissertation, Stanford University.
Kohlberg, E. and J-F. Mertens (1986) 'On the strategic stability of equilibria', *Econometrica*, **54**: 1003–1038.
Kohlleppel, L. (1983) 'Multidimensional market signalling', University of Bonn, mimeo.
Mailath, G. (1987) 'Incentive compatibility in signaling games with a continuum of types', *Econometrica*, **55**: 1349–1365.
Mailath, G., M. Okuno-Fujiwara and A. Postlewaite (1993) 'On belief based refinements in signaling games', *Journal of Economic Theory*, **60**: 241–276.
Maskin, E. and J. Tirole (1990) 'The principal-agent relationship with an informed principal: The case of private values', *Econometrica*, **58**: 379–409.
Maskin, E. and J. Tirole (1990) 'The principal-agent relationship with an informed principal: The case of common values', *Econometrica*, **60**: 1–42.
Milgrom, P. and J. Roberts (1982) 'Limit pricing and entry under incomplete information: An equilibrium analysis', *Econometrica*, **50**: 443–459.
Mirrlees, J.A. (1971) 'An exploration of the theory of optimum income taxation', *Review of Economic Studies*, **28**: 195–208.
Myerson, R. (1985) 'Mechanism design by an informed principal', *Econometrica*, **47**: 61–73.
Quinzii, M. and J.-C. Rochet (1985), 'Multidimensional signalling', *Journal of Mathematical Economics*, **14**: 261–284.
Rabin, M. (1990) 'Communication between rational agents', *Journal of Economic Theory*, **51**: 144–70.
Ramey, G. (1988) 'Intuitive signaling equilibria with multiple signals and a continuum of types', University of California at San Diego, mimeo.
Reinganum, J. and L. Wilde (1986) 'Settlement, litigation, and the allocation of litigation costs', *Rand Journal of Economics*, **17**: 557–566.
Riley, J. (1979) 'Informational equilibrium', *Econometrica*, **47**: 331–359.
Rothschild, M. and J.E. Stiglitz (1976) 'Equilibrium in competitive insurance markets: An essay on the economics of imperfect information', *Quarterly Journal of Economics*, **80**: 629–649.
Rubinstein, A. (1985), 'A bargaining model with incomplete information about time preferences', *Econometrica*, **53**: 1151–1172.
Seidman, D. (1992) 'Cheap talk games may have unique, informative equilibrium outcomes,' *Games and Economic Behavior*, **4**: 422–425.
Sobel, J. (1989) 'An analysis of discovery rules', *Law and Contemporary Problems*, **52**: 133–59.
Spence, A.M. (1974) *Market Signaling*. Cambridge, MA: Harvard University Press.
Stiglitz, J. (1977) 'Monopoly, non-linear pricing and imperfect information: The insurance market', *Review of Economic Studies*, **94**: 407–430.
Stiglitz, J. and A. Weiss (1990) 'Sorting out the differences between screening and signaling models', in: M. Bachrach et al., eds., *Oxford Mathematical Economics Seminar*, Twenty-fifth Anniversary Volume. Oxford: Oxford University Press.
Tirole, J. (1988) *The Theory of Industrial Organization*. Cambridge, MA: MIT Press.
Wärneyrd, K. (1993) 'Cheap talk, coordination, and evolutionary stability', *Games and Economic Behavior*, **5**: 532–546.
Wilson, C. (1977) 'A model of insurance markets with incomplete information', *Journal of Economic Theory*, **16**: 167–207.
Van Damme, E. (1987) 'Equilibria in noncooperative games', in: H.J.M. Peters and O.J. Vrieze, eds., *Surveys in Game Theory and Related Topics*. Amsterdam: Centre for Mathematics and Computer Science, pp. 1–35.
Zahavi, A. (1975) 'Mate selection – a selection for a handicap', *Journal of Theoretical Biology*, **53**: 205–14.
Zahavi, A. (1977) 'Reliability in communication systems and the evolution of altruism', in: B. Stonehouse and C. Perrins, eds., *Evolutionary Ecology*, London: Macmillan.

Chapter 26

MORAL HAZARD

PRAJIT K. DUTTA

Columbia University and University of Wisconsin

ROY RADNER*

Bell Laboratories and New York University

Contents

1.	Introduction	870
2.	The principal–agent model	871
	2.1. The static model	871
	2.2. The dynamic model	873
3.	Analyses of the static principal–agent model	874
4.	Analyses of the dynamic principal–agent model	878
	4.1. Second-best contracts	879
	4.2. Simple contracts	881
5.	Games of imperfect monitoring	886
	5.1. Partnership model	887
	5.2. Repeated games with imperfect monitoring	892
6.	Additional bibliographical notes	897
	References	900

*We thank the editors and two anonymous referees for helpful comments. The views expressed here are those of the authors and do not necessarily reflect the viewpoint of AT&T Bell Laboratories.

Handbook of Game Theory, Volume 2, Edited by R.J. Aumann and S. Hart
© *Elsevier Science B.V., 1994. All rights reserved*

1. Introduction

The owner of an enterprise wants to put it in the hands of a manager. The profits of the enterprise will depends both on the actions of the manager as well as the environment within which he operates. The owner cannot directly monitor the agent's action nor can he costlessly observe all relevant aspects of the environment. This situation may also last a number of successive periods. The owner and the manager will have to agree on how the manager is to be compensated, and the owner wants to pick a compensation mechanism that will motivate the manager to provide a good return on the owner's investment, net of the payments to the manager. This is the well-known "*principal–agent*" problem with *moral hazard*. Some other principal–agent relationships in economic life are: client–lawyer, customer–supplier, insurer–insured and regulator–public utility.[1]

The principal–agent relationship embodies a special form of moral hazard, which one might call "one-sided", but moral hazard can also be "many-sided". The paradigmatic model of many-sided moral hazard is the *partnership*, in which there are many agents but no principal. The output of the partnership depends jointly on the actions of the partners and on the stochastic environment; each partner observes only the output (and his own action) but not the actions of the other partners nor the environment. This engenders a free-rider problem. As in the case of principal–agent relationships, a partnership, too, may last many periods.[2]

In Section 2 we present the principal–agent model formally. Section 3 discusses some salient features of optimal principal–agent contracts when the relationship lasts a single period. The first main point to make here is that in a large class of cases an equilibrium in the one-period game is Pareto-inefficient. This is the well-known problem involved in providing a risk-averse agent insurance while simultaneously giving him the incentives to take, from the principal's perspective, appropriate actions. We also discuss, in this section, other properties of static contracts such as monotonicity of the agent's compensation in observed profits.

In Section 4 we turn to repeated moral hazard models. Section 4.1 discusses some known properties of intertemporal contracts; the main points here are that

---

[1] The insurer–insured relationship is the one that gave rise to the term "moral hazard" and the first formal economic analysis of moral hazard was probably given by Arrow (1963, 1965).

[2] More complex informational models can be formulated for both the principal–agent as well as the partnership framework; models in which some agents obtain (incomplete) information about the environment or the actions of others. We do not directly discuss these generalizations although many of the results that follow can be extended to these more complex settings (see also the further discussion in Section 6). Note too that we do not treat here an important class of principal–agent models, the "adverse selection" models. The distinction between moral hazard and adverse selection models is that in the former framework, the principal is assumed to know all relevant characteristics of the agent (i.e., to know his "type") but not to know what action the agent chooses whereas in the latter model the principal is assumed not to know some relevant characteristic of the agent although he is able to observe the agent's actions. (See Section 6 for a further discussion.)

an optimal contract will, typically, reward the agent on the basis of past performance as well as current profits. Furthermore, although a long-term contract allows better resolution of the incentives-insurance trade-off, in general, some of the inefficiency of static contracts will persist even when the principal–agent relationship is long-lived. However, if the principal and agent are very patient, then almost all inefficiency can, in fact, be resolved by long-term contracts – and, on occasion, simple long-term contracts. These results are discussed in Section 4.2.

Many-sided moral hazard is studied in Section 5. The static partnership model is discussed in Section 5.1. The main focus here is on the possible resolution of the free-rider problem when the partners are risk-neutral. We also discuss some properties of optimal sharing rules, such as monotonicity, and the effect of risk-aversion on partners' incentives. Again, in general, static partnership contracts are unable to generate efficiency. This motivates a discussion of repeated partnership models. Such a model is a special case of a repeated game with imperfect monitoring; indeed results for repeated partnerships can be derived more readily from studying this more general class of games. Hence, in Section 5.2 we present known results on the characterization of equilibria in repeated games with imperfect monitoring.

It should be noted that the principal–agent framework is in the spirit of mechanism design; the principal chooses a compensation scheme, i.e., chooses a game form in order to motivate the manager to take appropriate actions and thereby the principal maximizes his own equilibrium payoff. The static partnership model is similarly motivated; the partners' sharing rule is endogenous to the model. In contrast, one can take the compensation scheme or sharing rule as exogenously given, i.e., one can take the game form is given, and focus on the equilibria generated by this game form. In the second approach, therefore, a moral hazard or partnership model becomes a special case of a game with imperfect monitoring. This is the approach used in Section 5.2.

Section 6 brings together additional bibliographical notes and discusses some extensions of the models studied in this paper.

2. The principal–agent model

2.1. *The static model*

A static (or stage-game) principal–agent model is defined by the quintuple $(A, \varphi, \mathbf{G}, U, W)$. A is the set of actions that the agent can choose from. An action choice by the agent determines a distribution, $\varphi(a)$, over output (or profit) G; $G \in \mathbf{G}$. The agent's action is unobservable to the principal whereas the output is observable. The agent is paid by the principal on the basis of that which is observable; hence, the compensation depends only on the output and is denoted $\mathbf{I}(G) = I$. U will denote the utility function of the agent and its arguments are the action undertaken and the realized compensation; $U(a, I)$. Finally, the principal's payoff depends on his net return $G - I$ and is denoted $W(G - I)$. (Note that G and I are real-valued.)

The *maintained assumptions* will be:

(A1) There are only a finite number of possible outputs; $G_1, G_2, \ldots, G_n$.

(A2) The set of actions A is a compact subset of some Euclidean space.

(A3) The agent's utility function U is strictly increasing in I and the principal's payoff function W is also strictly increasing.

A compensation scheme for the agent will be denoted $I_1, \ldots I_n$. Furthermore, with some abuse of notation, we will write $\varphi_j(a)$ for the probability that the realized output is G_j, $j = 1, \ldots, n$, when the action taken is a.

The time structure is that of a two-move game. The principal moves first and announces the compensation function **I**. Then the agent chooses his action, after learning **I**. The expected utilities for principal and agent are, respectively, $\sum_j \varphi_j(a) W(G_j - I_j)$ and $\sum_j \varphi_j(a) U(a, I_j)$. The *principal–agent problem* is to find a solution to the following optimization exercise:

$$\max_{I_1, \ldots, I_n, \hat{a}} \sum_j \varphi_j(\hat{a}) W(G_j - I_j) \tag{2.1}$$

s.t. $$\sum_j \varphi_j(\hat{a}) U(\hat{a}, I_j) \geq \sum_j \varphi_j(a) U(a, I_j), \quad \forall a \in A, \tag{2.2}$$

$$\sum_j \varphi_j(\hat{a}) U(\hat{a}, I_j) \geq \bar{U}. \tag{2.3}$$

The constraint (2.2) is referred to as the *incentive constraint*; the agent will only take those actions that are in his best interest. Constraint (2.3) is called the *individual-rationality constraint*; the agent will accept an arrangement only if his expected utility from such an arrangement is at least as large as his outside option $\bar{U}$. The objective function, maximizing the principal's expected payoff, is, in part, a matter of convention. One interpretation of (2.1) is that there are many agents and only one principal, who consequently gets all the surplus, over and above the outside options of principal and agent, generated by the relationship.[3]

If there is a $\bar{U}$ such that $(a^*, \mathbf{I}^*)$ is a solution to the principal–agent problem, then $(a^*, \mathbf{I}^*)$ will be called a *second-best* solution. This terminology distinguishes $(a^*, \mathbf{I}^*)$ from a Pareto-optimal (or *first-best*) action–incentives pair that maximizes (2.1) subject only to the individual-rationality constraint (2.3).

---

[3] An alternative specification would be to maximize the agent's expected payoffs instead; in this case, the constraint (2.3) would be replaced by a constraint that guarantees the principal his outside option. Note furthermore the assumption, implicit in (2.1)–(2.3), that in the event of indifference the agent chooses the action which maximizes the principal's returns. This assumption is needed to ensure that the optimization problem has a solution. A common, albeit informal, justification for this assumption is that, for every $\varepsilon > 0$, there is a compensation scheme similar to the one under consideration in which the agent has a strict preference and which yields the principal a net profit within ε of the solution to (2.1)–(2.3).

2.2. The dynamic model

In a repeated principal–agent model, in each period $t = 0, 1, \ldots T$, the stage-game is played and the output observed by both principal and agent; denote the output realization and the compensation pair, $G(t)$ and $I(t)$ respectively. The relationship lasts for $T(\leqslant \infty)$ periods, where T may be endogenously determined. The public history at date t, that both principal and agent know, is $h(t) \equiv (G(0), I(0), \ldots G(t-1), I(t-1))$, whereas the private history of the agent is $h_\alpha(t) \equiv (a(0), G(0), I(0), \ldots a(t-1), G(t-1), I(t-1))$. A strategy for the principal is a sequence of maps $\sigma_p(t)$, where $\sigma_p(t)$ assigns to each public history, $h(t)$, a compensation function $\mathbf{I}(t)$. A strategy for the agent is a sequence of maps $\sigma_\alpha(t)$, where $\sigma_\alpha(t)$ assigns to each pair, a private history $h_\alpha(t)$ and the principal's compensation function $\mathbf{I}(t)$, an action $a(t)$. A strategy choice by the principal and agent induces, in the usual way, a distribution over the set of histories $(h(t), h_\alpha(t))$; the pair of strategy choices therefore generate expected payoffs for principal and agent in period t; denote these $W(t; \sigma_p, \sigma_\alpha)$ and $U(t; \sigma_p, \sigma_\alpha)$. Lifetime payoffs are evaluated under discount factors δ_p and δ_α, for principal and agent respectively, and equal $(1 - \delta_p) \sum_{t=0}^{T} \delta_p^t W(t; \sigma_p, \sigma_\alpha)$ and $(1 - \delta_\alpha) \sum_{t=0}^{T} \delta_\alpha^t U(t; \sigma_p, \sigma_\alpha)$. The *dynamic principal–agent problem* is:[4]

$$\max_{\sigma_p, \hat{\sigma}_a} (1 - \delta_p) \sum_{t=0}^{T} \delta_p^t W(t; \sigma_p, \hat{\sigma}_\alpha) \tag{2.4}$$

$$\text{s.t.} \quad (1 - \delta_\alpha) \sum_{t=0}^{T} \delta_\alpha^t U(t; \sigma_p, \hat{\sigma}_\alpha) \geqslant (1 - \delta_\alpha) \sum_{t=0}^{T} \delta_\alpha^t U(t; \sigma_p, \sigma_\alpha), \quad \forall \sigma_\alpha, \tag{2.5}$$

$$(1 - \delta_\alpha) \sum_{t=0}^{T} \delta_\alpha^t U(t; \sigma_p, \sigma_\alpha) \geqslant \bar{U}. \tag{2.6}$$

The incentive constraint is (2.5) whereas the individual-rationality constraint is (2.6). Implicit in the formulation of the dynamic principal–agent problem is the idea that principal and agent are bound to the arrangement for the contract length T. Such a commitment is not necessary if we require that (a) the continuations of σ_α must satisfy (2.5) and (2.6) after *all* private histories $h_\alpha(t)$ and principal's compensation choice $\mathbf{I}(t)$, and (b) that the continuations of σ_p must solve the optimization problem (2.4) after all public histories $h(t)$.

---

[4] In the specification that follows, we add the principal's (as well as the agent's) payoffs over the contract horizon $0, \ldots T$ only. If T is less than the working lifetime of principal and agent, then the correct specification would be to add payoffs over the (longer) working lifetime in each case. Implicit in (2.5)–(2.6) is the normalization that the agent's aggregate payoffs, after the current contract expires, are zero. The principal's payoffs have to include his profits from the employment of subsequent agents. It is straightforward to extend (2.4) to do that and in Section 3.2 we will, in fact, do so formally.

3. Analyses of the static principal–agent model

It is customary to assume that an agent, such as the manager of a firm, is more risk-averse than the principal, such as the shareholder(s). From a first-best perspective, this suggests an arrangement between principal and agent in which the former bears much of the risk, and indeed, if the principal is risk-neutral, bears all of the risk. However, since the agent's actions are unobservable, the provision of such insurance may remove incentives for the agent to take onerous, but profitable, actions that the principal prefers. The central issue consequently, in the design of optimal contracts under moral hazard, is how best to simultaneously resolve (possible) conflicts between insurance and incentive considerations.

To best understand the nature of the conflict imagine, first, that the agent is in fact risk-neutral. In this case first-best actions (and payoffs) can be attained as second-best outcomes, and in a very simple way. An effective arrangement is the following: the agent pays the principal a fixed fee, independent of the gross return, but gets to keep the entire gross return for himself. (The fixed fee can be interpreted as a "franchise fee.") This arrangement internalizes, for the agent, the incentives problem and leads to a choice of first-best actions. Since the agent is risk-neutral, bearing all of the risk imposes no additional burden on him.[5]

On the other hand, imagine that the agent is strictly risk-averse whereas the principal is risk-neutral. Without informational asymmetry, the first-best arrangement would require the principal to bear all of the risk (and pay the agent a constant compensation). However, such a compensation scheme only induces the agent to pick his most preferred action. If this is not a first-best action, then we can conclude that the second-best payoff for the principal is necessarily less than his first-best payoff. These ideas are formalized as:

Proposition 3.1. (i) *Suppose that $U(a, \cdot)$ exhibits risk-neutrality, for every $a \in A$ (and the principal is either risk-averse or risk-neutral). Let (a_F, I_F) be any first-best pair of action and incentive scheme. Then, there is a second-best contract (a^*, I^*) such that the expected payoffs of both principal and agent are identical under (a_F, I_F) and (a^*, I^*).*

(ii) *Suppose that $U(a, \cdot)$ exhibits strict risk-aversion for every $a \in A$, and furthermore, that the principal is risk-neutral. Suppose at every first-best action, a_F, (a) $\varphi_j(a_F) > 0$, $j = 1, \ldots n$, and (b) for every I' there is $a' \in A$ such that $U(a', I') > U(a_F, I')$. Then, the principal's expected payoffs in any solution to the principal–agent problem is strictly less than his expected first-best payoff.*

Proof. (i) Let (a_F, I_F) be a first-best pair of action and incentive scheme and let the average retained earnings for the principal be denoted $\bar{G} - \bar{I} \equiv \sum_j \varphi_j(a_F) \cdot (G_j - I_{jF})$. Consider the incentive scheme I^* in which the agent pays a fixed fee

---

[5] The above argument is valid regardless of whether the principal is risk-neutral or risk-averse.

$\bar{G} - \bar{I}$ to the principal, regardless of output. Since the agent is risk-neutral, his utility function is of the form, $U(a, I) = H(a) + K(a)I$. Simple substitution then establishes the fact that $U(a_F, I^*) = U(a_F, I_F)$. Hence, the new compensation scheme is individually rational for the agent. Moreover, since the principal is either risk-averse or risk-neutral, his payoff under this scheme is at least as large as his payoff in the first-best solution; $W(\bar{G} - \bar{I}) \geqslant \sum_j \varphi_j(a_F) W(G_j - I_{jF})$. The scheme is also incentive compatible for the action a_F. For suppose, to the contrary, that there is an action a' such that $H(a') + K(a')[\sum_j \varphi_j(a') G_j - (\bar{G} - \bar{I})] > H(a_F) + K(a_F)\bar{I}$. Then there is evidently a fixed fee $\bar{G} - \bar{I} + \varepsilon$, for some $\varepsilon > 0$, that if paid by the agent to the principal, constitutes an individually rational compensation scheme. Further, the principal now makes strictly more than his first-best payoff; and that is a contradiction.

(a_F, I^*) is a pair that satisfies constraints (2.2) and (2.3) and yields the principal at least as large a payoff as the first-best. Since, by definition, the second-best payoff cannot be any more than the first-best payoff, in particular the two payoffs are equal and equal to that under (a_F, I^*), $W(\bar{G} - \bar{I})$.[6]

(ii) Let (a^*, I^*) be a solution to the principal–agent problem. If this is also a solution to the first-best problem, then, given the hypothesis, $\varphi_j(a^*) > 0, j = 1, \ldots n$, and principal and agent attitudes to risk, it must be the case that $I_j^* = I_{j'}^* \equiv I^*$, for all j, j'. But then, by hypothesis, a^* is not an incentive-compatible action[7]. □

Proposition 3.1(ii) strongly suggests that whenever the agent is risk-averse, there will be some efficiency loss in that the principal will provide incomplete insurance, in order to maintain incentives. The results we now turn to provide some characterization of the exact trade–off between incentives and insurance in a second-best contract. The reader will see, however, that not too much can be said, in general, about the optimal contract. Part of the reason will be the fact that although, from an incentive standpoint, the principal would like to reward evidence of "good behavior" by the agent, such evidence is linked to observable outputs in a rather complex fashion.

Grossman and Hart (1983) introduced a device for analyzing principal–agent problems that we now discuss. Their approach is especially useful when the agent's preferences satisfy a separability property; $U(a, I) \equiv H(a) + V(I)$.[8] Suppose also that the principal is risk-neutral and the agent is risk-averse.[9] Now consider

---

[6] A corollary to the above arguments is clearly that if the principal is risk-averse, while the agent is risk-neutral, then the unique first- (and second-) best arrangement is for the agent to completely insure the principal.

[7] Whether or not there is always a solution to the principal–agent problem is an issue that has been discussed in the literature. Mirrlees (1974) gave an example in which the first-best payoff can be approximated arbitrarily closely but cannot actually be attained. Sufficient conditions for a solution to the principal–agent problem to exist are given, for example, by Grossman and Hart (1983).

[8] Grossman and Hart admit a somewhat more general specification; $U(a, I) = H(a) + K(a)V(I)$ where $K(a) > 0$. That specification is equivalent to the requirement that the agent's preferences over income lotteries be independent of his action. See Grossman and Hart (1983) for further details.

[9] Since Proposition 3.1 has shown that a risk-neutral agent can be straightforwardly induced to take first-best actions, for the rest of this section we will focus on the hypothesis that the agent is, in fact, risk-averse.

any action $a \in A$ and let $C(a)$ denote the minimum expected cost at which the principal can induce the agent to take this action, i.e.

$$C(a) \equiv \min_{v_1,\ldots v_n} \sum_j \varphi_j(a) V^{-1}(v_j) \tag{3.1}$$

$$\text{s.t.} \quad H(a) + \sum_j \varphi_j(a) v_j \geqslant H(a') + \sum_j \varphi_j(a') v_j, \quad \forall a', \tag{3.2}$$

$$H(a) + \sum_j \varphi_j(a) v_j \geqslant \bar{U}, \tag{3.3}$$

where $v_j \equiv V(I_j)$. (3.2) and (3.3) are simply the (rewritten) incentive and individual-rationality constraints and the point to note is that the incentive constraints are linear in the variables $v_1,\ldots v_n$. Furthermore, if V is concave, then the objective function is convex and hence we have a convex programming problem.[10] The full principal–agent problem then is to find an action that maximizes the net benefits to the principal, $\sum_j \varphi_j(a) G_j - C(a)$.

Although the (full) principal–agent problem is typically not convex, analysis of the cost-minimization problem alone can yield some useful necessary conditions for an optimal contract. For example, suppose that the set of actions is, in fact, finite. Then the Kuhn–Tucker conditions yield:[11]

$$[V'(I_j)]^{-1} = \lambda + \sum_{a' \neq a} \mu(a') \left(1 - \frac{\varphi_j(a')}{\varphi_j(a)}\right), \tag{3.4}$$

where $\lambda, \mu(a')$, are (non-negative) Lagrange multipliers associated with, respectively, the individual-rationality and incentive constraints (one for each $a' \neq a$). The interpretation of (3.4) is as follows: the agent is paid a base wage, λ, which is adjusted if the jth output is observed. In particular, if the incentive constraint for action a' is binding, $\mu(a') > 0$, then the adjustment is positive if and only if the jth output is more likely under the desired action a, than under a'.

One further question of interest is whether there are conditions under which the optimal contract is *monotonically increasing* in that it rewards higher outputs with larger compensations; if we adopt the convention that outputs are ordered so that $G_j \leqslant G_{j+1}$, the question is, (when) is $I_j \leqslant I_{j+1}$? This question makes sense when "higher" inputs do, in fact, make higher outputs more likely. So suppose that $A \subset \mathbb{R}$ (for example, the agent's actions are effort levels) and, to begin with, that $a' > a$ implies that the distribution function corresponding to a' first-order stochastically dominates that corresponding to a.

---

[10] The earlier literature on principal–agent models replaced the set of incentive constraints (2.8) by the single constraint that, when the compensation scheme $I_1,\ldots I_n$ is used, the agent satisfies his first-order conditions at the action a. That this procedure is, in general, invalid was first pointed out by Mirrlees (1975). One advantage of the Grossman and Hart approach is, of course, that it avoids this "first-order approach".

[11] The expression that follows makes sense, of course, only when $\varphi_j(a) > 0$ and V is differentiable.

Now although the first-order stochastic monotonicity condition does imply that higher outputs are more likely when the agent works harder, we cannot necessarily infer, from seeing a higher output, that greater effort was in fact expended. The agent's reward is conditioned on precisely this inference and since the inference may be non-monotone so might the compensation.[12] Milgrom (1981) introduced into the principal–agent literature the following stronger condition under which higher output does, in fact, signal greater effort by the agent:

Monotone likelihood ratio condition (MLRC). If $a' > a$, then the likelihood ratio $[\varphi_j(a')]/[\varphi_j(a)]$ is increasing in j.

Under MLRC, the optimal compensation scheme will indeed be monotonically increasing *provided* the principal does in fact want the agent to exert the greatest effort. This can be easily seen from (3.4); the right-hand side is increasing in the output level. Since V^{-1} is convex, this implies that v_j, and hence I_j, is increasing in j. If, however, the principal does not want the agent to exert the greatest effort, rewarding higher output provides the wrong incentives and hence, even with MLRC, the optimal compensation need not be monotone.[13] Mirrlees (1975) introduced the following condition that, together with MLRC, implies monotonicity [let $F(a)$ denote the distribution function corresponding to a]:

Concavity of the distribution function (CDF). For all a, a' and $\theta \in (0,1)$, $F(\theta a + (1-\theta)a')$ first-order stochastically dominates $\theta F(a) + (1-\theta)F(a')$.

It can be shown by standard arguments that, under CDF, the agent's expected payoffs are a concave function of his actions (for a fixed monotone compensation scheme). In turn this implies that whenever an action $\hat{a}$ yields the agent higher payoffs than any $a < \hat{a}$, then, in fact, it yields higher payoffs than all other actions (including $a > \hat{a}$). Formally, these ideas lead to:

Proposition 3.2 [Grossman and Hart (1983)]. *Assume that V is strictly concave and differentiable and that MLRC and CDF hold. Then a second-best incentive scheme $(I_1, \ldots I_n)$ satisfies $I_1 \leq I_2 \leq \cdots \leq I_n$.*

Proposition 3.2 shows that the sufficient conditions on the distribution functions for, what may be considered, an elementary property of the incentive scheme, monotonicity, are fairly stringent. Not surprisingly, more detailed properties, such

---

[12] For example, suppose that there are two actions, $a_1 > a_2$ and three outputs, G_j, $j=1,\ldots 3$, $G_j \leq G_{j+1}$. Suppose also that the probability of G_1 is positive under both actions but the probability of G_2 is zero when action a_1 is employed (but positive under a_2). It is obvious that if action a_1 is to be implemented, then the compensation, if G_2 is observed, must be the lowest possible. Here the posterior probabilty of a_1, given the higher output G_2, is smaller than the corresponding probabilty when the lowest output G_1 is observed.

[13] Note that if $\mu(a') > 0$, for some $a' > a$, then (3.4) shows that on account of this incentive constraint the right-hand side decreases with j.

as convexity, are even more difficult to establish. The arguments leading to the proposition have, we hope, given the reader an appreciation of why this should be the case, namely the subtleties involved in inverting observed outcomes into informational inferences.[14]

One other conclusion emerges from the principal–agent literature: optimal contracts will be, in general, quite delicately conditioned on the parameters of the problem. This can be appreciated even from an inspection of the first-order condition (3.4). This is also a corollary of the work of Holmstrom (1979) and Shavell (1979). These authors asked the question: if the principal has available informational signals other than output, (when) will the optimal compensation scheme be conditioned on such signals? They showed that whenever output is *not* a sufficient statistic for these additional signals, i.e. whenever these signals do yield additional information about the agent's action, they should be contracted upon. Since a principal, typically, has many sources of information in addition to output, such as evidence from monitoring the agent or the performance of agents who manage related activities, these results suggest that such information should be used; in turn, this points towards quite complex optimal incentive schemes.

However, in reality contracts tend to be much simpler than those suggested by the above results. To explain this simplicity is clearly the biggest challenge for the theory in this area. Various authors have suggested that the simplicity of observable schemes can be attributed to some combination of: (a) the costs of writing and verifying complex schemes, (b) the fact that the principal needs to design a scheme that will work well in a variety of circumstances and under the care of many different agents and (c) the long-term nature of many observable incentive schemes. Of these explanations it is only (c) that has been explored at any length. Those results will be presented in the next section within our discussion of dynamic principal–agent models.

4. Analyses of the dynamic principal–agent model

In this section we turn to a discussion of repeated moral hazard. There are at least two reasons to examine the nature of long-term arrangements between principal and agent. The first is that many principal–agent relationships, such as that between a firm and its manager or that between insurer and insured or that between client and lawyer/doctor are, in fact, long-term. Indeed, observed contracts often exploit the potential of a long-term relationship; in many cases the contractual relationship continues only if the two parties have fulfilled prespecified obligations and met predesignated standards. It is clearly a matter of interest then to investigate how

---

[14]Grossman and Hart (1983) do establish certain other results on monotonicity and convexity of the optimal compensation scheme. They also show that the results can be tightened quite sharply when the agent has available to him only two actions.

such observed long-term contractual arrangements resolve the trade-off between insurance and incentives that bedevils static contracts.

A second reason to analyze repeated moral hazard is that there are theoretical reasons to believe that repetition does, in fact, introduce a rich set of incentives that are absent in the static model. Repetition introduces the possibility of offering the agent intertemporal insurance, which is desirable given his aversion to risk, without (completely) destroying his incentives to act faithfully on the principal's behalf. The exact mechanisms through which insurance and incentives can be simultaneously addressed will become clear as we discuss the available results. In Section 4.1 we discuss characterizations of the second-best contract at fixed discount factors. Subsequently, in Section 4.2 we discuss the asymptotic case where the discount factors of principal and agent tend to one.

4.1. Second-best contracts

Lambert (1983) and Rogerson (1985a) have established necessary conditions for a second-best contract. We report here the result of Rogerson; the result is a condition that bears a family resemblance to the well-known Ramsey–Euler condition from optimal growth theory. It says that the principal will smooth the agent's utilities across time periods in such a fashion as to equate his own marginal utility in the current period to his expected marginal utility in the next. We also present the proof of this result since it illustrates the richer incentives engendered by repeating the principal–agent relationship.[15]

Recall the notation for repeated moral hazard models from Section 2.2. A public (respectively, private) history of observable outputs and compensations (respectively, outputs, compensations and actions) up to but not including period t is denoted $h(t)$ [Respectively, $h_a(t)$]. Denote the output that is realized in period t by G_j. Let the period t compensation paid by the principal, after the public history $h(t)$ and then the observation of G_j, be denoted I_j. After observing the private history $h_a(t)$ and the output/compensation realized in period t, G_j/I_j, $j=1,\ldots n$, the agent takes an action in period $t+1$; denote this action a_j. Denote the output that is realized in period $t+1$ (as a consequence of the agent's action a_j) G_k, $k=1,\ldots n$. Finally denote the compensation paid to the agent in period $t+1$ when this output is observed I_{jk}, $j=1,\ldots n$, $k=1,\ldots n$.

Proposition 4.1 [Rogerson (1985a)]. *Suppose that the principal is risk-neutral and the agent's utility function is separable in action and income. Let (σ_p, σ_a) be a second-best contract. After every history $(h(t), h_a(t))$, the actions taken by the agent*

---

[15] This proof of the result is due to James Mirrlees.

and the compensation paid by the principal must be such that

$$[V'(I_j)]^{-1} = \frac{\delta_p}{\delta_\alpha} \sum_{k=1}^{n} \varphi_k(a_j)[V'(I_{jk})]^{-1}, \quad j = 1, \ldots n. \tag{4.1}$$

Proof. Pick any history pair $(h(t), h_\alpha(t))$ in the play of $(\sigma_p, \sigma_\alpha)$. As before let $v_j \equiv V(I_j)$. Construct a new incentive scheme σ_p^* that differs from σ_p only after $(h(t), h_\alpha(t))$ and then too in the following special way: $v_j = v_j^*$, $v_{jk} = v_{jk}^*$ for all k and $\mathbf{j} \neq j$, but $v_j^* = v_j - y$, $v_{jk}^* = v_{jk} + y/\delta_\alpha$ where y lies in any small interval around zero. In words, in the contract σ_p^*, after the history $(h(t), G_j)$, the principal offers a utility "smoothing" of y between periods t and $t+1$.

It is straightforward to check, given the additive separability of the agent's preferences, that the new scheme continues to have a best response of σ_α, the agent's utility is unchanged (and therefore, the scheme is individually rational). Since $(\sigma_p, \sigma_\alpha)$ is a solution to the principal–agent problem, σ_p is, in fact, the least costly scheme for the principal that implements σ_α [a la Grossman and Hart (1983)]. In particular, $y = 0$ must solve the principal's cost minimization exercise along this history. The first-order condition for that to be the case is easily verified to be (4.1).[16] □

Since the principal can be equivalently imagined to be providing the agent monetary compensation, I_j, or the utility associated with such compensation v_j, $V^{-1}(v)$ can be thought to be the principal's "utility function". Equation (4.1), and the proof of the proposition, then says that the principal will maintain intertemporal incentives and provide insurance so as to equate his (expected) marginal utilities across periods.

An immediate corollary of (4.1) is that second-best compensation schemes will be, in general, history-dependent; the compensation paid in the current period will depend not just on the observed current output, but also on past observations of output. To see this note that if I_{jk}, the compensation in period $t+1$, were independent of period t output, $I_{jk} = I_{\mathbf{j}k}$ for $j \neq \mathbf{j}$, then the right-hand side of (4.1) is itself independent of j and hence so must the left-hand side be independent of j. If V is strictly concave this can be true only if $I_j = I_\mathbf{j}$ for $j \neq \mathbf{j}$. But we know that a fixed compensation provides an agent with perverse incentives, from the principal's viewpoint.[17] History dependence in the second-best contract is also quite intuitive; by conditioning future payoffs on current output, and varying these

---

[16] In the above argument it was necessary for the construction of the incentive scheme σ_p^* that the principal be able to offer a compensation strictly lower than $\min(I_j, I_{jk})$, $j = 1, \ldots n$, $k = 1, \ldots n$. This, in turn, is possible to do whenever there is unbounded liability which we have allowed. If we restrict the compensations to be at least as large as some lower bound $\underline{I}$, then the argument would require the additional condition that $\min(I_j, I_{jk}) > \underline{I}$.

[17] The result, that second-best contracts will be history-dependent, was also obtained by Lambert (1983).

payoffs appropriately in the observed output, the principal adds a dimension of incentives that are absent in static contracts (which only allow for variations across current payments).

An unfortunate implication of history dependence is that the optimal contract will be very complex, conditioning as it ought to on various elements of past outputs. Such complexities, as we argued above, fly in the face of reality. An important question then is whether there are environments in which optimal contracts are, in fact, simple in demonstrable ways. Holmstrom and Milgrom (1987) have shown that if the preferences of both principal and agent are multiplicatively separable across time, and if each period's utility is representable by a CARA function, then the optimal contract is particularly simple; the agent performs the same task throughout and his compensation is only based on current output.[18] Since such simplification is to be greatly desired, an avenue to pursue would be to examine the robustness of their result within a larger class of "reasonable preferences."

4.2. Simple contracts

The second-best contracts studied in the previous subsection had two shortcomings: not all of the inefficiency due to moral hazard is resolved even with long-term contracts and furthermore, the best resolution of inefficiency required delicate conditioning on observable variables. In this subsection we report some results that remedy these shortcomings. The price that has to be paid is that the results require both principal and agent to be very patient.

The general intuition that explains why efficiency gains are possible in a repeated moral hazard setting is similar to that which underlies the possibility of efficiency gains in any repeated game with imperfect monitoring. Since this is the subject of Section 5.2, we restrict ourselves, for now, to a brief remark. The lifetime payoffs of the agent [see (2.4)] can be decomposed into an immediate compensation and a "promised" future reward. The agent's incentives are affected by variations in each of these components and when the agent is very patient, variations in future payoffs are (relatively) the more important determinant of the agent's incentives. A long-term perspective allows principal and agent to focus on these dynamic incentives.

A more specific intuition arises from the fact that the repetition of the relationship gives the principal an opportunity to observe the results of the agent's actions over a number of periods and obtain a more precise inference about the likelihood

---

[18] Fudenberg et al. (1990) have shown the result to also hold with additively separable preferences and CARA utility, under some additional restrictions. In this context also see Fellingham et al. (1985) who derive first-order conditions like (4.1) for alternative specifications of separability in preferences. They then show that utility functions obeying CARA and/or risk-neutrality satisfy these first-order conditions.

that the agent used an appropriate action. The repetition also allows the principal opportunity to "punish" the agent for perceived departures from the appropriate action. Finally, the fact that the agent's actions in any one period can be made to depend on the outcomes in a number of previous periods provides the principal with an indirect means to insure the agent against random fluctuations in the output that are not due to fluctuations in the agent's actions.

We now turn to a class of simple incentive schemes called *bankruptcy schemes*. These were introduced and analyzed by Radner (1986b); subsequently Dutta and Radner (1994) established some further properties of these schemes.

For the sake of concreteness, in describing a bankruptcy scheme, we will refer to the principal (respectively, the agent) as the owner (respectively, the manager). Suppose the owner pays the manager a fixed compensation per period, say w, as long as the manager's performance is "satisfactory" in a way that we define shortly; thereafter, the manager is fired and the owner hires a new manager. Satisfactory performance is defined as maintaining a positive "cash reserve", where the cash reserve is determined recursively as follows:

$$Y_0 = y$$
$$Y_t = Y_{t-1} + G_t - r, \quad t > 0. \tag{4.2}$$

The numbers y, r and w are parameters of the owner's strategy and are assumed to be positive.

The interpretation of a bankruptcy scheme is the following: the manager is given an initial cash reserve equal to y. In each period the manager must pay the owner a fixed "return", equal to r. Any excess of the actual return over r is added to the cash reserve, and any deficit is subtracted from it. The manager is declared "bankrupt" the first period, if any, in which the cash reserve becomes zero or negative, and the manager is immediately fired. Note that the cash reserve can also be thought of as an accounting fiction, or "score"; the results do not change materially under this interpretation.

It is clear that bankruptcy schemes have some of the stylized features of observable contracts that employ the threat of dismissal as an incentive device and use a simple statistic of past performance to determine when an agent is dismissed. Many managerial compensation packages have a similar structure; evaluations may be based on an industry-average of profits. Insurance contracts in which full indemnity coverage is provided only if the number of past claims is no larger than a prespecified number is a second example.

The principal's strategy is very simple; it involves a choice of the triple (y, w, r). The principal is assumed to be able to commit to a bankruptcy scheme. A strategy of the agent, say σ_α, specifies the action to be chosen after every history $h_\alpha(t)$ – and the agent makes each period's choice from a compact set A. Suppose that both principal and agent are infinitely-lived and suppose also that their discount factors are the same, i.e. $\delta_p = \delta_\alpha \equiv \delta$. Let $T(\sigma_\alpha)$ denote the time period at which the agent goes bankrupt; note that $T(\sigma_\alpha)$ is a random variable whose distribution is determined by the agent's strategy σ_α as well as the level of the initial cash reserve y and the required average rate of return r. Furthermore, $T(\sigma_\alpha)$ may take the value infinity.

The manager's payoffs from a bankruptcy contract are denoted $\mathbf{U}(\sigma_\alpha; y, w, r)$: $\mathbf{U}(\sigma_\alpha; y, w, r) \equiv (1-\delta)\sum_{t=0}^{T(\sigma_\alpha)} \delta^t U(a(t), w)$.[19] In order to derive the owner's payoffs we shall suppose that each successive manager uses the same strategy. This assumption is justified if successive managers are identical in their characteristics; the assumption then follows from the principle of optimality.[20] Denote the owner's payoffs $\mathbf{W}(y, w, r; \sigma_\alpha)$. Then

$$\mathbf{W}(y, w, r; \sigma_\alpha) = (1-\delta)E\sum_{t=0}^{T(\sigma_\alpha)} \delta^t[r - w - (1-\delta)y]$$
$$+ E\delta^{T(\sigma_\alpha)}[(1-\delta)y + \mathbf{W}(y, w, r; \sigma_\alpha)]. \tag{4.3}$$

Collecting terms in (4.3) we get

$$\mathbf{W}(y, w, r; \sigma_\alpha) = r - w - \frac{(1-\delta)y}{1 - E\delta^{T(\sigma_\alpha)}}. \tag{4.4}$$

The form of the principal's payoffs, (4.4), is very intuitive. Regardless of which generation of agent is currently employed, the principal always gets per period returns of $r - w$. Every time an agent goes bankrupt, however, the principal incurs the cost of setting up a new agent with an initial cash reserve of y. These expenses, evaluated according to the discount factor δ, equal $(1-\delta)y/1 - E\delta^{T(\sigma_\alpha)}$; note that as $\delta \to 1$, this cost converges to $y/ET(\sigma_\alpha)$, i.e. the cash cost divided by the frequency with which, on average, this expenditure is incurred.[21]

The dynamic principal-agent problem, (2.2)-(2.4) can then be restated as

$$\max_{(y,w,r)} \mathbf{W}(y, w, r; \hat{\sigma}_\alpha) \tag{4.5}$$

$$\text{s.t.} \quad \mathbf{U}(\hat{\sigma}_\alpha; y, w, r) \geq \mathbf{U}(\sigma_\alpha; y, w, r), \quad \forall \sigma_\alpha, \tag{4.6}$$

$$\mathbf{U}(\hat{\sigma}_a; y, w, r) \geq \bar{U}. \tag{4.7}$$

Suppose that the first-best solution to the static model is the pair (a_F, w_F), where the agent takes the action a_F and receives the (constant) compensation w_F. Let the principal's gross expected payoff be denoted r_F; $r_F = \sum_j \varphi_j(a_F)G_j$. Since the dynamic model is simply a repetition of the static model, this is also the dynamic first-best solution.

---

[19] Implicit in this specification is a normalization which sets the agent's post contract utility level at zero.
[20] The results presented here can be extended, with some effort, to the case where successive managers have different characteristics.
[21] In our treatment of the cash cost we have made the implicit assumption that the principal can borrow at the same rate as that at which he discounts the future.

We now show that there is a bankruptcy contract in which the payoffs of both principal and agent are arbitrarily close to their first-best payoffs, provided the discount factor is close to 1. In this contract, $w = w_F$, $r = r_F - \varepsilon/2$, for a small $\varepsilon > 0$, and the initial cash reserve is chosen to be "large".

Proposition 4.2 [Radner (1986b)]. *For every $\varepsilon > 0$, there is $\delta(\varepsilon) < 1$ such that for all $\delta \geqslant \delta(\varepsilon)$, there is a bankruptcy contract $(y(\delta), w_F, r_F - \varepsilon/2)$ with the property that whenever the agent chooses his strategy optimally, the expected payoffs for both principal and agent are within ε of the first-best payoffs. It follows that the corresponding second-best contract has this property as well (even if it is not a bankruptcy contract).*[22]

Proof. Faced with a bankruptcy contract of the form $(y(\delta), w_F, r_F - \varepsilon/2)$, one strategy that the agent can employ is to always pick the first-best action a_F. Therefore,

$$U(\hat{\sigma}_a; y(\delta), w, r) \geqslant U(a_F, w_F)(1 - E\delta^{T(F)}), \qquad (4.8)$$

where $T(F)$ is the time at which the agent goes bankrupt when his strategy is to use the constant action a_F. As $\delta \uparrow 1$, $(1 - E\delta^{T(F)})$ converges to Prob. $(T(F) = \infty)$. Since the expected output, when the agent employs the action a_F, is r_F and the amount that the agent is required to pay out is only $r_F - \varepsilon/2$, the random walk that the agent controls is a process with positive drift. Consequently, Prob. $(T(F) = \infty) > 0$, and indeed can be made as close to 1 as desired by taking the initial cash reserve, $y(\delta)$, sufficiently large [see, e.g., Spitzer (1976, pp. 217–218)]. From (4.8) it is then clear that the agent's payoffs are close to the first-best payoffs whenever δ is large.

The principal's payoff will now be shown to be close to his first-best payoffs as well. From the derivation of the principal's payoffs, (4.4), it is evident that a sufficient condition for this to be the case is that the (appropriately discounted) expected cash outlay per period, $(1-\delta)y/1 - E\delta^{T(\sigma_a)}$, be close to zero at high discount factors [or that the representative agent's tenure, $ET(\sigma_a)$, be close to infinity]. We demonstrate that whenever the agent plays a best response such is, in fact, a consequence. Write U^0 for $\max_a U(a, w_F)$. Since the agent's post-bankruptcy utility level has been normalized to 0, the natural assumption is $U^0 > 0$. Note that

$$U^0(1 - E\delta^{T(\hat{\sigma}(\delta))}) \geqslant U(\hat{\sigma}(\delta); y, w, r), \qquad (4.9)$$

where $T(\hat{\sigma}(\delta))$ is the time of bankruptcy under the optimal strategy $\hat{\sigma}(\delta)$. Hence,

$$U^0(1 - E\delta^{T(\hat{\sigma}(\delta))}) \geqslant U(a_F, w_F)(1 - E\delta^{T(F)})$$

---

[22]Using a continuous time formulation for the principal–agent model, Dutta and Radner (1994) are able to, in fact, give an upper bound on the extent of efficiency loss from the optimal bankruptcy contract, i.e., they are able to give a rate of convergence to efficiency as $\delta \to 1$.

from which it follows that

$$(1 - E\delta^{T(\hat{\sigma}(\delta))}) \geq c(1 - E\delta^{T(F)}), \tag{4.10}$$

where $c \equiv U(a_F, w_F)/U^0$. Substituting (4.10) into the principal's payoffs, (4.4), we get the desired conclusion; the principal's payoffs are close to his first-best payoffs, $r_F - w_F$, provided his discount factor is close to 1. The proposition is proved.[23] □

In a (constant wage) bankruptcy scheme the principal can extend increasing levels of insurance to the agent, as $\delta \to 1$, by specifying larger and larger levels of the initial cash reserve $y(\delta)$. The reason that this gives a patient agent the incentive to take actions close to the first-best is suggested by some results in permanent income theory. Yaari (1976) has shown that, for some specifications of bankruptcy, a patient risk-averse agent whose income fluctuates, but who has opportunity to save, would find it optimal to consume his expected income every period; i.e., would want and be able to smooth consumption completely. A bankruptcy scheme can be interpreted as *forced* consumption smoothing with the principal acting as a bank; an almost patient agent would like to (almost) follow such a strategy anyway.[24,25]

The first study of simple contracts to sustain asymptotic efficiency, and indeed the first analyses of repeated moral hazard, were Radner (1981) and Rubinstein (1979). Radner (1981) showed that for sufficiently long but finite principal–agent games, with no discounting, one can sustain approximate efficiency by means of approximate equilibria. Rubinstein showed in an example how to sustain exact efficiency in an infinitely repeated situation with no discounting. For the case of discounting, Radner (1985) showed that approximate efficiency can be sustained, even without precommitment by the principal, by use of *review strategies*. Review strategies are a richer version of bankruptcy schemes. In these schemes the principal holds the agent to a similar performance standard, maintaining an acceptable average rate of return, but (a) reviews the agent periodically (instead of every period) and (b) in the event of the agent failing to meet the standard, "punishes" him for a length of time (instead of severing the relationship forever). After the punishment phase, the arrangement reverts to the normal review phase. The insurance-incentive trade-off in these schemes is similar to those under bankruptcy schemes.

---

[23] We have not shown that an optimal bankruptcy contract, i.e., a solution of (4.5)–(4.7), exists. A standard argument can be developed to do so [for details, see Dutta and Radner (1992)].

[24] There are some delicate issues that we gloss over; the Yaari model is a pure consumption model (whereas our agent works as well) and bankruptcy as defined in the finite-horizon Yaari model has no immediate analog in the infinite-horizon framework adopted here.

[25] That allowing the agent to save opens up self-insurance possibilities in a repeated moral hazard model, has been argued recently by a number of authors such as Allen (1985), Malcomson and Spinnewyn (1988) and Fudenberg et al. (1990). In particular, the last paper shows that, even if the agent runs a franchise, and is exposed to all short-term risk, he can guarantee himself an average utility level close to the first-best.

Radner (1986b) introduced the concept of bankruptcy strategies for the principal, which were described above, and showed that they yield efficient payoffs in the limit as the principal's and agent's discount factors go to 1. Dutta and Radner (1994), in a continuous time formulation, provide a characterization of the optimal contract, within the class of bankruptcy contracts, and establish a lower bound on the rate at which principal–agent values must go to efficiency as $\delta \to 1$.

Up to this point we have assumed that the principal can precommit himself to a particular strategy.[26] Although precommitments can be found in some principal–agent relationship (e.g., customer–supplier, client–broker), it may not be a satisfactory description of many other such relationships. This issue is particularly problematic in repeated moral hazard contracts since at some point the principal has an incentive to renege on his commitment [if his continuation strategy does not solve (4.5)–(4.7) at that history].[27] For example, in a bankruptcy contract the principal has an incentive to renege after the agent, by virtue of either hard work or luck, has built up a large case reserve (and consequently will "coast" temporarily).

A bankruptcy (or review) strategy can be modified to ensure that the principal has no incentives to renege, i.e., that an equilibrium is perfect. One way to do so would be to modify the second feature of review strategies which was described above; an agent is never dismissed but principal and agent temporarily initiate a punishment phase whenever *either* the agent does not perform satisfactorily *or* the principal reneges on his scheme.[28] We shall not present those results here since in Section 5.2 we discuss the general issue of (perfect) folk theorems in games with imperfect monitoring.

5. Games of imperfect monitoring

In the principal–agent model of Sections 2–4 there is only one agent whose actions directly affect gross return; the moral hazard is due, therefore, to the unobservability of this agent's actions. In this section we analyze moral hazard issues that arise when there are multiple agents who affect gross returns and whose individual actions are hidden from each other.

The paradigmatic model of many-sided moral hazard is the *partnership* model, in which there is no principal but there are several agents – or partners – who jointly own the productive asset. Any given formula for sharing the return – or output – determines a game; the partners are typically presumed to choose their

---

[26] The agent need not however commit to his strategy. Although the incentive constraint, (4.6) may suggest that the agent is committed to his period 0 strategy choice, standard dynamic optimality arguments show that in fact all continuations of his strategy are themselves optimal.

[27] Note that a (limited) precommitment by the principal is also present in static moral hazard contracts; the principal has to abide by his announcement of the incentive scheme after the output consequences of the agent's action are revealed.

[28] The standard which the agent is held to, in such a strategy, has to be chosen somewhat carefully. For details on the construction, see Radner (1985).

actions in a self-interested way. The equilibria of the static partnership model are discussed in Section 5.1.

It is not very difficult to see from the above discussion that many-sided moral hazard is an example of the more general class of *games with imperfect monitoring*; indeed, for some of the results that follow it is more instructive to take this general perspective. So Section 5.2 will, in fact, introduce the general framework for such games, present results from *repeated games* with imperfect monitoring and discuss their implication for the repeated partnership model.

5.1. Partnership model

A *static* (or stage-game) partnership model is defined by a quintuple $(A_i, \varphi, \mathbf{G}, S_i, U_i; i = 1, \ldots m)$; i is the index for a partner, there are m such partners and each partner picks actions – e.g., inputs – from a set A_i. Let an action profile be denoted a; $a \equiv (a_1, \ldots a_m)$. The partners' action profile determines a distribution φ over the set of possible outputs $\mathbf{G}$. The m-tuple $S = (S_1, \ldots S_m)$ is the sharing rule; partner i's share of the total output is $S_i(G)$. A sharing rule must satisfy the *balanced budget* requirement:

$$\sum_i S_i(G) = G, \quad \text{for all } G. \tag{5.1}$$

The appropriate versions of the assumptions (A1)–(A3) will continue to hold. In other words, we will assume that the range of $\mathbf{G}$ is finite, with $\varphi_j(a)$ denoting the probability that the realized output is G_j, $j = 1, \ldots n$. Furthermore, each A_i is compact; indeed in the literature, the partners' action sets A_i have been assumed either to be finite or a (compact) continuum. For the results of this subsection, we will assume the latter; in particular, A_i will be an interval of the real line.[29] Given a sharing rule S and the actions of all the partners, partner i's expected utility is $EU_i(S_i(G), a_i)$. Note that this expected utility depends on the actions chosen by the others only through the effect of such actions on the distribution of the output. Finally, in addition to assuming that U_i is strictly increasing in the ith partner's share, we will also assume that φ_j and $U_i(S_i(G), \cdot)$ are differentiable functions.

An m-tuple of inputs is a *Nash equilibrium* of the partnership game if no partner can increase his own payoff by unilaterally changing his input. An *efficient* m-tuple of inputs is one for which no other feasible input tuple yields each partner at least as much expected utility and yields one partner strictly more.

Note that since each partner only observes the total output, their individual compensations can, at most, depend on this realized output. This is precisely what creates a potential *free-rider* problem; each partner's input generates a positive

---

[29] All of the results discussed in this subsection continue to hold when A_i is a set with a finite number of elements.

externality for the other partners and, especially since inputs are unobservable, each partner may therefore have an incentive to provide too little of his input. Two questions can then be posed: (i) The normative one: is there a' sharing rule under which the free-rider problem can be resolved in that the equilibria of the corresponding game are efficient? (This is the subject of Section 5.1). (ii) The positive question: how much of inefficiency is caused by the free-rider problem if the sharing rule is fixed ex ante? (This is the subject of Section 5.2).

We begin the discussion of partnership models with the case of risk-neutral partners. Recall that with one-sided moral hazard, there are efficient incentive-compatible arrangements between principal and agent when the latter is risk-neutral even in the static game. The first question of interest is whether this result can be generalized, i.e., are there sharing rules for risk-neutral partners such that the Nash equilibria of the resulting partnership game are efficient?

Let $U(S_i(G), a_i) \equiv S_i(G) - Q_i(a_i)$. As is well-known, risk neutrality in this case implies that utility is transferable. Hence, the efficiency problem can be written as

$$\max_a \sum_j G_j \varphi_j(a) - \sum_i Q_i(a_i). \tag{5.2}$$

Suppose that $\hat{a}$ is an interior solution to the efficiency problem (5.2). The question we are studying can be precisely stated as: is there a sharing rule S that satisfies the budget-balancing condition (5.1) and has the property that

$$\hat{a}_i \in \operatorname{argmax} \sum_j S_i(G_j) \varphi_j(a_i, \hat{a}_{-i}) - Q_i(a_i), \quad \forall i. \tag{5.3}$$

An early paper by Holmstrom (1982) suggested an inevitable conflict between budget balance and efficiency; (5.3) can be satisfied only if (5.1) is sacrificed by allowing a surplus in some states of the world $[\sum_i S_i(G_j) < G_j,$ for some $j]$. If it is indeed the case that a residual claimant – or principal – is always required to ensure efficiency in a partnership model, then we could correctly conclude that there is an advantage to an organization with separation between owner and management. As it turns out, the environments in which efficiency and budget balance are incompatible are limited although they contain the (important) symmetric case where each partner's effect on output is identical.

Suppose the distribution of output is affected by the partners' inputs only through some *aggregate variable*; i.e., there are (differentiable) functions $\theta: A \to \mathbb{R}$ and $\xi_j: \mathbb{R} \to [0, 1]$, such that for all j[30]

$$\varphi_j(a) = \xi_j(\theta(a)). \tag{5.4}$$

Proposition 5.1. *Suppose the aggregate effect condition (5.4) holds and suppose further that $\partial \theta / \partial a_i \neq 0$, for all a, i. Then there does not exist any sharing rules that*

---

[30] The condition (5.4) can be shown to be equivalent to the following condition on the derivatives of the likelihood functions: for any pair of partners $(i, \mathbf{i})$ and any pair of outputs $(j, \mathbf{j})$ and for all action tuples a, $\varphi_{ji}(a)/\varphi_{\mathbf{j}i}(a) = \varphi_{j\mathbf{i}}(a)/\varphi_{\mathbf{j}\mathbf{i}}(a)$, where $\varphi_{ji}(a) \equiv \partial \varphi_j / \partial a_i$.

both balances the budget and under which an efficient input profile $\hat{a}$ is a Nash equilibrium of the partnership game.

Proof. Suppose to the contrary that there is such a sharing rule S. The first-order conditions for efficiency yield [from (5.2)]:

$$\sum_j G_j \cdot \xi'_j(\theta(\hat{a})) = Q'_i(\hat{a}_i)/\theta_i(\hat{a}_i), \quad \forall i, \tag{5.5}$$

where the notation is: $\xi'_j \equiv \partial \xi_j/\partial \theta(a)$, $Q'_i \equiv \partial Q_i/\partial a_i$ and $\theta_i \equiv \partial \theta/\partial a_i$. Since $\hat{a}$ is a Nash equilibrium under sharing rule S, the first-order condition for best response yields [from (5.3)]:

$$\sum_j S_i(G_j) \cdot \xi'_j(\theta(\hat{a})) = Q'_i(\hat{a}_i)/\theta_i(\hat{a}_i), \quad \forall i. \tag{5.6}$$

Note that the right-hand sides of (5.5) and (5.6) are identical. Summing the left-hand side of (5.5) over the index i yields $m \cdot \sum_j G_j \xi'_j(\theta(\hat{a}))$. A similar summation over the left-hand side of (5.6) yields, after invoking the budget balance condition (5.1), $\sum_j G_j \xi'_j(\theta(\hat{a}))$. Since $m > 1$, the two sums do not agree and we have a contradiction. □

Remark. When the output is deterministic, i.e., $G = \theta(a)$, the aggregate condition is automatically satisfied.[31] Indeed, this was precisely the case studied by Holmstrom (1982). Similarly, if there are only two outcomes the condition is automatically satisfied as well; this can be seen, for example, from the equivalent representation for this condition (see footnote 30), that the derivatives of the likelihood ratios are equal across partners. Finally, if the agents are symmetric in their effect on output, the aggregate condition is immediate as well.[32]

Williams and Radner (1989) were the first to point out that in order to resolve the organization problem, even with risk-neutral partners, it is necessary that there be some *asymmetry* in the effect that each partner's input has on the distribution of the aggregate output. The intuition is quite straightforward: if, at the margin, the effect of partner i's input is more important in the determination of output j than the determination of output k (and the converse is true for partner $\hat{i}$'s input vis-a-vis outputs k and j), then the share assigned to i for output j has relatively greater impact on his decisions than the share assigned for output k; a parallel argument works for partner $\hat{i}$ and outputs k and j. Consequently, shares can be assigned in order to give the partners' appropriate incentives. Indeed, Williams and

---

[31] In this case, ξ cannot, evidently, be a differentiable function. The appropriate modification of the proof of Proposition 5.1 is, however, immediate.

[32] The aggregate condition on output distribution is also at the heart of the Radner et al. (1986) example of a repeated game with discounting in which efficiency cannot be sustained as an equilibrium outcome. This was pointed out by Matsushima (1989a, b).

Radner (1989) show that generically, in the space of distributions, there are sharing rules that do resolve the free-rider problem *and* balance the budget.

Since the aggregate condition (5.2) is no longer being assumed, the first-order condition (5.6), for the efficient input profile $\hat{a}$ to be a Nash equilibrium, can be written as

$$\sum_j S_i(G_j) \cdot \varphi_{ji}(\hat{a}) = Q'_i(\hat{a}_i), \quad \forall i. \tag{5.7}$$

If we wish to design the sharing rule S so that $\hat{a}$ satisfies the first-order conditions for an equilibrium, then the mn unknowns, $[S_i(G_j)]$, must satisfy the $(m+n)$ equations implied by (5.7) and the budget balance condition (5.1). The basic lemma of Williams and Radner (1989), reported below, is that, generically in the data of the model, such a solution can be found if $n > 2$, and that in particular this can be done if $\hat{a}$ is an efficient vector of inputs. Of course, to complete the argument it remains to show that there are reasonable conditions under which a solution to the "first-order" problem is actually an equilibrium.

Theorem 5.2 [Williams and Radner (1989)]. (i) *When $n > 2$, there exists a solution to the first-order conditions, (5.7), and the budget balance conditions, (5.1), for each pair of distribution and utility functions $(\varphi_j, Q_i, j = 1, \ldots m, i = 1, \ldots n)$ in some open dense subset (in the Whitney C^1 topology) of the set of feasible problems.*

(ii) *Suppose that $m = 2$ and $n = 3$. Assume further that φ_j is first-order stochastically increasing in its arguments, for $j = 1, \ldots 3$. Then there exists a one-parameter solution to the problem of finding a sharing rule whose Nash equilibrium is efficient if the follwing two additional hypotheses are satisfied at the efficient input profile $\hat{a}$:*
(a) $\varphi_{11}(\hat{a})\varphi_{22}(\hat{a}) - \varphi_{21}(\hat{a})\varphi_{12}(\hat{a}) > 0$
(b) $\varphi_{21}(\cdot, \hat{a}_2)/\varphi_{11}(\cdot, \hat{a}_2)$ *is an increasing function whereas* $\varphi_{22}(\hat{a}_1, \cdot)/\varphi_{21}(\hat{a}_1, \cdot)$ *is a decreasing function.*

Other conditions for solutions to this problem in static partnership models have been presented by Legros and Matsushima (1991), Legros (1989) and Matsushima (1989b). All of these conditions resonate with the idea that "symmetry is detrimental to efficiency" in partnership models.[33]

A question of some interest is what properties will efficiency-inducing sharing rules possess. In particular, as in the case of principal–agent models, we can ask: will the sharing rules be *monotonically increasing* in output, i.e., will a higher output increase the share of *all* partners? Of course, such a question makes sense only when higher inputs do, in fact, make higher outputs more likely – i.e., $\varphi_j(\cdot)$ is

---

[33] Interestingly, the results from the static model, with risk-neutral partners, will turn out to be very helpful in the subsequent analysis of the repeated partnership model with general (possibly risk-averse) utility functions. This is because intertemporal expected utility will be seen to have a decomposition very similar to that between the monetary transfer and the input-contingent expected utility in the static risk-neutral case; this point will be clearer after our discussion in Section 5.2.

first-order stochastically increasing – since a higher output may then be taken as a signal of higher inputs and appropriately rewarded. It is easy to show, however, that such *monotonicity* is *incompatible with efficiency*. The intuition is straightforward: if all partners benefit from a higher output then the social benefit to any one partner increasing his input is greater than that partner's private benefit. However, in an equilibrium that is efficient, the private and social benefits have to be equal. This idea is formalized as:

Proposition 5.3. *Suppose that $\varphi_j(a_1,\ldots a_m)$ is strictly first-order stochastically increasing in its arguments. Let S be a sharing rule for which the first-best profile of inputs â is a Nash equilibrium. Then there is some partner, say i, whose share, S_i, does not always increase with output.*

Proof. Suppose, to the contrary, that the sharing rules, S_i, are increasing for all partners. The social marginal benefit to increasing partner **i**'s input is:

$$\sum_i \left[\sum_j S_i(G_j) \cdot \varphi_{ji}(\hat{a}) \right] - Q'_i(\hat{a}_i). \tag{5.8}$$

Since $\hat{a}_i$ is a best response, $\sum_j S_i(G_j) \cdot \varphi_{ji}(\hat{a})] = Q'_i(\hat{a}_i)$. Substituting this into (5.8) yields $\sum_{i \neq \mathbf{i}} \sum_j S_i(G_j) \cdot \varphi_{ji}(\hat{a})$; the assumption on first-order stochastic dominance implies that $\sum_j S_i(G_j) \cdot \varphi_{ji}(\hat{a}) > 0$, for all $i \neq \mathbf{i}$. Hence, social utility would be increased by expanding partner **i**'s input.[34] □

Remark. One immediate corollary of the proposition is that the proportional sharing rule, $S_i(G_j) = G_j/m$, does not solve the organization problem by inducing efficient provision of inputs.

If partners' utility functions exhibit risk-aversion there will not be, in general, a sharing rule that sustains the efficient outcome as an equilibrium. This is because efficiency arrangements in this case requires efficient risk-sharing as well as efficient provision of inputs. To see this note that an efficient solution to the partnership problem is given by any solution to the following: $\max_{a, S_i} \sum_i \lambda_i [\sum_j U_i(S_i(G_j), a_i) \cdot \varphi_j(a)]$, where $\lambda_i > 0, i = 1,\ldots m$. [There may also be efficient expected utility vectors corresponding to some choices of $(\lambda_1,\ldots \lambda_m)$ with some coordinates λ_i equal to zero.] A solution to the above maximization problem will typically involve not just the action profile, as with risk-neutrality, but also the sharing rules. Moreover, the rules that share risk efficiently may not be incentive-compatible. An alternative way of seeing this is to note that if the sharing rules are specified in the efficient solution then there is no further degree of freedom left to specify these rules such that they also satisfy the Nash equilibrium first-order conditions (5.7).

---

[34] It is obvious, from the proof, that the proposition is true as long as φ_j is first-order stochastically increasing in the input level of at least one partner.

There are several questions that remain open in the static partnership model. The first involves the characterization of solutions to the second-best problem; what are the properties of the most efficient Nash equilibrium in a partnership game (when the efficient solution is unattainable)? In particular, it may be fruitful to employ the Grossman and Hart (1983) cost-minimization approach to the partnership problem to investigate properties such as monotonicity and convexity for the second-best solution.[35]

A related question to ask is whether input profiles (and sharing rules) (arbitrarily) close to the efficient vector can, in fact, be implemented an incentive-compatible fashion, even if exact efficiency is unattainable. Recent developments in implementation theory have introduced weaker (and yet compelling) notions that involve implementability of profiles close to that desired; see Abreu and Matsushima (1992). This approach may even be able to dispense with the requirement that partners be asymmetric.[36]

We now turn to the general class of games with imperfect monitoring.

5.2. Repeated games with imperfect monitoring

A static (or stage) game with imperfect monitoring is defined by a triple $(A_i, \varphi_j, U_i; i = 1, \ldots m, j = 1, \ldots n)$; i is the index for a player and each player picks actions a_i from a finite set A_i. This choice is not observed by any player other than i. An action profile $a \equiv (a_1, \ldots a_m)$ induces a probability distribution on a public outcome $G_j, j = 1, \ldots n$; the probability that the outcome is G_j when the action profile a is chosen is denoted $\varphi_j(a)$. Each player's realized payoff depends on the public outcome and his own action but not on the actions of the other players; the payoff is denoted $U_i(G, a_i)$. We will allow players to pick mixed actions as well; denote a generic mixed action by α_i. For each profile of mixed actions $\alpha = (\alpha_1, \ldots \alpha_m)$, the conditional probability of public outcomes and the player's expected payoffs are computed in the obvious fashion. Abusing notation, we write $\varphi_j(\alpha)$ to be the probability of the outcome G_j under the mixed action profile α. It will be useful to denote player i's expected payoffs as $\Gamma_i(\alpha)$.

It is clear that a partnership model, with a fixed sharing rule, is an example of a game of imperfect monitoring. So also is the principal–agent model of Sections 2–4. Imagine that player 2 is the principal and his action is the choice of a compensation scheme for the agent. Since the agent actually moves after the principal – whereas in the above game, moves are simultaneous – a_1 now must be interpreted as a contingent effort rule that specifies the agent's effort for every compensation rule

---

[35] Mookherjee (1984) has studied the related problem in which there is a principal, in addition to the partners. He characterizes the second-best outcome from the principal's perspective.

[36] Legros (1989) shows that even in the deterministic partnership model (with risk-neutrality), ε-efficiency can be attained if partners are allowed to randomize in their choice of inputs.

that the principal could choose. The public outcome is then the realized output level plus the principal's compensation scheme.[37]

In a repeated game with imperfect monitoring, in each period $t = 0, 1, \ldots$, the stage game is played and the associated public outcome revealed. The public history at date t is $h(t) = (G(0), G(1), \ldots G(t-1))$ whereas the private history of player i is $h_i(t) = (a_i(0), G(0), a_i(1), G(1), \ldots a_i(t-1), G(t-1))$. A strategy for player i is a sequence of maps $\sigma(t)$, where $\sigma(t)$ assigns to each pair of public and private histories $(h(t), h_i(t))$ a mixed action $\alpha_i(t)$. A strategy profile induces, in the usual way, a distribution over the set of histories $(h(t), h_i(t))$ and hence an expected payoff for player i in the tth period; denote this $\Gamma_i(t)$. Lifetime payoffs are evaluated under a (common) discount factor δ (< 1) and equal $(1 - \delta) \sum_{t=0}^{\infty} \delta^t \Gamma_i(t)$.

Player i seeks to maximize his lifetime payoffs. We restrict attention to a subclass of Nash equilibria that have been called *perfect public equilibria* in the literature; a strategy profile $(\sigma_1, \ldots \sigma_m)$ is a perfect public equilibrium if, (a) for all time periods t and all players i, the continuation of σ_i after history $(h(t), h_i(t))$ only depends on the public history $h(t)$ and (b) the profile of continuation strategies constitute a Nash equilibrium after every history.[38] Suppose that the stage game has a Nash equilibrium; an infinite repetition of this stage-game equilibrium is an example of a perfect public equilibrium. Let V denote the set of payoff vectors corresponding to all perfect public equilibria in the repeated game.

5.2.1. A characterization of the set of equilibrium payoffs

In this subsection we will provide an informal discussion of the Abreu et al. (1986, 1990) recursive characterization of the equilibrium payoff set V. The heart of their analysis is to demonstrate that the set of equilibrium payoffs in repeated games has a Bellman-equation-like representation similar to the one exhibited by the value function in dynamic programming.

Suppose, to begin with, that we have a perfect public equilibrium profile σ^*. Such an equilibrium can be decomposed into (a) an action profile in period zero, say $\alpha^*(0)$, and (b) an expected continuation payoff (or "promised future payoff") profile, $v^j(1)$, $j = 1, \ldots n$, that is contingent on the public outcome G_j realized in period zero. Since σ^* is an equilibrium it follows that: (i) $\alpha^*(0)$ must satisfy the incentive constraint that no player can unilaterally improve his payoffs given the

---

[37] Fudenberg et al. (1989), who suggest the above interpretation of the principal–agent model, show that several other models can also be encompassed in the current framework. In particular, the oligopoly models of Porter (1983) and Green and Porter (1984) are easily accommodated.

[38] Note that a player is not restricted to choosing a strategy in which he can only condition on the public history. If every other player but i chooses such a strategy, elementary dynamic programming arguments can be used to show that player i cannot, in his best response problem, do any better by choosing a strategy that conditions on his private history as well. A second point to note is that were we to restrict attention to pure strategies only, then without any loss of generality we could in fact restrict players to choosing strategies which only condition on public histories [for this and related arguments see Abreu et al. (1990)].

twin expectations of other players' actions in that period, $\alpha^*_{-i}(0)$, and the continuation payoffs, $v^j(1)$; (ii) the continuation payoffs must themselves be drawn from the set of equilibrium payoffs V. Moreover, an identical argument is true for every equilibrium strategy profile and after all histories, i.e., an equilibrium in the repeated game is a sequence of incentive-compatible "static" equilibria.

Now consider an arbitrary set of continuation payoffs $W \subset \mathbb{R}^m$; these need not be equilibrium payoffs. Define an action prifile $\hat{\alpha}$ to be enforceable, with respect to W, if there are payoff profiles $w^j \in W$, $j = 1, \ldots n$ with the property that $\hat{\alpha}$ is a Nash equilbrium of the "static" game with payoffs $(1 - \delta)\Gamma!_i(\alpha) + \delta E w_i(\alpha)$. Let $B(W)$ be the set of Nash equilibrium payoffs to these "static" games (with all possible continuation payoffs being drawn from W). If a bounded set W has the property that $W \subset B(W)$, (and Abreu, Pearce and Stachetti call such a set *self-generating*) then it can be shown that all payoffs in $B(W)$ are actually repeated-game equilibrium payoffs, i.e., $B(W) \subset V$.[39] In other words a sequence of static Nash equilibria, all of whose payoffs are self-referential in the manner described above, is a perfect equilibrium of the repeated game.

More formally let us define, for any set $W \subset \mathbb{R}^m$:

$$B(W) \equiv \left\{ w \in \mathbb{R}^m : \exists w^j \in \mathbb{R}^m, \quad j = 1, \ldots n, \text{ and } \exists \alpha \text{ s.t.} \right.$$

$$w_i = (1 - \delta)\Gamma_i(\alpha) + \delta \sum_j w_i^j \varphi_j(\alpha) \tag{5.9}$$

$$\left. (1 - \delta)\Gamma_i(\alpha) + \delta \sum_j w_i^j \varphi_j(\alpha) \geq (1 - \delta)\Gamma_i(a_i, \alpha_{-i}) + \delta \sum_j w_i^j \varphi_j(a_i, \alpha_{-i}) \quad \forall i, a_i \right\}. \tag{5.10}$$

Theorem 5.4 [Abreu, Pearce and Stachetti (1990)]. (i) (*Sufficiency*) *Every bounded self-generating set is a subset of the equilibrium payoffs set; if W is bounded and $W \subset B(W)$ then $B(W) \subset V$.*

(ii) (*Necessity*) *The equilibrium payoffs set V is the largest self-generating set among the class of bounded self-generating sets; $V = B(V)$.*

The recursive approach has two useful consequences. The sufficiency characterization, part (i), says that if a subset of feasible payoffs can be shown to be self-generating, then all of its elements are equilibrium payoffs; this (constructive) approach can be used to provide upper bound on the difference between second-

---

[39] The argument is as follows: by definition, if $\hat{w}$ is in $B(W)$, then it can be decomposed into an action profile $\hat{\alpha}(0)$ and "continuation" payoffs $\hat{w}(1)$ where $\hat{\alpha}(0)$ is a Nash equilibrium in the "static" game with payoffs $(1 - \delta)\Gamma_i(\alpha) + \delta E \hat{w}_i(\alpha)$. Since $\hat{w}(1) \in W \subset B(W)$, it can also be similarly decomposed. In other words there is a strategy $\hat{\sigma}$, which can be deduced from these arguments, with the property that no one-shot deviation against it is profitable for any player. The unimprovability principle of discounted dynamic programming then implies that there are, in fact, no profitable deviations against $\hat{\sigma}$.

best and efficient payoffs. The constructive approach is used in proving the folk theorem that we discuss shortly.

A second (inductive) approach can be employed to determine properties of the second-best equilibrium. In this approach one conjectures that the equilibrium payoff set has the properties one seeks to establish and then demonstrates that such properties are, in fact, maintained under the recursion. Abreu et al. (1986, 1990) have utilized this approach to provide conditions on the primitives under which the equilibrium payoff set is compact, convex and monotonically increasing in the discount factor.

5.2.2. A folk theorem with imperfect monitoring

We turn now to the second-best problem: how large is the inefficiency caused by free-riding in a game of imperfect monitoring? The repeated perspective allows the design of a richer set of incentives. This is immediate from the incentive constraint, (5.10), above; by way of different specifications of the promised future payoff, w_i^j, there is considerable room to fine-tune current incentives. However, for the future to have significant bearing on the players' decisions today it must be the case that there is sufficient weight attached to the future; indeed, the folk theorem, which answers the second-best question, is, in fact, an asymptotic result that obtains for players with δ close to 1.

There is a formal similarity between the payoffs to risk-neutral players in the static partnership model and the intertemporal decomposition of payoffs in the repeated game, (5.9). In the static model, the player's (risk-neutral) payoffs are $\sum_j S_i(G_j)\varphi_j(\alpha) - Q_i(\alpha_i)$. The intertemporal payoff is $(1-\delta)\Gamma_i(\alpha) + \delta \sum_j w_i^j \varphi_j(\alpha)$, where w_i^j can be interpreted as player i's "share" of future payoffs resulting from an output G_j.[40] The difference in the analyses is that the future payoffs w_i^j, have to be self-generated whereas the static shares $S_i(G_j)$, have to satisfy budget balance, $\sum_i S_i(G_j) = G_j$.

It turns out, however, that the analogy between the two models goes a little further still. This is because the folk-theorem proof techniques of Fudenberg et al. (1989) and Matsushima (1989a) critically employ a construction in which actions are enforced by continuation payoffs w_i^j that are restricted to lie on a hyperplane, i.e., are such that $\sum_i w_i^j = 0$. This is, of course, a restatement of the budget balance condition, after relabelling variables. This explains why one hypothesis of the folk theorem below is an asymmetry condition like the ones that were used in solving the static risk-neutral incentives problem:

Pairwise full rank. The stage game satisfies the pairwise full rank condition if,

---

[40] Indeed it is easy to check that in all of the analyses of the static risk-neutral case, the own action-contingent utility, $Q_i(\alpha_i)$, could be replaced with the exact analogue in (5.9), $\Gamma_i(\alpha) \equiv E[U_i(G, a_i)/\alpha]$, without changing any of the results. Hence, the correspondence, between the intertemporal (possibly risk-averse) payoffs and the static risk-neutral payoffs, is exact.

for all pairs of players $(i, \mathbf{i}, i \neq \mathbf{i})$ there exists a profile α such that the matrix

$$\{\varphi_j(a_i, \alpha_{-i}), \varphi_j(a_\mathbf{i}, \alpha_{-\mathbf{i}})\}$$

with rows corresponding to the elements of $A_i \times A_\mathbf{i}$ and columns corresponding to the outcomes $G_j, j = 1, \ldots n$, has rank $|A_i| + |A_\mathbf{i}| - 1$.[41]

Theorem 5.5 [Fudenberg, Levine and Maskin (1989)]. *Suppose the stage game satisfies the pairwise full rank condition and, additionally, the following two hypotheses*:

(i) *For all players i and all pure action profiles $\hat{a}$, the $|A_i|$ vectors, $\{\varphi_j(a_i, \hat{a}_{-i}), j = 1, \ldots n\}$, are linearly independent.*

(ii) *The set of individually rational payoff vectors, say F^* has dimension equal to the number of players.*[42]

Then, for every closed set W in the relative interior of F^* there is a $\underline{\delta} < 1$ such that for all $\delta > \underline{\delta}$, every payoff in W is a perfect public equilibrium payoff.

Remarks. (1) An obvious implication of the above result is that, as $\delta \to 1$, any corresponding sequence of second-best equilibrium payoffs is asymptotically efficient.

(2) In the absence of the pairwise full rank condition, asymptotic efficiency may fail to obtain; Radner et al. (1986) contains the appropriate example. In this example, the aggregate condition of Section 5.2 [condition (5.4)] holds and hence, for reasons identical to the static risk-neutral case, the efficient solution cannot be sustained as equilibrium behavior.

(3) Condition (i) is required to ensure that player i, when called upon to play action $\hat{a}_i$ cannot play a more profitable action $\tilde{a}_i$ whose outcome consequences are identical to those of $\hat{a}_i$. Condition (ii) is required to ensure that there are feasible asymmetric lifetime payoffs; when a deviation by player i (respectively $\mathbf{i}$) is inferred, there exists a continuation strategy σ^i whose payoffs are smaller than the payoffs to some other continuation strategy $\sigma^\mathbf{i}$ (and vice versa for player $\mathbf{i}$).[43]

(4) Since the principal–agent model of Sections 2–4 is a special case of the game with imperfect monitoring, Theorem 5.5 also yields a folk theorem, and asymptotic efficiency, in that model. However, the informational requirements of this result are evidently more stringent than those employed by Radner (1985) to prove asymptotic efficiency in the principal–agent model. Restricted to the principal–agent

---

[41]This is really a full rank condition since the row vectors must always admit at least one linear dependence. Also, a necessary condition for the pairwise full rank condition to hold is clearly that the number of outcomes $n \geq |A_i| + |A_\mathbf{i}| - 1$.

[42]Note that mixed strategies ae admissible and hence an individually rational payoff vector is one whose components dominate the mixed strategy minimax for each player.

[43]A similar condition is also required in repeated games with perfect monitoring; see Fudenberg and Maskin (1986). Recent work [Abreu et al. (1992)] has shown that, under perfect monitoring, the full-dimensionality assumption can be replaced by the weaker requirement that players' preferences not be representable by an identical ordering over mixed action profiles. Whether full-dimensionality can be similarly weakened in the imperfect monitoring case remains an open question.

model, Theorem 5.5 can, however, be proved unde weaker hypotheses, and Radner's result can be generalized; see Fudenberg et al. (1989) for details.

To summarize, under certain conditions, repetition of a partnership allows the design of intertemporal incentives such that the free-rider problem can be asymptotically resolved by patient partners. Of course, the curse of the folk theorem, Theorem 5.5, is that it proves that a lot of other, less attractive, arrangements can also be dynamically sustained.

In many partnerships there are variables other than just the partners' actions that determine the output, as for example the usage of (commonly owned) capital stock; this is an additional source of information. One question of interest, for both static and especially repeated partnership models, is whether, and how much, such additional information alleviates the free-rider problem. A second open question is whether bounds can be derived for the rate of convergence to efficiency as the discount factor goes to one (in a spirit similar to the Dutta and Radner (1994) exercise for the principal–agent model).

6. Additional bibliographical notes

Notes on Section 3. In the field of economics, the first formal treatment of the principal–agent relationship and the phenomenon of moral hazard was probably given by Arrow (1963, 1965), although a paper by Simon (1953) was an early forerunner of the principal–agent literature.[44] Early work on one-sided moral hazard was done by Wilson (1969), Spence and Zeckhauser (1971) and Ross (1973). James Mirrlees contributed early, and sophisticated, analyses of the problem; much of his work is unpublished [but see Mirrlees (1974, 1976)]. Holmstrom (1979) and Shavell (1979) investigated conditions under which it is beneficial for the principal to monitor the agent, or use any other sources of information about the agent's performance in writing the optimal contract. In addition to these papers other characterizations of the second-best contract, all of which employ the first-order approach, include Harris and Raviv (1979) and Stiglitz (1983). [For analyses that provide sufficient conditions under which the first-order approach is valid, see Rogerson (1985b) and Jewitt (1988)]. The paper by Grossman and Hart (1983) provides a particularly thorough and systematic treatment of the one-period model.

The static principal–agent model has been widely applied in economics. As we have indicated earlier, the phenomenon, and indeed the term itself, came from the

---

[44]The recognition of incentive problems is of much older vintage however. In an often quoted passage from the "Wealth of Nations" Adam Smith says, "The directors of such companies, being the managers rather of other peoples' money than their own, it cannot well be expected, that they should watch over it with the same anxious vigilance with which the partners in a private co-partnery frequently watch over their own... Negligence and profusion therefore must always prevail in the management of the affairs of such a company." [See Smith (1937).]

study of insurance markets. One other early application of the theory has been to agrarian markets in developing economies; a principal question here is to understand the prevalence, and uniformity, of sharecropping contracts. An influential paper is Stiglitz (1974) that has subsequently been extended in several directions by, for example, Braverman and Stiglitz (1982); for a recent survey of this literature, see Singh (1989). Other applications of the theory include managerial incentives to (a) invest capital in productive activities, rather than perquisites [Grossman and Hart (1982)], (b) to invest in human capital, [Holmstrom and Ricart i Costa (1986)] and (c) to obtain information about and invest in risky assets, [Lambert (1986)].

Three topics in static moral hazard that we have not touched upon are: (a) incentive issues when moral hazard is confounded with informational asymmetries due to adverse selection; see, for example, Foster and Wan (1984) who investigate involuntary unemployment due to such a mix of asymmetries, and defense contracting issues as studied by Baron and Besanko (1987) and McAfee and McMillan (1986); (b) the general equilibrium consequences of informational asymmetries, a topic that has been studied in different contexts by Joseph Stiglitz and his co-authors; see, for example, Arnott and Stiglitz (1986) and Greenwald and Stiglitz (1986) and (c) the implications of contract renegotiation. The last topic asks the question, (when) will principal and agent wish to write a new contract to replace the current one and has been recently addressed by a number of authors including Fudenberg and Tirole (1990). The survey of principal–agent models by Hart and Holmstrom (1986) is a good overview of the literature; furthermore, it expands on certain other themes that we have not been able to address.

Notes on Section 4. Lambert (1983) derives a characterization of dynamic second-best contracts that also yields the history dependence implied by Rogerson's result; he takes, however, a first-order approach to the problem. Spear and Srivastava (1988) employ the methods of Abreu et al. (1990) to derive further characterizations of the optimal compensation scheme, such as monotonicity in output.

The second strand in the literature on dynamic principal–agent contracts has explored the implications of simple contracts that condition on history in a parsimonious fashion. Relevant papers here are Radner (1981, 1985, 1986b), Rubinstein (1979), Rubinstein and Yaari (1983), and Dutta and Radner (1992). These papers have been reviewed in some detail in Section 4.2.

Recently a number of papers have investigated the consequences of allowing the agent to borrow and lend and thereby provide himself with self-insurance. Indeed if the agent is able to transact at the same interest rates as the principal, an assumption that is plausible if capital markets are perfect (but only then), there exist simple output contingent schemes (that look a lot like franchises) which approximate efficiency. Papers in this area include Allen (1985), Malcomson and Spinnewyn (1988), and Fudenberg et al. (1990).

In the study of labor contracts a number of authors have investigated some simple history-dependent incentive schemes under which an employee cannot be compensated on the basis of observed output but rather has to be paid a fixed wage; the employee may, however, be fired in the event that shirking is detected.[45] Shapiro and Stiglitz (1984) show that involuntary unemployment is necessary, and will emerge, in the operation of such incentive schemes. An application of these ideas to explain the existence of dual rural labor markets in agrarian economies can be found in Easwaran and Kotwal (1985).

Notes on Section 5. Static partnership models were first studied formally by Holmstrom (1982) – see also the less formal discussion in Alchian and Demsetz (1972). For characterizations of conditions under which the first-best is sustainable as an equilibrium by risk-neutral partners, see, in addition to the Williams and Radner (1989) paper that we have discussed, Legros (1989), Matsushima (1989b) and Legros and Matsushima (1991). For a discussion of the case of risk-averse partners see Rasmussen (1987).

Mookherjee (1984) generalized the Grossman and Hart (1983) approach to single-sided moral hazard problems to characterize the second-best contract when there is a principal and several agents (or partners). (His framework covers both the case of a partnership, where production is joint, as well as the case of independent production). He derived an optimality condition that is the analog of condition (3.4) above and used this to investigate conditions under which (a) an agent's compensation should be independent of other agents' output, and (b) agents' compensations should be based solely on their "rank" (an ordinal measure of relative output). The attainability of first-best outcomes through rank order tournaments has also received extensive treatment in the context of labor contracts; see Lazear and Rosen (1981) for the first treatment and subsequent analyses by Green and Stokey (1983) and Nalebuff and Stiglitz (1983).

Radner (1986a) was the first paper to study repeated partnerships; in his model partners do not discount the future but rather employ the long-run average criterion to evaluate lifetime utility. This paper showed that the efficient expected utility vectors *can* be sustained as a perfect equilibrium of a repeated partnership game (even under risk-aversion) for a "large" class of partnership models.[46] Subsequent work, which has incorporated discounting of future payoffs, has included Radner et al. (1986), Abreu et al. (1991) and Radner and Rustichini (1989). Radner et al. (1986) gave an example in which equilibrium payoffs for the repeated game with discounting are uniformly bounded away from one-period efficiency for all discount

---

[45] These contracts therefore bear a family resemblence to the bankruptcy schemes we have discussed in this paper; the one difference is that in a bankruptcy scheme observed output is utilized in deciding whether or not to terminate an agent's contract whereas in the papers referred to here it is usually assumed that shirking can be directly observed (and penalized).

[46] The exact condition that needs to be satisfied is as follows: fix a sharing rule S and suppose $\hat{a}$ is the (efficient) input profile that is to be sustained. Then there exist positive constants K_i such that $[EU_i(S_i(G), \hat{a}) - EU_i(S_i(G), \hat{a}_{-i}, a_i)] + K_i[E(G|\hat{a}_{-i}, a_i) - E(G|\hat{a})] \leq 0$, for all a_i, where the expectations are taken under the input profiles $\hat{a}$ and $(\hat{a}_{-i}, a_i)$.

factors strictly less than one.[47] A model formally similar to a repeated partnership is that of a oligopoly with unobserved quantity choices by each firm; this model was studied by Porter (1983), Green and Porter (1984) and Abreu et al. (1986).

Abreu et al. (1986, 1990) contain the analyses, reported in Section 5.2, that characterize the equilibrium payoff set in repeated games with discounting and imperfect monitoring. Fudenberg et al. (1989), and Matsushima (1989a) have employed the sufficiency part of the characterization to prove that efficiency is sustainable, even with many-sided moral hazard, for a large class of repeated games with imperfect monitoring. These results are important in that they make clear the conditions needed to give agents appropriate dynamic incentives in order to sustain efficiency.

As noted in Section 1, models of adverse selection and misrepresentation have not been discussed in the present article. For this topic the reader is referred to Melumad and Reichelstien (1989) and the references cited therein. For a recent survey of the related topic of incentive compatibility and the revelation of preferences for public goods, see Groves and Ledyard (1987). An elementary exposition of the problem of misrepresentation and the Groves–Vickrey–Clark mechanism can be found in Radner (1987), and further work on incentive compatibility appears in the volume edited by Groves et al. (1987).

References

Abreu, D., P.K. Dutta and L. Smith (1992) 'The Folk Theorem for Discounted Repeated Games: A NEU Condition', *Econometrica*, forthcoming.

Abreu, D. and H. Matsushima (1992) 'Virtual Implementation in Iteratively Undominated Strategies: Complete Information', *Econometrica*, **60**: 993–1008.

Abreu, P. Milgrom and D. Pearce (1991) 'Information and Timing in Repeated Patnerships', *Econometrica*, **59**.

Abreu, D., D. Pearce and E. Stachetti (1986) 'Optimal Cartel Monitoring with Imperfect Information', *Journal of Economic Theory*, **39**: 251–269.

Abreu, D., D. Pearce and E. Stachetti (1990) 'Towards a Theory of Discounted Repeated Games with Imperfect Monitoring', *Econometrica*, **58**: 1041–1063.

Alchian, A. and H. Demsetz (1972) 'Production, Information Costs and Economic Organizations', *American Economic Review*, **62**: 777–795.

Allen, F. (1985) 'Repeated Principal Agent Relationships with Lending and Borrowing', *Economic Letters*, **17**: 27–31.

Arnott, R. and J. Stiglitz (1986) 'Labor Turnover, Wage Structures and Moral Hazard: The Inefficiency of Competitive Markets', *Journal of Labor Economics*, **3**: 434–462.

---

[47] As discussed above there is a formal similarity between the two models, static partnerships in which the sharing rule is endogenous and a discounted repeated partnership with a fixed sharing rule (but endogenous future compensation). In particular in both of these cases an ability to treat partners asymmetrically is essential to the sustainability of efficient outcomes. The Radner, Myerson and Maskin example restricts itself to the aggregate condition (5.4), much as the Holmstrom analysis did in the static partnership context (and furthermore only considers a symmetric sharing rule). Hence, in both cases efficiency cannot be sustained. The undiscounted case is different in that the asymmetric treatment of partners is inessential to the sustainability of efficient outcomes.

Arrow, K.J. (1963) 'Uncertainty and the Welfare Economics of Medical Care', *American Economic Review*, **53**: 941–973.
Arrow, K.J. (1965) 'Aspects of the Theory of Risk-Bearing', Yrjo Johansonian Saatio', Lecture 3, Helsinki.
Baron, D. and D. Besanko (1987) 'Monitoring, Moral Hazard, Asymmetric Information and Risk-Sharing in Procurement Contracting', *Rand Journal of Economics*, **18**: 509–532.
Braverman, A. and J. Stiglitz (1982) 'Sharecropping and the Interlinking of Agrarian Markets', *American Economic Review*, **72**: 695–715.
Dutta, P.K. and R. Radner (1992) 'Optimal Principal Agent Contracts for a Class of Incentive Schemes: A Characterization and the Rate of Approach to Efficiency', *Economic Theory*, **4**: 483–503.
Easwaran, M. and A. Kotwal (1985) 'A Theory of Two-Tier Labor Markets in Agrarian Economies', *American Economic Review*, **75**: 162–177.
Fellingham, J., D. Newman and Y. Suh (1985) 'Contracts Without Memory in Multiperiod Agency Models', *Journal of Economic Theory*, **37**: 340–355.
Foster, J. and H. Wan (1984) 'Involuntary unemployment as a Principal-Agent Equilibrium', *American Economic Review*, **74**: 476–484.
Fudenberg, D., B. Holmstrom and P. Milgrom (1990) 'Short-Term Contracts and Long-Term Agency Relationships', *Journal of Economic Theory*, **51**: 1–31.
Fundenberg, D., D. Levine and E. Maskin (1989) 'The Folk Theorem with Imperfect Public Information', MIT, mimeo.
Fudenberg, D. and E. Maskin (1986) 'The Folk Theorem in Repeated Games with Discounting or with Incomplete Information', *Econometrica*, **52**: 975–994.
Fudenberg, D. and J. Tirole (1990) 'Moral Hazard and Renegotiation in Agency Contracts', *Econometrica*, **58**: 1279–1301.
Green, E. and R. Porter (1984) 'Noncooperative Collusion Under Imperfect Price Information', *Econometrica*, **52**: 87–100.
Green, J. and N. Stokey (1983) 'A Comparison of Tournaments and Contracts', *Journal of Political Economy*, **91**: 349–364.
Greenwald, B. and J. Stiglitz (1986) 'Externalities in Economies with Imperfect Information and Incomplete Markets', *Quarterly Journal of Economics*, **101**: 229–264.
Grossman, S.J. and O.D. Hart (1982) 'Corporate Financial Structure and Managerial Incentives', in: J. McCall, ed., *The Economics of Information and Uncertainty*. Chicago: University of Chicago Press.
Grossman, S.J. and O.D. Hart (1983) 'An Analysis of the Principal–Agent Problem', *Econometrica*, **51**: 7–45.
Groves, T. and J. Ledyard (1987) 'Incentive Compatibility since 1972', in: T. Groves R. Radner and S. Reiter, eds., *Information, Incentives and Economic Mechanisms*. Minnesota: University of Minnesota Press.
Groves, T., R. Radner and S. Reiter (1987) *Information, Incentives and Economic Mechanisms*. Minnesota: University of Minnesota Press.
Harris, M. and A. Raviv (1979) 'Optimal Incentive Contracts with Imperfect Information', *Journal of Economic Theory*, **20**: 231–259.
Hart, O.D. and B. Holmstrom (1986) 'The Theory of Contracts', in: T. Bewley, ed., *Advances in Economic Theory*. Cambridge: Cambridge University Press.
Holmstrom, B. (1979) 'Moral Hazard and Observability', *Bell Journal of Economics*, **10**: 74–91.
Holmstrom, B. (1982) 'Moral Hazard in Teams', *Bell Journal of Economics*, **13**: 324–340.
Holmstrom, B. and P. Milgrom (1987) 'Aggregation and Linearity in the Provision of Intertemporal Incentives', *Econometrica*, **55**: 303–329.
Holmstrom, B. and J. Ricart i Costa (1986) 'Managerial Incentives and Capital Management', *Quarterly Journal of Economics*, **101**: 835–860.
Jewitt, I. (1988) 'Justifying the First-Order Approach to Principal–Agent Problems', *Econometrica*, **56**: 1177–1190.
Lambert, R. (1983) 'Long-Term Contracts and Moral Hazard', *Bell Journal of Economics*, **14**: 441–452.
Lambert, R. (1986) 'Executive Effort and the Selection of Risky Projects', *Rand Journal of Economics*, **16**: 77–88.
Lazear, E. and S. Rosen (1981) 'Rank Order Tournaments as Optimal Labor Contracts', *Journal of Political Economy*, **89**: 841–864.
Legros, P. (1989) 'Efficiency and Stability in Partnerships', Ph.D Dissertation, California Institute of Technology.

Legros, P. and H. Matsushima (1991) 'Efficiency in Partnerships', *Journal of Economic Theory*, **55**: 296–322.
Malcomson, J. and F. Spinnewyn (1988) 'The Multiperiod Principal–Agent Problem', *Review of Economic Studies*, **55**: 391–408.
Matsushima, H. (1989a) 'Efficiency in Repeated Games with Imperfect Monitoring', *Journal of Economic Theory*, **48**: 428–442.
Matsushima, H. (1989b) 'Necessary and Sufficient Condition for the Existence of Penalty Rules with Budget Balancing', Institute of Socio–Economic Planning, University of Tsukuba, mimeo.
McAfee, R. and J. McMillan (1986) 'Bidding for Contracts: A Principal–Agent Analysis', *Rand Journal of Economics*, **17**: 326–338.
Melumad, N. and S. Reichelstein (1989) 'Value of Communication in Agencies', *Journal of Economic Theory*, **47**: 334–368.
Milgrom, P. (1981) 'Good News and Bad News: Representation Theorems and Applications', *Bell Journal of Economics*, **12**: 380–391.
Mirrlees, J. (1974) 'Note on Welfare Economics, Information and Uncertainty', in: M. Balch, D. McFadden and S. Wu, eds., *Essays on Economic Behavior Under Uncertainty*. Amsterdam: North-Holland.
Mirrlees, J. (1975) 'The Theory of Moral Hazard and Unobservable Behavior', Nuffield College, Oxford University, mimeo.
Mirrlees, J. (1976) 'The Optimal Structure of Incentives and Authority Within an Organization', *Bell Journal of Economics*, **7**: 5–31.
Mookherjee, D. (1984) 'Optimal Incentive Schemes with Many Agents', *Review of Economic Studies*, **51**: 433–446.
Nalebuff, B. and J. Stiglitz (1983) 'Prizes and Incentives: Towards a General Theory of Compensation and Competition', *Bell Journal of Economics*, **13**: 21–43.
Porter, R. (1983) 'Optimal Cartel Trigger Price Strategies', *Journal of Economic Theory*, **29**: 313–338.
Radner, R. (1981) 'Monitoring Cooperative Agreements in a Repeated Principal–Agent Relationship', *Econometrica*, **49**: 1127–1148.
Radner, R. (1985) 'Repeated Principal–Agent Games with Discounting', *Econometrica*, **53**: 1173–1198.
Radner, R. (1986a) 'Repeated Partnership Games with Imperfect Monitoring and No Discounting', *Review of Economic Studies*, **53**: 43–57.
Radner, R. (1986b) 'Repeated Moral Hazard with Low Discount Rates', in: W.P. Heller, R. Starr and D. Starett, eds., *Uncertainty, Information and Communication: Essays in Honor of Kenneth Arrow*, Vol. 3. Cambridge: Cambridge University Press, pp. 25–64.
Radner, R. (1987) 'Decentralization and Incentives', in: T. Groves, R. Radner and S. Reiter, eds., *Information, Incentives and Economic Mechanisms*. Minnesuta: University of Minnesota Press.
Radner, R., R. Myerson and E. Maskin (1986) 'An Example of a Repeated Partnership game with Discounting and with Uniformly Inefficient Equilibria', *Review of Economic Studies*, **53**: 59–69.
Radner, R. and A. Rustichini (1989) 'The Design and Sharing of Performance Rules for a Partnership in Continuous Times', AT&T Bell Laboratories, Murray Hill, NJ, mimeo.
Rasmussen, E. (1987) 'Moral Hazard in Risk-Averse Teams', *Rand Journal of Economics*, **18**: 428–435.
Rogerson, W. (1985a) 'Repeated Moral Hazard', *Econometrica*, **53**: 69–76.
Rogerson, W. (1985b) 'The First-Order Approach to Principal–Agent Problems', *Econometrica*, **53**: 1357–1368.
Ross, S. (1973) 'The Economic Theory of Agency: The Principal's Problem', *American Economic Review*, **63**: 134–139.
Rubinstein, A. (1979) 'Offenses That May have Been Committed by Accident – an Optimal Policy of Retribution', in: S. Brams, A. Schotter and G. Schrodiauer, eds., *Applied Game Theory*. Wurzburg: Physica–Verlag.
Rubinstein, A. and M. Yaari (1983) 'Repeated Insurance Contracts and Moral Hazard', *Journal of Economic Theory*, **30**: 74–97.
Shapiro, C. and J. Stiglitz (1984) 'Equilibrium Unemployment as a Worker Discipline Device', *American Economic Review*, **74**: 433–444.
Shavell, S. (1979) 'Risk Sharing and Incentives in the Principal and Agent Relationship', *Bell Journal of Economics*, **10**: 55–73.
Simon, H. (1953) 'A Formal Theory of the Employment Relationship', *Econometrica*, **19**: 293–305.

Singh, N. (1989) 'Theories of Sharecropping', in: P. Bardhan, ed., *The Economic Theory of Agrarian Institutions.* Oxford: Oxford University Press.
Smith, A. (1937) in: E. Cannan, ed., *An Inquiry into the Nature and Causes of the Wealth of Nations.* New York: Modern Library, p. 700.
Spear, S. and S. Srivastava (1988) 'On Repeated Moral Hazard with Discounting', *Review of Economic Studies,* **55**.
Spence, A.M. and R. Zeckhauser (1971) 'Insurance, Information and Individual Action', *American Economic Review,* **61**: 380–387.
Spitzer, F. (1976) *Principles of Random Walk,* 2nd. edn New York: Springer-Verlag.
Stiglitz, J. (1974) 'Incentives and Risk-Sharing in Sharecropping', *Review of Economic Studies,* **41**: 219–255.
Stiglitz, J. (1983) 'Risk, Incentives and the Pure Theory of Moral Hazard', *The Geneva Papers on Risk and Insurance,* **8**: 4–33.
Williams, S. and R. Radner (1989) 'Efficiency in Partnerships when the Joint Output is Uncertain, mimeo, Northwestern University'.
Wilson, R. (1969) 'The Theory of Syndicates', *Econometrica,* **36**: 119–132.
Yaari, M. (1976) 'A Law of Large Numbers in the Theory of Consumption Choice Under Uncertainty, *Journal of Economic Theory,* **12**: 202–217.

Chapter 27

SEARCH

JOHN McMILLAN and MICHAEL ROTHSCHILD*
University of California at San Diego

Contents

1.	Search theory	906
2.	Stopping rules	907
3.	Price dispersion	912
4.	Search and bargaining	916
5.	Conclusion	921
References		922

---

*We thank Robert Aumann, Donald Deere, Sergiu Hart, Yong-Gwan Kim, Eric Maskin, Peter Morgan, Jennifer Reinganum, Louis Wilde, and a referee for comments and advice. For research support, McMillan thanks the National Science Foundation (SES 8721124), and Rothschild thanks the Deutsche Forschungsgemeinschaft, Gottfried-Wilhelm-Leibniz-Förderpreis and the National Science Foundation (SES 857711).

Handbook of Game Theory, Volume 2, Edited by R.J. Aumann and S. Hart
© *Elsevier Science B.V., 1994. All rights reserved*

1. Search theory

"One should hardly have to tell academicians that information is a valuable resource: knowledge *is* power. And yet it occupies a slum dwelling in the town of economics." So wrote George Stigler in his 1961 paper introducing search theory. Things have changed. The erstwhile slum-dweller has prospered: information issues are now to be found at the most elite and prestigious of economics addresses. Search theory has provided a simple and remarkably robust laboratory which economic theorists have used to examine a wide variety of questions about the acquisition of information. Early work on search, inspired by the seminal work of Stigler (1961), modeled the individual's searching decisions and drew inferences about the value of information and the nature of frictional unemployment. More recent work, building on sequential-bargaining analysis (reviewed in the Chapter by Binmore, Osborne, and Rubinstein in this Handbook), has focused on the interactions among searching agents and has deepened our understanding of the nature and meaning of competition.

Section 2 of this survey analyzes the classical search problem: the optimal search rule for an individual who can, for a fixed and constant cost, take a random sample from a distribution $F(\ )$ of economic opportunities. This standard formulation of the search problem raises several important questions. The opportunity to search from $F(\ )$ is of economic value; what determines who gets this opportunity? To the best of our knowledge, economic theorists have addressed this problem only tangentially: several papers [Lucas and Prescott (1974) and Mortenson (1976) are examples] have discussed how agents allocate themselves across different markets where each market is, from the point of view of the agent, characterized by a distribution $F(\ )$ of economic opportunities; David (1973) has used a similar model to analyze labor migration.

The bulk of recent work on search has focused on two other questions. The first is about the determinants of the distribution $F(\ )$. The second concerns the rules under which search and sampling take place and the way these affect the division of surplus between the different sides of a market. A simple example underscores both of these questions. Consider the market for a single good that is sold at several stores. Assume that all stores have identical cost structures and produce the item at a net cost of $1; all buyers place the same value, $2, on the item and incur of cost of k to visit any store. When a potential customer walks into a store, therefore, there is a potential surplus of $1 to be split between them. How will this surplus be divided? Diamond (1971) analyzed this problem, using the following argument. Suppose that the rules of the game are such that stores quote prices to buyers (that is, each seller is a Stackelberg leader with respect to the buyers). Then buyers decide whether or not to accept this offer. If they decline the offer, then they walk out of the store and search again. Diamond argued that, when buyers' search costs are positive, the only equilibrium of this process has all stores charging $2 and appropriating all the surplus. The monopoly price rules no matter how many sellers there are. The argument is simple. Suppose, to the contrary, that there is an equilibrium in which some store is charging less than $2. Let p be the lowest price charged by any store. A store charging p can increase its profits by raising its price slightly; if it raises its price by less than k, no customer who enters the store will refuse to buy. A customer who walks out must pay a search cost of k; he will save less than k by leaving and searching again. Thus, if the store raises its price by this small amount it will make the same number of sales. However, its profits from each sale will increase. It follows that there is no equilibrium in which any store charges less than $2. This argument is quite general: it continues to hold (with obvious modifications) if customers differ in their valuations and in their search costs, as long as all customers have search costs bounded away from zero. Also, the argument is symmetric: if we allowed customers rather than stores to make the first offer, gave stores no choice but to accept or reject this offer, and postulated that it cost stores something (in inventory costs) to wait for another customer then the market would clear at a price of $1.

How can a nondegenerate price distribution $F(\ )$ be sustained? Are there models of markets in which distributions of economic opportunities are part of a Nash equilibrium? At first blush it is difficult to understand how this could come to be; it is clear that if agents on one side of a market are the same, they will in general face the same problem and solve it in the same way. Thus identical stores will charge the same prices of all their customers and the distribution $F(\ )$ will be degenerate, as in Diamond's example. A nondegenerate distribution can be part of an equilibrium only if it somehow turns out that profit is maximized at several different prices or if agents are different. Section 3 below surveys some attempts to solve this problem. This work establishes that nondegenerate distributions can be part of an equilibrium in many models in which all participants behave optimally.

Diamond's example makes two further points. First, the details of who gets to make an offer matters; and, second, to analyze these situations completely bargaining theory is needed. In Section 4 we discuss how search theory and bargaining theory have been used to analyze these questions. The marriage of search theory and bargaining theory has produced some of the more interesting recent economic theory. This research has illuminated such diverse and important topics as the nature of the competitive mechanism and the possible impact of externalities and multiple equilibria on macroeconomic performance.

2. Stopping rules

The classical search problem is as follows. By paying a fixed cost, k, the searcher gains the right to take a random sample (of size one) from a distribution $F(\ )$. The objects drawn from $F(\ )$ – we shall call them opportunities – are of value to the searcher. Typical examples are a job (with a given expected lifetime income),[1] the right to buy some desired object at a given price, or – as will be discussed in some detail below – the opportunity to strike a bargain with someone who has something the searcher wants. The value of opportunities is denominated in the same units as the cost of search, k. These units can be in monetary or utility terms; since most search theory assumes constant marginal utility, however, there is little practical difference between the two approaches.[2]

After each draw the searcher has a choice: he can keep what he has drawn or he can pay k and draw another opportunity from the distribution $F(\ )$. It is clear from this description that the searcher will reap a profit consisting of the opportunity he eventually accepts minus the search costs he pays. This profit need not be positive; if all opportunities have negative value (as they will when the searcher is seeking to buy something at the lowest possible cost and value is denominated in monetary units) he will suffer a loss. The profit the searcher earns is a random variable whose value depends both on the actual draws he gets from the distribution $F(\ )$ and on his decisions to accept or reject particular opportunities. The strategy he uses will determine the expected value of his profit. An optimal decision rule[3] maximizes the expected value of the profit.

Let V^* be the expected value of the searcher's profit if he follows an optimal strategy. Clearly the searcher should never accept an opportunity with value less than V^*. If he rejects the opportunity he is in the same situation as a searcher who is starting anew: he can expect to make a profit of V^*. The timing is such that search costs are incurred immediately, but any benefits from search are received in the next period; and the searcher discounts future gains using the discount factor δ. Thus V^* must satisfy

$$V^* = \delta \int_{-\infty}^{\infty} \max[y, V^*] \, dF(y) - k. \tag{1}$$

The searcher follows a *reservation-value* rule: he accepts all offers greater than or equal to V^* and rejects all those less than V^*. The reservation property of the optimal search rule is a consequence of the stationarity of the search problem. We assume that the searcher who discards an opportunity and starts searching again is in exactly the same position as he was before he started searching. If the searcher's situation changes, then he will change his search behavior. The searcher's situation

---

[1] The job-search literature is surveyed by Mortenson (1986).
[2] Exceptions include Danforth (1979), Hall et al. (1979), and Manning and Morgan (1982).
[3] It can be shown that an optimal rule exists under very mild assumptions [see De Groot (1970), Kohn and Shavell (1974), Lippman and McCall (1976, 1981), McCall (1970)]. If, however, $F(\ )$ has no finite moments, an optimal decision rule does not exist. The reservation-price rule has been extended to the case where the searcher can take more than one observation at a time, choosing his sample size [Benhabib and Bull (1983), Gal et al. (1981), Harrison and Morgan (1990), Morgan (1983, 1986), Morgan and Manning (1985)]; and to the case of search across several goods [Anglin (1990), Burdett and Maleug (1981), Carlson and McAfee (1984), Vishwanath (1988), Weitzman (1979)].

can change either because the distribution $F(\ )$ changes,[4] or because the cost of search changes.[5]

It is instructive to use the fact that the optimal search strategy has the reservation property to calculate V^*. If the searcher sets a reservation value of x, then the value of searching is

$$V(x) = \delta\left[F(x)V(x) + \int_x^\infty y\,dF(y)\right] - k. \tag{2}$$

This equation states that if the searcher sets a reservation value of x, the expected profits from searching are equal to the discounted value of next period's expected benefits minus the cost of search. With probability $F(x)$ the searcher will reject the opportunity and be back where he started – with an expected value of $V(x)$. With probability $1 - F(x)$ he will have chosen a value from the upper tail of the distribution $F(\ )$. The expected value of this choice is $[\int_x^\infty y\,dF(y)]/[1 - F(x)]$. From equation (2),

$$V(x) = \frac{\delta \int_x^\infty y\,dF(y) - k}{1 - \delta F(x)}. \tag{3}$$

Let x^* maximize $V(x)$. Then $V(x^*) = V^*$. It is easy to check (by differentiating Eq. (3) and setting the derivative to zero) that $x^* = V^*$ as claimed and to derive an implicit equation for x^*:

$$x^* = \frac{\delta \int_{x^*}^\infty y\,dF(y) - k}{1 - \delta F(x^*)}. \tag{4}$$

This equation has a simple interpretation: V^* or x^* represents the expected discounted value of net benefits.[6] It is useful to rewrite Eq. (4) as

$$x^* = \delta\left[x^* + \int_{x^*}^\infty (y - x^*)\,dF(y) - k\right]. \tag{5}$$

---

[4] The distribution $F(\ )$ could change for a number of reasons. Some examples: (i) if search is modeled as being without replacement, so that the searcher discards opportunities as search progresses, the distribution of remaining opportunities changes [Rosenfeld and Shapiro (1981)]; (ii) if the searcher does not know the distribution from which he is drawing, then sampling yields information about it and his subjective estimate of $F(\ )$ changes [Bikhchandani and Sharma (1989), Burdett and Vishwanath (1988), Christensen (1986), Kohn and Shavell (1974), Morgan (1985), Rosenfeld and Shapiro (1981), Rothschild (1974a)]; and (iii) in job-search models, the searcher ages as search progresses, so that lifetime wages decline [Diamond and Rothschild (1989, pp. 450–454), Groneau (1971)].

[5] Having search costs increase is a way of introducing changing marginal utility into the analysis. In the job-search literature it is reasonable to assume that as searchers have been unemployed longer, the cost of search increases. Sometimes the cost of search can be directly affected by government policy – as in the length of time the government chooses to pay unemployment benefits [Burdett (1978)].

[6] Note that, since $\sum_{i=1}^\infty [\sum_{j=0}^i \delta^j F(x)^j][1 - F(x)] = [1 - \delta F(x)]^{-1}$, the expected discounted value of the total cost of searching is $k/[1 - \delta F(x)]$.

This equation also has an attractive interpretation. Suppose the searcher has just drawn an opportunity worth x^*. Then the left-hand side of (5) is what he gets if he stops searching and keeps x^*. The right-hand side is his expected discounted profit if he searches exactly one more time; if on that search he gets something better than x^*, he keeps it; otherwise he stops searching and keeps x^*. Equation (5) states that the expected discounted values of these two strategies are the same. We may summarize the discussion as follows:

Theorem. *If it costs k per unit to pick a draw from a distribution F (with finite first and second moments) then the optimal search rule is a reservation-value rule, where the reservation value x^* satisfies Eq. (4). The gain to the searcher from following this optimal strategy is given by Eq. (3), with $x = x^*$.*

One important property of the optimal search rule is that it is *myopic* in the following sense: the searcher who follows a reservation price strategy will never decide to accept an opportunity he has once rejected. Thus, in deciding whether or not to accept an opportunity he need only consider whether he wants it now; he need not worry that he might at some later date find it attractive. In technical terms the optimal search strategy is the same whether or not recall is permitted. Myopic search rules are attractive to economic theorists because if agents follow them economic behavior can be simply modeled. A market in which prospective buyers retain the right to take up offers previously made is much more complicated than a market in which all offers are made on a take-it-or-leave-it basis. If search is myopic this complexity need not be modeled; searchers would not avail themselves of the opportunity to take up offers once spurned. The tractability and plausibility of the bargaining models to be described in Section 4 depends on agents' myopic search rules.

The myopic property is distinct from the reservation-value property. A stationary reservation-value rule is myopic. However, it is easy to construct examples where the optimal search rule is myopic but does not have the reservation-value property [see Rothschild (1974a)]. Similarly, if the searcher follows a reservation-value rule but the value is falling (as will be the case if the cost of searching is rising), then the optimal search rule will not be myopic. Stationary optimal search rules are myopic; if a searcher's situation does not change as he searches, his decision rule will not change. Stationarity is not necessary for the myopic property, however: Rosenfeld and Shapiro (1981) give examples of optimal search rules which are myopic but not stationary (because the searcher is learning about the distribution of values as he searches).

Two important comparative statics results can be derived from Eq. (4) or Eq. (5). First, $dx^*/dk < 0$ and $dx^*/d\delta > 0$: as the cost of search or the cost of time increases the searcher becomes less picky (and the value of the entire search enterprise declines). Second, because a mean-preserving spread of $F(\ )$ will increase $\int_x^\infty y\, dF(y)$, as the distribution $F(\ )$ becomes more dispersed [in the sense of Rothschild and Stiglitz (1970)], V^* and x^* increase.

Many studies of search adopt a slightly richer formulation of the searcher's problem. Suppose that instead of paying a fixed cost k for an opportunity to draw from the distribution $F(\ )$, the searcher looks for opportunities with an intensity which he controls. Specifically, suppose that opportunities arrive as a Poisson process with an arrival rate of s and that someone who searches with an intensity of s for a length of time T incurs search costs of $k(s)T$. The Poisson assumption (which preserves stationarity in what is now a continuous time setup) means that in a short time interval, Δ, the probability that precisely one opportunity arrives is $s\Delta + o(\Delta)$.

We use the same argument that led to Eq. (2) to analyze this model. Assume that the time interval Δ is so small that we may ignore that $o(\Delta)$ possibility that more than one opportunity arrives in Δ. When the searcher controls both the intensity with which he searches and the reservation value, the probability that a searcher will accept an offer in an interval Δ is approximately $s\Delta[1 - F(x)]$, or the probability that he gets an opportunity times the probability that the offer is accepted. Thus, if the searcher follows a policy with values s and x, his expected gain is $V(s, x)$ where $V(s, x)$ satisfies (with r denoting the searcher's discount rate)

$$V(s, x) \approx e^{-r\Delta}\left([1 - s\Delta + s\Delta F(x)]V(s, x) + s\Delta \int_x^\infty y\, dF(y)\right) - k(s)\Delta. \qquad (6)$$

The interpretation of Eq. (6) is exactly the same as that of Eq. (2). With probability $s\Delta[1 - F(x)]$ the searcher will accept an offer with expected value $[\int_x^\infty y\, dF(y)]/[1 - F(x)]$; with complementary probability we will reject the offer and start searching again. Whether the searcher accepts an offer or not, he incurs search costs of $k(s)\Delta$ in searching over the interval. Once again, we can solve for $V(s, x)$. Approximating $e^{-r\Delta}$ by $1 - r\Delta$, and discarding all terms involving Δ^2, we obtain after simplification

$$V(s, x) = \frac{s \int_x^\infty y\, dF(y) - k(s)}{r + s[1 - F(x)]}, \qquad (7)$$

which has the same interpretation[7] as Eq. (4). Again, it is easy to see that if x^* is chosen to maximize $V(s, x)$, then $V(s, x^*) = x^*$. If s^* is chosen to maximize $V(s, x^*)$

---

[7] Note again that the expected discounted costs of searching are

$$\int_0^\infty k(s) \int_0^t e^{-ru}\, du\, s[1 - F(x)] e^{-s[1 - F(x)]t}\, dt = \frac{k(s)}{r + s[1 - F(x)]}.$$

then

$$k'(s) = \int_{x^*}^{\infty} (y - x^*) \, dF(y); \tag{8}$$

search intensity is set so that the marginal cost of increasing search intensity is equal to the expected improvement from an additional draw. It is straightforward to rewrite Eq. (7) in a form similar to Eq. (5):

$$rV(s, x^*) = s \int_{x^*}^{\infty} (y - x^*) \, dF(y) - k(s). \tag{9}$$

We may summarize this discussion as follows:

Theorem. *If opportunities arrive as a Poisson process with arrival rate s, where k(s) is the cost per unit time of generating search opportunities with intensity s, then a reservation-value rule is optimal, and the optimal intensity, reservation price, and value satisfy Eq. (7), (8), and (9).*

3. Price dispersion

The classical law of one price says that frictionless competition will force identical items to be sold at the same price; but in fact homogeneous goods often sell for widely varying prices.[8] Can search costs result in an equilibrium with dispersed prices? As we have seen, in the Diamond (1971) model with identical buyers and sellers, there is no price dispersion in equilibrium; all sellers charge the monopoly price. To generate a price distribution, there must be some heterogeneity among sellers and/or buyers. The sources of price dispersion include differences in the sellers' production costs [Bénabou (1988a), Bester (1988a), Carlson and McAfee (1983), MacMinn (1980), McAfee and McMillan (1988), Reinganum (1979)]; differences in buyers' search costs [Axell (1977), Bénabou (1988a), MacMinn (1980), Rob (1985), Stahl (1989), Stiglitz (1987), von zur Muehlen (1980)]; differences in buyers' beliefs about the price distribution [Rothschild (1974b)] or preferences [Diamond (1987)]; the use by buyers of nonsequential search strategies [Braverman (1980), Burdett and Judd (1983), Chan and Leland (1982), Gale (1988), Hey (1974), Sadanand and Wilde (1982), Salop and Stiglitz (1976), Schwartz and Wilde (1985), Wilde (1977, 1987), Wilde and Schwartz (1979)]; the repetitiveness of purchases and resultant customer loyalties [Bénabou (1988a), Cressy (1983), McMillan and Morgan (1988), Rosenthal (1982), Sutton (1980)]; the use by sellers of mixed strategies, varying their prices over time [Shilony (1977), Varian (1980)]; stockpiling by buyers [Bucovetsky (1983), Salop and Stiglitz (1982)]; and advertising by sellers

---

[8]Carlson and Pescatrice (1980), Dahlby and West (1986), Pratt et al. (1979), Stigler (1961).

that randomly reaches buyers, inducing differences in buyers' information about prices [Butters (1977)].

Most extant models of price dispersion assume away much of the complexity of the bargaining between buyer and seller by assuming that only the buyers search and that the "bargaining" consists of the seller making a take-it-or-leave-it price offer to any buyer who has found him. Thus the sellers play a Nash noncooperative game among themselves and a Stackelberg game against the buyers, who take the prices they find as given.

One way of having buyers differ in what they know is to posit advertising by sellers that reaches buyers randomly. In the seminal model of Butters (1977), buyers demand one unit up to a maximum price of a. All sellers have the same constant average production cost c. For an advertising cost of b a seller reaches a single buyer, but the seller cannot target a particular buyer; each buyer has an equal chance of receiving a message. Buyers do no search for themselves, but just purchase from the lowest-price seller whose message they have received. The main result is that in equilibrium all prices strictly between $c+b$ and a are advertised. To establish this, suppose there were an interval (p^-, p^+) within $(c+b, a)$ for which no prices were advertised. Then a seller advertising p^- could increase his profit per sale by $p^+ - p^-$ with no loss in sales by raising his price to p^+; this could not be a Nash equilibrium. In particular, if the highest price advertised were less than a then the highest-price seller could increase profit by raising his price to a. Finally, suppose the lowest price were strictly higher than $c+b$. Then the lowest-priced seller could get an extra sale for sure by sending an additional message, and strictly positive profit from this sale. (Since the seller is an infinitesimal part of the market, reaching a buyer he has already reached is a probability-zero event.) This is inconsistent with equilibrium. Hence the advertising induces price dispersion.

Butters's model shows that, for price dispersion to exist when sellers are identical, buyers must differ in what they know at the time they purchase. If buyers can search for themselves across sellers, then after they have finished acquiring information, different buyers must have found different best prices. For competition among sellers not to eliminate price dispersion when sellers have identical production costs, there must be some buyers who know only one price at the time of purchase [as observed by Wilde (1977)]. If all buyers know at least two prices, price differences are not consistent with Nash equilibrium; one of the identical sellers could change his price so as to attract more buyers and increase his profits. On the other hand, there must be some buyers who know two prices for there to be price dispersion; if all buyers could learn only one price the equilibrium would be at the monopoly price. Burdett and Judd (1983) give a model of information acquisition that leaves some buyers knowing only one price, and some knowing two. All sellers have the same constant production cost, and all buyers have the same search cost. Buyers are assumed not to use a sequential sampling rule but instead to use a fixed-sample-size rule: they commit in advance of sampling to a sample size. At any equilibrium, there are always some buyers who sample once, and there may be some who sample twice. To establish this, note that, if all buyers took two or more samples, price would be driven down to marginal cost; but once this happened it would be in no buyer's interest to take more than one sample. Hence equilibrium requires that some buyers search only once. Since all search costs are identical, either all buyers observe the same number of prices or they are indifferent between observing n and $n+1$ for some integer n; but we have established that n must equal one. Hence samples of at most two are taken. If the search cost is high enough, each searcher samples only one seller, and the only equilibrium is at the monopoly price. But it can be shown [see Burdett and Judd (1983) for the algebra] that for low enough search costs there exist equilibria in which some searchers sample two sellers, and therefore prices are dispersed. Related models that work by having some buyers sample only one seller while other buyers sample more than one seller include Braverman (1980), Chan and Leland (1982), Sadanand and Wilde (1982), Salop and Stiglitz (1977), Schwartz and Wilde (1979, 1985), and Wilde and Schwartz (1979).

If the sellers are not identical, then a dispersion of prices might arise simply as a reflection of the dispersion of production costs. To exemplify how the heterogeneity of sellers' production costs can create price dispersion, consider the model of Reinganum (1979). There are many sellers each with constant average cost c drawn from a distribution $G(c)$ with support $[c_0, c_1]$, with $c_0 < c_1 \leqslant (c_0 e)/(1+e)$. There are many identical buyers with the indirect utility function $U(P, p) + W$, where W represents wealth, p the price of the searched-for good, and P the vector of (nonstochastic) prices of all other goods. Let $q(p)$ represent the corresponding demand curve and assume that it has constant elasticity $e < -1$. All buyers know the prevailing distribution of prices $F(p)$ on $[p_0, p_1]$. The search cost is k per sample. As in Section 2, a buyer searches until he finds a price at least as low as his reservation price. The value of searching when the reservation price is p_r is

$$V(p_r) = [1 - F(p_r)]V(p_r) + \int_{p_0}^{p_r} U(P, p)\,dF(p) - k. \tag{10}$$

This is analogous to Eq. (2).[9] After solving Eq. (10) for $V(p_r)$, differentiating, and putting the result equal to zero, we find that the optimal reservation price p_r^* is defined, analogously to Eq. (4), by

$$\int_{p_0}^{p_r^*} [U(P, p) - U(P, p_r^*)]\,dF(p) = k. \tag{11}$$

The quantity demanded by any buyer at any price p is $q(p)$ if $p \leqslant p_r^*$ and zero otherwise. An equilibrium consists of a reservation price p_r^* and a price distribution $F^*(p)$ such that (a) given F^*, buyers choose their optimal reservation price p_r^*; and (b) given p_r^*, the sellers collectively generate F^* by each setting the price that

---

[9] But (16) differs from (2) in that the buyer's utility is not linear in price; lower draws from the distribution are better, whereas in (2) high draws are desired; and the discount factor δ is set equal to one.

maximizes profit. The result to be established is that there exists an equilibrium with price dispersion; that is, $dF^*(p) > 0$ on $[p_0^*, p_1^*]$, with $p_1^* > p_0^*$. The equilibrium price distribution is constructed as follows. Let $F(p) = G[p(1 + e)/e]$ on $[p_0, p_1] = [c_0 e/(1 + e), c_1 e/(1 + e)]$. Then there is a unique reservation price p_r^* defined by Eq. (11), which can be shown to satisfy $p_0 < p_r^* \leqslant p_1$. A seller maximizing profits will set his price $p \leqslant p_r^*$ since otherwise profit is zero. Given this, any buyer will buy from the first seller he visits. All sellers look alike to a buyer, so λ, the ratio of buyers to sellers, is the expected number of buyers who draw their first sample from any particular seller. A seller with cost c chooses a price to maximize his expected profit, which is $(p - c)q(p)\lambda$ if $p \leqslant p_r^*$ and is zero if $p > p_r^*$. The price p^* that maximizes profit is given by the marginal-revenue-equals-marginal-cost condition $p^*(1 + e)/e = c$, provided $p^* \leqslant p_r^*$; and $p^* = p_r^*$ otherwise. Since $p_r^* > p_0 \geqslant c_1$, all sellers make positive profit, so none exits. The resulting prices range from $p_0^* = c_0 e/(1 + e)$ to $p_1^* = p_r^* \leqslant c_1 e/(1 + e)$; and the induced price distribution is $F^*(p) = G[p(1 + e)/e]$, but with a mass point at p_r^*. It remains to show that p_r^*, defined to be the reservation price for F, remains the reservation price for F^*. But F and F^* coincide except at p_r^*, where F^* has a mass point. At p_r^* the integrand in Eq. (11) is zero, so the equation defining the reservation price remains the same when F^* replaces F. Hence the price distribution induced by the sellers' cost distribution is consistent with optimal search by buyers.

As in the Diamond (1971) monopoly-price model, in Reinganum's model the buyers' search costs turn each seller into a local monopolist. But because the sellers have different costs and the buyers have downward-sloping demands, the sellers exercise their monopoly power by setting different prices, creating an equilibrium price distribution. Perversely for a model of search, no search actually takes place: each buyer purchases from the first seller he encounters. Bénabou (1988a) shows, however, that generalizing the model by giving buyers a distribution of search costs results in active search by the buyers.

All of the price-dispersion models so far discussed are static. In the model of McMillan and Morgan (1988), buyers purchase the good in each period over an infinite time horizon. A buyer need pay no search cost to return to the seller he patronized in the previous period, but it is costly to learn the price of any other seller. As a result of the repetitiveness of purchases, a seller's customers rationally form implicit contracts with the seller to purchase repeatedly from him; and the seller reciprocates by implicitly contracting not to vary his price over time. If a seller raises his price, he induces his marginal customers (those with a reservation price just above his former price) to search for an alternative seller. But if he lowers his price then, on the Nash assumption that only he has changed his price, he gets no new customers, as all the other buyers remain with their accustomed seller. Thus the price elasticity of a seller's demand is higher for price rises than price cuts; the demand curve facing each seller is kinked. Because the corresponding marginal-revenue curve is discontinuous, many prices are consistent with optimizing behavior; in particular, there can be price-dispersed equilibria.

4. Search and bargaining

The parties to a negotiation are usually able, with effort and some luck, to develop alternative opportunities. Common-sense advice for a bargainer is to work on improving his alternatives, for the better his fallback, the stronger his bargaining position. In this section, we expound a model that corroborates this common-sense view.

Examining the bargaining foundations of search theory, or adding search to bargaining theory, eliminates some unsatisfactory features of both search theory and bargaining theory. Bargaining theory traditionally assumes exchanges are idiosyncratic: a single potential buyer faces a single potential seller. In some applications this is the appropriate assumption (for example, a negotiation between an insurance assessor and a claimant); but in most economic negotiations alternative trading partners can be searched for. Search theory, on the other hand, usually posits an extreme outcome to the bargaining between buyer and seller: after a searching buyer has located a seller, the seller makes a take-it-or-leave-it price offer [as in the Diamond (1971) model and the price-dispersion models discussed above, for example]. Such offers are not subgame perfect; if the buyer rejects the offer it is not in the seller's interest to refuse to bargain. The implicit assumption, therefore, is that the seller has the ability to make commitments. In some applications this is appropriate: a store might refuse to bargain with one customer because bargaining would destroy its ability to make credible take-it-or-leave-it offers to other customers and therefore reduce the profits it extracts from the other customers. But in many applications it is unrealistic to assume that one of the agents can capture most of the gains from trade.[10]

The line of research that assumes that both sellers and buyers incur search costs in finding potential trading partners and investigates the bargaining between any matched buyer and seller was begun by Diamond, Maskin, and Mortenson.[11] Diamond (1981, 1982) and Mortenson (1982a,b) represent the bargaining process

---

[10] In the model of McAfee and McMillan (1988), it is the searching buyer, rather than the nonsearching seller, who makes commitments. Each potential seller has private information about his own costs, and extracts rents from this information by quoting a price strictly above his cost. The buyer/searcher optimally announces not one but two reservation prices, p^* and p^{**}, with $p^{**} > p^*$. The buyer immediately purchases if he finds a seller asking a price less than the lower reservation price p^*; and he immediately rejects any seller asking a price above the higher reservation price p^{**}. Any seller quoting a price between the two reservation prices is put on a list and asked to wait. If the buyer ends up sampling all of the potential sellers without getting a price quote below p^*, he returns to, and purchases from, the cheapest of the listed sellers. If the buyer receives no quotes below p^{**}, he makes no purchase. Related analyses include Bester (1988a, b), Daughety and Reinganum (1991), Fudenberg et al. (1987), and Riley and Zeckhauser (1983).

[11] Diamond (1981, 1982, 1984a, b), Diamond and Maskin (1979, 1981), Mortenson (1976, 1982a, b; 1988). Other contributors to this line of research include Bester (1988a), Binmore and Herrero (1988a, b), Deere (1988), Gale (1986, 1987), Howitt (1985), Howitt and McAfee (1987, 1988), Iwai (1988a, b), McLennan and Sonnenschein (1991), McKenna (1988), Peters (1991), Pissarides (1985), Rosenthal and Landau (1981), and Rubinstein and Wolinsky (1985).

by the cooperative Nash bargaining model (see the Chapter in this Handbook by Thomson).[12] The most detailed investigation of the strategic interactions between search and bargaining is that of Wolinsky (1987), based on noncooperative bargaining concepts; we shall follow Wolinsky's analysis here.

If neither bargainer had an alternative trading partner, sequential bargaining would divide equally the surplus being bargained over, provided buyer and seller were equally patient and the time between offers was short (see Chapter 7 of this Handbook by Binmore, Osborne, and Rubinstein). The existence of alternatives, however, means that the negotiated division of the surplus depends upon the bargainers' search capabilities. The buyer's ability to find an alternative seller increases the share of the surplus he can negotiate; and the seller's ability to find an attractive alternative reduces the buyer's negotiated share. To make this idea precise, consider a market in which many potential buyers and sellers search for each other. When a pair meets, they learn that the difference between the seller's and the buyer's valuation is m. Ex ante, m is viewed as a random draw from a continuously differentiable distribution F with support $[0, m^+]$. Let the subscript 1 denote sellers and 2 buyers. After meeting, a pair can either negotiate a price and leave the market, or continue to search. As in the second model in Section 2 above, search takes place in continuous time; a searcher chooses his search intensity s_i, $i = 1, 2$, where s_i is the probability that, within a unit interval of time, the searcher initiates a contact with an agent of the opposite type. Search activity costs the searcher $k_i(s_i)$ per unit of time, where $k_i(s_i)$ is an increasing, concave function. Let s_i^0 represent the probability per unit of time that the type i agent is contacted as a result of the search activity on the other side of the market: if there are N_i searchers of type i, $i = 1, 2$, then $s_i^0 = (N_j/N_i)s_j$, $j \neq i$. Thus, given that there are many buyers and sellers, the probability that a type i agent is matched is $(s_i + s_i^0)$ per unit of time. There is a common rate of time preference r. Assume a steady state: the (exogenous) rate of entry of new searchers equals the (endogenous) rate of exit as agreements are made. Equilibrium in this game is characterized by an agreement $(w_1(m), w_2(m))$ that divides the surplus from the match, m. If one, and only one, of the players could make commitments (as the seller was assumed to be able to do in the price-dispersion models of the previous section), he would get the whole surplus, m. Let us require instead that both bargainers behave in a subgame-perfect way. The derivation of the perfect equilibrium combines the techniques of optimal search decisions, given in Section 2 above, with the backward-induction method of finding perfect equilibrium in a bargaining model (exposited in the Chapter by Binmore, Osborne, and Rubinstein). We shall give an outline; for the details, see Wolinsky (1987).

First, consider search strategies. An agent of type i chooses a search intensity s_i and a constant reservation value m (i.e., he takes the first opportunity whose

---

[12] Diamond and Maskin (1979) and Diamond (1984a) assume the bargaining results in equal division of the surplus, which would be the Nash solution if the bargainers had identical disagreement payoffs.

joint surplus exceeds m).[13] Let $V_i(s_i, m; s_i^0)$ represent the present discounted value of this search policy. Then V_i satisfies a modified version of Eq. (9):

$$rV_i(s_i, m; s_i^0) = (s_i + s_i^0) \int_m^{m^+} [w_i(y) - V_i(s_i, m; s_i^0)] \, dF(y) - k_i(s_i). \tag{12}$$

A searcher can be in one of two different situations: either he is unmatched, or he has already found a partner and is in the process of negotiating. Consider the former. Define

$$V_i^* \equiv V_i(s_i^*, m^*; s_i^0) = \max_{s_i, m} V_i(s_i, m; s_i^0). \tag{13}$$

Here s_i^* and m^* characterize optimal search by an agent who has not yet found a match, given s_i^0, $w_i(m)$, and $F(\ )$. A match will be made only if $m \geq V_1^* + V_2^*$; otherwise the two agents are better off continuing to search for partners. A searcher who is already matched, on the other hand, has a simpler decision than one who is still looking for a partner. His reservation value is fixed at the surplus he has already found, m (which must exceed m^* or he would have rejected it). His only choice is the rate of search, s_i^{**}, which he sets to equate the marginal benefit in terms of expected improved match to the marginal cost, as in Eq. (8):

$$\int_m^{m^+} [w_i(x) - w_i(m)] \, dF(x) = k_i'(s_i). \tag{14}$$

Since $m > m^*$, this implies slower search: $s_i^{**} < s_i^*$. The corresponding expected return is $V_i^{**} = V_i(s_i^{**}, m, s_i^0)$.

The main result from this analysis shows how the two search decisions shape the bargaining agreement. Recall that the limiting perfect equilibrium in a sequential bargaining model is the same as the Nash cooperative solution, provided the disagreement point for the Nash solution is appropriately chosen (see Chapter 7 of this Handbook by Binmore, Osborne, and Rubinstein). That is, the spoils are divided so as to maximize the product of the two bargainers' net gains from the agreement. The division of the surplus at a perfect equilibrium of this search-and-bargaining game, when the time between offers goes to zero, is given by

$$w_i(m) = (\tfrac{1}{2}[m + d_i(m) - d_j(m)]), \quad i \neq j = 1, 2, \tag{15}$$

where $d_i(m)$, $i = 1, 2$, is a weighted average of the returns from the two kinds of search:

$$d_i(m) = \alpha_i V_i^{**} + (1 - \alpha_i) V_i^*, \tag{16}$$

where, in turn, the weights reflect the search rates, the discount rate, and the

---

[13] The assumption of a large number of searchers of either type means that we need not consider the interactions among the search decisions of different agents of a given type. On such interactions when the number of searchers is small, see Reinganum (1982, 1983).

probability of finding a better match:

$$\alpha_i = \frac{r + (s_i + s_i^0)[1 - F(m)]}{r + (s_i + s_i^0 + s_j + s_j^0)[1 - F(m)]}, \quad i \neq j = 1, 2. \tag{17}$$

Equation (15) defines the Nash cooperative solution for dividing the sum m with the disagreement point (d_1, d_2) [since (15) maximizes the Nash product $(w_1 - d_1) \cdot (w_2 - d_2)$ subject to full division of the surplus, $w_1 + w_2 = m$]. Equation (16) defines this disagreement point to be the weighted average of V_i^* (i.e., the return agent i expects if the negotiations break down because agent j finds a better match, so that agent i must search for another partner) and V_i^{**} (i.e., the return agent i expects if he searches while negotiating with his current partner). To summarize:

Theorem. *Consider a market in which many potential buyers and sellers search for trading partners. The difference between a buyer's and a seller's valuation is m, which is a random draw from a continuously differentiable distribution F. The searchers have a common rate of time preference r. Both buyer and seller can choose their search intensities. Let s_i, $i = 1, 2$, represent the probability (chosen by the searcher of type i) that within a unit interval of time the searcher finds an agent of the opposite type; and let s_i^0 represent the probability that within a unit interval of time an agent of type i is contacted as a result of search activity on the other side of the market. V_i^* is the return to search by a searcher who has not yet found a match; V_i^{**} is the return to further search by a searcher who has already found a potential trading partner (both are defined above). The perfect equilibrium (as the time between offers approaches zero) of the search-and-bargaining game is defined by Eq. (15), (16), and (17).*

Thus an agent's bargaining power is affected by search in two ways.[14] First, the ability to search while negotiating effectively reduces the cost to the agent of delaying agreement. The agent who has the lower cost of delay tends to get the better of the bargain, because of his negotiating partner's impatience to settle. Second, the cost of search for a new partner determines the losses from breakdown. The agent who would suffer less should the negotiations break down is in the better bargaining position. Hence the bargaining solution depends on both the opportunities available in the event of breakdown, and the opportunities for search during the bargaining. It pays a bargainer to have low search costs.[15]

---

[14] In equilibrium, agreement is reached immediately a buyer and seller meet (provided m is large enough), because the bargainers foresee these losses from protracted negotiations: there is no time for search during the negotiations. But the implicit, and credible, threat of search shapes the terms of the agreement.

[15] The foregoing analysis assumes that a matched pair know exactly the size of the surplus to be divided between them. For the beginnings of an analysis of the case in which information about the surplus is imperfect, see Bester (1988b), Deere (1988), Rosenthal and Landau (1981), and Wolinsky (1990).

Increasing the search activity of the agents on one side of the market would reduce the expected time without a match of each agent on the other side of the market; in a job-market interpretation of the model, this corresponds to reducing unemployment. Thus, as noted by Deere (1987), Diamond (1981, 1982), Mortenson (1976, 1982a, b), and Wilde (1977), there is an externality: the searchers fail to take into account the benefits their search decisions convey to agents on the other side of the market, and too little search takes place.[16] Diamond (1981, 1982, 1984a, b), Howitt (1985), Howitt and McAfee (1987, 1988), and Pissarides (1985) have used this externality as the driving force in macroeconomic models with some Keynesian properties. Because there is too little job search and recruiting, equilibrium occurs at an inefficiently low level of aggregate activity. If workers search more, the returns to firms' search increase; firms then increase their search, which in turn raises the return to workers' search. This feedback effect generates multiple equilibria; there is a role for macroeconomic policy to push the economy away from Pareto-dominated equilibria.

As the transactional frictions (the cost of search and/or the discounting of future returns) go to zero, does the equilibrium of the search-and-bargaining game converge on the Walrasian equilibrium? In other words, can perfectly competitive, price-taking behavior be given a rigorous game-theoretic foundation? Rubinstein and Wolinsky (1985), Gale (1986; 1987), McLennan and Sonnenschein (1991), Bester (1988a), Binmore and Herrero (1988a, b), and Peters (1991) investigate this question and, under varying assumptions, obtain both affirmative and negative answers; the limiting behavior of search-and-bargaining models turn out to be surprisingly subtle. Rubinstein and Wolinsky (1985) simplify the above model by assuming the value of a match, denoted m above, is the same for all pairs of agents; the agents cannot control their rate of search; and an agent always drops an existing partner as soon as he finds a new one. The perfect equilibrium of this model is determined, as above, by each agent's risk of losing his partner while bargaining. In a steady state in which the inflow of new agents equals the outflow of successfully matched agents, the limiting perfect equilibrium (as the discount rate approaches one) divides the surplus in the same ratio as the number of agents of each type in the market at any point in time, $N_1:N_2$; in particular, the side of the market that has the fewer participants does not get all of the surplus. Is this contrary to the Walrasian model? Binmore and Herrero (1988a, b) and Gale (1986, 1987) argue that it is not, on the grounds that, in this dynamic model, supply and demand should be defined in terms not of stocks of agents present in the market but of flows of agents. Since the inflow of agents is, by the steady-state assumption, equal to the outflow the market never actually clears, so any division of the surplus is consistent with Walrasian equilibrium. A model in which the steady-state assumption is dropped, so that the agents who leave the market upon being successfully matched are not

---

[16] In the terminology of Milgrom and Roberts (1990), this game has the property of *supermodularity*: one agent's increasing the level of his activity increases the other agents' returns to their activity.

replaced by new entrants, has a frictionless limit equilibrium in which the market clears in the Walrasian sense and the short side of the market receives all of the surplus.[17]

5. Conclusion

The preceding sections have traced the development of the main ideas of search theory.[18] We saw in Section 4 that game theory provides an excellent framework for integrating the theory of search with the theory of bargaining. For the case in which two parties meet and find that a mutually beneficial exchange is possible, but that they can continue searching for other trading partners if a deal is not consummated, Wolinsky (1987) has shown that search theory can be married to bargaining theory in a virtually seamless way. The two theories fit together as well and as easily as hypothesis testing and Bayesian decision theory. One of the more fruitful offspring of this marriage is a satisfactory answer to a question which most other models of search behavior leaves open. In the conventional model of a buyer searching for a purchase, the distribution of offers from which the buyer selects is a given. Clearly, it is worth something to a seller to have potential buyers include that seller's price offer in the distribution from which a buyer is sampling. Since having a place in this distribution is worth something, a natural question is: how is this good allocated; how do sellers become part of the sample from which a buyer selects? In the conventional search literature, the advertising model of Butters (1977) attempts to answer that question. Bargaining theory provides an easy answer. When both sides of the market are modeled symmetrically, then it is clear that what matters to a market participant is making an encounter with someone on the other side of the market. This is determined by the intensity with which a market participant searches (and by the search intensity of those on the other side of the market.) While this merger of the two theories is very neat, it does leave open some questions. Bargaining theory can tell us how, if two parties haggle, a deal is struck, and what the terms of the agreement will be. It does not have much to say, for a given market, about whether haggling is the rule of the day or whether sales are made at posted (and inflexible) prices. An open and fertile area for future research would seem to be the use of search theory and bargaining theory to put a sharper focus on the conventional microeconomic explanations of why different markets operate under different rules.

Search theory provides a toolkit for analyzing information acquisition. Much of modern game theory and economic theory is concerned with the effects of

---

[17] For more on the literature on the bargaining foundations of perfect competition, see Osborne and Rubinstein (1990) and Wilson (1987), as well as the Chapter by Binmore, Osborne, and Rubinstein.

[18] Other surveys of search theory, emphasizing different aspects of the theory, include Burdett (1989), Hey (1979, Chs. 11, 14), Lippman and McCall (1976, 1981), McKenna (1987a, b), Mortensen (1986), Rothschild (1973), and Stiglitz (1988).

informational asymmetries (see, for example, the Chapters in this Handbook by Geanakoplos (Ch. 40), Wilson (Chs. 8 and 10, Volume I), and Kreps ad Sobel (Ch. 25). Usually in such models the distribution of information – who knows what – is taken as given. In some applications this is natural: for example, individuals' tastes are inherently unobservable. In other applications, however, it might be feasible for an individual to reduce his informational disadvantage: for example, if the private information is about production costs, then it is reasonable to suppose that the initially uninformed individual could, at some cost, reduce or even eliminate the informational asymmetry. The techniques of search theory might be used in future research to endogenize the information structure in asymmetric-information models.[19]

References

Albrecht, J.W., B. Axell and H. Lang (1986) 'General equilibrium wage and price distributions', *Quarterly Journal of Economics*, **101**: 687–706.
Anglin, P.M. (1988) 'The sensitivity of consumer search to wages', *Economics Letters*, **28**: 209–213.
Anglin, P.M. (1990) 'Disjoint search for the prices of two goods consumed jointly', *International Economic Review*, **31**: 383–408.
Anglin, P.M. and M.R. Baye (1987) 'Information, multiprice search, and cost-of-living index theory', *Journal of Political Economy*, **95**: 1179–1195.
Anglin, P.M. and M.R. Baye (1988) 'Information gathering and cost of living differences among searchers', *Economics Letters*, **28**: 247–250.
Axell, B. (1974) 'Price dispersion and information', *Swedish Journal of Economics*, **76**: 77–98.
Axell, B. (1977) 'Search market equilibrium', *Scandinavian Journal of Economics*, **79**: 20–40.
Bagwell, K. and M. Peters (1987) 'Dynamic monopoly power when search is costly', Discussion Paper No. 772, Northwestern University.
Bagwell, K. and G. Ramey (1992) 'The Diamond paradox: A dynamic resolution', Discussion Paper No. 92–45, University of California, San Diego.
Bagwell, K. and G. Ramey (1994) 'Coordination economies, advertising and search behavior in retail markets', *American Economic Review*, **87**: 498–517.
Bénabou, R. (1988a) 'Search market equilibrium, bilateral heterogeneity and repeat purchases', Discussion Paper No. 8806, CEPREMAP.
Bénabou, R. (1988b) 'Search, price setting, and inflation', *Review of Economic Studies*, **55**: 353–376.
Bénabou, R. (1992) 'Inflation and efficiency in search markets', *Review of Economic Studies*, **59**: 299–330.
Bénabou, R. and R. Gertner (1990) 'The informativeness of prices: Search with learning and inflation uncertainty', Working Paper No. 555, MIT.
Benhabib, J. and C. Bull (1983) 'Job search: The choice of intensity', *Journal of Political Economy*, **91**: 747–764.
Bester, H. (1988a) 'Bargaining, search costs and equilibrium price distributions', *Review of Economic Studies*, **55**: 201–214.
Bester, H. (1988b) 'Qualitative uncertainty in a market with bilateral trading', *Scandanavian Journal of Economics*, **90**: 415–434.
Bester, H. (1989) 'Noncooperative bargaining and spatial competition', *Econometrica*, **57**: 97–113.
Bikhchandani, S. and S. Sharma (1989) 'Optimal search with learning', Working Paper #89–30, UCLA.
Binmore, K. and M.J. Herrero (1988a) 'Matching and bargaining in dynamic markets', *Review of Economic Studies*, **55**: 17–32.

---

[19] One such issue – bidders at an auction strategically acquiring information about the item's value before bidding – has been modeled by hausch and Li (1990), Lee (1984, 1985), Matthews (1984), and Milgrom (1981).

Binmore, K. and M.J. Herrero (1988b) 'Security equilibrium,' *Review of Economic Studies*, **55**: 33–48.
Braverman, A. (1980) 'Consumer search and alternative market equilibria', *Review of Economic Studies*, **47**: 487–502.
Bucovetsky, S. (1983) 'Price dispersion and stockpiling by consumers,' *Review of Economic Studies*, **50**: 443–465.
Burdett, K. (1978) 'A theory of employee job search and quit rates', *American Economic Review*, **68**: 212–220.
Burdett, K. (1989) 'Search market models: A survey', University of Essex mimeo.
Burdett, K. and K.L. Judd (1983) 'Equilibrium price distributions', *Econometrica*, **51**: 955–970.
Burdett, K. and D.A. Malueg (1981) 'The theory of search for several goods', *Journal of Economic Theory*, **24**: 362–376.
Burdett, K. and T. Vishwanath (1988) 'Declining reservation wages and learning', *Review of Economic Studies*, **55**: 655–666.
Butters, G.R. (1977) 'Equilibrium distributions of sales and advertising prices', *Review of Economic Studies*, **44**: 465–491. Reprinted in Diamond and Rothschild (1989).
Carlson, J. and R.P. McAfee (1983) 'Discrete equilibrium price dispersion', *Journal of Political Economy*, **91**: 480–493.
Carlson, J. and R.P. McAfee (1984) 'Joint search for several goods', *Journal of Economic Theory*, **32**: 337–345.
Carlson, J. and D.R. Pescatrice (1980) 'Persistent price distributions', *Journal of Economics and Business*, **33**: 21–27.
Chan, Y.-S. and H. Leland (1982) 'Prices and qualities in markets with costly information', *Review of Economic Studies*, **49**: 499–516.
Christensen, R. (1986), "Finite stopping in sequential sampling without recall from a Dirichlet process', *Annals of Statistics*, **14**: 275—282.
Cressy, R.C. (1983) 'Goodwill, intertemporal price dependence and the repurchase decision', *Economic Journal*, **93**: 847–861.
Dahlby, B. and D.S. West (1986) 'Price dispersion in an automobile insurance market', *Journal of Political Economy*, **94**: 418–438.
Danforth, J. (1979) 'On the role of consumption and decreasing absolute risk aversion in the theory of job search,' in: S.A. Lippman and J.J. McCall, eds., *Studies in the Economics of Search*. Amsterdam: North-Holland.
Daughety, A. (1992) 'A model of search and shopping by homogeneous customers without price pre-commitment by firms,' *Journal of Economics and Management Strategy*, **1**: 455–473.
Daughety, A.F. and J.F. Reinganum (1992) 'Search equilibrium with endogenous recall', *Rand Journal of Economics*, **23**: 184–202.
Daughety, A. and J.F. Reinganum (1991) 'Endogenous availability in search equilibrium', *Rand Journal of Economics*, **22**: 287–306.
David, P.A. (1973) 'Fortune, risk, and the micro-economics of migration', in: P.A. David and M.W. Reder, eds., *Nations and Households in Economic Growth: Essays in Honor of Moses Abromovitz*. New York: Academic Press.
Deere, D.R. (1987) 'Labor turnover, job-specific skills, and efficiency in a search model', *Quarterly Journal of Economics*, **102**: 815–833.
Deere, D.R. (1988) 'Bilateral trading as an efficient auction over time', *Journal of Political Economy*, **96**: 100–115.
De Groot, M. (1970) *Optimal statistical decision*. New York: McGraw-Hill.
Diamond, P.A. (1971) 'A model of price adjustment', *Journal of Economic Theory*, **3**: 156–168.
Diamond, P.A. (1981) 'Mobility costs, frictional unemployment, and efficiency', *Journal of Political Economy*, **89**: 798–812.
Diamond, P.A. (1982) 'Wage determination and efficiency in search equilibrium', *Review of Economic Studies*, **49**: 217–228.
Diamond, P.A. (1984a) 'Money in search equilibrium', *Econometrica*, **52**: 1–20.
Diamond, P.A. (1984b) *A Search Theoretic Approach to the Micro Foundations of Macroeconomics*. Cambridge: MIT Press.
Diamond, P.A. (1987) 'Consumer differences and prices in a search model', *Quarterly Journal of Economics*, **12**: 429–436.
Diamond, P.A. and E. Maskin (1979) 'An equilibrium analysis of search and breach of contract, I: Steady states', *Bell Journal of Economics*, **10**: 282–316.

Diamond, P.A. and E. Maskin (1981) 'An equilibrium analysis of search and breach of contract, II: A non-steady state example', *Journal of Economic Theory*, **25**: 165–195.
Diamond, P.A. and M. Rothschild, eds. (1989) *Uncertainty in Economics*. New York: Academic Press, revised edition.
Evenson, R.E. and Y. Kislev (1976) 'A stochastic model of applied research', *Journal of Political Economy*, **84**: 265–281.
Fudenberg, D., D.K. Levine, and J. Tirole (1987) 'Incomplete information bargaining with outside opportunities', *Quarterly Journal of Economics*, **102**: 37–50.
Gabszewicz, J. and P.G. Garella (1986) 'Subjective price search and price competition', *International Journal of Industrial Organisation*, **4**: 305–316.
Gal, S., M. Landsberger and B. Levykson (1981) 'A compound strategy for search in the labor market', *International Economic Review*, **22**: 597–608.
Gale, D. (1986) 'Bargaining and competition, Part I: Characterization; Part II: Existence', *Econometrica*, **54**: 785–818.
Gale, D. (1987) 'Limit theorems for markets with sequential bargaining', *Journal of Economic Theory*, **43**: 20–54.
Gale, D. (1988) 'Price setting and competition in a simple duopoly model', *Quarterly Journal of Economics*, **103**: 729–739.
Gastwirth, J.L. (1976) 'On probabilistic models of consumer search for information', *Quarterly Journal of Economics*, **90**: 38–50.
Groneau, R. (1971) 'Information and frictional unemployment', *American Economic Review*, **61**: 290–301.
Hall, J., S.A. Lippman and J.J. McCall (1979) 'Expected utility maximizing job search', in: S.A. Lippman and J.J. McCall, eds., *Studies in the Economics of Search*. Amsterdam: North-Holland.
Harrison, G.W. and P. Morgan (1990) 'Search intensity in experiments', *Economic Journal*, **100**: 478–486.
Hausch, D.B. and L. Li (1990) 'A common value auction model with endogenous entry and information acquisition', University of Wisconsin, mimeo.
Hey, J.D. (1974) 'Price adjustment in an atomistic market', *Journal of Economic Theory*, **8**: 483–499.
Hey, J.D. (1979) *Uncertainty in Microeconomics*. New York: New York University Press.
Howitt, P. (1985) 'Transaction costs in the theory of unemployment', *American Economic Review*, **75**: 88–100.
Howitt, P. and R.P. McAfee (1987) 'Costly search and recruiting', *International Economic Review*, **28**: 89–107.
Howitt, P. and R.P. McAfee (1988) 'Stability of equilibria with aggregate externalities', *Quarterly Journal of Economics*, **103**: 261–277.
Iwai, K. (1988a) 'The evolution of money: A search-theoretic foundation of monetary economics', CARESS Working Paper #88-03, University of Pennsylvania.
Iwai, K. (1988b) 'Fiat money and aggregate demand management in a search model of decentralized exchange', CARESS Working Paper #88-16, University of Pennsylvania.
Karnai, E. and A. Schwartz (1977) 'Search theory: The case of search with uncertain recall', *Journal of Economic Theory*, **16**: 38–52.
Kohn, M. and S. Shavell (1974) 'The theory of search', *Journal of Economic Theory* **9**: 93–123.
Landsberger, M. and D. Peled (1977) 'Duration of offers, price structure, and the gain from search', *Journal of Economic Theory*, **16**: 17–37.
Lee, T. (1984) 'Incomplete information, high-low bidding, and public information in first-price auctions', *Management Science*, **30**: 1490–1496.
Lee, T. (1985) 'Competition and information acquisition in first-price auctions', *Economics Letters*, **18**: 129–132.
Lippman, S.A. and J.J. McCall (1976) 'The economics of job search: A survey', *Economic Inquiry*, **14**: 155–89 and 347–368.
Lippman, S.A. and J.J. McCall, eds. (1979) *Studies in the Economics of Search*. Amsterdam: North-Holland
Lippman, S.A. and J.J. McCall (1981) 'The economics of uncertainty: Selected topics and probabilistic methods', in: K.J. Arrow and M.D. Intriligator, eds., *Handbook of Mathematical Economics*, Vol. 1. Amsterdam: North-Holland.
Lucas, R. and E. Prescott (1974) 'Equilibrium search and unemployment', *Journal of Economic Theory*, **7**: 188–209. Reprinted in Diamond and Rothschild (1989).
MacMinn, R. (1980) 'Search and market equilibrium', *Journal of Political Economy*, **88**: 308–327.

Manning, R. (1976) 'Information and sellers' behavior', *Australian Economic Papers*, **15**: 308–321.
Manning, R. (1989a) 'Budget-constrained search', Discussion Paper No. 8904, State University of New York, Buffalo.
Manning, R. (1989b) 'Search while consuming', *Economic Studies Quarterly*, **40**: 97–108.
Manning, R. and P.B. Morgan (1982) 'Search and consumer theory', *Review of Economic Studies*, **49**: 203–216.
Matthews, S.A. (1984) 'Information acquisition in discriminatory auctions', in: M. Boyer and R. Kihlstrom, eds., *Bayesian Models in Economic Theory*. Amsterdam: North-Holland.
McAfee, R.P. and J. McMillan (1988) 'Search mechanisms', *Journal of Economic Theory*, **44**: 99–123.
McCall, J.J. (1970) 'The economics of information and job search', *Quarterly Journal of Economics*, **84**: 113–126.
McKenna, C.J. (1987a) 'Models of search market equilibrium', in: J. Hey and P. Lambert, eds., *Surveys in the Economics of Uncertainty*. London: Basil Blackwell.
McKenna, C.J. (1987b) 'Theories of individual search behaviour', in: J. Hey and P. Lambert, eds., *Surveys in the Economics of Uncertainty*. London: Basil Blackwell.
McKenna, C.J. (1988) 'Bargaining and strategic search', *Economics Letters*, **28**: 129–134.
McLennan, A. and H. Sonnenschein (1991) 'Sequential bargaining as a noncooperative foundation for perfect competition', *Econometrica*, **59**: 1395–1424.
McMillan, J. and P.B. Morgan (1988) 'Price dispersion, price flexibility, and repeated purchasing', *Canadian Journal of Economics*, **21**: 883–902.
Milgrom, P.R. (1981) 'Rational expectations, information acquisition, and competitive bidding', *Econometrica*, **49**: 921–943.
Milgrom, P.R. and J. Roberts (1990) 'Rationalizability, learning and equilibrium in games with strategic complementarities', *Econometrica*, **58**: 1255–1278.
Morgan, P.B. (1983) 'Search and optimal sample sizes', *Review of Economic Studies*, **50**: 659–675.
Morgan, P.B. (1985) 'Distributions of the duration and value of job search with learning', *Econometrica*, **53**: 1191–1232.
Morgan, P.B. (1986) 'A note on 'job search: The choice of intensity', *Journal of Political Economy*, **94**: 439–442.
Morgan, P.B. and R. Manning (1985) 'Optimal search', *Econometrica*, **53**: 923–944.
Mortenson, D.T. (1976) 'Job matching under imperfect information', in: O. Ashenfelter and J. Blum, eds., *Evaluating the Labor-Market Effects of Social Programs*. Industrial Relations Program, Princeton University.
Mortenson, D.T. (1982a) 'The matching process as a noncooperative bargaining game', in: J.J. McCall, ed., *The Economics of Information and Uncertainty*. Chicago: University of Chicago Press.
Mortenson, D.T. (1982b) 'Property rights and efficiency in mating, racing, and related games', *American Economic Review*, **72**: 968–980.
Mortenson, D.T. (1986) 'Job search and labor market analysis', in: O. Ashenfelter and R. Layard, eds., *Handbook of Labor Economics*. Amsterdam: North-Holland.
Mortenson, D.T. (1988) 'Matching Finding a partner for life or otherwise', *American Journal of Sociology*, **94**: S215–S240.
Nelson, P. (1970) 'Information and consumer behavior', *Journal of Political Economy*, **78**: 311–329.
Nelson, P. (1974) 'Advertising as information', *Journal of Political Economy*, **82**: 729–754.
Osborne, M.J. and Rubinstein, A. (1990) *Bargaining and Markets*. San Diego: Academic Press.
Peters, M. (1991) 'Ex ante price offers in matching games: Non-steady states', *Econometrica*, **59**: 1425–1454.
Pissarides, C.A. (1985) 'Taxes, subsidies and equilibrium unemployment', *Review of Economic Studies*, **52**: 121–133.
Pratt, J., D. Wise and R. Zeckhauser (1979) 'Price differences in almost competitive markets', *Quarterly Journal of Economics*, **93**: 189–212.
Reinganum, J.F. (1979) 'A simple model of equilibrium price dispersion', *Journal of Political Economy*, **87**: 851–858.
Reinganum, J.F. (1982) 'Strategic search theory', *International Economic Review*, **23**: 1–18.
Reinganum, J.F. (1983) 'Nash equilibrium search for the best alternative', *Journal of Economic Theory*, **30**: 139–152.
Riley, J.G. and J. Zeckhauser (1983) 'Optimal selling strategies: When to haggle, when to hold firm', *Quarterly Journal of Economics*, **98**: 267–289.

Rob, R. (1985) 'Equilibrium price distributions', *Review of Economic Studies*, **52**: 487–504.
Robert, J. and D.O. Stahl II (1993) 'Informative price advertising in a sequential search model', *Econometrica*, **61**: 657–686.
Rosenfeld, D.B. and R.D. Shapiro (1981) 'Optimal adaptive price search', *Journal of Economic Theory*, **25**: 1–20.
Rosenthal, R.W. (1982) 'A dynamic model of duopoly with customer loyalties', *Journal of Economic Theory*, **27**: 69–76.
Rosenthal, R.W. and H.J. Landau (1981) 'Repeated bargaining with opportunities for learning', *Journal of Mathematical Sociology*, **8**: 61–74.
Rothschild, M. (1973) 'Models of market organization with imperfect information: A survey', *Journal of Political Economy*, **81**: 1283–1308. Reprinted in Diamond and Rothschild (1989).
Rothschild, M. (1974a) 'Searching for the lowest price when the distribution of prices is unknown', *Journal of Political Economy*, **82**: 689–711. Reprinted in Diamond and Rothschild (1989).
Rothschild, M. (1974b) 'A two-armed bandit theory of market pricing', *Journal of Economic Theory*, **9**: 185–202.
Rothschild, M., and J.E. Stiglitz (1970) 'Increasing risk: A definition', *Journal of Economic Theory*, **2**: 225–243. Reprinted in Diamond and Rothschild (1989).
Rubinstein, A. and A. Wolinsky, (1985) 'Equilibrium in a market with sequential bargaining', *Econometrica*, **53**: 1133–1151.
Sadanand, A. and L.L. Wilde (1982) 'A generalized model of pricing for homogeneous goods under imperfect information', *Review of Economic Studies*, **49**: 229–240.
Salop, S. and J.E. Stiglitz (1977) 'Bargains and ripoffs: A model of monopolistic competition', *Review of Economic Studies*, **44**: 493–510.
Salop, S. and J.E. Stiglitz (1982) 'The theory of sales: A simple model of equilibrium price dispersion with many identical agents', *American Economic Review*, **72**: 1121–1130.
Schwartz, A. and L.L. Wilde (1979) 'Intervening in markets on the basis of imperfect information: A legal and economic analysis', *Pennsylvania Law Review*, **127**: 630–682.
Schwartz, A. and L.L. Wilde (1982a) 'Competitive equilibria in markets for heterogeneous goods under imperfect information', *Bell Journal of Economics*, **12**: 181–193.
Schwartz, A. and L.L. Wilde (1982b) 'Imperfect information, monopolistic competition, and public policy', *American Economic Review: Papers and Proceedings*, **72**: 18–24.
Schwartz, A. and L.L. Wilde (1985) 'Product quality and imperfect information', *Review of Economic Studies*, **52**: 251–262.
Shilony, Y. (1977) 'Mixed pricing in oligopoly', *Journal of Economic Theory*, **14**: 373–388.
Stahl, D.O. II (1989) 'Oligopolistic pricing with sequential consumer search', *American Economic Review*, **79**: 700–712.
Stigler, G.J. (1961) 'The economics of information', *Journal of Political Economy*, **69**: 213–225.
Stigler, G.J. (1962) 'Information in the labor market', *Journal of Political Economy*, **70**: 94–104.
Stiglitz, J.E. (1979) 'Equilibrium in product markets with imperfect information', *American Economic Review: Papers and Proceedings*, **69**: 339–345.
Stiglitz, J.E. (1987) 'Competition and the number of firms in a market: Are duopolies more competitive than atomistic firms?' *Journal of Political Economy*, **95**: 1041–1061.
Stiglitz, J.E. (1988) 'Imperfect information in the product market', in: R. Schmalensee and R. Willing, eds., *Handbook of Industrial Organization*, Vol. 1. Amsterdam: North-Holland.
Sutton, J. (1980) 'A model of stochastic equilibrium in a quasi-competitive industry', *Review of Economic Studies*, **47**: 705–721.
Talman, G. (1992) 'Search from an unknown distribution: An explicit solution', *Journal of Economic Theory*, **57**: 141–157.
Varian, H.R. (1980) 'A model of sales', *American Economic Review*, **70**: 651–659.
Veendorp, E.C. (1984) 'Sequential search without reservation price', *Economics Letters* **6**: 53–58.
Vishwanath, T. (1988) 'Parallel search and information gathering', *American Economic Review: Papers and Proceedings*, **78**: 110–116.
Von zur Muehlen, P. (1980) 'Monopolistic competition and sequential search', *Journal of Economic Dynamics and Control*, **2**: 257–281.
Weitzman, M.L. (1979) 'Optimal search for the best alternative', *Econometrica*, **47**: 641–655.
Wernerfelt, B. (1988) 'General equilibrium with real time search in labor and product markets', *Journal of Political Economy*, **96**: 821–831.

Wilde, L.L. (1977) 'Labor market equilibrium under nonsequential search', *Journal of Economic Theory*, **16**: 373–393.
Wilde, L.L. (1980) 'On the formal theory of inspection and evaluation in product markets', *Econometrica*, **48**: 1265–1280.
Wilde, L.L. (1985) 'Equilibrium search models as simultaneous move games', Social Science Working Paper No. 584, California Institute of Technology.
Wilde, L.L. (1987) 'Comparison shopping as a simultaneous move game', California Institute of Technology, mimeo.
Wilde, L.L. and A. Schwartz (1979) 'Equilibrium comparison shopping', *Review of Economic Studies*, **46**: 543–554.
Wilson, R. (1987) 'Game-theoretic analyses of trading processes', in: T. Bewley, ed., *Advances in Economic Theory: Fifth World Congress*. Cambridge: Cambridge University Press.
Wolinsky, A. (1987) 'Matching, search, and bargaining', *Journal of Economic Theory* **32**: 311–333.
Wolinsky, A. (1988) 'Dynamic markets with competitive bidding', *Review of Economic Studies*, **55**: 71–84.
Wolinsky, A. (1990) 'Information revelation in a market with pairwise meetings', *Econometrica*, **58**, 1–24.

Chapter 28

GAME THEORY AND EVOLUTIONARY BIOLOGY

PETER HAMMERSTEIN

Max-Planck-Institut für Verhaltensphysiologie

REINHARD SELTEN

University of Bonn

Contents

1.	Introduction	931
2.	Conceptual background	932
	2.1. Evolutionary stability	932
	2.2. The Darwinian view of natural selection	933
	2.3. Payoffs	934
	2.4. Game theory and population genetics	935
	2.5. Players	936
	2.6. Symmetry	936
3.	Symmetric two-person games	937
	3.1. Definitions and notation	937
	3.2. The Hawk–Dove game	937
	3.3. Evolutionary stability	938
	3.4. Properties of evolutionarily stable strategies	940
4.	Playing the field	942
5.	Dynamic foundations	948
	5.1. Replicator dynamics	948
	5.2. Disequilibrium results	951
	5.3. A look at population genetics	952
6.	Asymmetric conflicts	962

*We are grateful to Olof Leimar, Sido Mylius, Rolf Weinzierl, Franjo Weissing, and an anonymous referee who all helped us with their critical comments. We also thank the Institute for Advanced Study Berlin for supporting the final revision of this paper.

Handbook of Game Theory, Volume 2, Edited by R.J. Aumann and S. Hart
© *Elsevier Science B.V., 1994. All rights reserved*

7.	Extensive two-person games	965
	7.1. Extensive games	965
	7.2. Symmetric extensive games	966
	7.3. Evolutionary stability	968
	7.4. Image confrontation and detachment	969
	7.5. Decomposition	970
8.	Biological applications	971
	8.1. Basic questions about animal contest behavior	972
	8.2. Asymmetric animal contests	974
	8.3. War of attrition, assessment, and signalling	978
	8.4. The evolution of cooperation	980
	8.5. The great variety of biological games	983
References		987

1. Introduction

The subject matter of evolutionary game theory is the analysis of conflict and cooperation in animals and plants. Originally game theory was developed as a theory of human strategic behavior based on an idealized picture of rational decision making. Evolutionary game theory does not rely on rationality assumptions but on the idea that the Darwinian process of natural selection drives organisms towards the optimization of reproductive success.

A seminal paper by Maynard Smith and Price (1973) is the starting point of evolutionary game theory but there are some forerunners. Fisher (1930, 1958) already used a game-theoretic argument in his sex ratio theory. Hamilton (1967) in a related special context conceived the notion of an unbeatable strategy. An unbeatable strategy can be described as a symmetric equilibrium strategy of a symmetric game. Trivers (1971) referred to supergame theory when he introduced the concept of reciprocal altruism. However, the efforts of the forerunners remained isolated whereas the conceptual innovation by Maynard Smith and Price immediately generated a flow of successful elaborations and applications. The book of Maynard Smith (1982a) summarizes the results of the initial development of the field. In the beginning there was little interaction between biologists and game theorists but nowadays the concept of an evolutionarily stable strategy and its mathematical exploration has been integrated into the broader field of non-cooperative game theory. An excellent overview concerning mathematical results is given by van Damme (1987) in Chapter 9 of his book on stability and perfection of Nash equilibria. Another overview paper with a similar orientation is due to Bomze (1986). However, it must be emphasized that the reader who is interested in substantial development, biological application, and conceptual discussion must turn to the biological literature which will be reviewed in Section 8.

The interpretation of game models in biology on the one hand and in economics and the social sciences on the other hand is fundamentally different. Therefore, it is necessary to clarify the conceptual background of evolutionary game theory. This will be done in the next section. We then proceed to introduce the mathematical definition of evolutionary stability for bimatrix games in Section 3; important properties of evolutionarily stable strategies will be discussed there. In Section 4 we shall consider situations in which the members of a population are not involved in pairwise conflicts but in a global competition among all members of the population. Such situations are often described by the words "playing the field". The mathematical definition o evolutionary stability for models of this kind will be introduced and its properties will be discussed. Section 5 deals with the dynamic foundations of evolutionary stability; most of the results covered concern a simple system of asexual reproduction called replicator dynamics; some remarks will be made about dynamic population genetics models of sexual reproduction. Section

6 presents two-sided asymmetric conflicts. It is first shown how asymmetric conflicts can be imbedded in symmetric games. A class of game models with incomplete information will be examined in which animals can find themselves in different roles such as owner and intruder in a territorial conflict. If the roles on both sides are always different, then an evolutionarily stable strategy must be pure. Section 7 is devoted to evolutionary stability in extensive games. problems arise with the usual normal form definition of an evolutionarily stable strategy. A concept which is better adapted to the extensive form will be defined and its properties will be discussed. In the last section some remarks will be made on applications and their impact on current biological throught.

2. Conceptual background

In biology strategies are considered to be inherited programs which control the individual's behavior. Typically one looks at a population of members of the same species who interact generation after generation in game situations of the same type. Again and again the joint action of mutation and selection replaces strategies by others with a higher reproductive success. This dynamic process may or may not reach a stable equilibrium. Most of evolutionary game theory focuses attention on those cases where stable equilibrium is reached. However, the dynamics of evolutionary processes in disequilibrium is also an active area of research (see Section 5).

2.1. Evolutionary stability

In their seminal paper John Maynard Smith and George R. Price (1973) introduced the notion of an evolutionarily stable strategy which has become the central equilibrium concept of evolutionary game theory. Consider a population in which all members play the same strategy. Assume that in this population a mutant arises who plays a different strategy. Suppose that initially only a very small fraction of the population plays the mutant strategy. The strategy played by the vast majority of the population is stable against the mutant strategy if in this situation the mutant strategy has the lower reproductive success. This has the consequence that the mutant strategy is selected against and eventually vanishes from the population. A strategy is called *evolutionarily stable* if it is stable, in the sense just explained, against any mutant which may arise.

A population state is *monomorphic* if every member uses the same strategy and *polymorphic* if more than one strategy is present. A mixed strategy has a monomorphic and a polymorphic interpretation. On the one hand we may think of a monomorphic population state in which every individual plays this mixed strategy.

On the other hand a mixed strategy can also be interpreted as a description of a polymorphic population state in which only pure strategies occur; in this picture the probabilities of the mixed strategy describe the relative frequencies of the pure strategies.

The explanation of evolutionary stability given above is monomorphic in the sense that it refers to the dynamic stability of a monomorphic population state against the invasion of mutants. A similar idea can be applied to a polymorphic population state described by a mixed strategy. In this polymorphic interpretation a potential mutant strategy is a pure strategy not represented in the population state. Stability of a polymorphic state requires not only stability against the invasion of mutants but also against small perturbations of the relative frequencies already present in the population.

Biologists are reluctant to relinquish the intuitive concept of evolutionary stability to a general mathematical definition since they feel that the great variety of naturally occurring selection regimes require an openness with respect to formalization. Therefore they do not always use the term evolutionarily stable strategy in the exact sense of the definition prevailing in the formal literature [Maynard Smith and Price (1973), Maynard Smith (1982)]. This definition and its connections to the intuitive notion of evolutionary stability will be introduced in Sections 3 and 4.

2.2. The Darwinian view of natural selection

Darwin's theory of natural selection is the basis of evolutionary game theory. A common misunderstanding of the Darwinian view is that natural selection optimizes the welfare of the species. In the past even eminent biologists explained phenomena of animal interaction by vaguely defined benefits to the species. It is not clear what the welfare of the species should be. Is it the number of individuals, the total biomass, or the expected survival of the species in the long run? Even if a reasonable measure of this type could be defined it is not clear how the interaction among species should result in its optimization.

The dynamics of selection among individuals within the same species is much quicker than the process which creates new species and eliminates others. This is due to the fact that the life span of an individual is negligibly short in comparison to that of the species. An adiabatic approximation seems to be justified. For the purpose of the investigation of species interaction equilibrium within the species can be assumed to prevail. This shows that the process of individual selection within the species is the more basic one which must be fully understood before the effects of species interaction can be explored. Today most biologists agree that explanations on the basis of individual selection among members of the same species are much more fruitful than arguments relying on species benefits [Maynard Smith (1976)].

In the 1960s a theory of group selection was proposed [(Wynne-Edwards (1962)] which maintains that evolution may favor the development of traits like restraint in reproduction which are favorable for a local group within a structured population even if they diminish the reproductive success of the individual. It must be emphasized that theoretical explanations of group selection can be constructed on the basis of individual selection. Inasfar as such explanations are offered the idea of group selection is not in contradiction to the usual Darwinian view. However, the debate on group selection has shown that extreme parameter constellations are needed in theoretical models in order to produce the phenomenon [Levins (1970), Boorman and Levitt (1972, 1973), Maynard Smith (1976), see also Grafen (1984) for recent discussions on the term group selection].

Only very few empirical cases of group selection are documented in the literature, e.g. the case of myxomatosis (a disease of rabbits in Australia). A quicker growth rate within the infected rabbit is advantageous for the individual parasite but bad for the group of parasites in the same animal since a shorter life span of the rabbit decreases the opportunities for infection of other rabbits [Maynard Smith (1989)].

The species and the group are too high levels of aggregation for the study of conflict in animals and plants. It is more fruitful to look at the individual as the unit of natural selection. Often an even more reductionist view is proposed in the literature; the gene rather than the individual is looked upon as the basic unit of natural selection [e.g. Williams (1966), Dawkins (1976, 1982)]. It must be admitted that some phenomena require an explanation in terms of genes which pursue their own interest to the disadvantage of the individual. For example, a gene may find a way to influence the process of meiosis in its favor; this process determines which of two genes of a pair of chromosomes in a parent is contributed to an egg or sperm. However, in the absence of strong hints in this direction one usually does not look for such effects in the explanation of morphological and behavioral traits in animals or plants. The research experience shows that the individual as the level of aggregation is a reasonable simplification. The significance of morphological and behavioral traits for the survival of the individual exerts strong pressure against disfunctional results of gene competition within the individual.

2.3. Payoffs

Payoffs in biological games are in terms of fitness, a measure of reproductive success. In many cases the fitness of an individual can be describd as the expected number of offspring. However, it is sometimes necessary to use a more refined definition of fitness. For example, in models for the determination of the sex ratio among offspring it is necessary to look more than one generation ahead and to count grandchildren instead of children [Fisher (1958), Maynard Smith (1982a)]. In models involving decisions on whether offspring should be born earlier or later in the lifetime of the mother it may be necessary to weigh earlier offspring more heavily than later ones. Under conditions of exteme variations of the environment which affect all individuals with the same strategy in the same way the expected logarithm of the number of offspring may be a better definition of fitness than the usual one [Gillespie (1977)].

The significance of the fitness concept lies in its ability to connect short run reproductive success with long run equilibrium properties. Darwinian theory is not tautological. It does not say that those survive who survive. Instead of this it derives the structure of long-run equilibrium from the way in which short-term reproductive success measured by fitness depends on the environment and the population state. However, as it has been explained above, different types of processes of natural selection may require different ways of making the intuitive concept of reproductive success more precise.

2.4. Game theory and population genetics

Biologists speak of frequency-dependent selection if the fitness of a type depends on the frequency distribution over types in the population. This does not necessarily mean that several types must be present at equilibrium. Frequency-dependent selection has been discussed in the biological literature long before the rise of evolutionary game theory. Game-theoretic problems in biology can be looked upon as topics of frequency-dependent selection and therefore some biologists feel that game theory does not add anything new to population genetics. However, it must be emphasized that the typical population genetics treatment of frequency-dependent selection focuses on the genetic mechanism of inheritance and avoids the description of complex strategic interaction. Contrary to this the models of evolutionary game theory ignore the intricacies of genetic mechanisms and focus on the structure of strategic interaction.

The empirical investigator who wants to model strategic phenomena in nature usually has little information on the exact way in which the relevant traits are inherited. Therefore game models are better adapted to the needs of empirical research in sociobiology and behavioral ecology than dynamic models in population genetics theory. Of course the treatment of problems in the foundation of evolutionary game theory may require a basis in population genetics. However, in applications it is often preferable to ignore foundational problems even if they are not yet completely solved.

In biology the word genotype refers to a description of the exact genetic structure of an individual whereas the term phenotype is used for the system of morphological and behavioral traits of an individual. Many genotypes may result in the same phenotype. The models of evolutionary game theory are called phenotypical since they focus on phenotypes rather than genotypes.

2.5. Players

The biological interpretation of game situations emphasizes strategies rather than players. If one looks at strategic interactions within a population it is important to know the relative frequencies of actions, it is less interesting to know which member plays which strategy. Therefore, the question who are the players is rarely discussed in the biological literature.

It seems to be adequate to think of a "player" as a randomly selected animal. There are two ways of elaborating this idea. Suppose that there are N animals in the population. We imagine that the game is played by N players who are randomly assigned to the N animals. Each player has equal chances to become each one of the N animals. We call this the "many-player interpretation".

Another interpretation is based on the idea that there are only a small number of players, for example 2, which are assigned to the two roles (e.g. owner and intruder) in a conflict at random. Both have the same chance to be in each of both roles. Moreover, there may be a universe of possible conflicts from which one is chosen with the appropriate probability. There may be incomplete information in the sense that the players do not know exactly which conflict has been selected when they have to make their decision. We call this the "few-player interpretation".

The few-player interpretation can be extended to conflicts involving more than two animals. The number of animals in the conflict may even vary between certain limits. In such cases the number n of players is the maximal number of animals involved in a conflict and in any particular conflict involving m animals m players are chosen at random and randomly assigned to the animals.

2.6. Symmetry

In principle evolutionary game theory deals only with fully symmetric games. Asymmetric conflicts are imbedded in symmetric games where each player has the same chance to be on each side of the conflict. Strategies are programs for any conceivable situation. Therefore one does not have to distinguish between different types of players. One might think that it is necessary to distinguish, for example, between male and female strategies. But apart from the exceptional case of sex-linked inheritance, one can say that males carry the genetic information for female behavior and vice versa.

The mathematical definition of evolutionary stability refers to symmetric games only. Since asymmetric conflicts can be imbedded in symmetric games, this is no obstacle for the treatment of asymmetric conflicts. In the biological literature we often find a direct treatment of asymmetric conflicts without any reference to the symmetrization which is implicitly used in the application of the notion of evolutionary stability to such situations.

3. Symmetric two-person games

3.1. Definitions and notation

A symmetric two-person game $G = (S, E)$ consists of a finite nonempty *pure strategy set* S and a *payoff function* E which assigns a real number $E(s, t)$ to every pair (s, t) of pure strategies in S. The number $E(s, t)$ is interpreted as the payoff obtained by a player who plays s against an opponent who plays t.

A *mixed strategy* q is a probability distribution over S. The probability assigned to a pure strategy s is denoted by $q(s)$. The set of all mixed strategies is denoted by Q. The payoff function E is extended in the usual way to pairs of mixed strategies (p, q).

A *best reply* to q is a strategy r which maximizes $E(\cdot, q)$ over Q. An *equilibrium point* is a pair (p, q) with the property that p and q are best replies to each other. A symmetric equilibrium point is an equilibrium point of the form (p, p).

A strategy r is a *strict best reply* to a strategy q if it is the only best reply to q. A strict best reply must be a pure strategy. An equilibrium point (p, q) is called strict if p and q are strict best replies to each other.

3.2. The Hawk–Dove game

Figure 1 represents a version of the famous Hawk–Dove game [Maynard Smith and Price (1973)]. The words Hawk and Dove refer to the character of the two strategies H and D and have political rather than biological connotations. The game describes a conflict between two animals of the same species who are

	H	D
H	$\frac{1}{2}(V-W)$, $\frac{1}{2}(V-W)$	V , 0
D	0 , V	$\frac{1}{2}V$, $\frac{1}{2}V$

Figure 1. The Hawk–Dove game

competing for a resource, for example a piece of food. Strategy H has the meaning of serious aggression, whereas D indicates peaceful behavior. If both choose H, the animals fight until one of them is seriously injured. Both contestants have the same probability to win the fight. The damage W caused by a serious wound is assumed to be higher than the value V of the resource. If only one of the animals plays the aggressive strategy H, this animal will win the resource and the other will flee. If both choose D, then some kind of unspecified random mechanism, for example a ritual fight, decides who gains the resource. Again, both players have the same chance to win. The numbers V and W are measured as changes in fitness. This is the usual interpretation of payoffs in biological games.

The game has two strict equilibrium points, namely (H, D) and (D, H), and one equilibrium point in mixed strategies, namely (r, r) with $r(H) = V/W$.

Only the symmetric equilibrium point is biologically meaningful, since the animals do not have labels 1 and 2. They cannot choose strategies dependent on the player number. Of course, one could think of correlated equilibria [Aumann (1974)], and something similar is actually done in the biological literature [Maynard Smith and Parker (1976), Hammerstein (1981)]. Random events like "being there first" which have no influence on payoffs may be used to coordinate the actions of two opponents. However, in a biological game the correlating random event should be modelled explicitly, since it is an important part of the description of the phenomenon.

3.3. Evolutionary stability

Consider a large population in which a symmetric two-person game $G = (S, E)$ is played by randomly matched pairs of animals generation after generation. Let p be the strategy played by the vast majority of the population, and let r be the strategy of a mutant present in a small frequency. Both p and r can be pure or mixed. For the sake of simplicity we assume non-overlapping generations in the sense that animals live only for one reproductive season. This permits us to model the selection process as a difference equation.

Let x_t be the relative frequency of the mutant in season t. The *mean strategy* q_t of the population at time t is given by

$$q_t = (1 - x_t)p + x_t r.$$

Total fitness is determined as the sum of a basic fitness F and the payoff in the game. The mutant r has the total fitness $F + E(r, q_t)$ and the majority strategy p has the total fitness $F + E(p, q_t)$. The biological meaning of fitness is expressed by the mathematical assumption that the growth factors x_{t+1}/x_t and $(1 - x_{t+1})/(1 - x_t)$ of the mutant subpopulation and the majority are proportional with the same proportionality factor to total fitness of r and p, respectively. This yields the following difference equation:

$$x_{t+1} = \frac{F + E(r, q_t)}{F + E(q_t, q_t)} x_t, \quad \text{for } t = 0, 1, 2, \ldots. \tag{1}$$

This process of inheritance describes an asexual population with randomly matched conflicts. We want to examine under what conditions x_t converges to 0 for $t \to \infty$. For this purpose, we look at the difference $x_t - x_{t+1}$:

$$x_t - x_{t+1} = \frac{E(q_t, q_t) - E(r, q_t)}{F + E(q_t, q_t)} x_t.$$

Obviously x_{t+1} is smaller than x_t if and only if the numerator of the fraction on the right-hand side is positive. We have

$$E(q_t, q_t) = (1 - x_t) E(p, q_t) + x_t E(r, q_t)$$

and therefore

$$E(q_t, q_t) - E(r, q_t) = (1 - x_t)[E(p, q_t) - E(r, q_t)].$$

This shows that $x_{t+1} < x_t$ holds if and only if the expression in square brackets on the right-hand side is positive, or equivalently if and only if the following inequality holds:

$$(1 - x_t)[E(p, p) - E(r, p)] + x_t[E(p, r) - E(r, r)] > 0. \tag{2}$$

It can also be seen that $x_{t+1} > x_t$ holds if and only if the opposite inequality is true.

Assume $E(p, p) < E(r, p)$. Then the process (1) does not converge to zero, since for sufficiently small x_t the expression on the left-hand side of inequality (2) is negative.

Now, assume $E(p, p) > E(r, p)$. Then for sufficiently small x_t the left-hand side of (2) is positive. This shows that an $\varepsilon > 0$ exists such that for $x_0 < \varepsilon$ the process (1) converges to zero.

Now, consider the case $E(p, p) = E(r, p)$. In this case the process (1) converges to zero if and only if $E(p, r) > E(r, r)$.

We say that p is *stable against* r if for all sufficiently small positive x_0 the process (1) converges to zero. What are the properties of a strategy p which is stable against every other strategy r? Our case distinction shows that p is stable against every other strategy r if and only if it is an evolutionarily stable strategy in the sense of the following definition:

Definition 1. An *evolutionarily stable strategy* p of a symmetric two-person game $G = (S, E)$ is a (pure or mixed) strategy for G which satisfies the following two conditions (a) and (b):

(a) *Equilibrium condition*: (p, p) is an equilibrium point.
(b) *Stability condition*: Every best reply r to p which is different from p satisfies

the following inequality:

$$E(p,r) > E(r,r) \tag{3}$$

The abbreviation ESS is commonly used for the term evolutionarily stable strategy.

A best reply to p which is different from p is called an alternative best reply to p. Since the stability condition only concerns alternative best replies, p is always evolutionarily stable if (p,p) is a strict equilibrium point.

Vickers and Cannings (1987) have shown that for every evolutionarily stable strategy p an ε exists such that inequality (2) holds for $0 < x_t < \varepsilon$ and every $r \neq p$. Such a bound ε is called a *uniform invasion barrier*.

Result 1. For all r different from p the process (1) converges to zero for all sufficiently small x_0 if and only if p is an evolutionarily stable strategy. Moreover, for every evolutionarily stable strategy a uniform invasion barrier ε can be found with the property that process (1) converges to zero fo all $0 < x_0 < \varepsilon$ for all r different from p.

The uniformity result of Vickers and Cannings (1987) holds for finite symmetric two-person games only. They present a counterexample with a countable infinity of strategies.

Comment. Result 1 shows that at least one plausible selection process justifies Definition 1 as an asymptotically stable dynamic equilibrium. However, it must be emphasized that this process is only one of many possible selection models, some of which will be discussed in Section 5.

Process (1) relates to the monomorphic interpretation of evolutionary stability which looks at an equilibrium state where all members of the population use the same strategy. This monomorphic picture seems to be adequate for a wide range of biological applications. In Section 5 we shall also discuss a polymorphic justification of Definition 1.

3.4. Properties of evolutionarily stable strategies

An evolutionarily stable strategy may not exist for symmetric two-person games with more than two pure strategies. The standard example is the Rock–Scissors–Paper game [Maynard Smith (1982a)]. This lack of universal existence is no weakness of the concept in view of its interpretation as a stable equilibrium of a dynamic pocess. Dynamic systems do not always have stable equilibria.

A symmetric two-person game may have more than one evolutionarily stable strategy. This potential multiplicity of equilibria is no drawback of the concept, again in view of its dynamic interpretation. The history of the dynamic system decides which stable equilibrium is eached, if any equilibrium is reached at all.

The *carrier* of a mixed strategy p is the set of pure strategies s with $p(s) > 0$. The *extended carrier* of p contains in addition to this all pure best replies to p. The following two results are essentially due to Haigh (1975).

Result 2. Let p be an ESS of $G = (S, E)$. Then G has no ESS r whose carrier is contained in the extended carrier of p. [See Lemma 9.2.4. in van Damme (1987).]

Result 3. A symmetric two-person game $G = (S, E)$ has at most finitely many evolutionarily stable strategies.

It can be seen easily that Result 2 is due to the fact that r would violate the stability condition (3) for p. Result 3 is an immediate consequence of Result 2, since a finite game has only finitely many carriers.

An ESS p is called *regular* if $p(s) > 0$ holds for every pure best reply s to p. In other words, p is regular if its extended carrier coincides with its carrier. One meets irregular ESSs only exceptionally. Therefore, the special properties of regular ESSs are of considerable interest.

Regularity is connected to another property called essentiality. Roughly speaking, an ESS p of $G = (S, E)$ is *essential* if for every payoff function E_+ close to E the game $G_+ = (S, E_+)$ has an ESS p_+ near to p. In order to make this explanation more precise we define a distance of E and E_+ as the maximum of $|E(s, t) - E_+(s, t)|$ over all s and t in S. Similarly, we define a distance of p and p_+ as the maximum of $|p(s) - p_+(s)|$ over all s in S.

An ESS p of $G = (S, E)$ is *essential* if the following condition is satisfied. For every $\varepsilon > 0$ we can find a $\delta > 0$ such that every symmetric two-person game $G_+ = (S, E_+)$ with the property that the distance between E and E_+ is smaller than δ has an ESS p_+ whose distance from p is smaller than ε. The ESS p is strongly essential if not only an ESS but a regular ESS p_+ of this kind can be found for G_+.

The definition of essentiality is analogous to the definition of essentiality for equilibrium points introduced by Wu Wen-tsün and Jian Jia-he (1962).

Result 4. If p is a regular ESS of $G = (S, E)$ then p is a strongly essential ESS of G. [See Selten (1983), Lemma 9.]

An irregular ESS need not be essential. Examples can be found in the literature [Selten (1983), van Damme (1987)].

A *symmetric equilibrium strategy* is a strategy p with the property that (p, p) is an equilibrium point. Haigh (1975) has derived a useful criterion which permits to decide whether a symmetric equilibrium strategy is a regular ESS. In order to express this criterion we need some further notation. Consider a symmetric equilibrium strategy p and let $s_1, \ldots, s_n$ be the pure strategies in the carrier of p.

Define
$$a_{ij} = E(s_i, s_j) \tag{4}$$
for $i, j = 1, \ldots, n$. The payoffs a_{ij} form an $n \times n$ matrix:
$$A = (a_{ij}). \tag{5}$$
We call this matrix the carier matrix of p even if it does not only depend on p but also on the numbering of the pure strategies in the carrier. The definition must be understood relative to a fixed numbering.

Let $D = (d_{ij})$ be the following $n \times (n-1)$ matrix with
$$d_{ij} = \begin{cases} 1, & \text{for } i = j < n, \\ -1, & \text{for } i = n, \\ 0, & \text{else.} \end{cases}$$

We say that Haigh's criterion is satisfied for p if the matrix $D^T A D$ is negative quasi-definite. The upper index T indicates transposition. Whether Haigh's criterion is satisfied or not does not depend on the numbering of the pure strategies in the carrier.

Result 5. Let p be a symmetric equilibrium strategy of $G = (S, E)$ whose carrier coincides with its extended carrier. Then p is a regular ESS if and only if it satisfies Haigh's criterion.

The notion of an ESS can be connected to various refinement properties for equilibrium points. Discussions of these connections can be found in Bomze (1986) and van Damme (1987). Here, we mention only the following result:

Result 6. Let p be a regular ESS p of $G = (S, E)$. Then (p, p) is a proper and strictly perfect equilibrium point of G. [Bomze (1986)].

4. Playing the field

In biological applications one often meets situations in which the members of a population are not involved in pairwise conflicts but in a global competition among all members of the population. Such situations are often described as "playing the field". Hamilton and May (1977) modelled a population of plants which had to decide on the fraction of seeds dropped nearby. The other seeds are equipped with an umbrella-like organ called pappus and are blown away by the wind. Fitness is the expected number of seeds which succeed to grow up to a plant. The mortality of seeds blown away with the wind is higher than that of seeds dropped nearby, but in view of the increased competition among the plant's own offspring, it is disadvantageous to drop too many seeds in the immediate neighborhood. In this example every plant competes with all other plants in the population. It is not possible to isolate pairwise interactions.

In the following we shall explain the structure of playing-the-field models. Consider a population whose members have to select a strategy p out of a strategy set P. Typically P is a convex subset of an Euclidean space. In special cases P may be the set of mixed strategies in a finite symmetric game. In many applications, like the one mentioned above, P is a biological parameter to be chosen by the members of the population.

For the purpose of investigating monomorphic stability of a strategy p we have to look at bimorphic population states in which a mutant q is represented with a small relative frequency ε whereas strategy p is represented with the frequency $1-\varepsilon$. Formally a bimorphic population state z can be described as a triple $z = (p, q, \varepsilon)$ with the interpretation that in the population p is used with probability $1-\varepsilon$ and q is used with probability ε. In order to describe the structure of a playing-the-field model we introduce the notion of a population game.

Definition 2. A *population game* $G = (P, E)$ consists of a strategy set P and a payoff function E with the following properties:

(i) P is a nonempty compact convex subset of a Euclidean space.
(ii) The payoff function $E(r; p, q, \varepsilon)$ is defined for all $r \in P$ and all bimorphic population states (p, q, ε) with $p, q \in P$ and $0 \leq \varepsilon \leq 1$.
(iii) The payoff function E has the property that $E(r; p, q, 0)$ does not depend on q.

It is convenient to use the notation $E(r, p)$ for $E(r; p, q, 0)$. We call $E(\cdot, \cdot)$ the *short payoff function* of the population game $G = (P, E)$.

Regardless of q all triples $(p, q, 0)$ describe the same bimorphic population state where the whole population plays p. Therefore, condition (iii) must be imposed on E.

A strategy r is a *best reply* to p if it maximizes $E(\cdot, p)$ over P; if r is the only strategy in P with this property, r is called a *strict best reply* to p. A strategy p is a (*strict*) *symmetric equilibrium strategy* of $G = (P, E)$ if it is a (strict) best reply to itself.

The book by Maynard Smith (1982a) offers the following definition of evolutionary stability suggested by Hammerstein.

Definition 3. A strategy p for $G = (P, E)$ is *evolutionarily stable* if the following two conditions are satisfied:

(a) *Equilibrium condition*: p is a best reply to p.
(b) *Stability condition*: For every best reply q to p which is different from p an $\varepsilon_q > 0$ exists such that the inequality

$$E(p; p, q, \varepsilon) > E(q; p, q, \varepsilon) \tag{6}$$

holds for $0 < \varepsilon < \varepsilon_q$.

In this definition the invasion barrier ε_q depends on q. This permits the possibility that ε_q becomes arbitrarily small as q approaches the ESS p. One may argue that therefore the definition of evolutionary stability given above is too weak. The following definition requires a stronger stability property.

Definition 4. An ESS p of $G = (P, E)$ is called an *ESS with uniform invasion barrier* if an $\varepsilon_0 > 0$ exists such that (6) holds for all $q \in P$ with $q \neq p$ for $0 < \varepsilon < \varepsilon_0$.

It can be seen immediately that a strict symmetric equilibrium strategy is evolutionarily stable, if E is continuous in ε everywhere. Unfortunately, it is not necessarily true that a strict symmetric equilibrium strategy is an ESS with uniform invasion barrier even if E has strong differentiability properties [Crawford (1990a, b)].

Population games may have special properties present in some but not all applications. One such property which we call *state substitutability* can be expressed as follows:

$$E(r; p, q, \varepsilon) = E(r, (1 - \varepsilon)p + \varepsilon q), \tag{7}$$

for $r, p, q \in P$ and $0 \leq \varepsilon \leq 1$.

Under the condition of state substitutability the population game is adequately described by the strategy space P and the short payoff function $E(\cdot, \cdot)$. However, state substitutability cannot be expected if strategies are vectors of biological parameters like times spent in mate searching, foraging, or hiding. State substitutability may be present in examples where P is the set of all mixed strategies arising from a finite set of pure strategies. In such cases it is natural to assume that $E(r, q)$ is a linear function of the first component r, but even then it is not necessarily true that E is linear in the second component q. Animal conflicts involving many participants may easily lead to payoff functions which are high-order polynomials in the probabilities for pure strategies. Therefore, the following result obtained by Crawford (1990a, b) is of importance.

Result 7. Let $G = (P, E)$ be a population game with the following properties:

  (i) P is the convex hull of finitely many points (the pure strategies).
  (ii) State substitutability holds for E.
  (iii) E is linear in the first component.

Then an ESS p is an ESS with uniform invasion barrier if a neighborhood U of p exists such that for every $r \in P$ the payoff function $E(r, q)$ is continuous in q within U. Moreover, under the same conditions a strict symmetric equilibrium strategy is an ESS of G.

A result similar to the first part of Result 7 is implied by Corollary 39 in Bomze and Pötscher (1989). Many other useful mathematical findings concerning playing-the-field models can be found in this book.

We now turn our attention to population games without state substitutability and without linearity assumptions on the payoff function. We are interested in the question under which condition a strict symmetric equilibrium strategy is an ESS with uniform invasion barrier. As far as we know, this important question has not been investigated in the literature. Unfortunately, our attempt to fill this gap will require some technical detail.

Let p be a strict symmetric equilibrium strategy of $G = (P, E)$. In order to facilitate the statement of continuity and differentiability conditions to be imposed on E, we define a deviation payoff function $F(t_1, t_2, r, \varepsilon)$:

$$F(t_1, t_2, r, \varepsilon) = E((1-t_1)p + t_1 r; p, (1-t_2)p + t_2 r, \varepsilon), \tag{8}$$

for $0 \leq t_i \leq 1$ ($i = 1, 2$) and $0 \leq \varepsilon \leq 1$.

The deviation payoff function is a convenient tool for the examination of the question what happens if the mutant strategy r is shifted along the line $(1-t)p + tr$ in the direction of p. Consider a mutant present in the relative frequency ε whose strategy is $(1-t)p + tr$. Its payoff in this situation is $F(t, t, r, \varepsilon)$. The payoff for the strategy p is $F(0, t, r, \varepsilon)$. Define

$$D(t, r, \varepsilon) = F(0, t, r, \varepsilon) - F(t, t, r, \varepsilon). \tag{9}$$

The intuitive significance of (9) lies in the fact that the mutant is selected against if and only if the payoff difference $D(t, r, \varepsilon)$ is positive.

We shall use the symbols F_i and F_{ij} with $i, j = 1, 2$ in order to denote the first and second derivatives with respect to t_1 and t_2. The indices 1 and 2 indicate differentiation with respect to t_1 and t_2, respectively.

Result 8. Let $G = (P, E)$ be a population game and let p be a strict symmetric equilibrium strategy for G. Assume that the following conditions (i) and (ii) are satisfied for the deviation payoff function F defined by (8):

(i) The deviation payoff function F is twice differentiable with respect to t_1 and t_2. Moreover, F and its first and second derivatives with respect to t_1 and t_2 are jointly continuous in t_1, t_2, r, ε.

(ii) A closed subset R of the border of the strategy set P exists (possibly the whole border) with the following properties (a) and (b):

(a) Every $q \in P$ permits a representation of the form

$$q = (1-t)p + tr, \quad \text{with } r \in R.$$

(b) The set R has two closed subsets R_0 and R_1 whose union is R, such that conditions (10) and (11) are satisfied:

$$\left. \frac{d^2 E((1-t)p + tr, p)}{dt^2} \right|_{t=0} < 0 \quad \text{for } r \in R_0, \tag{10}$$

$$\left.\frac{dE((1-t)p + tr, p)}{dt}\right|_{t=0} < -\alpha \quad \text{for } r \in R_1, \tag{11}$$

where $\alpha > $ is a constant which does not depend on r.

Under these assumptions p is an ESS with uniform invasion barrier.

Proof. Consider the first and second derivatives $D'(t, r, \varepsilon)$ and $D''(t, r, \varepsilon)$ respectively of $D(t, r, \varepsilon)$ with respect to t:

$$D'(t, r, \varepsilon) = F_2(0, t, r, \varepsilon) - F_1(t, t, r, \varepsilon) - F_2(t, t, r, \varepsilon), \tag{12}$$

$$D''(t, r, \varepsilon) = F_{22}(0, t, r, \varepsilon) - F_{11}(t, t, r, \varepsilon) - F_{12}(t, t, r, \varepsilon) \tag{13}$$
$$\qquad - F_{21}(t, t, r, \varepsilon) - F_{22}(t, t, r, \varepsilon).$$

In view of (iii) in Definition 2 we have

$$F(t_1, 0, r, \varepsilon) = E((1-t_1)p + t_1 r; p, p, \varepsilon) = E((1-t_1)p + t_1 r, p). \tag{14}$$

Since p is a strict symmetric equilibrium strategy, this shows that we must have $F_1(0, 0, r, \varepsilon) \leq 0$. Consequently, in view of (12) we can conclude

$$D'(0, r, \varepsilon) \geq 0 \quad \text{for } 0 \leq \varepsilon \leq 1. \tag{15}$$

Equation (13) yields

$$D''(0, r, \varepsilon) = -F_{11}(0, 0, r, \varepsilon) - 2F_{12}(0, 0, r, \varepsilon).$$

In view of (8) we have $F_2(t_1, t_2, r, 0) = 0$. This yields $F_{12}(0, 0, r, 0) = 0$. With the help of (iii) in the definition of a population game we can conclude

$$D''(0, r, 0) = -\left.\frac{d^2 E((1-t)p + tr, p)}{dt^2}\right|_{t=0} > 0, \quad \text{for } r \in R_0. \tag{16}$$

Similarly, (11) together with (12) yields

$$D'(0, r, 0) = -\left.\frac{dE((1-t)p + tr, p)}{dt}\right|_{t=0} > \alpha, \quad \text{for } r \in R_1. \tag{17}$$

Define

$$f(\mu) = \min_{\substack{0 \leq t \leq \mu \\ 0 \leq \varepsilon \leq \mu \\ r \in R_0}} D''(t, r, \varepsilon),$$

$$g(\mu) = \min_{\substack{0 \leq t \leq \mu \\ 0 \leq \varepsilon \leq \mu \\ r \in R_1}} D'(t, r, \varepsilon),$$

$$h(\mu) = \min_{\substack{\mu \leq t \leq 1 \\ 0 \leq \varepsilon \leq \mu \\ r \in R}} D(t, r, \varepsilon).$$

The functions $f(\mu), g(\mu)$, and $h(\mu)$ are continuous; this follows from the fact that the set of all triples (t, r, ε) over which the minimization is extended is compact, and that the function to be minimized is continuous by assumption in all three cases. We have $f(0) > 0$ in view of (16), and $g(0) > 0$ in view of (17). Moreover, we have

$$D(t, r, 0) = E(p, p) - E((1 - t)p + tr, p)$$

and therefore $h(\mu) > 0$ for sufficiently small $\mu > 0$, since p is a strict equilibrium point. Since f, g, and h are continuous, a number $\mu_0 > 0$ can be found such that for $0 \leq t \leq \mu_0$ and $0 \leq \varepsilon \leq \mu_0$ we have

$$D''(t, r, \varepsilon) > 0, \quad \text{for } r \in R_0,$$
$$D'(t, r, \varepsilon) > 0, \quad \text{for } r \in R_1,$$

and for $\mu_0 \leq t \leq 1$ and $0 \leq \varepsilon \leq \mu_0$ the following inequality holds:

$$D(t, r, \varepsilon) > 0$$

Together with (15) the last three inequalities show that μ_0 is a uniform invasion barrier. □

Remarks. Suppose that the strategy set P is one-dimensional. Then P has only two border points r_1 and r_2. Consider the case that $E(q, p)$ is differentiable with respect to q at (p, p). Then (10) is nothing else than the usual sufficient second-order condition for a maximum of $E(q, p)$ at $q = p$. It sometimes happens in applications [Selten and Shmida (1991)] that $E(q, p)$ has positive left derivative and negative right derivative with respect to q at $q = p$. In this case condition (11) is satisfied for r_1 and r_2. Result 8 is applicable here even if $E(q, p)$ is not differentiable with respect to q at $q = p$. It can be seen that (ii) in result 8 imposes only a mild restriction on the short payoff function E. The inequalities (10) and (11) with < replaced by ≤ are necessarily satisfied if p is a symmetric equilibrium strategy.

Eshel (1983) has defined the concept of a continuously stable ESS for the case of a one-dimensional strategy set [see also Taylor (1989)]. Roughly speaking, continuous stability means the following. A population which plays a strategy r slightly different from the ESS p is unstable in the sense that every strategy q sufficiently close to r can successfully invade if and only if q is between r and p. One may say that continuous stability enables the ESS to track small exogenous changes of the payoff function E. As Eshel has shown, continuous stability requires

$$E_{11}(q, p) + E_{12}(q, p) \leq 0,$$

if E is twice continuously differentiable with respect to q and p. Here, the indices 1 and 2 indicate differentiation with respect to the first and second argument, respectively. Obviously, this condition need not be satisfied for an ESS p for which the assumptions of Result 8 hold. Therefore, in the one-dimensional case an ESS with uniform invasion barrier need not be continuously stable.

Result 8 can be applied to models which specify only the short payoff function. It is reasonable to suppose that such models can be extended in a biologically meaningful way which satisfies condition (i) in Result 8. Condition (ii) concerns only the short payoff function.

Continuity and differentiability assumptions are natural for large populations. It is conceivable that the fitness function has a discontinuity at $\varepsilon = 0$; the situation of one mutant alone may be fundamentally different from that of a mutant in a small fraction of the population. However, in such cases the payoff for $\varepsilon = 0$ should be defined as the limit of the fitness for $\varepsilon \to 0$, since one is not really interested in the fitness of an isolated mutant.

It is doubtful whether the population game framework can be meaningfully applied to small finite populations. Mathematical definitions of evolutionary stability for such populations have been proposed in the literature [Schaffer (1988)]. However, these definitions do not adequately deal with the fact that in small finite populations [or, for that matter, in large populations – see Foster and Young (1990) and Young and Foster (1991)] the stochastic nature of the genetic mechanism cannot be neglected.

5. Dynamic foundations

Evolutionarily stable strategies are interpreted as stable results of processes of natural selection. Therefore, it is necessary to ask the question which dynamic models justify this interpretation. In Section 3.3 we have already discussed a very simple model of monomorphic stability. In the following we shall first discuss the replicator dynamics which is a model with continuous reproduction and exact asexual inheritance. Later we shall make some remarks on processes of natural selection in sexually reproducing populations without being exact in the description of all results.

We think that a distinction can be made between evolutionary game theory and its dynamic foundations. Therefore, our emphasis is not on the subject matter of this section. An important recent book by Hofbauer and Sigmund (1988) treats problems in evolutionary game theory and many other related subjects from the dynamic point of view. We recommend this book to the interested reader.

5.1. Replicator dynamics

The *replicator dynamics* has been introduced by Taylor and Jonker (1978). It describes the evolution of a polymorphic population state in a population whose members are involved in a conflict describd by a symmetric two-person game

$G = (S, E)$. Formally, the population state is represented by a mixed strategy q for G. The replicator dynamics is the following dynamic system:

$$\dot{q}(s) = q(s)[E(s, q) - E(q, q)], \quad \text{for all } s \in S. \tag{18}$$

As usually $\dot{q}(s)$ denotes the derivative of $q(s)$ with respect to time.

The replicator dynamics can be intuitively justified as follows. By a similar reasoning as in Section 3.3 concerning the process (1), in a discrete time model of non overlapping generations we obtain

$$q_{t+1}(s) = \frac{F + E(s, q_t)}{F + E(q_t, q_t)} q_t(s).$$

This yields

$$q_{t+1}(s) - q_t(s) = \frac{E(s, q_t) - E(q_t, q_t)}{F + E(q_t, q_t)} q_t(s).$$

If the changes from generation to generation are small, this difference equation can be approximated by the following differential equation:

$$\dot{q}(s) = \frac{E(s, q) - E(q, q)}{F + E(q, q)} q(s).$$

The denominator on the right-hand side is the same for all s and therefore does not influence the orbits. Therefore, this differential equation is equivalent to (18) as far as orbits are concerned.

A strategy q with $\dot{q}(s) = 0$ for all s in S is called a *dynamic equilibrium*. A dynamic equilibrium q is called *stable* if for any neighborhood U of q there exists a neighborhood V of q contained in U such that any trajectory which starts in V remains in U. A dynamic equilibrium q is called *asymptotically stable* if it is stable and if in addition to this there exists a neighborhood U of q such that any trajectory of (18) that starts in U converges to q.

It can be seen immediately that every pure strategy is a dynamic equilibrium of (18). Obviously, the property of being a dynamic equilibrium without any additional stability properties is of little significance.

A very important stability result has been obtained by Taylor and Jonker (1978) for the case of a regular ESS. Later the restriction of regularity has been removed by Hofbauer et al. (1979) and Zeeman (1981). This result is the following one:

Result 9. A strategy p for a symmetric two-person game $G = (S, E)$ is asymptotically stable with respect to the replicator dynamics if it is an ESS of G.

A strategy q can be asymptotically stable without being an ESS. Examples can be found in the literature [e.g. van Damme (1987), Weissing (1991)]. The question arises whether in view of such examples the ESS concept is a satisfactory static substitute for an explicit dynamic analysis. In order to examine this problem it is necessary to broaden the framework. We have to look at the possibility that not only pure but also mixed strategies can be represented as types in the population.

A nonempty finite subset $R = \{r_1, \ldots, r_n\}$ of the set of all mixed strategies Q will be called a *repertoire*. The pure or mixed strategies r_i are interpreted as the types which are present in the population. For $i = 1, \ldots, n$ the relative frequency of r_i will be denoted by x_i. For a population with repertoire R a *population state* is a vector $x = (x_1, \ldots, x_n)$. The symbol q_x will be used for the mean strategy of the population at state x. The replicator dynamics for a population with repertoire R can now be described as follows:

$$\dot{x}_i = x_i[E(r_i, q_x) - E(q_x, q_x)] \quad \text{for } i = 1, \ldots, n, \tag{19}$$

with

$$q_x = \sum_{i=1}^{n} x_i r_i. \tag{20}$$

Two population states x and x' which agree with respect to their mean strategy are phenotypically indistinguishable in the sense that in both population states the pure strategies are used with the same probability by a randomly selected individual. Therefore we call the set of all states x connected to the same mean strategy q the *phenotypical equivalence class* for q.

Consider a repertoire $R = \{r_1, \ldots, r_n\}$ for a symmetric two-person game $G = (S, E)$. Let x^j be the population state whose jth component is 1; this is the population state corresponding to r_j. We say that r is *phenotypically attractive in R* if r belongs to the convex hull of R and a neighborhood U of r exists such that every trajectory of (19) starting in U converges to the phenotypical equivalence class of r. This definition is due to Franz Weissing in his unpublished diploma thesis.

Result 10. A (pure or mixed) strategy p for a symmetric two-person game $G = (S, E)$ is stable and phenotypically attractive in every repertoire R whose convex hull contains p if and only if it is an ESS of G.

In a paper in preparation this result will be published by Weissing (1990). Cressman (1990) came independently to the same conclusion. He coined the notion of "strong stability" in order to give an elegant formulation of this result [see also Cressman and Dash (1991)].

Weissing proves in addition to Result 10 that a trajectory which starts near to the phenotypical equivalence class of the ESS ends in the phenotypical equivalence class at a point which, in a well defind sense, is nearer to the ESS than the starting point. This tendency towards monomorphism has already been observed in special examples in the literature [Hofbauer and Sigmund (1988)]. Weissing also shows that an asymptotically stable dynamic equilibrium p with respect to (18) is not stable with respect to (19) in a repertoire which contains p and an appropriately chosen mixed strategy q. This shows that an asymptotically stable dynamic equilibrium with respect to (18) which fails to be an ESS can be destabilized by mutations, whereas the same is not true for an ESS. All these results together provide a strong justification for the view that at least as far as the replicator dynamics is concerned, the ESS concept is a satisfactory static substitute for an explicit dynamic analysis. However, it must be kept in mind that this view is based on the originally intended interpretation of an ESS as a strategy which is stable against the invasion of mutations and not as a polymorphic equilibrium in a temporarily existing repertoire.

5.2. Disequilibrium results

We now turn our attention to disequilibrium properties of the replicator dynamics. Schuster et al. (1981) have derived a very interesting result which shows that under certain conditions a completely mixed symmetric equilibium strategy can be interpreted as a time average. An equilibrium strategy is called *completely mixed* if it assigns positive probabilities to all pure strategies s. The *omega-limit* of an orbit q_t is the set of all its accumulation points. With the help of these definitions we are now able to state the result of Schuster et al. [see also Hofbauer and Sigmund (1988) p. 136].

Result 11. Let $G = (S, E)$ be a symmetric two-person game with one and only one completely mixed equilibrium strategy p, and let q_t be an orbit starting in $t = 0$ whose omega-limit is in the interior of the set of mixed strategies Q. Then the following is true:

$$\lim_{T \to \infty} \frac{1}{T} \int_0^T q_t(s) \, dt = p(s), \quad \text{for all } s \text{ in } S. \tag{21}$$

Schuster et al. (1979) have introduced a very important concept which permits statements for certain types of disequilibrium behavior of dynamical systems. For the special case of the replicator dynamics (18) this concept of permanence is defined as follows. The system (18) is *permanent* if there exists a compact set K in the interior of the set Q of mixed strategies such that all orbits starting in the interior of Q end up in K.

Permanence means that none of the pure strategies in S will vanish in the population if initially all of them are represented with positive probabilities. The following result [Theorem 1 in Chapter 19.5 in Hofbauer and Sigmund (1988)] connects permanence to the existence of a completely mixed symmetric equilibrium strategy.

Result 12. If the system (18) is permanent then the game $G = (S, E)$ has one and only one completely mixed symmetric equilibrium strategy. Moreover, if (18) is permanent, equation (21) holds for every orbit q_t in the interior of Q.

In order to express the next result, we introduce some definitions. Consider a symmetric two-person game $G = (S, E)$. For every nonempty proper subset S' of S we define a restricted game $G' = (S', E')$. The payoff function E' as a function of pairs of pure strategies for G' is nothing else than the restriction of E to the set of these pairs. We call an equilibrium point of a restricted game $G' = (S', E')$ a *border pre-equilibrium* of G. The word border emphasizes the explicit exclusion of the case $S = S'$ in the definition of a restricted game. A border pre-equilibrium may or may ot be an equilibrium point of G. Note that the border pre-equilibria are the dynamic equilibria of (18) on the boundary of the mixed strategy set. The pure strategies are special border pre-equilibria. The following intriguing result is due to Jansen (1986) [see also Theorem 1 in Hofbauer and Sigmund (1988) p. 174].

Result 13. The replicator system (18) is permanent for a game $G = (S, E)$ if there exists a completely mixed strategy p for G such that

$$E(p, q) > E(q, q) \tag{22}$$

holds for all border pre-equilibria q of G.

Superficially inequality (22) looks like the stability condition (3) in the definition of an ESS. However, in (22) p is not necessarily a symmetric equilibrium strategy, and q is not necessarily an alternative best reply to p. A better interpretation of (22) focuses on the fact that all border pre-equilibria q are destabilized by the same completely mixed strategy p. We may say that Result 13 requires the existence of a completely mixed universal destabilizer of all border pre-equilibria.

Suppose that p is a completely mixed ESS. Then (22) holds for all border pre-equilibria in view of (3), since they are alternative best replies. The proof of Result 9 shows that a completely mixed ESS is globally stable. This is a special case of permanence. The significance of (22) lies in the fact that it also covers cases where the replicator dynamics (18) does not converge to an equilibrium. In particular, Result 13 is applicable to symmetric two-person games for which no ESS exists.

5.3. A look at population genetics

The replicator dynamics describes an asexual population, or more precisely a population in which, apart from mutations, genetically each individual is an exact copy of its parent. The question arises whether results about the replicator dynamics can be transferred to more complex patterns of inheritance. The investigation of such processes is the subject matter of population genetics. An introduction to population genetic models is beyond our scope. We shall only explain some game-theoretically interesting results in this area.

Hines and Bishop (1983, 1984a, b) have investigated the case of strategies controlled by one gene locus in a sexually reproducing diploid population. A gene locus is a place on the chromosome at which one of several different alleles of a gene can be located. The word *diploid* indicates that an individual carries each chromosome twice but with possibly different alleles at the same locus.

It has been shown by Hines and Bishop that an ESS has strong stability properties in their one-locus continuous selection model. However, they also point out that the set of all population mean strategies possible in the model is not necessarily convex. Therefore, the population mean strategy can be "trapped" in a pocket even if an ESS is feasible as a population mean strategy. The introduction of new mutant alleles, however, can change the shape of the set of feasible mean strategies. Here we shall not describe the results for one-locus models in detail. Instead of this we shall look at a discrete time two-locus model which contains a one-locus model as a special case.

We shall now describe a standard two-locus model for viability selection in a sexually reproducing diploid population with random mating. We first look at the case without game interaction in which fitnesses of genotypes are exogenous and constant. Viability essentially is the probability of survival of the carrier of a genotype. Viability selection means that effects of selection on fertility or mating success are not considered.

Let $A_1, \ldots, A_n$ be the possible alleles for a locus A, and $B_1, \ldots, B_m$ be the alleles for a locus B. For the sake of conveying a clear image we shall assume that both loci are linked which means that both are on the same chromosome. The case without linkage is nevertheless covered by the model as a special case. An individual carries pairs of chromosomes, therefore, a genotype can be expressed as a string of symbols of the form $A_i B_j / A_k B_l$. Here, A_i and B_j are the alleles for loci A and B on one chromosome, and A_k and B_l are the alleles on both loci on the other chromosome. Since chromosomes in the same pair are distinguished only by the alleles carried at their loci, $A_i B_j / A_k B_l$ and $A_k B_l / A_i B_j$ are not really different genotypes, even if they are formally different. Moreover, it is assumed that the effects of genotypes are *position-independent* in the sense that $A_i B_l / A_k B_j$ has the same fitness as $A_i B_j / A_k B_l$. The fitness of a genotype $A_i B_j / A_k B_l$ is denoted by w_{ijkl}. For the reasons explained above we have

$$w_{ijkl} = w_{ilkj} = w_{kjil} = w_{klij}.$$

An offspring has one chromosome of each of its parents in a pair of chromosomes. A chromosome received from a parent can be a result of recombination which means that the chromosomes of the parent have broken apart and patched together such that the chromosome transmitted to the offspring is composed of parts of both chromosomes of the parent. In this way genotype $A_i B_j / A_k B_l$ may transmit a chromosome A_iB_l to an offspring. This happens with probability $r \leq \frac{1}{2}$ called the *recombination* rate. The relative frequency of chromosomes with the same allele combination at time t is denoted by $x_{ij}(t)$. The model describes a population with nonoverlapping generations by a system of difference equations.

$$x_{ij}(t+1) = \frac{1}{W(t)}\left(r\sum_{kl} w_{ilkj}x_{il}(t)x_{kj}(t) + (1-r)\sum_{kl} w_{ijkl}x_{ij}(t)x_{kl}(t)\right],$$

for $i = 1,\ldots,n$ and $j = 1,\ldots,m$. (23)

Here, $W(t)$ is the mean fitness in the population:

$$W(t) = \sum_{ijkl} w_{ijkl}x_{ij}(t)x_{kl}(t).$$

The case that the two loci are on different chromosomes is covered by $r = \frac{1}{2}$. In model (23) the fitness w_{ijkl} of a genotype is constant. The selection described by the model is frequency-independent in the sense that fitnesses of genotypes do not depend on the frequencies $x_{ij}(t)$.

Moran (1964) has shown that in this model natural selection does not guarantee that the mean fitness of the population increases in time. It may even happen that the mean fitness of the population decreases until a minimum is reached. The same is true for the multilocus generalization of (23) and in many other population genetics models [Ewens (1968), Karlin (1975)]. Generally it cannot be expected even in a constant environment that the adaptation of genotype frequencies without any mutation converges to a population strategy which optimizes fitness.

The situation becomes even more difficult if game interaction is introduced into the picture. One cannot expect that the adaptation of genotype frequencies alone without any mutations moves results in convergence to an ESS unless strong assumptions are made. Suppose that the fitnesses w_{ijkl} depend on the strategic interaction in a symmetric two-person game $G = (S, E)$. Assume that a genotype A_iB_j/A_kB_l plays a mixed strategy u_{ijkl} in Q. In accordance with the analogous condition for the fitnesses w_{ijkl} assume

$$u_{ijkl} = u_{ilkj} = u_{kjil} = u_{klij}.$$

Define

$$w_{ijkl}(t) = F + E(u_{ijkl}, q_t),$$ (24)

with

$$q_t(s) = \sum_{ijkl} x_{ij}(t)x_{kl}(t)u_{ijkl}(s), \quad \text{for all } s \text{ in } S.$$ (25)

Here, q_t is the mean strategy of the population. If in (23) and in the definition of $W(t)$ the w_{ijkl} are replaced by $w_{ijkl}(t)$ we obtain a new system to which we shall refer as the *system (23) with frequency-dependent selection*.

This system has been investigated by Ilan Eshel and Marcus Feldman (1984).

It is very difficult to obtain any detailed information about its dynamic equilibria. Eshel and Feldman avoid this problem. They assume that originally only the alleles $A_1,\ldots,A_{n-1}$ are present at locus A. They look at a situation where a genotype frequency equilibrium has been reached and examine the effects of the introduction of a new mutant allele A_n at the locus A. Initially A_n appears in a low frequency at a population state near the equilibrium.

The ingenious idea to look at the effects of a mutant at a genotype frequency equilibrium is an important conceptual innovation. Eshel (1991) emphasizes that it is important to distinguish between the dynamics of genotype frequencies without mutation and the much slower process of gene substitution by mutation. The dynamics of genotype frequencies without mutation cannot be expected to converge towards a game equilibrium unless special conditions are satisfied. Evolutionary game theory is more concerned about equilibria which are not only stable with respect to the dynamics of genotype frequencies without mutation, but also with respect to gene substitution by mutation. This stability against mutations is often referred to as external stability [Liberman (1983)]. Surprisingly, the conceptual innovation by Eshel and Feldman (1984) also helps to overcome the analytical intractability of multilocus models, since their questions permit answers without the calculation of genotype frequencies at equilibrium.

A *population state* of model (23) is a vector $x=(x_{11},\ldots,x_{nm})$ whose components x_{ij} are the relative frequencies of the allele frequencies A_iB_j. The population mean strategy $q(x)$ is defined analogously to (25). For the purpose of examining qustions of external stability we shall consider the entrance of a mutant allele on locus A. We assume that originally only the alleles $A_1,\ldots,A_{n-1}$ are present. Later a mutant allele A_n enters the system. We assume that before the entrance of A_n the system has reached a dynamic equilibrium y with a population mean strategy p.

A particular specification of the frequency-dependent system (23) is described by a symmetric two-person game $G=(S,E)$ and a system of genotype strategies u_{ijkl}. It will be convenient to describe this system by two arrays, the *inside genotype strategy array*

$$U = (u_{ijkl}), \quad i,k = 1,\ldots,n-1, \quad j,l = 1,\ldots,m$$

and the *outside genotype strategy array*

$$U_n = (u_{njkl}), \quad k = 1,\ldots,n, \quad j,l = 1,\ldots,m.$$

The outside array contains all strategies of genotypes carrying the mutant allele (position independence requires $u_{njkl} = u_{kjnl}$). The inside array contains all strategies for genotypes without the mutant allele. A specific frequency-dependent system can now be described by a triple (G, U, U_n). We look at the numbers m,n and r as arbitrarily fixed with $n \geqslant 2$ and $m \geqslant 1$ and r in the closed interval between 0 and $\frac{1}{2}$.

We say that a sequence $q_0, q_1, \ldots$ of population mean strategies is *generated* by $x(0)$ in (G, U, U_n) if for $t = 0, 1, \ldots$ that strategy q_t is the population mean strategy of $x(t)$ in the sequence $x(0), x(1), \ldots$ satisfying (23) for this specification. For any two strategies p and q for G the symbol $|p-q|$ denotes the maximum of all absolute differences $|p(s)-q(s)|$ over $s \in S$. Similarly, for any population states x and y the symbol $|x-y|$ denotes the maximum of all absolute differences $|x_{ij} - y_{ij}|$. An *inside population state* is a population state $x = (x_{11}, \ldots, x_{nm})$ with

$$x_{nj} = 0 \quad \text{for } j = 1, \ldots, m.$$

With these notations and ways of speaking we are now ready to introduce our definition of external stability. This definition is similar to that of Liberman (1988) but with an important difference. Liberman looks at the stability of population states and requires convergence to the original inside population state y after the entrance of A_n in a population state near y. We think that Libermans definition is too restrictive. Therefore we require phenotypic attractiveness in the sense of Weissing (see Section 5.1) instead of convergence. The stability of the distribution of genotypes is less important than the stability of the population mean strategy.

Definition 5. Let $y = (y_{11}, \ldots, y_{n-1,m}, 0, \ldots, 0)$ be an inside population state. We say that y is *phenotypically externally stable* with respect to the game $G = (S, E)$ and the inside genotype strategy array U if for every U_n the specification (G, U, U_n) of (23) has the following property: For every $\varepsilon > 0$ a $\delta > 0$ can be found such that for every population state $x(0)$ with $|x(0) - y| < \delta$ the sequence of population mean strategies $q_0, q_1, \ldots$ generated by $x(0)$ satisfies two conditions (i) and (ii) with respect to the population mean strategy p of y:

(i) For $t = 0, 1, \ldots$ we have $|q_t - p| < \varepsilon$.
(ii) $\lim_{t \to \infty} q_t = p$.

Eshel and Feldman (1984) have developed useful methods for the analysis of the linearized model (23). However, as far as we can see their results do not imply necessary or sufficient conditions for external stability. Here we shall state a necessary condition for arbitrary inside population states and a necessary and sufficient condition for the case of a monomorphic population. The word monomorphic means that all genotypes not carrying the mutant allele play the same strategy. We shall not make use of the linearization technique of Eshel and Feldman (1984) and Liberman (1988).

Result 14. If an inside population state y is externally stable with respect to a game $G = (S, E)$ and an inside genotype strategy array U, then the population mean strategy p of y is a symmetric equilibrium strategy of G.

Proof. Assume that p fails to be a symmetric equilibrium strategy of G. Let s be a pure best reply to p. Consider a specification (G, U, U_n) of (23) with

$$u_{njkl} = s, \quad \text{for } j, l = 1, \ldots, m, \quad k = 1, \ldots, n.$$

Equations (24) yields

$$w_{njkl}(t) = F + E(s, q_t), \quad \text{for } j, l = 1, \ldots, m, \quad k = 1, \ldots, n.$$

For $i = n$ Eq. (23) assumes the following form:

$$x_{nj}(t+1) = \frac{F + E(s, q_t)}{F + E(q_t, q_t)} \sum_{kl} [rx_{nl}(t)x_{kj}(t) + (1-r)x_{nj}(t)x_{kl}(t)], \quad \text{for } j = 1, \ldots, m. \tag{26}$$

Define

$$a(t) = \sum_{i=1}^{n-1} \sum_{j=1}^{m} x_{ij}(t), \tag{27}$$

$$b(t) = \sum_{j=1}^{m} x_{nj}(t). \tag{28}$$

We call $a(t)$ the *inside part* of $x(t)$ and $b(t)$ the *outside part* of $x(t)$. In view of

$$\sum_{k=1}^{n} \sum_{l=1}^{m} x_{nl}(t)x_{kj}(t) = b(t) \sum_{k=1}^{n} x_{kj}(t)$$

and

$$\sum_{k=1}^{n} \sum_{l=1}^{m} x_{nj}(t)x_{kl}(t) = x_{nj}(t)$$

the summation over the Eqs. (26) for $j = 1, \ldots, m$ yields

$$b(t+1) = \frac{F + E(s, q_t)}{F + E(q_t, q_t)} b(t). \tag{29}$$

Since s is a best reply to p and p is not a best reply to p, we have

$$E(s, p) > E(p, p).$$

Therefore, we can find an $\varepsilon > 0$ such that $E(s, q) > E(p, q)$ holds for all q with $|p - q| < \varepsilon$. Consider a population state $x(0)$ with $b(0) > 0$ in the ε − neighborhood of p. We examine the process (23) starting at $x(0)$. In view of the continuity of $E(\cdot, \cdot)$ we can find a constant $g > 0$ such that

$$\frac{F + E(s, q_t)}{F + E(q_t, q_t)} > 1 + g$$

holds for $|q_t - p| < \varepsilon$. Therefore eventually q_t must leave the ε − neighborhood of p. This shows that for the ε under consideration no δ can be found which satisfies the requirement of Definition 5. □

The inside genotype strategy array $U = (u_{ijkl})$ with

$$u_{ijkl} = p, \quad \text{for } i, k = 1, \ldots, n-1, \quad j, l = 1, \ldots, m$$

is called the *strategic monomorphism* or shortly the *monomorphism* of p. A monomorphism U of a strategy p for $G = (S, E)$ is called *externally stable* if every inside population state y is externally stable with respect to G and U.

If only one allele is present on each locus one speaks of *fixation*. A monomorphism in our sense permits many alleles at each locus. The distinction is important since in the case of a monomorphism the entrance of a mutant may create a multitude of new genotypes with different strategies and not just two as in the case of fixation. Maynard Smith (1989) has pointed out that in the case of fixation the ESS-property is necessary and sufficient for external stability against mutants which are either recessive or dominant. The entrance of a mutant at fixation is in essence a one-locus problem. Contrary to this the entrance of a mutant at a monomorphism in a two-locus model cannot be reduced to a one-locus problem.

Result 15. Let p be a pure or mixed strategy of $G = (S, E)$. The monomorphism of p is externally stable if and only if p is an ESS of G.

Proof. Consider a specification (G, U, U_n) of (23), where U is the monomorphism of p. Let $x(0), x(1), \ldots$ be a sequence satisfying (23) for this specification. Let the inside part $a(t)$ and the outside part $b(t)$ be defined as in (27) and (28). A genotype $A_i B_j / A_k B_l$ is called *monomorphic* if we have $i < n$ and $k < n$. The joint relative frequency of all monomorphic genotypes at time t is $a^2(t)$. A genotype $A_i B_j / A_k B_l$ is called a *mutant heterozygote* if we have $i = n$ and $k < n$ or $i < n$ and $k = n$. A genotype $A_i B_j / A_k B_l$ is called a *mutant homozygote* if $i = k = n$ holds.

At time t the three classes of monomorphic genotypes, mutant heterozygotes, and mutant homozygotes have the relative frequencies $a^2(t)$, $2a(t)b(t)$, and $b^2(t)$, respectively. It is useful to look at the average strategies of the three classes. The average strategy of the monomorphic genotype is p. The average strategies u_t of the mutant heterozygotes and v_t of the mutant homozygotes are as follows:

$$u_t = \frac{1}{a(t)b(t)} \sum_{j=1}^{m} \sum_{k=1}^{n-1} \sum_{l=1}^{m} x_{nj}(t) x_{kl}(t) u_{njkl}, \tag{30}$$

$$v_t = \frac{1}{b^2(t)} \sum_{j=1}^{m} \sum_{l=1}^{m} x_{nj}(t) x_{nl}(t) u_{njnl}. \tag{31}$$

The alleles $A_1, \ldots, A_{n-1}$ are called *monomorphic*. We now look at the average strategy α_t of all monomorphic alleles and the average strategy β_t of the mutant allele A_n at time t.

$$\alpha_t = a(t)p + b(t)u_t, \tag{32}$$

$$\beta_t = a(t)u_t + b(t)v_t. \tag{33}$$

A monomorphic allele has the relative frequency $a(t)$ of being in a monomorphic genotype and the relative frequency $b(t)$ of being in a mutant heterozygote. Similarly, the mutant is in a mutant heterozygote with relative frequency $a(t)$ and in a mutant homozygote with frequency $b(t)$. The mean strategy q_t of the population satisfies the following equations:

$$q_t = a(t)\alpha_t + b(t)\beta_t, \tag{34}$$

$$q_t = a^2(t)p + 2a(t)b(t)u_t + b^2(t)v_t. \tag{35}$$

We now look at the relationship between $a(t)$ and $a(t+1)$. Equation (23) yields:

$$a(t+1) = \frac{F + E(p, q_t)}{F + E(q_t, q_t)} \sum_{i=1}^{n-1} \sum_{j=1}^{m} \sum_{k=1}^{n-1} \sum_{l=1}^{m} [r x_{il}(t) x_{kj}(t) + (1-r) x_{ij}(t) x_{kl}(t)]$$

$$+ \frac{1}{F + E(q_t, q_t)} \sum_{i=1}^{n-1} \sum_{j=1}^{m} \sum_{l=1}^{m} r[F + E(u_{ilnj}, q_t)] x_{il}(t) x_{nj}(t)$$

$$+ \frac{1}{F + E(q_t, q_t)} \sum_{i=1}^{n-1} \sum_{j=1}^{m} \sum_{l=1}^{m} (1-r)[F + E(u_{ijnl}, q_t)] x_{ij}(t) x_{nl}(t).$$

It can be seen without difficulty that this is equivalent to the following equation:

$$a(t+1) = \frac{1}{F + E(q_t, q_t)} \{a^2(t)[F + E(p, q_t)] + a(t)b(t)[F + E(u_t, q_t)]\}.$$

In view of the definition of α_t this yields

$$a(t+1) = \frac{F + E(\alpha_t, q_t)}{F + E(q_t, q_t)} a(t).$$

It follows by (34) that we have

$$\frac{F + E(\alpha_t, q_t)}{F + E(q_t, q_t)} = 1 + \frac{E(\alpha_t, q_t) - E(q_t, q_t)}{F + E(q_t, q_t)}$$

$$= 1 + \frac{b(t) E(\alpha_t, q_t) - b(t) E(\beta_t, q_t)}{F + E(q_t, q_t)}.$$

This yields (36). With the help of $a(t) + b(t) = 1$ we obtain a similar equation for $b(t)$.

$$a(t+1) = \left(1 + b(t) \frac{E(\alpha_t, q_t) - E(\beta_t, q_t)}{F + E(q_t, q_t)}\right) a(t), \tag{36}$$

$$b(t+1) = \left(1 + a(t) \frac{E(\beta_t, q_t) - E(\alpha_t, q_t)}{F + E(q_t, q_t)}\right) b(t). \tag{37}$$

Obviously, the difference $E(\alpha_t, q_t) - E(\beta_t, q_t)$ is decisive for the movement of $a(t)$ and $b(t)$. We now shall investigate this difference. For the sake of simplicity we drop t in $u_t, v_t, \alpha_t, \beta_t, q_t, a(t)$ and $b(t)$. It can be easily verified that the following is true:

$$E(\alpha, q) - E(\beta, q) = a^3 [E(p,p) - E(u,p)]$$
$$+ a^2 b \{2[E(p,u) - E(u,u)] + E(u,p) - E(v,p)\}$$
$$+ ab^2 \{2[E(u,u) - E(v,u)] + E(p,v) - E(u,v)\}$$
$$+ b^3 [E(u,v) - E(v,v)]. \tag{38}$$

We now prove the following *assertion*: If p is an ESS of G and $u \neq p$ or $v \neq p$ holds, then $E(\alpha, q) - E(\beta, q)$ is positive for all sufficiently small $b > 0$. – It is convenient to distinguish four cases:

(i) u is not a best reply to p. Then the first term in (38) is positive.

(ii) u is a best reply to p with $u \neq p$. Then the first term in (38) is zero and the second one is positive.

(iii) $u = p$ and v is not a best reply to p. Here, too the first term in (38) is zero and the second one is positive.

(iv) $u = p$ and v is a best reply to p. We must have $v \neq p$. The first three terms vanish and the fourth one is positive.

The discussion has shown that in all four cases $E(\alpha, q) - E(\beta, q)$ is positive for sufficiently small $b > 0$.

In the case $u = v = p$ we have $\alpha = \beta$. In this case the difference $E(\alpha, q) - E(\beta, q)$ vanishes. Consider a sequence $x(0), x(1), \ldots$ generated by the dynamic system. Assume that p is an ESS. If $b(0)$ is sufficiently small, then $b(t)$ is nonincreasing. The sequence of population mean strategies $q_0, q_1, \ldots$ generated by $x(0)$ remains in an ε-neighborhood of p with $\varepsilon > b(0)$. The sequence $q_0, q_1, \ldots$ must have an accumulation point q. Assume that q is different from p. In view of the continuity of (23) this is impossible, since for q_t sufficiently near to q the inside part $b(t)$ would have to decrease beyond the inside part b of q. Therefore, the sequence $q_0, q_1, \ldots$ converges to p. We have shown that every inside population state is externally stable if p is an ESS.

It remains to show that p is not externally stable if it is not an ESS. In view of Result 14 we can assume that p is a symmetric equilibrium strategy, but not an ESS. Let v be a best reply to p with

$$E(p, v) \leq E(v, v).$$

Since p is a symmetric equilibrium strategy, but not an ESS, such a strategy can be found. We look at the specification (G, U, U_n) with the following outside array:

$u_{njkl} = p$, for $j, l = 1, \ldots, m$ and $k = 1, \ldots, n-1$,

$u_{njnl} = v$, for $j, l = 1, \ldots, m$.

Consider a sequence $x(0), x(1), \ldots$ generated by the dynamic system. The mean strategies u_t and v_t do not depend on t. We always have $u_t = p$ and $v_t = v$. Assume $E(p, v) < E(v, v)$. In this case (38) shows that $E(\alpha_t, q_t) - E(\beta_t, q_t)$ is always negative.

In view of the continuity of (23) the sequence of the $b(t)$ always converges to 1 for $b(0) > 0$. In the case $E(p, v) = E(v, v)$ the difference $E(\alpha_t, q_t) - E(\beta_t, q_t)$ always vanishes and we have

$$q_t = [1 - b^2(0)]p + b^2(0)v, \quad \text{for } t = 0, 1, \ldots.$$

In both cases the sequence $q_0, q_1, \ldots$ of the population mean strategies does not converge to p, whenever $b(0) > 0$ holds. Therefore, p is not externally stable. □□

The proof of 14 reveals additional stability properties of a monomorphism whose phenotype is an ESS p. Consider a specification (G, U, U_n) of (23) and let U be the monomorphism of genotype p. We say that a population state c is *nearer* to the monomorphism U than a population state x' or that x' is *farther* from U than x if the outside part b of x is smaller than the outside part b' of x'. We say that a population state x is *shifted* towards the monomorphism if for every sequence $x(0), x(1), \ldots$ generated by (23) starting with $x(0) = x$ every $x(t)$ with $t = 1, 2, \ldots$ is nearer to the monomorphism than x; if for $x(0) = x$ every $x(t)$ with $t = 1, 2, \ldots$ is not farther away from the monomorphism than x we say that x is not *shifted away* from the monomorphism. An ε-neighborhood N_ε of an inside state y is called *drift resistant* if all population states $x \in N_\varepsilon$ with a population mean strategy different from p are shifted towards the monomorphism and no population state $x \in N_\varepsilon$ is shifted away from the monomorphism. An inside state y is called *drift resistant* if for some $\varepsilon > 0$ the ε-neighborhood N_ε of y is drift resistant. The monomorphism U is *drift resistant* if for every U_n every inside state y is drift resistant in (G, U, U_n).

Result 16. Let p be an ESS of G. Then the monomorphism U of phenotype p is drift resistant.

Proof. As we have seen in the proof of Result 15 the dynamics of (23) leads to Eq. (37) which together with (38) has the consequence that a population state x with a sufficiently small outside part b is not shifted farther away from the monomorphism and is shifted towards the monomorphism if its population mean strategy is different from p. □

Water resistant watches are not water proof. Similarly drift resistance does not offer an absolute protection against drift. A sequence of perturbances away from the monomorphism may lead to a population state outside the drift resistant ε-neighborhood. However, if perturbances are small relative to ε, this is improbable and it is highly probable that repeated perturbances will drive the mutant allele towards extinction. Of course, this is not true for the special case in which all genotypes carrying the mutant allele play the monomorphic ESS p. In this case the entrance of the mutant creates a new monomorphism, with one additional allele, which again will be drift resistant.

6. Asymmetric conflicts

Many conflicts modelled by biologists are asymmetric. For example, one may think of territorial conflicts where one of two animals is identified as the territory owner and the other one as the intruder (see Section 8.2). Other examples arise if the opponents differ in strength, sex, or age. Since a strategy is thought of as a program for all situations which may arise in the life of a random animal, it determines behavior for both sides of an asymmetric conflict. Therefore, in evolutionary game theory asymmetric conflicts are imbedded in symmetric games.

In the following we shall describe a class of models for asymmetric conflicts. Essentially the same class has first been examined by Selten (1980). In the models of this class the animals may have incomplete information about the conflict situation. We assume that an animal can find itself in a finite number of *states of information*. The set of all states of information is denoted by U. We also refer to the elements of U as *roles*. This use of the word role is based on applications in the biological literature on animal contests. As an example we may think of a strong intruder who faces a territory owner who may be strong or weak. On the one hand, the situation of the animal may be described as the role of a strong intruder and, on the other hand, it may be looked upon as the state of information in this role.

In each role u an animal has a nonempty, finite set C_u of choices. A *conflict situation* is a pair (u, v) of roles with the interpretation that one animal is in the role u, and the other in the role v. The game starts with a random move which selects a conflict situation (u, v) with probability w_{uv}. Then the players make their choices from the choice sets C_u and C_v respectively. Finally they receive a payoff which depends on the choices and the conflict situation. Consider a conflict situation (u, v) and let c_u and c_v be the choices of two opponents in the roles u and v respectively; under these conditions $h_{uv}(c_u, c_v)$ denotes the payoffs obtained by the player in the role u; for reasons of symmetry, the payoff of the player in the role of v is $h_{vu}(c_v, c_u)$.

We define an asymmetric conflict as a quadruple

$$M = (U, C, w, h). \tag{39}$$

Here, U, the *set of information states or roles*, is a nonempty finite set; C, the *choice set function*, assigns a nonempty finite choice set C_u to every information state u in U; the *basic distribution*, w, assigns a probability w_{uv} to every conflict situation (u, v); finally, h, the *payoff function*, assigns a payoff $h_{uv}(c_u, c_v)$ to every conflict situation (u, v) with $w_{uv} > 0$ together with two choices c_u in C_u and c_v in C_v. The probabilities w_{uv} sum up to 1 and have the symmetry property

$$w_{uv} = w_{vu}.$$

Formally the description of a model of type (39) is now complete. However, we would like to add that one may think of the payoffs of both opponents in a conflict situation (u, v) as a bimatrix game. Formally, it is not necessary to make this picture explicit, since the payoff for the role v in the conflict situation (u, v) is determined by h_{vu}.

A *pure strategy* s for a model of the type (39) is a function which assigns a choice c_u in C_u to every u in U. Let S be the set of all pure strategies.

From here we could directly proceed to the definition of a symmetric two-person game (S, E) based on the model. However, this approach meets serious difficulties. In order to explain these difficulties we look at an example.

Example. We consider a model with only two roles u and v, and $w_{uu} = w_{vv} = \frac{1}{2}$ and $C_u = C_v = \{H, D\}$. The payoff functions h_{uu} and h_{vv} are payoffs for Hawk–Dove games (see Figure 1) with different parameters W. We may think of (u, u) and (v, v) as different environmental conditions like rain and sunshine which influence the parameter W. The pure strategies for the game $G = (S, E)$ are HH, HD, DH, and DD, where the first symbol stands for the choice in u and the second symbol for the choice in v. Let p_1 be the ESS for the Hawk–Dove game played at (u, u) and p_2 be the ESS of the Hawk–Dove game played at (v, v). We assume that p_1 and p_2 are genuinely mixed. The obvious candidate for an ESS in this model is to play p_1 at u and p_2 at v. This behavior is realized by all mixed strategies q for G which satisfy the following equations:

$$q(HH) + q(HD) = p_1(H),$$

$$q(HH) + q(DH) = p_2(H).$$

It can be seen immediately that infinitely many mixed strategies q satisfy these equations. Therefore, no q of this kind can be an ESS of G, since all other strategies satisfying the two equations are alternative best replies which violate the stability condition (b) in the Definition 1 of evolutionary stability. Contrary to common sense the game G has no ESS.

The example shows that the description of behavior by mixed strategies introduces a spurious multiplicity of strategies. It is necessary to avoid this multiplicity. The concept of a behavior strategy achieves this purpose.

A *behavior strategy* b for a model of the type (39) assigns a probability distribution over the choice set C_u of u to every role u in U. A probability distribution over C_u is called a *local strategy* at u. The probability assigned b to a choice c in C_u is denoted by $b(c)$. The symbol B is used for the set of all behavior strategies b.

We now define an expected payoff $E(b, b')$ for every pair (b, b') of behavior strategies:

$$E(b, b') = \sum_{(u,v)} w_{uv} \sum_{c \in C_u} \sum_{c' \in C_v} b(c) b'(c') h_{uv}(c, c'). \tag{40}$$

The first summation is extended over all pairs (u, v) with $w_{uv} > 0$. The payoff $E(b, b')$ has the interpretation of the expected payoff of an individual who plays b in a population playing b'. We call

$$G_M = (B, E)$$

the *population game associated with* $M = (U, C, w, h)$.

$E(b, b')$ is a bilinear function of the probabilities $b(c)$ and $b'(c')$ assigned by the behavior strategies b and b' to choices. Therefore, the definition of evolutionary stability by two conditions analogous to those of Definition 1 is adequate. In the case of a bilinear payoff function E Definitions 1 and 3 are equivalent. A behavior strategy b is a best reply to a behavior strategy b' if it maximizes $E(\cdot, b')$ over B. A behavior strategy b^* for G_M is *evolutionarily stable* if the following conditions (a) and (b) are satisfied:

(a) *Equilibrium condition*: b^* is a best reply to b^*.

(b) *Stability condition*: Every best reply b to b^* which is different from b^* satisfies the following inequality:

$$E(b^*, b) > E(b, b) \tag{41}$$

An evolutionarily stable strategy b^* is called *strict* if b^* is the only best reply to b^*. It is clear that in this case b^* must be a pure strategy.

In many applications it never happens that two animals in the same role meet in a conflict situation. For example, in a territorial conflict between an intruder and a territory owner the roles of both opponents are always different, regardless of what other characteristics may enter the definition of a role. We say that $M(U, C, w, h)$ satisfies the condition of *role asymmetry* [information asymmetry in Selten (1980)] if the following is true:

$$w_{uu} = 0, \quad \text{for all } u \text{ in } U. \tag{42}$$

The following consequence of role asymmetry has been shown by Selten (1980).

Result 17. Let M be a model of the type (39) with role asymmetry and let $G_M = (B, E)$ be the associated population game. If b^* is an evolutionarily stable strategy for G_M, then b^* is a pure strategy and a strict ESS.

Sketch of the proof. If an alternative best reply is available, then one can find an alternative best reply b which deviates from b^* only in one role u_1. For this best reply b we have $E(b, b^*) = E(b^*, b^*)$. This is due to the fact that in a conflict situation (u_1, v) we have $u_1 \neq v$ in view of the role asymmetry assumption. Therefore, it never matters for a player of b whether an opponent plays b or b^*. The equality of $E(b, b^*)$ and $E(b^*, b^*)$ violates the stability condition (41). Therefore, the existence of an alternative best reply to b^* is excluded.

If the role asymmetry condition is not satisfied, an ESS can be genuinely mixed. This happens in the example given above. There the obvious candidate for an ESS corresponds to an ESS in G_M. This ESS is the behavior strategy which assigns the local strategies p_1 and p_2 to u and v, respectively.

A special case of the class of models of the type (39) has been examined by Hammerstein (1981). He considered a set U of the form

$$U = (u_1, \ldots, u_n, v_1, \ldots, v_n)$$

and a basic distribution w with $w(u, v) > 0$ for $u = u_i$ and $v = v_i$ ($i = 1, \ldots, n$) and $w(u, v) = 0$ for $u = u_i$ and $v = v_j$ with $i \neq j$. In this case an evolutionarily stable strategy induces strict pure equilibrium points on the n bimatrix games played in the conflict situations (u_i, v_i). In view of this fact it is justified to speak of a strict pure equilibrium point of an asymmetric bimatrix game as an evolutionarily stable strategy. One often finds this language used in the biological literature. The simplest example is the Hawk–Dove game of Figure 1 with the two roles "owner" and "intruder".

7. Extensive two-person games

Many animal conflicts have a sequential structure. For example, a contest may be structured as a sequence of a number of bouts. In order to describe complex sequential interactions one needs extensive games. It is not possible to replace the extensive game by its normal form in the search for evolutionarily stable strategies. As in the asymmetric animal conflicts in Section 6, the normal form usually has no genuinely mixed ESS, since infinitely many strategies correspond to the same behavior strategy. It may be possible to work with something akin to the agent normal form, but the extensive form has the advantage of easily identifiable substructures, such as subgames and truncations, which permit decompositions in the analysis of the game.

7.1. Extensive games

In this section we shall assume that the reader is familiar with basic definitions concerning games in extensive form. The results presented here have been derived by Selten (1983, 1988). Unfortunately, the first paper by Selten (1983) contains a serious mistake which invalidates several results concerning sufficient conditions for evolutionary stability in extensive two-person games. New sufficient conditions have been derived in Selten (1968).

Notation. The word *extensive game* will always refer to a finite two-person game with perfect recall. Moreover, it will be assumed that there are at least two choices at every information set. A game of this kind is described by a septuple

$$\Gamma = (K, P, U, C, p, h, h').$$

K is the *game tree*. The set of all endpoints of K is denoted by Z.

$P = (P_0, P_1, P_2)$ is the *player partition* which partitions the set of all decision points into a random decision set P_0 and decision sets P_1 and P_2 for players 1 and 2.

U is the *information partition*, a refinement of the player partition.

C is the *choice partition*, a partition of the set of alternatives (edges of the tree) into choices at information sets u in U. The set of all random choices is denoted by C_0. For $i = 1, 2$ the set of all choices for player i is denoted by C_i. The choice set at an information set u is denoted by C_u. The set of all choices on a play to an endpoint set is denoted by $C(z)$.

p is the *probability assignment* which assigns probabilities to random choices.

h and h' are the *payoff functions* of players 1 and 2, respectively which assign payoffs $h(z)$ and $h'(z)$ to endpoints z.

For every pair of behavior strategies (b, b') the associated payoffs for players 1 and 2 are denoted by $E(b, b')$ and $E'(b, b')$, respectively.

No other strategies than behavior strategies are admissible. Terms such as *best reply*, *equilibrium point*, etc. must be understood in this way. The probability assigned to a choice c by a behavior strategy b is denoted by $b(c)$.

We say that an information set u of player 1 is *blocked* by a behavior strategy b' of player 2 if u cannot be reached if b' is played. In games with perfect recall the probability distribution over vertices in an information set u of player 1 if u is reached depends only on the strategy b' of player 2. On this basis a *local payoff* $E_u(r_u, b, b')$ for a local strategy r_u at u if b and b' are played can be defined for every information set u of player 1 which is not blocked by b'. The local payoff is computed starting with the probability distribution over the vertices of u determined by b' under the assumption that at u the local strategy r_u is used and later b and b' are played.

A local *best reply* b_u at an information set u of player 1 to a pair of behavior strategies (b, b') such that u is not blocked by b' is a local strategy at u which maximizes player 1's local payoff $E_u(r_u, b, b')$. We say that b_u is a *strict local best reply* if b_u is the only best reply to (b, b') at u. In this case b_u must be a pure local strategy, or in other words a choice c at u.

7.2. Symmetric extensive games

Due to structural properties of game trees, extensive games with an inherent symmetry cannot be represented symmetrically by an extensive form. Thus two simultaneous choices have to be represented sequentially. One has to define what is meant by a symmetric two-person game.

Definition 6. A *symmetry* f of an extensive game $\Gamma = (K, P, U, C, p, h, h')$ is a mapping from the choice set C onto itself with the following properties (a)–(f):

(a) If $c \in C_0$, then $f(c) \in C_0$ and $p(f(c)) = p(c)$.
(b) If $c \in C_1$ then $f(c) \in C_2$.
(c) $f(f(c)) = c$ for every $c \in C$.
(d) For every $u \in U$ there is a $u' \in U$ such that for every choice c at u, the image $f(c)$ is a choice at u'. The notation $f(u)$ is used for this information set u'.
(e) For every endpoint $z \in Z$ there is a $z' \in Z$ with $f(C(z)) = C(z')$, where $f(C(z))$ is the set of all images of choices in $C(z)$. The notation $f(z)$ is used for this endpoint z'.
(f) $h(f(z)) = h'(z)$ and $h'(f(z)) = h(z)$.

A symmetry f induces a one-to-one mapping from the behavior strategies of player 1 onto the behavior strategies of player 2 and vice versa:

$b' = f(b)$, if $b'(f(c)) = b(c)$, for every $c \in C_1$,
$b = f(b')$, if $b' = f(b)$.

An extensive game may have more than one symmetry. In order to see this, consider a game Γ with a symmetry f. Let Γ_1 and Γ_2 be two copies of Γ, and let f_1 and f_2 be the symmetries corresponding to f in Γ_1 and Γ_2, respectively. Let Γ_3 be the game which begins with a random move which chooses one of both games Γ_1 and Γ_2 both with probability $\frac{1}{2}$. One symmetry of Γ_3 is composed of f_1 and f_2, and a second one maps a choice c_1 in Γ_1 on a choice c_2 in Γ_2 which corresponds to the same choice c in Γ as c_1 does.

In biological aplications there is always a natural symmetry inherent in the description of the situation. "Attack" corresponds to "attack", and "flee" corresponds

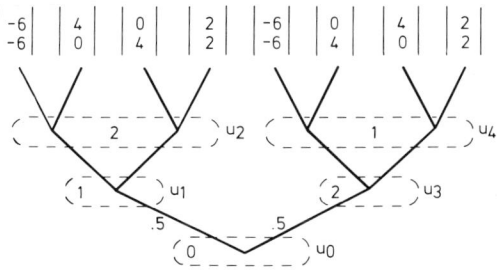

Figure 2. Example of a game with two symmetries. This game starts with a random move after which players 1 and 2 find themselves in a Hawk–Dove contest. Their left choices mean to play Hawk, right choices mean to play Dove. The initial random move can be *distinguishing* or *neutral* in the following sense. Suppose that the left random choice determines player 1 as the original owner of a disputed territory and the right random choice determines player 2 as the original owner. In this case the random move distinguishes the players so that they can make their behavior dependent on the roles "owner" and "intruder". On the other hand, suppose that the random move determines whether there is sunshine (left) or an overcast sky (right). This is the neutral case where nothing distinguishes player 1 and player 2. The two possible symmetries of this game specify whether the random move is distinguishing or neutral. If the symmetry maps the information set u_1 to u_3, it is distinguishing. If it maps u_1 to u_2, it is neutral [Selten (1983)].

to "flee" even if a formal symmetry may be possible which maps "attack" to "flee". Therefore, the description of the natural symmetry must be added to the extensive form in the evolutionary context (see Figure 2).

Definition 7. A *symmetric extensive game* is a pair (Γ, f), where Γ is an extensive game and f is a symmetry of Γ.

7.3. Evolutionary stability

A definition of evolutionary stability suggests itself which is the analogue of Definition 1. As we shall see later, this definition is much too restrictive and will have to be refined. Since it is the direct analogue of the usual ESS definition we call it a direct ESS.

Definition 8. Let (Γ, f) be a symmetric extensive game. A *direct ESS* for (Γ, f) is a behavior strategy b^* for player 1 in Γ with the following two properties (a) and (b):

(a) *Equilibrium condition*: $(b^*, f(b^*))$ is an equilibrium point of Γ.
(b) *Stability condition*: If b is a best reply to $f(b^*)$ which is different from b^*, then we have

$$E(b^*, f(b)) > E(b, f(b)).$$

A behavior strategy b^* which satisfies (a) but not necessarily (b) is called a *symmetric equilibrium strategy*. An equilibrium point of the form $(b^*, f(b^*))$ is called a *symmetric equilibrium point*.

The definition of a direct ESS b^* is very restrictive, since it implies that every information set u of Γ must be reached with positive probability by the equilibrium path generated by $(b^*, f(b^*))$ [Selten (1983) Lemma 2 and Theorem 2, p. 309]. This means that most biological extensive form models can be expected to have no direct ESS. Nevertheless, the concept of evolutionary stability can be defined in a reasonable way for extensive two-person games.

Since every information set is reached with positive probability by $(b^*, f(b^*))$, no information set u of player 1 is blocked by $f(b^*)$. Therefore, a local payoff $E_u(r_u, b^*, f(b^*))$ is defined at all information sets u of player 1 if b^* is a direct ESS of (Γ, f). We say that a direct ESS b^* is *regular* if at every information set u of player 1 the local strategy b_u^* assigned by b^* to u chooses every pure local best reply at u with positive probability.

The definition of perfectness [Selten (1975)] was based on the idea of mistakes which occur with small probabilities in the execution of a strategy. In the biological context it is very natural to expect such mistakes; rationality is not assumed and genetic programs are bound to fail occasionally for physiological reasons if no others. Therefore, it is justified to transfer the trembling hand approach to the definition of evolutionary stability in extensive games. However, in contrast to the definition of perfectness, it is here not required that every choice must be taken with a small positive probability in a perturbed game. The definition of a perturbed game used here permits zero probabilities for some or all choices. Therefore, the game itself is one of its perturbed games.

A *perturbance* α for (Γ, f) is a function which assigns a minimum probability $\alpha_c \geq 0$ to every choice c of player 1 and 2 in Γ such that (a) the choices at an information set u sum to less than 1 and (b) the equation $\alpha_c = \alpha_d$ always holds for $d = f(c)$. A *perturbed game* of (Γ, f) is a triple $\Gamma' = (\Gamma, f, \alpha)$ in which α is a perturbance for (Γ, f). In the perturbed game Γ' only those behavior strategies are admissible which respect the minimum probabilities of α in the sense $b(c) \geq \alpha_c$. Best replies in Γ' are maximizers of $E(\cdot, b')$ or $E(b, \cdot)$ within these constraints. The definition of a *direct ESS* for Γ' is analogous to Definition 6. A *regular direct ESS* of a perturbed game is also defined analogously to a regular direct ESS of the unperturbed game.

The maximum of all minimum probabilities assigned to choices by a perturbance α is denoted by $|\alpha|$. If b and b^* are two behavior strategies, then $|b - b^*|$ denotes the maximum of the absolute difference between the probabilities assigned by b and b^* to the same choice. With the help of these auxiliary definitions we can now give the definition of a (regular) limit ESS.

Definition 9. A behavior strategy b^* of player 1 for a symmetric extensive two-person game (Γ, f) is a *(regular) limit ESS* of (Γ, f) if for every $\varepsilon > 0$ at least one perturbed game $\Gamma' = (\Gamma, f, \alpha)$ with $|\alpha| < \varepsilon$ has a (regular) direct ESS b with $|b - b^*| < \varepsilon$.

Loosely speaking, a limit ESS b^* is a strategy which can be arbitrarily closely approximated by a direct ESS of a perturbed game. For a biological game model this means that in a slightly changed game model with small mistake probabilities for some choices a direct ESS close to b^* can be found.

Since the special case of a perturbance $\alpha_c = 0$ for all choices c of players 1 and 2 is not excluded by the definition of a perturbed game, a direct ESS is always a limit ESS.

7.4. Image confrontation and detachment

In the following we shall state a result which can be looked upon as the analogue of Result 17 for symmetric extensive games. Let (Γ, f) be a symmetric extensive game. We say that an information set u in Γ is *image confronted* if at least one play in Γ intersects both u and $f(u)$; otherwise u is called *image detached*. The following result can be found in Selten (1983):

Result 18. Let b^* be a limit ESS for a symmetric extensive game (Γ, f). Then the following is true:

(a) The pair $(b^*, f(b^*))$ is an equilibrium point of (Γ, f).
(b) If u is an image detached information set in (Γ, f), then the local strategy b_u^* is pure, or in other words $b(c) = 1$ holds for one choice c at u.

Result 18 narrows down the search for ESS candidates. In many models most information sets are image detached, since the opponents in a sequential conflict who are initially indistinguishable may quickly become distinguishable by the history of the game. Thus one of both animals may be identifiable as that one which attacked first; thereby all later information sets become image detached.

7.5. Decomposition

In the determination of subgame perfect equilibrium points one can replace a subgame by the payoff vector for one of its equilibrium points and then determine an equilibrium point of the truncation formed in this way. The equilibrium points of the subgame and the truncation together form an equilibrium point of the whole game. Unfortunately, a direct ESS or a limit ESS does not have the same decomposition properties as a subgame perfect equilibrium point. A direct ESS or a limit ESS cannot be characterized by purely local conditions. Counterexamples can be found in the literature (van Damme 1987, Selten 1988). Nevertheless, a limit ESS does have some decomposition properties which are useful in the analysis of extensive game models. In order to describe these decomposition properties we now introduce the definitions of an "upper layer" and an "abridgement".

We say that an information set v *preceeds* an information set u if at least one play intersects both information sets and if on every play which intersects both information sets the vertex in v comes before the vertex in u. An *upper layer* of a symmetric extensive two-person game (Γ, f) is a nonempty subset V of the set of all information sets U in Γ which satisfies the following conditions (i) and (ii):

(i) If $v \in V$ then $f(v) \in V$
(ii) If $v \in V$ preceeds $u \in U$ then $u \in V$.

An information set v in an upper layer V is called a starting set in V if no other information sets $u \in V$ preceed v. The vertices of starting sets in V are called starting points in V.

Let V be an upper layer of (Γ, f). For every strategy b of player 1 for Γ we construct a new symmetric two-person game called the *b-abridgement* of (Γ, f) with respect to V. The b-abridgement is a game (Γ_*, f_*) constructed as follows. All vertices and edges after the starting point of V are removed from the game tree. The starting points of V become end points. The payoffs at a starting point

x are the conditional payoff expectations in the original game if x is reached and b and $f(b)$ are played later on. The other specifications of (Γ_*, f_*) are obtained as restrictions of the corresponding specifications of (Γ, f) to the new game tree.

A strategy b of player 1 for (Γ, f) is called *quasi-strict* at an information set u if a number $\varepsilon > 0$ exists for which the following is true: If r is a strategy in the ε-neighborhood of b and if $f(r)$ does not block u, then the local strategy b_u which b assigns to u is a strong local best reply to r and $f(r)$ in Γ.

Obviously, b is quasi-strict at u if b_u is a strict best reply. Moreover, if b is quasi-strict at u, then b_u is a pure local strategy, or in other words a choice at u.

The following result is the Theorem 4 in Selten (1988).

Result 19. Let (Γ, f) be a symmetric extensive two-person game, let b be a strategy of player 1 for Γ and let V be an upper layer of (Γ, f). Moreover, let (Γ_*, f_*) be the b-abridgement of (Γ, f) with respect to V. Then b is a regular limit ESS of (Γ, f) if the following two conditions (i) and (ii) are satisfied:

(i) For every information set v of player 1 which belongs to V the following is true: b is quasi-strict at v in (Γ, f).

(ii) The strategy b_* induced on (Γ_*, f_*) by b is a regular limit ESS of (Γ_*, f_*).

Results 18 and 19 are useful tools for the analysis of evolutionary extensive game models. In many cases most of the information sets are image detached and the image detached information sets form an upper layer. In the beginning two animals involved in a conflict may be indistinguishable, but as soon as something happens which makes them distinguishable by the history of the conflict all later information sets become image detached. A many-period model with ritual fights and escalated conflicts [Selten (1983, 1988)] provides an example. Result 18 helps to find candidates for a regular limit ESS and Result 19 can be used in order to reach the conclusion that a particular candidate is in fact a regular limit ESS. Further necessary conditions concerning decompositions into subgames and truncations can be found in Selten (1983). However, the sufficient conditions stated there in connection to subgame–truncation decompositions are wrong.

8. Biological applications

Evolutionary game theory can be applied to an astonishingly wide range of problems in zoology and botany. Zoological applications deal, for example, with animal fighting, cooperation, communication, coexistence of alternative traits, mating systems, conflict between the sexes, offspring sex ratio, and the distribution of individuals in their habitats. botanical applications deal, for example, with seed dispersal, seed germination, root competition, nectar production, flower size, and sex allocation. In the following we shall briefly review the major applications of evolutionary game theory. It is not our intent to present the formal mathematical structure of any of the specific models but rather to emphasize the multitude of new insights biologists have gained from strategic analysis. The biological literature on evolutionary games is also reviewed in Maynard Smith (1982a), Riechert & Hammerstein (1983), Parker (1984), Hines (1987), and Vincent & Brown (1988).

8.1. Basic questions about animal contest behavior

Imagine the following general scenario. Two members of the same animal species are contesting a *resource*, such as food, a territory, or a mating partner. Each animal would increase its Darwinian fitness by obtaining this resource (the *value of winning*). The opponents could make use of dangerous weapons, such as horns, antlers, teeth, or claws. This would have negative effects on fitness (the *cost of escalation*). In this context behavioral biologists were puzzled by the following functional questions which have led to the emergence of evolutionary game theory:

Question 1. Why are animal contests often settled *conventionally*, without resort to injurious physical escalation; under what conditions can such conventional behavior evolve?

Question 2. How are conflicts resolved; what can decide a non-escalated contest?

Classical ethologists attempted to answer Question 1 by pointing out that it would act against the good of the species if conspecifics injured or killed each other in a contest. Based on this argument, Lorenz (1966) even talked of "behavioral analogies to morality" in animals. He postulated the fairly widespread existence of an innate inhibition which prevents animals from killing or injuring members of the same species.

However, these classical ideas are neither consistent with the methodological individualism of modern evolutionary theory, nor are they supported by the facts. Field studies have revealed the occurrence of fierce fighting and killing in many animal species when there are high rewards for winning. For example, Wilkinson and Shank (1977) report that in a Canadian musk ox population 5–10 percent of the adult bulls may incur lethal injuries from fighting during a single mating season. Hamilton (1979) describes battlefields of a similar kind for fig wasps, where in some figs more than half the males died from the consequences of inter-male combat. Furthermore, physical escalation is not restricted to male behavior. In the desert spider *Agelenopsis aperta* females often inflict lethal injury on female opponents when they fight over a territory of very high value [Hammerstein and Riechert (1988)].

Killing may also occur in situations with moderate or low rewards when it is cheap for one animal to deal another member of the same species the fatal blow. For example, in lions and in several primate species males commit infanticide by killing a nursing-female's offspring from previous mating with another male. This infanticide seems to be in the male's "selfish interest" because it shortens the period of time until the female becomes sexually receptive again (Hausfater and Hrdy 1984).

Another striking phenomenon is that males who compete for access to a mate may literally fight it out on the female's back. This can sometimes lead to the death of the female. In such a case the contesting males destroy the very mating partner they are competing for. This happens, for example, in the common toad [Davies and Halliday (1979)]. During the mating season the males locate themselves near those ponds where females will appear in order to spawn. The sex ratio at a given pond is highly male-biased. When a single male meets a single female he clings to her back as she continues her travel to the spawning site. Often additional males pounce at the pair and a struggle between the males starts on the female's back. Davies & Halliday describe a pond where more than 20 percent of the females carry the heavy load of three to six wrestling males. They also report that this can lead to the death of the female who incurs a risk of being drowned in the pond.

It is possible to unravel this peculiar behavior by looking at it strictly from the individual male's point of view. His individual benefit from interacting with a female would be to fertilize her eggs and thus to father her offspring. If the female gets drowned, there will be no such benefit. However, the same zero benefit from this female will occur if the male completely avoids wrestling in the presence of a competitor. Thus it can pay a male toad to expose the female to a small risk of death.

The overwhelming evidence for intra-specific violence has urged behavioral biologists to relinquish the Lorenzian idea of a widespread "inhibition against killing members of the same species" and of "behavioral analogies to morality". Furthermore, this evidence has largely contributed to the abolishment of the "species welfare paradigm". Modern explanations of contest behavior hinge on the question of how the behavior contributes to the individual's success rather than on how it affects the welfare of the species. These explanations relate the absence or occurrence of fierce fighting to costs and benefits in terms of fitness, and to biological constraints on the animals' strategic possibilities.

We emphasize that the change of paradigm from "species welfare" to "individual success" has paved the way for non-cooperative game theory in biology (Parker and Hammerstein 1985). The *Hawk–Dove* game in Figure 1 (Maynard Smith and Price 1973) should be regarded as the simplest model from which one can deduce that natural selection operating at the individual level may forcefully restrict the amount of aggression among members of the same species, and that this restriction of aggression should break down for a sufficiently high value of winning. In this sense the Hawk–Dove game provides a modern answer to Question 1, but obviously it does not give a realistic picture of true animal fighting.

The evolutionary analysis of the Hawk–Dove game was outlined in Section 3. Note, however, that we confined ourselves to the case where the game is played between genetically unrelated opponents. Grafen (1979) has analyzed the *Hawk–Dove game between relatives* (e.g. brothers or sisters). He showed that there are serious problems with adopting the well known biological concept of "inclusive fitness" in the context of evolutionary game theory [see also Hines & Maynard Smith (1979)]. The inclusive fitness concept was introduced by Hamilton (1964) in order to explain "altruistic behavior" in animals. It has had a major impact on the early development of sociobiology. Roughly speaking, *inclusive fitness* is a measure of reproductive success which includes in the calculation of reproductive output the effects an individual has on the number of offspring of its genetic relatives. These effects are weighted with a coefficient of relatedness. The inclusive fitness approach has been applied very successfully to a variety of biological problems in which no strategic interaction occus.

8.2. *Asymmetric animal contests*

We now turn to Question 2 about how conflict is resolved. It is typical for many animal contests that the opponents will differ in one or more aspect, e.g. size, age, sex, ownership status, etc. If such an asymmetry is discernible it may be taken as a cue whereby the contest is conventionally settled. This way of settling a dispute is analyzed in the theory of asymmetric contests (see Section 6 for the mathematical background).

It came as a surprise to biologists when Maynard Smith and Parker (1976) stated the following result about simple contest models with a single asymmetric aspect. A contest between an "owner" and an "intruder" can be settled by an evolutionarily stable "owner wins" convention even if ownership does not positively affect fighting ability or reward for winning (e.g. if ownership simply means prior presence at the territory). Selten (1980) and Hammerstein (1981) clarified the game-theoretic nature of this result. Hammerstein extended the idea by Maynard Smith and Parker to contests with several asymmetric aspects where, for example, the contest is one between an owner and an intruder who differ in size (strength). He showed that a *payoff irrelevant* asymmetric aspect may decide a contest even if from the beginning of the contest another asymmetric aspect is known to both opponents which is *payoff relevant* and which puts the conventional winner in an inferior strategic position. For example, if escalation is sufficiently costly, a weaker owner of a territory may conventionally win against a stronger intruder without having more to gain from winning.

This contrasts sharply with the views traditionally held in biology. Classical ethologists either thought they had to invoke a "home bias" in order to explain successful territorial defense, or they resorted to the well known logical fallacy that it would be more important to avoid losses (by defending the territory) than to make returns (by gaining access to the territory). A third classical attempt to explain the fighting success of territory holders against intruders had been based on the idea that the owner's previous effort already put into establishing and maintaining the territory would bias the value of winning in the owner's favor and thus create for him a higher incentive to fight. However, in view of evolutionary game theory modern biologists call this use of a "backward logic" the Concorde fallacy and use a "forward logic" instead. The value of winning is now defined as the individual's future benefits from winning.

The theory of asymmetric contests has an astonishingly wide range of applications in different parts of the animal kingdom, ranging from spiders and insects to birds and mammals [Hammerstein (1985)]. Many empirical studies demonstrate that asymmetries are decisive for conventional conflict resolution [e.g. Wilson (1961), Kummer et al. (1974), Davies (1978), Riechert (1978), Yasukawa and Bick (1983), Crespi (1986)]. These studies show that differences in ownership status, size, weight, age, and sex are used as the cues whereby contests are settled without major escalation. Some of the studies also provide evidence that the conventional settlement is based on the logic of deterrence and thus looks more peaceful than it really is (the animal in the winning role is ready to fight). This corresponds nicely to qualitative theoretical results about the structure of evolutionarily stable strategies for the asymmetric contest models mentioned above. More quantitative comparisons between theory and data are desirable, but they involve the intriguing technical problem of measuring game-theoretic payoffs in the field. Biological critics of evolutionary game theory have argued that it seems almost impossible to get empirical access to game-theoretic payoffs [see the open peer commentaries of a survey paper by Maynard Smith (1984)].

Question 3. Is it possible to determine game-theoretic payoffs in the field and to estimate the costs and benefits of fighting?

Despite the pessimistic views of some biologists, there is a positive answer to this question. Hammerstein and Riechert (1988) analyze contests between female territory owners and intruders of the funnel web spider *Agelenopsis aperta*. They use data [e.g. Riechert (1978, 1979, 1984)] from a long-term field study about demography, ecology, and behavior of these desert animals in order to estimate all *game payoffs* as *changes in the expected lifetime number of eggs* laid. Here the matter is complicated by the fact that the games over web sites take place long before eggs will be laid. Subsequent interactions occur, so that a spider who wins today may lose tomorrow against another intruder, and today's loser may win tomorrow against another owner.

In the major study area (a New mexico population), web site tenancy crucially affects survival probability, fighting ability, and the rate at which eggs are produced. The spiders are facing a harsh environment in which web sites are in short supply. Competition for sites causes a great number of territorial interactions. At an average web site which ensures a moderate food supply, a contest usually ends without leading into an injurious fight. For small differences in weight, the

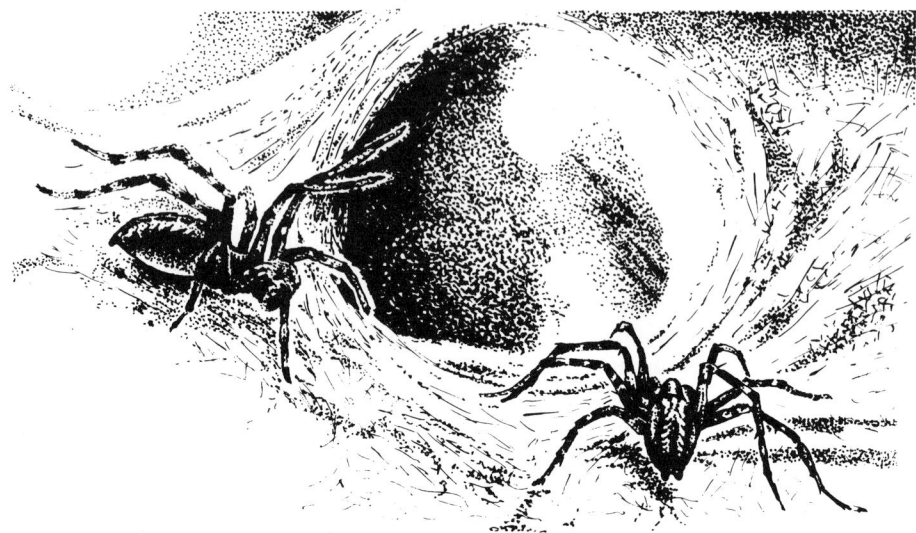

Figure 3. Fighting in the desert spider *Agelenopsis aperta*. The left animal is an intruder who challenges the owner of a territory. The contest takes place on the owner's funnel web. Typically the opponents do not touch each other and keep a minimum distance of approximately one inch. The web is used as a "scale" in order to determine relative body weight. The heavier spider wins conventionally if there is an important weight difference. Otherwise ownership is respected in most cases. (Drawing by D. Schmidl after a photograph by S.E. Riechert.)

owner–intruder asymmetry typically settles the dispute in favor of the owner. For great differences in weight, the weight asymmetry settles it in favor of the heavier spider. Apparently the information about relative weight is revealed when the spiders are shaking the web. This follows from the observation that in most contests the spiders do not even touch each others' bodies, and from the fact that vision is relatively poor in these creatures whose sensory systems are more specialized on dealing with vibratory and chemical information. Experiments with artificial weights glued on the spider's abdomen also support this view.

In order to examine carefully whether we understand the evolutionary logic of fighting in these desert animals it is clearly impotant to deal with Question 3 about measuring payoffs in the field. However, only the immediate consequences of a contest and not the game payoffs can be measured directly. For example, the outcome of a territorial fight between two spiders may be that one of the contestants has lost *two legs* during a vicious biting match, and that the opponent has gained the territory which yields a daily energy intake of *72 joule* per day. Here, we are dealing with very different currencies: how do joules compare to leg amputations?

The currency problem can be solved by converting legs and joules into the leading currency of evolutionary biology, i.e. fitness. This requires numerical simulations of future life histories for healthy and injured spiders with different starting conditions. In the study by Hammerstein and Riechert these simulations are based on data from 15 years of desert field work. The results are used to calculate, for example, the value of winning as the difference in expected future egg mass between a healthy owner and a healthy wanderer. The wanderer's expected egg mass can be interpreted as a biological *opportunity cost*.

A glance at the outcome of these calculation reveals an interesting aspect of this spider society. The estimated benefit of winning is 20 mg for an average site and 72 mg for an excellent site. These values differ by a factor of more than 3. In contrast, the immediate benefits in terms of daily weight gain are not even twice as good for excellent sites (5.1 versus 7.9 mg). This "multiplier effect" can be explained as follows. An individual who occupies an excellent site gains weight at a rate which is near the maximum for this population. As a result, an individual's weight relative to the population mean will increase as long as it inhabits the excellent site. This is important for remaining in possession of the site because weight counts in territorial defense. Therefore, an originally "richer" spider will have better future chances of maintaining its wealth than an otherwise equal "poorer" individual. This anthropomorphic feature of the spider society explains the multiplier effect.

The negative consequences of fighting can be calculated on the same scale as the value of winning (egg mass). Leg loss costs 13 mg and death costs 102 mg in the New Mexico population. Furthermore, the risk of injury is known for escalated fights. Using these data it can be shown that intruders would pay a high price for "breaking the rules" of this society by disregarding the status of an owner of similar weight. This cost is high enough to ensure the evolutionary stability of the "owner wins" convention.

In order to give a consistent picture of the spider territorial system the assumption had to be made that an individual spider has little information about how its weight compares to the rest of the population, and good informtion about its relative size in a given contest situation. Otherwise small spiders would have to engage in more fierce fights with small opponents because they would then be in a "now or never" situation. Grafen (1987) calls asymmetries *devisive* when they put animals in such now or never situations. This feature does not seem to prevail in the spider population because of the apparent lack of an individual's information about its "position" in the population weight distribution. At this point it should be emphasized that game-theoretic studies of field data often reveal details about information that is not available to an animal although it would be advantageous for the individual to have this information. Such constraints, for example, were also discovered in a famous field study about digger wasps [Brockmann et al. (1979)]. These wasps face a strategic decision to either work hard and digg a burrow or to save this effort and enter an existing burrow which may or may not be in use by a conspecific animal. Surprisingly, the wasps seem to be unable to distinguish between empty and occupied burrows although in view of the fighting risk it would be important to have this capability.

The analysis of spider behavior by Hammerstein and Riechert (1988) is based on a way of modelling called the *fixed background approach*. This means that payoffs for a given contest are calculated from field data about habitat structure, demography, and observed behavior of the population under investigation. The "fixed background" is a mathematical representation of biological facts about the existing population. In other theoretical studies that are more detached from empirical information one can also adopt the *variable background approach*. Here, the entire sequence of an individual's contests with varying opponents is modelled within a single large population game. Houston and McNamara (1988a, 1991) study the Hawk–Dove game from this point of view.

8.3. War of attrition, assessment, and signalling

The Hawk–Dove game and related models emphasize the problem of when an individual should initiate injurious biting or other forms of esclated fighting behavior. It is assumed that the onset of true aggression is a discrete step rather than a gradual transition (such a discrete step occurs, for example, in the spider example presented above). Maynard Smith (1974) introduced the "war of attrition" as a different model that captures ways of fighting in which costs arise in a smoother, gradual way. This modelling approach emphasizes the following problem:

Question 4. When should an animal stop a costly fighting activity?

Bishop and Cannings (1978a, b) analyzed the symmetric war of attrition and another variant of this game with random rewards about which the contestants have private information. Hammerstein and Parker (1981) analyzed the asymmetric war of attrition, i.e. a conflict where a role difference between two opponents is known to both of them with some small error. They showed that typically the benefit–cost ratio V/C should be decisive: the individual with the higher benefit–cost ratio should win (here, V is the value of winning, C is the constant time rate of cost, and both parameters may differ for different roles). This result has an important implication, namely that common-sense rules for conflict resolution are the only evolutionarily stable solutions in the asymmetric war of attrition model! Needless to say that this is in sharp contrast with the results obtained for models of the Hawk–Dove type (see previous section).

The original war of attrition model is lacking an important feature. Behavioral biologists find it hard to imagine a real animal contest in which no assessment of the opponent takes place during an ongoing aggressive interaction. This has led to an interesting refinement of the war of attrition model by Enquist and Leimar. They developed the "sequential assessment game" in which the opponents acquire information about their true fighting abilities throughout an agonistic interaction [Enquist and Leimar (1983, 1987, 1988), Leimar (1988)]. The beauty of the sequential assessment game is its potential for creating quantitative predictions that can be tested by empirical biologists. Enquist et al. (1990) demonstrate this for fighting in the cichlid fish *Nannacara anomala*, Leimar et al. (1991) for fighting in the bowl and doily spider *Frontinella pyramitela* [see also Austad (1983)].

The problem of assessment is closely related to the problem of signalling and communication in animal contests. When animals assess each other, it may pay an individual to convey information about strength by use of signals rather than by actual fighting [e.g. Clutton-Brock and Albon (1979)]. Evolutionary game theory has strongly influenced the biological study of signalling. It caused behavioral biologists to address the strategic aspects of animal communication.

Question 5. What is communicated by animal display behavior?

It seems obvious that animals should demonstrate strength or resource holding potential during an agonistic display. However, it is much less clear as to whether or not animals should communicate behavioral intentions, such as the intention to escalate. Maynard Smith (1982b) expressed the extreme view that the intention to escalate could not be communicated by a cheap signal, such as barking. He argued that if such a signal did indeed deter opponents it would be profitable to use this signal even in situations where no escalation is intended. As a result of this cheating the signal would then lose its meaning.

It was Maynard Smith's merit to bring this problem of cheating to the attention of behavioral biologists. However, he clearly pushed the cheating argument a little too far. Enquist (1985) provided the first biological model in which cheap signals do relate to intended fighting behavior at evolutionary equilibrium. As far as the empirical side of this problem is concerned, Caryl (1979) has reanalyzed a number of bird studies on aggresion in the light of Maynard Smith's idea. He concluded that the displays are far from being good predictors of physical attack. In contrast, Enquist et al. (1985) povide an example of aggressive behavior in fulmars, where the behavior shown by an individual at an early stage during a contest is related to this individual's persistence at a later stage. Andersson (1984) discusses the astonishingly high degree of interspecific variation found in threat displays.

So far we addressed the use of cheap signals in animal communication. More recent developments in biological signalling theory deal with the evolution of costly signals. Zahavi (1975, 1977, 1987) made the first major attempt to consider horns, antlers, and other structures as costly signals used in animal communication. He formulated the so-called handicap principle which roughly states that animals should demonstrate their quality to a rival or a potential mate by building costly morphological structures that in some sense handicap the signaller. The handicap principle was formulated in a verbal way by a field biologist. It attracted major criticism by Maynard Smith (1976, 1985) and other theoreticians who rejected Zahavi's idea on the ground of simple model considerations. However, recently

Pomiankowski (1987) and Grafen (1990a, b) used more intricate models from population genetics and game theory to re-establish Zahavi's principle. This shows nicely how biologists may sometimes learn more quickly about game rationality from studying real animals than by dealing with mathematical models.

8.4. The evolution of cooperation

There are a number of different ways (see Questions 6–9) in which cooperation between animals can be explained from an evolutionary point of view. Hamilton (1964) argued convincingly that altruistic behavior towards genetic relatives becomes understandable if one includes into the evolutionary measure of success (fitness) not only the altruist's own offspring but also the effect of the altruistic act on the receiver's reproductive output. The so-called *kin selection theory* weighs this effect on the receiver's offspring with the coefficient of genetical relatedness between donor and receiver. Hamilton's theory therefore takes indirect gene transmission via relatives into account.

Question 6. Is genetic relatedness an important key to the understanding of animal cooperation?

Because of its emphasis on genetic relatedness, kin selection theory does not attempt to explain cooperation between non-relatives. Except for this limitation in scope it has become the most successful theory about cooperation in animals. It is crucial, for example, for the understanding of castes and reproductive division of labor in highly social species of ants, bees, wasps, and termites. The remarkable complexity of social organization in ant colonies was recently surveyed by Hölldobler and Wilson (1990). Interestingly, many of the features of a typical ant colony are also found in a highly social mammal called the naked mole-rat [Sherman et al. (1991)]. Again, elementary kin selection theory rather than evolutionary game theory seems to provide the essential ideas for the understanding of sociality in this mammal. Kin selection also plays an important role in the explanation of helping and cooperative breeding in various bird species [Stacey and Koening (1990)].

Question 7. Does the (non-genetic) inheritance of resources or mating partners lead to animal cooperation?

Reyer (1990) described a cooperatively breeding bird population in which two kinds of cooperation occur that call for very different explanations. In the Pied Kingfisher, young male birds stay with their parents and help them to raise brothers and sisters instead of using the reproductive season for their own breeding attempts. This is explicable by straight forward kin selection arguments. In contrast, other young birds of the same species are also helpers but are not genetically related to the receivers of their help.

Reyer was able to show that the latter helpers through their help increased their future chances of becoming the breeding female's mating partner in the subsequent season. Note here that the mating market in Reyer's population is fairly asymmetrical. Female partners are scarce "resources". There is a game involved in the sense that it would be in the breeding pair's interest to impose a high work load on the helper whereas the helper would be better off by inheriting the female without incurring the cost of help. There seem to be a number of other biological examples where games of a similar kind are being played.

Question 8. Do animals cooperate because of partially overlapping interests?

Houston and Davies (1985) give an example of cooperation in birds where it may happen that two unrelated males care cooperatively for the same female's offspring. Here the point is that both have copulated with this female and both have a positive probability of being the father of the offspring they are raising. There is nevertheless a game-like conflict between the males as to how much each contributes to the joint care. Houston and Davies used ESS models to analyze the consequences of unequal mating success for the amount of male parental care. They also included the female as a third player in this game.

Question 9. Has evolution taught its organisms the logic of repeated games?

Trivers (1971) introduced the idea into evolutionary theory that a long-term benefit of an altruistic act may result from the fact that it will be reciprocated later by the receiver. He caused biologists to search for the phenomenon of reciprocal altruism. In his seminal paper, Trivers already referred to the game-theoretic literature on cooperation in a repeated game, such as the supergame of the prisoner's dilemma. After more than a decade of search for reciprocal altruism, however, it became clear that this search has not been very successful [Packer (1986)] except for the discovery of elaborate alliances in primates [e.g. Seyfart and Cheney (1984)].

In non-primates, only a few cases are known that have at least some flavor of reciprocity. Perhaps the most interesting example was provided by Wilkinson (1984) who studied vampire bats in Costa Rica. In the population under investigation, female vampire bats roost in groups of several individuals. They fly out every night searching for a blood meal. If they fail to obtain such a meal for two consecutive nights, death from starvation is very likely to occur. Females who return to the roost without a blood meal solicit help from other females by licking under their wings. If their begging is successful, blood will be donated by regurgitation. Wilkinson demonstrated that bats who had previously received help from a begging individual had a stronger tendency to provide help than others.

Although Trivers (1971) had already introduced the idea of cooperation in repeated games, this idea entered the biological literature for a second time when Axelrod and Hamilton (1981) analyzed the repeated Prisoner's dilemma game in the context of evolutionary game theory. They emphasize that cooperation could not very easily start to evolve in a population of non-cooperating animals because a threshold value for the proportion of cooperators in the population would have to be exceeded in order to make cooperation a successful evolutionary strategy. Here, kin selection may once again play an important role if one attempts to understand the initial spread of cooperation.

Axelrod and Hamilton also stated that "tit for tat" would be an evolutionarily stable strategy. This, however, is not quite correct because "tit for tat" fails to satisfy the second condition for evolutionary stability [Hammerstein (1984), Selten and Hammerstein (1984)]. Boyd and Richerson (1988) discussed further problems with the logic of biological cooperation in groups of more than two individuals. Nowak and Sigmund (1992) studied repeated conflict in heterogeneous populations.

Brown and Brown (1980) and Caraco and Brown (1986) interpreted the phenomenon of food sharing in communally breeding birds as cooperation based on the repeated structure of an evolutionary game. Their key argument is the following. When neighboring parents share the task of delivering food to their offspring, this may result in a fairly steady supply of nutrition even when an individual's daily foraging success varies a lot. This benefit from food sharing may be achievable in a game equilibrium, since foraging takes place repeatedly.

The idea that "tit for tat"-like cooperation might evolve caused Milinski (1987) to run a series of experiments in which he manipulated cooperation in sticklebacks. These fish swim towards a predator in order to inspect the enemy. Milinski studied pairwise predator inspection trips. Using a mirror, he replaced the partner of a stickleback by the animal's own mirror image. Depending on the mirror's angle, the artificial companion behaved in a more or less cooperative way by either following or swimming away. With a cooperative mirror image, the sticklebacks approached the predator more closely than with a non-cooperative mirror image.

At first glance this seems to be explicable in a fairly simple way through the effect of risk dilution, and there seems to be no need to invoke the far-sighted logic of repeated game interaction. However, Milinski offers a number of reasons that indicate the importance of the repeated structure [see also Milinski et al. (1990a, b)]. He created a lively discussion of the predator inspection phenomenon [e.g. Lazarus and Metcalfe (1990), Milinski (1990)] and caused other fish biologists to search for similar phenomena in their species [Dugatkin (1988, 1991), Dugatkin and Alfieri (1991)].

Fischer (1980, 1981, 1988) suggested another example of animal behavior that seemed to have some similarity with the "tit for tat" strategy. His study object is a fish called the black hamlet. This is a simultaneous hermaphrodite who produces sperm and eggs at the same time. When mating takes place between a pair of fish, each fish alternates several times between male and female role. Fischer argues that eggs are relatively few and expensive compared with sperm, and that, therefore, it is easy for a hamlet to get its own eggs fertilized by another fish. Thus a fish can use the eggs in its possession to trade them in for the opportunity to fertilize the eggs of another fish. The egg exchange takes place in a sequence containing several distinct spawning bouts.

To trade, or not to trade; that is the question Friedman and Hammerstein (1991) asked about the Hamlet's peculiar mating behavior. They modelled the conflict as a population game in which the search for other partners after desertion of a given partner is explicitly taken into account. It follows from their analysis that there is less similarity with a repeated prisoner's dilemma game and with "tit for tat" than Fischer (1988) had originally suggested. Friedman and Hammerstein show in particular that cheating does not pay in the Hamlet's egg trading, and that there is no scope for specialization on male function. Enquist and Leimar (1993) analyzed more generally the evolution of cooperation in mobile organisms.

8.5. The great variety of biological games

A final "tour d'horizon" will take us through a number of different biological applications. Evolutionary conflict seems to exist in almost any area of animal and plant behavior where individuals interact directly or indirectly. Consider, for example, the problem of parental care. Both father and mother of a young animal would benefit from offering parental care if this increased their offspring's chances of survival. However, when there is a cost involved in parental care, both father and mother would often be better off if they could transfer some or all of the work load to the other sex.

A similar point can be made about the problem of mate searching. Although both sexes have an interest in finding each other for the purpose of mating, an individual of either sex would benefit from saving the cost of mate searching. Once the sexes have found each other, there may be a conflict as to whether or not mating should take place. It may pay off for one sex to mate but not for the other.

The problem of conflict over parental investment was first described by Trivers (1972). However, through his way of looking at this problem he came close to committing the so-called Concorde fallacy [Dawkins and Carlisle (1976)]. Maynard Smith (1977) analyzed conflict over parental investment in the context of evolutionary game theory. Hammerstein (1981) provided a formal justification for his modelling approach. Grafen and Sibly (1978) further discussed the evolution of mate desertion. Mate searching games and conflict over mating were studied by Bengtsson (1978), Packer (1979), Parker (1979), and Hammerstein and Parker (1987).

Finding a mate can be facilitated if one sex calls in order to attract a mate. This is what some males – the callers – do, for example, in the natterjack toad. Other males in this species – the satellites – remain silent and try to intercept females as they approach the caller. If everybody is silent, there is a high incentive to behave as a caller. If everybody calls, some males have a high incentive to remain silent. Arak (1988) used game theory in order to model the evolution of this system.

Alternative male mating tactics are also found in various other parts of the animal kingdom. For example, in a population of the white-faced dragonfly males either defend territories on the pond or they act as transients in the vegetation surrounding the pond. Waltz and Wolf (1988) used ESS theory in order to understand the dragonfly mating strategies. Another dragonfly mating system was investigated in a similar spirit by Poethke and Kaiser (1987).

Before mating takes place a male may have to defend the female against other males who show an interest in the same partner (pre-copulatory mate guarding). Even after mating a male may have to defend the female against other males if a new mate would still be able to fertilize a considerable fraction of the eggs (post-copulatory mate guarding). These conflicts were analyzed by Grafen and Ridley (1983), Ridley (1983), and Yamamura (1986, 1987).

Related to the problem of mate guarding and access to a mate is the evolution of male armament. In many species males have a greater body size than females, and they are equipped with morphological structures that may serve as weapons in a fight. The evolution of such features resembles an arms race that can be treated as an evolutionary game [Parker (1983)]. However, the equilibrium approach of ESS theory may not always be appropriate [Maynard Smith and Brown (1986)] for this kind of problem.

Males use their weapons not only against their rivals. In several species of primates and in lions, males have a tendency to kill the offspring of a nursing female after they chased away the father of this offspring. This infant killing seems to considerably shorten the time span that elapses until the female becomes sexually receptive again. Hausfater et al. (1982), and Glass et al. (1985) argued that infant killing is an equilibrium strategy of an evolutionary game. One can look in a similar way at the phenomenon of siblicide and at other forms of sib competition [Dickins and Clark (1987), Parker et al. (1989)].

Fortunately, there are many cases in nature where less violent means than infant killing serve the purpose of gaining access to a mate. For example, the female may have to be "conquered" by donation of a nuptial gift that would resemble a box of chocolates in the human world. Parker and Simmons (1989) studied game-theoretic models of nuptial feeding in insects.

Yet another problem concerning the sexes is that of sex ratio among a female's offspring. Throughout the animal kingdom there is a strong tendency to produce both sexes in roughly equal numbers. This fact was already known to Darwin (1871) who reviewed sex ratio data for various groups of animals with little success in explaining the 1:1 property. It was indeed left to Fisher (1930, 1958) and to Shaw and Mohler (1953) to give a convincing explanation of this phenomenon. There is an implicit game-theoretic structure in Fisher's way of reasoning. This structure was revealed by Maynard Smith (1982a) who reformulated Fisher's thoughts in the formal framework of evolutionary game theory. Maynard Smith

(1980) also introduced a new theory of sexual investment in which the evolution of the sex ratio is subject to severe constraints. Reviews of sex ratio theory can be found in Charnov (1982), Karlin and Lessard (1986), and Bull and Charnov (1988).

When animals live in groups, such as a flock of birds, game-like situations arise from the problem of how individuals should avoid dangers caused by predators. There is an obvious conflict concerning the safest locations within the group. However, there is also an evolutionary conflict with regard to the individual rate of scanning for predators. Properties of this vigilance game were investigated by Pulliam et al. (1982), Parker and Hammerstein (1985), Motro (1989), Motro and Cohen (1989), Lima (1990), and McNamara and Houston (1992).

The distribution of foraging animals over a patchy habitat can be studied theoretically under the extreme assumption that each individual is *free* to choose its foraging patch and can change between patches at no cost. Fretwell and Lucas (1970) and Fretwell (1972) pointed out that *ideally* each animal should move to the patch were its success will be highest. When competitive ability is equal this should result in a distribution over patches with the property that (a) individuals in different patches have equal gains, and that (b) average success is the same for all patches. This pattern of habitat utilization is called the *ideal free distribution.*

A number of experiments [e.g. Milinski (1979), Harper (1982)] have shown that prediction (b) is supported by facts even when there are differences in competitive ability so that (a) does not hold. This has led to interesting theoretical work by Houston and McNamara (1988b) who enrich the game-theoretic analysis with additional methods from statistical mechanics [see also Parker and Sutherland (1986)]. A review of models in relation to the ideal free distribution is given by Milinski and Parker (1991). Yamamura and Tsuji (1987) examine the use of patchily distributed resources as a game that is closely related to standard models in optimal foraging theory.

Organisms are not always free to choose their location. An interesting example of unfree choice of location is that of seed dispersal. Here the parental plant can impose its "will" on the offspring by attaching morphological structures to it that will enforce migration. Hamilton and May (1977) used evolutionary game theory in order to discuss the question of how (enforced) seed dispersal is affected by intraspecific competition. They found that substantial dispersal should occur even when the habitat is homogeneous, constant, and saturated, and when there are high levels of seed mortality during dispersal. This is in sharp contrast with the view previously held in ecology that seed dispersal serves the main purpose of colonizing new empty patches. The seminal paper by Hamilton and May stimulated various authors to analyze dispersal as an evolutionary game [Comins et al. (1980), Comins (1982), Hastings (1983), Stenseth (1983), Levin et al. (1984), Holt (1985), Frank (1986), Lomnicki (1988), Taylor (1988), Cohen and Motro (1989)].

A somewhat related matter is the game between individuals of a parasitoid who have to decide where to lay their eggs. This game can lead to the phenomenon of superparasitism where the number of eggs laid in a host is far greater than the number of offspring that can emerge from this host. Oviposition in a previously parasitized host can be observed even when only one offspring will emerge from this host. Originally, biologists tended to interpret this phenomenon as a mistake made by the egg-laying parasitoid. However, with the help of evolutionary game theory superparasitism can be understood much more convincingly as an adaptive phenomenon [van Alphen and Visser (1990)].

Parental plants can impose their "will" not only on the spatial distribution of their offspring, but also on the temporal distribution of when the seeds should germinate; different degrees of encapsulation lead to different germination patterns. These patterns have to be adapted to the environment in competition with other members of the same species. Germination games and the problem of dormancy were investigated by Ellner (1985a, b; 1986, 1987).

There are more games played by plants. Desert shrubs use their root systems in order to extract water and nutrients from the soil. Near the surface, water is in short supply and there are many indications that neighboring plants compete for access to this scarce resource. Riechert and hammerstein (1983) modelled root competition as a population game and showed that the evolutionarily stable root strategy would not result in optimal resource exploitation at the species level. This result casts some doubt on classical ideas in ecology that are based on species welfare considerations.

The root game can be extended to a co-evolutionary case where plants of different species interact. Co-evolutionary games were first studied by Lawlor and Maynard Smith (1976). Riechert and Hammerstein (1983) introduced a modelling approach to co-evolution in which the game-theoretic nature of interspecific conflict is made more explicit. They analyzed the problem of root evolution under the simultaneous occurrence of intra- and interspecific competition for water. Their model shows that co-evolution should lead to the use of different soil strata by different species, and that similar species should show the phenomenon of character displacement in areas were they overlap. As a matter of fact, most communities of desert plants are stratified in this sense. A number of other papers relate co-evolution to the framework of evolutionary game theory [e.g. Eshel and Akin (1983), Brown and Vincent (1987), Eshel (1978), Leimar et al. (1986)].

So far we have discussed applications of evolutionary game theory to either zoology or botany. However, some ESS models relate to both disciplines. This is most obvious in the field of pollination biology. There is a game-like conflict between plants and insects where the plant uses its flower to advertise nectar. Once a pollinator visits the flower, the insect's pollination service is rendered regardless of whether or not nectar is found. Given that this is the case, why should a plant offer nectar at all and how much nectar should it provide? Cochran (1986) explained the existence of non-rewarding orchids using her own empirical work. In contrast, Selten and Shmida (1991) answered the question of why a population of rewarding plants can be in stable equilibrium.

References

Andersson, M. (1980) 'Why are there so many threat displays?', *Journal of Theoretical Biology*, **86**: 773–781.
Arak, A. (1988) 'Callers and satellites in the natterjack toad: evolutionarily stable decision rules', *Animal Behaviour*, **36**: 416–423.
Aumann, R.J. (1974) 'Subjectivity and correlation in randomized strategies', *Journal of Mathematical Economics*, **1**: 67–96.
Austad, S.N. (1983) 'A game theoretical interpretation of male combat in the bowl and doily spider (Frontinella pyramitela)', *Animal Behaviour*, **31**: 59–73.
Axelrod, R. and W.D. Hamilton (1981) 'The evolution of cooperation', *Science*, **211**: 1390–1396.
Bengtsson, B.O. (1978) 'Avoiding inbreeding: at what cost?', *Journal of Theoretical Biology*, **73**: 439–444.
Bishop, D.T. and C. Cannings (1978a) 'A generalized war of attrition', *Journal of Theoretical Biology*, **70**: 85–124.
Bishop, D.T. and C. Cannings (1978b) 'The war of attrition with random rewards', *Journal of Theoretical Biology*, **74**: 377–388.
Bomze, I.M. (1986) 'Non-cooperative 2-person games in biology: a classification', *International Journal of Game Theory*, **15**: 31–59.
Bomze, I.M. and B. Pötscher (1989) *Game theoretical foundations of evolutionary stability*. Berlin: Springer-Verlag.
Boorman, S.A. and P.R. Levitt (1972) 'A frequency-dependent natural selection model for the evolution of social cooperation networks', *Proceedings of the National Academy of Sciences, U.S.A.*, **69**: 2711–2713.
Boorman, S.A. and P.R. Levitt (1973) 'Group selection on the boundary of a stable population', *Theoretical Population Biology*, **4**: 85–128.
Boyd, R. and P.J. Richerson (1988) 'The evolution of reciprocity in sizable groups', *Journal of Theoretical Biology*, **132**: 337–356.
Brockmann, H.J., A. Grafen and R. Dawkins (1979) 'Evolutionarily stable nesting strategy in a digger wasp', *Journal of Theoretical Biology*, **77**: 473–496.
Brown, J.L. and E.R. Brown (1980) 'Reciprocal aid-giving in a communal bird', *Zeitschrift für Tierpsychologie*, **53**: 313–324.
Brown, J.S. and T.L. Vincent (1987) 'Coevolution as an evolutionary game', *Evolution*, **41**: 66–79.
Bull, J.J. and E.L. Charnov (1988) 'How fundamental are Fisherian sex ratios?', In: P.H. Harvey and L. Partridge, eds., *Oxford Surveys in Evolutionary Biology*. Vol. **5**: Oxford: Oxford University Press, pp. 96–135.
Caraco, T. and J.L. Brown (1986) 'A game between communal breeders: when is food sharing stable?', *Journal of Theoretical Biology*, **118**: 379–393.
Caryl, P.G. (1979) 'Communication by agonistic displays: what can games theory contribute to ethology?', *Behaviour*, **68**: 136–169.
Charnov, E.L. (1982) *The theory of sex allocation*. Princeton: Princeton University Press.
Clutton-Brock, T.H. and S.D. Albon (1979) The roaring of red deer and the evolution of honest advertisement. *Behaviour*, **69**: 145–170.
Cochran, M.E. (1986) *Consequences of pollination by chance in the pink lady's-slipper, Cypripedium acaule*, Ph.D. Dissertation, University of Tennessee, Knoxville.
Cohen, D. and U. Motro (1989) 'More on optimal rates of dispersal: taking into account the cost of the dispersal mechanism', *American Naturalist*, **134**: 659–663.
Comins, H.N. (1982) 'Evolutionarily stable strategies for dispersal in two dimensions', *Journal of Theoretical Biology*, **94**: 579–606.
Comins, H.N., W.D. Hamilton and R.M. May (1980) 'Evolutionarily stable dispersal strategies', *Journal of Theoretical Biology*, **82**: 205–230.
Crawford, V.P. (1990a) 'On the definition of an evolutionarily stable strategy in the "playing the field" model', *Journal of Theoretical Biology*, **143**: 269–273.
Crawford, V.P. (1990b) 'Nash equilibrium and evolutionary stability in large- and finite-population "playing the field" models', *Journal of Theoretical Biology*, **145**: 83–94.
Cressman, R. (1990) 'Strong stability and density-dependent evolutionarily stable strategies', *Journal of Theoretical Biology*, **145**: 319–330.

Cressman, R. and A.T. Dash (1991) 'Strong stability and evolutionarily stable strategies with two types of players', *Journal Mathematical Biology*, **30**: 89–99.
Crespi, B.J. (1986) 'Size assessment and alternative fighting tactics in Elaphrothrips tuberculatus (Insecta: Thysanoptera)', *Animal Behavior*, **34**: 1324–1335.
Darwin, C. (1871) *The descent of man, and selection in relation to sex.* London: John Murray.
Davies, N.B. (1978) 'Territorial defence in the speckled wood butterfly (Pararge aegeria): the resident always wins', *Animal Behaviour*, **26**: 138–147.
Davies, N.B. and T.R. Halliday (1979) 'Competitive mate searching in male common toads', *Bufo bufo Animal Behaviour*, **27**: 1253–1267.
Dawkins, R. (1976) *The selfish gene.* Oxford: Oxford University Press.
Dawkins, R. (1982) *The extended phenotype.* Oxford: Freeman.
Dawkins, R. and T.R. Carlisle (1976) 'Parental investment, mate desertion and a fallacy', *Nature, London*, **262**: 131–133.
Dickins, D.W. and R.A. Clark (1987) 'Games theory and siblicide in the Kittiwake gull, Rissa tridactyla', *Journal of Theoretical Biology*, **125**: 301–305.
Dugatkin, L.A. (1988) 'Do guppies play tit for tat during predator inspection visits?', *Behavioural Ecology and Sociobiology*, **25**: 395–399.
Dugatkin, L.A. (1991) 'Dynamics of the tit for tat strategy during predator inspection in the guppy (Poecilia reticulata)', *Behavioral Ecology and Sociobiology*, **29**: 127–132.
Dugatkin, L.A. and M. Alfieri (1991) 'Guppies and the tit for tat strategy: preference based on past interaction', *Behavioral Ecology of Sociobiology*, **28**: 243–246.
Ellner, S. (1985a) 'ESS germination strategies in randomly varying environments. II. Reciprocal yield law models', *Theoretical Population Biology*, **28**: 80–115.
Ellner, S. (1985b) 'ESS germination strategies in randomly varying environments. I. Logistic-type models', *Theoretical Population Biology*, **28**: 50–79.
Ellner, S. (1986) 'Germination dimorphisms and parent-offspring conflict in seed germination', *Journal of Theoretical Biology*, **123**: 173–185.
Ellner, S. (1987) 'Competition and dormancy: a reanalysis and review', *American Naturalist*, **130**: 798–803.
Enquist, M. (1985) 'Communication during aggressive interactions with particular reference to variation in choice of behaviour', *Animal Behaviour*, **33**: 1152–1161.
Enquist, M., E. Plane and J. Roed (1985) 'Aggressive communication in fulmars (Fulmarus glacialis) competing for food', *Animal Behaviour*, **33**: 1107–1120.
Enquist, M. and O. Leimar (1983) 'Evolution of fighting behaviour. Decision rules and assessment of relative strength', *Journal of Theoretical Biology*, **102**: 387–410.
Enquist, M. and O. Leimar (1987) 'Evolution of fighting behaviour: the effect of variation in resource value', *Journal of Theoretical Biology*, **127**: 187–205.
Enquist, M. and O. Leimar (1988) 'The evolution of fatal fighting', *Animal Behaviour*, **39**: 1–9.
Enquist, M. and O. Leimar (1993) 'The evolution of cooperation in mobile organisms', *Animal Behaviour*, **45**: 747–757.
Enquist, M., O. Leimar, T. Ljungberg, Y. Mallner and N. Segerdahl (1990) 'Test of the sequential assessment game I: fighting in the cichlid fish Nannacara anomala', *Animal Behaviour*, **40**: 1–14.
Eshel, I. (1978) 'On a prey-predator non-zero-sum game and the evolution of gregarious behavior of evasive prey', *American naturalist*, **112**: 787–795.
Eshel, I. (1983) 'Evolutionary and continuous stability', *Journal of Theoretical Biology*, **103**: 99–111.
Eshel, I. (1991) 'Game theory and population dynamics in complex genetical systems: the role of sex in short term and in long term evolution', in: R. Selten, ed., *Game Equilibrium Models. Vol. I, Evolution and Game Dynamics.* Berlin: Springer-Verlag, pp. 6–28.
Eshel, I. and E. Akin (1983) 'Coevolutionary instability of mixed Nash solutions', *Journal of Mathematical Biology*, **18**: 123–133.
Eshel, I. and M.W. Feldman (1984) 'Initial increase of new mutants and some continuity properties of ESS in two locus systems', *American Naturalist*, **124**: 631–640.
Ewens, W.J. (1968) 'A genetic model having complex linkage behavior', *Theoretical and Applied Genetics*, **38**: 140–143.
Fischer, E.A. (1980) 'The relationship between mating system and simultaneous hermaphroditism in the coral reef fish', Hypoplectrus nigricans (Serranidae). *Animal Behavior*, **28**: 620–633.

Fischer, E.A. (1981) 'Sexual allocation in a simultaneously hermaphrodite coral reef fish', *American Naturalist*, **117**: 64–82.
Fischer, E.A. (1988) 'Simultaneous hermaphroditism, tit for tat, and the evolutionary stability of social systems', *Ethology and Sociobiology*, **9**: 119–136.
Fisher, R.A. (1930, 1958) *The genetical theory of natural selection*, 1st edn. Oxford: Clarendon Press; 2nd revised edn. New York: Dover Publications.
Foster, D. and H.P. Young (1990) 'Stochastic evolutionary game dynamics', *Theoretical Population Biology*, **38**: 219–232.
Frank, S.A. (1986) 'Dispersal polymorphisms in subdivided populations', *Journal of Theoretical Biology*, **122**: 303–309.
Fretwell, S.D. (1972) *Seasonal environments*. Princeton: Princeton University Press.
Fretwell, S.D. and H.L. Lucas (1970) 'On territorial behaviour and other factors influencing habitat distribution in birds', *Acta Biotheoretical*, **19**: 16–36.
Friedman, J. and P. Hammerstein (1991) 'To trade, or not to trade; that is the question', in: R. Selten, ed., *Game Equilibrium Models*. Vol. I, *Evolution and Game Dynamics*. Berlin: Springer-Verlag, pp. 257–275.
Gillespie, J.H. (1977) 'Natural selection for variances in offspring numbers: a new evolutionary principle', *American Naturalist*, **111**: 1010–1014.
Glass, G.E., R.D. Holt and N.A. Slade (1985) 'Infanticide as an evolutionarily stable strategy', *Animal Behaviour*, **33**: 384–391.
Grafen, A. (1979) 'The hawk–dove game played between relatives', *Animal Behaviour*, **27**: 905–907.
Grafen, A. (1984) 'Natural selection, kin selection and group selection', in: J.R. Krebs and N.B. Davies, eds., *Behavioural Ecology: An Evolutionary Approach*, 2nd edn. Oxford: Blackwell Scientific Publication pp. 62–84.
Grafen, A. (1987) 'The logic of devisively asymmetric contests: respect for ownership and the desperado effect', *Animal Behaviour*, **35**: 462–467.
Grafen, A. (1990a) 'Sexual selection unhandicapped by the Fisher process', *Journal of Theoretical Biology*, **144**: 473–516.
Grafen, A. (1990b) 'Biological signals as handicaps', *Journal of Theoretical Biology*, **144**: 517–546.
Grafen A. and M. Ridley (1983) 'A model of mate guarding', *Journal of Theoretical Biology*, **102**: 549 567.
Grafen, A. and R.M. Sibly (1978) 'A model of mate desertion', *Animal Behaviour*, **26**: 645–652.
Haigh, J. (1975) 'Game theory and evolution', *Advances in Applied Probability*, **7**: 8–11.
Hamilton, W.D. (1964) 'The genetical evolution of social behaviour', I and II, *J. Theoretical Biology*, **7**: 1–16, 17–52.
Hamilton, W.D. (1967) 'Extraordinary sex ratios', *Science*, **156**: 477–488.
Hamilton, W.D. (1979) 'Wingless and fighting males in fig wasps and other insects'. In: M.S. Blum and N.A. Blum, eds., *Sexual Selection and Reproductive Competition in Insects*. London: Academic Press, pp. 167–220.
Hamilton, W.D. and R.M. May (1977) 'Dispersal in stable habitats', *Nature, London*, **269**: 578–581.
Hammerstein, P. (1981) 'The role of asymmetries in animal contests', *Animal Behaviour*, **29**: 193–205.
Hammerstein, P. (1984) 'The biological counterpart to non-cooperative game theory', *Nieuw Archief voor Wiskunde (4)*, **2**: 137–149.
Hammerstein, P. (1985) 'Spieltheorie und Kampfverhalten von Tieren', *Verhandlungen der Deutschen Zoologischen Gesellschaft*, **78**: 83–99.
Hammerstein, P. and G.A. Parker (1981) 'The asymmetric war of attrition', *Journal of Theoretical Biology*, **96**: 647–682.
Hammerstein, P. and G.A. Parker (1987) 'Sexual selection: games between the sexes', in: J.W. Bradbury and M.B. Andersson eds., *Sexual selection: testing the alternatives*. Chichester: Wiley, pp. 119–142.
Hammerstein, P. and S.E. Riechert (1988) 'Payoffs and strategies in territorial contests: ESS analysis of two ecotypes of the spider Agelenopsis aperta', *Evolutionary Ecology*, **2**: 115–138.
Harper, D.G.C. (1982) 'Competitive foraging in mallards: 'ideal free' ducks', *Animal Behavior*, **30**: 575–584.
Hastings, A. (1983) 'Can spatial variation alone lead to selection for dispersal?', *Theoretical Population Biology*, **24**: 244–251.
Hausfater, G., S. Aref and S.J. Cairns (1982) 'Infanticide as an alternative male reproductive strategy in langurs: a mathematical model', *Journal of Theoretical Biology*, **94**: 391–412.

Hausfater, G. and S.B. Hrdy, eds. (1984) *Infanticide: comparative and evolutionary perspectives*. New York: Aldine.
Hines, W.G.S. (1987) 'Evolutionary stable strategies: a review of basic theory', *Theoretical Population Biology*, **31**: 195–272.
Hines, W.G.S. and D.T. Bishop (1983) 'Evolutionarily stable strategies in diploid populations with general inheritance patterns', *Journal of Applied Probability*, **20**: 395–399.
Hines, W.G.S. and D.T. Bishop (1984a) 'Can and will a sexual diploid population attain an evolutionary stable strategy?', *Journal of Theoretical Biology*, **111**: 667–686.
Hines, W.G.S. and D.T. Bishop (1984b) 'On the local stability of an evolutionarily stable strategy in a diploid population', *Journal of Applied Probability*, **21**: 215–224.
Hines, W.G.S. and J. Maynard Smith (1979) 'Games between relatives', *Journal of Theoretical Biology*, **79**: 19–30.
Hölldobler, B. and E.O. Wilson (1990) *The ants*. Heidelberg: Springer-Verlag.
Hofbauer, J., P. Schuster and K. Sigmund (1979) 'A note on evolutionarily stable strategies and game dynamics', *Journal of Theoretical Biology*, **81**: 609–612.
Hofbauer, J. and K. Sigmund (1988) *The theory of evolution and dynamical systems*. Cambridge: Cambridge University Press.
Holt, R.D. (1985) 'Population dynamics in two-patch environments: some anomalous consequences of an optimal habitat distribution', *Theoretical of Population Biology*, **28**: 181–208.
Houston, A. (1983) 'Comments on "learning the evolutionarily stable strategy"', *Journal of Theoretical Biology*, **105**: 175–178.
Houston, A.I. and N.B. Davies (1985) 'The evolution of cooperation and life history in the Dunnock, Prunella modularis', in: R.M. Sibly and R.H. Smith, eds., *Behavioural Ecology: Ecological Consequences of Adaptive Behaviour*. Oxford: Blackwell Scientific publications, pp. 471–487.
Houston, A.I. and J. McNamara (1988a) 'Fighting for food: a dynamic version of the Hawk-Dove game', *Evolutionary Ecology*, **2**: 51–64.
Houston, A.I. and J. McNamara (1988b) 'The ideal free distribution when competitive abilities differ: an approach based on statistical mechanics', *Animal Behaviour*, **36**: 166–174.
Houston, A.I. and J. McNamara (1991) 'Evolutionarily stable strategies in the repeated Hawk-Dove game', *Behavioral Ecology*, **2**: 219–227.
Jansen, W. (1986) 'A permanence theorem for replicator and Lotka-Volterra systems', *Journal of Mathematical Biology*, **25**: 411–422.
Karlin, S. (1975) 'General two-locus selection models: some objectives, results and interpretations', *Theoretical Population Biology*, **7**: 364–398.
Karlin, S. and S. Lessard (1986) *Sex ratio evolution*. Princeton: Princeton University Press.
Kummer, H., W. Götz and W. Angst (1974) 'Triadic differentiation: an inhibitory process protecting pair bond in baboons', *Behaviour*, **49**: 62–87.
Lawlor, L.R. and J. Maynard Smith (1976) 'The coevolution and stability of competing species', *American Naturalist*, **110**: 79–99.
Lazarus, J. and N.B. Metcalfe (1990) 'Tit-for-tat cooperation in sticklebacks: a critique of Milinski', *Animal Behaviour*, **39**: 987–988.
Leimar, O. (1988) *Evolutionary analysis of animal fighting*. Doctoral dissertation, University of Stockholm.
Leimar, O., S.N. Austad and M. Enquist (1991) 'Test of the sequential assessment game II: fighting in the Bowl and doily spider Frontinella pyramitela', *Evolution*, **45**: 1862–1874.
Leimar, O., M. Enquist and B. Sillén Tullberg (1986) 'Evolutionary stability of aposematic coloration and prey unprofitability: a theoretical analysis', *American Naturalist*, **128**: 469–490.
Levin, S.A., D. Cohen and A. Hastings (1984) 'Dispersal strategies in patchy environments', *Theoretical Population Biology*, **26**: 165–191.
Levins, R. (1970) 'Extinction', in: M. Gerstenhaber, ed., *Some mathematical questions in biology*, Lectures on Mathematics in the Life Sciences, Vol. 2, American Mathematical Society, Providence, RI, pp. 77–107.
Liberman, U. (1988) 'External stability and ESS: criteria for initial increase of new mutant allele', *Journal of Mathematical Biology*, **26**: 477–485.
Lima, S.L. (1990) 'The influence of models on the interpretation of vigilance', in: M. Bekoff and D. Jamieson eds., *Interpretation and Explanation in the Study of Animal Behaviour: Comparative Perspectives*. Boulder, CO: Westview Press.

Lomnicki, A. (1988) *Population ecology of individuals*. Princeton: Princeton University Press.
Lorenz, K. (1966) *On aggression*. New York: Harcourt, Brace and World.
Maynard Smith, J. (1974) 'The theory of games and the evolution of animal conflicts', *Journal of Theoretical Biology*, **47**: 209–221.
Maynard Smith, J. (1976) 'Group selection', *The Quarterly Review of Biology*, **51**: 277–283.
Maynard Smith, J. (1977) 'Parental investment: a prospective analysis', *Animal Behaviour*, **25**: 1–9.
Maynard Smith, J. (1980) 'A new theory of sexual investment', *Behavioral Ecology and Sociobiology*, **7**: 247–251.
Maynard Smith, J. (1982a) *Evolution and the theory of games*. Cambridge: Cambridge University Press.
Maynard Smith, J. (1982b) 'Do animals convey information about their intentions?', *Journal of Theoretical Biology*, **97**: 1–5.
Maynard Smith, J. (1984) 'Game theory and the evolution of behaviour', *The Behavioral and Brain Sciences.*, **7**: 95–125.
Maynard Smith, J. (1989) *Evolutionary genetics*. Oxford: Oxford University Press.
Maynard Smith, J. and R.L.W. Brown (1986) 'Competition and body size', *Theoretical Population Biology*, **30**: 166–179.
Maynard Smith, J. and G.A. Parker (1976) 'The logic of asymmetric contests', *Animal Behaviour*, **24**: 159–175.
Maynard Smith, J. and G.R. Price (1973) 'The logic of animal conflict', *Nature, London*, **246**: 15–18.
McNamara, J. and A.I. Houston (1992) 'Evolutionarily stable levels of vigilance as a function of group size', *Animal Behaviour*, **43**, 641–658.
Milinski, M. (1979) 'An evolutionarily stable feeding strategy in sticklebacks', *Zeitschrift für Tierpsychologie*, **51**: 36–40.
Milinski, M. (1987) 'Tit for tat in sticklebacks and the evolution of cooperation', *Nature*, **325**: 433–435.
Milinski, M. (1990) 'No alternative to tit-for-tat cooperation in sticklebacks', *Animal Behaviour*, **39**: 989–999.
Milinski, M. and G.A. Parker (1991) 'Competition for resources', in: J.R. Krebs, and N.B. Davies, eds., *Behavioural Ecology. An Evolutionary Approach*, pp. 137–168.
Milinski, M., D. Pfluger, Külling and R. Kettler (1990a) 'Do sticklebacks cooperate repeatedly in reciprocal pairs?', *Behavioral Ecology and Sociobiology*, **27**: 17–21.
Milinski, M., D. Külling and R. Kettler (1990b) 'Tit for tat: sticklebacks (Gasterosteus aculeatus) "trusting" a cooperating partner', *Behavioural Ecology*, **1**: 7–11.
Moran, P.A.P. (1964) 'On the nonexistence of adaptive topographies', *Annals of Human Genetics*, **27**, 383–393.
Motro, U. (1989) 'Should a parasite expose itself? (Some theoretical aspects of begging and vigilance behaviour)', *Journal of Theoretical Biology*, **140**, 279–287.
Motro, U. and D. Cohen (1989) 'A note on vigilance behavior and stability against recognizable social parasites', *Journal of Theoretical Biology*, **136**: 21–25.
Nowak, M.A. and K. Sigmund (1992) 'Tit for tat in heterogeneous populations', *Nature*, **355**: 250–253.
Packer, C. (1979) 'Male dominance and reproductive activity in Papio anubis', *Animal Behaviour*, **27**: 37–45.
Packer, C. (1986) 'Whatever happened to reciprocal altruism?', *Tree*, **1**: 142–143.
Parker, G.A. (1974) 'Assessment strategy and the evolution of animal conflicts', *Journal of Theoretical Biology*, **47**: 223–243.
Parker, G.A. (1979) 'Sexual selection and sexual conflict', in: M.S. Blum and N.A. Blum, eds., *Sexual Selection and Reproductive Competition in Insects*. New York: Academic Press, pp. 123–166.
Parker, G.A. (1983) 'Arms races in evolution – an ESS to the opponent-independent costs game', *Journal of Theoretical Biology*, **101**: 619–648.
Parker, G.A. (1984) 'Evolutionarily stable strategies', in: J.R. Krebs and N.B. Davies, eds., *Behavioural Ecology: An Evolutionary Approach*, 2nd edn. Oxford. Blackwell Scientific Publications, pp. 30–61.
Parker, G.A. (1985) 'Population consequences of evolutionarily stable strategies', in: R.M. Sibly and R.H. Smith, eds., *Behavioural Ecology: Ecological Consequences of Adaptive Behaviour*. Oxford: Blackwell Scientific Publications, pp. 33–58.
Parker, G.A. and P. Hammerstein (1985) 'Game theory and animal behaviour', in: P.J. Greenwood, P.H. Harvey, and M. Slatkin, eds., *Evolution: Essays in honour of John Maynard Smith*. Cambridge: Cambridge University Press, pp. 73–94.
Parker, G.A., D.W. Mock and T.C. Lamey (1989) 'How selfish should stronger sibs be?', *American Naturalist*, **133**: 846–868.

Parker, G.A. and D.I. Rubinstein (1981) 'Role assessment, reserve strategy, and acquisition of information in asymmetrical animal conflicts', *Animal Behaviour*, **29**: 135–162.
Parker, G.A. and L.W. Simmons (1989) 'Nuptial feeding in insects: theoretical models of male and female interests', *Ethology*, **82**: 3–26.
Parker, G.A. and W.J. Sutherland (1986) 'Indeal free distributions when animals differ in competitive ability: phenotype-limited ideal free models', *Animal Behaviour*, **34**: 1222–1242.
Poethke, H.J. and H. Kaiser (1987) 'The territoriality threshold: a model for mutual avoidance in dragonfly mating systems', *Behavioral Ecology and Sociobiology*, **20**: 11–19.
Pulliam, H.R., G.H. Pyke and T. Caraco (1982) 'The scanning behavior of juncos: a game-theoretical approach', *Journal of Theoretical Biology*, **95**: 89–103.
Reyer, H.U. (1990) 'Pied Kingfishers: ecological causes and reproductive consequences of cooperative breeding', in: P.B. Stacey and W.D. Koenig, eds., *Cooperative breeding in birds*. Cambridge: Cambridge University Press, pp. 529–557.
Ridley, M. (1983) *The explanation of organic diversity*. Oxford: Clarendon Press.
Riechert, S.E. (1978) 'Games spiders play: behavioral variability in territorial disputes', *Behavioral Ecology and Sociobiology*, **3**: 135–162.
Riechert, S.E. (1979) 'Games spiders play II: resource assessment strategies', *Behavioral Ecology and Sociobiology*, **6**: 121–128.
Riechert, S.E. (1984) 'Games spiders play III: Cues underlying context associated changes in agonistic behavior', *Animal Behavior*, **32**: 1–15.
Riechert, S.E. and P. Hammerstein (1983) 'Game theory in the ecological context', *Annual Review of Ecology and Systematics*, **14**: 377–409.
Schaffer, M.E. (1988) 'Evolutionarily stable strategies for a finite population and a variable contest size. *Journal of Theoretical Biology*, **132**: 469–478.
Schuster, P., K. Sigmund, J. Hofbauer and R. Wolff (1981) 'Selfregulation of behavior in animal societies. Part I: Symmetric contests. II: Games between two populations with selfinteraction', *Biological Cybernetics*, **40**: 17–25.
Schuster, P., K. Sigmund and R. Wolff (1979) 'Dynamical systems under constraint organization. Part 3: Cooperative and competitive behavior of hypercycles', *Journal of Differential Equations*, **32**: 357–368.
Selten, R. (1975) 'Reexamination of the perfectness concept for equilibrium points in extensive games', *International Journal of Game Theory*, **4**: 25–55.
Selten, R. (1980) 'A note on evolutionarily stable strategies in asymmetric animal conflicts', *Journal of Theoretical Biology*, **84**: 93–101.
Selten, R. (1983) 'Evolutionary stability in extensive 2-person games', *Mathematical Social Sciences*, **5**: 269–363.
Selten, R. (1988) 'Evolutionary stability in extensive two-person games – correction and further development', *Mathematical Social Sciences*, **16**: 223–266.
Selten, R. and P. Hammerstein (1984) 'Gaps in harley's argument on evolutionarily stable learning rules and in the logic of "tit for tat"', *The Behavioral and Brain Sciences*.
Selten, R. and A. Shmida (1991) 'Pollinator foraging and flower competition in a game equilibrium model', in: R. Selten, ed., *Game Equilibrium Models. Vol. I, Evolution and Game Dynamics*. Berlin: Springer-Verlag, pp. 195–256.
Seyfart, R.M. and D.L. Cheney (1984) 'Grooming, alliances and reciprocal altruism in vervet monkeys', *Nature*, **308**: 541–543.
Shaw, R.F. and J.D. Mohler (1953) 'The selective advantage of the sex ratio', *American Naturalist*, **87**: 337–342.
Sherman, P.W., J.U.M. Jarvis and R.D. Alexander, eds. (1991) *The biology of the naked mole-rat*. Princeton: Princeton University Press.
Stacey, P.B. and W.D., Koenig, eds. (1990) *Cooperative breeding in birds*. Cambridge: Cambridge University Press.
Stenseth, N.C. (1983) 'Causes and consequences of dispersal in small mammals', in: I.R. Swingland and P.J. Greenwood, eds., *The Ecology of Animal Movement*. Oxford: Clarendon, pp. 63–101.
Sutherland, W.J. and G.A. parker (1985) 'Distribution of unequal competitors', in: R.M. Sibly and R.H. Smith, eds., *Behavioural Ecology: Ecological Consequences of Adaptive Behaviour*. Oxford: Blackwell Scientific Publications, pp. 255–273.
Taylor, P.D. (1988) 'An inclusive fitness model for dispersal of offspring', *Journal of Theoretical Biology*, **130**: 363–378.

Taylor, P.D. (1989) 'Evolutionary stability in one-parameter models under weak selection', *Theoretical Population Biology*, **36**: 125–143.
Taylor, P.D. and L.B. Jonker (1978) 'Evolutionarily stable strategies and game dynamics', *Mathematical Biosciences*, **40**: 145–156.
Trivers, R.L. (1971) 'The evolution of reciprocal altruism', *The Quarterly Review of Biology*, **46**: 35–57.
Trivers, R.L. (1972) 'Parental investment and sexual selection', in: B. Campbell, ed., *Sexual Selection and the Descent of Man*. Chicago: Aldine, pp. 139–179.
van Alphen, J.J.M. and M.E. Visser (1990) 'Superparasitism as an adaptive strategy', *Annual Review of Entomology*, **35**: 59–79.
van Damme, E. (1987) *Stability and perfection of Nash equilibria*. Berlin: Springer-Verlag.
Vickers, G.T. and C. Cannings (1987) 'On the definition of an evolutionarily stable strategy', *Journal of Theoretical Biology*, **129**: 349–353.
Vincent, T.L. and J.S. Brown (1988) 'The evolution of ESS theory', *Annual Review of Ecology and Systematics*, **19**: 423–443.
Waltz, E.C. and L.L. Wolf (1988) 'Alternative mating tactics in male white-faced dragonflies (Leucorrhinia intacta): plasticity of tactical options and consequences for reproductive success', *Evolutionary Ecology*, **2**: 205–231.
Weissing, F.J. (1990) *On the relation between evolutionary stability and discrete dynamic stability*. Manuscript.
Weissing, F.J. (1991) 'Evolutionary stability and dynamic stability in a class of evolutionary normal form games', in: R. Selten, ed., *Game Equilibrium Models I: Evolution and Game Dynamics*. Berlin: Springer-Verlag, pp. 29–97.
Wilkinson, G.S. (1984) 'Reciprocal food sharing in the vampire bat', *Nature*, **308**: 181–184.
Wilkinson, P.F. and C.C. Shank (1977) 'Rutting-fight mortality among musk oxen on Banks Island, Northwest Territories, Canada'. *Animal behavior*, **24**: 756–758.
Williams, G.C. (1966) *Adaptation and natural selection*. Princeton: Princeton University Press.
Wilson, F. (1961) 'Adult reproductive behaviour in Asolcus basalis (Hymenoptera: Selonidae)', *Australian Journal of Zoology*, **9**: 739–751.
Wu Wen-tsün and Jian Jia-he (1962) 'Essential equilibrium points of n-person non-cooperative games', *Scientia Sinica*, **11**: 1307–1322.
Wynne-Edwards, V.C. (1962) *Animal dispersion in relation to social behavior*. Edinburgh: Oliver and Boyd.
Yamamura, N. (1986) 'An evolutionarily stable strategy (ESS) model of postcopulatory guarding in insects', *Theoretical Population Biology*, **29**: 438–455.
Yamamura, N. (1987) 'A model on correlation between precopulatory guarding and short receptivity to copulation', *Journal of Theoretical Biology*, **127**: 171–180.
Yamamura, N. and N. Tsuji (1987) 'Optimal patch time under exploitative competition', *American Naturalist*, **129**: 553–567.
Yasukawa, K. and E.I. Bick (1983) 'Dominance hierarchies in dark-eyed juncos (Junco hyemalis): a test of a game-theory model', *Animal Behaviour*, **31**: 439–448.
Young, H.P. and D. Foster (1991) 'Cooperation in the short and in the long run', *Games and Economic Behaviour*, **3**: 145–156.
Zahavi, A. (1975) 'Mate selection – a selection for a handicap', *Journal of Theoretical Biology*, **53**: 205–214.
Zahavi, A. (1977) 'The cost of honesty (Further remarks on the handicap principle)', *Journal of Theoretical Biology*, **67**: 603–605.
Zahavi, A. (1987) 'The theory of signal detection and some of its implications', in: V.P. Delfino, ed., *International Symposium of Biological Evolution*. Bari: Adriatica Editrice.
Zeeman, E.C. (1981) 'Dynamics of the evolution of animal conflicts', *Journal of Theoretical Biology*, **89**: 249–270.

Chapter 29

GAME THEORY MODELS OF PEACE AND WAR

BARRY O'NEILL*

Yale University

Contents

1.	International relations game theory	996
	1.1. Game analyses of specific international situations	996
	1.2. The debate on realism and international cooperation	999
	1.3. International negotiations	1003
	1.4. Models of armsbuilding	1004
	1.5. Deterrence and signalling resolve	1006
	1.6. The myth that game theory shaped nuclear deterrence theory	1010
	1.7. First-strike stability and the outbreak of war	1013
	1.8. Escalation	1014
	1.9. Alliances	1016
	1.10. Arms control verification	1017
2.	Military game theory	1018
	2.1. Analyses of specific military situations	1019
	2.2. Early theoretical emphases	1021
	2.3. Missile attack and defence	1023
	2.4. Tactical air war models	1024
	2.5. Lanchester models with fire control	1025
	2.6. Search and ambush	1026
	2.7. Pursuit games	1026
	2.8. Measures of effectiveness of weapons	1027
	2.9. Command, control and communication	1027
	2.10. Sequential models of a nuclear war	1028
	2.11. Broad military doctrine	1029
3.	Comparisons and concluding remarks	1029
References		1031

*This work was done in part while I was a visiting fellow at the Program on International Security and Arms Control, Yale, and the Center for International Security and Arms Control, Stanford. It was supported by an SSRC/MacArthur Fellowship in Security. I would like to thank Bruce Blair, Jerry Bracken, Steve Brams, Bruce Bueno de Mesquita, George Downs, Joanne Gowa, David Lalman, Fred Mosteller, Standers Mac Lane, James Morrow, Barry Nalebuff, Michael Nicholson, Richard Snyder, James G. Taylor and John Tukey for their information and suggestions.

Handbook of Game Theory, Volume 2, Edited by R.J. Aumann and S. Hart
© *Elsevier Science B.V., 1994. All rights reserved*

Game theory's relevance to peace and war was controversial from the start. John McDonald (1949) praised the new discipline as "more avant-garde than Sartre, more subtle than a Jesuit," and quoted one Pentagon analyst, "We hope it will work, just as we hoped in 1942 that the atomic bomb would work." P.M.S. Blackett, who had guided British operational research during the war, was a sceptic: "I think the influence of game theory has been almost wholly detrimental." If it had practical relevance, he claimed, investors and card players would have snapped it up, but since they have not, "it is clearly useless for the much more complicated problems of war" (1961).

The debate has continued up to the present day, but it has been conducted mostly in the abstract, with critics trying to prove a priori that game theory is inapplicable. The present chapter surveys the work actually done on peace and war and uses it to rebut some common beliefs. It focuses on the subjects studied and the techniques of application rather than the mathematics. The international relations (IR) section aims to be comprehensive, and while this means including some obsolete or less-than-high-quality papers, it gives the reader an unfiltered account of what has been done. The military section is not comprehensive but includes the main subjects modelled so far, and notes the interaction of game models with new military strategy and technology.

1. International relations game theory

"Philosophical" discussions of game theory in IR are by Deutsch (1954, 1968), Waltz (1959), Quandt (1961), R. Snyder (1961), Boll (1966), Shubik (1968, 1991), Robinson (1970), Forward (1971), Rosenau (1971), Junne (1972), George and Smoke (1974), Plon (1976), Martin (1978), Wagner (1983), Maoz (1985), Snidal (1985a), Hardin (1986b), Larson (1987), Jervis (1988), O'Neill (1989b), Rapoport (1989), Hurwitz (1989), Gates and Humes (1990), Hollis and Smith (1990), Ludwig (1990), Bennett (1991), Abbey (1991), Riker (1992), Wagner (1992), Nicholson (1992), Morrow (1992b), and Leilzel (1993).

1.1. Game analyses of specific international situations

Compared to games in IR theory, studies of specific conflicts tend to use uninnovative mathematics, but they can be interesting when the authors are writing about their own countries and concerns. The goal is not prediction, of course, since the outcome is already recorded, but a game organizes the account of each nation's decisions and shows the essence of the conflict. The favorite subjects have been the 1962 Cuban Missile crisis, the United States/Soviet arms competition and the Middle East wars.

Economics game models involving goods or money allow assumptions about continuity, risk aversion, or conservation of the total after a division. Without the mathematical handle of an extensive commodity, IR theorists have often used 2 × 2 matrices and looked at only the ordinal properties of the payoffs to avoid having to justify specific values. Rapoport and Guyer's taxonomy (1966) of 78 ordinal games without tied payoffs has been used repeatedly. The most prominent study is by Synder and Diesing (1977) who set up 2 × 2 matrices for sixteen historical situations. Other examples are Maoz (1990) on Hitler's expansion, and on Israel/Syria interactions, Measor (1983), Boulding (1962), Nicholson (1970), and Brown (1986) on the arms race, Gigandet (1984) on German/Soviet relations between the World Wars, Dekmajian and Doran (1980) on the Middle East conflict, Ravid (1981) on the decision to mobilize in the 1973 War, Forward (1971) and Weres (1982) on the Cuban missile crisis, Levy (1985) on the war in Namibia, Satawedin (1984) on the Thai–American relations in the early 1960s, Schellenberg (1990) on the Berlin and Formosa Straits Crises of 1958, and Kirby (1988) and Call (1989) on antimissile systems. Scholars from South Korea and West Germany used matrix games to discuss reunifying their countries [Pak (1977), Gold (1973), Krelle 1972 see also Cook (1984)]. Hsieh (1990b) modelled the U.S. Iraq confrontation over Kuwait.

To find cardinal payoffs, some researchers have interviewed the real players and scaled their responses. Delver (1986, 1987, 1989) contacted senior Dutch naval officers and applied Dutch developments in bargaining theory. His interest was the effect of NATO maritime strategy on superpower relations. Lumsden (1973) interviewed Greek students about the Cyprus crisis, but was unable to talk with their Turkish counterparts. Plous (1985, 1987, 1988, 1993) sent questionnaires to all members of the US Senate, the Israeli Knesset, the Australian Parliament and the Politburo, asking them to estimate the benefits if each superpower reduced its nuclear arsenal. The Politburo was unresponsive in those pre-glasnost days, but thirty-two senators sent back data that would make good examples for introductory classes.

Two-by-two matrices often miss the conflict's essence. In the arms reduction questionnaire, for example, it is implausible that one side would be acting with no information about what the other did. Snyder and Diesing portray their situations as matrices, but their text often interprets the moves as sequential. Many applications should have considered games in extensive form or with incomplete information. One extensive form example on the Falklands/Malvinas War is by Sexton and Young (1985), who add an interesting analysis of the consequences of misperception. Young (1983) argues that the technique is no more complex but more adequate than 2 × 2 matrices. Johr (1971), a Swiss political scientist, uses the extensive form to discuss his country's strivings for neutrality in the face of Nazi Germany. With perfect information game trees, Joxe (1963) describes the Cuban Missile Crisis, O'Neill (1990c) examines Eisenhower's 1955 offer to Chiang Kai-shek to blockade China's coast, Hsieh (1990a) discusses Taiwan's contest with

China for diplomatic recognition from the rest of the world, and Bau (1991) models the changing relations of the two nations over forty years.

Game trees with perfect information are simpler models than they appear. Players are called on to make only a few of the possible payoff comparisons, so the modeller has to commit to few assumptions about utilities. For example, a two-person binary tree where each player makes two moves, has sixteen endpoints. A modeller ranking all outcomes for each player's preference would face 16!16! possibilities, equivalent in information to $\log_2(16!16!) = 89$ binary decisions on which of two outcomes is preferred. In fact by working backwards through the tree, only $8 + 4 + 2 + 1 = 15$ two-way decisions are enough to fix the solution. This is near the $\log_2(4!4!) = 9.2$ bits needed to fill in a 2×2 ordinal matrix, which has only half as many moves. Trees can be solved using very sparse information about the payoffs.

Another example of moving beyond 2×2 matrices is Zagare's $3 \times 3 \times 3$ game on the Vietnam war negotiations (1977). Incomplete information analyses include Guth (1986) on negotiations over an intermediate missile ban in Europe, Mishal et al. (1990) on the PLO/Israel conflict, and Niou (1992) on Taiwan signalling its resolve to the People's Republic. Wagner on the Cuban missile crisis (1989a) gives a mathematically significant model that also takes the facts seriously. Hashim (1984), at the University of Cairo, uses mostly three-person games to treat the Middle East conflict.

Responding to the limits of 2×2 games and Nash equilibria, some writers have devised their own solution theories, prominent ones being Howard's *metagames* (1971), *conflict analysis* [(Fraser and Hipel (1984), Fraser et al. (1987)], which are related to *hypergames* [Bennett and Dando (1982), and Brams and Hessel's *theory of moves* (1984) with its non-myopic equilibria. [For summaries and comparisons, see Hovi (1987), Powell (1988), Nicholson (1989), Zagare (1986) and Brams (1994).] Conflict analysis and hypergames were applied to the Suez crisis [Wright et al. (1980), Shupe et al. (1980)], the fall of France [Bennett and Dando (1970, 1977)], the 1979 Zimbabwe conflict [Kuhn et al. (1983)], the Cuban Missile Crisis [Fraser and Hipel (1982)], the 1973 Middle East War [Said and Hartley (1982)], the Falklands/Malvinas conflict [Hipel et al. (1988)], arms control verification [Cheon and Fraser (1987)], the nuclear arms race [Bennett and Dando (1982)], the Armenian/Azerbaijan conflict [Fraser et al. (1990)] and post-Cold War nuclear stability [Moore (1991)]. Some metagame analyses are by Richelson (1979) on nuclear strategy, Alexander (1976) on Northern Ireland, Benjamin (1981) on Angola, the Kurdish uprising and various other conflicts, Benjamin (1982) and Radford (1988) on the Iranian hostage crisis, and Howard (1968, 1972) on the Vietnam war, the testban negotiations and the Arab/Israeli conflict [see also Schelling (1968)]. With the theory of moves Leng (1988) studied crisis bargaining; Mor (1993), deception in the 1967 Middle East War; Brams studied the Cuban missile crisis (1985, 1994), the bombing of North Vietnam and the Middle East conflict (1994); Brams et al. (1994), international trade negotiations; Brams and

Mattli (1993) the Iranian Hostage Crisis; and Zagare the Middle East conflicts of 1967 and 1973 (1981, 1983), the Geneva conference of 1954 (1979), and the nuclear arms race (1987a). Brams (1976) compared standard and metagame analyses of the Cuban missile crisis, and Gekker (1989a) looked at the 1948 Berlin crisis through sequential equilibria and the theory of moves.

Advocates of these alternatives object to the mainstream theory for the wrong reasons, in my view, since the difficulties are treatable within it. Hypergame analyses involve one side misunderstanding the other's goals, as do incomplete information games. Metagame theory prescribes general policies of play, as do repeated games. Non-myopic equilibria deal with players who can switch back and forth in the cells of a matrix, like Danskin's matrix-differential games [Washburn (1977)] or some parts of repeated game theory. The theory of moves translates a 2×2 matrix into a particular extensive-form game and makes substantial assumptions about who can move at each point. These need to be justified within a given application but often they are not.

Advocates of these methods can point out that they are simple enough for decision makers to comprehend, unlike repeated game theory, so they can recommend moves in real decisions. To this end, Howard as well as Fraser and Hipel developed large interactive software programs, and government agencies have shown interest in such programs.

Only one other group has focused on methodology for converting the judgements of IR experts into games. The *Schwaghof Papers* [Selten et al. (1977)] summarize a 1976 Bonn conference where political analysts and game theorists met to model Middle East issues. Their method of *scenario bundles* estimated governments' goals, fears, strengths and plausible coalitions, and generated games of perfect information. Their first scenario in 1976 had Iraq tempted to invade Kuwait but deterred by the prospect of an Iran/Saudi/U.S. coalition and a Soviet withdrawal of support. In 1990 Iraq really invaded, and the common wisdom was that Saddam went ahead *in spite of* the Soviet Union's retirement as his patron. The model suggests the opposite possibility, that Soviet absence may have given Iraq less to lose. The little-known report is available from the first author at the University of Bonn.

1.2. The debate on realism and international cooperation

Turning to IR theory, a prevalent framework is "realism", sometimes refined as "neorealism" or "structural realism" [Keohane (1986a)]. It takes the main actors to be states, which pursue their interests within an anarchic world. To avoid the continuing danger of subjugation, they seek power, which in a competitive system is defined *relative to* other states and so can never be achieved mutually and finally. Significant cooperation is very difficult. The basic explanation of international events lies in the structure of this system, the resources held by its members and their alliances. Alternatives to realist theory allow more hope for cooperation, either through international institutions and interdependence (traditional idealism), a dominant power who keeps order (hegemonic stability theory), the common interests of subregions (regional integration theory), international institutions and standards of behavior (regime theory), or transnational class interests (Marxist structuralism). The more basic critique of postmodernism or reflectivism emphasizes that the actors' identities and interests are social creations.

Participants in the debate have increasingly used game models to state their arguments. Prisoner's Dilemma (PD) games have formalized a central issue between realists and regime theorists: How can cooperation develop in an anarchic world? Axelrod (1984) discussed cooperation in repeated PD with probabilistic termination, comparing it with war and competition between species. His ideas were connected to political realism by Stein (1982a, 1990), Keohane (1985, 1986b), Lipson (1984) and Grieco (1988, 1990). Informal work on game models of cooperation was done by Axelrod and Keohane (1985), Brams (1985), Jervis (1988), Larson (1987), Nicholson (1981), Snidal (1985a,b), Cudd (1990), Iida (1993b) and the authors in Oye's volume (1986). Taylor's work (1976, 1987), although not addressed to IR, has been influential. Snidal (1985b, 1991a) uses n-person and non-PD games to analyze international cooperation, McGinnis (1986) considers issue linkage, Sinn and Martinez Oliva (1988) policy coordination, Stein (1993) international dependence, and Pahre (1994) and Martin (1992b), the choice between bilateral and multilateral arrangements, all central ideas of regime theory.

Axelrod's "theory of cooperation" has triggered much interaction between mathematical and non-mathematical researchers, and produced papers of different degrees of formalization. It has influenced many non-formal policy writings on superpower dealings, e.g., Goldstein and Freeman (1990), Evangelista (1990), Weber (1991) and Bunn (1992). The simplified model was accessible and appeared just when the related political theory questions were prominent. Axelrod incorrectly believed that tit-for-tat was in effect a perfect equilibrium and he overlooked actual equilibria, but these errors were fortunate in a way since technicalities would have reduced his readership. An analysis of repeated PD with significant discounting turns out to be quite complicated [Stahl (1991)].

The study of cooperation connects with international institutions. One neglected area is game-theoretical analyses of voting to propose decision rules for international organizations, a rare example being Weber and Wiesmeth's paper (1994). Snidal and Kydd (1994) give a summary of applications of game theory to regime theory.

Relevant to the realist debate on cooperation are game models in international political economy, in the sense of international trade. Some studies interweave historical detail, conventional theory and simple matrix games. Conybeare's work (1984, 1987) separates trade wars into different 2×2 games with careful empirical comparison of the models. Josko de Gueron (1991) investigates Venezuela's foreign debt negotiations. Alt et al. (1988), Gowa (1986, 1989), Snidal (1985c), Krasner (1991) and Kroll (1992), Gowa and Mansfield (1993) analyze hegemony and regimes. In Iida's model (1991) the United States drastically cuts its contribution to the economic order to signal that it will no longer act as the hegemon. Wagner (1988a) uses Nash bargaining theory to analyze how economic interdependence can be converted to political influence, and Kivikari and Nurmi (1986) and Baldwin and Clarke (1987) model trade negotiations. A frequent theme has been the Organization of Petroleum Exporting Countries [Bird and Sampson (1976), Brown et al. (1976), Bird (1977), Sampson (1982)], often analyzed as a coalitional game.

A central realist concept is power. Selten and Mayberry (1968) postulate two countries who build weapons at linear costs, and share a prize according to the fraction that each holds of the total military force. The prize might involve political compliance from the adversary or third countries. Their model's form was identical to Tullock on rent-sharing twelve years later (1980). Harsanyi (1962) uses bargaining theory to define the notion of power, and compares the result with Robert Dahl's famous treatment. Brito and Intriligator (1977) have countries invest in weapons to increase their share of an international pie, as determined by a version of Nash bargaining theory, but their arsenals also increase the risk of war. Shapley and Aumann (1977) revise standard bargaining theory so that threats involve a non-refundable cost, e.g., the money spent on arms, and Don (1986) applies their model. Brito and Intriligator (1985) have countries that commit themselves to a retaliatory strategy to be implemented in the second period in order to induce a redistribution of goods in the first. In Garfinkel's model (1991) each side divides its resources between domestic consumption versus arms. She postulates that the resource quantities vary randomly and relates the state of the two economies to their armsbuilding. Matsubara (1989) gives a theoretical discussion of international power. Zellner (1962), Satawedin (1984) and Gillespie and Zinnes (1977) discuss large countries' attempts to gain influence in the Third World, the latter authors applying differential games. Mares (1988) discusses how a large power can force its will on a middle one, examining the United States versus Argentina in the Second World War and later Mexico. Gates (1989) looks at American attempts to prevent Pakistan from acquiring nuclear weapons, and Snidal (1990a) discusses how economic investment in the People's Republic of China by smaller but richer Taiwan, may aid the latter's security. Bueno de Mesquita (1990) and Bueno de Mesquita and Lalman (1990; 1992) examine how wars change hegemony in the international system, exemplified by the Seven Weeks War of 1866. Wagner (1989b) argues from bargaining theory that economic sanctions are ineffective at compelling a state to change its policy, and Tsebelis (1990) is similarly negative. Martin (1992a) uses 2×2 games to define varieties of international cooperation on sanctions. A subsequent model (1993) investigates influencing another state to join in sanctions against a third. Her approach represents a recent postitive trend of combining game models with empirical statistics or case studies [e.g., James and Harvey (1989, 1992), Bueno de Mesquita and Lalman (1992), Gates et al. (1992), Gowa and Mansfield (1993), Fearon (1994)].

Realist explanations of international events take states as having goals, but is this assumption justified? Kenneth Arrow devised his social choice theorem after Olaf Helmer, a colleague at the RAND Corporation, asked him whether strategic analysis was coherent in treating countries as single actors [Feiwel (1987) p. 193]. Thirty years later, Achen (1988) gave conditions on the decision-making process of a state's competing domestic groups such that the state would possess a utility function. Seo and Sakawa (1990) model domestic forces in a novel way, by assigning fuzzy preference orderings to states. Downs and Rocke (1991) discuss the agency problem of voters who want to prevent their leaders from using foreign interventions as a reelection tactic. McGinnis and Williams (1991, 1993) view correlated equilibria as a technique to link domestic politics and foreign policy. Whereas Achen's approach treated the entities within a country as unstructured, McGinnis and Williams suggest extending his idea by finding assumptions that yield a national utility function *and* reflect the structure of a government's decision-making institutions. Other game analyses involving domestic and international levels are by Bendor and Hammond (1992), Mayer (1992) and Lohmann (1993).

The IR mainstream tends to perceive game theory as innately realist-oriented in that it casts governments as the players and their choices as "rational". Rosenau (1971), for one, writes, "The external behavior called for in game-theoretical models, for example, presumes rational decision-makers who are impervious to the need to placate their domestic opponents or, indeed, to any influences other than the strategic requisites of responding to adversaries abroad". This is a misunderstanding. Many current game models fit this pattern, but this reflects realism's prevalence, not the theory's requirement, and other studies have used players within or across states. Intriligator and Brito (1984) suggest a coalitional game with six players: the governments, electorates and military-industrial complexes of the United States and the Soviet Union. Although the two military-industrial complexes cannot talk directly, they coordinate their activities toward their goals. In 1968, Eisner looked at Vietnam War policy, the players being North Vietnam, President Johnson, and the American public. Iida (1993a,c) and Mo (1990) formalize Putnam's notion of two-level games (1988), with negotiations conducted at the international level that must be ratified domestically. Bueno de Mesquita and Lalman (1990a), Ahn (1991) and Ahn and Bueno de Mesquita (1992) investigate the influence of domestic politics on crisis bargaining. Morrow (1991a) has the US administration seeking a militarily favorable arms agreement with the Soviet Union, but hoping to please the US electorate, which cares only about the economy and whether there is an agreement. Sandretto's players (1976) include multinational corporations, and he extends the ideas of Lenin and Rosa Luxemburg on how transnational control of the means of production influences world politics. These examples show that nothing precludes domestic or transnational organizations as players.

There is an irony in another misunderstanding, that game theory requires confidence in everyone's "rationality": the first significant IR game model, Schelling on pre-emptive instability (1958b, 1960), had each side worrying that the other would attack irrationally. The confusion derives from the ambiguous word "rationality", which perhaps could be dropped from our discourse, if it means that players' goals are always soberly considered and based on long-term self-centered "national interests"; game models do not assume that. O'Neill's NATO competitors in nuclear escalation (1990a) act from transitory emotions of anger and fear. It is not game theory that determines who are the players and what are their goals; that is up to the modeller. The essence of game theory is that each player considers the other's view of the situation in its own choice of strategy.

A cluster of papers debating a realist issue comprises Powell (19991a,b), Snidal (1990b, 1991a,c, 1993), Morrow (1990b), Niou and Ordeshook (1992), Grieco (1993) and Busch and Reinhardt (1993). They discuss whether pursuit of resources and power is absolute or relative, how immediate goals derive from long-term ones, and whether they preclude cooperation. Relevant also to realist theory, Niou and Ordeshook have developed a set of models of alliance behavior, regimes, the balance of power and national goals (1990b, 1990c, 1991, and others treated below in the alliance section).

Although reflectivist writers are usually hostile to game theory and regard it as the mathematical apology for realism, in fact the two approaches agree on the importance of socially-defined acts and subjective knowledge. Kratochwil (1989) uses matrix games in a complex account of the development of norms. My work on international symbols (O'Neill, 1994b) tries to define symbolic events in game terms and analyze symbolic arms contests, symbolic tension-provoking crisis events and symbolic leadership. Other models of honor, insults and challenges (O'Neill, 1991d, 1994b) emphasize the power of language acts to change players' goals.

1.3. International negotiations

The first applications to arms bargaining were mathematically auspicious. Herbert Scoville, Director of the U.S. Arms Control and Disarmament Agency, funded many prominent theorists to investigate the topic, and their papers pioneered the study of repeated games with incomplete information [Aumann et al. (1966, 1967, 1968)]. Saaty (1968) summarizes one of their models. Their work was too technical for political science, and the approach has been taken up only recently when Downs and Rocke (1987, 1990) studied arms control negotiations using repeated Prisoner's Dilemma and simulation, in an excellent interweaving of the factual and theoretical.

A fine conceptual use of sample 2 × 2 games for arms bargaining is Schelling's 1976 paper. When he revised it for the widely-read periodical *Daedalus* (1975), the editor insisted that he omit the matrices. The article has strongly influenced US thinking on arms control. No doubt policymakers found it easier to read in its bowdlerized form, but they missed the path that led Schelling to his ideas.

Guth and van Damme's interesting "Gorbi-games" (1991) analyze when arms control proposals show honorable intentions. Picard (1979), Makins (1985) and Bunn and Payne (1988) study strategic arms negotiations as elementary games, and Jonsson (1976) looks at the nuclear test ban. Hopmann (1978) uses Harsanyi's model of power to discuss conventional arms talks. Guth (1985, 1991) treats an issue that dogged NATO, that negotiating with a rival requires negotiating with one's allies. Brams (1990a) applies a model of sophisticated voting.

No studies have yet compared game models of bargaining with real international negotiations, probably because the earlier mathematical bargaining theory focused on axioms for the outcomes and not dynamics. Several applications to IR have suggested clever mechanisms to reach an accord. Kettele (1985) proposes that each side feed its best alternative to an agreement into a computer that simply announces whether any compromise is possible. A "go ahead" might prompt more effort to agree. Unaware of the theoretical literature, he did not ask whether each side would tell the computer the truth. Green and Laffont (1994) outline an interesting procedure in which the parties have only one agreement available, to sign or not sign. Each is unsure whether the agreement will help or hurt it, and has only partial knowledge about its own and the other's gains. In a series of simultaneous moves, each says "yes" or "no" until both accept together or time runs out. At the equilibrium each can infer some of the other's knowledge from the other's delays in accepting.

Other procedures to promote agreement include the use of coalitional Shapley values for the practice of European arms negotiators of meeting in caucuses before the plenary [O'Neill (1994a)]. Frey (1974) discusses an "insurance system" to convert PD traps of arms-building into mechanisms that prompt agreements to reduce arms. Vickrey (1978) applies the Clarke tax to induce governments to state their true interests, and Isard's (1965) technique uses metagames. Brams (1990b) puts his negotiating mechanisms in an international context. Singer (1963) and Salter (1986) expand an application originally proposed by Kuhn to make divide-and-choose symmetrical: each side separates its arms into n piles, and each chooses one of the other's piles to be dismantled. The goal is that neither can claim that it has had to disarm more, but what assumptions about how weapons holdings interact to produce security would guarantee a complaint-proof process?

1.4. Models of armsbuilding

When does armsbuilding dissuade the other from arming and when does it provoke? Jervis' answer (1978) involved the "security dilemma", in which each side tries to increase its own security, and thereby lowers the other's. If technology could provide a purely defensive weapon that could not be used for attack, the dilemma would disappear, but most weapons can be used for protection or aggression. Jervis modelled the security dilemma by the Stag Hunt game shown below, where the moves might be passing up some new military system or building it. Governments do end up at the poor equilibrium, each fearing the other will choose the second strategy. (Payoffs are ordinal with "4" the most preferred and the two equilibria are underlined.)

	Refrain	Build
Refrain	<u>4,4</u>	1,3
Build	3,1	<u>2,2</u>

Jervis' paper drew attention to the Stag Hunt, sometimes called Assurance or Reciprocity, which had been overshadowed by PD and Chicken. Sharp (1986), Snyder (1984) and Nicholson (1989) also discuss the Stag Hunt vis-a-vis the security dilemma. It is as old as PD: back in 1951 Raiffa had presented it to experimental subjects and discussed it in never-published parts of his Ph.D. thesis (1952). Sen (1967, 1973, 1974) had applied it in welfare economics, and Schelling (1960) combined it with incomplete information to model preemptive attacks, but otherwise it was ignored. Political scientists like Jervis saw its importance while game theorists were dismissing it as trivial on the grounds that one equilibrium is ideal for both players. Some recent theoretical papers have looked at it following Aumann's (1990) suggestion that the upper left equilibrium may not be self-enforcing [O'Neill (1991c)].

Avenhaus et al. (1991) let each government build offence- or defence-oriented weapons which the other will observe to judge aggressive intentions. My model [O'Neill (1993b)] relates the degree of offensive advantage and the tendency to military secrecy, another theme of the security dilemma literature. Kydd (1994) presents an interesting model where a player's probability distribution over the other's payoff for war is the former's private information.

Lewis Frye Richardson's differential equations for an arms race have fascinated so many scholars, it is not surprising that they have been incorporated into differential games [Deyer and Sen (1984), Gillespie and Zinnes (1977), Simaan and Cruz (1973, 1975a,b, 1977), van der Ploeg and de Zeeuw (1989, 1990), Gillespie et al. (1975, 1977), see also Case (1979), Moriarty (1984), Saaty (1968) and Amit and Halperin (1987)]. Melese and Michel (1989) analyze the closed-loop equilibrium of a differential game of arms racing, and show that reasonable assumptions about a player's payoff stream as a function of its own rate of buildup, the other's holdings, and a time discount factor, can lead to the traditional solutions of Richardson's equations.

A substantial literature addresses repeated PD models of arms races [Brams (1985), Downs et al. (1985), Lichbach (1988, 1989, 1990), Majeski (1984, 1986a,b), Schlosser and Wendroff (1988), McGinnis (1986, 1990), Stubbs (1988)]. "When do arms races lead to war?" is a fundamental question in the non-modelling literature, but it is hard to find a game model that relates mutual armsbuilding and the decision to attack, an exception being Powell (1993). Engelmann (1991) discusses how a country might change its preferences continuously from a PD game to achieve a cooperative outcome. Sanjian (1991) uses fuzzy games for the superpower competition through arms transfers to allies.

Another issue is whether to keep one's arsenal a secret [Brams (1985)]. Should you keep the adversary uncertain, or will your silence be seen as a token of weakness? In Nalebuff's model (1986), McGuire's (1966) and Hutchinson's (1992) secrecy is harmful, and in Sobel's (1988), extended by O'Neill (1993b), keeping secrets would help both, but is not an equilibrium. At the sensible equilibrium the two sides reveal their holdings, even though they would be better off by keeping them secret.

Sobel assumes two aggressive governments, each possessing a military strength known to itself and verifiably revealable to the other. Each decides whether to reveal, and then chooses either to attack or to adopt a defensive posture. If there is a war, whoever has the larger force wins, and losing is less harmful if the loser has a defensive posture. The payoffs to X decrease in this order: (1) X and Y attack, X wins; (2) X attacks and Y defends, X wins; (3) both defend, so there is no war; (4) X defends and Y attacks, X loses; and (5) X attacks and Y attacks, X loses. The payoffs to Y are analogous. A strategy for each must involve a cutoff such that a government will reveal its strength if and only if it is greater than the cutoff. However an equilibrium pair with a non-zero cutoff cannot exist, because a government just slightly below it will not want its strength estimated as that of the average government below. It will reveal its strength, and thus the set of keeping secrecy will peal away from the top end, like the market for lemons.

Some multistage games on the choice of guns or butter are by Brito and Interiligator (1985), Powell (1993), and John et al. (1993). A body of work not surveyed in detail here looks at strategies of weapons procurement [e.g., Laffont and Tirole (1993), Rogerson (1990)]. Also, I will only mention samples from the many experiments or simulations on game matrices representing the arms race. Alker and Hurwitz paper (n.d.) discusses PD experiments; Homer–Dixon and Oliveau's After-MAD game [Dewdney (1987)] is primarily educational; Alker et al. (1980) analyze a computer simulation on international cooperation based on repeated PD; experiments like those of Pilisuk (1984), Pilisuk and Rapoport (1963) and Plous (1987) interpret game moves as arming and disarming, escalating or deescalating; and Lindskold's experiment (1978) studies GRIT, the tit-for-tat tension-reducing procedure.

1.5. Deterrence and signalling resolve

The puzzle of deterrence has been how to make threats of retaliation credible. When they involve nuclear weapons, they are usually not worth carrying out. One example was NATO's traditional threat to use tactical nuclear arms in Europe if it was losing a conventional war. Far from saving Europe, this would have devastated it beyond any historical precedent. The difficulty is this: to trust deterrence is to believe (1) that the adversary will not act suicidally, and (2) that the adversary thinks that you might react suicidally. Game models did not cause this tension, they just made it more obvious.

One response is Chain Store-type models of "reputation", where the other holds some small credence that threatener would actually prefer to retaliate, and the threatener is careful not to extinguish this possibility as the game is repeated [Wilson (1989)]. However, the assumptions of repeated plays and evolving reputations do not fit nuclear crises. Issues change from one to the next, government leaders come and go, and in the case of the ultimate nuclear threat, the game will not be repeated. Security scholars have taken other approaches to the credibility problem, following especially Schelling (1960, 1966a, 1967). He gave two alternatives to brandishing threats of total destruction. The first is to perform a series of small escalations that gradually increase the pain suffered by both antagonists. The second method involves setting up a mechanism that repeatedly generates a possibility of total disaster, allowing the two sides to compete in risk-taking until one gives in, like two people roped together moving down a slope leading to a precipice. A threat to commit murder-and-sucide might be ignored, but one to incur a chance of it might work. These schemes assume too much information and control to really be implemented [O'Neill (1992)], but they became the paradigms in nuclear deterrence theory. For United States nuclear strategy they generated attempts to set up "limited nuclear options", and deploy weapons that fill in "gaps in the escalation ladder".

Powell (1987a,b, 1989a,b,c, 1990, 1994) makes Schelling's ideas mathematically rigorous with sequential games of incomplete information. A government's resolve can be high or low, and if it stays in the conflict, the other will raise its estimate of this resolve. The mainstream literature uses term "resolve" vaguely but Powell's models justify its definition as a ratio of certain utility increments, the ratio involved in determining who will prevail. He discusses how perceptions of resolve, not just actual resolve, influences the outcomes how the information held by the parties affects the likelihood of inadvertent war, and how a change in the size of the escalatory steps alters the likelihood of war. Nalebuff's related model (1986) has a continuum of escalatory moves.

The common concept of *showing* resolve is full of puzzles: Clearly the weaker you are the more important it is to show resolve, so why are aggressive actions not interpreted as signs of weakness [Jervis (1988)]? In many cases, weak and strong alike would make a display of resolve, perhaps because the action costs little, and because no action would be taken surely to mean weakness. But if the weak always act, what rational basis is there for inferring weakness from inaction? Inferences from moves no one would make are the domain of equilibrium refinement theory, and Nalebuff (1991) applies it in a simple model of promoting reputation. A situation presents you with the choice of Acting or Not Acting, and acting confers on you a benefit of x, known to all, and a cost of c, known only to you. Letting $c = c_A$ or c_N be everyone else's estimate of your cost following their observation of you choosing Act or Not Act, you will receive a "reputation for toughness" payoff of $a(1-c)$, where $a>0$. Thus, Act yields you $x-c+a(1-c_A)$ and Not Act, $a(1-c_N)$.

Nalebuff's structure is not strictly a game since one side simply observes, but it shows how different equilibrium theories restrict others' belief about your cost. Assuming that $x=\frac{1}{4}$, $a=\frac{1}{2}$, and that the audience holds a uniform distribution for c before they observe your move, one sequential equilibrium calls for Acting if $c<\frac{1}{2}$, for Not Acting otherwise, so those who can act cheaply will do so. In another, you should choose Act no matter what your type, since it is assumed that others are disposed to believe someone choosing Not Act has the maximum cost of 1. A further sequential equilibrium calls for Not Acting for all types, since it is assumed that if you were to Act, others would believe that your cost is 1, that your eagerness to prove yourself betrays weakness. The requirement of rationalizability rejects the equilibrium Not Act for any cost. The most that Act could benefit you through reputation is $\frac{1}{2}$, so if others see you Act, they should decide that your cost must be no higher than $\frac{1}{2}$, certainly not 1. A cost of $\frac{1}{2}$ is the same cost as they would estimate for you if you chose Not Act, so even when others' beliefs are worst for you, Acting does not change your reputation, so Act only if it is worth it disregarding reputation, i.e., if $c<\frac{1}{4}$. Nalebuff eliminates other equilibria using universally divine equilibria and perfect sequential equilibria. His approach adds to our understanding of both refinement theories and deterrence. Looking at a specific application, some government's attempt to signal, we can ask whether the refinement is reasonable to admit or reject some beliefs, and we are led to seek out historical evidence on how the target of the signal actually received it. The political science literature has mostly neglected the latter question, and the few who have examined it, e.g., Thies (1980) on U.S. signalling during the Vietnam war and George and Smoke (1974) on Chinese intervention in Korea, have not supported the elaborate schemes of military buildups and violence that governments have used as signals.

Referring to the 1990 war of insults between Bush and Saddam Hussein, I discuss displays of resolve, including a volatile "war of face", where the sides use amounts of their prestige as "bids" for a prize [O'Neill (1994b)]. The game is like a war of attrition with incomplete information about the values of the prize, except that only the loser must pay. Fearon (1992b,c) develops a model where the loser alone must pay, but has the outside option of going to war. Fearon (1990) and Brito and Intriligator (1987) present costly signalling models, and Banks (1990) gives a general theorem on when greater resolve leads to success in a crisis. O'Neill (1991d) suggests a mechanism behind large powers' preoccupation with commitment over issues of no innate importance, based on an analysis of the chivalric custom of challenges to honor in the Middle Ages.

Kilgour and Zagare (1991a,b, 1993) set up an incomplete information game to compare the conventional literature's concept of credibility with game theory's, and Kilgour (1991) uses it to theorize about whether democratic states are more warprone. Wagner (1988b, 1992) and Zagare (1990a) use sequential equilibria to answer those political scientists who have criticized deterrence theory from psychological and historical perspectives, an issue also addressed by Mor (1993). Bueno de Mesquita and Lalman (1990a,b) discuss how each side's incomplete information about the other's intentions can lead to a war desired by neither, and argue that the confirmation of deterrence theory may be in part an artifact. Quester (1982) uses matrices to categorize ways that wars can start, using historical examples. Stein (1980, 1982b) looks at various incidents from the viewpoint of conditional rewards and punishments, and at types of misperception. Wilson (1986) gives a repeated game model of attacks and counterattacks in which a sequential equilibrium directs each side to maintain a positive probability of striking at each trial. Zagare (1985, 1987b) axiomatizes the structure of mutual deterrence situations to derive that it is a PD game, but Brams and Kilgour (1988b) critique his assumptions and end up at Chicken.

To analyze partial credibility of a deterrent threat, O'Neill (1990a) posits a threatener whose probability of retaliation increases with the damage done by the adversary and decreases with the cost to the threatener of retaliating. The model assesses one justification put forward for "coupling", NATO's conscious attempt to ensure that a conventional war in Europe would trigger a global nuclear war. Wagner (1982) and Allan (1980, 1983) think of deterrence as negotiation over the boundaries of power, over what rivals may do without facing retaliation. Examining multi-sided threats, Scarf (1967) assumes that each government allocates some weapons specifically against each of the others, and that each configuration of all parties' allocations yields a vector of ordinal utilities. Adapting his n-person solution theory (1971), he states assumptions guaranteeing a core allocation of military threats. His theory is developed by Kalman (1971) and modified to a sequential game by Selten and Tietz (1972).

In O'Neill (1989a, 1994b) I extend Kull's "perception theory" explanation (1988) of why the superpowers built far more nuclear weapons than needed for deterrence and chose weapons designed for unrealistic counterforce wars. Each side knows that warfighting ability and superiority in numbers are irrelevant, and knows that the adversary knows this, but they are unsure that the adversary knows that they know it. Using Aumann's representation of limited common knowledge (1988), the model exhibits belief structures that induce governments to build militarily useless armaments.

Balas (1978) calculates the appropriate size of the U.S. Strategic Petroleum Reserve, proposed to reduce vulnerability to oil embargoes. Compared to a non-game model based on an exogenous likelihood of an embargo, his approach calls for smaller reserves, because it takes account that reserves tend to deter an embargo. Several authors look at deterrence and negotiation with "terrorists" [e.g., Sandler and Lapan (1988)].

Intriligator (1967) and Brito (1972) model a missile war as a differential game, with bounds on the rates of fire and the attrition of missiles and economic targets governed by differential equations. Intriligator's approach has been widely cited. It was extended by Chaatopadhyay (1969), and applied in many subsequent articles to derive a lower limit on armaments needed to guarantee deterrence.

Two French scholars, Bourqui (1984) and Rudnianski (1985, 1986), considered their country's worry, how a small nuclear power can deter a much larger one. Threats and deterrence in elementary games are treated also by Dacey (1987), Maoz and Fiesenthal (1987), Moulin (1981) Nalebuff (1988), Reisinger (1989), Schelling (1966b), Snyder (1971), Stefanek (1985), Nicholson (1988), Bueno de Mesquita and Lalman (1988, 1989), Fogarty (1990), Gates and Ostrom (1991), Langlois (1991a,c) Rudnianski (1991) and Cederman (1991). Some matrix games portray the ethical dilemmas of nuclear weapons [Hardin (1983, 1986a), Gauthier (1984), Woodward (1989)]. More general discussions of games and deterrence are by Nicholson (1990), Kugler and Zagare (1987, 1990), P. Bracken (1984), P. Bracken et al., (1984), and Gekker (1989b), and O'Neill (1989b) surveys the goals of game models of deterrence.

McGinnis (1992) critiques signalling applied to deterrence, and O'Neill (1992) suggests that game models are biassed towards armsbuilding in that they usually start after the crisis is in progress, and also ignore the promise half of deterrence: if you don't attack me, I won't attack you.

1.6. The myth that game theory shaped nuclear deterrence strategy

One myth about game models and deterrence is worth refuting in detail. It is that in the late 1940s and 1950s thinking on nuclear strategy was molded by game theory. By the end of the Cold War this claim was so widely believed that no evidence was needed to support it, and it is often used to dismiss the game theory approach as tried and rejected. Nye and Lynn-Jones (1988) write that as the field of strategic studies was developing, "Deterrence theory and game theory provided a powerful unifying framework for those central issues, but often at a cost of losing sight of the political and historical context" (p.6). The current view, they say, holds that "the abstract formulations of deterrence theory – often derived from game theory – disregard political realities" (p.11). Evangelista (1988, p.270) writes, "the early study of the postwar Soviet–American arms race was dominated by game theory and strategic rational-actor approaches". Horowitz (1970, p.270) states that game theory became the "operational codebook" of the Kennedy Administration. According to Steve Weber (1990), its mathematical elegance made it "almost irresistible" to those trying to untangle nuclear strategy, and it promoted a basic delusion of American policy, that the United States must match Soviet weapons type-for-type. In his highly-praised history of nuclear strategy, Freedman (1983, p.181) states, "for a time, until the mid-1960s, the employment of matrices was the *sine qua non* of a serious strategist".

In fact, with a couple of exceptions, substantial game modelling of international strategy started only in the later 1960s, after the tenets of nuclear strategy had already developed. Nye and Lynn-Jones give no supporting references, and when Freedman asserts that any serious strategist had used matrix games, can we believe that non-users Brodie, Kahn, Kaufmann, Kissinger or Wohlstetter were less than serious? Hedley Bull, who had associated with the American nuclear strategists, wrote (1967, p.601), "the great majority of civilian strategists have made no use at all of game theory and indeed would be at a loss to give any account of it". Licklider's survey of 177 non-governmental nuclear strategists found only two mathematicians (1971). Donald Brennan (1965), a prominent strategist at the Hudson Institute, estimated that his center's reports, classified and unclassified, might fill twenty feet of shelf space, but "the number of individual pages on which there is any discussion of concepts from game theory could be counted on one hand; there has been no use at all made of game theory in a formal or quantitative sense... gaming is much used by nuclear strategists, game theory is not".

Aiming to show game theory's influence, Freedman presents the 2×2 Chicken matrix as one used by strategists of the 1950s and early 1960s. Chicken did not appear until the mid-1960s, and it is ironical that its sources were not nuclear planners, but two noted peace activists, Bertrand Russell and Anatol Rapoport. In 1957 Russell used the teenage dare as an analogy for the arms race, and in 1965 Rapoport defined "Chicken" as the now-familiar 2×2 matrix, [O'Neill (1991c)]. Russell's highway metaphor was prominent in the early discussions of strategists Schelling, Kahn and others, but the matrix game had not yet been defined.

When one tracks down cited examples of game theory's influence, most fall into one or another type of non-application. Some employ wargaming or systems analysis, rather than game theory. Some deal with military tactics, like submarine search or fire control, rather than strategy and foreign policy. Some use parlor games like monopoly, chess or poker as metaphors for international affairs, but give no formal analysis. Some only explain the principles of mathematical games and hope that these might lead to a model [e.g., Morton Kaplan, (1957)]. Others construct matrices, but analyze them using only decision theory. In particular, Ellsberg (1961) and Snyder (1961), writing on deterrence, and Thomas Schelling (1958a, 1960) on threat credibility, calculated no game-theoretical equilibria. A decision-maker in their models maximized an expectation against *postulated* probabilities of the adversary's moves. It was just to avoid assuming such probability values that minimax strategies and equilibria were developed. The adversary's move probabilities are not assumed but *deduced* from the situation. George and Smoke's influential book explains Snyder's 1961 deterrence model in a "modified version" (1974, p.67). Referring back to the original, one finds that the modification was to drop Snyder's postulated probabilities of a Soviet attack, thereby turning his decision model into a game. Their treatment gives the impression that IR game theory began earlier than it really did.

Fred Kaplan's lengthy attempt (1983) to link nuclear strategy to game theory focuses on the RAND Corporation. He describes the enthusiasm of mathematician

John Williams to assemble a team there, but evidently those hopes never materialized; RAND's applications addressed mainly military tactics. The only arguable exceptions that have appeared were Schelling's preemptive instability model (1958a) written during a sabbatical year at RAND, and the borderline game model of Goldhamer, Marshall and Leites' secret report on deterrence (1959), which derived no optimal strategies. In Digby's insider history of strategic thought at RAND (1988, 1990), game theory is barely mentioned [see also Leonard (1991)].

A milder claim was that game theory set an attitude; strategists approached their problem with game theory principles in mind, by considering the adversary's perspective. As an example of game reasoning, Kaplan recounts Albert Wohlstetter's early 1950s argument that American bomber bases near the Soviet Union were vulnerable to surprise attack. This is no more than thinking from the enemy's viewpoint, which must be as old as warfare.

As a criterion for what counts as a "game model", one should expect at least a precise statement of assumptions and a derivation of the parties' best strategies, but there are almost none of these in the early nuclear strategy literature. The only such models published were Morton Kaplan's on deterrence (1958) and Schelling's use of 2×2 matrices to explicate trust and promises, and study first-strike instability, in *The Strategy of Conflict*. Schelling's book as a whole has been enormously influential, but its formal models had small effect and were left untouched by other authors for many years [e.g., Nicholson, 1972]. Schelling interspersed the models with shrewd observations about tacit bargaining and coordination in crises, but he did not derive the latter from any formal theory. His informal ideas were beyond extant mathematics and in fact the reverse happened: his ideas spurred some later mathematical developments [Crawford (1990)].

The claimed link between game theory and the bomb began in the late 1940s with suggestions that the new logic of strategy would reveal how to exploit the new weapon. It was bolstered by von Neumann's visible activity in defense matters, as an advisor to the US government on nuclear weapons and ICBMs [Heims (1980)], and later Chairman of the Atomic Energy Commission. The Hollywood movie *War Games* depicted a mathematical genius whose supercomputer controls the U.S. missile force and solves nuclear war as a poker game. Likewise Poundstone's *Prisoner's Dilemma* (1992) intersperses chapters on von Neumann's life, nuclear strategy and game theory basics. A reader might conclude that von Neumann connected the two activities, but nowhere does the author document von Neumann ever applying games to the Cold War. On the contrary, according to his colleague on the Air Force Scientific Advisory Board, Herbert York (1987, p.89), "although I frequently discussed strategic issues with him, I never once heard him use any of the jargon that game theorists and operations analysts are wont to use – 'zero-sum game', 'prisoner's dilemma' and the like – nor did he ever seem to use such notions in his own thinking about strategy... he always employed the same vocabulary and concepts that were common in the quality public press at the time". Morgenstern's writings on military matters included parlor game analogies and mathematical-sounding terms (e.g., 1959), but no models.

During the 1960s the debate on the validity of nuclear strategy reinforced the myth. The critics attacked game theory and nuclear strategy together, as acontextual, ahistorical, technocratic, positivistic and counter to common sense. Supporters saw game theory as right for the atomic era, when wars must be planned by deduction rather than based on historical cases. The antis opposed a method that apparently prescribes behavior in complex situations with no role for any experiences, intuitions or goals that cannot be formalized. They saw its mathematical character as a warning that the same technocrats that gave the world the nuclear arms race, now wanted control over decisions on war. The unfortunate name "game theory" suggested that international politics was a frivolous pastime played for the sake of winning, and the idea of a "solution" seemed arrogant. [Examples include Maccoby (1961), Aron (1962), and Green (1966) as attackers; Wohlstetter (1964) and Kaplan (1973) as defenders; and Schelling (1960) and Rapoport (1964, 1965, 1966) as reformers.]

The debate proceeded along these lines, but neither group pinpointed just how game theory had influenced nuclear thinking. The best critiques of strategic thought [Rapoport (1964), Green, 1966)] did not assert that game theory was widely used to formulate it. Rapoport's thesis was more subtle, that knowledge of formal conflict theory's existence creates an expectation that nuclear dilemmas can be solved by a priori reasoning. Many other writers, however, asserted a link without specifying just how game theory was involved [e.g., Zuckermann (1961), Horowitz (1962, 1963a,b, 1970), Heims (1980), Morris (1988)].

The notion that game theory has already been tried probably dampens current interest and provides an excuse to deemphasize it in graduate programs. Some appropriate mathematical techniques, like incomplete information, equilibrium selection in extensive games and limitations on common knowledge, were invented fairly recently, and have seen application only in the last few years.

1.7. First-strike stability and the outbreak of war

In his model of preemptive instability (1960), Schelling published the first non-trivial game application to international strategy, a full-fledged game of incomplete information. Starting with a Stag Hunt game, he assumed that each side assigns some probability that the other might attack irrationally, and traced the payoff-dominating equilibrium as this fear grows. Nicholson (1970) extended his model.

Game theory has contributed by sharpening existing informal concepts, sometimes suggesting a numerical measure. Examples are Harsanyi's definition of power (1962), Axelrod's formula for conflict of interest (1969), O'Neill's measure of advantage in a brinkmanship crisis (1991c), Powell's formula for resolve (1987a,b, 1990), and various measures for the values for the worths of weapons, discussed later. O'Neill offers a formula for degree of first-strike instability (1987), the temptation for each government to attack due to fear that the other is about to, fuelled by a military advantage from attacking first. The decision is cast as a Stag Hunt, and the measure is applied to the debate on space-based missile defences. Scheffran (1989) uses this formula as part of a computer simulation of arms-building dynamics, and Fichtner (1986a) critiques it. Bracken (1989b) proposes an alternative measure. Interest in missile defences in space has generated about three dozen quantitative papers asking whether these systems increase the temptation to strike [Oelrich and Bracken (1988)], and more recent work has analyzed the stability of arms reduction proposals [Kent and Thaler (1989)]. In many cases their authors were grappling with a basically game-theoretical problem, but were not ready to use the right technique, and built inconsistencies into their models [O'Neill (1987)].

Brown (1971) and Callaham (1982) offer early uses of game theory to analyze instability. In Nalebuff's model of first-strike instability (1989), government A holds a conditional probability function stating a likelihood that B will strike expressed as a function of B's assessed likelihood that A will strike. Some exogenous crisis event impels these beliefs up to an equilibrium. Powell (1989a) gives assumptions on a game of escalation will complete information such that there will be no temptation to strike, and Bracken and Shubik (1993a) extend his analysis. Franck's (1983) work is related. Wagner (1991) gives a model intended to rationalize US counterforce policy.

Responding to the depolarization of the world, Bracken and Shubik (1991, 1993b) and Best and Bracken (1993a,b) look at the war consequences of all possible coalitions of the nuclear powers in regard to preemptive stability, the later study including many classes of weapons, using very large and fast computers to solve game trees with tens of millions of nodes. They sometimes reproduce a result of the theory of truels, three-person duels, that the weakest party may hold the best position. A different approach is Avenhaus et al. (1993).

Greenberg (1982) represents the decision of how to launch a surprise attack, and O'Neill (1989c) models the defender's decision when to mobilize in the face of uncertain warning, using the stochastic processes methods on quickest detection. The goal is to produce a measure of the danger of a war through a false alarm, expressed as a function of the probability parameters of the warning systems.

1.8. Escalation

Two views of escalation regard it as either a tool that each side manipulates, or a self-propelling force that takes control of the players. Models with complete information and foresight tend to the former picture, but those that place limits on the governments' knowledge and judgement usually predict that they will be swept up into the spiral of moves and responses. An example of the former is the dollar auction. This simple bidding game resembles international conflict in that both the winner and the loser must forfeit their bids. O'Neill (1985, 1986, 1990b), Leininger (1989), Ponssard (1990) and Demange (1990) discuss the solution, in which one bids a certain amount determined by the sums of money in the pockets of each bidder, and the other resigns. The rule is unintuitive and counter to the experimental finding that people do bid up and up, but it shows one role of game theory, to establish what constitutes rational behavior within a set of assumptions, in order to identify a deviation to be explained.

Langlois (1988, 1989) develops a repeated continuous game model postulating limited foresight in escalation: each side acts not move by move, but chooses a reaction function, a rule for how strongly it will respond to the other side's aggressiveness. He derives conditions for an equilibrium pair of functions, investigates when a crisis would escalate or calm down, and argues for the repeated game approach in general. Langlois (1991b) discusses optimal threats during escalation. O'Neill (1988) portrays limited foresight using the artificial intelligence theory of game-playing heuristics, to show that under some conditions players fare worse when they consider more possible moves or look further down the tree. Zagare (1990b, 1992) gives some complete information games of escalation.

In a series of models, most collected in their book, Brams and Kilgour (1985, 1986b, 1987a,c,d, 1988a,b, 1994) ask how threats of retaliation can support a cooperative outcome. They investigate multistage games involving plays of continuous versions of a Chicken or Prisoner's Dilemma matrix, where players adopt degrees of cooperation, followed by possible retaliation. In an empirical test, James and Harvey (1989) compare the Brams and Kilgour model with several dozen incidents between the superpowers, and Gates et al. (1992) compare an extended deterrence model with military expenditure data. Uscher (1987) compares repeated PD and Chicken models of escalation. Powell (1987a,b; 1990) analyzes crisis bargaining in the context of his escalatory model, and Leng (1988) uses the theory of moves to discuss the learning of bargaining stances from the experience of past crises. Further studies on escalation are by Brown, et al. (1973) and Ackoff et al. (1968). Hewitt et al. (1991) generate predictions from a simple extensive game model, and test them on 826 cases over the last seventy years.

Morrow (1989b) considers bargaining and escalating as complementary moves during a crisis, and has each side learn about the other's capacity for war through its delay of an agreement or its willingness to fight a small conflict. In a follow-on paper (1992a), he asks whether each state can learn enough about the adversary to avoid the next crisis. A block to resolving a crisis is that any side initiating a compromise may be seen as weak. Morrow (1989a) looks for information conditions that lessen this problem, which is the diplomatic analogue of a military first-strike advantage.

1.9. Alliances

Game models of alliances have focused on four issues: who joins in coalitions, which alliance structures promote peace, who pays for the "public good" of defense and how much do they buy, and where do allies set the tradeoff between giving up their freedom of action and acquiring support from others.

Arguing informally from coalition-form game theory, William Riker (1962) suggested that a smallest winning coalition will be the one that forms. He discussed the implications for national and international affairs. In the theory of national politics his idea has been very influential, but has been followed up only sparingly in the international context [Siverson and McCarty (1978), Diskin and Mishal (1984)].

How will nations respond to an increase in one party's power? Will they "bandwagon," i.e., join the powerful country to be on the winning side, or will they "balance," form a countervailing alliance against the threat? Much of U.S. foreign policy has followed the notion that small countries will bandwagon. Arosalo (1971), Chatterjee (1975), Young (1976), Don (1986), Guner (1990, 1991), Maoz (1989a,b), Nicholson (1989), Luterbacher (1983), Hovi (1984), Yu (1984), Dudley and Robert (1992), Niou and Ordeshook (1986, 1987, 1989a,b) and Niou et al. (1989), Lieshout (1992), Linster (1993) and Smith (1993) have addressed this and related questions using mostly coalitional games, the last-mentioned authors stating a solution theory developed from the nucleolus and bargaining set. Their book operationalizes and tests the theory using large European power relations from 1871 to 1914. The recent formal literature on coalition formation exemplified in Roth's volume (1988) has not been applied to military alliances, nor to "peaceful alliances," like the signators of the treaty banning certain nuclear tests and the treaty against proliferation of nuclear weapons. Gardner and Guth's paper (1991) is the closest work.

Wagner (1987) addresses the traditional question of the optimal number of players to produce balance-of-power stability. Selten (1992) invented a board game that reproduces balance-of-power phenomena and is simple enough to be analyzed mathematically. Zinnes et al. (1978b) give a differential game model of n-nation stability. Kupchan (1988) represents relations within an alliance by several 2×2 matrix games, while Sharp (1986) and Maoz (1988) discuss the dilemma between gaining support through an alliance versus getting entangled in a partner's wars. O'Neill (1990a) uses games of costly signalling to clarify the idea of "reassurance" in alliances. Allies invest in militarily pointless arms to show each other that they are motivated to lend support in a crisis. Brams and Mor (1993) consider the victor in a war who may restore the standing of the loser in hopes of a future alliance.

Weber and Wiesmeth (1991a,b) apply the notion of egalitarian equivalent allocations, and Sandler and Murdoch (1990) set a Lindahl allocations model and a non-cooperative game model against actual contributions to NATO. Guth (1991) applies his resistance avoidance equilibrium refinement. Bruce (1990) gives the most fully game theoretical model of burdernsharing: unlike most economic models he includes the adversary's reactions as well as the allies'. Morrow (1990a, 1994) and Sorokin (1994) deal with the decision to give up one's autonomy in an alliance in return for a less than fully credible promise of help.

1.10. Arms control verification

Game-theoretical studies of verification divide into two groups. The first and earlier involves decisions about allocating inspection resources or a quota of inspections limited by treaty. The second ask whether to cheat and whether to accuse in the face of ambiguous evidence.

The first group, on how to allocate inspections, began in the early 1960s. In the nuclear test ban talks the American side claimed seven yearly on-site inspections were necessary, but the Soviets would not go above three, citing the danger of espionage. The negotiations stalled and a complete ban has still not been achieved. Maximizing the deterrent value of each inspection should please a government worried about either compliance or espionage, and this goal motivated the work of Davis (1963), Kuhn (1963), Dresher (1962) and Maschler (1966, 1967). Their inspector confronts a series of suspicious events, and each time decides whether to expend a visit from a quota, knowing that there will be fewer left for the next period. Rapoport (1966) gives a simple example. The problem is somewhat like a search game extended over time rather than space, and Kilgour (1993) in fact has the players select places, rather than times to violate. Extensions involve non-zero-sum payoffs [Avenhaus and von Stengel (1991)] or several illegal acts [von Stengel (1991)]. Maschler's papers showed the advantage to an inspector who can publicly precommit to a strategy, a theme developed by Avenhaus and Okada (1992), Avenhaus et al. (1991) and Avenhaus and von Stengel (1993). Moglewer (1973), Hopfinger (1975), Brams, et al. (1988) and Kilgour (1990) extend quota verification theory, and Filar (1983, 1984) and Filar and Schultz (1983) treat an on-site inspector who has different travel costs between different sites, these being known to invader who plans accordingly. Some recent work has examined the monitoring of materials in nuclear energy plants, in support of the Non-Proliferation Treaty [Avenhaus (1977, 1986), Avenhaus and Canty (1987), Avenhaus et al. (1987), Bierlein (1983), Zamir (1987), Avenhaus and Zamir (1988), Canty and Avenhaus (1991, Canty et al. (1991), Petschke (1992)].

A challenging puzzle from the first group has a violator choose a time to cheat on the $[0,1]$ interval, simultaneously with an inspector who allocates n inspection times on the interval. The inspector gets a free inspection at $t = 1$. The payoff to the violator is the time from the violation to the next inspection, while the inspector receives the negative of this. Diamond (1982) derives the mixed strategy solution, not at all trivial, where the draw of a single random variable fixes all n inspection times [see also von Stengel (1991)].

Almost all studies in the first group are zero-sum, like Diamond's game. At first it seems odd that a government would be happy to find that the other side has violated, but the model's viewpoint is not the inspecting government's but that of its verification organization. The latter's duty is to behave *as if* the game were zero-sum, even to assume that the other side will violate, as Diamond's game and others do. Verification allocations are to peace what military operations are to war, and are more oppositional than the grand strategy of the state.

The second group of verification studies takes the governmental point of view, with non-zero-sum payoffs and a decision of whether, rather than how, to violate and whether to accuse. Maschler's early paper (1963) is an example. Evidence is sometimes represented as a single indicator that suggests innocence or guilt [Frick (1976), Avenhaus and Frick (1983), Fichtner (1986b), Dacey (1979), Wittman (1989), Weissenberger (1991)], dichotomously or on a continuum of evidential strength [O'Neill (1991b)]. Brams, Kilgour and Davis have developed several models [Brams (1985), Brams and Davis (1987), Brams and Kilgour (1986a, 1987a), Kilgour and Brams (1992)], as has Chun (1989), Bellany (1982) and Avenhaus (1993). Weissenberger's model (1990) jointly determines the optimal treaty provisions and the verification system. O'Neill (1991b) proves the odd result that with more thorough verification, the inspector may be less certain about the other's guilt after observing guilty evidence. With better verification technology, the inspector's prior, before the evidence was that the other would not dare to violate, and this aspect can outweigh the greater strength of evidence of guilt from the verification scheme.

Most research has concentrated on these two areas, but many more verification issues are amenable to game theory. Future arms treaties will more often be multilateral and it would be impractical for each government to send inspectors to monitor every treaty partner, but relying on third parties introduces principal–agent issues. A second example involves the claim that as stockpiles are greatly reduced, verification must be more and more thorough since a small evasion will give a bigger advantage. How would a game model develop this idea? Another example arose from past treaties: Should you make an accusation that requires proof when that proof compromises intelligence sources needed to spot a future violation? A final topic involves a Soviet concern that was recurrent since the testban talks: how to balance between effective verification and military secrecy.

2. Military game theory

Progress in military game theory has gone mostly unknown beyond that community. While some studies are available in *Operations Research and Naval Research Logistics Quarterly*, many have been issued as reports from government laboratories, strategic studies institutes or private consulting firms. Much of the American work is not publicized or is stamped secret, sometimes placed in classified journals like *The Proceedings of the Military Operations Research Society Symposium* or *The Journal of Defense Research*. Secrecy obstructs the circulation of new ideas so even game theorists with a security clearance can be unaware of their colleagues' work.

Most of the research below is from the United States. I do not know how much has gone on elsewhere, hidden by classification. Some writers claimed that the Soviet Union was advanced in military game applications, but there seems to be no direct evidence about this either way. Some Soviet texts included military game theory [Ashkenazi (1961), Suzdal (1976), Dubin and Suzdal (1981)], but their treatment was in no way deeper or more practical than Western works (Rehm, n.d.). This review will concentrate on applications in the United States, where access to information is the freest.

The social value of military game theory has been controversial. To use mathematics to make killing more efficient strikes one as a travesty, and it also seems pointless, since each side's gains can usually be imitated by the adversary, with both ending up less secure. On the other hand, a logical analysis might counter governments' natural instincts to build every possible weapon in great numbers, since, for all anyone can say, it might confer some benefit some day. Analysis could prompt officials to think clearly about their goals and evaluate the means more objectively. One policy document that used game theory, Kent's Damage Limitation Study, had this effect, and helped to slow the arms race.

General discussions with military examples are given by Dresher (1961b, 1968), Shubik (1983, 1987), Thomas (1966), Finn and Kent (1985) and O'Neill (1993c). The NATO symposium proceedings edited by Mensch (1966) give a good sample of problems. Leonard (1992) recounts the history of military interest in game theory, and Mirowski (1992) argues, unconvincingly in my view, that military concerns have shaped game theory as a whole.

2.1. *Analyses of specific military situations*

In 1917 Thomas Edison came remarkably near to devising the first game model of a real military situation. His game was well-defined and analytically solvable, but practically minded as he was and lacking a solution theory, he had people gain experience playing it and he examined their strategies [Tidman (1984)]. The problem was how to get transport ships past German U-boats into British ports. He divided a chart of the harbor waters into squares, forty miles on a side, to approximate the area over which a submarine could spot the smoke rising from a ship's funnel. One player placed pegs in the squares corresponding to a path of entry, and the other player on a separate board placed pegs representing waiting submarines, defining a game of ambush [Ruckle (1983)]. He concluded that the last leg of the Atlantic crossing was not as dangerous as had been thought.

The next world war prompted a full solution theory of a game of anti-submarine patrols, possibly the first full game study on record of a real problem beyond board games and gambling. Philip Morse, who had overseen the wartime Antisubmarine Warfare Operations Research Group, posited a sub running a corridor. To recharge its batteries it must travel on the surface for some total time during its passage [Morse and Kimball (1946)]. The corridor widens and narrows, making anti-submarine detection less and more effective. Mixed strategies were calculated for how to surface and how to concentrate the search. The measure-theoretic foundations were complex and it took Blackwell (1954) to add the details. Allentuck (1968) believes that anti-submarine game theory was decisive in winning the Battle of the North Atlantic, but his evidence is indirect, and the theory most likely never guided a real decision [Morse (1977), Tidman (1984), Waddington (1973)].

Perhaps the most influential analysis of a specific situation was Goldhammer, Marshall and Leites' secret report on nuclear counterforce policy [1959; see also O'Neill (1989b)]. They polled fellow RAND strategists to get their estimates of Soviet utilities for various nuclear war outcomes, and described a tree of several stages. However they kept their tree unsolvable by omitting many payoffs and the probabilities of chance moves, leaving out in particular the utility for an American surrender. A more recent extensive form model on overall strategic policy is by Phillips and Hayes (1978).

Successfully capturing simplicity and reality is Hone's matrix analysis (1984) of the naval arms race during the years between the world wars. Each side could build ships emphasizing armor, firepower, speed or range, each quality being differentially effective against the others, as in Scissors, Paper and Stone. He applied Schelling's technique (1964) of categorizing payoffs as simply high, medium or low. [(A similar method had been devised independently for a secret 1954 British Army report on the optimum gun/armor balance in a tank, Shephard (1966)]. Schellenberg (1990) terms these *primitive* games and lists the 881 2×2 types.

One "historical" episode was analyzed as a game but has now been exposed as fabricated to preserve a military secret. The 1944 defeat of the German Seventh Army near Avranche had been attributed to the astute foresight of the American commander, but documents released later showed that the Allies had secretly intercepted Hilter's message to his general [Ravid (1990)]. The analyses of Haywood (1950, 1954), Brams (1975) and Schellenberg (1990) would be tidy applications if the event had ever happened.

Game theorists sometimes apologize for recommending mixed strategies, but Beresford and Preston (1955) cite how real combatants used them. In daily skirmishes between Malaysian insurgents and British overland convoys, the insurgents could either storm the convoy or snipe at it, and the British could adopt different defensive formations. Each day the British officer in charge hid a piece of grass in one hand or the other, and a comrade chose a hand to determine the move. It is not clear whether they generated their procedure themselves or owed it to von Neumann and Morgenstern.

2.2. Early theoretical emphases

In the late 1940s at the RAND Corporation, Lloyd Shapley, Richard Bellman, David Blackwell and others worked on the theory of *duels*. Two sides approach each other with bullets of specified accuracies. Each must decide when to fire: shooting early gives you a lower chance of hitting but waiting too long risks that your opponent shoots first and eliminates you. In other versions, the bullets are silent (the opponent does not know when one has been fired), or noisy (the opponent hears your shots, and will know when your ammunition is gone), or there is continuous fire, or several weapons on each side, or a time lag between firing and hitting. The pressing problem then was defence against atomic bombers: When should an interceptor open fire on a bomber, and vice versa [Thomas (1966)]? The destructiveness of atomic weapons justified zero-sum, non-repeated-game assumptions: the payoff was the probability of destroying the bomber, with no regard for the survival of the fighter. Dresher (1961b) gives the basic ideas and Kimmeldorf (1983) reviews recent duelling theory. A related application was the optimal detonation height of attacking warheads facing optimally detonating anti-missile interceptors [Kitchen (1962)].

Also close to duels were *games of missile launch timing*, investigated at RAND in the late 1950s. The first intercontinental missiles were raised out of protective underground silos or horizontal "coffins," prepared and launched. During the preparation they would be more vulnerable to attack. Several papers [Dresher (1957, 1961a, pp. 114–115), Johnson (1959), Dresher and Johnson (1961), Thomas (1966) and Brown's camouflaged piece (1957)] considered a missile base that chooses a series of firing times, versus an attacker who chooses the times for its warheads to arrive at the base. A warhead destroys a missile with a given probability if it is still in its silo, and with higher likelihood if it is outside in preparation for firing.

By 1962 the Strategic Air Command had begun deploying missiles fired directly out of silos, so the launch timing question lost some relevance. Now SAC began worrying about "pin-down". Soviet submarine missiles lacked the accuracy to destroy silos but they could explode their warheads above the North American continent in the flyout paths of US missiles. This would create an "X-ray fence" damaging the electronics of missiles flying through it, so US weapons would be stuck in their silos until Soviet ICBMs arrived. Keeping up a complete blockade would exceed Soviet resources, so the sub missiles would have to be used prudently. How should SAC choose an optimal fire-out strategy against an optimal Soviet fire-in strategy? For many years the game's solution guided actual plans for launching rates [Zimmerman (1988), p. 38].

Games of aiming and evasion [Isaacs (1954)] are still relevant today, and still largely unsolved. In the simplest version a bomber attacks a battleship that moves back and forth on a one-dimensional ocean. It turns and accelerates to its maximum velocity instantly. The plane drops a bomb in a chosen direction; the bomb reaches the surface after a time lag, destroying anything within a certain distance. The ship knows when the bomb is released but not where it is heading. Finding the approximate mixed strategy solutions is not too difficult: if the radius of destruction is small compared to the interval the ship can reach, the aim point and ship's position should be chosen from a uniform distribution. However a slight change makes the problem more interesting: suppose the ship does not know when the bomb is dropped. It must then make its position continuously unpredictable to the bomber, who is trying to guess where it will be a time lag from now [Washburn (1966, 1971)]. A discrete version where the ship moves on a uni-dimensional lattice yields some results when the duration of the bomb's fall equals one, two or three ship movements [Lee and Lee (1990, and their citations)]. The modern story would involve an intercontinental missile attacking a missile-carrying train or a submarine located by underwater sensors.

Several papers addressed the strategy of naval *mine-laying and detecting*. In Scheu's model (1968), the mine-layer leaves magnetic influence mines or time-actuated mines to destroy the searcher, and the searcher may postpone the search and the advance of ships until some of the mines have exploded. Another post-war focus involved *ordnance selection*, such as the armament of a fighter plane that trades off between cannons and machine guns. Fain and Phillips' working paper (1964) gives an example on the optimal mix of planes on an aircraft carrier.

Reconnaissance problems received much attention [Belzer (1949), Blackwell (1949), Bohnenblust et al. (1949), Danskin (1962a), Dresher (1957), Shapley (1949), Sherman (1949)] in regard to the value of acquiring information for a later attack, and the value of blocking the adversary's reconnaissance attempts. Tompkins' *Hotspot game* [Isbell and Marlow (1956)] is an elegant abstract problem suggested by military conflict, involving allocation of resources over time. Each side sends some integral number of resources to a "hotspot" where one or the other loses a unit with a probability determined by the two force sizes. When a unit disappears, either side sends more until one player's resources are exhausted.

A class of conflicts suggested by warfare is *Colonel Blotto*, where players simultaneously divide their resources among several small engagements. Borel stated a simple version in 1921, not long before he became France's Minister of the Navy. Each player divides one unit among three positions and whoever has assigned the greater amount to two out of the three positions, wins. Although the game is easy to state, some mixed strategy solutions found by Gross and Wagner (1950) have as their supports intricate two-dimensional Cantor sets. Soviet researchers contributed to the theory [Zauberman (1975)], which is reviewed by Washburn (1978).

Military applications of Blotto games turned out to be rarer than expected. There seem to be few real contexts where unforeseen numbers of forces from both sides show up at several locations. Two applications are Shubik and Weber (1979, 1982) on the defence of a network, in which they introduce payoffs that are not additive in the engagements, and Grotte and Brooks' (1983) measure of aircraft carrier presence. An application of Blotto-like games involves missile attack and defence, treated next.

2.3. Missile attack and defence

Missile defence theory has had an important influence on arms policy. Some work had been done at the RAND Corporation [Belzer et al. (1949), Dresher and Gross (1950), Gross (1950a,b)] but the most significant model has been Prim–Read theory, named for its two originators. In a simple version, an attacker sends missile warheads to destroy a group of fixed targets, and a defender tries to protect them using interceptors, which are themselves missiles. At the first stage the defender divides its interceptors among sites, and plans how many to send up from each targeted site against each incoming warhead. The attacker, knowing the allocation and the firing plans, divides its warheads among the sites, and launches the missiles in each group sequentially at their target. If an attacking warhead gets through, it destroys the site with certainty, but each interceptor succeeds only with a probability. There is no reallocation: if the attacker has destroyed a site, further weapons directed there cannot be sent elsewhere, nor can any interceptors be shifted. The payoff is the sum of the values of the target destroyed.

The recommended rule for allocating the defence interceptors, the Prim–Read deployment, typically does not divide the defences in proportion to the values of the targets, but induces the attacker to send warheads in roughly those proportions. A remarkable result is that defensive deployments exist that are best independent of the number of attacking weapons, so the defender need not know the opposing force size. The basics were laid out by Read (1957, 1961), and Karr (1981) derives some uniqueness and optimality results. Examples of modern applications are by Holmes (1982) and Bracken and Brooks (1983).

Prim–Read theory was used in a report that may have helped dampen the arms competition. In 1964, US Air Force General Glenn Kent completed his secret "Damage-Limitation Study" on how to make nuclear war less destructive. One proposal was to build a vast anti-missile system. With Prim–Read theory, Kent's group derived a supportable estimate of a defence's effectiveness, and the verdict was negative: an antimissile system would be expensive to build, and cheap to thwart by an adversary's increasing the attacking force. Kent's study bolstered the case for a ban on large missile defence systems [Kaplan (1983)], and in 1972 the two powers signed the Anti-Ballistic Missile Treaty. It continues in force today, probably the greatest success of nuclear arms control, a move to an equilibrium Pareto-superior to a more intense race in defensive and offensive weapons.

Few policy decisions are on record as influenced by game theory, but we have to separate innate reasons from institutional ones. Perhaps the opportunity was there to apply the theory, but there was no "inside client", no right person in place to carry it through. Kent had a reputation for innovativeness, had associated with Schelling at Harvard and had written on the mathematics of conflict (1963). When he gained a position of influence, he used the theory effectively.

Prim–Read theory in Kent's study assumes that the attacker knows the defender's firing schedule. This is questionable, but the opposite premise would require a mixed strategy solution. Mixed strategies are more complicated, and were alien to the operations researchers working on these problems, who were trained on optimization and programming. McGarvey (1987), however, investigated the case. Another branch of missile defence theory alters another part of Prim–Read, by assuming that the attacker is unaware of the defender's allocation. The attitude has been that Prim–Read theory treats attacks against population centers and the alternative branch deals with attacks on missile sites. Strauch's models (1965, 1967) also assumed perfectly functioning interceptors, and thus were formally Blotto games. Other versions [Matheson (1966, 1967)] had imperfect interception, so each side had a strategic decision beyond the Blotto allocation, of how many interceptors to send up against each attacking warhead, not knowing how many more would be descending later on the particular target. Unlike the Prim–Read case, a deployment that is ideal against an all-out attack may be poor against a smaller one, but Bracken et al. (1987) found "robust" defence allocations that for any attack level stay within a certain percentage of the optimum. Burr et al. (1985) find solutions without the approximation of continuously divisible missiles. Matlin (1970) reviewed missile defence models, about a dozen of which involved games, and Eckler and Burr (1972) wrote a thorough report on mathematical research up to that time. Many of the studies should have treated their situations as games, but used one-person optimization, and so were forced to assume inflexible or irrational behavior by the other player.

Ronald Reagan's vision of anti-missile satellites shielding an entire country bypassed the mathematically fertile study of allocation among defended sites. New game models appeared more slowly than one would expect from the bountiful funding available, but recent examples are by O'Meara and Soland (1988), as well as several authors they reference, on the coordination problem of guarding a large area. Another theme is the timing of the launch of heat decoys against a system that attacks missiles just as they rise from their silos.

2.4. *Tactical air war models*

Strategy for tactical air operations has preoccupied many military operations researchers and for good reason. Each side would be continually dividing its fighter-bombers among the tasks of attacking the adversary's ground forces, intercepting its aircraft and bombing its airfields. The two sides' decisions would be interdependent, of course, since one's best choice today depends on what it expects the other to do in the future. Through the Cold War there was concern that the aircraft allocation rule would play a large role in a conventional war in Europe, but dynamic analyses of the "conventional balance" sidestepped it. Some large computer models stressed air activity, some the ground war, often as a function of the sponsoring military service, but few addressed their interaction. An exception was the program TAC CONTENDER, forerunner of the current RAND program TAC SAGE [Hillstead (1986)], which was thought to find an optimal strategy, until Falk (1973) produced a counterexample.

Satisfactory tactical air models might bolster each side's feelings of security and prompt it to reduce its forces, but so far the problem has been solved only in simplified versions. Research began quite early [Giamboni et al (1951), and Berkovitz and Dresher later succeeded in solving a complicated sequential game of perfect information [(1959, 1960), Dresher (1961b)]. The dilemma is to make a problem realistic and computationally manageable, and there have been three approaches: Lagrange multipliers as in TAC CONTENDER, which provide only a sufficient condition for optimality; grid methods, applied by Dantzig [Control Analysis Corporation (1972)], with the same shortcoming; and solutions of trees of matrix games [Bracken et al. (1975), Schwartz (1979)], which are sure to be optimal but allow models of fewer stages.

2.5. Lanchester models with fire control

Lanchester's widely-known equations mean to describe the rates at which two armies in combat destroy each other. They assume a homogeneous force on each side. Extending them to a mixed force, like infantry and artillery, introduces the question of how to allocate each component against each of the adversary's components. Almost always the approach has been to divide the forces in some arbitrary proportion that stays fixed throughout the battle, but a clever opponent might beat this by concentrating forces first on one target and later on the other. Accordingly, one stream of work treats the allocation-of-fire decision as a differential game constrained by Lanchester's equations. In Weiss's fine initiating study (1957, 1959), each side could direct its artillery against the other's artillery or ground forces, but the ground forces attacked only each other. Whereas air war applications are usually discrete, with planes sent out daily, fire control games tend to be continuous. Taylor (1974) made Weiss's results more rigorous, and Kawara (1973), followed up by Taylor (1977, 1978) and Taylor and Brown (1978), adds an interesting variant in which the attacker's ground forces shoot while they move towards the defender's line, but with "area fire", not precisely aimed because they are moving. The artillery bombardment must cease when the two sides meet, so the each side tries for the highest ratio of its forces to the other's by that time. Moglewer and Payne (1970) treat fire control and resupply jointly [see also Sternberg (1971) and Isaacs' Game of Attack and Attrition (1965)]. Taylor (1983, Ch.8) gives an excellent summary of the field.

2.6. Search and ambush

In a *search game*, a Searcher tries to find a Hider within some time limit or in the shortest time. Sometime the Hider is moving, like a submarine, sometimes fixed, like a mine. As described earlier, the research began in World War II but has found few applicable results, perhaps because of the difficulty of incorporating two-or three-dimensional space. The bulk of applied search theory today is only "one-sided", postulating a non-intelligent evader located in different places with exogenous probabilities, but several dozen papers solving abstract games are listed by Dobbie (1968) and the Chudnovskys (1988). Gal's book (1980) finds some strategies that approach minimax as the search area becomes large with respect to the range of detection. Lalley and Robbins (1987) give examples of the peculiar hide-and-seek behavior that can arise as a minimax solution. One variant requires the evader to call at a port within a time interval [Washburn (1971)], and Dobbie (1968) suggested a game where both sides are searchers, a ship trying to rendezvous with its convoy. A problem apparently not yet investigated as a game allows the hider to become the seeker, such as a submarine versus a destroyer.

Related to search games are continuous *games of ambush*, which Ruckle investigates in his book (1983) and a series of articles in *Operations Research*. Usually the Ambusher does not move, so the game is easier to solve than a search. One or both players typically choose some geometrical shape. A simple example has the Evader travelling from one side of the unit square to the other, while the Ambusher selects a set of prescribed area in the square, and receives as payoff the length of the path that lies in the area. A less elegant but more practical example [Randall (1966)] involves anti-submarine barriers. Danskin on convoy routing (1962b) is also relevant.

Simple to state but difficult to solve is the *cookie-cutter game*. The target chooses a position within a circle of unit radius while the attacker chooses the center of a circle with given radius $r < 1$, and wins if that circle contains the target. The case of $r = 0.5$ is elegantly solvable [Gale and Glassey (1975)], but for r less than approximately 0.4757 little is known [Danskin (1990)].

2.7. Pursuit games

Pursuit problems [Isaacs (1951, 1965), Hayek (1980)], studied in the 1950s to improve defence against atomic bombers, sparked the whole area of differential games. Pursuit game theory has grown steadily and a recent bibliography [Rodin (1987)] lists hundreds of articles. Typically the Pursuer tries to come within capture distance of the Evader, but in some games the angle of motion is important, as in the case of maneuvering fighters. Solutions of practical pursuit problems are hard to obtain and hard to implement, since human behavior cannot be programmed second-by-second. Aerial dogfights involving complete turns are too complicated, and add the wrinkle that the Evader is trying to switch roles with the Pursuer. However, strategies optimal for sections of an engagement have been calculated, with the goal of abstracting general rules [Ho (1970)]. A more feasible application is the control of maneuvering nuclear warheads and interceptor missiles. The short engagement times and the interceptors' high accelerations mean less maneuvering, so their movements can be represented by simpler kinematic laws. Even though they can be analysed mathematically, the weapons are not in wide use; as of recent years the only maneuvering warheads facing antimissile interceptors were British missiles aimed at Moscow. Discussions of differential games in military operations research are given by Isaacs (1975), Ho and Olsder (1983) and Schirlitzki (1976).

2.8. Measures of effectiveness of weapons

An issue of practical interest is how to assign numbers to represent the military worths of weapons. The goal is to generate indices of overall strength, in order to assess the "military balance", to investigate doctrine that directs battle decisions of what should be "traded" for what, to score war games, or to set fair arms control agreements. Most non-game-theory procedures work from the bottom up, estimating the qualities of individual weapons based on their design features, then adding to evaluate the whole arsenal, but clearly additivity is a questionable assumption here since some weapons complement each other and others are substitutes. A game-theoretical approach might start with the benefits of having a certain arsenal and infer back to the worths of the components. Pugh and Mayberry (1973)[see also Pugh (1973) and Assistant Chief of Staff, USAF Studies and Analysis (1973)] treat the question of the proper objective function of war, following Nash's general bargaining model, to argue that each side will try for the most favorable negotiated settlement by conducting war as strictly competitive. Anderson (1976) critiques their approach, and discusses games where the payoffs are the ratios of military strengths. O'Neill (1991a) applies the Shapley value, regarding the weapons as the players in a coalitional game. The characteristic function displays non-monotonicity whenever the enemy's armaments join the coalition. Robinson's important papers (1993) link the traditional eigenvalue approach in which values are defined recursively as functions of the values of the adversary's weapons destroyed, with the values calculated from payoffs of a game. Both methods face the problem of non-uniqueness. These models do not lead to a measure for specific decisions, but do clarify the informal debate.

2.9. Command, control and communication

Some writers have analyzed the contest of a jammer versus a transmitter/receiver team. Fain's early report (1961) on the tradeoff between offensive forces and jamming units was belittled by the other conference participants, but more studies followed. Some discuss a jammer who chooses a signal with a limit on its bandwidth and power [Weiss and Schwartz (1985), Stark (1982), Helin (1983), Basar (1983), Basar and Wu (1985), Bansal and Basar (1989)]. Others posit a network where the transmitter can route the communication a certain way, and the jammer attacks certain nodes or links [Polydoros and Cheng (1987)]. Other game analyses have looked at the authentication of messages [Simmons (1981), Brickell (1984), and the identification of an aircraft as friend or foe [Bellman et al. (1949)]. These papers may be the tip of a larger classified literature.

McGuire (1958) assumed unreliable communication links connecting ICBMs and discussed the tradeoff between retaliating in error, e.g., in response to an accidental breakdown of a link, versus not retaliating after a real attack. For a given network, the attacker plans what to target and the defender plans what instructions for retaliation to issue to base commanders who find themselves incommunicado. Independently, Weiss (1983) considered missiles that retaliate against target of differing values, where each launch control officer may retarget based on partial knowledge of what other missiles have been destroyed. The retaliator issues plans for the missile commander, contingent on the latter's knowledge of what others remain, and the stiker decides what to attack. In Weiss' view, implementing the theory would bolster a government's deterrent threat without adding weapons that it could use in a first strike; in other words, it would ease the security dilemma. Notwithstanding, the Strategic Air Command and its successor STRATCOM did not transfer this kind of control down to the level of a missile base commander.

2.10. Sequential models of a nuclear war

"Nuclear exchange" models' goal has been to suggest the consequences of changing employment plans, adding new weapons or signing an arms agreement. Several models involve the attacker dividing its weapons between the adversary's population and retaliating missiles. The sides maximize some objective functions of the damage to each after a fixed number of attacks. Usually the solution method is a max–min technique, but one exception is Dalkey's study (1965). Early papers used Lagrange techniques on two-stage models and so guranteed only local optimality, but Bracken et al. (1977) found a method for the global optimum in a class of reasonable problems. Grotte's paper (1982) gives an example with many references. A model by Bracken (1989a) shows the philosophy behind this work, trying to add as much detail as possible. It has twelve stages, where each side allocates missiles among the roles of attacking population targets, attacking military targets and saving as reserve forces, and also decides how much of its missile defence system to hold back for use against future strikes. This discrete game of perfect information has 20 000 000 possible paths, and takes several minutes on a Cray computer.

The Arsenal Exchange Model is the program most used by US government agencies to investigate nuclear war strategies, and it is not game-theoretic. An attempt was made to include optimal choices by both sides, but the task turned out to be too complex, and it is still the user who chooses levels of attack and priorities of targets, whereupon a linear programming routine assigns weapons to targets.

2.11. Broad military doctrine

An army ceases to fight before its physical endurance is gone; its psychological breakpoint involves partly the mutual trust and expectations of the soldiers. No one seems to have applied the Stag Hunt, although Molander (1988) used the Prisoner's Dilemma to analyze the cohesion of society as a whole in wartime.

A unique study is by Niou and Ordeshook (1990a) on the military thinking of Sun Tzu. Many of his teachings, set down over two thousand years ago, can be reproduced formally, although he sometimes cut off the logic of think and doublethink at an arbitrary level; for example, he neglected the consequences if your opponent knows your opportunities for deception.

3. Comparisons and concluding remarks

Military game theory is structured more like a field of mathematics, each writer's theorems building on those of others. IR game models have tended to take their problems from the mainstream literature rather than each other, and so have touched on an immense number of questions, as this survey shows. Only recently have identifiable research streams of game models formed, such as PD models of cooperation, signalling analyses of deterrence in a crisis, and models of relative versus absolute gains as national interests.

The early theory of military tactics emphasized simple games with continuous strategy spaces, as appear in Karlin's (1959) and Dresher's (1961d) books. Now the growing practice is to approximate very detailed problems as finite and solve them by computer, but the old theorems were elegant and helped show a situation's strategic structure more than a giant linear program could. IR models, in contrast, are becoming more mathematically sophisticated. Most authors began as social scientists and learned the mathematics later on their own, so many of the earlier pieces surveyed here are at the level of "proto-game theory" [O'Neill (1989b)], using the concepts but no formal derivations, often setting up the game but not solving it. Now techniques like signalling models are percolating through from economics and the papers state their assumptions precisely and derive theorems.

While IR models have tried to clarify theoretical debates, military models have been used as sources of practical advice. The view has been that the models can never include enough factors to be precisely true, but they give general principles of good strategy [Dresher (1966)], resonant with the idea of military doctrine. The target-defence principle of Prim–Read theory, for example, says never assign defences that induce the adversary to assault a defended target while sending nothing against an undefended one. It is related to the "no soft-spot principle" in general studies of defence allocation. Tactical air war models continually suggest that the superior force should split its attack, while the inferior should concentrate on a place randomly chosen.

Like these examples of military principles, game models have impressed some important truths on the IR discipline: international struggles are not zero-sum, but justify cooperative acts; good intentions are not enough for cooperation, which can be undermined by the structure of the situation; building more weapons may increase one's danger of being attacked.

The reference section shows about twenty-five publications with models from the 1960s, sixty from the 1970s and over two hundred from the 1980s. The nuclear strategy debate of the 1950s and 1960s made the theory controversial and more prominent. Despite the accusation that game theory was a tool of the Cold War, many in the peace research community became users. After a lull in the early 1970s, the activity of Brams and his colleagues helped keep games visible to IR theorists. In the 1980s worries over Reagan's buildup and hawkish statements triggered a boom in security studies in general, and more interest in game models. The staying influence of Schelling's 1960 book, still required reading in most graduate programmes, has given game theory a presumption of importance, and Axelrod's work has promoted access to it by the non-mathematical majority. Articles are appearing more often in the main political science journals.

With the end of the Cold War, attention is moving to subjects like international political economy, alliances and verification, and new models are appearing at a steady or increasing rate. However the attitude of many political scientists is still hostile. The need for mathematical knowledge blocks progress: tens of thousands of undergraduates hear about the Prisoner's Dilemma each year, but those graduate students who want serious applications usually find no mentor in their department. Many IR scholars misunderstand game theory profoundly, rejecting it as the ultimate rational actor theory, repeatedly explaining its "innate limitations", or, like Blackett quoted above, castigating it because it cannot make precise numerical predictions, a test no other theoretical approach can pass.

How should we judge Blackett's point that if game theory were practical, practical people would be using it? He claims that if it does not apply to a simple context like gambling, it must be irrelevant to complicated international issues. In fact game theory clarifies international problems exactly because they are more complicated. Unlike card games, the rules of interaction are uncertain, the aims of the actors are debatable and even basic terms of discourse are obscure. What does it mean to "show resolve"? What constitutes "escalation"? What assumptions imply that cooperation will emerge from international anarchy? The contribution of game models is to sort out concepts and figure out what the game might be.

References

Abbey, R. (1991) 'Reading Between the Matrices; Conflicting Strategies in *The Strategy of Conflict*', Working Paper #9, York Centre for International and Strategic Studies, York University, Toronto.
Achen, C. (1988) 'A State with Bureaucratic Politics is Representable as a Unitary Rational Actor', American Political Science Association Meeting.
Ackoff, R. et al. (1968) *The Escalation and Deescalation of Conflict*, Vol. I. *Conflicts and Their Escalation: A Metagame Analysis*, Arms Control and Disarmament Agency Report ST-149.
Ahn, B. (1991) 'Domestic Politics, Rational Actors and Foreign War', Department of Political Science, University of Rochester, mimeo.
Ahn, B. and B. Bueno de Mesquita (1992) 'When Hawks are Doves and Doves are Hawks: Domestic Conditions and Foreign Conflict', Hoover Institution.
Alexander, J. (1976) 'A study of conflict in Northern Ireland, an application of metagame theory', *Journal of Peace Science*, 2: 113–134.
Alker, H., J. Bennett and D. Mefford (1980) 'Generalized precedence logics for resolving insecurity dilemmas', *International Interactions*, 7: 165–206.
Alker, H. and R. Hurwitz (n.d.) 'Resolving the Prisoner's Dilemma. Department of Political Science', Massachusetts Institute of Technology.
Allan, P. (1980) 'Dynamics of Bargaining in International Conflict – US/Soviet Relations during the Cold War, 1946–1963', Institut Universitaire des Hautes Etudes Internationales.
Allan, P. (1983) 'Bargaining power and game structures', in: *Crisis Bargaining and the Arms Race*. Cambridge, MA: Ballinger, Ch. 5.
Allentuck, A. (1968) 'Game theory and the Battle of the Atlantic', *Research Studies*, 36: 61–68.
Alt, J., R. Calvert and B. Humes (1988) 'Reputation and hegemonic stability: a game-theoretic analysis', *American Political Science Review*, 82: 445–466.
Amit, I. and A. Halperin (1987) 'The Optimal Accumulation of Power', Center for Military Analyses, Haifa, Israel.
Anderson, L. (1976) 'Antagonistic games', Paper P-1204, Institute for Defense Analyses, Arlington, VA.
Aron, R. (1962) *Paix et Guerre Entre les Nations*. [Peace and War among Nations.] Paris: Calmann–Levy.
Arosalo, U. (1971) 'Peliteoria ja kansainvaliset liittoutumat mallin ja empirian vertailua', ['Game theory and international alliances: a model and its empirical validity'.] *Politiikka* (Helsinki), 13: 115–128.
Ashkenazi, V. (1961) *Primenenye Teorii Igr v Veonnem Dele*. [Applications of Game Theory to Military Affairs.] Moscow: Soviet Radio.
Assistant Chief of Staff, United States Air Force Studies and Analysis (1973) 'Methodology for Use in Measuring the Effectiveness of General Purpose Forces (An Algorithm for Approximating the Game-theoretic Value of N-staged Games)', Washington.
Aumann, R.J. (1990) 'Nash equilibria are not self-enforcing', in: J.J. Gabszewicz, J.-F. Richard and L. Wolsey, eds., *Economic Decision Making: Games, Econometrics, and Optimization*, Amsterdam: Elsevier, pp. 201–206.
Aumann, R.J. (1992) 'Irrationality in Game Theory', in: P. Dasgupta, D. Gale, O. Hart and E. Maskin, eds., *Economic Analysis of Markets and Games*, Boston, MA: MIT Press, pp. 214–227.
Aumann, R., J. Harsanyi, M. Maschler, G. O'Brien, R. Radner and M. Shubik (1966) *Development of Utility Theory for Arms Control and Disarmament*, Mathematica Corporation, Arms Control and Disarmament Agency Report ST-80.
Aumann, R., J. Harsanyi, M. Maschler, J. Mayberry, H. Scarf, R. Selten and R. Stearns (1967) '*Models of the Gradual Reduction of Arms*', Mathematica Corporation, Arms Control and Disarmament Agency Report ST-116.
Aumann, R., J. Harsanyi, M. Maschler, J. Mayberry, R. Radner, H. Scarf, R. Selten and R. Stearns (1968) *The Indirect Measurement of Utility*, Mathematica Corporation, Arms Control and Disarmament Agency Report ST-143.
Avenhaus, R. (1977) *Material Accountability: Theory, Verification and Applications*. New York: Wiley.
Avenhaus, R. (1986) *Safeguard Systems Analysis*. New York: Plenum Press.
Avenhaus, R. (1993) 'Towards a theory of arms control and disarmament verification', in: R. Huber and R. Avenhaus, eds., *International stability in a Multipolar world: Issues and models for Analysis*. Baden-Baden: Nomos, pp. 269–282.
Avenhaus, R., and M. Canty (1987) 'Game Theoretical Analysis of Safeguards Effectiveness: Attribute Sampling', Spezielle Berichte der Kernforschungsanlage Juelich, Report 417.

Avenhaus, R., J. Fichtner and G. Vachon (1987) 'The Identification of Factors and Relationships Relevant to a Routine Random Inspection Procedure Under a Future Chemical Weapons Convention and the IAEA Experience', Arms Control and Disarmament Division, Department of External Affairs, Canada.

Avenhaus, R., and H. Frick (1983) 'Analyze von Fehlalarmen in Ueberwachungssystem mit Hilfe von Zweipersonen Nichtnullsummenspielen', ['Analysis of False Alarms in Inspection Systems with the Help of Two-person Non-zero-sum Games'.] *Operations Research Verfahren*, **16**: 629–639.

Avenhaus, R., W. Guth and R. Huber (1991) 'Implications of the Defence Efficiency Hypothesis for the Choice of Military Force Structures', Part I. 'Games with and without Complete Information about the Antagonist's Intentions, Part II. 'A Sequential Game Including the Possibility of Restructuring Forces', in: R. Selten, ed., *Game Equilibrium Models IV*. New York: Springer-Verlag, pp. 256–288 and 289–318.

Avenhaus, R., and A. Okada (1992) 'Statistical criteria for sequential inspector-leadership games', *Journal of the Operations Research Society of Japan*, **35**: 134–151.

Avenhaus, R., A. Okada and S. Zamir (1991) 'Inspector-leadership games with incomplete information', in: R. Selten, ed., *Game Equilibrium Models IV*. New York: Springer-Verlag, pp. 319–361.

Avenhaus, R., and B. von Stengel (1991) 'Non-zerosum Dresher inspection games', *Contributions to the 16th Symposium on Operations Research*, Trier, Germany.

Avenhaus, R., and B. von Stengel (1993) 'Verification of attributes and variables: perfect equilibria and inspector leadership', in: R. Huber and R. Avenhaus, eds., *International Stability in a Multipolar World: Issues and Models for Analysis*. Baden-Baden: NOMOS, pp. 296–317.

Avenhaus, R., B. von Stengel and S. Zamir (1993) 'A game-theoretic approach to multipolar stability', in: R. Huber and R. Avenhaus, eds., *International Stability in a Multipolar World: Issues and Models for Analysis*. Baden-Baden: NOMOS, pp. 145–154.

Avenhaus, R., and S. Zamir (1988) 'Safeguards Games with Applications to Material Control', University of Bielefeld, Zentrum fur Interdisziplinare Forschung Working Paper No. 12.

Axelrod, R. (1969) *Conflict of Interest*. Chicago: Markham.

Axelrod, R. (1984) *The Evolution of Cooperation*. New York: Basic Books.

Axelrod, R., and R. Keohane (1985) 'Achieving cooperation under anarchy', *World Politics*, **38**: 226–254.

Balas, E. (1978) 'Choosing the Overall Size of the Strategic Petroleum Reserve', Management Sciences Research Group, Carnegie-Mellon University, Energy Policy Modelling Conference, mimeo.

Baldwin, R., and R. Clarke (1987) 'Game-modeling multilateral trade negotiations', *Journal of Policy Modeling*, **9**: 257–284.

Banks, J. (1990) 'Equilibrium behavior in crisis bargaining games', *American Journal of Political Science*, **34**: 509–614.

Bansal, R., and T. Basar (1989) 'Communication games with partially soft power constraints', *Journal of Optimization Theory and Applications*, **6**: 329–346.

Basar, T. (1983) 'The Gaussian test channel with an intelligent jammer', *IEEE Transactions on Information Theory*, **29**: 152–157.

Basar, T., and Y. Wu (1985) 'Complete characterization of minimax and maximin encoder-decoder policies for communication channels with incomplete statistical descriptions', *IEEE Transactions on Information Theory*, **31**: 482–489.

Bau, T. (1991) 'Taipei-Peking interaction as a two-person conflict: a game theoretical analysis, 1949–1989', *Issues and Studies*, **27**: 72.

Bellany, I. (1982) 'An introduction to verification', *Arms Control*, **3**: 1–13.

Bellman, R., D. Blackwell and J. LaSalle (1949) 'Application of Theory of Games to Identification of Friend and Foe', RM-197, RAND Corporation, Santa Monica.

Belzer, R. (1949) 'Solution of a Special Reconnaissance Game', RM-203, RAND Corporation, Santa Monica.

Belzer, R., M. Dresher, and O. Helmer (1949) 'Local Defence of Targets of Equal Value', RM-319, RAND Corporation, Santa Monica.

Bendor, J., and T. Hammond (1992) 'Rethinking Allison's models', *American Political Science Review*, **86**: 301–322.

Benjamin, C. (1981) 'Developing a Game/Decision-theoretic Approach to Comparative Foreign Policy Analysis: Some Cases in Recent American Foreign Policy'. Ph.D. Thesis, University of Southern California.

Benjamin, C. (1982) 'The Iranian Hostage Negotiations: A Metagame Analysis of the Crisis', International Studies Association Meeting.

Bennett, P. (1991) 'Modelling complex conflicts: formalism or expertise', *Review of International Studies*, **17**: 349–364.

Bennett, P., and M. Dando (1970) 'Complex strategic analysis: a hypergame perspective of the fall of France', *Journal of the Operational Research Society*, **30**: 23–32.

Bennett, P., and M. Dando (1977) 'Fall Gelb and other games: a hypergame perspective of the fall of France', *Journal of the Conflict Research Society*, **1**: 1–32.

Bennett, P., and M. Dando (1982) 'The nuclear arms race: some reflections on routes to a safer world', in: M. Dando and B. Newman, eds., *Nuclear Deterrence: Policy Options for the 1980's*. London: Castell, pp. 177–192.

Beresford, R., and M. Preston (1955) 'A mixed strategy in action', *Operational Research Quarterly*, **6**: 173–175.

Berkovitz, L., and M. Dresher (1959) 'A game theory analysis of tactical air war', *Operations Research*, **7**: 599–620.

Berkovitz, L., and M. Dresher (1960) 'Allocation of two types of aircraft in tactical air war: a game-theoretical analysis', *Operations Research*, **8**: 694–706.

Best, M., and J. Bracken (1993a) 'First-strike stability in a multipolar world', in: R. Huber and R. Avenhaus, eds., *International Stability in a Multipolar World: Issues and Models for Analysis*. Baden-Baden: NOMOS, pp. 223–254.

Best, M. and J. Bracken (1993b) 'First-strike Stability in a Multipolar World: Three Independent Sides'. World Laboratory Conference, Erice, Italy.

Bierlein, D. (1983) 'Game-theoretical modelling of safeguards of different types of illegal activities', in: *Proceedings of the Fourth Formator Symposium on Mathematical Methods in the Analysis of Large-scale Systems*. Prague: Czechoslovakian Academy of Science.

Bird, C. (1977) 'An infinite player cooperative game theory model for the analysis of the effect of various factors on international policy-making', in: J. Gillespie and D. Zinnes, eds., *Mathematical Systems in International Relations Research*. New York: Praeger, pp. 396–430.

Bird, C., and M. Sampson (1976) 'A game-theory model of OPEC, oil consumers and oil companies emphasizing coalition formations', in: D. Zinnes and J. Gillespie eds., *Mathematical Models in International Relations*. New York: Praeger, pp. 376–395.

Blackett, P. (1961) 'Critique of some contemporary defense thinking', *Encounter*, **16**: 9–17.

Blackwell, D. (1949) 'Comparison of Reconnaissances', RM-241, RAND Corporation, Santa Monica.

Blackwell, D. (1954) 'A representation problem', *Proceedings of the American Mathematical Society*, **5**: 283–287.

Borel, E. (1921) 'La theorie du jeu et les equations integrales a noyau symetrique', *Comptes Rendus de l'Academie des Sciences*, **173**: 1304–1308. (1953) Trans. L. Savage, "Theory of play and integral equations with skew symmetric kernels", Econometrica, **21**: 97–100.

Bohnenblust, H., L. Shapley and S. Sherman (1949) 'Reconnaissance in Game Theory', RM-208, RAND Corporation, Santa Monica.

Boulding, K. (1962) *Conflict and Defence*. New York: Harper.

Bourqui, P.-F. (1984) 'Etude des moyens d'influence de l'etat faible sur l'etat fort par quatre structures de la theorie des jeux', [A study of how a weak state may influence a strong one: four structures from the theory of games.] In: *Application de La Theorie des Jeux a l'Etude des Relations Internationales*. Geneva: Institut Universitaire de Hautes Etudes Internationales.

Bracken, J. (1989a) 'A Multistage Strategic Nuclear Offense-defense Force Allocation Model', Military Operations Research Society Symposium.

Bracken, J. (1989b) 'A Measure of First-strike Instability', School of Organization and Management, Yale University, mimeo.

Bracken, J., and P. Brooks (1983) 'Attack and Defense of ICBMs Deceptively Based in a Number of Identical Areas', Institute for Defence Analysis Paper P-1730.

Bracken, J., P. Brooks and J. Falk (1987) 'Robust preallocated preferential defense', *Naval Research Logistics*, **34**: 1–22.

Bracken, J., J. Falk and A. Karr (1975) 'Two models for optimal allocation of aircraft sorties', *Operations Research*, **23**: 979–996.

Bracken, J., J. Falk and F. Miercort (1977) 'A strategic weapons exchange allocation model', *Operations Research*, **25**: 968–976.

Bracken, J., and M. Shubik (1991) 'Strategic offensive and defensive forces and worldwide coalition games', in: R. Avenhaus, H. Karkar and M. Rudnianski, eds., *Defense Decision Making: Analytical Support and Crisis Management*. New York: Springer-Verlag, pp. 169–185.
Bracken, J., and M. Shubik (1993a) 'Crisis stability games', *Naval Research Logistics*, **40**: 289–303.
Bracken, J., and M. Shubik (1993b) 'Worldwide nuclear coalition games: a valuation of strategic offensive and defensive forces', *Operations Research*, **41**: 655–668.
Bracken, P. (1984) 'Deterrence, gaming and game theory', *Orbis*, **27**: 790–802.
Bracken, P., M. Haviv, M. Shubik and U. Tulowitzki (1984) 'Nuclear Warfare, C^3I and First and Second-strike Scenarios', Cowles Paper 712, Yale University.
Brams, S. (1975) *Game Theory and Politics*. New York: Free Press.
Brams, S. (1976) *Paradoxes in Politics*. New York: Free Press.
Brams, S. (1985) *Superpower Games: Applying Game Theory to Superpower Conflict*. New Haven: Yale University Press.
Brams, S. (1990a) 'Power and instability in multilateral arms control negotiations', in: R. Huber, H. Linnenkamp and I. Scholich, eds., *Military Stability, Prerequisites and Analysis Requirements for Conventional Stability in Europe*. Baden-Baden: NOMOS, pp. 87–96. Also published (1992) as 'Puissance et instabilite dans les negotiations multilaterales sur la maitrice des armements', in: M. Rudnianski, ed., *L'Aide a la Decision dans la Crise Internationale*. Paris: Foundation pour les Etudes de Defense Nationale, pp. 213–225.
Brams, S. (1990b) *Negotiation Games: Applying Game Theory to Bargaining and Arbitration*. New York: Routledge.
Brams, S. (1994) *The Theory of Moves*. Cambridge, UK: Cambridge University Press.
Brams, S., and M. Davis (1987) 'The verification problem in arms control', in: C. Cioffi-Revilla, R. Merritt and D. Zinnes, eds., *Communication and Interactions in Global Politics*. Beverly Hills: Sage.
Brams, S., M. Davis and M. Kilgour (1991) 'Optimal cheating and inspection strategies under INF', 318–335 in: R. Avenhaus, H. Karkar and M. Rudnianski, eds. *Defense Decision Making: Analytical Support and Crisis Management*. New York: Springer-Verlag.
Brams, S., A. Doherty and M. Weidner (1994) 'Game theory and multilateral negotiations: the Single European Act and the Uruguay Round', in: W. Zartmann, ed., *Many Are Called But Few Choose: The Analysis of Multilateral Negotiations*. San Francisco: Jossey-Basse.
Brams, S., and M. Hessel (1984) 'Threat power in sequential games', *International Studies Quarterly*, **28**: 23–44.
Brams, S., and M. Kilgour (1985) 'Optimal deterrence', *Social Philosophy and Policy*, **3**: 118–135.
Brams, S., and M. Kilgour (1986a) 'Notes on arms control verification', 337–350 in: R. Avenhaus and R. Huber, eds. *Modelling and Analysis of Arms Control Problems*. Berlin: Springer-Verlag.
Brams, S., and M. Kilgour (1986b) 'Is nuclear deterrence rational?' *PS*, **19**: 645–651.
Brams, S., and M. Kilgour (1987a) 'Verification and stability', 193–213 in: A. Din, ed. *Arms and Artificial Intelligence*. Oxford: Oxford University Press.
Brams, S., and M. Kilgour (1987b) 'Threat escalation and crisis stability', *American Political Science Review*, **81**: 833–850.
Brams, S., and M. Kilgour (1987c) 'Optimal threats', *Operations Research*, **35**: 524–536.
Brams, S., and M. Kilgour (1987d) 'Winding down if preemption or escalation occurs', *Journal of Conflict Resolution*, **31**: 547–572.
Brams, S., and M. Kilgour (1988a) 'Deterrence versus defense: a game-theoretic analysis of Star Wars', *International Studies Quarterly*, **32**: 2–28.
Brams, S., and M. Kilgour (1988b) *Game Theory and National Security*. New York: Basil Blackwell.
Brams, S., and M. Kilgour (1994) 'Are crises rational? a game-theoretic analysis', in: M. Intriligator and U. Luterbacher, eds., *Cooperative Models in International Relations*, forthcoming.
Brams, S., and W. Mattli (1993) 'Theory of moves: overview and examples", *Conflict Management and Peace Science*, **12**: 1–39.
Brams, S., and B. Mor (1993) 'When is it rational to be magnanimous in victory?' *Rationality and Society*.
Brennan, D. (1965) 'Review of Anatol Rapoport's *Strategy and Conscience*', *Bulletin of the Atomic Scientists*, **21**: 25–30.
Brickell, E. (1984) 'A Few Results in Message Authentification', Sandia National Laboratories, Albuquerque, New Mexico.
Brito, D. (1972) 'A dynamic model of an arms race', *International Economic Reveiw*, **35**: 359–375.

Brito, D., and M. Intriligator (1977) 'Strategic weapons and the allocation of international rights', 199–200 in: Gillespie and D. Zinnes, eds., *Mathematical Systems in International Relations Research*, New York: Praeger.
Brito, D., and M. Intriligator (1985) 'Conflict, war and redistribution', *American Political Science Review*, **79**: 943–957.
Brito, D., and M. Intriligator (1987) 'Arms races and the outbreak of war: applications of principal-agent theory and asymmetric information', 104–121 in: C. Schmidt and F. Blackaby, eds. *Peace, Welfare, and Economic Analysis*. New York: St. Martin's Press.
Brown, M., J. Mearsheimer, and W. Petersen (1976) 'A Model of Cartel Formation', P-5708, RAND Corporation, Santa Monica.
Brown, R. (1957) 'The solution of a certain two-person zerosum game', *Operations Research*, **5**: 63–67.
Brown, S. (1986) 'The superpowers' dilemma', *Negotiation Journal*, **2**: 371–384.
Brown, T. (1971) 'Models of Strategic Stability', Southern California Arms Control and Foreign Policy Seminar, Los Angeles.
Brown, T., S. Johnson and M. Dresher (1973) 'Some Methods for the Study of Escalation', WN-8123, RAND Corporation, Santa Monica.
Bruce, N. (1990) Defence expenditures by countries in allied and adversarial relationship. *Defence Economics*, **1**: 179–195.
Bueno de Mesquita, B., and D. Lalman (1988) 'Arms races and the propensity for peace', *Synthese*, **76**: 263–284.
Bueno de Mesquita, B., and D. Lalman (1989) 'The road to war is strewn with peaceful intentions', 253–266 in: P. Ordeshook, ed. *Models of Strategic Choice in Politics*. Ann Arbor: University of Michigan Press.
Bueno de Mesquita, B. (1990) 'Pride of place: the origins of German hegemony', *World Politics*, **43**: 28–52.
Bueno de Mesquita, B., and D. Lalman (1990) 'Domestic opposition and foreign war', *American Political Science Review*, **84**: 747–766.
Bueno de Mesquita, B., and D. Lalman (1992) *War and Reason*. New Haven: Yale University Press.
Bull, H. (1966) 'International theory, the case for a classical approach', *World Politics*, **18**: 361–377.
Bull, H. (1967) 'Strategic studies and its critics', *World Politics*, **20**: 593–605.
Bunn, G. (1992) *Arms Control by Committee: Managing Negotiations with the Russians*. Stanford: Stanford University Press.
Bunn, G., and R. Payne (1988) 'Tit-for-tat and the negotiation of nuclear arms control', *Arms Control*, **9**: 207–283.
Burr, S., J. Falk and A. Karr (1985) Integer Prim-Read solutions to a class of targeting problems. *Operations Research*, **33**: 726–745.
Busch, M., and E. Reinhardt (1993) 'Nice strategies in a world of relative gains', *Journal of Conflict Resolution*, forthcoming.
Callaham, M. (1982) 'Ballistic Missile Defense Systems and Strategies'. Ph.D. Thesis, Carnegie-Mellon University.
Call, G. (1989) 'SDI: Arms Race Solution or Bargaining Chip', Amherst College, mimeo.
Canty, M., and R. Avenhaus (1991) 'Inspection Games over Time', Berichte des Forschungszentrums Julich No. Jul-2472.
Canty, M., R. Avenhaus, and B. von Stengel (1991) 'Sequential aspects of nuclear safeguards: interim inspections of direct use material', American Nuclear Society Conference, Albuquerque.
Case, J. (1979) *Economics and the Competitive Process*. New York: New York University.
Cederman, L. (1991) 'The Logic of Superpower Extended Deterrence', University of Michigan, mimeo.
Chaatopadhyay, R. (1969) 'Differential game-theoretic analysis of a problem of warfare', *Naval Research Logistics Quarterly*, **16**: 435–441.
Chatterjee, P. (1975) *Arms, Alliances and Stability*. New York: Halsted.
Cheon, W., and N. Fraser (1987) 'The Arms Control Game and the Role of Verification', Society for Political Psychology.
Chudnovsky, D., and G. Chudnovsky (1988) *Search Theory: Some Recent Developments*. New York: Dekker.
Chun, S. (1989) 'Verification of Compliance, Mathematical, Political and Economic Principles'. Ph.D. Thesis, University of Waterloo.
Control Analysis Corporation, Palo Alto (1972) 'Approximate Solutions of Multi-stage Network Games', mimeo.

Conybeare, J. (1984) 'Public goods, Prisoner's Dilemma and the international political economy', *International Studies Quarterly*, **28**: 5–22.
Conybeare, J. (1987) *Trade Wars*, New York: Columbia University Press.
Cook, D. (1984) 'Korean Unification: A Game-theoretical and Bargaining Analysis'. Master's thesis, Naval Postgraduate School, Monterey, California.
Crawford, V. (1990) 'Thomas Schelling and the Analysis of Strategic Behavior', University of California, San Diego, Department of Economics, Working Paper 90-11.
Cudd, A. (1990) 'Conventional foundationalism and the origin of norms', *Southern Journal of Philosophy*, **28**: 485–504.
Dacey, R. (1979) 'Detection and disarmament', *International Studies Quarterly*, **23**: 589–598.
Dacey, R. (1987) 'Ambiguous information and the manipulation of plays of the arms race game and the mutual deterrence game', 163–179 in: C. Cioffi-Revilla, R. Merritt and D. Zinnes, eds. *Interaction and Communication in Global Politics*. London: Sage.
Dalkey, N. (1965) 'Solvable nuclear war models', *Management Science*, **9**: 783–791.
Danskin, J. (1962a) 'A theory of reconnaissance, II', *Operations Research*, **10**: 300–309.
Danskin, J. (1962b) 'A two-person zerosum game model of a convoy routing', *Operations Research*, **10**: 774–785.
Danskin, J. (1990) 'On the cookie-cutter game: search and evasion on a disc', *Mathematics of Operations Research*, **15**: 573–596.
Davis, M. (1963) 'Verification of disarmament by inspection: a game-theoretic model', Econometric Research Program Memorandum 62, Princeton University.
Dekmajian, M., and G. Doron (1980) 'Changing patterns of equilibrium in the Arab/Israel conflict', *Conflict Management and Peace Science*, **5**: 41–54.
Delver, R. (1986) 'Sea Link '86 in Speltheoretisch perspectief', ['Sea Link '86 from a game-theoretical perspective'.] *Marineblad*, **96**: 388–395.
Delver, R. (1987) 'Over strategische conflictafwikkeling', ['On the resolution of strategic conflict'.] *Acta Politica*, **4**: 385–407.
Delver, R. (1989) 'Strategic Rivalry in Cooperative Games of Properly Mixed Interests', Royal Netherlands Naval Academy Report SS-88-1.
Demange, G. (1990) 'Equilibria of the Dollar Auction', International Conference on Game Theory, Stony Brook, N.Y.
Deutsch, K. (1954) 'Game theory and politics: some problems of application', *Canadian Journal of Economics and Political Science*, **20**: 76–83.
Deutsch, K. (1968) 'Threats and deterrence as mixed motive games', 124–130 in: *Analysis of International Relations*. Englewood Cliffs, N.J.: Prentice-Hall.
Dewdney, K. (1987) 'After-MAD, a game of nuclear strategy that ends in a Prisoner's Dilemma', *Scientific American*, **257**: 174–177.
Deyer, S., and S. Sen (1984) 'Optimal control and differential game model of military disputes in less developed countries', *Journal of Economic Dynamics and Control*, **7**: 153–169.
Diamond, H. (1982) 'Minimax policies for unobservable inspections', *Mathematics of Operations Research*, **7**: 139–153.
Digby, J. (1988) 'Operations research and systems analysis at RAND, (1948–1967)', *OR/MS Today*, **15**: 10–13.
Digby, J. (1990) 'Strategic Thought at RAND, (1948–1963). The Ideas, Their Origins, Their Fates', N-3096-RC, Rand Corporation, Santa Monica.
Diskin, A., and S. Mishal (1984) 'Coalition formation in the Arab world, an analytical perspective', *International Interactions*, **11**: 43–59.
Dobbie, J. (1968) 'A survey of search theory', *Operations Research*, **16**: 525–537.
Don, B. (1986) 'The Interalliance Game', in: *Allies and Adversaries, Policy Insights into Strategic Defense Relationships*. P-7242, RAND Corporation, Santa Monica, Ch. 4.
Downs, G., and D. Rocke (1987) 'Tacit bargaining and arms control', *World Politics*, **39**: 297–325.
Downs, G., and D. Rocke (1990) *Tacit Bargaining, Arms Races and Arms Control*. Ann Arbor: University of Michigan Press.
Downs, G., and D. Rocke (1991) 'Intervention, Escalation and Gambling for Resurrection', Princeton University.
Downs, G., D. Rocke and R. Siverson (1985) 'Arms races and cooperation', in: Oye, ed., pp. 118–146.

Dresher, M. (1957) 'Optimal Tactics in Delayed Firing: a Game-theoretical Analysis', RM-1886, RNAD Corporation, Santa Monica.
Dresher, M. (1961a) *Games of Strategy: Theory and Applications.* Englewood Cliffs, New Jersey: Prentice-Hall.
Dresher, M. (1961b) Some Military Applications of the Theory of Games. P-1849, RAND Corporation, Santa Monica. 597–604 in: *Proceedings of the Second International Conference on Operations Research.* New York: Wiley.
Dresher, M. (1962) 'The Sampling Inspector Problem in Arms Control Agreements', RM-2972, RAND Corporation, Santa Monica.
Dresher, M. (1966) 'Some principles of military operations derived by the theory of games', in: Mensch, ed., pp. 360–362.
Dresher, M. (1968) 'Mathematical Models of Conflict', 228–239 in: E. Quade, Boucher, eds. *Systems Analysis and Policy Planning: Applications in Defense.* New York: Elsevier.
Dresher, M., and O. Gross (1950) 'Local Defense of Targets of Equal Value: Extension of Results', RM-320, RAND Corporation, Santa Monica.
Dresher, M., and S. Johnson (1961) 'Optimal Timing in Missile Launching: A Game-theoretical Analysis', RM-2723, RAND Corporation, Santa Monica.
Dubin, G., and V. Suzdal (1981) *Vvedenie v Prikladnuu Teoria Igr.* [Introduction to Applied Game Theory.] Moscow: Nauka.
Dudley, L. and J. Robert (1992) 'A Non-cooperative Model of Alliances and Warfare', Universite de Montreal.
Eckler, R., and S. Burr (1972) *Mathematical Models of Target Coverage and Missile Allocation.* Alexandria, VA.: Military Operations Research Society.
Eisner, R. (1968) 'War and Peace: A New View of the Game', Department of Economics, Northwestern University.
Ellsberg, D. (1961) 'The crude analysis of strategic choices', *American Economic Review*, **51**: 472–478.
Engelmann, W. (1991) 'Conditions for Disarmament – Game Theoretic Models of Superpower Conflict', Working Paper 11, Interdisciplinary Research Group on Science, Technology and Security Policy, Technical University, Darmstadt, Germany.
Evangelista, M. (1988) *Innovation and the Arms Race.* Ithaca: Cornell University Press.
Evangelista, M. (1990) 'Cooperation theory and disarmament negotiations in the 1950s', *World Politics*, **42**: 502–528.
Fain, W. (1961) 'The role of communication in warfare', 565–578 in: *Proceedings of the Second International Conference on Operational Research.* New York: Wiley.
Fain, W., and Phillips, J. (1964) 'Application of game theory to real military decisions', Douglas Aircraft. Abridged in: Mensch (1966).
Falk, J. (1973) 'Remarks on Sequential Two-person Zero-sum Games and TAC CONTENDER', George Washington University Institute for Management Science and Engineering Program in Logistics.
Fearon, J. (1990) 'Deterrence and the Spiral Model, the Role of Costly Signalling in Crisis Bargaining', American Political Science Association Meeting, San Francisco.
Fearon, J. (1992b) 'Audience Costs, Learning and the Escalation of International Disputes', Department of Political Science, University of Chicago.
Fearon, J. (1992c) 'Threats to use force: the role of costly signals in international crises'. Ph.D. dissertation, University of California, Berkeley.
Fearon, J. (1994) 'Signalling versus the balance of power and interests', *Journal of Conflict Resolution*, **38**: 236–269.
Fichtner, J. (1986a) 'Zur Modellierung der mehrstufigen Abschreckung', Working Paper, Neubilberg.
Fichtner, J. (1986b) 'On solution concepts for solving two person games which model the verification problem in arms control', 421–441 in: R. Avenhaus, R. Huber and J. Kettelle, eds. *Modelling and Analysis in Arms Control.* Berlin: Springer-Verlag.
Feiwel, G., ed. (1987) *Arrow and the Ascent of Modern Economic Theory.* New York: New York University Press.
Filar, J. (1983) 'The Travelling Inspector Model', Technical Report 374, Department of Mathematical Sciences, The Johns Hopkins University.
Filar, J. (1984) Player aggregation in the travelling inspector model', *Proceedings of the IEEE Conference on Decision and Control*, **2**: 1192–1199.

Filar, J., and T. Schultz (1983) 'Interactive Solutions for the Travelling Inspector Model and Related Problems', Operations Research Group Report Series #83-06, Department of Mathematical Sciences, The Johns Hopkins Univeristy.
Finn, M., and G. Kent. (1985) 'Simple Analytical Solutions to Complex Military Problems', N-2211, RAND Corporation, Santa Monica.
Fogarty, T. (1990) 'Deterrence games and social choice', International Interactions, 15: 203–226. Reprinted, F. Zagare, ed., Modeling International Conflict. New York: Gordon and Breach, 1990.
Forward, N. (1971) The Field of Nations. New York: Little-Brown.
Franck, R. (1983) 'The Option of War and Arms Race Behavior: A Microeconomic Analysis'. Ph.D. Thesis, Harvard University.
Fraser, N., and K. Hipel (1982) 'Dynamic modeling of the Cuban Missile Crisis', Conflict Management and Peace Science, 6: 1–18.
Fraser, N., and K. Hipel (1984) Conflict Analysis, Models and Resolutions. New York: Elsevier.
Fraser, N., K. Hipel and J. Jaworsky (1990) 'A conflict analysis of the Armenian-Azerbaijani dispute', Journal of Conflict Resolution, 34: 652–677.
Fraser, N., C. Powell and C. Benjamin (1987) 'New methods for applying game theory to international conflict', International Studies Notes, 13: 9–18.
Freedman, L. (1983) The Evolution of Nuclear Strategy. London: MacMillan.
Frey, B. (1974) 'An insurance system for peace', Papers of the Peace Research Society International, 22: 111–128.
Frick, H. (1976) 'Spieltheoretische Behandlung mehrfacher Inventurprobleme'. Ph.D. Dissertation, Universitat Fridericiana Karlsruhe.
Gal, S. (1980) Search Games. New York: Academic Press.
Gale, D., and G. Glassey (1975) 'Elementary Problem #2468', American Mathematical Monthly, 82: 521–522.
Gardner, R. and W. Guth. (1991) 'Modelling alliance formation, a noncooperative approach', 210–228 in: R. Selten, ed., Game Equilibrium Models, IV. New York: Springer Verlag
Garfinkel, M. (1991) 'Arming as a strategic investment in a cooperative equilibrium', American Economic Review, 80: 50–68.
Gates, S. (1989) 'The Limits of Conditionality'. Ph.D. Thesis, University of Michigan.
Gates, S., and B. Humes (1990) 'Game Theoretical Models and International Relations', Michigan State University, mimeo.
Gates, S., and C. Ostrom, Jr. (1991) 'The Long Peace: A Game Theoretic Analysis', Department of Political Science, Michigan State University, mimeo.
Gates, S., C. Ostrom and S. Quinones (1992) 'Shortrun Reciprocity, Credibility and the "Long Peace": An Empirical Test of Some Game-theoretic Propositions', American Political Science Association, Chicago.
Gauthier, D. (1984) 'Deterrence, maximization and rationality', Ethics, 94: 474–495.
Gekker, R. (1989a) 'Subgame perfect equilibria and the Berlin crisis of 1948', York Centre for International and Strategic Studies, York University, mimeo.
Gekker, R. (1989b) 'The deterrence paradox revisited: a game-theoretic account', York Centre for International and Strategic Studies, York University, mimeo.
George, A., and R. Smoke (1974) Deterrence in American Foreign Policy. New York: Columbia.
Giamboni, L., A. Mengel and R. Dishington (1951) 'Simplified model of a symmetric tactical air war', RM-711, RAND Corporation, Santa Monica.
Gigandet, C. (1984) 'Les relations germano-sovietiques (1922–1941), essai d'application d'un modele de jeu a deux acteurs', ['German/Soviet relations, 1922–1941, an attempt to apply a two-person game theory model'.] in: Application de La Theorie des Jeux a l'Etude des Relations Internationales. Geneva: Institut Universitaire de Hautes Etudes Internationales.
Gillespie, J., and D. Zinnes (1977) 'Embedded game analysis of an international conflict', Behavioral Science, 22: 22–29.
Gillespie, J., D. Zinnes and G. Tahim (1975) 'Foreign assistance and the arms race, a differential game model', Papers of the Peace Science Society, 25: 35–51.
Gillespie, J., D. Zinnes and G. Tahim (1977) 'Deterrence as second attack capability: an optimal control and differential game model', 367–395 in: J. Gillespie and D. Zinnes, eds., Mathematical Systems in International Relations Research. New York: Praeger.

Gold, V. (1973) 'Spieltheorie und politische Realitat' (Game theory and political reality.) *Politische Studien*, **21**: 257–276.
Goldhamer, H., A. Marshall and N. Leites (1959) 'The Strategy and Deterrence of Total War, 1959–1961', RM-2301, RAND Corporation, Santa Monica.
Goldstein, J., and J. Freeman (1990) *Three-Way Street, Strategic Reciprocity in World Politics*. Chicago: University of Chicago Press.
Gowa, J. (1986) 'Anarchy, egoism and third images: the evolution of cooperation and international relations', *International Organization*, **40**: 176–177.
Gowa, J. (1989) 'Rational hegemons, excludable goods and small groups', *World Politics*, **41**: 307–324.
Gowa, J., and E. Mansfield (1993) 'Power politics and international trade', *American Political Science Review*, **87**: 408–420.
Green, J., and J. Laffont (1994) 'International Agreement Through Sequential Unanimity Games', in: M. Intriligator and U. Luterbacher, eds.
Green, P. (1966) *Deadly Logic*. Columbus: Ohio State University Press.
Greenberg, I. (1982) 'The role of deception in decision theory', *Journal of Conflict Resolution*, **26**: 139–156.
Grieco, J. (1988) 'Realist theory and the problem of international cooperation, analysis with an amended Prisoner's Dilemma model', *Journal of Politics*, **50**: 600–624.
Grieco, J. (1990) *Cooperation Among Nations*. Ithaca: Cornell University Press.
Grieco, J. (1993) 'The relative gains problem for international co-operation' *American Political Science Review*, **87**, 729–736.
Gross, O. (1950a) 'Local Defense of Targets of Equal Value: Completion of Results', RM-329, RAND Corporation, Santa Monica.
Gross, O. (1950b) '*n* Targets of differing Vulnerability with Attack Stronger Than Defense', RM-359, RAND Corporation, Santa Monica.
Gross, O., and R. Wagner (1950) 'A continuous Colonel Blotto game', RM-408, RAND Corporation, Santa Monica.
Grotte, J. (1982) 'An optimizing nuclear exchange model for the study of war and deterrence', *Operation Research*, **30**: 428–445.
Grotte, J., and P. Brooks (1983) 'Measuring naval presence using Blotto games', *International Journal of Game Theory*, **12**: 225–236.
Guner, S. (1990) 'A Game Theoretical Analysis of Alliance Formation and Dissolution; The United States, the Soviet Union and the People's Republic of China, 1949–1972', Institut Universitaire de Hautes Etudes Internationales, Geneva.
Guner, S. (1991) 'A Systemic Theory of Alliance Stability', University of Texas, Austin, mimeo.
Guth, W. (1985) 'An extensive game approach to model the nuclear deterrence debate', *Zeitschrift fur die gesamte Staatswissenschaft/Journal of Institutional and Theoretical Economics*, **14**: 525–538.
Guth, W. (1986) 'Deterrence and incomplete information: the game theory approach', in: R. Avenhaus, R. Huber and J. Kettelle, eds. *Modelling and Analysis in Arms Control*. New York: Springer-Verlag.
Guth, W. (1991) 'The stability of the Western defense alliance: a game-theoretical analysis', 229–255 in: R. Selten, ed. *Game Equilibrium Models* IV. Springer-Verlag.
Guth, W. and E. van Damme (1991) 'Gorbi-games: A game-theoretical analysis of disarmament campaigns and the defense-efficiency hypothesis', 215–241 in: R. Avenhaus, H. Karkar and M. Rudnianski, eds. *Defense Decision Making: Analytical Support and Crisis Management*. New York: Springer-Verlag.
Hardin, R. (1983) 'Unilateral versus mutual disarmament', *Philosophy and Public Affairs*, **12**: 236–254.
Hardin, R. (1986a) 'Risking Armageddon', 201–232 in: A. Cohen and S. Lee, eds. *Nuclear Weapons and the Future of Humanity: the Fundamental Questions*. New York: Rowman and Allanheld.
Hardin, R. (1986b) 'Winning versus doing well', 103–117 in: C. Kelleher, F. Kerr and G. Quester, eds., *Nuclear Deterrence, New Risks, New Opportunities*. London: Pergammon-Brassey.
Harsanyi, J. (1962) 'Measurement of social power, opportunity costs, and the theory of two-person bargaining games', *Behavioral Science*, **7**: 67–80.
Hashim, H. (1984) *Nazariyat al-Mubarayat wa-dawruha fi tahlil al-sira' at al-dawliyah ma'a al-tatbiq'ala al-sira' al'-Arabi al-Isra'ili*. (The Theory of Games and Its Role in International Relations, with Applications to the Arab/Israel Conflict.) Cairo: Cairo University.
Hayek, O. (1980) *Pursuit Games*. New York: Academic Press.

Haywood, O. (1950) 'Military decision and the mathematical theory of games', *Air University Quarterly Review*, **4**: 17–30.
Haywood, O. (1954) 'Military decisions and game theory', *Operations Research*, **2**: 365–385.
Heims, S. (1980) *John von Neumann and Norbert Weiner: From Mathematics to the Technology of Life and Death*. Cambridge: MIT Press.
Helin, P. (1983) 'Minimax Signal Design'. Ph.D. Thesis, University of California, Santa Barbara.
Hewitt, J., M. Boyer and J. Wilkenfeld (1991) 'Crisis Actors in Protracted Social Conflicts', Department of Political Science, University of Maryland, mimeo.
Hillstead, R. (1986) 'SAGE: An Algorithm for the Allocation of Resources in a Conflict Model', N-2533, Rand Corporation, Santa Monica.
Hipel, K., M. Wang and N. Fraser (1988) 'A hypergame analysis of the Falkland/Malvinas conflict', *International Studies Quarterly*, **32**: 335–358.
Ho, Y., ed. (1970) *First International Conference on the Theory and Applications of Differential Games*, Department of Engineering, Harvard University.
Ho, Y., and G. Olsder (1983) 'Differential games: concepts and applications', 127–185 in: M. Shubik, ed. *Mathematics of Conflict*. New York: North Holland.
Hollis, M., and S. Smith (1990) 'The games nations play', Ch. 6 in: *ibid.*, eds. *Explaining and Understanding International Relations*. Oxford: Oxford University Press.
Holmes, R. (1982) 'Optimization Methodology for Analysis of Defense of a Combined MX/MM Force', Lincoln Laboratory, Massachusetts Institute of Technology.
Hone, T. (1984) 'Game theory and symmetrical arms competition: the Dreadnought Race', *Journal of Strategic Studies*, **7**: 169–177.
Hopfinger, E. (1975) *Reliable Inspection Strategies: A Game Theoretical Analysis of Sampling Inspection Problems*. Meisenhaim am Glau: Hain.
Hopmann, T. (1978) 'Asymmetrical bargaining and the conference on security and cooperation in Europe', *International Organization*, **32**: 141–177.
Horowitz, I. (1962) 'Arms, policies and games', *American Scholar*, **31**: 94–107.
Horowitz, I. (1963a) *Games, Strategies and Peace*. Philadelphia: American Friends Service Committee.
Horowitz, I. (1963b) *The War Game: Studies of the New Civilian Militarists*. New York: Ballantine.
Horowitz, I. (1970) 'Deterrence games: from academic casebook to military codebook', 277–291 in: *The Structure of Conflict*. New York: Academic Press.
Hovi, J. (1984) 'Om bruk av spollteoretiske modeller med variable preferanser i studlet av internasjonal politikk', ('On the application of game-theoretical models with variable preferences to the study of international politics'.) *Internasjonal Politikk* (Oslo). **3**: 81–1.
Hovi, J. (1987) 'Probabilistic behavior and non-myopic equilibria in superpower games', *Journal of Peace Research*, **24**: 407–416.
Howard, N. (1968) 'Metagame analysis of the Viet Nam policy', *Papers of the Peace Research Society International*, **10**: 126–142.
Howard, N. (1971) *Paradoxes of Rationality: The Theory of Metagames and Political Behavior*. Cambridge: MIT Press.
Howard, N. (1972) 'The Arab-Israeli conflict, a metagame analysis done in 1970', *Papers of the Peace Research Society International*, **19**: 35–60.
Hsieh, J. (1990a) 'The Recognition Game: How Far Can the Republic of China Go'? National Chengchi University, Taipei, mimeo.
Hsieh, F. (1990b) 'Chan fu? Ho fu? Tsung po-yi li-lun k'an Chung-tung wei-chi', ['War or peace? A game-theoretical view of the Middle East Crisis'.] *Mei-kuo yueh-k'an* [American Monthly, Taipei], December: 9–13.
Hurwitz, R. (1989) 'Strategic and social fictions in the Prisoner's Dilemma', 113–134 in: J. Der Derian and M. Shapiro, eds. *International/Intertextual Relations*. Lexington, MA.: Lexington.
Hutchinson, H. (1992) 'Intelligence: escape from the Prisoner's Dilemma', *Intelligence and National Security*, **7**: 327–334.
Iida, K. (1991) 'Post-hegemonic Transition and "Overshooting": A Game-theoretic Analysis', Department of Politics, Princeton University, mimeo.
Iida, K. (1993a) When and how do domestic constraints matter?: Two-level games with uncertainty. Forthcoming, *Journal of Conflict Resolution*.
Iida, K. (1993b) "Analytic uncertainty and international cooperation" *International Studies Quarterly* **37**: 451–457.

Iida (1993c) Secret diplomacy and open economies endogenous surprise in two-level games mimeo. Princeton University.

Intriligator, M. (1967) 'Strategy in a Missile War, Targets and Rates of Fire', Security Studies Paper 10, University of California, Los Angeles.

Intriligator, M., and D. Brito (1984) 'A Game Theory Model of the Military Industrial Complex', Department of Economics, University of California, mimeo.

Intriligator, M. and U. Luterbacher, eds. (1994) *Cooperative Game Models in International Relations*. Boston: Kluwer.

Isaacs, R. (1951) 'Games of Pursuit', P-257, RAND Corporation, Santa Monica.

Isaacs, R. (1954) 'Games of Aiming and Evasion', RM-1385, RAND Corporation, Santa Monica.

Isaacs, R. (1965) *Differential Games*. New York: Wiley.

Isaacs, R. (1975) 'Theory and applications of differential games', in: J. Grotte, ed., NATO *Advanced Study Institute Series C*. 13. Dordrecht.

Isard, W. (1965) 'The Veto-incemax procedure potential for Viet Nam conflict resolution', *Papers of the Peace Research Society International*, **10**: 148–162.

Isbell, J., and Marlow, W. (1965) 'Attrition games', *Naval Research Logistics Quarterly*, **3**: 71–94.

James, P., and F. Harvey (1989) 'Threat escalation and crisis stability: superpower cases, 1948–1979', *Canadian Journal of Political Science*, **22**: 523–545.

James, P., and F. Harvey (1992) 'The most dangerous game: superpower rivalry in international crises, 1948–1985', *Journal of Politics*, **54**: 24–53.

Jervis, R. (1978) 'Cooperation under the security dilemma', *World Politics*, **30**: 167–214.

Jervis, R. (1988) 'Realism, game theory and cooperation', *World Politics*, **40**: 317–349.

John, A., R. Pecchenino and S. Schreft (1993) 'The macroeconomics of Dr. Strangelove', *American Economic Review*, **83**: 43–62.

Johnson, S. (1959) 'A Game Solution to a Missile-launch Scheduling Problem', RM-2398, RAND Corporation, Santa Monica.

Johr, M. (1971) 'Zur Strategie der Schweiz im Zweiten Weltkreig – Versuch einer Spieltheoretischen Answertung des Bonjour-Berichtes', ['On Switzerland's strategy in the Second World War – a game theory analysis of the Bonjour Report'.] *Wirtschaft und Recht* (Zurich), **24**: 13–59.

Jonsson, C. (1976) *The Soviet Union and the Test Ban: A Study in Soviet Negotiating Behavior*. Lund: Studenterlitteratur.

Josko de Gueron, E. (1991) 'Las Estructuras y los Resultados de la Negociacion: La Dueda Externa Venezolana', ['Bargaining Structures and Outcomes: Venezuela's Foreign Debt'.] Instituto de Estudios Politicos, Universidad Central de Venezuela.

Joxe, A. (1963) 'La Crise Cubaine de 1962, Elements Principaux de Decision, au Cours de la Crise "Chaude"', Document 3, EPHE Groupe d'Etudes Mathematiques des Problemes Politiques et Strategiques, Paris.

Junne, G. (1972) 'Spieltheorie in der Internationalen Politik; die Beschrenkte Rationalitat Strategischen Denkens', ['Game Theory in International Politics: The Limited Rationality of Strategic Throught'.] Dusseldorf: Bertelsmann Universitat-verlag.

Kalman, P. (1971) 'Applications of a theorem by Scarf on cooperative solutions to n-person games to problems of disarmament', *Journal of Political Economy*, **79**: 1388–1396.

Kaplan, F. (1983) *The Wizards of Armageddon*. New York: Simon and Schuster.

Kaplan, M. (1957) *System and Process in International Politics*. Wiley: New York.

Kaplan, M. (1958) 'The calculus of nuclear deterrence', *World Politics*, **11**: 20–43.

Kaplan, M. (1973) 'Strategy and morality', 13–38 in: M. Kaplan, ed. *Strategic Theory and its Moral Implications*. Chicago: Center for Policy Studies, University of Chicago.

Karlin, S. (1959) *Mathematical Methods and Theory in Games, Programming and Economics*. Vol. II. Reading, MA: Addison–Wesley.

Karr, A. (1981) 'Nationwide Defense against Nuclear Weapons: Properties of Prim–Read Deployments', P-1395, Institute for Defense Analyses, Alexandria, VA.

Kawara, Y. (1973) 'Allocation problem of fire support in combat as a differential game', *Operations Research*, **21**: 954–951.

Kent, G. (1963) 'On the Interaction of Opposing Forces under Possible Arms Agreements', Center for International Affairs, Harvard University.

Kent, G. (1964) 'Summary Study of the Strategic Offensive and Defensive Forces of the U.S. and USSR', United States Department of Defense.

Kent, G., and D. Thaler (1989) 'First Strike Instability: A Methodology for Evaluating Strategic Forces', R-3765, RAND Corporation, Santa Monica.
Keohane, R. (1985) *After Hegemony: Cooperation and Discord in the World Political Economy*. Princeton: Princeton University Press.
Keohane, R., ed. (1986a) *Neorealism and its Critics*. New York: Columbia.
Keohane, R. (1986b) 'Reciprocity in international relations', *International Organization*, **40**: 1–28.
Kettelle, J. (1985) 'A computerized third party', 347–367 in: R. Avenhaus, R. Huber and J. Kettelle, eds. *Modelling and Analysis in Arms Control*. New York: Springer-Verlag.
Kilgour, M. (1990) 'Optimal cheating and inspection strategies under a chemical weapons treaty', *INFOR*, **28**: 27–39.
Kilgour, M. (1991) 'Domestic political structure and war behavior: a game-theoretical approach', *Journal of Conflict Resolution*, **35**: 266–284.
Kilgour, M. (1993) 'Site Selection for on-site inspection', *Arms Control*, **13**: 439–462.
Kilgour, M., and S. Brams (1992) 'Putting the other side "on notice" can induce compliance in arms control', *Journal of Conflict Resolution*, **36**: 395–414.
Kilgour, M., and F. Zagare (1991a) 'Credibility, Uncertainty and Deterrence', *American Journal of Political Science*, **35**: 305–334.
Kilgour, M., and F. Zagare (1991b) 'Uncertainty and the Role of the Pawn in Extended Deterrence', Department of Mathematics, Wilfred Laurier University, mimeo.
Kilgour, M., and F. Zagare (1993) 'Asymmetric deterrence', *International Studies Quarterly*, **37**: 1–27.
Kimmeldorf, G. (1983) 'Duels, an overview', 55–71 in: M. Shubik, ed. *Mathematics of Conflict*. New York: North-Holland.
Kirby, D. (1988) 'SDI and Prisoner's Dilemma', *Parameters*, **18**: 88–96.
Kitchen, J. (1962) 'Optimal Altitude for Anti-missile Detonations', RM-3328, RAND Corporation, Santa Monica.
Kivikari, U., and H. Nurmi (1986) 'A game theory approach to potential problems of East-West trade', *Cooperation and Conflict*, **21**: 65–78.
Krasner, S. (1991) 'Global communications and national power: life on the Pareto frontier', *World Politics*, **43**: 336–366.
Kratochwil, F. (1989) *Rules, Norms and Decisions: On the Conditions of Practical and Legal Reasoning in International Relations and Domestic Affairs*. Cambridge: Cambridge University Press.
Krelle, W. (1972) 'Theoretical methods for decision making in foreign policy', *Europa-Archiv*, **27**: 387–398.
Kroll, J. (1992) 'Avenues of Commerce: Free Trade and Protectionism in International Politics', Dartmouth College, mimeo.
Kugler, J., and F. Zagare (1987) 'Risk, deterrence and war', 69–89 in: J. Kugler and F. Zagare, eds. *Exploring the Stability of Deterrence*. Denver: Lynne Reinner.
Kugler, J., and F. Zagare (1990) 'The long-term stability of deterrence', International Interactions, **15**: 255–278. Reprinted (1990) F. Zagare, ed., *Modeling International Conflict*. New York: Gordon and Breach.
Kuhn, H. (1963) 'Recursive inspection games', in: *Mathematica, Inc., The Application of Statistical Methodology to Arms Control and Disarmament*. United States Arms Control and Disarmament Agency Report ST-3.
Kuhn, J., K. Hipel and N. Fraser (1983) 'A coalition analysis algorithm with application to the Zimbabwe conflict', *IEEE Transactions on Systems, Man and Cybernetics*, **3**: 338–352.
Kull, S. (1988) *Minds at War*. New York: Basic.
Kupchan, C. (1988) 'NATO and the Persian Gulf: examining intra-alliance behavior', *International Organization*, **42**: 317–346.
Kydd, A. (1994) *Security and conflict*. Ph.D. Dissertation, University of Chicago.
Laffont, J. and J. Tirole (1993) *A Theory of Incentives in Procurement and Regulation*. Cambridge: MIT Press.
Lalley, S., and H. Robbins (1987) 'Asymptotically minimax stochastic search in the plane', *Proceedings of the National Academy of Sciences*, **84**: 2111–2112.
Langlois, J. (1988) 'On crisis instability and limited lookahead', San Francisco State University, mimeo.
Langlois, J. (1989) 'Modeling deterrence and international crises', *Journal of Conflict Resolution*, **33**: 67–83.
Langlois, J. (1991a) 'A Theory of Rational Expectations in Recurring International Disputes', International Conference on Game Theory, Florence.

Langlois, J. (1991b) 'Rational deterrence and crisis stability', *American Journal of Political Science*, **35**: 801–832.
Langlois, J. (1991c) 'Decision Systems Analysis of Conflict Management', San Francisco State University, mimeo.
Larson, D. (1987) 'Game theory and the psychology of reciprocity', American Political Science Association Annual Meeting.
Lee, L., and Lee, K. (1990) 'The discrete evasion game with three-move lags', *Naval Research Logistics*, **37**: 867–874.
Leininger, W. (1989) 'Escalation and cooperation in conflict situations: the dollar auction revisited', *Journal of Conflict Resolution*, **33**: 231–254.
Leng, R. (1988) 'Crisis learining games', *American Political Science Review*, **82**: 179–194.
Leonard, R. (1991), War as a 'simple economic problem; the rise of defense' 260–283 in C. Goodwin, ed. *Economics and National Security* Durham: Duke University Press.
Leonard, R. (1992) 'Creating a context for game theory', in R. Weintraub, ed. *Towards a History of Game Theory* Durham: Duke University Press.
Levy, M. (1985) 'Namibia as a Prisoner's Dilemma', *Journal of Conflict Resolution*, **29**: 581–604.
Lichbach, M. (1988) 'Giving Peace Time: Solving an Arms Race that is a Prisoner's Dilemma with Repeated Play', Working Paper, University of Illinois at Chicago.
Lichbach, M. (1989) 'Stability in Richardson's arms races and cooperation in Prisoner's Dilemma arms rivalries', *American Journal of Political Science*, **33**: 1016–1047.
Lichbach, M. (1990) 'When is an arms rivalry a Prisoner's Dilemma? Richardson's models and 2 × 2 Games', *Journal of Conflict Resolution*, **34**: 29–56.
Licklider, R. (1971) *The Private Nuclear Strategists*. Columbus: Ohio State University Press.
Lieshout, R. (1992) 'The theory of games and the formation of degensive alliances', *Acta Politica*, **27**: 111.
Leitzel, J. (1993) 'Game theory and national security: forward induction.' in J. Leitzel, ed. *Economics and National Security* Boulder: West view.
Lindskold, S. (1978) 'The GRIT proposal and the effects of conciliatory acts on conflict and cooperation', *Psychological Bulletin*, **85**: 772–795.
Linster, B. (1903) 'A rent-seeking model of international competition and alliances', *Defence Economics*, **4**, 213–226.
Lipson, C. (1984) 'International cooperation in economic and security affairs', *World Politics*, **37**: 1–23.
Lohmann, S. (1993) 'Electoral cycles and international policy cooperation', *European Economic Review*, forthcoming.
Ludwig, R. (1990) 'Peace Research: Developments in Post-World War II Thought'. Ph.D. Thesis, University of Michigan.
Lumsden, M. (1973) 'The Cyprus conflict as a Prisoner's Dilemma game', *Journal of Conflict Resolution*, **17**: 7–31.
Luterbacher, U. (1983) 'Coalition Formation in the Triad: The Case of US, USSR and China', Institut Universitaire de Hautes Etudes Internationales, Geneva.
Maccoby, M. (1961) 'The social psychology of deterrence', *Bulletin of the Atomic Scientists*, **17**: 278–281.
Majeski, S. (1984) 'Arms races as iterated Prisoner's Dilemma games', *Mathematics Social Sciences*, **7**: 253–266.
Majeski, S. (1986a) 'Mutual and unilateral cooperation in arms race settings', *International Interactions*, **12**: 343–361.
Majeski, S. (1986b) 'Technological innovation and cooperation in arms races', *International Studies Quarterly*, **30**: 176–191.
Makins, C. (1985) 'The superpower's dilemma: negotiating in the nuclear age', *Survival*, **27**: 169–178.
Maoz, Z. (1985) 'Decision-theoretic and game-theoretic models of international conflict', 76–109 in: U. Luterbacher and M. Ward, eds. *Dynamic Models on International Conflict*. Boulder: Lynne Reinner.
Maoz, Z. (1988) 'The Ally's Paradox', American Political Science Association Annual Meeting.
Maoz, Z. (1989a) *Paradoxes of War: The Art of National Self-entrapment*. Boston: Allen and Unwin.
Maoz, Z. (1989b) 'National preferences, international structures and balance-of-power politics', International Studies Association Meeting.
Maoz, Z. (1990) *National Choices and International Processes*. Cambridge: Cambridge University Press.
Maoz, Z., and D. Fiesenthal (1987) 'Self-binding commitments, the inducement of trust, social choice and the theory of interntional cooperation', *International Studies Quarterly*, **31**: 177–220.

Mares, D. (1988) 'Middle powers and regional hagemony: to challenge or acquiesce in hegemonic enforcement', *International Studies Quarterly*, **32**: 453–471.
Martin, B. (1978) 'The selective usefulness of game theory', *Social Studies of Science*, **8**: 85–110.
Martin, L. (1992a) 'Coercive Cooperation: Explaining Multilateral Economic Sanctions', Princeton: Princeton University Press.
Martin, L. (1992b) 'Interests, power and multilateralism', *International Organization*, **46**: 765–792.
Martin, L. (1993) 'Credibility, costs and institutions', *World Politics*, **45**: 406–432.
Maschler, M. (1963) 'A non-zerosum game related to a test ban treaty', Arms Control and Disarmament Agency, Paper 6098-018.
Maschler, M. (1966) 'A price leadership model for solving the inspector's non-constant sum game', *Naval Research Logistics Quarterly*, **13**: 11–33.
Maschler, M. (1967) 'The inspector's non-constant sum game: its dependence on a system of detectors', *Naval Research Logistics Quarterly*, **14**: 275–290.
Matheson, J. (1966) 'Preferential strategies', AR 66-2, Analytical Services, Inc. Arlington, VA.
Matheson, J. (1967) 'Preferential strategies with incomplete information', AR 67-1, Analytical Services, Inc., Arlington, VA.
Matlin, S. (1970) 'A review of the literature on the missile allocation problem', *Operations Research*, **18**: 334–373.
Matsubara, N. (1989) 'Conflict and limits of power', *Journal of Conflict Resolution*, **33**: 113–141.
Mayer, F. (1992) 'Managing domestic differences in international negotiations: the strategic use of internal side payments', *International Organization*, **46**: 793–818.
McDonald, J. (1949) 'A theory of strategy', *Fortune*. **39**: June, 100–110. Reprinted (1966) 1–22. in: D. Mulvihill, ed. *Guide to the Quantitative Age*. New York: Holt, Rinehart and Winston.
McGarvey. D. (1987) 'Mixed Strategy Solutions for Prim-Read Defenses', P-7302, RAND Corporation, Santa Monica.
McGinnis, M. (1986) 'Issue linkage and the evolution of international cooperation', *Journal of Conflict Resolution*, **30**: 141–170.
McGinnis, M. (1990) 'Limits to cooperation: iterated, graduated games and the arms race', *International Interactions*, **16**: 271–293
McGinnis, M. (1992) 'Bridging or broadening the gap? A comment on Wagner's Rationality and Misperception in Deterrence', *Journal of Theoretical Politics*, **4**: 443–457.
McGinnis, M., and J. Williams. (1991) 'Configurations of cooperation: correlated equilibria in co-ordination and iterated Prisoner's Dilemma games', Peace Science Association, Ann Arbor.
McGinnis, M., and J. Williams. (1993) 'Policy uncertainty in two-level games: examples of correlated equilibria', *International Studies Quarterly*, **37**: 29–54.
McGuire, C. (1958) 'Reliability and Vulnerability of a Retaliatory Force under Control Systems that Provide for Decentralized Action', D-5527, RAND Corporation, Santa Monica.
McGuire, M. (1966) 'The arms race: an interaction process', in: *Secrecy and the Arms Race*. Cambridge: Harvard University Press, Ch. 5.
Measor, N. (1983) 'Game theory and the nuclear arms race', 132-156 in: N. Blake and K. Pole, eds. *Dangers of Deterrence: Philosophers on Nuclear Strategy*. London: Routledge and Kegan Paul.
Melese F., and P. Michel (1989) 'A Differential Game Formulation of Richardson's Arms Model', Naval Postgraduate School, Monterey, California.
Mensch, A., ed. (1966) *Theory of Games: Techniques and Applications*. New York: American Elsevier.
Mirowski, P. (1992) 'When games grow deadly serious: the military influence on the evolution of game theory', in: R. Weintraub, ed. *Toward a history of game theory*. Durham: Duke University Press.
Mishal, S., D. Schmeidler and I. Sened (1990) 'Israel and the PLO: a game with differential information', 336-357 in: T. Ichiishi, A. Neyman and Y. Tauman, eds. *Game Theory and Applications*. New York: Academic Press.
Mo, J. (1990) 'International Bargaining and Domestic Political Competition', Stanford University, mimeo.
Moglewer, S. (1973) 'A Game-theoretical Approach to Auditing Inspection', Paper 6165, McDonnell-Douglas Corporation, Long Beach, CA.
Moglewer S., and C. Payne. (1970) 'A game theory approach to logistics allocation', *Naval Research Logistics Quarterly*, **17**: 87–97.
Molander, P. (1988) *Society at War: Planning Perspectives*. Stockholm: Foersvarets Forskningsanstalt.
Mor, B. (1993) *Decision and Interaction in Crisis*. Wesport, Conn.: Greenwood.

Moore, J. (1991) 'Nuclear Stability in the Post-Cold War Era'. Ph.D. Dissertation, Massachusetts Institute of Technology.
Morgenstern, O. (1959) *The Question of National Defense*. New York: Random House.
Moriarty, G. (1984) 'Differential game theory applied to a model of the arms race', *IEEE Technology and Society Magazine*, **3**: 10-17.
Morris, C. (1988) *Iron Destinies, Lost Opportunities: The Arms Race between the USA and the USSR, 1945–1987*. New York: Harper and Row.
Morrow, J. (1989a) 'Bargaining in repeated crises: a limited information model', 207–228 in: P. Ordeshook, ed. *Models of Strategic Choice in Politics*. Ann Arbor: University of Michigan Press.
Morrow, J. (1989b) 'Capabilities, uncertainty and resolve: a limited information model of crisis bargaining', *American Journal of Political Science*, **33**: 941–972.
Morrow, J. (1990a) 'Alliance Credibility and Recontracting, the Case of US-ROC Relations', International Conference on Game Theory and ROC Foreign Policy, Taipei.
Morrow, J. (1990b) 'Modelling Intenational Regimes', Hoover Institution.
Morrow, J. (1991a) 'Electoral and congressional incentives and arms control', *Journal of Conflict Resolution*, **35**: 245–265.
Morrow, J. (1992a) 'Signaling difficulties with linkage in crisis bargaining', *International Studies Quarterly*, **36**: 153–172. Also in: M. Intriligator and U. Luterbacher, eds. (1994).
Morrow, J. (1992b) 'Sorting through the Wealth of Notions? Progress and Prospects after Ten Years of Rational Choice Formal Models in International Politics', American Political Science Association, Chicago.
Morrow, J. (1994) 'Alliances, credibility and peacetime costs', *Journal of Conflict Resolution*, **38**: 220–297.
Morse, P. (1977) *At the Beginnings: A Physicist's Life*. Cambridge, Mass: MIT Press.
Morse, P., and G. Kimball (1946) *Methods of Operations Research*. U.S. Navy Operational Research Group Report 54. Reprinted (1956), Cambridge: MIT Press.
Moulin, H. (1981) 'Deterrence and cooperation, a classification of two-person games', *European Economic Review*, **15**: 179–183.
Nalebuff, B. (1988) 'Minimal nuclear deterrence', *Journal of Conflict Resolution*, **32**: 411–425.
Nalebuff, B. (1986) 'Brinkmanship and nuclear deterrence: the neutrality of escalation', *Conflict Management and Peace Science*, **9**: 19–30.
Nalebuff, B. (1989) 'A New Measure of Crisis Instability', School of Organization and Management, Yale University, mimeo.
Nalebuff, B. (1991) 'Rational deterrence in an imperfect world', *World Politics*, **43**: 313–335. Also in: M. Intriligator and U. Luterbacher, eds. (1994).
Nicholson, M. (1970) *Conflict Analysis*. London: English University Press.
Nicholson, M. (1972) 'Uncertainty and crisis behaviour: an illustration of conflict and peace research', 237–265 in: C. Carter and J. Ford, eds. *Uncertainty and Expectations in Economics*. Oxford: Blackwell.
Nicholson, M. (1981) 'Cooperation, anarchy and "random" change', 173–188 in: B. Buzan and B. Jones, eds. *Change and the Study of International Relations*. New York: St. Martin's.
Nicholson, M. (1988) 'Quick and Slow Deterrence Games: An Application to Technological Arms Races', American Political Science Association Annual Meeting.
Nicholson, M. (1989) *Formal Theories of Interntional Relations*. Cambridge: Cambridge University Press.
Nicholson, M. (1990) 'Deterrence, uncertainty and cooperative behavior', *International Interactions*, **15**: 227–242. Reprinted (1990), F. Zagare, ed., *Modeling International Conflict*. New York: Gordon and Breach.
Nicholson, M. (1992) *Rationality and the Analysis of International Conflict*. New York: Cambridge University Press.
Niou, E. (1992) 'An analysis of the Republic of China's security issues', *Issues and Studies*, **28**: 82–95.
Niou, E. and P. Ordeshook (1986) 'A theory of the balance of power in international systems', *Journal of Conflict Resolution*, **30**: 685–715
Niou, E, and P. Ordeshook (1987) 'Preventive war and the balance of power: a game-theoretic analysis', *Journal of Conflict Resolution*. **31**: 387–419.
Niou, E. and P. Ordeshook (1989a) 'The geographical imperatives of the balance of power in three-country systems', *Mathematics and Computer Modelling*, **12**: 519–531.

Niou, E., and P. Ordeshook (1989b) 'Stability in international systems and the costs of war', 229–251 in: Ordeshook, ed.
Niou, E. and P. Ordeshook (1990a) 'A Game-theoretical interpretation of Sun Tzu's *The Art of War*', Social Science Working Paper 738, California Institute of Technology.
Niou, E. and P. Ordeshook (1990b) 'The Prospects for a New International Order Based on Collective Security and Taiwan's Role in this Order', International Conference on Game Theory and ROC Foreign Policy, Taipei.
Niou, E. and P. Ordeshook (1990c) 'Stability in anarchic interntional systems', *American Political Science Review*, **84**: 1207–1234.
Niou, E. and P. Ordeshook (1991) 'Realism versus neorealism: a game-theoretical formulation', *American Journal of Political Science*, **35**: 481–511.
Niou, E., and P. Ordeshook (1992) 'Less Filling, Tastes Great: The Realist-Neoliberal Devate', Duke University Program in Political Economy Working Paper 174.
Niou, E., P. Ordeshook and G. Rose (1989) *The Balance of Power: Stability in International Systems.* New York: Cambridge University Press.
Nye, J. and S. Lynn-Jones (1988) 'International security studies', *International Security*, **12**: 5–27.
Oelrich, I. and J. Bracken (1988) 'A Comparison and Analysis of Strategic Defense Transition Stability Models', Institute for Defense Analysis Paper P-2145.
O'Meara, N. and R. Soland (1988) 'Sensitivity of expected survival value to the attacker's perceptions of defense capabilities'. P-7484, RAND Corporation, Santa Monica.
O'Neill, B. (1986) 'International escalation and the dollar auction', *Journal of Conflict Resolution*, **30**: 33–50. Abridged (1985), 220–227 in: U. Luterbacher and M. Ward, eds. *Dynamic Models on International Conflict*. Boulder: Lynne Reinner.
O'Neill, B. (1987) 'A measure for crisis instability, with an application to space-based anti-missile systems', *Journal of Conflict Resolution*, **31**: 631–672.
O'Neill, B. (1988) 'Two Dogmas of Strategic Analysis', International Political Science Association Meeting.
O'Neill, B. (1989a) 'Rationalist Alternatives to Deterrence Theory', International Studies Association Meeting.
O'Neill, B. (1989b) Game theory and the study of the deterrence of war. 134–156 in: P. Stern, R. Axelrod, R. Jervis, and Roy Radner, eds. *Perspectives on Deterrence*. New York: Oxford University Press.
O'Neill, B. (1989c) 'The Decision to Mobilize Based on Imperfect Warning', Colloquium, Association pour la Reflexion sur les Enseignements Scientifiques Appliques a la Defense, Paris.
O'Neill, B. (1990a) The Intermediate Nuclear Forces: an analysis of coupling and reassurance. *International Interactions*, **15**: 329–347. Reprinted, F. Zagare, ed., *Modeling International Conflict*. New York: Gordon and Breach.
O'Neill, B. (1990b) Competitive moves in bargaining: delays, threats, escalations and ultimata. in: Peyton Young, ed. *Negotiation Analysis*. Ann Arbor: University of Michigan Press.
O'Neill, B. (1990c) 'Game Theory, Its Premises and Promises', Conference on Game Theory and Republic of China Foreign Policy, Taipei.
O'Neill, B. (1991a) 'How to Measure Military Worth (At Least in Theory)', Working Paper #7, York University Centre for International and Strategic Studies.
O'Neill, B. (1991b) 'Why a Better Verification Scheme Can Give More Ambiguous Evidence', Working Paper #8, York University Centre for International and Strategic Studies.
O'Neill, B. (1991c) 'A Credibility Measure for the Outcomes of Chicken and the Stag Hunt', York University, mimeo.
O'Neill, B. (1991d) 'The strategy of challenges: two beheading games in medieval literature', 124–149 in: R. Selten, ed. *Game Equilibrium Models IV*. Springer-Verlag.
O'Neill, B. (1992) 'Are game models of deterrence biassed towards arms-building?' *Journal of Theoretical Politics*, **4**: 459–477.
O'Neill, B. (1993a) 'Game theory sources for international relations specialists', in: M. Intriligator and Urs Luterbacher, eds., *Cooperative Game Models in International Relations*., forthcoming.
O'Neill, B. (1993b) '*The Strategy of Secrets*', Yale University, mimeo.
O'Neill, B. (1993c) 'Operations research and strategic nuclear war', *International Military Defense Encyclopedia*. Pergammon–Brassey.

O'Neill, B. (1994a) 'A negotiating problem from the Vienna arms talks', in: M. Intriligator and U. Luterbacher, eds., forthcoming.
O'Neill, B. (1994b) *'Symbols, Honor and War'*, Yale University, mimeo.
Oye, K., ed. (1986) *Cooperation Under Anarchy.* Princeton: Princeton University Press. Special issue of *World Politics*, 1985.
Pahre, R. (1994) 'Multilateral cooperation in an iterated prisoners' dilemma', *Journal of Conflict Resolution*, **38**: 326–352.
Pak, S. (1977) *Nam-pukhan Kwan'gye e Chogyong Hal Kakchong Keim Iren Model Ui Kaebal Yon gu.* [The Development of Various Game Theory Models for Application to the South Korea/North Korea Relationship.] Seoul: Department of Policy Planning, Ministry of Terriotrial Reunification.
Petschke, C. (1992) *Spieltheoretische Analyse einer Klasse von mehrstufigen Uberwachungssytem mit Fehlalarmen.* [Game Theoretic Analysis of a Class of Multistage Safeguards Systems with False Alarms.] Frankfurt am Main: Hain.
Phillips, W., and R. Hayes (1978) 'Linking forecasting to policy planning: an application of the theory of non-cooperative games', 39–62 in: W. Hollist, ed. *Exploring Competitive Arms Processes.* New York: Dekker.
Picard, L. (1979) 'The use of game theory to explain the SALT process', *Policy Studies Journal*, **8**: 120–126.
Pilisuk, M. (1984) 'Experimenting with the arms race', *Journal of Conflict Resolution*, **28**: 296–315.
Pilisuk, M. and A. Rapoport (1963) 'A non-zerosum game model of some disarmament problems', *Papers of the Peace Research Society International*, **1**: 57–78.
Plon, M. (1976) *La Theorie des Jeux: Une Politique Imaginaire.* [The Theory of Games: A Fantasy in Political Analysis] Paris: Maspero.
Plous, S. (1985) 'Perceptual illusions and military realities: a social psychological analysis of the nuclear arms race', *Journal of Conflict Resolution*, **29**: 363–389.
Plous, S. (1987) 'Perceptual illusions and military realities: results from a computer-generated arms race', *Journal of Conflict Resolution*, **31**: 5–33.
Plous, S. (1988) 'Modeling the nuclear arms race as a perceptual dilemma', *Philosophy and Public Affairs*, **17**: 44–53.
Plous, S. (1993) 'The nuclear arms race: prisoner's dilemma or perceptual dilemma?' *Journal of Peace Research*, **30**: 163–179.
Polydoros, A. and U. Cheng (1987) 'Topology-selective Jamming of Certain Code-Division Multiple-Access Monohop Networks', Axiomatix, Los Angeles.
Ponssard, J. (1990) 'A Note on Forward Induction and Escalation Games with Perfect Information', Laboratoire d'Econometrie, L'Ecole Polytechnique.
Poundstone, W. (1992) *Prisoner's Dilemma.* New York: Doubleday.
Powell, C. (1988) "Problems and breakdowns of rationality in foreign policy and conflict analysis", *Peace and Change*, **13**: 65–94.
Powell, R. (1987a) 'Nuclear brinkmanship with two-sided incomplete information', *American Political Science Review*, **82**: 155–178.
Powell, R. (1987b) 'Crisis bargaining, escalation and MAD', *American Political Science Review*, **81**: 717–736.
Powell, R. (1989a) 'Crisis stability in the nuclear age', *American Political Science Review*, **83**: 61–76.
Powell, R. (1989b) 'Nuclear deterrence and the strategy of limited retaliation', *American Political Science Review*, **83**: 503–520.
Powell, R. (1989c) 'The dynamics of longer brinkmanship crises', 151–175 in: P. Ordeshook, ed. *Models of Strategic Choice in Politics.* Ann Arbor: University of Michigan Press.
Powell, R. (1990) *Nuclear Deterrence Theory: The Search for Credibility.* Cambridge University Press.
Powell, R. (1991a) 'Absolute and relative gains in international relations theory', *American Political Science Review*, **85**: 1303–1320.
Powell, R. (1991b) 'The Pursuit of Power and Plenty', Department of Political Science, University of California, Berkeley, mimeo.
Powell, R. (1993) 'Guns, butter and anarchy', *American Political Science Review*, **87**: 115–132.
Powell, R. (1994) 'Limited nuclear options and the dynamics of escalation', in: M. Intriligator and U. Luterbacher, eds.
Pugh, G. (1973) 'Theory of measures of effectiveness for general purpose military forces: Part II, Lagrange dynamic programming in time-sequential combat games', *Operations Research*, **21**: 886–906.

Pugh, G. and J. Mayberry (1973) 'Theory of measures of effectiveness for general-purpose military forces: Part I, A zero-sum payoff appropriate for evaluating combat strategies', *Operations Research*, **21**: 867–885.
Putnam, R. (1988) 'Diplomacy and domestic politics: The logic of two-level games', *International Organization*, **42**: 427–460.
Quandt, R. (1961) 'On the use of game models in theories of international relations', 69–76 in: Klaus Knorr and Sidney Verba, eds. *The International System, Theoretical Essays, World Politics*, **14**: 69–76.
Quester, G. (1982) 'The six causes of war', *Jerusalem Journal of International Relations*, **6**: 1–23.
Radford, K. (1988) 'Analysis of an international conflict: the US hostages in Iran', *Peace and Change*, **13**: 132–144.
Randall, R. (1966) 'A game-theoretic model of a submarine barrier to detect transitory submarines', 333–354 in: Mensch, ed.
Raiffa, H. (1952) Arbitration Schemes for Generalized Two-person Games. Ph.D. thesis. University of Michigan.
Rapoport, A. (1964) *Strategy and Conscience*. New York: Harper and Row.
Rapoport, A. (1965) 'Chicken a la Kahn', *Virginia Quarterly Review*, **41**: 370–389.
Rapoport, A. (1966) *Two-person Game Theory: The Essential Ideas*. Ann Arbor: University of Michigan Press.
Rapoport, A. (1989) *The Origins of Violence: Approaches to the Study of Conflict*. New York: Paragon.
Rapoport, A. and M. Guyer (1966) 'A taxonomy of 2 × 2 games', *General Systems*, **11**: 203–214.
Ravid, I. (1981) 'To Mobilize or Not to Mobilize, October 1973', Center for Military Analyses Report 81/27, Haifa, Israel.
Ravid, I. (1990) 'Military decision, game theory and intelligence: an anecodote', *Operations Research*, **38**: 260–264.
Read, T. (1957) 'Tactics and deployment for anti-missile defenses', Bell Telephone Laboratories, Whippany, N.J.
Read, T. (1961) 'Strategy for active defense', *American Economic Review*, **51**: 465–471.
Rehm, A. (n.d.) 'Soviet Military Operations Research', United States Central Intelligence Agency.
Reisinger, W. (1989) 'Research note: 2 × 2 games of commitment in world politics', *International Studies Quarterly*, **33**: 111–118.
Richelson, J. (1979) 'Soviet strategic posture and limited nuclear operations: a metagame analysis', *Journal of Conflict Resolution*, **23**: 326–336.
Riker, W. (1962) *The Theory of Political Coalitions*. New Haven: Yale University Press.
Riker, W. (1992) 'The entry of game theory into political science' *History of Political Economy*, **24**: 207–234.
Robinson, S. (1993) 'Shadow prices for measures of effectiveness; I Linear Model; II General Model', Operations Research, **31**: 518–535, 536–548.
Robinson, T. (1970) 'Game Theory and Politics, Recent Soviet Views', RM-5839, RAND Corporations, Santa Monica.
Rodin, E. (1987) 'A pursuit–evasion bibliography, version 1', *computers and Mathematics with Applications*, **13**: 275–340. Also published (1987) in: Y. Yavin and M. Pachter, eds., *Pursuit-Evasion Differential Games*. Elmsford, NY: Pergamon.
Rogerson, W. (1990) 'Quality and quantity in military procurement', *American Economic Review*, **80**: 83–92.
Rosenau, J. (1971) *The Scientific Study of Foreign Policy*. New York: Free Press.
Roth, A. (1988) *The Shapley Value*. New York: Cambridge University Press.
Ruckle, W. (1983) *Geometric Games and Their Applications*. Boston: Pitman.
Rudnianski, M. (1985) 'Une analyze de la notion de dissuasion: calculs strategiques au dela des valeurs-seuils', ("An analysis of the concept of deterrence: calculation of strategic thresholds".) *Strategique*, **26**: 159–175.
Rudnianski, M. (1986) 'Une approche de quelques problems strategiques par la theorie des jeux, Premiere partie: les differentes formes de dissuasion', ['An approach to some strategic problems by the theory of games. First part: different types of deterrence'] *L'Armement*, **1**: 129–160.
Rudnianski, M. (1991) 'Deterrence typology and nuclear stability: a game theoretical approach', 137–168 in: R. Avenhaus, H. Karkar and M. Rudnianski, eds. *Defense Decision Making: Analytical Support and Crisis Management*. New York: Springer-Verlag.

Saaty, T. (1968) *Mathematical Models of Arms Control and Disarmament – Application of Mathematical Structures in Politics*. New York: Wiley.
Said, A. and D. Hartley (1982) 'A hypergame approach to crisis decision-making: the 1973 Middle East War', *Journal of the Operational Research Society*, **33**: 937–948.
Salter, S. (1986) 'Stopping the arms race: a modest proposal', *Issues in Science and Technology*, **2**: 75–82.
Sampson, M. (1982) 'Some necessary conditions for international policy coordination', *Journal of Conflict Resolution*, **26**: 359–384.
Sandler, T. and H. Lapan (1988) 'The calculus of dissent: analysis of terrorists' choice of targets', *Synthese*, **76**: 245–262.
Sandler, T. and J. Murdoch (1990) 'Nash-Cournot or Lindahl behavior? An empirical test for the NATO allies', *Quarterly Journal of Economics*, **105**: 875–894.
Sandretto, R. (1976) *'Inegalites Transnationales: Une Application de la Theorie des Jeux'*, *Aix-Marseilles*: Centre National De la Recherche Scientifique.
Sanjian, G. (1991) 'A Fuzzy Game Model of US/USSR Arms Trade Decision-making', Bucknell University, mimeo.
Satawedin, D. (1984) The Bargaining Power of a Small State Thai–American Alliance during the Laotian Crisis, 1959–1962. Ph.D. thesis, Northern Illinois University.
Scarf, H. (1967) 'Exploration of a stability concept for armed nations by means of a game-theoretical model', in: Aumann *et al.*
Scarf, H. (1971) 'On the existence of a cooperative solution for a general calss of n-person games', *Journal of Economic Theory*, **3**: 169–181.
Scheffran, J. (1989) 'Strategic Defense, Disarmament and Stability'. Doctoral Thesis, Department of Physics, University of Marburg, Germany.
Schellenberg, J. (1990) *Primitive Games*. Boulder: Westview.
Schelling, T. (1958a) 'The strategy of conflict, a prospectus for a reorientation of game theory', *Journal of Conflict Resolution*, **2**: 203–264.
Schelling, T. (1958b) 'The Reciprocal Fear of Surprise Attack', P-1342, RAND Corporation, Santa Monica.
Schelling, T. (1960) *The Strategy of Conflict*. Cambridge: Harvard University Press.
Schelling, T. (1964) 'Assumptions about enemy behavior', in: E. Quade, ed., *Analysis for Military Decisions*. R-387, RAND Corporation, Santa Monica. Chicago: Rand-McNally, pp. 199–216.
Schelling, T. (1966a) *Arms of Influence*. New Haven: Yale University Press.
Schelling, T. (1966b) 'Uncertainty, brinkmanship and the game of Chicken', in: K. Archibald, ed., *Strategic Interaction and Conflict*. Berkeley: Institute of International Studies, pp. 74–87.
Schelling, T. (1967) 'The strategy of inflicting costs', *Issues in Defense Economics*. New York: National Bureau of Economic Research.
Schelling, T. (1968) 'Notes on policies, games, metagames and Viet Nam', *Papers of the Peace Research Society International*, **10**: 134–147.
Schelling, T. (1976) 'A framework for the evaluation of arms control proposals', in: M. Pfaff, ed., *Frontiers in Social Thought: Essays in Honor of Kenneth E. Boulding*. New York: North-Holland, pp. 283–305. Reprinted (1984) in: *Choice and Consequence*. Cambridge: Harvard university Press, pp. 243–267. Revised (1975) in: Daedalus, **104**: 187–200.
Scheu, J. (1968) 'A mine-clearing game', in: *Seventh Symposium on Gaming*. Pacific Grove, CA, Chs. 1–5.
Schirlitzki, J. (1976) 'Differentialspiele – eine analytische Methode der militarischen Systemanalyse', ['Differential games – an analytical method for military systems analysis'.] 65–98 in: R. Huber, R. Niemeyer and H. Hofmann, eds., *Operational Research Games for Defense*. Munich: Oldenbourg.
Schlosser, J. and B. Wendroff (1988) 'Iterated Games and Arms Race Models', Los Alamos National Laboratory.
Schwartz, E. (1979) 'An improved computational procedure for optimal allocation of aircraft sorties', *Operations Research*, **27**: 621–627.
Selten, R. (1992) 'Balance of power in parlor game', 150–209 in: R. Selten, ed. *Game Equilibrium Models IV*. New York: Springer-Verlag.
Selten, R. and J. Mayberry (1968) 'Application of bargaining I-games to Cold War competition', 278–297 in: Aumann, *et al.*

Selten, R., R. Stoecker and R. Avenhaus. (1977) 'Research Conference on Strategic Decision analysis Focusing on the Persian Gulf (The Schwaghof Papers)', Report, Verein zur Forderung der Systemanalyse der Arms Control. Bonn.

Selten, R. and R. Tietz. (1972) 'Security equilibria', 103–122, 185–207 in: R. Rosecrance, ed. *The Future of the International Strategic System*. San Francisco: Chandler. Also in: Aumann et al. (1967).

Sen, A. (1967) 'Isolation, assurance and the social rate of discount', *Quarterly Journal of Economics*, **81**: 112–124.

Sen, A. (1973) *On Economic Inequality*. Oxford: Oxford University Press.

Sen, A. (1974) 'Choice, orderings and morality', 54–67 in: S. Korner, ed., *Practical Reason*.

Seo, F. and M. Sakawa (1990) 'A game theoretic approach with risk assessment for international conflict solving', *IEEE Transactions on Systems, Man and Cybernetics*, **20**: 141–148.

Sexton, T. and D. Young (1985) 'Game tree analysis of international crises', *Journal of Policy Analysis and Management*, **4**: 354–370.

Shapley, L. (1949) 'A Tactical Reconnaissance Model', RM-205, RAND Corporation, Santa Monica.

Shapley, L. and R. Aumann (1977) 'Long-term Competition', RAND Corporation, Santa Monica.

Sharp, J. (1986) 'Alliance Security Dilemmas', Center for European Studies, Harvard University, mimeo.

Shephard, R. (1966) 'Some reasons for present limitations in the application of the theory of games to army problems', 373–388 in: Mensch, ed.

Sherman, S. (1949) 'Total Reconnaissance with Total Countermeasures', RM-202, RAND Corporation, Santa Monica.

Shubik M. (1968) 'On the study of disarmament and escalation', *Journal of Conflict Resolution*. **12**: 83–101.

Shubik, M. (1983) 'Game theory, the language of strategy', 1–28 in: M. Shubik, ed., *Mathematics of Conflict*. Amsterdam: Elsevier.

Shubik, M. (1987) 'The uses, value and limitations of game-theoretic methods in defense analysis', 53–84 in: C. Schmidt and F. Blackaby, eds. *Peace, Defense and Economic Analysis*. New York: St. M.'s Press.

Shubik, M. (1991) 'Models of Strategic Behavior and Nuclear Deterrence', International Security and Arms Control Program, Report 6.

Shubik, M., and R. Weber (1979) 'Systems defense games, Colonel Blotto, command and control', *Naval Research Logistics Quarterly*, **28**: 281–287.

Shubik, M. and R. Weber (1982) 'Competitive valuation of cooperative games', Yale University.

Shupe, M., W. Wright, K. Hipel and N. Fraser (1980) 'Nationalization of the Suez Canal, a hypergame analysis', *Journal of Conflict Resolution*, **24**: 477–493.

Simann, M. and J. Cruz (1973) 'A multi-stage game formulation of the arms race and its relationship to Richardson's model', *Modelling and Simulation*, **4**: 149–153.

Simaan, M., and J. Cruz (1975a) 'Formulation of Richardson's model of the arms race from a differential game viewpoint', *Review of Economic Studies*, **42**: 62–77.

Simaan, M., and J. Cruz (1975b) 'Nash equilibrium strategies for the problem of armament races and control', *Management Science*, **22**: 96–105.

Simaan, M. and J. Cruz (1977) 'Equilibrium concepts for arms race problems', 342–366 in: J. Gillespie and D. Zinnes, eds. *Mathematical Systems in International Relations Research*. New York: Praeger.

Simmons, G. (1981) 'A Game Theory Model of Digital Message Authentication', Sandia National Laboratories.

Singer, E. (1963) 'A bargaining model for international negotiations', *Journal of Conflict Resolution*, **7**: 21–25.

Sinn, S. and J. Martinez Oliva (1988) 'The game theoretic approach to international policy coordination: assessing the role of targets', *Review of World Economics*, **124**: 252–269.

Siverson, R. and C. McCarty (1978) 'War, implementation costs and minimum winning coalitions', *International Interactions*, **5**: 31–42.

Smith, A. (1993) Alliance formation and war, mimeo. University of Rochester.

Snidal, D. (1985a) 'The game theory of international politics', *World Politics*, **38**: 25–57.

Snidal, D. (1985b) 'Coordination versus Prisoner's Dilemma, implications for international cooperation and regimes', *American Political Science Review*, **79**: 923–942.

Snidal, D. (1985c) 'The limits of hegemonic stability theory', *International Organization*, **39**: 579–613.

Snidal, D. (1990a) 'A problem of assurance: The R.O.C. and Taiwanese Investment in the P.R.C.', International Conference on Game Theory and ROC Foreign Policy, Taipei.

Snidal, D. (1991a) 'International cooperation among relative gains maximizers', *International Studies Quarterly*, **35**: 387–402. Also in: U. Luterbacher and M. Intriligator, eds. (1994).
Snidal, D. (1991c) 'Relative gains and the pattern of international cooperation', *American Political Science Review*, **85**: 701–726.
Snidal, D. (1993) 'The relative gains problem for international cooperation'. *American Political Science Review*, **87**: 738–743.
Snidal, D. and A. Kydd (1994) 'The use of game theory in the explanation of international regimes.' in V. Rittberger, ed. *Regime Theory in International Relations.* New York: Oxford University Press.
Snyder, G. (1961) *Deterrence and Defense.* Princeton: Princeton University Press.
Snyder, G. (1971) 'Prisoner's Dilemma and Chicken models in international politics', *International Studies Quarterly*, **15**: 66–103.
Snyder, G. (1984) 'The security dilemma in alliance politics' *World Politics*, **59**: 461–495.
Snyder, G. and P. Diesing (1977) *Conflict Among Nations.* Princeton: Princeton University Press.
Snyder, R. (1961) 'Game theory and the analysis of political behavior', 381–390 in: J. Rosenau, ed. *International Politics and Foreign Policy.* New York: Free Press.
Sobel, J. (1988) 'When and How to Communicate with the Enemy', University of California, San Diego, mimeo.
Sorokin, G. (1994) 'Alliance formation and general deterrence', *Journal of Conflict Resolution*, **38**: 298–325.
Stahl, D. (1991) 'The graph of the Prisoners' Dilemma supergame payoffs as a function of the discount factor', *Games and Economic Behavior*, **3**: 368–384.
Stark, W. (1982) 'Coding for Frequency-hopped Spread Spectrum Channels with Partial-band Interference', University of Illinois, Coordinated Science Laboratory.
Stefanek, J. (1985) 'Deterrence and Threats', Systems Research Group, Polish Academy of Sciences, Warsaw.
Stein, A. (1980) 'The politics of linkage', *World Politics*, **33**: 62–81.
Stein, A. (1982a) 'Coordination and collaboration regimes in an anarchic world', *International Organization*, **36**: 299–324.
Stein, A. (1982b) 'When misperception matters', *World Politics*, **35**: 505–526.
Stein, A. (1990) *Why Nations Cooperate.* Ithaca: Cornell University Press.
Stein, A. (1993) 'Interdependence in international relations', in: J. Husbands and P. Stern, eds., *Behavior, Society and International Conflict. Vol. III.* New York: Oxford University Press.
Stenberg, S. (1971) 'Development of Optimal Allocation Strategies in Heterogeneous Lanchester-type Processes', Report TR 71-1, Systems Research Laboratory, University of Michigan.
Strauch, R. (1965) 'A Preliminary Treatment of Mobile SLBM Defense: A Game-theoretical Analysis', RM-4439, RAND Corporation, Santa Monica.
Strauch, R. (1967) '"Shell Game" Aspects of Mobile Terminal ABM systems', RM-5474, RAND Corporation, Santa Monica.
Stubbs, E. (1988) 'Cost–benefit and Cost–effectiveness in a Game-theoretical Model of Arms Race Instabilities', Political Science Department, Harvard University, mimeo.
Suzdal, V. (1976) *Teoria Igr dla Flota.* [Theory of Games for the Navy.] Moscow: Voyenizdat. Translation, USAF Foreign Technology Division. FTD-ID(RS)T-0978-78.
Taylor, J. (1974) 'Some differential games of tactical interest', *Operations Research*, **22**: 304–317.
Taylor, J. (1977) 'Determining the class of payoffs that yield force-level-independent optimal fire-support strategies', *Operations Research*, **25**: 506–516.
Taylor, J. (1978) 'Differential-game examination of optimal time-sequential fire-support strategies', *Naval Research Logistics Quarterly*, **25**: 323–355.
Taylor, J. (1983) *Lanchester Models of Combat.* Alexandria, VA: Military Operations Research Society.
Taylor, J., and G. Brown (1978) 'An examination of the effects of the criterion functional on optimal fire-support strategies', *Naval Research Logistics Quarterly*, **25**: 183–211.
Taylor, M. (1976) *Anarchy and Cooperation.* New York: Wiley.
Taylor, M. (1987) *The Possibility of Cooperation: Studies in Rationality and Social Change.* Cambridge: Cambridge University Press.
Thies, W. (1980) *When Governments Collide.* Boulder: University of Colorado Press.
Thomas, C. (1966) 'Some past applications of game theory to problems of the United States Air Force', 250–267 in: A. Mensch, ed. *Theory of Games: Techniques and Applications*, New York: American Elsevier.

Tidman, R. (1984) *The Operations Evaluations Group: A History of Naval Operations Analysis*. Annapolis: Naval Institute Press.

Tsebelis, G. (1990) 'Are sanctions effective? A game-theoretic analysis', *Journal of Confict Resolution*, **34**: 3–28.

Tullock, G. (1980) 'Effecient rent-seeking', 97–112 in: J. Buchanan, R. Tollison and Gordon Tullock, eds., *Toward a Theory of the Rent-Seeking Society*. Texas A&M University Press.

Uscher, A. (1987) 'On Thermonuclear Peace: The Tightrope of Deterrence'. Ph.D. Thesis, George Washington University.

van der Ploeg, F. and A. de Zeeuw (1989) 'Conflict over arms accumulation in market and command economies', in: ibid, ed. *Dynamic Policy Games in Economics*. Amsterdam: North-Holland.

van der Ploeg, F. and A. de Zeeuw (1990) 'Perfect equilibrium in a model of competitive arms accumulation', *International Economic Review*, **31**: 131–146.

Vickrey, W. (1978) 'Application of demand-revealing procedures to international disputes', *Papers of the Peace Research Society International*, **28**: 91–100.

von Stengel, B. (1991) 'Recursive Inspection Games', Paper S-9106, Institut fur Angewandte Systemforschung und Operations Research, Universitat der Bundeswehr, Munich.

Waddington, C. (1973) *O.R. in World War II: Operations Against the U-Boat*. London: Ekel.

Wagner, H. (1982) 'Deterrence and bargaining', *Journal of Conflict Resolution*, **26**: 329–358.

Wagner, H. (1983) 'The theory of games and the problem of international cooperation', *American Political Science Review*, **77**: 330–346.

Wagner, H. (1987) 'The theory of games and the balance of power', *World Politics*, **38**: 546–576.

Wagner, H. (1988a) 'Economic interdependence, bargaining power and political influence', *International Organization*, **42**: 461–483.

Wagner, H. (1988b) 'Reputation and the Credibility of Military Threats: Rational Choice versus Psychology', American Political Science Association Annual Meeting.

Wagner, H. (1989a) 'Uncertainty, rational learning, and bargaining in the Cuban Missile Crisis', 177–205 in: P. Ordeshook, ed. *Models of Strategic Choice in Politics*. Ann Arbor: University of Michigan Press.

Wagner, H. (1989b) 'Economic interdependence, bargaining power and political influence', *International Organization*, **42**: 461–482.

Wagner, H. (1991) 'Nuclear deterrence, counterforce strategies and the incentive to strike first', *American Political Science Review*, **85**: 727–749.

Wagner, H. (1992) 'Rationality and misperception in deterrence theory', *Journal of Theoretical Politics*, **4**: 115–141.

Waltz, K. (1959) *Man, the State and War*. New York: Columbia.

Washburn, A. (1966) 'Infinite-track Approximation to the Rail-mobile System', Boeing Corp, Coordination Sheet 2-1301-0-008.

Washburn, A. (1971) 'An Introduction to Evasion Games', Naval Postgraduate School, mimeo.

Washburn, A. (1977) 'A computation method for matrix differential games', *International Journal of Game Theory*, **5**: 1–10.

Washburn, A. (1978) 'An Introduction to Blotto Games', Naval Postgraduate School, Monterey, California, mimeo.

Weber, S. and H. Wiesmeth (1991a) 'Economic models of NATO', *Journal of Public Economics*, **46**: 181–193.

Weber, S. and H. Wiesmeth (1991b) 'Burden-sharing in NATO: an economic analysis', 83–95 in: R. Avenhaus, H. Karkar and M. Rudnianski, eds. *Defense Decision Making: Analytical Support and Crisis Management*. New York: Springer-Verlag.

Weber, S. and H. Wiesmeth (1994) 'An Economic Analysis of Cooperation in the European Community', in: M. Intriligator and U. Luterbacher, eds.

Weber, S. (1990) 'Cooperation and interdependence', *Daedalus*, **120**: 183–202.

Weber, S. (1991) *Cooperation and Discord in US-Soviet Arms Control*. Princeton: Princeton University Press.

Weiss, H. (1957) 'Lanchester-type models of warfare', in: M. Daws, R. Eddvier and T. Page, eds., *Proceedings of the First International Conference on Operational Research*. Baltimore: ORSA.

Weiss, H. (1959) 'Some differential games of tactical interest and the value of a supporting weapon system', *Operations Research*, **7**: 180–196.

Weiss, M. (1983) 'Two Game-theoretical Studies of Military Command and Control'. Ph.D. thesis, Department of Industrial Engineering and Management Sciences, Northwestern University.

Weiss, M., and S. Schwartz (1985) 'Optimal Minimax Jamming and Detection of Radar Signals', Princeton University Information Sciences and Systems Laboratory, mimeo.

Weissenberger, S. (1990) 'Deterrence and the Design of Treaty Verification Systems', Lawrence Livermore National Laboratory, UCRL-JC-105810.

Weissenberger, S. (1991) 'Treaty Verification with an Uncertain Partner', Lawrence Livermore National Laboratory, UCRL-JC-105885.

Weres, L. (1982) *Teoria Gier w Amerykanskiej Nauce o Stosunkach Miedzynarodowych.* [Game Theory in American Research on International Relations.] Poznan, Poland: Instytut Zachodni.

Wilson, R. (1986) 'A Note on Deterrence', School of Business, Stanford University, mimeo.

Wilson, R. (1989) Deterrence and oligopolistic competition. 157–190 in: P. Stern, R. Axelrod, R. Jervis, and Roy Radner, eds. *Perspectives on Deterrence.* New York: Oxford University Press.

Wittman, D. (1989) Arms control verification and other games involving imperfect detection. *American Political Science Review*, **83**: 923–945.

Wohlstetter, A. (1964) 'Sin and games in America', 209–225 in: M. Shubik, ed. *Game Theory and Related Approaches to Social Behavior.* Huntington, NY: Krieger.

Woodward, P. (1989) 'The "game" of nuclear strategy: Kavka on strategic defense', *Ethics*, **99**: 673–691.

Wright, W., M. Shupe, N. Fraser and K. Hipel (1980) 'A conflict analysis of the Suez Invasion of 1956', *Conflict Management and Peace Science*, **5**: 27–40.

York, H. (1987) *Making Weapons, Talking Peace.* New York: Basic Books.

Young, D. (1983) 'Game Tree Analysis of Public Policy Decisions: The Case of the Falkland Islands', Public Choice Society Annual Meeting.

Young, O. (1976) 'Rationality, coalition formation and international relations', 223–245 in: S. Benn and G. Mortimore, eds., *Rationality and the Social Sciences.* London: Routledge and Kegan Paul.

Yu, P. (1984) *A Strategic Model for Chinese Checkers: Power and Exchange in Beijing's Interactions with Washington and Moscow.* New York: Peter Lang.

Zagare, F. (1977) 'A game-theoretic analysis of the Vietnam negotiations: preferences and strategies, 1968–1973', *Journal of Conflict Resolution*, **21**: 663–684.

Zagare, F. (1979) 'The Geneva Conference of 1954, a case of tacit deception', *International Studies Quarterly*, **23**: 390–411.

Zagare, F. (1981) 'Nonmyopic equilibria and the Middle East Crisis of 1967', *Conflict Management and Peace Science*, **5**: 139–162.

Zagare, F. (1983) 'A game-theoretic evaluation of the cease-fire alert decision of 1973', *Journal of Peace Research*, **20**: 73–86.

Zagare, F. (1985) 'Toward a reconciliation of game theory and the theory of mutual deterrence', *International Studies Quarterly*, **29**: 155–170.

Zagare, F. (1986) 'Recent advances in game theory and political science', 60–90 in: S. Long, ed. *Annual Review of Political Science.* Norwood, NJ: Ablex.

Zagare, F. (1987a) 'A stability analysis of the US-USSR strategic relationship', 123–149 in: J. Kugler and F. Zagare, eds. *Exploring the Stability of Deterrence.* Denver: Lynne Reinner.

Zagare, F. (1987b) *The Dynamics of Deterrence.* Chicago: University of Chicago Press.

Zagare, F. (1990a) 'Rationality and deterrence', *World Politics*, **42**: 238–260.

Zagare, F. (1990b) 'The dynamics of escalation', *Information and Decision Technologies*, **16**: 249–261.

Zagare, F. (1992) 'NATO, rational escalation and flexible response', *Journal of Peace Research*, **29**: 435–454.

Zamir, S. (1987) 'A Two-period Material Safeguards Game', Department of Economics, University of Pittsburgh.

Zauberman, A. (1975) *Differential Games and Other Game Theoretical Topics in Soviet Literature.* New York: New York University Press.

Zellner, A. (1962) 'War and peace, a fantasy in game theory', *Journal of Conflict Resolution*, **6**: 39–41.

Zimmerman, C. (1988) *Insider at SAC: Operations Analysis Under General LeMay.* Manhattan, Kansas: Sunflower University Press.

Zinnes, D., J. Gillespie, P. Schrodt, G. Tahim and R. Rubison (1978a) 'Arms and aid, a differential game analysis', 17–38 in: W. Hollist, ed. *Exploring Competitive Arms Processes.* New York: Marcel Dekker.

Zinnes, D., J. Gillespie and G. Tahim (1978b) 'A formal analysis of some issues in the balance of power', *International Studies Quarterly*, **22**: 323–356.

Zuckermann, S. (1961) 'Judgement and control in modern warfare', *Foreign Affairs*, **40**: 196–212.

Chapter 30

VOTING PROCEDURES

STEVEN J. BRAMS*

New York University

Contents

1.	Introduction	1056
2.	Elections and democracy	1057
3.	Sincerity, strategyproofness, and efficacy	1058
	3.1. Voter preferences and dominance	1059
	3.2. Dominance between strategies	1061
	3.3. Admissible strategies	1063
	3.4. Sincere voting and strategyproofness	1067
	3.5. Efficacy	1069
4.	Condorcet properties	1069
	4.1. Dichotomous preferences	1070
	4.2. Runoff systems	1072
	4.3. Condorcet possibility and guarantee theorems	1073
5.	Preferential voting and proportional representation	1075
	5.1. The Hare system of single transferable vote (STV)	1075
	5.2. The Borda count	1078
	5.3. Cumulative voting	1080
	5.4. Additional-member systems	1081
6.	Conclusions	1084
	References	1085

*The financial support of the National Science Foundation under grant SES-871537 is gratefully acknowledged. I thank Philip D. Straffin, Jr., an anonymous referee, and the editors for valuable comments and suggestions.

Handbook of Game Theory, Volume 2, Edited by R.J. Aumann and S. Hart
© *Elsevier Science B.V., 1994. All rights reserved*

1. Introduction

The cornerstone of a democracy is fair and periodic elections. The voting procedures used to elect candidates in large part determine whether elections are considered fair [Fishburn 1983, Nurmi 1983, Dummett 1984)]. By *procedures* I mean the rules that govern how votes in an election are aggregated and how a winner or winners are determined.

Thus, I exclude from consideration in this chapter the means by which a person becomes an official candidate, which may of course be biased against certain individuals and even prevent their candidacies. Similarly, I ignore the possibility of ballot-stuffing and other forms of election fraud, which may turn elections based on ostensibly fair voting procedures into sham contests, and factors related to turnout.

On the other hand, strategies that *voters* employ to try to effect better outcomes – as long they are allowed by the rules of a voting procedure – will be analyzed in this chapter. However, campaign strategies that *candidates* employ to win elections in certain kinds of contests will not be studied. Just as voting rules can be exploited by voters, campaigns enable candidates to try to abet their chances of winning – or, on occasion, satisfy other goals – by selecting strategies (e.g., negative campaigning) that capitalize on certain features in a campaign [Riker (1989)].

Readers interested in spatial models of voting, the simplest of which assume that candidates can choose points along a single dimension – often interpreted as positions on a left-right issue – are discussed in several recent works, including Enelow and Hinich (1984) and Cox (1984, 1985, 1987), wherein the effects of different voting procedures on candidate strategies are compared. Models of resource-allocation strategies in U.S. presidential and other elections are analyzed in Brams and Davis (1973, 1974, 1982), Lake (1979), and Snyder (1989).

Voting procedures, of course, not only affect campaign strategies and electoral outcomes but also influence strategies and outcomes in legislatures, councils, and other voting bodies, wherein the alternatives that are voted on are not candidates but bills and resolutions. The study of *legislative* voting procedures, which include both voting rules and rules that give special status to such figures as chairs who can break ties – whose prerogatives may not always redound to their advantage [Brams, Felsenthal and Maoz (1986, 1987) – was pioneered by Farquharson (1969); an overview of work in this field is given in the Chapter on Social Choice, Chapter 31.

I will focus in this chapter on how decision theory, and to a less extent game theory [Peleg (1984)], can be used to illuminate the choice of better and worse strategies under different voting procedures. Some attention will also be given to the nonstrategic properties of these procedures. Moreover, I will concentrate on *practical* voting procedures that have actually been used in elections, especially procedures that do not require voters to rank candidates.

A common framework for comparing such procedures will be developed in some detail in this chapter. The most prominent ranking procedures, and methods of proportional representation, will also be considered; more general theoretical analysis can be found in Chapter 31.

An underlying assumption of the subsequent analysis is that voters are rational with respect to certain goals [Brams (1985)]. Not only do they prefer better to worse outcomes, but they also may act with knowledge that other actors will behave so as to maximize achievement of their own goals, which may or may not be in conflict with theirs.

The strategic calculations that voters make result in outcomes that can be evaluated according to different normative criteria. I shall discuss some of the most important criteria that have been used to assess the quality of outcomes in voting bodies.

Judgments about better and worse voting procedures can be made on the basis of these criteria. In offering such judgments, I shall try to show how the strategic analysis of voting procedures can aid in the selection of procedures that seem best equipped to meet certain needs. To the degree that these procedures are perceived as fair and thereby minimize divisive and unnecessary conflict, then decision theory, game theory, and social theory can contribute both to the enhanced understanding and, when applied, the better functioning of democratic government.

2. Elections and democracy

The stirrings of democracy go back at least to the Bible, but elections, as we know them today, did not occur in biblical times. Saul, the first Israelite king, apparently surfaced because he was "an excellent young man; no one among the Israelites was handsomer than he; he was a head taller than any of the people" [1 Samuel 9:2; translations from *The Torah* (1967)]. Previously, when the Israelites were clamoring for a king "like all other nations" [1 Samuel 8:5], God had told Samuel, the first prophet and judge of Israel, that "it is not us that they (the Israelites) have rejected; it is Me they have rejected as their king" [1 Samuel 8:7]. Grudgingly, God then instructed Samuel to "heed their demands and appoint a king for them" [1 Samuel 8:22], which is perhaps the first divine concession to democratic principles.

Certain choices by lot are also reported in the Bible, but only later in Greece and Rome were more systematic procedures developed for the election of political leaders [Stavely (1972)]. More important, the first philosophical writings on democracy and elections can be found at this time. In his *Politics*, for example, Aristotle gave considerable attention to better and worse forms of government, including representative democracies. In *The Republic*, Plato, speaking for Socrates, did not advance a democratic credo but did expound on the proper training of individuals for public office, recommending – among other things – the extended study of both poetry and mathematics.

Political theorists since the Greeks have concentrated on explicating ideals of democracy, such as liberty and equality, and given relatively little attention to its methods. Here I shall reverse the usual priorities, which reflects what Riker (1982) calls the *liberal* interpretation of democracy – that its crucial feature is to enable citizens, through voting, to control public officials by means of periodic elections.

Social choice theory over the past forty years has convincingly established that elections may produce social choices that lack internal consistency, independent of how they were achieved. For example, because of a paradox of voting, which Arrow (1963, 1951) cast in much more general form in his famous impossibility theorem, preferences may be cyclical, rendering the very notion of *the* popular will a contradiction in terms (an example of cyclical preferences will be given in Section 4.1).

Even if preferences do not cycle, the "popular will" may be difficult to decipher, mainly because no voting procedure can unerringly ferret out what people want. For one thing, they may not reveal their true preferences for strategic reasons; for another, even true preferences may be distorted (in some sense) by the aggregation procedure. In subsequent sections of this chapter, I shall indicate why the truthful revelation of preferences may not be rational as well as some of the biases of different methods of aggregation.

Precisely because all procedures are manipulable, biased in some direction and to some extent, or otherwise deficient, they require careful analysis if both their weaknesses and their strengths are to be fairly assessed. They *are* the method of liberal democracy and, therefore, are essential to its functioning, haphazard as the outcomes they produce may seem [Miller (1983)]. If they are variously effective in promoting democracy's ideals, as discussed by political philosophers from time immemorial to the present, it behooves us to compare them so that more informed and better choices can be made.

3. Sincerity, strategyproofness, and efficacy

The focus of this and the next two sections will be on voting procedures in which voters can vote for but not rank candidates.[1] In this section some basic ideas about voting that will be helpful in distinguishing better from worse strategies will be introduced. By "better" I mean roughly those strategies that a voter would seriously consider in deciding for whom to vote. Crucial to this determination will be eliminating those "worse" strategies that a voter would never consider because they are "dominated" – under no circumstances would they yield a better outcome

---

[1] The analysis in these sections is adapted from Brams and Fishburn (1983, chs. 2, 3, and 5).

than some other strategies, and sometimes they would give unequivocally worse outcomes. Those strategies which are not dominated will be called "admissible."

In general, a voter has more than one admissible strategy, so his or her problem is to choose from among them. Those admissible strategies which are "sincere," and more honestly reflect a voter's preferences in a sense to be made precise later, will be distinguished from those that are not under single-ballot voting systems that do not require voters to rank candidates (runoff systems will be considered in Section 4.1). Different voting systems themselves will be characterized as sincere and, more stringently, "strategyproof," for different configurations of voter preferences. Finally, the "efficacy" of strategies under different voting systems will be briefly discussed.

3.1. Voter preferences and dominance

Denote individual candidates by small letters $a, b, c, \ldots$, and subsets of candidates by $A, B, C, \ldots$. For any two subsets A and B, $A \cup B$, or the union of A and B, is the set of all candidates who are in A or in B. $A \setminus B$, or the difference of A and B, is the set of all candidates who are in A and not in B.

A voter's *strict preference relation* on the candidates will be denoted by P, so that aPb abbreviated ab, means that the voter definitely prefers a to b. Similarly, R will denote a voter's *nonstrict preference relation* on the candidates, so that aRb means that the voter likes a at least as much b. Alternatively, aRb means that the voter either strictly prefers a to b or is indifferent between a and b.

It will be assumed that each voter's preferences are *connected*: he or she has a definite preference order on the candidates so that, for any two candidates, a voter prefers one to the other or is indifferent between them. Also, a voter's preference and indifference relations on all candidates are assumed to be *transitive*: if the voter strictly prefers a to b and b to c, he or she will strictly prefer a to c, and similarly for nonstrict preference and indifference relations.

Given connectivity and transitivity, the set of all candidates can be partitioned into nonempty subsets, say $A_1, A_2, \ldots, A_n$, for a given P so that the voter is indifferent among all candidates within each A_i and strictly prefers every candidate in A_i to every candidate in A_j if $i < j$ – that is, he or she ranks all the "indifferent" candidates in A_i higher than the "indifferent" candidates in A_j.

According to this designation, A_1 is the voter's subset of most-preferred candidates, and A_n is the voter's subset of least-preferred candidates. If the voter is indifferent among all candidates, then A_1 and A_n are the same, but otherwise A_1 and A_n are disjoint. The following comprehensive definition introduces a number of terms that will be used in this and later sections:

Definition 1. Suppose P partitions the set of all candidates into $n \geqslant 1$ nonempty subsets $A_1, A_2, \ldots, A_n$, so that the voter is indifferent among all candidates within each A_i and has aPb when a is a member of A_i and b is a member of A_j iff (if and only if) $i < j$. Then P is *unconcerned* iff $n = 1$; P is *dichotomous* iff $n = 2$; P is *trichotomous* iff $n = 3$; and P is *multichotomous* iff $n \geqslant 4$. In addition, a subset of candidates B is *high for P* iff whenever it contains a candidate in A_j it contains all (more-preferred) candidates in A_i for every $i < j$; and B is *low for P* iff whenever it contains a candidate in A_i it contains all (less-preferred) candidates in A_j for every $j < i$.

A voter who has an unconcerned P will be referred to as an *unconcerned voter* since he or she is indifferent among all candidates. If P is unconcerned, then every subset of candidates is both high and low for P.

When P is concerned — that is, when P is either dichotomous, trichotomous, or multichotomous — exactly two subsets of candidates are *both* high and low for P, namely the empty set and the set of all candidates. To illustrate these concepts, if P is trichotomous on a set of five candidates $\{a, b, c, d, e\}$ — say, with $A_1 = \{a\}$, $A_2 = \{b, c\}$, and $A_3 = \{d, e\}$ — then P has six high subsets in addition to the empty and whole sets, namely $\{a\}$, $\{a, b\}$, $\{a, c\}$, $\{a, b, c\}$, $\{a, b, c, d\}$, and $\{a, b, c, e\}$; and P has six low subsets in addition to the empty set and whole set, namely $\{e\}$, $\{d\}$, $\{e, d\}$, $\{c, d, e\}$, $\{b, d, e\}$, and $\{b, c, d, e\}$. In all cases, the number of high subsets is equal to the number of low subsets since a subset is high iff its complement (i.e., all other candidates) is low.

The characterizations of dominance and admissible strategies to be developed in Sections 3.2 and 3.3 depend not only on the relations P and R as applied to individual candidates but also on extensions of these relations to subsets of candidates. Thus, APB means that the voter prefers outcome A to outcome B, and ARB means that he or she finds A at least as good as B.

When A and B are one-candidate subsets, say $A = \{a\}$ and $B = \{b\}$, it is natural to assume that APB iff aPb, and that ARB iff aRb. Assume also that, for any nonempty A and B, APB and BRA cannot both hold. In addition, the following will be assumed for all candidates a and b and for all subsets of candidates A, B, and C.

Assumption P. If aPb, then $\{a\} P \{a, b\}$ and $\{a, b\} P \{b\}$.

Assumption R. If $A \cup B$ and $B \cup C$ are not empty and if aRb, bRc, and aRc for all a belonging to A, b belonging to B, and c belonging to C, then $(A \cup B) R (B \cup C)$.

Assumption P asserts that if candidate a is preferred to candidate b, then outcome $\{a\}$ is preferred to the tied outcome $\{a, b\}$, which in turn is preferred to $\{b\}$. This seems quite reasonable, regardless of how the tie between a and b might be broken when $\{a, b\}$ occurs, if the voter believes that a and b each has a positive probability of being elected when the two are tied after the initial ballot. Similarly, assumption R says that if everything in A is at least as good as everything in B and C, and if everything in B is at least as good as everything in C, then outcome $A \cup B$ will be at least as good as outcome $B \cup C$.

3.2. Dominance between strategies

Define a *strategy* to be any subset of candidates. Then *choosing* a strategy means voting for all candidates in the subset. For identification purposes, let S and T rather than $A, B, C, \ldots$, identify strategies. A voter uses S if he or she votes for each candidate in S and no candidate not in S.

A strategy is *feasible* for a particular voting system iff it is permitted by that system. The abstention strategy, which is the empty subset of candidates, is assumed always to be feasible. For every other strategy, a voter's ballot is counted iff he or she uses a feasible strategy.

The notion of admissible strategies to be developed in Section 3.3 depends on feasibility and on dominance. Roughly speaking, strategy S dominates strategy T for a particular voter if he or she likes the outcome of S as much as the outcome of T in every possible circumstance, and strictly prefers the outcome of S to the outcome of T in at least one circumstance.

To define dominance precisely, define a *contingency* as a function f that assigns a nonnegative integer to each candidate. A contingency is interpreted as specifying the number of votes each candidate receives from all voters *other than* the voter for whom dominance is being defined.

Call the latter voter the *focal voter*. Given a contingency f and a strategy S for this voter, let $F(S,f)$ denote the *outcome* of the vote, That is, $F(S,f)$ is the subset of candidates who have the greatest vote total under F and S. For any candidate a and strategy S, let $S(a) = 1$ if a belongs to S, with $S(a) = 0$ otherwise. Then, with $f(a)$ the integer assigned by contingency f to candidate a, a belongs to $F(S,f)$ iff $f(a) + S(a) \geq f(b) + S(b)$ for all candidates $b \neq a$. That is, a necessary and sufficient condition for candidate a to be contained in the outcome is that he or she receives at least as many votes from all voters (including the focal voter) as does every other candidate.

One of the main tasks of the subsequent analysis is to determine strategies for a voter that lead to outcomes he or she most prefers. Although different strategies may be preferred under different contingencies, some strategies are uniformly as good as, or better than, other strategies, regardless of the contingency. That is, one strategy may dominate another strategy.

Definition 2. Given the strict and nonstrict preference relations P and R for a voter, strategy S *dominates* strategy T, or S dom T for this voter, iff $F(S,f) R F(T,f)$ for all possible contingencies f, and $F(S,f) P F(T,f)$ for at least one contingency.

This definition does not require S and T to be feasible strategies and it is therefore applicable to all nonranking voting systems. Feasibility will enter the analysis explicitly through the definition of admissibility in Section 3.3.

Assumption R implies that an unconcerned voter (Definition 1) will be indifferent among all outcomes as well as among all individual candidates. Because Definition

2 requires that $F(S,f) P F(T,f)$ for some contingency f, it follows that no strategy is dominated for a voter with preference order P if P is unconcerned. The following theorem characterizes dominance between strategies for all concerned P.

Theorem 1 (Dominance). *Suppose P is concerned and Assumptions P and R hold. Then S dom T for P iff $S \neq T$, $S \setminus T$ is high for P, $T \setminus S$ is low for P, and neither $S \setminus T$ nor $T \setminus S$ is the set of all candidates.*

Proofs of Theorem 1 and other theorems and corollaries in Section 3 are given in Brams and Fishburn (1978). The intuitive reasoning underlying Theorem 1 is that because dominance is based on all contingencies, and the focal voter votes for all candidates in $S \cap T$ when he or she uses either S or T, S dominates T for P iff $S \setminus T$ dominates $T \setminus S$ for P. That is, dominance shows up in the nonoverlapping candidates, with those in S being high and those in T being low.

Although Theorem 1 is predicated on Assumptions P and R, the necessary and sufficient conditions for S dom T do not explicitly use the P and R relations on the outcomes. That is, dominance between strategies can be determined completely on the basis of a voter's strict preference relation P over the individual candidates. This greatly simplifies the identification of dominated strategies for a voter.

For example, if the set of candidates is $\{a,b,c\}$, and P is trichotomous with a preferred to b and b preferred to c, then Theorem 1 says that strategy $\{a\}$, under which the voter votes only for his or her most-preferred candidate, dominates strategies $\{c\}$, $\{a,c\}$, $\{b,c\}$, $\{a,b,c\}$, and the abstention strategy. Moreover, these are the only strategies that $\{a\}$ dominates, whereas $\{a,b\}$ dominates these strategies and $\{b\}$ in addition.

To illustrate the applicability of Theorem 1 to specific voting systems, define *plurality voting* to be a system in which a voter can vote for only one candidate or abstain (i.e., a feasible strategy is to vote for any subset containing exactly one candidate or none). Define *approval voting* to be a system in which a voter can vote for any number of candidates (i.e., a feasible strategy is *any* subset of candidates); this is a relatively new system that has been axiomatized in two different ways by Fishburn (1978a).

Under approval voting, Theorem 1 says that if voters consider voting for their second choice b, then they should also vote for their first choice a since the latter strategy is as good as, and sometimes better than, the strategy of voting for b alone. However, under plurality voting, a vote for b alone could be a voter's best strategy since in this case $\{b\}$ is not dominated by any other feasible strategy. As will be shown in Section 3.3 (see Definition 4), strategy $\{b\}$ is "admissible" for plurality voting but "inadmissible" for approval voting.

It can be demonstrated by exhaustive enumeration that in none of the sixteen contingencies in this example does strategy $\{b\}$ induce a better outcome for a voter with preference scale abc than does $\{a,b\}$ under approval voting [Brams (1978, pp. 199–202), Brams (1983, pp. 38–41)]. Fortunately, Theorem 1 relieves one of the necessity of checking each and every contingency, which becomes virtually impossible if there are more than three candidates, to establish which strategies are dominated. If a voter has information about the preferences of other voters – even in aggregated form (e.g., from a poll) – he or she may be able to narrow down the undominated choices [Brams (1982b), Brams and Fishburn (1983, ch. 7), Merrill (1988, ch. 6)], which will be characterized next in terms of admissibility.

3.3. Admissible strategies

In this section a theorem will be presented that characterizes all admissible strategies for every concerned P and for all nonranking voting systems. Admissible strategies under different simple voting systems will then be compared.

To begin the analysis, two definitions are needed. The first makes the concept of a nonranking voting system precise by tying it to the numbers of candidates that voters are allowed to vote for (e.g., one, two, one or two, etc.) in order that their ballots be considered legal.

Definition 3. Suppose there are m candidates. Then a *nonranking voting system* is a nonempty subset s of $\{1, 2, \ldots, m-1\}$.

As noted in Section 3.2, abstention – voting for no candidates – will be considered feasible for all voting systems. Since a vote for all candidates is tantamount to an abstention insofar as the determination of the outcome of a vote is concerned, m is not included as a possible number in s in Definition 3. This will not present problems with the later analysis of admissibility since, when it is allowed, a vote fo all m candidate (like an abstention) is dominated by some other feasible strategy whenever P is concerned.

According to Definition 3, plurality voting is system $\{1\}$: a voter is allowed to vote for only one candidate. Under system $\{2\}$, each voter must vote for exactly two candidates for his or her ballot to be legal. System $\{1, 2\}$ allows a voter to vote for either one or two candidates. System $\{1, 2, \ldots, m-1\}$ is approval voting.

For any subset A of candidates, let $|A|$ denote the *number* of candidates in A. Then strategy S is *feasible* for system s iff either S is the abstention strategy or $|S|$ belongs to s. This characterization of feasibility is consistent with Definition 3, for it says that feasible voting strategies are exactly those that are permitted by the system – vote for none, or vote for only numbers of candidates contained in s.

Next, to provide a formal definition of admissibility, feasibility and dominance are combined.[2]

---

[2]Carter (1990) has proposed a somewhat different definition of admissibility as well as sincerity (see Section 3.4).

Definition 4. Strategy S is *admissible* for system s and preference order P iff S is feasible for s and there is no strategy T that is also feasible for s and has T dom S for P.

Definition 4 suggests that, because of the feasibility requirement, a strategy feasible for each of two voting systems may be admissible for one system but inadmissible for the other. Indeed, in the example in Section 3.2 in which P was abc, it was noted that strategy $\{b\}$ is admissible for plurality voting but inadmissible for approval voting.

Before examining these specific systems in more detail, it is useful to have a theorem that characterizes all admissible strategies for every voting system and every concerned preference order P. This is so because the analysis that follows is based on the assumption that nonabstaining voters use only admissible strategies.

To facilitate the statement of the admissibility theorem, let

$M(P) = A_1$, the subset of most-preferred candidates under P,
$L(P) = A_n$, the subset of least-preferred candidates under P,

where A_1 and A_n are as given in Definition 1. The admissibility theorem below appears complex, but as later corollaries will make clear, it is not difficult to apply to particular voting systems. Moreover, comparisons among several systems will show that they possess striking differences that bear on their strategic manipulability.

Theorem 2 (Admissibility). *Suppose preference order P is concerned and Assumptions P and R hold. Then strategy S is admissible for system s and preference order P iff S is feasible for s and either C1 or C2 (or both) holds:*

C1: Every candidate in $M(P)$ is in S, and it is impossible to divide S into two nonempty subsets S_1 and S_2 such that S_1 is feasible for s and S_2 is low for P.

C2: No candidate in $L(P)$ is in S, and there is no nonempty subset A of candidates disjoint from S such that $A \cup S$ is feasible for s and A is high for P.

Since the abstention strategy satisfies neither C1 nor C2, it is never admissible for a concerned voter. Because abstention is inadmissible in the formal sense, a vote for all m candidates must likewise be inadmissible if it is permitted. Thus, though the abstention and "vote for all" strategies are assumed to be feasible, they can be omitted from the formal analysis since they are always inadmissible for a concerned voter.

When Theorem 2 is applied to approval voting, the following result is obtained.

Corollary 1. *Strategy S is admissible for approval voting and a concerned P iff S contains all candidates in $M(P)$ and none in $L(P)$.*

Hence, concerned voters use one of their admissible strategies under approval voting iff they vote for every one of their most-preferred candidates and do not vote for any of their least-preferred candidates. Thus, if $m = 4$ and a voter has preference order $abcd$, then his or her admissible strategies are $\{a\}$, $\{a,b\}$, $\{a,c\}$ and $\{a,b,c\}$.

Corollary 2. *A voter has a unique admissible (or dominant) strategy under approval voting iff his or her preference order P is dichotomous. This dominant strategy is the voter's subset of most-preferred candidates.*

Thus, if a voter has dichotomous preferences $(ab)(cd)$ – with parentheses denoting indifference subsets – then $\{a,b\}$ is his or her dominant strategy.

It is instructive to compare approval voting with two other systems to illustrate the numbers of admissible strategies they offer different types of voters. The first system is plurality voting. The second system is $s = \{1, m-1\}$, in which a voter can vote for either one, or all but one, candidate, is called *negative voting*. It was proposed by Boehm (1976) and has been analyzed in Brams (1977, 1978, 1983); an extension to allow a third option of abstention has been proposed by Felsenthal (1989), but this is not a nonranking voting system.

Under negative voting, each voter is allowed to cast one vote. This vote can be either for or against a candidate. A *for* vote adds one point to the candidate's score, and an *against* vote subtracts one point from the candidate's score. The outcome of a negative voting ballot is the subset of candidates with the largest *net vote* total (algebraic sum of for and against points), which may, of course, be negative.

Since a vote against a candidate has the same effect as a vote for every other candidate, the negative voting system is tantamount to system $\{1, m-1\}$. When $m = 3$, this system is equivalent to approval voting, because a vote against one candidate has the same effect as approval votes for each of the other two. But for $m > 3$, negative voting has fewer feasible strategies than approval voting.

This argument can be made more precise. If there are $m > 3$ candidates, under approval voting a voter can cast either an approval vote or no vote for each candidate, giving each voter 2^m feasible voting strategies. However, since an abstention (vote for none) has the same net effect as a vote for all candidates, the number of effectively different choices is $2^m - 1$. By contrast, plurality voting allows $m + 1$ different choices (a vote for one of the m candidates or an abstention), and various other systems allow between $m + 1$ and $2^m - 1$ different strategies. Negative voting, for example, allows a voter $2m + 1$ strategies, because he or she can vote for or against each of the m candidates or abstain. If $m = 4$, there are nine feasible strategies, whereas under approval voting a voter would have 15 feasible strategies.

The following corollaries of Theorem 2 identify the admissible strategies of a concerned voter under plurality and negative voting. Recall that $L(P)$ is the voter's subset of least-preferred candidates. In Corollary 4, $\bar{a}$ denotes the strategy in which the voter votes for all candidates other than candidate a (or casts a vote against candidate a).

Corollary 3. *Strategy $\{a\}$ is admissible for plurality voting and concerned P iff a is not in L(P).*

Corollary 4. *Suppose $m > 3$. Then* (i) *strategy $\{a\}$ is admissible for negative voting and concerned P iff the voter strictly prefers a to at least two other candidates, and* (ii) *strategy $\bar{a}$ is admissible for negative voting and concerned P iff the voter strictly prefers at least two other candidates to a.*

Corollaries 1, 3, and 4, which provide necessary and sufficient conditions for admissible strategies under three different simple voting systems, can be used to identify and compare sets of admissible strategies for various preference orders of voters for the candidates. For example, given the trichotomous preference order $(ab)cd$ for the set of four candidates $\{a,b,c,d\}$, the sets of admissible strategies for each of the systems is:
  (1) *Approval*: $\{a,b\}$, $\{a,b,c\}$. These are the only feasible strategies that contain all the voter's most-preferred, and none of his or her least-preferred, candidates.
  (2) *Plurality*: $\{a\}$, $\{b\}$, $\{c\}$. These are the only feasible strategies that do not contain the voter's least-preferred candidates.
  (3) *Negative*: $\{a\}$, $\{b\}$, $\bar{c}, \bar{d}$. These are the only feasible strategies in which the voter strictly prefers the candidate to at least two others, or strictly prefers at least two other candidates to the barred candidate.

The numbers of admissible strategies for all concerned P orders for four candidates are shown in Table 1. It is clear that the relative numbers of admissible strategies for the three systems are very sensitive to the specific form of P. For example, approval voting offers voters more admissible strategies than the other systems when P is $a(bc)d$ but fewer when P is $(ab)cd$. Hence, although the number of feasible strategies increases exponentially with m under approval voting (but not the other systems), the number of admissible strategies under approval voting

Table 1
Numbers of admissible voting strategies for three voting systems with four candidates.

Concerned preference order		Number of admissible strategies for		
		Apporval voting	Negative voting	Plurality voting
Dichotomous	$a(bcd)$	1	1	1
	$(abc)d$	1	1	3
	$(ab)(cd)$	1	4	2
Trichotomous	$(ab)cd$	2	4	3
	$ab(cd)$	2	4	2
	$a(bc)d$	4	2	3
Multichotomous	$abcd$	4	4	3

is comparable to that of the other systems and so should not overwhelm the voter with a wealth of viable options.

3.4. Sincere voting and strategyproofness

To facilitate a general comparison of nonranking voting systems in terms of their ability to elicit the honest or true preferences of voters, the following notions of sincere strategies and strategyproofness are helpful.

Definition 5. Let P be a concerned preference order on the candidates. Then strategy S is *sincere* for P iff S is high for P; voting system s is *sincere* for P iff all admissible strategies for s and P are sincere; and voting system s is *strategyproof* for P iff exactly one strategy is admissible for s and P (in which case this strategy must be sincere).

Sincere strategies are essentially strategies that directly reflect the true preferences of a voter, i.e., that do not report preferences "falsely." For example, if P is $abcd$, then $\{a,c\}$ is not sincere since a and c are not the voter's two most-preferred candidates. Because a democratic voting system should base the winner of an election on the true preferences of the voters, sincere strategies are of obvious importance to such systems.

They are also important to individual voters, for if a system is sincere, voters will always vote for *all* candidates ranked above the lowest-ranked candidates included in their chosen admissible strategies. To illustrate when this proposition is not true, if P is $abcd$, $\{a,c\}$ is admissible under approval voting but obviously not sincere since this strategy involves voting for candidate c without also voting for preferred candidate b.

Using Corollaries 1, 3, and 4, one can easily verify that, for the seven prototype preference orders for four candidates given in Table 1, approval voting is sincere in six cases (only $abcd$ is excluded), negative voting is sincere in four cases, and plurality voting is sincere in only the first three cases. In fact, it is no accident that approval voting is "more sincere" than both of the other systems given in Table 1; as the following theorem demonstrates, approval voting is the uniquely most sincere system of all the nonranking voting systems described by Definition 3.

Theorem 3. *If P is dichotomous, then every voting system s is sincere for P. If P is trichotomous, then approval voting is sincere for P, and this is the only system that is sincere for every trichotomous P. If P is multichotomous, then no voting system is sincere for P.*

No system is sincere when P is multichotomous because, for every s and every P with four or more indifference subsets A_i, there is an admissible strategy that is not sincere. When there are relatively few candidates in a race, however, it is reasonable to expect that many voters will have dichotomous or trichotomous preference orders. Indeed, Theorem 3 says that when voters do not (or cannot) make finer distinctions, approval voting is the most sincere of all nonranking voting systems, and this result can be extended to voters with multichotomous preferences [Fishburn (1978b)].

Even if a voting system is sincere for P, however, it is not strategyproof for P if it allows more than one admissible strategy. Although manipulation of election outcomes by strategic voting in multicandidate elections is an old subject, only since the pioneering work of Gibbard (1973) and Satterthwaite (1975) has it been established that virtually every type of reasonable election method is vulnerable to strategic voting.

Like sincerity, strategyproofness seems a desirable property for a voting system to process. If voters have only one admissible strategy, they will never have an incentive to deviate from it even if they know the result of voting by all the other voters. For this is a dominant strategy, so whatever contingency arises, a voter cannot be hurt, and may be helped, by choosing it.

Sincerity, on the other hand, does not imply such stability but instead asserts that whatever admissible strategy a voter chooses, whenever it includes voting for some candidate, it also includes voting for all candidates preferred to him or her. In effect, a voting system is sincere if it does not force voters to "sacrifice," for strategic reasons, more-preferred for less-preferred candidates.

Because the demands of strategyproofness are more stringent than those for sincere voting, the circumstances that imply strategyproofness are less likely to obtain than the circumstances that imply sincerity. Nevertheless, as with sincerity, the approval voting system is the uniquely most strategyproof of all systems covered by Definition 3.

Theorem 4. *If P is dichotomous, then approval voting is strategyproof for P, and this is the only system s that is strategyproof for every dichotomous P. If P is trichotomous or multichotomous, then no voting system is strategyproof for P.*

Theorem 3 and 4 provide strong support for approval voting based on the criteria of sincerity and strategyproofness, which can be extended to committees [(Fishburn (1981)]. Yet, the limitations of these results should not be forgotten: strategyproofness depends entirely on the preferences of all voters' being dichotomous; sincerity extends to trichotomous preferences, but it is a weaker criterion of nonmanipulativeness than strategyproofness.[3]

---

[3] For other concepts of manipulability, see Brams and Zagare (1977, 1981) and Chamberlin (1986); the manipulability of approval voting has provoked exchanges between Niemi (1984) and Brams and Fishburn (1985), and between Saari and Van Newenhizen (1988) and Brams, Fishburn and Merrill (1988). Different strategic properties of voting systems are compared in Merrill (1981, 1982, 1988), Nurmi (1984, 1987) and Saari (1994).

3.5. Efficacy

There is not space here to define formally the "efficacy" of a voting strategy, which is a measure of that strategy's ability, on the average, to change the outcome of an election from what it would be if the voter in question abstained [Fishburn and Brams (1981b, c), Brams and Fishburn (1983)]. In large electorates, the most efficacious approval voting strategies are for a focal voter to vote for either the top one or top two candidates in three-candidate contests, approximately the top half of all candidates in more populous contests, given that all possible ways that other voters can vote are equiprobable and ties are broken randomly. When cardinal utilities are associated with the preferences of a voter, the voter's utility-maximizing strategy in large electorates is to vote for all candidates whose utilities exceed his or her average utility over all the candidates. Hoffman (1982, 1983) and Merrill (1979, 1981, 1982, 1988) have independently derived similar results; in doing so, they consider goals other than expected-utility maximization.

A voter's utility-maximizing strategy can lead to substantially different expected-utility gains, depending on his or her utilities for the various candidates. However, it can be shown that plurality voting gains are even more disparate – despite the fact that every voter is restricted to voting for just one candidate [Fishburn and Brams (1981b, c, 1983), Rapoport and Felsenthal (1990)] – so approval voting is more equitable in the sense of minimizing differences among voters.

As a case in point, plurality voting affords a dichotomous voter who equally likes four candidates but despises the fifth in a five-candidate race little opportunity to express his or her preferences, compared with the voter who greatly prefers one candidate to all the others. Approval voting, on the other hand, is equitable to both – allowing the first voter to vote for his or her top four choices, the second to vote for his or her top choice – despite the extreme differences in each's utilities for the candidates. In general, not only is a voter able to be more efficacious under approval than plurality voting, but he or she cannot suffer as severe deprivations, or utility losses, under the former procedure.

4. Condorcet properties

When there are three or more candidates, and only one is to be elected to a specified office, there are widespread differences of opinion as to how the winner should be determined. These differences are perhaps best epitomized by Condorcet's (1785) criticism of Borda's (1781) method, which is recounted by Black (1958), who also provides an historical treatment and analysis of other proposed voting methods. Borda recommended that in an election among three candidates, the winner be the candidate with the greatest point total when awards, say, of two points, one point, and zero points are made, respectively, to each voter's most-preferred, next most-preferred, and least-preferred candidates. Condorcet argued to the contrary that the winner ought to be the candidate who is preferred by a simple majority of voters to *each* of the other candidates in pairwise contests – provided that such a majority candidate exists – and showed that Borda's method can elect a candidate other than the majority candidate.

Although many analysts accept Condorcet's criterion, it leaves open the question of which candidate should win when there is no majority candidate – that is, when the so-called paradox of voting occurs (see Section 4.1 for a simple example). In a series of books, Straffin (1980), Riker (1982), Schwartz (1986), Nurmi (1987), and Merrill (1988) have reviewed a number of methods for determining a winner from voters' preferences when there is no such candidate (called a "Condorcet candidate") and conclude that, although some methods are better than others – in the sense of satisfying properties considered desirable – there is no obviously best method.

The admissibility results of Section 3.3 will be used to determine the abilities of various nonranking systems to elect a strict (single) Condorcet candidate when one exists. Runoff voting systems will then be briefly compared with single-ballot systems. After investigating the existence of admissible and sincere strategies that elect the Condorcet candidate, cases in which the Condorcet candidate is invariably elected, whatever admissible strategies voters use, will be considered.

4.1. Dichotomous preferences

Let V denote a finite list of preference orders for the candidates, with each order in the list representing the preferences of a particular voter. Clearly, each allowable preference order may appear more than once in V, or not at all. The *Condorcet candidates* with respect to any given V are the candidates in the set $\mathrm{Con}(V)$, where

$\mathrm{Con}(V) = \{$all candidates a such that, for each candidate $b \neq a$, at least as many orders in V have a preferred to b as have b preferred to $a\}$.

Thus, candidate a is in $\mathrm{Con}(V)$ iff at least as many voters prefer a to b as prefer b to a for each b other than a.

When there is a paradox of voting that includes all candidates, $\mathrm{Con}(V)$ will be empty, which occurs, for example, if $V = (abc, bca, cab)$. Then a majority of two voters prefers a to b, a different majority b to c, and a still different majority c to a. Thus, while a defeats b, and b defeats c, c defeats a, making the majorities cyclical. Because each candidate can be defeated by one other, there is no candidate who does at least as well against all other candidates as they against him or her. Other voting paradoxes are described in Fishburn (1974), and an extension of Condorcet's criterion to committees is given in Gehrlein (1985).

Condorcet's basic criterion asserts that candidate a wins the election when $\mathrm{Con}(V)$ contains only a and no other candidate. But because it is possible for more than one candidate to be in Con(V), as when the same number of voters prefer a to b as prefer b to a, it is useful to extend Condorcet's rule to assert that if Con(V) is not empty, then *some* candidate in Con(V) wins the election. In particular, if every candidate who is in the outcome from the ballot is in Con(V), provided it is not empty, the extended rule will apply.

It has been shown by Inada (1964) that if all preference orders in V are dichotomous, then Con(V) is not empty. Using the previous results, which presume Assumptions P and R (Section 3.1), I shall show more than this, namely that the use of admissible strategies under approval voting when preferences are dichotomous always yields Con(V) as the outcome. Moreover, for any *other* nonranking voting system s, the use of admissible strategies, given dichotomous preferences for all voters, may give an outcome that contains no candidate in Con(V).

The following formulation will be used to express these results more rigorously:

Definition 5. For any finite list V of preference orders for the candidates, and for any voting system s as identified in Definition 3, let $V(s)$ be the set of all functions that assign an admissible strategy to each of the terms in V. For each function α in $V(s)$, let $F(\alpha)$ be the outcome (the set of candidates with the greatest vote total) when every voter uses the admissible strategy that is assigned to his or her preference order by α.

As an illustration, assume $V = (abc, abc, c(ab))$, consisting of two voters who prefer a to b to c and one voter who is indifferent between a and b but prefers c to both a and b. If s is plurality voting, then $V(s)$ contains $2 \times 2 = 4$ functions since each of the first two voters has two admissible strategies, and the third voter has one admissible strategy according to Corollary 2. The outcomes for the four α functions are $\{a\}$, $\{a,b,c\}$, $\{a,b,c\}$, and $\{b\}$. $F(\alpha) = \{a,b,c\}$, for example, if α assigns strategy $\{a\}$ to the first voter and strategy $\{b\}$ to the second voter; the third voter would necessarily choose strategy $\{c\}$ because it is his or her only admissible strategy.

Theorem 5. *Suppose all preference orders in V are dichotomous and s is the approval voting system. Then $F(\alpha) = $ Con(V) for every α belonging to $V(s)$.*

In other words, if all voters have dichotomous preferences, their use of admissible strategies under approval voting invariably yields Con(V) as the outcome.

To show how the dichotomous-preference situation differs under plurality voting, suppose that the candidate set is $\{a,b,c\}$ and there are $2n+1$ voters in V: One is $a(bc)$, with the order $2n$ orders divided evenly between $b(ac)$ and $(ac)b$. Then Con(V) = $\{a\}$. However, if as few as two of the n voters who have the order $(ac)b$ vote for c rather than a under plurality voting, then $F(\alpha) = b$, and $F(\alpha)$ and Con(V) are therefore disjoint. The following theorem shows that a similar result holds for every system other than approval voting.

Theorem 6. *Suppose s is a voting system as described in Definition 3 but not approval voting. Then there exists a V consisting entirely of dichotomous preference orders and an α belonging to $V(s)$ such that no candidate in $F(\alpha)$ is in $\text{Con}(V)$.*

In contrast to the definitive picture obtained for dichotomous preferences, comparisons among approval voting and other single-ballot systems are less clear-cut when some voters divide the candidates into more than two indifference sets. Recent reviews of the literature by Merrill (1988) and Nurmi (1987), based primarily on computer simulations [see also Bordley (1983, 1985), Chamberlin and Cohen (1978), Chamberlin and Featherston (1986), Fishburn and Gehrlein (1976, 1977, 1982), Nurmi (1988)], suggest that approval voting is generally as good or better than other nonranking systems, particularly plurality voting [Nurmi and Uusi-Heikkilä (1985), Felsenthal and Maoz (1988)], in electing Condorcet candidates. Indeed, it compares favorably with most ranking systems not only in terms of its Condorcet efficiency [(Merrill (1985)] but also in terms of the "social utility" of the candidates it elects [Weber (1977), Merrill (1984, 1988)].

4.2. Runoff systems

All the theoretical results discussed so far pertain only to nonranking voting systems in which the election outcome is decided by a single ballot. But there are more complicated nonranking voting systems, the most prominent being runoff systems. These systems consist of voting procedures whose first ballots are similar to the ballots of the single-ballot nonranking systems; the second or runoff ballot is a simple-majority ballot between the two candidates who receive the most first-ballot votes.

Several important differences between single-ballot and runoff systems are worth mentioning here. First, runoff systems are more costly than single-ballot systems since they require two elections. Second, historically it is almost always true that fewer voters go to the polls to vote in the second election than the first. Third, runoff systems encourage more candidates to enter [Wright and Riker (1989)] and provide voters with more opportunities for strategic manipulation. In particular, runoff systems may encourage some voters to vote for candidates on the first ballot only because they could be beaten by the voter's favorite (whom the voter thinks does not need help to get on the ballot) in a runoff. Since the degree of manipulability in this sense is presumably correlated with the number of candidates a person can vote for on the first ballot, the plurality runoff system would appear to be the least manipulable, the approval runoff system the most manipulable. In fact, it can be shown that coupling approval voting to a runoff system procedure a combination that is *never* strategyproof [Fishburn and Brams (1981a), Brams and Fishburn (1983)].

But it is runoff plurality voting that is by far the most common multiballot election system in use today. This system is used extensively in U.S. state and local elections, particularly in party primaries; its main rationale is to provide a device to ensure the election of the strongest candidate should he or she not triumph in the first plurality contest. However, if "strongest" is defined to be the Condorcet candidate (if one exists), this claim on behalf of the runoff is dubious, particularly when compared with single-ballot approval voting.

Generally speaking, runoff systems are less sincere than single-ballot systems, with runoff plurality less sincere than runoff approval because it requires a more restrictive condition when preferences are dichotomous. Thus, under single-ballot and runoff systems, approval voting is more sincere than plurality voting.

Neither runoff approval nor plurality voting is strategyproof, even when preferences are dichotomous. Runoff approval voting, especially, is subject to severe manipulation effects and is even more manipulable when the preferences of voters are not dichotomous. For example, if a plurality of voters has preference order abc and is fairly sure that a would beat c but lose to b in a runoff, these voters may well vote for a and c on the first ballot in an attempt to engineer a runoff between a and c. Other examples, and a more precise statement of general results, is given in Fishburn and Brams (1981a) and Brams and Fishburn (1983).

4.3. Condorcet possibility and guarantee theorems

Consider whether a Condorcet candidate x can be elected under different voting systems when all voters use sincere strategies:

Theorem 7. *Under both single-ballot and runoff systems, sincere strategies can always elect Condorcet candidate x under approval voting but not necessarily under other nonranking voting systems.*

This is not to say, however, that all sincere strategies guarantee x's election, even under approval voting. Occasionally, however, it is possible to make this guarantee, as the following example for three candidates $\{x, a, b\}$ illustrates:

1 voter has preference order xab;
1 voter has preference order $(ax)b$;
1 voter has preference order $(bx)a$.

Candidate x is the Condorcet candidate because x is preferred by two voters to a and two voters to b. Under single-ballot approval voting, the only admissible strategies of the second two voters is to vote for their two preferred candidates. The first voter has two sincere admissible strategies: vote for his or her most-preferred, or two most-preferred, candidate(s). Whichever strategy the first voter uses, x wins the election. Hence, single-ballot approval voting *must* elect x when voters use sincere admissible strategies.

Now consider single-ballot plurality voting, and runoff plurality and runoff approval voting as well. Under these three other systems, the following strategies of each voter are both admissible and sincere: the first voter votes for x, the second voter for a, and the third voter for b. These strategies do not guarantee the election of x under plurality voting, nor do they guarantee the election of x under either runoff system since the runoff pair could be $\{a, b\}$. Therefore, single-ballot approval voting is the only one of the four systems to guarantee the election of Condorcet candidate x when voters use sincere admissible strategies.

A second example demonstrates the proposition that when all voters use admissible (though not necessarily sincere) strategies, a Condorcet candidate's election may again be ensured only under single-ballot approval voting:

2 voters have preference order $(xa)b$;
1 voter has preference order xba;
1 voter has preference order bxa.

Once again, x is the Condorcet candidate, and single-ballot approval voting ensures his or her election no matter what admissible strategies voters choose. (In this example, as in the previous one with only three candidates, all admissible strategies are necessarily sincere under single-ballot approval voting since no voter's preference order can be multichotomous.) On the other hand, if the second two voters voted for b under single-ballot plurality voting (one would have to vote insincerely), x would not be ensured election.

Should this occur, and should the first two voters also vote for a, neither runoff system would elect x either. Thus, not only may approval voting be the only system to guarantee the election of a Condorcet candidate when voters are restricted to sincere admissible strategies, but it also may be the only system to provide this guarantee when voters are less constrained and can use any admissible strategies.

In generalizing these results, assume that voters use admissible, but not necessarily sincere admissible, strategies. Moreover, to simplify matters, only the four systems previously discussed will be analyzed, namely single-ballot and runoff plurality voting and single-ballot and runoff approval voting.

In the preceding examples, x was a *strict* Condorcet candidate – x did not share his or her status with any other candidate. This is not a coincidence since, as shown by the following theorem, x must be a strict Condorcet candidate if some system – single-ballot or runoff – guarantees his or her election when voters use admissible strategies.

Theorem 8. *Suppose that regardless of which system is used, all voters use admissible strategies on the only (or first) ballot and, in the case of a runoff system, each voter votes on the runoff ballot iff he is not indifferent between the candidates on that ballot. Then the four systems can be ordered in terms of the following chain of* *implications for any m > 3, where ⇒ is the logical relation "implies."*

[*x must be elected under runoff plurality voting*]
⇒ [*x must be elected under runoff approval voting*]
⇒ [*x must be elected under single-ballot plurality voting*]
⇒ [*x must be elected under single-ballot approval voting*]
⇒ [*x must be a strict Condorcet candidate*].

Since examples can be constructed to show that the converse implications are false for some m, the ability of a system to guarantee the election of a strict Condorcet candidate under the conditions of Theorem 8 is highest for single-ballot approval voting, next highest for single-ballot plurality voting, third highest for runoff approval voting, and lowest – in fact, nonexistent – for runoff plurality voting.

The ability of the different systems to elect, or guarantee the election of, Condorcet candidates decreases in going from single-ballot runoff systems, and from approval voting to plurality voting. Because single-ballot approval voting also encourages the use of sincere strategies, which systems with more complicated decision rules generally do not [Merrill and Nagle (1987), Merrill (1988)], voters need not resort to manipulative strategies to elect Condorcet candidates under this system.

5. Preferential voting and proportional representation

There are a host of voting procedures under which voters either can rank candidates in order of their preferences or allocate different numbers of votes to them. Unfortunately, there is no common framework for comparing these systems, although certain properties that they satisfy, and paradoxes to which they are vulnerable, can be identified. I shall describe the most important of these systems and indicate briefly the rationales underlying each, giving special attention to the proportional representation of different factions or interests. Then I shall offer some comparisons, based on different criteria, of both ranking and nonranking voting procedures.

5.1. The Hare system of single transferable vote (STV)

First proposed by Thomas Hare in England and Carl George Andrae in Denmark in the 1850s, STV has been adopted throughout the world. It is used to elect public officials in such countries as Australia (where it is called the "alternative vote"), Malta, the Republic of Ireland, and Northern Ireland; in local elections in Cambridge, MA, and in local school board elections in New York City, and in numerous private organizations. John Stuart Mill (1862) placed it "among the greatest improvements yet made in the theory and practice of government."

Although STV violates some desirable properties of voting systems [Kelly (1987)], it has strengths as a system of proportional representation. In particular, minorities can elect a number of candidates roughly proportional to their numbers in the electorate if they rank them high. Also, if one's vote does not help elect a first choice, it can still count for lower choices.

To describe how STV works and also illustrate two properties that it fails to satisfy, consider the following examples [Brams (1982a), Brams and Fishburn (1984d)]. The first shows that STV is vulnerable to "truncation of preferences" when two out of four candidates are to be elected, the second that it is also vulnerable to "nonmonotonicity" when there is one candidate to be elected and there is no transfer of so-called surplus votes.

Example 1. Assume that there are three classes of voters who rank the set of four candidates $\{x, a, b, c\}$ as follows:

I. 6 voters: $xabc$,
II. 6 voters: $xbca$,
III. 5 voters: $xcab$.

Assume also that two of the set of four candidates are to be elected, and a candidate must receive a quota of 6 votes to be elected on any round. A "quota" is defined as $[n/(m+1)] + 1$, where n is the number of voters and m is the number of candidates to be elected.

It is standard procedure to drop any fraction that results from the calculation of the quota, so the quota actually used is $q = [[n/(m+1)] + 1]$, the integer part of the number in the brackets. The integer q is the smallest integer that makes it impossible to elect more than m candidates by first-place votes on the first round. Since $q = 6$ and there are 17 voters in the example, at most two candidates can attain the quota on the first round (18 voters would be required for three candidates to get 6 first-place votes each). In fact, what happens is as follows:

First round: x receives 17 out of 17 first-place votes and is elected.

Second round: There is a "surplus" of 11 votes (above $q = 6$) that are transferred in the proportions 6:6:5 to the second choices (a, b, and c, respectively) of the three classes of voters. Since these transfers do not result in at least $q = 6$ for any of the remaining candidates (3.9, 3.9, and 3.2 for a, b, and c, respectively), the candidate with the fewest (transferred) votes (i.e., c) is eliminated under the rules of STV. The supporters of c (class III) transfer their 3.2 votes to their next-highest choice (i.e., a), giving a more than a quota of 7.1. Thus, a is the second candidate elected. Hence, the two winners are $\{x, a\}$.

Now assume two of the six class II voters indicate x is their first choice, but they do not indicate a second or third choice. The new results are:

First round: Same as earlier.

Second round: There is a "surplus" of 11 votes (above $q = 6$) that are transferred in the proportions 6:4:2:5 to the second choices, if any (a, b, no second choice, and c, respectively) of the voters. (The two class II voters do not have their votes transferred to any of the remaining candidates because they indicated no second choice.) Since these transfers do not result in at least $q = 6$ for any of the remaining candidates (3.9, 2.6, and 3.2 for a, b, and c, respectively), the candidate with the fewest (transferred) votes (i.e., b) is eliminated. The supporters of b (four voters in class II) transfer their 2.6 votes to their next-highest choice (i.e., c), giving c 5.8, less than the quota of 6. Because a has fewer (transferred) votes (3.9), a is eliminated, and c is the second candidate elected. Hence, the set of winners is $\{x, c\}$.

Observe that the two class II voters who ranked only x first induced a better social choice for themselves by truncating their ballot ranking of candidates. Thus, it may be advantageous not to rank all candidates in order of preference on one's ballot, contrary to a claim made by a mathematical society that "there is no tactical advantage to be gained by marking few candidates" [Brams 1982a]. Put another way, one may do better under the STV preferential system by *not* expressing preferences – at least beyond first choices.

The reason for this in the example is the two class II voters, by not ranking bca after x, prevent b's being paired against a (their last choice) on the second round, wherein a beats b. Instead, c (their next-last choice) is paired against a and beats him or her, which is better for the class II voters.

Lest one think that an advantage gained by truncation requires the allocation of surplus votes, I next give an example in which only one candidate is to be elected, so the election procedure progressively eliminates candidates until one remaining candidate has a simple majority. This example illustrates a new and potentially more serious problem with STV than its manipulability due to preference truncation, which I shall illustrate first.

Example 2. Assume there are four candidates, with 21 voters in the following four ranking groups:

I. 7 voters: *abcd*
II. 6 voters: *bacd*
III. 5 voters: *cbad*
IV. 3 voters: *dcba*

Because no candidate has a simple majority of $q = 11$ first-place votes, the lowest first-choice candidate, d, is eliminated on the first round, and class IV's 3 second-place votes go to c, giving c 8 votes. Because none of the remaining candidates has a majority at this point, b, with the new lowest total of 6 votes, is eliminated next, and b's second-place votes go to a, who is elected with a total of 13 votes.

Next assume the three class IV voters rank only d as their first choice. Then d is still eliminated on the first round, but since the class IV voters did not indicate a second choice, no votes are transferred. Now, however, c is the new lowest candidate, with 5 votes; c's elimination results in the transfer of his or her supporters' votes to b, who is elected with 11 votes. Because the class IV voters prefer b to a, it is in their interest not to rank candidates below d to induce a better outcome for themselves, again illustrating the truncation problem.

It is true that under STV a first choice can never be hurt by ranking a second choice, a second choice by ranking a third choice, etc., because the higher choices are eliminated before the lower choices can affect them. However, lower choices can affect the order of elimination and, hence, transfer of votes. Consequently, a higher choice (e.g., second) can influence whether a lower choice (e.g., third or fourth) is elected.

I wish to make clear that I am not suggesting that voters would routinely make the strategic calculations implicit in these examples. These calculations are not only rather complex but also might be neutralized by counterstrategic calculations of other voters. Rather, I am saying that to rank *all* candidates for whom one has preferences is not always rational under STV, despite the fact that it is a ranking procedure. Interestingly, STV's manipulability in this regard bears on its ability to elect Condorcet candidates [Fishburn and Brams (1984)].

Example 2 illustrates a potentially more serious problem with STV: raising a candidate in one's preference order can actually hurt that candidate, which is called *nonmonotonicity* [Smith (1973), Doron and Kronick (1977), Fishburn (1982), Bolger (1985)]. Thus, if the three class IV voters raise a from fourth to first place in their rankings – without changing the ordering of the other three candidates – b is elected rather than a. This is indeed perverse: a loses when he or she moves up in the rankings of some voters and thereby receives more first-place votes. Equally strange, candidates may be helped under STV if voters do not show up to vote for them at all, which has been called the "no-show paradox" [Fishburn and Brams (1983), Moulin (1988), Ray (1986), Holzman (1988, 1989)] and is related to the Condorcet properties of a voting system.

The fact that more first-place votes (or even no votes) can hurt rather than help a candidate violates what arguably is a fundamental democratic ethic. It is worth noting that for the original preferences in Example 2, b is the Condorcet candidate, but a is elected under STV. Hence, STV also does not ensure the election of Condorcet candidates.

5.2. The Borda count

Under the Borda count, points are assigned to candidates such that the lowest-ranked candidate of each voter receives 0 points, the next-lowest 1 point, and so on up to the highest-ranked candidate, who receives $m-1$ votes if there are m candidates. Points for each candidate are summed across all voters, and the candidate with the most points is the winner. To the best of my knowledge, the Borda count and similar scoring methods [Young (1975)] are not used to elect candidates in any public elections but are widely used in private organizations.

Like STV, the Borda count may not elect the Condorcet candidate [Colman and Pountney (1978)], as illustrated by the case of three voters with preference order *abc* and two voters with preference order *bca*. Under the Borda count, *a* receives 6 points, *b* 7 points, and *c* 2 points, making *b* the Borda winner; yet *a* is the Condorcet candidate.

On the other hand, the Borda count would elect the Condorcet candidate (*b*) in Example 2 in the preceding section. This is because *b* occupies the highest position *on the average* in the rankings of the four sets of voters. Specifically, *b* ranks second in the preference order of 18 voters, third in the order of 3 voters, giving *b* an average ranking of 2.14, which is higher (i.e., closer to 1) than *a*'s average ranking of 2.19 as well as the rankings of *c* and *d*. Having the highest average ranking is indicative of being broadly acceptable to voters, unlike Condorcet candidate *a* in the preceding paragraph, who is the last choice of two of the five voters.

The appeal of the Borda count in finding broadly acceptable candidates is similar to that of approval voting, except that the Borda count is almost certainly more subject to manipulation than approval voting. Consider again the example of the three voters with preference order *abc* and two voters with preference order *bca*. Recognizing the vulnerability of their first choice, *a*, under Borda, the three *abc* voters might insincerely rank the candidates *acb*, maximizing the difference between their first choice (*a*) and *a*'s closest competitor (*b*). This would make *a* the winner under Borda.

In general, voters can gain under the Borda count by ranking the most serious rival of their favorite last in order to lower his or her point total [Ludwin (1978)]. This strategy is relatively easy to effectuate, unlike a manipulative strategy under STV that requires estimating who is likely to be eliminated, and in what order, so as to be able to exploit STV's dependence on sequential eliminations and transfers.

The vulnerability of the Borda count to manipulation led Borda to exclaim, "My scheme is intended only for honest men!" [Black (1958, p. 238)]. Recently Nurmi (1984, 1987) showed that the Borda count, like STV, is vulnerable to preference truncation, giving voters an incentive not to rank all candidates in certain situations. However, Chamberlin and Courant (1983) contend that the Borda count would give effective voice to different interests in a representative assembly, if not always ensure their proportional representation.

Another type of paradox that afflicts the Borda count and related point-assignment systems involves manipulability by changing the agenda. For example, the introduction of a new candidate, who cannot win – and, consequently, would appear irrelevant – can completely reverse the point-total order of the old candidates, even though there are no changes in the voter's rankings of these candidates [Fishburn (1974)]. Thus, in the example below, the last-place finisher among three candidates (a, with 6 votes) jumps to first place (with 13 votes) when "irrelevant" candidate x is introduced, illustrating the extreme sensitivity of the Borda count to apparently irrelevant alternatives:

3 voters: cba	$c = 8$	3 voters: $cbax$	$a = 13$
2 voters: acb	$b = 7$	2 voters: $axcb$	$b = 12$
2 voters: bac	$a = 6$	2 voters: $baxc$	$c = 11$
			$x = 6$

Clearly, it would be in the interest of a's supporters to encourage x to enter simply to reverse the order of finish.

5.3. *Cumulative voting*

Cumulative voting is a voting system in which each voter is given a fixed number of votes to distribute among one or more candidates. This allows voters to express their *intensities* of preference rather than simply to rank candidates, as under STV and the Borda count. It is a system of proportional representation in which minorities can ensure their approximate proportional representation by concentrating their votes on a subset of candidates commensurate with their size in the electorate.

To illustrate this system and the calculation of optimal strategies under it, assume that there is a single minority position among the electorate favored by one-third of the voters and a majority position favored by the remaining two-thirds. Assume further that the electorate comprises 300 voters, and a simple majority is required to elect a six-member governing body.

If each voter has six votes to cast for as many as six candidates, and if each of the 100 voters in the minority casts three votes each for only two candidates, these voters can ensure the election of these two candidates no matter what the 200 voters in the majority do. For each of these two minority candidates will get a total of 300 (100×3) votes, whereas the two-thirds majority, with a total of 1200 (200×6) votes to allocate, can at best match this sum for its four candidates ($1200/4 = 300$).

If the two-thirds majority instructs its supporters to distribute their votes equally among five candidates ($1200/5 = 240$), it will not match the vote totals of the two minority candidates (300) but can still ensure the election of four (of its five) candidates – and possibly get its fifth candidate elected if the minority puts up three candidates and instructs its supporters to distribute their votes equally among the three (giving each $600/3 = 200$ votes).

Against these strategies of either the majority (support five candidates) or the minority (support two candidates), it is easy to show that neither side can improve its position. To elect five (instead of four) candidates with 301 votes each, the majority would need 1505 instead of 1200 votes, holding constant the 600 votes of the minority; similarly, for the minority to elect three (instead of two) candidates with 241 votes each, it would need 723 instead of 600 votes, holding constant the 1200 votes of the majority.

It is evident that the optimal strategy for the leaders of both the majority and minority is to instruct their members to allocate their votes as equally as possible among a certain number of candidates. The number of candidates they should support for the elected body should be proportionally about equal to the number of their supporters in the electorate (if known).

Any deviation from this strategy – for example, by putting up a full slate of candidates and not instructing supporters to vote for only some on this slate – offers the other side an opportunity to capture more than its proportional "share" of the seats. Patently, good planning and disciplined supporters are required to be effective under this system.

A systematic analysis of optimal strategies under cumulative voting is given in Brams (1975). These strategies are compared with strategies actually adopted by the Democratic and Republican parties in elections for the Illinois General Assembly, where cumulative voting was used until 1982. This system has been used in elections for some corporate boards of directors. In 1987 cumulative voting was adopted by two cities in the United States (Alamorgordo, NM, and Peoria, IL), and other small cities more recently, to satisfy court requirements of minority representation in municipal elections.

5.4. Additional-member systems

In most parliamentary democracies, it is not candidates who run for office but political parties that put up lists of candidates. Under party-list voting, voters vote for the parties, which receive representation in a parliament proportional to the total numbers of votes that they receive. Usually there is a threshold, such as 5 percent, which a party must exceed in order to gain any seats in the parliament.

This is a rather straightforward means of ensuring the proportional representation (PR) of parties that surpass the threshold. More interesting are systems in which some legislators are elected from districts, but new members may be added to the legislature to ensure, insofar as possible, that the parties underrepresented on the basis of their national-vote proportions gain additional seats.

Denmark and Sweden, for example, use total votes, summed over each party's district candidates, as the basis for allocating additional seats. In elections to

Germany's Bundestag and Iceland's Parliament, voters vote twice, once for district representatives and once for a party. Half of the Bundestag is chosen from party lists, on the basis of the national party vote, with adjustments made to the district results so as to ensure the approximate proportional representation of parties. In 1993, Italy adopted a similar system for its chamber of deputies.

In Puerto Rico, no fixed number of seats is added unless the largest party in one house of its bicameral legislature wins more than two-thirds of the seats in district elections. When this happens, that house can be increased by as much as one-third to ameliorate underrepresentation of minority parties.

To offer some insight into an important strategic feature of additional-member systems, assume, as in Puerto Rico, that additional members can be added to a legislature to adjust for underrepresentation, but this number is variable. More specifically, assume a voting system, called *adjusted district voting*, or ADV [Brams and Fishburn (1984b,c)], that is characterized by the following four assumptions:

(1) There is a jurisdiction divided into equal-size districts, each of which elects a single representative to a legislature.

(2) There are two main factions in the jurisdiction, one majority and one minority, whose size can be determined. For example, if the factions are represented by political parties, their respective sizes can be determined by the votes that each party's candidates, summed across all districts, receive in the jurisdiction.

(3) The legislature consists of all representatives who win in the districts *plus* the largest vote-getters among the losers – necessary to achieve PR – if PR is not realized in the district elections. Typically, this adjustment would involve adding minority-faction candidates, who lose in the district races, to the legislature, so that it mirrors the majority–minority breakdown in the electorate as closely as possible.

(4) The size of the legislature would be *variable*, with a lower bound equal to the number of districts (if no adjustment is necessary to achieve PR), and an upper bound equal to twice the number of districts (if a nearly 50-percent minority wins no district seats).

As an example of ADV, suppose that there are eight districts in a jurisdiction. If there is an 80-percent majority and a 20-percent minority, the majority is likely to win all the seats unless there is an extreme concentration of the minority in one or two districts.

Suppose the minority wins no seats. Then its two biggest vote-getters could be given two "extra" seats to provide it with representation of 20 percent in a body of ten members, exactly its proportion in the electorate.

Now suppose the minority wins one seat, which would provide it with representation of $\frac{1}{8} \approx 13$ percent. It it were given an extra seat, its representation would rise to $\frac{2}{9} \approx 22$ percent, which would be closer to its 20-percent proportion in the electorate. However, assume that the addition of extra seats can never make the minority's proportion in the legislature exceed its proportion in the electorate.

Paradoxically, the minority would benefit by winning no seats and then being granted two extra seats to bring its proportion up to exactly 20 percent. To prevent a minority from benefitting by *losing* in district elections, assume the following *no-benefit constraint*: the allocation of extra seats to the minority can never give it a greater proportion in the legislature than it would obtain had it won more district elections.

How would this constraint work in the example? If the minority won no seats in the district elections, then the addition of two extra seats would give it $\frac{2}{10}$ representation in the legislature, exactly its proportion in the electorate. But I just showed that if the minority had won exactly one seat, it would *not* be entitled to an extra seat – and $\frac{2}{9}$ representation in the legislature – because this proportion exceeds its 20 percent proportion in the electorate. Hence, its representation would remain at $\frac{1}{8}$ if it won in exactly one district.

Because $\frac{2}{10} > \frac{1}{8}$, the no-benefit constraint prevents the minority from gaining two extra seats if it wins no district seats initially. Instead, it would be entitled in this case to only one extra seat, because the next-highest ratio below $\frac{2}{10}$ is $\frac{1}{9}$; since $\frac{1}{9} < \frac{1}{8}$, the no-benefit constraint is satisfied.

But $\frac{1}{9} \approx 11$ percent is only about half of the minority's 20-percent proportion in the electorate. In fact, one can prove in the general case that the constraint may prevent a minority from receiving up to about half of the extra seats it would be entitled to – on the basis of its national vote total – were the constraint *not* operative and it could therefore get up to this proportion (e.g., 2 out of 10 seats in the example) in the legislature [Brams and Fishburn (1984b)].

The constraint may be interpreted as a kind of "strategyproofness" feature of ADV: It makes it unprofitable for a minority party deliberately to lose in a district election in order to do better after the adjustment that gives it extra seats. (Note that this notion of strategyproofness is different from that given for nonranking systems in Section 3.)

But strategyproofness, in precluding any possible advantage that might accrue to the minority from throwing a district election, has a price. As the example demonstrates, it may severely restrict the ability of ADV to satisfy PR, giving rise to the following dilemma: Under ADV, one cannot guarantee a close correspondence between a party's proportion in the electorate and its representation in the legislature if one insists on the no-benefit constraint; dropping it allows one to approximate PR, but this may give an incentive to the minority party to lose in certain district contests in order to do better after the adjustment.

It is worth pointing out that the "second chance" for minority candidates afforded by ADV would encourage them to run in the first place, because even if most or all of them are defeated in the district races, their biggest vote-getters would still get a chance at the (possibly) extra seats in the second stage. But these extra seats might be cut by up to a factor of two from the minority's proportion in the electorate should one want to motivate district races with the no-benefit constraint. Indeed, Spafford (1980, p. 393), anticipating this dilemma, recommended that only an (unspecified) fraction of the seats that the minority is entitled to be allotted to it in the adjustment phase to give it "some incentive to take the single-member contests seriously,..., though that of course would be giving up strict PR."

6. Conclusions

There is no perfect voting precedure [Niemi and Riker (1976), Fishburn (1984), Nurmi (1986)]. But some procedures are clearly superior to others in satisfying certain criteria. Among nonranking voting systems, approval voting distinguishes itself as more sincere, strategyproof, and likely to elect Condorcet candidates (if they exist) than other systems, including plurality voting and plurality voting with a runoff. Its recent adoption by a number of professional societies – including the Institute of Management Science [Fishburn and Little (1988)], the Mathematical Association of America [Brams (1988)], the American Statistical Association [Brams and Fishburn (1988)], and the Institute of Electrical and Electronics Engineers [Brams and Nagel (1991)], with a combined membership of over 400 000 [Brams and Fishburn (1992a)] – suggests that its simplicity as well as its desirable properties augur a bright future for its more widespread use, including its possible adoption in public elections [Brams (1993a)]. Indeed, bills have been introduced in several U.S. state legislatures for its enactment in such elections, and its consideration has been urged in such countries as Finland [Anckar (1984)] and New Zealand [Nagel (1987)].[4]

Although ranking systems, notably STV, have been used in public elections to ensure proportional representation of different parties in legislatures,[5] the vulnerability of STV to preference truncation and the no-show paradox illustrates its manipulability, and its nonmonotonicity casts doubt upon its democratic character. In particular, it seems bizarre that voters may prevent a candidate from winning by raising him or her in their rankings.

While the Borda count is monotonic, it is easier to manipulate than STV. It is difficult to calculate the impact of insincere voting on sequential eliminations and transfers under STV, but under the Borda count the strategy of ranking the most serious opponent of one's favorite candidate last is a transparent way of diminishing a rival's chances. Also, the introduction of a new and seemingly irrelevant candidate,

---

[4] Other empirical analyses of approval voting, and the likely effects it and other systems would have in actual elections, can be found in Brams and Fishburn (1981, 1982, 1983), De Maio and Muzzio (1981), De Maio et al. (1983, 1986), Fenster (1983), Chamberlin et al. (1984), Cox (1984), Nagel (1984), Niemi and Bartels (1984), Fishburn (1986), Lines (1986), Felsenthal et al. (1986), Koc (1988), Merrill (1988), and Brams and Merrill (1994). Contrasting views on approval voting's probable empirical effects can be found in the exchange between Arrington and Brenner (1984) and Brams and Fishburn (1984a).

[5] If voters are permitted to vote for more than one party in party-list systems, seats might be proportionally allocated on the basis of approval votes, although this usage raises certain problems (Brams (1987), Chamberlin (1985)]; for other PR schemes, see Rapoport et al. (1988a,b), Brams and Fishburn (1992b, 1993b), and Brams (1990).

as I illustrated, can have a topsy-turvy effect, moving a last-place candidate into first place and vice-versa.

Additional-member systems, and specifically ADV that results in a variable-size legislature, provide a mechanism for approximating PR in a legislature without the nonmonotoncity of STV or the manipulability of the Borda count. Cumulative voting also offers a means for parties to ensure their proportional representation, but it requires considerable organizational efforts on the part of parties. In the face of uncertainty about their level of support in the electorate, party leaders may well make nonoptimal choices about how many candidates their supporters should concentrate their votes on, which weakens the argument that cumulative voting can in practice guarantee PR. But the no-benefit constraint on allocation of additional seats to underrepresented parties under ADV – in order to rob them of the incentive to throw district races – also vitiates fully satisfying PR, underscoring the difficulties of satisfying a number of desiderata.

An understanding of these difficulties, and possible trade-offs that must be made, facilitates the selection of procedures to meet certain needs. Over the last forty years the explosion of results in social choice theory, and the burgeoning decision-theoretic and game-theoretic analyses of different voting systems, not only enhance one's theoretical understanding of the foundations of social choice but also contribute to the better design of practical voting procedures that satisfy criteria one deems important.

References

Anckar, D. (1984) 'Presidential Elections in Finland: A Plea For Approval Voting', *Electoral Studies*, **3**: 125–138.
Arrington, T.S. and S. Brenner (1984) 'Another Look at Approval Voting', *Polity*, **17**: 118–134.
Arrow, K.J. (1963) *Social choice and individual values*, 2nd edn. (1st edn., 1951). New Haven: Yale University Press.
Black, D. (1958) *The theory of committees and elections*. Cambridge: Cambridge University Press.
Boehm, G.A.W. (1976) 'One Fervent Voting against Wintergreen', mimeographed.
Bolger, E.M. (1985) 'Monotonicity and Other Paradoxes in Some Proportional Representation Schemes', *SIAM Journal on Algebraic and Discrete Methods*, **6**: 283–291.
Borda, Jean-Charles de (1781) 'Mémoire sur les élections au scrutin', *Histoire de l'Académie Royale des Sciences*, Paris.
Bordley, R.F. (1983) 'A Pragmatic Method for Evaluating Election Schemes through Simulation', *American Political Science Review*, **77**: 123–141.
Bordley, R.F. (1985) 'Systems Simulation: Comparing Different Decision Rules', *Behavioral Science*, **30**: 230–239.
Brams, S.J. (1975) *Game theory and politics*. New York: Free Press.
Brams, S.J. (1977) 'When Is It Advantageous to Cast a Negative Vote?', in: R. Henn and O. Moeschlin, eds., *Mathematical Economics and Game Theory: Essays in Honor of Oskar Morgenstern*, Lecture Notes in Economics and Mathematical Systems, Vol. 141. Berlin: Springer-Verlag.
Brams, S.J. (1978) *The presidential election game*. New Haven: Yale University Press.
Brams, S.J. (1982a) 'The AMS Nomination Procedure Is Vulnerable to Truncation of Preferences', *Notices of the American Mathematical Society*, **29**: 136–138.
Brams, S.J. (1982b) 'Strategic Information and Voting Behavior', *Society*, **19**: 4–11.

Brams, S.J. (1983) 'Comparison Voting', in: S.J. Brams, W.F. Lucas, and P.D. Straffin, Jr., eds., *Modules in applied mathematics: political and related models*, Vol. 2. New York: Springer-Verlag.
Brams, S.J. (1985) *Rational Politics: Decisions, Games, and Strategy*. Washington, DC: CQ Press.
Brams, S.J. (1987) 'Approval Voting and Proportional Representation', mimeographed.
Brams, S.J. (1988) 'MAA Elections Produce Decisive Winners', *Focus: The Newsletter of the Mathematical Association of America*, **8**: 1–2.
Brams, S.J. (1993a) 'Approval Voting and the Good Society', *PEGS Newsletter*, **3**: 10, 14.
Brams, S.J. (1990) 'Constrained Approval Voting: A Voting System to Elect a Governing Board', *Interfaces*, **20**: 65–80.
Brams, S.J. and M.D. Davis (1973) 'Models of Resource Allocation in Presiental Campaigning: Implications for Democratic Representation', *Annals of the New York Academy of Sciences* (L. Papayanopoulos, ed., *Democratic Representation and Apportionment: Quantitative Methods, Measures, and Criteria*), **219**: 105–123.
Brams, S.J. and M.D. Davis (1974) 'The 3/2's Rule in Presidential Campaigning', *American Political Science Review*, **68**: 113–134.
Brams, S.J. and M.D. Davis (1982) 'Optimal Resource Allocation in Presidential Primaries', *Mathematical Social Sciences*, **3**: 373–388.
Brams, S.J., D.S. Felsenthal and Z. Maoz (1986) 'New Chairman Paradoxes', in: A. Diekmann and P. Mitter, eds., *Paradoxical effects of social behavior: essays in honor of Anatol Rapoport*. Heidelberg: Physica-Verlag.
Brams, S.J., D.S. Felsenthal and Z. Maoz (1987) 'Chairman paradoxes under approval voting', in: G. Eberlein and H. Berghel, eds., *Theory and decision: essays in honor of Werner Leinfellner*. Dordrecht: D. Reidel.
Brams, S.J. and P.C. Fishburn (1978) 'Approval Voting', *American Political Science Review*, **72**: 831–847.
Brams, S.J. and P.C. Fishburn (1981) 'Reconstructing Voting Processes: The 1976 House Majority Leader Election under Present and Alternative Rules', *Political Methodology*, **7**: 95–108.
Brams, S.J. and P.C. Fishburn (1982) 'Deducing Preferences and Choices in the 1980 Election', *Electoral Studies*, **1**: 39–62.
Brams, S.J. and P.C. Fishburn (1983) *Approval voting*. Cambridge: Birkhäuser Boston.
Brams, S.J. and P.C. Fishburn (1984a) 'A Careful Look at Another Look at Approval Voting', *Polity*, **17**: 135–143.
Brams, S.J. and P.C. Fishburn (1984b) 'A note on variable-size legislatures to achieve proportional representation', in: A. Lijphart and B. Grofman, eds., *Choosing an electoral system: issues and alternatives*. New York: Praeger.
Brams, S.J. and P.C. Fishburn (1984c) 'Proportional Representation in Variable-Size Electorates', *Social Choice and Welfare*, **1**: 397–410.
Brams, S.J. and P.C. Fishburn (1984d) 'Some logical defects of the single transferable vote', in: A. Lijphart and B. Grofman, eds., *Choosing an electoral system: issues and alternatives*. New York: Praeger.
Brams, S.J. and P.C. Fishburn (1985) 'Comment on 'The Problem of Strategic Voting under Approval Voting'', *American Political Science Review*, **79**: 816–818.
Brams, S.J. and P.C. Fishburn (1988) 'Does Approval Voting Elect the Lowest Common Denominator'?, *PS: Political Science & Politics*, **21**: 277–284.
Brams, S.J., P.C. Fishburn and S. Merrill, III (1988) 'The Responsiveness of Approval Voting: Comments on Saari and Van Newenhizen "and" Rejoinder to Saari and Van Newenhizen', *Public Choice*, **59**: 121–131 and 149.
Brams, S.J. and J.H. Nagel (1991) 'Approval Voting in Practice', *Public Choice* **71**: 1–17
Brams S.J. and P.C. Fishburn (1992a) 'Approval Voting in Scientific and Engineering Societies', *Group Decision and Negotiation*, **1**: 35–50.
Brams, S.J. and P.C. Fishburn (1992b) 'Coalition Voting', in: P.E. Johnson, ed., *Mathematical and Computer Modelling (Formal Models of Politics II: Mathematical and Computer Modelling)*, **16**: 15–26.
Brams, S.J. and P.C. Fishburn (1993b) 'Yes-No Voting', *Social Choice and Welfare*, **10**: 35–50.
Brams, S.J. and S. Merrill, III (1994) 'Would Ross Perot Have Won the 1994 Presidential Election under Approval Voting?', *PS: Political Science and Politics* **27**: 39–44.
Brams, S.J. and F.C. Zagare (1977) 'Deception in Simple Voting Games', *Social Science Research*, **6**: 257–272.
Brams, S.J. and F.C. Zagare (1981) 'Double Deception: Two against One in Three-Person Games', *Theory and Decision*, **13**: 81–90.

Carter, C. (1990) 'Admissible and sincere strategies under approval voting', *Public Choice*, **64**: 1–20.
Chamberlin, J.R. (1985) 'Committee selection methods and representative deliberations', mimeographed, University of Michigan.
Chamberlin, John R. (1986) 'Discovering Manipulated Social Choices: The Coincidence of Cycles and Manipulated Outcomes', *Public Choice*, **51**: 295–313.
Chamberlin, J.R. and M.D. Cohen (1978) 'Toward Applicable Social Choice Theory: A Comparison of Social Choice Functions under Spatial Model Assumptions', *American Political Science Review*, **72**: 1341–1356.
Chamberlin, J.R. and P.N. Courant (1983) 'Representative Deliberations and Representative Decisions: Proportional Representation and the Borda Rule', *American Political Science Review*, **77**: 718–733.
Chamberlin, J.R., J.L. Cohen, and C.H. Coombs (1984) 'Social Choice Observed: Five Presidential Elections of the American Psychological Association', *Journal of Politics*, **46**: 479–502.
Chamberlin, J.R. and F. Featherston (1986) 'Selecting a Voting System', *Journal of Politics*, **48**: 347–369.
Colman, A. and I. Pountney (1978) 'Borda's Voting Paradox: Theoretical Likelihood and Electoral Occurrences', *Behavioral Science*, **23**: 15–20.
Condorcet, Marquis de (1785) *Essai sur l'application de l'analyse à la probabilité des decisions rendues à la pluralité des voix*. Paris.
Cox, Gary W. (1984) 'Strategic Electoral Choice in Multi-Member Districts: Approval Voting in Practice', *American Journal of Political Science*, **28**: 722–734.
Cox, Gary W. (1985) 'Electoral equilibrium under Approval Voting', *American Journal of Political Science*, **29**: 112–118.
Cox, Gary W. (1987) 'Electoral Equilibrium under Alternative Voting Institutions', *American Journal of Political Science*, **31**: 82–108.
De Maio, G. and D. Muzzio (1981) 'The 1980 Election and Approval Voting', *Presidental Studies Quarterly*, **9**: 341–363.
De Maio, G., D. Muzzio, and G. Sharrard (1983) 'Approval Voting: The Empirical Evidence', *American Politics Quarterly*, **11**: 365–374.
De Maio, G., D. Muzzio and G. Sharrard (1986) 'Mapping Candidate Systems via Approval Voting', *Western Political Quarterly*, **39**: 663–674.
Doron, G. and R. Kronick (1977) 'Single Transferable Vote: An Example of a Perverse Social Choice Function, *American Journal of Political Science*, **21**: 303–311.
Dummett, M. (1984) *Voting procedures*. Oxford: Oxford University Press.
Enelow, J.M. and M.J. Hinich (1984) *The spatial theory of election competition: an introduction*. Cambridge: Cambridge University Press.
Farquharson, R. (1969) *Theory of voting*. New Haven: Yale University Press.
Felsenthal, D.S. (1989) 'On Combining Approval with Disapproval Voting', *Behavioral Science*, **34**: 53–60.
Felsenthal, D.S. and Z. Maoz (1988) 'A Comparative Analysis of Sincere and Sophisticated Voting under the Plurality and Approval Procedures', *Behavioral Science*, **33**: 116–130.
Felsenthal, D.S., Z. Maoz and A. Rapport (1986) 'Comparing Voting System in Genuine Elections: Approval-Plurality versus Selection-Plurality', *Social Behavior*, **1**: 41–53.
Fenster, M.J. (1983) 'Approval Voting: Do Moderates Gain?', *Political Methodology*, **9**: 355–376.
Fishburn, P.C. (1974) 'Paradoxes of Voting', *American Political Science Review*, **68**: 537–546.
Fishburn, P.C. (1978a) 'Axioms for Approval Voting: Direct Proof', *Journal of Economic Theory*, **19**: 180–185.
Fishburn, P.C. (1978b) 'A Strategic Analysis of Nonranked Voting Systems', SIAM *Journal on Applied Mathematics*, **35**: 488–495.
Fishburn, P.C. (1981) 'An Analysis of Simple Voting Systems for Electing Committees', SIAM *Journal on Applied Mathematics*, **41**: 499–502.
Fishburn, P.C. (1982) 'Monotonicity Paradoxes in the Theory of Elections', *Discrete Applied Mathematics*, **4**: 119–134.
Fishburn, P.C. (1983) 'Dimensions of Election Procedures: Analyses and Comparisons', *Theory and Decision*, **15**: 371–397.
Fishburn, P.C. (1984) 'Discrete Mathematics in Voting and Group Choice', SIAM *Journal on Algebraic and Discrete Methods*, **5**: 263–275.
Fishburn, P.C. (1986) 'Empirical Comparisons of Voting Procedures', *Behavioral Science*, **31**: 82–88.

Fishburn, P.C. and S.J. Brams (1981a) 'Approval Voting, Condorcet's Principle, and Runoff Elections', *Public Choice*, **36**: 89–114.

Fishburn, P.C. and S.J. Brams (1981b) 'Efficacy, Power, and Equity under Approval Voting', *Public Choice*, **37**: 425–434.

Fishburn, P.C. and S.J. Brams (1981c) 'Expected Utility and Approval Voting', *Behavioral Science*, **26**: 136–142.

Fishburn, P.C. and S.J. Brams (1983) 'Paradoxes of Preferential Voting', *Mathematics Magazine*, **56**: 207–214.

Fishburn, P.C. and Brams, S.J. (1984) 'Manipulability of Voting by Sincere Truncation of Preferences', *Public Choice*, **44**: 397–410.

Fishburn, P.C. and W.V. Gehrlein (1976) 'An Analysis of Simple Two-stage Voting Systems', *Behavioral Science*, **21**: 1–12.

Fishburn, P.C. and W.V. Gehrlein (1977) 'An Analysis of Voting Procedures with Nonranked Voting', *Behavioral Science*, **22**: 178–185.

Fishburn, P.C. and W.V. Gehrlein (1982) 'Majority Efficiencies for Simple Voting Procedures: Summary and Interpretation', *Theory and Decision*, **14**: 141–153.

Fishburn, P.C. and J.C.D. Little (1988) 'An Experiment in Approval Voting', *Management Science*, **34**: 555–568.

Gehrlein, W.V. (1985) 'The Condorcet Criterion and Committee Selection', *Mathematical Social Sciences*, **10**: 199–209.

Gibbard, A. (1973) 'Manipulation of Voting Schemes: A General Result', *Econometrica*, **41**: 587–601.

Hoffman, D.T. (1982) 'A Model for Strategic Voting', *SIAM Journal on Applied Mathematics*, **42**: 751–761.

Hoffman, D.T. (1983) 'Relative Efficiency of Voting Systems', SIAM *Journal on Applied Mathematics*, **43**:1213–1219.

Holzman, Ron (1988/1989) 'To Vote or Not to Vote: What Is the Quota?', *Discrete Applied Mathematics*, **22**: 133–141.

Inada, K. (1964) 'A Note on the Simple Majority Decision Rule', *Econometrica*, **32**: 525–531.

Kelly, J.S. (1987) *Social choice theory: an introduction*. New York: Springer-Verlag.

Koc, E.W. (1988) 'An Experimental Examination of Approval Voting under Alternative Ballot Conditions', *Polity*, **20**: 688–704.

Lake, M. (1979) 'A new campaign resource allocation model', in: S.J. Brams, A. Schotter and G. Schwödiauer, eds., *Applied game theory: proceedings of a conference at the Institute for Advanced Studies*, Vienna, June 13–16, 1978. Würzburg: Physica-Verlag.

Lines, Marji (1986) 'Approval Voting and Strategy Analysis: A Venetian Example', *Theory and Decision*, **20**: 155–172.

Ludwin, W.G. (1978) 'Strategic Voting and the Borda Method', *Public Choice*, **33**: 85–90.

Merrill, S. (1979) 'Approval Voting: A "Best Buy" Method for Multicandidate Elections?', *Mathematics Magazine*, **52**: 98–102.

Merrill, S. (1981) 'Strategic Decisions under One-Stage Multicandidate Voting Systems', *Public Choice*, **36**: 115–134.

Merrill, S. (1982) 'Strategic voting in multicandidate elections under uncertainty and under risk', in: M. Holler, ed., *Power, voting, and voting power*. Würzburg: Physica-Verlag.

Merrill, S., III (1984) 'A Comparison of Efficiency of Multicandidate Electoral Systems', *American Journal of Political Science*, **28**: 23–48.

Merrill, S., III (1985) 'A Statistical Model for Condorcet Efficiency Using Simulation under Spatial Model Assumptions', *Public Choice*, **47**: 389–403.

Merrill, S., III (1988) *Making multicandidate elections more democratic*. Princeton: Princeton University Press.

Merrill, S., III, and J. Nagel (1987) 'The Effect of Approval Balloting on Strategic Voting under Alternative Decision Rules', *American Political Science Review*, **81**: 509–524.

Miller, N.R. (1983) 'Pluralism and Social Choice', *American Political Science Review*, **77**: 734–747.

Moulin, Hervé (1988) 'Condorcet's Principle Implies the No Show Paradox', *Journal of Economic Theory*, **45**: 53–64.

Nagel, J. (1984) 'A Debut for Approval Voting', *PS*, **17**: 62–65.

Nagel, J. (1987) 'The Approval Ballot as a Possible Component of Electoral Reform in New Zealand', *Political Science*, **39**: 70–79.

Niemi, R.G. (1984) 'The Problem of Strategic Voting under Approval Voting', *American Political Science Review* **78**: 952–958.
Niemi, R.G. and L.M. Bartels (1984) 'The Responsiveness of Approval Voting to Political Circumstances', *PS*, **17**: 571–577.
Niemi, R.G. and W.H. Riker (1976) 'The Choice of Voting Systems', *Scientific American*, **234**: 21–27.
Nurmi, H. (1983) 'Voting Procedures: A Summary Analysis', *Bristish Journal of Political Science*, **13**: 181–208.
Nurmi, H. (1984) 'On the Strategic Properties of Some Modern Methods of Group Decision Making', *Behavioral Science*, **29**: 248–257.
Nurmi, H. (1986) 'Mathematical Models of Elections and Their Relevance for Institutional Design', *Electoral Studies*, **5**: 167–182.
Nurmi, H. (1987) *Comparing voting systems*. Dordrecht: D. Reidel.
Nurmi, H. (1988) 'Discrepancies in the Outcomes Resulting from Different Voting Schemes', *Theory and Decision*, **25**: 193–208.
Nurmi, H. and Y. Uusi-Heikkilä (1985) 'Computer Simulations of Approval and Plurality Voting: The Frequency of Weak Pareto Violations and Condorcet Loser Choices in Impartial Cultures', *European Journal of Political Economy*, **2/1**: 47–59.
Peleg, Bezalel (1984) *Game-theoretical analysis of voting in committees*. Cambridge: Cambridge University Press.
Rapoport, A. and D.S. Felsenthal (1990) 'Efficacy in small electorates under plurality and approval voting', *Public Choice*, **64**: 57–71.
Rapoport, A., D.S. Felsenthal and Z. Maoz (1988a) 'Mircocosms and Macrocosms: Seat Allocation in Proportional Representation Systems', *Theory and Decision*, **24**: 11–33.
Rapoport, A., D.S. Felsenthal and Z. Maoz (1988b) 'Proportional Representation in Israel's General Federation of Labor: An Empirical Evaluation of a New Scheme', *Public Choice*, **59**: 151–165.
Ray, D. (1986) 'On the Practical Possibility of a 'No Show Paradox' under the Single Transferable Vote', *Mathematical Social Sciences*, **11**: 183–189.
Riker, W.H. (1982) *Liberalism against populism: a confrontation between the theory of democracy and the theory of social choice*. San Francisco: Freeman.
Riker, W.H. (1989) 'Why negative campaigning is rational: the rhetoric of the ratification campaign of 1787–1788', mimeographed, University of Rochester.
Saari, D.G. (1994) *Geometry of Voting*. New York: Springer-Verlag.
Saari, D.G. and J. Van Newenhizen (1988) 'The Problem of Indeterminacy in Approval, Multiple, and Truncated Voting Systems "and" Is Approval Voting an 'Unmitigated Evil'?: A Response to Brams, Fishburn, and Merrill', *Public Choice*, **59**: 101–120 and 133–147.
Satterthwaite, M.A. (1975) 'Strategy-Proofness and Arrow's Conditions: Existence and Correspondence Theorems for Voting Procedures and Social Welfare Functions', *Journal of Economic Theory*, **10**: 187–218.
Schwartz, Thomas (1986) *The Logic of Collective Choice*. New York: Columbia University Press.
Smith, J.H. (1973) 'Aggregation of Preferences with Variable Electorate', *Econometrica*, **47**: 1113–1127.
Snyder, J.M. (1989) 'Election Goals and the Allocation of Political Resources', *Econometrica*, **57**: 637–660.
Spafford, D. (1980) 'Book Review', *Candian Journal of Political Science*, **11**: 392–393.
Stavely, E.S. (1972) *Greek and roman voting and elections*. Ithaca: Cornell University Press.
Straffin, P.D., Jr. (1980) *Topics in the theory of voting*. Cambridge: Birkhäuser Boston.
The Torah: *The Five Books of Moses* (1987) 2nd edn. Philadelphia: Jewish Publication Society.
Weber, R.J. (1977) 'Comparison of Voting Systems', Cowles Foundation Discussion Paper No. *498A*, Yale University.
Wright, S.G. and W.H. Riker (1989) 'Plurality and Runoff Systems and Numbers of Candidates', *Public Choice*, **60**: 155–175.
Young, H.P. (1975) 'Social Choice Scoring Functions', SIAM *Journal on Applied Mathematics*, **28**: 824–838.

Chapter 31

SOCIAL CHOICE

HERVÉ MOULIN*

Duke University

Contents

0.	Introduction	1092
1.	Aggregation of preferences	1094
	1.1. Voting rules and social welfare preorderings	1094
	1.2. Independence of irrelevant alternatives and Arrow's theorem	1095
	1.3. Social welfare preorderings on the single-peaked domain	1096
	1.4. Acyclic social welfare	1098
2.	Strategyproof voting	1101
	2.1. Voting rules and game forms	1101
	2.2. Restricted domain of preferences: equivalence of strategyproofness and of IIA	1103
	2.3. Other restricted domains	1104
3.	Sophisticated voting	1105
	3.1. An example	1105
	3.2. Dominance-solvable game forms	1106
	3.3. Voting by successive vetos	1109
4.	Voting by self-enforcing agreements	1110
	4.1. Majority versus minority	1110
	4.2. Strong equilibrium in voting by successive veto	1113
	4.3. Strong monotonicity and implementation	1116
	4.4. Effectivity functions	1118
5.	Probabilistic voting	1121
References		1123

*The comments of an anonymous referee are gratefully acknowledged.

Handbook of Game Theory, Volume 2, Edited by R.J. Aumann and S. Hart
© *Elsevier Science B.V., 1994. All rights reserved*

0. Introduction

Social choice theory started in Arrow's seminal book [Arrow (1963)] as a tool to explore fundamental issues of welfare economics. After nearly forty years of research its influence is considerable in several branches of public economics, like the theory of inequality indexes, taxation, the models of public choice, of planning, and the general question of incentive compatibility. Sen (1986) provides an excellent survey of the non-strategic aspects of the theory.

The core of the theory deals with abstract public decision making, best described as a voting situation: a single outcome (candidate) must be picked out of a given issue (the issue is the set of eligible candidates), based on the sole data of the voters' ordinal preferences. No compensation, monetary or else, is available to balance out the unavoidable disagreements about the elected outcome. The ordinal voting model is not the only context to which the social choice methodology applies (see below at the end of this introduction); yet it encompasses the bulk of the literature.

In the late 1960s interest grew for the strategical properties of preference aggregation in general and of voting rules in particular. Although the pioneer work by Farqharson (1969) is not phrased in the game theoretical language, it is pervaded with strategic arguments. Soon thereafter the question was posed whether or not existed a strategyproof voting rule (namely one where casting a sincere ballot is a dominant strategy for every voter at every preference profile – or equivalently, the sincere ballot is a best response no matter how other players vote) that would not amount to give full decision power to a dictator-voter. It was negatively answered by Gibbard (1973) and Satterthwaite (1975) (in two almost simultaneous papers) who formulated and proved a beautifully simple result (Theorem 3 below). This theorem had a profound influence on the then emerging literature on incentive compatibility and mechanism design. It also had some feedback into the seminal social choice results because its proof is technically equivalent to Arrow's.

The next step in the strategic analysis of social choice was provided by the implementation concept. After choosing an equilibrium concept, we say that a social choice function (voting rule) – or social choice correspondence – is implementable if there exists a game form of which the equilibrium outcome (or outcomes) is (are) the one(s) selected by our social choice function.

The implementation concept is the organizing theme of this Chapter. We provide, without proofs, the most significant (at least in this author's view) results of the strategic theory of social choice in the ordinal framework of voting. Systematical presentations of this theory can be found in at least three books, namely Moulin (1983), Peleg (1984) and Abdou and Keiding (1991).

In order to offer a self-contained exposition with rigorous statements of deep results, we had to limit the scope of the chapter in at least two ways. First of all, only the strategic aspects of social choice are discussed. Of course, Arrow's theorem cannot be ignored because of its connection with simple games (Lemma 1) and with strategyproofness (Theorem 4). But many important results on familiar voting rules such as the Borda rule, general scoring methods as well as approval voting have been ignored (on these Chapter 30 is a useful starting point). Moreover, the whole cardinal approach (including collective utility functions) is ignored, despite the obvious connection with axiomatic bargaining (Chapter 35).

Implementation theory has a dual origin. One source of inspiration is the strategic theory of voting, as discussed here. The other source stems from the work of Hurwicz (1973) and could be called the economic tradition. It is ignored in the current chapter (and forms the heart of the chapter on 'Implementation' in Volume III of this Handbook), which brings the second key limitation of its scope. Generally speaking, the implementation theory developed by economists is potentially broader in its applications (so as to include the exchange and distribution of private goods alongside voting and the provision of public goods) but the strategic mechanisms it uncovers are less easy to interpret, and less applicable than, say, voting by veto.

Nevertheless many of the recent and most general results of implementation theory are directly inspired by Maskin's theorem on implementation in Nash equilibrium (Lemma 7 and Theorem 8) and are of some interest for voting rules as well: the most significant contributions are quoted in Sections 4.3 and 4.4. An excellent survey of that literature is Moore (1992).

In Section 1 we discuss aggregation of individual preferences into a social welfare ordering. The axiom of independence of irrelevant alternatives (IIA) is presented and Arrow's impossibility result in stated (Theorem 1). We propose two ways of overcoming the impossibility. One is to restrict the preferences to be single-peaked, in which case majority voting yields a satisfactory aggregation. The other is to require that the social welfare relation be only acyclic: Nakamura's theorem (Theorem 2) then sets narrow limits to the decisiveness of society's preferences.

Section 2 is devoted to the Gibbard-Satterthwaite impossibility result (Theorem 3) and its relation to Arrow's. A result of Blair and Muller (1983) shows the connection between the IIA axiom and strategyproofness on any restricted domain (Theorem 4).

Sophisticated voting is the subject of Section 3: this is how Farqharson (1969) called the successive elimination of dominated voting strategies. Game forms where this process leads to a unique outcome [called dominance-solvable by Moulin (1979)] are numerous: a rich family of examples consists of taking the subgame perfect equilibrium of game trees with perfect information (Theorem 5). The particularly important example of voting by veto is discussed in some detail.

In Section 4 we analyze voting in strong and Nash equilibrium. We start with the voting by veto example where the strong equilibrium outcomes are consistent with the sophisticated ones (Theorem 6). Then we discuss implementation by strong and Nash equilibrium of arbitrary choice correspondences. Implementability in Nash equilibrium is almost characterized by the strong monotonicity property

(Theorem 8). Implementability in strong equilibrium, on the other hand, relies on the concept of effectivity functions (Theorems 9 and 10).

Our last Section 5 presents the main result about probabilistic voting, that is to say voting rules of which the outcome is a lottery over the deterministic candidates. We give the characterization due to Gibbard (1978) and Barbera (1979) of those probabilistic voting rules that are together strategyproof, anonymous and neutral (Theorem 11).

1. Aggregation of preferences

1.1. Voting rules and social welfare preorderings

The formal study of voting rules initiated by Borda (1781) and Condorcet (1985) precedes social choice theory by almost two centuries (Arrow's seminal work appeared first in 1950). Yet the filiation is clear.

Both objects, a voting rule and a social welfare preordering, are mappings acting on the same data, namely preference profiles. A preference profile tells us in detail the opinion of each (voting) agent about all feasible outcomes (candidates) – a formal definition follows –. The voting rule, then, aggregates the information contained in the preference profile into a single choice (it elects a single candidate) whereas a social welfare preordering computes from the profile a full-fledged preordering of all feasible outcomes, and interprets this preordering as the compromise opinion upon which society will act.

The social welfare preordering is a more general object than a voting rule: it induces the voting methods electing a top outcome of society's preordering (this holds true if the set of outcomes – the issue – is finite, or if it is compact and if the social preordering is continuous). Hence the discussion of strategic properties of voting is relevant for preference aggregation as well. See in particular Theorem 4 in Section 2, linking explicitly strategyproofness of voting methods to the IIA property for social welfare preorderings.

Despite the close formal relation between the two objects, they have been generally studied by two fairly disjoint groups of scholars. Voting methods is a central topics of mathematical politics, as Chapter 30 makes clear. Aggregation of preferences, on the other hand, primarily concerns welfare economists, for whom the set of outcomes represents any collection of purely public decisions to which some kind of social welfare index must be attached. On this see the classical book by Sen (1970).

Binary choice (namely the choice between exactly two candidates) is the only case where voting rules and social welfare preorderings are effectively the same thing. Ordinary majority voting is, then, the unambiguously best method (the winner is socially preferred to the loser). This common sense intuition admits a simple and elegant formulation due to May (1952). Consider the following three axioms. *Anonymity* says that each voter's opinion should be equally important (one man, one vote). *Neutrality* says that no candidate should be a priori discriminated against. *Monotonicity* says that more supporters for a candidate should not jeopardize its election. These three properties will be defined precisely in Section 1.4 below. May's theorem says that majority voting is the only method satisfying these three axioms. It is easy to prove [see e.g. Moulin (1988) Theorem 11.1].

The difficulty of preference aggregation is that the majority rule has no easy generalization when three outcomes or more are at stake.

Let A denote the set of outcomes (candidates) not necessarily finite. Each agent (voter) has preferences on A represented by an ordering u_i of A: it is a complete, transitive and asymmetric relation on A. We denote by $L(A)$ the set of orderings of A. Indifferences in individual opinions are excluded throughout, but only for the sake of simplicity of exposition. All results stated below have some kind of generalization when individual preferences are *preorderings* on A (complete and transitive).

A *preference profile* is $u = (u_i)_{i \in N}$ where N is the given (finite) society (set of concerned agents) and u_i is agent i's preference. Thus $L(A)^N$ is the set of preference profiles. A *social welfare preordering* is a mapping R from preference profiles into a (social) preordering $R(u)$ of A. Thus $aR(u)b$ means that the social welfare is not lower at a than at b. Note that $R(u)$ conveys only ordinal information: we do not give any cardinal meaning to the notion of social welfare. Moreover social indifferences are allowed.

1.2. Independence of irrelevant alternatives and Arrow's theorem

The independence of irrelevant alternatives axiom (in short IIA) is an informational requirement: comparing the social welfare of any two outcomes a, b should depend upon individual agents' opinions about a versus b, and about those opinions only. Thus the relative ranking of a versus any other (irrelevant) outcome c, or that of b versus c in anybody's preference, should not influence the social comparison of a versus b. This decentralization property (w.r.t. candidates) is very appealing in the voting context: when the choice is between a and b we can compare these two on their own merits.

Notation. If u is a profile ($u \in L(A)^N$) and a, b are two distinct elements of A:

$$N(u, a, b) = \{i \in N / u_i(a) > u_i(b)\}.$$

Notice that agents' preferences are written as utility functions for notational convenience only (yet if the set A is infinite, a preference in $L(A)$ may not be formally representable by a cardinal utility function!).

We say that the social welfare ordering R satisfies *Arrow's IIA axiom* if we have for all a, b in A and u, v in $L(A)^N$:

$$\{N(u, a, b) = N(v, a, b)\} \Rightarrow \{aR(u)b \Leftrightarrow aR(v)b\}.$$

We show in Section 2 (Theorem 4) that violation of IIA is equivalent to the possibility of strategically manipulating the choice mechanism deduced from the social welfare preordering R. Thus the IIA axiom can be viewed as a strategyproofness requirement as well.

Theorem 1 [Arrow (1963)]. *Suppose A contains three outcomes or more. Say that R is a social welfare preordering (in short SWO) satisfying the unanimity axiom:*
 Unanimity: for all u in $L(A)^N$ and all a, b in A, $\{N(u,a,b) = N\} \Rightarrow \{aP(u)b\}$ where $P(u)$ is the strict preference relation associated with $R(u)$.
Then R satisfies the IIA axiom if and only if it is dictatorial:
 Dictatorship: there is an agent i such that $R(u) = u_i$ for all u in $L(A)^N$.

The theorem generalizes to SWOs that do not even satisfy unanimity. Wilson (1972) proves that if R is onto $L(A)$ and satisfies IIA, it must be either dictatorial or anti-dictatorial (there is an agent i such that $R(u)$ is precisely the opposite of u_i).

All ordinary SWOs satisfy unanimity and are not dictatorial. Hence they violate the IIA axiom. This is easily checked for the familiar *Borda ranking*: each voter ranks the p candidates, thus giving zero point to the one ranked last, one point to the one ranked next to last, and so on, up to $(p-1)$ points to his top candidate. The social preordering simply compares the total scores of each candidate.

On the other hand, consider the majority relation: a is preferred to b (respectively indifferent to b) if more voters (respectively the same number of voters) prefer a to b than b to a [$|N(u,a,b)| > |N(u,b,a)|$, respectively $|N(u,a,b)| = |N(u,b,a)|$. This relation $M(u)$ satisfies the IIA axiom, and unanimity. The trouble, of course, is that it is not a preordering of A: it may violate transitivity (an occurrence known as the Condorcet paradox).

1.3. Social welfare preorderings on the single-peaked domain

The simplest way to overcome Arrow's impossibility result is to restrict the domain of feasible individual preferences from $L(A)$ to a subset D. In the voting context this raises a normative issue (should not voters have the inalienable right to profess *any* opinion about the candidates?) but from a positive point of view, the subdomain D may accurately represent some special feature shared by everyone's preference. Think of the assumptions on preferences for consumption goods that economists routinely make.

The most successful restrictive assumption on preferences is single-peakedness. Say that A is finite and linearly ordered:

$$A = \{a_1, a_2, \ldots, a_p\}. \tag{1}$$

The preference u is *single-peaked* (relative to the above ordering of A) if there exists an outcome a^* (the peak of u) such that u is increasing before a^* and decreasing after a^*:

$$a^* = a_{k^*}, \quad 1 \leqslant k < l \leqslant k^* \Rightarrow u(a_k) < u(a_l),$$
$$k^* \leqslant k < l \leqslant p \Rightarrow u(a_l) < u(a_k).$$

Denote by $SP(A)$ the subset of $L(A)$ made up of single-peaked preferences [relative to the given ordering (1) of A].

Black (1958) was the first to observe that the majority relations is transitive over the single-peaked domain. Suppose that N contains an *odd* number of agents and consider a single-peaked preference profile u is $SP(A)^N$. Then the majority relation $M(u)$ defined as follows:

$$aM(u)b \Leftrightarrow |N(u, a, b)| \geqslant |N(u, b, a)|$$

is an ordering of $A[M(u)$ is in $L(A)]$. Its peak a is the unique Condorcet winner of profile u (a defeats each and every other candidate in binary majority duels).

When N contains an *even* number of candidates, the majority relation is only quasi-transitive (its strict component is transitive) and only the existence of a *weak* Condorcet winner is guaranteed [outcome a is a weak Condorcet winner if for all outcomes b we have $|N(u, a, b)| \geqslant |N(u, b, a)|$]. In general, for all N (whether $|N|$ is odd or even), there exist many social welfare preorderings satisfying the IIA axiom and unanimity, yet not dictatorial. Indeed there are several SWOs on the domain $SP(A)$ satisfying IIA, unanimity and anonymity (no discrimination among voters). An example is the leftist majority relation $M_-(u)$, namely the variant of the majority relation that break ties in favor of the smallest element according to the ordering (1): $a_k M_-(u) a_l \Leftrightarrow \{|N(u, a_k, a_l)| > |N(u, a_l, a_k)|$ or $|N(u, a_k, a_l)| = |N(u, a_l, a_k)|$ and $k < l)\}$. Note that, characteristically, anonymity is preserved here at the cost of neutrality (the SWO *is* somewhat biased in favor of lowest ranked outcomes).

A social welfare preordering such as M_- satisfies a number of desirable properties that we list below:

(i) It aggregates all profiles of single-peaked preferences into a single-peaked social ordering [it maps $SP(A)^N$ into $SP(A)$].

(ii) It satisfies IIA and is monotonic: see in Section 1.4 the definition of monotonicity as property (2). (By Theorem 4 in Section 2, the voting rule induced by M_- over any subset of A is thus strategyproof).

(iii) It is unanimous and anonymous.

All social welfare orderings on SPA satisfying properties (i)–(iii) have been characterized [Moulin (1988) Theorem 11.6, see also Moulin (1984)]. They are all derived from the ordinary majority relation by (a) choosing $(n-1)$ fixed ballots in $SP(A)$, called phantom voters, and (b) computing the ordinary majority relation of the $(2n-1)$ preference profile combining the n true voters and the $(n-1)$ phantom voters.

Remark 1. Single-peakedness is not the only domain restriction allowing for the existence of a non-dictatorial SWO satisfying IIA and unanimity. But it is the only such restriction known so far where the structure of those SWOs is well understood. Abstract results characterizing the "successful" domain restrictions are proposed by Kalai and Muller (1977), Blair and Muller (1983) among others. See the review of this literature in Muller and Satterthwaite (1985).

1.4. Acyclic social welfare

When we need to aggregate preferences over the full domain $L(A)^N$ (or even its extension to preorderings), the only way out of Arrow's impossibility result is to impose a weaker rationality requirement upon the social preference relation $R(u)$. The weakest such requirement is acyclicity.

Denote by $P(u)$ the strict component of $R(u)$. We want to interpret the statement $aP(u)b$ as "outcome a is strictly preferred to b." This excludes b from being chosen. Therefore, if the strict relation $P(u)$ had a cycle, such as $aP(u)b$, $bP(u)c$, $cP(u)a$, we would not be able to choose any outcome in $\{a,b,c\}$.

We say that a complete binary relation R on A is *acyclic* if its strict component P (aPb iff $\{aRb$ and no $bRa\}$) has no cycles. There is no sequence $a_1, \ldots, a_K$ in A such that

$$a_k P a_{k-1} \text{ all } k = 2, \ldots, K; \; a_1 P a_K.$$

When a relation is acyclic, out of every finite subset B of A we can pick at least one maximal element a (namely aRb for all b in B).

An *acyclic social welfare* is a mapping R from $L(A)^N$ into the set of acyclic complete relations on A. Many acyclic social welfares satisfy the IIA axiom as well as Unanimity and yet are not dictatorial. A complete characterization of those is fairly complicated. Partial results were obtained by Blau and Deb (1977) and Blair and Pollak (1982). Nakamura (1978) was the first to associate a simple game with an acyclic social welfare: this insight is an essential ingredient of the strategic approach to social choice. It turns our that under the two additional properties of monotonicity and neutrality, the acyclic social welfares satisfying IIA and unanimity are completely described by means of a simple game.

Monotonicity. For all a in A, all u,v in $L(A)^N$, *if* u and v coincide on $A\setminus\{a\}$ and for all b and all agent i, $\{u_i(a) > u_i(b) \Rightarrow v_i(a) > v_i(b)\}$ *then* we have

$$\text{for all } b \quad \{aR(u)b \Rightarrow aR(v)b\}. \tag{2}$$

In words if the relative position of a improves upon from u to v (while the relative position of other outcomes is unaffected), then the relative position of a improves as well from $R(u)$ to $R(v)$.

Ch. 31: Social Choice

Neutrality. For any permutation σ of A, and any profile u in $L(A)^N$, denote by u_i^σ the permuted preference: $u_i^\sigma(a) = u_i(\sigma(a))$. Then we have

$$aR(u^\sigma)b \Leftrightarrow \sigma(a)R(u)\sigma(b).$$

Neutrality simply rules out a priori discrimination among outcomes on the basis of their name. Note that Sen (1970) calls neutrality the combination of the above property with Arrow's IIA.

A *simple game* (see also Chapter 36) is a collection W of coalitions (subsets) of N such that

$$T \in W, T \subset T' \Rightarrow T' \in W \quad \text{for all } T, T' \subseteq N.$$

A simple game is *proper* if it satisfies

$$T \in W \Rightarrow N \setminus T \notin W \quad \text{for all } T \subseteq N.$$

In a proper simple game, we may interpret W as the set of *winning* coalitions, namely coalitions having full control of the collective decision.

Lemma 1. *Let R be an acyclic social welfare on A, N satisfying (a) the IIA axiom, (b) neutrality, (c) monotonicity. Then there exists a proper simple game W that exactly represents R in the following sense:*

$$\text{for all } a, b \in A, u \in L(A)^N: aR(u)b \Leftrightarrow N(u, b, a) \notin W. \tag{3}$$

To prove Lemma 1, one simply takes property (3) as the definition of the simple game W. Note that the strict preference $P(u)$ is easily written in terms of W:

$$aP(u)b \Leftrightarrow N(u, a, b) \in W.$$

Thus society prefers a to b iff the set of agents preferring a to b is a winning coalition. In this case we say that a *dominates* b at profile u via the game W.

The interesting question is whether the converse of Lemma 1 holds as well. Given a proper simple game, property (3) surely defines a complete binary relation $R(u)$ that is monotonic and neutral. Moreover it satisfies IIA since $aR(u)b$ depends upon $N(u, a, b)$ only. But it is not always the case that $R(u)$ is acyclic.

Nakamura's theorem gives a very simple characterization of those proper simple games for which the binary relation (3) is acyclic. This latter property is equivalent to the non-emptiness of the core of the dominance relation of W at every profile

$$a \in \text{Core}(W, u) \Leftrightarrow \{\text{there is no outcome } b \text{ dominating } a \text{ at profile } u\}.$$

Theorem 2 [Nakamura (1975)]. *Given is the society N. To every proper simple game W, attach its Nakamura number $v(W)$*

$$v(W) = +\infty \quad \text{if} \bigcap_{T \in W} T \neq \emptyset,$$

$$v(W) = \inf\{|W_0|/W_0 \subset W, \bigcap_{T \in W_0} T = \emptyset\}.$$

Given a finite set A of outcomes, the relation R(u) defined by (3) is acyclic for all preference profiles u if and only if A contains fewer outcomes than v(W):

$$\{R(u) \text{ is acyclic for all } u\} \Leftrightarrow \{|A| < v(W)\}.$$

Suppose that at some profile u the relation $P(u)$ [the strict component of $R(u)$] has a cycle $a_1, \ldots, a_K$:

$$a_{k+1} P(u) a_k \text{ all } k = 1, \ldots, K-1; \; a_1 P(u) a_K.$$

Then each coalition $T_k = N(u, a_{k+1}, a_k)$, $k = 1, \ldots, K-1$, as well as $T_K = N(u, a_1, a_K)$, is in W and their intersection must be empty, otherwise some agent would have cyclic preferences. Therefore the Nakamura number $v(W)$ is at most K. As A contains at least K distinct outcomes, this proves "half" of the Theorem.

For many simple games the Nakamura number is hard to compute, but in some cases it is not. Consider for example an *anonymous* simple game W, namely a quota game W_q:

$$T \in W \Leftrightarrow |T| \geq |q|,$$

where the quota q is such that $|N|/2 < q \leq |N|$ (in order that W be proper). It is easy to compute the Nakamura number of W_q:

$$v(W_q) = \left\lceil \frac{n}{n-q} \right\rceil,$$

where $n = |N|$ and $[z]$ is the smallest integer greater than or equal to z.

Thus the quota-game W_q yields an acyclic relation on A if and only if

$$|A| < \left\lceil \frac{n}{n-q} \right\rceil \Leftrightarrow q > n\left(\frac{|A|-1}{|A|}\right).$$

This says that with four outcomes ($|A| = 4$) we need a quota strictly greater than $\frac{3}{4}$ (76 voters out of 100) otherwise the dominance relation may cycle. With ten outcomes we need 91 voters out of 100. The corresponding relation $R(u)$ will declare many pairs of outcomes indifferent, hence its core is likely to be a big subset of A. Thus the acyclic social welfare ends up largely undecisive.

Note that the characterization of acyclic social welfare satisfying IIA, anonymity (instead of neutrality) and monotonicity can be given as well [see Moulin (1985)] and yields a similar undecisiveness result.

2. Strategyproof voting

2.1. Voting rules and game forms

Given the society N and the set A of outcomes a *voting rule* is a (single-valued) mapping from the set $L(A)^N$ of preference profiles into A. A game form endows each agent with an arbitrary message space (also called strategy space) within which this agent freely picks one.

Definition 1. Given A, the set of outcomes, and N, the set of agents, a game form g is an $(N+1)$-tuple $g = (X_i, i \in N; \pi)$, where:

(a) X_i is the strategy set (or message space) of agent i, and
(b) π is a (single-valued) mapping from $X_N = \prod_{i \in N} X_i$ into A.

The mapping π is the decision rule: if for all i agent i chooses strategy x_i, the overall strategy N-tuple is denoted $x = (x_i)_{i \in N}$ and the decision rule forces the outcome $\pi(x) \in A$.

Thus a game form is a more general object than a voting rule inasmuch as the message space does not necessarily coincide with the set of individual preferences. A game form with message space $L(A)$ for every agent is simply a voting rule. Lemma 2 below reduces the search of strategyproof game forms to that of strategyproof voting rules. The full generality of the game form concept, which allows in particular the message space to be bigger than $L(A)$, is needed when we explore a more complex strategic behavior than the dominating strategy equilibrium. See Sections 3 and 4 below.

Given a game form $g = (X_i, i \in N, \pi)$, we associate to every preferences profile $u = (u_i)_{i \in N}$ the normal form game $g(u) = (X_i, u_i \circ \pi, i \in N)$, where agent i's strategy is x_i ans his utility level is $u_i(\pi(x))$. This game reflects the interdependence of the individual agents' opinions (utility) and their strategic abilities (agent i is free to send any message within X_i).

In this section we focus on those game forms such that for all profiles u, every agent has a straightforward non-cooperative strategy whether or not he knows of the other agents' preference orderings. This is captured by the notion of dominant strategy.

Definition 2. Given A and N and a game form g we say that g is *strategyproof* if for every agent i there exists a mapping from $L(A)$ into A_i denoted $u_i \to x_i(u_i)$ such that the following holds true:

$$\forall u_i \in L(A), \forall x_{-i} \in X_{-i}, \forall y_i \in X_i : u_i(\pi(y_i, x_{-i})) \leq u_i(\pi(x_i(u_i), x_{-i})). \quad (4)$$

Given a voting rule S, we say that S is *strategyproof* if for every agent i we have

$$\forall u_i \in L(A), \forall u_{-i} \in L(A)^{N \setminus \{i\}}, \forall v_i \in L(A) : u_i(S(v_i, u_{-i})) \leq u_i(S(u_i, u_{-i})). \quad (5)$$

Property (4) says that if agent i's utility is u_i then strategy $x_i(u_i)$ is a best response to every possible strategic behavior x_{-1} of the other agents, in short $x_i(u_i)$ is a dominant strategy. Notice that $x_i(u_i)$ is a decentralized behavior by agent i, who can simply ignore the utility of the other agents. Even if agent i happens to know of other agents' preferences and/or strategic choices, this information is useless as long as he plays non-cooperatively (it may happen, however, that the dominant strategy equilibrium is cooperatively unstable, i.e. Pareto-inferior – the Prisoners Dilemma is an example).

In the context of voting rules, strategyproofness means that telling the truth is a dominant strategy at every profile and for all agents: hence we expect direct revelation of their preferences by the non-cooperative agents involved in a strategy-proof direct game form.

The following result, a particular case of the "revelation principle", states that as far as strategyproofness is concerned, it is enough to look at voting rules.

Lemma 2. *Leg g be a strategyproof game form (g.f.). For all $i \in N$ and all $u_i \in L(A)$ let us denote by $D_i(u_i) \subset X_i$ the set of agent i's dominant strategies*

$$\{x_i \in D_i(u_i)\} \Leftrightarrow \{\forall x_{-i} \in X_{-i} \forall y_i \in X_i u_i(\pi(y_i, x_{-i})) \leq u_i(\pi(x_i, x_{-i}))\}.$$

Then for all profiles $u \in L(A)^N$ the set $\pi(D_i(u_i), i \in N)$ is a singleton and defines a strategyproof voting rule.

Let S be a voting rule (a single-valued mapping from $L(A)^N$ into A). We say that S satisfies *citizen sovereignty* if $S(L(A)^N) = A$, i.e. if no outcome is a priori excluded by S. It is a very mild property, even weaker than the *unanimity* condition [namely if a is on top of u_i for all $i \in N$, then $S(u) = a$].

Theorem 3 [Gibbard (1973), Satterthwaite (1975)]. *Let society N be finite and A (not necessarily finite) contain at least three distinct outcomes. Then a voting rule S satisfying citizen sovereignty is strategyproof if and only if it is dictatorial: there is an agent i (the dictator) whose top outcome is always elected*

$$\text{for all } u \in L(A)^N \quad S(u) = \text{top}(u_i).$$

The proof of Theorem 3 is technically equivalent to that of Arrow's theorem (Theorem 1). The most natural (although not the quickest) method of proving Theorem 3 is to check that strategyproofness implies strong monotonicity (Definition 8), and then to use Theorem 7 (in Section 4.3 below). See Moulin (1983) or Peleg (1984). A short proof is given by Barbera (1983). A more intuitive proof under additional topological assumptions is in Barbera and Peleg (1990). The latter proof technique has many applications to economic environments.

For binary choices ($|A| = 2$) strategyproofness is equivalent to monotonicity and is therefore satisfied by many non-dictatorial voting rules. As soon as three outcomes are on stage we cannot find a reasonable strategyproof voting rule. This suggests two lines of investigation. In the first one we insist on the strategyproofness requirement, that is we want our mechanisms to allow "pure" decentralization of the decision process, requiring that an agent's optimal strategy is unambiguous, even if he ignores the other preferences, and is still unaffected if this agent happens to know the preferences of some among his fellow agents. Then, by the Gibbard–Satterthwaite theorem we must *restrict the domain* of feasible preferences. This line will be explored in the next section.

Another way of escaping the Gibbard–Satterthwaite result is to weaken the equilibrium concept: not demanding that a dominant strategy equilibrium exists for all profiles still leaves room for patterns of behavior that are, to a large extent, non-cooperatively decentralized (see Section 3). Or we can take a cooperative view of the decision-making mechanism so that different equilibrium concepts are in order (Section 4).

2.2. Restricted domain of preferences: equivalence of strategyproofness and of IIA

A restricted domain is a subset D of $L(A)$. We assume that every agent's preferences vary in D. For any social welfare ordering defined on D, Arrow's IIA axiom is equivalent to a set of strategyproofness conditions. This simple result underlines the implicitly strategic character of Arrow's axiom. It demonstrates that Arrow's impossibility theorem (Theorem 1) and Gibbard–Satterthwaite's theorem (Theorem 3) are related.

By definition a voting rule on D is a mapping from D^N into A. The definition of strategyproofness extends straightforwardly (no misreport of preferences in D is profitable).

Theorem 4 [Blair and Muller (1983)] *Given are the set of A of outcomes and N of agents. Let $D \subset L(A)$ be a restricted domain, and $u \to R(u)$ be a social welfare ordering on D [for each $u \in D^N$, $R(u)$ is an element of $L(A)$]. To R associate the following family of voting rules (one for each issue $B \subseteq A$):*

$$\text{fix } B \subseteq A: \text{ for all } u \in D^N \quad S(u, B) = \max_B R(u),$$

then the two following statements are equivalent:

(i) *R is monotonic and satisfies Arrow's IIA,*
(ii) *for all fixed issue $B \subset A$, the voting rule $S(\cdot, B)$ is strategyproof on D^N.*

Monotonicity has been defined in Section 1 for acyclic social welfare. The same definition applies here. To give the intuition for the proof suppose that $a = S(u, B)$, $b = S(u', B)$ where u and u' are two preference profiles differing only in agent 1's preference. If a and b are different, we must have (by definition of S) $aP(u)b$ and $bP(u')a$.

But the difference between $N(u, a, b)$ and $N(u', a, b)$ can only come from agent 1. So if $u_1(a) < u_1(b)$ we must have $u'_1(a) > u'_1(b)$ and thus a contradiction of the monotonicity of R. We conclude that $u_1(a) < u_1(b)$ is impossible, so that agent 1 cannot manipulate S afterall.

When property (i) in Theorem 4 is satisfied, the voting rules $S(\cdot, B)$ are robust even against manipulations by coalitions. Given B, a profile u in D^N and a coalition $T \subseteq N$, there is no joint misreport v_T in D^T such that:

$$S(u, B) = a, S(v_T, u_{N \setminus T}) = b \text{ and } u_i(b) > u_i(a) \text{ for all } i \in T.$$

Notice that coalitional strategyproofness implies Pareto optimality if $S(\cdot, B)$ satisfies citizen-sovereignty (take $T = N$).

A prominent application of Theorem 4 is to the domain $SP(A)$ of single-peaked preferences (w.r.t. a given ordering of A) see Section 1.3. When N has an odd number of agents we know that the majority relation M is a social welfare ordering satisfying IIA and monotonicity. Thus the corresponding Condorcet winner voting rule (namely the top outcome of the majority relation; it is simply the median peak) is (coalitionally) strategyproof on the single-peaked domain).

All strategyproof voting rules on $SP(A)^N$ (whether $|N|$ is odd or even) have been characterized [Moulin (1980a) Barbera and Jackson (1992)]. They are all derived by (a) choosing $(n-1)$ fixed outcomes (corresponding to the top outcomes of $(n-1)$ phantom voters) and (b) computing the Condorcet winner of the $(2n-1)$-profile with the n true voters and $(n-1)$ phantoms voters whose peaks are fixed in (a). Notice the similarity with the generalized majority relations described at the end of Section 1.3.

A systematical, albeit abstract, characterization of all domains described in Theorem 4 is given by Kalaï and Muller (1977).

2.3. Other restricted domains

If the domain of individual preferences allows for Arrowian social welfare ordering, we can find strategyproof voting rules as well (Theorem 4). However, there are some restricted domains where non-dictatorial strategyproof voting rules exist whereas Arrow's impossibility result prevails (so that these voting rules do not pick the top candidate of an Arrowian social welfare ordering). An example of such a domain is the set of single-peaked preferences on a tree [Demange (1982), see also Moulin (1988, Sect. 10.2)]. Another example comes from the problem of electing a committee, namely a subset of a given pool of candidates. Assuming that utility for committees is separably additive in candidates, Barbera et al. (1991) show that voting by quotas is strategyproof.

Several definitions of the single-peaked domain to a multi-dimensional space of outcomes have been proposed: Border and Jordan (1983) work with separable quadratic preferences, whereas Barbera et al. (1993a) use a lattice structure on the set of outcomes, without requiring separability. Their concept of single-peakedness will no doubt lead to many applications. For both definitions, the one-dimensional characterization of strategyproof voting rules extends to the multi-dimensional case.

Further variants and generalizations of the Gibbard–Satterthwaite theorem to decomposable sets of outcomes (the set A is a Cartesian product) are in Laffond (1980), Chichilnisky and Heal (1981), Peters et al. (1991, 1992), Le Breton and Sen (1992), Le Breton and Weymark (1993). See also Barbera et al. (1993b) for a case where the set A is only a subset of a Cartesian product.

Chichilnisky has proposed a topological model of social choice, where the domain of preferences is a connected topological space and the aggregation (or voting) rule is continuous. Impossibility results analogous to Arrow's and Gibbard–Satterthwaite's obtain if and only if the space of preferences is not contractible [Chichilnisky (1983) is a good survey]. Many of the preference domains familiar in economic and political theory (e.g., single-peaked preferences, or monotonic preferences over the positive orthant) are contractible, so it follows that these domains do admit continuous social choice rules.

3. Sophisticated voting

3.1. An example

In this section we analyze the non-cooperative properties of some game forms. The key concept is that of dominance-solvable game forms, that corresponds to the subgame perfect equilibrium of (prefect information) game trees. It turns out that quite a few interesting game forms have such an equilibrium at all profiles. They usually have a sequential definition (so that their message space is much bigger than a simple report of one's preferences).

In his pioneer work on strategic voting, Farqharson (1969) introduced the notion of sophisticated voting, later generalized to arbitrary game forms by the concept of dominance solvability [Moulin (1979)]. We give the intuition for the subsequent definitions in an example due to Farqharson.

Example. *The Chair's Paradox.* There are three voters $\{1, 2, 3\}$ and three candidates $\{a, b, c\}$. The rule is plurality voting. Agent 1, the chair, breaks ties: if every voter proposes a different name, voter 1's candidate passes. The preferences display a Condorcet cycle:

1	2	3
a	c	b
b	a	c
c	b	a

Players know each and every opinion. For instance I know your preferences; you know that I know it; I know that you know that I know it, and so on.... To prevent the formation of coalitions (which are indeed tempting since any two voters control the outcome if they vote together), our voters are isolated in three different rooms where they must cast "independent" votes x_i, $i = 1, 2, 3$. What candidate is likely to win the election?

Consider the decision problem of voter 1 (the chair). No matter what the other two decided, voting a is optimal for him: indeed if they cast the same vote ($x_2 = x_3$), player 1's vote will not affect the outcome so $x_1 = a$ is as good as anything else. On the other hand if 2 and 3 disagree ($x_2 \neq x_3$) then player 1's vote is the final outcome, so he had better vote for a.

This argument does not apply to player 2: sure enough he would never vote $x_2 = b$ (his worst candidate) yet sometimes (e.g., when $x_1 = b$, $x_3 = c$) he should support c and some other times (e.g., when $x_1 = b$, $x_3 = a$) he should support a instead. By the same token player 3 cannot a priori (i.e., even before he has any expectations about what the others are doing) decide whether to support b or c; he can only discard $x_3 = a$.

To make up his mind player 3 uses his information about other players' preferences: he knows that player 1 is going to support a, and player 2 will support c or a. Since none of them will support b, this candidate cannot pass: given this it would be foolish to support b (since the "real" choice is a majority vote between a and c) hence player 3 decides to support c.

Of course, player 2 figures out this argument (he knows the preference profile as well) hence expects 1 to support a and 3 to support c. So he supports c as well, and the outcome is c after all. The privilege of the chair turns out to his disadvantage!

3.2. Dominance-solvable game forms

Definition 3. *Undominated strategy.* Given A and N, both finite, let $g = (X_i, i \in N; \pi)$ be a game form and $u \in L(A)^N$ be a fixed profile. For any subsets $T_i \subseteq X_i$, $i \in N$, we denote by $\Delta_j(u_j; Y_N)$ the set of agent j's undominated strategies when the strategy spaces are restricted to Y_i, $i \in N$. Thus, x_j belongs to $\Delta_j(u_j; Y_N)$ if and only if

$x_j \in Y_j$ and for no $y_j \in Y_j$,
$\forall x_{-j} \in Y_{-j}: u_j(\pi(x_j, x_{-j})) \leq u_j(\pi(y_j, x_{-j}))$,
$\exists x_{-j} \in Y_{-j}: u_j(\pi(x_j, x_{-j})) < u_j(\pi(y_j, x_{-j}))$.

Definition 4. Given a game form g and a profile $u \in L(A)^N$ the *successive elimination of dominated strategies* is the following decreasing sequence:

$$X_j^t, j \in N, t = 1, 2, \ldots,$$
$$X_j^0 = X_j; X_j^{t+1} = \Delta_j(u_j; X_N^t) \subseteq X_j^t.$$

We say that g is *dominance-solvable at* u if there is an integer t such that $\pi(X_N^t)$ is a singleton, denoted $S(u)$. We say that g is *dominance-solvable* if it is so at every profile. In that case we say that g *sophisticatedly implements* the voting rule S.

The behavioral assumptions underlying the concept of sophisticated voting are complete information (every agent is aware of the whole profile) and non-cooperation. A dominance-solvable game form is a decentralization device to the extent that it selects an unambiguous Nash equilibrium, without requiring explicit coordination of the agents. This contrasts sharply with implementation in Nash equilibrium (Section 4.3) where the choice of a particular equilibrium rests on preplay communication.

To understand the link between dominance solvability and subgame perfect equilibrium, we need to restrict attention to game trees, namely extensive form games with perfect information where to each terminal node is attached a candidate (an element of A).

Indeed a large family of dominance solvable forms is generated by finite game trees. In the above tree a candidate (a, b, c, d, and e) is attached to each terminal node; to each non-terminal node is attached a game form to select one of the successor nodes (i.e., one immediately below on the tree). For instance on the tree of the figure, at node m a first game form g_m decides to go left (n) or right (n'), and so on....

Theorem 5 [Moulin (1979), Gretlein (1983), Rochet (1980)]. *For any finite tree, the above construction (attach a candidate to every terminal node and to every non-terminal node attach a game form to select one of the successor nodes) defines*

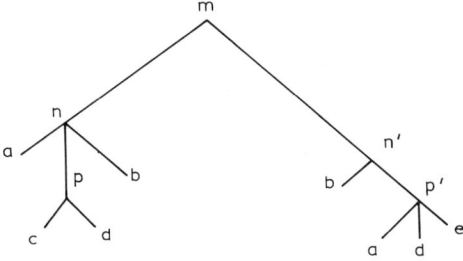

Figure 1

a dominance-solvable game form provided each game form is strategyproof. Moreover, for every profile u in $L(A)^N$ the sophisticated equilibrium outcome $S(u)$ is also the subgame perfect equilibrium outcome of the corresponding extensive game.

The intuition of the proof is simple. Fix a profile and consider a game form attached to a node followed by terminal nodes only (every tree has at least one such node: for instance p and p' on the figure). In this g.f. each voter has a dominant strategy. Delete every other strategy as dominated hence making the node p in effect a terminal one to which is attached the equilibrium candidate of the game form g_p. Repeat this process until only one candidate remains, which is the sophisticated equilibrium outcome. But this algorithm is just the backward induction algorithm of subgame perfection.

Thus dominance-solvable game forms a rich class of decision-making procedures (allowing great flexibility in the distribution of power) and therefore implement a great variety of voting rules. By weakening strategyproofness into dominance solvability we convert the Gibbard–Satterthwaite impossibility theorem into a wide possibility result.

In the game trees described in Theorem 5, we must attach a strategyproof game form to each non-terminal node. From Theorem 4 we know that there are essentially two kinds of strategyproof g.f.s. First we have dictatorial g.f.s. whereby some agent is in full control of the decision at each non-terminal node. Of course we can nominate different dictators at different nodes so that the overall distribution of power is fairly equitable. The prominent example in this first kind of game trees is voting by veto in which the next section is devoted.

The second class of strategyproof g.f.s. are binary. Suppose that our tree is binary (each non-terminal node is followed by exactly two nodes). Then the decision at any non-terminal node can be taken by majority vote (or by any such monotonic rule). This leads to the second important class of dominance-solvable game forms, namely voting by successive majority votes. References on these rules include McKelvey and Niemi (1978), Moulin (1979), Miller (1977, 1980), Shepsle and Weingast (1984).

An open problem. From Definition 4, a dominance-solvable game form *implements* a certain voting rule S. In other words if voters have complete information about their mutual preferences and behave non-cooperatively, our prediction is that at profile u the candidate actually elected will be $S(u)$. From the social planner's point of view, S is the relevant output that he can force upon the agents by imposing the corresponding game form g. *Question*: which voting rules S are implementable by some g.f., however complicated?

Little is known to answer this question. Moulin (1983) (Theorem 4, p. 106) gives a necessary condition on S for implementability: this condition implies for instance that Borda's voting rule is not implementable. Golberg and Gourvitch (1986) give an example of a voting rule implementable by a game form, but not by a game tree. Hence the sequential voting methods described in Theorem 5 are not enough to implement all that can be implemented.

3.3. Voting by successive vetos

Given the finite set A with p outcomes, and N with n voters, consider a sequence in N with length $(p-1)$ (so σ is a mapping from $\{1,\ldots,p-1\}$ into N).

The voting by veto game form g_σ works as follows: the agents take turns as σ prescribes to veto one different outcome at a time [agent $\sigma(1)$ goes first, next $\sigma(2)$ – who could be the same agent – and so on]. The elected outcome is the (unique) veto-free outcome after exhaustion of σ.

For instance suppose $n=2$, $p=7$. Then the sequence $\sigma = \{121212\}$ corresponds to alternating vetos between the two players starting with agent 1.

To the game form g_σ and a profile u in $L(A)^N$, we associate the algorithm of sincere successive veto.

a_1 is the worst outcome of $u_{\sigma(1)}$ within A,
a_k is the worst outcome of $u_{\sigma(k)}$ within $A \setminus \{a_1,\ldots,a_{k-1}\}$,
for all $k = 2,\ldots,p-1$.

The elected outcome is denoted $G_\sigma(u)$:

$$G_\sigma(u) = A \setminus \{a_1,\ldots,a_{p-1}\}.$$

Lemma 3 [Mueller (1978), Moulin (1981a)]. *The game form g_σ is dominance solvable. It implements the voting rule $G_{\bar{\sigma}}$ where $\bar{\sigma}$ is the reverse sequence from σ:*

$$\bar{\sigma}(1) = \sigma(p-1); \cdots \bar{\sigma}(k) = \sigma(p-k) \quad \text{all } k = 1,\ldots,p-1.$$

More precisely given a profile u, let $b_1,\ldots,b_{p-1}$ the algorithm of sincere successive veto in $g_{\bar{\sigma}}$. Then at the sophisticated equilibrium of g_σ at u, player $\sigma(1) = \bar{\sigma}(p-1)$ starts by vetoing b_{p-1}; next player $\sigma(2) = \bar{\sigma}(p-2)$ vetos $b_{p-2}; \cdots$; next player $\sigma(k)$ vetos $b_{p-k}; \cdots$; finally player $\sigma(p-1)$ vetos b_1.

We illustrate the result by two exmples. Our first example involves five candidates and four players endowed with one veto right apiece. The profile is:

Voter	1	2	3	4
	a	d	b	c
	c	e	c	e
	b	b	e	d
	d	c	a	b
	e	a	d	a

Given the ordering $\sigma = \{1234\}$, sincere successive veto elects c after eliminating successively e, a, d and b (since player 4 is left with $\{b,c\}$). Yet given the ordering $\bar{\sigma} = \{4321\}$, sincere successive veto elects b after eliminating successively a, d, c (since player 2 is left with $\{b,c,e\}$), and e. Thus in g_σ the outcomes of sincere voting is c while that of sophisticated voting is b. In this example, no matter how we choose σ, the two outcomes of sincere and sophisticated voting differ, and are b, c or c, b, respectively.

For a second example, consider voting by alternating veto by two agents among seven candidates ($\sigma = \{121212\}$) when the preference profile is

Voter	1	2
	a	g
	b	f
	c	d
	d	e
	e	c
	f	a
	g	b

There we have $G_\sigma(u) = A \setminus \{g, b, f, a, e, c\} = d$ and $G_{\bar{\sigma}}(u) = A \setminus \{b, g, a, f, c, e\} = d$ so both the sincere and sophisticated outcome coincide (although the corresponding strategies differ widely). Also it does not matter who starts the elimination first (since $\{212121\}$ is just $\bar{\sigma}$). The voting by veto algorithm yields a neutral but not anonymous voting rule. Under alternating veto, however, anonymity is "almost" satisfied and the equivalence of the sincere and sophisticated algorithms is a general feature [see Moulin (1981a)]. For certain cardinalities of n and p, it is possible to adapt the alternating veto algorithm so as to make its outcome exactly anonymous [see Moulin (1980b)].

4. Voting by self-enforcing agreements

By self-enforcing agreements, we mean here strong equilibrium and Nash equilibrium, where the agents use pre-play communication to select one particular equilibrium.

4.1. Majority versus minority

We consider an electorate small enough to allow cooperative behavior, through the formation of coalitions of voters who jointly decide how to cast their ballots.

The natural interpretation is thus a committee, whose members trust each other enough to cooperate [see Peleg (1978)].

We look for voting rules (or, in general, game forms) where the cooperative behavior (i) has some stable outcome(s) no matter what the profile of individual preferences, and (ii) where the cooperatively stable outcomes are easily identifiable.

The simplest cooperative stability concept is Nash equilibrium (interpreted as a self-enforcing agreement). In most familiar voting rules, we have a host of Nash equilibrium outcomes at every profile. In plurality voting, for instance, every candidate is a Nash equilibrium outcome when all voters agree to cast their vote for this very candidate. By contrast if the stability concept is that of strong equilibrium (see Definition 5 below), for most familiar voting rules there exist preference profiles at which a strong equilibrium does not exist. For instance with plurality, a strong equilibrium exists if and only if a Condorcet winner exists.

Hence two symmetrical problems: when we use Nash equilibrium we need to knock-out equilibrium outcomes (see Theorem 8 below) whereas with strong equilibrium (see Section 4.4 below) we need to make sure that an equilibrium always exist.

It turns out that no matter what equilibrium concept we use (strong equilibrium, Nash equilibrium, or core) there is no non-dictatorial voting rule that would bring exactly one equilibrium outcome at every profile. This impossibility result (Theorem 7 below) is technically equivalent to Arrow's and Gibbard–Satterthwaite's (Theorem 3).

Thus a fundamental alternative. On the one hand we can work with voting rules that select an unambiguous cooperative outcome at some profiles but fail to have any cooperatively stable outcome at some other profiles. Condorcet consistent voting methods (where a Condorcet winner, if one exists is elected) are of this type: when there is a Condorcet winner, it is the unique core outcome of the corresponding voting games; but if the Condorcet paradox occurs, we have cooperative unstability.

On the other hand we can use voting games where a cooperatively stable outcome exists always: voting by veto methods are the main example. But in this case cooperative behavior is genuinely undetermined at many profiles: when many outcomes are cooperatively stable there is much room for further negotiations and bargaining within the committee.

These two options have long been present in political theory (in a less abstract formulation). The pluralist tradition [see the article by Miller (1983) for historical references] gathers most supporters of the majority principle. It postulates that the opinion of any majority of voters should prevail. The normative tradition (with such diverse supporters as the philosopher Rousseau, the utopian political thinker Proudhon, as well as some social choice theorists) argues on the contrary that the majority principle in effect makes an homogeneous coalition controlling 50% of the electorate an actual dictator, allowing it to ignore the opinion of the antagonistic minority which may be as large as 49%.

Formally, the remedy to the tyranny of the majority is the *minority principle* requiring that *all* coalitions, however small, should be given some fraction of the decision power. We illustrate the principle by a simple example involving ten voters and eight outcomes. Take a profile where two homogeneous antagonistic coalitions are opposed:

Number of voters	6	4
	a	h
	b	g
	c	f
	d	e
	e	d
	f	c
	g	b
	h	a

While the Condorcet winner is a (the favorite of the majority), the minority principle allows the four-voter coalitions to oppose any three outcomes, hence a, b, c, while the six-voter coalitions can oppose (veto) any four outcomes, hence e, f, g, h. Thus d emerges as the fair compromise: the majority opinion is more influential but not tyrannical.

Formally, consider an election involving n voters and p candidates. The minority principle [Moulin (1981b)] says that a coalition T of size t, $1 \leq t \leq n$, should have the power to veto $v(t)$ candidates, where

$v(t)$ is the greatest integer smaller than $p \cdot t / n$.

Thus if the voters in T coordinate their strategies, they can eliminate any subset with $v(t)$ – or less – candidates.

The conflict of the two thesis (majority or minority principle) is even more striking at those profiles where there is no Condorcet winner. For the normativist, if the voting rule actually endows any majority with full decision power, then Condorcet cycles generates a cooperative unstability threatening the very roots of social consensus. By all means the social rules should avoid such "explosive" configurations, by bringing about within the decision rule itself some stable compromises upon which the collective tensions will come at a rest.

But the pluralists deny that cooperative stability is at all desirable. They argue that Condorcet cycles are the best protection for minorities, who can always live in the hope of belonging tomorrow to the ever changing ruling majority [see Miller (1983)].

The formal analysis of cooperative voting is at the heart of this important debate.

4.2. Strong equilibrium in voting by successive veto

Notation. For a given coalition T we denote by X_T the Cartesian product of the strategy sets X_i, $i \in T$, with current element x_T. We call x_T joint strategy of coalition T.

Definition 5. Given are A, the set of outcomes, with size p, N the society and a game form $g = (X_i, i \in N; \pi)$. We say that the strategy n-tuple $x \in X_N$ is a strong equilibrium of g at profile $u \in L(A)^N$, if for all coalition $T \subseteq N$ there is no joint strategy $y_T \in X_T$ such that

$$\text{for all } i \in T \quad u_i(\pi(y_T, x_{N/T})) > u_i(\pi(x)). \tag{6}$$

In words no coalition of agents can jointly change their messages so as to force the election of an outcome preferred by all deviating agents. This is the most demanding stability concept when coalitions can form: it implies in particular the Nash equilibrium property [apply (6) to single agent coalitions] and Pareto optimality [apply (6) to $T = N$]. In general normal form games, strong equilibria rarely exist. Yet in the voting context a variety of game forms guarantee existence of a strong equilibrium at every profile. The main example is voting by successive veto.

Using the notation of Section 3.3 above, we pick a game form g_σ derived from a $(p-1)$ sequence σ in N. To σ we associate the *veto power* β_i of each and every agent: $\beta_i = |\sigma^{-1}(i)|$ is the number of veto decisions that agent i must take. By construction $\sum_{i \in N} \beta_i = p - 1$.

Lemma 4 [Moulin (1982)]. *Fix a profile $u \in L(A)^N$. Consider an outcome $a \in A$ such that $A/\{a\}$ can be partitioned as*

$$A/\{a\} = B_1 \cup B_2 \cup \cdots \cup B_n,$$

where for all $i \in N$

- *the set B_i has cardinality β_i*
- *agent i prefers a to B_i: for all $b \in B$, $u_i(a) > u_i(b)$* $\biggr\}$ (7)

Then any strategy n-tuple where for all i, agent i decides to eliminate all outcomes in B_i no matter what, is a strong equilibrium of g_σ at u with associated outcome a.

This result is typical of the strong equilibrium behavior. The agents (cooperatively) share the task of eliminating $A/\{a\}$. By choosing properly the subsets B_i (namely making sure that $u_i(a) > u_i(B_i)$) they rule out profitable deviations. To see this, consider a deviation by coalition T. As long as the agents in $N \setminus T$ stick to the above strategies, for any j outside T the outcomes in B_j will be eliminated. Thus the outcome when T deviates is either a (in which case the deviation is useless) or belongs to B_i for some i in T (in which case the deviation hurts agent i).

Of course Lemma 4 is of any use only if we can characterize the outcomes in A such that $A/\{a\}$ can be partitioned as in (7).

Definition 6. Fix a distribution β of *veto power*: for each $i \in N$, β_i is a non-negative integer, and $\sum_{i \in N} \beta_i = p - 1$. Given a profile $u \in L(A)^N$, define the β *veto core* to be the set of those outcomes $a \in A$ such that

$$\text{for all } T \subset N: \sum_{i \in T} \beta_i + |\Pr(T, a, u)| \leq p - 1, \tag{8}$$

where $\Pr(T, a, u)$ contains the outcomes unanimously preferred by all agents in T to a:

$$\Pr(T, a, u) = \{b \in A / \text{for all } i \in T: u_i(b) > u_i(a)\}$$

We shall denote by $C_\beta(u)$ the β veto core at profile u.

Suppose condition (8) fails for some coalition T:

$$\sum_{i \in T} \beta_i + |\Pr(T, a, u)| \geq p \Leftrightarrow \sum_{i \in T} \beta_i \geq |A/\Pr(T, a, u)|.$$

Then, by joining forces (namely pooling their veto powers) the agents in T can together veto the whole subset $A/\Pr(T, a, u)$, hence forcing the final outcome within $\Pr(T, a, u)$, to their unanimous benefit. The next result is central to the analysis of voting by successive veto.

Lemma 5 [Moulin and Peleg (1982, Theorem 4.13)]. *Fix a distribution β of veto power, and a profile $u \in L(A)^N$. For any outcome $a \in A$, the three following statements are equivalent:*

  (i) *The set $A/\{a\}$ can be partitioned as $B_1 \cup \cdots \cup B_n$, where B_i, $i \in N$ satisfy (7).*
  (ii) *a belongs to the β veto core at u: $a \in C_\beta(u)$.*
  (iii) *For some $(p-1)$ sequence σ with distribution of veto power β, a is the outcome of the algorithm of sincere successive veto (see Section 3.3 above).*
  $a = G_\sigma(u)$ *for some σ such that:* $|\sigma^{-1}(i)| = \beta_i$, *all* $i \in N$.

An important consequence of Lemma 5 is the nonemptyness of the β veto core at every profile u. Next we have the complete characterization of the strong equilibrium outcomes of g_σ.

Theorem 6 [Moulin and Peleg (1982)]. *Given a profile u, the set of strong equilibrium outcomes of g_σ is nonempty and equals the β veto core $C_\beta(u)$.*

A nice feature of voting by successive veto is that the set of strong equilibrium outcomes forget about the particular sequence σ, retaining only the allocation of veto power. Thus, if $p = 11$, $n = 5$, and $\beta_1 = \cdots = \beta_5 = 2$, the β veto core is a perfectly anonymous (and neutral) choice correspondence, even though any associated game form g_σ must violate anonymity (since someone must start).

Another remarkable property is uncovered by Lemma 5: every algorithm of sincere successive veto ends up in the corresponding veto core, and all elements of the veto core can be reached in this way. Thus the sophisticated equilibrium outcomes also cover the veto core (by virtue of Lemma 3). This consistency of non-cooperative and cooperative behavior at every profile is quite unique among the familiar voting games (I do not claim nor venture to conjecture that it is a characteristic property of voting by veto).

We give a few examples. First take $n = 3$, $p = 4$, so that the anonymous distribution of veto power is $\beta_i = 1$, $i = 1, 2, 3$.
Consider this profile:

u_1	u_2	u_3
d	a	a
c	d	c
b	b	b
a	c	d

Here a is the Condorcet winner (the Borda winner as well), whereas the veto core is $\{b\}$ since outcomes a, c and d are eliminated respectively by agents 1, 2 and 3 in any algorithm of sincere veto. Indeed b is the only candidate that no voter regards as extremist. Note that b is the Condorcet loser (the Borda loser as well).

Next, we go back to the two agents profile considered at the very end of Section 3.3.

1	2
a	e
b	d
c	c
d	g
e	f
f	a
g	b

Taking $\beta_1 = \beta_2 = 3$ we have the veto core $\{c, d\}$. To achieve c by sincere successive veto we must take a sequence like $\sigma = \{222111\}$. To achieve d we can use the alternating sequences $\sigma = \{121212\}$ or $\{212121\}$ (as we did in Section 3.2).

More details on the strategic properties of voting by veto can be found in Moulin (1983, Ch. 6).

4.3. Strong monotonicity and implementation

Voting by successive veto are not the only game forms with at least one strong equilibrium outcome at every profile. Actually we are more interested in the strong equilibrium outcomes(s) than in the game form themselves.

Definition 7. Given are A, N and a *multi-valued* voting rule S: to each profile, S associates a nonempty choice set $S(u) \subseteq A$. We say that S is *implementable in strong equilibrium* (respectively *in Nash equilibrium*) if there is a game form g, of which the strong equilibrium (respectively Nash equilibrium) outcomes at every profile coincide with the choice set suggested by S:

for all $u \in L(A)^N$: $\pi(SE(u)) = S(u)$,

where $SE(u) \subseteq X_N$ is the set of strong equilibrium outcomes at u of the game form $g = (X_i, i \in N, \pi)$. [Respectively for all $u \in L(A)^N$, $\pi(NE(u)) = S(u)$ where $NE(u) \subseteq X_N$ is the set of Nash equilibrium outcomes of g at u.]

We will use the following notations: given two profiles u, v and outcome a we say that *a keeps or improves its relative position from u to v* if

$u_i(a) > u_i(b) \Rightarrow v_i(a) > v_i(b)$ for all i, all $b \neq a$.

Next we say that v *is deduced from* u *by lifting* a if

(i) u and v coincide on $A/\{a\}$: $u_i(b) > u_i(c) \Leftrightarrow v_i(b) > v_i(c)$ for all i, all $b, c \neq a$,
(ii) a keeps or improves its relative position for u to v,
(iii) $v \neq u$: hence the position of a improves for at least one agent.

Definition 8 [Maskin (1977)]. A multi-valued voting rule S is *strongly monotonic* if we have for all profiles u, v and outcome a: $\{a \in S(u)$ *and a keeps or improve its relative position from u to v*$\} \Rightarrow \{a \in S(v)\}$

Lemma 6 [Peleg (1984)]. *A multi-valued voting rule S is strongly monotonic if and only if for all u, v and all a, we have*

$\{v$ *is deduced from u by lifting* $a\} \Rightarrow \{S(v) \subset S(u) \cup \{a\}\}$.

If a voting rule is strongly monotonic, a push up to any candidate can help election of this very candidate, or bar some candidates previously elected; but it cannot allow the election of *another* candidate. In short, campaigning for one candidate can only help that very candidate. However natural this property is, most of the usual voting rules do violate it: Borda's rule, almost all scoring methods [see Theorem 2.3.22 in Peleg (1984) for details], as well as all Condorcet consistent rules.

A trivial example of a strongly monotonic voting rule is the Pareto correspondence: $a \in S(u)$ iff {no outcome b is Pareto-superior to a}. Indeed we decide whether or not a is in the choice set simply by checking its relative position to other outcomes. A more interesting example is the veto core associated with some veto power β (Definition 6). Indeed, we decide whether or not a is in the choice set by checking inequalities (8). If a keeps or improve its relative position from u to v, then $\Pr(T, a, v) \subseteq \Pr(T, a, u)$ so (8) are preserved.

We turn now to the main impossibility result for strong monotonicity.

Theorem 7 [Muller and Satterthwaite (1977)]. *Suppose A contains three outcomes or more, and S is a single-valued voting rule (satisfying citizen-sovereignty). Then S is strongly monotonic iff S is dictatorial (for some $i^* \in N$, $S(u) = \text{top}(u_i^*)$ for all u).*

From this result a proof of Theorem 3 is easily derived.

Lemma 7 (Maskin (1977, 1979)]. *If the multi-valued voting rule S is implementable in strong equilibrium or in Nash equilibrium, then S is strongly monotonic.*

Proof. Say that S is implementable in Nash equilibrium by g. Fix two profiles u, v and an outcome a satisfying the premises of Definition 8. Let x be a Nash equilibrium of g at u that elects a. Consider a deviation from x by agent i resulitng in the election of outcome b. By the equilibrium property of x, we have $u_i(b) \leq u_i(a)$. Because a keeps or improve its relative position from u to v, we have $v_i(b) \leq v_i(a)$ as well. Thus x is still a Nash equilibrium of g at v: as g implements S this implies $a \in S(v)$ as was to be proved. The proof for the strong equilibrium case is similar. □□

Combining Theorem 7 and Lemma 7 we derive the fundamental undeterminacy of implementable voting rules: the only single-valued voting rules implementable in strong equilibrium and/or Nash equilibrium are dictatorial. Interestingly, the converse of Lemma 7 is almost true for implementability in Nash equilibrium (but not in strong equilibrium).

Theorem 8. *Suppose that we have three players or more. Let S be a strongly monotonic multi-valued voting rule. Then S is implementable in Nash equilibrium if it satisfies at least one of the following properties:*

(i) S is neutral (no discrimination among candidates).

(ii) S gives no veto power to any agent: for all $i \in N$ all profile u and all outcome a:

$$\{a = \text{top}(u_j) \text{ for all } j \neq i\} \Rightarrow \{a \in S(u)\}$$

(iii) S is inclusion minimal strongly monotonic:

$$\{\emptyset \neq S'(u) \subseteq S(u) \text{ for all } u, \text{ and } S' \text{ strongly monotonic}\} \Rightarrow S' = S.$$

Maskin (1977) proved that a strongly monotonic rule is implementable if it satisfies (ii). Repullo (1987) gives a simplified proof of this result. Moulin (1983) (Theorem 11, p. 202) proves that it is implementable if it satisfies (i).

The proof of (iii) follows from a general characterization result by Danilov (1992a). Danilov strengthens Maskin's monotonicity property and obtains a necessary and sufficient condition for Nash implementability when there are three players or more.

Among strongly monotonic voting rules the inclusion minimal ones are the most deterministic. An example of an inclusion minimal strongly monotonic rule is the β-veto core C_β of Definition (6). It is implementable in strong equilibrium as well (see at the end of the next section). Theorem 8 virtually closes the question of implementation by Nash equilibrium in the voting content.

The game forms used to prove Theorem 8 are unappealing because they ask that each voter announces the entire preference profile and an equilibrium results only if those announcements coincide.

In the more general context of implementation in economic environments (see Introduction) the techniques of Maskin's theorem can be adapted to show that virtually all (multi-valued or single-valued) voting rules are implementable by subgame perfect equilibrium [Moore and Repullo (1988), Abreu and Sen (1990)] or by Nash equilibrium using only undominated strategies [Palfrey and Srivastava (1991), Jackson (1992), Jackson et al. (1992), Sjöström (1993b)]. Related results in the context of implementation by Nash equilibrium are in Moore and Repullo (1992), Dutta and Sen (1991b), Sjöström (1992), Yamato (1993).

To understand equally well implementation by strong equilibrium, we need to introduce the concept of effectivity functions.

4.4. Effectivity functions

Definition 9 [Moulin and Peleg (1982)]. Given A and N, both finite, an effectivity function is a binary relation defined on (nonempty) coalitions of N and (nonempty) subsets of A and denoted eff. Thus eff is a subset of $2^N \times 2^A$: we write T eff B and say that coalition T is effective for B, if (and only if) the set eff contains (T, B). The relation eff satisfies two monotonicity properties w.r.t. coalitions and subsets of outcomes, as well as two boundary conditions:

$\{T \text{ eff } B \text{ and } T \subset T'\} \Rightarrow \{T' \text{ eff } B\}$,
$\{T \text{ eff } B \text{ and } B \subset B'\} \Rightarrow \{T \text{ eff } B'\}$,
$T \text{ eff } A$ for all (nonempty) T,
$N \text{ eff } B$ for all (nonempty) B.

The interpretation of T eff B is that the agents in T can, by some unspecified cooperative action, force the final outcome within the set B — they can veto the subset A/B —.

A simple game (Section 1.4 above) is a particular case of effectivity function whereby a winning coalition ($T \in W$) is effective for any (nonempty) subset B of A (in particular it can force the choice of any outcome a in A), whereas a non-winning coalition is effective for A only (namely it is unable to veto any outcome whatsoever).

To any game form $g = (X_i, i \in N; \pi)$ is associated the effectivity function whereby coalition T is effective for B if (any only if) there is a joint strategy x_T in X_T such that

$$\pi(x_T, x_{N \setminus T}) \in B \text{ for all } x_{N \setminus T} \in X_{N \setminus T}.$$

The same definition allows us to associate an effectivity function to every voting rule by viewing the latter as a particular game form.

Consider for instance a vector $\beta = (\beta_i)_{i \in N}$ of veto powers as in Section 4.2 above: each β_i is an integer and $\sum_{i \in N} \beta_i = p - 1$. To β corresponds the following effectivity function:

$$T \text{ eff}_\beta B \Leftrightarrow \sum_{i \in T} \beta_i + |B| \geq n. \tag{9}$$

Here T is effective for B if the total veto power of coalition T is at least the size of $N \setminus B$. Clearly eff_β is the effevtivity function associated with any voting by successive veto game form g_σ with veto power β [$\beta_i = |\sigma^{-1}(i)|$ for all i].

There is a natural generalization of the above effectivity function eff_β, called an *additive* effectivity function. Pick two probability distributions α and β on A and N respectively such that

$$\alpha_a > 0 \text{ for all } a, \sum_A \alpha_a = 1,$$

$$\beta_i > 0 \text{ for all } i, \sum_N \beta_i = 1.$$

Then define an effectivity function as follows:

$$T \text{ eff}_{\alpha, \beta} B \Leftrightarrow \sum_T \beta_i + \sum_B \alpha_a > 1.$$

The idea is that β_i is agent i's veto endowment and α_a is outcome a's veto price: coalition T is effective for B iff its total veto endowment exceeds the price of $N \setminus B$. The family of additive effectivity functions is derived from a class of game forms generalizing voting by successive vetoes [see Moulin (1983, Chs. 6, 7 for details)].

Our next definition translates the notion of core into the effectivity function language. Notations as in Definition 6.

Definition 10. Given are an effectivity function eff on A, N and a profile u in $L(A)^N$. We say that outcome a is in the core of eff u if no coalition is effective for the subset of outcomes that it unanimously prefers to a:

there is no $T \subseteq N$: $T \text{ eff } \Pr(T, a, u)$

We denote by Core(eff, u) the (possibly empty) core of eff at u. We say that the effectivity function eff is *stable* if its core is nonempty at all profiles in $L(A)^N$.

When the effectivity function is just a simple game, its stability is completely characterized by its Nakamura number (see Theorem 2 above). No such compact characterization is available for effectivity function (even for neutral ones). However we know of a very powerful sufficient condition for stability.

Definition 11. An effectivity function is *convex* if for every T_1, T_2 in N, B_1, B_2 in A we have

$$\{T_1 \text{ eff } B_1 \text{ and } T_2 \text{ eff } B_2\} \Rightarrow \{T_1 \cup T_2 \text{ eff } B_1 \cap B_2 \text{ and/or } T_1 \cap T_2 \text{ eff } B_1 \cup B_2\}.$$

This definition is related to the ordinal concept of convexity for NTU cooperative games.

Observe that an additive effectivity function is convex. Indeed suppose that $T_1 \cup T_2$ is not effective for $B_1 \cap B_2$ nor is $T_1 \cap T_2$ effective for $B_1 \cup B_2$:

$$\beta(T_1 \cup T_2) + \alpha(B_1 \cap B_2) \leq 1,$$
$$\beta(T_1 \cap T_2) + \alpha(B_1 \cup B_2) \leq 1.$$

(with the convention $\beta(T) = \sum_T \beta_i$). Summing up and regrouping, this implies

$$\beta(T_1) + \beta(T_2) + \alpha(B_1) + \alpha(B_2) \leq 2.$$

On the other hand $T_i \text{ eff } B_i$ $i = 1, 2$ reads $\beta(T_i) + \alpha(B_i) > 1$ for $i = 1, 2$. Thus our next result implies in particular that every additive effectivity function is stable.

Theorem 9 [Peleg (1984)]. *A convex effectivity function is stable.*

Peleg's original proof was shortened by Ichiishi (1985) (who proves, however, a slightly weaker result). Keiding (1985) gives a necessary and sufficient condition for stability in terms of an "acyclicity" property. A corollary of Keiding's theorem is a direct combinatorial proof of Peleg's theorem.

We are now ready to state the general results about implementation in strong equilibrium. Say that an effectivity function eff is *maximal* if we have for all T, B

$$T \text{ eff } B \Leftrightarrow No\{(N \setminus T) \text{ eff}(A \setminus B)\}.$$

Clearly for a stable effectivity function the implication $\Rightarrow$ holds true, but not necessarily the converse one.

In the following result, we assume that the multi-valued voting rule S satisfies citizen-sovereignty: for all outcome a there is a profile u such that $S(u) = \{a\}$.

Theorem 10 [Moulin and Peleg (1982)]. *Given are A and N, both finite*

(a) Say that S is a multi-valued voting rule implementable by strong equilibrium (Definition 7). Then its effectivity function eff_S is stable and maximal, and S is contained

in its core:

for all $u \in L(A)^N$: $S(u) \subseteq \text{Core}(\text{eff}_S, u)$.

(b) *Conversely, let eff be a stable and maximal effectivity function. Then its core defines a voting correspondence implementable by strong equilibrium.*

Danilov (1992b) proposes in the voting context a necessary and sufficient condition for implementability in strong equilibrium. In the general context of economic environment, Dutta and Sen (1991a) give another such condition.

Theorem 10 implies that the core correspondence of stable and maximal effectivity functions are the inclusion *maximal* correspondences implementable by strong equilibrium. When this core correspondence happens to be also inclusion minimal strongly monotonic [as is the case for the "integer" effectivity function (9)], we conclude that the core is the *only* implementable correspondence yielding this distribution of veto power.

The concept of effectivity functions proves useful to formulate the idea of core in several models of public decision-making. See the surveys by Ichiishi (1986), and Abdou and Keiding (1991).

5. Probabilistic voting

In this section we distinguish more carefully an ordinal preference u in $L(A)$ from its cardinal representations. To every preference u in $L(A)$ we associate the set $C(u)$ of its cardinal representation. Thus $C(u)$ is the subset of R^A made up of those vectors V such that

[for all a, b in A: $V(a) > V(b)$ iff a is preferred to b at u].

Definition 12. Given A and N, both finite, a probabilistic voting rule S is a (single-valued) mapping from $L(A)^N$ into $P(A)$, the set of probability distribution over A. Thus S is described by p mappings S_a, $a \in A$ where $S_a(u)$ is the probability that a is elected at profile u:

$$S_a(u) \geq 0, \sum_{a \in A} S_a(u) = 1 \text{ for all } u \in L(A)^N.$$

Notice that individual messages are purely deterministic, even though collective decision is random.

Definition 13. The probabilistic voting rule S is strategyproof if for all profile u and all cardinal representation V of this profile

$V_i \in C(u_i)$ for all i,

then the following inequality holds true:

$$\sum_{a\in A} V_i(a)S_a(u) \geq \sum_{a\in A} V_i(a)S_a(u'_i, u_i), \quad \text{all } i\in N$$
$$\text{all } u'_i \in L(A)$$

Notice that the cardinal representation U_i of u_i cannot be part of agent i's message: he is only allowed to reveal his (ordinal) ordering of the deterministic outcomes. It is only natural to enlarge the class of probabilistic voting rules so as to include any mapping S from $(R^A)^N$ into $P(A)$ and state the strategyproofness property as follows:

$$\sum_{a\in A} V_i(a)S_a(V) \geq \sum_{a\in A} V_i(a)S_a(V'_i, V_i) \quad \text{all } i, V, V'_i.$$

Unfortunately, the much bigger set of such strategyproof cardinal probabilistic voting rules is unknown.

An example of strategyproof probabilistic voting rule is *random dictator*: an agent is drawn at random (with uniform probability) and he gets to choose the (deterministic) outcome as pleases him.

Gibbard (1978) gives a (complicated) characterization of all strategyproof voting rule in the sense of the above defintion. Hylland (1980) shows that if a probabilistic voting rule is strategyproof and selects an outcome which is ex post Pareto-optimal (that is to say the deterministic outcome eventually selected is not Parento-inferior in the ordinal preference profile) then this voting rule is undistinguishable from a random dictator rule. This can be deduced also from Barbera's theorem below.

This last result characterizes all strategyproof (probabilistic) voting rules that, in addition, are anonymous and neutral. We need a couple of definitions to introduce the result. First take a vector of scores such that

$$0 \leq s_0 \leq \cdots \leq s_{p-1} \text{ and } s_0 + \cdots + s_{p-1} = \frac{1}{n}$$

and define the (probabilistic) scoring voting rule S^s as follows:

$$S^s_a(u) = \sum s_{v_i(a)} \quad \text{all } a, u,$$

where v_i is this element in $C(u_i)$ with range $0, \ldots, p-1$.

Next take a vector of weights t such that

$$0 \leq t_0 \leq \cdots \leq t_n; \; t_l + t_{n-l} = \frac{2}{p(p-1)} \quad \text{all } l = 0, \ldots, n$$

and define the (probabilistic) supporting size voting rule S^t as follows

$$S^t_a(u) = \sum_{b\neq a} t_{\tau(,a,b)} \quad \text{all } a, u,$$

where $\tau(u, a, b) = |N(u, a, b)|$ is the number of agents preferring a to b.

Theorem 11 [Barbera (1979)]. *Let S be an anonymous and neutral probabilistic voting rule. The two following statements are equivalent:*
  (i) *S is strategyproof (Definition 13).*
  (ii) *S is a convex combination of one (probabilistic) scoring rule and one (probabilistic) supporting size rule.*

Note that the proof of Theorem 11 relies heavily on Gibbard's characterization result [Gibbard (1978)]. So it is probably fair to call this elegant result the Gibbard–Barbera theorem.

In the context of probabilistic voting, Abreu and Sen (1991) have shown that all social choice functions (even single-valued) are implementable by Nash equilibrium with an arbitrary small probabilistic error. See also Abreu and Matsushima (1989) for the case of sophisticated implementation, and Sjöström (1993a) for the case of implementation by trembling hand perfect equilibrium.

References

Abdou, J. and H. Keiding (1991) *Effectivity functions in social choice*. Dordrecht: Kluwer Academic Publishers.
Abreu, D. and H. Matsushima (1989) 'Virtual implementation in iteratively undominated strategies: complete information', mimeo.
Abreu, D. and A. Sen (1990) 'Subgame perfect implementation: a necessary and almost sufficient condition', *Journal of Economic Theory*, **50**: 285–299.
Abreu, D. and A. Sen (1991) 'Virtual implementation in Nash equilibrium', *Econometrica*, **59**: 997–1021.
Arrow, K. (1963) *Social choice and individual values*, 2nd edn. New York: Wiley, 1st edn. 1951.
Barbera, S. (1979) 'Majority and positional voting in a probabilistic framework', *Review of Economic s Studies*, **46**: 389–397.
Barbera, S. (1983) 'Strategy-proofness and pivotal voters: a direct proof of the Gibbard–Satterthwaite theorem', *International Economic Review*, **24**: 413–417.
Barbera, S., F. Gül and E. Stachetti (1993) 'Generalized median voter schemes and committees', *Journal of Economic Theory*, **61**(2): 262–289.
Barbera, S. and M. Jackson (1992) 'A characterization of strategy-proof social choice functions for economies pure public goods', *Social Choice and Welfare*, forthcoming.
Barbera, S., J. Massò and A. Neme (1993) 'Voting under constraints,' mimeo, Universitat Autònoma de Barcelona.
Barbera, S. and B. Peleg (1990) 'Strategy-proof voting schemes with continuous preferences', *Social Choice and Welfare*, **7**: 31–38.
Barbera, S., H. Sonnenschein and L. Zhou (1991) 'Voting by committees', *Econometrica*, **59**: 595–609.
Black, D. (1958) *The theory of committees and elections*, Cambridge University Press, Cambridge.
Blair, D. and E. Muller (1983) 'Essential aggregation procedures on restricted domains of preferences', *Journal of Economic Theory*, **30**, 34–53.
Blair, D. and R. Pollak (1982) 'Acyclic collective choice rules', *Econometrica*, **50**: 4: 931–943.
Blau, J. and R. Deb (1977) 'Social decision functions and the veto', *Econometrica*, **45**: 871–879.
Borda, J.C. (1781) 'Mémoire sur les élections au scrutin', *Histoire de l'Académie Royale des Sciences*, Paris.
Border K. and J. Jordan (1983) 'Straightforward elections, unanimiy and phantom agents', *Review of Economic Studies*, **50**: 153–170.
Chichilnisky, G. and Heal G. (1981) 'Incentive Compatibility and Local Simplicity', mimeo, University of Essex.
Chichilnisky, G. (1983) 'Social Choice and Game Theory: recent results with a topological approach', in: P. Pattanaik and M. Salles, ed., *Social Choice and Welfare*. Amsterdam: North-Holland.

Condorcet, Marquis de (1985) 'Essai sur l'application de l'analyse à la probabilité des décisions rendues à la pluralité des voix', Paris.

Danilov, V. (1992a) 'Implementation via Nash equilibrium', *Econometrica*, **60**: 43–56.

Danilov, V. (1992b) 'On strongly implementable social choice correspondences', presented at the Meeting of the Social Choice and Welfare Society, Caen, June 1992.

Demange, G. (1982) 'Single peaked orders on a tree', *Mathematical Social Sciences*, **3**(4): 389–396.

Dutta, B. and A. Sen (1991a) 'Implementation under strong equilibrium: a complete characterization', *Journal of Mathematical Economics*, **20**: 49–68.

Dutta, B. and A. Sen (1991b) 'A necessary and sufficient condition for two-person Nash implementation', *Review of Economic Studies*, **58**: 121–128.

Farqharson, R. (1969) *Theory of voting*, Yale University Press, New Haven.

Gibbard, A. (1973) 'Manipulation of voting schemes: a general result', *Econometrica*, **41**: 587–601.

Gibbard, A. (1978) 'Straightforwardness of game forms with lotteries as outcomes', *Econometrica*, **46**: 595–614.

Golberg, A. and Gourvitch, V. (1986) 'Secret and extensive dominance solvable veto voting schemes', mimeo, Academy of Sciences, Moscow.

Gretlein, R. (1983) 'Dominance elimination procedures on finite alternative games', *International Journal of Game Theory*, **12**(2): 107–114.

Hurwicz, L. (1973) 'The design of mechanisms for resource allocation', *American Economic Review*, **63**: 1–30.

Hylland, A. (1980) 'Strategyproofness of voting procedures with lotteries as outcomes and infinite sets of strategies', mimeo, University of Oslo.

Ichiishi, T. (1986) 'The effectivity function approach to the core', in: W. Hildenbrand and A. MasColell, ed., *Contributions to Mathematical Economics* (in honor of Gérard Debreu). Amsterdam: North-Holland, Ch. 15.

Ichiishi, T. (1985) 'On Peleg's theorem for stability of convex effectivity functions', mimeo, University of Iowa.

Jackson, M. (1992) 'Implementation in undominated strategies: a look at bounded mechanisms', *Review of Economic Studies*, **59**: 757–775.

Jackson, M., T. Palfrey and S. Srivastava (1992) 'Undominated Nash implementation in bounded mechanisms, forthcoming', *Games and Economic Behavior*.

Kalai, E. and E. Muller (1977) 'Characterization of domains admitting nondictatorial social welfare functions and nonmanipulable voting procedures', *Journal of Economic Theory*, **16**: 457–469.

Keiding, H. (1985) 'Necessary and sufficient conditions for stability of effectivity functions', *International Journal of Game Theory*, **14**(2): 93–102.

Laffond, G. (1980) 'Révélation des préférences et utilités unimodales, mimeo, Laboratoire d'Econométrie', Conservatoire National des Arts et Métiers.

Le Breton, M. et Sen, A. (1992) 'Strategyproof social choice functions over product domains with unconditional preferences, mimeo', Indian Statistical Institute.

Le Breton, M. and Weymark, J. (1993) 'Strategyproof social choice with continuous separable preferences, mimeo', University of British Columbia.

Maskin, E. (1977) 'Nash equilibrium and welfare optimality', mimeo, Massachussetts, Institute of Technology.

Maskin, E. (1979) 'Implementation and strong Nash equilibrium', in: J.J. Laffont, ed., *Aggregation and Revelation of Preferences*. Amsterdam: North-Holland.

May, K. (1952) 'A set of independent necessary and sufficient conditions for simple majority decision', *Econometrica*, **20**: 680–684.

McKelvey, R. and R. Niemi (1978) 'A multi-stage game representation of sophisticated voting for binary procedures', *Journal of Economic Theory*, **18**: 1–22.

Miller, N. (1977) 'Graph-theoretical approaches to the theory of voting', *American Journal of Political Science*, **21**: 769–803.

Miller, N. (1980) 'A new solution set for tournaments and majority voting: further graph-theoretical approaches to the theory of voting', *American Journal of Political Science*, **24**: 68–69.

Miller, N. (1983) 'Pluralism and social choice', *American Political Science Review*, **77**: 734–735.

Moore, J. (1992) 'Implementation in economic environments with complete information', in: J.J. Laffont, ed., *Advances in Economic Theory*. Cambridge: Cambridge University Press.

Moore, J. and R. Repullo (1988) 'Subgame perfect implementation', *Econometrica*, **56**(5): 1191–1220.

Moore, J. and R. Repullo (1992) 'Nash implementation: a full characterization', *Econometrica*, **58**: 1083–1099.
Moulin, H. (1979) 'Dominance-solvable voting schemes', *Econometrica*, **47**: 1137–1351.
Moulin, H. (1980a) 'On strategyproofness and single peakedness', *Public Choice*, **35**: 437–455.
Moulin, H. (1980b) 'Implementing efficient, anonymous and neutral social choice functions', *Journal of Mathematical Economics*, **7**: 249–269.
Moulin, H. (1981a) 'Prudence versus sophistication in voting strategy', *Journal of Economic Theory*, **24**(3): 398–412.
Moulin, H. (1981b) 'The proportional veto principle', *Review of Economic Studies*, **48**: 407–416.
Moulin, H. (1982) 'Voting with proportional veto power', *Econometrica*, **50**(1): 145–160.
Moulin, H. (1983) *The strategy of social choice*, advanced textbook in Economics. Amsterdam: North-Holland.
Moulin, H. (1984) 'Generalized Condorcet winners for single peaked and single plateau preferences', *Social Choice and Welfare*, **1**: 127–147.
Moulin, H. (1985) 'From social welfare orderings to acyclic aggregation of preferences', *Mathematical Social Sciences*, **9**: 1–17.
Moulin, H. (1988) *Axioms of cooperative decision-making*, Cambridge University Press, Cambridge.
Moulin, H. and B. Peleg (1982) 'Core of effectivity functions and implementation theory', *Journal of Mathematical Economics*, **10**: 115–145.
Mueller, D. (1978) Voting by veto, *Journal of Public Economics*, **10**(1): 57–76.
Muller, E. and M. Satterthwaite (1977) 'The equivalence of strong positive association and strategyproofness', *Journal of Economic Theory*, **14**: 412–418.
Muller, E. and M. Satterthwaite (1985) 'Strategyproofness: the existence of dominant strategy mechanisms', in: Hurwicz et al. eds., *Social Goals and Social Organization*. Cambridge: Cambridge University Press.
Nakamura, K. (1975) 'The core of a simple game with ordinal preferences', *International Journal of Game Theory*, **4**: 95–104.
Nakamura, K. (1978) 'Necessary and sufficient condition on the existence of a class of social choice functions', *The Economic Studies Quarterly*, **29**(3): 259–267.
Palfrey, T. and S. Srivastava (1991) 'Nash implementation using undominated strategies', *Econometrica*, **59**: 479–501.
Peleg, B. (1978) 'Consistent voting systems', *Econometrica*, **46**: 153–161.
Peleg, B. (1984) *Game theoretic analysis of voting in committees*, Cambridge University Press, Cambridge.
Peters, H., H. van der Stel, and T. Storcken (1991a) 'Generalized median solutions, strategy-proofness and strictly convex norms', Reports in Operations Research and Systems Theory No. M 91-11, Department of Mathematics, University of Limburg.
Peters, H., H. van der Stel, and T. Storcken (1992) 'Pareto optimality, anonymity, and strategy-proofness in location problems', *International Journal of Game Theory* **21**: 221–235.
Repullo, R. (1987) 'A simple proof of Maskin's theorem on Nash implementation', *Social Choice and Welfare*, **4**(1): 39–41.
Rochet, J.C. (1980) 'Selection of a unique equilibrium payoff for extensive games with perfect information', mimeo, Université de Paris 9.
Satterthwaite, M.A. (1975) 'Strategyproofness and Arrow's conditions: existence and correspondence theorems for voting procedures and social welfare functions', *Journal of Economic Theory*, **10**: 198–217.
Sen, A.K. (1970) *Collective Choice and Social Welfare* Holden Day, San Francisco.
Sen, A.K. (1986) 'Social choice theory', in: K. Arrow and M. Intriligator eds., *Handbook of Mathematical Economics*. Amsterdam: North-Holland.
Shepsle, K. and B. Weingast (1984) 'Uncovered sets and sophisticated voting outcomes with implications for agenda institutions', *American Journal of Political Science*, **28**: 49–74.
Sjöstrom, T. (1992) 'On the necessary and sufficient conditions for Nash implementation', *Social Choice and Welfare*, forthcoming.
Sjöstrom, T. (1993a) 'Implementation in perfect equilibria', *Social Choice and Welfare*, **10**(1): 97–106.
Sjöstrom, T. (1993b) 'Implementation in undominated Nash equilibrium without integer games', *Games and Economic Behaviour*, forthcoming.
Wilson, R. (1972) 'Social choice without the Pareto principle', *Journal of Economic Theory*, **5**: 478–486.
Yamato, T. (1993) 'Double implementation in Nash and undominated Nash equilibria', mimeo.

Chapter 32

POWER AND STABILITY IN POLITICS

PHILIP D. STRAFFIN Jr.

Beloit College

Contents

1.	The Shapley–Shubik and Banzhaf power indices	1128
2.	Structural applications of the power indices	1130
3.	Comparison of the power indices	1133
4.	Dynamic applications of the power indices	1138
5.	A spatial index of voting power	1140
6.	Spatial models of voting outcomes	1144
Bibliography		1150

"Political science, as an empirical discipline, is the study of the shaping and sharing of power" – Lasswell and Kaplan, *Power and Society*.

In this chapter we will treat applications of cooperative game theory to political science. Our focus will be on the idea of power. We will start with the use of the Shapley and Banzhaf values for simple games to measure the power of political actors in voting situations, with a number of illustrative applications. The applications will point to the need to compare these two widely used power indices, and we will give two characterizations of them. We will then consider several ways in which political dynamics might be analyzed in terms of power. Finally, we will consider the role of political ideology and show how it can be modeled geometrically. For these spatial models of voting, we will consider both the problem of measuring the power of voters, and the problem of prescribing rational voting outcomes. For other directions in which game theory has influenced political science, see Chapters 30 (Voting procedures) and 31 (Social choice).

1. The Shapley–Shubik and Banzhaf power indices

A voting situation can be modeled as a cooperative game in characteristic function form in which we assign the value 1 to any coalition which can pass a bill, and 0 to any coalition which cannot. The resulting game is known as a *simple game* [Shapley (1962)]. We call the coalitions which can pass bills *winning coalitions*, and note that the game is completely determined by its set of winning coalitions.

Definition. A simple game (or voting game) is a set N and a collection $\mathscr{W}$ of subsets of N such that

 (i) $\emptyset \notin \mathscr{W}$.
 (ii) $N \in \mathscr{W}$.
 (iii) $S \in \mathscr{W}$ and $S \subseteq T \Rightarrow T \in \mathscr{W}$.

A coalition S is called a *minimal winning coalition* if $S \in \mathscr{W}$, but no proper subset of S is in $\mathscr{W}$. The "monotonicity" property (iii) implies that the winning coalitions of any simple game can be described as the supersets of its minimal winning coalitions. A simple game is *proper* if $S \in \mathscr{W} \Rightarrow N \setminus S \notin \mathscr{W}$, or equivalently if there are not two disjoint winning coalitions. A simple game is *strong* if $S \notin \mathscr{W} \Rightarrow N \setminus S \in \mathscr{W}$, so that "deadlocks" are not possible.

A common form of simple game is a *weighted voting game* $[q; w_1, w_2, \ldots, w_n]$. Here there are n voters, the ith voter casts w_i votes, and q is the *quota* of votes needed to pass a bill. In other words, $S \in \mathscr{W} \Leftrightarrow \sum_{i \in S} w_i \geq q$. A weighted voting game will be proper if $q > w/2$, where w is the sum of the w_i. It will also be strong if w

is an odd integer and $q = (w + 1)/2$. Weighted voting games appear in many contexts: stockholder voting in corporations, New York State county boards, United Nations agencies, the United States Electoral College, and multi-party legislatures where parties engage in bloc voting. See Lucas (1983) for many examples. On the other hand, not all politically important simple games are weighted voting games. A familiar example is the United States legislative scheme, in which a winning coalition must contain the President and a majority of both the Senate and the House of Representatives, or two-thirds of both the Senate and the House.

Given a simple game, we desire a measure of the power of individual voters in the game. Shapley and Shubik (1954) proposed using the Shapley value as such a measure:

"There is a group of individuals all willing to vote for some bill. They vote in order. As soon as enough members have voted for it, it is declared passed, and the member who voted last is given credit for having passed it. Let us choose the voting order of members randomly. Then we may compute how often a given individual is *pivotal*. This latter number serves to give us our index." [Shapley and Shubik (1954)]

In other words, the *Shapley–Shubik power index* of voter i is

$$\phi_i = \frac{\text{the number of orders in which voter } i \text{ is pivotal}}{n!},$$

where $n!$ is the total number of possible orderings of the voters. For example, in the weighted voting game

[7; 4, 3, 2, 1]
 A B C D

the orderings, with pivotal voters underlined, are

A<u>B</u>CD A<u>B</u>DC AC<u>B</u>D AC<u>D</u>B AD<u>B</u>C AD<u>C</u>B B<u>A</u>CD B<u>A</u>DC BC<u>A</u>D BCD<u>A</u>
BD<u>A</u>C BDC<u>A</u> CA<u>B</u>D CA<u>D</u>B CB<u>A</u>D CBD<u>A</u> CD<u>A</u>B CDB<u>A</u> D<u>A</u>BC D<u>A</u>CB
DB<u>A</u>C DBC<u>A</u> DC<u>A</u>B DCB<u>A</u>

and the Shapley–Shubik power indices are

$$\phi_A = \tfrac{14}{24} \quad \phi_B = \tfrac{6}{24} \quad \phi_C = \phi_D = \tfrac{2}{24}.$$

To get a combinatorial formula for the Shapley–Shubik power index, define i to be a *swing* voter for coalition S if $S \in \mathcal{W}$, $S \setminus \{i\} \notin \mathcal{W}$. Then

$$\phi_i = \sum_{\substack{i \text{ swings} \\ \text{for } S}} \frac{(s-1)!(n-s)!}{n!},$$

where $s = |S|$, the number of voters in S. This follows from the observation that voter i is pivotal for an ordering if and only if i is a swing voter for the coalition S of i and all voters who precede i. There are $(s-1)!$ ways in which the voters before i could be ordered, and $(n-s)!$ ways in which the voters who follow i could be ordered.

A second game theoretic power index was proposed by Banzhaf (1965):

"The appropriate measure of a legislator's power is simply the number of different situations in which he is able to determine the outcome. More explicitly, in a case in which there are n legislators, each acting independently and each capable of influencing the outcome only by means of his votes, the ratio of the power of legislator X to the power of legislator Y is the same as the ratio of the number of possible voting combinations of the entire legislature in which X can alter the outcome by changing his vote, to the number of combinations in which Y can alter the outcome by changing his vote." [Banzhaf (1965)].

In other words, voter i's power should be proportional to the number of coalitions for which i is a swing voter. It is convenient to divide this number by the total number of coalitions containing voter i, obtaining the *unnormalized Banzhaf index*

$$\beta'_i = \frac{\text{number of swings for voter } i}{2^{n-1}}.$$

The standard Banzhaf index β is this index normalized to make the indices of all voters add to 1. In the weighted voting game above the winning coalitions, with swing voters underlined, are

$$\underline{AB} \quad \underline{ABC} \quad \underline{ABD} \quad \underline{ACD} \quad \underline{A}BCD$$

and the Banzhaf indices are

$$\beta'_A = \tfrac{5}{8} \quad \beta'_B = \tfrac{3}{8} \quad \beta'_C = \beta'_D = \tfrac{1}{8}$$
$$\beta_A = \tfrac{5}{10} \quad \beta_B = \tfrac{3}{10} \quad \beta_C = \beta_D = \tfrac{1}{10}.$$

Although the Shapley–Shubik and Banzhaf indices were first proposed in the second half of the twentieth century, Riker (1986) has found a similar combinatorial discussion of voting power in the works of Luther Martin, a delegate to the United States Constitution Convention in 1787. For methods of computing the Shapley–Shubik and Banzhaf indices, see Lucas (1983).

2. Structural applications of the power indices

Nassau County Board, 1964. This example prompted Banzhaf's original investigation of voting power (1965). The players were the county board representatives from

Hempstead #1, Hempstead #2, North Hempstead, Oyster Bay, Glen Cove and Long Beach. The weighted voting game was

[58; 31, 31, 21, 28, 2, 2].
   H1 H2 N   O   G L

Notice that whether a bill passes or not is completely determined by voters H1, H2, and O: if two of them vote for the bill it will pass, while if two of them vote against the bill it will fail. Voters N, G and L are *dummies* in this game. They can never affect the outcome by their votes. The Shapley–Shubik power indices are

$$\phi_{H1} = \phi_{H2} = \phi_O = \tfrac{1}{3} \quad \text{and} \quad \phi_N = \phi_G = \phi_L = 0$$

and the Banzhaf indices are the same.

United Nations Security Council. Shapley and Shubik (1954) analyzed the Security Council, which then had eleven members. A winning coalition needed seven members, including all five of the Council's permanent members, who each had veto power proposed actions. Denote the players by PPPPPNNNNNN, where we will not distinguish among permanent members, or among non-permanent members, since they played symmetric roles. There were $\binom{11}{5} = 462$ possible orderings. Of these, a non-permanent member pivots only in an ordering which looks like (PPPPPN)N(NNNN), where the parenthesis notation simply means that the pivotal N is preceded by five Ps and one N, and followed by four Ns. The number of such orderings is $\binom{6}{1}\binom{4}{0} = 6$. Hence the six non-permanent members together held only $6/462 = 0.013$ of the Shapley–Shubik power, with the remaining 0.987 held by the five permanent members.

In 1965 the Security Council was expanded to include ten non-permanent and five permanent members. A winning coalition now needs nine members, still including all five permanent members. There are now $\binom{15}{5} = 3003$ orderings of PPPPPNNNNNNNNNN, with a non-permanent member pivoting in orderings which look like (PPPPPNNN)N(NNNNNN), of which there are $\binom{8}{3}\binom{6}{0} = 56$. Hence the expansion increased the proportion of Shapley–Shubik power held by the non-permanent members just slightly, to $56/3003 = 0.019$.

Slightly harder calculations for the Banzhaf index give the total proportion of power held by the non-permanent members as 0.095 before 1965 and 0.165 after 1965. See Brams (1975). Notice that the difference between the two power indices is significant in this case.

Council of the European Economic Community. Table 1 gives the Shapley–Shubik indices of representatives in the Council of the EEC in 1958, and after the expansion in 1973. The Banzhaf indices are similar – see Brams (1976). The figures illustrate several interesting phenomena. First, note that Luxembourg was a dummy in the 1958 Council. It was not in 1973, although new members had joined and Luxembourg's share of the votes *decreased* from 1958 to 1973. Brams (1976) calls

Table 1
Power indices for the Council of the European Economic Community

Member	1958		1973	
	Weight	Shapley–Shubik index	Weight	Shapley–Shubik index
France	4	0.233	10	0.179
Germany	4	0.233	10	0.179
Italy	4	0.233	10	0.179
Belgium	2	0.150	5	0.081
Netherlands	2	0.150	5	0.081
Luxembourg	1	0.000	2	0.010
Denmark	–	–	3	0.057
Ireland	–	–	3	0.057
United Kingdom	–	–	10	0.179
Quota	12 of 17		41 of 58	

this the Paradox of New Members: adding new members to a voting body may actually increase the voting power of some old members. See Brams and Affuso (1985) for other voting power paradoxes illustrated by the addition of new members to the EEC in the 1980s.

United States Electoral College. In the United States Electoral College, the Presidential electors from each state have traditionally voted as a bloc, although they are not constitutionally required to do so. Hence we can view the Electoral College as a weighted voting game with the states as players. Since electors are not assigned in proportion to population, it is of interest to know whether voting power is roughly proportional to population. In particular, is there a systematic bias in favor of either large states or small states? Banzhaf (1968), Mann and Shapley (1962), Owen (1975) and Rabinowitz and MacDonald (1986) have used the Banzhaf and Shapley–Shubik power indices to study this question. The Shapley–Shubik index indicates that power is roughly proportional to the number of electoral votes (not population), with a slight bias in favor of large states. The large state bias is more extreme with the Banzhaf index. We will consider Rabinowitz and MacDonald's work in Section 5.

Canadian Constitutional amendment scheme. An interesting example of a voting game which is not a weighted voting game was a Canadian Constitutional amendment scheme proposed in the Victoria Conference of 1971. An amendment would have to be approved by Ontario, Quebec, two of the four Atlantic provinces (New Brunswick, Nova Scotia, Prince Edward Island and Newfoundland), and either British Columbia and two prairie provinces (Alberta, Saskatchewan and Manitoba) or all three prairie provinces. Table 2 shows the Shapley–Shubik and Banzhaf indices for this scheme, from Miller (1973) and Straffin (1977a). First of

Table 2
Power indices for Canadian Constitutional amendment schemes

Province	1980 Population (%)	Victoria scheme (1971)		Adopted scheme (1982)	
		Shapley–Shubik	Banzhaf	Shapley–Shubik	Banzhaf
Ontario	35.53	31.55	21.78	14.44	12.34
Quebec	26.52	31.55	21.78	12.86	11.32
British Columbia	11.31	12.50	16.34	10.28	10.31
Alberta	9.22	4.17	5.45	9.09	9.54
Manitoba	4.23	4.17	5.45	9.09	9.54
Saskatchewan	3.99	4.17	5.45	9.09	9.54
Nova Scotia	3.49	2.98	5.94	9.09	9.54
New Brunswick	2.87	2.98	5.94	8.69	9.29
Newfoundland	2.34	2.98	5.94	8.69	9.29
Prince Edward Island	0.51	2.98	5.94	8.69	9.29

all, notice how well Shapley–Shubik power approximates provincial populations, even though the scheme was not constructed with knowledge of the index. Second, notice that the Banzhaf index gives quite different results, even differing in the *order* of power: it says that the Atlantic provinces are more powerful than the prairie provinces, while the Shapley–Shubik index says the opposite.

Unfortunately, this clever and equitable scheme was not adopted. The last two columns of the table give the Shapley–Shubik and Banzhaf indices for the considerably more egalitarian scheme of Canada's Constitution Act, as approved in 1982 [Kilgour (1983), Kilgour and Levesque (1984)].

3. Comparison of the power indices

We have seen that the Shapley–Shubik and Banzhaf indices can give quite different results. The Banzhaf index, for example, gave considerably more power to the non-permanent members of the U.N. Security Council, and to the smaller Canadian provinces. The asymptotic behavior of the two indices for large voting bodies is also very different. Consider three examples:

(1) In the game [5; 3, 1, 1, 1, 1, 1, 1] call the large voter X. Then $\phi_X = 0.429$, while $\beta_X = 0.455$. However, if we keep X having one-third of the votes and keep the quota at a majority, but let the number of small voters go to infinity, then $\phi_X \to \frac{1}{2}$, while $\beta_X \to 1$.[1]

---

[1] Intuitively, what is happening here is that for the Shapley–Shubik index, X will pivot if he votes after $\frac{1}{4}$ of the small voters and before $\frac{3}{4}$ of them, i.e. half the time. For the Banzhaf index, X will be a swing voter for any coalition with between $\frac{1}{2}$ and $\frac{5}{6}$ of the votes, whereas a small voter will be a swing only for coalitions with "exactly half" of the votes. The latter are very rare compared to the former.

(2) In the game [5; 3, 2, 1, 1, 1, 1] call the large voters X and Y. Then $\phi_X = 0.400$ and $\phi_Y = 0.200$, while $\beta_X = 0.393$ and $\beta_Y = 0.179$. However, if we keep X having one-third of the votes, Y having two-ninths, and the quota at a majority, but let the number of small voters go to infinity, then $\phi_X \to 0.391$ and $\phi_Y \to 0.141$, while $\beta_X \to 1$ and $\beta_Y \to 0$.[2]

(3) In their original paper Shapley and Shubik (1954) considered a legislative scheme in which a bill must be approved by a president P and by a majority in each of a three-member senate (SSS) and a five-member house (HHHHH). Shapley–Shubik power indices are $\phi_P = 0.381$, $\phi_S = 0.107$, $\phi_H = 0.059$. Banzhaf indices are $\beta_P = 0.228$, $\beta_S = 0.114$, $\beta_H = 0.086$. The discrepancy grows as the size of the senate and house grows. For the United States legislative scheme with a senate of 101 (including the vice-president) and house of 435, we have $\phi_P = 0.5000$, $\phi_S = 0.0025$, $\phi_H = 0.0006$ but $\beta_P = 0.0313$, $\beta_S = 0.0031$, $\beta_H = 0.0015$. The difference in the power of the president is particularly striking.

Given these differences, it is important to characterize the Shapley–Shubik and Banzhaf indices well enough to understand which index would be most appropriate in which situations. We will consider here two approaches: an axiomatic characterization due to Dubey [Dubey (1975) and Dubey and Shapley (1979)], and a probability characterization due to Straffin (1977a).

First, some terminology. A *power index* is a function K which assigns to each player i in a simple game G a real number $K_i(G) \geq 0$. Recall that a voter i is a *dummy* in a simple game $(N, \mathcal{W})$ if $S \in \mathcal{W}$ implies $S \setminus \{i\} \in \mathcal{W}$. The *unanimity game* U_N with player set N is the game whose only winning coalition is the grand coalition N of all players. Finally, given two simple games $G_1 = (N_1, \mathcal{W}_1)$ and $G_2 = (N_2, \mathcal{W}_2)$, with N_1 and N_2 not necessarily disjoint

Definition. $G_1 \wedge G_2$ is the simple game with $N = N_1 \cup N_2$ and $S \in \mathcal{W}$ if and only if $S \cap N_1 \in \mathcal{W}_1$ and $S \cap N_2 \in \mathcal{W}_2$.

Definition. $G_1 \vee G_2$ is the simple game with $N = N_1 \cup N_2$ and $S \in \mathcal{W}$ if and only if $S \cap N_1 \in \mathcal{W}_1$ or $S \cap N_2 \in \mathcal{W}_2$.

Thus to win in $G_1 \wedge G_2$ a coalition must win in both G_1 and G_2, whereas to win in $G_1 \vee G_2$ it must win either in G_1 or G_2.

Consider the following axioms for a power index K:

Axiom 1. $K_i(G) = 0$ if and only if i is a dummy in G.

---

[2] For the "oceanic game" where the number of small voters becomes a continuum, the limit values are attained exactly both here and in (1). See Milnor and Shapley (1978), Dubey and Shapley (1979), Straffin (1983).

Axiom 2. $K_i(G_1) + K_i(G_2) = K_i(G_1 \wedge G_2) + K_i(G_1 \vee G_2)$.

Axiom 3. If i is a player in U_N, then $K_i(U_N) = 1/|N|$.

Axiom 3'. If i is a player in U_N, then $K_i(U_N) = 1/2^{|N|-1}$.

It is clear that the Shapley–Shubik index satisfies Axioms 1 and 3, and the unnormalized Banzhaf index satisfies Axioms 1 and 3'. Both indices also satisfy Axiom 2. (The normalized Banzhaf index β does *not* satisfy Axiom 2, which is why β' is more convenient.) Dubey proved:

Theorem 3.1 [Dubey (1975)]. *The Shapley–Shubik index is the unique power index which satisfies Axioms 1, 2 and 3.*

Theorem 3.2 [Dubey and Shapley (1979)]. *The unnormalized Banzhaf index is the unique power index which satisfies Axioms 1, 2 and 3'.*

Proofs. We will show that a power index satisfying Axioms 1 and 2 is determined by its values on unanimity games. Note that any simple game G can be written as $G = U_{S_1} \vee U_{S_2} \vee \cdots \vee U_{S_m}$, where $S_1, \ldots, S_m$ are the minimal winning coalitions of G. Our proof is by induction on m, the number of minimal winning coalitions. If $m = 1$, G is a unanimity game with dummies and K is determined. If $m > 1$, then by Axiom 2 we have

$$K_i(G) = K_i(U_{S_1}) + K_i(U_{S_2} \vee \cdots \vee U_{S_m}) - K_i(U_{S_1} \wedge (U_{S_2} \vee \cdots \vee U_{S_m}))$$
$$= K_i(U_{S_1}) + K_i(U_{S_2} \vee \cdots \vee U_{S_m}) - K_i(U_{S_1 \cup S_2} \vee \cdots \vee U_{S_1 \cup S_m})$$

and the games on the right have fewer minimal winning coalitions than G, so that K_i is determined inductively. □

For example, consider the game

$$G = [7;\ 4,\ 3,\ 2,\ 1]$$
$$\text{A}\ \ \text{B}\ \ \text{C}\ \ \text{D}$$

of Section 1.

The minimal winning coalitions are AB and ACD. Hence

$$K_i(G) = K_i(U_{AB}) + K_i(U_{ACD}) - K_i(U_{ABCD})$$

so that

$$\phi(G) = (\tfrac{1}{2}, \tfrac{1}{2}, 0, 0) + (\tfrac{1}{3}, 0, \tfrac{1}{3}, \tfrac{1}{3}) - (\tfrac{1}{4}, \tfrac{1}{4}, \tfrac{1}{4}, \tfrac{1}{4})$$
$$= (\tfrac{7}{12}, \tfrac{3}{12}, \tfrac{1}{12}, \tfrac{1}{12})$$

and

$$\beta'(G) = (\tfrac{1}{2},\tfrac{1}{2},0,0) + (\tfrac{1}{4},0,\tfrac{1}{4},\tfrac{1}{4}) - (\tfrac{1}{8},\tfrac{1}{8},\tfrac{1}{8},\tfrac{1}{8})$$
$$= (\tfrac{5}{8},\tfrac{3}{8},\tfrac{1}{8},\tfrac{1}{8}).$$

Because Axiom 2 plays such a powerful role in this characterization of the Shapley–Shubik and Banzhaf indices, using the characterization to understand the difference between the indices requires having a good intuitive feel for why Axiom 2 should hold. However, we can at least say, looking at Axioms 3 and 3′, that which of the two indices you use might depend on how powerful you think players are in unanimity games of different sizes. Adding a new player to a unanimity game U_N cuts β' of each old player by a factor of two, but only lowers ϕ of each old player from $1/|N|$ to $1/(|N|+1)$.

For an alternative characterization of ϕ and β' in probabilistic terms, consider a voting model in which each voter i's probability p_i of voting "yes" on a bill is a random variable. Each voter asks the *question of individual effect*: What is the probability that my vote will make a difference to the outcome? That is, what is the probability that other voters will cast their votes for a bill in such a way that the bill will pass if I vote for it, but fail if I vote against it?

Of course, the answer to this question depends not only on the voting game, but also on the probability distributions of the p_i's. Two possible assumptions are the

Independence assumption. Each p_i is chosen independently from the uniform distribution on $[0,1]$.

Homogeneity assumption. A random variable p is chosen from the uniform distribution on $[0,1]$, and $p_i = p$ for all i.

We then have [Straffin (1977a)]

Theorem 3.3. *The answer to voter i's question of individual effect under the independence assumption is β'_i.*

Theorem 3.4. *The answer to voter i's question of individual effect under the homogeneity assumption is ϕ_i.*

Proofs. The answer to voter i's question of individual effect is the probability that a bill passes if we assume i votes for it, but would fail if i voted against it. This is exactly the probability that i will be a swing for the coalition S of "yes" voters, assuming i votes yes. This probability is

$$\sum_{\substack{i \text{ swings} \\ \text{for } S}} \left(\prod_{j \in S-i} p_j\right)\left(\prod_{j \in N-S}(1-p_j)\right). \tag{1}$$

If we set all $p_j = p$ and take the average value of (1) over p in $[0, 1]$ we get

$$\int_0^1 \sum_{\substack{i \text{ swings} \\ \text{for } S}} p^{s-1}(1-p)^{n-s} \, dp = \sum_{\substack{i \text{ swings} \\ \text{for } S}} \int_0^1 p^{s-1}(1-p)^{n-s} \, dp$$

$$= \sum_{\substack{i \text{ swings} \\ \text{for } S}} \frac{(s-1)!(n-s)!}{n!} = \phi_i.$$

In the penultimate step we used the "beta function identity" $\int_0^1 x^a(1-x)^b \, dx = a!b!/(a+b+1)!$. This proves Theorem 3.4.

On the other hand, averaging as each p_j ranges independently over $[0, 1]$ is equivalent to setting each $p_j = \frac{1}{2}$. If we do this for (1) we get

$$\sum_{\substack{i \text{ swings} \\ \text{for } S}} \frac{1}{2^{s-1}} \frac{1}{2^{n-s}} = \sum_{\substack{i \text{ swings} \\ \text{for } S}} \frac{1}{2^{n-1}} = \frac{\text{number of swings for } i}{2^{n-1}} = \beta_i'.$$

This proves Theorem 3.3. □

One way to interpret these theorems practically is to think of p_i as the "acceptability" of a given bill to voter i. The independence assumption says that the acceptability of a bill to voter i is independent of its acceptability to any other voter j. Under the homogeneity assumption the acceptability of any given bill is the same to all voters: voters judge bills by common standards. If we believe that voters in a certain body have such common standards, the Shapley–Shubik index is applicable; if we believe voters behave independently, the Banzhaf index is the instrument of choice. For the United States legislative scheme, the Shapley–Shubik index might be most appropriate, while for the diverse Canadian provinces, we might prefer the Banzhaf index. The model also gives insight into the strange behavior of the Banzhaf index in examples 1 and 2 at the beginning of this section. If a large number of small voters vote independently with p_j chosen from the uniform distribution on $[0, 1]$, the law of large numbers says that about half of them will vote yes. In that case, the outcome will be determined by how the largest voter X votes.

Alternatively, it is possible to use this model to define "partial homogeneity assumptions" tailored to particular voting situations. Certain groups of voters would choose their p_i homogeneously, independent from other groups of voters. We can even build in an elementary form of ideological opposition by having some groups of voters choose p_i opposite to other groups ($p_i = 1 - p_j$). See Straffin (1977a) and Straffin et al. (1981) for examples. In Section 5 we will see another, more powerful approach to measuring voting power when ideological considerations are important.

4. Dynamic applications of the power indices

If politics is the shaping of power, political actors might act to increase their power, and the rational choice assumption that they do so might have some explanatory efficacy in political dynamics. We will consider three possible situations of this type.

First, notice that voters can sometimes increase their power by forming coalitions. For example, consider [3; 1, 1, 1, 1, 1]. If the first three voters form a coalition and act as a voting bloc, the resulting game is [3; 3, 1, 1], in which the bloc has all the voting power. Each of the voters in the bloc has increased his share of the power from $\frac{1}{5}$ to $\frac{1}{3}$. Strategy of this kind – that voters should strive to form a minimal winning coalition to maximize their power – has been formalized by Riker as the *size principle* [Riker (1962), Riker and Ordeshook (1973)].

However, forming coalitions does not always increase power, since other players may also be forming coalitions. For example, if both the first two voters and the last two voters in the five person majority game form coalitions, the resulting game is [3; 2, 1, 2]. Each of the two-voter blocs has power $\frac{1}{3}$, which is less than the $\frac{2}{5}$ its members had before. An example of this phenomenon is considered in [Straffin (1977b)]. The county board of Rock County, Wisconsin is a 40-player majority game. There are two large cities in the county, Beloit with 11 board members, and Janesville with 14 board members. Bloc voting by board members from the two cities has not emerged, but has occasionally been urged upon the board members from Beloit. Indeed, if the board members from Beloit did form a bloc, it would have (by the Shapley–Shubik index) $\frac{11}{30} = 0.367$ of the power, instead of the $\frac{11}{40} = 0.275$ its members have without bloc voting. However, if the Janesville board members also formed a bloc in response to Beloit bloc voting, the resulting game would be

[21; 11, 14, 1, ···, 1],
     B   J   15 others

and in this game $\phi_B = 0.180$, with Beloit board members considerably *worse* off than without bloc voting. Perhaps the Beloit board members are being politically canny in resisting urgings to vote as a bloc. Straffin (1977b) gives general conditions for when bloc voting by each of two blocs helps both blocs, helps one bloc but hurts the other, or hurts both blocs.

If one way for two voters to change their power is to form a coalition and agree to cast their votes together, another way is to *quarrel*, i.e. "agree" to cast their votes differently. Kilgour (1974) first analyzed the power effects of quarreling voters. For a simple example, consider

[3; 2, 1, 1].
  A B C

To calculate the Shapley–Shubik indices we list the orderings and underline the pivot in each:

$$\underline{AB}C^* \quad A\underline{C}B \quad B\underline{A}C^* \quad BC\underline{A}^{*+} \quad C\underline{A}B \quad CB\underline{A}^{*+}$$

so that $\phi_A = \frac{2}{3}$ and $\phi_B = \phi_C = \frac{1}{6}$. Now suppose A and B quarrel. Kilgour interprets this to mean that orderings which have both A and B at or before the pivot are no longer feasible. With the orderings marked by the asterisk ruled out, and the remaining orderings considered equally likely, the Shapley–Shubik indices become $\phi_A^Q = \frac{1}{2}$, $\phi_B^Q = 0$, and $\phi_C^Q = \frac{1}{2}$. Both A and B have lost power because of their quarrel. The Banzhaf index, with the quarreling interpretation that winning coalitions which contain both A and B are not feasible, gives a similar result.

If it were always true that quarreling members lose power, we would have support for the moral dictum that nastiness can only hurt you. However, in some situations quarreling can *increase* the power of the quarrelers. If B and C quarrel in the above example, the orderings marked by + are infeasible and the resulting Shapley–Shubik indices are $\phi_A^Q = \frac{1}{2}$, $\phi_B^Q = \phi_C^Q = \frac{1}{4}$. B and C have gained power at the expense of A. Once again, the Banzhaf index gives a similar result. Brams (1976) called this the "paradox of quarreling members," and pointed out that

> It would appear, therefore, that power considerations – independently of ideological considerations – may inspire conflicts among members of a voting body simply because such conflicts enhance the quarreling members' voting power. [Brams (1976, p. 190)].

For other possibilities and examples, see Straffin (1983).

Brams and Riker (1972) and Straffin (1977c) have suggested that power dynamics might be a way to explain the onset of a "bandwagon effect" at a critical point in the building of rival coalitions. As an example, consider the 1940 Republican Presidential nominating convention in the United States. On the fifth ballot Wendell Willkie had 429 votes, Robert Taft 377, and the remaining 194 of the 1000 votes were split among other candidates. Suppose we model this situation as

$$[501; \quad 429, \quad 377, \quad 1, \quad \cdots, \quad 1].$$
$$ W \quad\ \ T \quad\ \ 194\ \text{Others}$$

The Shapley–Shubik power indices work out to be $\phi_W = 0.4023$, $\phi_T = 0.1356$ and $\phi_O = 0.0024$. Consider a single "O" deciding whether to remain uncommitted to either major candidate, or to support one of them. If she commits to Taft, the game becomes

$$[501; \quad 429, \quad 378, \quad 1, \quad \cdots, \quad 1],$$
$$ W \quad\ \ T \quad\ \ 193\ \text{O's}$$

with $\phi_W = 0.3999$, $\phi_T = 0.1370$. The commitment to T has raised T's power by 0.0014, which is less than the power our O would retain by remaining uncommitted. If power is thought of as a marketable commodity, which can be paid for by concessions or political promises, commitment to T is a losing proposition. On the other hand, commitment to W would produce

[501; 430, 377, 1, ⋯, 1]
     W   T   193 O's

with $\phi_W = 0.4064$, $\phi_T = 0.1333$. This commitment raises W's power by 0.0041, considerably more than O's uncommitted power. Hence our focal O might be wise to bargain for commitment to W on the next ballot. Since this is true for *all* O's, we might predict a bandwagon rush to commit to W. Willkie did win handily on the next ballot.

In general, as long as rival blocs are fairly evenly matched and far from the winning quota, an uncommitted voter has more power than that voter would add to either bloc by joining that bloc. However, as the blocs slowly build strength, there comes a time when it is power-advantageous for any uncommitted voter (hence all uncommitted voters) to join the larger bloc. The precise form for this bandwagon threshold for large games is derived in Straffin (1977c), where other examples from U.S. nominating conventions are considered. The claim, of course, is not that delegates to nominating conventions calculate Shapley–Shubik or Banzhaf power indices and make decisions accordingly, but that a well-developed political sense might be tuned to shifts of power in a form not unrelated to what these indices measure.

5. A spatial index of voting power

The Shapley–Shubik index measures the probability that a voter will be pivotal as a coalition forms, assuming that all orders of coalition formation are equally likely. It is clear that in any actual voting body, all orders of coalition formation are not equally likely. One major distorting influence is the factor of ideology, which has traditionally been modeled geometrically. A familiar example is the one-dimensional left–right liberal–conservative spatial model of voter ideology. For a more sophisticated model, we could place voters in a higher dimensional *ideological space*, where distance between voters represents ideological difference. Owen (1971) and Shapley (1977) proposed adapting the Shapley–Shubik index to such spatially placed voters.

Figure 1 shows a two-dimensional placement of the voters A, B, C, D, E in a symmetric majority game [3; 1, 1, 1, 1, 1]. The Shapley–Shubik index, ignoring ideology, would give each voter $\frac{1}{5}$ of the voting power. On the other hand, common political wisdom would expect the "centrist" voter D to have more power than the other voters. In this context Shapley pictures a forming coalition as a line [in the general k-dimensional case, a $(k-1)$-dimensional hyperplane] sweeping through the plane in a fixed direction given by some unit vector *u*, picking up voters as it moves. The vector *u* could be thought of as describing the "ideological direction"

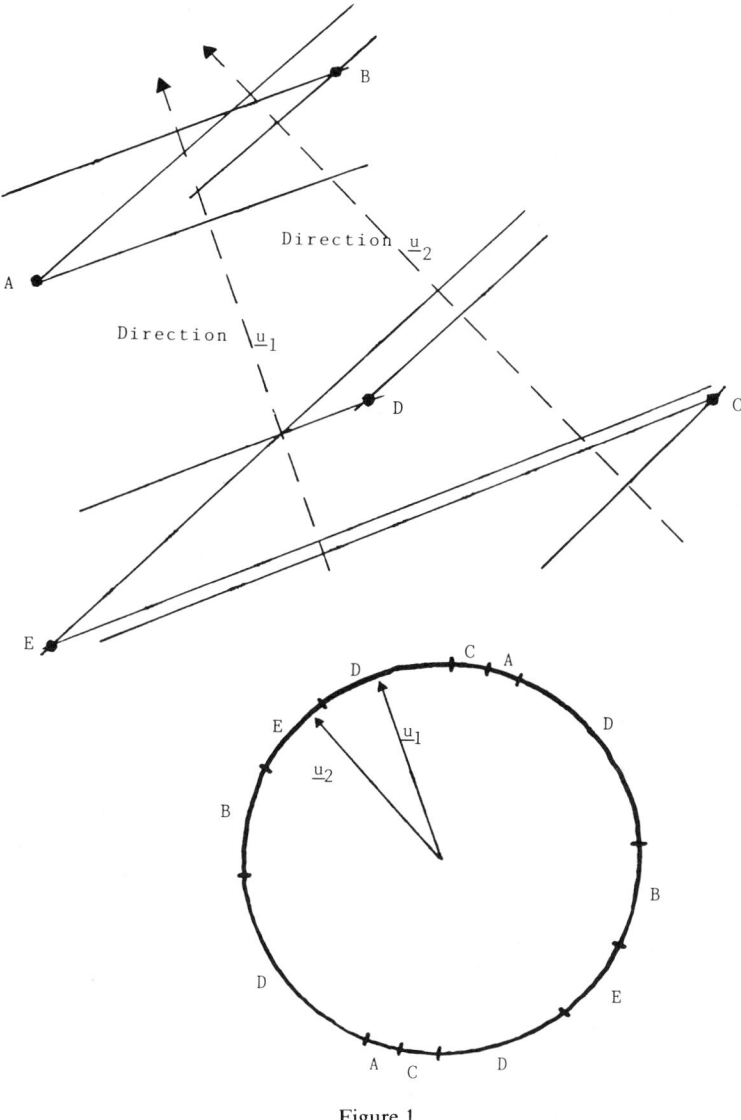

Figure 1

of the bill under consideration – the particular mix of ideological factors embodied in the bill. In Figure 1, a coalition forming in direction u_1 picks up voters in order CEDAB, with D as pivot, while a coalition forming in direction u_2 picks up voters in order CDEBA with E as pivot. In this way, each unit vector, hence each point on the unit circle [in general the unit $(k-1)$-sphere] is labeled by a voter. Voter

i's *Shapley–Owen power index* ψ_i is the proportion of the unit circle labeled by that voter. In other words, it is probability that voter i is pivotal, assuming that all *directions* of coalition formation are equally likely. In the example of Figure 1 the indices are

$$\psi_A = 20°/360° = 0.06, \quad \psi_B = 62°/360° = 0.17, \quad \psi_C = 22°/360° = 0.06,$$
$$\psi_D = 204°/360° = 0.57, \quad \psi_E = 52°/360° = 0.14.$$

Voter D is indeed most powerful.

In two dimensions, the Shapley–Owen index is efficiently calculable by a *rotation algorithm*, which I will describe for the case of a simple majority game in which the total number of votes is odd (although the algorithm is more generally applicable). Define a *median line* to be a line L in the plane R^2 such that each open half plane H_1 and H_2 of $R^2 \setminus L$ contains less than a majority of votes. It follows that

(i) every median line contains at least one voter point, and each closed half plane $H_1 \cup L$ and $H_2 \cup L$ contains a majority of votes,
(ii) there is exactly one median line in each direction, and
(iii) voter i's Shapley–Owen power index is the sum of the angles swept out by median lines passing through point i, divided by 180° (thinking of lines as undirected).

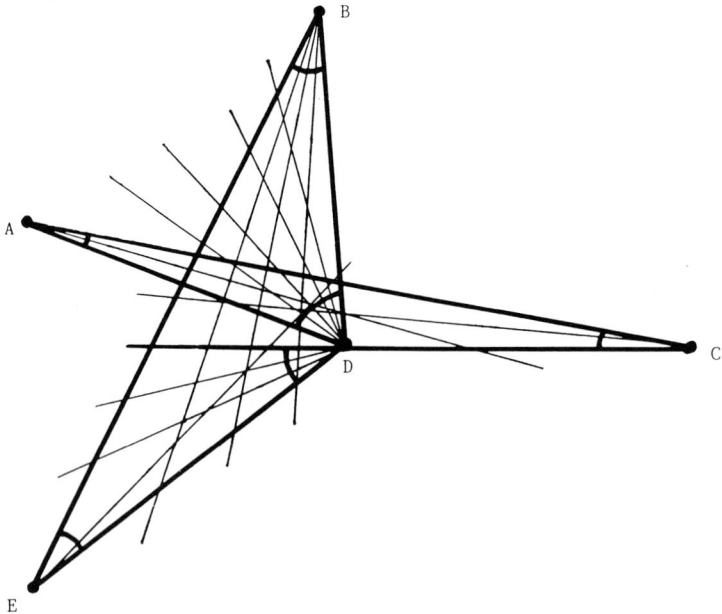

Figure 2

Hence to calculate the indices, start with any median line, say one passing through A in Figure 2. Imagine rotating it counterclockwise about A until it passes through another voter point (C in this example), assigning the angle of rotation to voter A. As we rotate past AC, the median line switches to passing through C. Continue rotating, always assigning the angle of rotation to the voter through whose point the median line passes, until after 180° rotation the line returns to its original position. In practice, of course, the algorithm is discrete, since the median line can only change its voter point at those directions, finite in number, when it passes through two or more voter points. The Shapley–Owen indices of the voters in our example are the angles marked in Figure 2, divided by 180°.

Figure 3 shows a two-dimensional spatial plot of states in the United States Electoral College, from Rabinowitz and Macdonald (1986). It was constructed from a principal component analysis of Presidential election data in the period 1944–1980.

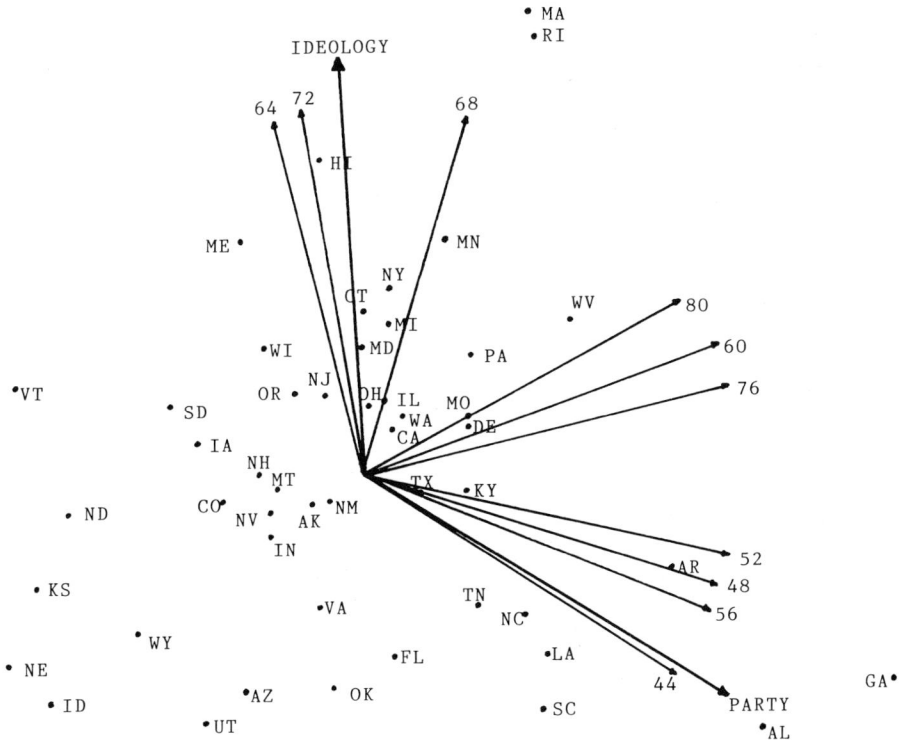

Figure 3

The first component (vertical, explaining 48% of the variance) is identifiable as liberal–conservative. The second component (horizontal, explaining an additional 31% of the variance) has to do with traditional Republican–Democratic party identification. Since the third component explains only an additional 7% of the variance, the two-dimensional spatial plot captures most of the ideological information in the election data. Notice that the analysis also gives the projection of the ten Presidential election axes onto the principal component plane.

In calculating the Shapley–Owen indices of the states in Figure 3, Rabinowitz and Macdonald noted that the directions for all of the Presidential elections 1964–1980 lie within a sector of about 90°. Hence to get a measure of electoral power applicable to the modern period, they used only directions in this sector. The results are shown in Table 3. The modified Shapley–Owen indices are given in column 3, and are translated into effective electoral votes in column 5. Column 6 gives the classical Shapley–Shubik indices in the same form. The difference between these figures, in column 7, is a measure of the extent to which a state's ideological position increases or decreases its effective electoral power. Notice that the states which benefit most heavily in proportion to their number of votes – California, Texas, Illinois, Ohio and Washington – are all centrally located in Figure 3. States which lose most because of ideology are the outlying states – Massachusetts, Rhode Island, Idaho, Nebraska and Utah.

Rapoport and Golan (1985) carried out an interesting comparative study of the Shapley–Shubik, Banzhaf, and Deegan–Packel (1979) power indices and their corresponding spatial generalizations. They computed all six indices for parties in the Israeli Knesset in 1981, and compared the results to the perceived distribution of power as rated by three groups of political experts. The spatial corrections improved the match (as measured by mean absolute deviation) between the power indices and perceived power, for all three indices. The resulting match was quite good for Banzhaf, not quite as good for Shapley–Owen, and poor for Deegan–Packel.

6. Spatial models of voting outcomes

When we represent voters by points in an ideological space, we can also represent voting alternatives by points in the same space. For example, the two dimensions of R^2 as an ideological space might represent yearly expenditures for military purposes and for social welfare purposes. A voter is represented by his *ideal point*, whose coordinates are the amounts he would most like to see spent for these two purposes. A budget bill is represented by the point whose coordinates are its expenditures for the two purposes. The assumption is that in a vote between two rival budget bills each voter, at least when voting is sincere, will vote for the bill closest to his ideal point. It is most common to use the standard Euclidean measure of distance, although in some situations other distance measures might be more appropriate.

Table 3
Power in the modern sector [Rabinowitz and Macdonald (1986)]

Rank	State	Power (%)[a]	Electoral votes	Power electoral votes[b]	Shapley electoral votes	Power–Shapley[c]
1	California	12.02	47	64.64	49.91	+14.74
2	Texas	6.99	29	37.60	29.60	+8.00
3	New York	6.70	36	36.02	37.29	−1.27
4	Illinois	5.71	24	30.74	24.25	+6.50
5	Ohio	5.47	23	29.42	23.19	+6.23
6	Pennsylvania	4.96	25	26.68	25.31	+1.37
7	Michigan	3.97	20	21.37	20.05	+1.33
8	New Jersey	3.66	16	19.68	15.92	+3.77
9	Florida	3.36	21	18.09	21.09	−3.00
10	North Carolina	2.53	13	13.63	12.86	+0.77
11	Missouri	2.41	11	12.95	10.84	+2.11
12	Wisconsin	2.34	11	12.57	10.84	+1.73
13	Washington	2.33	10	12.52	9.84	+2.68
14	Tennessee	2.22	11	11.93	10.84	+1.08
15	Indiana	2.19	12	11.78	11.85	−0.07
16	Maryland	2.16	10	11.62	9.84	+1.78
17	Kentucky	2.02	9	10.86	8.84	+2.02
18	Virginia	1.91	12	10.29	11.85	+1.56
19	Louisiana	1.85	10	9.95	9.84	−0.11
20	Connecticut	1.60	8	8.62	7.84	+0.78
21	Iowa	1.56	8	8.37	7.84	+0.52
22	Oregon	1.53	7	8.22	6.85	+1.37
23	Colorado	1.48	8	7.93	7.84	+0.10
24	Georgia	1.47	12	7.89	11.85	−3.96
25	Minnesota	1.32	10	7.10	9.84	−2.74
26	South Carolina	1.26	8	6.78	7.84	−1.06
27	Alabama	1.26	9	6.77	8.84	−2.07
28	Arkansas	1.07	6	5.77	5.86	−0.09
29	New Mexico	1.06	5	5.69	4.87	+0.82
30	Oklahoma	0.95	8	5.08	7.84	−2.76
31	West Virginia	0.85	6	4.57	5.86	−1.29
32	New Hampshire	0.83	4	4.45	3.89	+0.56
33	Montana	0.82	4	4.39	3.89	+0.50
34	Mississippi	0.78	7	4.17	6.85	−2.68
35	Nevada	0.76	4	4.06	3.89	+0.17
36	Maine	0.73	4	3.91	3.89	+0.02
37	Delaware	0.65	3	3.51	2.91	+0.60
38	Kansas	0.63	7	3.41	6.85	−3.44
39	Alaska	0.62	3	3.34	2.91	+0.43
40	Arizona	0.62	7	3.33	6.85	−3.52
41	South Dakota	0.57	3	3.07	2.91	+0.16
42	Hawaii	0.52	4	2.77	3.89	−1.12
43	Vermont	0.44	3	2.39	2.91	−0.53
44	North Dakota	0.37	3	2.00	2.91	−0.92
45	Massachusetts	0.33	13	1.78	12.86	−11.08
46	Utah	0.32	5	1.75	4.87	−3.12
47	Wyoming	0.27	3	1.46	2.91	−1.46
48	Nebraska	0.26	5	1.42	4.87	−3.45
49	Idaho	0.19	4	1.02	3.89	−2.87
50	Rhode Island	0.12	4	0.64	3.89	−3.25

[a] Percentage of times the state occupies the pivotal position.
[b] Percentage of times the state occupies the pivotal position multiplied by the total number of electoral votes (538).
[c] Difference between power measures.

Suppose voters in a voting game are positioned in R^k and consider points in R^k as alternatives from which they are to make a collective choice. Could we say from game theoretical considerations which point should be the outcome or, if that is asking too much, at least specify a set of points within which the outcome should lie? From among a number of proposed answers to this question, we will consider three. For simplicity of presentation, we will assume that $k \leqslant 2$, and that the voting game is a simple majority game with an odd number of voters.

The core

Definition. An alternative x is in the *core* of a spatial voting game if and only if there is no alternative y such that y defeats x.

If the voting game is one-dimensional, then for a simple majority game with an odd number of voters, the core consists of exactly one point, the ideal point of the median voter. Hence in particular the core is always non-empty. For higher dimensions, the core will be non-empty if and only if the voter ideal points satisfy a restrictive symmetry condition first given by Plott (1967). To derive this condition for $k = 2$, recall the idea of a median line from the last section. Notice that if L is a median line and x is a point not on L, then x will lose under majority voting to x', which is the reflection of x across L, moved slightly towards L. This is because x' will be preferred to x by all voters in $L \cup H_2$, where H_2 is the half plane containing x', and this is a majority of voters. Since x can be beaten if there is some median line which does not contain x, an alternative x can only be in the core if it lies on all median lines. We have proved

Theorem 6.1a [Davis et al. (1972), Feld and Grofman (1987)]. *The core of a simple majority game in R^2 will be non-empty if and only if all median lines pass through a single point. If they do, that point, which must be a voter ideal point, is the unique point in the core.*

A bit of geometry shows that this is equivalent to

Theorem 6.1b [Plott (1967)]. *The core of a simple majority game in R^2 will be non-empty if and only if the voter ideal points all lie on a collection of lines L_i such that*

  (i) *the lines L_i all pass through the ideal point of one voter A, and*
  (ii) *A is the median voter on each of the lines L_i.*

This condition is clearly structurally unstable – small perturbations of voter ideal points will destroy it. Hence

Theorem 6.1c. *Generically, simple majority games in dimensions two or higher have empty core.*

The top cycle set. If the core is empty, then no alternative can directly defeat all other alternatives. However, we might consider the set of alternatives x which at least can defeat all other alternatives in a finite number of steps.

Definition. An alternative x is in the *top cycle set* of a spatial voting game if for any other alternative y, there is a finite sequence of alternatives $x = x_0, x_1, \ldots, x_m = y$ such that x_i defeats x_{i+1} for all $i = 0, \ldots, m - 1$.

It is surprising how large the top cycle set is:

Theorem 6.2 [McKelvey (1976)]. *If the core of a simple majority game in R^k is empty, then the top cycle set is all of R^k.*

To see what this theorem means, suppose we have three voters A, B, C as in Figure 4, and suppose we start with any alternative w. Choose any other point in the plane, perhaps one far away like z. McKelvey's theorem says that there must be a chain of alternatives starting with w and ending with z, such that each alternative in the chain is preferred to the preceding alternative by a majority of the voters. If we control the voting agenda and present the alternatives in the

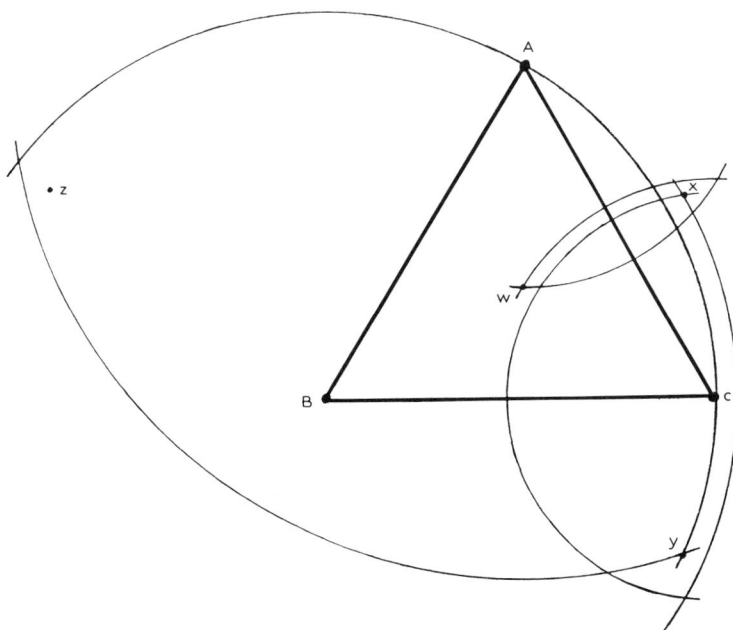

Figure 4

chain in order, we can get our small society to accept z, or any other alternative in the plane. The agenda controller, in the generic case where the core is empty, has complete power.

Figure 4 illustrates how to construct a McKelvey chain for this example. Notice that AB, AC and BC are all median lines. If we reflect w across AC and move it in slightly to get x, then x beats w. Now reflect x across BC and move it in slightly to get y, which beats x. Now reflect y across AB and move it in slightly to get z. We have found the required chain w, x, y, z.

This simple example has in it the proof of the general theorem for $k = 2$. For first suppose that, in this example, we were trying to get to some other point t. Just keep repeating the three reflections in order, obtaining points x_1, y_1, z_1, $x_2, y_2, z_2, \ldots$ as far away from A, B and C as we please. When we get some z_m far enough away beyond t, t will be unanimously preferred to z_m and we will have our chain. Second, suppose we have any configuration of any number of voters, but empty core. Then by Theorem 6.1a there must be three median lines which do not pass through a common point, and hence bound a triangle. Let this triangle play the role of ABC in the above argument.

McKelvey's result, with its implications of the inherent instability of majority rule and the power of agenda control, has generated a large literature. For a thoughtful discussion, see Riker (1980).

The strong point. If the core is empty, then every alternative is defeated by some other alternative. However, we might look for the alternative which is defeated by the fewest possible other alternatives.

Definition. For a point x in R^k, **def**(x) is the set of all points in R^k which defeat x.

Definition. The *strong point* of a simple majority game in R^k is the point x in R^k for which k-volume of **def**(x) is as small as possible. [Grofman et al. (1987)].

It is easy to show that the strong point exists and is unique. However, it might seem that the strong point would be difficult to compute. This is not true, at least for $k = 2$, by a recent result of Owen and Shapley, which connects voting outcomes to voting power.

Theorem 6.3 [Owen and Shapley (1989)]. *The strong point in a two-dimensional majority voting game is the weighted average of the voter ideal points, where the weights are the Shapley–Owen power indices of the voters.*

In other words, the strong point is located exactly at the "center of power". I refer you to Owen and Shapley (1989) for the general proof of this theorem, but I would like to illustrate in a simple case the elegant geometric insight on which it is based. In Figure 5 we would like to find a formula for the area of **def**(X),

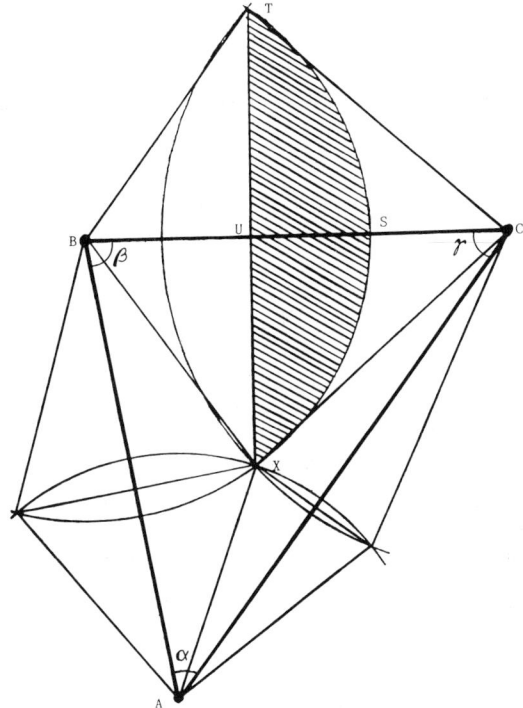

Figure 5

which in this case is the union of six segments of circles, one of which, $XSTU$, is shaded in the figure. The area of this segment is twice the difference between the circular sector BSX and the triangle BUX. Adding six such areas together, we get

$$\text{Area of } \mathbf{def}(X) = 2\left(\frac{\alpha}{2}\overline{AX}^2 + \frac{\beta}{2}\overline{BX}^2 + \frac{\gamma}{2}\overline{CX}^2 - \triangle ABC\right),$$

where α, β and γ are measured in radians.

Since the area of triangle ABC is a constant not depending on X, to minimize the area of $\mathbf{def}(X)$ we must choose X to minimize $\alpha\overline{AX}^2 + \beta\overline{BX}^2 + \gamma\overline{CX}^2$. It is well known that this weighted sum of squared distances is minimized by taking

$$X = \frac{\alpha A + \beta B + \gamma C}{\alpha + \beta + \gamma} = \frac{\alpha}{\pi}A + \frac{\beta}{\pi}B + \frac{\gamma}{\pi}C.$$

These coefficients of A, B and C are exactly the Shapley–Owen power indices of voters A, B and C.

The conclusion of Theorem 6.3 does not, unfortunately, hold in dimensions $k > 2$. For other solution concepts for the problem of voting outcomes, and for a general introduction to other spatial voting ideas, see Enelow and Hinich (1984). For a survey of coalitional questions in spatial voting, see Straffin and Grofman (1984).

Bibliography

Brams, S.J. (1975) *Game Theory and Politics*. New York: Free Press.
Brams, S.J. (1976) *Paradoxes in Politics*. New York: Free Press.
Brams, S.J. and P. Affuso (1985) 'New paradoxes of voting power on the EC Council of Ministers', Electoral Studies, **4**: 135–139 & 290.
Brams, S.J. and W.H. Riker (1972) 'Models of coalition formation in voting bodies', in: Herndon and Bernd, eds., *Mathematical Applications in Political Science* **VI**. Charlottesville: University of Virginia Press.
Banzhaf, J. (1965) 'Weighted voting doesn't work: a mathematical analysis', *Rutgers Law Review*, **19**: 317–343.
Banzhaf, J. (1968) 'One man, 3.312 votes: a mathematical analysis of the electoral college', *Villanova Law Review*, **13**: 304–332.
Davis, O., M. Degroot and M. Hinich (1972) 'Social preference orderings and majority rule', *Econometrica*, **40**: 147–157.
Deegan, J. and E. Packel (1979) 'A new index of power for simple n-person games', *International Journal of Game Theory*, **7**: 113–123.
Dubey, P. (1975) 'On the uniqueness of the Shapley value', *International Journal of Game Theory*, **4**: 131–139.
Dubey, P. and L.S. Shapley (1979) 'Mathematical properties of the Banzhaf index', *Mathematics of Operations Research*, **4**: 99–131.
Enelow, J. and M. Hinich (1984) *'The Spatial Theory of Voting: An Introduction*. Cambridge: Cambridge University Press.
Feld, S. and B. Grofman (1987) 'Necessary and sufficient conditions for a majority winner in n-dimensional spatial voting games: an intuitive geometric approach', *American Journal of Political Science*, **31**: 709–728.
Grofman, B., G. Owen, N. Noviello and G. Glazer (1987) 'Stability and centrality of legislative choice in the spatial context', *American Political Science Review*, **81**: 539–552.
Kilgour, D.M. (1974) 'A Shapley value analysis for cooperative games with quarreling', in Rapoport, A., *Game Theory as a Theory of Conflict Resolution*. Dordrecht: Reidel.
Kilgour, D.M. (1983) 'A formal analysis of the amending formula of Canada's Constitution Act, 1982', *Canadian Journal of Political Science*, **16**: 771–777.
Kilgour, D.M. and T.J. Levesque (1984) 'The Canadian constitutional amending formula: bargaining in the past and future', *Public Choice*, **44**: 457–480.
Lucas, W.F. (1983) 'Measuring power in weighted voting systems', in: S.J. Brams, W. Lucas and P.D. Straffin, eds., *Political and Related Models*. New York: Springer Verlag.
Mann, I. and L.S. Shapley (1962) 'The a priori voting strength of the electoral college', in Shubik, M., *Game Theory and Related Approaches to Social Behavior*. New York: Wiley.
McKelvey, R. (1976) 'Intransitivities in multidimensional voting models and some implications for agenda control', *Journal of Economic Theory*, **12**: 472–482.
Miller, D. (1973) 'A Shapley value analysis of the proposed Canadian constitutional amendment scheme', *Canadian Journal of Political Science*, **4**: 140–143.
Milnor, J. and L.S. Shapley (1978) 'Values of large games II: oceanic games', *Mathematics of Operations Research*, **3**: 290–307.
Owen, G. (1971) 'Political games', *Naval Research Logistics Quarterly*, **18**: 345–355.
Owen, G. (1975) 'Evaluation of a Presidential election game', *American Political Science Review*, **69**: 947–953.

Owen, G. and L.S. Shapley (1989) 'Optimal location of candidates in ideological space', *International Journal of Game Theory*, **18**: 339–356.
Ordeshook, P. (1978) *Game Theory and Political Science*. New York: New York University Press.
Plott, C. (1967) 'A notion of equilibrium and its possibility under majority rule', *American Economic Review*, **57**: 787–806.
Rabinowitz, G. and S. Macdonald (1986) 'The power of the states in U.S. Presidential elections', *American Political Science Review*, **80**: 65–87.
Rapoport, A. and E. Golan (1985) 'Assessment of political power in the Israeli Knesset', *American Political Science Review*, **79**: 673–692.
Riker, W.H. (1962) *The Theory of Political Coalitions*. New Haven: Yale University Press.
Riker, W.H. (1980) 'Implications from the disequilibrium of majority rule for the study of institutions', *American Political Science Review*, **74**: 432–446.
Riker, W.H. (1986) 'The first power index', *Social Choice and Welfare*, **3**: 293–295.
Riker, W.H. and P. Ordeshook (1973) *An Introduction to Positive Political Theory*. Englewood Cliffs: Prentice Hall.
Roth, A. (1988) 'The Shapley Value: Essays in Honor of Lloyd S. Shapley'. Cambridge: Cambridge University Press.
Shapley, L.S. (1962) 'Simple games: an outline of the descriptive theory', *Behavioral Science*, **7**: 59–66.
Shapley, L.S. (1977) 'A comparison of power indices and a non-symmetric generelization', RAND Paper P-5872, Rand Corporation Santa Monica.
Shapley, L.S. and M. Shubik (1954) 'A method for evaluating the distribution of power in a committee system', *American Political Science Review*, **48**: 787–792.
Straffin, P.D. (1977a) 'Homogeneity, independence and power indices', *Public Choice*, **30**: 107–118.
Straffin, P.D. (1977b) 'The power of voting blocs: an example', *Mathematics Magazine*, **50**: 22–24.
Straffin, P.D. (1977c) 'The bandwagon curve', *American Journal of Political Science*, **21**: 695–709.
Straffin, P.D. (1980) *Topics in the Theory of Voting*. Boston: Birkhauser.
Straffin, P.D. (1983) 'Power indices in politics', in: S.J. Brams, W. Lucas and P.D. Straffin, eds., *Political and Related Models*. New York: Springer Verlag.
Straffin, P.D., M. Davis and S.J. Brams (1981) 'Power and satisfaction in an ideologically divided voting body', in: M. Holler ed., *Power, Voting and Voting Power*. Wurzburg: Physica Verlag.
Straffin, P.D. and B. Grofman (1984) 'Parliamentary coalitions: a tour of models', *Mathematics Magazine*, **57**: 259–274.

Chapter 33

GAME THEORY AND PUBLIC ECONOMICS

MORDECAI KURZ*

Stanford University

Contents

1.	Introduction	1154
2.	Allocation of pure public goods	1155
	2.1. Pure public goods: Lindahl equilibrium and the core	1156
	2.2. Decentralization in public goods economies: are public goods large or small?	1161
	2.3. The core of the second best tax game	1165
	2.4. Shapley value of public goods games	1168
3.	Externalities, coalition structures and local public goods	1174
	3.1. On the core of games with externalities and public goods	1174
	3.2. Equilibria and cores of local public goods economies	1177
	3.3. Second best tax game revisited	1182
4.	Power and redistribution	1183
5.	Some final reflections	1189
	References	1190

*This work was supported by National Science Foundation Grant IRI-8814953 at the Institute for Mathematical Studies in the Social Science, Stanford University, Stanford, CA. The author expresses his gratitude to Daniel Klein, Myrna Wooders and Suzanne Scotchmer for helpful comments on an earlier draft.

Handbook of Game Theory, Volume 2, Edited by R.J. Aumann and S. Hart
© Elsevier Science B.V., 1994. All rights reserved

1. Introduction

This paper explores those areas of public economics where game theory has had a constructive contribution. I note first that the field of "public economics" did not exist as a coherent field until the late 1960s. Up to that time the field consisted of two separate areas of study, neither one of which was pursued with great vitality in the post World War II period. One was welfare economics including welfare measurements, externalities, and efficiency properties of alternative fiscal systems. The other was applied public finance which examined the structure of public budgets and expenditures including the incidence of taxes. Public economics has been profoundly altered and unified by the systematic introduction of the general equilibrium framework and game theoretic tools. Since my goal is to review these areas of the field where game theory has been effectively used, almost all the contributions I review were written after 1970. In those papers reviewed I will highlight the main results aiming to give the reader a broad perspective on the main ideas involved. This task is made especially difficult because of the great diversity in the formulation of problems in this area. This point needs an explanation. Virtually all the problems of public economics which I review will involve the consideration of public goods, externalities in consumption or production, redistribution of resources and the effects of the distribution of power on public decisions. As distinct from the unified, comprehensive structure of a private goods economy, an economy with a public sector is not a well defined object. Rather, there is a complex family of models of such economies and these vary dramatically both in the generality of their formulation as well as the specific aspects of the public sector which are under study. This means that the examination of this field requires the study of diverse models which vary in their basic structure and/or in their underlying assumptions. Consequently, each section of this paper will present its own prototype model and expose the basic results relative to this model.

A secondary practical consequence of the diversity in the field is that I cannot hope to review in this paper all the important work that has been done. My limited objective is to select a sample of representative results that will give the reader a comprehensive picture of how game theory is being used in this field of study. This is obviously a subjective matter. I perceive it to be an effective means of exposing the main results without meaning to evaluate contributors in terms of importance. I must therefore tender my apologies to any scholar whose work has not been adequately reviewed here. Given the extremely large number of papers available and the limited space I have, this apology must be widely extended.

The layout of the paper follows a simple structure: Section 2 surveys the problem of allocating pure public goods under conditions of complete information. Section 3 analyzes the controversial issue of local public goods. Section 4 covers the effect of power on the tax structure and income redistribution. Some final reflections will conclude the paper.

2. Allocation of pure public goods

We often think of a "public good" as a good whose use can be shared by many agents. However, even this basic concept is subject to the diversity of modeling mentioned in the introduction. To clarify some *basic terminology* let $N = \{1, 2, \ldots, n\}$ be the set of consumers, let X^i be the consumption set of i and let x^i be a vector of consumption by agent i. Following the notation of Milleron (1972), I will sometimes find it useful to decompose x^i as follows:

$$x^i = (x_L^i, x_Q^i) \in X^i \subseteq R^{l+q},$$

where $x_L^i = $ a vector of l private goods consumed, $x_Q^i = $ a vector of q public goods consumed. Denoting the supply of private goods to be allocated among the n consumers by z_L then feasibility requires

$$\sum_{i=1}^{n} x_L^i \leq z_L. \qquad (1)$$

However, for public goods this is not ture. Pure public goods are often identified by the feasibility condition

$$x_Q^i \leq z_Q \quad \text{all } i = 1, 2, \ldots, n, \qquad (2)$$

where z_Q is the supply of these public goods. One must note, however, that condition (2) contains a few important assumptions:

(i) *No exclusion.* The good in question is like radio waves and any consumer may enjoy all of the available supply; thus $x_Q^i = z_Q$ is feasible.

(ii) *Free disposal.* Each consumer may turn off the radio and not use all of z_Q. However, if z_Q is air pollution and the consumer has no ability to exclude himself then $x_Q^i < z_Q$ is not feasible. Note that "no exclusion" and "free disposal" are often intimately related.

(iii) *No congestion.* If z_Q is, say, the area of public beach available for use then the condition $x_Q^i \leq z_Q$ presumes that only i is on the beach. If other people are also on the beach causing congestion then the amount of free beach used by i may be written in the form

$$x_Q^i \leq z_Q g(x_Q^1, x_Q^2, \ldots, x_Q^n), \qquad (3)$$

where $g(\cdot)$ measures congestion.

This discussion reveals that a pure public good without exclusion, with free disposal, and without congestion induces a feasibility condition like (2). This also provides a hint of some of the subtle issues of modeling economies with public goods.

The term *local* public goods as distinct from pure public goods will be employed extensively. It refers to public goods with exclusion, but the exclusion is almost always associated with the spatial character of the good. Thus the main street of a town is a local public good since exclusion takes place effectively by the requirement that the user be present in the town in order to use this public good. Naturally, local public goods may or may not be subject to congestion and they may be further excluded within the locality (e.g. the municipal swimming pool).

I turn now to the theory of Lindahl equilibrium and the core.

2.1. Pure public goods: Lindahl equilibrium and the core

The presence of public goods and taxes raises the immediate question of how a competitive economy would function under such conditions. Samuelson (1954) was the first to note that a Pareto optimum with public goods would call for the qualization of the marginal rate of transformation of a public good into private goods with the *sum* of the private marginal rates of substitution. It was also reasonably well understood at that time that market externalities can be "corrected" with the proper set of taxes and subsidies. The essential contribution of Foley (1967, 1970) was to integrate these ideas into a general equilibrium formulation. Foley (1967) introduced the concept of *a public competitive equilibrium* and studied its efficiency properties. He then reformulated the concept of Lindahl equilibrium and showed that it is in the core of a certain game. Foley's work was extended and generalized in many directions and these extensions will be briefly reviewed later. Foley's formulation is simple and enables a direct exposition of the main ideas and for this reason I will follow it here.

The real line is R and the m-dimensional Euclidean space is R^m. If x and y are in R^m, I write $x \geq y$ if $x^i \geq y^i$ for all i, $x > y$ if $x \geq y$ and $x \neq y$ and $x \gg y$ if $x^i > y^i$ all i. The set of all $x \in R^m$ such that $x \geq 0$ is denoted by R^m_+.

The economy has l private goods and q public goods. There is a set $N = \{1, 2, \ldots, n\}$ of consumers but only a single aggregate enterprise. The economy $\mathscr{E}$ is composed of

(i) n consumption sets $X^i \subseteq R^{l+q}$. Each consumer is endowed with an initial vector ω_L^i of private goods. $\omega_L = \sum_{i=1}^n \omega_L^i$,

(ii) n complete and transitive preference orderings $\underset{i}{\succcurlyeq}$ on X^i,

(iii) an aggregate production set $Y \subset R^{l+q}$. A production program is written $y = (y_L, y_Q)$ where y_L is the *net* output of private goods and y_Q is the output of public goods. No endowment of public goods is allowed.

(iv) ownership shares θ^i. $\sum_{i=1}^n \theta^i = 1$. θ^i denotes the ownership of consumer i in the profits of the aggregate enterprise.

A program $(x^{1*}, x^{2*}, \ldots, x^{n*}, y^*)$ is attainable for $\mathscr{E}$ if

(a) for every i, $x^{i*} \in X^i$,
(b) $y^* \in Y$,
(c) $\sum_{i=1}^{n} x_L^{i*} = y_L^* + \sum_{i=1}^{n} \omega_L^i$,
(d) $x_Q^{i*} = y_Q^k$, $i = 1, 2, \ldots, n$.

Foley (1970) proposed a set of assumption about X^i, $\underset{i}{\succcurlyeq}$ and Y which are familiar from the theory of general equilibrium. These have been weakened in many directions and this will be briefly discussed later. Unfortunately the weaker assumptions tend to be rather complicated and for this reason we state mainly the simpler assumption of Foley:

A.1. For each i, $X^i \subset R^{l+q}$ is bounded from below, closed, convex and has an interior in the private goods subspace. The endowment ω_L^i is in this interior.

A.2. For each i, $\underset{i}{\succcurlyeq}$ is convex, continuous and monotonic (increasing in x_{Lj}^i and x_{Qj}^i).

A.3. Y is a closed convex cone with the following four properties:
  (a) $0 \in Y$,
  (b) *private goods are net inputs*: if $0 \neq (x_L, x_Q) \in Y$ then $x_{Lj} < 0$ for some j,
  (c) *public goods are not needed as production inputs*: if $(x_L, x_Q) \in Y$ and

$$\bar{x}_{Qj} = x_{Qj} \quad \text{for } j \text{ with } x_{Qj} \geq 0,$$
$$\bar{x}_{Qj} = 0 \quad \text{for } j \text{ with } x_{Qj} < 0,$$

  then $(x_L, \bar{x}_Q) \in Y$,
  (d) *possibility of public production*: for all $j = 1, 2, \ldots, n$ there exists $(x_L, x_Q) \in Y$ with $x_{Qj} > 0$.

I can now introduce the first competitive concept with a public sector. *A public competitive equilibrium (with lump-sum taxes)* is an attainable program for $\mathscr{E}$, a price system $p^* \in R^{l+q}$ and a system of taxes $(t_1^*, t_2^*, \ldots, t_n^*)$ with $p_Q^* y_Q^* = \sum_{i=1}^{n} t_i^*$ such that

(a) $p^* y^* \geq p^* y$ for all $y \in Y$,
(b) for each consumer i the vector x_L^i satisfies

$$p_L^* x_L^i = p_L^* \omega_L^i + \theta^i p^* y^* - t_i^*$$

and given y_Q^* the vector x_L^{i*} is a minimizer of $p_L^* x_L^i$ in the set

$$\{x_L^i | (x_L^i, y_Q^*) \in X^i, (x_L^i, y_Q^*) \underset{i}{\succcurlyeq} (x_L^{i*}, y_Q^*)\},$$

(c) there is no other governmental proposal $(\bar{y}_Q, \bar{t}_1, \bar{t}_2, \ldots, \bar{t}_n)$ at prices p^* with the property that for every i there exists $\bar{x}_L^i$ with $(\bar{x}_L^i, \bar{y}_Q) \underset{i}{\succcurlyeq} (x_L^{i*}, y_Q^*)$ with at least one strict preference and $p_L^*(\bar{x}_L^i - x_L^{i*}) \leq t_i^* - \bar{t}_i$ for all $i \in N$.

In a public competitive equilibrium the production sector is profit maximizing and the consumers are optimizing given prices, the profit level of the production sector and lump-sum taxes. To interpret the behavior of government is more complex. The government evaluates alternative policies of public goods production and tax schedules; in equilibrium it cannot find a welfare improving program. However, the concept of "welfare improving" is delicate. Note that a consumer enjoying a bundle (x_L^i, y_Q) pays the sum of expenditures on private goods and his lump-sum tax, that is: $p_L^* x_L^i + t_i$. A welfare improving proposal is one which all consumers prefer and which, in addition, does not cost more than the original one. That is, for all i

$$(\bar{x}_L^i, \bar{y}_Q) \succsim_i (x_L^{i*}, y_Q^*),$$

$$p_L^* \bar{x}_L^i + \bar{t}_i \leq p_L^* x_L^{i*} + t_i^*,$$

with strict inequality for some i. Without using any of the specialized assumptions we now have:

Theorem 2.1A [Foley (1967)]. *Any attainable program satisfying conditions* (a) *and* (c) *of the definition of a public competitive equilibrium is Pareto optimal. In particular, any public competitive equilibrium (with lump-sum taxes) is Pareto optimal.*

The converse of Theorem 2.1A is much more complicated. However, utilizing assumptions A.1–A.3 we have:

Theorem 2.1B [Foley (1970)]. *Let* $(x^{1*}, x^{2*}, \ldots, x^{n*}, y^*)$ *be a Pareto-optimal program. Under assumptions* A.1–A.3 *above, for any distribution of the initial resources* $(\omega_L^1, \omega_L^2, \ldots, \omega_L^n)$ *and any system of shares* $(\theta^1, \theta^2, \ldots, \theta^n)$ *there exists a price system* $p^* \in R^{l+q}$ *and lump-sum taxes* $(t_1^*, \ldots, t_n^*)$ *such that the given program, together with* p^* *and* $(t_1^*, \ldots, t_n^*)$, *constitutes a public competitive equilibrium (with lump-sum taxes).*

The reader may note that one could have defined a public competitive equilibrium relative to other tax systems. In fact, Foley (1967) proved the existence and optimality of such equilibria with taxes which are proportional to private wealth (i.e: tax of $i = t_i p_L \omega_L^i$). Moreover, in a series of papers Greenberg (1975, 1977, 1983) shows that public competitive equilibria exist relative to continuous tax schemes under very general conditions of the public sector including externalities in production and consumption. It is then clear that due to the possibility of many alternative tax systems there are too many public competitive equilibria. The consideration of Lindahl equilibrium and the core of the associated public goods game is an attempt to define stronger kinds of outcomes. The idea of a Lindahl equilibrium [due to Lindahl (1967) but known to have been developed as early as 1919] is to replace the low dimensional tax structure $t \in R^n$ with a complex system of *individualized* public goods prices related to all agents

$(p_Q^1, p_Q^2, \ldots, p_Q^n) \in R^{nq}$. (4)

In an extension of the above model to an economy with m firms indexed by $f = 1, 2, \ldots, m$, prices of public goods will be individualized to firms as well and thus the complete system becomes

$(p_Q^1, p_Q^2, \ldots, p_Q^n, p_{Qf}^1, \ldots, p_{Qf}^m) \in R^{(n+m)q}$.

To explain Lindahl's idea (for the case of a single firm) let p_L^* be the price vector (in R^l) of private goods and let π be the profits of the production sector. The income W^i of consumer i is defined by

$$W^i = p_L^* \omega_L^i + \theta^i \pi.$$ (5)

The consumer is then visualized as "purchasing" with his income a vector x_L^i of private goods and a vector x_Q^i of public goods. With this construction, consumers form demand functions for private *as well as public goods* by selecting consumption vectors $x^i = (x_L^i, x_Q^i)$ which are maximal with respect to $\underset{i}{\succcurlyeq}$ in the budget sets

$$\{x^i \in X^i | p_L^* x_L^i + p_Q^{i*} x_Q^i \leqslant W^i\}.$$ (6)

The aggregate production sector acts as a profit maximizing firm. The profit π is defined by

$$\pi = p_L^* y_L + \left(\sum_{i=1}^n p_Q^{i*} \right) y_Q.$$ (7)

With these defined we can now introduce the following concept:

A *Lindahl equilibrium* is an attainable program $\{(x_L^{i*}, x_Q^{i*}) \; i = 1, \ldots, n; y^*\}$ and a price system $p^* = (p_L^*, p_Q^{1*}, \ldots, p_Q^{n*}) \; p^* \in R^{l+nq}$ such that

(a) y^* maximizes π as defined in (7) over $y \in Y$. The maximized value π^* defines W^{i*} in (5),
(b) for each i, if $\tilde{x}^i \underset{i}{\succ} x^{i*}$ then

$$p_L^* \tilde{x}_L^i + p_Q^{i*} \tilde{x}_Q^i > p_L^* x_L^{i*} + p_Q^{i*} x_Q^{i*} = W^{i*},$$

(c) $p^* \neq 0$.

Theorem 2.1C [Foley (1970)]. *Under assumptions A.1–A.3 there exists a Lindahl equilibrium.*

Theorems 2.1B and 2.1C can be proved under much more general conditions. The reader may consult the excellent survey of Milleron (1972) to see that the following extensions of the theorems are possible: (i) public goods are allowed to be a mixture of pure public goods and excludable public goods (subject to free disposal); (ii) monotonicity of preferences is relaxed and public "bads" are allowed; (iii) multiple competitive firms may be introduced to produce private and public goods; (iv) the production technologies may not be cones allowing positive profits in production; (v) firms may use public goods as a production input; (vi) an initial endowment of public goods may be introduced. These extensions involve substantial complexity which we avoid here. The central feature of the theory remains intact: a Lindahl equilibrium is a public competitive equilibrium with a price structure replacing taxes and a profit maximizing behavior by the production firms replacing the program evaluation function of the public sector in a public competitive equilibrium. It is because of these "competitive" features that the early developers of the theory hoped that it may provide a natural extension of the equilibrium of the Arrow–Debreu economy and provide a unified theory of value for economies with public goods and externalities. These hopes did not materialize as will be explained below when I present the critique of the concept of Lindahl equilibrium. Here I mention the fact that Lindahl prices were criticized since there is no obvious mechanism to establish them as a basis for allocation. Researchers thus searched for a game theoretical formulation of the public goods problem with solutions that implement "reasonable" allocations. Lindahl allocations, being first best, clearly remain a reference relative to which other candidate solutions must be evaluated. This brings us to the consideration of the *core* of the associated public goods game.

Given an initial distribution of endowments ω_L^i of private goods, a coalition S *can ensure itself* of a program $\{(x_L^i, x_Q^i), i \in S; (y_L, y_Q)\}$ if

(i) $(x_L^i, x_Q^i) \in X^i$ all $i \in S$,

(ii) $(y_L, y_Q) \in Y$,

(iii) $\sum_{i \in S} x_L^i = y_L + \sum_{i \in S} \omega_L^i$,

(iv) $x_Q^i = y_Q$.

A consumption program $\{(x_L^{i*}, x_Q^{i*}), i = 1, 2, \ldots, n\}$ *is blocked by a coalition* S if S can ensure itself a program $\{\hat{x}_L^i, \hat{x}_Q^i), i \in S; (\hat{y}_L, \hat{y}_Q)\}$ such that $(\hat{x}_L^i, \hat{x}_Q^i) \succ_i (x_L^{i*}, x_Q^{i*})$ all $i \in S$.

Theorem 2.1D [Foley (1970)]. *Under assumptions A.1–A.3, any Lindahl equilibrium is in the core.*

It is again to be noted that Theorem 2.1D may be proved under weaker conditions than those stated. Milleron (1972) proves such a result but he does assume Y to be a cone and, in addition, he permits all private and public goods to be freely disposable.

I turn now to the crucial question of the behavior of the core in a large economy. The supposition of Lindahl equilibrium being the public goods economy counterpart to the Walrasian equilibrium was formulated as the question whether the core of the public goods economy (properly enlarged) will converge to the set of

Lindahl equilibria. This proved to be false. Muench (1972) gave a counterexample of a non-pathological continuum economy for which the core and the set of Lindahl equilibria do not coincide. Milleron (1972) provides a simple example of an economy with one private good and one public good. The replicated economy has a unique Lindahl equilibrium allocation but the core does not shrink to it. Champsaur et al. (1975) present a model of an economy with a single private good and postulate mild assumptions on preferences (weak monotonicity) and production (no free production). Under these conditions the authors show that when the economy is replicated the core does not shrink at all. Also they present a simple example in which the core actually expands! The conclusion of this research is that the limit theorems of the core of markets with private goods do not extend to economies with public goods. Moreover, some authors like Rosenthal (1971) and Champsaur et al. (1975) express serious doubts whether core-like notions are useful to explain the behavior of large economies with public goods and externalities.

2.2. Decentralization in public goods economies: are public goods large or small?

My objective in this section is to discuss some questions which are essential to an equilibrium theory with public goods. I will develop these through a critical evaluation of the concept of Lindahl equilibrium. This will bring the discussion back to the asymptotic behavior of the core in public goods economies and will enable me to raise a fundamental question regarding the modeling of public goods economies.

One must recognize first that Lindahl equilibrium have properties which are crucially important to any equilibrium theory of public goods economies: they exist under standard assumptions, they are Pareto optimal (thus "first" best) and when Y is a cone with free disposal they are also in the core of the associated market game. Unfortunately, the discussion above revealed the fact that the core may not converge to the set of Lindahl equilibria when the number of agents is increased. This can be interpreted to mean that in a Lindahl equilibrium an agent does not have a negligible effect on prices no matter how large the economy is and hence agents have the incentive to act strategically. This has two interpretations. First, in an economy with public goods it may pay to form coalition structures and to induce bargaining among stable coalitions. This immediately opens the door for other solution concepts, as well as other formulations of public goods games, which allow richer strategy spaces. Such formulations are examined later in this paper.

The second interpretation relates to the fact that the preferences of the consumers are not observed and since there is no mechanism to force them to reveal their demands they are likely to attempt to "free ride" by distorting their announced preferences. This means that Lindahl equilibria are not "incentive compatible".

An alternate way of seeing the same phenomenon is to note that Lindahl prices are not functioning in true "markets" since each "market" has only one agent on the buying or selling side. Thus, no matter how large the economy is, an agent remains large in the market in which his personalized price is determined and hence he has an incentive to act in a non-competitive manner. Considerations of incentive compatibility force us to recognize that first best may not be feasible. I will consider explicitly in Section 2.3 a second best tax game which is used to study the financing of the production of public goods.

The importance of the concept of Lindahl equilibrium is that it formulates precisely the requirements for a full decentralization of public decision making with pure public goods. The broad critique of this solution concept amounts to the perplexing but deep conclusion that a full decentralization of the first best solution of the public goods allocation problem is not generally achievable. I am stressing this basic point here since it will resurface in the discussion on local public goods and Tiebout equilibrium. Since this is one important reason why game theoretic tools are effectively used in the study of public economics, I will further elaborate on this point. Recall that due to the important function of the government in a public competitive equilibrium, it is not a fully decentralized allocation mechanism. We can think of such mechanisms as *quasi-decentralized*. Recall further that the central difference between a Lindahl equilibrium and a public competitive equilibrium has to do with the role of the government in the equilibrium. Given a set of Lindahl equilibrium prices the government has no function to perform. On the other hand, in a public competitive equilibrium a government agency is required to take prices as given *and evaluate the desirability and cost of alternative packages of public goods and taxes*. Out of equilibrium the government can find a package which will dominate. This quasi-decentralized concept of equilibrium has been the dominant market equilibrium concept used in almost all papers studying equilibria with public goods since the early 1970s. The rule of conduct proposed for the government in part (c) of the definition of public competitive equilibrium is due to Foley (1967). Since then many other rules have been proposed depending upon the nature of public goods, their spatial character and the externalities generated. It is important to note, however, that all these different quasi-decentralized equilibria share one critical flaw with Lindahl equilibria: they are not incentive compatible since they require the government to *know* the preferences of the consumers. This fact explains why one finds in the literature references to the individualized taxes of a public competitive equilibrium as "Lindahl taxes".

The persistent difficulties with the incentive compatibility of quasi-decentralized equilibrium concepts is only one side of the problem involved. The mirror image of the same problem is the irregular behavior of the core and other solutions of public goods games as the economies increase in size. The lack of convergence of the core to any competitive solution in the examples provided above should appear a bit striking. This appears to contrast results of models (to be reviewed later) in which the cores of large economies with externalities and/or public goods do get close to an equilibrium or an "approximate" equilibrium of a corresponding market. [See for example: Shapley and Shubik (1966), Wooders (1978, 1980), Scotchmer and Wooders (1987) and many others.] It will thus be useful to examine again the examples cited above in order to understand their significance.

In the example of Muench (1972) the agents are represented as a continuum in the space $T = [0, 1]$. The utility functions of the agents are defined over pairs (x, y) where x and y are real numbers; x denotes the amount of the private good and y is the amount of the public good. The production set is taken as

$$F = \{(x, y) \in R^2 \mid x \leqslant 0,\ y \geqslant 0,\ y + ax \leqslant 0\} \tag{8}$$

and a feasible allocation for a coalition S is a measurable function $z(t)$ and a real number y such that

(i) $x = \int_S [z(t) - \omega(t)]\, dt$,

(ii) $(x, y) \in F$.

Note first that for any coalition S the "size" of the public good is large compared to the endowment of an agent. If preferences are monotonic in public goods then all consumers want more public goods but the *per capita cost of producing a given amount of public goods is falling with the size of coalitions*. This makes small coalitions relatively ineffective.

Milleron (1972) recognizes the difficulties inherent in using a measure space of consumers. Intending to take the asymptotic approach he notes that using the Debreu–Scarf approach would lead to an infinite number of agents with an infinite production of public goods (since initial resources will rise without bound). Citing a discussion with Aumann he suggests that preferences should change with size and he gives an interpretation which amounts to the recognition of congestion. Milleron selects the opposite procedure of letting the initial endowment of each agent be ω_L/n (where n is the replica) but the utility function is assumed to be $u(x_L, x_Q) = \log(nx_L) + \log x_Q$. A constant returns to scale production technology is assumed: $y_L + y_Q = 0$, $y_L \leqslant 0$. Given the initial endowment assumed, each replica has a constant aggregate endowment ω_L and in a Lindahl equilibrium a constant fraction of it is dedicated to the production of public goods. Thus here again the "size" of each agent measured by his endowment and his importance relative to the size of the economy *falls* at the rate of $1/n$ whereas the size of the public good in the economy remains the same.

The Champsaur et al. (1975) theorem about the non-contraction of the core also arises in a similar model in which the endowment and the importance of each agent fall at the rate of $1/n$; the per capita cost of producing the public good also continues to fall with the size of the economy. The size of the public good relative to the size of each agent rises without bound.

It seems to me that apart from the examples of non-producible public goods like "the U.S. Constitution" or the "political safety" which the constitution produces, one is hard pressed to come up with important examples of producible pure public goods (to which the theory at hand applies) which do not experience congestion with increased use. The cost of national defense increases with the size of the economy and even the classical example of public television suffers from congestion due to the limited range of any television station. In practically any case of producible public goods (e.g. roads, bridges, schools, libraries, etc.) there are some limitations of space, fixed capital or other resources such that as the number of users increases the cost of providing the public good increases and/or the quality of the good declines due to increased congestion. If one accepts this view as a basis for a modeling strategy then the pure public good with no exclusion and no congestion (i.e.: a "large" public good) is a useful concept only for an economy of a fixed size or one that varies within a narrow range of sizes. On the other hand we know of many "small" public goods: the municipal swimming pool or the Golden Gate Bridge have finite optimal number of users and would thus be negligible relative to the size of an expanding economy. To some extent it makes sense to *define* a local public good to be *a public good confined to a given locality and having a finite optimal number of users*. Consideration of an *expanding* local public goods economy raises immediate problems: given that space is a fixed factor how do you replicate a local public goods economy? One sensible modeling strategy is to assume a fixed number of "localities" and expand the size of the population in each locality but require that all local public goods be subject to congestion and have an optimal size. Thus, as the economy expands a point is reached where a second swimming pool or a second bridge are needed. This means that *in a replica economy the number of operating units of each local public good must also be replicated*!

The modeling of public goods economies will not be complete without the consideration of such services as the Environmental Protection Agency or the Social Security Administration. However *these public goods are in fact public services and their cost of production is proportional to the population which uses them*. I use the term "small" to refer to local public goods *and* public services and the term "large" to refer to pure public goods!

This discussion amounts to the view that if a researcher is interested in the decentralization of public decisions and the asymptotic behavior of the core, his model of the public sector should contain only public goods with (local) optimal size and public services. *Within such models the large, pure, public good is a fiction which should be employed with care since it could distort the conclusions*. The model of pure public goods is useful only to handle an economy of fixed size in which a choice among public goods is made. If a city with 100 000 people is to choose between libraries and parks a model of pure public goods with a continuum of players may be appropriate.

Given the classification proposed here and the implied modeling strategy it is not unreasonable to expect that public goods economies with "small" or local public goods would have either exact or approximate quasi-decentralizable allocations and such equilibria would have entirely sensible interpretation including competition among the different providers of "small" public goods. Such economies are likely to have nonempty cores or approximate cores when they are replicated and such cores may have regular asymptotic properties. This is the essential idea developed in the work of Shapley, Shubik, Wooders and others reported in Sections 3.1 and 3.2 below. On the other hand the results cited in the previous section suggest that an economy with "large" public goods may not have quasi-decentralizable equilibria and, in any event, such equilibria do not have a sensible meaning. Moreover the associated games may have an expanding number of players but empty cores and even if the cores are not empty they are not likely to have sensible asymptotic properties.

2.3. The core of the second best tax game

The theory of second best introduces the "incentive compatibility" condition which specifies that tax rates cannot depend upon the unobserved preferences. A second best tax equilibrium can be thought of as a public competitive equilibrium where the tax scheme is restricted to depend upon what is observable. The nature of such equilibria is well understood [see for example Greenberg (1975, 1977), Guesnerie (1979), Fuchs and Guesnerie (1983)] and I shall not explore it here. To study the core of the associated game I follow the simple model of Guesnerie and Oddou (1979, 1981).

The authors construct a cooperative game (N, V) in which there is one pure public good and one private good. Using the notation of Section 2.1 above $N = \{1, 2, \ldots, n\}$ and the utility function of consumer i is denoted by $u_i(x, y)$; it is monotonic and strictly quasi-concave. Consumers have initial endowments $\omega_i > 0$ of the private good and no endowment of the public good. Production technology is constant and returns to scale: it transforms one unit of the private good into one unit of the public good. The tax structure is restricted to be linear with a positive tax rate $t \in [0, 1]$ and applies equally to all. Thus the tax payment equals $t(\sum_{i=1}^{n} \omega_i)$ which is then used to produce $y = t(\sum_{i=1}^{n} \omega_i)$ units of the public good. This implies $x^i = (1 - t)\omega_i$.

To formulate the tax game Guesnerie and Oddou (1981) makes the standard assumptions used in Section 2.1 above: each coalition can use the production technology and does not assume that it will benefit from any public good produced by its complement. This is the MaxMin construction of the characteristic function. Guesnerie and Oddou also assume that each coalition can obtain resources from its members only by taxing all of them at the same rate. With these assumptions they define

$$V(S) = \left\{ \alpha = (\alpha^1, \ldots, \alpha^n) \in R^n \,\middle|\, \begin{array}{l} \exists t \in [0, 1] \text{ such that } \forall i \in S \\ \alpha^i \leq u_i((1-t)\omega_i, t \sum_{i=1}^{n} \omega_i) \end{array} \right\}. \tag{9}$$

Note the crucial role played by the requirement that a coalition can allocate resources only via a tax system with the tax rate equalized among all members. This "second best" feature suggests that the game may not be superadditive and that the core may be empty. However, if the core is empty there may be a stable arrangement in which N will be broken up into a *structure* $\mathscr{S} = (S_1, S_2, \ldots, S_m)$ which is a partition of N. In $\mathscr{S}$ each coalition establishes its own tax rate and its own production of the public good. The economy becomes a local public goods economy.

An *efficient outcome* of the game is a vector $\bar{u} \in R_+^n$ such that

(i) there is a structure $\mathscr{S} = (S_k)_{k \in M}$ for which $\bar{u} \in \bigcap_{k \in M} V(S_k)$ $M = \{1, 2, \ldots, m\}$,

(ii) for all structures $\mathscr{S}' = (S_j)_{j \in L}$, $\{u \in R_+^n, u > \bar{u}\} \cap \left\{ \bigcap_{j \in L} V(S_j) \right\} = \varnothing$[1]

$L = \{1, 2, \ldots, l\}$.

An *efficient structure* $\mathscr{S} = (S_k)_{k \in M}$ is a structure such that there exists an efficient outcome $\bar{u}$ satisfying $\bar{u} \in \bigcap_{k \in M} V(S_k)$. A *universally efficient* structure is a structure such that if $\bar{u}$ is any efficient outcome then $\bar{u} \in \bigcap_{k \in M} V(S_k)$.

The central issue of concern now is the nonemptiness of the core of (N, V) and the universal efficiency of N. Guesnerie and Oddou (1981) introduced the concept of *existence of bilateral merging agreement* which I will now define. To do this let

$$U_i(t, S) \equiv u_i((1-t)\omega_i, t \sum_{i \in S} \omega_i), \tag{10a}$$
$$U_i^*(S) \equiv \underset{t \in [0,1]}{\text{Max }} u_i(t, S). \tag{10b}$$

Condition EBMA for coalition S: Existence of bilateral merging agreement for coalition S requires that for all coalitions R and T such that $R \cup T = S$, $R \cap T = \varnothing$, for all $i \in R$ and $j \in T$ there is a $t \in [0, 1]$ such that

(a) $U_i^*(R) \leq U_i(t, S)$,
(b) $U_j^*(T) \leq U_j(t, S)$.

Guesnerie and Oddou (1981) interpret the condition to mean that if $i \in R$ and $j \in T$ were "dictators" and could merge R and T into a single coalition $S = R \cup T$ they would be able to find a tax rate such that both will be as well off as they were in S.

I turn now to the results which clarify some of the concepts involved here.

Theorem 2.3A [Guesnerie and Oddu (1981)]. *The tax game V is superadditive if and only if condition EBMA holds for all coalitions S.*

---

[1] $u > \bar{u}$ means $u_i \geq \bar{u}_i$ all i, $u_k > \bar{u}_k$ some k.

Theorem 2.3B [Guesnerie and Oddou (1981)]. *The grand coalition N constitutes a universally efficient structure if and only if condition EBMA holds for N.*

As for the nonemptiness of the core of (N, V), Guesnerie and Oddou provide the following result:

Theorem 2.3C [Guesnerie and Oddou (1981)]. *If the grand coalition N constitutes a universally efficient structure then the core of the tax game (N, V) is not empty.*

As a corollary of Theorems 2.3A, 2.3B and 2.3C one notes that the core of (N, V) is not empty either when V is superadditive or when EBMA holds for the grand coalition N.

A natural question arises with respect to the conditions that would ensure that (N, V) is balanced. Such conditions would provide alternative ways of proving the nonemptiness of the core of (N, V). Greenberg and Weber (1986) contributed to this discussion by proposing a criterion that is equivalent to balancedness. In doing so they were able to weaken a bit the conditions used by Guesnerie and Oddou (1981) as follows:

$u_i(x, y)$ is quasi-concave utility function
(rather than strictly quasi-concave), (11a)

$\omega_i \geq 0$ all i (rather than $\omega_i > 0$). (11b)

Theorem 2.3D [Greenberg and Weber (1986)]. *Under the weaker conditions (11a), (11b) the tax game (N, V) is balanced (hence its core is not empty) if and only if condition EMBA holds for N.*

In considering solutions to supplement the core Guesnerie and Oddou (1979) propose a different solution which they call C-stability. *A C-stable solution* is a vector $u^* \in R^n$ of utilities such that

(i) There is a structure $\mathscr{S} = (S_k)_{k \in M}$ which can achieve u^* thus $u^* \in \bigcap_{k \in M} V(S_k)$.

(ii) There is no other structure $\mathscr{S}'$ with $S \in \mathscr{S}'$ such that S can improve upon u^* for its members; that is

$V(S) \cap \{u \in R^n, u \gg u^*\} = \emptyset.$

A structure $\mathscr{S}$ which can achieve a C-stable solution u^* is called a *stable structure*.

An examination of stable structures will be made in connection with the study of local public goods. Here I list a few results:

(1) A C-stable solution is an efficient outcome.
(2) The core of (N, V) is a set of C-stable solutions.
(3) If N is universally efficient then any C-stable solution is necessarily in the core of (N, V).

(4) The set N constitute a stable structure if and only if the core of (N, V) is not empty.

It is clear that when V is not superadditive the grand coalition is not necessarily a universally efficient structure. Moreover, in this case the core of (N, V) may be empty and therefore the set of C-stable solutions appear as a natural generalization. It is entirely proper to think of the C-stable solutions as the core of a game in which agents propose *both allocations u and coalition structures* $\mathscr{S}$ which can achieve u. I return to this subject in Section 3.3.

2.4. Shapley value of public goods games

In this section I review examples of pure public goods games (without exclusion) where the central focus shifts from decentralization of public decisions to the examination of the strategic behavior of individuals and coalitions. Aumann, Kurz and Neyman (1983, 1987) (in short AKN) consider an economy with a continuum of consumers and with initial resources represented by the integrable function $e(t)$: $T \to E^l_+$ where T is the set of consumers. The initial resources can be used only to produce public goods and *public goods are the only desirable goods*; hence the economy is modeled as a *pure public goods* economy. This formulation is clearly in sharp contrast to the type of economies considered so far in this survey. AKN further assume that consumers may have voting rights represented by a measure v on T. It is *not* assumed that all voters have the same weight but rather that the non-negative voting measure v is arbitrary with $v(T) = 1$. In AKN (1983) two games are investigated: one with simple majority voting and one without voting; the value outcomes of these two are then compared.

The main objective of the AKN analysis is to study the effect of the voting measures on the outcomes. Thus they formulate two specific questions:

(a) Given the institution of voting how does a *change in the voting measure* alter the value outcome?
(b) Comparing societies with and without voting, how do the value outcomes *differ* in the two societies?

The first question is asked *within the context of the institution of voting*. Thus think of a public goods economy with two public goods: libraries and television. Assume that there are two types of consumers: those who like only books and those who like only television. Now consider two circumstances: in the first one the voting measure is uniform on T (thus all voters have equal weight). In the second all the weights are given to the television lovers and no voting weight is given to the book lovers. How would these two societies differ in their allocation of public goods?

To understand the second question note that there is a fundamental difference between the strategic games *with* the institution of voting and without it. With voting a winning coalition can become dictatorial: it can select any strategy from the strategy space of the majority whereas the minority may be restricted to a narrow set of options. In fact, in the voting game of AKN (1983) the minority is confined to one strategy only: do nothing and contribute nothing.[2] The *non-voting game* is a very straightforward strategic game: a coalition S uses its resources $e(S)$ to produce a vector x of public goods whereas the coalition $T\setminus S$ produces y. Since these are economies without exclusion all consumers enjoy $(x + y)$. The second question essentially investigates what happens to the value outcome of this, rather simple game, if we add to it the institution of voting giving any majority the power it did not have before and drastically restricting the options of the minority.

The first question is investigated in AKN (1987) whereas the second in AKN (1983). I will present the results in this order. The notation I use is close to the one used by AKN in both papers. A nonatomic *public goods economy* consists of

(i) A measure space $(T, \mathscr{F}, \mu)$ (T is the space of agents or players, $\mathscr{F}$ the family of coalitions, and μ the *population measure*); AKN assume that $\mu(T) = 1$ and that μ is σ-additive, non-atomic and non-negative.
(ii) Positive integers l (the number of different kinds of resources) and m (the number of different kinds of public goods).
(iii) A correspondence G from R^l_+ to R^m_+ (the *production correspondence*).
(iv) For each t in T, a number $e(t)$ of R^l_+ [$e(t)\mu(dt)$ is dt's endowment of resources].
(v) For each t in T, a function $u_t: R^m_+ \to R$ (dt's von Neumann–Morgenstern utility).
(vi) A σ-additive, non-atomic, non-negative measure v on $(T, \mathscr{F})$ (*the voting measure*); assume $v(T) = 1$.

Note that the total endowment of a coalition S – its input into the production technology if it wishes to produce public goods by itself – is $\int_S e(t)\mu(dt)$; for simplicity, this vector is sometimes denoted $e(S)$. A public goods bundle is called *jointly producible* if it is in $G(e(T))$, i.e., can be produced by all of society.

AKN assume that the measurable space $(T, \mathscr{F})$ is isomorphic[3] to the unit interval $[0, 1]$ with the Borel sets. They also assume:

Assumption 1. $u_t(y)$ is Borel measurable simultaneously in t and y, continuous in y for each fixed t, and bounded uniformly in t and y.

Assumption 2. G has compact and nonempty values.

---

[2] This assumption can be relaxed by allowing the minority a richer strategy space as long as "contributing noting" (i.e.: $x = 0$) is feasible and is an optimal threat for the minority. See on this, Section 4 of this chapter.

[3] An isomorphism is a one-to-one correspondence that is measurable in both directions.

I turn now to describe a family of public goods games analyzed by AKN. Recall that a strategic game with player space $(T, \mathcal{F}, \mu)$ is defined by specifying, for each coalition S, a set X^s of strategies, and for each pair (σ, τ) of strategies belonging respectively to a coalition S and its complement $T \backslash S$, a payoff function $H^s_{\sigma\tau}$ from T to R.

In formally defining public goods games, one describes pure strategies only; but it is to be understood that arbitrary mixtures of pure strategies are also available to the players. The pure strategies of the game will have a natural Borel structure, and mixed strategies should be understood as random variables whose values are pure strategies.

As explained earlier AKN (1987) consider a class of public goods economies in which all the specifications except for the voting measure v are fixed. They assume that the set X^S_v of pure strategies of a coalition S in the economy with voting measure v is a compact metric space, such that

$$S \supset U \text{ implies } X^S_v \supset X^U_v \tag{12a}$$

and for any voting measure η

$$(v(S) - \tfrac{1}{2})(\eta(S) - \tfrac{1}{2}) > 0 \text{ implies } X^S_v = X^S_\eta. \tag{12b}$$

The reader should note that an important difference between AKN (1983) and AKN (1987) is found in the specification of the role of voting. In the voting game AKN (1983) specify exactly what the majority and minority can do whereas in the general strategic public goods games under consideration here, only the mild conditions (12a) and (12b) are specified. Now, from (12b) it follows that X^T_v is independent of v so that $X^T_v = X^T$; and from (12a) it follows that $X^S_v \subset X^T$ for all v, i.e., X^T contains all strategies of all coalitions. Now AKN postulate that there exists a continuous function that associates with each pair (σ, τ) in $X^T \times X^T$ a public goods bundle $y(\sigma, \tau)$ in R^m_+. Intuitively, if a coalition S chooses σ and its complement $T \backslash S$ chooses τ, then the public goods bundle produced is $y(\sigma, \tau)$. Finally, define

$$H^S_{\sigma\tau}(t) = u_t(y(\sigma, \tau)) \tag{13}$$

for each S, v, σ in X^S_v, and τ in $X^{T \backslash S}_v$.

Note that the feasible public goods bundles – those that can actually arise as outcomes of a public goods game – are contained in a compact set [the image of $X^T \times X^T$ under the mapping $(\sigma, \tau) \to y(\sigma, \tau)$], and hence constitute a bounded set.

The solution concept adopted by AKN (1983, 1987) is the *asymptotic value*, which is an analogue of the finite-game Shapley value for game with a continuum of players, obtained by taking limits of finite approximations. Let Γ be a public goods game. A comparison function is a non-negative valued μ-integrable function $\lambda(S) = \int_S \lambda(t) \mu(dt)$. A *value outcome* in Γ is then a random bundle of public goods associated with the Harsanyi–Shapley NTU value based on φ; i.e., a random variable y with values in $G(e(T))$, for which there exists a comparison function λ

such that the Harsanyi coalitional form v_λ^T of the game $\lambda \Gamma$ is defined and has an asymptotic value, and

$$(\varphi v_\lambda^\Gamma)(S) = \int_S Eu_t(\underset{\sim}{y}) \lambda(\mathrm{d}t) \quad \text{for all } S \in \mathscr{F},$$

where $Eu_t(\underset{\sim}{y})$ is the expected utility of $\underset{\sim}{y}$.

Theorem 2.4A. *In any public goods game, the value outcomes are indepenent of the voting measure.*

Theorem 2.4A is a rather surprising result; for the example of the television and book lovers it says that it does not matter if all the voting weights are given to the book lovers or to the television lovers, the value outcome is the same! To put it in perspective, note that the power and influence of every individual and coalition consists of two components: first, the material endowment which is the *economic* resources of the individual or coalition and second, the voting weight which is the *political* resources of the individual or coalition. What the theorem says is that the economic resoruces are the dicisive component in the determination of value outcomes whereas changes in the political resources have no effect on the value outcome.

The surprising nature of Theorem 2.4A arises from two sources. First, on the formal level, Aumann and Kurz (1977a, b, 1978) developed a similar model for private goods economies and their conclusion is drastically different: the value allocation of a private goods economy is very sensitive to the voting measure.[4] Second, Theorem 2.4A seems to contradict our casual, common-sense, view of the political process which suggests that those who have the vote will get their way. Theorem 2.4A insists that public goods are drastically different from private goods and that the bargaining process envisioned by value theory makes the "common-sense" view a bit simplistic. An alternative political vision suggests that even when majorities and governments change, actual policies change much more slowly and often only in response to a perceived *common view* that a change is needed rather than the view of a specific majority in power.

It may be of some interest to provide an intuitive explanation of the process which one may imagine to take place in the calculations of a value allocation. Think of any majority coalition that decides on an allocation of public goods. The minority may approach any member to switch his vote in exchange for a side payment. Such a member reasons that since he is "small" he is not essential to the majority and would be responsive to "sell" his vote to the minority. Since public goods are not excluded a member who "sells" his vote retains the benefits both from the public goods which the majority produces as well as from the side payments he received from the minority for his vote. In equilibrium, competition

---
[4] This work is reviewed below, Section 4.

forces the value of a vote to zero. Note that when exclusion is possible then a defection may be extremely costly: a member who switches his vote will lose the right to enjoy the public goods produced by the majority and the voting measure will have an important effect on the value outcome. It is therefore very interesting that the "free rider" problem which is so central to the problem of allocating and financing public goods, is the essential cause of why the value outcome of public goods game is insensitive to the voting measure. I turn now to the second question.

In order to study the impact of the entire institution of voting AKN (1983) compare the value outcome of a voting vs. non-voting game. The *non-atomic public goods economy* is defined in exactly the same way as above but AKN (1983) add three more assumptions to the two assumptions specified above. The additional assumptions are:

Assumption 3. If $x \leqslant y$, then $G(x) \subset G(y)$ and $u_t(x) \leqslant u_t(y)$ for all t.

Assumption 4. $0 \in G(0)$.

Assumption 5. Either (i) u^t is C^1 (continuously differentiable) on R_+^m and the derivatives $\partial u_t / \partial y^j$ are strictly positive and uniformly bounded, or (ii) there are only finitely many different utility functions u_t.

Assumption 3 may be called "*monotonicity of production and utility*" or "*free disposal of resources and of public goods*"; it excludes the case of public "bads". Assumption 4 says that the technology is capable of producing nothing from nothing. In Assumption 5, AKN assume that either the utility functions are smooth, or that there are only finitely many "utility types" (though perhaps a continuum of "endowment types"). This is, essentially, a technical assumption.

Turning now to the two public goods games, their strategy spaces are simplified. More specifically, in the non-voting game, a pure strategy for a coalition S is simply a member x of $G(e(S))$, i.e., a choice of a public goods bundle which can be produced from the total resource bundle $e(S)$. If S has chosen $x \in G(e(S))$ and $T \setminus S$ has chosen $y \in G(e(T \setminus S))$, then the payoff to any t is $u_t(x + y)$.

In the voting game, a strategy for a coalition S in the majority $(v(S) > \frac{1}{2})$ is again a member x of $G(e(S))$. Minority coalitions $(v(S) < \frac{1}{2})$ *have only one strategy* (essentially "doing nothing"). If a majority coalition S chooses $x \in G(e(S))$ and $T \setminus S$ chooses its single strategy (as it must), then the payoff to any t is $u_t(x)$. The definition of strategies and payoffs for coalitions with exactly half the vote is not important, as these coalitions play practically no role in the analysis.

Note that the set of feasible public goods bundles – those that can actually arise as outcomes of one of the games – is precisely the compact set $G(e(T))$.

Under these conditions AKN (1983) show

Theorem 2.4B. *The voting game has the same value outcomes as the non-voting game.*

It is no surprise that Theorem 2.4A follows from Theorem 2.4B; the latter uses additional assumptions. However, as explained earlier, Theorem 2.4B provides an answer to question 2 while Theorem 2.4A answers question 1. To understand the subtle difference between Theorems 2.4A and 2.4B consider two examples where *the first theorem is true* whereas *the second theorem fails.*

Example 1. Public bads. This is the case in which some commodity is undesirable (e.g.: air pollution) but the amount of that commodity which any coalition S can produce depends only on $e(S)$. Monotonicity (Assumption 3) is violated in this case; however, the condition is not utilized in the proof of Theorem 2.4A and hence this theorem remains true. AKN (1983) show (Example 6, pp. 687–688) that if public bads are present Theorem 2.4B fails.

Example 2. Modify the voting game so that the majority can expropriate the resources of the minority against its will. In addition, provide the minority the additional strategic options of destroying any part of its endowment of resources. However, assume now that one commodity is like "land" and it cannot be destroyed (if a commodity is labor then the prohibition of slavery ensures the minority the right to destroy their labor by not working). In this example it is not feasible for the minority to contribute nothing in face of an expropriation by the majority; under all circumstances the majority will expropriate the undestroyed "land" owned by the minority. Note that there is nothing in this example to violate the conditions used in the proof of Theorem 2.4A and this theorem remains true. The example alters the strategic option specified for the minority (to be able to contribute "nothing") and AKN (1983) show (pp. 688–689) that Theorem 2.4B fails in this case as well.

To understand the common ground of these examples note that the Harsanyi–Shapley value entails an extensive amount of bargaining between each coalition S and its complement $T\setminus S$. In the non-voting game under the postulated assumptions each coalition has an optimal threat which is "to contribute nothing". This is exactly the strategy available to the minority in the voting game. In both examples 1 and 2 above this relationship breaks down. In Example 1 a small coalition can produce public bads and hence in a non-voting game it has a threat against the majority which is more effective than the "do nothing" strategy which a dictatorial majority imposes on the minority in the voting game. In Example 2 a small coalition in the non-voting game can still contribute nothing (by not using any of their resources) while this option is not available in the voting game where the majority can expropriate the non-destroyed part of the resources of the minority. In summary, under the assumptions of AKN (1983) the optimal threat of any coalition is the same in the voting and the non-voting games. In the two examples provided this equalization breaks down and for this reason *the mere introduction of the institution of voting* changes the balance of power between small and large coalitions.

3. Externalities, coalition structures and local public goods

This section reports important ideas which are currently being developed. However, it overlaps significantly with Chapter 37 on 'Coalitional structures' by J. Greenberg. Since Greenberg's chapter is more specialized I will confine my exposition of the overlap to the essential ideas only.

3.1. On the core of games with externalities and public goods

It was recognized early in the 1960s that non-convexity of preferences presents no significant problem for the existence of reasonable solutions to private goods markets as long as the economy is "large". For finite but large private economies Shapley and Shubik (1966) introduce the concept of weak ε-core.[5] When utility is transferable it is the set of all payoff vectors $(\alpha_1, \alpha_2, \ldots, \alpha_n)$ such that $\sum_{i \in S} \alpha_i \geqslant v(S) - \varepsilon |S|$ for all $S \subset N$. For the standard private goods exchange economy with l commodities and where all the consumers have the same utility function $U(x)$ and where R^l_+ is the commodity space, Shapley and Shubik propose the following condition:

There exists a *linear* function $L(x)$ and a *continuous* function $K(x)$ such that

$$K(x) \leqslant U(x) \leqslant L(x) \quad \text{all } x \in R^l_+. \tag{14}$$

Under this condition Shapley and Shubik (1966) prove that for any $\varepsilon > 0$ there exists a sufficiently large replica of the market such that the weak ε-core of the large economy is nonempty. Results of similar nature were derived by Kannai (1969, 1970), Hildenbrand et al. (1973) and others. In this development, replication tends to convexify the game.

Shubik and Wooders (1983a) introduce the idea of *approximate cores* to extend the notion of quasi-cores[6] to economies with generalized non-convexities and public goods. The conditions of Shubik and Wooders (1983a) can be viewed as "balancing" the core of the replica game as the number of replications increases and this enables them to prove the nonemptiness of the approximate core. I will provide here a *brief* formal statement of the development of Shubik and Wooders (1983a).

Start with (A, V), a game without side payments where A is a finite set of players and V is the game correspondence from the set of nonempty subsets of A into subsets of R^A. Now introduce the definition

$$V^p(S) = \left\{ u \in R^S \middle| \begin{array}{l} \text{for some } u' \in V(S), u \text{ is} \\ \text{the projection of } u' \text{ on } R^S \end{array} \right\},$$

---

[5] In terminology adopted later the weak ε-core became simply an ε-core.
[6] Shapley and Shubik (1966) introduce also the concept of *strong* ε-core to identify those payoff vectors $(\alpha_1, \ldots, \alpha_n)$ such that $\sum_{i \in S} \alpha_i \geqslant v(S) - \varepsilon$ for all $S \subset N$. The weak and strong ε cores are called quasi-cores.

where R^S is the subspace of R^A associated with S. Let $\tilde{V}$ be the *balanced cover* of V and $\tilde{V}^c$ be the *comprehensive cover of $\tilde{V}$*. Assume that the set $\{1,\ldots,T\}$ of integers is a fixed set denoting the T *types of players*. A payoff u is said to satisfy the *equal treatment property* if players of the same type receive the same payoff.

A sequence $(A_r, V_r)_{r=1}^\infty$ of games is required to satisfy the condition $A_r \subset A_{r+1}$ for all r. Each A_r has the composition

$$A_r = \{(t,q) | t \in \{1,\ldots,T\}, q \in \{1,\ldots,r\}\}$$

and u^{tq} denotes the payoff of player (t,q) in (A_r, V_r). A sequence $(A_r, V_r)_{r=1}^\infty$ is said to be *a sequence of replica games* if

(1) for each r and each t, all players of type t of the rth game are substitutes,
(2) for any r' and r'', $r' < r''$ and $S \subset A_{r'}$ imply $V_{r'}^p(S) \subset V_{r''}^p(S)$ (i.e. the achievable outcomes of coalition S does not decrease with r).

A sequence $(A_r, V_r)_{r=1}^\infty$ is said to be *per capita bounded* if there is a constant K such that for all r and all equal treatment payoffs u in $V_r(A_r)$ one has

$$u^{tq} \leq K. \tag{15}$$

This condition is somewhat peculiar since, for a fixed game, this is not much more than a normalization. Note however that with externalities and congestion the utility function of players in replica r may depend upon the total population rT as is the case in a pure public goods economy with congestion. Thus, in a sequence of replica games we really have a *sequence* of utility functions and (15) requires a *uniform normalization* but this is hardly a restriction. Note that in Section 2.2 I presented an example of a sequence of economies due to Milleron (1972). The example intended to show nonempty cores of the replica economy which contain the Lindahl equilibrium but do not converge to it. In that example the utility functions are given by

$$u(n, x_L, x_Q) = \log n x_L + \log x_Q,$$

which are clearly unbounded. The example specifies the endowment to be $(1/n)\omega_L$. Since the main issue here is how to model the replication of a public goods economy, the question of boundedness appears to be technical; one can reformulate Milleron's example with bounded utility functions. I turn now to the approximate cores.

A sequence $(A_r, V_r)_{r=1}^\infty$ of replica games is said to have a nonempty *strong approximate core* if given any $\varepsilon > 0$, there is an r^* large enough such that for all $r \geq r^*$ the ε-core of (A_r, V_r) is nonempty. The sequence is said to have a nonempty *weak approximate core* if given any $\varepsilon > 0$ and any $\lambda > 0$, there is an r^* large enough such that for each $r \geq r^*$, there exists some equal treatment vector $\tilde{u}$ in the ε-core of $(A_r, \tilde{V}^c)$ and some $u \in V_r(A_r)$ such that

$$|\{(t,q) \in A_r | \tilde{u}^{tq} \neq u^{tq}\}| < \lambda |A_r|.$$

Given the above Shubik and Wooders (1983a) show the following:

Theorem 3.1A [Shubik and Wooders (1983a)]. *Let $(A_r, V_r)_{r=1}^{\infty}$ be a sequence of superadditive, per capita bounded replica games. Then the weak approximate core is nonempty. There exists a subsequence r_k such that $(A_{r_k}, V_{r_k})_{k=1}^{\infty}$ has a nonempty strong approximate core.*

Theorem 3.1A represents an interesting line of research. It is important to see, however, that the proposed approximation will prove useful only if broad applications can be demonstrated making use of the theorem. Other papers utilizing this approach employed a richer structure which resulted in sharper results, particularly with respect to the games $(A_r, \tilde{V}_r^C)$. One example of this is Wooders (1983) (see Theorem 3). To present this result I need a bit more terminology. Thus, let $S \subset A_r$ be a coalition. The *profile of S* is denoted by $\rho(S) = (s_1, \ldots, s_T) \in R^T$ where s_j is the number of players of type j in S. A sequence $(A_r, V_r)_{r=1}^{\infty}$ is said to satisfy the condition of *minimum efficient scale of coalitions* if there is r^* such that for all $r > r^*$, given $x \in \tilde{V}_r$, there is a balanced collection β of subsets of A_r with the properties:

(1) $\rho(S) \leq \rho(A_{r^*})$ all $S \in \beta$,
(2) $u \in \bigcap_{S \in \beta} V_r(S)$.

In Milleron's example coalitions do not have a minimum efficient scale and this is the crucial reason for his conclusion; it has nothing to do with the per capita boundedness.

A game (A, V) satisfies *the condition of quasi-transferable utilities* if given any $S \subset A$, if u is on the boundary of $V^p(S)$ but there is no $u' \in V^p(S)$ with $u' \gg u$, then $V^p(S) \cap \{u' \in R^S | u' \geq u\} = u$. A sequence $(A_r, V_r)_{r=1}^{\infty}$ satisfies this assumption if each game in the sequence does. Wooders (1983) then shows

Theorem 3.1B [Wooders (1983)]. *Let $(A_r, V_r)_{r=1}^{\infty}$ be a sequence of superadditive replica games satisfying the conditions of minimum efficient scale of coalitions with bound r^* and the condition of quasi-transferable utilities. Then, for any $r \geq r^*$ the core of $(A_r, \tilde{V}_r)$ is nonempty and if x is a payoff in the core, then x has the equal treatment property.*

If a sequence of games satisfies the conditions of Theorem 3.1B one concludes that the sequence has a nonempty weak approximate core. Moreover, the reference allocation $\tilde{u}$ is not in the (equal treatment) ε-core of $(A_r, \tilde{V}_r^C)$ but rather in the (equal treatment) core of these games [i.e. $(A_r, \tilde{V}_r^C)$].

The developments in this section obviously relate to my discussion, in Section 2.2, of modeling small or large public goods in an expanding economy. Condition (14) of Shapley and Shubik (1966) as well as the conditions specifying the existence of a finite minimum efficient or optimal size of coalitions which is independent of the size of the entire economy, say that any increasing returns to size would ultimately be exhausted and either constant or decreasing returns to size will set in. For public goods economies these conditions mean that as the economy expands public goods become "small" relative to the economy. In Section 2.2, I concentrated on the *asymptotic behavior* of the core whereas here the central question is the *existence* of an approximate or an exact core for large economies, Other papers that contribute to this discussion include, for example, Wooders (1978, 1980, 1981, 1983), Shubik and Wooders (1983b, 1986) and Wooders and Zame (1984).

3.2. Equilibria and cores of local public goods economies

The issue of equilibria in local public goods economies has been controversial. This theory is very close to the theory of "clubs" and both generated a substantial literature with diverse and often contradictory concepts, assumptions and conclusions.

"Local public goods" are generally defined as public goods with exclusion. In most instances the theory suggests that the set of consumers (or players) be partitioned where members of each set in the partition are associated with such objects as jurisdictions, communities, locations etc. all of which have a spatial character. Note, however, that from the game theoretic viewpoint a partition of the players into "communities" has the same formal structure as the formation of a coalition structure; in the development below it would be proper to think of a collection of communities as a coalition structure. Most writers assume that local public goods are exclusive to a given locality without "spillovers" (or externalities across locations) which represent utility or production interactions across communities. There are some cases [e.g. Greenberg (1977)] in which cross jurisdiction public goods externalities are permitted and studied. Local public goods are studied either with or without congestion. Without congestion local public goods are experienced by any member of the community and the level of "consumption" is independent of the number of users. With congestion the number of users may influence both the *cost* as well as the *quality* of the public goods provided. With congestion, the theory of local public goods is similar to the theory of "clubs".

Some of the ambiguity and confusion may be traced back to the original paper by Tiebout (1956). He proposed a vague concept of an equilibrium in which optimizing communities offer a multitude of bundles of public goods and taxes; the maximizing consumers take prices and taxes as given and select both the optimal bundle of private goods *as well as the locality* that offers the optimal mix of local public goods and taxes. Tiebout proposed that such an equilibrium will attain a Pareto optimum and will thus solve the problem of optimal allocation of resources to public goods through a market decentralized procedure.

Unfortunately, Tiebout's paper constituted an extremely vague and imprecise set of statements. Without specifying what optimization criterion should be adopted by a community he then assumes that each community has an optimal size [see Tiebout (1956) p. 419] and that communities are in competition. Tiebout left obscure the issue whether he conjectured that an "equilibrium" (or perhaps what we would formulate as a core) exists for any finite economy or only for large economies with many communities and with great variability among communities [see Tiebout (1956) pp. 418 and 421]. This leaves the door wide open for an approximate equilibrium or an ε-core. The outcome of these ambiguities is a large number of different interpretations of what constitutes a "Tiebout equilibrium". At present there does not appear to be a consensus on a definition that would be generally accepted.

Given the controversial nature and the unsettled state of the theory of equilibrium with local public goods the limited space available here makes it impossible for me to sort out all the different points of view. Moreover, since the focus of my exposition is on game theoretic issues, this controversy is not central to the development here: my coverage will thus be selective. First I want to *briefly* review models which extend the concepts of "competitive public equilibrium" and "Lindahl equilibrium" to economies with local public goods. Next I will examine some results related to the core of the market games associated with these economies. Finally I hope to integrate the discussion with the general questions of convergence and nonemptiness of the core of the expanding public good economies. This will also require an integration of the discussion of strategies of modeling economies with public goods.

The early literature extended the equilibrium concepts of Section 2 to the case of local public goods by considering a fixed set of jurisdictions and a fixed allocation of consumers to jurisdictions. Examples of such contributions include Ellickson (1973), Richter (1974), and Greenberg (1977, 1983). The general result is that such equilibria exist and, subject to conditions similar to those presented in Section 2, the equilibria are Pareto optimal in the restricted sense that no dominating allocation exists given the distribution of consumers to the given "localities" or coalitional structure. On the other hand for the more interesting case of optimal selection of location by the consumers, which is the heart of the Tiebout hypothesis, a long list of theorems and counter examples were discovered. There are examples of equilibria which are not Pareto optimal; of sensible economies where equilibria do not exist and reasonable models for which the corresponding games have empty cores. Such examples may be found in Pauly (1967, 1970), Ellickson (1973), Westhoff (1977), Johnson (1977), Bewley (1981), Greenberg (1983) and many others.

The following example [Bewley (1981), Example 3.1] provides a well known explanation of why equilibrium may not be Pareto optimal in a small economy. There are two identical consumers and two regions. There is only one public good and one private good called "labor". The endowment of each consumer consists of 1 unit of labor. The utility function in either region is $u(l, y) = y$ where l is leisure and y is the amount of the public good provided in the region (i.e. utility does not depend upon leisure). Production possibilities are expressed by $y_j \leq L_j \; j = 1, 2$

where L_j is the amount of labor input. It is easy to see that the following configuration is an equilibrium: the price of labor and the price of the public good is 1; the tax in both regions is 1 (thus the endowment is taxed away); one consumer lives in each region and each region provides one unit of public goods. Each local government acts to maximize the welfare of its own citizen by providing the maximal amount of the desired good. Yet this is not a Pareto-optimal allocation since the two consumers would do better if they coordinate their decisions and lived together in one region in which two units of the public good be provided. This classical "matching" or "coordination" problem arises since the economy is small and the public good has no optimal size.

The possibility of an empty core of a game with local public goods should also be clear from a similar and familiar problem with the core of games with coalition structures. Since I discuss such a game in the section on 'second best' (Section 3.3) let me briefly review the point here. Let (N, V) be a game without transferable utility. Postulate that for any given coalition structure $\mathcal{S} = \{S_1, S_2, \ldots, S_K\}$, which is a partition of N, there exists a feasible set $Y(\mathcal{S})$ of utility allocations to the players. A pair $(\mathcal{S}, u)$ with $u \in Y(\mathcal{S})$ is *blocked* if there is $S \subset N$ with $v \in V(S)$ such that $v_i > u_i$ for all $i \in S$. The *core of the coalition structure game* is the set of all pairs $(\mathcal{S}, u)$ which are unblocked. Guesnerie and Oddou (1979, 1981) call this core the *C-stable solution*. When the grand coalition has the capacity to form coalition structures so that allocations may be attained through all such structures then $Y(\mathcal{S}) \subseteq V(N)$ for all $\mathcal{S}$ and (N, V) is superadditive. There are many interesting situations where this is not possible in which case the game may not be superadditive. In any event, it is well known that such games may have empty cores (see Section 3.3 below) and the local public goods economy, where the regional configuration is the coalition structure, is an example.

The paper by Bewley (1981) presents a long list of counterexamples for the existence or optimality of equilibrium in economies with public goods. Most of the examples are for finite economies and some for a continuum of consumers and a finite number of public goods. All of Bewley's examples are compelling but hardly surprising or unfamiliar. Bewley then considers the case of *pure public service* where the cost of using a public good is proportional to the number of users.[7] For an economy with a continuum of consumers, a finite number of regions, a finite number of pure public services and profit maximizing government he proves the existence and the optimality of a Tiebout-like equilibrium. Bewley's main point in proving these theorems is to argue that the case of pure public services makes the public goods economy essentially a private goods economy. He then concludes that the existence and optimality of a Tiebout-like equilibrium can be established only when the public goods economy has been stripped of its

---

[7] In the example given above the production possibilities of local public goods was $y \leq L$. With pure public service you specify the production possibilities by $ky \leq L$ where k is the number of users. This is equivalent to proportional congestion in production, not utility.

essential public goods properties. I have no difficulty in accepting Bewley's answer but have substantial difficulties with his question and the model used to analyze it.

In Section 2.2 I have taken the view that almost all local producible public goods suffer from congestion in use and therefore must have a finite optimal (or near-optimal) size which, in most instances, is independent of the size of the population of the economy. This means that as the economy expands the optimal size of any public good becomes small relative to the economy. It is thus far from obvious how to model an expanding or replica economy with public goods. However, if one assumes a finite number of regions then a sensible way to model a large economy with local public goods is to have large regions each with *multiple* roads, *multiple* bridges, *multiple* schools, *multiple* swimming pools and *multiple* other public goods, where each unit of such public goods (perhaps individuals), is small relatively to the large economy. It does not take too many counterexamples to see that due to indivisibilities and other matching problems any finite economy will have a Tiebout-like equilibrium only by chance and its core may very well be empty. This raises the natural interest in an approximation theory that seeks to establish the existence of an "approximate" equilibrium with near optimal properties and nonempty approximate core. Does the approximation improve as the size of the economy increase? If yes, how good is the approximation of any finite economy? These become the central questions to be investigated.

The model with a continuum of consumers, a finite number of jurisdictions and a *fixed* finite quantity of public goods is not an approximation of the economy I outlined above. Moreover, the usefulness of a model with a continuum of consumers to examine the issues raised above is yet to be demonstrated. I think that the asymptotic approach is superior since it focuses on the nature of the approximation.

The papers by Wooders (1978, 1981) present a model of local public goods economies allowing for congestion and assuming exclusion. Wooders proves the existence of an exact equilibrium, its Pareto optimality and the nonemptiness of the core of the associated game. Wooders assumes, however, that each public good has an interval of optimal sizes (i.e. the set of optimal sizes contains at least two consecutive integers or two relatively prime integers). In addition the economy is assumed to have finite number of types of consumers and *sufficiently large number of replications so that the matching problem is exactly resolved*. Other specialized assumptions are made including one public good and one private good. In Wooders (1980) most of the specialized assumptions are dropped except that it is still assumed that *the economy has only one public good and one private good*. A state of the economy is an allocation of consumers over jurisdictions and an allocation of private and public goods which is feasible. Wooders (1980) proposes to apply a Shapley–Shubik (1966) like notion of an ε-core to the economy with public goods. A *Tiebout ε-equilibrium* is defined as a stable of the economy, a system of individualized taxes, a price for the private good and a set of prices for the public good in each jurisdiction such that (i) markets clear, (ii) each jurisdiction maximizes profits in the production of its goods, (iii) the budget constraint of each consumer is satisfied and if a subset S of consumers finds a preferred allocation $(\hat{x}, \hat{y})$ in any jurisdiction k it is more expensive within $\varepsilon|S|$, i.e.:

$$\sum_{j \in S} (\hat{x}^j - \omega^j) + q_k \hat{y} > \pi_k - \varepsilon|S|,$$

where q_k is the price of the public good in jurisdiction k and π_k is the profits from production in jurisdiction k, (iv) the budget of all jurisdictions combined is balanced.

Space does not permit me to spell out a long list of technical conditions on utilities and production made by Wooders (1980). I only mention the key assumption that the optimal size of every public good may depend upon taxes but is uniformly smaller than m which is the number of consumers in the economy without replication. With r replications the size of the population is rm but the optimal size does not depend upon r.

Theorem 3.2A [Wooders (1980)]. *Given any $\varepsilon_0 > 0$ there exists an r_0 such that if $r \geq r_0$ the ε_0-core of the economy is not empty.*

Theorem 3.2B [Wooders (1980)]. *Every ε-equilibrium allocation is in the ε-core.*

Theorem 3.2C [Wooders (1980)]. *If the production set in each jurisdiction is a convex cone then given any $\varepsilon_0 > 0$, there exists an r_0 such that if $r \geq r_0$ an ε_0-equilibrium exists.*

Many features of the Wooders (1980) model (e.g. one public good and one private good, Lindahl taxation) drastically reduce the applicability of these results.[8] There are, however, three methodological points which I wish to be highlighted:

(1) The model defines a local public good so that its use entails congestion and it has an optimal size. Replication of the economy requries building parallel facilities and this means the replication of local public goods.

(2) Given the externalities involved one seeks only ε-equilibrium and ε-cores for any finite economy.

(3) The approximation improves with size but no results on the quality of the approximation are given.

It is clear that this conception of a "large" economy with public goods stands in sharp contrast to the formulation of the problem by Bewley (1981). This leaves a wide but important, open issue for the reader to reflect upon.

---

[8] She recognizes this as well on page 1482, Section 5.

3.3. Second best tax game revisited

In Section 2.3 I presented a model due to Guesnerie and Oddou (1979, 1981). Recall that when the tax game (N, V) defined in (10) is not superadditive, Guesnerie and Oddou introduced, as a candidate for a solution, the notion of *a C-stable solution* which is the same as the *core of the coalition structure game* defined in Section 3.2. It is important to understand why the tax game may not be superadditive. Recall that the rules of the game require that if any coalition is formed it must impose a uniform tax rate on all its members. This restriction may be so drastic that if the diversity among players is sufficiently great it may pay for coalitions to break away, impose their own separate tax rates and produce their own separate bundles of public goods. Thus $V(N)$ may be dominated by an outcome which results from the formation of a coalition structure $(S_1, S_2, \ldots, S_K)$ and thus (N, V) is not superadditive. Guesnerie and Oddou (1979) showed that if $n \leq 3$ the core of the coalition structure game is not empty. Since there is no general existence theorem of stable coalition structures, the question was whether the special structure of the tax game will lead to such a theorem. Weber and Zamir (1985) provided a counterexample in which the core of the coalition structure game associated with the tax game is empty. This negative conclusion provides one more indication of the complexity of the problem of externalities, coalities structures and local public goods.

On the positive side Greenberg and Weber (1986) introduced the notion of *consecutive games*. To define it call S a *consecutive coalition* if for any three players i, j, k the conditions $i < j < k$ and $i \in S$, $k \in S$ imply $j \in S$. The consecutive structure of S, denoted by $C(S)$, is the unique partition of S into consecutive coalitions which are maximal with respect to set inclusion. Now let (N, V) be any game. The associated consecutive game (N, V_c) is defined by

$$V_c(S) = \bigcap_{T \in C(S)} V(T).$$

Greenberg and Weber (1986) then prove

Theorem 3.3A. *Let (N, V) be a game. Then the core of the coalition structure game associated with the consecutive game (N, V_c) is nonempty.*

With the aid of this theorem Greenberg and Weber consider the special case where the utility functions of the players are of the form

$$u^i(x, y) = f(y) + g_i(x), \quad i = 1, \ldots, n,$$

where y is the amount of the public good and x the amount of the private good. In addition, they consider two specifications:

$$g_i(x) = \alpha_i x + \beta_i, \quad i = 1, \ldots, n, \tag{16a}$$

$g_i(x) = g(x)$ all i and $g(x)$ is concave or convex. (16b)

Under either of the specifications (16a) or (16b) they show that the core of the coalition structure game associated with the tax game [i.e. (N, V) where V is defined in (10)] is nonempty. Although instructive the cases (16a) and (16b) are rather special. This research area is clearly wide open.

4. Power and redistribution

The question of income and wealth distribution has always been a central problem for economic theory. Given a set of initial property rights of agents, the theory of general competitive equilibrium provides a positive theory for the distribution of income which emerge with prices. The theory does not provide an explanation for the extensive amount of income *redistribution* which is carried out through the public sector. The traditional approach to this question is to formulate the problem of redistribution as an activity that would be a consequence of the formation of some "social welfare function" and public policy "objectives". In this context redistribution is viewed as one more public good which arises as a consequence of the externalities in utility among different members of society. The plain facts are, however, that many recipients of education support are children of the middle or upper income families; that many of the supported farmers are very well off and some are powerful and wealthy corporations; that industries which have received extensive public support includes industries like oil and coal, lumber, steel and others; that many of the veterans and of the retirees are very well off and some are outright wealthy.

In this section I present a drastically different view which explains the redistribution phenomenon as an outcome of a complex balance of power in society. The ideas originate in the work of Aumann and Kurz (1977a, b, 1978). Some subsequent contributions include Osborne (1979, 1984), Kurz (1977, 1980), Gardner (1981), Imai (1983) and Peck (1986). Aumann and Kurz (1977a) explicitly state that they do not reject the social welfare function approach to the process of public policy formation. They state that there are public policy issues with respect to which a consensus may be reached and in that case the public sector becomes an instrument to carry out the will of society. This, according to Aumann and Kurz, does not handle some of the very hard questions of redistribution which may not be based on a consensus.

Aumann and Kurz aim to obtain a characterization of the income distribution in the economic-political equilibrium which they study. However, they investigate two different approaches. In the first, called the "*commodity redistribution*" approach they study the entire economic-political system as a single integrated game; the solution which they adopt is the NTU Harsanyi-Shapley value. This procedure is analogous to the treatment of the voting public goods games in Section 2.4

above. In the second approach, called the *"income redistribution"* approach, the economic side of the model – consumption and exchange – is assumed to take place in a normal competitive environment relative to endogeneously determined prices. The political side of the society – represented by the redistribution of income – is assumed governed by game theoretic considerations; again by the Harsanyi–Shapley NTU value. Thus, apart from the desire to characterize the equilibrium distribution of income, Aumann and Kurz have a second central objective: to study the *existence* of equilibrium commodity allocations in each of the models based on the two approaches outlined above and the *relationship* between the two implied commodity allocations. I turn now to the formal description of the models and the results.

A *market M* consists of

(i) A measurable space $(T, \mathscr{F})$ (the space of agents) together with a σ-additive, non-negative measure μ on $\mathscr{F}$ with $\mu(T) = 1$ (*the population measure*).
(ii) The non-negative orthant R^l_+ – called the consumption set – of a Euclidean space R^l (l represents the number of different commodities in the market).
(iii) An integrable function e from T to R^l_+ (the *endowment* function or initial allocation).
(iv) For each t in T, a function u_t on R^l_+ (the utility function of t).

A market is called finite if T is *finite*, and *non-atomic* if μ is non-atomic. Aumann and Kurz assume that every market is either finite or non-atomic. They also assume (as in Section 2.4 above) that the measureable space $(T, \mathscr{F})$ is finite or isomorphic to the unit interval $[0, 1]$ with the Borel sets. The origin of R^l as well as the number zero, are denoted 0. A function u on R^l_+ is *increasing* if $x \geqslant y$ implies $u(x) > u(y)$. The partial derivative $\partial u/\partial^i$ of a function u on R^l_+ is denoted u^i, and the gradient $(u^1, \ldots, u^l)$ is denoted u'. The following assumptions are made:

(A.1) For each t, u is *increasing*, *concave*, and *continuous* on R^l_+.
(A.2) $u_t(0) = 0$.
(A.3) $u_t(x)$ is simultaneously measurable[9] in x and t.
(A.4) $\int e > 0$.
(A.5) For each t and i, the partial derivative $u^i_t(x)$ exists and is continuous at each x in R^l_+ with $x^i > 0$.

A market is called *bounded* if
(A.6) u_t is uniformly bounded, i.e.

$$\text{Sup } \{u_t(x): t \in T, x \in R^l_+\} < \infty$$

and

(A.7) $u_t(1, \ldots, 1)$ is uniformly positive, i.e.

$$\text{Inf } \{u_t(1, \ldots, 1): t \in T\} > 0.$$

---
[9] I.e. measurable in the product σ-field $\mathscr{B} \times \mathscr{F}$ where $\mathscr{B}$ is the Borel σ-field on R^l_+.

Finally a market is called *trivial* if $|T| = 2$ and one of them has an endowment vector equal to 0.

As in Section 2.4 $e(S)$ stands for $\int_S e$. An *S-allocation* is a measurable function $x(t)$ from S to R^l_+ with $\int_S x = e(S)$. An *allocation* is a T-allocation. If p is a price vector and $x \in R^l$ then $\sum_{i=1}^{l} p^i x^i$ is denoted px.

Given a market M, I turn now to the *commodity redistribution game $\Gamma(M)$* which is described as follows: T, $\mathcal{F}$, and μ are as in the market M. As for the strategy spaces and payoff functions, Aumann and Kurz do not describe these fully, because that would lead to irrelevant complications; they do make three assumptions about them, which suffice to characterize completly the associated Harsanyi coalitional form and its values.

The first of the three assumptions is:

(A.8) If $\mu(S) > \frac{1}{2}$, then for each S-allocation x there is a strategy σ of S such that for each strategy τ of $T \setminus S$,

$$h^s_{\sigma\tau}(t) \begin{cases} \geq u_t(x(t)), & t \in S, \\ = 0, & t \in S. \end{cases}$$

This means that a coalition in the majority can force every member outside of it down to the zero level, while reallocating to itself its initial bundle in any way it pleases.

Next, they assume

(A.9) If $\mu(S) \geq \frac{1}{2}$, then there is a strategy τ of $T \setminus S$ such that for each strategy σ of S, there is an S-allocation x such that

$$h^s_{\sigma\tau}(t) \leq u_t(x(t)), \quad t \in S.$$

This means that a coalition in the minority can prevent the majority from making use of any endowment other than its own (the majority's). Finally, they assume

(A.10) If $\mu(S) = \frac{1}{2}$, then for each S-allocation x there is a strategy σ of S such that for each strategy τ of $T \setminus S$ there is a $T \setminus S$-allocation y such that

$$h^s_{\sigma\tau}(t) \begin{cases} \geq u_t(x(t)), & t \in S, \\ \leq u_t(y(t)), & t \in T \setminus S. \end{cases}$$

This simply means that if neither S nor its complement are in the majority, then each side can divide its endowment in any way it pleases, while at the same time not giving anything to the other side.

As stated earlier the solution concept adopted by Aumann and Kurz is the Harsanyi–Shapley NTU value [for details see Aumann and Kurz (1977b) pp. 186–192]. Very important for the analysis are the allocations x such that $u(x)$ is a value of $\Gamma(M)$. These are the commodity redistributions which can be observed in any practical situation. They are called by Aumann and Kurz *the commodity tax allocations for M*.

I turn now to the third component which is *income redistribution*. Its essential feature is that given any market price p society can impose a tax on income which is the monetary worth of the endowment vector. Given a price vector p, define the *indirect utility function* u_t^p of trader dt to be the function from R to itself given by

$$u_t^p(y) = \max\{u_t(x): x \in R_+^l \text{ and } px \leq y\}. \tag{17}$$

Intuitively, $u_t^p(y)$ is the highest utility dt can attain by buying goods at prices p with a maximum expenditure of y.

If all traders are assured that they can always trade at the fixed prices p, then the given l-good economy becomes an economy with only one good, namely money. In this economy the initial endowments are $pe(t)$, and the utility functions are u_t^p; this may be analyzed as *a redistribution game with only one commodity*, and this analysis yields a certain taxation–redistribution system.

Taxation and redistribution in this one-good "money" economy, as well as the ordinary incentives for trading, will in general create a situation in which at prices p, the supply and demand in the original l-good economy are out of balance. One therefore would like to know whether there exists a price vector p such that if the above procedure is carried out, supply and demand for each of the l goods will match after the taxation and redistribution are carried out, where the same price vector p is used both in assessing the endowment for purposes of taxation and in trading after taxes have been collected. Such a p, together with the resulting tax scheme, is called a *"competitive tax equilibrium."*

Formally, given a market M and a price vector p, define a market M^p, called *the market derived from M at prices p* (or simply the *derived market*) as follows: $(T, \mathscr{F}, \mu)$ is as in M; the number of commodities is 1; the initial allocation is pe; and the utility functions are the indirect utilities u_t^p defined in (17). Aumann and Kurz show that M^p does indeed satisfy all the conditions required of markets as defined earlier. *A competitive tax equilibrium in M* is a pair consisting of an allocation x and a price vector p such that

$$x(t) \text{ a.e. maximizes } u_t \text{ over } \{x \in R_+^l, px \leq px(t)\}, \tag{18a}$$

and

$$px \text{ is a commodity tax allocation in } M^p. \tag{18b}$$

If (x, p) is a competitive tax equilibrium, then x will be called an *income tax allocation*. Note that in M^p there is just one "commodity," namely money, and that the quantities $px(t)$ appearing in (18b) are in units of that "commodity." Note also that in any market in which $l = 1$, i.e. in which there is just one commodity, the commodity and income tax allocations are the same. Hence when $l = 1$, one may refer simply to *tax allocations*.

I turn now to the three basic results.

Theorem 4A (Existence). *Every non-trivial bounded market has a commodity tax allocation and an income tax allocation.*

Theorem 4B (Equivalence). *In a non-atomic bounded market, the commodity tax allocations coincide with the income tax allocations.*

The fundamental importance of Theorems 4A and 4B is that they demonstrate the possible compatibility between the operations of a competitive, decentralized market economy (in private goods) and a highly strategic behavior on the political-public sector level. Theorem 4B also shows that the complex social bargaining about final commodity vectors is not necessary: the same set of outcomes will be reached if the social bargaining takes place with respect to monetary "income" using competitive market prices. Aumann and Kurz (1977b) show that Theorem 4B fails if T is finite and this means that equivalence is approximately correct only in large markets. The final result is as follows:

Theorem 4C (Characterization). *A non-atomic bounded market with a single commodity ($l = 1$) has a unique tax allocation x. This allocation is characterized by a.e. $x(t) > 0$ and*

$$x(t) + \frac{u_t(x(t))}{u'_t(x(t))} = e(t) + \int \frac{u(x)}{u'(x)}. \tag{19}$$

Aumann and Kurz (1977a) provide a detailed analysis of the implications of Theorem 4C. Here is a brief summary.

Implication 1. Theorem 4C establishes a monotonic increasing relation between net income $x(t)$ and gross income $e(t)$. Moreover, let

$$c = \int \frac{u(x)}{u'(x)}$$

and call c the tax credit. Also let

$$\eta(t) = \frac{u'_t(x(t))}{u_t(x(t))} x(t).$$

Then one can calculate

$$x(t) = \frac{\eta(t)}{1 + \eta(t)} [c + e(t)].$$

This means that if an agent has no income [i.e. $e(t) = 0$] then he receives a *support* of size $[\eta(t)/(1 + \eta(t))]c$. As his income increases the agent pays a positive marginal income tax $M(t)$ until the support is exhausted, at which time he begins to pay a positive amount of total taxes.

In Section 2.4 I noted that in this family of strategic voting games an agent has two endowments: his *economic* endowment in the form of the initial resources at his disposal and his *political* endowment in the form of his voting weight. Implication 1 reveals that even when $e(t) = 0$, it is the value of his political endowment that ensures the agent some support. This is a clear form in which the institution of democracy has a profound effect on the redistribution of private goods.

Implication 2. If u_t is twice continuously differentiable then the marginal tax rate $M(t)$ is defined by

$$M(t) = 1 - \frac{1}{2 + (-u_t'' u_t / u_t'^2)}$$

and $\frac{1}{2} \leq M(t) < 1$, i.e.: the marginal income tax is between 50% and 100% but the tax structure is progressive, regressive or neutral.

Implication 3. The size of the tax paid by agent t is $e(t) - x(t)$ and a quick calculation shows that

$$t\text{'s income tax} = \frac{u_t(x(t))}{u_t'(x(t))} - c.$$

Aumann and Kurz (1977b) call the term $u(x)/u'(x)$ *the fear of ruin at* x. They also call the inverse u'/u the *boldness at* x. To understand these terms suppose that t is considering a bet in which he risks his entire fortune x against a possible gain of a small amount h. The probability q of ruin would have to be very small in order for him to be indifferent between such a bet and retaining his current fortune. Moreover, the more unwilling he is to risk ruin, the smaller q will be. Thus q is an inverse measure of t's aversion to risking ruin, and a direct measure of boldness; obviously q tends to 0 as the potential winnings h shrink. Observe that the boldness is *the probability of ruin per dollar of potential winnings for small potential winnings*, i.e., it is the limit of q/h as $h \to 0$. To see this, note that for indifference we must have

$$u(x) = (1-q)u(x+h) + qu(0) = (1-q)u(x+h).$$

Hence

$$\frac{q}{h} = \frac{[u(x+h) - u(x)]/h}{u(x+h)}$$

and as $h \to 0$, this tends to u'/u. In conclusion *the tax equals the fear of ruin at the net income, less a constant tax credit*. Thus the more fearful a person, the higher he may expect his tax to be.

The basic model described here was extended in a few directions. Osborne (1979, 1984) investigated the sensitivity of the model to the composition of the endowment and to alteration in the retaliatory ability of the minority. He shows that if the endowment vector $e(t)$ is composed of some goods which cannot be destroyed (like "land") the tax rate on them will be higher than the tax rate on those components of the endowment which can escape taxation (by destruction). Osborne also shows that the basic result of Aumann and Kurz, (i.e.: the emergence of an endogenous income or wealth tax) remains essentially unaltered when the relative power of the majority and the minority is altered. However, the value allocations are sensitive to this distribution of power. Kurz (1977, 1980) consider the possibility that agents in the Aumann and Kurz model may attempt to distort their preferences. Under such circumstances the outcome would be a linear income tax with a marginal tax rate of 50%. Gardner (1981) introduces into the measure μ an atomic player with a veto power. It is interpreted as a "collegial polity" whose vote is needed in order to exercise the power of the state. The rest of the voters are represented by a continuum. Using the assumption that both the atom and a positive measure of the other players are risk neutral Gardner is able to characterize the value allocations; he shows that in their basic structure they are similar to the result in (19) [see Gardner (1981), Proposition 3, p. 364].

The Lindahl and Tiebout programs for the allocation and financing decisions of the public sector stand in sharp contrast to the Aumann and Kurz agenda. Both Lindahl and Tiebout searched for a way of extending the operation of the private market economy to the public sphere. Aumann and Kurz proposed to draw a sharp distinction between the competitive nature of the private economy and the highly strategic character of the public sphere. Aumann and Kurz sought to show that this distinction does not prevent the development of an integrated and coherent view of the total economy. My aim in this paper was to highlight these contrasting approaches to the analysis of the public sector.

5. Some final reflections

This paper reviewed some areas of public economics where the theory of games has made important contributions. Game theory has been successfully employed in economics whenever the standard, private goods, competitive model did not provide an adequate framework for analysis. Public economics is one of these cases. I have reviewed in this paper developments and results in a long list of specific areas of public economics. In closing I want to mention a line of research which developed in recent years. To begin with, recall that we have seen again and again in this paper that any decision in the public sphere encounters great difficulties with the incentives of individuals and public agents to do what is socially desirable. These problems arise from many reasons: consumers may distort their preference when asked to participate in financing decisions of public goods; agents may misrepresent their private information which may be needed for equilibrium redistribution of income; managers of public agencies may have no incentive to carry out efficiently the production decisions of the public sector and others. These type of questions were only partly reviewed in this paper and the reader may consult the paper by L. Hurwicz on 'implementation' (this volume) for more details. I would like to note that important progress has been made in analyzing these type of incentive problems within the framework of game theoretic models. More specifically, the process of public decisions is formulated as a game (called "a game form") where each agent may reveal voluntarily any part of the private information at his disposal. A strategy by each player induces a collective action. Thus the set of *incentive compatible* outcomes are those outcomes which are obtained in some equilibrium of the game [for some details see for example Hurwicz (1972, 1979), Hammond (1979), Dasgupta et al. (1979), Maskin (1980)]. Various concepts of equilibria (e.g. Nash equilibrium, dominant strategy Nash equilibrium, etc.) and their refinement have been examined in the development of these ideas.

From the perspective of this paper it is noteworthy how well the game theoretic concepts and the economic theoretic ideas were integrated in this field of research. This integration is so complete that the standard tools of non-cooperative game theory provided both the *language* as well as the *scientific framework* within which the theory of incentive compatible implementation was developed. This provides a good example how game theory comes into essential use in economics when the standard competitive model is inadequate.

References

Aumann, R.J., R.J. Gardner and R.W. Rosenthal (1977) 'Core and Value for A Public-Goods Economy: An Example', *Journal of Economic Theory*, **5**: 363–365.
Aumann, R.J. and M. Kurz (1977a) 'Power and Taxes', *Econometrica*, **45**: 1137–1161.
Aumann, R.J. and M. Kurz (1977b) 'Power and Taxes in a Multicommodity Economy', *Israel Journal of Mathematics*, **27**: 185–234.
Aumann, R.J. and M. Kurz (1978) 'Taxes and Power in a Multicommodity Economy (Updated)', *Journal of Public Economics*, **9**: 139–161.
Aumann, R.J., M. Kurz and A. Neyman [1983] 'Voting for Public Goods', *Review of Economic Studies*, **50**: 677–693.
Aumann, R.J., M. Kurz and A. Neyman (1987) 'Power and Public Goods', *Journal of Economic Theory*, **42**: 108–127.
Bewley, T. (1981) 'A Critique of Tiebout's Theory of Local Public Expenditures', *Econometrica*, **49**: 713–741.
Champsaur, P. (1975) 'How to Share the Cost of a Public Good', *International Journal of Game Theory*, **4**: 113–129.
Champsaur, P., D.J. Roberts and R.W. Rosenthal (1975) 'On Cores in Economies with Public Goods', *International Economic Review*, **16**: 751–764.
Dasgupta, P., P. Hammond and E. Maskin (1979) 'The Implementation of Social Choice Rules: Some General Results', *Review of Economic Studies*, **46**: 185–211.
Foley, D.K. (1967) 'Resource Allocation and the Public Sector', *Yale Economic Essays*, Spring, 43–98.
Foley, D.K. (1970) "Lindahl's Solution and the Core of an Economy with Public Goods', *Econometrica*, **38**: 66–72.
Fuchs, G. and R. Guesnerie [1983] 'Structure of Tax Equilibria', *Econometrica*, **51**: 403–434.
Gardner R. (1981) 'Wealth and Power in a Collegial Polity', *Journal of Economic Theory*, **25**: 353–366.
Greenberg, J. (1975) 'Efficiency of Tax Systems Financing Public Goods in General Equilibrium Analysis', *Journal of Economic Theory*, **11**: 168–195.
Greenberg, J. (1977) 'Existence of an Equilibrium with Arbitrary Tax Schemes for Financing Local Public Goods', *Journal of Economic Theory*, **16**: 137–150.
Greenberg, J. (1983) 'Local Public Goods with Mobility: Existence and Optimality of a General Equilibrium', *Journal of Economic Theory*, **30**: 17–33.

Greenberg, J. and S. Weber (1986) 'Strong Tiebout Equilibrium Under Restricted Preferences Domain', *Journal of Economic Theory*, **38**: 101–117.

Guesnerie, R. (1975) 'Production of the Public Sector and Taxation in a Simple Second-Best Model', *Journal of Economic Theory*, **10**: 127–156.

Guesnerie, R. (1979) 'Financing Public Goods with Commodity Taxes: The Tax Reform Viewpoint', *Econometrica*, **47**: 393–421.

Guesnerie, R. and C. Oddou (1979) 'On Economic Games which are not Necessarily Superadditive. Solution Concepts and Application to a Local Public Problem with Few Agents', *Economic Letters*, **3**: 301–306.

Guesnerie, R. and C. Oddou (1981) 'Second Best Taxation as a Game', *Journal of Economic Theory*, **25**: 66–91.

Hammond, P.J. (1979) 'Straightforward Individual Incentive Compatibility in Large Economies', *Review of Economic Studies*, **46**: 263–282.

Hildenbrand, W., D. Schmeidler and S. Zamir (1973) 'Existence of Approximate Equilibria and Cores', *Econometrica*, **41**: 1159–1166.

Hurwicz, L. (1972) 'On Informationally Decentralized Systems', in: C.B. McGuire and R. Radner, eds. *Decision and organization*. Amsterdam: North-Holland.

Hurwicz, L. (1979) "Outcome Functions Yielding Walrasian and Lindahl Allocations at Nash Equilibrium Points', *Review of Economic Studies*, **46**: 217–226.

Imai, H. (1983) 'Voting, Bargaining, and Factor Income Distribution', *Journal of Mathematical Economics*, **11**: 211–233.

Ito, Y. and M. Kaneko (1981) 'Ratio Equilibrium in an Economy with Externalities', *Zeischrift für Nationalökonomie, Journal of Economics*, Springer-Verlag, **41**: 279–294.

Johnson, Michael (1977) 'A Pure Theory of Local Public Goods', Ph.D. Dissertation, Northwestern University, Evanston, IL.

Kaneko, M. and M.H. Wooders (1982) 'Cores of Partitioning Games', *Mathematical Social Sciences*, **3**: 313–327.

Kannai, Y. (1969) 'Countably Additive Measure in Cores of Games', *Journal of Mathematical Analysis and Applications*, **27**: 227–240.

Kannai, Y. (1970) 'Continuity Properties of the Core of a Market', *Econometrica*, **38**: 791–815.

Kurz, M. (1977) 'Distortion of Preferences, Income Distribution and the Case for a Linear Income Tax', *Journal of Economic Theory*, **14**: 291–298.

Kurz, M. (1980) 'Income Distribution and the Distortion of Preferences: The l Commodity Case', *Journal of Economic Theory*, **22**: 99–107.

Lindahl, E. (1967) 'Just Taxation: a Positive Solution', in: R.A. Musgrave and A.T. Peacock, eds., *Classics in the theory of public finance*. London: MacMillan.

Maskin, E. (1980) 'On First-Best Taxation', in: D. Collard, R. Lecomber and M. Slater, eds., *Income distribution: the limits to redistribution*. Bristol, England: John Wright and Sons.

McGuire, M. (1974) 'Group Segregation and Optimal Jurisdictions', *Journal of Political Economy*, **82**: 112–132.

Milleron, J.C. (1972) 'Theory of Value with Public Goods: A Survey Article', *Journal of Economic Theory*, **5**: 419–477.

Muench, T.J. (1972) 'The Core and the Lindahl Equilibrium of an Economy with Public Goods: an Example', *Journal of Economic Theory*, **4**: 241–255.

Osborne, M.J. (1979) 'An Analysis of Power in Exchange Economies', Technical Report No. 291, Economic Series, Institute for Mathematical Studies in the Social Sciences, Stanford University.

Osborne, M.J. (1984) 'Why do Some Goods Bear Higher Taxes Than Others?', *Journal of Economic Theory*, **32**: 301–316.

Pauly, M. (1967) 'Clubs, Commonality, and the Core: An Integration of Game Theory and the Theory of Public Goods', *Economica*, **34**: 314–324.

Pauly, M. (1970) 'Optimality, 'Public' Goods, and Local Governments: A General Theoretical Analysis', *Journal of Political Economy*, **78**: 572–585.

Pauly, M. (1976) 'A Model of Local Government Expenditure and Tax Capitalization', *Journal of Public Economics*, **6**: 231–242.

Peck, R. (1986) 'Power and Linear income Taxes: An Example', *Econometrica*, **54**: 87–94.

Richter, D. (1974) 'The Core of a public Goods Economy', *International Economic Review*, **15**: 131–142.

Rosenmuller, J. (1981) 'Values of Non-sidepayment Games and Their Application in the Theory of

Public Goods', in: *Essays in Game Theory and mathematical Economics in Honor of Oskar Morgenstern*, Vol. 4 of *Gesellschaft, Recht Wirtschaft*. Mannheim, Vienna and Zurich: Bibliographisches Institut, pp. 111–29.

Rosenmuller, J. (1982) 'On Values, Location Conflicts, and Public Goods', in: M. Deistler, E. Furst and G. Schwodianer, eds., *Proceedings of the Oskar Morgenstern Symposium at Vienna*. Vienna: Physica-Verlag, pp. 74–100.

Rosenthal, R.W. (1971) 'External Economies and Cores', *Journal of Economic Theory*, **3**: 182–188.

Samuelson, P.A. (1954) 'The pure Theory of Public Expenditures', *Review of Economics and Statistics*, **36**: 387–389.

Scotchmer, S. and M.H. Wooders (1987) 'Competitive Equilibrium and the core in Economies with Anonymous Crowding', *Journal of public Economics*, **34**: 159–174.

Shapley, L.S. and M. Shubik (1966) 'Quasi-Cores in a Monetary Economy with Nonconvex Preferences', *Econometrica*, **34**: 805–827.

Shapley, L.S. and H.E. Scarf (1974) 'On Cores and Indivisibilities', *Journal of Mathematical Economics*, **1**: 23–28.

Shubik, M. and M.H. Wooders (1983a) 'Approximate cores of Replica Games and Economies. Part I: Replica Games, Externalities, and Approximate Cores', *Mathematical Social Sciences*, **6**: 27–48.

Shubik, M. and M.H. Wooders (1983b) 'Approximate Cores of Replica Games and Economies. Part II: Set-up Costs and Firm Formation in Coalition production Economies', *Mathematical Social Sciences*, **6**: 285–306.

Shubik, M. and M.H. Wooders (1986) 'Near-Market and Market Games', *The Economic Studies Quarterly*, **37**: 289–299.

Tiebout, M. (1956) 'A pure Theory of Local Expenditures', *Journal of Political Economy*, **64**: 413–424.

Weber, S. and S. Zamir (1985) 'Proportional Taxation; Non Existence of Stable Structures in Economy with a Public Good', *Journal of Economic Theory*, **35**: 107–114.

Westhoff, F. (1977) 'Existence of Equilibria in Economies with a Local Public Good', *Journal of Economic Theory*, **14**: 84–112.

Wooders, M.H. (1978) 'Equilibria, the Core, and Jurisdiction Structures in Economies with a Local Public Good', *Journal of Economic Theory*, **18**: 328–348.

Wooders, M.H. (1980) 'The Tiebout Hypothesis: Near Optimality in Local Public Good Economies', *Econometrica*, **48**: 1467–1485.

Wooders, M.H. (1981) 'Equilibria, the Core, and Jurisdiction Structures in Economies with a Local Public Good: A Correction', *Journal of Economic Theory*, **25**: 144–151.

Wooders, M.H. (1983) 'The Epsilon Core of A Large Replica Game' *Journal of Mathematical Economics*, **11**: 277–300.

Wooders, M.H. and W.R. Zame (1984) 'Approximate Cores of Large Games', *Econometrica*, **52**: 1327–1349.

Zeckhauser, R. and M. Weinstein (1974) 'The Topology of Pareto Optimal Regions of Public Goods', *Econometrica*, **42**: 643–666.

Chapter 34

COST ALLOCATION

H.P. YOUNG*

Johns Hopkins University

Contents

1.	Introduction	1194
2.	An illustrative example	1195
3.	The cooperative game model	1197
4.	The Tennessee Valley Authority	1198
5.	Equitable core solutions	1203
6.	A Swedish municipal cost-sharing problem	1206
7.	Monotonicity	1209
8.	Decomposition into cost elements	1211
9.	The Shapley value	1213
10.	Weighted Shapley values	1215
11.	Cost allocation in the firm	1217
12.	Cost allocation with variable output	1219
13.	Ramsey prices	1219
14.	Aumann–Shapley prices	1220
15.	Adjustment of supply and demand	1223
16.	Monotonicity of Aumann–Shapley prices	1223
17.	Equity and competitive entry	1224
18.	Incentives	1225
19.	Conclusion	1228
Bibliography		1230

*This work was supported in part by the National Science Foundation, Grant SES 8319530. The author thanks R.J. Aumann, M.L. Balinski, S. Hart, H. Moulin, L. Mirman, A. Roth, and W.W. Sharkey for constructive comments on an earlier draft of the manuscript.

Handbook of Game Theory, Volume 2, Edited by R.J. Aumann and S. Hart
© *Elsevier Science B.V., 1994. All rights reserved*

1. Introduction

Organizations of all kinds allocate common costs. Manufacturing companies allocate overhead expenses among various products and divisions. Telephone companies allocate the cost of switching facilities and lines among different types of calls. Universities allocate computing costs among different departments. Cost allocation is also practiced by public agencies. Aviation authorities set landing fees for aircraft based on their size. Highway departments determine road taxes for different classes of vehicles according to the amount of wear and tear they cause to the roadways. Regulatory commissions set rates for electricity, water, and other utilities based on the costs of providing these services. Cost allocation is even found in voluntary forms of organization. When two doctors share an office, for example, they need to divide the cost of office space, medical equipment, and secretarial help. If several municipalities use a common water supply system, they must reach an agreement on how to share the costs of building and operating it. When the members of NATO cooperate on common defense, they need to determine how to share the burden. The common feature in all of these examples is that prices are not determined externally by market forces, but are set internally by mutual agreement or administrative decision.

While cost allocation is an interesting accounting problem, however, it is not clear that it has much to do with game theory. What does the division of defense costs in NATO or overhead costs in General Motors have in common with dividing the spoils of a game? The answer is that cost allocation *is* a kind of game in which costs (and benefits) are shared among different parts of an organization. The organization wants an allocation mechanism that is efficient, equitable, and provides appropriate incentives to its various parts. Cooperative game theory provides the tools for analyzing these issues. Moreover, cooperative game theory and cost allocation are closely intertwined in practice. Some of the central ideas in cooperative game theory, such as the core, were prefigured in the early theoretical literature on cost allocation. Others, such as the Shapley value, have long been used implicitly by some organizations. Like Moliere's M. Jourdain, who was delighted to hear that he had been speaking prose all his life, there are people who use game theory all the time without ever suspecting it.

This chapter provides an overview of the game theoretic literature on cost allocation. The aim of the chapter is two-fold. First, it provides a concrete motivation for some of the central solution concepts in cooperative game theory. Axioms and conditions that are usually presented in an abstract setting often seem more compelling when interpreted in the cost allocation framework. Second, cost allocation is a practical problem in which the salience of the solution depends on contextual and institutional details. Thus the second objective of the chapter is to illustrate various ways of *modelling* a cost allocation situation.

2. An illustrative example

Consider the following simple example. Two nearby towns are considering whether to build a joint water distribution system. Town A could build a facility for itself at a cost of $11 million, while town B could build a facility at a cost of $7 million. If they cooperate, however, then they can build a facility serving both communities at a cost of $15 million. (See Figure 1.) Clearly it makes sense to cooperate since they can jointly save $3 million. Cooperation will only occur, however, if they can agree on how to divide the charges.

One solution that springs to mind is to share the costs equally – $7.5 million for each. The argument for equal division is that each town has equal power to enter into a contract, so each should shoulder an equal burden. This argument is plausible if the towns are of about the same size, but otherwise it is suspect. Suppose, for example, that town A has 36 000 residents and town B has 12 000 residents. Equal division between the towns would imply that each resident of A pays only *one-third* as much as each resident of B, even though they are served by the same system. This hardly seems fair, and one can imagine that town B will not agree to it. A more plausible solution would be to divide the costs equally among the *persons* rather

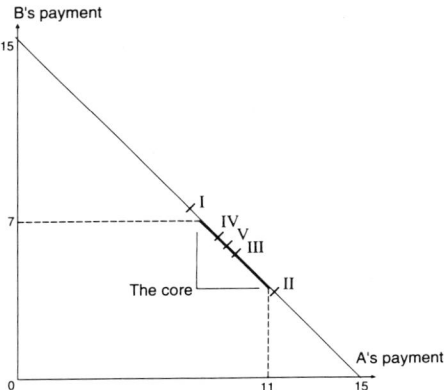

Figure 1. The core of the cost-sharing game.

Table 1
Five cost allocations for two towns

	Town A	Town B
I. Equal division of costs between towns	7.50	7.50
II. Equal division of costs among persons	11.25	3.75
III. Equal division of savings between towns	9.50	5.50
IV. Equal division of savings among persons	8.75	6.25
V. Savings (and costs) proportional to opportunity costs	9.17	5.83

than the towns. This results in a charge of $312.50 per capita, and altogether the citizens of town A pay $11.25 million while the citizens of town B pay $3.75 million (see Table 1).

Unfortunately, neither of these proposals takes into account the opportunity costs of the parties. B is not likely to agree to equal division, because $7.5 million exceeds the cost of building its own system. Similarly, A is not likely to agree to equal division per capita, since $11.25 exceeds the cost of building its own system. Thus the equity issue is complicated by the need to give the parties an incentive to cooperate. Without such incentives, cooperation will probably not occur and the outcome will be inefficient. Thus we see that the three major themes of cost allocation – efficiency, equity, and incentives – are closely intertwined.

Let us consider the incentives issue first. The simplest way to ensure that the parties have an incentive to cooperate is to focus on the amounts that they *save* rather than on the amounts that they *pay*. Three solutions now suggest themselves. One is to divide the $3 million in savings equally among the towns. In this case town A would pay $11 − 1.5 = $9.5 million and town B would pay $7 − 1.5 = $5.5 million. A second, and perhaps more plausible, solution is to divide the savings equally among the residents. Thus everyone would save $62.50, and the total cost assessments would be $8.75 million for town A and $6.25 million for town B. Yet a third solution would be to allocate the savings in proportion to each town's opportunity cost. This yields a payment of $9.17 million for A and $5.83 million for B. (Note that this is the same thing as allocating total cost in proportion to each town's opportunity cost.)

All three of these allocations give the parties an incentive to cooperate, because each realizes positive savings. Indeed, any solution in which A pays at most $11 million and B pays at most $7 million creates no *dis*incentive to cooperation. The set of all such solutions is known as the *core* of the cost-sharing game, a concept that will be defined more generally in Section 4 below. In the present case the core is the line segment shown in Figure 1.

This example illustrates several points. First, there is no completely obvious answer to the cost allocation problem even in apparently simple cases. Second, the problem cannot be avoided: costs must be allocated for the organization to be viable. Third, there is no external market mechanism that does the job. One might try to mimic the market by setting price equal to marginal cost, but this does not work either. In the preceding example, the marginal cost of including town A is $15 − 7 = \$8$ million (the difference between the cost of the project with A and the cost of the project without A), while the marginal cost of including town B is $15 − 11 = \$4$ million. Thus the sum of the marginal costs does not equal total cost – indeed it does not even *cover* total cost. Hence we must find some other means to justify a solution. This is where ideas of equity come to the fore: they are the instruments that the participants use to reach a joint decision. Equity principles, in other words, are not merely normative or philosophical concepts. Like other kinds of norms, they play a crucial economic role by coordinating players' expectations, without which joint gains cannot be realized.

3. The cooperative game model

Let us now formulate the problem in more general terms. Let $N = \{1, 2, \ldots, n\}$ be a set of *projects, products, or services* that can be provided jointly or severally by some organization. Let $c(i)$ be the cost of providing i by itself, and for each subset $S \subseteq N$, let $c(S)$ be the cost of providing the items in S jointly. By convention, $c(\phi) = 0$. The function c is called a *discrete cost function* or sometimes a *cost-sharing game*. An *allocation* is a vector $(x_1, \ldots, x_n)$ such that $\sum x_i = c(N)$, where x_i is the amount charged to project i. A *cost allocation method* is a function $\phi(c)$ that associates a unique allocation to every cost-sharing game.

In some contexts it is natural to interpret $c(S)$ as the *least costly* way of carrying out the projects in S. Suppose, for example, that each project involves providing a given amount of computing capability for each department in a university. Given a subset of departments $S, c(S)$ is the cost of the most economical system that provides the required level of computing for all the members of S. This might mean that each department is served by a separate system, or that certain groups of departments are served by a common system while others are served separately, and so forth. In other words, the cost function describes the *cost* of the most economical way of combining activities, it does not describe the physical *structure* of the system. If the cost function is interpreted in this way, then for any partition of a subset of projects into two disjoint subsets S' and S'', we have

$$c(S' \cup S'') \leqslant c(S') + c(S''). \tag{1}$$

This property is known as *subadditivity*.

A second natural property of a cost function is that costs increase the more projects there are, that is, $c(S) \leqslant c(S')$ for all $S \subseteq S'$. Such a cost function is *monotonic*. Neither monotonicity nor subadditivity will be assumed in subsequent results unless we specifically say so.

The reason for carrying out projects separately rather than jointly is that it generates cost savings. Hence it often makes sense to focus attention on the cost-savings directly. For each subset S of projects the potential *cost saving* is

$$v(S) = \sum_{i \in S} c(i) - c(S). \tag{2}$$

The function v is called the *cost-savings game*.

If c is subadditive, then v is nonnegative and monotone increasing in S. Indeed, for every S and $i \notin S$, subadditivity implies that $c(S+i) \leq c(S) + c(i)$, from which it follows that $v(S) \leq v(S+i)$. (We write "$S+i$" instead of "$S \cup \{i\}$".) Since $v(\phi) = 0$, it follows that v is nonnegative and monotonic. It also follows that $v(N)$ is the largest among all $v(S)$, so from a purely formal point of view, N is the efficient set of projects to undertake. This assumption will be implicit throughout the remainder of the chapter. We do not, however, need to assume subadditivity and monotonicity of the cost function in order to prove many of the theorems quoted below, and we shall not assume them unless explicitly noted.

4. The Tennessee Valley Authority

The Tennessee Valley Authority was a major regional development project created by an act of Congress in the 1930s to stimulate economic activity in the mid-southern United States. The goal was to construct a series of dams and reservoirs along the Tennessee River to generate hydroelectric power, control flooding, and improve navigational and recreational uses of the waterway. Economists charged with analyzing the costs and benefits of this project observed that there is no completely obvious way to attribute costs to these purposes, because the system is designed to satisfy all of them simultaneously. The concepts that they devised to deal with this problem foreshadow modern ideas in game theory, and one of the formulas they suggested has since become (after minor modifications) the standard method for allocating the cost of multi-purpose reservoirs.

Table 2
Cost function for navigation (1), flood control (2), and power (3), in thousands of dollars

Subsets S	Cost $c(S)$
$\emptyset$	0
$\{1\}$	163 520
$\{2\}$	140 826
$\{3\}$	250 096
$\{1,2\}$	301 607
$\{1,3\}$	378 821
$\{2,3\}$	367 370
$\{1,2,3\}$	412 584

Table 2 shows the cost function for the TVA case as analyzed by Ransmeier (1942, p. 329). There are three purposes: navigation (1), flood control (2), and power (3).

Ransmeier (1942, p. 220) suggested the following criteria for a cost allocation formula:

> The method should have a reasonable logical basis... It should not result in charging any objective with a greater investment than would suffice for its development at an alternate single-purpose site. Finally, it should not charge any two or more objectives with a greater investment than would suffice for alternate dual or multiple purpose development.

In terms of the joint cost function $c(S)$ these requirements state that, if x_i is the charge to purpose i, then the following inequality should hold for every subset S of purposes (including singletons),

$$x(S) \leq c(S), \tag{3}$$

where $x(S) = \sum_{i \in S} x_i$. Condition (3) is known as the *stand-alone cost test*. Its rationale is evident: if cooperation among the parties is voluntary, then self-interest dictates that no participant – or group of participants – be charged more than their stand-alone (opportunity) cost. Otherwise they would have no incentive to agree to the proposed allocation.

A related principle known as the "incremental cost test" states that no project should be charged *less* than the marginal cost of including it. In Table 2, for example, the cost of including project 1 at the margin is

$$c(1,2,3) - c(2,3) = 45\,214.$$

In general, the *incremental* or *marginal cost* of a subset S is defined to be $c(N) - c(N-S)$, and the *incremental cost test* requires that the allocation $x \in R^N$ satisfy

$$x(S) \geq c(N) - c(N-S) \quad \text{for all } S \subseteq N. \tag{4}$$

Whereas (3) provides incentives for voluntary cooperation, (4) arises from considerations of equity. For, if (4) were violated for some S, then it could be said that the coalition $N - S$ is *subsidizing* S. In other words, even if there is no need to give the parties an incentive to cooperate, there is still an argument for a core allocation on equity grounds[1].

It is easily seen that conditions (3) and (4) are equivalent given that costs are allocated exactly, namely,

$$x(N) = c(N). \tag{5}$$

---

[1] This idea has been extensively discussed in the literature on public utility pricing. See for example Faulhaber (1975), Sharkey (1982a, b, 1985), and Zajac (1993).

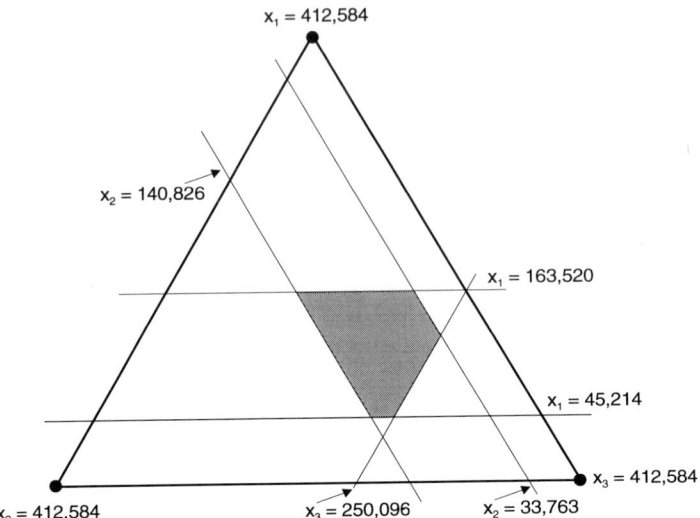

Figure 2. The core of the TVA cost-sharing game.

The *core* of the cost-sharing game c, written Core (c), is the set of all allocations $x \in R^N$ such that (3) and (5) [equivalently (4) and (5)] hold for all $S \subseteq N$.

The core of the TVA cost game is illustrated in Figure 2. The top vertex x_1 represents the situation where all costs are allocated to purpose 1 (navigation); the right-hand vertex allocates all costs to purpose 3 (power), and the left-hand vertex allocates all costs to purpose 2 (flood control). Each point in the triangle represents a division of the $412 584 among the three purposes.

In this example the core is fairly large, because there are strongly increasing returns to scale. This is due to the fact that the marginal cost of building a higher dam (to serve more purposes) decreases with the height of the dam.

The core is a closed, compact, convex subset of R^N. It may, however, be empty. Consider the cost function

$$c(1) = c(2) = c(3) = 6$$
$$c(1,2) = c(1,3) = c(2,3) = 7$$
$$c(1,2,3) = 11.$$

If x is in the core, then

$$x_1 + x_2 \leqslant 7, x_1 + x_3 \leqslant 7, x_2 + x_3 \leqslant 7.$$

However, the sum of these inequalities yields $2(x_1 + x_2 + x_3) \leqslant 21$, which contradicts the break-even requirement $x_1 + x_2 + x_3 = 11$. Hence the core is empty. Furthermore this is true even though the cost function is both subadditive and monotonic. A sufficient condition that the core be nonempty is that the cost function exhibit increasing returns to scale, as we shall show in Section 9.

Table 3
The ACA method applied to the TVA data

	Purpose			Total
	1	2	3	
1. Alternate cost $c(i)$	163 520	140 826	250 096	554 442
2. Separable cost $c(N) - c(N - i)$	147 630	136 179	213 393	497 202
3. Alternate cost avoided (1-2)	15 890	4 647	36 704	57 240
4. Allocation of nonsep. cost	4 941	1 445	11 412	17 798
5. Allocation (2 + 4)	152 571	137 624	224 805	515 000

Although the Tennessee Valley Authority did not adopt a formal method for allocating costs, they took as a basis an approach known as the "alternative justifiable expenditure method" and then rounded off the results according to "judgment."[2] This has become, after further refinements, the principal textbook method used by civil engineers to allocate the costs of multi-purpose reservoirs, and is known as the "separable costs remaining benefits method" [James and Lee (1971)]. We shall now describe a simple version of this method.

Given a cost-sharing game c, define the separable cost of a purpose $i \in N$ to be its marginal cost

$$s_i = c(N) - c(N - i).$$

The *alternate cost* for i is the stand-alone cost $c(i)$. The difference between the alternate cost and the separable cost is the *alternate cost avoided*

$$r_i = c(i) - s_i.$$

The *alternate cost avoided method* (ACA) assigns costs according to the formula

$$x_i = s_i + [r_i/r(N)][c(N) - s(N)].^3 \qquad (6)$$

In other words, each project pays its separable cost and the "nonseparable cost" $c(N) - s(N)$ is allocated in proportion to the numbers r_i. The implicit assumption here is that all $r_i \geq 0$, which is the case if c is subadditive. Table 3 illustrates the calculation for the TVA cost data.

The ACA method can be given a more succinct and intuitive formulation in terms of the cost-savings game. For each project $i \in N$ define i's *marginal savings*

---

[2] As Ransmeier remarks, "there is little to recommend the pure judgement method for allocation. In many regards it resembles what Professor Lewis has called the 'trance method' of utility valuation." (1942, p.342).

[3] The more sophisticated *separable costs remaining benefits method* (SCRB) incorporates benefits as follows. Let $b(i)$ be the benefit from undertaking project i by itself. Then the *maximum justifiable expenditure* for i is min $\{b(i), c(i)\}$ and r_i in formula (6) is defined to be $r_i = \min\{b(i), c(i)\} - s_i$. [James and Lee (1971)].

to be

$$v^i(N) = v(N) - v(N-i). \tag{7}$$

Given that i is charged x_i, the resulting savings are $y_i = c(i) - x_i$. A simple manipulation of (6) shows that the ACA imputes savings according to the formula

$$y_i = \left[v^i(N) / \sum_{j \in N} [v^j(N)] \right] v(N). \tag{8}$$

In other words, the ACA allocates cost savings in proportion to each project's marginal contribution to savings [Straffin and Heaney (1981)]. This solution was proposed independently in the game theory literature as a means of minimizing players' "propensity to disrupt" the solution [Gately (1974), Littlechild and Vaidya (1976), Charnes et al. (1979)].

There is no reason to think that the ACA method yields a solution in the core, and indeed it does not in general (see Section 6 below for an example). However, when there are at most three projects, and the cost function is subadditive, then it is in the core provided that the core is nonempty. Indeed we can show more. Define the *semicore* of a cost function to be the set of all allocations x such that,

$$\text{for every } i, \quad x_i \leqslant c(i), \quad x(N-i) \leqslant c(N-i), \quad \text{and } x(N) = c(N). \tag{9}$$

In the case of two or three projects the semicore is clearly the same as the core.

Theorem 1. *If c is subadditive, then the alternate cost avoided method is in the semicore whenever the latter is nonempty.*

Proof. It is easiest to work in terms of the cost-savings game v. By assumption c is subadditive, so v is nonnegative and monotone. By assumption, c has a nonempty semicore, so there is an allocation of savings y such that

$$v^i(N) \geqslant y_i \geqslant v(i) = 0 \quad \text{for every } i \in N.$$

Hence

$$y(N) = v(N) \leqslant \sum_{j \in N} v^j(N). \tag{10}$$

The ACA allocation is defined by

$$y_i^* = v^i(N) \left[v(N) / \sum_{j \in N} v^j(N) \right].$$

From this and (10) we conclude that

$$0 \leqslant y_i^* \leqslant v^i(N). \tag{11}$$

Letting $x_i^* = c(i) - y_i^*$ it follows that x^* is in the semicore of c. □

5. Equitable core solutions

Core allocations provide incentives for cooperation. They are also *fair* in the sense that no subgroup subsidizes any other. If core allocations exist, however, there are usually an uncountable number of them. Which is most equitable? Here we shall suggest one answer to this question.

Let us begin by noticing that in the case of two projects the natural solution is to choose the *midpoint* of the core. This solution treats the two projects equally in the sense that they save equal amounts relative to their opportunity costs. There is another way of justifying this answer however. Consider a cost allocation situation in which some costs can be attributed directly to particular projects. In the TVA case, for example, the cost of the generators is directly attributable to hydropower generation, the cost of constructing levees is directly attributable to controlling flooding, and the cost of building locks is directly attributable to navigation. Sometimes the distinction between direct and joint costs is not so clear, however. Deepening the river channel probably benefits navigation most, but it has a favorable impact on flood control and hydropower generation as well.

From a formal point of view, we say that a cost function c *decomposes into direct costs* $d = (d_1, d_2, \ldots, d_n)$ and *joint costs* c^* if c can be written in the form

$$c(S) = d(S) + c^*(S) \quad \text{for every } S \subseteq N. \tag{12}$$

A cost allocation method ϕ is *invariant in direct costs* if whenever c satisfies (12), then

$$\phi(c) = d + \phi(c^*), \tag{13}$$

that is, $\phi_i(c) = d_i + \phi_i(c^*)$ for every i.

A cost allocation method is *symmetric* if it is invariant under any renaming of the projects. In other words, given any cost function c on N, and any permutation π of N, if we define the cost function πc such that $\pi c(\pi S) = c(S)$, then

$$\phi_{\pi(i)}(\pi c) = \phi_i(c) \quad \text{for every } i \in N. \tag{14}$$

Theorem 2. *If ϕ is symmetric and invariant in direct costs, then for any two-project cost function c defined on $N = \{1, 2\}$,*

$$\phi_i(c) = c(i) - s/2 \quad \text{where } s = c(1) + c(2) - c(1, 2). \tag{15}$$

The proof of this result is straightforward and is left to the reader.

The method defined by (15) is called the *standard two-project solution*. When total savings s are nonnegative (which is true if c is subadditive), the standard solution is simply the midpoint of the core. Thus Theorem 2 provides an axiomatic justification of a solution that is intuitively appealing on equity grounds.

When there are more than two projects, it seems natural to apply the same idea and divide the cost-savings equally among the various projects. In the TVA case, this yields the cost allocation (116 234, 93 540, 202 810) which is in the core. But

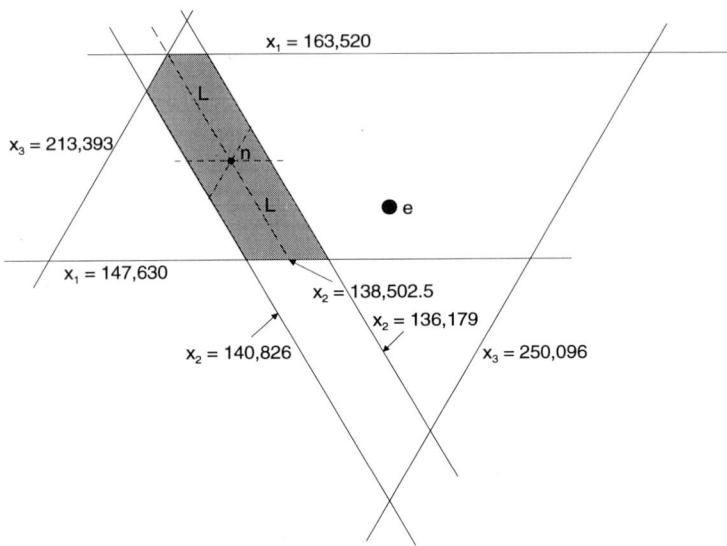

Figure 3. The core of the modified TVA cost game.

there are many situations where the equal savings allocation is *not* in the core. Consider the following variation of the TVA data: total costs are $515 000 000 and all other costs remain the same as before. The core of this modified TVA game is shown in Figure 3, and the equal division of savings (e) is not in it. Hence we must find some other principle for determining an equitable allocation in the core.

Consider the following approach: instead of insisting that all projects be treated equally, let us treat all projects – and all subgroups of projects – *as nearly equally as possible*. By this we mean the following. Let c be a cost function and let x be an allocation of total cost $c(N)$. Each subset of projects $S \subseteq N$ realizes cost savings equal to $e(S, x) = c(S) - x(S)$. We shall say that the set S is strictly *better off* than the set T if $e(S, x) > e(T, x)$. A natural criterion of equity is to *maximize the position of the least well-off subset*, that is, to find an allocation x that maximizes $\min_S e(S, x)$. This is known as the *maximin criterion*.[4] A maximin allocation is found by solving the linear programming problem

$$\max \varepsilon$$
subject to $e(S, x) \geq \varepsilon$ for all S, $\phi \subset S \subset N$
and $x(N) = c(N)$. (16)

The set of all solutions to x (16) is called the *least core* of c.

If there is more than one allocation in the least core, we may whittle it down further by extending the maximin criterion as follows. Order the numbers $e(S, x)$,

---

[4]This idea is central to John Rawls's *A Theory of Justice* (1971).

$\phi \subset S \subset N$, from lowest to highest, and denote this vector of dimension $(2^n - 2)$ by $\theta(x)$. The *prenucleolus* is the allocation x that maximizes $\theta(x)$ lexicographically, that is, if y is any other cost allocation and k is the first index such that $\theta_k(x) \neq \theta_k(y)$, then $\theta_k(x) > \theta_k(y)$ [Schmeidler (1969)].[5]

The prenucleolus occupies a central position in the core in the sense that the minimal distance from any boundary is as large as possible. In Figure 3 it is the point labelled "n", with coordinates $x_1 = \$155\,367.2, x_2 = \$138\,502.5, x_3 = \$221\,130.25$. (Note that this is *not* the center of gravity.)

We claim that the prenucleolus is a natural extension of the standard two-project solution. Let us begin by observing that, when there are just two projects, the prenucleolus agrees with the standard two-project solution. Indeed, in this case there are just two proper subsets $\{1\}$ and $\{2\}$, and the allocation that maximizes the smaller of $c(1) - x_1$ and $c(2) - x_2$ is clearly the one such that $c(1) - x_1 = c(2) - x_2$.

When there are more than two projects, the prenucleolus generalizes the standard two-project solution in a more subtle way. Imagine that each project is represented by an agent, and that they have reached a preliminary agreement on how to split the costs. It is natural for each subgroup of agents to ask whether they fairly divide the cost allocated to them as a subgroup. To illustrate this idea, consider Figure 3 and suppose that the agents are considering the division $n = (155\,367.2, 138\,502.5, 221\,130.25)$. Let us restrict attention to agents 1 and 3. If they view agent 2's allocation as being fixed at $\$138\,502.50$, then they have $\$376\,497.50$ to divide between them. The range of possible divisions that lie within the core is represented by the dotted line segment labelled L. In effect, L is the core of a smaller or "reduced" game on the two-player set $\{1,3\}$ that results when 2's allocation is held fixed. Now observe that (n_1, n_3) is the midpoint of this segment. In other words, it is the standard two-project solution of the reduced game. Moreover the figure shows that the prenucleolus n bisects *each* of the three line segments through n in which the charge to one agent is held fixed. This observation holds in general: *the prenucleolus is the standard two-project solution when restricted to each pair of agents*. This motivates the following definition. Let c be any cost function defined on the set of projects N, and let x be any allocation of $c(N)$. For each proper subset T of N, define the *reduced cost function* $c_{T,x}$ as follows:

$$c_{T,x}(S) = \min_{S' \subseteq N-T} \{c(S \cup S') - x(S')\} \quad \text{if } \phi \subset S \subset T,$$

$$c_{T,x}(T) = x(T),$$

$$c_{T,x}(\phi) = 0. \tag{17}$$

A cost allocation method ϕ is *consistent* if for every N, every cost function c on

---

[5]If x' and x'' are distinct allocations that maximize $\theta(\cdot)$ lexicographically, then it is relatively easy to show that $\theta(x'/2 + x''/2)$ is strictly larger lexicographically than both $\theta(x')$ and $\theta(x'')$. Hence θ has a unique maximum. The lexicographic criterion has been proposed as a general principle of justice by Sen (1970).

N, and every proper subset T of N,

$$\phi(c) = x \Rightarrow \phi(c_{T,x}) = x_T. \tag{18}$$

If (18) holds for every subset T of cardinality two, then ϕ is *pairwise consistent*. Note that this definition applies to all cost games c, whether or not they have a nonempty core.[6]

Consistency is an extremely general principle of fair division which says that if an allocation is fair, then every subgroup of claimants should agree that they share the amount allotted to them fairly. This idea has been applied to a wide variety of allocation problems, including the apportionment of representation [Balinski and Young (1982)], bankruptcy rules [Aumann and Maschler (1985)], surplus sharing rules [Moulin (1985)], bargaining problems [Harsanyi (1959), Lensberg (1985, 1987, 1988)], taxation [Young (1988)], and economic exchange [Thomson (1988)]. For reviews of this literature see Thomson (1990) and Young (1994).

To state the major result of this section we need one more condition. A cost allocation rule is *homogeneous* if for every cost function c and every positive scale factor λ, $\phi(\lambda c) = \lambda \phi(c)$.

Theorem 3 [Sobolev (1975)]. *The prenucleolus is the unique cost allocation method that is symmetric, invariant in direct costs, homogeneous, and consistent.*

We remark that it will not suffice here to assume pairwise consistency instead of consistency. Indeed, it can be shown that, *for any cost function c, the set of all allocations that are pairwise consistent with the standard two-project solution constitutes the prekernel of c* [Peleg (1986)]. The prekernel contains the prenucleolus but possibly other points as well. Hence pairwise consistency with the standard solution does not identify a unique cost allocation.

6. A Swedish municipal cost-sharing problem

In this section we analyze an actual example that illustrates some of the practical problems that arise when we apply the theory developed above. The Skåne region of southern Sweden consists of eighteen municipalities, the most populous of which is the city of Malmö (see Figure 4). In the 1940s several of them, including Malmö, banded together to form a regional water supply utility known as the Sydvatten (South Water) Company. As water demands have grown, the Company has been under increasing pressure to increase long-run supply and incorporate outlying municipalities into the system. In the late 1970s a group from the International Institute for Applied Systems Analysis, including this author, were invited to

---

[6]For an alternative definition of the reduced game see Hart and Mas-Colell (1989).

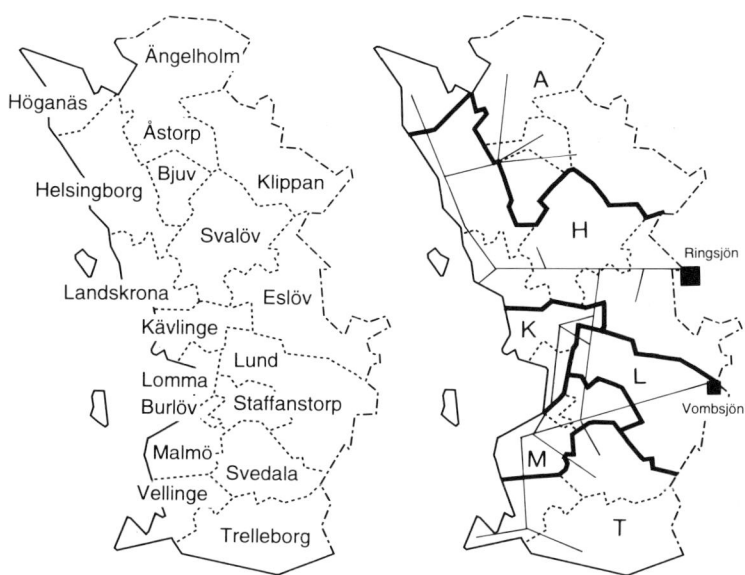

Figure 4. The region of Skåne, Sweden and its partition into groups of municipalities.

analyze how the cost of expanding the system should be allocated among the various townships, We took the system as it existed in 1970 and asked how much it would cost to expand it in order to serve the water demands projected by 1980.

In theory we should have estimated the system expansion cost for each of the $2^{18} = 262\,144$ possible subgroups, but this was clearly infeasible. To simplify the problem, we noted that the municipalities fall into natural groups based on past associations, geographical proximity, and existing water transmission systems. This led us to group the eighteen municipalities into the six units shown in Figure 4. We treated these groups as single actors in developing the cost function. Of course, once a cost allocation among the six groups is determined, a second-stage allocation must be carried out within each subgroup. This raises some interesting modelling issues that will be discussed in Section 10 below. Here we shall concentrate on the problem of allocating costs among the six units in the aggregated cost-sharing game.

One of the first problems that arises in defining the cost function is how to distinguish between direct costs and joint costs. Within each municipality, for example, a local distribution network is required no matter where the water comes from, so one might suppose that this is a direct cost. However, in some cases the water delivered by the regional supply network must first be pumped up to a reservoir before being distributed within the municipality. The cost of these pumping facilities depends on the pressure at which the water is delivered by the regional system. Hence the cost of the local distribution facilities is *not* completely independent of the method by which the water is supplied. This illustrates why the borderline between direct and joint costs is somewhat fuzzy, and why it is important to use a method that does not depend on where the line is drawn, i.e., to use a method that is invariant in direct costs.

For each subset of the six units, we estimated the cost of expanding the system to serve the members of this subset using standard engineering formulas, and the result is shown in Table 4. [See Young et al. (1982) for details.]

Note the following qualitative features of the cost function. Even though L is close to the two major sources of supply (lakes Ringsjön and Vombsjön), it has a high stand-alone cost because it does not have rights to withdraw from these sources. Hence we should expect L's charge to be fairly high. By contrast, H and M have relatively low stand-alone costs that can be reduced even further by including other municipalities in the joint scheme. However, the system owned by H (Ringsjön) has a higher incremental capacity than the one owned by M (Vombsjön). Hence the incremental cost of including other municipalities in a coalition with M is higher than the incremental cost of including them in a coalition with H. In effect, H has more to offer its partners than M does, and this should be reflected in the cost allocation.

Table 4
Costs of alternative supply systems, in millions of Swedish crowns. Coalitions are separated by commas if there are no economies of scale from integrating them into a single system, that is, we write S, S' if $c(S) + c(S') = c(S \cup S')$

Group	Total cost	Group	Total cost	Group	Total cost
A	21.95	AHK	40.74	AHKL	48.95
H	17.08	AHL	43.22	AHKM	60.25
K	10.91	AH, M	55.50	AHK, T	62.72
L	15.88	AH, T	56.67	AHL, M	64.03
M	20.81	A, K, L	48.74	AHL, T	65.20
T	21.98	A, KM	53.40	AH, MT	74.10
		A, K, T	54.84	A, K, LM	63.96
AH	34.69	A, LM	53.05	A, K, L, T	70.72
A, K	32.86	A, L, T	59.81	A, K, MT	72.27
A, L	37.83	A, MT	51.36	A, LMT	73.41
A, M	42.76	HKL	27.26	HKL, M	48.07
A, T	43.93	HKM	42.55	HKL, T	49.24
HK	22.96	HK, T	44.94	HKMT	59.35
HL	25.00	HL, M	45.81	HLMT	64.41
H, M	37.89	HL, T	46.98	KLMT	56.61
H, T	39.06	H, MT	56.49	AHKL, T	70.93
K, L	26.79	K, LM	42.01	AHKLM	69.76
KM	31.45	K, L, T	48.77	AKHMT	77.42
K, T	32.89	K, MT	50.32	AHLMT	83.00
LM	31.10	LMT	51.46	AKLMT	73.97
L, T	37.86			HKLMT	66.46
MT	39.41				
				AHKLMT	83.82

Table 5
Cost allocation of 83.82 million Swedish crowns by four methods

	A	H	K	L	M	T
Stand-alone cost	21.95	17.08	10.91	15.88	20.81	21.98
Prop. to pop.	10.13	21.00	3.19	8.22	34.22	7.07
Prop. to demand	13.07	16.01	7.30	6.87	28.48	12.08
ACA	19.54	13.28	5.62	10.90	16.66	17.82
Prenucleolus	20.35	12.06	5.00	8.61	18.32	19.49

Table 5 shows cost allocations by four different methods. The first two are "naive" solutions that allocate costs in proportion to population and water demand respectively. The third is the standard engineering approach described in Section 4 (the ACA method), and the last is the prenucleolus. Note that both of the proportional methods charge some participant more than its stand-alone cost. Allocation by demand penalizes M, while allocation by population penalizes both H and M. These two units have large populations but they have ready access to the major sources of supply, hence their stand-alone costs are low. A and T, by contrast, are not very populous but are remote from the sources and have high stand-alone costs. Hence they are favored by the proportional methods. Indeed, neither of these methods charges A and T even the *marginal* cost of including them.

The ACA method is apparently more reasonable because it does not charge any unit more than its stand-alone cost. Nevertheless it fails to be in the core, which is nonempty. H, K, and L can provide water for themselves at a cost of 27.26 million Swedish crowns, but the ACA method charges them a total of 29.80 million Swedish crowns. In effect they are subsidizing the other participants. The prenucleolus, by contrast, is in the core and is therefore subsidy-free.

7. Monotonicity

Up to this point we have implicitly assumed that all cost information is in hand, and the agents need only reach agreement on the final allocation. In practice, however, the parties may need to make an agreement before the actual costs are known. They may be able to *estimate* the total cost to be divided (and hence their prospective shares), but in reality they are committing themselves to a *rule* for allocating cost rather than to a single cost allocation. This has significant implications for the type of rule that they are likely to agree to. In particular, if total cost is higher than anticipated, it would be unreasonable for anyone's charge to go down. If cost is lower than anticipated, it would be unreasonable for anyone's charge to go up. Formally, an allocation rule ϕ is *monotonic in the aggregate* if for any set of projects N, and any two cost functions c and c' on N

$$c'(N) \geq c(N) \quad \text{and } c'(S) = c(S) \quad \text{for all } S \subset N$$
$$\text{implies } \phi_i(c') \geq \phi_i(c) \text{ for all } i \in N. \tag{19}$$

Table 6
Allocation of a cost overrun of 5 million Swedish crowns by two methods

	A	H	K	L	M	T
Prenucleolus	0.41	1.19	−0.49	1.19	0.84	0.84
ACA	1.88	0.91	−0.16	0.07	0.65	0.65

This concept was first formulated for cooperative games by Megiddo (1974). It is obvious that any method based on a proportional criterion is monotonic in the aggregate, but such methods fail to be in the core. The alternate cost avoided method is neither in the core nor is it monotonic in the aggregate. For example, if the total cost of the Swedish system increases by 5 million crowns to 87.82, the ACA method charges K less than before (see Table 6). The prenucleolus is in the core (when the core is nonempty) but it is also not monotonic in the aggregate, as Table 6 shows.

The question naturally arises whether *any* core method is monotonic in the aggregate. The answer is affirmative. Consider the following variation of the prenucleolus. Given a cost function c and an allocation x, define the *per capita savings* of the proper subset S to be $d(x,S) = (c(S) - x(S))/|S|$. Order the $2^n - 2$ numbers $d(x,S)$ from lowest to highest and let the resulting vector be $\gamma(x)$. The *per capita prenucleolus* is the unique allocation that lexicographically maximizes $\gamma(x)$ [Grotte (1970)].[7] It may be shown that the per capita prenucleolus is monotonic in the aggregate and in the core whenever the core is nonempty. Moreover, it allocates any increase in total cost in a natural way: the increase is split equally among the participants [Young et al. (1982)]. In these two respects the per capita prenucleolus performs better than the prenucleolus, although it is less satisfactory in that it fails to be consistent.

There is a natural generalization of monotonicity, however, that both of these methods fail to satisfy. We say that the cost allocation method ϕ is *coalitionally monotonic* if an increase in the cost of any particular coalition implies, *ceteris paribus*, no decrease in the allocation to any member of that coalition. That is, for every set of projects N, every two cost functions c, c' on N, and every $T \subseteq N$,

$$c'(T) \geq c(T) \quad \text{and} \quad c'(S) = c(S) \quad \text{for all } S \neq T$$
$$\text{implies } \phi_i(c') \geq \phi_i(c) \quad \text{for all } i \in T. \tag{20}$$

It is readily verified that (20) is equivalent to the following definition: ϕ is *coalitionally monotonic* if for every N, every two cost functions c' and c on N, and every $i \in N$,

if $c'(S) \geq c(S)$ for all S containing i and $c'(S) = c(S)$ for all S not containing i, then $\phi_i(c') \geq \phi_i(c)$.

---
[7]Grotte (1970) uses the term "normalized nucleolus" instead of "per capita nucleolus".

The following "impossibility" theorem shows that coalitional monotonicity is incompatible with staying in the core.

Theorem 4 [Young (1985a)]. *For $|N| \geq 5$ there exists no core allocation method that is coalitionally monotonic.*

Proof. Consider the cost function c defined on $N = \{1, 2, 3, 4, 5\}$ as follows:

$$c(S_1) = c(3,5) = 3, \quad c(S_2) = c(1,2,3) = 3,$$
$$c(S_3) = c(1,3,4) = 9, \quad c(S_4) = c(2,4,5) = 9,$$
$$c(S_5) = c(1,2,4,5) = 9, \quad c(S_6) = c(1,2,3,4,5) = 11.$$

For $S \neq S_1, \ldots, S_5, S_6, \phi$, define

$$c(S) = \min_k \{c(S_k): S \subseteq S_k\}.$$

If x is in the core of c, then

$$x(S_k) \leq c(S_k) \quad \text{for } 1 \leq k \leq 5. \tag{21}$$

Adding the five inequalities defined by (21) we deduce that $3x(N) \leq 33$, whence $x(N) \leq 11$. But $x(N) = 11$ because x is an allocation. Hence, all inequalities in (21) must be equalities. These have the unique solution $x = (0, 1, 2, 7, 1)$, which constitutes the core of c.

Now consider the game c', which is identical to c except that $c'(S_5) = c'(S_6) = 12$. A similar argument shows that the unique core element of this game is $x' = (3, 0, 0, 6, 3)$. Thus the allocation to both 2 and 4 decreases even though the cost of some of the sets containing 2 and 4 monotonically increases. This shows that no core allocation procedure is monotonic for $|N| = 5$, and by extension for $|N| \geq 5$. □

8. Decomposition into cost elements

We now turn to a class of situations that calls for a different approach. Consider four homeowners who want to connect their houses to a trunk power line (see Figure 5). The cost of each segment of the line is proportional to its length, and a segment costs the same amount whether it serves some or all of the houses. Thus the cost of segment OA is the same whether it carries power to house A alone or to A plus all of the houses more distant than A, and so forth.

If the homeowners do not cooperate they can always build parallel lines along the routes shown, but this would clearly be wasteful. The efficient strategy is to construct exactly four segments OA, AB, BC, and BD and to share them. But what is a reasonable way to divide the cost?

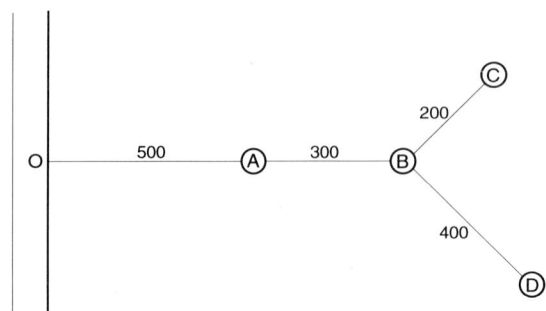

Figure 5. Cost of connecting four houses to an existing trunk power line.

The answer is transparent. Since everyone uses the segment OA, its cost should be divided equally among all four homeowners. Similarly, the cost of segment AB would be divided equally among B, C, D, the cost of BC should be borne exclusively by C and the cost of BD exclusively by D. The resulting cost allocation is shown in Table 7.

Let us now generalize this idea. Suppose that a project consists of m distinct components or *cost elements*. Let $C_\alpha \geq 0$ be the cost of component $\alpha, \alpha = 1, 2, \ldots, m$. Denote the set of potential beneficiaries by $N = \{1, 2, \ldots, n\}$. For each cost element α, let $N_\alpha \subseteq N$ be the set of parties who use α. Thus the stand-alone cost of each subset $S \subseteq N$ is

$$c(S) = \sum_{N_\alpha \cap S \neq \phi} C_\alpha. \tag{22}$$

A cost function that satisfies (22) *decomposes* into nonnegative cost elements. *The decomposition principle states that when a cost function decomposes, the solution is to divide each cost element equally among those who use it and sum the results.*

It is worth noting that the decomposition principle yields an allocation that is in the core. Indeed, $c(S)$ is the sum of the cost elements used by members of S, but the charge for any given element is divided equally among all users, some of

Table 7
Decomposition of electrical line costs

Cost elements	Homes				Segment cost
	A	B	C	D	
OA	125	125	125	125	500
AB		100	100	100	300
BC			200		200
BD				400	400
Charge	125	225	425	625	1400

Table 8
Aircraft landings, runway costs, and charges at Birmingham airport, 1968–69

Aircraft type	No. landings	Total cost*	Shapley value
Fokker Friendship 27	42	65 899	4.86
Viscount 800	9555	76 725	5.66
Hawker Siddeley Trident	288	95 200	10.30
Britannia	303	97 200	10.85
Caravelle VIR	151	97 436	10.92
BAC 111 (500)	1315	98 142	11.13
Vanguard 953	505	102 496	13.40
Comet 4B	1128	104 849	15.07
Britannia 300	151	113 322	44.80
Corvair Corronado	112	115 440	60.61
Boeing 707	22	117 676	162.24

*Total cost of serving this type of plane and all smaller planes.

which may not be in S. Hence the members of S are collectively not charged *more* than $c(S)$.

As a second application of the decomposition principle consider the problem of setting landing fees for different types of planes using an airport [Littlechild and Thompson (1977)]. Assume that the landing fees must cover the cost of building and maintaining the runways, and that runways must be longer (and therefore more expensive) the larger the planes are. To be specific, let there be m different types of aircraft that use the airport. Order them according to the length of runway that they need: type 1 needs a short runway, type 2 needs a somewhat longer runway, and so forth. Schematically we can think of the runway as being divided into m sections. The first section is used by all planes, the second is used by all but the smallest planes, the third by all but the smallest two types of planes, and so forth.

Let the annualized cost of section α be $c_\alpha, \alpha = 1, 2, \ldots, m$. Let n_α be the number of landings by planes of type α in a given year, let N_α be the *set* of all such landings, and let $N = \cup N_\alpha$. Then the cost function takes the form

$$c(S) = \sum_{N_\alpha \cap S \neq \phi} c_\alpha,$$

so it is decomposable. Table 8 shows cost and landing data for Birmingham airport in 1968/69, as reported by Littlechild and Thompson (1977), and the charges using the decomposition principle.

9. The Shapley value

The decomposition principle involves three distinct ideas. The first is that everyone who uses a given cost element should be charged equally for it. The second is that those who do not use a cost element should not be charged for it. The third is that the results of different cost allocations can be added together. We shall now show how these ideas can be extended to cost functions that do *not* necessarily decompose into nonnegative cost elements.

Fix a set of projects N and let ϕ be a cost allocation rule defined for every cost function c on N. The notion that everyone who uses a cost element should be charged equally for it is captured by symmetry (see p. 1203). The idea that someone should not be charged for a cost element he does not use generalizes as follows. Say that project i is a *dummy* if $c(S+i) = c(S)$ for every subset S not containing i. It is natural to require that the charge to a dummy is equal to zero.

Finally, suppose that costs can be broken down into different categories, say operating cost and capital cost. In other words, suppose that there exist cost functions c' and c'' such that

$$c(S) = c'(S) + c''(S) \quad \text{for every} \quad S \subseteq N.$$

The rule ϕ' is *additive* if

$$\phi(c) = \phi(c') + \phi(c'').$$

Theorem 5. [Shapley (1953a, b)]. *For each fixed N there exists a unique cost allocation rule ϕ defined for all cost functions c on N that is symmetric, charges dummies nothing, and is additive, namely the Shapley value*

$$\phi_i(c) = \sum_{S \subseteq N-i} \frac{|S|!(|N-S|-1)!}{|N|!} [c(S+i) - c(S)].$$

When the cost function decomposes into cost elements, it may be checked that the Shapley value gives the same answer as the decomposition principle. In the more general case the Shapley value may be calculated as follows. Think of the projects as being added one at a time in some arbitrary order $R = i_1, i_2, \ldots, i_n$. The *cost contribution of project $i = i_k$ relative to the order R* is

$$\gamma_i(R) = c(i_1, i_2, \ldots, i_k) - c(i_1, i_2, \ldots, i_{k-1}).$$

It is straightforward to check that the Shapley value for i is just the average of $\gamma_i(R)$ over all $n!$ orderings R.

When the cost function decomposes into distinct cost elements, the Shapley value is in the core, as we have already noted. Even when the game does not decompose, the Shapley value may be in the core provided that the core is large enough. In the TVA game, for example, the Shapley value is (117 829, 100 756.5, 193 998.5), which is comfortably inside the core. There are perfectly plausible examples, however, where the cost function has a nonempty core and the Shapley value fails to be in it. If total cost for the TVA were 515 000, for example (see

Figure 3) the Shapley value would be

(151 967 2/3,   134 895 1/6,   228 137 1/6).

This is not in the core because the total charges for projects 1 and 3 come to 380 104 5/6, which exceeds the stand-alone cost $c(1,3) = 378\,821$.

There is, however, a natural condition under which the Shapley value is in the core – namely, if the marginal cost of including any given project decreases the more projects there are. In other words, the Shapley value is in the core provide there are increasing (or at least not decreasing) returns to scale. To make this idea precise, consider a cost function c on N. For each $i \in N$ and $S \subseteq N$, i's *marginal cost contribution relative to S* is

$$c^i(S) = \begin{cases} c(S) - c(S - i), & \text{if } i \in S, \\ c(S + i) - c(S), & \text{if } i \notin S. \end{cases} \qquad (23)$$

The function $c^i(S)$ is called the *derivative* of c with respect to i. The cost function is *concave* if $c^i(S)$ is a nonincreasing function of S for every i, that is, if $c^i(S) \geqslant c^i(S')$ whenever $S \subset S' \subseteq N$.[8]

Theorem 6 [Shapley (1971)]. *The core of every concave cost function is nonempty and contains the Shapley value.*

10. Weighted Shapley values

Of all the properties that characterize the Shapley value, symmetry seems to be the most innocuous. Yet from a modelling point of view this assumption is perhaps the trickiest, because it calls for a judgment about what should be treated equally. Consider again the problem of allocating water supply costs among two towns A and B as discussed in Section 2. The Shapley value assigns the cost savings ($3 million) equally between them. Yet if the towns have very different populations, this solution might be quite inappropriate. This example illustrates why the symmetry axiom is not plausible when the partners or projects differ in some respect *other than cost* that we feel has a bearing on the allocation.

Let us define the *cost objects* to be the things we think deserve equal treatment provided that they contribute equally to cost. They are the "elementary particles" of the system. In the municipal cost-sharing case, for example, the objects might be the *towns* or the *persons* or (conceivably) the *gallons of water* used. To apply

---

[8] An equivalent condition is that c be submodular, that is, for any $S, S' \subseteq N$, $c(S \cup S') + c(S \cap S') \leqslant c(S) + c(S')$ for all $S, S' \subseteq N$.

the Shapley value we would then compute the cost function for all subsets of the cost objects.

If the objects are very numerous, however, this approach is impractical. Simplifying assumptions must be made. We could assume, for instance, that serving any part of a town costs the same amount as serving the whole town. Thus, if persons are the cost objects, and N^* is the set of all 48 000 persons, we would define the cost function c^* on N^* as follows:

$$c^*(S) = 11 \text{ if } S \subseteq A, \quad c^*(S) = 7 \text{ if } S \subseteq B, \quad c^*(S) = 15 \text{ otherwise.}$$

This game has the feature that it is composed of distinct "blocks" or "families." More generally, given a cost function c^* defined on a set N^*, a *family* of c^* is a nonempty subset $S \subseteq N^*$ such that

$$\text{for every } T \subseteq N, \quad S \cap T \neq \phi \quad \text{implies } c^*(T \cup S) = c^*(T). \tag{24}$$

In other words, S is a family if we incur the full cost of serving S whenever we have to serve at least one member of S.

Consider any partition of N^* into families $S_1, S_2, \ldots, S_m$, some or all of which may be singletons. Let w_j be the number of persons in family j, $1 \leq j \leq m$. Define the *aggregated cost function* c on the index set $M = \{1, 2, \ldots, m\}$ as follows:

$$c(T) = c^*\left(\bigcup_{j \in T} S_j\right), \quad T \subseteq M.$$

Consider all $m!$ orderings of the m families. For any such ordering $R = (i(1), i(2), \ldots, i(m))$ define the *probability of* R *with respect to* w to be

$$P_w(R) = \prod_{k=1}^{m} \left(w_{i(k)} \bigg/ \sum_{j=k}^{m} w_{i(j)}\right).$$

One way of constructing such a probability distribution over orderings is as follows: Add one family at a time, where the probability that a family is chosen equals its weight divided by the weight of all partnerships remaining to be chosen.

The Shapley value of c^* may now be computed in two steps. First we compute the expected marginal contribution to cost of each family j over all orderings. In other words, for each ordering R, let $\gamma_j(R)$ be the difference between the cost of family j together with all its predecessors in R, and the cost of all j's predecessors in R (excluding j). Define

$$(\phi_w)_j(c) = \sum_R P_w(R) \gamma_j(R). \tag{25}$$

The function ϕ_w is called the *weighted Shapley value* of c with weights w [see Loehman and Whinston (1976), Shapley (1981), Kalai and Samet (1987)]. It can be shown that the Shapley value of individual project i in the *original* cost function c^* is just the weighted Shapley value of the family to which i belongs, divided by the number of members in that family, that is,

$$\text{for all } i \in S_j, \quad \phi_i(c^*) = (\phi_w)_j(c)/w_j.$$

The question now arises whether we can justify the weighted Shapley value from first principles without the symmetry axiom. Here is one such axiomatization. [For alternative approaches see Loehman and Whinston (1976), Shapley (1981).] Fix the set N, and consider a cost allocation method ϕ defined for all cost functions c on N. The method ϕ is *positive* if whenever c is monotonic [$S \subseteq S'$ implies $c(S) \leq c(S')$] and contains no dummy players, then $\phi_i(c) > 0$ for all i.

For every $S \subseteq N$, let u_S be the cost function such that $u_S(T) = 1$ whenever $T \cap S \neq \emptyset$ and $u_S(T) = 0$ whenever $T \cap S = \emptyset$. The method ϕ is *family consistent* if for any family S in c, $\phi_i(c) = \phi_i(\gamma u_S)$, where $\gamma = \sum_{i \in S} \phi_i(c)$. Family consistency means that the total charge to any family is divided among its members just as it would be if this were the unique maximal family and all other players were dummies.

Theorem 7 [Kalai and Samet (1987)]. *A cost allocation method ϕ on N is additive, positive, family consistent, and charges dummies nothing if and only if there is a vector of weights $w \in \mathbb{R}_{++}$ such that $\phi = \phi_w$.* ∎

For a review of the literature on weighted Shapley values see Kalai and Samet (1988).

11. Cost allocation in the firm

We shift our attention now to firms that want to allocate common costs among various product lines or divisions. The reason for allocating such costs is to make division managers sensitive to the burden that they are placing on shared facilities. Ideally, therefore, the allocation should provide an incentive for managers to operate more efficiently, thereby reducing the incremental burden they place on these facilities.

This problem can be modelled using cooperative game theory as follows. [This approach was pioneered by Shubik (1962)]. Let N denote a set of n divisions or product lines in a firm, each of which is represented by a division manager. For each $S \subseteq N$, let $c(S)$ represent the overhead cost that the divisions in S would incur if they were in business by themselves, that is, if the firm were stripped of the other divisions. Suppose, for example, that the firm consists of two divisions that share warehouse space for their products. We assume that the volume of business is fixed, and the issue is how to divide the cost of the warehouse among the divisions given the volume of business that they do. (The case where volume varies in response to price will be taken up in the next section.) The cost function might be as follows:

$$c(\emptyset) = 0, \quad c(1) = 40, \quad c(2) = 60, \quad c(1,2) = 75.$$

A customary solution under these circumstances is to divide total cost in proportion to some rough measure of each division's usage of the facility, such as stand-alone costs. In this case the allocation would be $x_1 = 30$, $x_2 = 45$.

Suppose now that division manager 1 institutes a policy that reduces his demands on warehouse space, so that the cost of all coalitions containing 1 decreases. We would certainly want this change to be reflected in a reduced charge for division 1, for otherwise there would be no incentive for the division manager to undertake the cost-cutting moves. This condition translates into the requirement that the cost allocation rule be *coalitionally monotonic* in the sense of (20). It is easy to see that allocating in proportion to stand-alone cost has this property (assuming that stand-alone costs are positive), as do a number of other natural methods, such as dividing cost savings equally.

We know from Theorem 4, however, that coalitional monotonicity is inconsistent with staying in the core when $|N| \geq 5$. Thus in more complicated examples we are faced with a choice: we can adopt an allocation rule that gives managers appropriate incentives to cut costs, or that gives them an incentive to cooperate, but not both. One could argue that in the present context the incentives created by monotonicity are more compelling than the incentives embodied in the core. The reason is that a firm is not a voluntary association of divisions that must be given an incentive to remain together (which is the argument for a core allocation), rather, the problem for the firm is to send the right signals to divisions to get them to act efficiently.

When we probe a bit deeper, however, we see that coalitional monotonicity is not sufficient to create the cost-cutting incentives that the firm would like. It encourages cost-cutting moves by some divisions *provided* that there are no offsetting cost increases by other divisions. Yet this is exactly what might happen in practice. Suppose, for example, that division 1 reduces the demands that it places on warehouse space by 1 unit, and simultaneously division 2 introduces wasteful policies that increase demands on space by 11 units. The new cost function is as follows:

$$c(\emptyset) = 0, \quad c(1) = 39, \quad c(2) = 71, \quad c(1,2) = 85.$$

The principle of coalitional monotonicity does not apply here because the efficiency losses due to division 2 offset the efficiency gains by division 1. Nevertheless it seems clear that division 1 should be charged less than before and that division 2 should be charged more than before. The reason is that the *marginal* cost contribution of division 1 has decreased, whereas the marginal cost contribution of 2 has increased. Indeed, the partial derivative $c^1(S)$ has decreased by one unit for every coalition S, while $c^2(S)$ has increased by 11 units for every coalition. Allocation in proportion to stand-alone cost, however, yields the new charges $x_1 = 30.1$, $x_2 = 54.9$. In other words, division 1 is penalized for backsliding by division 2. This seems unreasonable.

Ch. 34: Cost Allocation   1219

To generalize these observations, fix a firm with a set of divisions N. We shall say that a cost allocation rule ϕ is *strongly monotonic* (does not create perverse incentives) if for every two cost functions c and C on N, and for every $i \in N$,

$$c^i(S) \leq C^i(S) \quad \text{for all } S \subseteq N \quad \text{implies } \phi_i(c) \leq \phi_i(C).$$

The reader may check that strong monotonicity implies coalitional monotonicity, but not vice versa.

Theorem 8 [Young (1985a)]. *For every set N the Shapley value is the unique cost allocation method that is symmetric and strongly monotonic.*

12. Cost allocation with variable output

We turn now to the problem of allocating joint costs when output can vary. Let there be n goods that are jointly produced by a firm, and let $C(q)$ be the joint cost of producing the bundle $q = (q_1, \ldots, q_n)$, where $q_i \geq 0$ is the quantity of good i. We shall assume that $C(0) = 0$ and that C has continuous first partial derivatives on the domain R^n_+ (one-sided on the boundary). (The latter condition can be relaxed slightly as will be indicated below.)

Given a target level of production $q^* > 0$, the goal is to find unit prices $p = (p_1, \ldots, p_n)$ such that costs are exactly covered:

$$\sum_{i=1}^{n} p_i q_i^* = C(q^*). \tag{26}$$

This condition is known as the *break-even* or *zero-profit* constraint. Normally, C is defined to include the cost of capital, so (26) means that there is no profit after stockholders are allowed a normal return on their investment.

13. Ramsey prices

The traditional approach to allocating costs in this setting is to consider how demands adjust in response to the cost of the goods. In other words, we interpret the allocated cost of product i as a published price, and consumers demand any amount they wish at that price (there is no rationing). An example would be a publicly regulated utility that sets prices to cover total cost, and supplies whatever the market demands at these prices.

To simplify matters, we shall assume that demands for the products are independent. Let $q_i = Q_i(p_i)$ be the amount demanded of product i when prices are set at p_i. The inverse demand function Q_i is assumed to be strictly monotone decreasing in p_i and continuously differentiable. In a competitive setting the firm would produce quantities that clear the market, and price would equal marginal cost. As we have already noted, however, this will not work in a regulated setting because marginal cost pricing will not normally cause the firm to break even. Indeed, many public utilities have the property that marginal costs *decrease* as quantity increases, in which case marginal cost pricing does not even *cover* total cost.

The solution proposed by Ramsey (1927) is to determine a price–quantity pair (p,q) that *maximizes consumer surplus subject to the break-even constraint*, that is, which maximizes

$$S(q) = \sum_{i=1}^{n} \int_0^{q_i} Q_i^{-1}(t)\,dt - C(q), \quad \text{subject to} \quad \sum_{i=1}^{n} p_i q_i - C(q) = 0. \tag{27}$$

This is a standard exercise in constrained optimization. Form the Lagrangian

$$L(q) = S(q) + \lambda \left(\sum_{i=1}^{n} q_i Q_i^{-1}(q_i) - C(q) \right).$$

Let $p_i = Q_i^{-1}(q_i)$ and $c_i(q) = \partial C(q)/\partial q_i$. A necessary condition that the pair (p,q) be optimal is that there exist a real number λ such that for all i

$$p_i - c_i(q) + \lambda \left(p_i + q_i \frac{\partial p_i}{\partial q_i} - c_i(q) \right) = 0. \tag{28}$$

Let $\mu_i = -(p_i/q_i)(\partial q_i/\partial p_i)$ be the demand elasticity for i at q. Then (28) is equivalent to

$$(p_i - c_i)/p_i = \lambda/(1+\lambda)\mu_i. \tag{29}$$

This is the *Ramsey formula* and prices satisfying it are known as *Ramsey prices*.[9] Their essential property is that *the percentage difference between price and marginal cost for each good is inversely proportional to the elasticity of demand for that good*. As W. Arthur Lewis (1949, p. 21) put it:

> The principle is... that those who cannot escape must make the largest contribution to indivisible cost, and those to whom the commodity does not matter much may escape. The man who has to cross Dupuit's bridge to see his dying father is mulcted thoroughly; the man who wishes only to see the scenery on the other side gets off lightly.

14. Aumann–Shapley prices

A drawback of Ramsey pricing is that it is highly sensitive to demand elasticities, which in practice may not be known with much accuracy. In this section we

---

[9]Ramsey first proposed this approach as a way of setting optimal tax rates on consumer goods (Ramsey, 1927). Later it was applied to public utility pricing by Manne (1952), Baumol and Bradford (1970), Boiteux (1971) and others.

examine an alternative approach that does not rely heavily on an analysis of demands. We posit instead that the quantities to be produced q^* are given exogenously (perhaps as the result of a back-of-the-envelope demand analysis), and the object is to allocate the cost $C(q^*)$ fairly among the n products. In this set-up a *cost allocation method* is a function $f(C, q^*) = p$ where p is a nonnegative vector of prices satisfying $\sum p_i q_i^* = C(q^*)$.

Many of the concepts introduced for discrete cost functions carry over to this case. Suppose, for example, that C is the sum of two cost functions, say capital cost C' and operating cost C''. The cost allocation method f is *additive* if

$$C(q) = C'(q) + C''(q) \quad \text{for all } q \leqslant q^* \text{ implies } f(C, q^*) = f(C', q^*) + f(C'', q^*).$$

Let us now consider the analog of symmetry. We want to say that when two products look alike from the standpoint of costs, their prices should be equal. To see why this issue is not quite as straightforward as it first appears, consider a refinery that makes gasoline for the U.S. market (q_1) and gasoline for the British market (q_2). The quantity q_1 is expressed in U.S. gallons and the quantity q_2 in Imperial gallons; otherwise they are the same. Thus the cost function takes the form $C(q_1, q_2) = C'(q)$ where $q = 0.833 q_1 + q_2$ is the total quantity in Imperial gallons. Note that the cost function C is not symmetric in q_1 and q_2; nevertheless 1 and 2 are essentially the same products. In this situation it is natural to require that the prices satisfy $p_1 = 0.833 p_2$.

More generally, we say that f is *weakly aggregation invariant* [Billera and Heath (1982)] if, for every C, C' and q^*,

$$C(q_1, q_2, \ldots, q_n) = C'(\sum \lambda_i q_i) \quad \text{for all } q \leqslant q^*$$
$$\text{implies } f_i(C, q^*) = \lambda_i f(C', \sum \lambda_i q_i^*).$$

The example of U.S. and Imperial gallons may seem a bit contrived. Consider, however, the following situation. A refinery blends m grades of petroleum distillate to make n grades of gasoline for sale at the pump. Assume that the cost of blending is negligible. Let $C(y_1, y_2, \ldots, y_m)$ be the joint cost of producing the m refinery grades in amounts $y_1, y_2, \ldots, y_m$. Suppose that one unit of blend j uses a_{ij} units of grade i, $1 \leqslant i \leqslant m$, $1 \leqslant j \leqslant n$. Let $x = (x_1, x_2, \ldots, x_n)$ be the amounts produced of the various blends. The joint cost of producing x is $C(Ax)$.

The cost allocation method f is *aggregation invariant* [Young (1985b)] if, for every $m \times n$ nonnegative matrix A, and every target level of production $x^* > 0$ such that $Ax^* > 0$,

$$f(C(Ax^*), x^*) = f(C, Ax^*) A.[10]$$

---

[10] Several weaker forms of this axiom have been proposed in the literature. Suppose, for example, that A is a square diagonal matrix, that is, $a_{ii} > 0$ for all i and $a_{ij} = 0$ for all $i \neq j$. Thus each product is simply rescaled by a positive factor. Then the axiom says that the prices should be scaled accordingly. This is the *rescaling axiom* [Mirman and Tauman (1982a)]. Another variation is the following. Suppose that the n products can be divided into m disjoint, nonempty subgroups $S_1, S_2, \ldots, S_m$ such that the products within any subgroup are equivalent for cost purposes. By this we mean that the cost of

A third natural condition on the allocation rule is that, if costs are increasing (or at least nondecreasing) in every product, then all prices should be nonnegative. Formally, we say that the cost allocation method f is *nonnegative* if, for every cost function C and target $q^* > 0$,

$$C(q) \leq C(q') \quad \text{for all} \quad 0 \leq q < q' \leq q^* \text{ implies } f(C, q^*) \geq 0. \tag{30}$$

Theorem 9 [Billera and Heath (1982), Mirman and Tauman (1982a)].[11] *There exists a unique cost allocation method f that is additive, weakly aggregation invariant, and nonnegative, namely*

$$\forall i, \quad p_i = f_i(C, q^*) = \int_0^1 (\partial C(tq^*)/\partial q_i) \, dt. \tag{31}$$

In other words, the price of each product is its marginal cost *averaged* over all vectors $tq^*: 0 \leq t \leq 1$ that define the ray from 0 to q^*. These are known as *Aumann–Shapley* (AS) *prices* and are based on the Aumann–Shapley value for nonatomic games [Aumann and Shapley (1974)].

We remark that marginal cost pricing has all of these properties except that it fails to satisfy the break-even requirement. (In the special case where the cost function exhibits constant returns to scale, of course, the two methods are identical.) Samet and Tauman (1982) characterize marginal cost pricing axiomatically by dropping the break-even requirement and strengthening nonnegativity to require that $f(C, q^*) \geq 0$ whenever C is nondecreasing in a neighborhood of q^*.

Billera et al. (1978) describe how the AS method was used to price telephone calls at Cornell University. The university, like other large organizations, can buy telephone service in bulk from the telephone company at reduced rates. Two types of contracts are offered for each class of service. The first type of contract requires that the university buy a large amount of calling time at a fixed price, and any amount of time exceeding this quota is charged at a small incremental cost. The second type of contract calls for the university to buy a relatively small amount of time at a lower fixed price, and the incremental cost for calls over the quota is higher. There are seven classes of service according to the destination of the call: five classes of WATS lines, overseas (FX) lines, and ordinary direct distance dialing (DDD). The university buys contracts for each class of service based on its estimate of expected demand. Calls are then broken down into types according to three

---

producing any quantities $x_1, \ldots, x_n$ can be written in the form $C(\sum_{i \in S_1} x_i, \sum_{i \in S_2} x_i, \ldots, \sum_{i \in S_m} x_i)$, for some cost function C defined on m variables. In this case the A-matrix consists of zeros and ones, and the sum of the rows is $(1, 1, \ldots, 1)$. This version of aggregation invariance was proposed by Mirman and Neyman (1983), who dubbed it "consistency". (Note that this use of the term consistency is not the same as in Section 5). This condition is very natural, and says that splitting a product into several equivalent products does not effectively change their prices. When the A-matrix is a *single* row vector of ones, the condition is called *weak consistency* [Mirman and Tauman (1982a)].

[11] Mirman and Tauman (1982a) prove this result under the assumptions of rescaling and weak consistency instead of weak aggregation invariance.

criteria: the time of day the call is placed, the day on which it is placed (business or nonbusiness), and the type of line along which it is routed (five types of WATS, FX, or DDD). Time of day is determined by the hour in which the call begins: midnight to 1 A.M., 1–2 A.M., and so forth. Thus there are $n = 24 \times 7 \times 2 = 338$ types of "products." Quantities are measured in minutes. For each combination of products demanded $q = q_1, q_2, \ldots, q_n$, the least cost $C(q)$ of meeting these demands can be computed using an optimization routine. The cost is then allocated using Aumann–Shapley prices, that is, for each demand vector q^* the unit price is computed according to formula (31), which determines the rate for each type of call.[12]

15. Adjustment of supply and demand

When a regulated firm uses Aumann–Shapley prices, it is natural to ask whether there exists a level of production q^* such that supply equals demand, i.e., such that markets clear. The answer is affirmative under fairly innocuous assumptions on the demand and cost functions. Assume that: (i) there are m consumers $j = 1, 2, \ldots, m$ and that each has a utility $u_j(q)$ for bundles $q = (q_1, q_2, \ldots, q_n)$ that is continuous, quasi-concave, and monotonically increasing in q; (ii) each consumer j has an initial money budget equal to $b_j > 0$. The cost function is assumed to satisfy the following conditions: (iii) there are no fixed costs [$C(0) = 0$]; (iv) $C(q)$ is continuous and nondecreasing; (v) $\partial C/\partial q_i$ is continuously differentiable except for at most a finite number of points on the ray $\{tq: 0 \leq t \leq q^*\}$; moreover, $\partial C/\partial q_i$ is continuous for all q such that $q_i = 0$ except perhaps when $q = 0$.

Theorem 10 [Mirman and Tauman (1982a, b)]. *Under assumptions (i)–(v) there exists a level of output q^* such that Aumann–Shapley prices clear the market, that is, there exists a distribution of q^* among the m consumers $q^1, q^2, \ldots, q^m$ such that, at the AS prices p, q^j maximizes $u_j(q)$ among all bundles $q \geq 0$ such that $p \cdot q \leq b_j$.*

For related results see Mirman and Tauman (1981, 1982a, b) Boes and Tillmann (1983), Dierker et al. (1985), and Boehm (1985). An excellent survey of this literature is given by Tauman (1988).

16. Monotonicity of Aumann–Shapley prices

Aumann–Shapley prices have an important property that is analogous to strong monotonicity in discrete cost-sharing games. Consider a decentralized firm in which

---

[12] The cost function computed in this manner is not differentiable, i.e., there may be abrupt changes in slope for neighboring configurations of demands. It may be proved, however, that the characterization of Aumann–Shapley prices in Theorem 9 also holds on the larger domain of cost functions for which the partial derivatives exist almost everywhere on the diagonal and are integrable.

each of the n product lines is supervised by a division manager. Corporate headquarters wants a cost accounting scheme that encourages managers to innovate and reduce costs. Suppose that the cost function in period 1 is C and the cost function in period 2 is C'. We can say that i's contribution to costs decreases if $\partial C'(q)/\partial q_i \leq \partial C(q)/\partial q_i$ for all $q \leq q^*$. A cost allocation method f is *strongly monotonic* if whenever i's contribution to cost decreases, then i's unit price does not increase, that is, if $f_i(C', q^*) \leq f_i(C, q^*)$.

Theorem 11 [Young (1985b)]. *For every set N there is a unique cost allocation method that is aggregation invariant and strongly monotonic, namely, Aumann–Shapley pricing.*[13]

17. Equity and competitive entry

Cost allocation not only provides internal signals that guide the firm's operations, it may also be a response to external competitive pressures. As before, we consider a firm that produces n products and whose cost of production is given by $C(q)$, where $C(0) = 0$ and $C(q)$ is continuous and nondecreasing in q. The firm is said to be a *natural monopoly* if the cost function is subadditive:

for all $q, q' \geq 0$, $\quad C(q) + C(q') \geq C(q + q')$.

(This is the analog of condition (1) for discrete cost functions.) Typical examples are firms that rely on a fixed distribution network (subways, natural gas pipelines, electric power grids, telephone lines), and for this reason their prices are often regulated by the state. If such a firm is subjected to potential competition from firms that can enter its market, however, then it may be motivated to regulate its own prices in order to deter entry. Under certain conditions this leads to a price structure that can be justified on grounds of equity as well. This is the subject of *contestable market theory*.[14]

The *core* of the cost function C for a given level of production q^* is the set of all price vectors $p = p_1, p_2, \ldots, p_n$ such that

$\sum p_i q_i^* = C(q^*)$ and $p \cdot q \leq C(q)$ whenever $0 \leq q \leq q^*$.

Let $q = Q(p)$ be the inverse demand function for the firm's products. A price vector p is *anonymously equitable* if p is in the core of C given $q^* = Q(p)$.

Consider a firm that is currently charging prices p^* and is subject to competitive entry. For the prices to be sustainable, revenue must cover cost given the demand at these prices, that is,

$p^* \cdot q^* \geq C(q^*)$ when $q^* = Q(p^*)$.

---

[13] Monderer and Neyman (1988) show that the result holds if we replace aggregation invariance by the weaker conditions of rescaling and consistency (see footnote 10).
[14] See Baumol et al. (1977), Panzar and Willig (1977), Sharkey and Telser (1978), Baumol et al. (1982).

If p^* is not in the core, then there exists some bundle $q \leq q^*$ such that $p^* \cdot q > C(q)$. This means, however, that another firm can profitably enter the market. The new firm can undercut the old firm's prices and capture the portion q of the old firm's market; moreover it can choose these prices so that $p \cdot q \geq C(q)$. (We assume the new firm has the same production technology, and hence the same cost function, as the old firm.) Thus to deter entry the old firm must choose prices p^* that are in the core.

In fact, entry deterrence requires more than being in the core. To see why, consider a subset of products S, and let $Q_S(p_S, p^*_{N-S})$ be the inverse demand function for S when the entering firm charges prices p_S and the original firm charges prices p^*_{N-S} for the other products. The entering firm can undercut p^* on some subset S and make a profit unless it is the case that

for all $S \subseteq N$, $p_S \leq p^*_S$ and $q_S \leq Q_S(p_S, p^*_{N-S})$ implies $p_S q_S \leq C(q_S, 0_{N-S})$. (32)

A vector of prices p^* that satisfies (32) is said to be *sustainable*. Sustainable prices have the property that no entrant can anticipate positive profits by entering the market and undercutting these prices. This means, in particular, that sustainable prices yield zero profits, that is, costs are covered exactly.

We now examine conditions under which AS prices are sustainable. The cost function C exhibits *cost complementarity* if it is twice differentiable and all second-order partial derivatives are nonincreasing functions of q:

$$\frac{\partial^2 C(q)}{\partial q_i \partial q_j} \leq 0 \quad \text{for all } i, j.$$

The inverse demand function $Q(p)$ satisfies *weak gross substitutability* if, for every i, Q_i is differentiable, and $\partial Q_i / \partial p_j \geq 0$ for every distinct i and j. Q is *inelastic below* p^* if

$$\frac{\partial Q_i(p)/Q_i(p)}{\partial p_i/p_i} \geq -1 \quad \text{for every } i \text{ and all } p \leq p^*.$$

Theorem 12 [Mirman, Tauman, and Zang (1985a)]. *If C satisfies cost complementarity and $Q(p)$ is upper semicontinuous, then there exists an AS vector p^* that is in the core of C given $q^* = Q(p^*)$. Moreover, if Q satisfies weak gross substitutability and is inelastic below p^* then p^* is sustainable.*

18. Incentives

One of the reasons why firms allocate joint costs is to provide their divisions with incentives to operate more efficiently. This problem can be modelled in a variety of ways. On the one hand we may think of the cost allocation mechanism as an incentive to change the cost function itself (i.e., to innovate). This issue was discussed in Section 16, where we showed that it leads to AS prices. In this section we take a somewhat different view of the problem. Let us think of the firm as being composed of n divisions that use inputs provided by the center. Suppose, for simplicity, that each division i uses one input, and that the cost of jointly producing these inputs is $C(q_1, q_2, \ldots, q_n)$. Each division has a technology for converting q_i into a marketable product, but this technology is unknown to the center and cannot even be observed *ex post*. Let $r_i(q_i)$ be the *maximum revenue* that division i can generate using the input q_i and its most efficient technology. Assume that the revenue generated by each division is independent of the revenue generated by the other divisions. The *firm's* objective is to maximize net profits

$$\max_q F(q) = \sum r_i(q_i) - C(q). \tag{33}$$

Since the true value of r_i is known only to division i, the firm cannot solve the profit maximization problem posed by (33). Instead, it would like to design a cost allocation scheme that will give each division the incentive to "do the right thing". Specifically we imagine the following sequence of events. First each division sends a message $m_i(q_i)$ to the center about what its revenue function is. The message may or may not be true. Based on the vector of messages $m = (m_1, m_2, \ldots, m_n)$, the center determines the quantities of inputs $q_i(m)$ to provide and allocates the costs according to some scheme $t = g(m, C)$, where t_i is the total amount that the division is assessed for using the input q_i (i.e., t_i/q_i is its unit "transfer" price). The combination of choices $(q(m), g(m, C))$ is called a *cost allocation mechanism*. Note that the function g depends on the messages as well as on the cost function C, so it is more general than the cost allocation methods discussed in earlier sections. Note also that the cost assessment does not depend on the divisional revenues, which are assumed to be unobservable by the center.

We impose the following requirements on the cost allocation mechanism. First, the assessments $t = g(m, C)$ should exactly equal the center's costs:

$$\sum g_i(m, C) = C(q(m)). \tag{34}$$

Second, the center chooses the quantities that would maximize profit assuming the reported revenue functions are accurate:

$$q(m) = \text{argmax} \sum m_i(q_i) - C(q). \tag{35}$$

Each division has an incentive to reveal its true revenue function provided that reporting some other message would never yield a higher profit, that is, if reporting the true revenue function is a dominant strategy. In this case the cost allocation mechanism $(q(m), g(m, C))$ is *incentive-compatible*:

for every i and every m, $r_i(q_i(r_i, m_{-i})) - g_i((r_i, m_{-i}), C) \geq r_i(q_i(m)) - g_i(m, C)$.
$$\tag{36}$$

Theorem 13 [Green and Laffont (1977), Hurwicz (1981), Walker (1978)]. *There exists no cost allocation mechanism $(q(m), g(m, C))$ that, for all cost functions C and all revenue functions r_i, allocates costs exactly (34), is efficient (35), and incentive-compatible (36).*

One can obtain more positive results by weakening the conditions of the theorem. For example, we can devise mechanisms that are efficient and incentive-compatible, though they may not allocate costs exactly. A particularly simple example is the following. For each vector of messages m let

$$q(m_{-i}) = \underset{q}{\operatorname{argmax}} \sum_{j \neq i} m_j(q_j) - C(q). \tag{37}$$

Thus $q(m_{-i})$ is the production plan the center would adopt if i's message (and revenue) is ignored. Let

$$P_i(m) = \sum_{j \neq i} m_j(q_j(m)) - C(q(m)),$$

where $q(m)$ is defined as in (35), and let

$$P_i(m_{-i}) = \sum_{j \neq i} m_j(q_j(m_{-i})) - C(q(m_{-i})).$$

$P_i(m)$ is the profit from adopting the optimal production plan based on *all* messages but not taking into account i's reported revenue, while $P_i(m_{-i})$ is the profit if we ignore both i's message and its revenue. Define the following cost allocation mechanism: $q(m)$ maximizes $\sum m_i(q_i) - C(q)$ and

$$g_i(m, C) = P_i(m_{-i}) - P_i(m). \tag{38}$$

This is known as the *Groves mechanism*.

Theorem 14 [Groves (1973, 1985)]. *The Groves mechanism is incentive-compatible and efficient.*

It may be shown, moreover, that any mechanism that is incentive-compatible and efficient is equivalent to a cost allocation mechanism such that $q(m)$ maximizes $\sum m_i(q_i) - C(q)$ and $g_i(m, C) = A_i(m_{-i}) - P_i(m)$, where A_i is any function of the messages that does not depend on i's message [Green and Laffont (1977)].

Under more specialized assumptions on the cost and revenue functions we can obtain more positive results. Consider the following situation. Each division is required to meet some exogenously given demand or target q_i^0 that is unknown to the center. The division can buy some or all of the required input from the center, say q_i, and make up the deficit $q_i^0 - q_i$ by some other (perhaps more expensive) means. (Alternatively we may think of the division as incurring a penalty for not meeting the target.)

Let $(q_i^0 - q_i)_+$ denote the larger of $q_i^0 - q_i$ and 0. Assume that the cost of covering the shortfall is linear:

$$c_i(q_i) = a_i(q_i^0 - q_i)_+ \quad \text{where } a_i > 0.$$

The division receives a fixed revenue r_i^0 from selling the target amount q_i^0. Assume that the center knows each division's unit cost a_i but not the values of the targets. Consider the following mechanism. Each division i reports a target $m_i \geq 0$ to the center. The center then chooses the efficient amount of inputs to supply assuming that the numbers m_i are true. In other words, the center chooses $q(m)$ to minimize total revealed cost:

$$q(m) = \operatorname{argmin} \left[a_i(m_i - q_i)_+ + C(q) \right]. \tag{39}$$

Let the center assign a nonnegative weight λ_i to each division, where $\sum \lambda_i = 1$, and define the cost allocation scheme by

$$g_i(m, C) = a_i q_i(m) - \lambda_i \left[\sum a_j q_j(m) - C(q(m)) \right]. \tag{40}$$

In other words, each division is charged the amount that it saves by receiving $q_i(m)$ from the center, minus the fraction λ_i of the joint savings. Notice that if $q_i(m) = 0$, then division i's charge is zero or negative. This case arises when i's unit cost is lower than the center's marginal cost of producing q_i.

The cost allocation scheme g is *individually rational* if $g_i(m, C) \leq a_i q_i^0$ for all i, that is, if no division is charged more than the cost of providing the good on its own.

Theorem 15 [Schmeidler and Tauman (1994)]. *The mechanism described by (39) and (40) is efficient, incentive-compatible, individually rational, and allocates costs exactly.*

Generalizations of this result to nonlinear divisional cost functions are discussed by Schmeidler and Tauman (1994). It is also possible to implement cost allocations via mechanisms that rely on other notions of equilibrium, e.g., Nash equilibrium, strong equilibrium or dominance-solvable equilibrium. For examples of this literature see Young (1980, 1985c, Ch. 1), Jackson and Moulin (1992), and Moulin and Schenker (1992).

19. Conclusion

In this chapter we have examined how cooperative game theory can be used to justify various methods for allocating common costs. As we have repeatedly emphasized, cost allocation is not merely an exercise in mathematics, but a practical problem that calls for translating institutional constraints and objectives into mathematical language.

Among the practical issues that need to be considered are the following. What are the relevant units on which costs are assessed – persons, aircraft landings, towns, divisions in a firm, quantities of product consumed? This is a nontrivial issue because it amounts to a decision about what is to be treated equally for purposes of the cost allocation. A second practical issue concerns the amount of available information. In allocating the cost of water service between two towns, for example, it is unrealistic to compute the cost of serving all possible subsets of individuals. Instead we would probably compute the cost of serving the two towns together and apart, and then allocate the cost savings by some method that is weighted by population. Another type of limited information concerns levels of demand. In theory, we might want to estimate the cost of different quantities of water supply, as well as the demands for service as a function of price. Ramsey pricing requires such an analysis. Yet in most cases this approach is infeasible. Moreover, such an approach ignores a key institutional constraint, namely the need to allocate costs so that both towns have an incentive to accept. (Ramsey prices need not be in the core of the cost-sharing game.)

This brings us to the third modelling problem, which is to identify the purpose of the cost allocation exercise. Broadly speaking there are three objectives: the allocation decision should be *efficient*, it should be *equitable*, and it should create appropriate *incentives* for various parts of the organization. These objectives are closely intertwined. Moreover, their interpretation depends on the institutional context. In allocating water supply costs among municipalities, for example, efficiency calls for meeting fixed demands at least cost. An efficient solution will not be voluntarily chosen, however, unless the cost allocation provides an incentive for all subgroups to participate. This implies that the allocation lie in the core. This condition is not sufficient for cooperation, however, because the parties still need to coordinate on a *particular solution* in the core. In this they are guided by principles of equity. In other words, equity promotes efficient solutions because it helps the participants realize the potential gains from cooperation.

Cost allocation in the firm raises a somewhat different set of issues. Efficiency is still a central concern, of course, but creating voluntary cooperation among the various units or divisions is not, because they are already bound together in a single organization. Incentives are still important for two reasons, however. First, the cost allocation mechanism sends price signals within the firm that affects the decisions of its divisions, and therefore the efficiency of the outcome. Second, it creates external signals to potential competitors who may be poised to enter the market. For prices to be sustainable (i.e., to deter entry) they need to lie in the core; indeed that must satisfy a somewhat stronger condition than being in the core. Thus incentive considerations prompted by external market forces are closely related to incentives that arise from the need for cooperation.

If the firm has full information on both costs and demands, and is constrained to break even, then the efficient (second-best) solution is given by Ramsey pricing. This solution may or may not be sustainable. If demand data is not known but the cost function is, there is a good case for using Aumann–Shapley pricing, since this is essentially the only method that rewards innovations that reduce marginal costs. Under certain conditions Aumann–Shapley prices are also sustainable. If key aspects of the cost structure are known only to the divisions, however, then it is impossible to design a general cost allocation mechanism that implements an efficient outcome in dominant strategies and fully allocates costs. More positive results are attainable for particular classes of cost functions, and for mechanisms that rely on weaker forms of equilibrium.

We conclude from this discussion is that there is no single, all-purpose solution to the cost allocation problem. Which method suits best depends on context, organizational goals, and the amount of information available. We also conclude, however, that cost allocation is a significant real-world problem that helps motivate the central concepts in cooperative game theory, and to which the theory brings important and unexpected insights.

Bibliography

Aumann, R.J. and M. Maschler, M. (1985) 'Game theoretic analysis of a bankruptcy problem from the Talmud', *Journal of Economic Theory*, **36**: 195–213.
Aumann, R.J. and L.S. Shapley (1974) *Values of Non-Atomic Games*. Princeton, NJ: Princeton University Press.
Balachandran, B.V. and R.T.S. Ramakrishnan (1981) 'Joint cost allocation: a unified approach', *Accounting Review*, **56**: 85–96.
Balinski, M.L. and F.M. Sand (1985) 'Auctioning landing rights at congested airports', in: H.P. Young, ed., *Cost Allocation: Methods, Principles, Applications*. Amsterdam: North-Holland.
Balinski, M.L. and H.P. Young (1982) *Fair Representation*. New Haven, CT: Yale University Press.
Banker, R.D. (1981) 'Equity considerations in traditional full cost allocation practices: an axiomatic perspective', in: S. Moriarity, ed., *Joint Cost Allocations*. Norman, OK: University of Oklahoma.
Baumol, W.J., E.E. Bailey and R.D. Willig (1977) 'Weak Invisible Hand Theorems on the Sustainability of Multiproduct Natural Monopoly'. *American Economic Review*, **67**: 350–365.
Baumol, W. and D. Bradford (1970) 'Optimal departures from marginal cost pricing', *American Economic Review*, **60**: 265–83.
Baumol, W.J., J.C. Panzar and R.D. Willig (1982) *Contestable Markets and the Theory of Industry Structure*. New York: Harcourt Brace Jovanovich.
Bentz, W.F. (1981) Comments on R.E. Verrechia's 'A question of equity: use of the Shapley value to allocate state and local income and franchise taxes', in: S. Moriarity, ed., *Joint Cost Allocations*. Norman, OK: University of Oklahoma.
Biddle, G.C. (1981) Comments on J.S. Demski's 'Cost allocation games', in: S. Moriarity, ed., *Joint Cost Allocations*. Norman, OK: University of Oklahoma.
Biddle, G.C. and R. Steinberg (1984) 'Allocation of joint and common costs', *Journal of Accounting Literature*, **3**: 1–45.
Biddle, G.C. and R. Steinberg (1985) 'Common cost allocation in the firm', in: H.P. Young, ed., *Cost Allocation: Methods, Principles, Applications*. Amsterdam: North Holland.
Billera, L.J. and D.C. Heath (1982) 'Allocation of shared costs: a set of axioms yielding a unique procedure', *Mathematics of Operations Research*, **7**: 32–39.
Billera, L.J., D.C. Heath and J. Ranaan (1978) 'Internal telephone billing rates – a novel application of non-atomic game theory', *Operations Research*, **26**: 956–965.
Billera, L.J. and D.C. Heath, and R.E. Verrechia (1981) 'A unique procedure for allocating common costs from a production process', *Journal of Accounting Research*, **19**: 185–96.
Bird, C.G. (1976) 'On cost allocation for a spanning tree: a game theoretic approach', *Networks*, **6**: 335–350.

Boatsman, J.R., D.R. Hansen and J.I. Kimbrell (1981) 'A rationale and some evidence supporting an altenative to the simple Shapley value', in: S. Moriarity, ed., *Joint Cost Allocations*. Norman, OK: University of Oklahoma.

Boehm, V. (1985) 'Existence and Optimality of Equilibria with Price Regulation', mimeo, Universität Mannheim.

Boes, D. and G. Tillmann (1983) 'Cost-axiomatic regulatory pricing', *Journal of Public Economics*, **22**: 243–256.

Bogardi, I. and F. Szidarovsky (1976) 'Application of game theory to water management', *Applied Mathematical modelling*, **1**: 11–20.

Boiteux, M. (1971) 'On the management of public monopolies subject to budgetary constraints', *Journal of Economic Theory*, **3**: 219–40.

Brief, R.P. and J. Owen (1968) 'A least squares allocation model', *Journal of Accounting Research*, **6**: 193–198.

Callen, J.L. (1978) 'Financial cost allocations: a game theoretic approach', *Accounting Review*, **53**: 303–308.

Champsaur, P. (1975) 'How to share the cost of a public good', *International Journal of Game Theory*, **4**: 113–129.

Charnes, A., D. Bres, D. Eckels, S. Hills, R. Lydens, J. Rousseau, K. Russell and M. Shoeman (1979) 'Costs and their assessment to users of a medical library', in: S. Brams, G. Schwodiauer and A. Schotter, eds., *Applied Game Theory*. Wurzburg: Physica-Verlag.

Clark, J.M. (1923) *Studies in the Economics of Overhead Costs*. Chicago: University of Chicago Press.

Claus, A. and D.J. Kleitman (1973) 'Cost allocation for a spanning tree', *Netwoks*, **3**: 289–304.

Coase, R.H. (1946) 'The marginal cost controversy', *Econometrica*, **13**: 169–182.

Curien, N. (1985) 'Cost allocation and pricing policy: the case of French telecommunications', in: H.P. Young, ed., *Cost Allocation: Methods, Principles, Applications*. Amsterdam: North Holland.

Demski, J.S. (1981) 'Cost allocation games', in: S. Moriarity, ed., *Joint Cost Allocations*. Norman, OK: University of Oklahoma.

Dierker, E., R. Guesnerie, and W. Neuefeind (1985) 'General equilibrium when some firms follow special pricing rules', *Econometrica*, **53**: 1369–1393.

Dopuch, N. (1981) 'Some perspectives on cost allocations', in: S. Moriarity, ed., *Joint Cost Allocations*. Norman, OK: University of Oklahoma.

Eckel, L.G. (1976) 'Arbitrary and incorrigible allocations', *Accounting Review*, **51**: 764–777.

Faulhaber, G. (1975) 'Cross-subsidization: pricing in public enterprises', *American Economic Review*, **65**: 966–977.

Faulhaber, G. and S.B. Levinson (1981) 'Subsidy-Free Prices and Anonymous Equity'. *American Economic Review*, **71**: 1083–1091.

Fishburn, P.C. and H.O. Pollak (1983) 'Fixed route cost allocation', *American Mathematical Monthly*, **90**: 366–378.

Gangolly, J.S. (1981) 'PM joint cost allocation: independent cost proportional scheme (JCPS) and its properties', *Journal of Accounting Research*, **19**: 299–312.

Gately, D. (1974) 'Sharing the gains from regional cooperation: a game theoretic application to planning investment in electric power', *International Economic Review*, **15**: 195–208.

Giglio, R.J. and R. Wrightington (1972) 'Methods for apportioning costs among participants in a regional system', *Water Resources Research*, **8**: 1133–1144.

Gillies, D.B. (1953) *Some Theorems on n-Person Games*. PhD Thesis, Department of Mathematics, Princeton University.

Granot, D. and F. Granot (1992) 'On some network flow games', *Mathematics of Operations Research*, **17**: 792–841.

Granot, D. and G. Huberman (1981) 'On minimum cost spanning tree games', *Mathematical Programming*, **21**: 1–18.

Granot D. and G. Huberman (1984) 'On the core and nucleolus of minimum cost spanning tree games', *Mathematical Programming*, **29**: 323–347.

Green, J. and J.-J. Laffont (1977) 'Characterizations of Satisfactory Mechanisms for the Revelation of Preferences for Public Goods'. *Econometrica*, **45**: 427–38.

Grotte, J.H. (1970) 'Computation of and Observations on the Nucleolus and the Central Games', MSc Thesis, Department of Operations Research, Cornell University.

Groves, T. (1973) 'Incentives in Teams'. *Econometrica*, **41**: 617–663.

Groves, T. (1985) 'The impossibility of incentive-compatible and efficient full cost allocation schemes', in: H.P. Young, ed., *Cost Allocation: Methods, Principles, Applications*. Amsterdam: North Holland.
Groves, T. and M. Loeb (1979) 'Incentives in a divisionalized firm', *Management Science*, **25**: 221–230.
Hamlen, S.S. and W.A. Hamlen (1981) 'The concept of fairness in the choice of joint cost allocation', in: S. Moriarity, ed., *Joint Cost Allocations*. Norman, OK: University of Oklahoma.
Hamlen, S.S., W.A. Hamlen and J.T. Tschirhart (1977) 'The use of core theory in evaluating joint cost allocation schemes', *Accounting Review*, **52**: 616–627.
Hamlen, S.S., W.A. Hamlen and J.T. Tschirhart (1980) 'The use of the generalized Shapley value in joint cost allocation', *Accounting Review*, **55**: 269–287.
Harsanyi, J.C. (1959) 'A bargaining model for the cooperative *n*-person game', in: A.W. Tucker and D. Luce, eds., *Contributions to the Theory of Games*, Vol. IV. Annals of Mathematics Studies No. 40. Princeton, NJ: Princeton University Press.
Hart, S. and A. Mas-Colell (1989) 'Potential, value and consistency', *Econometrica*, **57**: 589–614.
Heaney, J.P. (1979) 'Efficiency/equity analysis of environmental problems – a game theoretic perspective', in: S. Brams, G. Schwodiauer, and A. Schotter, eds., *Applied Game Theory*. Wurzburg: Physica-Verlag.
Hogarty, T. (1970) 'The profitability of corporate mergers', *Journal of Business*, **43**: 317–327.
Hughes, J.S. and J.H. Scheiner (1980) 'Efficiency properties of mutually satisfactory cost allocations', *Accounting Review*, **55**: 85–95.
Hurwicz, L. (1981) 'On the Incentive Problem in the Design of Non-Wasteful Resource Allocation Mechanisms', in: N. Assorodobraj-Kula et al. eds., *Studies in Economic Theory and Practice*. Amsterdam: North-Holland.
Jackson, M. and H. Moulin (1992) 'Implementing a Public Project and Distributing Its Cost', *Journal of Economic Theory*, **57**: 125–140.
James, L.D. and R.R. Lee (1971) *Economics of Water Resources Planning*. New York: McGraw-Hill.
Jensen, D.L. (1977) 'A class of mutually satisfactory allocations', *Accounting Review*, **52**: 842–856.
Kalai, E. and D. Samet (1987) 'On weighted Shapley values', *International Journal of Game Theory*, **16**: 205–222.
Kalai, E. and D. Samet (1988) 'Weighted Shapley values', in: A. Roth, ed., *The Shapley Value*. New York: Cambridge University Press.
Lensberg, T. (1985) 'Bargaining and fair allocation', in: H.P. Young, (ed.), *Cost Allocation: Methods, Principles, Applications*. Amsterdam: North Holland.
Lensberg, T. (1987) 'Stability and Collective Rationality', *Econometrica*, **55**: 935–962.
Lensberg, T. (1988) 'Stability and the Nash solution', *Journal of Economic Theory*, **45**: 330–341.
Lewis, W.A. (1949) *Overhead Cost*. London: Allen and Unwin.
Littlechild, S.C. (1970) 'Marginal cost pricing with joint costs', *Economic Journal*, **80**: 323–335.
Littlechild, S.C. (1975) 'Common costs, fixed charges, clubs and games', *Review of Economic Studies*, **42**: 117–124.
Littlechild, S.C. and G. Owen (1973) 'A simple expression for the Shapley value in a special case', *Management Science*, **20**: 370–2.
Littlechild, S.C. and G. Owen (1974) 'A further note on the nucleolus of the 'airport game'', *International Journal of Game Theory*, **5**: 91–95.
Littlechild, S.C. and Thompson, G.F. (1977) 'Aircraft landing fees: a game theory approach', *Bell Journal of Economics*, **8**: 186–204.
Littlechild, S.C. and K.C. Vaidya (1976) 'The propensity to disrupt and the disruption nucleolus of a characteristic function game', *International Journal of Game Theory*, **5**: 151–161.
Loehman, E.T., E. Orlando, J.T. Tschirhart and A.B. Whinston (1979) 'Cost allocation for a regional waste-water treatment system', *Water Resources Research*, **15**: 193–202.
Loehman, E.T., D. Pingry and A.B. Whinston (1973) 'Cost allocation for regional pollution treatment systems', in: J.R. Conner and E.T. Loehman, eds., *Economics and Decision Making for Environmental Quality*. Gainesville: University of Florida Press.
Loehman, E.T. and A.B. Whinston (1971) 'A new theory of pricing and decision-making for public investment', *Bell Journal of Economics and Management Science*, **2**: 606–625.
Loehman, E.T. and A.B. Whinston (1974) 'An axiomatic approach to cost allocation for public investment', *Public Finance Quarterly*, **2**: 236–251.
Loehman, E.T. and A.B. Whinston (1976) 'A generalized cost allocation scheme', in: S. Lin, ed., *Theory and Measurement of Economic Externalities*. New York: Academic Press.
Louderback, J.G. (1976) 'Another approach to allocating joint costs: a comment', *Accounting Review*, **51**: 683–685.

Loughlin, J.C. (1977) 'The efficiency and equity of cost allocation methods for multipurpose water projects', *Water Resources Research*, **13**: 8–14.
Lucas, W.F. (1981) 'Applications of cooperative games to equitable allocations', in: W.F. Lucas, ed., *Game Theory and Its Applications. Proceedings of Symposia in Applied Mathematics 24*. Providence, RI: American Mathematical Society.
Manes, R.P. and V.L. Smith (1965) 'Economic joint cost theory and accounting practice', *Accounting Review*, **40**: 31–35.
Manes, R.P. and R. Verrecchia (1982) 'A new proposal for setting intra-company transfer prices', *Accounting and Business Research*, **13**: 77–104.
Mayer, H. (1923) 'Zurechnung', in: *Handworterbuch der Staatswissenschaftten*, Vol. 8, 4th Ed. Jena, pp. 1206–1228.
Manne, A. (1952) 'Multi-purpose public enterprises – criteria for pricing', *Economica*, NS, **19**: 322–6.
Megiddo, N. (1974) 'On the nonmonotonicity of the bargaining set, the kernel, and the nucleolus of a game', *SIAM Journal on Applied Mathematics*, **27**: 355–8.
Megiddo, N. (1978a) 'Cost allocation for Steiner trees', *Networks*, **8**: 1–6.
Megiddo, N. (1978b) 'Computational complexity of the game theory approach to cost allocation for a tree', *Mathematics of Operations Research*, **3**: 189–195.
Mirman, L.J. and A. Neyman (1983) 'Diagonality of cost allocation prices', *Mathematics of Operations Research*, **9**: 66–74.
Mirman, L.J., D. Samet and Y. Tauman (1983) 'An axiomatic approach to the allocation of a fixed cost through prices', *Bell Journal of Economics*, **14**: 139–151.
Mirman, L.J. and Y. Tauman (1981) 'Valeur de Shapley et repartition equitable des couts de production', *Les cahiers du Seminaire d'Econometrie*, **23**: 121–151.
Mirman, L.J. and Y. Tauman (1982a) 'Demand compatible equitable cost sharing prices', *Mathematics of Operations Research*, **7**: 40–56.
Mirman, L.J. and Y. Tauman (1982b) 'The continuity of the Aumann–Shapley price mechanism', *Journal of Mathematical Economics*, **9**: 235–249.
Mirman, L.J., Y. Tauman and I. Zang (1985a) 'Supportability, sustainability and subsidy free prices', *The Rand Journal of Economics*, **16**: 114–126.
Mirman, L.J., Y. Tauman, and I. Zang (1985b) 'On the use of game-theoretic methods in cost accounting', in: H.P. Young, ed., *Cost Allocation: Methods, Principles, Applications*. Amsterdam: North-Holland.
Monderer, D. and A. Neyman (1988) 'Values of smooth nonatomic games: the method of multilinear extension', in: A. Roth, ed., *The Shapley Value*. New York: Cambridge University Press.
Moriarity, S. (1975) 'Another approach to allocating joint costs', *Accounting Review*, **50**: 791–795.
Moriarity, S., ed. (1981) *Joint Cost Allocations*. Norman, OK: University of Oklahoma Press.
Moriarity, S. (1981b) 'Some rationales for cost allocations', in: S. S. Moriarity ed., *Joint Cost Allocations*. Norman, OK: University of Oklahoma Press.
Mossin, J. (1968) 'Merger agreements: some game theoretic considerations', *Journal of Business*, **41**: 460–471.
Moulin, H. (1985) 'The separability axiom and equal sharing methods', *Journal of Economic Theory*, **36**: 120–148.
Moulin, H. and S. Schenker (1992) 'Serial cost sharing', *Econometrica*, **60**: 1009–1037.
Okada, N. (1985) 'Cost allocation in multipurpose reservoir development: the Japanese experience', in: H.P. Young, ed., *Cost Allocation: Methods, Principles, Applications*. Amsterdam: North Holland.
Panzar, J.C. and R.D. Willig (1977) 'Free Entry and the Sustainability of Natural Monopoly'. *Bell Journal of Economics*, **8**: 1–22.
Parker, T. (1927) 'Allocation of the Tennessee Valley Authority projects', *Transactions of the American Society of Civil Engineers*, **108**: 174–187.
Peleg, B. (1986) 'On the reduced game property and its converse', *International Journal of Game Theory*, **15**: 187–200.
Pigou, A.C. (1913) 'Railway rates and joint cost', *Quarterly Journal of Economics*, **27**: 536–538.
Ramsey, F. (1927) 'A contribution to the theory of taxation', *Economic Journal*, **37**: 47–61.
Ransmeier, J.S. (1942) 'The Tennessee Valley Authority: A Case Study in the Economics of Multiple Purpose Stream Planning', Nashville, TN: Vanderbilt University Press.
Rawls, J. (1971) *A Theory of Justice*. Cambridge, MA: Harvard University Press.
Roth, A.E. and R.E. Verrecchia (1979) 'The Shapley value as applied to cost allocation: a reinterpretation', *Journal of Accounting Research*, **17**: 295–303.

Samet, D. and Y. Tauman (1982) 'The determination of marginal cost prices under a set of axioms', *Econometrica*, **50**: 895–909.
Samet, D., Y. Tauman and I. Zang (1983) 'An application of the Aumann–Shapley prices to the transportation problem', *Mathematics of Operations Research*, **10**: 25–42.
Schmeidler, D. (1969) 'The nucleolus of a characteristic function game', *SIAM Journal on Applied Mathematics*, **17**: 1163–70.
Schmeidler, P. and Y. Tauman (1994) 'Incentive-compatible cost allocation mechanisms', *Journal of Economic Theory*, **63**: 189–207.
Sen, A. (1970) *Collective Choice and Social Welfare*. San Francisco: Holden–Day.
Shapley, L.S. (1953a) 'Additive and Non-Additive Set Functions', Ph.D. Thesis, Princeton University.
Shapley, L.S. (1953b) 'A value for n-person games', in: H.W. Kuhn and A.W. Tucker, eds., *Contributions to the Theory of Games*, **Vol. II**. Annals of Mathematics Studies No. 28. Princeton, NJ: Princeton University Press.
Shapley, L.S. (1971) 'Cores of convex games', *International Journal of Game Theory*, **1**: 11–26.
Shapley, L.S. (1981) 'Comments on R.D. Banker's 'Equity considerations in traditional full cost allocation practices: an axiomatic perspective', in: S. Moriarity, ed., *Joint Cost Allocations*. Norman, OK: University of Oklahoma Press.
Shapley, L.S. and M. Shubik (1969) 'On the core of an economic system with externalities', *American Economic Review*, **59**: 337–362.
Sharkey, W.W. (1982a) *The Theory of Natural Monopoly*. London: Cambridge University Press.
Sharkey, W.W. (1982b) 'Suggestions for a game-theoretic approach to public utility pricing and cost allocation', *Bell Journal of Economics*, **13**: 57–68.
Sharkey, W.W. (1985) 'Economic and game-theoretic issues associated with cost allocation in a telecommunications network', in: H.P. Young, ed., *Cost Allocation: Methods, Principles, Applications*. Amsterdam: North-Holland.
Sharkey, W.W. and L.G. Telser (1978) 'Supportable Cost Functions for the Multiproduct Firm'. *Journal of Economic Theory*, **18**: 23–37.
Shubik, M. (1962) 'Incentives, decentralized control, the assignment of joint costs and internal pricing', *Management Science*, **8**: 325–43.
Shubik, M. (1985) 'The cooperative form, the value, and the allocation of joint costs and benefits', in: H.P. Young, ed., *Cost Allocation: Methods, Principles, Applications*. Amsterdam: North Holland.
Sobolev, A.I. (1975) 'Characterization of the principle of optimality for cooperative games through functional equations', in: N.N. Voroby'ev, ed., *Mathematical Methods in the Social*, Vipusk 6, Vilnius, USSR, pp. 92–151.
Sorenson, J.R., J.T. Tschirhart and A.B. Whinston (1976) 'A game theoretic approach to peak load pricing', *Bell Journal of Economics and Management Science*, **7**: 497–520.
Sorenson, J.R., J.T. Tschirhart and A.B. Whinston (1978a) 'Private good clubs and the core', *Journal of Public Economics*, **10**: 77–95.
Sorenson, J.R., J.T. Tschirhart and A.B. Whinston (1978b) 'A theory of pricing under decreasing costs', *American Economic Review*, **68**: 614–624.
Straffin, P. and J.P. Heaney (1981) 'Game theory and the Tennessee Valley Authority', *International Journal of Game Theory*, **10**: 35–43.
Suzuki, M. and M. Nakamaya (1976) 'The cost assignment of the cooperative water resource development: a game theoretical approach', *Management Science*, **22**: 1081–1086.
Tauman, Y. (1988) 'The Aumann–Shapley Prices: A Survey', in: A. Roth, ed., *The Shapley Value.* New York: Cambridge University Press.
Tennessee Valley Authority (1938) 'Allocation of Investment in Norris, Wheeler, and Wilson Projects', U.S. House of Representatives Document No 709, Congress Third Session, Washington, DC: U.S. Government Printing Office.
Thomas, A.L. (1969) 'The Allocation Problem in Financial Accounting', *Studies in Accounting Research No. 3*, American Accounting Association.
Thomas, A.L. (1971) 'Useful arbitrary allocations', *Accounting Review*, **46**: 472–479.
Thomas, A.L. (1974) 'The Allocation Problem', Part Two, *Studies in Accounting Research No. 9*, American Accounting Association.
Thomas, A.L. (1978) 'Arbitrary and incorrigible allocations: a comment', *Accounting Review*, **53**: 263–269.
Thomas, A.L. (1980) *A Behavioral Analysis of Joint-Cost Allocation and Transfer Pricing*. Champaign, IL: Stipes Publishing Company.

Thomson, W. (1988) 'A study of choice correspondences in economies with a variable number of agents', *Journal of Economic Theory*, **46**: 237–254.
Thomson, W. (1990) 'The consistency principle', in: T. Ichiishi, A. Neyman, and Y. Tauman, eds., *Game Theory and Applications*. New York: Academic Press.
Verrecchia, R.E. (1981) 'A question of equity: use of the Shapley value to allocate state and local income and franchise taxes', in S. Moriarity, ed., *Joint Cost Allocations*. Norman, OK: University of Oklahoma Press.
Verrecchia, R.E. (1982) 'An analysis of two cost allocation cases', *Accounting Review*, **57**: 579–593.
von Neumann, J. and O. Morgenstern (1944) *Theory of Games and Economic Behavior*. Princeton, NJ: Princeton University Press.
Walker, M. (1978) 'A Note on the Characterization of Mechanisms for the Revelation of Preferences.' *Econometrica*, **46**: 147–152.
Young, H.P. (1979) 'Exploitable surplus in n-person games', in: S. Brams, G. Schwodiauer and A. Schotter eds., *Applied Game Theory*. Wurzburg: Physica-Verlag.
Young, H.P. (1980) 'Cost Allocation and Demand Revelation in Public Enterprises', Working Paper WP-80-130, International Institute for Applied Systems Analysis, Laxenburg, Austria.
Young, H.P. (1985a) 'Monotonic solutions of cooperative games', *International Journal of Game Theory*, **14**: 65–72.
Young, H.P. (1985b) 'Producer incentives in cost allocation', *Econometrica*, **53**: 757–765.
Young, H.P., ed. (1985c) *Cost Allocation: Methods, Principles, Applications*. Amsterdam: North-Holland.
Young, H.P. (1988) 'Distributive justice in taxation', *Journal of Economic Theory*, **44**: 321–335.
Young, H.P. (1994) *Equity in Theory and Practice*. Princeton, NJ: Princeton University Press.
Young, H.P., N. Okada and T. Hashimoto (1982) 'Cost allocation in water resources development', *Water Resources Research*, **18**: 463–75.
Zajac, E.E. (1985) 'Perceived economic justice: the case of public utility regulation', in: H.P. Young, ed., *Cost Allocation: Methods, Principles, Applications*. Amsterdam: North-Holland.
Zajac, E. (1993) *Fairness of Efficiency: An Introduction to Public Utility Pricing*, 2nd edn. Cambridge, MA: Ballinger.
Zimmerman, J.L. (1979) 'The costs and benefits of cost allocations', *Accounting Review*, **54**: 504–521.

Chapter 35

COOPERATIVE MODELS OF BARGAINING

WILLIAM THOMSON*

University of Rochester

Contents

1.	Introduction	1238
2.	Domains. Solutions	1240
3.	The main characterizations	1245
	3.1. The Nash solution	1245
	3.2. The Kalai–Smorodinsky solution	1249
	3.3. The Egalitarian solution	1251
4.	Other properties. The role of the feasible set	1254
	4.1. Midpoint domination	1254
	4.2. Invariance	1254
	4.3. Independence and monotonicity	1256
	4.4. Uncertain feasible set	1258
5.	Other properties. The role of the disagreement point	1260
	5.1. Disagreement point monotonicity	1261
	5.2. Uncertain disagreement point	1262
	5.3. Risk-sensitivity	1264
6.	Variable number of agents	1265
	6.1. Consistency and the Nash solution	1266
	6.2. Population monotonicity and the Kalai–Smorodinsky solution	1268
	6.3. Population monotonicity and the Egalitarian solution	1270
	6.4. Other implications of consistency and population monotonicity	1271
	6.5. Opportunities and guarantees	1271
	6.6. Replication and juxtaposition	1273
7.	Applications to economics	1274
8.	Strategic considerations	1275
Bibliography		1277

*I am grateful to the National Science Foundation for its support under grant No. SES 8809822, to H. Moulin, a referee, and the editors for their advice, and particularly to Youngsub Chun for his numerous and detailed comments. All remaining errors are mine.

Handbook of Game Theory, Volume 2, Edited by R.J. Aumann and S. Hart
© *Elsevier Science B.V., 1994. All rights reserved*

1. Introduction

The axiomatic theory of bargaining originated in a fundamental paper by J.F. Nash (1950). There, Nash introduced an idealized representation of the *bargaining problem* and developed a methodology that gave the hope that the undeterminateness of the terms of bargaining that had been noted by Edgeworth (1881) could be resolved.

The canonical bargaining problem is that faced by management and labor in the division of a firm's profit. Another example concerns the specification of the terms of trade among trading partners.

The formal and abstract model is as follows: Two agents have access to any of the alternatives in some set, called the feasible set. Their preferences over these alternatives differ. If they agree on a particular alternative, that is what they get. Otherwise, they end up at a prespecified alternative in the feasible set, called the disagreement point. Both the feasible set and the disagreement point are given in utility space. Let they be given by S and d respectively. Nash's objective was to develop a theory that would help predict the compromise the agents would reach. He specified a natural class of bargaining problems to which he confined his analysis, and he defined a *solution*, to be a rule that associates with each problem (S, d) in the class a point of S, to be interpreted as this compromise. He formulated a list of properties, or *axioms*, that he thought solution should satisfy, and established the existence of a unique solution satisfying all the axioms. It is after this first *axiomatic characterization of a solution* that much of the subsequent work has been modelled.

Alternatively, solutions are meant to produce the recommendation that an impartial arbitrator would make. There, the axioms may embody normative objectives of fairness.

Although criticisms were raised early on against some of the properties Nash used, the solution he identified, now called the Nash solution, was often regarded as *the* solution to the bargaining problem until the mid-seventies. Then, other solutions were introduced and given appealing characterizations, and the theory expanded in several directions. Systematic investigations of the way in which solutions could, or should, depend on the various features of the problems to be solved, were undertaken. For instance, the crucial axiom on which Nash had based his characterization requires that the solution outcome be unaffected by certain contractions of the feasible set, corresponding to the elimination of some of the options initially available. But, is this independence fully justified? Often not. A more detailed analysis of the kinds of transformations to which a problem can be subjected led to the formulation of other conditions. In some cases, it seems quite natural that the compromise be allowed to move in a certain direction, and perhaps be required to move, in response to particular changes in the configuration of the options available.

The other parameters entering in the description of the problem may change too. An improvement in the fallback position of an agent, reflected in an increase in his coordinate of the disagreement point, should probably help him. Is it actually the case for the solutions usually discussed? This improvement, if it does occur, will be at a cost to the other agents. How should this cost be distributed among them?

The feasible set may be subject to uncertainty. Then, how should the agents be affected by it? And, what should the consequences of uncertainty in the disagreement point be? How should solutions respond to changes in the risk attitude of agents? Is it preferable to face an agent who is more, or less, risk-averse?

The set of agents involved in the bargaining may itself change. Imagine some agents to withdraw their claims. If this affects the set of options available to the remaining agents favorably, it is natural to require that each of them be affected positively. Conversely, when the arrival of additional agents implies a reduction in the options available to the agents initially present, should not they all be negatively affected? And, if some of the agents leave the scene with their payoffs, or promise of payoffs, should not the situation, when reevaluated from the viewpoint of the agents left behind, be thought equivalent to the initial situation? If yes, then each of them should still be attributed the very same payoff as before. If not, renegotiations will be necessary that will greatly undermine the significance of any agreement.

What is the connection between the abstract models with which the theory of bargaining deals and more concretely specified economic models on the one hand, and strategic models on the other? How helpful is the theory of bargaining to the understanding of these two classes of problems?

These are a sample of the issues that we will discuss in this review. It would of course be surprising if the various angles from which we will attack them all led to the same solution. However, and in spite of the large number of intuitively appealing solutions that have been defined in the literature, three solutions (and variants) will pass a significantly greater number of the tests that we will formulate than all of the others. They are Nash's original solution, which selects the point of S at which the product of utility gains from d is maximal, a solution due to Kalai and Smorodinsky (1975), which selects the point of S at which utility gains from d are proportional to their maximal possible values within the set of feasible points dominating d, and the solution that simply equates utility gains from d, the Egalitarian solution. In contexts where interpersonal comparisons of utility would be inappropriate or impossible, the first two would remain the only reasonable candidates. We find this conclusion to be quite remarkable.

An earlier survey is Roth (1979c). Partial surveys are Schmitz (1977), Kalai (1985), Thomson (1985a), and Peters (1987a). Thomson and Lensberg (1989) analyze the case of a variable number of agents. Peters (1992) and Thomson (1994) are detailed accounts.

2. Domains. Solutions

A union contract is up for renewal; management and labor have to agree on a division of the firm's income; failure to agree results in a strike. This is an example of the sort of conflicts that we will analyze. We will consider the following abstract formulation: An n-person bargaining problem, or simply a *problem*, is a pair (S, d) where S is a subset of the n-dimensional Euclidean space, and d is a point of S. Let Σ_d^n be the class of problems such that [Figure 1(a)]:

(i) S is convex, bounded, and closed (it contains its boundary).
(ii) There is at least one point of S strictly dominating d.

Each point of S gives the utility levels, measured in some von Neumann–Morgenstern scales, reached by the agents through the choice of one of the alternatives, or randomization among those alternatives, available to them. Convexity of S is due to the possibility of randomization; boundedness holds if utilities are bounded; closedness is assumed for mathematical convenience. The existence of at least one $x \in S$ with $x > d$ is postulated to avoid the somewhat degenerate case when only some of the agents stand to gain from the agreement.[1] In addition, we will usually assume that

(iii) (S, d) is *d-comprehensive*: If $x \in S$ and $x \geqslant y \geqslant d$, then $y \in S$.

The property of (S, d) follows from the natural assumption that utility is freely disposable (above d). It is sometimes useful to consider problems satisfying the slightly stronger condition that the part of their boundary that dominates d does not contain a segment parallel to an axis. Along that part of the boundary of such a *strictly d-comprehensive* problem, "utility transfers" from one agent to another

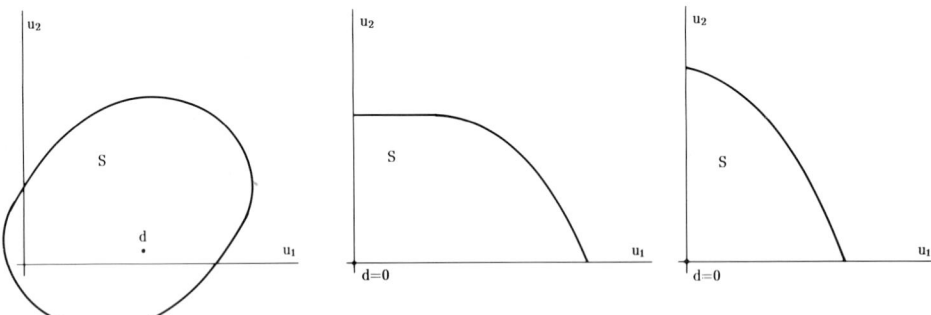

Figure 1. Various classes of bargaining problems: (a) An element of Σ_d^2. (b) An element of Σ_0^2. (c) A strictly comprehensive element of Σ_0^2.

---

[1] Vector inequalities: given $x, x' \in \mathbb{R}^n$, $x \geqslant x'$ means $x_i \geqslant x'_i$ for all i; $x \geq x'$ means $x \geqslant x'$ and $x \neq x'$; $x > x'$ means $x_i > x'_i$ for all i.

are always possible. Let $\partial S = \{x \in S \mid \nexists x' \in S \text{ with } x' > x\}$ be the undominated boundary of S.

In most of the existing theory the choice of the zero of the utility scales is assumed not to matter, and for convenience, we choose scales so that $d = 0$ and we ignore d in the notation altogether. However, in some sections, the disagreement point plays a central role; it is then explicitly reintroduced. When $d = 0$, we simply say that a problem is comprehensive instead of d-comprehensive. Finally, and in addition, we often require that

(iv) $S \subset \mathbb{R}^n_+$.

Indeed, an argument can be made that alternatives at which any agent receives less than what he is guaranteed at $d = 0$ should play no role in the determination of the compromise. (This requirement is formally stated later on.)

In summary, we usually deal with the class Σ^n_0 of problems S as represented in Figure 1(b) and (c) (the problem of Figure 1c is strictly comprehensive, whereas that of Figure 1b is only comprehensive since its boundary contains a non-degenerate horizontal segment.) We only occasionally consider *degenerate* problems, that is, problems whose feasible set contains no point strictly dominating the disagreement point.

Sometimes, we assume that utility can be disposed of in *any* amount: if $x \in S$ then any $y \in \mathbb{R}^n$ with $y \leqslant x$ is also in S. We denote by $\Sigma^n_{d,-}$ and $\Sigma^n_{0,-}$, the classes of such *fully comprehensive problems* corresponding to Σ^n_d and Σ^n_0.

The class of games analyzed here can usefully be distinguished from the class of "games in coalitional form" (in which a feasible set is specified for each group of agents, see Chapters 12-14, 17, 36, 37), and from various classes of economic and strategic models (in the former, some economic data are preserved, such as endowments, technology; see Chapter 7; in the latter, a set of actions available to each agent is specified, each agent being assumed to choose his action so as to bring about the outcome he prefers; most of the chapters in this handbook follow the strategic approach). In Section 7 and 8, we briefly show how the abstract model relates to economic and strategic models.

A *solution* defined on some domain of problems associates with each element (S, d) of the domain a unique point of S interpreted as a prediction, or a recommendation, for that problem.

Given $A \subset \mathbb{R}^n_+$, cch$\{A\}$ denotes the "convex and comprehensive hull" of A: it is the smallest convex and comprehensive subset of $\mathbb{R}^n_+$ containing A. If $x, y \in \mathbb{R}^n_+$, we write cch$\{x, y\}$ instead of cch$\{\{x, y\}\}$. $\Delta^{n-1} = \{x \in R^n_+ \mid \Sigma x_i = 1\}$ is the $(n-1)$-dimensional unit simplex.

Other classes of problems have been discussed in the literature in particular non-convex problems and problems that are unbounded below. In some studies, no disagreement point is given [Harsanyi (1955), Myerson (1977), Thomson (1981c)]. In others, an additional reference point is specified; if it is in S, it can be interpreted as a status quo [Brito et al. (1977) choose it on the boundary of S], or as a first step towards the final compromise [Gupta and Livne (1988)]; if it is outside of S, it represents a vector of claims [Chun and Thomson (1988), Bossert (1992a, b, 1993) Herrero (1993), Herrero and Marco (1993), Marco (1994a, b). See also Conley, McLean and Wilkie (1994) who apply the technique of bargaining theory to multi-objective programming.] Another extension of the model is proposed by Klemisch-Ahlert (1993). Some authors have considered multivalued solutions [Thomson (1981a), Peters et al. (1993)], and others, probabilistic solutions [Peters and Tijs (1984b)].

Three solutions play the central role in the theory as it appears today. We introduce them first, but we also present several others so as to show how rich

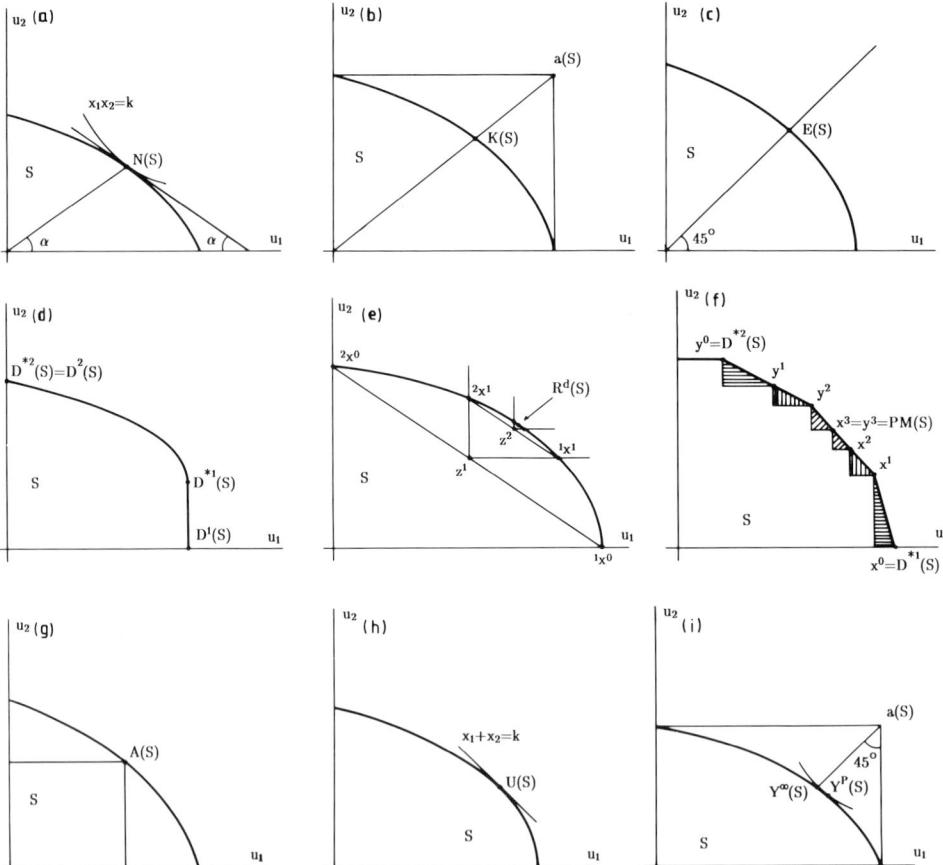

Figure 2. Examples of solutions. (a) The Nash solution. (b) The Kalai–Smorodinsky solution. (c) The Egalitarian solution. (d) The Dictatorial solutions. (e) The discrete Raiffa solution. (f) The Perles–Maschler solution. (g) The Equal Area solution. (h) The Utilitarian solution. (i) The Yu solutions.

and varied the class of available solutions is. Their definitions, as well as the axioms to follow shortly, are stated for an arbitrary $S \in \Sigma_0^n$, [or $(S, d) \in \Sigma_d^n$].

For the best-known solution, introduced by Nash (1950), the compromise is obtained by maximizing the product of utility gains from the disagreement point.

Nash solution, N [Figure 2(a)]: $N(S)$ is the maximizer of the product $\prod x_i$ over S. [$N(S, d)$ is the maximizer of $\prod(x_i - d_i)$ for $x \in S$ with $x \geq d$.]

The Kalai–Smorodinsky solution sets utility gains from the disagreement point proportional to the agents' most optimistic expectations. For each agent, this is defined as the highest utility he can attain in the feasible set subject to the constraint that no agent should receive less than his coordinate of the disagreement point.

Kalai–Smorodinsky solution, K [Figure 2(b)]: $K(S)$ is the maximal point of S on the segment connecting the origin to $a(S)$, *the ideal point of S*, defined by $a_i(S) \equiv \max\{x_i | x \in S\}$ for all i. [$K(S, d)$ is the maximal point of S on the segment connecting d to $a(S, d)$ where $a_i(S, d) \equiv \max\{x_i | x \in S, x \geq d\}$ for all i.]

The idea of equal gains is central to many theories of social choice and welfare economics. Here, it leads to the following solution:

Egalitarian solution, E [Figure 2(c)]: $E(S)$ is the maximal point of S of equal coordinates. [$E_i(S, d) - d_i = E_j(S, d) - d_j$ for all i, j.]

The next solutions are extreme examples of solutions favoring one agent at the expense of the others. They occur naturally in the construction of other solutions and sometimes are useful indicators of the strength of some proposed list of axioms (just as they do in Arrow-type social choice).

Dictatorial solutions, D^i and D^{*i} [Figure 2(d)]: $D^i(S)$ is the maximal point x of S with $x_j = 0$ for all $j \neq i$. [Similarly, $D_j^i(S, d) = d_j$ for all $j \neq i$.] If $n = 2$, $D^{*i}(S)$ is the point of $PO(S) \equiv \{x \in S | \nexists x' \in S \text{ with } x' \geq x\}$ of maximal ith coordinate.

If S is strictly comprehensive, $D^i(S) = D^{*i}(S)$. If $n > 2$, the maximizer of x_i in $PO(S)$ may not be unique, and some rule has to be formulated to break possible ties. A lexicographic operation is often suggested.

The next two solutions are representatives of an interesting family of solutions based on processes of balanced concessions: agents work their way from their preferred alternatives (the dictatorial solution outcomes) to a final position by moving from compromise to compromise.

The (discrete) Raiffa solution, R^d [Figure 2(e)]: $R^d(S)$ is the limit point of the sequence $\{z^t\}$ defined by: $x^{i0} = D^i(S)$ for all i; for all $t \in \mathbb{N}, z^t = \Sigma x^{i(t-1)}/n$, and $x^{it} \in WPO(S) \equiv \{x \in S | \nexists x' \in S \text{ with } x' > x\}$ is such that $x_j^{it} = z_j^t$ for all $j \neq i$. [On Σ_d^n, start from the point $D^i(S, d)$ instead of the $D^i(S)$.

A continuous version of the solution is obtained by having $z(t)$ move at time t in the direction of $\Sigma x^i(t)/n$ where $x^i(t) \in WPO(S)$ is such that $x_j^i(t) = z_j(t)$ for all $j \neq i$].

Perles–Maschler solution, *PM* [Figure 2(f)]: For $n = 2$. If ∂S is polygonal, $PM(S)$ is the common limit point of the sequences $\{x^\nu\}, \{y^t\}$, defined by: $x^0 = D^{*1}(S)$, $y^0 = D^{*2}(S)$; for each $t \in \mathbb{N}$, x^t, $y^t \in PO(S)$ are such that $x_1^t \geq y_1^t$, the segments $[x^{t-1}, x^t]$, $[y^{t-1}, y^t]$ are contained in $PO(S)$ and the products $|(x_1^{t-1} - x_1^t)(x_2^{t-1} - x_2^t)|$ and $|(y_1^{t-1} - y_1^t)(y_2^{t-1} - y_2^t)|$ are equal and maximal. [Equality of the products implies that the triangles of Figure 2(f) are matched in pairs of equal areas.] If ∂S is not polygonal, $PM(S)$ is defined by approximating S by a sequence of polygonal problems and taking the limit of the associated solution outcomes. [On Σ_d^n, start from the $D^i(S, d)$ instead of the $D^i(S)$.]

The solution can be given the following equivalent definition when ∂S is smooth. Consider two points moving along ∂S from $D^{*1}(S)$ and $D^{*2}(S)$ so that the product of the components of their velocity vectors in the u_1 and u_2 directions remain constant: the two points will meet at $PM(S)$. The differential system describing this movement can be generalized to arbitrary n; it generates n paths on the boundary of ∂S that meet in one point that can be taken as the desired compromise.

The next solution exemplifies a family of solutions for which compromises are evaluated globally. Some notion of the sacrifice made by each agent at each proposed alternative is formulated and the compromise is chosen for which these sacrifices are equal. In a finite model, a natural way to measure the sacrifice made by an agent at an alternative would be simply to count the alternatives that the agent would have preferred to it. Given the structure of the set of alternatives in the model under investigation, evaluating sacrifices by areas is appealing.

Equal Area solution, *A* [Figure 2(g)]: For $n = 2$. $A(S)$ is the point $x \in PO(S)$ such that the area of S to the right of the vertical line through x is equal to the area of S above the horizontal line through x. (There are several possible generalizations for $n \geq 3$. On Σ_d^n, ignore points that do not dominate d.)

The next solution has played a major role in other contexts. It needs no introduction.

Utilitarian solution, *U* [Figure 2(h)]: $U(S)$ [or $U(S, d)$] is a maximizer in $x \in S$ of Σx_i.

This solution presents some difficulties here. First, the maximizer may not be unique. To circumvent this difficulty a tie-breaking rule has to be specified; for $n = 2$ it is perhaps most natural to select the midpoint of the segment of maximizers (if $n > 2$, this rule can be generalized in several different ways). A second difficulty is that as defined here for Σ_d^n, the solution does not depend on d. A partial remedy is to search for a maximizer of Σx_i among the points of S that dominate d. In spite of these limitations, the Utilitarian solution is often advocated. In some situations, it has the merit of being a useful limit case of sequences of solutions that are free of the limitations.

Agents cannot simultaneously obtain their preferred outcomes. An intuitively appealing idea is to try to get as close as possible to satisfying everyone, that is, to come as close as possible to what we have called the ideal point. The following one-parameter family of solutions reflects the flexibility that exists in measuring how close two points are from each other.

Yu solutions, Y^p [Figure 2(i)]: Given $p \in]1, \infty[$, $Y^p(S)$ is the point of S for which the p-distance to the ideal point of S, $(\Sigma |a_i(S) - x_i|^p)^{1/p}$, is minimal. [On Σ_d^n, use $a(S, d)$ instead of $a(S)$.]

Versions of the Kalai–Smorodinsky solution appear in Raiffa (1953), Crott (1971), Butrim (1976), and the first axiomatization is in Kalai and Smorodinsky (1975). A number of variants have been discussed, in particular by Rosenthal (1976, 1978) Kalai and Rosenthal (1978) and Salonen (1985, 1987). The Egalitarian solution cannot be traced to a particular source but Egalitarian notions are certainly very old. The Equal Area solution is analyzed in Dekel (1982), Ritz (1985), Anbarci (1988), Anbarci and Bigelow (1988) and Calvo (1989); the Yu solutions in Yu (1973) and Freimer and Yu (1976); the Raiffa solution in Raiffa (1953) and Luce and Raiffa (1957). The member of the Yu family obtained for $p = 2$ is advocated by Salukvadze (1971a, b). The extension of the Yu solutions to $p = \infty$ is to maximize $\min\{|a_i(S) - x_i|\}$ in $x \in S$ but this may not yield a unique outcome except for $n = 2$. For the general case, Chun (1988a) proposes, and axiomatizes, the Equal Loss solution, the selection from $Y^\infty(S)$ that picks the maximal point of S such that $a_i(S) - x_i = a_j(S) - x_j$ for all i, j. The solution is further studied by Herrero and Marco (1993). The Utilitarian solution dates back to the mid-19th century. The two-person Perles–Maschler solution appears in Perles–Maschler (1981) and its n-person extension in Kohlberg et al. (1983), and Calvo and Gutiérrez (1993).

3. The main characterizations

Here we present the classic characterizations of the three solutions that occupy center stage in the theory as it stands today.

3.1. The Nash solution

We start with Nash's fundamental contribution. Nash considered the following axioms, the first one of which is a standard condition: all gains from cooperation should be exhausted.

Pareto-optimality: $F(S) \in PO(S) \equiv \{x \in S | \nexists x' \in S \text{ with } x' \geq x\}$.

The second axiom says that if the agents cannot be differentiated on the basis of the information contained in the mathematical description of S, then the solution should treat them the same.

Symmetry: If S is invariant under all exchanges of agents, $F_i(S) = F_j(S)$ for all i, j.

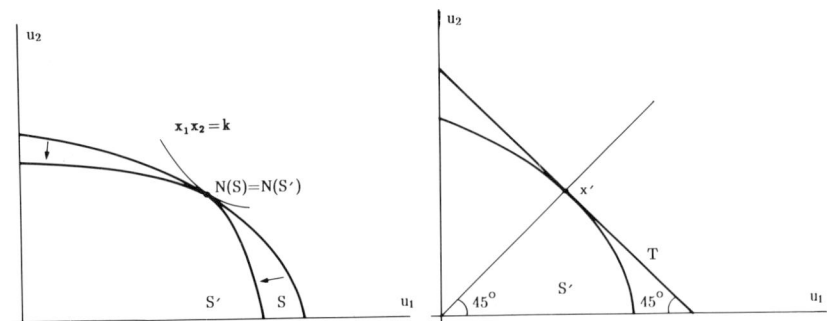

Figure 3. The Nash solution. (a) The Nash solution satisfies *contraction independence*. (b) Characterization of the Nash solution on the basis of *contraction independence*.

This axiom applies to problems that are "fully symmetric", that is, are invariant under all permutations of agents. But some problems may only exhibit partial symmetries, which one may want solutions to respect. A more general requirement, which we will also use, is that solutions be independent of the name of agents.

Anonymity: Let $\pi: \{1,\ldots,n\} \to \{1,\ldots,n\}$ be a bijection. Given $x \in \mathbb{R}^n$, let $\tilde{\pi}(x) \equiv (x_{\pi(1)}, \ldots, x_{\pi(n)})$ and $\tilde{\pi}(s) \equiv \{x' \in \mathbb{R}^n | \exists x \in S \text{ with } x' = \tilde{\pi}(x)\}$. Then, $F(\tilde{\pi}(S)) = \tilde{\pi}(F(S))$.

Next, remembering that von Neumann–Morgenstern utilities are unique only up to a positive affine transformation, we require that the solution should be independent of which particular members in the families of utility functions representing the agents' preferences are chosen to describe the problem.

Let $\Lambda_0^n: \mathbb{R}^n \to \mathbb{R}^n$ be the class of independent person by person, positive, linear transformations ("scale transformations"): $\lambda \in \Lambda_0^n$ if there is $a \in \mathbb{R}_{++}^n$ such that for all $x \in \mathbb{R}^n$, $\lambda(x) = (a_1 x_1, \ldots, a_n x_n)$. Given $\lambda \in \Lambda_0^n$ and $S \subset \mathbb{R}^n$, $\lambda(S) \equiv \{x' \in \mathbb{R}^n | \exists x \in S \text{ with } x' = \lambda(x)\}$.

Scale invariance: $\lambda(F(S)) = F(\lambda(S))$.

Finally, we require that if an alternative is judged to be the best compromise for some problem, then it should still be judged best for any subproblem that contains it. This condition can also be thought of as a requirement of informational simplicity: a proposed compromise is evaluated only on the basis of information about the shape of the feasible set in a neighborhood of itself.

Contraction independence:[2] If $S' \subseteq S$ and $F(S) \in S'$, then $F(S') = F(S)$.

In the proof of our first result we use the fact that for $n = 2$, if $x = N(S)$, then S has at x a line of support whose slope is the negative of the slope of the line connecting x to the origin (Figure 2a).

---

[2] This condition is commonly called "independence of irrelevant alternatives".

Theorem 1 [Nash (1950)]. *The Nash solution is the only solution on Σ_0^n satisfying Pareto-optimality, symmetry, scale invariance, and contraction independence.*

Proof (for $n = 2$). It is easy to verify that N satisfies the four axioms (that N satisfies *contraction independence* is illustrated in Figure 3(a). Conversely, let F be a solution on Σ_0^2 satisfying the four axioms. To show that $F = N$, let $S \in \Sigma_0^2$ be given and let $x \equiv N(S)$. Let $\lambda \in \Lambda_0^2$ be such that $x' \equiv \lambda(x)$ be on the $45°$ line. Such λ exists since $x > 0$, as is easily checked. The problem $S' \equiv \lambda(S)$ is supported at x' by a line of slope-1 [Figure 3(b)]. Let $T \equiv \{y \in \mathbb{R}_+^2 \mid \Sigma y_i \leqslant \Sigma x_i'\}$. The problem T is symmetric and $x' \in PO(T)$. By *Pareto-optimality* and *symmetry*, $F(T) = x'$. Clearly, $S' \subseteq T$ and $x' \in S'$, so that by *contraction independence* $F(S') = x'$. The desired conclusion follows by *scale invariance*. □

No axiom is universal applicable. This is certainly the case of Nash's axioms and each of them has been the object of some criticism. For instance, to the extent that the theory is intended to predict how real-world conflicts are resolved, *Pareto-optimality* is certainly not always desirable, since such conflicts often result in dominated compromises. Likewise, we might want to take into account differences between agents pertaining to aspects of the environment that are not explicitly modelled, and differentiate among them even though they enter symmetrically in the problem at hand; then, we violate *symmetry*. *Scale invariance* prevents basing compromises on interpersonal comparisons of utility, but such comparisons are made in a wide variety of situations. Finally, if the contraction described in the hypothesis of *contraction independence* is skewed against a particular agent, why should the compromise be prevented from moving against him? In fact, it seems that solutions should in general be allowed to be responsive to changes in the geometry of S, at least to its main features. It is precisely considerations of this kind that underlie the characterizations of the Kalai–Smorodinsky and Egalitarian solutions reviewed later.

Nash's theorem has been considerably refined by subsequent writers. Without *Pareto-optimality*, only one other solution becomes admissible: it is the trivial *disagreement solution*, which associates with every problem its disagreement point, here the origin [Roth (1977a, 1980)]. Dropping *symmetry*, we obtain the following family: given $\alpha \in \Delta^{n-1}$, the *weighted Nash solution* with weights α is defined by maximizing over S the product $\prod x_i^{\alpha_i}$ [Harsanyi and Selten (1972)]; the Dictatorial solutions and some generalizations [Peters (1986b, 1987b)] also become admissible. Without *scale invariance*, many other solutions, such as the Egalitarian solution, are permitted.

The same is true if *contraction independence* is dropped; however, let us assume that a function is available that summarizes the main features of each problem into a *reference point* to which agents find it natural to compare the proposed compromise in order to evaluate it. By replacing in *contraction independence* the hypothesis of identical disagreement points (implicit in our choice of domains) by the hypothesis of identical reference points, variants of the Nash solution, defined by maximizing the product of utility gains from that reference point, can be obtained under weak assumptions on the reference function. [Roth (1977b), Thomson (1981a)].

Contraction independence bears a close relation to the axioms of revealed preference of demand theory [Lensberg (1987), Peters and Wakker (1991), Bossert (1992b)]. An extension of the Nash solution to the domain of non-convex problems and a characterization appear in Conley and Wilkie (1991b). Non-convex problems are also discussed in Herrero (1989). A characterization without the expected utility hypothesis is due to Rubinstein et al. (1992).

We close this section with the statement of a few interesting properties satisfied by the Nash solution (and by many others as well). The first one is a consequence of our choice of domains: the Nash solution outcome always weakly dominates the disagreement point, here the origin. On Σ_d^n, the property would of course not necessarily be satisfied, so we write it for that domain.

Individual rationality: $F(S,d) \in I(S,d) \equiv \{x \in S | x \geqslant d\}$.

In fact, the Nash solution (and again many others) satisfies the following stronger condition: all agents should strictly gain from the compromise. The Dictatorial and Utilitarian solutions do not satisfy the property.

Strong individual rationality: $F(S,d) > d$.

The requirement that the compromise depend only on $I(S,d)$ is implicitly made in much of the literature. If it is strongly believed that no alternative at which one or more of the agents receives less than his disagreement utility should be selected, it seems natural to go further and require of the solution that it be unaffected by the elimination of these alternatives.

Independence of non-individually rational alternatives: $F(S,d) = F(I(S,d),d)$.

Most solutions satisfy this requirement. A solution that does not, although it satisfies *strong individual rationality*, is the Kalai–Rosenthal solution, which picks the maximal point of S on the segment connecting d to the point $b(S)$ defined by $b_i(S) = \max\{x_i | x \in S\}$ for all i.

Another property of interest is that small changes in problems do not lead to wildly different solution outcomes. Small perturbations in the feasible set, small errors in the way it is described, small errors in the calculation of the utilities achieved by the agents at the feasible alternatives; or conversely, improvements in the description of the alternatives available, or in the measurements of utilities, should not have dramatic effects on payoffs.

Continuity: If $S^v \to S$ in the Hausdorff topology, and $d^v \to d$, then $F(S^v, d^v) \to F(S,d)$.

All of the solutions of Section 2 satisfy *continuity*, except for the Dictatorial solutions D^{*i} and the Utilitarian and Perles–Maschler solutions (the tie-breaking rules necessary to obtain single-valuedness of the Utilitarian solutions are responsible for the violations).

Other continuity notions are formulated and studied by Jansen and Tijs (1983). A property related to *continuity*, which takes into account closeness of Pareto optimal boundaries, is used by Peters (1986a) and Livne (1987a). Salonen (1992, 1993) studies alternative definitions of *continuity* for unbounded problems.

3.2. The Kalai–Smorodinsky solution

We now turn to the second one of our three central solutions, the Kalai–Smorodinsky solution. Just like the Egalitarian solution, examined last, the appeal of this solution lies mainly in its monotonicity properties. Here, we will require that an expansion of the feasible set "in a direction favorable to a particular agent" always benefits him: one way to formalize the notion of an expansion favorable to an agent is to say that the range of utility levels attainable by agent j ($j \neq i$) remains the same as S expands to S', while for each such level, the maximal utility level attainable by agent i increases. Recall that $a_i(S) \equiv \max\{x_i | x \in S\}$.

Individual monotonicity (for $n = 2$): If $S' \supseteq S$, and $a_j(S') = a_j(S)$ for $j \neq i$, then $F_i(S') \geq F_i(S)$.

By simply replacing *contraction independence* by *individual monotonicity*, in the list of axioms shown earlier to characterize the Nash solution we obtain a characterization of the Kalai–Smorodinsky solution.

Theorem 2 [Kalai and Smorodinsky (1975)]. *The Kalai–Smorodinsky solution is the only solution on Σ_0^2 satisfying Pareto-optimality, symmetry, scale invariance, and individual monotonicity.*

Proof. It is clear that K satisfies the four axioms [that K satisfies *individual monotonicity* is illustrated in Figure 4(a)]. Conversely, let F be a solution on Σ_0^2

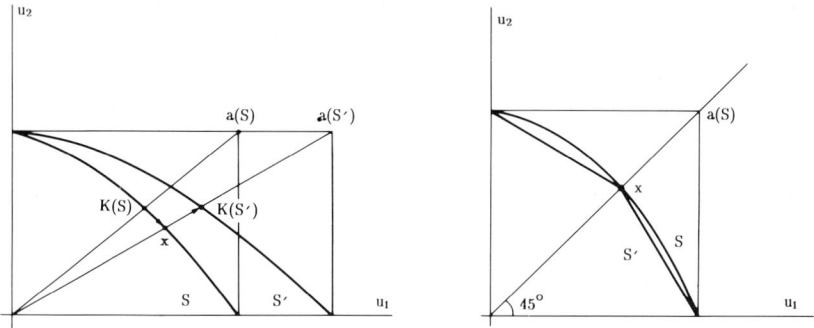

Figure 4. The Kalai–Smorodinsky solution. (a) The solution satisfies *individual monotonicity*. (b) Characterization of the solution on the basis of *individual monotonicity* (Theorem 2).

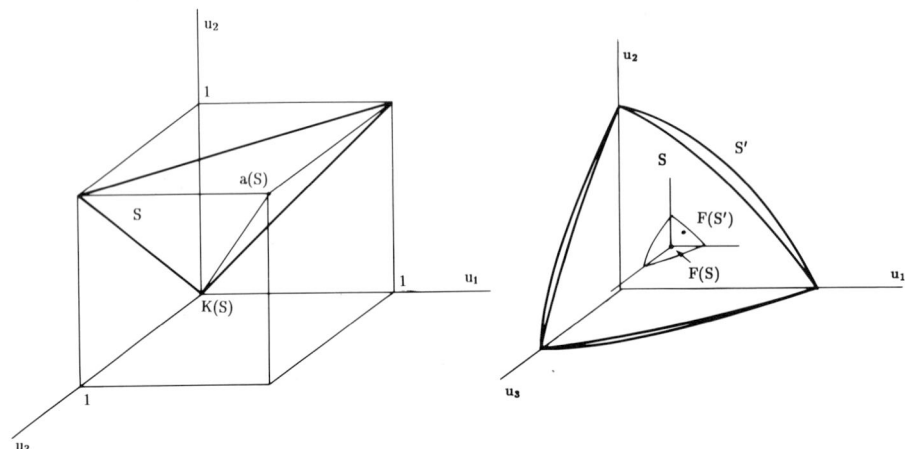

Figure 5. (a) A difficulty with the Kalai–Smorodinsky solution for $n \geq 2$. If S is not comprehensive, $K(S)$ may be strictly dominated by all points of S. (b) The axiom of *restricted monotonicity*: An expansion of the feasible set leaving unaffected the ideal point benefits all agents.

satisfying the four axioms. To see that $F = K$, let $S \in \Sigma_0^2$ be given. By *scale invariance*, we can assume that $a(S)$ has equal coordinates [Figure 4(b)]. This implies that $x \equiv K(S)$ itself has equal coordinates. Then let $S' \equiv \text{cch}\{(a_1(S), 0), x, (0, a_2(S))\}$. The problem S' is symmetric and $x \in PO(S')$ so that by *Pareto-optimality* and *symmetry*, $F(S') = x$. By *individual monotonicity* applied twice, we conclude that $F(S) \geq x$, and since $x \in PO(S)$, that $F(S) = x = K(S)$. □

Before presenting variants of this theorem, we first note several difficulties concerning the possible generalization of the Kalai–Smorodinsky solution itself to classes of not necessarily comprehensive n-person problems for $n > 2$. On such domains the solution often fails to yield Pareto-optimal points, as shown by the example $S = \text{convex hull}\{(0, 0, 0), (1, 1, 0), (0, 1, 1)\}$ of Figure 5(a): there $K(S)(= (0, 0, 0))$ is in fact dominated by all points of S [Roth (1979d)]. However, by requiring comprehensiveness of the admissible problems, the solution satisfies the following natural weakening of *Pareto-optimality*:

Weak Pareto-optimality: $F(S) \in WPO(S) \equiv \{x \in S | \not\exists x' \in S, x' > x\}$.

The other difficulty in extending Theorem 2 to $n > 2$ is that there are several ways of generalizing *individual monotonicity* to that case, not all of which permit the result to go through. One possibility is simply to write "for all $j \neq i$" in the earlier statement. Another is to consider expansions that leave the ideal point unchanged [Figure 5(b), Roth (1979d), Thomson (1980)]. This prevents the skewed expansions that were permitted by *individual monotonicity*. Under such "balanced" expansions, it becomes natural that all agents benefit:

Restricted monotonicity: If $S' \supseteq S$ and $a(S') = a(S)$, then $F(S') \geq F(S)$.

To emphasize the importance of comprehensiveness, we note that *weak Pareto-optimality*, *symmetry*, and *restricted monotonicity* are incompatible if that assumption is not imposed [Roth (1979d)].

A lexicographic (see Section 3.3) extension of K that satisfies *Pareto-optimality* has been characterized by Imai (1983). Deleting *Pareto-optimality* from Theorem 2, a large family of solutions becomes admissible. Without *symmetry*, the following generalizations are permitted: Given $\alpha \in \Delta^{n-1}$, the *weighted Kalai–Smorodinsky solution with weights* α, K^α: $K^\alpha(S)$ is the maximal point of S in the direction of the α-weighted ideal point $a^\alpha(S) \equiv (\alpha_1 a_1(S), \ldots, \alpha_n a_n(S))$. These solutions satisfy *weak Pareto-optimality* (but not *Pareto-optimality*, even if $n = 2$). There are other solutions satisfying only *weak Pareto-optimality*, *scale invariance*, and *individual monotonicity*; they are normalized versions of the "Monotone Path solutions", discussed below in connection with the Egalitarian solution [Peters and Tijs (1984a, 1985b)]. Salonen (1985, 1987) characterizes two variants of the Kalai–Smorodinsky solution. These results, as well as the characterization by Kalai and Rosenthal (1978) of their variant of the solution, and the characterization by Chun (1988a) of the Equal Loss solution, are also close in spirit to Theorem 2. Anant et al. (1990) and Conley and Wilkie (1991) discuss the Kalai–Smorodinsky solution in the context of non-convex games.

3.3. The Egalitarian solution

The Egalitarian solution performs the best from the viewpoint of monotonicity and the characterization that we offer is based on this fact. The monotonicity condition that we use is that all agents should benefit from *any* expansion of opportunities; this is irrespective of whether the expansion may be biased in favor of one of them, (for instance, as described in the hypotheses of *individual monotonicity*). Of course, if that is the case, nothing prevents the solution outcome from "moving more" is favor of that agent. The price paid by requiring this strong monotonicity is that the resulting solution involves interpersonal comparisons of utility (it violates *scale invariance*). Note also that it satisfies *weak Pareto-optimality* only, although $E(S) \in PO(S)$ for all strictly comprehensive S.

Strong monotonicity: If $S' \supseteq S$, then $F(S') \geqslant F(S)$.

Theorem 3 [Kalai (1977)]. *The Egalitarian solution is the only solution on Σ_0^n satisfying weak Pareto-optimality, symmetry, and strong monotonicity.*

Proof (for $n = 2$). Clearly, E satisfies the three axioms. Conversely, to see that if a solution F on Σ_0^2 satisfies the three axioms, then $F = E$, let $S \in \Sigma_0^2$ be given, $x \equiv E(S)$, and $S' \equiv \text{cch}\{x\}$ [Figure 6(a)]. By *weak Pareto-optimality* and *symmetry*, $F(S') = x$. Since $S \supseteq S'$, *strong monotonicity* implies $F(S) \geqslant x$. Note that $x \in WPO(S)$. If, in fact, $x \in PO(S)$, we are done. Otherwise, we suppose by contradiction that

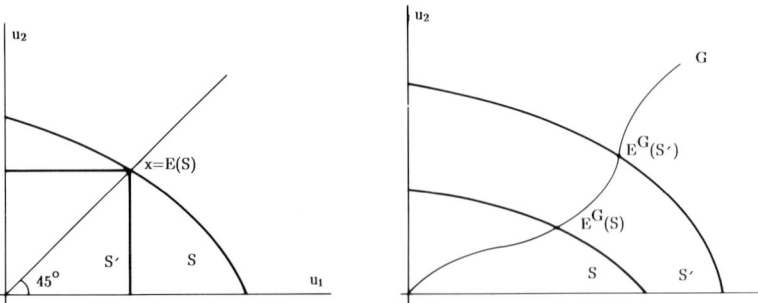

Figure 6. Egalitarian and Monotone Path solutions. (a) Characterization of the Egalitarian solution on the basis of strong monotonicity (Theorem 3). (b) Monotone Path solutions.

$F(S) \neq E(S)$ and we construct a strictly comprehensive problem S' that includes S, and such that the common value of the coordinates of $E(S')$ is smaller than $\max_i F_i(S)$. The proof concludes by applying *strong monotonicity* to the pair S, S'. □

It is obvious that comprehensiveness of S is needed to obtain weak Pareto-optimality of $E(S)$ even if $n = 2$. Moreover, without comprehensiveness, *weak Pareto-optimality* and *strong monotonicity* are incompatible [Luce and Raiffa (1957)].

Deleting *weak Pareto-optimality* from Theorem 3, we obtain solutions defined as follows: given $k \in [0, 1]$, $E^k(S) = kE(S)$. However, there are other solutions satisfying *symmetry* and *strong monotonicity* [Roth (1979a,b)]. Without *symmetry* the following solutions become admissible: Given $\alpha \in \Delta^{n-1}$, the *weighted Egalitarian solution with weights* α, E^α selects the maximal point of S in the direction α [Kalai (1977)]. *Weak Pareto-optimality* and *strong monotonicity* essentially characterize the following more general class: given a strictly monotone path G in $\mathbb{R}^n_+$, the *Monotone Path solution relative to* G, E^G chooses the maximal point of S along G [Figure 6(b), Thomson and Myerson (1980)]. For another characterization of the Egalitarian solution, see Myerson (1977). For a derivation of the solution without expected utility, see Valenciano and Zarzuelo (1993).

It is clear that *strong monotonicity* is crucial in Theorem 3 and that without it, a very large class of solutions would become admissible. However, this axiom can be replaced by another interesting condition [Kalai (1977b)]: imagine that opportunities expand over time, say from S to S'. The axiom states that $F(S')$ can be indifferently computed in one step, ignoring the initial problem S altogether, or in two steps, by first solving S and then taking $F(S)$ as starting point for the distribution of the gains made possible by the new opportunities.

Decomposability: If $S' \supseteq S$ and $S'' \equiv \{x'' \in \mathbb{R}^n_+ \mid \exists x' \in S' \text{ such that } x' = x'' + F(S)\} \in \Sigma^n_0$, then $F(S') = F(S) + F(S'')$.

The weakening of *decomposability* obtained by restricting its applications to cases where $F(S)$ is proportional to $F(S'')$ can be used together with *Pareto-optimality*, *symmetry*, *independence of non-individually rational alternatives*, the requirement that the solution depend only on the individually rational part of the feasible set, *scale invariance*, and *continuity* to characterize the Nash solution [Chun (1988b)]. For a characterization of the Nash solution based on yet another decomposability axiom, see Ponsati and Watson (1994).

As already noted, the Egalitarian solution does not satisfy *Pareto-optimality*, but there is a natural extension of the solution that does. It is obtained by a lexicographic operation of a sort that is familiar is social choice and game theory. Given $z \in \mathbb{R}^n$, let $\tilde{z} \in \mathbb{R}^n$ denote the vector obtained from z by writing its coordinates in increasing order. Given $x, y \in \mathbb{R}^n$, x is *lexicographically greater than* y if $\tilde{x}_1 > \tilde{y}_1$ or $[\tilde{x}_1 = \tilde{y}_1$ and $\tilde{x}_2 > \tilde{y}_2]$, or, more generally, for some $k \in \{1, \ldots, n-1\}$, $[\tilde{x}_1 = \tilde{y}_1, \ldots, \tilde{x}_k = \tilde{y}_k$, and $\tilde{x}_{k+1} > \tilde{y}_{k+1}]$. Now, given $S \in \Sigma_0^n$, its *Lexicographic Egalitarian solution* outcome $E^L(S)$, is the point of S that is lexicographically maximal. It can be reached by the following simple operation (Figure 7): let x^1 be the maximal point of S with equal coordinates [this is $E(S)$]; if $x^1 \in PO(S)$, then $x^1 = E^L(S)$; if not, identify the greatest subset of the agents whose utilities can be simultaneously increased from x^1 without hurting the remaining agents. Let x^2 be the maximal point of S at which these agents experience equal gains. Repeat this operation from x^2 to obtain x^3, etc., ..., until a point of $PO(S)$ is obtained.

The algorithm produces a well-defined solution satisfying *Pareto-optimality* even on the class of problems that are not necessarily comprehensive. Given a problem in that class, apply the algorithm to its comprehensive hull and note that taking the comprehensive hull of a problem does not affect its set of Pareto-optimal points. Problems in Σ_d^n can of course be easily accommodated. A version of the

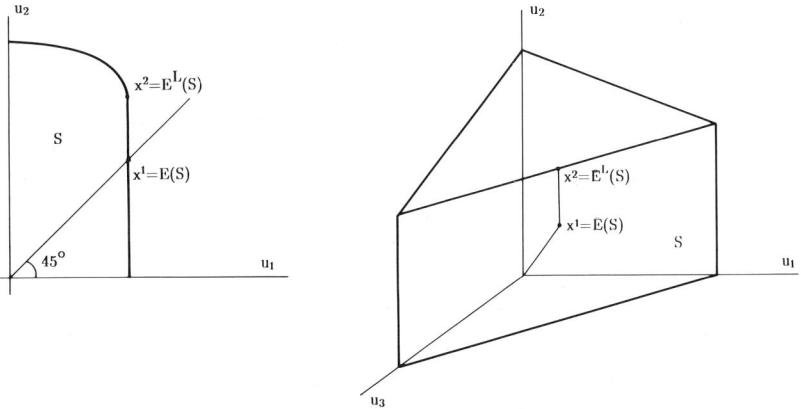

Figure 7. The Lexicographic Egalitarian solution for two examples. In each case, the solution outcome is reached in two steps. (a) A two-person example. (b) A three-person example.

Kalai–Smorodinsky solution that satisfies *Pareto-optimality* on Σ_d^n for all n can be defined in a similar way.

For characterizations of E^L based on monotonicity considerations, see Imai (1983) and Chun and Peters (1988). Lexicographic extensions of the Monotone Path solutions are defined, and characterized by similar techniques for $n=2$, by Chun and Peters (1989a). For parallel extensions, and characterizations thereof, of the Equal Loss solution, see Chun and Peters (1991).

4. Other properties. The role of the feasible set

Here, we change our focus, concentrating on properties of solutions. For many of them, we are far from fully understanding their implications, but taken together they constitute an extensive battery of tests to which solutions can be usefully subjected when they have to be evaluated.

4.1. Midpoint domination

A minimal amount of cooperation among the agents should allow them to do at least as well as the average of their preferred positions. This average corresponds to the often observed tossing of the coin to determine which one of two agents will be given the choice of an alternative when no easy agreement on a deterministic outcome is obtained. Accordingly, consider the following two requirements [Sobel (1981), Salonen (1985), respectively], which correspond to two natural definitions of "preferred positions".

Midpoint domination: $F(S) \geqslant [\Sigma D^i(S)]/n$.

Strong midpoint domination: $F(S) \geqslant [\Sigma D^{*i}(S)]/n$.

Many solutions satisfy *midpoint domination*. Notable exceptions are the Egalitarian and Utilitarian solutions (of course, this should not be a surprise since the point that is to be dominated is defined in a scale-invariant way); yet we have (compare with Theorem 1):

Theorem 4 [Moulin (1983)]. *The Nash solution is the only solution on Σ_0^n satisfying midpoint domination and contraction independence.*

Few solutions satisfy *strong midpoint domination* [the Perles–Maschler solution does however; Salonen (1985) defines a version of the Kalai–Smorodinsky solution that does too].

4.2. Invariance

The theory presented so far is a cardinal theory, in that it depends on utility functions, but the extent of this dependence varies, as we have seen. Are there

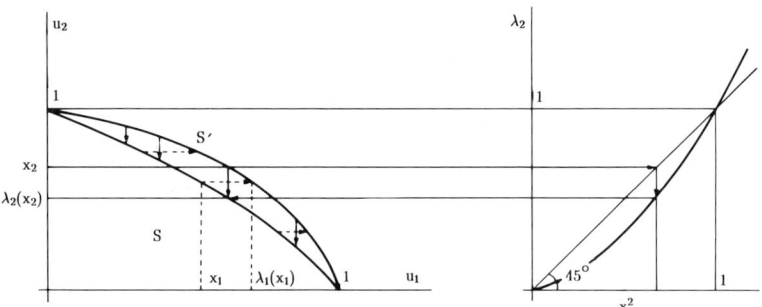

Figure 8. Strong individual rationality and ordinal invariance are incompatible on $\tilde{\Sigma}_0^2$ (Theorem 5). (a) S is globally invariant under the composition of the two transformations defined by the horizontal and the vertical arrows respectively. (b) An explicit construction of the transformation to which agent 2's utility is subjected.

solutions that are invariant under *all* monotone increasing, and independent agent by agent, transformations of utilities, i.e., solutions that depend only on ordinal preferences? The answer depends on the number of agents. Perhaps surprisingly, it is negative for $n = 2$ but not for all $n > 2$.

Let $\tilde{\Lambda}_0^n$ be the class of these transformations: $\lambda \in \tilde{\Lambda}_0^n$ if for each i, there is a continuous and monotone increasing function $\lambda_i: \mathbb{R} \to \mathbb{R}$ such that given $x \in \mathbb{R}^n$, $\lambda(x) = (\lambda_1(x_1), \ldots, \lambda_n(x_n))$. Since convexity of S is not preserved under transformations in $\tilde{\Lambda}_0^n$, it is natural to turn our attention to the domain $\tilde{\Sigma}_0^n$ obtained from Σ_0^n by dropping this requirement.

Ordinal invariance: For all $\lambda \in \tilde{\Lambda}_0^n, F(\lambda(S)) = \lambda(F(S))$.

Theorem 5 [Shapley (1969), Roth (1979c)]. *There is no solution on $\tilde{\Sigma}_0^2$ satisfying strong individual rationality and ordinal invariance.*

Proof. Let F be a solution on $\tilde{\Sigma}_0^2$ satisfying *ordinal invariance* and let S and S' be as in Figure 8(a). Let λ_1 and λ_2 be the two transformations from $[0, 1]$ to $[0, 1]$ defined by following the horizontal and vertical arrows of Figure 8(a) respectively. [The graph of λ_2 is given in Figure 8(b); for instance, $\lambda_2(x_2)$, the image of x_2 under λ_2, is obtained by following the arrows from Figure 8(a) to 8(b)]. Note that the problem S is globally invariant under the transformation $\lambda \equiv (\lambda_1, \lambda_2) \in \tilde{\Sigma}_0^2$, with only three fixed points, the origin and the endpoints of $PO(S)$. Since none of these points is positive, F does not satisfy *strong individual rationality*. □

Theorem 6 [Shapley (1984), Shubik (1982)]. *There are solutions on the subclass of $\tilde{\Sigma}_0^3$ of strictly comprehensive problems satisfying Pareto-optimality and ordinal invariance.*

Proof. Given $S \in \tilde{\Sigma}_0^3$, let $F(S)$ be the limit point of the sequence $\{x^t\}$ where x^1 is the point of intersection of $PO(S)$ with $\mathbb{R}^{\{1,2\}}$ such that the arrows of Figure 9(a)

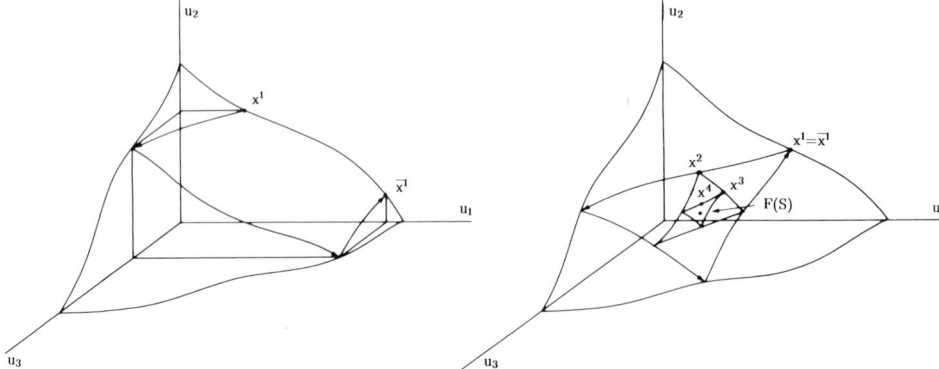

Figure 9. A solution on $\tilde{\Sigma}_0^3$ satisfying *Pareto-optimality, strong individual rationality,* and *ordinal invariance.* (a) The fixed point argument defining x^1. (b) The solution outcome of S is the limit point of the sequence $\{x^t\}$.

lead back to x^1; x^2 is the point of $PO(S)$ such that $x_2^2 = x_2^1$ and a similarly defined sequence of arrows leads back to x^2; this operation being repeated forever [Figure 9(b)]. The solution F satisfies *ordinal invariance* since at each step, only operations that are invariant under ordinal transformations are performed. □

There are other solutions satisfying these properties and yet other such solutions on the class of smooth problems [Shapley (1984)].

In light of the negative result of Theorem 5, it is natural to look for a weaker invariance condition. Instead of allowing the utility transformations to be independent across agents, require now that they be the same for all agents:

Weak ordinal invariance: For all $\lambda \in \tilde{\Lambda}_\cap^n$ such that $\lambda_i = \lambda_j$ for all i, j, $F(\lambda(S)) = \lambda(F(S))$.

This is a significantly weaker requirement than *ordinal invariance*. Indeed, we have:

Theorem 7 [Roth (1979c), Nielsen (1983)]. *The Lexicographic Egalitarian solution is the only solution on the subclass of $\tilde{\Sigma}_0^2$ of problems whose Pareto-optimal boundary is a connected set to satisfy weak Pareto-optimality, symmetry, contraction independence, and weak ordinal invariance.*

4.3. Independence and monotonicity

Here we formulate a variety of conditions describing how solutions should respond to changes in the geometry of S. An important motivation for the search for alternatives to the monotonicity conditions used in the previous pages is that these conditions pertain to transformations that are not defined with respect to the compromise initially chosen.

One of the most important conditions we have seen is *contraction independence*. A significantly weaker condition which applies only when the solution outcome of the initial problem is the only Pareto-optimal point of the final problem is:

Weak contraction independence: If $S' = \text{cch}\{F(S)\}$, then $F(S) = F(S')$.

Dual conditions to *contraction independence* and *weak contraction independence*, requiring invariance of the solution outcome under expansions of S, provided it remains feasible, have also been considered. Useful variants of these conditions are obtained by restricting their application to smooth problems. The Nash and Utilitarian solutions can be characterized with the help of such conditions [Thomson (1981b, c)]. The smoothness restriction means that utility transfers are possible at the same rate in both directions along the boundary of S. Suppose S is not smooth at $F(S)$. Then, one could not eliminate the possibility that an agent who had been willing to concede along ∂S up to $F(S)$ might have been willing to concede further if the same rate at which utility could be transferred from him to the other agents had been available. It is then natural to think of the compromises $F(S)$ as somewhat artificial and to exclude such situations from the range of applicability of the axiom. A number of other conditions that explicitly exclude kinks or corners have been formulated [Chun and Peters (1988, 1989a), Peters (1986a), Chun and Thomson (1990c)]. For a characterization of the Nash solution based on yet another expansion axiom, see Anbarci (1991).

A difficulty with the two monotonicity properties used earlier, *individual monotonicity* and *strong monotonicity*, as well as with the independence conditions, is that they preclude the solution from being sensitive to certain changes in S that intuitively seem quite relevant. What would be desirable are conditions pertaining to changes in S that are defined *relative to the compromise initially established*. Consider the next conditions [Thomson and Myerson (1980)], written for $n = 2$, which involve "twisting" the boundary of a problem around its solution outcome, only "adding", or only "substracting", alternatives on one side of the solution outcome (Figure 10).

Twisting: If $x \in S' \setminus S$ implies $[x_i \geq F_i(S) \text{ and } x_j \leq F_j(S)]$ and $x \in S \setminus S'$ implies $[x_i \leq F_i(S) \text{ and } x_j \geq F_j(S)]$, then $F_i(S') \geq F_i(S)$.

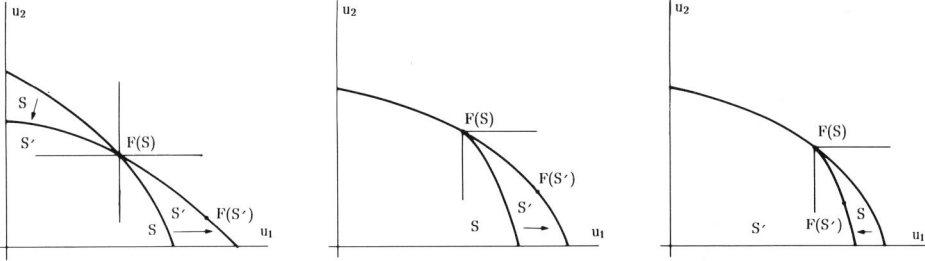

Figure 10. Three monotonicity conditions. (a) Twisting. (b) Adding. (c) Cutting.

Adding: If $S' \supset S$, and $x \in S' \setminus S$ implies $[x_i \geq F_i(S)$ and $x_i \leq F_j(S)]$, then $F_i(S') \geq F_i(S)$.

Cutting: If $S' \subset S$, and $x \in S \setminus S'$ implies $[x_i \geq F_i(S)$ and $x_j \leq F_j(S)]$, then $F_i(S') \leq F_i(S)$.

The main solutions satisfy *twisting*, which is easily seen to imply *adding* and *cutting*. However, the Perles–Maschler solution does not even satisfy *adding*. *Twisting* is crucial to understanding the responsiveness of solutions to changes in agents' risk aversion (Section 5.3).

Finally, we have the following strong axiom of solidarity. Independently of how the feasible set changes, all agents gain together or all agents lose together. No assumptions are made on the way S relates to S'.

Domination: Either $F(S') \geq F(S)$ or $F(S) \geq F(S')$.

A number of interesting relations exist between all of these conditions. In light of *weak Pareto-optimality* and *continuity*, *domination* and *strong monotonicity* are equivalent [Thomson and Myerson (1980)], and so are *adding* and *cutting* [Livne (1986a)]. *Contraction independence* implies *twisting* and so do *Pareto-optimality* and *individual monotonicity* together [Thomson and Myerson (1980)]. Many solutions (Nash, Perles–Maschler, Equal Area) satisfy *Pareto-optimality* and *twisting* but not *individual monotonicity*. Finally, *weak Pareto-optimality*, *symmetry*, *scale invariance* and *twisting* together imply *midpoint domination* [Livne (1986a)].

The axioms *twisting*, *individual monotonicity*, *adding* and *cutting* can be extended to the n-person case in a number of different ways.

4.4. Uncertain feasible set

Suppose that bargaining takes place today but that the feasible set will be known only tomorrow: It may be S^1 or S^2 with equal probabilities. Let F be a candidate solution. Then the vector of expected utilities today from waiting until the uncertainty is resolved is $x^1 \equiv [F(S^1) + F(S^2)]/2$ whereas solving the "expected problem" $(S^1 + S^2)/2$ produces $F[(S^1 + S^2)/2]$. Since x^1 is in general not Pareto-optimal in $(S^1 + S^2)/2$, it would be preferable for the agents to reach a compromise today. A necessary condition for this is that both benefit from early agreement. Let us then require of F that it gives all agents the incentive to solve the problem today: x^1 should dominate $F[(S^1 + S^2)/2]$. Slightly more generally, and to accommodate situations when S^1 and S^2 occur with unequal probabilities, we formulate:

Concavity: For all $\lambda \in [0, 1]$, $F(\lambda S^1 + (1 - \lambda)S^2) \geq \lambda F(S^1) + (1 - \lambda)F(S^2)$.

Alternatively, we could imagine that the feasible set is the result of the addition of two component problems and require that both agents benefit from looking at the situation globally, instead of solving each of the two problems separately and adding up the resulting payoffs.

Super-additivity: $F(S^1 + S^2) \geq F(S^1) + F(S^2)$.

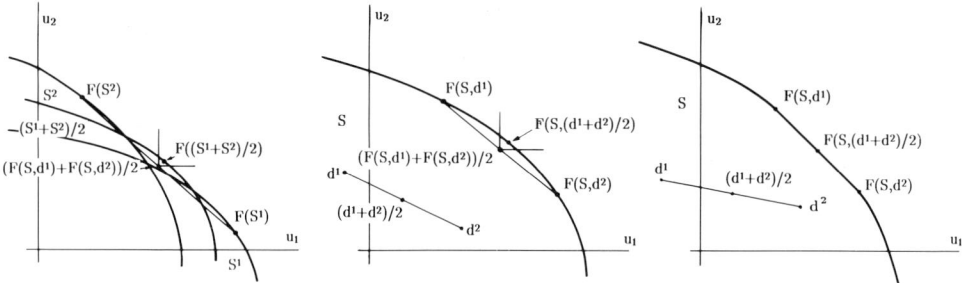

Figure 11. Concavity conditions. (a) (*Feasible set*) *concavity*: the solution outcome of the average problem $(S^1 + S^2)/2$ dominates the average $(F(S^1) + F(S^2))/2$ of the solution outcomes of the two component problems S^1 and S^2. (b) *Disagreement point concavity*: the solution outcome of the average problem $(S, (d^1 + d^2)/2)$ dominates the average $(F(S, d^1) + F(S, d^2))/2$ of the solution outcomes of the two component problems (S, d^1) and (S, d^2). (c) *Weak disagreement point concavity*: this is the weakening of *disagreement point concavity* obtained by limiting its application to situations where the boundary of S is linear between the solution outcomes of the two components problems, and smooth at these two points.

Neither the Nash nor Kalai–Smorodinsky solution satisfies these conditions, but the Egalitarian solution does. Are the conditions compatible with *scale invariance*? Yes. However, only one *scale invariant* solution satisfies them together with a few other standard requirements. Let Γ_0^2 designate the class of problems satisfying all the properties required of the elements of Σ_0^2, but violating the requirement that there exists $x \in S$ with $x > 0$.

Theorem 8 [Perles and Maschler (1981)]: *The Perles–Maschler solution is the only solution on $\Sigma_0^2 \cup \Gamma_0^2$ satisfying Pareto-optimality, symmetry, scale invariance, super-additivity, and to be continuous on the subclass of Σ_0^2 of strictly comprehensive problems.*

Deleting *Pareto-optimality* from Theorem 8, the solutions PM^λ defined by $PM^\lambda(S) \equiv \lambda PM(S)$ for $\lambda \in [0, 1]$ become admissible. Without *symmetry*, we obtain a two-parameter family [Maschler and Perles (1981)]. *Continuity* is indispensable [Maschler and Perles (1981)] and so are *scale invariance* (consider E) and obviously *super-additivity*. Theorem 8 does not extend to $n > 2$: In fact, *Pareto-optimality*, *symmetry*, *scale invariance*, and *super-additivity* are incompatible on Σ_0^3 [Perles (1981)].

Deleting *scale invariance* from Theorem 8 is particular interesting: then, a joint characterization of Egalitarianism and Utilitarianism on the domain $\Sigma_{0,-}^n$ can be obtained (note however the change of domains). In fact, *super-additivity* can be replaced by the following strong condition, which says that agents are *indifferent* between solving problems separately or consolidating them into a single problem and solving that problem.

Linearity: $F(S^1 + S^2) = F(S^1) + F(S^2)$.

Theorem 9 [Myerson (1981)]. *The Egalitarian and Utilitarian solutions are the only solutions on $\Sigma_{0,-}^n$ satisfying weak Pareto-optimality, symmetry, contraction independence, and concavity. The Utilitarian solutions are the only solutions on $\Sigma_{0,-}^n$ satisfying Pareto-optimality, symmetry, and linearity. In each of these statements, the Utilitarian solutions are covered if appropriate tie-breaking rules are applied.*

On the domain $\Sigma_{0,-}^2$, the following weakening of *linearity* (and *super-additivity*) is compatible with *scale invariance*. It involves a smoothness restriction whose significance has been discussed earlier (Section 4.4).

Weak linearity: If $F(S^1) + F(S^2) \in PO(S^1 + S^2)$ and ∂S^1 and ∂S^2 are smooth at $F(S^1)$ and $F(S^2)$ respectively, then $F(S^1 + S^2) = F(S^1) + F(S^2)$.

Theorem 10 [Peters (1986a)]. *The weighted Nash solutions are the only solutions on $\Sigma_{d,-}^2$ satisfying Pareto-optimality, strong individual rationality, scale invariance, continuity, and weak linearity.*

The Nash solution can be characterized by an alternative weakening of *linearity* [Chun (1988b)]. Randomizations between all the points of S and its ideal point, and all the points of S and its solution outcome, have been considered by Livne (1988, 1989a,b). He based on these operations the formulation of invariance conditions which he then used to characterize the Kalai–Smorodinsky and continuous Raiffa solutions.

To complete this section, we note that instead of considering the "addition" of two problems we could formulate a notion of "multiplication," and require the invariance of solutions under this operation. The resulting requirement leads to a characterization of the Nash solution.

Given $x, y \in \mathbb{R}_+^2$, let $x * y \equiv (x_1 y_1, x_2 y_2)$; given $S, T \in \Sigma_0^2$, let $S * T \equiv \{z \in \mathbb{R}_+^2 \mid z = x * y$ for some $x \in S$ and $y \in T\}$. The domain Σ_0^2 is not closed under the operation $*$, which explains the form of the condition stated next.

Separability: If $S * T \in \Sigma_0^2$, then $F(S * T) = F(S) * F(T)$.

Theorem 11 [Binmore (1984)]. *The Nash solution is the only solution on Σ_0^2 satisfying Pareto-optimality, symmetry, and separability.*

5. Other properties. The role of the disagreement point

In our exposition so far, we have ignored the disagreement point altogether. Here, we analyze its role in detail, and, for that purpose, we reintroduce it in the notation: a bargaining problem is now a pair $(S; d)$ as originally specified in Section 2. First we consider increases in one of the coordinates of the disagreement point; then, situations when it is uncertain. In each case, we study how responsive solutions

Ch. 35: Cooperative Models of Bargaining

are to these changes. The solutions that will play the main role here are the Nash and Egalitarian solutions, and generalizations of the Egalitarian solution. We also study how solutions respond to changes in the agents' attitude toward risk.

5.1. Disagreement point monotonicity

We first formulate monotonicity properties of solutions with respect to changes in d [Thomson (1987a)]. To that end, fix S. If agent i's fallback position improves while the fallback position of the others do not change, it is natural to expect that he will (weakly) gain [Figure 12(a)]. An agent who has less to lose from failure to reach an agreement should be in a better position to make claims.

Disagreement point monotonicity: If $d'_i \geqslant d_i$ and for all $j \neq i$, $d'_j = d_j$, then $F_i(S, d') \geqslant F_i(S, d)$.

This property is satisfied by all of the solutions that we have encountered. Even the Perles–Maschler solution, which is very poorly behaved with respect to changes in the feasible set, as we saw earlier, satisfies this requirement.

A stronger condition which is of greatest relevance for solutions that are intended as normative prescriptions is that under the same hypotheses as *disagreement point monotonicity*, not only $F_i(S, d') \geqslant F_i(S, d)$ but in addition for all $j \neq i$, $F_j(S, d') \leqslant F_j(S, d)$. The gain achieved by agent i should be at the expense (in the weak sense) of all the other agents [Figure 12(b)]. For a solution that select Pareto-optimal compromises, the gain to agent i has to be accompanied by a loss to at least one agent $j \neq i$. One could argue that an increase in some agent k's payoff would unjustifiably further increase the negative impact of the change in d_i on all agents $j, j \notin \{i, k\}$. (Of course, this is a property that is interesting only if $n \geqslant 3$.) Most

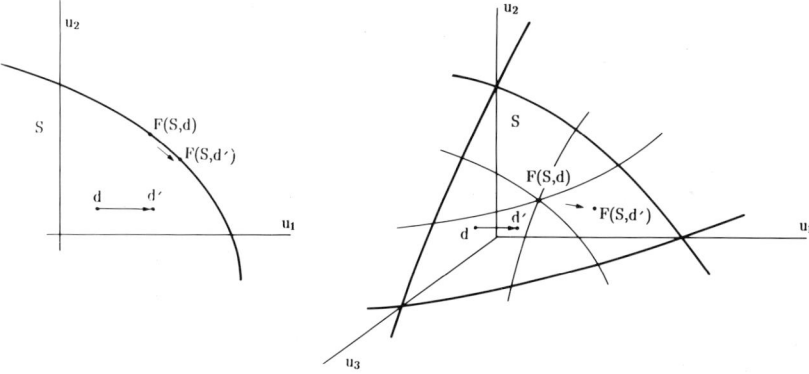

Figure 12. Conditions of monotonicity with respect to the disagreement point. (a) Weak disagreement point monotonicity for $n = 2$, an increase in the first coordinate of the disagreement point benefits agent 1. (b) Strong disagreement point monotonicity for $n = 3$, an increase in the first coordinate of the disagreement point benefits agent 1 at the expense of both other agents.

solutions, in particular the Nash and Kalai–Smorodinsky solutions and their variants, violate it. However, the Egalitarian solution does satisfy the property and so do the Monotone Path solutions.

Livne (1989b) shows that the continuous Raiffa solution satisfies a strengthening of *disagreement point monotonicity*. Bossert (1990a) bases a characterization of the Egalitarian solution on a related condition.

5.2. Uncertain disagreement point

Next, we imagine that there is uncertainty about the disagreement point. Recall that earlier we considered uncertainty in the feasible set but in practice, the disagreement point may be uncertain just as well. Suppose, to illustrate, that the disagreement point will take one of two positions d^1 and d^2 with equal probabilities and that this uncertainty will be resolved tomorrow. Waiting until tomorrow and solving then whatever problem has come up results in the expected payoff vector today $x^1 \equiv [F(S, d^1) + F(S, d^2)]/2$, which is typically Pareto-dominated in S. Taking as new disagreement point the expected cost of conflict and solving the problem $(S, (d^1 + d^2)/2)$ results in the payoffs $x^2 \equiv F(S, (d^1 + d^2)/2)$. If $x^1 \leqslant x^2$, the agents will agree to solve the problem today. If neither x_1 dominates x_2 nor x_2 dominates x_1, their incentives to wait will conflict. The following requirement prevents any such conflict:

Disagreement point concavity: For all $\lambda \in [0, 1]$, $F(S, \lambda d^1 + (1 - \lambda)d^2) \geqslant \lambda F(S, d^1) + (1 - \lambda)F(S, d^2)$.

Of all the solutions seen so far, only the weighted Egalitarian solutions satisfy this requirement. It is indeed very strong, as indicated by the next result, which is a characterization of a family of solutions that further generalize the Egalitarian solution: given δ, a continuous function from the class of n-person fully comprehensive feasible sets into Δ^{n-1}, and a problem $(S, d) \in \Sigma^n_{d,-}$ the *Directional solution relative to* δ, E^δ, selects the point $E^\delta(S, d)$ that is the maximal point of S of the form $d + t\delta(S)$, for $t \in \mathbb{R}_+$.

Theorem 12 [Chun and Thomson (1990a, b)]. *The Directional solutions are the only solutions on $\Sigma^n_{d,-}$ satisfying weak Pareto-optimality, individual rationality, continuity, and disagreement point concavity.*

This result is somewhat of a disappointment since it says that *disagreement point concavity* is incompatible with full optimality, and permits *scale invariance* only when $\delta(S)$ is a unit vector (then, the resulting Directional solution is a Dictatorial solution). The following weakening of *disagreement point concavity* allows recovering full optimality and *scale invariance*.

Weak disagreement point concavity: If $[F(S,d^1), F(S,d^2)] \subset PO(S)$ and $PO(S)$ is smooth at $F(S,d^1)$ and $F(S,d^2)$, then for all $\lambda \in [0,1]$, $F(S, \lambda d^1 + (1-\lambda)d^2) = \lambda F(S,d^1) + (1-\lambda)F(S,d^2)$.

The boundary of S is linear between $F(S,d^1)$ and $F(S,d^2)$ and it seems natural to require that the solution respond linearly to linear movements of d between d^1 and d^2. This "partial" linearity of the solution is required however only when neither compromise is at a kink of ∂S. Indeed, an agent who had been willing to trade off his utility against some other agent's utility up to such a point might have been willing to concede further. No additional move has taken place because of the sudden change in the rates at which utility can be transferred. One can therefore argue that the initial compromise is a little artifical and the smoothness requirement is intended to exclude these situations from the domain of applicability of the axiom. This sort of argument is made earlier in Section 4.3.

Theorem 13 [Chun and Thomson (1990c)]. *The Nash solution is the only solution on $\Sigma_{d,-}^n$ satisfying Pareto-optimality, independence of non-individually rational alternatives, symmetry, scale invariance, continuity, and weak disagreement point concavity.*

A condition related to *weak disagreement point concavity* says that a move of the disagreement point in the direction of the desired compromise does not call for a revision of this compromise.

Star-shaped inverse: $F(S, \lambda d + (1-\lambda)F(S,d)) = F(S,d)$ for all $\lambda \in \,]0,1]$.

Theorem 14 [Peters and van Damme (1991)]. *The weighted Nash solutions are the only solutions on $\Sigma_{d,-}^n$ satisfying strong individual rationality, independence of non-individually rational alternatives, scale invariance, disagreement point continuity, and star-shaped inverse.*

Several conditions related to the above three have been explored. Chun (1987b) shows that a requirement of *disagreement point quasi-concavity* can be used to characterize a family of solutions that further generalize the Directional solutions. He also establishes a characterization of the Lexicographic Egalitarian solution [Chun (1989)]. Characterizations of the Kalai–Rosenthal solution are given in Peters (1986c) and Chun (1990). Finally, the continuous Raiffa solution for $n=2$ is characterized by Livne (1989a), and Peters and van Damme (1991). They use the fact that for this solution the set of disagreement points leading to the same compromise for each fixed S is a curve with differentiability, and certain monotonicity, properties. Livne (1988) considers situations where the disagreement point is also subject to uncertainty but information can be obtained about it, and he characterizes a version of the Nash solution

5.3. Risk-sensitivity

Here we investigate how solutions respond to changes in the agents' risk-aversion. Other things being equal, is it preferable to face a more risk-averse opponent? To study this issue we need explicitly to introduce the set of underlying physical alternatives. Let C be a set of *certain options* and L the set of *lotteries* over C. Given two von Neumann–Morgenstern utility functions u_i and $u_i': L \to \mathbb{R}$, u_i' is *more risk-averse than* u_i if they represent the same ordering on C and for all $c \in C$, the set of lotteries that are u_i'-preferred to c is contained in the set of lotteries that are u_i-preferred to c. If $u_i(C)$ is an interval, this implies that there is an increasing concave function $k: u_i(C) \to \mathbb{R}$ such that $u_i' = k(u_i)$. An *n-person concrete problem* is a list (C, e, u), where C is as above, $e \in C$, and $u = (u_1, \ldots, u_n)$ is a list of von Neumann–Morgenstern utility functions defined over C. The *abstract problem associated with* (C, e, u) is the pair $(S, d) \equiv (\{u(l) | l \in L\}, u(e))$.

The first property we formulate focuses on the agent whose risk-aversion changes. According to his old preferences, does he necessarily lose when his risk-aversion increases?

Risk-sensitivity: Given (C, e, u) and (C', e', u'), which differ only in that u_i' is more risk-averse than u_i, and such that the associated problems $(S, d), (S', d')$ belong to Σ_d^n, we have $F_i(S, d) \geq u_i(l')$, where $u'(l') = F(S', d')$.

In the formulation of the next property, the focus is on the agents whose preferences are kept fixed. It says that all of them benefit from the increase in some agent's risk-aversion.

Strong risk-sensitivity: Under the same hypotheses as *risk-sensitivity*, $F_i(S, d) \geq u_i(l')$ and in addition, $F_j(S, d) \leq u_j(l')$ for all $j \neq i$.

The concrete problem (C, e, u) *is basic* if the associated abstract problem (S, d) satisfies $PO(S) \subset u(C)$. Let $B(\mathscr{C}_d^n)$ be the class of basic problems. If (C, e, u) is basic and u_i' is more risk-averse than u_i, then (C, e, u_i', u_{-i}) also is basic.

Theorem 15 [Kihlstrom, Roth and Schmeidler (1981), Nielsen (1984)]. *The Nash solution satisfies risk-sensitivity on* $B(\mathscr{C}_d^n)$ *but it does not satisfy strong risk-sensitivity. The Kalai–Smorodinsky solution satisfies strong risk sensitivity on* $B(\mathscr{C}_0^n)$.

There is an important logical relation between *risk-sensitivity* and *scale invariance*.

Theorem 16 [Kihlstrom, Roth and Schmeidler (1981)]. *If a solution on* $B(\mathscr{C}_d^2)$ *satisfies Pareto-optimality and risk sensitivity, then it satisfies scale invariance. If a solution on* $B(\mathscr{C}_0^n)$ *satisfies Pareto-optimality and strong risk sensitivity, then it satisfies scale invariance.*

For $n = 2$, interesting relations exist between *risk sensitivity* and *twisting* [Tijs and Peters (1985)] and between *risk sensitivity* and *midpoint domination* [Sobel (1981)].

Further results appear in de Koster et al. (1983), Peters (1987a), Peters and Tijs (1981, 1983, 1985a), Tijs and Peters (1985) and Klemisch–Ahlert (1992a).

For the class of non-basic problems, two cases should be distinguished. If the disagreement point is the image of one of the basic alternatives, what matters is whether the solution is appropriately responsive to changes in the disagreement point.

Theorem 17 [Based on Roth and Rothblum (1982) and Thomson (1987a)]. *Suppose $C = \{c_1, c_2, e\}$, and F is a solution on Σ_d^2 satisfying Pareto-optimality, scale invariance and disagreement point monotonicity. Then, if u_i is replaced by a more risk-averse utility u'_i, agent j gains if $u_i(l) \geq \min\{u_i(c_1), u_i(c_2)\}$ and not otherwise.*

The *n*-person case is studied by Roth (1988). Situations when the disagreement point is obtained as a lottery are considered by Safra et al. (1990). An application to insurance contracts appears in Kihlstrom and Roth (1982).

6. Variable number of agents

Most of the axiomatic theory of bargaining has been written under the assumption of a fixed number of agents. Recently, however, the model has been enriched by allowing the number of agents to vary. Axioms specifying how solutions could, or should, respond to such changes have been formulated and new characterizations of the main solutions as well as of new solutions generalizing them have been developed. A detailed account of these developments can be found in Thomson and Lensberg (1989).

In order to accommodate a variable population, the model itself has to be generalized. There is now an infinite set of "potential agents", indexed by the positive integers. Any finite group may be involved in a problem. Let $\mathcal{P}$ be the set of all such groups. Given $Q \in \mathcal{P}$, $\mathbb{R}^Q$ is the utility space pertaining to that group, and Σ_0^Q the class of subsets of $\mathbb{R}_+^Q$ satisfying all of the assumptions imposed earlier on the elements of Σ_0^n. Let $\Sigma_0 \equiv \cup \Sigma_0^Q$. A *solution* is a function F defined on Σ_0 which associates with every $Q \in \mathcal{P}$ and every $S \in \Sigma_0^Q$ a point of S. All of the axioms stated earlier for solutions defined on Σ_0^n can be reformulated so as to apply to this more general notion by simply writing that they hold for every $P \in \mathcal{P}$. As an illustration, the optimality axiom is written as:

Pareto-Optimality: For each $Q \in \mathcal{P}$ and $S \in \Sigma_0^Q$, $F(S) \in PO(S)$.

This is simply a restatement of our earlier axiom of *Pareto-optimality* for each group separately. To distinguish the axiom from its fixed population counterpart, we will capitalize it. We will similarly capitalize all axioms in this section.

Our next axiom, *Anonymity*, is also worth stating explicitly: it says that the solution should be invariant not only under exchanges of the names of the agents in each given group, but also under replacement of some of its members by other agents.

Anonymity: Given $P, P' \in \mathscr{P}$ with $|P| = |P'|$, $S \in \Sigma_0^P$ and $S' \in \Sigma_0^{P'}$, if there exists a bijection $\gamma: P \to P'$ such that $S' = \{x' \in \mathbb{R}^{P'} | \exists x \in S \text{ with } x'_i = x_{\gamma(i)} \forall i \in P\}$, then $F_i(S') = F_{\gamma(i)}(S)$ for all $i \in P$.

Two conditions specifically concerned with the way solutions respond to changes in the number of agents have been central to the developments reported in the section. One is an independence axiom, and the other a monotonicity axiom. They have led to characterizations of the Nash, Kalai–Smorodinsky and Egalitarian solution. We will take these solutions in that order.

Notation: Given $P, Q \in \mathscr{P}$ with $P \subset Q$ and $x \in \mathbb{R}^Q$, x_P denotes its projection on $\mathbb{R}^P$. Similarly, if $A \subseteq \mathbb{R}^Q$, A_P denotes its projection on $\mathbb{R}^P$.

6.1. Consistency and the Nash solution

We start with the independence axiom. Informally, it says that the desirability of a compromise should be unaffected by the departure of some of the agents with their payoffs. To be precise, given $Q \in \mathscr{P}$ and $T \in \Sigma_0^Q$, consider some point $x \in T$ as the candidate compromise for T. For x to be acceptable to all agents, it should be acceptable to all subgroups of Q. Assume that it has been accepted by the subgroup P' and let us imagine its members leaving the scene with the understanding

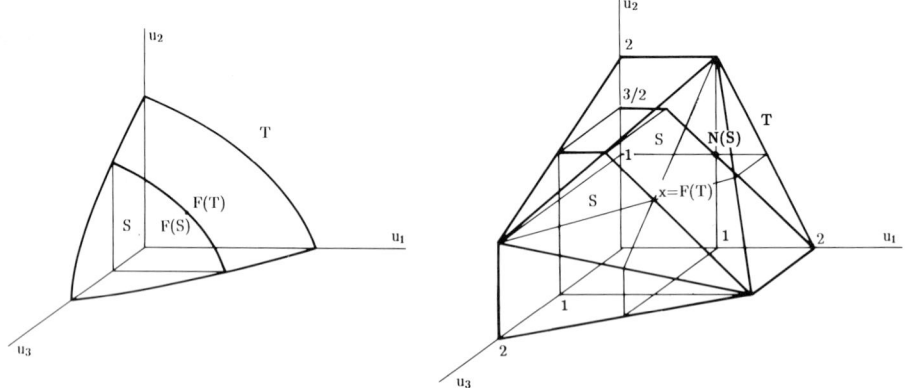

Figure 13. Consistency and the Nash solution. (a) The axiom of Consistency: the solution outcome of the "slice" of T by a plane parallel to the coordinate subspace relative to agents 1 and 2 through the solution outcome of T, $F(T)$, coincides with the restriction of $F(T)$ to that subspace. (b) Characterization of the Nash solution (Theorem 18).

that they will indeed receive their payoffs $x_{P'}$. Now, let us reevaluate the situation from the viewpoint of the group $P = Q \setminus P'$ of remaining agents. It is natural to think as the set $\{y \in \mathbb{R}^P | (y, x_{Q \setminus P}) \in T\}$ consisting of points of T at which the agents in P' receive the payoffs $x_{P'}$, as the feasible set for P. Let us denote it $t_P^x(T)$. Geometrically, $t_P^x(T)$ is the "slice" of T through x by a plane parallel to the coordinate subspace relative to the group P. If this set is a well-defined member of Σ_0^P, does the solution recommend the utilities x_P? If yes, and if this coincidence always occurs, the solution is *Consistent* [Figure 13(a)].

Consistency: Given $P, Q \in \mathcal{P}$ with $P \subset Q$, if $S \in \Sigma_0^P$ and $T \in \Sigma_0^Q$ are such that $S = t_P^x(T)$, where $x = F(T)$, then $x_P = F(S)$.

Consistency is satisfied by the Nash solution [Harsanyi (1959)] but not by the Kalai–Smorodinsky solution nor by the Egalitarian solution. Violations are usual for the Kalai–Smorodinsky solution but rare for the Egalitarian solution; indeed, on the class of strictly comprehensive problems, the Egalitarian solution does satisfy the condition, and if this restriction is not imposed, it still satisfies the slightly weaker condition obtained by requiring $x_P \leqslant F(S)$ instead of $x_P = F(S)$. Let us call this weaker condition *Weak Consistency*. The Lexicographic Egalitarian solution satisfies *Consistency*.

By substituting this condition for *contraction independence* in Nash's classic theorem (Theorem 1) we obtain another characterization of the solution.

Theorem 18 [Lensberg (1988)]. *The Nash solution is the only solution on Σ_0 satisfying Pareto-Optimality, Anonymity, Scale Invariance, and Consistency.*

Proof [Figure 13(b)]. It is straightforward to see that N satisfies the four axioms. Conversely, let F be a solution on Σ_0 satisfying the four axioms. We only show that F coincides with N on Σ_0^P if $|P| = 2$. Let $S \in \Sigma_0^P$ be given. By *Scale Invariance*, we can assume that S is normalized so that $N(S) = (1, 1)$.

In a first step, we assume that $PO(S) \supset [(\frac{3}{2}, \frac{1}{2}), (\frac{1}{2}, \frac{3}{2})]$. Let $Q \in \mathcal{P}$ with $P \subset Q$ and $|Q| = 3$ be given. Without loss of generality, we take $P = \{1, 2\}$ and $Q = \{1, 2, 3\}$. [In Figure 13(b), $S \equiv \text{cch}\{(2, 0), (\frac{1}{2}, \frac{3}{2})\}$]. Now, we translate S by the third unit vector, we replicate the result twice by having agents 2, 3 and 1, and then agents 3, 1 and 2 play the roles of agents 1, 2 and 3 respectively; finally, we define $T \in \Sigma_0^Q$ to be the convex and comprehensive hull of the three sets so obtained. Since $T = \text{cch}\{(1, 2, 0), (0, 1, 2), (2, 0, 1)\}$ is invariant under a rotation of the agents, by *Anonymity*, $F(T)$ has equal coordinates, and by *Pareto-Optimality*, $F(T) = (1, 1, 1)$. But, since $t_P^{(1, 1, 1)}(T) = S$, *Consistency* gives $F(S) = (1, 1) = N(S)$, and we are done.

In a second step, we only assume that $PO(S)$ contains a non-degenerate segment centered at $N(S)$. Then, we may have to introduce more than one additional agent and repeat the same construction by replicating the problem faced by agents 1 and 2 many times. If the order of replication is sufficiently large, the resulting T

is indeed such that $t_P^{(1,\ldots,1)}(T) = S$ and we conclude as before. If S does not contain a non-degenerate segment centered at $N(S)$, a continuity argument is required. □

The above proof requires having access to groups of arbitrarily large cardinalities, but the Nash solution can still be characterized when the number of potential agents is bounded above, by adding *Continuity* [Lensberg (1988)]. Unfortunately, two problems may be close in the Hausdorff topology and yet sections of those problems through two points that are close by, parallel to a given coordinate subspace, may not be close to each other: Hausdorff continuity ignores slices whereas slices are central to *Consistency*. A weaker notion of continuity recognizing this possibility can however be used to obtain a characterization of the Nash solution, even if the number of potential agents is bounded above [Thomson (1985b)]. Just as in the classic characterization of the Nash solution, *Pareto-Optimality* turns out to play a very minor role here: without it, the only additional admissible solution is the disagreement solution [Lensberg and Thomson (1988)].

Deleting *Symmetry* and *Scale Invariance* from Theorem 18, the following solutions become admissible: For each $i \in \mathbb{N}$, let $f_i: \mathbb{R}_+ \to \mathbb{R}$ be an increasing function such that for each $P \in \mathscr{P}$, the function $f^P: \mathbb{R}_+^P \to \mathbb{R}$ defined by $f^P(x) = \Sigma_{i \in P} f_i(x)$ be strictly quasi-concave. Then, given $P \in \mathscr{P}$ and $S \in \Sigma_0^P$, $F^f(S) \equiv \operatorname{argmax}\{f^P(x) | x \in S\}$. These *separable additive* solutions F^f are the only ones to satisfy *Pareto-Optimality Continuity* and *Consistency* [Lensberg (1987), Young (1988) proves a variant of this result].

6.2. Population monotonicity and the Kalai–Smorodinsky solution

Instead of allowing some of the agents to depart with their payoffs, we will now imagine them to leave empty-handed, without their departure affecting the oppor-

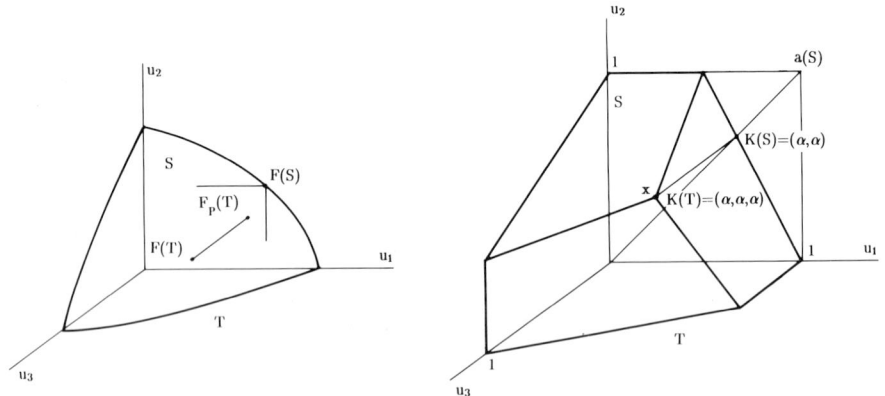

Figure 14. Population Monotonicity and the Kalai–Smorodinsky solution. (a) The axiom of Population Monotonicity: the projection of the solution outcome of the problem T onto the coordinate subspace pertaining to agents 1 and 2 is dominated by the solution outcome of the intersection of T with that coordinate subspace. (b) Characterization of the Kalai–Smorodinsky solution (Theorem 19) on the basis of Population Monotonicity.

tunities of the agents that remain. Do all of these remaining agents gain? If yes, the solution will be said to satisfy *Population Monotonicity*. Conversely and somewhat more concretely, think of a group of agents $P \in \mathscr{P}$ dividing goods on which they have equal rights. Then new agents come in, who are recognized to have equally valid rights. This requires that the goods be redivided. *Population Monotonicity* says that none of the agents initially present should gain. Geometrically [Figure 14(a)], this means that the projection of the solution outcome of the problem involving the large group onto the coordinate subspace pertaining to the smaller group of remaining agents is Pareto-dominated by the solution outcome of the intersection of the large problem with that subspace.

Population Monotonicity: Given $P, Q \in \mathscr{P}$ with $P \subset Q$, if $S \in \Sigma_0^P$ and $T \in \Sigma_0^Q$ are such that $S = T_P$, then $F(S) \geqslant F_P(T)$.

The Nash solution does not satisfy this requirement but both the Kalai–Smorodinsky and Egalitarian solutions do: In fact, characterizations of these two solutions can be obtained with the help of this condition.

Theorem 19 [Thomson (1983c)]. *The Kalai–Smorodinsky solution is the only solution on Σ_0 satisfying Weak Pareto-Optimality, Anonymity, Scale Invariance, Continuity, and Population Monotonicity.*

Proof [Figure 14(b)]. It is straightforward to see that K satisfies the five axioms. Conversely, let F be a solution on Σ_0 satisfying the five axioms. We only show that F coincides with K on Σ_0^P if $|P| = 2$. So let $S \in \Sigma_0^P$ be given. By *Scale Invariance*, we can assume that S is normalized so that $a(S)$ has equal coordinates. Let $Q \in \mathscr{P}$ with $P \subset Q$ and $|Q| = 3$ be given. Without loss of generality, we take $P = \{1, 2\}$ and $Q = \{1, 2, 3\}$. (In the figure $S = \text{cch}\{(1, 0), (\frac{1}{2}, 1)\}$ so that $a(S) = (1, 1)$.) Now, we construct $T \in \Sigma_0^Q$ by replicating S in the coordinates subspaces $\mathbb{R}^{\{2,3\}}$ and $\mathbb{R}^{\{3,1\}}$, and taking the comprehensive hull of S, its two replicas and the point $x \in \mathbb{R}^Q$ of coordinates all equal to the common value of the coordinates of $K(S)$. Since T is invariant under a rotation of the agents and $x \in PO(T)$, it follows from *Anonymity* and *Weak Pareto-Optimality* that $x = F(T)$. Now, note that $T_P = S$ and $x_P = K(S)$ so that by *Population Monotonicity*, $F(S) \geqslant K(S)$. Since $|P| = 2$, $K(S) \in PO(S)$ and equality holds.

To prove that F and K coincide for problems of cardinality greater than 2, one has to introduce more agents and *Continuity* becomes necessary. □

Solutions in the spirit of the solutions E^α described after Theorem 20 below satisfy all of the axioms of Theorem 19 except for *Weak Pareto-Optimality*. Without *Anonymity*, we obtain certain generalizations of the Weighted Kalai–Smorodinsky solutions [Thomson (1983a)]. For a clarification of the role of *Scale Invariance*, see Theorem 20.

6.3. Population monotonicity and the Egalitarian solution

All of the axioms used in the next theorem have already been discussed. Note that the theorem differs from the previous one only in that *Contraction Independence* is used instead of *Scale Invariance*.

Theorem 20 [Thomson (1983d)]. *The Egalitarian solution is the only solution on Σ_0 satisfying Pareto-Optimality, Symmetry, Contraction Independence, Continuity, and Population Monotonicity.*

Proof. It is easy to verify that E satisfies the five axioms [see Figure 15(a) for *Population Monotonicity*]. Conversely, let F be a solution on Σ_0 satisfying the five axioms. To see that $F = E$, let $P \in \mathcal{P}$ and $S \in \Sigma_0^P$ be given. Without loss of generality, suppose $E(S) = (1, \ldots, 1)$ and let $\beta \equiv \max\{\sum_{i \in P} x_i | x \in S\}$. Now, let $Q \in \mathcal{P}$ be such that $P \subset Q$ and $|Q| \geqslant \beta + 1$; finally, let $T \in \Sigma_0^Q$ be defined by $T \equiv \{x \in \mathbb{R}_+^Q | \sum_{i \in Q} x_i \leqslant |Q|\}$. [In Figure 15(b), $P = \{1, 2\}$ and $Q = \{1, 2, 3\}$.] By *Weak Pareto-Optimality* and *Symmetry*, $F(T) = (1, \ldots, 1)$. Now, let $T' \equiv \text{cch}\{S, F(T)\}$. Since $T' \subset T$ and $F(T) \in T'$, it follows from *Contraction Independence* that $F(T') = F(T)$. Now, $T'_P = S$, so that by *Population Monotonicity*, $F(S) = F_P(T') = E(S)$. If $E(S) \in PO(S)$ we are done. Otherwise we conclude by *Continuity*. □

Without *Weak Pareto-Optimality*, the following family of *truncated Egalitarian solutions* become admissible: let $\alpha \equiv \{\alpha^P | P \in \mathcal{P}\}$ be a list of non-negative numbers such that for all $P, Q \in \mathcal{P}$ with $P \subset Q$, $\alpha^P \geqslant \alpha^Q$; then, given $P \in \mathcal{P}$ and $S \in \Sigma_0^P$, let $E^\alpha(S) \equiv \alpha^P(1, \ldots, 1)$ if these points belongs to S and $E^\alpha(S) = E(S)$ otherwise [Thomson (1984b)]. The Monotone Path solutions encountered earlier, appropriately gene-

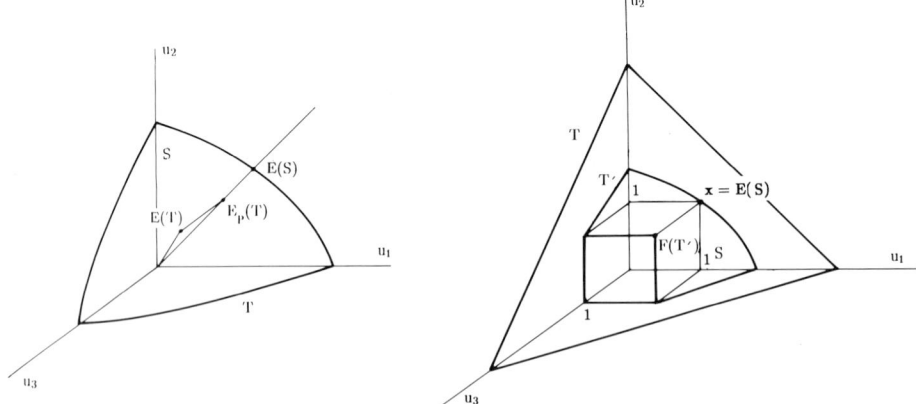

Figure 15. Population Monotonicity and the Egalitarian solution. (a) The Egalitarian solution satisfies Population Monotonicity. (b) Characterization of the Egalitarian solution on the basis of Population Monotonicity (Theorem 20).

ralized, satisfy all the axioms of Theorem 20, except for *Symmetry*: let $G \equiv \{G^P | P \in \mathscr{P}\}$ be a list of monotone paths such that $G^P \subset \mathbb{R}^P_+$ for all $P \in \mathscr{P}$ and for all $P, Q \in \mathscr{P}$ with $P \subset Q$, the projection of G^Q onto $\mathbb{R}^P$ be contained in G^P. Then, given $P \in \mathscr{P}$ and $S \in \Sigma_0^P$, $E^G(S)$ is the maximal point of S along the path G^P [Thomson (1983a, 1984b)].

6.4. Other implications of consistency and population monotonicity

The next result involves considerations of both *Consistency* and *Population Monotonicity*.

Theorem 21 [Thomson (1984c)]. *The Egalitarian solution is the only solution on Σ_0 satisfying Weak Pareto-Optimality, Symmetry, Continuity, Population Monotonicity and Weak Consistency.*

In order to recover full optimality, the extension of *individual monotonicity* to the variable population case can be used.

Theorem 22 [Lensberg (1985a)(1985b)]. *The Lexicographic Egalitarian solution is the only solution on Σ_0 satisfying Pareto-Optimality, Symmetry, Individual Monotonicity, and Consistency.*

6.5. Opportunities and guarantees

Consider a solution F satisfying *Weak Pareto-Optimality*. When new agents come in without opportunities enlarging, as described in the hypotheses of *Population Monotonicity*, one of the agents originally present will lose.

We propose here a way of quantifying these losses and of ranking solutions on the basis of the extent to which they prevent agents from losing too much.

Formally, let $P, Q \in \mathscr{P}$ with $P \subset Q$, $S \in \Sigma_0^P$, and $T \in \Sigma_0^Q$ with $S = T_P$. Given $i \in P$, consider the ratio $F_i(T)/F_i(S)$ of agent i's final to initial utilities: let $\alpha_F^{(i,P,Q)} \in \mathbb{R}$ be the greatest number α such that $F_i(T)/F_i(S) > \alpha$ for all S, T as just described. This is the *guarantee offered to i by F when he is initially part of P and P expands to Q*: agent i's final utility is guaranteed to be at least $\alpha_F^{(i,P,Q)}$ times his initial utility. If F satisfies *Anonymity*, then the number depends only on the cardinalities of P and $Q \setminus P$, denoted m and n respectively, and we can write it as α_F^{mn}:

$$\alpha_F^{mn} \equiv \inf \left\{ \frac{F_i(T)}{F_i(S)} \,\middle|\, S \in \Sigma_0^P, T \in \Sigma_0^Q, P \subset Q, S = T_P, |P| = m, |Q \setminus P| = n \right\}.$$

We call the list $\alpha_F \equiv \{\alpha_F^{mn} | m, n \in \mathbb{N}\}$ the *guarantee structure of F*.

We now proceed to compare solutions on the basis of their guarantee structures. Solutions offering greater guarantees are of course preferable. The next theorem says that the Kalai–Smorodinsky is the best from the viewpoint of guarantees. In particular, it is strictly better than the Nash solution.

Theorem 23 [Thomson and Lensberg (1983)]. *The guarantee structure α_K of the Kalai–Smorodinsky solution is given by $\alpha_K^{mn} = 1/(n+1)$ for all $m, n \in \mathbb{N}$. If F satisfies Weak Pareto-Optimality and Anonymity, then $\alpha_K \geqslant \alpha_F$. In particular, $\alpha_K \geqslant \alpha_N$.*

Note that solutions could be compared in other ways. In particular, protecting individuals may be costly to the group to which they belong. To analyze the trade-off between protection of individuals and protection of groups, we introduce the coefficient

$$\beta_F^{mn} \equiv \inf\{\Sigma_{i\in P}[F_i(T)/F_i(S)] \,|\, S \in \Sigma_0^P, T \in \Sigma_0^Q, P \subset Q, S = T_P, |P| = m, |Q \setminus P| = n\},$$

and we define $\beta_F \equiv \{\beta_F^{mn} | m, n \in \mathbb{N}\}$ as the *collective guarantee structure of F*. Using this notion, we find that our earlier ranking of the Kalai–Smorodinsky and Nash solutions is reversed.

Theorem 24 [Thomson (1983b)]. *The collective guarantee structure β_N of the Nash solution is given by $\beta_N^{mn} = n/(n+1)$ for all $m, n \in \mathbb{N}$. If F satisfies Weak Pareto-Optimality and Anonymity, then $\beta_N \geqslant \beta_F$. In particular, $\beta_N \geqslant \beta_K$.*

Theorem 23 says that the Kalai–Smorodinsky solution is best in a large class of solutions. However, it is not the only one to offer maximal guarantees and to satisfy *Scale Invariance and Continuity* [Thomson and Lensberg (1983)]. Similarly, the Nash solution is not the only one to offer maximal collective guarantees and to satisfy *Scale Invariance and Continuity* [Thomson (1983b)].

Solutions can alternatively be compared on the basis of the opportunities for gains that they offer to individuals (and to groups). Solutions that limit the extent to which individuals (or groups) can gain in spite of the fact that there may be more agents around while opportunities have not enlarged, may be deemed preferable. Once again, the Kalai–Smorodinsky solution performs better than any solution satisfying *Weak Pareto-Optimality* and *Anonymity* when the focus is on a single individual, but the Nash solution is preferable when groups are considered. However, the rankings obtained here are less discriminating [Thomson (1987b)].

Finally, we compare agent i's percentage loss $F_i(T)/F_i(S)$ to agent j's percentage loss $F_j(T)/F_j(S)$, where both i and j are part of the initial group P: let

$$\varepsilon_F^{mn} \equiv \inf\left\{\frac{F_j(T)/F_j(S)}{F_i(T)/F_i(S)} \,\bigg|\, S \in \Sigma_0^P, T \in \Sigma_0^Q, P \subset Q, S = T_P, |P| = m, |Q \setminus P| = n\right\},$$

and $\varepsilon_F \equiv \{\varepsilon_F^{mn} | (m, n) \in (\mathbb{N} \setminus 1) \times \mathbb{N}\}$. Here, we would of course prefer solutions that prevent agents from being too differently affected. Again, the Kalai–Smorodinsky solution performs the best from this viewpoint.

Theorem 25 [Chun and Thomson (1989)]. *The relative guarantee structures ε_K and ε_E of the Kalai–Smorodinsky and Egalitarian solutions are given by $\varepsilon_K^{mn} = \varepsilon_E^{mn} = 1$ for all $(m,n) \in (\mathbb{N}\setminus 1) \times \mathbb{N}$. The Kalai–Smorodinsky solution is the only solution on Σ_0 to satisfy Weak Pareto-Optimality, Anonymity, Scale Invariance and to offer maximal relative guarantees. The Egalitarian solution is the only solution on Σ_0 to satisfy Weak Pareto-Optimality, Anonymity, Contraction Independence and to offer maximal relative guarantees.*

6.6. Replication and juxtaposition

Now, we consider the somewhat more special situations where the preferences of the new agents are required to bear some simple relation to those of the agents originally present, such as when they are exactly opposed or exactly in agreement. There are several ways in which opposition or agreement of preferences can be formalized. And to each such formulation corresponds a natural way of writing that a solution respects the special structure of preferences.

Given a group P of agents facing the problem $S \in \Sigma_0^P$, introduce for each $i \in P, n_i$ additional agents "of the same type" as i and let Q be the enlarged group. Given any group P' with the same composition as P [we write comp(P') = comp(P)], define the problem $S^{P'}$ faced by P' to be the copy of S in $\mathbb{R}^{P'}$ obtained by having each member of P' play the role played in S by the agent in P of whose type he is. Then, to construct the problem T faced by Q, we consider two extreme possibilities. One case formalizes a situation of maximal compatibility of interests among all the agents of a given type:

$$S^{\max} \equiv \cap \{S^{P'} \times \mathbb{R}^{Q\setminus P'} | P' \subset Q, \text{comp}(P') = \text{comp}(P)\}.$$

The other formalizes the opposite; a situation of minimal compatibility of

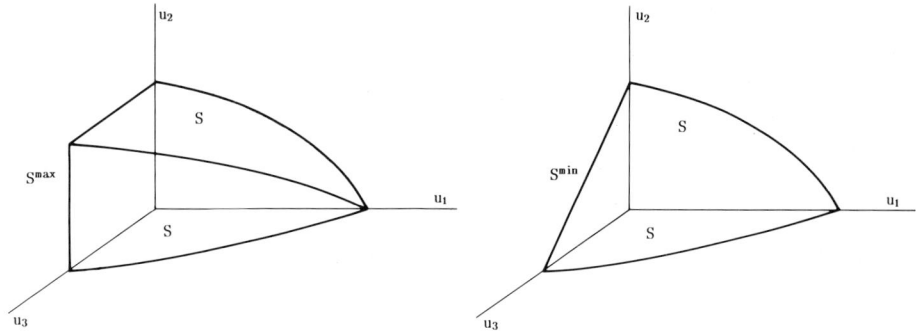

Figure 16. Two notions of replication. (a) Maximal compatibility of interest. (b) Minimal compatibility of interest.

interests,

$$S^{\min} \equiv \operatorname{cch}\{S^{P'} | P' \subset Q, \operatorname{comp}(P') = \operatorname{comp}(P)\}.$$

These two notions are illustrated in Figure 16 for an initial group of 2 agents (agents 1 and 2) and one additional agent (agent 3) being introduced to replicate agent 2.

Theorem 26 [based on Kalai (1977a)]. *In $S^{\max}$, all of the agents of a given type receive what the agent they are replicating receives in S if either the Kalai–Smorodinsky solution or the Egalitarian solution is used. However, if the Nash solution is used, all of the agents of a given type receive what the agent they are replicating would have received in S under the application of the weighted Nash solution with weights proportional to the orders of replication of the different types.*

Theorem 27 [Thomson (1984a, 1986)]. *In $S^{\min}$, the sum of what the agents of a given type receive under the replication of the Nash, Kalai–Smorodinsky, and Egalitarian solutions, is equal to what the agent they are replicating receives in S under the application of the corresponding weighted solution for weights proportional to the orders of replication.*

7. Applications to economics

Solutions to abstract bargaining problems, most notably the Nash solution, have been used to solve concrete economic problems, such as management-labor conflicts, on numerous occasions; in such applications, S is the image in utility space of the possible divisions of a firm's profit, and d the image of a strike. Problems of fair division have also been analyzed in that way; given some bundle of infinitely divisible goods $\Omega \in \mathbb{R}_+^l$, S is the image in utility space of the set of possible distributions of Ω, and d is the image of the 0 allocation (perhaps, of equal division). Alternatively, each agent may start out with a share of Ω, his individual endowment, and choosing d to be the image of the initial allocation may be more appropriate.

Under standard assumptions on utility functions, the resulting problem (S, d) satisfies the properties typically required of admissible problems in the axiomatic theory of bargaining. Conversely, given $S \in \Sigma_0^n$, it is possible to find exchange economies whose associated feasible set is S [Billera and Bixby (1973)].

When concrete information about the physical alternatives is available, it is natural to use it in the formulation of properties of solutions. For instance, expansions in the feasible set may be the result of increases in resources or improvements in technologies. The counterpart of *strong monotonicity*, (which says that such an expansion would benefit all agents) would be that all agents benefit from greater resources or better technologies. How well-behaved are solutions on this domain? The answer is that when there is only one good, solutions are better behaved than on abstract domains, but as soon as the number of goods is greater than 1, the same behavior should be expected of solutions on both domains [Chun and Thomson (1988a)].

The axiomatic study of solutions to concrete allocation problems is currently an active area of research. Many of the axioms that have been found useful in the abstract theory of bargaining have now been transposed for this domain and their implications analyzed. Early results along those lines are characterizations of the Walrasian solution [Binmore (1987)] and of Egalitarian-type solutions [Roemer (1986a,b, 1988) and Nieto (1992)]. For a recent contribution, see Klemisch-Ahlert and Peters (1994).

8. Strategic considerations

Analyzing a problem (S, d) as a strategic game requires additional structure: strategy spaces and an outcome function have somehow to be associated with (S, d). This can be done in a variety of ways. We limit ourselves to describing formulations that remain close to the abstract model of the axiomatic theory and this brief section is only meant to facilitate the transition to chapters in this Handbook devoted to strategic models.

Consider the following game: each agent demands a utility level for himself; the outcome is the vector of demands if it is in S and d otherwise. The set of Nash (1953) equilibrium outcomes of this *game of demands* is $PO(S) \cap I(S, d)$ (to which should be added d if $PO(S) \cap I(S, d) = WPO(S) \cap I(S, d)$), a typically large set, so that this approach does not help in reducing the set of outcomes significantly. However, if S is known only approximately (replace its characteristic function by a *smooth* function), then as the degree of approximation increases, the set of equilibrium outcomes of the resulting *smoothed game of demands* shrinks to $N(S, d)$ [Nash (1950), Harsanyi (1956), Zeuthen (1930), Crawford (1980), Anbar and Kalai (1978), Binmore (1987), Anbarci (1992, 1993a), Calvo and Gutiérrez (1994)].

If bargaining takes place over time, agents take time to prepare and communicate proposals, and the worth of an agreement reached in the future is discounted, a *sequential game of demands* results. Its equilibria (here some perfection notion has to be used) can be characterized in terms of the weighted Nash solutions when the time period becomes small: it is $N^\delta(S, d)$ where δ is a vector related in a simple way to the agents' discount rates [Rubinstein (1982), Binmore (1987), see Chapter 7 for an extensive analysis of this model. Livne (1987) contains an axiomatic treatment].

Imagine now that agents have to justify their demands: there is a family $\mathscr{F}$ of "resonable" solutions such that agent i can demand $\bar{u}_i$ only if $\bar{u}_i = F_i(S, d)$ for some $F \in \mathscr{F}$. Then strategies are in fact elements of $\mathscr{F}$. Let F^1 and F^2 be the strategies chosen by agents 1 and 2. If $F^1(S, d)$ and $F^2(S, d)$ differ, eliminate from S all alternatives at which agent 1 gets more than $F_1^1(S,d)$ and agent 2 gets more than $F_2^2(S,d)$; one could argue that the truncated set S^1 is the relevant set over which to bargain; so repeat the procedure: compute $F^1(S^1,d)$ and $F^2(S^1,d)\ldots$ If, as $v \to \infty$, $F^1(S^v,d)$ and $F^2(S^v,d)$ converge to a common point, take that as the solution outcome of this induced *game of solutions*. For natural families $\mathscr{F}$, convergence does occur for all F^1 and $F^2 \in \mathscr{F}$, and the only equilibrium outcome of the game so defined is $N(S,d)$ [van Damme (1986); Chun (1984) studies a variant of the procedure].

Thinking now of solutions as normative criteria, note that in order to compute the desired outcomes, the utility functions of the agents will be necessary. Since these functions are typically unobservable, there arises the issue of *manipulation*. To the procedure is associated a game of misrepresentation, where strategies are utility functions. What are its equilibria? In the game so associated with the Nash solution when applied to a one-dimensional division problem, each agent has a dominant strategy, which is to pretend that his utility function is linear. The resulting outcome is equal division [Crawford and Varian (1979)]. If there is more than one good and preferences are known ordinally, a dominant strategy for each agent is a least concave representation of his preferences [Kannai (1977)]. When there are an increasing number of agents, only one of whom manipulates, the gain that he can achieve by manipulation does not go to zero although the impact on each of the others vanishes; only the first of these conclusions holds, however, when it is the Kalai-Smorodinsky solution that is being used [Thomson (1994)]. In the multi-commodity case, Walrasian allocations are obtained at equilibria, but there are others [Sobel (1981), Thomson (1984d)].

Rather than feeding in agents' utility functions directly, one could of course think of games explicitly designed so as to take strategic behavior into account. Supposing that some solution has been selected as embodying society's objectives, does there exist a game whose equilibrium outcome always yields the desired utility allocations? If yes, the solution is *implementable*. The Kalai-Smorodinsky solution is implementable by stage games [Moulin (1984)]. Other implementation results are from Anbarci (1990) and Bossert and Tan (1992). Howard (1992) establishes the implementability of the Nash solution.

List of solutions

Dictatorial
Egalitarian
Equal Loss
Equal Area
Kalai-Rosenthal
Kalai-Smorodinsky
Lexicographic Egalitarian

Nash
Perles-Maschler
Raiffa
Utilitarian
Yu

List of axioms

- adding
- anonymity
- concavity
- consistency
- continuity
- contraction independence
- converse consistency
- cutting
- decomposability
- disagreement point concavity
- disagreement point monotonicity
- domination
- independence of non individually rational alternatives
- individual monotonicity
- individual rationality
- linearity
- midpoint domination
- ordinal invariance
- pareto-optimality
- population monotonicity
- restricted monotonicity
- risk sensitivity
- scale invariance
- separability
- star-shaped inverse
- strong individual rationality
- strong midpoint domination
- strong monotonicity
- strong disagreement point monotonicity
- strong risk sensitivity
- superadditivity
- symmetry
- twisting
- weak contraction independence
- weak disagreement point
- weak linearity
- weak ordinal invariance
- weak pareto-optimality

Bibliography

Anant, T.C.A., K. Basu and B. Mukherji (1990) 'Bargaining Without Convexity: Generalizing the Kalai–Smorodinsky solution', *Economics Letters*, **33**: 115–119.

Anbar, D. and E. Kalai (1978) 'A One-Shot Bargaining Problem', *International Journal of Game Theory*, **7**: 13–18.

Anbarci, N. (1988) 'The Essays in Axiomatic and Non-Cooperative Bargaining Theory', Ph.D. Thesis, University of Iowa.

Anbarci, N. (1989) 'The Kalai–Smorodinsky Solution with Time Preferences', *Economics Letters*, **31**: 5–7.

Anbarci, N. (1991) 'The Nash Solution and Relevant Expansions', *Economics Letters*, **36**: 137–140.

Anbarci, N. (1992) 'Final-offer arbitration and Nash's demand game', University of Buffalo Discussion Paper.

Anbarci, N. (1993a) 'Bargaining with finite number of alternatives: strategic and axiomatic approaches', University of Buffalo Discussion Paper.

Anbarci, N. (1993b) 'Non-Cooperative Foundations of the Area Monotonic Solution', *Quarterly Journal of Economics*, **108**: 245–258.

Anbarci, N. and F. Bigelow (1988) 'The Area Monotonic Solution to the Cooperative Bargaining Problem', Working Paper, University of Missouri; *Mathematical Social Sciences*, forthcoming.

Anbarci, N. and F. Bigelow (1993) 'Non-dictatorial, Pareto-monotonic, Cooperative Bargaining: an Impossibility Result', *European Journal of Political Economy*, **9**: 551–558.

Anbarci, N. and G. Yi (1992) 'A Meta-Allocation Mechanism in Cooperative Bargaining', *Economics Letters*, **20**: 176–179.

Arrow, K. (1965) *Aspects of the Theory of Risk-Bearing*, Yrjö Jahnsson Foundation, Helsinki.

Billera, L.F. and R.E. Bixby (1973) 'A Characterization of Pareto Surface', *Proceedings of the American Mathematical Society*, **41**: 261–267.

Binmore, K.G. (1984) 'Bargaining Conventions', *International Journal of Game Theory*, **13**: 193–200.

Binmore, K.G. (1987) 'Nash Bargaining Theory I, II, III', in: K.G. Binmore and P. Dasgupta, eds., *The Economics of Bargaining*. Basil Blackwell.

Blackorby, C., W. Bossert and D. Donaldson (1992) 'Generalized Ginis and Cooperative Bargaining Solutions', University of Waterloo Working Paper 9205, forthcoming in *Econometrica*.

Bossert, W. (1990) 'Disagreement Point Monotonicity, Transfer Responsiveness, and the Egalitarian Bargaining Solution', mimeo.

Bossert, W. (1992a) 'Monotonic Solutions for Bargaining Problems with Claims', *Economics Letters*, **39**: 395–399.

Bossert, W. (1992b) 'Rationalizable Two-Person Bargaining Solutions', University of Waterloo mimeo.

Bossert, W. (1993) 'An alternative solution to two-person bargaining problems with claims', *Mathematical Social Sciences*, **25**: 205–220.

Bossert, W., E. Nosal and V. Sadanand (1992) 'Bargaining under Uncertainty and the Monotonic Path Solutions', University of Waterloo mimeo.

Bossert, W. and G. Tan (1992) 'A Strategic Justification of the Egalitarian Bargaining Solution', University of Waterloo Working paper 92–12.

Brito, D.L., A.M. Buoncristiani and M.D. Intriligator (1977) 'A New Approach to Nash's Bargaining Problem', *Econometrica*, **45**: 1163–1172.

Brunner, J.K. (1992) 'Bargaining with Reasonable Aspirations', University of Linz mimeo.

Butrim, B.I. (1976) 'A Modified Solution of the Bargaining Problem', (in Russian), *Zhurnal Vychistitel'noi Matematiki Matematicheskoi Fiziki*, **16**: 340–350.

Calvo, E. (1989) 'Solucion de Alternativea No Dominadas Iguales', University of Bilbao Doctoral Dissertation.

Calvo, E. and E. Gutiérrez (1993) 'Extension of the Perles–Maschler Solution to N-Person Bargaining Games', University of Bilbao Discussion Paper.

Calvo, E. and E. Gutiérrez (1994) 'Comparison of bargaining solutions by means of altruism indexes', University of Bilbao Discussion Paper.

Cao, X. (1981) 'Preference Functions and Bargaining Solutions', *IEEE* 164–171.

Chun, Y. (1984) 'Note on 'The Nash Bargaining Solution is Optimal'', University of Rochester, Mimeo.

Chun, Y. (1986) 'The Solidarity Axiom for Quasi-Linear Social Choice Problems', *Social Choice and Welfare*, **3**: 297–310.

Chun, Y. (1987a) 'The Role of Uncertain Disagreement Points in 2-Person Bargaining', Discussion Paper No. 87–17, Southern Illinois University.

Chun, Y. (1987b) 'Axioms Concerning Uncertain Disagreement Points for 2-Person Bargaining Problems', Working Paper No. 99, University of Rochester.

Chun, Y. (1988a) 'The Equal-Loss Principle for Bargaining Problems', *Economics Letters*, **26**: 103–106.

Chun, Y. (1988b) 'Nash Solution and Timing of Bargaining', *Economics Letters*, **28**: 27–31.

Chun, Y. (1989) 'Lexicographic Egalitarian Solution and Uncertainty in the Disagreement Point', *Zeitschrift für Operations Research*, **33**: 259–306.

Chun, Y. (1990) 'Minimal Cooperation in Bargaining', *Economic Letters*, **34**: 311–316.

Chun, Y. (1991) 'Transfer Paradox and Bargaining Solutions', Seoul National University Mimeo.

Chun, Y. and H.J.M. Peters (1988) 'The Lexicographic Egalitarian Solution', *Cahiers du CERO*, **30**: 149–156.

Chun, Y. and H.J.M. Peters (1989) 'Lexicographic Monotone Path Solutions', *O.R. Spektrum*, **11**: 43–47.

Chun, Y. and H.J.M. Peters (1991) 'The Lexicographic Equal-Loss Solution', *Mathematical Social Science*, **22**: 151–161.

Chun, Y. and W. Thomson (1988) 'Monotonicity Properties of Bargaining Solutions When Applied to Economics', *Mathematical Social Sciences*, **15**: 11–27.

Chun, Y. and W. Thomson (1989) 'Bargaining Solutions and Stability of Groups', *Mathematical Social Sciences*, **17**: 285–295.

Chun, Y. and W. Thomson (1990a) 'Bargaining Problems with Uncertain Disagreement Points, *Econometrica*, **33**: 29–33.

Chun, Y. and W. Thomson (1990b) 'Egalitarian Solution and Uncertain Disagreement Points', *Economics Letters*, **33**: 29–33.

Chun, Y. and W. Thomson (1990c) 'Nash Solution and Uncertain Disagreement Points', *Games and Economic Behavior*, **2**: 213–223.

Chun, Y. and W. Thomson (1992) 'Bargaining Problems With Claims', *Mathematical Social Sciences*, **24**: 19–33.

Conley, J., R. McLean and S. Wilkie (1994) 'The Duality of Bargaining Theory and Multi-Objective Programming, and Generalized Choice Problems', Bellcore mimeo.
Conley, J. and S. Wilkie (1991a) 'Bargaining Theory Without Convexity', *Economics Letters*, **36**: 365–369.
Conley, J. and S. Wilkie (1991b) 'An Extension of the Nash Bargaining Solution to Non-Convex Problems: Characterization and Implementation', University of Illinois mimeo.
Crawford, V. (1980) 'A Note on the Zeuthen-Harsanyi Theory of Bargaining', *Journal of Conflict Resolution*, **24**: 525–535.
Crawford, V. and H. Varian (1979) 'Distortion of Preferences and the Nash Theory of Bargaining', *Economics Letters*, **3**: 203–206.
Crott, H.W. (1971) Experimentelle Untersuchung zum Verhandlungs verhaltern in Kooperativen Spielen, *Zeitschrift für Sozialpsychologie*, **2**: 61–74.
van Damme, E. (1986) 'The Nash Bargaining Solution is Optimal', *Journal of Economic Theory*, **38**: 78–100.
Dasgupta, P. and E. Maskin (1988) 'Bargaining and Destructive Powers', mimeo.
Dekel, E. (1982) Harvard University, untitled mimeo.
Edgeworth, F.Y. (1881) *Mathematical Physics*. London: C. Kegan Paul and Co.
Freimer M. and P.L. Yu (1976) 'Some New Results on Compromise Solutions for Group Decision Problems', *Management Science*, **22**: 688–693.
Furth, D. (1990) 'Solving Bargaining Games by Differential Equations', *Mathematics of Operations Research*, **15**: 724–735.
Gaertner, W. and M. Klemisch-Ahlert (1991) 'Gauthier's Approach to Distributive Justice and Other Bargaining Solutions', in: P. Vallentyne, ed., *Contractarianism and Rational Choice*. Cambridge University Press, pp. 162–176.
Green, J. (1983) 'A Theory of Bargaining with Monetary Transfers', Harvard Institute of Economic Research Discussion Paper.
Green, J. and J.J. Laffont (1988) 'Contract Renegotiation and the Underinvestment Effect', Harvard University Mimeo.
Gupta, S. and Z.A. Livne (1988) 'Resolving a Conflict Situation with a Reference Outcome: An Axiomatic Model', *Management Science*, **34**: 1303–1314.
Harsanyi, J.C. (1955) 'Cardinal Welfare, Individualistic Ethics, and Interpersonal Comparisons of Utility', *Journal of Political Economy*, **63**: 309–321.
Harsanyi, J.C. (1956) 'Approaches to the Bargaining Problem Before and After the Theory of Games: A Critical Discussion of Zeuthen's, Hicks', and Nash's Theories', *Econometrica*, **24**: 144–157.
Harsanyi, J.C. (1958) 'Notes on the Bargaining Problem', *Southern Economic Journal*, **24**: 471–476.
Harsanyi, J.C. (1959) 'A Bargaining Model for the Cooperative n-Person Game', in: A.W. Tucker and R.d. Luce, eds., *Contributions to the Theory of Games*, IV, Annals of Mathematical Studies No. 40. Princeton, NJ: Princeton University Press.
Harsanyi, J.C. (1961) 'On the Rationality Postulates Underlying the Theory of Cooperative Games', *Journal of Conflict Resolution*, **5**: 179–196.
Harsanyi, J.C. (1963) 'A Simplified Bargaining Model for the *n*-person Cooperative Game', *International Economic Review*, **4**: 194–220.
Harsanyi, J.C. (1965) 'Bargaining and Conflict Situations in the Light of a New Approach to Game Theory', *American Economic Review*, **55**: 447–457.
Harsanyi, J.C. (1977) *Rational Behavior and Bargaining Equilibrium in Games and Social Situations*. Cambridge: Cambridge University Press.
Harsanyi, J.C. and R. Selten (1972) 'A Generalized Nash Solution for Two-Person Bargaining Games with Incomplete Information', *Management Science*, **18**: 80–106.
Heckathorn, D. (1980) 'A Unified Model for Bargaining and Conflict', *Behavioral Science*, **25**: 261–284.
Heckathorn, D. and R. Carlson (1983) 'A Possibility Result Concerning *n*-Person Bargaining Games', University of Missouri Technical Report 6.
Herrero, C. (1993) 'Endogenous Reference Points and the Adjusted Proportional Solution for Bargaining Problems with Claims', University of Alicante mimeo.
Herrero, C. and M.C. Marco (1993) 'Rational Equal-Loss Solutions for Bargaining Problems', *Mathematical Social Sciences*, **26**: 273–286.
Herrero, M.J. (1989) 'The Nash Program: Non-Convex Bargaining Problems', *Journal of Economic Theory*, **49**: 266–277.

Howe, R.E. (1987) 'Sections and Extensions of Concave Functions', *Journal of Mathematical Economics*, **16**: 53–64.

Huttel, G. and W.F. Richter (1980) 'A Note on "an Impossibility Result Concerning n-Person Bargaining Games', University of Bielefeld Working Paper 95.

Imai, H. (1983) 'Individual Monotonicity and Lexicographic Maxmin Solution', *Econometrica*, **51**: 389–401, erratum in *Econometrica*, **51**: 1603.

Iturbe, I. and J. Nieto (1992) 'Stable and Consistent Solutions on Economic Environments', mimeo.

Jansen, M.J.M. and S.H. Tijs (1983) 'Continuity of Bargaining Solutions', *International Journal of Game Theory*, **12**: 91–105.

Kalai, E. (1977a) 'Nonsymmetric Nash Solutions and Replications of Two-Person Bargaining', *International Journal of Game Theory*, **6**: 129–133.

Kalai, E. (1977b) 'Proportional Solutions to Bargaining Situations: Interpersonal Utility Comparisons', *Econometrica*, **45**: 1623–1630.

Kalai, E. (1985) 'Solutions to the Bargaining Problem', in: L. Hurwicz, D. Schmeidler and H. Sonnenschein, Eds, *Social Goals and Social Organization*, Essays in memory of E. Pazner. Cambridge University Press, pp. 77–105.

Kalai, E. and R.W. Rosenthal (1978) 'Arbitration of Two-Party Disputes Under Ignorance', *International Journal of Game Theory*, **7**: 65–72.

Kalai, E. and M. Smorodinsky (1975) 'Other Solutions to Nash's Bargaining Problem', *Econometrica*, **43**: 513–518.

Kannai, Y. (1977) 'Concaviability and Construction of Concave Utility Functions', *Journal of Mathematical Economics*, **4**: 1–56.

Kihlstrom, R.E. and A.E. Roth (1982) 'Risk Aversion and the Negotiation of Insurance Contracts', *Journal of Risk and Insurance*, **49**: 372–387.

Kihlstrom, R.E., A.E. Roth and D. Schmeidler (1981) 'Risk Aversion and Solutions to Nash's Bargaining Problem', in: O. Moeschlin and D. Pallaschke, eds., *Game Theory and Mathematical Economics*. North-Holland, pp. 65–71.

Klemisch-Ahlert, M. (1991) 'Independence of the Status Quo? A Weak and a Strong Impossibility Result for Social Decisions by Bargaining', *Journal of Economics*, **53**: 83–93.

Klemisch-Ahlert, M. (1992a) 'Distributive Justice and Risk Sensitivity', *Theory and Decision*, **32**: 302–318.

Klemisch-Ahlert, M. (1992b) 'Distributive Effects Implied by the Path Dependence of the Nash Bargaining Solution', University of Osnabruck mimeo.

Klemisch-Ahlert, M. (1993) 'Bargaining when Redistribution is Possible', University of Osnabrück Discussion Paper.

Kohlberg, E., M. Maschler and M.A. Perles (1983) 'Generalizing the Super-Additive Solution for Multiperson Unanimity Games', Discussion Paper.

de Koster, R., H.J.M. Peters, S.H. Tijs and P. Wakker (1983) 'Risk Sensitivity, Independence of Irrelevant Alternatives and Continuity of Bargaining Solutions', *Mathematical Social Sciences*, **4**: 295–300.

Lahiri, S. (1990) 'Threat Bargaining Games with a Variable Population', *International Journal of Game Theory*, **19**: 91–100.

Lensberg, T. (1985a) 'Stability, Collective Choice and Separable Welfare', Ph.D. Dissertation, Norwegian School of Economics and Business Administration, Bergen, Norway.

Lensberg, T. (1985b) 'Bargaining and Fair Allocation', in: H.P. Young, ed., *Cost Allocation: Methods, Principles, Applications*. North-Holland, pp. 101–116.

Lensberg, T. (1987) 'Stability and Collective Rationality', *Econometrica*, **55**: 935–961.

Lensberg, T. (1988) 'Stability and the Nash Solution', *Journal of Economic Theory*, **45**: 330–341.

Lensberg, T. and W. Thomson (1988) 'Characterizing the Nash Bargaining Solution Without Pareto-Optimality', *Social Choice and Welfare*, **5**: 247–259.

Livne, Z.A. (1986) 'The Bargaining Problem: Axioms Concerning Changes in the Conflict Point', *Economics Letters*, **21**: 131–134.

Livne, Z.A. (1987) 'Bargaining Over the Division of a Shrinking Pie: An Axiomatic Approach', *International Journal of Game Theory*, **16**: 223–242.

Livne, Z.A. (1988) 'The Bargaining Problem with an Uncertain Conflict Outcome', *Mathematical Social Sciences*, **15**: 287–302.

Livne, Z.A. (1989a) 'Axiomatic Characterizations of the Raiffa and the Kalai–Smorodinsky Solutions to the Bargaining Problem', *Operations Research*, **37**: 972–980.

Livne, Z.A. (1989b) 'On the Status Quo Sets Induced by the Raiffa Solution to the Two-Person Bargaining Problem', *Mathematics of Operations Research*, **14**: 688–692.

Luce, R.D. and H. Raiffa (1957) *Games and Decisions: Introduction and Critical Survey*. New York: Wiley.

Marco, M.C. (1993a) 'Efficient Solutions for Bargaining Problems with Claims', University of Alicante Discussion Paper.

Marco, M.C. (1993b) 'An Alternative Characterization of the Extended Claim-Egalitarian Solution', University of Alicante Discussion Paper.

Maschler, M. and M.A. Perles (1981) 'The Present Status of the Super-Additive Solution', in: R. Aumann et al. eds., *Essays in Game Theory and Mathematical Economics in honor of Oskar Morgenstern*, pp. 103–110.

McLennan, A. (198·) 'A General Noncooperative Theory of Bargaining', University of Toronto Discussion Paper.

Moulin, H. (1978) 'Implementing Efficient, Anonymous and Neutral Social Choice Functions', Mimeo.

Moulin, H. (1983) 'Le Choix Social Utilitariste', Ecole Polytechnique DP.

Moulin, H. (1984) 'Implementing the Kalai–Smorodinsky Bargaining Solution', *Journal of Economic Theory*, **33**: 32–45.

Moulin, H. (1988) *Axioms of Cooperative Decision Making*, Cambridge University Press.

Myerson, R.B. (1977) 'Two-Person Bargaining Problems and Comparable Utility', *Econometrica*, **45**: 1631–1637.

Myerson, R.B. (1981) 'Utilitarianism, Egalitarianism, and the Timing Effect in Social Choice Problems', *Econometrica*, **49**: 883–897.

Nash, J.F. (1950) 'The Bargaining Problem', *Econometrica*, **28**: 155–162.

Nash, J.F. (1953) 'Two-Person Cooperative Games', *Econometrica*, **21**: 129–140.

Nash, J.F. (1961) 'Noncooperative Games', *Annals of Mathematics*, **54**: 286–295.

Nielsen, L.T. (1983) 'Ordinal Interpersonal Comparisons in Bargaining', *Econometrica*, **51**: 219–221.

Nielsen, L.T. (1984) 'Risk Sensitivity in Bargaining with More Than Two Participants', *Journal of Economic Theory*, **32**: 371–376.

Nieto, J. (1992) 'The Lexicographic Egalitarian Solution on Economic Environments', *Social Choice and Welfare*, **9**: 203–212.

O'Neill, B. (1981) 'Comparison of Bargaining Solutions, Utilitarianism and the Minmax Rule by Their Effectiveness', Discussion Paper, Northwestern University.

Perles, M.A. (1982) 'Nonexistence of Super-Additive Solutions for 3-person Games', *International Journal of Game Theory*, **11**: 151–161.

Perles, M.A. (1994a) 'Non-Symmetric Super-Additive Solutions', forthcoming.

Perles, M.A. (1994b) 'Super-Additive Solutions Without the Continuity Axiom', forthcoming.

Perles, M.A. and M. Maschler (1981) 'A Super-Additive Solution for the Nash Bargaining Game', *International Journal of Game Theory*, **10**: 163–193.

Peters, H.J.M. (1986a) 'Simultaneity of Issues and Additivity in Bargaining', *Econometrica*, **54**: 153–169.

Peters, H.J.M. (1986b) *Bargaining Game Theory*, Ph.D. Thesis, Maastricht, The Nertherlands.

Peters, H.J.M. (1987a) 'Some Axiomatic Aspects of Bargaining', in: J.H.P. Paelinck and P.H. Vossen, eds., *Axiomatics and Pragmatics of Conflict Analysis*. Aldershot, UK: Gower Press, pp. 112–141.

Peters, H.J.M. (1987b) 'Nonsymmetric Nash Bargaining Solutions', in: H.J.M. Peters and O.J. Vrieze, eds., *Surveys in Game Theory and Related Topics*, CWI Tract 39, Amsterdam.

Peters, H.J.M. (1992) *Axiomatic Bargaining Game Theory*, Kluwer Academic Press.

Peters, H.J.M. and E. van Damme (1991) 'Characterizing the Nash and Raiffa Solutions by Disagreement Point Axioms', *Mathematics of Operations Research*, **16**: 447–461.

Peters, H.J.M. and M. Klemisch-Ahlert (1994) 'An Impossibility Result Concerning Distributive Justice in Axiomatic Bargaining', University of Limburg Discussion Paper.

Peters, H.J.M. and S. Tijs (1981) 'Risk Sensitivity of Bargaining Solutions', *Methods of Operations Research*, **44**: 409–420.

Peters, H.J.M. and S. Tijs (1983) 'Risk Properties of n-Person Bargaining Solutions', Discussion Paper 8323, Nijmegen University.

Peters, H.J.M. and S. Tijs (1984a) 'Individually Monotonic Bargaining Solutions for n-Person Bargaining Games', *Methods of Operations Research*, **51**: 377–384.

Peters, H.J.M. and S. Tijs (1984b) 'Probabilistic Bargaining Solutions', *Operations Research Proceedings*. Berlin: Springer-Verlag, pp. 548–556.

Peters, H.J.M. and S. Tijs (1985a) 'Risk Aversion in n-Person Bargaining', *Theory and Decision*, **18**: 47–72.

Peters, H.J.M. and S. Tijs (1985b) 'Characterization of All Individually Monotonic Bargaining Solutions', *International Journal of Game Theory*, **14**: 219–228.
Peters, H.J.M., S. Tijs and R. de Koster (1983) 'Solutions and Multisolutions for Bargaining Games', *Methods of Operations Research*, **46**: 465–476.
Peters, H.J.M. and P. Wakker (1991) 'Independence of Irrelevant Alternatives and Revealed Group Preferences'. *Econometrica*, **59**: 1781–1801.
Ponsati, C. and J. Watson (1994) 'Multiple Issue Bargaining and Axiomatic Solutions', Stanford University mimeo.
Pratt, J.W. (1964) 'Risk Aversion in the Small and the Large', *Econometrica*, **32**, 122–136.
Raiffa, H. (1953) 'Arbitration Schemes for Generalized Two-Person Games', in: H.W. Kuhn and A.W. Tucker, eds., *Contributions to the Theory of Games II*, Annals of Mathematics Studies, No 28. Princeton: Princeton University Press, pp. 361–387.
Ritz, Z. (1985) 'An Equal Sacrifice Solution to Nash's Bargaining Problem', University of Illinois, mimeo.
Roemer, J. (1986a) 'Equality of Resources Implies Equality of Welfare', *Quarterly Journal of Economics*, **101**: 751–784.
Roemer, J. (1986b) 'The Mismarriage of Bargaining Theory and Distributive Justice', *Ethics*, **97**: 88–110.
Roemer, J. (1988) 'Axiomatic Bargaining Theory on Economic Environments', *Journal of Economic Theory*, **45**: 1–31.
Roemer, J. (1990) 'Welfarism and Axiomatic Bargaining Theory', *Recherches Economiques de Louvain*, **56**: 287–301.
Rosenthal, R.W. (1976) 'An Arbitration Model for Normal-Form Games', *Mathematics of Operations Research*, **1**: 82–88.
Rosenthal, R.W. (1978) 'Arbitration of Two-Party Disputes Under Uncertainty', *Review of Economic Studies*, **45**: 595–604.
Roth, A.E. (1977a) 'Individual Rationality and Nash's Solution to the Bargaining Problem', *Mathematics of Operations Research*, **2**: 64–65.
Roth, A.E. (1977b) 'Independence of Irrelevant Alternatives, and Solutions to Nash's Bargaining Problem', *Journal of Economic Theory*, **16**: 247–251.
Roth, A.E. (1978) 'The Nash Solution and the Utility of Bargaining', *Econometrica*, **46**: 587–594, 983.
Roth, A.E. (1979a) 'Proportional Solutions to the Bargaining Problem', *Econometrica*, **47**: 775–778.
Roth, A.E. (1979b) 'Interpersonal Comparisons and Equal Gains in Bargaining', Mimeo.
Roth, A.E. (1979c) *Axiomatic Models of Bargaining*. Berlin and New York: Springer-Verlag, No. 170.
Roth, A.E. (1979d) 'An Impossibility Result Concerning n-Person Bargaining Games', *International Journal of Game Theory*, **8**: 129–132.
Roth, A.E. (1980) 'The Nash Solution as a Model of Rational Bargaining', in: A.V. Fiacco and K.O. Kortanek, eds., *Extremal Methods and Systems Analysis*, Springer-Verlag.
Roth, A.E. (1988) 'Risk Aversion in Multi-Person Bargaining Over Risky Agreements', University of Pittsburg, mimeo.
Roth, A.E. and U. Rothblum (1982) 'Risk Aversion and Nash's Solution for Bargaining Games with Risky Outcomes', *Econometrica*, **50**: 639–647.
Rubinstein, A. (1982) 'Perfect equilibrium in a Bargaining Model', *Econometrica*, **50**: 97–109.
Rubinstein, A, Z. Safra and W. Thomson (1992) 'On the Interpretation of the Nash Solution and its Extension to Non-expected Utility Preferences', *Econometrica*, **60**: 1172–1196.
Safra, Z., L. Zhou and I. Zilcha (1990) 'Risk Aversion in the Nash Bargaining Problem With Risky Outcomes and Risky Disagreement Points', *Econometrica*, **58**: 961–965.
Safra, Z. and I. Zilcha (1991) 'Bargaining Solutions without the Expected Utility Hypothesis', Tel Aviv University mimeo.
Salonen, H. (1985) 'A Solution for Two-Person Bargaining Problems', *Social Choice and Welfare*, **2**: 139–146.
Salonen, H. (1987) 'Partially Monotonic Bargaining Solutions', *Social Choice and Welfare*, **4**: 1–8.
Salonen, H. (1992) 'A Note on Continuity of Bargaining Solutions', University of Oulu mimeo.
Salonen, H. (1993) 'Egalitarian Solutions for N-Person Bargaining Games', University of Oulu Discussion Paper.
Salukvadze, M.E. (1971a) 'Optimization of Vector Functionals I. The Programming of Optimal Trajectories', *Automation and Remote Control*, **32**: 1169–1178.
Salukvadze, M.E. (1971b) 'Optimization of Vector Functionals II. The Analytic Constructions of Optimal Controls', *Automation and Remote Control*, **32**: 1347–1357.

Schelling, T.C. (1959) 'For the Abandonment of Symmetry in Game Theory', *Review of Economics and Statistics*, **41**: 213–224.

Schmitz, N. (1977) 'Two-Person Bargaining Without Threats: A Review Note', *Proceedings of the OR Symposium*, Aachen **29**, pp. 517–533.

Segal, U. (1980) 'The Monotonic Solution for the Bargaining Problem: A Note', Hebrew University of Jerusalem Mimeo.

Shapley, L.S. (1953) 'A Value for N-Person Games', in: H. Kuhn and A.W. Tucker, eds., *Contributions to the Theory of Games, II*. Princeton University Press, pp. 307–317.

Shapley, L.S. (1969) 'Utility Comparison and the Theory of Games', in: G. Th. Guilbaud, ed., *La Decision*, Editions du CNRS, Paris, pp. 251–263.

Shapley, L.S. (1984) Oral Presentation at the I.M.A., Minneapolis.

Shubik, M. (1982) *Game Theory in the Social Sciences*, MIT University Press.

Sobel, J. (1981) 'Distortion of Utilities and the Bargaining Problem', *Econometrica*, **49**: 597–619.

Thomson, W. (1980) "Two Characterizations of the Raiffa Solution', *Economics Letters*, **6**: 225–231.

Thomson, W. (1981a) 'A Class of Solutions to Bargaining Problems', *Journal of Economic Theory*, **25**: 431–441.

Thomson, W. (1981b) 'Independence of Irrelevant Expansions', *International Journal of Game Theory*, **10**: 107–114.

Thomson, W. (1981c) "Nash's Bargaining Solution and Utilitarian Choice Rules', *Econometrica*, **49**: 535–538.

Thomson, W. (1983a) 'Truncated Egalitarian and Monotone Path Solutions', Discussion Paper (January), University of Minnesota.

Thomson, W. (1983b) 'Collective Guarantee Structures', *Economics Letters*, **11**: 63–68.

Thomson, W. (1983c) 'The Fair Division of a Fixed Supply Among a Growing Population', *Mathematics of Operations Research*, **8**: 319–326.

Thomson, W. (1983d) 'Problems of Fair Division and the Egalitarian Principle', *Journal of Economic Theory*, **31**: 211–226.

Thomson, W. (1984a) 'Two Aspects of the Axiomatic Theory of Bargaining', Discussion Paper, No. 6, University of Rochester.

Thomson, W. (1984b) 'Truncated Egalitarian Solutions', *Social Choice and Welfare*, **1**: 25–32.

Thomson, W. (1984c) 'Monotonicity, Stability and Egalitarianism', *Mathematical Social Sciences*, **8**: 15–28.

Thomson, W. (1984d) 'The manipulability of resource allocation mechanisms', *Review of Economic Studies*, **51**: 447–460.

Thomson, W. (1985a) 'Axiomatic Theory of Bargaining with a Variable Population: A Survey of Recent Results', in: A.E. Roth, ed., *Game Theoretic Models of Bargaining*. Cambridge University Press, pp. 233–258.

Thomson, W. (1985b) 'On the Nash Bargaining Solution', University of Rochester, mimeo.

Thomson, W. (1986) 'Replication Invariance of Bargaining Solutions', *International Journal of Game Theory*, **15**: 59–63.

Thomson, W. (1987a) 'Monotonicity of Bargaining Solutions with Respect to the Disagreement Point', *Journal of Economic Theory*, **42**: 50–58.

Thomson, W. (1987b) 'Individual and Collective Opportunities', *International Journal of Game Theory*, **16**: 245–252.

Thomson, W. (1990) 'The Consistency Principle in Economics and Game Theory', in: T. Ichiishi, A. Neyman and Y. Tauman, eds., *Game Theory and Applications*. Academic Press, pp. 187–215.

Thomson, W. (1994) *Bargaining Theory: The Axiomatic Approach*, Academic Press, forthcoming.

Thomson, W. and T. Lensberg (1983) 'Guarantee Structures for Problems of Fair Division', *Mathematical Social Sciences*, **4**: 205–218.

Thomson, W. and T. Lensberg (1989) *The Theory of Bargaining with a Variable Number of Agents*, Cambridge University Press.

Thomson, W. and R.B. Myerson (1980) 'Monotonicity and Independence Axioms', *International Journal of Game Theory*, **9**: 37–49.

Tijs S. and H.J.M. Peters (1985) 'Risk Sensitivity and Related Properties for Bargaining Solutions', in: A.E. Roth, ed., *Game Theory Models of Bargaining*. Cambridge University Press, pp. 215–231.

Valenciano, F. and J.M. Zarzuelo (1993) 'On the Interpretation of Nonsymmetric Bargaining Solutions and their Extension to Non-Expected Utility Preferences', University of Bilbao Discussion Paper.

Wakker, P., H. Peters, and T. van Riel (1986) 'Comparisons of Risk Aversion, with an Application to Bargaining', *Methods of Operations Research*, **54**: 307–320.
Young, P. (1988) 'Consistent Solutions to the Bargaining Problem', University of Maryland, mimeo.
Yu, P.L. (1973) 'A Class of Solutions for Group Decision Problems', *Management Science*, **19**: 936–946.
Zeuthen, F. (1930) *Problems of Monopoly and Economic Welfare*, London: G. Routledge.

Chapter 36

GAMES IN COALITIONAL FORM

ROBERT J. WEBER

Northwestern University

Contents

1.	Introduction	1286
2.	Basic definitions	1286
3.	Side payments and transferable utility	1287
4.	Strategic equivalence of games	1288
5.	Properties of games	1290
6.	Balanced games and market games	1293
7.	Imputations and domination	1294
8.	Covers of games, and domination equivalence	1295
9.	Extensions of games	1295
10.	Contractions, restrictions, reductions, and expansions	1297
11.	Cooperative games without transferable utility	1298
12.	Two notions of "effectiveness"	1299
13.	Partition games	1301
14.	Games with infinitely many players	1301
15.	Summary	1302
Bibliography		1302

1. Introduction

The ability of the players in a game to cooperate appears in several guises. It may be that they are allowed to communicate, and hence are able to correlate their strategy choices. In addition, they may be allowed to make binding commitments before or during the play of the game. These commitments may even extend to post-play behavior, as when several players agree to redistribute their final payoffs via side payments.

When there are only two players in a game, each faces an essentially dichotomous decision: to cooperate, or not to cooperate. But situations involving more than two players are qualitatively different. Not only must a player decide which of many coalitions to join, he also faces uncertainty concerning the extent to which the players outside of his coalition will coordinate their actions.

Questions concerning coalition formation can be explored in the context of extensive-form or strategic-form games. However, such approaches require a formal model of the cooperative moves which the players are able to make. An alternative approach is to sacrifice the elaborate strategic structure of the game, in an attempt to isolate coalitional considerations. It is this approach which we set forth in the chapter. For each coalition, the various rewards that its players can obtain through cooperation are specified. It is assumed that disjoint coalitions can select their players' rewards independently. Although not all cooperative games fit this format well, we shall see that many important classes of games can be conveniently represented in this manner.[1]

2. Basic definitions

Let $N = \{1, 2, \ldots, n\}$ be a set of players. Any subset of N is a *coalition*, and the collection of 2^n coalitions of N is denoted by 2^N. A *coalitional*[2] *function* $v: 2^N \to \mathbb{R}$ is a real-valued function which assigns a "worth" $v(S)$ to each coalition S, and which satisfies $v(\emptyset) = 0$. It is frequently assumed that the coalitional function is expressed in units of an infinitely divisible commodity which "stores" utility, and which can be transferred without loss between players. The single number $v(S)$ indicates the total amount that the players of S can jointly guarantee themselves.

---

[1]Shapley and Shubik use the term "c-game" to describe a game which is adequately represented by its coalitional form [see Shubik (1982), pp. 130–131]. While no formal definition is given, constant-sum strategic-form games, and games of orthogonal coalitions (such as pure exchange economies), are generally regarded to be c-games.

[2]The function v, which "characterizes" the potential of each coalition, has been traditionally referred to as the "characteristic function" of the game. However, the term "characteristic function" already has different meanings in several branches of mathematics. In this chapter, at the suggestion of the editors, the more descriptive "coalitional function" is used instead.

The coalitional function is often called, simply, a *game* (with transferable utility) *in coalitional form*.

Example 1. Side-payment market games. Consider a collection $N = \{1, 2, \ldots, n\}$ of traders. They participate in a market encompassing trade in m commodities. Any m-vector in the space $\mathbb{R}_+^m$ represents a bundle of commodities. Each trader i in N brings an initial endowment $\omega^i \in \mathbb{R}_+^m$ to the marketplace. Furthermore, each trader has a continuous, concave utility function $u_i: \mathbb{R}_+^m \to \mathbb{R}$ which measures the worth, to him, of any bundle of commodities. The triple $(N, \{\omega^i\}, \{u_i\})$ is called a market (or, more formally, a *pure exchange economy*). We assume that the utility functions of the traders are normalized with respect to a transferable commodity which can be linearly separated in each trader's preferences. (This assumption is discussed in the next section.)

If any particular coalition S forms, its members can pool their endowments, forming a total supply $\omega(S) \equiv \sum_{i \in S} \omega^i$. This supply can then be reallocated as a collection $\{a^i : i \in S\}$ of bundles, such that each $a^i \in \mathbb{R}_+^m$ and $a(S) = \omega(S)$. Such a reallocation is of total value $\sum_{i \in S} u_i(a^i)$ to the coalition, and this value can be arbitrarily distributed among the players of S through a series of side payments (in the transferable commodity). Since the individual utility functions are continuous, the total value function is a continuous function over the compact set of potential reallocations, and therefore attains a maximum. We call this maximum $v(S)$, the worth of the coalition. The coalitional worths determine a game in coalitional form, known as a *market game*, which corresponds in a natural way to the original market. Notice that the correspondence is strengthened by the fact that the players outside of the coalition S have no effect on the total value achievable by the players of S.

3. Side payments and transferable utility

The definition in the previous section assumes that there exists a medium through which the players can make side payments which transfer utility. Exactly what does this assumption entail?

Utility functions, representing the players' preferences, are determined only up to positive affine transformations. That is, if the function u represents an individual's preferences over an outcome space, and if $a > 0$ and b are any real numbers, then the function $a \cdot u + b$ will also serve to represent his preferences. Therefore, in representing the preferences of the n players in a game, there are in general $2n$ parameters which may be chosen arbitrarily.

Assume, in the setting of Example 1, that there exists some commodity – without loss of generality, the mth – such that the utility functions representing the players' preferences are of the form $u_i(x_1, \ldots, x_m) = \hat{u}_i(x_1, \ldots, x_{m-1}) + c_i x_m$, where each c_i is a positive constant. In this case, we say that the players' preferences are linearly separable with respect to that commodity. Transfers of the mth commodity between players will correspond to a linear transfer of utility. If we choose the utility functions $(1/c_i) \cdot u_i$ to represent the preferences of the players, then we establish a common utility scale in which the mth commodity plays the role of numeraire.

Throughout our discussions of games with transferable utility, we will assume the existence of a commodity in which side payments can be made, and we will further assume that each player holds a quantity of this commodity that is sufficient for any relevant side payments. (These assumptions will apply even in settings where specific commodities are not readily apparent.) The coalitional function will be assumed to be expressed in units of this commodity; therefore, the division of the amount $v(S)$ among the players in S can be thought of simultaneously as a redistribution of the side-payment commodity, and as a reallocation of utility. One important point should be appreciated: In choosing a common utility normalization for the players' preferences, we are *not* making direct interpersonal comparisons of utility.

Example 2. Constant-sum strategic-form games. Assume that the players in N are playing a strategic-form game, in which $K_1, \ldots, K_n$ are the players' finite pure-strategy spaces and $P_1, \ldots, P_n$ are the corresponding payoff functions. Assume that for every n-tuple of strategies $(k_1, \ldots, k_n)$, the total payoff $\sum_{i \in N} P_i(k_1, \ldots, k_n)$ is constant; such a game is said to be *constant-sum*. For any coalition S, let M^S be the set of all probability distributions over the product space $\prod_{i \in S} K_i$. Each element of M^S is an S-correlated strategy.

Fix a nonempty coalition S, with nonempty complement $\bar{S}$. We define a two-person strategic-form game played between S and $\bar{S}$. The pure-strategy spaces in this game are $\prod_{i \in S} K_i$ and $\prod_{j \in \bar{S}} K_j$, and the mixed-strategy spaces are M^S and $M^{\bar{S}}$. The expected payoff functions E^S and $E^{\bar{S}}$ are the sums of the components of the original expected payoff functions corresponding, respectively, to the players of S and $\bar{S}$.

In the game just described, the interests of the complementary coalitions are directly opposed. It follows from the minimax theorem that the game has an optimal strategy pair $(m^S, m^{\bar{S}})$, and that the "values" $v(S) = E^S(m^S, m^{\bar{S}})$ and $v(\bar{S}) = E^{\bar{S}}(m^S, m^{\bar{S}})$ are well-defined. By analyzing all 2^{n-1} of the two-coalition games, and taking $v(\emptyset) = 0$ and $v(N) = \sum_{i \in N} P_i(k_1, \ldots, k_n)$, we eventually determine the coalitional-form game associated with the original constant-sum strategic-form game. Notice that the appropriate definition of the coalitional function would not be so clear if the strategic-form game were not constant-sum. In such a case, the interests of complementary coalitions need not be directly opposed, and therefore no uniquely determined "value" need exist for the game between opposing coalitions.

4. Strategic equivalence of games

Since the choice of a unit of measurement for the side-payment commodity may be made arbitrarily, the "transferable utility" assumption actually consumes only

$n-1$ of the $2n$ degrees of freedom in the utility representation of an n-person game. A common (across players) scale factor, and the origins (zero-points) of the players' utility functions, may still be chosen freely.

Example 3. General-sum strategic-form games. Consider the situation treated in Example 2, but without the assumption that the strategic-form game under consideration is constant-sum. From a conservative perspective, coalition S can guarantee itself at most

$$v(S) = \max_{m^S \in M^S} \min_{m^{\bar{S}} \in M^{\bar{S}}} E^S(m^S, m^{\bar{S}}).$$

Von Neumann and Morgenstern (1944) suggested that the coalitional function associated with a general-sum strategic-form game be derived in this manner.

An alternative approach, which takes into account the possibility that S and $\bar{S}$ might have some common interests, was subsequently suggested by Harsanyi (1959). The two-coalition game may be viewed as a two-person bargaining game. For such games, the variable-threats version of the Nash solution yields a unique payoff pair. This pair of payoffs is used define $h(S)$ and $h(\bar{S})$ in the *modified coalitional function h* associated with the strategic-form game.

The Nash solution of a variable-threats game is particularly easy to obtain when the game permits side payments (and the utility functions of the players are normalized so that utility is transferable through side payments in a one-to-one ratio). Let

$$A = \max_{m^N \in M^N} E^N(m^N),$$

the maximum total payoff attainable by the players of the game through cooperation. For any coalition S, let

$$B_S = \max_{m^S \in M^S} \min_{m^{\bar{S}} \in M^{\bar{S}}} [E^S(m^S, m^{\bar{S}}) - E^{\bar{S}}(m^S, m^{\bar{S}})].$$

The $h(S) = (A + B_S)/2$, and $h(\bar{S}) = (A - B_S)/2$. It is easily checked that h is well-defined, i.e., that for all S, $B_S = -B_{\bar{S}}$. Similarly, it is simple to show that $h(S) \geq v(S)$ for every S. When the strategic-form game is constant-sum, the original and modified coalitional functions coincide.

Let v be an n-person game. For any constants $a > 0$ and $b_1, \ldots, b_n$, the game w defined for all coalitions S by

$$w(S) = a \cdot \left(v(S) - \sum_{i \in S} b_i \right)$$

is *strategically equivalent* to v. Motivation for this definition is provided by Example 3: If v is the coalitional function (classical or modified) associated with the strategic-form game in which the players' payoff functions are $P_1, \ldots, P_n$, then w is the coalitional function associated with the corresponding game with payoff functions $a(P_1 - b_1), \ldots, a(P_n - b_n)$; in both games, the players' strategic motivations are the same.

Assume that $v(N) > \sum_{i \in N} v(i)$; a game with this property is said to be *essential*. An essential game is one in which the players have something to gain through group cooperation. We shall assume that all games under consideration are essential, unless we make specific mention to the contrary. The $(0,1)$-*normalization* of v is the game w defined for all coalitions S by

$$w(S) = \frac{v(S) - \sum_{i \in S} v(i)}{v(N) - \sum_{i \in N} v(i)}.$$

Note that $w(i) = 0$ for all i in N, and $w(N) = 1$. The $(0, 1)$-normalization consumes the remaining $n + 1$ degrees of freedom in our derivation of the coalitional form: Each equivalence class of essential strategically-equivalent games contains a unique game in $(0, 1)$-normalization.

Other normalizations have been used on occasion. The early work of von Neumann and Morgenstern was focused on the coalitional form of constant-sum games, and they worked primarily with the $(-1, 0)$-normalization in which each $w(i) = -1$ and $w(N) = 0$.

5. Properties of games

A game v is *superadditive* if, for all disjoint coalitions S and T, $v(S \cup T) \geq v(S) + v(T)$. In superadditive games, the merger of disjoint coalitions can only improve their prospects. The games derived from markets and from strategic-form games in Examples 1–3 are all superadditive. Furthermore, the property of superadditivity is a property of equivalence classes of games: If a game is superadditive, all strategically equivalent games are, too.

If $v(S) \geq v(T)$ for all $S \supset T$, then v is *monotonic*. Monotonicity is not a particularly natural concept, since every monotonic game is strategically equivalent to a non-monotonic one (the converse is also true). An alternative definition, which is preserved under normalization and which more closely captures the property that "larger is better", is that a game is *zero-monotonic* if its $(0, 1)$-normalization is monotonic. Superadditivity implies zero-monotonicity, but is not implied by it.

A stronger notion than superadditivity is that of convexity: A game v is *convex* if for all coalitions S and T, $v(S \cup T) + v(S \cap T) \geq v(S) + v(T)$. An equivalent definition is that for any player i, and any coalitions $S \supset T$ not containing i, $v(S \cup i) - v(S) \geq v(T \cup i) - v(T)$. This has been termed the "snowball" property: The larger a coalition becomes, the greater is the marginal contribution of new members.

A game v is *constant-sum* if $v(S) + v(\bar{S}) = v(N)$ for all complementary coalitions S and $\bar{S}$. The coalitional function derived from a constant-sum strategic-form game (Example 2) is itself constant-sum; the modified coalitional function of any strategic-form game (Example 3) is also constant-sum.

The *dual* of a game v is the game v' defined by $v'(S) = v(N) - v(\bar{S})$ for all $S \subset N$. As $v(S)$ represents the payoff that can be "claimed" by the coalition S, so $v'(S)$ represents the share of the payoff which can be generated by the coalition N which cannot be claimed by the complement of S. Clearly, the only self-dual games are those which are constant-sum.

Example 4. Simple games. A *simple game* v is a monotonic (0, 1)-valued coalitional function on the coalitions of N. We often think of a simple game as representing the decision rule of a political body: Coalitions S for which $v(S) = 1$ are said to be *winning*, and those for which $v(S) = 0$ are said to be *losing*. Every simple game is fully characterized by its set of minimal winning coalitions.

The dual v' of the simple game v is also a simple game, which can be viewed as representing the "blocking" rule of a political body: The winning coalitions in v' are precisely those whose complements are losing coalitions in v.

A simple game v is *proper* if the complement of every winning coalition is losing: This is equivalent to v being superadditive, and is clearly a desirable property of a group decision rule. (If the rule were improper, two disjoint coalitions could pass contradictory resolutions.) A game is *strong* if the complement of every losing coalition is a winning coalition. A *decisive* simple game is both proper and strong. This is equivalent to the game being constant-sum, and hence self-dual.

The *weighted voting game* $[q; w_1, \ldots, w_n]$ is the simple game in which a coalition S is winning if and only if $\sum_{i \in S} w_i \geq q$, i.e., if the "weights" of the players in S add up to at least the "quota". Legislative bodies in which legislators represent districts of unequal populations are frequently organized as weighted voting games. The United Nations Security Council can pass a resolution only with the support of all five permanent members, together with at least four of the ten non-permanent members. This rule can be represented by the weighted voting game [39; 7,7,7,7,7, 1,1,1,1,1,1,1,1,1,1]. The game is proper, but not decisive, and hence deadlocks can occur. The nine-player weighted voting game [4; 1,1,1,1,1,1,1,1,1] represents the rule used by the U.S. Supreme Court in granting a writ of a certiorari (that is, in accepting a case to be heard by the full court). Note that this game is not proper, but that, since only a single action can be adopted, the improperness does not create any practical difficulties.

Both the Security Council and Supreme Court games are *homogeneous*, i.e., each has a weighted representation in which all minimal winning coalitions have total weight equal to the quota. There are seven decisive simple games with five players or less (ignoring symmetries and dummy players), and all are homogeneous. These are the games [1; 1], [2; 1,1,1], [3; 2,1,1,1], [3; 1,1,1,1,1], [4; 2,2,1,1,1], [4; 3,1,1,1,1], and [5; 3,2,2,1,1]. There are 23 decisive six-player games, eight of which are homogeneous: [4; 2,1,1,1,1,1], [5; 3,2,1,1,1,1], [5; 4,1,1,1,1,1], [6; 3,3,2,1,1,1], [6; 4,2,2,1,1,1], [7; 4,3,3,1,1,1], [7; 5,2,2,2,1,1], and [8; 5,3,3,2,1,1]. Six others are weighted voting games with no homogeneous representations. [5; 2,2,2,1,1,1], [6; 3,2,2,2,1,1], [7; 3,3,2,2,2,1], [7; 4,3,2,2,1,1], [8; 4,3,3,2,2,1], and [9; 5,4,3,2,2,1]. (In the first game, for example, since 123 and 124 are minimal winning coalitions, any homogeneous representation must assign equal weight to players 3 and 4. But since 135 wins and 145 loses, the weights of players 3 and 4 must differ). Finally, nine have no weighted voting representations at all: An example is the game with minimal winning coalitions 12, 134, 234, 135, 146, 236, and 245. (In any representation, the total weight of the winning coalitions 135 and 146 would have to be at least twice the quota. The total weight of the losing coalitions 136 and 145 would be the same, but would have to be less than twice the quota.) Von Neumann and Morgenstern (1944, sections 50–53) and Gurk and Isbell (1959) give these and other examples.

Finding a weighted voting representation for a simple game is equivalent to solving a system of linear inequalities. For decisive homogeneous games, the system has only a one-dimensional family of solutions. The nucleolus (one type of co-operative "solution" for games in coalitional form) lies in this family [Peleg (1968)], and a von Neumann–Morgenstern stable set (another type of solution) can be constructed from the representation (see also Chapters 17 and 18 of volume I of this Handbook). Ostmann (1987) has shown that, for non-decisive homogeneous games as well, the unique minimal representation with integer weights is the homogeneous representation.

The *counting vector* of an n-player simple game is the n-vector $(c_1,\ldots,c_n)$ in which c_i is the number of winning coalitions containing player i. No two distinct weighted voting games have identical counting vectors [Lapidot (1972)]. Einy and Lehrer (1989) have extended this result to provide a full characterization of those simple games which can be represented as weighted voting games.

Consider two simple games v and w with disjoint player sets. The *sum* $v \oplus w$ is the simple game on the union of the player sets in which a coalition wins if and only if it contains a winning coalition from at least one of the original games; the *product* $v \otimes w$ is the simple game on the union of the player sets in which a coalition wins if and only if it contains winning coalitions from both of the original games. Typically, sums are not proper, and products are not strong.

If v is a simple game with player set $N = \{1,\ldots,n\}$, and $w_1,\ldots,w_n$ are simple games with disjoint player sets, the *composition* $v[w_1,\ldots,w_n]$ has as its player set the union of the player sets of $w_1,\ldots,w_n$; a coalition S of $v[w_1,\ldots,w_n]$ is winning if $\{i \in N : S$ contains a winning coalition of $w_i\}$ is a winning coalition of v. (Both sums and products are special types of compositions.) The U.S. Presidential election is an example of a composition: The voters in the 50 states and the District of Columbia engage in simple majority games to determine the winning candidate in each of the 51 regions, and then the regions "play" a weighted voting game (with weights roughly proportional to their populations) in order to select the President.

The analysis of complex political games is sometimes simplified by representing the games as compositions, and then analyzing each of the components separately. Shapley (1959) provides examples of this approach, and Owen (1959) generalizes the approach to non-simple games.

The set of all games (essential and inessential) with player set N can be viewed as a $(2^n - 1)$-dimensional vector space. For any nonempty $S \subset N$, let v_S be the game defined by $v_S(T) = 1$ if $T \supset S$, and $v_S(T) = 0$ otherwise. Each game v_S is a simple game with a single minimal winning coalition. The collection $\{v_S\}$ is a basis for the vector space of games on N, and each game v admits a unique representation

$$v = \sum_{S \subset N} \left(\sum_{T \subset S} (-1)^{|S|-|T|} v(T) \right) v_S.$$

This representation finds use in the study of the valuation of games.

6. Balanced games and market games

We have previously seen how a game can be associated with any market. A natural, complementary question is: What games arise from markets? That is, what games are market games? In this section, we will completely answer this question (for games with transferable utility).

Consider an economy in which the only available commodities are the labors of the individual players. We represent this economy by assuming that there is an n-dimensional commodity space, and that the initial endowment of each trader is one unit of his own labor. Define e^S to be the n-dimensional vector with $e_i^S = 1$ if i is in S, and $e_i^S = 0$ otherwise. Then the initial endowments are $\omega^i = e^i$. Assume that each coalition S has available to it a productive activity which requires equal amounts of labor from all of its members. The output attainable by S from an input of e^S is denoted by $v(S)$. If, however, each player i contributes only x_i units of his labor to the activity, then only $(\min_{i \in S} x_i) \cdot v(S)$ units of output can be produced. If a trader holds a bundle $x = (x_1, \ldots, x_n)$, his utility is simply the maximum output he can attain from that bundle:

$$u_i(x) = u(x) \equiv \max \sum \gamma_T v(T),$$

where the maximum is taken over all $\{\gamma_T\}_{T \subset N}$ such that $\sum \gamma_T e^T = x$ and all $\gamma_T \geq 0$. (Such a collection $\{\gamma_T\}_{T \subset N}$ is said to be x-balanced.) Note that all players have the same utility function u, and that u is continuous, concave, and homogeneous of degree 1, i.e., for all $t > 0$, $u(tx) = t \cdot u(x)$. The triple $(N, \{\omega^i\}, \{u\})$ is the *direct market* associated with the coalitional function v.

Let $\bar{v}$ be the coalitional function of the game which arises from the direct market associated with v. By definition (Example 1),

$$\bar{v}(S) = \max_{\Sigma x^i = e^S} \sum u(x^i) = u(e^S);$$

the second equality follows from the homogeneity and concavity of u. Furthermore,

$$u(e^S) = \max_{\Sigma \gamma_T e^T = e^S} \sum \gamma_T v(T) \geq v(S).$$

(That is, each coalition is "worth" in the game $\bar{v}$ at least what it can produce on its own.)

Viewing v as a coalitional function, we say that the game v is *totally balanced* if $v = \bar{v}$, i.e., if for all S and all e^S-balanced collections $\{\gamma_T\}_{T \subset S}$, $v(S) \geq \sum \gamma_T v(T)$. We have just shown that every totally balanced game is a market game, i.e., is generated by its associated direct market. The converse is also true.

Theorem [Shapley and Shubik (1969)]: *A game is a market game if and only if it is totally balanced.*

Proof. It remains only to show that every market game is totally balanced. Let $(N, \{\omega^i\}, \{u_i\})$ by any market and let v be the associated market game. Consider any coalition S and a collection $\{x^{i,T}\}_{i \in T \subset S}$ of allocations such that for each T, $v(T) = \sum_{i \in T} u_i(x^{i,T})$ and $\sum_{i \in T} x^{i,T} = \omega(T)$. For any set $\{\gamma_T\}_{T \subset S}$ of e^S-balanced weights,

$$\sum_{i \in S} \sum_{T \ni i} \gamma_T x^{i,T} = \sum_{T \subset S} \gamma_T \sum_{i \in T} x^{i,T} = \sum_{T \subset S} \gamma_T \omega(T) = \omega(S).$$

Therefore,

$$\sum_{T \subset S} \gamma_T v(T) = \sum_{T \subset S} \gamma_T \left(\sum_{i \in T} u_i(x^{i,T}) \right) = \sum_{i \in S} \sum_{T \ni i} \gamma_T u_i(x^{i,T}) \leq \sum_{i \in S} u_i \left(\sum_{T \ni i} \gamma_T x^{i,T} \right).$$

The inequality follows from the concavity of the functions u_i. The last expression corresponds to the total utility achieved by S through a particular reallocation of $\omega(S)$, and hence is at most $v(S)$, the maximum total utility achievable through any reallocation. □

7. Imputations and domination

One reason for studying games in coalitional form is to attempt to predict (or explain) the divisions of resources upon which the players might agree. Any such division "imputes" a payoff to each player. Formally, the set of *imputations* of the game v is

$$A(v) = \{x \in \mathbb{R}^N | x(N) = v(N), \text{ and } x_i \geq v(i) \text{ for all } i \in N\}.$$

(For any coalition S, we write $x(S) = \sum_{i \in S} x_i$.) An imputation x divides the gain available to the coalition N among the players, while giving each player at least as much as he can guarantee himself.

An imputation x *dominates* another imputation y if there is some coalition S for which, for all i in S, $x_i > y_i$, and $x(S) \leq v(S)$. Basically, this requires that every player in S prefers x to y, and that the players of S together claim no more than their coalitonal worth $v(S)$ in the dominating imputation. Solution concepts such as the "core" and "von Neumann–Morgenstern stable sets" are based on this dominance relation.

Two games are *domination-equivalent* if there is a one-to-one mapping between their imputation sets which preserves the dominance relation. It is trivial to note that strategically-equivalent games are domination-equivalent. In the next section, we will see that domination-equivalence yields strictly larger equivalence classes of games than does strategic-equivalence.

8. Covers of games, and domination equivalence

A *cover* of a game is just what the name implies: another game which attributes at least as great a worth to every coalition. For example, the modified coalitional function of a strategic-form game is a constant-sum cover of the classical coalitional function.

Two particular covers of a game are of interest. Our interest in them is inspired by the fact that, if the cover assigns the same worth to the coalition N as does the original game, then certain properties of the original game will be preserved in the cover.

Let v be any game. The *superadditive cover* of v is the game v^{sa} defined for all coalitions S by

$$v^{sa}(S) = \max \sum_{T \in \mathscr{P}} v(T),$$

where the maximum is taken over all partitions $\mathscr{P}$ of S. It is easily seen that v^{sa} is the minimal superadditive game which covers v.

The *balanced cover* of v was defined in the previous section: It is the game $\bar{v}$ defined for all coalitions S by

$$\bar{v}(S) = \max \sum \gamma_T v(T),$$

where the maximum is taken over all e^S-balanced collections $\{\gamma_T\}$ of weights. The balanced cover of v is the minimal totally-balanced game which covers v.

Theorem. *Let v be any game.*
 (a) *If $v^{sa}(N) = v(N)$, then v^{sa} is domination-equivalent to v.*
 (b) *If $\bar{v}(N) = v(N)$, then $\bar{v}$ is domination-equivalent to v.*

9. Extensions of games

At times it is useful to view the coalitional function as representing the amount attainable by a coalition when its members participate fully in a cooperative venture in which the remaining players are not at all involved. When taking this view, we might wish to extend the coalitional function, to indicate what can be achieved when several players participate only partially in a coalitional activity. Formally, an *extension* of a coalitional function v is a real-valued function f defined on the unit n-cube, such that $f(e^S) = v(S)$ for every coalition S. For example, in Section 6 the utility function u in the direct market derived from a coalitional function v was an extension of the balanced cover of the coalitional function.

A slightly different construction yields an extension of any game. The *balanced extension* of v is the function f^b defined by

$$f^b(x) = \max \sum \gamma_T v(T),$$

where the maximum is taken over all x-balanced collections $\{\gamma_T\}$ such that $\Sigma \gamma_T = 1$. (The final constraint ensures that the extension coincides with the original coalitional function at the vertices of the unit n-cube). The graph of the balanced extension is the intersection of the graphs of all linear functions which are at least as great as the coalitional function at all vertices of the unit n-cube; it follows that the balanced extension is piecewise-linear and concave.

With respect to the balanced extension, we say that a coalition S *sees* another coalition $T \subset S$ (and that T sees S) if the line connecting the points $(e^S, v(S))$ and $(e^T, v(T))$ lies on the upper boundary of the graph of the extension. Many interesting classes of games can be characterized by the "seeing" relation. For example, a game is totally balanced if every coalition can see the empty coalition; a game is convex if every coalition can see all of its subsets and supersets.

The *core* of a game v is the set of imputations x for which $x(S) \geq v(S)$ for all coalitions S. [If $v^{sa}(N) = v(N)$, this is precisely the set of undominated imputations.] Each imputation in the core corresponds to a supporting hyperplane of the graph of the balanced extension which passes through both the origin and $(e^N, v(N))$. Hence, the core is nonempty if and only if N sees the empty coalition. A game is *exact* if for every coalition S there is an imputation x in the core of the game satisfying $x(S) = v(S)$; clearly, a game is exact if and only if every coalition sees both N and the empty coalition.

A *symmetric game* v is characterized by a sequence of $n+1$ numbers, $(0, v_1, \ldots, v_n)$, where for any coalition S, $v(S) = v_{|S|}$. The coalitional function of a symmetric game can be graphically represented in two dimensions by the function f on $\{0, 1, \ldots, n\}$ for which $f(k) = v_k$. This function can be extended to a piecewise-linear function on the interval $[0, n]$ by taking the extension to be linear on each interval $[k, k+1]$; the graph of this function is called the *line-graph* of the symmetric game. A "seeing" relation can be defined in this context: The integer s "sees" the integer t if the line segment connecting (s, v_s) and (t, v_t) lies on or above the line-graph. The two notions of "seeing" that we have just defined are related in a natural way: For any symmetric game, S sees T if and only if $|S|$ sees $|T|$.

The balanced extension of a game captures the idea of fractional player participation in coalition activities. Another approach to the extension of a coalitional function is to consider probabilistic participation. The *multilinear extension* of a game v is the function f^* defined for all x in the unit n-cube by

$$f^*(x) = \sum_{S \subset N} \prod_{i \in S} x_i \prod_{j \notin S} (1 - x_j) v(S).$$

If each player i in the game has probability x_i of being available for participation in a coalitional activity, then $f^*(x)$ is the expected coalitional worth of the coalition which actually arises. There are $2^n - 1$ different nonempty products of variables in f^*; since the coalitional function takes $2^n - 1$ nontrivial values, it is not surprising that f^* is the unique multilinear function extending v. The multilinear extension is of importance when studying the valuation of games; see Owen (1972).

10. Contractions, restrictions, reductions, and expansions

One method for defining or studying solution concepts for games is to examine the relationship between solutions of closely related games with differing player sets.

The *contraction* of a game v with respect to a coalition $S \subset N$ is the game w with player set $(N \setminus S) \cup \{S\}$ defined by $w(T) = v(T)$ and $w(T \cup \{S\}) = v(T \cup S)$ for all $T \subset N \setminus S$; the players in S have been replaced by a "syndicate," and are either all present in, or all absent from, any coalition in the game w. Two particularly natural types of syndicates are partnerships and families: A *partnership* in v is a coalition S for which $v(T) = v(T \setminus S)$ whenever $T \cap S \neq S$ (the presence of members of the partnership adds nothing to the worth of a coalition unless all of the partnership members are present), and a *family* in v is a coalition S for which $v(T) = v(T \cup S)$ whenever $T \cap S$ is nonempty (the presence of any member of the family is equivalent to the presence of the entire family). Each individual player of a game is both a partnership and a family; the contraction of a game with respect to an individual player is isomorphic to the original game. Kalai and Samet (1988) have used contractions with respect to both partnerships and families to characterize the "weighted Shapley values" of games.

The *restriction* of a game v to a coalition S is the game $v|_S$ with player set S defined by $v|_S(T) = v(T)$ for all $T \subset S$. A *reduction* of v is a game w with player set S such that $w(T) = v(T \cup \bar{S}) - g(\bar{S}, v|_{T \cup \bar{S}})$ for all $T \subset S$. The reduction function g represents a prespecified "withdrawal" from the game $v|_{T \cup \bar{S}}$ by the coalition $\bar{S}$; in the reduced game, each coalition has the cooperation of $\bar{S}$ potentially available, but must first allocate to the players of $\bar{S}$ their joint withdrawal amount. One simple reduction function which has been used in a variety of settings is $g(\bar{S}, v|_{T \cup \bar{S}}) = v(\bar{S})$. In studies of the "consistency" of solution concepts which select a single imputation $x(v)$ as the "solution" of every game v, the reduction function $g(\bar{S}, v|_{T \cup \bar{S}}) = \sum_{i \in \bar{S}} x_i(v|_{T \cup \bar{S}})$ has been employed; see, for example, Hart and Mas-Colell (1989).

The *constant-sum expansion* v^{cs} of a game v results from the introduction of a fictitious $(n+1)$st player who "fills out" the game; $v^{cs}(S) = v(S)$ for all $S \subset N$, and

$v^{cs}(S \cup \{n+1\}) = v(N) - v(\bar{S})$. Some of the early research on von Neumann–Morgenstern stable sets sought relationships between "solutions" to the constant-sum expansion and to the original game. The *dummy expansion* v^d of v results from the introduction of a player who makes no relevant contribution to any coalition: $v^d(S) = v(S \cap N)$ for all $S \subset N \cup \{n+1\}$. (Note that the restriction of either expansion to the original player set yields the original game.) In applications of cooperative game theory, it may be difficult to initially distinguish between relevant and irrelevant players. Solution theories which yield analogous results on v and v^d lessen the importance of such distinctions.

11. Cooperative games without transferable utility

When it is for some reason impossible for the players of a game to linearly transfer utility amongst themselves, the coalitional function as developed in the preceding sections fails to represent adequately the various outcomes attainable through cooperation. In such a case, we instead represent the game by describing for each coalition S the set of all utility S-tuples to which the players in S can reasonably aspire.

Example 5. Non-side-payment market games. Consider a pure exchange economy, as presented in Example 1, but assume that side payments between traders are not possible. The *attainable set* of coalition S is defined to be

$$V(S) = \{x \in \mathbb{R}^S: \text{for some } \{y^i\}_{i \in S} \text{ with } y(S) = \omega(S), \quad x_i \leq u_i(y^i) \text{ for all } i \in S\}.$$

This is the set of all outcomes attainable by the traders in S through a reallocation of their initial holdings. (We assume that utility is freely disposable, and therefore any S-vector which is smaller in all components than another S-vector attainable by the coalition is also attainable.) The correspondence assigning to each coalition its attainable set is the non-side-payment market game associated with the original market. Notice that the continuity and concavity of the traders' utility functions imply that each $V(S)$ is closed and convex.

In general, a *non-transferable utility game* (NTU-game) V is a correspondence which assigns to each coalition S a closed, convex subset $V(S)$ of $\mathbb{R}^S$. For technical reasons, it is usually assumed that for each S there is a strictly positive S-vector a, and a real number b, such that for all x in $V(S)$, $a \cdot x \leq b$. In essence, this assumption says that no player can receive a "very high" payoff in an outcome in $V(S)$, unless at least one other player receives a "very low" payoff; clearly this is a reasonable assumption in most of the real-world situations one would find oneself modeling.

Each transferable-utility game v is naturally associated with an NTU-game V for which $V(S) = \{x \in \mathbb{R}^S: x(S) \leq v(S)\}$.

Ch. 36: Games in Coalitional Form   1299

Example 5. Non-side-payment market games (continued). An NTU-game V is totally balanced if for all coalitions T, and all e^T-balanced collections $\{\gamma_S\}$,

$$\sum \gamma_S V(S) \subset V(T),$$

where the sum of sets is understood to consist of all sums of vectors from the sets. The game V is *totally balanced with slack* if for all coalitions T, and all e^T-balanced collections $\{\gamma_S\}$ with $\gamma_T = 0$,

$$\sum \gamma_S V(S) \subset \text{Int } V(T),$$

where Int $V(T)$ denotes the relative interior of $V(T)$. A set $A \subset \mathbb{R}^S$ *generates* another set B if $B = \{x \in \mathbb{R}^S : \text{for some } y \in A, x \leqslant y\}$.

Billera and Bixby (1973) showed that every set generated by a compact, convex set is the attainable set of some market. They also showed that every market game is totally balanced, and that if V is totally balanced and every $V(T)$ is a polyhedron, then V is a market game. Subsequently, Mas-Colell (1975) showed that every game which is totally balanced with slack is also a market game. A full analogue to the theorem of Section 6 (which characterizes all transferable-utility games which are market games) is still unknown.

If the utility functions of the traders in a market are only assumed to be quasiconcave, then even a characterization of all possible attainable sets is lacking. Weber (1981) presents partial results, and surveys a number of related issues.

12. Two notions of "effectiveness"

One of our goals in developing the coalitional form is to be able to study the prospects of various coalitions in a cooperative strategic-form game. In Example 3, we saw how two different coalitional forms could be derived from any strategic-form game in which side payments were possible. For two-player strategic-form games, $V(i)$ can be defined as (all points less than or equal to) i's maxmin value, the value to him of the zero-sum strategic-form game based on his own payoff matrix. This is a natural definition, since that value is simultaneously the greatest amount he can assure himself in the game, and the least amount to which he can be held by the other player.

In strategic-form games involving three or more players, it is no longer true, in general, that the payoff vectors a coalition can assure itself are precisely those which it cannot be denied.

Example 6. A three-person game [Aumann (1961)]. Consider the following three-player strategic-form game:

$$\text{I, II} \begin{bmatrix} 1, -1, \text{ any} & 0, 0, \text{ any} \\ 0, 0, \text{ any} & -1, 1, \text{ any} \end{bmatrix} \begin{matrix} \text{III} \\ \end{matrix}$$

What does it mean to say that $S = \{I, II\}$ can guarantee itself the payoff pair (x_1, x_2)? A natural definition is that there is a correlated strategy for S which simultaneously yields I an expected payoff of at least x_1, and II an expected payoff of at least x_2, no matter what strategy is adopted by III.

In general, we define

$$V_\alpha(S) = \{x \in \mathbb{R}^S : \text{there is an } m^S \in M^S \text{ such that for}$$
$$\text{every } m^{\bar{S}} \in M^{\bar{S}}, E^i(m^S, m^{\bar{S}}) \geq x_i \text{ for all } i \in S\}.$$

The correspondence V_α is called the α-*effectiveness* (*coalitional*) *form* of the cooperative strategic-form game.

In Example 6, the strategies available to S are mixtures $(p, 1-p)$, and to $\bar{S}$ are mixtures $(q, 1-q)$, with expected payoff $(pq - (1-p)(1-q), (1-p)(1-q) - pq) = (q - (1-p), (1-p) - q)$ to S. Neither I nor II can be guaranteed an expected payoff greater than zero: For any p, $q = 0$ holds I to $-(1-p)$, and $q = 1$ holds II to $-p$. In general, $(p, 1-p)$ guarantees at most $(-(1-p), -p)$.

The union of the sets of payoff vectors guaranteed by each strategy available to S provides an equivalent definition:

$$V_\alpha(S) = \bigcup_{m^S \in M^S} \left\{ x \in \mathbb{R}^S : x_i \leq \min_{m^{\bar{S}} \in M^{\bar{S}}} E^i(m^S, m^{\bar{S}}) \text{ for all } i \in S \right\}.$$

In Example 6, it is simple to see that $V_\alpha(S) = \{x \in \mathbb{R}^2 : x_1 \leq 0, x_2 \leq 0, x_1 + x_2 \leq -1\}$.

What payoff vectors can S not be denied? A definition analogous to that given above is

$$V_\beta(S) = \{x \in \mathbb{R}^S : \text{for every } m^{\bar{S}} \in M^{\bar{S}}, \text{ there is an } m^S \in M^S$$
$$\text{such that } E^i(m^S, m^{\bar{S}}) \geq x_i \text{ for all } i \in S\}.$$
$$= \bigcup_{m^{\bar{S}} \in M^{\bar{S}}} \{x \in \mathbb{R}^S : \text{for some } m^S \in M^S, x_i \leq E^i(m^S, m^{\bar{S}}) \text{ for all } i \in S\}.$$

The correspondence V_β is called the β-*effectiveness* (*coalitional*) *form* of the cooperative strategic-form game. In Example 6, $V_\beta(S) = \{x \in \mathbb{R}^2 : x_1 \leq 0, x_2 \leq 0\}$.

In general, $V_\alpha(S) \subset V_\beta(S)$, and equality need not hold if S consists of more than one player. This is simply a consequence of the fact that the minimax theorem does not hold for games with vector-valued payoffs.

As previously noted, any game in which utility is linearly transferable can be viewed as an NTU-game. In this case, the α-effectiveness and β-effectiveness forms coincide: $V_\alpha(S) = V_\beta(S) = \{x \in \mathbb{R}^S : x(S) \leq v(S)\}$.

Which of the two forms is the appropriate representation of the original game? One must base a decision on the context in which one is working. It is V_β which emerges as relevant when one wishes to relate the coalitional function to the "cooperative" equilibria of a repeatedly-played strategic-form game [Aumann (1961)].

13. Partition games

At times, the prospects of a coalition S depend on the manner in which the players in $\bar{S}$ divide themselves into coalitions. Let π be the set of all partitions of N. A *game in partition function form* specifies a coalitional worth $v(\mathscr{P}, S)$ for every partition $\mathscr{P}$ in π and every coalition S which is an element of $\mathscr{P}$. From this a coalitional function v is defined by $v(S) = \min_{\mathscr{P} \ni S} v(\mathscr{P}, S)$. A $\mathscr{P}$-imputation is an n-vector x satisfying $x_i \geq v(i)$ for every player i, and $x(N) = \sum_{S \in \mathscr{P}} v(\mathscr{P}, S)$; a $\mathscr{P}$-imputation x dominates a $\mathscr{P}'$-imputation y with respect to a coalition $S \in \mathscr{P}$ if $x^S > y^S$ and $x(S) \leq v(S)$. Much of the classical analysis of games in coalitional form has been extended to games in partition function form. [An early reference is Thrall and Lucas (1963).]

At other times, the ability of players to communicate amongst themselves is restricted. Communication restraints may be superimposed upon the coalitional function. A *game with cooperation structure* is a coalitional game v together with a cooperation structure $\mathscr{P}$ which associates with each coalition S a partition $\mathscr{P}(S)$ of S. The game $v^{\mathscr{P}}$ is defined by $v^{\mathscr{P}}(S) = \sum_{S_j \in \mathscr{P}(S)} v(S_j)$ for all coalitions S. A system of two-way communication linkages between players can be represented by a graph g with vertices corresponding to the players in N; the cooperation structure $\mathscr{P}^g$ associates with each S the partition which is induced by the connected components of the subgraph of g restricted to S. Aumann and Myerson (1988) have used this formulation to study the endogenous formation of communication links in a game.

14. Games with infinitely many players

When games have a large number of potential participants, it is sometimes mathematically convenient to represent the player set by either a countably infinite set, or by a continuum. Indeed, the space of "all" finite games may be modeled by assuming a countably infinite universe of players, and defining a game as a coalitional function on this infinite set. A *carrier* of a game v is a coalition S for which $v(T) = v(T \cap S)$ for all coalitions T; games with finite carriers correspond directly to finite games.

An extensive theory of games with a continuum of players has been developed. Measurability considerations usually require that the player set bring with it a set of permissible coalitions; hence the players and coalitions together are described by a pair $(I, \mathscr{C})$, where I is an uncountable set, and $\mathscr{C}$ is a σ-field of subsets of I. A game in coalitional form is a function $v: \mathscr{C} \to \mathbb{R}$ for which $v(\emptyset) = 0$. [An example of such a game is a real-valued function f of a vector of countably-additive, totally-finite, signed measures on $\mathscr{C}$, where $f(0,\ldots,0) = 0$.] Imputations are bounded, finitely-additive signed measures on $\mathscr{C}$. An *atom* of v is a coalition $S \in \mathscr{C}$ for which $v(S) > 0$ yet for every coalition $T \subset S$, either $v(T) = 0$ or $v(S \setminus T) = 0$. A *nonatomic* game is a game without atoms; much of the study of games representing large economies has been focused on properties of nonatomic games. [Aumann and Shapley (1974) provide a thorough treatment of the theory of nonatomic games.]

15. Summary

This chapter has had two principal purposes. One was to develop the idea of the "coalitional function" of a cooperative game as a means of abstracting from given settings (pure exchange economies, strategic-form games, political games, and the like) the possibilities available to the various coalitions through player cooperation. The other was to present several classes of games (market games, simple games, convex games, symmetric games) on which attention will be focused in subsequent chapters, and to provide several tools (covers and extensions, contractions and reductions, and the like) which will be useful in further analysis.

Bibliography

Auman, Robert J. (1961) 'The Core of a Cooperative Game without Side Payments', *Transactions of the American Mathematical Society*, **48**: 539–552.

Aumann, Robert J. and Roger B. Myerson (1988) 'Endogenous Formation of Links between Players and of Coalitions: An Application of the Shapley Value', in: A.E. Roth, ed., *The Shapley Value: Essays in Honor of Lloyd S. Shapley*. Cambridge University Press, pp. 175–191.

Aumann, Robert J. and Lloyd S. Shapley (1974) *Values of Non-Atomic Games*. Princeton University Press.

Billera, L.J. and R.E. Bixby (1973) 'A characterization of Polyhedral Market Games', *International Journal of Game Theory*, **2**: 253–261.

Einy, E. and E. Lehrer (1989) 'Regular Simple Games', *International Journal of Game Theory*, **18**: 195–207.

Gurk, H.M. and J.R. Isbell (1959) 'Simple Solutions', A.W. Tucker and R.D. Luce, eds., *Contributions to the Theory of Games*, Vol. 4. Princeton University Press, pp. 247–265.

Hart, Sergiu and Andreu Mas-Colell (1989) 'Potential, Value and Consistency', *Econometrica*, **57**: 589–614.

Harsanyi, John C. (1959) 'A Bargaining Model for the Cooperative n-Person Game', in: A.W. Tucker and R.D. Luce eds., *Contributions to the Theory of Games*, Vol. 4. Princeton University Press, 324–356.

Kalai, Ehud and Dov Samet (1988) 'Weighted Shapley Values', in: A.E. Roth, ed., *The Shapley Value: Essays in Honor of Lloyd S. Shapley*. Cambridge University Press, pp. 83–99.

Lapidot, E. (1972) 'The Counting Vector of a Simple Game', *Proceedings of the American Mathematical Society*, **31**: 228–231.

Mas-Colell, Andreu (1975) 'A Further Result on the Representation of Games by Markets', *Journal of Economic Theory*, **10**: 117–122.

von Neumann, John and Oskar Morgenstern (1944) *Theory of Games and Economic Behavior*. Princeton University Press.

Ostmann, A. (1987) 'On the Minimal Representation of Homogeneous Games', *International Journal of Game Theory*, **16**: 69–81.

Owen, Guillermo (1959) 'Tensor Composition of Non-Negative Games', M. Dresher, L.S. Shapley and A.W. Tucker, eds., *Advances in Game Theory*. Princeton University Press, pp. 307–326.

Owen, Guillermo (1972) 'Multilinear Extensions of Games', *Management Science*, **18**: 64–79.

Peleg, Bezalel (1967) 'On Weights of Constant-Sum Majority Games', *SIAM Journal of Applied Mathematics*, **16**, 527–532.
Shapley, Lloyd S. (1959) 'Solutions of Compound Simple Games', M. Dresher, L.S. Shapley, and A.W. Tucker, eds., *Advances in Game Theory*. Princeton University Press, pp. 267–305.
Shapley, Lloyd S. (1962) 'Simple Games: An Outline of the Descriptive Theory', *Behavioral Science*, **7**: 59–66.
Shapley, Lloyd S. and Martin Shubik (1969) 'On Market Games', *Journal of Economic Theory*, **1**: 9–25.
Shubik, Martin (1982) *Game Theory in the Social Sciences*. MIT Press.
Thrall, Robert M. and William F. Lucas (1963) 'n-Person Games in Partition Function Form', *Naval Research Logistics Quarterly*, **10**: 281–298.
Weber, Robert J. (1981) 'Attainable Sets of Markets: An Overview', In: M. Avriel, S. Schaible and W.T. Ziemba, eds., *Generalized Concavity in Optimization and Economics*. Academic Press, pp. 613–625.

Chapter 37

COALITION STRUCTURES

JOSEPH GREENBERG*

McGill University and C.R.D.E. Montreal

Contents

1.	Introduction	1306
2.	A general framework	1307
3.	Why do coalitions form?	1308
4.	Feasible outcomes for a fixed coalition structure	1311
5.	Individually stable coalition structures	1313
6.	Coalition structure core	1316
7.	Political equilibrium	1319
8.	Extensions of the Shapley value	1322
9.	Abstract systems	1324
10.	Experimental work	1327
11.	Noncooperative models of coalition formation	1329
References		1332

*I wish to thank Amnon Rapoport, Benyamin Shitovitz, Yves Sprumont, Abraham Subotnik, Shlomo Weber, and Licun Xue for valuable comments. Financial support from the Research Councils of Canada [both the Natural Sciences and Engineering (NSERC), and the Social Sciences and Humanities (SSHRC)] and from Quebec's Fonds FCAR is gratefully acknowledged.

Handbook of Game Theory, Volume 2, Edited by R.J. Aumann and S. Hart
© *Elsevier Science B.V., 1994. All rights reserved*

1. Introduction

Our social life is conducted almost entirely within the structure of groups of agents: individual consumers are households or families; individual producers are firms which are large coalitions of owners of different factors of production; workers are organized in trade unions or professional associations; public goods are produced within a complex coalition structure of Federal, State, and local jurisdictions; we organize and govern ourselves in political parties; and we socialize within a network of formal and informal social clubs.

The founders of game theory, von Neumann and Morgenstern recognized the central role and prominence of coalition formation. Indeed, most of their seminal work [von Neumann and Morgenstern (1944)] is devoted to a formal analysis of this very topic.[1] The purpose of this chapter is to survey some of the developments in this area, focusing mainly on social environments in which individuals choose to partition themselves into mutually exclusive and exhaustive coalitions. Naturally, it is impossible to encompass every scholarly work that deals with coalition structures, and what follows is, of necessity, a discussion of selected papers reflecting my personal taste and interests. Even then, in view of the scope of this subject, I can give only the broad conceptual framework and some idea of the formal results and applications that were obtained.

Following von Neumann and Morgenstern, the study of stable coalition structures, or more generally of coalition formation, was conducted mainly within the framework of games in coalitional form (see Definition 2.1). Unfortunately, this framework[2] became "unfashionable" during the last two decades, and, as a result, too little intellectual effort was devoted to the study of coalition formation. It is, therefore, encouraging to witness the revival, motivated by providing "noncooperative foundations for cooperative behavior", of the investigation of this important topic. The emerging literature (see Section 11) is a "hybrid", employing non-explicitly modelled "communications among players" (e.g., "cheap talk" and "renegotiation") within the framework of extensive or strategic form games.

My own view is that the consequences of allowing individuals to operate in groups should be investigated irrespective of the way (type of "game") the social environment is described. The theory of social situations [Greenberg (1990)] offers such an integrative approach. A unified treatment of cooperative and noncooperative social environments is not only "aesthetically" appealing, but is also mutually beneficial. For example, the basic premise of most solution concepts in games in

---

[1] Although von Neumann and Morgenstern do not explicitly consider coalition structures, much of their interpretive discussion of stable sets centers around which coalitions will form. [Aumann and Myerson, 1988, p. 176].

[2] It is noteworthy that it is von Neumann and Morgenstern who introduced both the strategic and the extensive form games. They, however, chose to devote most of their effort to studying ("stable sets" in) coalitional form games.

coalitional form is "blocking", i.e., the ability to use "credible threats". A necessary (but perhaps not sufficient) condition for a coalition to credibly object to a proposed payoff is that it can ("on its own") render each of its members better off. As was to be discovered many years later, the issue of credible threats is equally relevant for a single individual, giving rise to subgame perfection [Selten (1975)] and the various refinements of Nash equilibria. (See Chapters 5 and 6.)

The theory of social situations also highlights and amends the fact that none of the three game forms provides the essential information concerning the precise bargaining process, and in particular, the way in which a coalition can use its feasible set of actions. Even more disturbing is the fact that the meaning of "forming a coalition" is not entirely clear. Is it a binding commitment of the players to remain and never leave a coalition once it forms? Or is it merely a "declaration of intent" which can be revised, and if so, then under what conditions? That is, under what circumstances can a coalition form or dissolve? Who has property rights over a coalition?

Such fundamental open questions clearly demonstrate that more work should and can be done within the important area of coalition formation.

2. A general framework

Following von Neumann and Morgenstern, the study of stable coalition structures, or more generally of coalition formation, was conducted mainly within the framework of games in coalitional form. Recall the following definitions.

Definition 2.1. An *n-person game in coalitional form* (or *a game in characteristic function form*, or a *cooperative game*) is a pair (N, V), where $N = \{1, 2, \ldots, n\}$ is the set of *players* and for each coalition[3] S, $V(S)$ is a subset of R^N denoting the set of vectors of utility levels that members of S can attain on their own. (The payoffs of nonmembers are irrelevant, hence arbitrary.) The mapping V is called the *characteristic function*. The game (N, V) is called a *transferable utility (TU) game*, or a *game with side payments* if for every $S \subset N$, there exists a nonnegative scalar $v(S) \in R$, called *the worth of coalition* S, such that $x \in V(S)$ if and only if $x^i \geq 0$ for all $i \in S$, and $\Sigma_{i \in S} x^i \leq v(S)$.

Definition 2.2. Let (N, V) be an *n*-person game in coalitional form. *A coalition structure* (c.s.) is a partition $B = \{B_1, B_2, \ldots, B_K\}$ of the n players, i.e., $\cup B_k = N$ and for all $h \neq k$, $B_h \cap B_k = \varnothing$.

In most economic and political situations, however, the characteristic function is not given a priori. Rather, it is derived from the utilities individuals have over the actions taken by the coalitions to which they belong. (The nature of an "action" depends on the specific model considered. An action might be a production plan;

---

[3] A coalition is a nonempty subset of N.

a distribution of wealth or resources among the members of the coalition; a choice of location, price or quality; a tax rate imposed in order to finance the production of public goods; voting for a particular political candidate, etc.) For a coalition S, let $A(S)$ denote the set of "*S-feasible actions*". That is, $A(S)$ consists of all the actions S can take if and when it forms. [Interpreting the set $A(S)$ as the production possibilities of coalition S, yields the notion of "coalition production economies". See, e.g., Boehm (1973, 1974), Oddou (1972), and Sondermann (1974).]

The utility level enjoyed by individual i when he[4] belongs to coalition S and the S-feasible action $a \in A(S)$ is chosen, is given by $u^i(a, S)$. [The dependence of u^i on S, i.e., on the identity (or the number) of the members in i's coalition, was termed by Drèze and Greenberg (1980) the *hedonic aspect*.] The derived characteristic function of the game (N, V) associated with this society is given by

$$V(S) \equiv \{x \in R^N | \text{ there exists } a \in A(S) \text{ s.t. } x^i \leq u^i(a, S) \text{ for all } i \in S\}.$$

Although this framework is sufficiently general to encompass many contributions to the theory of coalition formation, it is not fully satisfactory. Most importantly, it ignores the possibility of externalities. The actions available to a coalition are assumed to be independent of the actions chosen by nonmembers. Indeed, $A(S)$ is defined without any reference to the actions players in $N \setminus S$ might adopt. Moreover, it is (at least implicitly) assumed that the utility levels members of S can attain, represented by the set $V(S)$, are also independent of the actions chosen by $N \setminus S$. In the presence of externalities the characteristic function is no longer unambiguous. In particular, what a coalition can guarantee its members might differ from what it cannot be prevented from achieving, leading, respectively, to the "α and β theories" [See Shubik (1984) on "c-games", and also Section 11.] Unfortunately, the more general formulation, Thrall and Lucas' (1963) "games in partition form", yields only a very limited set of results.

3. Why do coalitions form?

Among the very first questions that come to mind when analyzing coalition structures is why indeed do individuals form coalitions, rather than form one large group consisting of the entire society ("the grand coalition", N)? As Aumann and Drèze (1974), who were among the first to investigate explicitly, within the framework of TU games, solution concepts for coalition structures, wrote (p. 233):

> "After all, superadditivity[5] is intuitively rather compelling; why should not disjoint coalitions when acting together get at least as much as they can when acting separately?".

---

[4] Throughout the essay, I use the pronoun "he" to mean "she or he", etc.

[5] A game (N, V) is *superadditive* (relative to N) if for any two disjoint coalitions S and T (satisfying $S \cup T = N$), $x \in V(S) \cap V(T)$ implies $x \in V(S \cup T)$. In particular, a TU game is superadditive (relative to N) if for any two disjoint coalitions S and T (satisfying $S \cup T = N$), $V(S \cup T) \geq V(S) + V(T)$.

Aumann and Dréze provide several reasons for the existence of social environments that give rise to games that are not superadditive. The first and most straightforward is the possible "inherent" inefficiency of the grand coalition:

"Acting together may be difficult, costly, or illegal, or the players may, for various "personal" reasons, not wish to do so." [Aumann and Dréze (1974, p. 233)]

Thus, for example, the underlying technology might not be superadditive: The marginal productivity (of labor or management) might be negative. Alternatively, because of congestion in either production or consumption, or more generally, due to the hedonic aspect, the utility enjoyed by individuals might diminish as the size of the coalition to which they belong increases.

There are other less obvious but equally important reasons why a game might not be superadditive. These include: (i) normative considerations, and (ii) information (observability) imperfections. Consider, for example, the social norm of "equal treatment" or "nondiscrimination" within a coalition. Then, a game which is originally superadditive might become subadditive if this condition is imposed. Indeed, even in TU games, superadditivity does not imply that the per capita worth is an increasing function of the size of the coalition.

The second source of subadditivity is related to "moral hazard": the inability to monitor (or observe) players' (workers') performance might induce them to take actions (such as exert effort) which lead to inefficient or suboptimal outcomes.

"Acting together may change the nature of the game. For example, if two independent farmers were to merge their activities and share the proceeds, both of them might work with less care and energy; the resulting output might be less than under independent operations, in spite of a possibly more efficient division of labor." [Aumann and Dréze (1974, p. 233)]

Economies with local public goods involve both of these constraints. The lack of complete information gives rise to the well-known free rider problem, and the way in which taxes are levied involves normative (welfare) considerations. The following proportional taxation game [Guesnerie and Oddou (1981)] illustrates these two points. The economy consists of n individuals, a single private good, α, and a single public good, β. Individual $i \in N$ is endowed with $\omega^i \in R_+$ units of the private good (and with no public good), and his preferences are represented by the utility function $u^i(\alpha, \beta): R_+^2 \to R_+$. (No hedonic aspect is present.) The public good is produced according to the production function: $\beta = f(\alpha)$. By the very nature of (pure local) public goods, the most efficient partition is the grand coalition, i.e., $B = \{N\}$. But because of either lack of information (on the endowments or, more obviously, on the preferences) or due to ethical considerations (such as proportional to wealth contributions towards the production of the public good) it might be impossible to attain this "first best" solution. Guesnerie and Oddou require that whenever a coalition is formed it has to impose a uniform tax rate on all of its members. The resulting "indirect utility function" of individual i, who belongs to coalition S in which the tax rate is t, $0 \leqslant t \leqslant 1$, is given by

$$\varphi^i(t, S) \equiv u^i((1-t)\omega^i, f(t\Sigma_{i \in S}\omega^i)).$$

Observe that although no hedonic aspect is assumed on the individuals' utility functions $u^i, i \in N$, this aspect manifests itself in the indirect utility functions φ^i, $i \in N$. The associated game (N, V) is then

$$V(S) \equiv \{x \in R^N_+ \mid \text{there exists } t \in [0, 1] \text{ s.t. } x^i \leqslant \varphi^i(t, S) \text{ for all } i \in S\}.$$

It is obvious that if the individuals' preferences for the public good, β, and hence for the tax rates, are sufficiently different, this "second best" game (N, V) will not be superadditive. In this case, individuals will form different jurisdictions, each levying a different tax rate. [See Guesnerie and Oddou (1988), Greenberg and Weber (1982), and Weber and Zamir (1985).]

Labor-managed economies are, at least formally, very similar to economies with local public goods: Substitute "firms" and "working conditions" for "local communities" and "local public goods", respectively. Thus, labor-managed economies serve as another example for situations that give rise to nonsuperadditive games: First, there might be a technological "optimal size" of the firm. In addition, because of either information or ethical constraints, profits might be required to be equally or proportionally shared among the workers. It is then most likely that the more able workers would prefer to establish their own firm. [See, e.g., Farrell and Scotchmer (1988) and Greenberg (1979).]

Following Owen (1977), Hart and Kurz (1983; see Section 8) offer an additional and quite different reason for coalitions to form. Using the TU framework, they assume that the game is superadditive and that society operates efficiently. Hence, it is the grand coalition that will actually and eventually form. The formation of coalitions, in their framework, is a strategic act used by the players in order to increase their share of the total social "pie". They wrote (p. 1048):

> "Thus, we assume as a postulate that society as a whole operates efficiently; the problem we address here is how are the benefits distributed among the participants. With this view in mind, coalitions do not form in order to obtain their "worth" and then "leave" the game. But rather, they "stay" in the game and bargain as a unit with all other players."

Finally, it is interesting to note that some empirical data suggest that even when the actual game being played is superadditive, the grand coalition does not always form, not even in three-person superadditive TU games. [See, e.g., Medlin (1976)].

4. Feasible outcomes for a fixed coalition structure

It is doubtful that the process of coalition formation can be separated from the disbursement of payoffs. Typically, players do not first join a coalition and then negotiate their payoffs; rather, both coalitions and payoffs are determined simultaneously.

> "If the behavioral scientist thinks about decision-making in conflict situations in the model suggested by N-person game theory, he will focus on two fundamental questions: (1) Which coalitions are likely to form? (2) How will the members of a coalition apportion their joint payoff? It is noteworthy that N-person game theory has for the most part been concerned with the second of these questions, i.e., the question of apportionment... The behavioral scientist, on the other hand, may be more fundamentally interested in the first of the two questions." [Anatol Rapoport (1970, p. 286)].

Indeed, the early works on coalition structures (c.s.) avoided, possibly for methodological reasons, the issue of which c.s. are likely to form, and focused instead on the disbursement of payoffs for a fixed c.s. It turns out that even within this partial analysis it is not clear what is a feasible (let alone a "stable") payoff for a given partition. Aumann and Dréze (1974) suggest that the set of *feasible payoffs for a coalition structure B*, be given by

$$X(B) \equiv \cap \{x \in V(S) | S \in B\},$$

or, in their framework of TU games,

$$X(B) \equiv \{x \in R^N | \text{ for all } S \in B, \Sigma_{i \in S} x^i \leq v(S)\}.$$

That is, Aumann and Dréze insist that each coalition should distribute among its members the total payoff accruing to that coalition, and impose this feasibility condition on all the solution concepts they define (namely, the core, the Shapley value, the Nucleolus, von Neumann and Morgenstern solutions, the bargaining set, and the kernel). They stress this point, and write

> "*the major novel element* introduced by the coalition structure B lies in the conditions $x(B_k) = v(B_k)$, which constrain the solution to allocate exactly among the members of each coalition the total payoff of that coalition." [p. 231, my emphasis.]

Requiring that the solution for a given c.s. B belong to $X(B)$ implies that each coalition in B is an autarchy. In particular, no transfers among coalitions are allowed. This restriction might seem, at least at first sight, quite sensible, and indeed, it is employed in most works in this area. But, as the following example [Dréze and Greenberg (1980)] shows, this restriction might have some serious, and perhaps unexpected and unappealing, consequences.

Consider a society with three individuals, i.e., $N = \{1, 2, 3\}$, and assume that every pair of individuals can produce one unit of a (single) commodity. That is, the production technology yields the following sets of S-feasible allocations:

$A(S) = \{a \in R_+^S \mid \Sigma_{i \in S} a^i = 1\}$, if $|S| = 2$,

$A(S) = 0$, otherwise.

The preferences of player i depend on a^i – the quantity of the output he consumes, and S – the (number of) individuals belonging to his coalition. Specifically, for every $S \subset N$ and $i \in S$,

$u^i(a^i, S) = 2a^i$, if $S = \{i\}$,

$u^i(a^i, S) = a^i$, otherwise.

Let W (respectively, U) denote the set of utility levels that can be obtained if transfers among coalitions are (respectively, are not) allowed. Denoting by $a(B)$ the total output that can be produced by all the coalitions in B, we have

$W \equiv \{x \in R^N \mid$ there exist B and $a \in R^N$ s.t. for all $i \in S \in B$,
$x^i \leq u^i(a^i, S)$ and $\Sigma_{i \in N} a^i \leq a(B)\}$,

$U \equiv \{x \in R^N \mid$ there exist B and $a \in R^N$ s.t. for all $i \in S \in B$,
$x^i \leq u^i(a^i, S)$, and for all $S \in B$, $a^S \in A(S)\}$.

Thus, U is the union of $X(B)$ over all coalition structures B. We shall now see that in this example the utility levels that can be obtained if transfers among coalitions are allowed (i.e., the elements of W) strictly Pareto dominate the utility levels that can be obtained when such transfers are prohibited (i.e., the elements of U).

First, note that if B is not of the form $(\{i, j\}, \{k\})$, then $a(B) = 0$, implying $X(B) = \{(0, 0, 0)\}$, yielding the payoff $(0, 0, 0) \in U$. Consider the coalition structure $(\{1, 2\}, \{3\})$ and the distribution of $\frac{1}{3}$ of the commodity to each of the three individuals. (This is attainable if the coalition $\{1, 2\}$ produces 1 unit, and transfers, for free, $\frac{1}{3}$ to player 3.) The resulting vector of utility levels is $(\frac{1}{3}, \frac{1}{3}, \frac{2}{3}) \in W$, which strictly Pareto dominates $(0, 0, 0)$. Next, consider $x \in X(B)$, where B is of the form $(\{i, j\}, \{k\})$. Since the game is symmetric, without loss of generality, assume that $B = (\{1, 2\}, \{3\})$ and that $x_1 > 0$. Consider the coalition structure $B^* \equiv (\{1\}, \{2, 3\})$ and the allocation $a^* \equiv (\frac{3}{4} x_1, x_2 + \frac{1}{8} x_1, \frac{1}{8} x_1)$ which is feasible when transfers (of $\frac{3}{4} x_1$ from $\{2, 3\}$ to $\{1\}$) among the coalitions in B^* are allowed. The resulting vector of utility levels is $y \equiv (\frac{3}{2} x_1, x_2 + \frac{1}{8} x_1, \frac{1}{8} x_1) \in W$, which strictly Pareto dominates the vector of utility levels x.

Thus, for every coalition structure B and every $x \in X(B)$ there exist another coalition structure B^* and a transfer scheme among coalitions in B^* which yield each of the three players a higher utility level than the one he enjoys under x. Hence, *transfers may be necessary to attain Pareto optimality!* (This property could be understood as meaning that "altruism" may be required to achieve Pareto optimality.) It is evident from the above example that this phenomenon is robust, and that quite restrictive assumptions seem required to rule it out [see Greenberg (1980)].

Aumann and Dréze's feasibility restriction that payoffs belong to $X(B)$ is inadequate also in the Owen (1977) and in the Hart and Kurz (1983) models, where the raison d'etre for a coalition S to form is that its members try to receive more than $v(S)$ – the worth of S. Their approach

"differs from the Aumann and Dréze (1974) approach where each coalition $B_k \in B$ gets only its worth [i.e., $v(B_k)$]. The idea is that coalitions form not in order to get their worth, *but to be in a better position when bargaining with the others on how to divide the maximal amount available* [i.e., the worth of the grand coalition, which for superadditive games is no less than $\sum_k v(B_k)$]. It is assumed that the amount $v(N)$ will be distributed among the players and thus that all collusions and group formations are done with this in mind." [Kurz (1988, p. 160)]

Thus, even the appropriate definition of the basic notion of feasible payoffs for a given coalition structure is by no means straightforward. (Although, as stated above, most papers on coalition structures adopt Aumann and Dréze's restriction.) A much more complicated question, to which we now turn, is which coalition structures (and payoff configurations) are "stable" and hence more likely to prevail.

5. Individually stable coalition structures

It is useful, if only as a first step towards a more general stability analysis, to consider models in which only a single individual is allowed to join one of the coalitions in the existing structure. This restriction, which combines the notions of ψ-stability [Luce and Raiffa (1957)] and Nash equilibrium, seems particularly plausible whenever the individual is small relative to the size of the existing coalitions or whenever high transaction costs (due to lack of information and communication among the agents) are involved in forming a new coalition. Thus, for example, an unsatisfied professor (worker) considers moving to another existing University (firm) rather than establishing a new school (becoming an entrepreneur). Similarly, an individual may decide to change his sports club or the magazine to which he subscribes, rather than start a new one.

Perhaps the most immediate notion within this framework is that of an individually stable equilibrium (i.s.e.) [Greenberg (1977), and Dréze and Greenberg (1980)]. In an i.s.e. no individual can change coalitions in a way that is beneficial to himself and to all the members of the coalition which he joins. Since an individual can always decide to form his own coalition, we shall require that an i.s.e. be individually rational. Formally,

Definition 5.1. Let (N, V) be a game in coalitional form. The pair (x, B), where B is a partition and $x \in X(B)$ is an *individually stable equilibrium* (*i.s.e.*) if there exists no $S \in B \cup \{\emptyset\}$ and $i \in N \setminus S$ such that x belongs to the interior of the set $V(S \cup \{i\})$.

Clearly, if the game is superadditive then an i.s.e. always exists. Indeed, every Pareto optimal and individually rational payoff x together with the coalition structure $B = \{N\}$, constitutes an i.s.e. But in nonsuperadditive games it might be impossible to satisfy even this weak notion of stability. For example, it is easily verified that there is no i.s.e. in the TU game (N, v), where $N = \{1, 2, 3\}$ and

$$v(\{i\}) = 1, \quad v(S) = 4 \text{ if } |S| = 2, \quad \text{and } v(N) = 0.$$

The two main conceptual differences between an i.s.e. and the (noncooperative) notion of Nash equilibrium are that, in contrast to i.s.e., in a Nash equilibrium a player (i) takes the choices made by all other players as given, and (ii) can deviate unilaterally, i.e., he need not get the consent of the members of the coalition he wishes to join. Note that while (ii) facilitates mobility, (i), in contrast, makes mobility more difficult. This is because an existing coalition cannot revise its (strategy) choice in order to attract ("desirable") players.

For games with a continuum of (atomless) players, however, it is possible to establish the existence of an i.s.e. by proving the existence of a Nash equilibrium. Using this observation, Westhoff (1977) established the existence of an i.s.e. for a class of economies with a continuum of players, a single private and a single public good. Each consumer chooses among a finite number of communities, and each community finances its production of the public good using proportional income tax levied on its residents. An important novelty of this model is that the tax rate is determined by simple majority rule. Under some restrictive assumptions (notably that individuals can be ranked unidimensionally according to their marginal rates of substitution between the private and the public good) Westhoff proves the existence of a ("stratified") i.s.e.

Greenberg (1983) established the existence of an i.s.e. for a general class of economies with externalities and spillover effects both in consumption and in production. He considered local public goods economies with a continuum of players (each belonging to one of a finite number of types), a finite number of private goods and a finite number of public goods. Individuals can be partitioned into an exogenously given number of communities, with each community financing the production of its public goods by means of some (continuous) tax scheme that it chooses. At the prevailing equilibrium competitive prices, no individual would find it beneficial to migrate from his community to another one.

Other related works on the existence of i.s.e. in economies with local public goods, include: Denzau and Parks (1983), Dunz (1989), Epple et al. (1984), Greenberg and Shitovitz (1988), Konishi (1993), Nechyba (1993), Richter (1982), and Rose-Ackerman (1979). A recent interesting application of the notion of i.s.e.

is offered by Dutta and Suzumura (1993) who study the set of individually stable coalition structures in a two stage (initial investment and subsequent output) R&D game.

In any realistic model, however, the assumption that there is a continuum of individuals might be unacceptable. Thus, indivisibilities can be regarded as inherent to the analysis of coalition formation. Hence, if we wish to guarantee the existence of "individually stable partitions", the mere ability of a player to improve his payoff by joining another coalition (that is willing to accept him) cannot, by itself, be sufficient for him to do so; additional requirements might be called for. The classical modification of the notion of i.s.e. is the bargaining set and its variants (see Chapter 41). Drèze and Greenberg (1980) offered another modification – the notion of "*individually stable contractual equilibrium*" (i.s.c.e.) which requires that a player could only change coalitions when that move is beneficial to himself, to all the members of the coalition which he joins, and also to all the members of the coalition which he leaves. They show that an i.s.c.e. always exists, and moreover, it is possible to design a dynamic process that yields, in a finite number of steps, an i.s.c.e. Thus, contracts (preventing an individual from leaving a coalition if the remaining members object) and transfers (enabling the compensation of the remaining members by either positive or negative amount) together generate "individual stability" of efficient partitions.

The two equilibrium concepts, i.s.e. and i.s.c.e., may be linked to alternative institutional arrangements sometimes encountered in the real world. Thus, university departments may be viewed as coalitions of individuals (professors) who are allowed to move when they receive an attractive offer (i.e., when the move is beneficial to the individual and to the department he joins, no matter whether the department he leaves loses or gains...). In contrast, soccer teams typically "own" their players, who are not allowed to move to another team (coalition) unless a proper compensation is made.

Stability when only single players contemplate deviations emerges naturally in the analysis of cartels. A cartel can be regarded as a partition of firms into two coalitions: Those belonging to the cartel and those outside. Consider an economy in which a subset of firms form a cartel which acts as a price leader, and firms outside the cartel, the fringe, follow the imposed price. As is well-known, the cartel is in general not stable in the sense that it pays for a single firm to leave the cartel and join the fringe:

"The major difficulty in forming a merger is that it is more profitable to be outside a merger than to be a participant. The outsider sells at the same price but at the much larger output at which marginal cost equals price. Hence, the promoter of merger is likely to receive much encouragement from each firm – almost every encouragement, in fact, except participation." [Stigler (1950, 25–26)]

At least in a finite economy, it is reasonable to assume that each firm realizes that its decision to break away from the cartel and join the competitive fringe will affect the market structure. It is then possible that the fall in price due to the increased overall competition in the market leads to a fall in profit for the individual firm contemplating a move out of the cartel, which goes beyond the advantage of joining the competitive fringe. A cartel is stable if and only if firms inside do not find it desirable to exit and firms outside do not find it desirable to enter. In deciding the desirability of a move each firm hypothesizes that no other firm will change its strategy concerning membership in the cartel. Furthermore, firms are assumed to have exact knowledge of the dependence of profits at equilibrium on the size of the cartel. Under some restrictive assumptions, (such as identical firms), it is possible to establish the existence of stable cartels. [See, e.g., D'Aspremont et al. (1983) and Donsimoni et al. (1986).]

6. Coalition structure core

The most commonly used stability concept is the coalition structure core, which extends individual stability to group stability.

Definition 6.1. Let (N, V) be an n-person game in coalitional form. The *coalition structure B is stable* if its core is nonempty, i.e., if there exists a feasible payoff $x \in X(B)$ which does not belong to the interior of $V(S)$, for any $S \subset N$. (That is, x cannot be blocked.) The game (N, V) has *a coalition structure core* if there exists at least one partition which is stable.

Clearly, the game (N, V) has a coalition structure core if and only if its superadditive cover game, $(N, \hat{V})$, has a nonempty core. (Recall that the *superadditive cover game* is the game $(N, \hat{V})$, where

$$\hat{V}(S) \equiv \cup \{X(B^S) | B^S \text{ is a partition of } S\} \times R^{N \setminus S}.$$

By Scarf's (1967) theorem, therefore, a sufficient condition for the game (N, V) to have a coalition structure core is that its superadditive cover game is (Scarf) balanced. [Applications of this observation include: Boehm (1973, 1974), Greenberg (1979), Ichiishi (1983), Oddou (1972), and Sondermann 1974).] The interesting question here is which social environments lead to games that have a coalition structure core.

Wooders (1983) used the fact that the balancing weights are rational numbers (and hence have a common denominator) to show that balancedness is equivalent to superadditivity of a sufficiently large replica game. It follows, that if the game is superadditive, if per capita payoffs are bounded (essentially, that only "small" coalitions are effective), and if every player has many clones, then it has a coalition structure core.

Perhaps the best known example of social environments whose associated games have a coalition structure core is the central assignment game (in particular the "marriage game"). The basic result [Gale and Shapley (1962)], is that the game describing the following situation has a coalition structure core: the set of players is partitioned into two disjoint groups. Each member of one group has to be matched with one member (or more) of the second. The fact that such games have a coalition structure core implies that it is possible to pair the players in such a way that there are no two players that prefer to be together over their current partners. (See Chapter 33).

Shaked (1982) provides another example of a social environment which admits a stable partition. Shaked considers a market game with a continuum of players, given by the set $T = [0, 1]$, which is partitioned into m subsets, $\{T^h\}$, $h = 1, 2, \ldots, m$. All players that belong to T^h are said to be of type h. Let μ denote the Lebesgue measure on T. For a coalition[6] S, let $\alpha^h(S)$ denote the percentage of players of type h in S, i.e., $\alpha^h(S) \equiv \mu(S \cap T^h)/\mu(S)$, and let $\alpha(S)$ denote the normalized profile of S, i.e., $\alpha(S) \equiv (\alpha^1(S), \alpha^2(S), \ldots, \alpha^m(S))$. Assume that when S forms it cannot discriminate among its members that belong to the same type, i.e., $A(S)$ is the set of all equal treatment S-feasible allocations. The utility level that a player of type h attains when he belongs to S where the S-feasible allocation is $a(S)$, is given by $u^h(\alpha(S), a(S))$. Shaked's key assumption is that the hedonic aspect enters only through the profile, $\alpha(S)$, which, like the (equal treatment) allocation $a(S)$, does not depend on the size of S. That is, if S and $\hat{S}$ are such that, for some number $k > 0$, $\mu(S \cap T^h) = k\mu(\hat{S} \cap T^h)$ for all h, $h = 1, 2, \ldots, m$, then $A(S) = A(\hat{S})$, and for all $a \in A(S)$, $u^h(\alpha(S), a) = u^h(\alpha(\hat{S}), a)$, $h = 1, 2, \ldots, m$. Under these assumptions Shaked proves that if for every type h, $u^h(\alpha, a)$ is continuous and quasi-concave in the allocation a, then the associated game has a coalition structure core.

Greenberg and Weber (1986) pointed out another class of games with a coalition structure core, which they called consecutive games. Consecutive games have the property that players can be "numbered" (named, ordered) in such a way that a group of individuals can (strictly) benefit from cooperation only if it is consecutive, i.e., if all the individuals that lie in between the "smallest" and the "largest" players in that group (according to the chosen ordering) are also members of it. Formally,

Defintion 6.2. Let $N = \{1, 2, \ldots, n\}$, and let $S \subset N$. Coalition S is called *consecutive* if for any three players i, j, k, where $i < j < k$, if $i, k \in S$ then $j \in S$. The *consecutive structure of S*, denoted $C(S)$, is the (unique) partition of S into consecutive coalitions which are maximal with respect to set inclusion. Equivalently, it is the (unique) partition of S into consecutive coalitions that contains the minimal number of coalitions. [Thus, $C(S) = \{S\}$ if and only if S is consecutive.] Let (N, V) be a game.

---

[6] A coalition is a non-null measurable subset of T.

The *consecutive game* (N, V_c), generated by (N, V) is given by:

$$V_c(S) = \cap \{V(T) | T \in C(S)\}.$$

i.e., $x \in R^N$ belongs to $V_c(S)$ if $x \in V(T)$ for every (consecutive) coalition $T \in C(S)$.

An immediate example of a consecutive game is the one derived from the following social environment. Consider n landlords who are located along a line (a river). The profits a coalition (of landlords) can engender depend on the distance it can transport some commodities. Assume that without the consent of all landlords who own property between locations i and k, the transportation between these locations cannot be carried out. Then, only consecutive coalitions generate value added. For consecutive games we have the following result which is valid also for games with an infinite number of players [see Greenberg and Weber (1993b)].

Theorem 6.3. *Let (N, V) be a game. Then, the consecutive game (N, V_c) has a coalition structure core. Moreover, there exists a payoff x in the coalition structure core of (N, V_c) such that for every $T = \{1, 2, \ldots, t\}$, $t \leq n$, x^T belongs to the coalition structure core of the game (T, V^{T_c}), where for all $S \subset T$, $V^{T_c}(S)$ is the projection of $V_c(S)$ on R^T.*

Theorem 6.3 can be used to establish the existence of a coalition structure core also in nonconsecutive games with the property that if a payoff x is blocked, then it is also blocked by a consecutive coalition. Greenberg and Weber (1986) used this observation to prove the existence of a "strong Tiebout equilibrium" for a large class of economies with a single private good and several public goods. They showed that in such economies, individuals can be partitioned into K jurisdictions. These jurisdictions finance the production of their (local) public goods by either equal share or proportional taxation, in such a way that there exists no group of discontented consumer-voters who could improve their situation by establishing their own jurisdiction (in which they produce a vector of public goods and share its production costs according to the existing tax scheme). Moreover, no single individual would find it beneficial to leave the jurisdiction in which he currently resides and join any of the other existing jurisdictions.

Another application of Theorem 6.3 was offered by Demange and Henriet (1991). They considered an economy with two commodities: money and a single consumption good, α, whose quality, location, or price, might vary. Let A denote the possible specifications of α. The production technology of α, for a given attribute $a \in A$, is assumed to exhibit increasing returns to scale. Demange and Henriet show that under some mild assumptions on the preferences and the technology, the associated game has a coalition structure core. A similar result holds also for a model with a continuum of individuals. That is, there exists a list $\{(a_h, S_h)\}_{h=1,2,\ldots,H}$ (called a *sustainable configuration*) where (i) $\{S_h\}$ is a partition of N, (ii) for all h, $a_h \in A$, (iii) no individual can be made better off by joining another "coalition-firm" (a_t, S_t), (iv) every coalition-firm (a_h, S_h) makes nonnegative profits, and (v) no offer $a \in A$ can attract a group of individuals S such that (a, S) yields nonnegative profits. In particular, therefore, the coalition structure $B = \{S_h\}_{h=1,2,...,H}$ is stable.

Farrell and Scotchmer (1988) analyze "partnership formation", where people who differ in ability can get together to exploit economies of scale, but must share their output equally within groups. (The two examples they cite are information sharing in salmon fishing and law partnerships.) Specifically, they consider a game (N, V) where for all $S \subset N$, $V(S)$ contains a single outcome, $u(S)$, and, in addition, players "rank the coalitions" in the same way, i.e., whenever $i, j \in S \cap T$ we have that $u^i(S) > u^i(T)$ if and only if $u^j(S) > u^j(T)$. Farrell and Scotchmer show that such games have a coalition structure core, and moreover, coalitions in stable partitions are consecutive (when players are ordered according to their ability). They also prove that, generically, there is a unique stable partition.

Demange (1990) generalizes the notion of consecutive coalitions and Theorem 6.3 from the (unidimensional) line to a tree configuration, T, where a coalition is effective if the vertices that correspond to its members belong to a path in T.

7. Political equilibrium

Like firms and local jurisdictions, political parties can also be regarded as coalitions: each party (which is characterized by its political position) is identified with the set of voters supporting it. Multiparty equilibrium is, then, a stable partition of the set of voters among the different parties.

A fairly standard setting of the multiparty contest is the following: Society is represented by a finite set $N = \{1, 2, ..., n\}$ of n voters. Voter $i \in N$ has a preference relation over a finite set of alternatives (political positions), Ω, represented by the utility function $u^i: \Omega \to R$. Since Ω is finite, the alternatives in Ω can be ordered (assigned numbers) so that for every two alternatives, $a, b \in \Omega$, either $a \geq b$ or $b \geq a$. It is frequently assumed that preferences are *single-peaked*, i.e., Ω can be ranked in such a way that for each voter $i \in N$, there exists a unique alternative $q(i) \in \Omega$, called i's peak, such that for every two alternatives, $a, b \in \Omega$, if either $q(i) \geq a > b$ or $q(i) \leq a < b$, then $u^i(a) > u^i(b)$.

There exist, in general, two broad classes of electoral systems. The first, called the *fixed standard method*, sets down a standard, or criterion. Those candidates who surpass it are deemed elected, and those who fail to achieve it are defeated. A specific example is the uniform-quota method. Under this arrangement, a standard is provided exogenously, and party seats, district delegations size, and even the size of the legislature are determined endogenously. The second class of electoral system, operating according to the *fixed number method*, is much more common. According to this method, each electoral district is allotted a fixed number of seats (K), and parties or candidates are awarded these seats on the basis of relative performance at the polls. Of course, $K = 1$ describes the familiar "first-past-the-post" plurality method common in Anglo-American elections. In contrast to the fixed-standard method, here the number of elected officials is fixed exogenously, and the standard required for election is determined endogenously.

Consider first the fixed standard model. Let the exogenously given positive integer m, $m \leq n$, represent the least amount of votes required for one seat in the Parliament. Each running candidate adopts a political position in Ω, with which he is identified. Greenberg and Weber (1985) define an *m-equilibrium* to be a set of alternatives such that when the voters are faced with the choice among these alternatives, and when each voter (sincerely) votes for the alternative he prefers best in this set, each alternative receives at least m votes and, moreover, no new (potential) candidate can attract m voters by offering another alternative, in addition to those offered in the m-equilibrium. In order to define formally an m-equilibrium we first need

Definition 7.1. Let $A \subset \Omega$. The *support of alternative $a \in \Omega$ given A*, denoted $S(a; A)$, is the set of all individuals who strictly prefer alternative a over any other alternative in A. That is,

$$S(a; A) \equiv \{i \in N \mid u^i(a) > u^i(b) \text{ for all } b \in A, b \neq a\}.$$

A collection $A \subset \Omega$ is an m-equilibrium if the support of each alternative in A consists of at least m individuals (thus each elected candidate receives at least m votes), whereas no other alternative is supported (given A) by m or more voters (thus, no potential candidate can guarantee himself a seat). Formally,

Definition 7.2. A set of alternatives A is called an *m-equilibrium* if, for all $a \in \Omega$, $|S(a; A)| \geq m$ if and only if $a \in A$.

Greenberg and Weber (1985) proved

Theorem 7.3. *For every (finite) society N, (finite) set of alternatives Ω and quota m, $0 < m \leq n$, there exists an m-equilibrium.*

The game (N, V) associated with the above model is given by:

$$V(S) \equiv \{x \in R^N \mid \exists a \in \Omega \text{ for all } i \in S \text{ s.t. } x^i = u^i(a)\}, \quad \text{if } |S| \geq m,$$
$$V(S) = \{x \in R^N \mid x^i = 0 \text{ for all } i \in S\}, \quad \text{otherwise.}$$

The reader can easily verify that every m-equilibrium belongs to the coalition structure core of (N, V), but not vice versa. Thus, by Theorem 7.3, (N, V) has a coalition structure core. It is noteworthy that (N, V) need not be (Scarf) balanced (which might account for the relative complexity of the proof of Theorem 7.3). Greenberg and Weber (1993a) showed that their result is valid also for the more general case where the quota m is replaced with any set of winning coalitions. The more general equilibrium notion allows for both free entry and free mobility. The paper also studies its relation to strong and coalition-proof Nash equilibria (see Section 11). Two recent applications of Greenberg and Weber (1993a) are Deb, Weber, and Winter (1993) which focuses on simple quota games, and Le Breton and Weber (1993), which extends the analysis to a multidimensional set of alternatives but restricts the set of effective coalitions to consist only of pairs of players.

For the fixed number (say, K) method, Greenberg and Shepsle (1987) suggested the following equilibrium concept.

Definition 7.4. Let K be a positive integer. A set of alternatives A, $A \subset \Omega$, is a K-equilibrium if (i) A contains exactly K alternatives, and (ii) for all $b \in \Omega \setminus A$, $|S(b; A \cup \{b\})| \leqslant |S(a; A \cup \{b\})|$ for all $a \in A$.

Condition (i) requires that the number of candidates elected coincide with the exogenously given size K of the legislature. Condition (ii) guarantees that no potential candidate will find it worthwhile to offer some other available alternative $b \in \Omega \setminus A$, because the number of voters who would then support him will not exceed the number of voters who would support any of the K candidates already running for office. Indeed, condition (ii) asserts that the measure of the support of each of the original running candidates will rank (weakly) among the top K. Unfortunately,

Theorem 7.5. *For any given $K \geqslant 2$, there exist societies for which there is no K-equilibrium.*

Theorem 7.5 remains true even if voters have Euclidean preferences, i.e., for all $i \in N$ and all $a \in \Omega$, $u^i(a) = -|q(i) - a|$. Weber (1990a), [see also Weber (1990b)], has considerably strengthened Theorem 7.5 by proving

Theorem 7.6. *Define $\varphi(2) \equiv 7$, and for $K \geqslant 3$, $\varphi(K) = 2K + 1$. A necessary and sufficient condition for every society with n voters to have a K-equilibrium is that $n \leqslant \varphi(K)$.*

Another class of games in which much of the formal analysis of political games was done is that of simple games, where

Definition 7.7. A *simple game* is a pair (N, W), where W is the set of all *winning coalitions*. A simple game is *monotonic* if any coalition that contains a winning coalition is winning; *proper* if $S \in W$ implies $N \setminus S \notin W$; and *strong* if $S \notin W$ implies $N \setminus S \in W$. A (strong) *weighted majority game* is a (strong) simple game for which there exists a quota $q > 0$ and nonnegative weights $w^1, w^2, \ldots, w^n$, such that $S \in W$ if and only if $\Sigma_{i \in S} w^i \geqslant q$.

Peleg (1980, 1981) studied the formation of coalitions in simple games. To this end he introduced the notion of a dominant player. Roughly, a player is dominant if he holds a "strict majority" within a winning coalition. Formally,

Definition 7.8. Let (N, W) be a proper and monotonic simple game. A player i is *dominant* if there exists a coalition $S \in W$ which i *dominates*, i.e., such that $i \in S$ and for every $T \subset N$ with $T \cap S = \emptyset$, we have that $T \cup [S \setminus \{i\}] \in W$ implies $[T \cup \{i\}] \in W$, and there exists $T^* \in W$ such that $T^* \setminus \{i\} \notin W$.

Assume that the game (N, W) admits a single dominant player, i_0. Peleg studied the winning coalition that will result if i_0 is given a mandate to form a coalition. In particular, he analyzed the consequences of the following hypotheses:

(1) If $S \in W$ forms, then i_0 dominates S. That is, the dominant player forms a coalition which he dominates.

(2) If $S \in W$ forms, then $i_0 \in S$ and for every $i \in S \setminus i_0$, $S \setminus \{i\} \in W$. That is, the dominant player belongs to the coalition that forms, and, moreover, no other member of that coalition is essential for the coalition to stay in power.

(3) The dominant player tries to form that coalition in W which maximizes his Shapley value.

(4) The dominant player tries to form that coalition in W which maximizes his payoff according to the Nucleolus.

Among the interesting theoretic results, Peleg proved that if (N, W) is either a proper weighted majority game, or a monotonic game which contains veto players [i.e., where $\cap \{S | S \in W\} \neq \emptyset$], then (N, W) admits a single dominant player. In addition, these two papers contain real life data on coalition formation in town councils in Israel and in European parliaments (see Section 10).

8. Extensions of the Shapley value

Aumann and Dréze (1974) defined the notion of the *Shapley value for a given c.s. B, or the B-value*, by extending, in a rather natural way, the axioms that define the Shapley value (for the grand coalition N) for TU games. Their main result (Theorem 3, p. 220) is that the B-value for player $i \in B_k \in B$ coincides with i's Shapley value in the subgame (B_k, V), [where the set of players is B_k, and the worth of a coalition $S \subset B_k$ is $v(S)$]. It follows that the B-value always belongs to $X(B)$, i.e., the total amount distributed by the B-value to members of B_k is $v(B_k)$.

This same value arises in a more general and rather different framework. Myerson (1977) introduced the notion of a "cooperation structure", which is a graph whose set of vertices is the set of players, and an arc between two vertices means that the two corresponding players can "communicate". A coalition structure B is modelled by a graph where two vertices are connected if and only if the two corresponding players belong to the same coalition, $B_k \in B$. Myerson extended the Shapley value to arbitrary cooperation structures, G. The Myerson value, $\mu(G, V)$, is defined as follows: For $S \subset N$, let $S(G)$ be the partition of S induced by G. Given the cooperation structure G and the TU game (N, V), define the TU game (N, V_G) where $V_G(S) \equiv \Sigma_{T \in S(G)} V(T)$. Then, $\mu(G, V)$ is the Shapley value of the game (N, V_G). Myerson showed that if the cooperation graph represents a c.s. B, then the extended Shapley value coincides with the B-value. Aumann and Myerson (1988) used this framework to study, in weighted majority games, the very important and difficult subject of dynamic formation of coalitions.

Owen (1977, 1986) extended the Shapley value to games with prior coalition structures that serve as pressure groups. The Owen value for a TU game (N, V) and a coalition structure B assigns each player his expected marginal contribution with respect to the uniform probability distribution over all orders in which players of the same coalition appear successively. Winter (1992) showed that the difference between the Aumann and Dréze's B-value and the Owen's value can be attributed to the different definitions of the reduced games used for the axiomatic characterization of these values.

Following Owen, in Hart and Kurz (1983) coalitions form only for the sake of bargaining, realizing that the grand coalition (which is the most efficient partition) will eventually form. Given a c.s. B, the negotiation process among the players is carried out in two stages. In the first, each coalition acts as one unit, and the negotiations among these "augmented players" determine the worth of the coalitions (which might differ from the one given by v). The second stage involves the bargaining of the players within each coalition $B_k \in B$, over the pie that B_k received in the first stage. An important feature of the Hart and Kurz model is that in both stages, the same solution concept is employed, namely, the Shapley value. This two-stage negotiation process yields a new solution concept, called the *Coalitional Shapley value* (*CS-value*), which assigns to every c.s. B a payoff vector $\varphi(B)$ in R^N, such that (by the efficiency axiom) $\Sigma_{i \in N} \varphi^i(B) = v(N)$.

It turns out that the CS-value coincides with Owen's (1977, 1986) solution concept. But, in contrast to Owen (who extended the Shapley value to games in which the players are organized in a fixed, given c.s.), Hart and Kurz use the CS-value in order to determine which c.s. will form. (This fact accounts for the difference between the sets of axioms employed in the two works, that give rise to the same concept.) Based on the CS-value, Hart and Kurz define a *stable c.s.* as one in which no group of players would like to change, taking into account that the payoff distribution in a c.s. B is $\varphi(B)$. This stability concept is closely related to the notions of strong Nash equilibrium and the α and β cores. Specifically, depending on the reaction of other players, Hart and Kurz suggest two notions of stability, called γ and δ. In the γ-model, it is assumed that a coalition which is abandoned by some of its members breaks apart into singletons; in the δ-model, the remaining players form one (smaller) coalition.

Hart and Kurz (1984) characterize the set of stable c.s. for all three-person games, all symmetric four-player games, and all Apex games. Unfortunately, as they show, there exist (monotone and superadditive) games which admit no stable coalition structure (in either the γ or the δ model).

Winter (1989) offers a new "value" for level structures of cooperation. A level structure is a hierarchy of nested coalition structures each representing a cooperation on a different level of commitment. A level structure value is then defined as an operator of pairs of level structures and TU games, and is characterized axiomatically.

9. Abstract systems

The notions of an abstract system and an abstract stable set, due to von Neumann and Morgenstern, provide a very general approach to the study of social environments. In this section I review some applications of these notions within the framework of cooperative games. But, as Greenberg (1990) shows, many game-theoretic solution concepts, for all three types of games, can be derived from the "optimistic stable standard of behavior" for the corresponding "social situation", and hence these solution concepts can be defined as abstract stable sets for the associated abstract systems. This representation not only offers a unified treatment of cooperative and noncooperative games[1], but also delineates the underlying negotiation process, information, and legal institutions. I believe the potential of this framework for the analysis of coalition formation has not yet been sufficiently exploited.

Definition 9.1. A von Neumann and Morgenstern *abstract system* is a pair $(D, \angle)$ where D is a nonempty set and $\angle$ is a *dominance relation*: for $a, b \in D$, $a \angle b$ is interpreted to mean that b dominates a. The *dominion* of $a \in D$, denoted $\Delta(a)$, consists of all elements in D that a dominates, according to the dominance relation $\angle$. That is,

$$\Delta(a) = \{b \in D \mid b \angle a\}.$$

For a subset $K \subset D$, the dominion of K, denoted $\Delta(K)$, is the set

$$\Delta(K) = \cup \{\Delta(a) \mid a \in K\}.$$

Definition 9.2. Let $(D, \angle)$ be an abstract system. The set $K \subset D$ is the *abstract core* for the system $(D, \angle)$ if $K = D \setminus \Delta(D)$.

An element $x \in D$ belongs to the abstract core if and only if no other element in D dominates it. Thus, the abstract core is the set of maximal elements (with respect

---

[1] Section 11 includes several applications of this framework to coalition formation in noncooperative games.

to the dominance relation $\angle$). von Neumann and Morgenstern suggested the following more sophisticated solution concept. An element $x \in D$ belongs to an abstract stable set K if and only if no other element in K dominates it. Formally,

Definition 9.3. Let $(D, \angle)$ be an abstract system. A set $K \subset D$ is a von Neumann and Morgenstern *abstract stable set* for the system $(D, \angle)$ if $K = D \setminus \Delta(K)$.

The condition that $K \subset D \setminus \Delta(K)$, i.e., that if $x, y \in K$ then x does not dominate y, is called *internal stability*. The reverse condition that $K \supset D \setminus \Delta(K)$ is called *external stability*. (When applied to games in coalitional form, the above two general notions yield the familiar concepts of the core and the von Neumann and Morgenstern solution.)

Shenoy (1979), who was the first to use this general framework for the study of coalition formation, introduced the concept of the *dynamic solution*. This notion is closely related to von Neumann and Morgenstern abstract stable set for the transitive closure of the original dominance relation $\angle$, and to Kalai et al.'s (1976) "R-admissible set". It captures the dynamic aspect of the negotiations: when players consider a particular element in D, they consider all those elements in D that can be reached by *successive* (a chain of) dominations. Shenoy then studied the abstract core and the dynamic solution for the following three abstract systems. The abstract set D in the first two systems that Shenoy considered consists of the set of all possible coalition structures. The difference between the first two abstract systems lies in the dominance relation. Under one system, the partition B_2 dominates the partition B_1 if B_2 contains a coalition S such that *any* feasible payoff in B_1 is S-dominated by a payoff that is feasible for B_2, i.e.,

$B_1 \angle B_2$ if and only if there exists $S \in B_2$ s.t.
$[x \in X(B_1) \Rightarrow \exists y \in X(B_2)$ with $y^i > x^i$ for all $i \in S]$.

Shenoy (1978) showed that this dominance relation yields, in simple games, very similar predictions on coalition formation to those of Caplow's (1956, 1959) and Gamson's (1961) descriptive sociological theories.

The above dominance relation requires that for any proposed payoff in $X(B_1)$, the objecting coalition, S, can find a payoff in $V(S)$, [or, equivalently, in $X(B_2)$], which benefits all members of S. A stronger requirement is that S has to choose, in advance, a payoff in $V(S)$ which is superior (for its members) to any feasible payoff in the existing c.s. B_1. That is, the partition B_2 dominates B_1 if B_2 contains a coalition S and a fixed payoff $y \in X(B_2)$ which S-dominates every payoff that is feasible for B_1, i.e.,

$B_1 \angle B_2$ if and only if there exist $S \in B_2$ and $y \in X(B_2)$
s.t. $y^i > x^i$ for all $i \in S$ and all $x \in X(B_1)$.

Since elements of D are partitions, the above two abstract systems describe social environments in which coalitions first form, and only then is the disbursement of payoffs within each coalition determined. In contrast, the third abstract system suggested by Shenoy requires that both the partition and the corresponding payoff be determined simultaneously. Specifically, the abstract set is

$D^* \equiv \{(B, x) | B \text{ is a partition and } x \in X(B)\}$.

As elements of D^* already include payoffs, the natural dominance relation here is

$(B_1, x) \angle (B_2, y)$ iff $\exists S \in B_2$ s.t. $y^i > x^i$ for all $i \in S$.

It is easy to see that the abstract core of this third system yields the coalition structure core.

There are, of course, many other interesting abstract systems that can be associated with a given game (N, V). And, it seems most likely that the von Neumann and Morgenstern abstract stable sets for such systems will yield many illuminating results. Moreover, abstract stable sets may also prove useful in studying more general cooperation structures. Clearly, allowing for coalition formation need not result in coalition structures; in general, individuals may (simultaneously) belong to more than one coalition.

"Coalition structures, however, are not rich enough adequately to capture the subtleties of negotiation frameworks. For example, diplomatic relations between countries or governments need not be transitive and, therefore, cannot be adequately represented by a partition; thus both Syria and Israel have diplomatic relations with the United States but not with each other." [Aumann and Myerson (1988, p. 177)]

A cooperation structure which is closely related to that of an abstract stable set is McKelvey et al.'s (1978) notion of the "competitive solution". The competitive solution attempts to predict the coalitions that will form and the payoffs that will result in spatial voting games. McKelvey et al. assumed that potential coalitions bid for their members in a competitive environment via the proposals they offer. Given that several coalitions are attempting to form simultaneously, each coalition must, if possible, bid efficiently by appropriately rewarding its "critical" members. More specifically, let D be the set of *proposals*, i.e.,

$D = \{(S, x) | S \subset N \text{ and } x \in V(S) \cap V(N)\}$.

For two proposals (S, x) and (T, y) in $\cdot D$, the dominance relation is given by

$(S, x) \angle (T, y)$ if $y^i > x^i$ for all $i \in S \cap T$.

For any two coalitions S and T, the set $S \cap T$ consists of the *pivotal*, or *critical* players between S and T. Thus, $(S, x) \angle (T, y)$ if the pivotal players between S and T are better off under the proposal (T, y) than they are under (S, x). $K \subset D$ is a *competitive solution* if K is an abstract stable set such that each coalition represented in K can have exactly one proposal, i.e., if $(S, a) \in K$, then $(S, b) \in K$ implies $a = b$.

Clearly, if $x \in \text{Core}(N, V)$ then (N, x) belongs to every competitive solution. McKelvey et al. prove that many simple games admit a competitive solution, K, with the property that $(S, x) \in K$ implies that S is a winning coalition and x belongs to a ("main-simple") von Neumann and Morgenstern set. While the competitive solution seems to be an appealing concept, theorems establishing existence and uniqueness await proofs or (perhaps more probably) counterexamples.

A related notion is that of "aspiration" [see, e.g., Albers (1974), Bennett (1983), Bennett and Wooders (1979), Bennett and Zame (1988), and Selten (1981)]. Specifically, assume that prior to forming coalitions, every player declares a "reservation price" (in terms of his utility) for his participation in a coalition. A "price vector" $x \in R^N$ is an *aspiration* if the set of all the coalitions that can "afford" its players (at the declared prices) has the property that each player belongs to at least one coalition in that collection, and, in addition, x cannot be blocked. That is, for all $i \in N$ there exists $S \subset N$ such that $x \in V(S)$, and there exist no $T \subset N$ and $y \in V(T)$ such that $y^i > x^i$ for all $i \in T$. As is easily verified, every core allocation in the balanced cover game is an aspiration. By Scarf's (1967) theorem, therefore, the set of aspirations is nonempty. Clearly, the set of coalitions that support an aspiration is not a partition. Thus, despite its appeal, an aspiration fails to predict which coalitions will form, and moreover, it ignores the possibility that players who are left out will decide to lower their reservation price. [For more details concerning the aspiration set and its variants, see Bennett (1991)].

Other works that consider cooperative structures (in which players may belong to more than one coalition and the set of coalitions that can possibly form might be restricted) include: Auman and Myerson (1988), Gilles (1987, 1990), Gilles et al. (1989), Greenberg and Weber (1983), Kirman (1983), Kirman et al. (1986), Myerson (1977), Owen (1986), and Winter (1989).

10. Experimental work

Experimentation and studies of empirical data can prove useful in two ways: In determining which of the competing solution concepts have better predictive power, and, equally important, in potentially uncovering regularities which have not been incorporated by the existing models in game theory.

> "Psychologists speak of the "coalition structures" in families. The formal political process exhibits coalitions in specific situations (e.g., nominating conventions) with the precision which leaves no doubt. Large areas of international politics can likewise be described in terms of coalitions, forming and dissolving. Coalitions, then, offer a body of fairly clear data. The empirically minded behavioral scientist understandably is attracted to it. His task is to describe patterns of coalition formation systematically, so as to draw respectably general conclusions." [Anatol Rapoport (1970, p. 287)]

There is a large body of rapidly growing experimental works that study the formation of coalition structures, and test the descriptive power of different solution concepts. Unfortunately, these works do not arrive at clear conclusions or unified results. Most of the experiments were conducted with different purposes in mind, and they vary drastically from one another in experimental design, rules of the game, properties of the characteristic function, instructions, and population of subjects. Moreover, the evaluation of the findings is difficult because of several methodological problems such as controlling the motivation of the subjects, making sure they realize (and are not just told) the precise game they are playing, and the statistical techniques for comparing the performance of different solution concepts.

Moreover, most of the experiments have involved a very small (mainly 3–5) number of players, and have focused on particular types of games such as: Apex games [see e.g., Funk et al. (1980), Horowitz and Rapoport (1974), Kahan and Rapoport (1984, especially Table 13.1), and Selten and Schaster (1968)]; spatial majority games [see, e.g., Beryl et al. (1976), Eavy and Miller (1984), Fiorina and Plott (1978), and McKelvey and Ordeshook (1981, 1984)] and games with a veto player [see, e.g., Michener et al. (1976), Michener et al. (1975), Michener and Sakurai (1976), Nunnigham and Roth (1977, 1978, 1980), Nunnigham and Szwajkowski (1979), and Rapoport and Kahan (1979)].

Among the books that report on experimental works concerning coalition formation are: Kahan and Rapoport (1984); Part II of Rapoport (1970); Rapoport (1990); Sauermann (1978); and Part V of Tietz et al. (1988).

The only general result that one can draw from the experimental works is that none of the theoretical solution concepts fully accounts for the data. This "bad news" immediately raises the question of whether game theory is a descriptive or a prescriptive discipline. Even if, as might well be the case, it is the latter, experiments can play an important role. It might be possible to test for the appeal of a solution concept by running experiments in which one or some of the subjects are "planted experts" (social scientists or game theorists), each trying to convince the other players to agree on a particular solution concept. Among the many difficulties in designing such experiments is controlling for the convincing abilities of the "planted experts" (a problem present in every experiment that involves negotiations).

Political elections provide a useful source of empirical data for the study of coalition formation. DeSwaan (1973) used data from parliamentary systems to test coalition theories in political science. In particular, Section 5 of Chapter 4 presents two theories predicting that only consecutive coalitions will form. [As Axelrod (1970, p. 169) writes: "a coalition consisting of adjacent parties tends to have relatively low dispersion and thus low conflict of interest....".] The first theory is "Leisersen' minimal range theory", and the second is "Axelrod's closed minimal range theory". Data pertaining to these (and other) theories appear in Part II of DeSwaan's book. Peleg (1981) applied his "dominant player" theory (see Section 7) to real life data from several European parliaments and town councils in Israel. Among his findings are the following: 80% of the assemblies of 9 democracies considered in DeSwaan (1973) were "dominated", and 80% of the coalitions in the dominated assemblies contained a dominant player. Town councils in Israel exhibit a similar pattern: 54 out of 78 were "dominated", and 43 of the 54 dominated coalitions contained a dominant player. It is important to note that this framework allows analysis of coalition formation without direct determination of the associated payoffs. Rapoport and Weg (1986) borrow from both DeSwaan's and Peleg's works. They postulate that the dominant player (party) is given the mandate to form a coalition, and that this player chooses to form the coalition with minimal "ideological distances" (in some politically interpretable multidimensional space). Data from the 1981 Israeli Knesset elections are used to test the model.

11. Noncooperative models of coalition formation

During the last two decades research in game theory has focused on noncooperative games where the ruling solution concept is some variant of Nash equilibrium. A recent revival of interest in coalition formation has led to the reexamination of this topic using the framework of noncooperative games. In fact, von Neumann and Morgenstern (1944) themselves offer such a model where the strategy of each player consists of choosing the coalition he wishes to join, and a coalition S forms if and only if every member of S chooses the strategy "I wish to belong to S".

Aumann (1967) introduced coalition formation to strategic form games by converting them into coalitional form games, and then employing the notion of the core. The difficulty is that, as was noted in Section 2, coalitional games cannot capture externalities which are, in turn, inherent to strategic form games. Aumann considered two extreme assumptions that coalitions make concerning the behavior (choice of strategies) of nonmembers, yielding the notions of "the α and the β core".

Aumann (1959) offered another approach to the analysis of coalition formation within noncooperative environments. Remaining within the framework of strategic form games, Aumann suggested the notion of "strong Nash equilibrium", where coalitions form in order to correlate the strategies of their members. However, this notion involves, at least implicitly, the assumption that cooperation necessarily requires that players be able to sign "binding agreements". (Players have to follow the strategies they have agreed upon, even if some of them, in turn, might profit by deviating.)

But, coalitions can and do form even in the absence of (binding) contracts. This obvious observation was, at long last, recently recognized. The notion of "coalition proof Nash equilibrium" [CPNE Bernheim et al. (1987), Bernheim and Whinston (1987)], for example, involves "self-enforcing" agreements among the members of a coalition. Greenberg (1990) characterized this notion as the unique "optimistic stable standard of behavior", and hence, as a von Neumann and Morgenstern abstract stable set. Because this characterization is circular (Bernheim et al.'s is recursive) it enables us to extend the definition of CPNE to games with an infinite number of players. An interesting application of CPNE for such games was recently offered by Alesina and Rosenthal (1993) who analyzed CPNE in voting games with a continuum of players (voters) where policy choices depend not only on the executive branch but also on the composition of the legislature.

The characterization of CPNE as an optimistic stable standard of behavior also highlights the negotiation process that underlies this notion, namely, that only subcoalitions can further deviate. [The same is true for the notion of the core; see Greenberg (1990, Chapter 6.)] This opened the door to study abstract stable sets where the dominance relation allows for a subset, T, of a deviating coalition, S, to approach and attract a coalition Q, $Q \subset N \setminus S$, in order to jointly further deviate. Whether players in Q will join T depends, of course, on the information they have concerning the actions the other players will take. If this information is common knowledge, the resulting situation is the "coalitional contingent threats" (Greenberg 1990). If, on the other hand, previous agreements are not common knowledge, then in order for players in Q to agree to join T, they have to know what agreement was reached (previously and secretly) by S. Chakravorti and Kahn (1993) suggest that members of Q will join T only if any action of $T \cup Q$ that might decrease the welfare of a member of Q will also decrease that of a member of T (who is aware of the agreement reached by S). The abstract stable set for this system yields the notion of "universal CPNE". Chakravorti and Sharkey (1993) study the abstract stable set that results when each player of a deviating coalition may hold different beliefs regarding the strategies employed by the other players. Clearly, there are many other interesting negotiation processes, reflected by the "dominance relation", that are worth investigation (e.g., Ray and Vohra, 1992, and Kahn and Mookherjee, 1993).

The set of feasible outcomes in the "coalition proof Nash situation", whose optimistic stable standard of behavior characterizes CPNE, is the Cartesian product of the players' strategy sets. Recently, Ray (1993) replaced this set by the set of probability distributions over the Cartesian product of the players' strategy sets, and derived the notion of "ex-post strong correlated equilibrium" from the optimistic stable standard of behavior for the associated situation. Ray assumes that when forming a coalition, members share their information. The other extreme assumption, that (almost) no private information is shared, leads to Moreno and Wooders' (1993) notion of "coalition-proof equilibrium", (which is, in fact, a coalition proof ex-ante correlated equilibrium.) Einy and Peleg (1993) study correlated equilibrium when members of a coalition may exchange only partial (incentive-compatible) information, yielding the notion of "coalition-proof communication equilibrium".

The abundance of the different solution concepts demonstrates that the phrase "forming a coalition" does not have a unique interpretation, especially in games with imperfect information. In particular, the meaning of forming a coalition depends on the legal institutions that are available, which are not, but arguably ought to be, part of the description of the game.

The growing, and related, literature on "renegotiation-proofness", and "preplay-communications" in repeated games is moving in the same desirable direction: it analyzes formation of coalitions in "noncooperative" games, and in the absence of binding agreements. [See, e.g., Asheim (1991), Benoit and Krishna (1993), Bergin and Macleod (1993), DeMarzo (1992), Farrell (1983, 1990), Farrell and Maskin (1990), Matthews et al. (1991), Pearce (1987), and van Damme (1989).]

Coalition formation was recently introduced also to extensive form games, which, in contrast to games in strategic form, capture "sequential bargaining". Many of these works, however, involve only two players (see Chapter 16). In the last few years there is a growing literature on the "implementation" of cooperative solution concepts (mainly the core and the Shapley value) by (stationary) subgame perfect equilibria. [See, e.g., Bloch (1992), Gul (1989), Hart and Mas-Colell (1992), Moldovanu and Winter (1991, 1992), Perry and Reny (1992), and Winter (1993a, b).] Moldovanu (1990) studied noncooperative sequential bargaining based on a game in coalitional form. He showed that if this game is balanced then the set of payoffs which are supported by a coalition proof Nash equilibrium for the extensive form game coincides with the core of the original game in coalitional form.

While the general trend of investigating coalition formation and cooperation within the framework of noncooperative games is commendable, the insight derived from such "implementations" is perhaps limited. After all, both the core and the Shapley value stand on extremely sound foundations, and the fact that they can be supported by subgame perfect equilibrium in some specific game tree, while interesting, contributes little to their acceptability as solution concepts. Moreover, as the following two quotes suggest, it is doubtful that extensive form games offer a natural (or even an aceeptable) framework for the study of coalition formation.

"...even if the theory of noncooperative games were in a completely satisfactory state, there appear to be difficulties in connection with the reduction of cooperative games to noncooperative games. It is extremely difficult in practice to introduce into the cooperative games the moves corresponding to negotiations in a way which will reflect all the infinite variety permissible in the cooperative game, and to do this without giving one player an artificial advantage (because of his having the first chance to make an offer, let us say)." [McKinsey (1952, p. 359)]

"In much of actual bargaining and negotiation, communication about contingent behavior is in words or gestures, sometimes with and sometimes without contracts or binding agreements. A major difficulty in applying game theory to the study of bargaining or negotiation is that the theory is not designed to deal with words and gestures – especially when they are deliberately ambiguous – as moves. Verbal sallies pose two unresolved problems in game theoretic modeling: (1) how to code words, (2) how to describe the degree of commitment." [Shubik (1984, p. 293)]

The question then arises, which of the three types of games, and which of the multitude of solution concepts, is most appropriate for the analysis of coalition formation? It is important to note that seemingly different notions are, in fact, closely related. This is most clearly illustrated by the theory of social situations [Greenberg (1990)]. Integrating the representation of cooperative and non-cooperative social environments relates many of the currently disparate solution concepts and highlights the precise negotiation processes and behavioral assumptions that underlie them. This is demonstrated, for example, by Xue's (1993) study of farsighted coalitional stability. Farsightedness is reflected by the way individuals and coalitions view the consequences of their actions, and hence the opportunities that are available to them. Different degrees of farsightedness can, therefore, be captured by different social situations. Xue (1993) analyzes two such situations. In the first, individuals consider all "eventually improving" outcomes. The resulting optimistic and conservative stable standards of behavior turn out to be closely related to Harsanyi's (1974) "strictly stable sets", and to Chwe's (1993) "largest consistent set", respectively. Xue analyzes another situation where no binding agreements can be made. The resulting optimistic and conservative stable standards of behavior characterize the set of self-enforcing agreements among farsighted individuals and predict the coalitions that are likely to form.

I hope that this section as well as the rest of this survey suggests that, despite the interesting results that have already been obtained, doing research on coalition formation is both (farsighted) individually rational and socially desirable.

References

Albers, W. (1974) 'Zwei losungkonzepte fur kooperative Mehrpersonspiele, die auf Anspruchnisveaus der spieler Basieren', OR-Verfarhen, 1–13.
Alesina, A. and H. Rosenthal (1993) 'A Theory of Divided Government', mimeo, Harvard University.
Asheim, G. (1991) 'Extending renegotiation proofness to infinite horizon games', *Games and Economic Behavior*, **3**: 278–294.
Aumann, R. (1959) 'Acceptable points in general cooperative n-person games', *Annals of Mathematics Studies*, **40**: 287–324.
Aumann, R. (1967) 'A survey of cooperative games without side payments', M. Shubik, ed., *Essays in Mathematical Economics*. Princeton University Press, pp. 3–27.
Aumann, R. and J. Dréze (1974) "Cooperative games with coalition structures', International Journal of Game Theory, 217–237.
Aumann, R. and R. Myerson (1988) 'Endogenous formation of links between players and of coalitions: An application of the Shapley value', in: A. Roth, ed., *The Shapley value: Essays in honor of L.S. Shapley*, Cambridge University Press.
Axelrod, R. (1970) *Conflict of Interest*. Chicago: Markham.
Bennett, E. (1983) 'The aspiration approach to predicting coalition formation and payoff distribution in sidepayments games", *International Journal of Game Theory*, **12**: 1–28.
Bennett, E. (1991) 'Three approaches to predicting coalition formation in side payments games', in *Game Equilibrium Models*, R. Selten (ed.), Springer-Verlag, Heidelberg.
Bennett, E. and M. Wooders (1979) 'Income distribution and firm formation', *Journal of Comparative Economics*, **3**, 304–317.
Bennett, E. and W. Zame (1988) 'Bargaining in cooperative games', *International Journal of Game Theory*, **17**: 279–300.

Benoit, J.P. and V. Krishna (1993) 'Renegotiation in finitely repeated games', *Econometrica*, **61**: 303–323.
Bergin, J. and W.B. Macleod (1993) 'Efficiency and renegotiation in repeated games', *Journal of Economic Theory*, to appear.
Bernheim, B.D., B. Peleg and M. Whinston (1987) 'Coalition proof Nash equilibria, I: Concepts', *Journal of Economic Theory*, **42**: 1–12.
Bernheim, B.D. and M. Whinston (1987) 'Coalition proof Nash equilibria, II: Applications', *Journal of Economic Theory*, **42**: 13–29.
Beryl, J., R.D. McKelvey, P.C. Ordeshook and M.D. Wiener (1976) 'An experimental test of the core in a simple N-person, cooperative, nonsidepayment game', *Journal of Conflict Resolution*, **20**: 453–479.
Bloch, F. (1992) 'Sequential Formation of Coalitions with Fixed Payoff Division', mimeo, Brown University.
Boehm, V. (1973) 'Firms and market equilibria in a private ownership economy', *Zeitschrift für Nationalekonomie*, **33**: 87–102.
Boehm, V. (1974) 'The limit of the core of an economy with production', *International Economic Review*, **15**: 143–148.
Buchanan, J. (1965) 'An economic theory of clubs', *Economica*, 1–14.
Caplow, T. (1956) 'A theory of coalitions in the triad', *American Sociological Review*, **21**: 489–493.
Caplow, T. (1959) 'Further development of a theory of coalitions in the triad', *American Journal of Sociology*, **64**: 488–493.
Caplow, T. (1968) *Two against one: Coalitions in triads*, Englewood Cliffs, NJ: Prentice-Hall.
Chakravorti, B. and C. Kahn (1993) 'Universal coalition-proof equilibrium: Concepts and applications', mimeo, Bellcore.
Chakravorti, B. and W. Sharkey (1993) 'Consistency, un-common knowledge and coalition proofness', mimeo, Bellcore.
Charnes, A. and S.C. Littlechild (1975) 'On the formation of unions in n-person games', *Journal of Economic Theory*, **10**: 386–402.
Chwe, M. (1993) 'Farsighted Coalitional Stability', *Journal of Economic Theory*, to appear.
Cooper, T. (1986) 'Most-favored-customer pricing and tacit collusion', *Rand Journal of Economics*, **17**, 377–388.
Cross, J. (1967) 'Some theoretical characteristics of economic and political coalitions', *Journal of Conflict Resolution*, **11**: 184–195.
Dutta, B. and K. Suzumura (1993) 'On the Sustainability of Collaborative R&D Through Private Incentives', The Institute of Economic Research Histotsubashi Discussion Paper.
D'Aspremont, C., A. Jacquemin, J.J. Gabszewicz and J.A. Weymark (1983) 'On the stability of collusive price leadership', *Canadian Journal of Economics*, **16**: 17–25.
Deb, R., S. Weber and E. Winter (1993) 'An extension of the Nakamura Theorem to Coalition Structures', SMU Discussion Paper.
Demange, G. (1990) 'Intermediate Preferences and Stable Coalition Structures', DELTA EHESS Discussion Paper.
Demange, G. and D. Henriet (1991) 'Sustainable oligopolies', *Journal of Economic Theory*, **54**, 417–428.
DeMarzo, P.M. (1992) 'Coalitions, Leadership and Social Norms: The Power of Suggestion in Games", *Games and Economic Behavior*, **4**: 72–100.
Denzau, A.T. and R.P. Parks (1983) 'Existence of voting market equilibria', *Journal of Economic Theory*, **30**, 243–265.
DeSwaan, A. (1973) *Coalition theories and cabinet formations*, San Franscisco: Jossey-Bass, Inc.
Donsimoni, M., Economides N.S. and H.M. Polemarchakis (1986) 'Stable cartels', *International Economic Review*, **27**, 317–327.
Dréze, J. and J. Greenberg (1980) 'Hedonic coalitions: Optimality and Stability', *Econometrica*, **48**: 987–1003.
Dunz, K. (1989) 'Some comments on majority rule equilibria in local public good economies', *Journal of Economic Theory*, **47**, 228–234.
Eavy, C.L. and G.J. Miller (1984) 'Bureaucratic agenda control: Imposition or bargaining', *American Political Science Review*, **78**, 719–733.
Einy, E. and B. Peleg (1993) 'Coalition-Proof Communication Equilibrium', mimeo, Hebrew University of Jerusalem.

Epple, D., R. Filimon and T. Romer (1984) 'Equilibrium among local jurisdictions: Towards an integrated treatment of voting and residential choice", *Journal of Public Economics*, **24**, 281–308.
Farrell, J. (1983) 'Credible repeated game equilibria', mimeo, Berkeley University.
Farrell, J. (1990) 'Meaning and credibility in cheap talk games', in *Mathematical models in economics*, M. Dempster (ed.), Oxford University Press.
Farrell, J. and E. Maskin (1990) 'Renegotiation in repeated games', *Games and Economic Behavior*, **10**: 329–360.
Farrell, J. and S. Scotchmer (1988) 'Partnerships', *Quarterly Journal of Economics*, **103**: 279–297.
Fiorina, M. and C. Plott (1978) 'Committee decisions under majority rule', *American Political Science Review*, **72**, 575–598.
Funk, S.G., Rapoport, A. and Kahan, J.P. (1980) 'Quota vs. positional power in four-personal Apex games', *Journal of Experimental Social Psychology*, **16**, 77–93.
Gale, D. and L. Shapley (1962) 'College admission and stability of marriage' *American Mathematical Monthly*, **69**: 9–15.
Gamson, W.A. (1961) 'A theory of coalition formation', *American Sociological Review*, **26**: 373–382.
Gilles, R.P. (1987) 'Economies with coalitional structures and core-like equilibrium concepts', mimeo, Tilburg University.
Gilles, R.P. (1990) 'Core and equilibria of socially structured economies: The modeling of social constraints in economic behavior", Tilburg University.
Gilles, R.P., P.H.M. Ruys and S. Jilin (1989) 'On the existence of networks in relational models', mimeo, Tilburg University.
Greenberg, J. (1977) 'Pure and local public goods: A game-theoretic approach', in *Public Finance*, A, Sandmo (ed.), Lexington, MA: Heath and Co.
Greenberg, J. (1979) 'Existence and optimality of equilibrium in labor managed economies', *Review of Economic Studies*, **46**: 419–433.
Greenberg, J. (1980) 'Beneficial altruism', *Journal of Economic Theory*, **22**: 12–22.
Greenberg, J. (1983) 'Local public goods with mobility: Existence and optimality of general equilibrium', *Journal of Economic Theory*, **30**: 17–33.
Greenberg, J. (1990) *The theory of social situations: An alternative game theoretic approach*, Cambridge University Press.
Greenberg, J. and K. Shepsle (1987) 'Multiparty competition with entry: The effect of electoral rewards on candidate behavior and equilibrium', *American Political Science Review*, **81**: 525–537.
Greenberg. J. and B. Shitovitz (1988) 'Consistent voting rules for competitive local public good economies', *Journal of Economic Theory*, **46**, 223–236.
Greenberg, J. and S. Weber (1982) 'The equivalence of superadditivity and balancedness in the proportional tax game', *Economics Letters*, **9**: 113–117.
Greenberg, J. and S. Weber (1983) 'A core equivalence theorem with an arbitrary communication structure', *Journal of Mathematical Economics*, **11**: 43–55.
Greenberg, J. and S. Weber (1985) 'Multiparty Equilibria under proportional representation', *American Political Science Review*, **79**: 693–703.
Greenberg, J. and S. Weber (1986) 'Strong Tiebout equilibrium under restricted preferences domain', *Journal of Economic Theory*, **38**: 101–117.
Greenberg, J. and S. Weber (1993a) 'Stable coalition structures with unidimensional set of alternatives', *Journal of Economic Theory*, **60**, 62–82.
Greenberg, J. and S. Weber (1993b) 'Stable coalition structures in consecutive games', in: *Frontiers of Game Theory*. Binmore, Kirman and Tani (eds.), MIT Press.
Guesnerie, R. and C. Oddou (1981) 'Second best taxation as a game', *Journal of Economic Theory*, **25**: 67–91.
Guesnerie, R. and C. Oddou (1988) 'Increasing returns to size and their limits', *Scandinavian Journal of Economics*, **90**: 259–273.
Gul, F. (1989) 'Bargaining foundations of Shapley value', *Econometrica*, **57**, 81–95.
Harsanyi, J. (1974) 'An equilibrium-point interpretation of stable sets and a proposed alternative definition', *Management Science*, **20**: 1472–1495.
Hart, S. and M. Kurz (1983) 'Endogenous formation of coalitions', *Econometrica*, **51**, 1047–1064.
Hart, S. and M. Kurz (1984) 'Stable coalition structures', in *Coalitions and Collective Action*, M.T. Holler (ed.) Physica-Verlag, pp. 236–258.
Hart, S. and A. Mas-Colell (1992) 'N-person noncooperative bargaining', mimeo, Center for Rationality and Interactive Decision Theory.

Horowitz, A.D. and Rapoport, A. (1974) 'Test of the Kernel and two bargaining set models in four and five person games', in *Game theory as a theory of conflict resolution*, A. Rapoport (ed.) Dordrecht, the Netherlands: D. Reidel.
Ichiishi, T. (1983) *Game theory for economic analysis*. New York: Academic Press.
Kahan, J.P. and Rapoport, A. (1984) *Theories of coalition formation*. Hillsdale, NJ: Erlbaum.
Kalai, E., E. Pazner and D. Schmeidler (1976) 'Collective choice correspondences as admissible outcomes of social bargaining processes', *Econometrica*, **44**: 233–240.
Kaneko M. (1982) 'The central assignment game and the assignment markets', *Journal of Mathematical Economics*, **10**: 205–232.
Kaneko, M. and M. Wooders (1982) 'Cores of partitioning games', *Mathematical Social Sciences*, **2**: 313–327.
Kahn, C. and D. Mookherjee (1993) 'Coalition-Proof Equilibrium in an Adverse Selection Insurance Market, mimeo, University of Illinois.
Kirman, A. (1983) 'Communication in markets: A suggested approach', *Economics Letters*, **12**: 101–108.
Kirman, A., C. Oddou and S. Weber (1986) 'Stochastic communication and coalition formation', *Econometrica*, **54**: 129–138.
Konishi, H. (1993) 'Voting with ballots and feet: existence of equilibrium in a local public good economy', mimeo, University of Rochester.
Kurz, M. (1988) 'Coalitional value', in *The Shapley value: Essays in honor of L.S. Shapley*, A. Roth (ed.), Cambridge University Press, pp. 155–173.
Le Breton, M. and S. Weber (1993) 'Stable Coalition Structures and the Principle of Optimal Partitioning', to appear in *Social Choice and Welfare*, Barnett, Moulin, Salles, and Schofield eds., Cambridge University Press.
Lucas, W.F. and J.C. Maceli (1978) 'Discrete partition function games', in *Game Theory and Political Science*, P. Ordeshook (ed.), New York, pp. 191–213.
Luce, R.D. and H. Raiffa (1957) *Games and Decisions*. New York: Wiley.
Macleod, B. (1985) 'On adjustment costs and the stability of equilibria', *Review of Economic Studies*, **52**: 575–591.
Maschler, M. (1978) 'Playing an n-person game: An experiment' in *Coalition forming behavior*, H. Sauermann (ed.), Tubingen: J.C.B. Mohr.
Matthews, S., M. Okuno-Fujiwara and A. Postlewaite (1991) 'Refining cheap talk equilibria', *Journal of Economic Theory*, **55**, 247–273.
McGuire, M. (1974) 'Group segregation and optimal jurisdictions', *Journal of Political Economy*, **82**: 112–132.
McKelvey, R.D. and P.C. Ordeshook (1981) 'Experiments on the core: Some disconcerting results' *Journal of Conflict Resolution*, **25**, 709–724.
McKelvey, R.D. and P.C. Ordeshook (1984) 'The influence of committee procedures on outcomes: Some experimental evidence', *Journal of Politics*, **46**: 182–205.
McKelvey, R.D., P.C. Ordeshook and M.D. Wiener (1978) 'The competitive solution for N-person games without transferable utility, with an application to committee games', *American Political Science Review*, **72**: 599–615.
McKinsey, J. (1952) *Introduction to the theory of games*. New York: McGraw-Hill.
Medlin, S.M. (1976) 'Effects of grand coalition payoffs on coalition formation in three-person games', *Behavioral Science*, **21**: 48–61.
Michener, H.A., J.A. Fleishman and J.J. Vaska (1976) 'A test of the bargaining theory of coalition formation in four-person groups', *Journal of Personality and Social Psychology*, **34**, 1114–1126.
Michener, H.A., J.A. Fleishman, J.J. Vaske and G.R. Statza (1975) 'Minimum resource and pivotal power theories: A competitive test in four-person coalitional situations', *Journal of Conclict Resolution*, **19**: 89–107.
Michener, H.A. and M.M. Sakurai (1976) 'A search note on the predictive adequacy of the kernel', *Journal of Conflict Resolution*, **20**, 129–142.
Moldovanu, B. (1990) 'Sequential bargaining, cooperative games, and coalition-proofness', mimeo, Bonn University.
Moldovanu, B. and E. Winter (1990) 'Order independent equilibria', mimeo, Center for Rationality and Interactive Decision Theory.
Moldovanu, B. and E. Winter (1992) 'Core implementation and increasing returns for cooperation', mimeo, Center for Rationality and Interactive Decision Theory.
Moreno, D. and J. Wooders (1993) 'Coalition-Proof Equilibrium', mimeo, University of Arizona.

Myerson, R. (1977) 'Graphs and cooperation in games', *Mathematics of Operations Research*, **2**, 225–229.
Myerson, R. (1980) 'Conference structures and fair allocation rules', *International Journal of Game Theory*, **9**: 169–182.
Nechyba, T. (1993) 'Hierarchical public good economies: Existence of equilibrium, stratification, and results', mimeo, University of Rochester.
Nunnigham, J.K. and A.E. Roth (1977) 'The effects of communication and information availability in an experimental study of a three-person game', *Management Science*, **23**, 1336–1348.
Nunnigham, J.K. and A.E. Roth (1978) 'Large group bargaining in a characteristic function game', *Journal of Conflict Resolution*, **22**, 299–317.
Nunnigham, J.K. and A.E. Roth (1980) 'Effects of group size and communication availability on coalition bargaining in a veto game', *Journal of Personality and Social Psychology*, **39**: 92–103.
Nunnigham, J.K. and E. Szwajkowski (1979) 'Coalition bargaining in four games that include a veto player", Journal of Personality and Social Psychology, **37**, 1933–1946.
Oddou, C. (1972) 'Coalition production economies with productive factors', CORE D.P. 7231.
Owen, G. (1977) 'Values of games with a priori unions', in *Essays in Mathematical Economics and Game Theory*, R. Hein and O. Moeschlin (ed.), New York: Springer-Verlag, pp. 76–88.
Owen, G. (1986) 'Values of graph-restricted games', *Siam Journal of Discrete Methods*, **7**, 210–220.
Pearce, D. (1987) 'Renegotiation proof equilibria: Collective rationality and intertemporal cooperation', mimeo, Yale University.
Peleg, B. (1980) 'A theory of coalition formation in committees', *Journal of Mathematical Economics*, **7**, 115–134.
Peleg, B. (1981) 'Coalition formation in simple games with dominant players', *International Journal of Game Theory*, **10**: 11–33.
Perry, M. and P. Reny (1992) 'A noncooperative view of coalition formation and the core', mimeo, University of Western Ontario.
Postlewaite, A. and R. Rosenthal (1974) 'Disadvantageous syndicates', *Journal of Economic Theory*, **10**: 324–326.
Rapoport, A. (1970) *N-person game theory*, Ann Arbor, Mi: University of Michigan Press.
Rapoport, A. (1987) 'Comparison of theories for disbursements of coalition values', *Theory and decision*, **22**, 13–48.
Rapoport, A. (1990) *Experimental studies of interactive decision*, Dordrecht, the Netherlands: Kluwer Academic Press.
Rapoport, A. and J.P. Kahan (1979) 'Standards of fairness in 4-person monopolistic cooperative games', in *Applied game theory*, S.J. Brams, A Schotter, and G. Schwediaur (eds.) Würzburg: Physica-Verlag.
Rapoport, A. and J.P. Kahan (1984) 'Coalition formation in a five-person market game', *Management Science*, **30**: 326–343.
Rapoport, A. and E. Weg (1986) 'Dominated, connected, and tight coalitions in the Israel Knesset', *American Journal of Political Science*, **30**, 577–596.
Ray, D. and R. Vohra (1992) 'Binding Agreements', mimeo, Brown University.
Ray, I. (1993) 'Coalition-Proof Correlated Equilibrium', C.O.R.E. Discussion Paper.
Richter, D.K. (1982) 'Weakly democratic regular tax equilibria in a local public goods economy with perfect consumer mobility', *Journal of Economic Theory*, **27**: 137–162.
Riker, W. (1962) *The theory of political coalitions*, Yale University Press.
Riker, W. (1967) 'Bargaining in a three person game', *American Political Science Review*, **61**, 642–656.
Riker, W. and W.J. Zavoina (1970) 'Rational behavior in politics: Evidence from a three person game', *American Political Science Review*, **64**: 48–60.
Rose-Ackerman, S. (1979) 'Market models of local government: Exit, voting, and the land market', *Journal of Urban Economics*, **6**: 319–337.
Salant, S.W., S. Switzer and R.J. Reynolds (1993) 'Losses from horizontal merger: The effects of an exogenous change in industry structure on Cournot-Nash equilibrium', *Quarterly Journal of Economics*, **97**: 185–199.
Sauermann, H. (1978) 'Coalition forming behavior', Tubingen: J.C.B. Mohr (P. Siebeck).
Scarf, H. (1967) 'The core of an N-person game', *Econometrica*, **35**: 50–69.
Selten, R. (1972) 'Equal share analysis of characteristic function experiments', in *Beitrage zur experimentallen wirtschaftsforschung*, H. Sauermann, (ed.) Vol. III, Tubingen.
Selten, R. (1975) 'A reexamination of the perfectness concept for equilibrium points in extensive form games', *International Journal of Game Theory*, **4**, 25–55.

Selten, R. (1981) 'A noncooperative model of characteristic function bargaining' in Essays in Game Theory and Mathematical Economics in Honor of O. Morgenstern. V. Bohem and H. Nachtjamp (eds.), Bibliographisches Institut Mannheim, Wien-Zürich.
Selten, R. and K.G. Schaster (1968) 'Psychological variables and coalition-forming behavior' in *Risk and Uncertainty*, K. Borch and J. Mossin (eds.) London: McMillan, pp. 221–245.
Shaked, A. (1982) 'Human environment as a local public good', *Journal of Mathematical Economics*, **10**: 275–283.
Shapley, L. (1953) 'A value for n-person games', in *Contributions to the Theory of Games*, Kuhn and Tucker (ed.), 307–317.
Shapley, L. and M. Shubik (1975) 'Competitive outcomes in the cores of market games', *International Journal of Game Theory*, **4**: 229–327.
Shenoy, P.P. (1978) 'On coalition formation in simple games: A mathematical analysis of Caplow's and Bamson's theories', *Journal of Mathematical Psychology*, **18**: 177–194.
Shenoy, P.P. (1979) 'On coalition formation: A game-theoretic approach', *International Journal of Game Theory*, **8**, 133–164.
Shubik, M. (1984) *Game theory in the social sciences: Concepts and solutions*, Cambridge, MA: MIT Press.
Sondermann, D. (1974) 'Economies of scale and equilibria in coalition production economies', *Journal of Economic Theory*, **10**: 259–291.
Stigler, G.J. (1950) 'Monopoly and oligopoly by merger', *American Economic Review, Papers and Proceedings*, 23–34.
Tadenuma, K. (1988) 'An axiomatization of the core of coalition structures in NTU games', mimeo.
Thrall, R.M. and W.F. Lucas (1963) 'N-person games in partition function form', *Naval Research and Logistics*, **10**: 281–297.
Tietz, R., W. Albers and R. Selten (eds.) (1988) *Bounded rational behavior in experimental games and markets*, Berlin: Springer-Verlag.
van Damme, E. (1989) 'Renegotiation proof equilibria in repeated games', *Journal of Economic Theory*, **47**, 206–207.
von Neumann, J. and O. Morgenstern (1944) *Theory of games and economic behavior*, Princeton University Press.
Weber, S. (1990a) 'Existence of a fixed number equilibrium in a multiparty electoral system', *Mathematical Social Sciences*, **20**: 115–130.
Weber, S. (1990b) 'Entry detterence and balance of power', mimeo, York University.
Weber, S. and S. Zamir (1985) 'Proportional taxation: Nonexistence of stable structures in an economy with a public good', *Journal of Economic Theory*, **35**, 178–185.
Westhoff, F. (1977) 'Existence of equilibria with a local public good', *Journal of Economic Literature*, **15**: 84–112.
Winter, E. (1989) 'A value for cooperative games with levels structure of cooperation', *International Journal of Game Theory*, **18**, 227–242.
Winter, E. (1992) 'The consistency and the potential for values of games with coalition structure', *Games and Economic Behavior*, **4**, 132–144.
Winter, E. (1993a) 'Noncooperative bargaining in natural monopolies', *Journal of Economic Theory*, to appear.
Winter, E. (1993b) 'The demand commitment bargaining and the snowballing cooperation', *Economic Theory*, to appear.
Wooders, M. (1983) 'The Epsilon core of a large replica game', *Journal of Mathematical Economics*, **11**, 277–300.
Xue, L. (1993) 'Farsighted Optimistic and Conservative Coalitional Stability', mimeo, McGill University.

Chapter 38

GAME-THEORETIC ASPECTS OF COMPUTING

NATHAN LINIAL*

The Hebrew University of Jerusalem

Contents

1.	Introduction	1340
	1.1. Distributed processing and fault tolerance	1340
	1.2. What about the other three main issues?	1342
	1.3. Acknowledgments	1343
	1.4. Notations	1344
2.	Byzantine Agreement	1344
	2.1. Deterministic algorithms for Byzantine Agreement	1346
	2.2. Randomization in Byzantine Agreements	1350
3.	Fault-tolerant computation under secure communication	1352
	3.1. Background in computational complexity	1357
	3.2. Tools of modern cryptography	1360
	3.3. Protocols for secure collective computation	1368
4.	Fault-tolerant computation – The general case	1372
	4.1. Influence in simple games	1373
	4.2. Symmetric simple games	1379
	4.3. General perfect-information coin-flipping games	1381
	4.4 Quo vadis?	1384
5.	More points of interest	1385
	5.1. Efficient computation of game-theoretic parameters	1385
	5.2. Games and logic in computer science	1386
	5.3. The complexity of specific games	1387
	5.4. Game-theoretic consideration in computational complexity	1388
	5.5. Random number generation as games	1388
	5.6. Amortized cost and the quality of on-line decisions	1390
References		1391

---

*Work supported in part by the Foundation of Basic Research in the Israel Academy of Sciences. Part of this work was done while the author was visiting IBM Research – Almaden and Stanford University.

Handbook of Game Theory, Volume 2, Edited by R.J. Aumann and S. Hart
© Elsevier Science B.V., 1994. All rights reserved

1. Introduction

Computers may interact in a great many ways. A parallel computer consists of a group of processors which *cooperate* in order to solve large-scale computational problems. Computers *compete* against each other in chess tournaments and serve investment firms in their battle at the stock-exchange. But much more complex types of interaction do come up – in large computer networks, when some of the components fail, a scenario builds up, which involves both cooperation and conflict. The main focus of this article is on protocols allowing the well-functioning parts of such a large and complex system to carry out their work despite the failure of others. Many deep and interesting results on such problems have been discovered by computer scientists in recent years, the incorporation of which into game theory can greatly enrich this field.

Since we are not aware of previous attempts at a systematic study of the interface between game theory and theoretical computer science, we start with a list of what we consider to be the most outstanding issues and problems of common interest to game theory and theoretical computer science.

(1) How does the outcome of a given game depend on the players' computational power?

(2) Classify basic game-theoretic parameters, such as values, optimal strategies, equilibria, as being easy, hard or even impossible to compute. Wherever possible, develop efficient algorithms to this end.

(3) Theories of fault-tolerant computing: Consider a situation where a computational task is to be performed by many computers which have to cooperate for this purpose. When some of the computers malfunction, a conflict builds up. This situation generates a number of new game-theoretic problems.

(4) The theories of parallel and distributed computing concern efficient cooperation between computers. However, game-theoretic concepts are useful even in sequential computing.

For example, in the analysis of algorithms, one often considers an adversary who selects the (worst-case) input. It also often helps to view the same computer at two different instances as two cooperating players.

This article focuses on the third item in this list. We start with some background on fault-tolerant computing and go on to make some brief remarks on the other three items in this list.

1.1. *Distributed processing and fault tolerance*

A *distributed system* consists of a large number of loosely coupled computing devices (*processors*) which together perform some computation. Processors can communicate by sending messages to each other. Two concrete examples the reader may keep in mind, are large computer networks and neural systems (being thought of as networks consisting of many neurons). The most obvious difficulty is that data is available locally to individual processors, while a solution for the computational problem usually reflects a global condition. But distributed systems need also cope with serious problems of *reliability*. In such a large and complex setting components may be expected to fail: A data link between processors may cease passing data, introduce errors in messages or just operate too slowly or too fast. Processors may malfunction as well, and the reader can certainly imagine the resulting difficulties.

These phenomena create situations where failing components of the network may be viewed as playing against its correctly functioning parts. It is important to notice that the identity of failing components is not necessarily known to the reliable ones, and still the computational process is expected to run its correct course, as long as the number of failing components is not too large. Consequently, and unlike most game-theoretic scenarios, reliable or good players follow a predetermined pattern of behavior (dictated by the computer program they run) while bad players may, in the worst case, deviate from the planned procedure in an arbitrary fashion.

This is, then, the point of departure for most of the present article. How can n parties perform various computational tasks together if faults may occur? We present three different theories that address this problem: Section 2 concerns the Byzantine Agreement problem which is the oldest of the three and calls for processors to consistently agree on a value, rather a simple task. The main issues in this area are already fairly well understood. No previous background is needed to read this section.

A much more ambitious undertaking is surveyed in Section 3. Drawing heavily on developments in computational complexity and cryptography, one studies the extent to which computations can be carried out in a distributed environment in a reliable and secure way. We want to guarantee that the computation comes out correct despite sabotage by malfunctioning processors. Moreover, no classified information should leak to failing processors. Surprisingly satisfactory computational protocols can be obtained, under one of the following plausible assumptions concerning the means available to good players for hiding information from the bad ones. Either one postulates that communication channels are secure and cannot be eavesdropped, or that bad players are computationally restricted, and so cannot decipher the encrypted message sent by good players. This is still an active area of research. All necessary background in computational complexity and cryptography is included in the section.

Can anything be saved, barring any means for hiding information? This is the main issue in Section 4. It is no longer possible to warrant absolutely correct and leak-free computation, so only quantitative statements can be made. Namely, bad players are assumed to have some desirable outcome in mind. They can influence the protocol so as to make this outcome more likely, and the question is to find protocols which reduce their influence to a minimum. The notion of influence introduced in this section is, of course, conceptually related to the various measures developed in game theory to gauge the power of coalitions in games. Of all three, this theory is still at its earliest stage of development. This section is largely self-contained, except for a short part where (elementary) harmonic analysis is invoked.

Most of the research surveyed here has been developed from a theoretical standpoint, though some of it did already find practical application. A major problem in applying game theory to real life is the notorious difficulty in predicting the behavior of human players. This sharply contrasts with the behavior of a fault-free computer program, which at any given situation is completely predictable. This difference makes much of the foundational or axiomatic complications of game theory irrelevant to the study of interactions between computers. Two caveats are in order here: While it is true that to find out the behavior of a program on a *given* input, all we need to do is run the program on the input, there are numerous natural questions concerning a program's behavior on *unknown, general* inputs which cannot be answered. (Technically, we are referring to the fact that many such questions are *undecidable*.) This difficulty is perhaps best illustrated by the well known fact that the *halting problem* is undecidable [e.g. Lewis and Papadimifriou (1981)]. Namely, there is no algorithm, which can tell whether a given computer program halts on each and every possible input. (Or whether there is an input causing the program to loop forever). So, the statement that fault-free computer programs behave in a predictable way is to be taken with a grain of salt. Also, numerous models for the behavior of failing components have been considered, some of which are mentioned in the present article, but this aspect of the theory is not yet exhausted.

Most of the work done in this area supposes that the worst set of failures takes place. Game theorists would probably favor stochastic models of failure. Indeed, there is a great need for interesting theories on fault-tolerant computing where failures occur at random.

1.2. What about the other three main issues?

Of all four issues highlighted in the first paragraph we feel that the first one, i.e., how does a player's computational power affect his performance in various games, is conceptually and intellectually the most intriguing. Some initial work in this direction has already been performed, mostly by game theorists, see surveys by Sorin (Chapter 4) and Kalai (1990). It is important to observe that the scope of this problem is broad enough to include most of modern cryptography. The recent revolution in cryptography, begun with Diffie and Hellman's (1976), starts from the following simple, yet fundamental observation: In the study of secure communication, one should not look for *absolute* measures of security, but rather consider how secure a communication system is *against a given class of code-breakers*. The cryptanalyst's only relevant parameter is his computational power, the most interesting case being the one where a randomized polynomial-time algorithm is used in an attempt to break a communication system (all these notions are explained in Section 3). Thus we see a situation of conflict whose outcomes can be well understood, given the participants' computational power. A closer look into cryptography will probably provide useful leads to studying the advantage to computationally powerful players in other game playing.

Unfortunately, we have too little to report on the second issue – computational complexity of game-theoretic parameters. Some work in this direction is reviewed in Section 5. A systematic study of the computational complexity of game-theoretic parameters is a worthwhile project which will certainly benefit both areas. A historical example may help make the point: Combinatorics has been greatly enriched as a result of the search for efficient algorithms to compute, or approximate classical combinatorial parameters. In the reverse direction, combinatorial techniques are the basis for much of the development in theoretical computer science. It seems safe to predict that a similar happy marriage between game theory and theoretical computer science is possible, too.

Some of the existing theory pertaining to problem (4) is sketched in Section 5.4. Results in this direction are still few and isolated. Even the contour lines of this area have not been sketched yet.

1.3. Acknowledgments

This article is by no means a comprehensive survey of all subjects of common interest to game theory and theoretical computer science, and no claim for completeness is made. I offer my sincere apologies to all those whose work did not receive proper mention. My hope is that this survey will encourage others to write longer, more extensive articles and books where this injustice can be amended. I also had to give up some precision to allow for the coverage of more material. Each section starts with an informal description of the subject-matter and becomes more accurate as discussion develops. In many instances important material is missing and there is no substitute for studying the original papers.

I have received generous help from friends and colleagues, in preparing this paper, which I happily acknowledge: Section 3 could not have been written without many discussions with Shafi Goldwasser. Oded Goldreich's survey article (1988) was of great help for that chapter, and should be read by anyone interested in pursuing the subject further. Oded's remarks on an earlier draft were most beneficial. Explanations by Amotz Bar-Noy and the survey by Benny Chor and Cynthia Dwork (1989) were very useful in preparing Section 2.

Helpful comments on contents, presentation and references to the literature were made by Yuri Gurevich, Christos Papadimitriou, Moshe Vardi, Yoram Moses,

Michael Ben-Or, Daphne Koller, Jeff Rosenschein and Anna Karlin and are greatefully acknowledged.

1.4. Notations

Most of our terminology is rather standard. Theoretical computer scientists are very eloquent in discussing asymptotics, and some of their notations may not be common outside their own circles. As usual, if f, g are real functions on the positive integers, we say that $f = O(g)$, if there is a constant $C > 0$ such that $f(n) < Cg(n)$ for all large enough n. We say that $f = \Omega(g)$ if there is $C > 0$ such that $f(n) > Cg(n)$ for all large enough n. The notation $f = \Theta(g)$ says that f and g have the same growth rate, i.e., both $f = O(g)$ and $f = \Omega(g)$ hold. If for all $\varepsilon > 0$ and all large enough n there holds $f(n) < \varepsilon g(n)$, we say that $f = o(g)$. The set $\{1,\ldots,n\}$ is sometimes denoted $[n]$.

2. Byzantine Agreement

One of the earliest general problems in the theory of distributed processing, posed in Lamport et al. (1982), was how to establish consensus in a network of processors where faults may occur. The basic question of this type came to be known as the "Byzantine Agreement" problem. It has many variants, of which we explain one. We start with an informal discussion, then go into more technical details.

An overall description of the Byzantine Agreement problem which game theorists may find natural is: How to establish *common knowledge* in the absence of a mechanism which can guarantee reliable information passing? In known examples, where a problem is solved by turning some fact into common knowledge, e.g., the betraying wives puzzle [see Moses et al. (1986) for an amusing discussion of a number of such examples] there is an instance where all parties involved are situated in one location, the pertinent fact is being announced, thus becoming common knowledge. If information passes only through private communication lines, and moreover some participants try to avoid the establishment of common knowledge, is common knowledge achievable?

In a more formal setting, the Byzantine Agreement problem is defined by two parameters: n, the total number of players, and $t \leq n$, an upper bound on the number of "bad" players. Each player (a computer, or processor) has two special memory cells, marked *input* and *outcome*. As the game starts every input cell contains either zero or one (not necessarily the same value in all input cells). We look for computer programs to be installed in the processors so that after they are run and communicate with each other, all "good" processors will have the same value written in their outcome cells. To rule out the trivial program which always outputs 1, say, we require that if all input values are the same, then this is the value that has to appear in all outcome cells. The set of programs installed in the processors along with the rules of how they may communicate is a *protocol* in computer science jargon. The difficulty is that up to t of the processors may fail to follow the protocol. Their deviation from the protocol can be as destructive as possible, and in particular, there is no restriction on the messages they send. Bad players' actions are coordinated, and may depend on data available to all other bad players.

The Byzantine Agreement problem comes in many different flavors, largely depending on which data is available to the bad players at various times. In this survey we consider only the case where the bad players have complete knowledge of each other's information and nothing else. Another factor needed to completely specify the model is timing. Here we assume synchronous action: there is an external clock common to all processors. At each tick of the clock, which we call a *round*, all processors pass their messages. Internal computations are instantaneous. For a comprehensive discussion of other models see Chor and Dwork (1989).

A protocol *achieves Byzantine Agreement* (or *consensus*) in this model, if the following conditions hold when all "good" players (i.e., those following the protocol) halt:

Agreement. All good players have the same value written in their outcome cells.

Validity. If all inputs equal v, then all good outcome cells contain v.

A host of questions naturally come up, of which we mention a few:

- Given n, what is the largest t for which Byzantine Agreement can be achieved?
- When achievable, how costly is Byzantine Agreement: How many rounds of communication are needed? How long need the messages be? What computational power need the processors have?
- The above description of the problem implicitly assumes that every two processors are linked and can exchange messages. How will the results change under different modes of communication, e.g., if processors can send messages only to their neighbors in some (communication) graph?
- What if simultaneity of action is not guaranteed, and processors communicate via an asynchronous message passing mechanism? Can consensus still be achieved?
- Does randomization help? Are there any advantages to be gained by letting processors flip coins? I.e., can agreement be established more rapidly, with less computation, or against a larger number of bad players? Are there any gains in relaxing the correctness conditions and only asking that they hold with high probability and not with certainty?
- How crucial is the complete freedom granted to bad processors? For example, if failure can only mean that at some point a player stops playing, does that make consensus easier to achieve?

2.1. Deterministic algorithms for Byzantine Agreement

The earliest papers in this area where many of the problems were posed and some of the basic questions answered are Pease et al. (1980) and Lamport et al. (1982). A large number of papers followed, some of which we now briefly summarize along with review, commentary and a few proofs.

Theorem 2.1. (i) *A Byzantine Agreement (BA) among n players of which t are bad can be achieved iff $n \geq 3t + 1$.*
  (ii) *BA can be established in $t + 1$ rounds, but not in fewer.*
  (iii) *If G is the communication graph, then BA may be achieved iff G is $(2t + 1)$-connected.*

We start with a presentation of a protocol to establish BA. In the search for such protocols, a plausible idea is to delay all decisions to the very end. That is, investigate protocols which consist of two phases: information dispersal and decision-making. The most obvious approach to information dispersal is to have all players tell everyone else all they know. At round 1 of such protocols, player x is to tell his initial value to all other players. In the second round, x should tell all other players what player y told him on round 1, for each $y \neq x$, and so on. This is repeated a certain number of times, and then all players decide their output bit and halt. It turns out that consensus can indeed be established this way, though such a protocol is excessively wasteful is usage of communication links. We save our brief discussion of establishing BA *efficiently* for later.

Lamport, Pease and Shostak (1980) gave the first BA protocol. It follows the pattern just sketched: All information is dispersed for $t + 1$ rounds and then decisions are made, in a way we soon explain. They also showed that this protocol establishes BA if $n > 3t$. Our presentation of the proof essentially follows Bar-Noy et al. (1987). Explanations provided by Yoram Moses on this subject are gratefully acknowledged.

To describe the decision rule, which is, of course, the heart of the protocol, some notation is needed. We will be dealing with sequences or *strings* whose elements are names of players. Λ is the empty sequence, and a string consisting of a single term y is denoted y. The sequence obtained by appending player y to the string σ is denoted σy. By abuse of language we write $s \in \sigma$ to indicate that s is one of the names appearing in σ.

A typical message arriving at a player x during data dispersal has the form: "I, y_{i_k}, say that $y_{i_{k-1}}$ says that $y_{i_{k-2}}$ says that... that y_{i_1} says that his input is v". Player x encodes such data by means of the function V_x, defined as follows: $V_x(\Lambda)$ is x's input, $V_x(y_5, y_3)$ is the input value of y_5 as reported to x by y_3, and in general, letting σ be the sequence $y_{i_1}, \ldots, t_{i_k}$, the aforementioned message is encoded by setting $V_x(\sigma) = v$. If the above message repeats the name of any player more than once, it is ignored. Consequently, in what follows *only* those σ will be considered where *no player's name is repeated*. No further mention of this assumption is made.

The two properties of V_x that we need are

If x is a good player, then $\mathrm{input}(x) = V_x(\Lambda)$.

If x, y are good players, then $V_x(\sigma y) = V_y(\sigma)$.

The first statement is part of the definition. The second one says that a good player y correctly reports to (good) x the value of $V_y(\sigma)$ and x keeps that information via his data storage function V_x. If either x or y are bad, nothing can be said about $V_x(\sigma y)$.

The interesting part of the protocol is the way in which each player x determines his outcome bit based on the messages he saw. The goal is roughly to have the outcome bit represent the majority of the input bits to the good players. This simple plan can be easily thwarted by the bad players; e.g., if the input bits to the good players are evenly distributed, then a single bad player who sends conflicting messages to the good ones can fail this strategy. Instead, an approximate version of this idea is employed. Player x associates a bit $W_x(\sigma)$ with every sequence-with-no-repeats σ of *length* $|\sigma| \leq t+1$. The definition is

$$|\sigma| = t+1 \Rightarrow W_x(\sigma) = V_x(\sigma)$$

and for shorter σ

$$W_x(\sigma) = \mathrm{majority}\{W_x(\sigma y)\}$$

the majority being over all y not appearing in σ. Finally

$$\mathrm{outcome}(x) = W_x(\Lambda).$$

The validity of this protocol is established in two steps:

Proposition 2.1. If x and y are good players, then

$$W_x(\sigma y) = V_x(\sigma y) = V_y(\sigma).$$

Proof. The second equality is one of the two basic properties of the function V, quoted above, so it is only the first equality that has to be proved. If $|\sigma y| = t+1$ it follows from the definition of W in that case. For shorter σ we apply decreasing induction on $|\sigma|$: For a good player z the induction hypothesis implies $W_x(\sigma yz) = V_z(\sigma y) = V_y(\sigma)$. But

$$W_x(\sigma y) = \mathrm{majority}\{W_x(\sigma yx)\}$$

over all $s \notin \sigma y$ (good or bad). Note that most such s are good players: even if all players in σ are good, the number of bad $s \notin \sigma y$ is $\leq t$ and the number of such good s is $\geq n - |\sigma y| - t$, but

$$n - |\sigma y| - t > t,$$

because $|\sigma y| \leq t$, and by assumption $n \geq 3t + 1$ holds. It follows that most terms inside the majority operator equal $V_y(\sigma)$ (s behaves like z) and so $W_x(\sigma y) = V_y(\sigma)$ as claimed. □

Observe that this already establishes the *validity* requirement in the definition of BA: If all inputs are v, apply the lemma with $\sigma = \Lambda$ and conclude that if x, y are good, $W_x(y) = v$. In evaluating outcome$(x) = W_x(\Lambda)$ most terms in the majority thus equal v, so outcome$(x) = v$ for all good x, as needed.

For the second proposition define σ to be *closed* if either (i) its last element is the name of a good player, or (ii) it cannot be extended to a sequence of length $t + 1$ using only names of bad players. This condition obviously implies that Λ is closed. Note that if σ is closed and it ends with a bad name, then σs is closed for any $s \notin \sigma$: if s is good, this is implied by (i) and otherwise by (ii).

Proposition 2.2. If σ is closed and x, y are good, then $W_x(\sigma) = W_y(\sigma)$.

Proof. We first observe that the previous proposition implies the present one for closed strings which terminate with a good element. For if $\sigma = \mu z$ with z good, then both $W_x(\mu z)$ and $W_y(\mu z)$ equal $V_z(\mu)$. For closed strings ending with a bad player, we again apply decreasing induction on $|\sigma|$. If $|\sigma| = t + 1$, and its last element is bad, σ not closed, so assume $|\sigma| \leq t$. Since σ's last element is bad, all its extensions are closed, so by induction, $W_x(\sigma s) = W_y(\sigma s)$ for all s. Therefore

$$W_x(\sigma) = \text{majority}\{W_x(\sigma s)\} = \text{majority}\{W_y(\sigma s)\} = W_y(\sigma). \quad \square$$

Since Λ is closed, outcome$(x) = W_x(\Lambda)$ is the same for all good players x, as required by the *agreement* part of BA.

This simple algorithm has a serious drawback from a computational viewpoint, in that it requires a large amount of data to flow [$\Omega(n^t)$ to be more accurate]. This difficulty has been recently removed by Garay and Moses (1993), following earlier work in Moses and Waartz (1988) and Berman and Garay (1991):

Theorem 2.2. *There is an n-player BA protocol for $t < n/3$, which runs for $t + 1$ rounds and sends a polynomial (in n) number of information bits.*

The proof is based on a much more efficient realization of the previous algorithm, starting from the observation that many of the messages transmitted there are, in fact, redundant. This algorithm is complicated and will not be reviewed here.

A protocol to establish BA on a $(2t + 1)$-connected graph is based on the classical theorem of Menger (1927) which states that in a $(2t + 1)$-connected graph, there exist $2t + 1$ disjoint paths between every pair of vertices. Fix such a system of paths for every pair of vertices x, y in the communication graph and simulate the previous algorithm as follows: Whenever x sends a message to y in that protocol, send a copy of the message along each of the paths. Since no more than t of the paths contain a bad player, most copies will arrive at y intact and the correct message can be thus identified. New and interesting problems come up in trying to establish consensus in networks, especially if the underlying graph has a bounded degree, see Dwork et al. (1988).

Let us turn now to some of the impossibility claims made in the theorem. That for $n=3$ and $t=1$ BA cannot be achieved, means that for any three-players' protocol, there is a choice of a bad player and a strategy for him that prevents BA. The original proof of impossibility in this area were geared at supplying such a strategy for each possible protocol and tended to be quite cumbersome. Fischer, Lynch and Merritt (1986) found a unified way to generate such a strategy, and their method provides short and elegant proofs for many impossibility results in the field, including the fact that $t+1$ rounds are required, and the necessity of high connectivity in general networks.

Here, then, is Fischer et al.'s (1986) proof that BA is not achievable for $n=3, t=1$. The proof that in general, $n > 3t$ is a necessary condition for achieving BA, follows the same pattern, and will not be described here.

A protocol Q which achieves BA for $n=3, t=1$ consists of six computer programs $P_{i,j}(i=1,2,3; j=0,1)$ where $P_{i,j}$ determines the steps of player i on input j. We are going to develop a strategy for a bad player by observing an "experimental" protocol, $\hat{Q}$, in which all six programs are connected as in Figure 1 and run subject to a slight modification, explained below.

In $\hat{Q}$ all programs run properly and perform all their instructions, but since there are six programs involved rather than the usual three, we should specify how instructions in the program are to be interpreted. The following example should make the conversion rule clear. Suppose that $P_{3,1}$, the program for player 3 on input 1, calls at some point for a certain message M to be sent to player 2. In Figure 1, we find $P_{2,1}$ adjacent to our $P_{3,1}$, and so message M is sent to $P_{2,1}$. All this activity takes place in parallel, and so if, say, $P_{3,0}$ happens to be calling at the same time, for M' to be sent to player 2, then $P_{3,0}$ in $\hat{Q}$ will send M' to $P_{2,0}$, his neighbor in Figure 1. Observe that Figure 1 is arranged so that $P_{1,j}$ has one neighbor $P_{2,\alpha}$, and one $P_{3,\beta}$ etc., so this rule can be followed, and our experimental $\hat{Q}$ can be performed. $\hat{Q}$ terminates when all six programs halt. A strategy for a bad player in Q can be devised, by inspecting the messages sent in runs of $\hat{Q}$.

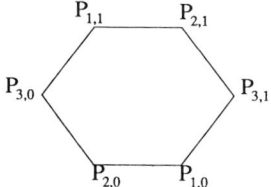

Figure 1. Protocol $\hat{Q}$.

We claim that if indeed, the programs $P_{i,j}$ always establish BA, then by the end of $\hat{Q}$ each $P_{i,j}$ reaches outcome j. E.g., suppose for contradiction, that $P_{3,1}$ does not reach 1. Then, we can exhibit a run of Q which fails to achieve BA, contrary to our assumption. Let Q run with good players 2 and 3 holding input 1 and a faulty player 1. Player 1 makes sure players 2 and 3 follow the same steps as $P_{2,1}$ and $P_{3,1}$ in $\hat{Q}$. He achieves this by sending player 2 the same messages $P_{1,1}$ sends to $P_{2,1}$ in $\hat{Q}$, and to 3 what $P_{1,0}$ sends $P_{3,1}$ there. This puts players 2, 3 in exactly the same position as are $P_{2,1}, P_{3,1}$ in $\hat{Q}$, since they have the same input and the same exchanges with the neighbors. In particular player 3 fails to reach the outcome 1, and BA is not reached, a contradiction.

But now we can take advantage of $P_{1,0}$ deciding 0 and $P_{3,1}$ deciding 1 in $\hat{Q}$. Run Q with good players 1 and 3 holding inputs 0 and 1 respectively and a bad player 2. Let player 2 send player 1 the messages sent by $P_{2,0}$ to $P_{1,0}$ in $\hat{Q}$ and to player 3 those $P_{2,1}$ sends to $P_{3,1}$ there. This forces player 1 to a 0 outcome and player 3 to a 1 outcome, so no consensus is achieved, a contradiction.

Observe the similarity between the mapping used in this proof and *covering maps* in topology. Indeed, further fascinating connections between topology and the impossibility of various tasks in asynchronous distributed computation have been recently discovered, see Borowski and Gafni (1993), Saks and Zaharoglou (1993), Herlihy and Shavit (1993).

The last remark of this section is that the assumed synchronization among processors is crucial for establishing BA. Fischer, Lynch and Patterson (1985) showed that if message passing is completely asynchronous then consensus cannot be achieved even in the presence of a *single* bad player. In fact, this player need not even be very malicious, and it suffices for him to stop participating in the protocol at some cleverly selected critical point. For many other results in the same spirit see Fischer's survey (1983).

2.2. Randomization in Byzantine Agreements

As in many other computational problems, randomization may help in establishing Byzantine Agreements. Compared with the deterministic situation, as summarized in Theorem 2.1, it turns out that a randomized algorithm can establish BA in expected time far below $t+1$. In fact the expected time can be bounded by a constant independent of n and t. It also helps defeating asynchrony, as BA may be achieved by a randomized *asynchronous* algorithm. The requirement that fewer than a third of the players may be bad cannot be relaxed, though, see Graham and Yao (1989) for a recent discussion of this issue. Goldreich and Petrank (1990) show that these advantages can be achieved without introducing any probability that the protocol does not terminate, i.e., the best of the deterministic and randomized worlds can be gotten.

We briefly describe the first two algorithms in this area, those of Ben-Or (1983) and Rabin (1983) and then explain how these ideas lead to more recent developments. For simplicity, consider Ben-Or's algorithm for small t and in a synchronous set-up: Each processor has a variable called *current value* which is initialized to the input value. At each round every processor is to send its current value to all other participants. Subsequently, each processor has n bits to consider: its own current bit and the $n-1$ values related to him by the rest. Three situations may come up: *high*, where at least $n/2 + 2t$ of these bits have the same value, say b. In this case the player sets its outcome bit to b, it announces its decision on the coming round and then retires. *Medium*: the more frequent bit occurs between $n/2 + t$ and $n/2 + 2t$ times. This bit now becomes the current value. In the *low* case the current value is set according to a coin flip. The following observations establish the *validity* and run time of this algorithm:

- If all input values are the same, then the algorithm terminates in one round.
- If in a certain round some good processor is in *high*, then every other good processor is either in *high* and commits on the same bit b or it is in *medium*, making b his current value, and so the algorithm terminates successfully in the next step.

Thus the expected run time of this algorithm is determined by how quickly "high" occurs. Now if $t = O(\sqrt{n})$ the expected number of rounds until some good processor goes into *high* is constant. (We have to wait for a random variable distributed according to a binomial law to exceed its expectation by a constant number of standard deviations.) Therefore for such values of t this is already a randomized BA algorithm with a constant expected run time. Is there such an algorithm also for larger t? Rabin's algorithm (found at about the same time as Ben-Or's) can be explained by modifying the previous algorithm as follows: If it were true that all the good processors at the *low* situation were flipping one and the same *collective* coin, then there would be no delay at all. This can be brought about by postulating the existence of a trusted referee who generates such global coin flips. Moreover, this scheme is seen to work in constant expected time even for $t = \Omega(n)$. However, the presence of a trusted referee makes the whole BA problem meaningless. Rabin observed that it suffices to sample a random source *before* the protocol is started, and reveal these random bits as needed. Again, revealing should take place without any help from an external trusted party. Such a mechanism for "breaking a secret bit into shares" and then revealing it when needed, is supplied by Shamir's secret sharing protocol (1979) to be described in Section 3.2.5. An important feature of this protocol is that no t of the players can gain any information about such a bit, while any $t + 1$ can find it out with certainty. Given this protocol, the only problematic part in Rabin's protocol is the assumption that a trusted source of random bits is available prior to its run. This difficulty has been later removed by Feldman and Micali (1988) who show:

Theorem 2.3 [Feldman and Micali (1988)]. *There is a randomized Byzantine Agreement algorithm with a constant expected run time.*

This algorithm may be perceived as performing Rabin's algorithm without a trusted dealer. Feldman and Micali achieve it by means of an elaborate *election* mechanism, whereby a somewhat-random player is elected by an approximate majority rule. Two important ingredients in this protocol, which show up again in Section 3 are a *commitment* mechanism and the *verifiable* secret-sharing method of Ben-Or et al. (1988), which is an improvement of Shamir's secret-sharing mechanism (1979).

3. Fault-tolerant computation under secure communication

The Byzantine Agreement problem of the previous section is but one instance of the following general problems: How can the action of n parties (players, processors etc.) be coordinated in the absence of a trusted coordinator? How well can they perform various tasks which require cooperation when some of them may fail to perform their roles? An interesting class of examples the reader may want to consider, is that of playing certain games without a trusted referee. If all parties involved are reliable, then no problem arises, since any player may act as a coordinator. Are there any means of coordination when reliability is not granted? This study raises many technical as well as conceptual difficulties, so we begin with a few examples to guide the reader's intuition. Our final goal in this section is to survey the articles of Goldreich et al. (1987) and Ben-Or et al. (1988), however, these papers build on many earlier interesting developments through which we first go. Space limitations prohibit proper mention and credit to the investigators who made significant contributions to the field. One should not fail, however, to mention Yao's contributions, in particular Yao (1979) as well as Yao (1982) where the two-player situation was resolved.

Models. To pose concrete questions in this area, we need to make some assumptions on the behavior of bad players and on the means available to good players for concealing information. Keeping our discussion at an intuitive level, let us say this: bad players are either assumed to be *curious* or *malicious*. Curious players do perform their part in protocol, but try and extract as much side information as possible from exchanges between themselves and the good players, thus raising the difficulty of avoiding *information leaks*. Malicious players will do anything (as do bad players in the Byzantine Agreement problem). This assumption places the focus on *correct* computation in the presence of faults.

There are two models for how good players hide information:

Information-theoretic. Secure communications lines exist between every two parties. No third party can gain any information by eavesdropping messages sent on such a line.

Alternatively, one postulates:

Bounded computational resources (or cryptographic set-up). Participants are assumed to have a restricted computational power.

Here, then, are some concrete examples of our general problem.

- Secure message passing: Consider a three parties' protocol, where Sender needs to relay a single bit to Receiver without Eavesdropper's gaining any information. This special case of our problem is, clearly, a most fundamental problem of cryptography. It is trivial, of course, if secure channels are provided, but is interesting if the parties have a limited computational power. Stated otherwise, it asks whether bounds on computational power allow simulating a secure communication channel.
- The "and" function: Two players have a single input bit each, and they need to compute the conjunction (logical "and") of these two bits. This extremely simple task brings up some of the care that has to be taken in the formal statement of our definitions: Notice that a player whose input bit is 1 will know at the game's end the other player's input bit, while a player whose input bit is 0, knows that the conjunction of the two bits is 0. This observation should be taken into account in making the formal definition of "no information leaks are allowed". Secure channels do not help in this problem, but this task can be performed in the cryptographic set-up.
- The millionaires' problem: Two (curious) millionaires would like to compare their wealth, without revealing any information about how much money they have. In other words: is there a protocol which allows players P_1 and P_2 to find out which of the integers x_1, x_2 is bigger, without P_1 finding out any other information about x_2, and vice versa? Again, this is only interesting in a cryptographic set-up and can be done if there is a commonly known upper bound on both x_1, x_2.
- Game playing without a referee: Can noncooperative games be played without a trusted referee? The definition of a correlated equilibrium in a noncooperative game depends on a random choice made by Fortune. Barany and Füredi (1983) show how $n \geq 4$ players $P_1, \ldots, P_n$ may safely play any noncooperative game in the absence of a referee, even if one of the players is trying to cheat. In our terminology the problem may be stated as follows: Let μ be a probability distribution over $A = A_1 \times \cdots \times A_n$ where A_i are finite sets. It is required to sample $a \in A$ according to μ in such a way that player P_i knows only the ith coordinate of a but attains no information at all on a's other coordinates. It is moreover

required that a player deviating from the protocol cannot bias the outcome and will be caught cheating with probability 1. As we shall see in this section, the above results can be strengthened, and the same will hold even if as many as $[(n-1)/3]$ players deviate from the rules of the game.
- Secure voting: Consider n voters, each of whom casts a yes/no vote on some issue. We would like to have a protocol which allows every player have exactly one vote and which eventually will inform all players on the outcome, while revaling nothing on votes of any individuals or subsets of players. Alternatively, we may ask that the tally be made known to all players. Or, we could ask that every female voter be told about the outcome among male voters or any other such task. As long as the number of bad players does not exceed a certain fraction of the total n, every such goal can be achieved.

Remark 3.1. Secure communication lines, may at first sight, seem superior to restrictions on faulty players' computational power. After all, what can cryptography buy us other than the freedom to communicate without being eavesdropped? This is, however, incorrect as indicated by the examples involving only two players.

Remark 3.2. Most of our discussion concerns protocols that allow players to safely compute a function, or a set of functions together. The example of non-cooperative game playing without a trusted referee, brings up a more general situation where with every list $x_1,\ldots,x_n$ of inputs to the players a probability space is associated, from which the players are to sample together a point z, and have player P_i find out a function $f_i(z)$, and nothing else. [When their goal is to compute a function f this space consists of the single point $f(x_1,\ldots,x_n)$. In the correlated equilibrium problem, $f_i(z)$ is the i-th coordinate of z.] The methods developed here are powerful enough to perform this more general task, in a private and fault-tolerant way. The changes required in the discussion are minimal and the reader can either supply them, or refer to the original papers.

Let us go into more details on the two main models for bad players' behavior:

(1) Unreliable participants are assumed to follow the protocol properly, except they store in memory all the messages they (legally) saw during the run of the protocol. Once computation is over, bad players join forces to extract as much information as possible from their records of the run. Bad players behaving in this way are sometimes said to be *curious*. There are certain quantities which they can infer, regardless of the protocol. Consider for example, five players each holding a single bit who compute a majority vote on their bits, and suppose that the input bits are $x_1 = x_4 = 0$, and $x_2 = x_3 = x_5 = 1$. If players 1 and 4 convene after finding out that *majority* = 1, they can deduce that all other x_i are 1. In saying that *no information leaks* in a protocol, it is always meant that only quantities such as this are computable by the unreliable players.

(2) A more demanding model assumes nothing with regard to the behavior of the bad players (as in the Byzantine Agreement problem). Such behavior is described as *malicious* or *Byzantine*. It is no longer enough to avoid the leak of classified information to the hands of bad players as in (1), for even the *correctness* of the computation is in jeopardy. What, for example, if we try to compute $f(x_1,\ldots,x_n)$ and player i refuses to reveal x_i which is known only to him? Or if, when supposed to compute a quantity depending on x_i and tell the outcome to other players, he intentionally sends an incorrect value, hoping no one will be able to detect it, since x_i is known only to him? If we momentarily relax the requirement for no information leak, then correctness can, at least in principle, be achieved through the following *commitment* mechanism: At the beginning of the protocol each player P_i puts a note in an envelope containing his value x_i. Then, all envelopes are publicly opened, and $f(x_1,\ldots,x_n)$ is evaluated. There are preassigned *default values* $z_1,\ldots,z_n$, and if player P_i fails to write a valid x_i on his note, then x_i is taken to be z_i. Protocols described in this section perform just that, without using physical envelopes. Given appropriate means for concealing information, as well as an upper bound on the number of faults, it is possible to compute both correctly and without leaking information.

A complete description of a problem in this class consists of:

- A specification of the task which is to be performed together.
- An upper bound t for the number of unreliable players out of a total of n.
- The assumed nature of faulty players: curious or malicious.
- The countermeasures available: Either secure communication lines or a bound on faulty players' computational power.

A secure communication line is a very intuitive concept, though the technical definition is not entirely straightforward. To explain the notion of restricted computational power and how this assumption allows sending encrypted messages, the fundamentals of the theory of computing and modern cryptography have to be reviewed, as we later do. It is a key idea in cryptography that the security of a communication channel should not be quantified in absolute terms but rather expressed in terms of the computational power of the would-be code-breaker. Similar ideas have already been considered in game theory literature (e.g., bounded recall, bounded rationality etc.) but still seem far from being exhausted, see surveys by Kalai (1990) and Sorin (Chapter 4).

The main result of this section is that if the number t of unreliable players is properly bounded, then in both modes of failure (curious, malicious) and both safety guarantees (secure lines, restricted computational power) correct and leak-free computation is possible. A more detailed annotated statement of the main results surveyed in this section follows:

(1) Given are functions $f; f_1,\ldots,f_n$ in n variables, and n processors $P_1,\ldots,P_n$ which communicate via secure channels and where each P_i holds input x_i known

only to it. There is a protocol which, on termination, supplies P_i with the value $f_i(x_1,\ldots,x_n)$. The protocol is leak-free against any proper minority of curious players. I.e., for any $S \subseteq \{1,\ldots,n\}$ with $|S| \leq [(n-1)/2]$, every quantity computable based on the set of messages passed to any $P_j(j \in S)$ can also be computed given only the x_j and $f_j(x_1,\ldots,x_n)$ for $j \in S$.

Also, the protocol computes correctly in the presence of any coalition S of $\leq$ $[(n-1)/3]$ malicious players. That is, the protocol computes a value $f(y_1,\ldots,y_n)$ so that $y_i = x_i$ for all $i \notin S$ and where the $y_j(j \in S)$ are chosen independent of the values of $x_i(i \notin S)$.

[Explanation: This definition captures the following intuition: in trying to compute $f(x_1,\ldots,x_n)$ there is no way to guarantee that unreliable players supply their correct input values. The best that can be hoped for is that they be made to *commit* on certain input values, chosen independent of the inputs to good players, and that after such a commitment stage, the computation of f proceeds correctly and without leaking information. In case a bad player refused to supply an input, his will be substituted for by some default value.]

The same results hold if the functions f_j are replaced by probability distributions and instead of computing the functions we need to sample according to these distributions, see Remark 3.2. The bounds in the theorem are tight.

(2) Assume that one-way trapdoor permutations exist.

(This is an unproven statement in computational complexity theory which is very plausible, and whose proof is far beyond the power of current methods. Almost nothing can be presently proved in cryptography without this, or similar, unproven hypothesis. We assume that 1:1 functions exist which are not hard to compute, but whose inverse requires infeasible amount of computation. See Section 3.2.2 for more.)

Modify the situation in (1) as follows: Channels are *not* secure, but processors are probabilistic, polynomial-time Turing Machines (this is theoretical computer scientists' way of saying that they are capable of any *feasible* computation, but nothing beyond). Similar conclusions hold with the bound $[(n-1)/2]$ and $[(n-1)/3]$ replaced by $n-1$ and $[(n-1)/2]$ respectively. Again the bounds are tight and the results hold also for sampling distributions rather than for evaluation of functions.

Remark 3.3. Certain (but not all) randomized algorithms have the disadvantage that they may fail due to a "statistical accident". As long as the probability of such failure is exponentially small in n, the number of players, *it will henceforth be ignored*.

For the reader's orientation, let us point out that in terms of performing various computational tasks, BA theory is completely subsumed by the present one. For example, the first part of Theorem 2.1 is contained in Theorem 3.5 below. In terms of efficiency, the present state of the theory leaves much to be desired. For more on efficiency see Section 3.3.1.

3.1. Background in computational complexity

3.1.1. Turing Machines and the classes **P** and **NP**

At this point of our discussion a (very brief) survey of the fundamental concepts of computational complexity theory is in order, see Garey and Johnson (1979) or Lewis and Papadimitriou (1981) for a thorough introduction to this area. A reader having some familiarity with computer science may safely skip this section. What we need now is a mathematically well-defined notion of a computer. This area is made up of two theories meant to answer some very fundamental questions:

Computability theory. What functions, relations etc. can be mechanically computed?

Complexity theory. Among computable functions which ones are easy and which are hard to compute?

What is needed is a definition powerful enough, to encompass all possible theoretical and physical models of computers. Also, a computer's performance certainly depends on a variety of physical parameters, such as speed, memory size, its set of basic commands, memory access time etc. Our definitions should be flexible enough so the dichotomy between easy and hard to compute functions does not depend on such technicalities. The most standard definition is that of a Turing Machine which we momentarily review. See Remark 3.4 below for why this choice seems justified.

Computability theory flourished 40–60 years ago and most of its big problems have been resolved. Complexity theory, on the other hand, is one of the most active and important areas of present study in computer science. The dichotomy between hard and easy computational tasks is usually made according to whether or not the problem is solvable in time bounded by a polynomial in the input length (a concept to be elaborated on in the coming paragraphs). As it turns out, this criterion for classifying problems as easy or hard does not depend on the choice of the model, and the classification induced by Turing Machines is the same as given by any other standard model.

Let us review the key features common to any computer the reader may be familiar with. It has a *control unit* which at each time resides in one of a (finite) number of *states*. There would also be *memory* and *input/output* devices. In the formal definition, memory and input/output units are materialized by an unbounded *tape* which initially contains the input to be processed. The tape may also be used for temporary storage, and will finally contain the output given by the machine. What the reader knows as the *computer program* is formally captured by the *transition function* which dictates what the new control state is, given the current state and the current symbol being read off the tape. Note that our definition differs from a regular computer, looking like a *special-purpose* computer, capable of performing only a single function. A computer which uses no software and where all behavior is determined by the hardware (i.e., the transition function). Indeed one of the early conceptual achievements of computability theory was the observation that the definition of a Turing Machine is flexible enough to allow the construction of a particular Turing Machine ("The Universal Turing Machine") that can be *programmed*. Thus a Turing Machine is a satisfactory mathematical abstraction of a general-purpose computer. Let us emphasize that the Universal Turing Machine does not require to be separately defined and is just an instance of a Turing Machine, according to the general definition.

We now turn to the formal definition. A Turing Machine M consists of:

- A finite set K of *states*, one of which, s, is called an *initial* state, and an extra *halting* state h, not in K.
- A *transition function* δ from $K \times \{0, 1, \#\}$ to $(K \cup \{h\}) \times \{0, 1, \#, L, R\}$, $\{\#$ is to be thought of as the blank symbol.)

The machine M operates on an *input* x, which is a finite string of zeros and ones whose *length* is denoted by $|x|$. The input is augmented on the left and on the right by an infinite sequence of #s. The augmented sequences is called the *tape*, which may be modified during M's run. A *run* formally depicts how a computation of M evolves. It is a sequence, each element of which consists of (M's *current*) state, a tape and a symbol in it (currently *scanned* by M.) The run starts with M in state s and scanning the leftmost bit of the input. If M is in state q, currently scanning a and $\delta(q, a) = (p, b)$, then M's next state is p. Now if $b = R$ (respectively L), the right (left) neighbor of the currently scanned symbol is scanned next, while if b is $0, 1$ or $\#$, the tape is modified, the scanned a is replaced by b, which is the next scanned symbol. The process terminates when M first enters state h, and the number of steps until termination is called the *running time* (of M on x). At termination there is a shortest segment of the tape where 0 and 1 may be found, all other symbols being $\#$. This finite segment is called the *output* of M when run on x. The length of the interval including those positions which are ever scanned during the run is called the *space* (utilized by M when run on x.) Sometimes there is need to talk about machines with more than one input or more than one output, but conceptually the definitions are the same.

A *decision problem* Q is specified by a *language* L i.e., a set of finite strings of zeros and ones (the strings are finite, but L itself is usually infinite). The output of M on input x should be either 1 or 0, according to whether x is in L or not. The main concern of computational complexity is the running time of M on inputs x *as a function of their length*. The "dogma" of this area says that Q is feasibly solvable if there is an M which solves the decision problem in time which is bounded above by some polynomial function of the length $|x|$. In such a case Q (or L) is said to be in the class **P**, (it is *solvable in polynomial time*).

One also makes similar definitions for *randomized* Turing Machines. Those make their transition based on coin-flipping. (Each current state and current scanned symbol, define a probability distribution on the next step. The new state, whether to move left or right, or to modify the tape are decided at random according to this distribution.) Running time becomes a random variable in this case, and there is room for more than one definition for "random polynomial time". The simplest definition requires the expected running time to be bounded above by a polynomial in the input length, see Johnson (1990) for more details.

There is another, less obvious, complexity measure for problems, leading to the definition of the class **NP**: Even if Q is not known to be solvable in polynomial time, how hard is checking the validity of a proposed solution to the problem? The difference is analogous to the one between an author of a paper and a referee. Even if you cannot solve a problem, still you may be able to check the correctness of a proposed solution. A language L is said to be in **NP** (the corresponding decision problem is solvable in *nondeterministic polynomial time*) if there is a Turing Machine M with two inputs x and y such that for every $x \in L$ there is a y such that after time polynomial in $|x|$, M outputs 1, but if $x \notin L$, then no such y exists. The string y can be thought of as a "proof" or *witness* as the usual term is, for x's membership in L. Extensive research in classifying natural computational problems yielded a list of a few thousand problems which are in **NP** and are *complete* in it, namely, none of these problems is known to be in **P**, but if any one of them is in **P**, then *all* of them are. This is usually taken as evidence that indeed **P** and **NP** are different. Whether or not **P** and **NP** are equal is generally considered the most important question in theoretical computer science. The standard text on the subject is Garey and Johnson (1979).

Example 3.1. Graph 3-colorability: The following language is **NP**-complete: It consists of all finite graphs G whose set of vertices $V = V(G)$ can be partitioned into three parts $V = \bigcup_{i=1}^{3} V_i$ such that if two vertices x, y belong to the same V_i, then there is no edge between them.

Remark 3.4. We close this section by mentioning the *Church-Turing Thesis* which says that all computable functions are computable by Turing Machines. This is not a mathematical statement, of course, but rather a summary of a period of research in the 1930s where many alternative models for computing devices were proposed and studied. Many of these models turned out to be equivalent to Turing Machines, but none were found capable of performing any computations beyond the power of Turing Machines. This point will again come up later in the discussion.

Algebraic circuits are among the other computing devices alluded to above. An algebraic circuit (over a field **F**) is, by definition, a directed acyclic graph with four types of nodes: input nodes, an output node, constant nodes and gate nodes. Input nodes have only outgoing edges, the output node has only one incoming edge, constant nodes contain fixed field elements, and only have outgoing edges. Gate nodes are either + gates or × gates. Such a circuit performs algebraic computations over **F** in the obvious way. Each input node is assigned a field element and each gate node performs its operation (addition or multiplication) on the values carried by its incoming edges. Finally, the value of the computed function resides in the output node.

Boolean circuits are similar, except they operate on "true" and "false" values and their gates perform logical "and", "or" and "not" operations.

We shall later interpret Church's Thesis as saying that any computable function is computable by some algebraic circuit. Such circuits can clearly simulate Boolean circuits, which are the model of choice in computational complexity.

3.2. Tools of modern cryptography

The reader may be aware of the transition that the field of cryptology has been undergoing in recent years, a development that was widely recorded in the general press. It became a legitimate area of open academic study which fruitfully interacts with number theory and computational complexity. Modern cryptography is not restricted to the design of codes for secure data transmission and code-breaking as had been the case since antiquity. It is also concerned with the design of *cryptographic protocols*. The problem in general is how to perform various transactions between mutually untrusting parties, for example how can you safely sign agreements over the telephone, or take a vote in a reliable way over a communication channel or any other similar transactions of commercial or military importance. See Rivest (1990) and Goldreich (1989) for recent surveys.

Such problems certainly are in the scope of game theory. It is therefore surprising that a formal connection between game theory and cryptography is still lacking.

In the coming paragraphs we briefly review some of the important developments in this area, in particular those which are required for game playing without a trusted referee and reliable computation in the presence of faults. It is crucial to develop this theory in a rigorous way, but lack of space precludes a complete discussion. The explanations concentrate more, therefore, on general ideas than on precise details and the interested reader will have to consult the original papers for full details.

3.2.1. Notions of security

Any decent theory of cryptography must start by defining what it means for a communication system to be secure. A conceptually important observation is that one should strive for a definition of the security of a cryptosystem *against a given class of code-breakers*. The most reasonable assumption would be that the cryptanalyst can perform any "feasible" computation. The accepted view is that "feasible" computation means those carried by a randomized polynomial-time Turing Machine, and so this is the assumption we adopt.

There is a variety of notions of security that are being considered in the literature, but the general idea is one and the same: *Whatever a randomized polynomial-time Turing Machine can compute, given a ciphertext, should also be computable without it*. In other words, the ciphertext should be indistinguishable from an encryption of purely random bits to such a machine. Among the key notions in attaining this end are those of a one-way function and of hard-core bits which are introduced now.

3.2.2. One-way functions and their applications

Most computer scientists believe that **P $\neq$ NP**, or that "the author's job is harder than that of the referee". However, this important problem is still very far from being resolved. If this plausible conjecture fails, then there is no hope for a theory of cryptography is any sense known to us. In fact, to lay the foundations for modern cryptography, researchers had to recourse to even more powerful hypothesis. At the early days of the subject, those were usually assumptions on the intractability of certain computational problems from number theory. For example, factoring numbers, checking whether an integer is a quadratic residue modulo a nonprime integer N and so on.

An ongoing research effort is being made to weaken, or eliminate the unproven assumption required, and in particular to substitute "concrete" unproven hypotheses by more general, abstract counterparts. At present, large parts of known cryptography are based on a single unproven assumption, viz., that *one-way functions exist*. This idea originated with Diffie and Hellman (1976) in a paper that marks the beginning of modern cryptography. Its significance for cryptographic protocols was realized shortly after, in Goldwasser and Micali (1984). Intuitively, a polynomial-time computable function f from binary strings to binary strings, is one-way, if any polynomial-time Turing Machine which tries to invert it must have a very small probability of success. (The probability space in which this experiment takes place is defined by x being selected uniformly, and by the random steps taken by M). Specifically, let M be a probabilistic, polynomial-time Turing Machine and let n be an integer. Consider the following experiment where a string x of length n is selected uniformly, at random, and M is run on input $f(x)$ (but see Remark 3.5 for some pedantic technicality). This run is considered a success, if the output string y has length n and $f(y) = f(x)$. The function f is said to be (*strongly*) *one-way* if every M has only a negligible chance of success, i.e., for every $c > 0$, if n is large enough, then the probability of success is smaller than n^{-c}.

- Certain restricted classes of one-way functions are sometimes required. Since we do not develop this theory in full, not all these classes are mentioned here, so when encountered in a theorem, the reader can safely ignore phrases like "non-uniform one-way functions" etc. These should be considered technicalities that are required for the precise statement of the theorem. For example, sometimes we need f to be a permutation, i.e., a 1:1 mapping between two sets of equal cardinality. Another class of one-way functions which come up sometime, are those having the trapdoor property. Rather than formally introducing this class, let us present the canonical example:

Example 3.2. There is a randomized polynomial-time algorithm [Lenstra and Lenstra Jr. (1990) and the references therein] to test whether a given number is prime or not. The expected run time of this algorithm on an n digit number is polynomial in n. By the prime number theorem [and in fact, even by older estimates of Tchebyshef, e.g. Hardy and Wright (1979)], about one out of any n integers with n digits is prime. It is therefore, possible to generate, in expected time polynomial in n, two primes p, q with n digits each. Let $N = pq$, and for $0 \leqslant x < N$, let $f(x, N) = (y, N)$ with $0 \leqslant y < N$ congruent to x^2 modulo N. It is not hard to see that any algorithm to invert f (i.e., take square roots mod N) can be used to factor N. Now, no polynomial-time algorithms are known for factoring integers, and if, indeed, factoring cannot be done in polynomial time, then this f is one-way. Additionally, it is *trapdoor*, in the sense that there is a short piece of information, namely the primes p and q given which, f may be inverted in polynomial time.

Example 3.3. Another example of a function which is generally believed to be one-way is the discrete log. It is a basic fact from algebra that the multiplicative group of a finite field is cyclic. Namely, there is a *generator* g such that every nonzero field element x has a unique representation of the form $x = g^k$ for some integer k. This k is the *discrete log* of x (to base g). While it is easy to evaluate g^k, given g, k, it is believed to be computationally infeasible to find k given x and g. More on computation in finite fields and computational number theory, can be found in Lenstra and Lenstra Jr. (1990).

Remark 3.5. Technically we need to assume in the definition of one-way functions, that M is run on $f(x)$ concatenated with a string of n 1s, to allow M run in time polynomial in n, because the length $|f(x)|$ may be much smaller than $n = |x|$.

The applications of one-way functions to cryptography are numerous. In the present discussion they are employed for creating "envelopes". The idea is to enable two players A and B to perform the following task in secure message passing: A has a message m to transmit to B. An encrypted version $C(m)$ of this message is to be sent to B, so that the following two conditions are met: Given only $C(m)$, B cannot discover any information about m. Secondly, A is *committed* to the message m, in the following sense: at a later stage, A can prove to B that indeed m is the message sent, but B is able to call A's bluff in case A tries to prove the false claim that the message sent is any $m' \neq m$.

To create envelopes, one more ingredient is required basides one-way functions, namely, *hard-core bits*. The idea is this: a one-way function is easy to compute, and hard to invert. But while the definition requires that inverting f on x be hard, it does not rule out the possibility that we do gain *some* information on $f^{-1}(x)$. Hard-core bits b map the domain of f into 0, 1 so that for every z in the range, b is constant on $f^{-1}(z)$. The condition is that given $f(x)$, it is hard to compute $b(x)$. For example, we may be claiming that not only is it hard to find x given $f(x)$, it is not even feasible to tell whether, the seventh bit of x is zero or one. Technically, let f be a function from binary strings to binary strings, (to be thought of as a one-way function). A polynomial-time computable Boolean function b which is constant of $f^{-1}(z)$ for all z, [e.g., $b(x)$ may be the seventh bit of x] is said to be a *hard-core bit for f* if for every randomized polynomial-time Turing Machine M, any $c > 0$ and a large enough n, the probability that $M(f(x)) = b(x)$ is smaller than $\frac{1}{2} + n^{-c}$, where $n = |x|$, the probability being over M's random steps. [I.e., if you are asked for the seventh bit of x, given $f(x)$ and your computational power is restricted, you might as well guess by tossing a coin]. It would make sense to assume that if f is one-way, there should be certain predicates b on x that cannot be computed in polynomial time with any advantage, given $f(x)$. The following theorem shows almost this. It is shown that any one-way function f can be slightly modified to yield another one-way function $\tilde{f}$ for which there is a hard-core bit. The concatenation of two bit strings $\hat{x}, \hat{y}$ is denoted $\hat{x}\hat{y}$, and if they have the same length, then $\langle \hat{x}, \hat{y} \rangle$ is their inner product (as vectors over the field of two elements)

Theorem 3.1 [Goldreich–Levin (1989)]. *Let $f(\cdot)$ be a one-way function and let $\tilde{f}(\hat{x}, \hat{y}) = f(\hat{x})\hat{y}$, then $\tilde{f}$ is one-way and $b(\hat{x}, \hat{y}) = \langle \hat{x}, \hat{y} \rangle$ is a hard-core bit for it.*

In other words, if I show you $f(x)$ and y, and ask you for the inner product of x and y, you might as well flip a coin to guess an answer. Any significant increase in the probability of success over $\frac{1}{2}$ requires super-polynomial computations. We will not say much about the proof, except explain, that given a program which can tell the inner product of x and y, with rate of success bounded away from $\frac{1}{2}$, such a program can be utilized to invert f on a nonnegligible fraction of its range. A recent proof of this theorem using harmonic analysis (as in the proof of Theorem 4.4) can be found in Kushilevitz and Mansour (1991).

Given this theorem, envelopes are easily constructed: Suppose f is a one-way function for which b is a hard-core bit. We make the extra simplifying assumption that f is one-to-one [this assumption may be relaxed, see Naor (1991)]. To "place a bit β in an envelope" do as follows: Pick a random x in the domain of f for which $b(x) = \beta$, (such an x may not be obtained on first try, but on average two random tries will do) and let the receiver know the value of $f(x)$. When the time to reveal β comes, tell the receiver the value of x. The following properties of the protocol are easy to verify:

- The receiver's chance of finding β before x is revealed, is only $\frac{1}{2} + o(1)$, because b is hard-core.
- The sender cannot make false claims on β's value because f is one-to-one.
- Finally, the process is computationally feasible, since f and b are polynomial-time computable.

3.2.3. Interactive proofs

Let us return for a moment to the definition of the class **NP**. The situation can be depicted as a two-person game, between a *Prover P* and a *Verifier V*, who wants to solve a decision problem: a language L and a binary string x are specified and the question is whether x belongs to L. The Verifier is computationally bounded, being a probabilistic polynomial-time Turing Machine, and so, possibly incapable of solving the problem on his own. The Prover is computationally unrestricted, and so if x is in L, he can find that out and even discover a witness y for x's membership in L, with $|y|$ polynomial in $|x|$. The Prover then shows y to V, who can verify in polynomial time that indeed $x \in L$.

If V trusts P, and is willing to take any statement from P on faith, then of course P's unbounded computational power is at V's disposal. But what if P is not trustworthy and V demands to see proofs that he can *verify* at least? We know that this can be done for any $L \in \mathbf{NP}$, but can P convince V about the truth of claims whose computational complexity is higher than **NP**? The immediate answer is negative. For example, permitting V to ask P more queries or even allowing for a series of exchanges would not help, if V is assumed to play deterministically. For then, the all-powerful Prover P could figure out all of V's messages ahead of time and supply all his replies at once. Things do change, though, if we deviate from this traditional scenario on two accounts:

- Allow V to flip coins (which P cannot guess in advance).
- Have V settle for something less than a proof, and be content with statistical evidence for the correctness of the claim $x \in L$. Still, V should be able to set the statistical significance of the test arbitrarily high.

That the statistical significance of the test may be as high as V wishes, has two consequences: Only with a preassigned small probability will Prover P fail to convince Verifier V that x is in L in case it is. At the same time V will almost never be convinced that x is in L if this is not the case, for *any adversarial strategy* the prover may choose. Again, with V trusting P, the interactive proof system becomes as powerful as Prover P. It is therefore only interesting to think of a *cheating* Prover who may try and convince V of false claims, but should, however, almost surely fail. This theory has been initiated in Goldwasser et al. (1988) and Babai (1986), and some of it is briefly surveyed here.

Technically, one defines the notion of a pair of interacting randomized Turing Machines, which captures the intuition of two players sending messages to each other, by the end of which exchange V outputs either 0 or 1. Two such machines P and V constitute an *interactive proof system* for a language L if:

- P almost surely convinces V that x is in L in case it is: For every $c > 0$ there is a polynomial $p(\cdot)$ such that on input $x \in L$ machine V runs for time at most $p(|x|)$ and (correctly) outputs 1 with probability at least $1 - |x|^{-c}$.
- *No one* has a significant chance of convincing V that x is in L in case it is not in L: For every $c > 0$ there is a polynomial $p(\cdot)$ such that for every Turing Machine P' interacting with V, and every $x \notin L$, V runs for time at most $p(|x|)$ and (correctly) outputs 0 with probability at least $1 - |x|^{-c}$.

(Think of P' as an impostor who tries to convince V that $x \in L$ when indeed $x \notin L$.)

If such P and V exist, we say that L is in **IP** (*there is an interactive proof system for L.*) Are interactive proof systems more powerful than **NP**? That is, are there languages which are in **IP** but not in **NP**? A series of recent papers indicate that apparently this is the case. The complexity class **PSPACE** includes all the languages L for which membership of x can be decided using *space* which is only polynomial in $|x|$. **PSPACE** can be easily shown to contain **NP**, and there is mounting evidence that it is in fact a much bigger class. Note that a Turing Machine which is given space s can run for time *exponential* in s.

In any event, the power of interactive proof is now known, viz.:

Theorem 3.2 [Lund et al. (1990), Shamir (1990)]

IP = PSPACE.

This fact may have some interesting implications for game-playing and the complexity of finding good strategies, since it is known that many computational problems associated with various board games are complete for the class **PSPACE**, see Garey and Johnson (1979).

3.2.4. Zero-knowledge proofs

In a conversation where Prover convinces Verifier that $x \in L$, what other information does P supply V with? In the cryptographic context it is often desirable for P to convince V of some fact (usually of the form "x is in L" for some $L \in \mathbf{NP}$) while supplying as little side information as possible. Are there languages L for which this can be done with *absolutely no* additional information passing to V other than the mere fact that $x \in L$? The (surprising) answer is that such languages exist, assuming the existence of one-way functions. Such L are said to have a *zero-knowledge* interactive proof system.

A moment's thought will convince the reader that the very definition of this new concept is far from obvious. The intuition is that once V is told that "x is in L" he should be able to reconstruct possible exchanges between himself and P

leading to this conclusion. More precisely, given an interactive proof system (P, V) for a language L and given an input x, there is a probability distribution on all exchanges between P and V. Suppose there is a polynomial-time probabilistic Turing Machine M, which on input x and a bit b (to be thought of as a shorthand for the statement "x is in L") randomly generates possible exchanges between P and V. Suppose moreover, that if indeed x is in L, then the distribution generated by M is identical with the one actually taking place between P and V. In this case we say that (P, V) constitute a *zero-knowledge* interactive proof system for L and L is said to be in the class **ZK**.

Since both V and M are probabilistic polynomial-time, we may as well merge them. Thus, given a correct, positive answer to the decision problem "is x in L?", the Verifier can randomly generate conversations between itself and the Prover, and the exchanges are generated according to the genuine distribution. Moreover, V does that *on its own*. This can surely be described as V having learnt only the answer to that decision problem and nothing else.

Zero-knowledge comes in a number of flavors. In the most important versions, one does not postulate that the conversations generated by V are identical with the correct distribution, only that it is *indistinguishable* from it. We do not go into this, only mention that indistinguishability is defined in the spirit of Section 3.2.1. Goldreich (1988) contains a more detailed (yet less uptodate) discussion.

We would like to show now that:

Theorem 3.3. *If one-way functions exist, then the language of all 3-colorable graphs is in* **ZK**.

Since 3-colorability is an **NP**-complete problem there follows:

Corollary 3.1. *If one-way functions exist, then*

NP $\subseteq$ **ZK**.

Together with the method that proved the equality between **IP** and **PSAPCE** (Theorem 3.2) even more can be said:

Corollary 3.2. *If one-way functions exist, then*

ZK = **IP**(= **PSPACE**).

Following is an informal description of the zero-knowledge interactive proof for graphs' 3-colorability. Given a 3-colorable graph G on a set of vertices $X = \{x_1, \ldots, x_n\}$, the Prover finds a 3-coloring of G, i.e., a partition (X_1, X_2, X_3) of X so that if x, y belong to the same X_i, there is no edge between them. If $x \in X_t$ this is denoted by $\phi(x) = t$. The Prover P also chooses a random permutation π on $\{1, 2, 3\}$ and, using the method of "envelopes" (Section 3.2.2) it relays to V the sequence $\pi(\phi(x_i)), i = 1, \ldots, n$. The Verifier V then picks an edge $e = [x_i, x_j]$ in the graph, and in return P exposes $\pi(\phi(x_i))$ and $\pi(\phi(x_j))$ which V checks to be distinct. Note that if G is 3-colorable, the above procedure will always succeed. If it is not, then no matter which ϕ is chosen by V, it is almost sure that P, choosing the inspected edge at random, will detect an edge whose two vertices are colored the same after, say, $10n^3$ tries. (There can be at most $\binom{n}{2}$ edges if the graph has n vertices and the previous claim follows immediately from the law of large numbers.)

It is intuitively clear that V gains no extra information from this exchange, other than the fact that G is 3-colorable. Technically, here is how the zero-knowledge property of this protocol is established: A Turing Machine which is input the graph G and is given the information that G is 3-colorable can simulate the exchange between P and V as follows: Select an edge in G, say $[x_i, x_j]$ at random (this simulates P's step). Select two random numbers from $\{1, 2, 3\}$ and associate one with x_i and the other with x_j, for V's response. It is easily verified that the resulting exchanges are distributed in exactly the same way as those between the real P and V.

3.2.5. Secret-sharing

This idea was first introduced by A. Shamir (1979). Informally, the simplest form of the question is: One would like to deal pieces, or shares, of a secret among n players so that a coalition of any r of them can together reconstruct it. However, no set of $r - 1$ or fewer players can find out anything about the secret. A little more formally, there is a secret s and each player is to receive a piece of information s_i so that s is reconstructable from any set of r of the s_i, but no set of $r - 1$ or less of the s_i supplies any information about s. Here is Shamir's solution to this problem.

Fix a finite field $\mathbf{F}$ with $\geq n$ elements and let $s \in \mathbf{F}$ be the secret. Pick a polynomial $f = f(x)$ of degree at most $r - 1$, from $\mathbf{F}[x]$, the ring of polynomials in x over $\mathbf{F}$ where f's free term is s and all other coefficients are selected from F, uniformly and independently, at random.

Associate with every player P_i a field element α_i (distinct players are mapped to distinct field elements). Player P_i is dealt the value $f(\alpha_i)$, for $i = 1, \ldots, n$. Note that any r of the elements $f(\alpha_i)$ uniquely define the polynomial f, and so, in particular the secret coefficient s. On the other hand, any proper subset of these field elements yields no information on s which could be any field element with uniform probability distribution.

The performance of Shamir's scheme depends on proper behavior of all parties involved. If the dealer deviates from the protocol or if the players do not show the share of the secret which they actually obtained, then the scheme is deemed useless. Chor, Goldwasser, Micali and Awerbuch (1985) proposed a more sophisticated *verifiable* secret-sharing wherein each player can prove to the rest that he followed the protocol, again without supplying any extra information. Bivariate, rather than univariate polynomials have to be used for this more sophisticated scheme. The basic idea is for the dealer to choose a random bivariate polynomial $f(x, y)$ subject to the conditions that $f(0,0) = s$, the secret, and that f has degree $\leq r - 1$ in both x and y. Again player P_i is associated with field element α_i and its share of the secret is the pair of univariate polynomials $f(\alpha_i, y)$ and $f(x, \alpha_i)$. A key step in verifying the validity of the shares is that players P_i and P_j compare their views of $f(\alpha_i, \alpha_j)$, which is included in both their shares. Mismatches are used in detecting faulty dealers and faulty share holders. The details can be found in Ben-Or et al. (1988).

3.3. Protocols for secure collective computation

Given a set of n players and an additional trusted party, there are various goals that can be achieved in terms of correct, reliable and leak-free computation. In fact, all they need to do is relay their input values to that party who can compute any functions of these inputs and communicate to every player any data desired. Speaking in broad terms, our goal here is to provide a protocol for the n parties to achieve all these tasks in the absence of a trusted party. The two most interesting instances of this general plan are:

Privacy. Consider a protocol for computing n functions $f_1, \ldots, f_n$ of n variables by n parties where originally party i holds x_i, the value of the ith variable and where by the protocol's end it gets to know $f_i(x_1, \ldots, x_n) (1 \leq i \leq n)$. The protocol is *t-private* if every quantity which is computable from the information viewed throughout the protocol's run by any coalition of $\leq t$ players, is also computable from their own inputs and outputs.

Fault tolerance. It is *t-resilient* if for every coalition S of no more than t parties, the protocol computes a value $f(y_1, \ldots, y_n)$ so that $y_i = x_i$ for all $i \notin S$ and so that $y_j (j \in S)$ are chosen independent on the value of $x_i (i \notin S)$.

Here are the results which hold under the assumption that bad players are computationally restricted.

Theorem 3.4. [Goldreich et al. (1987)]. *Assuming the existence of (non-uniform) trapdoor one-way functions, every function of n variables can be computed by n probabilistic polynomial-time Turing Machines, in an $(n-1)$-private, $[(n-1)/2]$-resilient protocol. The quantities $n-1$ and $[(n-1)/2]$ are best possible.*

Non-experts can safely ignore the phrase "non-uniform" in the theorem.
In the information-theoretic setting the results are summarized in:

Theorem 3.5. [Ben-Or et al. (1988), also Chaum et al. (1988)]. *Every function of n variables can be computed by n processors which communicate via secure channels in an $[(n-1)/2]$-private way. Similarly, a protocol exists which is both $[(n-1)/3]$-private and $[(n-1)/3]$-resilient. The bounds are tight.*

Remark 3.6. The same results hold, even if the players have to jointly sample a probability space determined by their inputs, see Remark 3.2.

Having reviewed all the necessary background, we are now ready to explain how the algorithms work. There are four protocols to describe, according to the model (cryptographic or information-theoretic) and whether bad players are assumed curious or malicious. All four protocols follow one general pattern, which is explained below. We review the solution for the (information-theoretic, curious) case in reasonable detail and then indicate how it is modified to deal with the other three situations.

The problem becomes more structured, when rather than dealing with a general function f the discussion focuses on a *circuit* which computes it (see Remark 3.4). This causes no loss of generality, since circuits can simulate Turing Machines. An idea originating in Yao (1982), and further developed in Goldreich et al. (1987) is for the players to collectively follow the computation carried out by the circuit moving from one gate to the next, but where each of the partial values computed in the circuit (values carried on the circuit's wires) is encoded as a secret shared by the players. To implement this idea one needs to be able to:

- Assign input values to the variables in a shared way.
- Perform the elementary field operations on values which are kept as shared secrets. The outcome should again be kept as a shared secret.
- If, at the computation's end, player P is to possess a certain value computed throughout, then all shares of this secret are to be handed to him by all other players.

The second item is the most interesting and will be our focus here.

In the case under consideration (information-theoretic, curious) this part is carried out as follows: Secrets are shared using Shamir's method, and we need to be able to add and to multiply field elements which are kept as secrets shared by all players. Addition of shared secrets poses no problem: if the secrets s_1, s_2 are encoded via the degree $\leq t$ polynomials f_1, f_2 (i.e., the free term is s_i and all other t coefficients are drawn from a uniform probability distribution over the field) then the secret $s = s_1 + s_2$ may be shared via the polynomial $f = f_1 + f_2$. Note that f constitutes a valid encoding of the secret s. Moreover, player P_i who holds the shares $f_1(\alpha_i)$ and $f_2(\alpha_i)$ of s_1, s_2 respectively, can compute its share in s by adding the above shares. Thus addition may be performed even without any communication among the players. The same holds also for multiplication by a scalar.

Multiplication of secrets creates more complications: While it is certainly possible to multiply the shares in just the same way to have $g = f_1 f_2$ encode the secret $s_1 s_2$, the polynomial g does not qualify as an encoding. Its degree may rise to $2t$, and also note that to qualify for a shared secret, all non-free terms of the polynomial have to be drawn from a uniform probability distribution. This requirement need not hold here, e.g., because g cannot be irreducible, except in the rare event that one of the f_i is constant. These two problems are resolved in two steps: A re-randomization, and a degree reduction. To re-randomize, every participant chooses a random polynomial q_i of degree $2t$ with a zero free term and deals it among all players. If g is replaced by $g + \sum q_i$, then no set of t players may gain any information on the first t coefficients. (This is a slight modification of the basic argument showing the validity of Shamir's secret-sharing method.) This polynomial has the correct free term, all other coefficients are uniformly, randomly distributed, so the only problem left is with the degree being (probably) too high, which thus needs to be reduced. This will be achieved by truncating the high-order terms in g.

Let h be the polynomial obtained by deleting all the terms in g of degree exceeding t. If $\boldsymbol{a}$ (respectively $\boldsymbol{b}$) is the vector whose ith coordinate is $g(\alpha_i)$ [respectively $h(\alpha_i)$] then there is a matrix C depending only on the α_i such that $\boldsymbol{b} = \boldsymbol{a}C$. This operation, and thus a degree reduction, may be performed in a shared way, as follows. Each P_j knows $g(\alpha_j)$ and for every i we need to compute $h(\alpha_i) = \sum_j c_{i,j} g(\alpha_j)$ and inform P_i. Now, P_j computes $c_{i,j} g(\alpha_j)$ and deals it as a shared secret among all players. Everyone sums his shares for $c_{i,j} g(\alpha_j)$ over all j, thus obtaining his share of $h(\alpha_i) = \sum c_{i,j} g(\alpha_j)$, which presently becomes a shared secret. (Recall that if s_v are secrets, and s_v^μ is the share of s_v held by player P_μ, then his share of $\sum s_v$ is $\sum s_v^\mu$.) Each player passes to P_i the share of $h(\alpha_i)$ he is holding, so now P_i can reconstruct the actual $h(\alpha_i)$. Now, the free term of g which is the same as the free term of h, is kept as a shared secret, as needed.

This establishes the $[(n-1)/2]$-privacy part of Theorem 3.5. The condition $n > 2t$ was required both at the multiplication step and for re-randomization.

That more curious players cannot be tolerated by any protocol follows from the impossibility of computing the logical "or" function by two players. See also Chor and Kushilevitz (1991).

We just saw how $[(n-1)/2]$ curious players may be handled, but how can a protocol tolerate $[(n-1)/3]$ *malicious* participants? We consider $n = 3t + 1$ players of which t may fail, and try to maintain the spirit of the previous protocol. This creates the need for a secret-sharing mechanism where t shares of a secret reveal nothing of the secret and where even if t out of n shares are either altered or missing, the secret can be recovered.

Notice the following simple fact about polynomials: Let f be a degree d polynomial over a (finite, but large enough) field F, and assume you are being told f's value at $3d + 1$ points. However, this data may be in error for as many as d of the data points. You can still recover f, despite the wrong data, as follows: Let the $3d+1$ points be $x_i \in F$ and the purported values be $y_i \in F$, so that $f(x_i) = y_i$ for at least $2d+1$ of the indices i. Suppose that g, h are two degree d polynomials, each satisfying at least $2d+1$ equalities of the form $g(x_i) = y_i$ and $h(x_i) = y_i$, respectively. This implies that for at least $2(2d+1) - (3d+1) = d+1$ indices, there holds $g(x_i) = h(x_i) = y_i$. But g, h have degree $\leq d$, so they must coincide (and be equal to f).

This implies that for $n \geq 3t+1$, even if as many as t of the shares are corrupted, the secret can be recovered. Consequently, the previous protocol can be performed successfully, if the following step is added: Whenever a player is receiving the n shares of a secret that he is supposed to know, he assumes that as many as t of them might be corrupted or missing. The previous remark implies that he can reconstruct the secret, nevertheless.

Incidently, a clever choice of the field elements α_i allows this recovery to be done in a computationally efficient way. Namely, if the field F has a primitive nth root of unity ω, i.e., $\omega^n = 1$ and $\omega^k \neq 1$ for $1 \leq k < n$, let $\alpha_i = \omega^i$ ($1 \leq i \leq n$). The efficient recovery mechanism is based on the discrete Fourier transform [e.g., Aho et al. (1974)], and is also related to the theory of error correcting codes [Mac Williams and Sloane (1977)].

Besides the corruption of shares, there is also a possibility that bad players who are supposed to share information with others will fail to do so. This difficulty is countered by using a verifiable secret-sharing mechanism, briefly described in Section 3.2.5.

No more than $[(n-1)/3]$ malicious participants can be handled by any protocol. This follows from the fact (Theorem 2.1) that Byzantine Agreement is achievable only if $n \geq 3t+1$.

Our last two comments are that if players are not restricted to communicate via the secure two-party lines but can also *broadcast* messages then resiliency can be increased from $[(n-1)/3]$ to $[(n-1)/2]$ [Rabin and Ben-Or (1989)]. Also, it is not a coincidence that our computations were always over a finite field. The situation when the domain is infinite is discussed in Chor et al. (1990).

We have reversed the historical order here, since Theorem 3.4, the solution in the cryptographic set-up was obtained [in Goldreich et al. (1987)] prior to the information-theoretic Theorem 3.5 [Ben-Or et al. (1988), Chaum et al. (1988)], which, in turn followed [Yao (1982)] when the two-player problem was solved. In hindsight, however, having presented the solution for the information-theoretic case, there is an easy way to describe the cryptographic solution. Note that in each step of the above protocol one of the players tries to convince another player of a certain statement such as "the value of a polynomial Φ at a point x is so and so". All these statements clearly belong to **NP** and so by Corollary 3.1 can be proved in a zero knowledge interactive protocol. Thus if the players are randomized polynomial-time processors, and if one-way functions exist, then the protocol may be run step by step safely. Namely, replace each step in the previous protocol where player P_i relays a message to P_j, using their secure communication line, by an interaction where P_i proves to P_j in zero knowledge, the appropriate statement. This proves Theorem 3.4.

3.3.1. Efficiency issues

While our motivating questions are now answered, there is much left to be desired in terms of efficiency. The protocols presented here allow us to transform any given protocol in distributed processing into one which cannot be disrupted by any set of faulty processors, as long as they are not too numerous. However, the transformation from the original protocol to the safe one creates a large overhead in processing time. Even if this transformation must, in general, be costly, could we supply similar, only more efficient results for specific computational tasks? In particular, in distributed computing there are many operations that have to be run in a network for purposes of maintenance [e.g. Afek et al. (1987)]. It would be interesting to find ways in which such tasks can be performed both safely and efficiently.

4. Fault-tolerant computation – The general case

Protocols presented in the previous two sections allow a set of players, some of whom may malfunction, to perform various computational tasks without disruption. These protocols require (i) an upper bound on the number of bad players and that (ii) good players can hide information, either through use of secure communication lines, or by exploiting the bad players' restricted computational power. Is fault-tolerant computation at all possible if (ii) is removed? We lower our expectations somewhat: rather than *absolute* results guaranteeing that computations always turn out correctly, *quantitative* statements are sought. It is assumed that the bad coalition favors a certain outcome for the protocol, and faulty players play an optimal strategy to maximize the probability of this outcome. The influence of a coalition towards a particular outcome is the largest increase in this outcome's probability the coalition can achieve. Protocols are sought to minimize all influences. The two tasks to be most extensively considered here are: flipping a coin and electing one of the players. In coin-flipping we try to make both outcomes as equally likely as possible, even though the bad players may favor, say, the outcome "heads". In the election game, the goal is to make it most likely for a good player to be elected, despite efforts by the bad players to have one of them elected.

The critical role of *timing* in such problems, is best illustrated, perhaps, through a simple example, viz. the *parity* function. Let all x_i take values 0 or 1, and let $f(x_1, \ldots, x_n) = 0$ or 1 depending on the parity of $\sum x_i$. Suppose that the bad coalition would like f to take the value 1. If all players are good and supply their correct inputs, and if the x_i are generated by independent coin flips, then f equals 1 with probability $\frac{1}{2}$. Can a bad coalition boost this probability up? That depends critically on the order in which the x_i are announced. If they are supplied simultaneously by all players, then even as many as $n-1$ bad players can exert no bias on f. If, however, the bad players let the good ones move first, then even a *single* bad player can decide f's value with certainty. Simultaneous action cannot be guaranteed in the presence of bad players, and so the relevant assumption is that good players move first.

The n-player computation of parity is an instance of the first class of problems considered in this section: Let f be an n-variable Boolean function, and let the assignment x_i to f's ith variable be known only to player P_i. Consider a coalition S whose goal is to have f take some desired value. All players not in S disclose the assignments to their variables, after which players in S together select values to be announced as assignments to their variables. Clearly, even after all non-S players announce their input values, f may still be undetermined, a fact which gives S an advantage. The probability of f remaining undetermined is defined as the *influence* of S on f.

The above discussion is a "one-shot" protocol for computing f. The most general mechanism for computing a function without hiding information can, as usual, be described in a tree form. Again we consider a coalition trying to make f take on a certain value and define influences accordingly (see Section 4.3).

Influences are the main subject of the current chapter, both for Boolean functions (which for a game theorist are *simple games*) and in general perfect information games. Many quantities are defined in game theory to measure the power of coalitions in various games and decision processes. Influences can be added to this long list and may be expected to come up in other game-theoretic problems. Specifically, one instance of our notions of influence turns out to coincide with the *Banzhaf–Coleman power index* [Owen (1982)].

Here are some of the main results under review. We will freely interchange terminologies between Boolean functions and simple games, and correspondingly sets of variables and coalitions. Our first result is stated both ways.

- For every n-variable Boolean function there is a set of $o(n)$ variables whose influence is $1 - o(1)$. [In every n-player simple game there is a coalition of $o(n)$ players whose influence is $1 - o(1)$.]
- There are n-player simple games where coalitions with fewer that $n/\log^2 n$ players have influence $o(1)$.
- In every perfect-information, n-player protocol to flip a coin, and for every $n \geqslant k \geqslant 1$, there is a k-player coalition with influence $> c(k/n)$ (where $c > 0$ is an absolute constant).
- There is an n-player perfect-information coin-flipping game, where the influence of every coalition of $\leqslant n/4$ players does not exceed 0.9.

4.1. Influence in simple games

We are now ready for formal definitions, after which some concrete examples are considered. Let $f: \{0, 1\}^n \to \{0, 1\}$ be a Boolean function and let $S \subseteq \{1, \ldots, n\}$. To define the influence of S on f, assign $0, 1$ values to all variables x_i not in S, where the assignments are chosen uniformly and independently at random. The probability that f remains undetermined after this partial assignment of values is the influence of S on f. In fact, notice that this partial assignment falls in one of three categories: either it determines that $f = 0$ regardless of the x_i for $i \in S$, or $f = 1$ must hold, or f remains undetermined. Let q_0, q_1 and q_2 be the probabilities of these three events. As we said, q_2 is called the *influence of S on f*, and is denoted $I_f(S)$.

There are two other quantities to mention. If $p_0 = p_0(f)$ is the probability that $f = 0$ when *all* its variables are assigned random values and $p_1 = 1 - p_0$, define the *influence of S on f towards zero*

$$I_f^0(S) = q_0 + q_2 - p_0.$$

(Explanation: All x_i with $i \in S$ are assigned after all other x_j are revealed, and are chosen so as to assure f equals 0, if possible. Therefore, the probability that f does evaluate to zero is $q_0 + q_2$. The excess of this probability over p_0 measures the advantage gained towards setting f to 0.) Similarly:

$$I_f^1(S) = q_1 + q_2 - p_1$$

is the *influence of S on f towards one*. It is pleasing to notice that the influence of S on f is the sum of the two

$$I_f(S) = q_2 = I_f^0(S) + I_f^1(S),$$

since $q_0 + q_1 + q_2 = p_0 + p_1 = 1$.

For $S = \{i\}$ a singleton, it is easily checked that $I^0(i) = I^1(i) = \frac{1}{2}I(i)$, and $I(i)$ is the same as the Banzhaf–Coleman power index. The reader should note that the subscript f was omitted and i was written instead of $\{i\}$, for typographic convenience. This shorthand writing will be used throughout.

Let us consider a few examples, all with $p_0 = p_1 = \frac{1}{2}$.

Example 4.1 (Dictatorship). This is the Boolean function $f(x_1, \ldots, x_n) = x_1$. It is easily verified that $I(1) = 1$, $I^0(1) = I^1(1) = \frac{1}{2}$, and that a set S has zero influence unless $1 \in S$.

Example 4.2 (Majority). For convenience, assume $n = 2m + 1$ is odd, and consider the influence of singletons first. The probability for the majority vote to be undetermined when only i's vote is unavailable is $I(i) = 2^{-2m}\binom{2m}{m} = \Theta(n^{-1/2})$. More generally, if S has $2s + 1$ elements, then

$$I(S) = \sum_{m+s \geq j \geq m-s} \binom{2m}{j} \Big/ 2^{2m-2s}.$$

This expression can be estimated using the normal distribution approximation to the binomial one. In particular, if $|S| = n^{1/2+o(1)}$ the influence of S equals $1 - o(1)$. That is, coalitions of such size have almost sure control over the outcome. This point will be elaborated on in the sequel.

The next example plays a central role in this area:

Example 4.3 (Tribes). Variables are partitioned into blocks ("tribes") of size b, to be determined soon and f equals 1 iff there is a tribe in which all variables equal 1. To guarantee $p_0 = p_1 = \frac{1}{2}$ the following condition must hold:

$$(1 - (\tfrac{1}{2})^b)^{n/b} = \tfrac{1}{2},$$

the solution of which is $b = \log n - \log\log n + \Theta(1)$. (In fact, the solution b is not an integer, so we round this number and make a corresponding slight change in f to guarantee $p_0 = p_1 = \frac{1}{2}$.) To figure out the influence of a single variable i, note that f is undefined iff all variables in i's tribe are 1 while no other tribe is a unanimous 1. The probability of the conjunction of these events is

$$I(i) = (\tfrac{1}{2})^{b-1}(1 - (\tfrac{1}{2})^b)^{n/b-1} = \Theta\left(\frac{\log n}{n}\right).$$

As for larger sets, there are sets of size $O(\log n)$ with influence close to $\frac{1}{2}$, e.g., a whole tribe, and no significantly smaller sets have influence bounded away from zero. It is also not hard to check that sets of variables with influence $1 - o(1)$ need to have $\Omega(n/\log n)$ variables and this is achievable. In fact, a set which meets all tribes and contains one complete tribe has influence 1.

We are looking for the best results of the following type: For every n-variable Boolean function of expected value $1 > p_1 > 0$, there is set of k variables of influence at least $J(n, k, p_1)$.

Already the case of $k = 1$, the best lower bound on the influence of a single player, turned out to be a challenging problem. An old result of S. Hart (1976) translated to the present context says, for example, that if $p_1 = \frac{1}{2}$ then $\sum_{x\in[n]} I_f(x) \geq 1$. The full statement of Hart's theorem supplies the best possible lower bound on the sum of influences for all $0 \leq p_1 \leq 1$. In combinatorial literature [e.g., Bollobas (1986)] this result is referred to as the *edge isoperimetric inequality* for the cube. That the bound is tight for $p_1 = \frac{1}{2}$ is exhibited by the dictatorship game. Consequently, in this case there is always a player with influence at least $1/n$. On the other hand, in all the above mentioned examples there is always a player with influence $\geq (c \log n)/n$. This indeed is the correct lower bound, as shown by Kahn, Kalai and Linial (1988).

Theorem 4.1. *For every Boolean function f on n variables (an n-player simple game) with $p_1(f) = \frac{1}{2}$, there is a variable (a player) with influence at least $(c \log n)/n$ where $c > 0$ is an absolute constant. This bound is tight except for the numerical value of c.*

Before sketching the proof let us comment on some of its extensions and corollaries. The vector $I(x), x \in [n]$ of influences of individual players, may be considered in various L^p norms. The edge isoperimetric inequality yields the optimal lower bound for L^1 while the above theorem answers the same question for L^∞. Can anything be said in other norms? As it turns out, the *tribes* function yields the best (asymptotic) result for all L^p, $p > 1 + o(1)$. The L^2 case is stated for all $0 < p_1 < 1$ in the following theorem:

Theorem 4.2 [Kahn, Kalai and Linial (1988)]. *For f as above, and $p_1 = p_1(f) \leq \frac{1}{2}$,*

$$\sum_{x \in [n]} I(x)^2 \geq c p_1^2 \frac{\log^2 n}{n},$$

where $c > 0$ is an absolute constant. The inequality is tight, except for the numerical value of c.

By repeated application of this theorem one concludes the interesting result that *there is always a coalition of negligible cardinality and almost absolute influence*:

Corollary 4.1. *For any simple game f with $p_1(f)$ bounded away from zero and one, there is a coalition of $o(n)$ players whose influence is $1 - o(1)$. In fact, for every $\omega = \omega(n)$ which tends to infinity with n there is such a coalition of size $n \cdot \omega(n) / \log n$. The result is optimal except for the ω term.*

We have established the existence of small coalitions whose influence is almost 1. Lowering the threshold, it is interesting to establish the existence of (possibly even smaller) coalitions whose influence is only bounded away from zero. While the last two bounds are tight for *tribes*, this game does have coalitions of size $O(\log n)$ with a constant influence. This is a very small cardinality, compared e.g., with the majority function, where cardinality $n^{1/2 + o(1)}$ is required to bound the influence away from zero. The situation is not completely clarified at the time of writing, though the gap is not nearly as big. Ajtai and Linial (1993) have an example of a Boolean function where sets of $o(n/\log^2 n)$ variables have $o(1)$ influence. It is plausible that the truth is closer to $n/\log n$. Following is an outline of this example.

Example 4.4. The example is obtained via a probabilistic argument, [for a reference on this method in general see Alon and Spencer (1992)]. First introduce the following modification of *tribes*: Reduce tribe sizes from $\log n - \log \log n + c_1$, to $\log n - 2\log \log n + c$, thus changing $p_0(f)$ from $\frac{1}{2}$ to $\Theta(1/n)$ (c, c_1 are absolute constants). Also introduce, for each variable i, a bit $\beta_i \in \{0, 1\}$ and define f to be 1 iff there is a tribe B such that for every $i \in B$ the corresponding input bit x_i equals β_i. This is essentially the same example as *tribes*, except for the slight change of parameters and the fact that some coordinates (where $\beta_i = 0$) are switched.

Now generate at random $m = \Theta(n)$ partitions $\pi^{(j)}(j = 1, \ldots, m)$ of the variables into blocks of the above size, as well as m corresponding binary vectors $\beta^{(j)}(j = 1, \ldots, m)$. Partition $\pi^{(j)}$ and vector $\beta^{(j)}$ define a function $f^{(j)}$ as above. The function considered is $f = \wedge_j f^{(j)}$. It can be shown that with positive probability $p_0(f)$ is very close to $\frac{1}{2}$ and that no set of size $o(n/\log^2 n)$ can have influence bounded away from zero. Some small modification of this f gives the desired example.

Let us turn to a sketch of the proof of Theorem 4.1. The reader is assumed to have some familiarity with the basics of harmonic analysis as can be found e.g., in Dym and McKean (1972). Identify $\{0, 1\}^n$ with the Abelian group $\mathbf{Z}_2^n$, so it makes sense to consider the Fourier transform of Boolean functions. It is also convenient to identify $\{0, 1\}^n$ with the collection of all subsets of $[n]$. The 2^n characters of this group are parameterized by all $S \subseteq [n]$ as follows:

$$u_S(T) = (-1)^{|S \cap T|}$$

and f is expressed as

$$f(T) = \sum_S \alpha_S u_S(T).$$

(α_S is usually denoted $\hat{f}_S$, the Fourier transform of f at S.) The easiest way to see the connection between influences and Fourier coefficients is to notice that if f is a monotone Boolean function (i.e., a monotone simple game), then the influence of player j is given by

$$I_f(j) = \alpha_{\{j\}}$$

and there is no loss of generality in considering only the monotone case because of the following:

Proposition 4.1 [Ben-Or and Linial (1989)]. *For every Boolean function g there is a monotone function h such that $p_1(h) = p_1(g)$ and so that for every $S \subseteq [n]$ and $\beta \in \{0, 1\}$, there holds*

$$I_h^\beta(S) \leqslant I_g^\beta(S).$$

So to show the existence of an influential variable one needs to prove lower bounds on certain Fourier coefficients of f. Monotonicity is not used in the present proof, but see Theorem 4.13. First one considers "directional derivatives of f:"

$$f^j(T) = f(T) - f(T \oplus \{j\}),$$

where $\oplus$ is mod 2 addition. Note that $I(j)$ is proportional to the fraction of the T for which $f^j(T) \neq 0$.

Since f^j takes only the values $-1, 0, 1$ the square of its L^2 norm is proportional to $I(j)$. Now Parseval's identity enables us to express the L^2 norm of a function in terms of its Fourier coefficients. At the same time there is a simple relationship between the coefficients of f^j and those of f. As a consequence an expression is obtained for $I(j)$ in terms of f's Fourier expansion:

$$I(j) = 4 \sum_{j \in S} \alpha_S^2,$$

and summing over all j,

$$\sum I(j) = 4 \sum_{|S|} |S| \alpha_S^2.$$

But Parseval's identity also gives

$$\sum \alpha_S^2 = \|f\|_2^2 = p_1(f)$$

and so the $I(j)$ can be small only if most of $\sum \alpha_S^2$ comes from sets S of bounded cardinality. The main part of the proof is to show that this cannot happen. In fact a similar statement is proved with regards to the functions f^j. The only two properties available about these functions is that their range is $\{-1, 0, 1\}$ and (arguing by contradiction) their supports are small. The problem is how to express the first property in analytical terms and the answer is that having $\{-1, 0, 1\}$ for range means that all L^p norms of such a function are easy to express. The issue then becomes to show that under some restrictions on L^p norms, it is impossible for almost all of the L^2 norm to come from coefficients α_S for S of bounded size.

In his work on the sharpest form of some classical inequalities in harmonic analysis Beckner (1975) considers the *two point space* $X = \{-1, 1\}$ and operators from $L^p(X)$ to $L^q(X)$. This is a two-parameter class of operators and Beckner finds tight bounds on their norms. To state his results, note that any function on the two point space is linear.

Lemma 4.1. *Consider the operator $T_1: L^p(X) \to L^2(X)$ which maps the function $a + bx$ to $a + \varepsilon bx$. If $p \leq 1 + \varepsilon^2$ then the norm of T_1 is 1.*

Another lemma of Beckner deals with similar problems concerning X^n. The space X^n is naturally identified with the n-dimensional cube and the operator $T = T_1^n$ is uniquely specified via knowledge of $T(u_S)$ for all S, since the characters span the whole space of real functions on the cube. As it turns out

$$T(u_S) = \varepsilon^{|S|} u_S,$$

and the lemma is:

Lemma 4.2. *If $p \leq 1 + \varepsilon^2$ then $\|T\| = 1$.*

The conclusion that is useful for our proof is given in the following lemma:

Lemma 4.3. *Let $g: \{0, 1\}^n \to \{-1, 0, 1\}$ be a function whose Fourier expansion is $g = \sum \gamma_S u_S$ and let g be nonzero on exactly $t \cdot 2^n$ points $(0 \leq t \leq 1)$. Then for every*

$0 \leqslant \delta \leqslant 1$

$$t^{2/(1+\delta)} \geqslant \sum \delta^{|S|} \gamma_S^2.$$

This last lemma is applied to $g = f^j$ and the resulting inequalities can be combined to derive the theorem.

4.1.1. The non-Boolean case

Even one-shot protocols for collective coin flipping need not be confined to a Boolean function, since players' input to the process might well contain more than a single bit. So, we may consider a function $f: X^n \to \{-1, 1\}$ where X is any probability space. Each player P_i samples a point $x_i \in X$, and the collective coin flip is $f(x_1, \ldots, x_n)$. (In the previous discussion we had $X = \{0, 1\}$, both points having probability $\frac{1}{2}$). The phenomena established in Theorems 4.1 and 4.2 have been recently extended to this more general situation, as we now explain. The definition of influence is easy to extend: let $S \subseteq [n]$ and assign all $x_i (i \notin S)$ values independently and according to the probability distribution on X. The probability that this partial assignment leaves f undetermined is $I_f(S)$, the influence of S on f. Though the same phenomena come up in the more general situation, the previous proof techniques breaks down, so new ideas have to be introduced. It is interesting to comment that this time Proposition 4.1 is being employed.

Theorem 4.3 [Bourgain, Kahn, Kalai, Katznelson and Linial (1992)]. *Let X be a probability space, and let $f: X^n \to \{0, 1\}$ have expectation $E(f) = \frac{1}{2}$. Then there is a variable with influence at least $(c \log n)/n$ where $c > 0$ is an absolute constant. The result is optimal except for the value of the constant c.*

The old theorem of Loomis and Whitney (1949) plays here the same role as does Hart's result in the case $X = \{0, 1\}$, providing the best L^1 estimate for the vector of influences. In this case the extremal example consists of $X = [0, 1]$, the unit interval with the function $f(x_1, \ldots, x_n)$ which equals 1 iff at least one of the x_i exceeds $2^{-1/n}$.

4.2. Symmetric simple games

A simple game f is said to be *symmetric* if $f(\bar{x}) = \overline{f(x)}$ for every input x where bar stands for Boolean negation. In other words, reversal of each individual vote reverses the overall decision. Clearly $p_1(f) = \frac{1}{2}$ holds for symmetric games. Evaluation of Boolean functions can be thought of as a voting scheme, and from this perspective, it is natural to require f to be symmetric. The same questions posed about general simple games may be asked in this context, though the results are more fragmentary.

The known results are summarized in the following theorems:

Theorem 4.4 [Feldman (1986), Dubiner (1986)]. *There is a symmetric n-player simple game where each player x has influence $I_f(x) = O(\log n/n)$.*

Theorem 4.5 [Ben-Or and Linial (1989)]. *There is a symmetric n-player game where a coalition having influence ε must have cardinality $\geq \varepsilon n^\alpha$, where $\alpha = \log 2/\log 3 = 0.63...$*

These theorems are derived from the following two examples:

Example 4.15 [**The circle game** (Feldman, Dubiner)]. Assume n to be odd, and imagine the variables being arranged around a circle. Consider any assignment for the variables of f and let $\lambda_1 \geq \lambda_2 \geq \ldots$ (respectively $\mu_1 \geq \mu_2 \ldots$) be the lengths of consecutive strings of zeros (respectively ones) in the input. The value of f is determined by a lexicographic comparison of these two sequences, i.e., let j be the smallest index for which $\lambda_j \neq \mu_j$. If $\lambda_j > \mu_j$, then $f = 0$, otherwise $f = 1$. Since n is odd, such a j exists and f is well-defined.

The analysis which shows that individual influences are $O(\log n/n)$ is not hard and is omitted. The reader looking to reconstruct a proof should notice that almost surely λ_1, μ_1, the longest consecutive blocks are equal to $\Theta(\log n)$.

Another remark concerning this game is that while it is asymptotically optimal with respect to minimizing individual influences it is not particularly efficient in bounding influences of larger coalitions. A coalition of size $O(\sqrt{n})$ with influence 1 can be constructed as follows: it has $\sqrt{n}$ members equally spaced on the circle, plus a consecutive segment of length $\sqrt{n}$. In this respect this example is not even as good as majority voting.

In the next example individual players have influence $O(n^{-\alpha})$, which is inferior to the circle game. However, it yields interesting bounds for the influences of larger coalitions.

Example 4.5 (Iterated majority of three). In this example it is convenient to assume that n is a power of 3. The set of variables is partitioned into three "committees" of equal size. Each committee is divided into three "sub-committees" which are themselves partitioned into three etc. until the level of individual variables is reached. The overall decision is taken by a simple majority vote among the three committees. Within each committee, decision is made by majority vote of subcommittees etc. Analysis of this game [Ben-Or and Linial (1989)] shows that a coalition of size $k \leq n^\alpha$ has influence $O(k/n^\alpha)$ where $\alpha = \log 2/\log 3 = 0.63...$

It is not clear whether symmetric games exist where individual influences are only $O(\log n/n)$ and at the same time, a size much larger that $\sqrt{n}$ is needed to achieve constant influence. Also, does this class have examples where constant influence can be achieved only by coalitions of size $\Omega(n^{1-\varepsilon})$ for all $\varepsilon > 0$? Not much is known at the moment.

4.3. General perfect-information coin-flipping games

It is not hard to think of more elaborate mechanisms for generating a random bit other than the one-shot games considered so far. Would it not be possible to find perfect-information games which are less amenable to bias than one-shot games? Indeed this is possible, but even in the most general perfect-information games there are influential coalitions. An exact description of the games under consideration has to come in first. This definition is rather cumbersome, but luckily, each such game can be approximated arbitrarily well by much more structured games. A *standard* coin-flipping game G is a binary tree each of whose internal nodes in *controlled* by one of the players and where the leaves are marked by zero or one. The game is started at the root. At each step the player which controls the present node has to flip a coin and accordingly the game proceeds to either the right or left son. When a leaf is reached the game terminates and the outcome is that leaf's label. The interested reader can find the general definition as well as the (easy) proof that there is no loss of generality in dealing with standard games in Ben-Or and Linial (1989). Most of the subsequent discussion concerns standard games and the reader should have no difficulty adapting it where necessary to, say, trees which are not binary or other slight modifications.

With the experience gained on one-shot games the definition of influence should be easy to figure out: Let $p_0 = p_0(G)$ be the probability that the outcome of G is zero (assuming all players abide by the rules) and let $p_1 = 1 - p_0$. Now suppose that a coalition $S \subseteq [n]$ tries to bias the outcome and maximize the probability of a zero outcome. So when a node controlled by a player from S is reached, it is not by a coin flip that the next step is determined, but rather according to an optimal, mutually coordinated strategy. The new probability for zero outcome is denoted $p_0(G) + I_G^0(S)$ and this additional term is called *the influence of S on G towards zero*. The corresponding term with zero replaced by one is also defined. In keeping with simple games we define the *influence of S on G* as $I_G(S) = I_G^0(S) + I_G^1(S)$.

Now that the necessary technicalities have been dealt with, we can come down to business and state the results.

Theorem 4.6 [Ben-Or and Linial (1989), Alon and Naor (1990), Boppana and Narayanan (1993)]. *Let G be an n-player coin-flipping game with $p_1(G)$ bounded away from zero and one and let $1 \leqslant k \leqslant n$, then*

$$\text{Average}(I_G^1(S)) > c\frac{k}{n},$$

where the average is over all coalitions S of k players and where c is an absolute constant.

For every $0 \leq \gamma < \frac{1}{2}$ and every large enough integer n, there is a game G such that if $|S| < \gamma n$ then

$$I_G(S) = O(|S|/n).$$

The lower bound on the average influence was shown in Ben-Or and Linial (1989) and the existence of a hard-to-bias protocol is form Alon and Naor (1990). The validity of this protocol for all $\gamma < \frac{1}{2}$ (which is essentially best possible) is from Boppana and Narayanan (1993b).

Proof (Sketch for the lower bound). Consider the case $k=1$ first. With every node v of the tree is associated a vector $(q_0^{(v)}, q_1^{(v)}, \ldots, q_n^{(v)})$ where $q_0^{(v)}$ is the fraction of zero leaves in the subtree rooted at v and $q_i^{(v)}$ is the probability of a zero outcome assuming i plays an optimal strategy to this end and all other players play at random. It turns out that the following inequality holds at every node v of the tree:

$$\prod_{i \neq 0} q_i^{(v)} \geq (q_0^{(v)})^{n-1}.$$

This inequality is proved by induction on the height of v in the tree. Let us skip this part and show only how to derive the theorem from it, and then proceed with a few extra remarks. When v is the root, $q_0 = p_0 = \frac{1}{2}$ and so

$$\frac{1}{n}\sum q_j \geq (\prod q_j)^{1/n} \geq (\tfrac{1}{2})^{(n-1)/n} = \frac{1}{2} + \frac{c}{n} + O(n^{-2}),$$

which proves the first part of the theorem for $k=1$. The general case is proved similarly.

A curious fact about the inequality $\prod_{i \neq 0} q_i^{(v)} \geq (q_0^{(v)})^{n-1}$ is that it is inhomogeneous (comparing degrees on both its sides). It can be shown, however, to completely characterize the attainable q-vectors. An analogous inequality appears in Loomis and Whitney (1949) (see comment following Theorem 4.3). □

Before we explain how the upper bound is proved, let us look at an example (which, incidentally, is not in standard form).

Example 4.6 (Baton-passing game). This game seems to have been suggested by a number of people. Its first analysis appears in Saks (1989) and further results were obtained in Ajtai and Linial (1993). The game starts with player 1 "holding the baton". At each round, the player holding the baton has to randomly select a player among those who had not previously held it and pass it on to that player. The last player to hold the baton is said to be *elected*. The elected player flips a coin which is the outcome of the game. Consider now a situation where this game is played by $n = s + t$ players where s abides by the rules and the complementary coalition B of t tries to bring about some desired outcome. Clearly, B can bias the outcome only if one of its members is elected. It is easily verified that the best strategy towards this end is for members of B to always pass the baton to a player outside B.

Let $f(s, t)$ denote the probability for a member of B being elected, given that the first player to hold the baton is not in B. It is easily seen that f is defined by the recurrence

$$f(s, t) = \frac{s}{s+t} f(s-1, t) + \frac{t}{s+t} f(s-1, t-1)$$

and the initial conditions $f(s, s+1) = 0$, and $f(s, 1) = 1/(s+1)$. The analysis of this game in Ajtai and Linial (1993) shows the second part of Theorem 4.6 for coalitions of size $O(n/\log n)$.

We now turn to some brief remarks on the protocol of Alon and Naor which is also an election game. If the elected player belongs to the bad coalition B we say that B *won*. If B's probability of winning (when B plays as well as possible and all others play at random) is smaller than some $c < 1$, the game is said to be *immune* against B. The Alon and Naor (1990) protocol is immune against any coalition of $n/3$ players or less. Consider a probability space $\mathcal{T} = \mathcal{T}_d$ consisting of all depth d complete binary trees, whose nodes are labeled by players' names, selected uniformly at random. Each such tree corresponds to an election protocol, where the label of an internal node is the name of the player who controls it. If the game terminates at a certain leaf, the player whose name marks that leaf is elected.

They show that a tree sampled from this space has a positive probability to be immune against all coalitions of $n/3$ players or less. The line of reasoning is this: Fix a bad coalition B, and consider a random variable $X = X_B^d$ which on every labeled tree of depth d evaluates to B's winning probability. If we can get tight enough bounds on X's expectation and variance, then we can conclude that with probability $> 1 - \delta$, a tree drawn from $\mathcal{T}$ is immune against B, for some small $\delta > 0$. If, moreover $\delta < 1/\binom{n}{3}$, then there is positive probability for a randomly drawn tree in $\mathcal{T}$ to be immune against *all* coalitions of $n/3$ members, i.e., this establishes the existence of the desired protocol. To estimate X_B^d's expectation and variance, consider what happens upon reaching a node v in the tree. If v is controlled by a good player, B's chance of winning is the average over the left and right subtrees rooted at v. If v is controlled by a bad player, this probability is the maximum over the two subtrees. However, the probabilities of winning for either subtree are themselves given by random variables distributed as X_B^r, for some $r < d$. Thus the key step in the proof is provided by the following lemma in probability:

Proposition 4.2. *Let Y and Z be independent random variables with equal expectations $E = E(Y) = E(Z)$ and equal variances $\sigma^2 = \text{Var}(T) = \text{Var}(Z)$. Let X be the random variable which equals $(Y + Z)/2$ with probability $1 - \varepsilon$ and $\max\{Y, Z\}$ with probability ε for some $0 \leqslant \varepsilon \leqslant 1$. Then $E(X) \leqslant E + \varepsilon\sigma/\sqrt{2}$ and $\text{Var}(X) \leqslant (1 + 3\varepsilon)\sigma^2/2$.*

This proof does not, however, tell us how to concretely construct such a game. This is a problem encountered often in trying to convert probabilistic arguments into algorithms. Alon and Naor do supply a somewhat weaker explicit construction for such game trees, based on the baton-passing game and Bracha's (1985) scheme from the theory of Byzantine Agreement, as well as some other steps of common use in derandomization arguments. This more complex protocol will not be reviewed here. The analysis in Boppana and Narayanan (1993a,b) is based on examining higher moments of the same random variables.

4.4. Quo vadis?

4.4.1. Fault-tolerant computation with full information?

We return to the following fundamental question: What are the limitations of fault-tolerant computing in the absence of secure communication and barring any restrictions on the computational power of bad players? The only attempt at the problem in this generality has so far been made by Goldreich, Goldwasser and Linial (1991). They introduce the following example:

Example 4.7 (Equality game). Two players draw a random number in the range $\{1, \ldots, n\}$. They compute a function f which equals 1, if the two numbers agree and is 0 otherwise. The expectation of f is, clearly $1/n$. However, if they compute f by revealing their inputs in turn, then the player who moves second can determine the outcome at will, by pretending to have an appropriate input. Consider, in contrast, the following protocol: player 1 is to report the remainder of his input mod $\sqrt{n}$, then player 2 is to report his input, and then 1 tells all his input. Here, even if one of the players cheats, trying to increase the expectation of the outcome, he can only boost it to $1/\sqrt{n}$.

A slight modification of the lower bound in Theorem 4.6 shows this protocol to be best possible. In Goldreich et al. (1991) the problem of fault-tolerant computing by two players is solved. The previous example turns out to be typical, in the following sense: Consider two parties trying to compute any function of two variables. By alternately revealing appropriate parts of their inputs, the function can be computed so that even if one of the players is trying to bias the outcome, his advantage would only very slightly exceed the above mentioned lower bound.

Computations by more than two players, bring about new complications. Already simple examples show that the satisfactory answer obtained for two players is incorrect for 3 players or more. In fact, there is not even a conjecture as to what the answer is general is. On the positive side Goldreich et al. (1991) show a protocol for computing functions of n variables, which for almost all functions reduces the bias to the lower bound in Theorem 4.6. Similar results are shown also for some specific functions. Most notably they show how n parties can sample from any probability distribution with minimum bias.

4.4.2. Efficient coin flipping

Alon and Naor's coin-flipping protocol reviewed above, takes time proportional to n, the number of players (and even slightly more in the protocol's explicit version). This raises the question whether, perhaps, a faster coin-flipping, or election is possible in a full-information game, even if a constant fraction of the players are bad. Cooper and Linial (1993) have recently shown that indeed these tasks can be performed in time only $O(\log^c n)$ for some constant c. Their current best protocol works with $c = 17$.

There is no reason known to us why the same performance will not be achievable in constant time. Such a protocol will greatly improve on Feldman and Micali's (1988) protocol, and would constitute a critical step in constructing protocols for fault-tolerant and efficient computation without secure communication.

5. More points of interest

This section contains a brief description of various problems and results where consideration from both theoretical computer science and game theory come to bear. In some instances the interaction between the two areas is very clear, in others it is more obscure, and the discussion should be taken as an invitation for researchers to try and find out whether a connection can be made. We tried to provide references wherever we are aware of a more complete discussion.

5.1. Efficient computation of game-theoretic parameters

Many of the quantities and parameters studied in game theory can, at least in principle, be computed or approximated. Are there efficient algorithms to compute these quantities? Obvious examples are optimal or near-optimal strategies for various games, states of equilibrium, values and other related quantities. Some recent work on this subject can be found in Koller and Megiddo (1992 and to appear) and the references therein. They consider the computational complexity of finding max-min strategies for two-person zero-sum games. Most instances of this problem are shown NP–complete, the most notable exception being that they provide an algorithm for a game given in extensive form, whose run time is polynomial in the size of the tree. See also Futia (1977) and Deng and Papadimitriou (1987). The case may be made to include much of the work on linear programming under this heading, because of the well-known connection between LP and zero-sum games [e.g., Owen (1982)]. Significant progress has been made in the design and analysis of algorithms for LP in the past few years [e.g., Megiddo (1987)]. Among the important advances are the development of efficient internal-point algorithms, better understanding of the average-case behavior of simplex-type algorithms, and the introduction of randomized algorithms to this area. These developments may help in designing better algorithms for a variety of computational problems from game theory.

A particularly fascinating problem in this area is that of computing the value of a stochastic game [Condon (1989)]. It is quite easy to show that this problem is in **NP** $\cap$ co $-$ **NP**, that is, both upper and lower bounds for the value can be established using short proofs. However, no polynomial-time algorithm is known, and the computational complexity of this problem is still unknown. See also Section 5.3.

5.2. *Games and logic in computer science*

This subject has its roots in the connection between games, descriptive set theory and logic. Mosschovakis (1989) and Hodges (1985) are two texts in this general area. Similar considerations turn out to be useful in some areas of theoretical computer science. For example Rabin's work (1969) on the decidability of the second-order theory of the infinite binary tree was shown by Gurevich and Harrington (1982) to be best viewed from a game-theoretic point of view. The method of Fraisse–Ehrenfeucht games [Ehrenfeucht (1961)] from mathematical logic turns out to be useful in delineating the expressive power of languages used in computer science. An interesting recent example for this approach is Ajtai and Fagin (1990) where a variation of a Fraisse–Ehrenfeucht game is used to show that connectivity problems for directed graphs are harder than for graphs, in a certain technical sense. Games of similar character were applied also in setting up the connection between classes of computational complexity and logical theories as in Papadimitriou's (1985) "games against nature", which can also be viewed as a forerunner to the theory of interactive proof system (see Section 3.2.3). A systematic discussion of this connection can be found in Condon's thesis (1988).

The notion of *Common Knowledge* which seems to be of interest to economists and game theorists received a good deal of attention in computer science circles as well. Formal theories were set up to capture this notion and decision procedures for the various logical theories which emerge were developed. For survey see the forthcoming book [Fagin et al. (to appear)] as well as the chapter by Geanakopolos in this Handbook.

5.3. The complexity of specific games

In the mid-70s a large amount of research was dedicated to studying the computational complexity of playing specific board games. For many combinatorial games (e.g., go, chess) it was shown that it is **NP**-hard or even **PSPACE**-hard to find an optimal strategy (technically, one defines a generalized version of chess played on an $n \times n$ board and the complexity is evaluated in terms of n.) While complexity of optimal play is already fairly well-understood, there is still much room for results on the complexity of finding *near*-optimal strategies. The standard reference on **NP**-completeness is Garey and Johnson (1979), which lists many of the known results; Berlekamp et al. (1982) is an encyclopedia of combinatorial games where complexity issues are considered as well. See also Simon and Schaeffer (Chapter 1).

Much attention was also paid to specific games as actually played by computers, traditional games such as chess, checkers and go as well as more modern ones. Games played by computers motivated a sizable amount of research in Artificial Intelligence (AI), with machines being constructed to play specific games and some methodologies developed to evaluate general game trees. Most notable among those in Knuth and Moore's α–β pruning method (1975) for efficient evaluation of game trees. This algorithm had many offsprings and ramifications, see e.g., Pearl's book (1984). Tarsi (1983) and Saks and Wigderson (1985) consider efficient randomized algorithms for evaluating game trees.

Having mentioned AI, let us briefly point out some interesting connections between this discipline and game theory, both in situations where a number of "intelligent" agents interact as well as in the design of a single agent. In the multi-agent situation, most of the work of date has focused on analyzing protocols for interactions among agents, and the performance of these protocols in various cooperative and competitive scenarios [e.g., Rosenschein and Genesereth (1988), Zlotkin and Rosenschein (1991), Krans and Wilkenfeld (1991)]. In the single-agent case as well, AI researchers have begun to use the tools of decision theory and game theory. Traditionally, AI theorists modeled objects such as an agent's goals and plans using mathematical logic, and attempted to define a prescriptive theory of action using the same tool. An alternative approach, placing more emphasis on decision-theoretic considerations (so as to ensure reasoned choice among alternative strategies) is represented, e.g., in Feldman and Sproull (1977), Horvitz (1988), Doyle (1990), Dean and Boddy (1988), and the book of Russell and Wefald (1991).

Computers playing classical games are already a field in itself with special-purpose computers and specialized periodicals. Two nontraditional games that may be worth mentioning are Diplomacy, for which a program was developed [Kraus et al. (1991)] which negotiates with the other players, besides taking the regular moves. Axelrod's experiments with computer programs playing the *Prisoner's Dilemma* are documented in Axelrod (1984); for a more recent survey see Axelrod and Diou (1988). In the context of Prisoner's Dilemma and repeated games, the idea of using bounded rationality to resolve the ensuing paradoxes was raised and considered extensively, see Kalai (1990) and Sorin (Chapter 4) for surveys. This approach ties naturally with methods of computer science where bounds on computational resources are a main concern.

More on this subject can be found in the survey of Simon and Schaeffer (Chapter 1).

5.4. Game-theoretic consideration in computational complexity

Interesting and largely unexplored connections with game theory arise in complexity theory. Various measures for the computational complexity of functions are considered in this area some of which are defined in terms of certain cooperative games. For example, let $f = f(x, y)$ be a function from n-bit strings to $\{0, 1\}$. The *communication complexity* of f is defined [Yao (1979)] as follows: Player 1 holds x and player 2 holds y and they want to compute $f(x, y)$ by sending messages to each other over a communication channel. This can be done e.g., by player 1 sending x over to player 2 who calculates $f(x, y)$ and sends the outcome over, for a total cost of $n + 1$ bits transmitted. For certain functions f, more efficient algorithms may be devised. For example, if $f(x, y)$ is the parity of $x + y$, it is enough for the first player to transmit a single bit indicating the parity of x, whence the second player can compute f. The smallest upper bound that can be achieved by the best algorithm is called f's communication complexity. Randomized and non-deterministic variations are being considered too. This is an important concept which comes up in many studies of complexity, but no concrete connection with game theory has been set up yet. A comprehensive survey article by Lovasz (1990) is highly recommended. The *depth* of a Boolean circuit (see Remark 3.4) is the length of a longest directed path from an input gate to an output gate in the underlying directed graph. (See Boppana and Sipser (1990) for an updated survey of circuit complexity). Given a function g, it is of interest to find the least possible depth of a circuit computing that function. It turns out that the least possible depth can be expressed in terms, of a cooperative two-person game, somewhat like the communication complexity game. This point of view enabled Karchmer and Wigderson (1990) to solve some interesting problems in circuit complexity.

It should be mentioned that in many search problems, one establishes a lower bound for time complexity by assuming that answers to the queries the algorithm makes, are supplied by a computationally unbounded adversary. The only restriction the adversary obeys is to never give contradictory answers. This standard point of view appears e.g., throughout Knuth's books (1973).

5.5. Random number generation as games

In recent years randomization has been playing a central role in the theory of efficient algorithms. It was Rabin's (1976) insight that randomized computation is a methodology in its own right, and this is by now widely realized. As observed by Megiddo (1989) the relationship between randomized and deterministic algorithms resembles that between mixed and pure strategies, though some interesting differences are pointed out as well. An easy but important result of Yao (1977) is based on viewing the run time of a randomized algorithm as the value of a game between the algorithm and an adversary designing a hard instance for it to run on. The min-max theorem implies that it is possible to replace the analysis of a randomized algorithm running on a hard instance by a discussion where firstly one side selects a probability distribution on problem instances and then the other devises a deterministic algorithm to have a low expected run time on inputs sampled from this distribution. This easy but very useful observation can sometimes simplify the analysis of randomized algorithms.

Randomized computing requires random bit generators, a need which creates new problems. The theory of pseudo-random bit generators was revolutionized by the recent advents in cryptography. Knuth (1973) explains the classical approach to pseudo random number generation. Goldreich's review (1988) explains the new point of view, see also Section 3.2 and Goldreich (1989), Rivest (1990). The property required from a random number generator is that no randomized polynomial-time Turing Machine can distinguish the generator's output from truly random numbers. A fundamental problem in this area has been recently resolved, by Impagliazzo, Levin and Luby (1989) and Hastad (1990) who showed:

Theorem 5.1 [Impagliazzo et al. (1989), Hastad (1990)]. *If one-way functions exist, then pseudo-random number generators exist.*

A traditional approach to random bit generation is to sample a physical process which "seems random enough". There are no theorems, of course, guaranteeing a truly random behavior of such sources. Consequently, the view developed of an imperfect source of randomness as an *adversary* who tries to supply the sampling program with "bad" random bits. This adversary cannot behave in a completely deterministic way, though, and the program should be smart enough to take advantage of the random aspects of the source's behavior and extract the randomness needed for its proper operation. U. Vazirani's thesis (1986) contains some of the fundamental results along this line of research. Further developments can be found in Chor and Geréb-Grauss (1988) and Lichtenstein et al. (1989) where possible connections with control theory are pointed out. Some typical results are: It is impossible to "improve the quality" of individual bits sampled from such an adversary source [Santha and U. Vazirani (1986)]. On the other hand the amount of randomness inherent in the source suffices for running randomized algorithms [U.V. Vazirani and V.V. Vazirani (1985)]. A survey of this area may be found in Ben-Or et al. (1987). The most recent progress in the area of "amplifying randomness" can be found in Zuckerman (1991) and the references therein.

5.6. *Amortized cost and the quality of on-line decisions*

A common problem which arises in many situations in economics is that decisions made at present have both an immediate impact, as well as consequences for the future. For example, the choice between saving plans of varying yields and time spans affects our situation at future points when interest rates may change. The economists' standard answer to problems such as this is to devise a statistical model for the future and invoke methods of Markov decision processes.

There is a class of examples from data structure theory, which prompted a different, more conservative, approach to similar problems in computer science. The aim is to develop decision-making strategies which are *guaranteed* to perform well, regardless of the upcoming sequence of events, and avoiding any statistical assumptions. See Section 1.1 where a similar difference in approach between computer scientists and game theorists manifests itself. Analysis of some problems in economics along the same line of thought have been performed e.g., in Cover (1988), Foster and Vorha (1987) and the survey of Mahmond (1984).

Let us present one of the canonical examples in this area: Consider a filing cabinet where many documents are kept. They are kept in no particular order, so when the need for a certain document arises, it has to be searched for sequentially starting at the top of the list. The cost of searching is proportional to the item's ordinal number in the cabinet at that time. To anticipate future requests, items may be permuted at any time, for a unit cost per any swap performed. The *move to front* strategy always locates the requested item and then moves it to the front of the cabinet. Given a sequence of requests, it is possible to find the optimal sequence of permutations which minimizes the overall cost (of searching and permuting). It is surprising, perhaps, that the cost incurred by the "move to front" strategy is at most twice this optimum. Notice that the optimum is achieved based on a complete knowledge of the sequence of requests, while the heuristic operates only on local information, never anticipating the future, and no statistical assumptions are made on the stream of requests.

Sleator and Tarjan's (1985) may be the earliest paper where this paradigm is presented, and McGeoch and Sleator (1991) is a collection of some recent articles in this area. In Borodin et al. (1987) a general framework for dealing with questions like this is laid out. The role of randomization in this area is elucidated in Ben-David et al. (1990), using game-theoretic considerations. Answering a question of Manasse et al. (1990), recently Fiat, Rabani and Ravid (1990) proved the *k-server* theorem: A player controls k servers which he can move around a metric space (X, d). At the beginning of the game all servers are located at a fixed origin point in X. At each step, an adversary makes a *request*, i.e., he selects a point p of X. The player has to move one of its servers to p, and it is up to him which of the k servers to move. (One may think of some kind of a service company operating k service vehicles and responding to calls of customers around the country). The cost incurred by the player at this step is the distance between that server's old location and p.

Fiat, A., Y. Rabani and Y. Ravid (1990) 'Competitive k-serve algorithms', *31st Symposium on the Foundations of Computer Science*, 454–463.

Fischer, M.J. (1983) 'The concensus problem in unreliable distributed systems (a brief survey)', *Technical Report*, Department of Computer Science, Yule University.

Fischer, M.J., N.A. Lynch and M. Merritt (1986) 'Easy impossibility proofs for distributed consensus problems', *Distributed Computing*, **1**: 26–39.

Fischer, M.J., N.A. Lynch and M. Patterson (1985) 'Impossibility of distributed concensus with one faulty processor', *Journal of ACM*, **32**: 374–382.

Foster, D.P. and R.V. Vorha (1987) On combining forecasts, Technical Report, Ohio State University.

Futia, C. (1977) 'The complexity of economic decision rules', *Journal of Mathematical Economics*, **4**.

Garay, J.A. and Y. Moses (1993) 'Fully polynomial Byzantine Agreement in $t+1$ rounds', *Proceedings of the 25th ACM Symposium on Theory of Computing*, 31–41.

Garey, M.R. and D.S. Johnson (1979) *Computers and interactability – a guide to the theory of NP-completeness*. Freeman.

Geanakopolos, J. Common knowledge, Chapter 40, this Handbook.

Genesereth, M.R. and J.S. Rosenschein (1988) 'Cooperation without communication', in: (A.H. Bond and L. Gasser, eds.), *Readings in distributed artificial Intelligence*. San Mateo, CA: Morgan Kaufmann Publishers, Inc., pp. 227–234.

Gilboa, I. and E. Zemel (1989) 'Nash and correlated equilibria: Some complexity considerations', *Games and Economic Behavior*, **1**.

Goldreich, O. (1988) 'Randomness, interactive proofs and zero-knowledge – A survey', in: (R. Herken ed.), *The universal Turing Machine, a half century survey*. Hamburg Berlin: Kammerer and Unverzagt, 377–405.

Goldreich, O. (1989) *Foundations of cryptography*, Class Notes, Computer Science Department, Technion, Haifa, Israel.

Goldreich, O., S. Goldwasser and N. Linial (1991) 'Fault-tolerant computation in the full information model', *32nd Symposium on the Foundations of Computer Science*, 447–457.

Goldreich, O. and L.A. Levin (1989) 'A hard-core predicate for all one-way functions', *Proceedings of the 21st ACM Symposium on Theory of Computing*, 25–32.

Goldreich, O., S. Micali and A. Wigderson (1991) 'Proofs that yield nothing but their validity, or all languages in *NP* have zero-knowledge proofs', *Journal of ACM*, **38**: 691–729; *27th Symposium on the Foundations of Computer Science* (1986) 174–187.

Goldreich, O., S. Micali and A. Wigderson (1987) 'How to play any mental game, or, a completeness theorem for protocols with honest majority', *Proceedings of the 19th ACM Symposium on Theory of Computing*, 218–229.

Goldreich, O. and E. Petrank (1990) 'The best of both worlds: guaranteeing termination in fast randomized Byzantine Agreement protocols', *Information Processing Letters*, **36**: 45–49.

Goldwasser, S. and S. Micali (1984) 'Probabilistic encryption', *Journal on Computer and System Science*, **28**: 270–299.

Goldwasser, S., S. Micali and C. Rackoff (1988) 'The knowledge complexity of interactive proof systems', *SIAM Journal on Computing*, **18**: 186–208.

Graham, R.L. and A.C. Yao (1989) 'On the improbability of reaching Byzantine Agreements', *Proceedings of the 21st ACM Symposium on Theory of Computing*, 467–478.

Gurevich, Y. and L. Harrington (1982) 'Trees, automata and games', *Proceedings of the 14th ACM Symposium on Theory of Computing*, 60–65.

Halpern, J.Y. and Y. Moses (1990) 'Knowledge and common knowledge in a distributed environment', *Journal of ACM*, **37**: 549–587.

Hardy, G.H. and E.M. Wright (1979) *An introduction to the theory of numbers*, 5th edn. Oxford University Press.

Hart, S. (1976) 'A note on the edges of the *n*-cube', *Discrete Mathematics*, **14**: 157–163.

Hastad, J. (1990) 'Pseudo-random generators under uniform assumptions', *Proceedings of the 22nd ACM Symposium on Theory of Computing*, 395–404.

Herlihy, M. and N. Shavit (1993) 'The asynchronous computability theorem for *t*-resilient tasks', *Proceedings of the 25th ACM Symposium on Theory of Computing*, 111–120.

Hodges, W. (1985) *Building models by games*, Cambridge University Press.

Horvitz, E.J. (1988) 'Reasoning under varying and uncertain resource constraints', in: *Proceedings of the Seventh National Conference on Artificial Intelligence*, 111–116.

Impagliazzo, R., L.A. Levin and M. Luby (1989) 'Pseudo-random generation from one-way functions', *Proceedings of the 21st ACM Symposium on Theory of Computing*, 12–24.
Johnson, D.S. (1990) 'A catalog of commplexity classes', in: (J. van Leeuwen ed.), *Handbook of theoretical computer science*. Vol. A, The MIT Press/Elsevier, 67–161.
Kahn, J., G. Kalai and N. Linial (1988) 'The influence of variables on Boolean functions', *29th Symposium on the Foundations of Computer Science*, 68–80.
Kalai, E. (1990) 'Bounded rationality and strategic complexity in repeated games', in: (T. Ichiishi, A. Neyman and Y. Tauman eds.), *Game theory and applications*. Academic Press, 131–157.
Karchmer, M. and A. Wigderson (1990) 'Monotone circuits for connectivity require superlogarithmic depth', *SIAM Journal of Discrete Mathmatics*, **3**: 255–265.
Knuth, D. (1973) *The art of computer programming* (3 vols.), Addison–Wesley.
Knuth, D.E. and R.N. Moore (1975) 'An analysis of the alpha–beta pruning', *Artificial Intelligence*, **6**: 293–322.
Koller, D. and N. Megiddo (1992) 'The complexity of two-person zero-sum games in extensive form', *Games and Economic Behavior*, **4**: 528–552.
Koller, D. and N. Megiddo (1991) 'Finding mixed strategies with small supports in extensive form games', Research Report RJ 8380, IBM Almaden Research Center. International Journal of Game Theory, to appear.
Kraus, S., E. Ephrati and D. Lehmann (1991) 'Negotiation in a non-cooperative environment', *Journal of Experimental and Theoretical Artificial Intelligence*, **4**: 255–282.
Kraus, S. and J. Wilkenfeld (1991) 'Negotiations over time in a multi-agent environment: Preliminary report', in *Proceedings of the Twelfth International Joint Conference on Artificial Intelligence*, 56–61.
Kushilevitz, E. and Y. Mansour (1991) 'Learning decision trees using the Fourier spectrum', *Proceedings of the 23rd Annual ACM Symposium on Theory of Computing*, 455–464.
Lamport, L., R. Shostak and M. Pease (1982) 'The Byzantine generals problem', *ACM TOPLAS*, **4**: 382–401.
Lenstra, A.K. and H.W. Lenstra Jr. (1990) 'Algorithms in number theory', in: (J. van Leeuwen ed.), *Handbook of theoretical computer science*. Vol. A, The MIT Press/Elsevier, 673–715.
Lewis, H.R. and C.H. Papadimitriou (1981) *Elements of the theory of computation*, Prentice Hall.
Lichtenstein, D., N. Linial and M. Saks (1989) 'Some extremal problems arising from discrete control processes', *Combinatorica*, **9**: 269–287.
Loomis, L. and H. Whitney (1949) 'An inequality related to the isoperimetric inequality', *Bulletin of the American mathematical Society*, **55**: 961–962.
Lovasz, L. (1990) 'Communication complexity: A survey', in: (B. Korte, L. Lovasz, H.J. Prömel and A. Schrijver eds.) *Paths, flows and VLSI layout*. Springer, 235–265.
Lund, C., L. Fortnow, H. Karloff and N. Nisan (1992) 'Algebraic methods for interactive proof systems', *Journal of ACM*, **39**(4): 855–868. Proceedings of the 31st Annual IEEE Symposium on Foundations of Computer Science, 16–25.
MacWilliams, J. and N. Sloane (1977) *The theory of error correcting codes*, North-Holland.
McGeoch, L.A. and D.D. Sleator (1991) *Workshop on On-Line Algorithms*, DIMACS Series in Discrete Mathematics and Theoretical Computer Science. Vol. 7, American Mathematical Society, Providence.
Mahmond, E. (1984) 'Accuracy in forecasting: A survey', *J. Forecasting*, **3**: 139–159.
Manasse, M.S., L.A. McGeoch and D.D. Sleator (1990) 'Competitive algorithms for server problems', *J. Algorithms*, **11**: 208–230.
Megiddo, N. (1987) 'On the complexity of linear programming', in: T. Bewley, ed., *Advances in economic theory fifth world congress*, Cambridge: Cambridge University Press, pp. 225–268.
Megiddo, N. (1989) 'On probabilistic machines, bounded rationality, and average-case complexity', Research Report RJ 7039, IBM Almaden Research Center, San Jose, California.
Megiddo, N. and A. Wigderson (1986) 'On play by means of computing machines', in: J.Y. Halpern, ed., *Theoretical aspects of reasoning about knowledge*. Los Altos, CA: Morgan Kaufmann Publishers, Inc., 259–274.
Menger, K. (1927) 'Zur allgemeinen Kurventheorie', *Fundamenta Mathematica*, **10**: 95–115.
Moses, Y., D. Dolev and J.Y. Halpern (1986) 'Cheating husbands and other stories: a case study of knowledge, action, and communication', *Distributed Computing*, **1**: 167–176.
Moses, Y. and O. Waartz (1988) 'Coordinated transversal: $(t+1)$-round Byzantine Agreement in polynomial time', *Proceedings of the 29th Annual IEEE Symposium on Foundations of Computer Science*, 246–255.

Mosschovakis, Y.N. (1980) *Descriptive set theory*. Amsterdam: North-Holland.
Naor, M. (1991) 'Bit commitment using pseudorandomness', *Journal of Cryptology*, **4**(2), 151–158.
Owen, G. (1982) *Game theory*, 2nd ed. New York: Academic Press.
Papadimitriou, C. (1985) 'Games against nature', *Journal of Compouting System Science*, **31**: 288–301.
Pearl, J. (1984) *Heuristics: intelligent search strategies for computer solving problems*, Addison–Wesley.
Pease, M., R. Shostak and L. Lamport (1980) 'Reaching agreement in the presence of faults', *Journal of ACM*, **27**: 228–234.
Rabin, M.O. (1969) 'Decidability of second-order theories and automata on infinite trees', *Transactions of the AMS*, **141**: 1–35.
Rabin, M.O. (1976) 'Probabilistic algorithms', in: (J. Traub ed.) *Algorithms and complexity*. Academic Press, 21–39.
Rabin, M.O. (1983) 'Randomized Byzantine generals', *Proceedings of the 24th Annual IEEE Symposium on Foundations of Computer Science*, 403–409.
Rabin, T. and M. Ben-or (1989) 'Verifiable secret sharing and multiparty protocols with honest majority', *Proceedings of the 21st Annual ACM Symposium on Theory of Computing*, 73–85.
Rivest, R. (1990) 'Cryptography', in: (J. van Leeuwen ed.), *Handbook of theoretical computer science*. Vol. A, The MIT Press/Elsevier, 719–755.
Rosenschein, J.S. and M.R. Genesereth (1988) 'Deals among rational agents', in: (B.A. Huberman ed.), *The ecology of computation*, North-Holland, 117–132.
Russell, S. and E. Wefald (1991) *Do the right thing*, MIT Press.
Saks, M. (1989) 'A robust noncryptographic protocol for collective coin flipping', *SIAM Journal of Discrete Mathematics*, **2**: 240–244.
Saks, M. and A. Wigderson (1985) 'Probabilistic decision trees and the complexity of evaluating game trees', *Proceedings of the 26th Annual IEEE Symposium on Foundations of Computer Science*, 29–38.
Saks, M. and F. Zaharoglou (1993) 'Wait-free k-set agreement is impossible: The topology of public knowledge', *Proceedings of the 25th ACM Symposium on Theory of Computing*, 101–110.
Santha, M. and U.V. Vazirani (1986) 'Generating quasi-random sequences from semi-random sources', *Journal of Computer and System Science*, **33**: 75–87.
Shamir, A. (1979) 'How to share a secret', *Communications of the ACM*, **22**: 612–613.
Shamir, A. (1990) **IP = PSPACE**, *Proceedings of the 31st Annual IEEE Symposium on Foundations of Computer Science*, 11–15.
Simon, H.A. and J. Schaeffer, The game of chess, Chapter 1, this Handbook.
Sleator, D. and R.E. Tarjan (1985) 'Amortized efficiency of list update and paging rules', *Communication of ACM*, **28**: 202–208.
Sorin, S. Repeated games with complete information, Chapter 4, this Handbook.
Tarsi, M. (1983) 'Optimal search on some game trees', *Journal of ACM*, **30**: 389–396.
Vazirani, U.V. (1986) 'Randomness, adversaries and computation', PhD Thesis, U. C. Berkeley.
Vazirani, U.V. and V.V. Vazirani (1985) 'Random polynomial time is equal to semi-random polynomial time', *Proceedings of the 26th Annual IEEE Symposium on Foundations of Computer Science*, 417–428.
Yao, A.C. (1977) 'Probabilistic computation: Towards a unified measure of complexity', *Proceedings of the 18th Annual IEEE Symposium on Foundations of Computer Science*, 222–227.
Yao, A.C. (1979) 'Some complexity questions related to distributed computing', *Proceedings of the 11th Annual ACM Symposium on Theory of Computing*, 209–213.
Yao, A.C. (1982) 'Protocols for secure communication', *Proceedings of the 23rd Annual IEEE Symposium on Foundations of Computer Science*, 160–164.
Yao, A.C. (1986) 'How to generate and exchange secrets', *Proceedings of the 27th IEEE Symposium on the Foundations of Computer Science*, 162–167.
Zlotkin, G. and J.S. Rosenschein (1991) 'Incomplete information and deception in multiagent negotiation', *Proceedings of the Twelfth International Joint Conference on Artificial Intelligence*, Sydney, Australia, pp. 225–231.
Zuckerman, D. (1991) 'Simulating BPP using general weak random source', *Proceedings of the 32nd IEEE Symposium on the Foundations of Computer Science*, 79–89.

Chapter 39

UTILITY AND SUBJECTIVE PROBABILITY

PETER C. FISHBURN

AT&T Bell Laboratories

Contents

1.	Introduction	1398
2.	Historical sketch	1400
3.	Ordinal utility and comparable differences	1401
4.	Expected utility and linear generalizations	1407
5.	Nonlinear utility	1411
6.	Subjective probability	1413
7.	Expected utility and subjective probability	1416
8.	Generalizations and alternatives	1420
References		1423

Handbook of Game Theory, Volume 2, Edited by R.J. Aumann and S. Hart
© *Elsevier Science B.V., 1994. All rights reserved*

1. Introduction

Utility and subjective probability involve the systematic study of people's preferences and beliefs, including quantitative representations of preference and belief that facilitate analyses of problems in decision making and choice behavior. Its history spans more than 250 years and is due mainly to people in economics, mathematics, statistics and psychology.

Utility theory is primarily concerned with properties of a binary relation $>$ on a set X, where X could be a set of commodity bundles, decision alternatives, monetary gambles, n-tuples of pure strategies, and so forth, and $x > y$ is interpreted as: *x is preferred to y*. Its classic representation is

$$\forall x, y \in X, x > y \Leftrightarrow u(x) > u(y),$$

where u is a real valued *utility* function on X that preserves $>$ in the indicated manner. Suppose a friend invites you to dinner tomorrow night and tells you that the entree will be steak or lobster or trout. If you prefer lobster to trout, trout to steak, and lobster to steak, your preferences can be presented by the utilities

$$u(\text{lobster}) = 2, u(\text{trout}) = 1, u(\text{steak}) = 0,$$

or by any other three numbers that satisfy $u(\text{lobster}) > u(\text{trout}) > u(\text{steak})$.

Subjective probability focuses on properties of a binary relation $>$ on a *Boolean algebra* $\mathcal{A}$ of subsets of a state space S ($S \in \mathcal{A}$; $A \cup B \in \mathcal{A}$ whenever $A, B \in \mathcal{A}$; A^c, the *complement* $S \setminus A$ of A in S, is in $\mathcal{A}$ whenever $A \in \mathcal{A}$), $A > B$ is interpreted as: the uncertain event A *is more probable than* the uncertain event B. Its classic representation is

$$\forall A, B \in \mathcal{A}, A > B \Leftrightarrow \pi(A) > \pi(B),$$

where π is a *probability measure* on $\mathcal{A}$, i.e., $\pi: \mathcal{A} \to [0, 1]$, $\pi(S) = 1$, and $\pi(A \cup B) = \pi(A) + \pi(B)$ whenever A and B are disjoint events in $\mathcal{A}$. Suppose tomorrow's weather will be either sunny all day (s), cloudy all day with possible rain (c), or a mixture of the two (m). With state space $S = \{s, c, m\}$, if you believe that s is more probable than m, m is more probable than c, and s is *less* probable than c or m jointly, and if your other comparisons are "consistent" with these, then all possible comparisons between events in S can be represented by a probability measure whose atomic probabilities satisfy

$$\pi(m) + \pi(c) > \pi(s) > \pi(m) > \pi(c) > 0$$

along with $\pi(s) + \pi(m) + \pi(c) = 1$.

Utility and subjective probability come together in Savage's (1954) representation for preference between *acts* $f, g, \ldots$ in the set $F = X^S$ of all mappings from S into an outcome set X:

$$\forall f, g \in F, f > g \Leftrightarrow \int_S u(f(s))\, d\pi(s) > \int_S u(g(s))\, d\pi(s).$$

When xAy denotes the act that yields outcome x if A obtains and y otherwise, $A > B$ in Savage's model corresponds to $xAy > xBy$ whenever $x > y$, i.e., you would rather bet on A than B for a preferred outcome. Suppose a three-horse race among a, b and c is about to be run, and you plan to buy a $2 win ticket on one of them. Assuming that the payoffs per dollar bet are known, your net profit matrix is as follows:

acts	states		
	a wins	b wins	c wins
bet on a	$3	−$2	−$2
bet on b	−$2	$1	−$2
bet on c	−$2	−$2	$14

Horse b is the strong favorite, and c is a long shot. If you personally prefer betting on a to c to b, then you can determine an increasing utility function on $\{-\$2, \$1, \$3, \$14\}$ and a probability distribution on the states that satisfies Savage's model, i.e., for which

$$u(\$3)\pi(a) + u(-\$2)[1-\pi(a)] > u(\$14)\pi(c) + u(-\$2)[1-\pi(c)]$$
$$> u(\$1)\pi(b) + u(-\$2)[1-\pi(b)].$$

Since $u(-\$2) = 0$ can be fixed, the requirements for the model come down to

$u(\$3)\pi(a) > u(\$14)\pi(c) > u(\$1)\pi(b)$,
$u(\$14) > u(\$3) > u(\$1) > 0$,
$\pi \geq 0$ and $\pi(a) + \pi(b) + \pi(c) = 1$.

The preceding discussion illustrates the *representational theory of measurement* that we emphasize throughout the chapter. Roughly speaking, the representational theory is concerned with one or more relations and operations on a set and with mappings into numerical structures (u, π, etc.) that preserve these qualitative relations and operations. It was formalized by Scott and Suppes (1958) and is followed extensively in recent works on the theory of measurement [Pfanzagl (1968), Fishburn (1970a, 1982a, 1988b), Krantz et al. (1971), Roberts (1979), Narens (1985), Suppes et al. (1989), Luce et al. (1990)].

The next section presents a brief historical sketch of our subject. Sections 3 through 5 focus on utility in the absence of subjective probability although numerical probabilities are heavily involved in the expected utility theory of Section 4 and its generalizations in Section 5. Subjective probability in the absence of utility is reviewed in Section 6 and is then merged with utility in our discussion of Savage's theory and its generalizations in Sections 7 and 8.

2. Historical sketch

More than 250 years ago, Daniel Bernoulli (1738) argued that a person's subjective value of wealth does not increase linearly in the amount but rather increases at a decreasing rate. This notion of diminishing marginal utility, applied in various ways to commodity bundles $(x_1, x_2, \ldots, x_n)$ in $\mathbb{R}^{+n}$, such as $u(x_1, \ldots, x_n) = u_1(x_1) + \cdots + u_n(x_n)$ with each u_i increasing and concave, became a centerpiece of the riskless theory of consumer economics during the second half of the nineteenth century. Major contributors included Gossen (1854), Jevons (1871), Menger (1871), Walras (1874) and Marshall (1890).

During this period, utility (subjective value, moral worth, psychic satisfaction) was often viewed "as a psychological entity measurable in its own right" [Strotz (1953, p. 84)], and the extent to which utility was precisely measurable was actively debated. The measurability issue waned under the ordinalist revolution of Edgeworth (1881), Fisher (1892), Pareto (1906) and Slutsky (1915), who maintained that utility represented only a person's preference order over commodity bundles or alternative riskless futures, so gradations in utility apart from its ordering are meaningless. Additional history of this era is provided by Samuelson (1947), Stigler (1950) and Kauder (1965).

A modest revival in measurable or "cardinal" utility [Hicks and Allen (1934), Fishburn (1976a)] was generated by Frisch (1926), Lange (1934) and Alt (1936), who axiomatized notions of comparable preference differences. They argued that we do in fact make strength-of-preference comparisons all the time which can be represented as $(x, y) >^* (z, w) \Leftrightarrow u(x) - u(y) > u(z) - u(w)$, where $(x, y) >^* (z, w)$ signifies that your intensity of preference for x over y (e.g., \$100 over \$0) exceeds your intensity of preference for z over w (e.g., \$210 over \$100).

Bernoulli (1738), motivated by his cousin Nicholas Bernoulli's St. Petersburg paradox [Menger (1967), Samuelson (1977)], which observes that most people in possession of an option with an infinite expected return (you win $\$2^n$ if the first head in a succession of coin tosses occurs at the nth toss) would sell it for a small sum (e.g., \$25), combined his riskless utility measure with outcome probability to resolve the paradox and other risky choices in which a person prefers an option with a lower expected return. His fundamental principle asserts that a person evaluates a lottery with probability p_i for outcome x_i, $\sum p_i = 1$, by its *expected utility* $\sum p_i u(x_i)$ rather than its expected return (or wealth level) $\sum p_i x_i$, and will choose the lottery with the highest expected utility.

The notion of expected utility was precisely axiomatized more than two centuries later by von Neumann and Morgenstern (1944). However, their interpretation differs radically from Bernoulli's since he starts with a riskless intensive utility function while they begin with axioms for the behavior of a preference relation $>$ on a set of risky options. Their utility function is based on comparisons between risky options and involves no overt notion of comparable preference differences [Luce and Raiffa (1957), Baumol (1958), Fishburn (1989b)].

A few years later, Savage (1954) presented axioms for $>$ on a set F of uncertain acts that led to the joint derivation of outcome utility and subjective probability for the model introduced in Section 1. His work is closely related to Ramsey's (1931) sketch of a theory of subjective expected utility, and it was strongly motivated by von Neumann and Morgenstern's derivation of expected utility without subjective probability and by de Finetti's pioneering investigations (1931, 1937) in subjective probability.

As in earlier axiomatizations of subjective or intuitive probability by Bernstein (1917), de Finetti (1931) and Koopman (1940), Savage assumes in effect that for every n the state space S can be partitioned into n equally likely events. This forces S to be infinite. Kraft, Pratt and Seidenberg (1959) later showed how to axiomatize subjective probability for finite S and in the process opened the way for many other axiomatizations involving finite structures.

Each of the main topics mentioned above has been extended and generalized in various ways that we shall describe in ensuing sections.

3. Ordinal utility and comparable differences

The formal development of utility and subjective probability is based on the notion of a binary relation. A *binary relation* R on a set X is a subset of $X \times X$. We write xRy to mean the same thing as $(x, y) \in R$, and take $\text{not}(xRy)$ as equivalent to $(x, y) \notin R$. A binary relation R on X is

reflexive if $\forall x \in X$, xRx,
symmetric if $\forall x, y \in X$, $xRy \Rightarrow yRx$,
asymmetric if $\forall x, y \in X$, $xRy \Rightarrow \text{not}(yRx)$,
transitive if $\forall x, y, z \in X$, $(xRy, yRz) \Rightarrow xRz$,
negatively transitive if $\forall x, y, z \in X$, $xRz \Rightarrow (xRy \text{ or } yRz)$,
complete if $\forall x, y \in X$, $x \neq y \Rightarrow (xRy \text{ or } yRx)$.

Special types of binary relations are identified by conjunctions of properties as follows:

equivalence: reflexive, symmetric and transitive;
linear order: asymmetric, transitive and complete;
weak order: asymmetric and negatively transitive;
partial order: asymmetric and transitive.

When E on X is an equivalence relation, it partitions X into mutually disjoint subsets called *equivalence classes*, such that x and y are in the same class if and only if xEy. The partition of X determined by E is denoted by X/E. Thus $X/E = \{\{y \in X: yEx\}: x \in X\}$.

Let $>$ denote an asymmetric binary relation on X, and define its *symmetric complement* $\sim$ by

$$x \sim y \quad \text{if not}(x > y) \text{ and not}(y > x),$$

When $>$ denotes *is preferred to*, $\sim$ is an *indifference* relation (or, sometimes, *incomparability* when $x \neq y$); when $>$ is a comparative probability relation *is more probable than*, $\sim$ signifies equal likelihood or, sometimes, incomparability [Keynes (1921)]. The union of $>$ and $\sim$, written as $\gtrsim$, means *is preferred or indifferent to* or *is at least as probable as*. (Many writers prefer to start with $\gtrsim$ as a reflexive relation. They then define $x > y$ by $[x \gtrsim y, \text{not}(y \gtrsim x)]$, $x \sim y$ by $(x \gtrsim y, y \gtrsim x)$, and x incomparable to y by $[\text{not}(x \gtrsim y), \text{not}(y \gtrsim x)]$). We use the former approach throughout this chapter.)

If $>$ is a linear order then $x \sim y \Leftrightarrow x = y$, and if $>$ is a weak order then $\sim$ is an equivalence whose *indifference classes* in $X/\sim$ can contain many pairs. In both cases, $>$, $\sim$ and $\gtrsim$ are transitive. When $>$ is a partial order, $\sim$ need not be transitive (as when $x \sim y$, $y \sim z$ and $x > z$) and hence may fail to be an equivalence.

The following theorem, due essentially to Cantor (1895), tells precisely when a linearly ordered set $(X, >)$ is isomorphically embeddable in the ordered real numbers. We say that $Y \subseteq X$ is $>$-*order dense* in X if, whenever $x > y$ and $x, y \in X \setminus Y$, there is a $z \in Y$ such that $x > z$ and $z > y$.

Theorem 3.1. *There is a one-to-one map $u: X \to \mathbb{R}$ such that*

$$\forall x, y \in X, \quad x > y \Leftrightarrow u(x) > u(y),$$

if and only if $>$ on X is a linear order and some countable subset of X is $>$-order dense in X.

If X is countable then the order denseness condition holds automatically, and with X enumerated as $x_1, x_2, x_3, \ldots$ it suffices to define u for the representation by

$$u(x_i) = \sum_{\{j: x_i > x_j\}} 2^{-j}$$

when $>$ is linear since the sums converge and $u(x_k) \geq u(x_h) + 2^{-h}$ if $x_k > x_h$. Proofs for uncountable X appear in Fishburn (1970a) and Krantz et al. (1971). An example of a linearly ordered set for which the order denseness condition does not hold is $\mathbb{R}^2$ with $>$ defined lexicographically by

$$(x_1, x_2) > (y_1, y_2) \quad \text{if } x_1 > y_1 \text{ or } \{x_1 = y_1, x_2 > y_2\}.$$

The following result is based on the fact that if $>$ on X is a weak order then

$>'$ defined on $X/\sim$ by $a>'b$ if $x>y$ for some (hence for all) $x\in a$ and $y\in b$ is a linear order on the indifference classes.

Corollary 3.1. *There is a $u: X \to \mathbb{R}$ such that, $\forall x, y \in X, x > y \Leftrightarrow u(x) > u(y)$, if and only if $>$ on X is a weak order and some countable subset of $X/\sim$ is $>'$-order dense in $X/\sim$.*

Weakly ordered preferences have been widely assumed in economic theory [Samuelson (1947), Stigler (1950), Debreu (1959)] where one often presumes that X is the nonnegative orthant $\mathbb{R}^{+n}$ of a finite-dimensional Euclidean space, preference increases along rays out from the origin, and the indifference classes or isoutility contours are smooth hypercurves concave away from the origin. The continuity of u implied by smoothness, and conditions of semicontinuity for u are discussed by Wold (1943), Wold and Jureen (1953), Yokoyama (1956), Debreu (1959, 1964), Newman and Read (1961), Eilenberg (1941), Rader (1963) and Herden (1989), among others. For example, if X is a topological space with topology $\mathcal{T}$ and the representation of Corollary 3.1 holds, then it holds for some continuous u if and only if the preferred-to-y and less-preferred-than-y sets

$$\{x \in X : x > y\} \text{ and } \{x \in X : y > x\}$$

are in $\mathcal{T}$ for every $y \in X$.

If you like your coffee black, you will probably be indifferent between x and $x + 1$ grains of sugar in your coffee for $x = 0, 1, 2, \ldots$, but will surely prefer 0 grains to 1000 grains so that your indifference relation is not transitive. Early discussants of partially ordered preferences and nontransitive indifference include Georgescu–Roegen (1936, 1958) and Armstrong (1939, 1948, 1950), with subsequent contributions by Luce (1956), Aumann (1962), Richter (1966), Fishburn (1970a, 1985a), Chipman (1971), and Hurwicz and Richter (1971), among others. Fishburn (1970b) surveys nontransitive indifference in preference theory. When partially ordered preferences are represented by a single functional, i.e., a real valued function, we lose the ability to represent $\sim$ precisely but can capture the *strong indifference* relation $\approx$ defined by

$$x \approx y \quad \text{if } \forall z \in X, x \sim z \Leftrightarrow y \sim z,$$

with $\approx$ an equivalence. We define $>''$ on $X/\approx$ by $a >'' b$ if $x > y$ for some $x \in a$ and $y \in b$. The following, from Fishburn (1970a), is a specialization of a theorem in Richter (1966).

Theorem 3.2. *If $>$ on X is a partial order and $X/\approx$ includes a countable subset that is $>''$-order dense in $X/\approx$, then there is a $u: X \to \mathbb{R}$ such that*

$$\forall x, y \in X, x > y \Rightarrow u(x) > u(y),$$
$$\forall x, y \in X, x \approx y \Rightarrow u(x) = u(y).$$

If a second functional is introduced, special types of partial orders have ⇔ representations. We say that $>$ on X is an *interval order* [Wiener (1914), Fishburn (1970a)] if it is asymmetric and satisfies

$$\forall x, y, a, b \in X, (x > a, y > b) \Rightarrow (x > b \text{ or } y > a),$$

and that it is *semiorder* [Armstrong (1939, 1948), Luce (1956)] if, in addition,

$$\forall x, y, z, a \in X, (x > y, y > z) \Rightarrow (x > a \text{ or } a > z).$$

Both conditions reflect assumptions about thresholds of discriminability in judgment. The coffee/sugar example stated earlier is a good candidate for a semiorder.

Theorem 3.3. *Suppose X is countable. Then $>$ on X is an interval order if and only if there are $u, \rho: X \to \mathbb{R}$ with $\rho \geq 0$ such that*

$$\forall x, y \in X, x > y \Leftrightarrow u(x) > u(y) + \rho(y),$$

and $>$ is a semiorder if and only if the same representation holds along with

$$\forall x, y \in X, u(x) < u(y) \Leftrightarrow u(x) + \rho(x) < u(y) + \rho(y).$$

Moreover, if X is finite and $>$ is a semiorder, then ρ can be taken to be constant.

The final assertion was first proved by Scott and Suppes (1958), and a complete proof is given in Fishburn (1985a). Generalizations for uncountable X are discussed in Fishburn (1985a), Doignon et al. (1984) and Bridges (1983, 1986), and an analysis of countable semiorders appears in Manders (1981). Aspects of families of semiorders and interval orders are analyzed by Roberts (1971), Cozzens and Roberts (1982), Doignon (1984, 1987) and Doigon et al. (1986).

A somewhat different approach to preference that takes a person's choices rather than avowed preferences as basic is the *revealed preference* approach pioneered by Samuelson (1938) and Houthakker (1950). Houthakker (1961) and Sen (1977) review various facets of revealed preferences, and Chipman et al. (1971) contains important papers on the topic. Sonnenschein (1971) and subsequent contributions by Mas-Colell (1974), Shafer and Sonnenschein (1975), and others, pursue a related approach in which transitivity is replaced by convexity assumptions.

To illustrate the revealed preference approach, suppose X is finite and C is a *choice function* defined on all nonempty subsets of X that satisfies

$$\emptyset \neq C(A) \subseteq A.$$

Roughly speaking, $C(A)$ identifies the most desirable elements of A. One condition for C [Arrow (1959)] is: for all nonempty subsets A and B of X,

$$A \subset B \text{ and } A \cap C(B) \neq \emptyset \Rightarrow C(A) = A \cap C(B).$$

This says that if some choice from the superset B of A is in A, then the choice set for A is precisely the set of elements from A that are in the choice set for B. It is not hard to show that it holds if and only if there is a weak order $>$ on X such that, for all nonempty $A \subseteq X$,

$$C(A) = \{x \in A : y > x \text{ for no } y \in A\}.$$

Many investigators address issues of interdependence, separability and independence among preferences for different factors when X is a product set $X_1 \times X_2 \times \cdots \times X_n$. Fisher (1892), other ordinalistists mentioned above, and Fishburn (1972a) discuss interdependent preferences in the multiattribute case. Lexicographic utilities, which involve an importance ordering on the X_i themselves, are examined by Georgescu-Roegen (1958), Chipman (1960), Fishburn (1974, 1980a) and Luce (1978). For vectors $(a_1, a_2, \ldots, a_n)$ and $(b_1, b_2, \ldots, b_n)$ of real numbers, we define the *lexicographic order* $>_L$ on $\mathbb{R}^n$ by

$$(a_1, \ldots, a_n) >_L (b_1, \ldots, b_n) \quad \text{if } a_i \neq b_i \text{ for some } i, \text{ and } a_i > b_i \text{ for the smallest such } i.$$

The preceding references discuss various types of preference structures for which preferences cannot be represented in the usual manner by a single functional but can be represented by utility vectors ordered lexicographically.

Additive utilities, which concern the decomposition $u(x_1, x_2, \ldots, x_n) = u_1(x_1) + u_2(x_2) + \cdots + u_n(x_n)$ and related expressions, are considered by Debreu (1960), Luce and Tukey (1964), Scott (1964), Aumann (1964), Adams (1965), Gorman (1968), Fishburn (1970a), Krantz et al. (1971), Debreu and Koopmans (1982) and Bell (1987). Investigations involving time preferences, in which i indexes time, include Koopmans (1960), Koopmans et al. (1964), Diamond (1965), Burness (1973, 1976), Fishburn (1978a) and Fishburn and Rubinstein (1982). Related works that address aspects of time preference and dynamic consistency include Strotz (1956), Pollak (1968, 1976), Peleg and Yaari (1973) and Hammond (1976a,b). We say a bit more about dynamic consistency at the end of Section 7.

Our next theorem [Debreu (1960)] illustrates additive utilities. We assume $X = X_1 \times \cdots \times X_n$ with each X_i a connected and separable topological space and let $\mathcal{T}$ be their product topology for X. Factor X_i is *essential* if $x > y$ for some $x, y \in X$ for which $x_j = y_j$ for all $j \neq i$.

Theorem 3.4. *Suppose $n \geq 3$, at least three X_i are essential, $>$ on X is a weak order, $\{x \in X : x > y\}$ and $\{x \in X : y > x\}$ are in $\mathcal{T}$ for every $y \in X$, and, for all $x, y, z, w \in X$, if $\{x_i, z_i\} = \{y_i, w_i\}$ for all i and $x \geq y$, then not $(z > w)$. Then there are continuous $u_i : X_i \to \mathbb{R}$ such that*

$$\forall x, y \in X, x > y \Leftrightarrow \sum_{i=1}^{n} u_i(x_i) > \sum_{i=1}^{n} u_i(y_i).$$

Moreover, v_i satisfy this representation in place of the u_i if and only if $(v_1, \ldots, v_n) = (\alpha u_1 + \beta_1, \ldots, \alpha u_n + \beta_n)$ for real numbers $\alpha > 0$ and $\beta_1, \ldots, \beta_n$.

The major changes for additivity when X is finite [Aumann (1964), Scott (1964),

Adams (1965), Fishburn (1970a)] are that a more general independence condition [than $\cdots x \gtrsim y \Rightarrow \text{not}(z > w)$] is needed and we lose the nice uniqueness property for the u_i at the end of Theorem 3.4.

One can also consider additivity for $X = X_1 \times \cdots \times X_n$ when it is not assumed that preferences are transitive. A useful model in this case is

$$x > y \Leftrightarrow \sum_{i=1}^{n} \phi_i(x_i, y_i) > 0,$$

where ϕ_i maps $X_i \times X_i$ into $\mathbb{R}$. Vind (1991) axiomatizes this for $n \geq 4$, and Fishburn (1991) discusses its axiomatizations for various cases of $n \geq 2$ under the restriction that each ϕ_i is *skew-symmetric*, which means that

$$\forall a, b \in X_i, \phi_i(a, b) + \phi_i(b, a) = 0.$$

The preceding theorems and discussion involve only simple preference comparisons on X. In contrast, comparable preference differences via Frisch (1926) and others use a quaternary relation $>^*$ on X which we write as $(x, y) >^* (z, w)$ as in the preceding section. In the following correspondent to Theorem 3.4 we let $\mathcal{T}$ be the product topology for $X \times X$ with X assumed to be a connected and separable topological space. Also, $(x, y) \sim^* (z, w)$ if neither $(x, y) >^* (z, w)$ nor $(z, w) >^* (x, y)$, and $\gtrsim^*$ is the union of $>^*$ and $\sim^*$.

Theorem 3.5. *Suppose* $\{(x, y) \in X \times X : (x, y) >^* (z, w)\}$ *and* $\{(x, y) \in X \times X : (z, w) >^* (x, y)\}$ *are in* $\mathcal{T}$ *for every* $(z, w) \in X \times X$, *and, for all* $x, x', x'', w, w', w'' \in X$, *if* x, x', x'', w, w', w'' *is a permutation of* y, y', y'', z, z', z'' *and* $(x, y) \gtrsim^* (z, w)$ *and* $(x', y') \gtrsim^* (z', w')$ *then not* $[(x'', y'') >^* (z'', w'')]$. *Then there is a continuous* $u: X \to \mathbb{R}$ *such that*

$$\forall x, y, z, w \in X, (x, y) >^* (z, w) \Leftrightarrow u(x) - u(y) > u(z) - u(w).$$

Moreover, v satisfies this representation in place of u if and only if $v = \alpha u + \beta$ for real numbers $\alpha > 0$ and β.

The final part of Theorem 3.5 is often abbreviated by saying that u is *unique up to a positive affine transformation*. The theorem is noted in Fishburn (1970a, p. 84) and is essentially due to Debreu (1960). Other infinite-X theorems for comparable differences appear in Suppes and Winet (1955), Scott and Suppes (1958), Chipman (1960), Suppes and Zinnes (1963), Pfanzagl (1959) and Krantz et al. (1971). Related theorems for finite X without the nice uniqueness conclusions are in Fishburn (1970a), and uniqueness up to a positive affine transformation with X finite is discussed in Fishburn et al. (1988). Fishburn (1986a) considers the generalized representation $(x, y) >^* (z, w) \Leftrightarrow \phi(x, y) > \phi(z, w)$ in which ϕ is a *skew-symmetric* $[\phi(y, x) = -\phi(x, y)]$ functional on $X \times X$. This generalization makes no transitivity assumption about simple preference comparisons on X, which might be cyclic. In contrast, if we define $>$ on X from $>^*$ in the natural way as $x > y$ if $(x, y) >^* (y, y)$, then the representation of Theorem 3.5 implies that $>$ on X is a weak order.

The revealed-by-choice preference approach of Samuelson (1938) and others is generalized to probabilistic choice models in a number of studies. For example, the theories of Debreu (1958, 1960) and Suppes (1961) make assumptions about binary choice probabilities $p(x, y)$ – that x will be chosen when a choice is required between x and y – which imply $u: X \to \mathbb{R}$ for which

$$p(x, y) > p(z, w) \Leftrightarrow u(x) - u(y) > u(z) - u(w),$$

and Luce's (1959) axiom for the probability $p(x, Y)$ that x will be chosen from Y included in finite X yields u unique up to multiplication by a nonzero constant such that, when $0 < p(x, Y) < 1$,

$$p(x, Y) = u(x) / \sum_{y \in Y} u(y).$$

The books by Luce (1959) and Restle (1961) and surveys by Becker et al. (1963) and Luce and Suppes (1965) provide basic coverage. A sampling of papers on binary choice probabilities, more general choice probabilities, and so-called random or stochastic utility models is Luce (1958), Marschak (1960), Marley (1965, 1968), Tversky (1972a, 1972b), Fishburn (1973a), Corbin and Marley (1974), Morgan (1974), Sattath and Tversky (1976), Yellott (1977), Manski (1977), Strauss (1979), Dagsvik (1983), Machina (1985), Tutz (1986), and Barbera and Pattanaik (1986). Recent contributions that investigate conditions on binary choice probabilities, such as the triangle condition

$$p(x, y) + p(y, z) \geq p(x, z),$$

that are needed for p to be induced by a probability distribution on the set of all linear orders on X, include Campello de Souza (1983), Cohen and Falmagne (1990), Fishburn and Falmagne (1989) and Gilboa (1990).

4. Expected utility and linear generalizations

We now return to simple preference comparisons to consider the approach of von Neumann and Morgenstern (1944) to preference between risky options encoded as probability distributions on outcomes or other elements of interest such as n-tuples of pure strategies. As mentioned earlier, their approach differs radically from Bernoulli's (1738) and its assessment of outcome utility by riskless comparisons of preference differences.

Throughout the rest of the chapter, P denotes a convex set of probability measures defined on a Boolean algebra $\mathcal{H}$ of subsets of a set X. *Convexity* means that $\lambda p + (1 - \lambda)q$, with value $\lambda p(A) + (1 - \lambda)q(A)$ for each $A \in \mathcal{H}$, is in P whenever $p, q \in P$ and $0 \leq \lambda \leq 1$. For example, if $p(A) = 0.3$, $q(A) = 0.2$, and $\lambda = \frac{1}{2}$, then $(\lambda p + (1 - \lambda)q)(A) = 0.25$. The von Neumann–Morgenstern theory also applies to more general convex sets, called mixture sets [Herstein and Milnor (1953), Fishburn (1970a)], but P will serve for most applications.

The von Neumann–Morgenstern axioms in the form used by Jensen (1967) are: for all $p, q, r \in P$ and all $0 < \lambda < 1$,

A1. $>$ *on P is a weak order,*

A2. $p > q \Rightarrow \lambda p + (1 - \lambda)r > \lambda q + (1 - \lambda)r$,

A3. $(p > q, q > r) \Rightarrow \alpha p + (1 - \alpha)r > q > \beta p + (1 - \beta)r \quad$ *for some* $\alpha, \beta \in (0, 1)$.

A2 is an independence axiom asserting preservation of $>$ under similar convex combinations: if you prefer \$2000 to a 50–50 gamble for \$1000 or \$3500 then, with $r(\$40\,000) = 1$ and $\lambda = 0.1$, you prefer a gamble with probabilities 0.1 and 0.9 for \$2000 and \$40 000 respectively, to a gamble that has probabilities 0.05, 0.05 and 0.9 for \$1000, \$3500 and \$40 000 respectively. A3 is a continuity or "Archimedean" condition that ensures the existence of real valued as opposed to vector valued or nonstandard utilities [Hausner (1954), Chipman (1960), Fishburn (1974, 1982a), Skala (1975), Blume et al. (1989)]. The last cited reference includes a discussion of the significance of lexicographic choice for game theory.

We say that $u: P \to \mathbb{R}$ is *linear* if

$$\forall p, q \in P, \forall \lambda < 0 < 1, u(\lambda p + (1 - \lambda)q) = \lambda u(p) + (1 - \lambda)u(q).$$

Theorem 4.1. *There is a linear functional u on P such that*

$$\forall p, q \in P, p > q \Leftrightarrow u(p) > u(q),$$

if and only if A1, A2 *and* A3 *hold. Moreover, such a u is unique up to a positive affine transformation.*

Proofs are given in Jensen (1967) and Fishburn (1970a, 1982a, 1988b), and alternative axioms for the linear representation are described by Marschak (1950), Friedman and Savage (1952), Herstein and Milnor (1953) and Luce and Raiffa (1957) among others.

Suppose $\mathcal{H}$ contains every singleton $\{x\}$. We then extend u to X by $u(x) = u(p)$ when $p(\{x\}) = 1$ and use induction with linearity to obtain the *expected utility* form

$$u(p) = \sum_X p(x)u(x)$$

for all *simple* measures ($p(A) = 1$ for some finite $A \in \mathcal{H}$) in P. Extensions of this form to $u(p) = \int u(x)\, \mathrm{d}p(x)$ for more general measures are axiomatized by Blackwell and Girshick (1954), Arrow (1958), Fishburn (1967, 1970a, 1975a), DeGroot (1970) and Ledyard (1971). The most important axioms used in the extensions, which may or may not force u to be bounded, are dominance axioms. An example is $[p(A) = 1, x > q$ for all $x \in A] \Rightarrow p \gtrsim q$. This says that if x is preferred to q for every x in a set on which p has probability 1, then p as a whole is preferred or indifferent to q.

The context in which X is an interval of monetary outcomes has been intensely studied with regard to risk attitudes [Friedman and Savage (1948), Markowitz (1952), Pratt (1964), Arrow (1974), Ross (1981), Machina and Neilson (1987)] and stochastic dominance [Karamata (1932), Hardy et al. (1934), Quirk and Saposnik (1962), Fishburn (1964, 1980b), Hadar and Russell (1969, 1971), Hanoch and Levy (1969), Whitmore (1970), Rothschild and Stiglitz (1970, 1971), Fishburn and Vickson (1978), Whitmore and Findlay (1978), Bawa (1982)]. Following Pratt and Arrow, increasing u on X is said to be *risk averse* (*risk neutral, risk seeking*) on an interval if any nondegenerate distribution on the interval is less preferred than (indifferent to, preferred to) the expected monetary value of the distribution. If u is twice differentiable, these alternatives are equivalent to the second derivative being negative (zero, positive) throughout the interval. For *changes* to present wealth, many people tend to be risk averse is gains but risk seeking in losses [Fishburn and Kochenberger (1979), Kahneman and Tversky (1979), Schoemaker (1980)]: but see Hershey and Schoemaker (1980) and Cohen et al. (1985) to the contrary.

A typical result for first ($>_1$) and second ($>_2$) degree stochastic dominance relates the first and second cumulatives of simple measures on X, defined by $p^1(x) = \sum_{y \leqslant x} p(y)$ and $p^2(x) = \int_{-\infty}^{x} (y) \, dy$, to classes of utility functions on X. Define

$$p >_i q \quad \text{if } p \neq q \text{ and } p^i(x) \leqslant q^i(x) \quad \text{for all } x \in X.$$

An equivalent way of stating first-degree stochastic dominance is that $p >_1 q$ if, for every x, p has as large a probability as q for doing better than x, and, for some x, the probability of getting an outcome greater than x is larger for p than for q.

Theorem 4.2. *Suppose X is an interval in $\mathbb{R}$, and p and q are simple measures. Then $p >_1 q$ if and only if $\sum p(x)u(x) > \sum q(x)u(x)$ for all increasing u on X, and $p >_2 q$ if and only if $\sum p(x)u(x) > \sum q(x)u(x)$ for all increasing and concave (risk averse) u on X.*

Axioms for the continuity of u on X in the $\mathbb{R}$ case are noted by Foldes (1972) and Grandmont (1972). I am not aware of similar studies for derivatives.

When $X = X_1 \times X_2 \times \cdots \times X_n$ in the von Neumann–Morgenstern model, special conditions lead to additive [Fishburn (1965), Pollak (1967)], multiplicative [Pollak (1967), Keeney (1968)], or other decompositions [Farquhar (1975), Fishburn and Farquhar (1982), Bell (1986)] for u on X or on P. Much of this is surveyed in Keeney and Raiffa (1976), Farquhar (1977, 1978) and Fishburn (1977, 1978b). Special considerations of i as a time index, including times at which uncertainties are resolved, are discussed in Drèze and Modigliani (1972), Spence and Zeckhauser (1972), Nachman (1975), Keeney and Raiffa (1976), Meyer (1977), Kreps and Porteus (1978, 1979), Fishburn and Rubinstein (1982), and Jones and Ostroy (1984).

The simplest decompositional forms for the multiattribute case are noted in our next theorem.

Theorem 4.3. *Suppose $X = X_1 \times X_2$, P is the set of simple probability distributions on X, linear u on P satisfies the representation of Theorem 4.1, and $u(x) = u(p)$ when*

$p(x) = 1$. Then there are $u_i \colon X_i \to \mathbb{R}$ such that

$$\forall x = (x_1, x_2) \in X, u(x_1, x_2) = u_1(x_1) + u_2(x_2)$$

if and only if $p \sim q$ whenever the marginal distributions of p and q on X_i are identical for $i = 1, 2$. And there are $u_1 \colon X_1 \to \mathbb{R}$ and $f, g \colon X_2 \to \mathbb{R}$ with $f > 0$ such that

$$\forall x \in X, u(x_1, x_2) = f(x_2)u_1(x_1) + g(x_2),$$

if and only if the conditional of $>$ on the distributions defined on X_1 for a fixed $x_2 \in X_2$ does not depend on the fixed level of X_2.

A generalization of Theorem 4.1 motivated by n-person games (X_i = pure strategy set for i, P_i = mixed strategy set for i) replaces P by $P_1 \times P_2 \times \cdots \times P_n$, where P_i is the set of simple probability distributions on X_i. A distribution p on $X = X_1 \times \cdots \times X_n$ is then *congruent with* the P_i if and only if there are $p_i \in P_i$ for each i such that $p(x) = p_1(x_1) \cdots p_n(x_n)$ for all $x \in X$. If we restrict a player's preference relation $>$ to $P_1 \times \cdots \times P_n$, or to the congruent distributions on X, then the axioms of Theorem 4.1 can be generalized [Fishburn (1976b, 1985a)] to yield $p > q \Leftrightarrow u(p) > u(q)$ for congruent distributions with $u(p) = (p_1, \ldots, p_n)$ given by the expected utility form

$$u(p_1, \ldots, p_n) = \sum_X p_1(x_1) \cdots p_n(x_n) u(x_1, \ldots, x_n).$$

Other generalizations of Theorem 4.1 remain in our original P context and weaken one or more of the von Neumann–Morgenstern axioms. Aumann (1962) and Fishburn (1971b, 1972b, 1982a) axiomatize partially ordered linear utility by modifying A1–A3. The following is from Fishburn (1982a).

Theorem 4.4. *Suppose $>$ on P is a partial order and, for all $p, q, r, s \in P$ and all $0 < \lambda < 1$:*

A2'. $(p > q, r > s) \Rightarrow \lambda p + (1 - \lambda)r > \lambda q + (1 - \lambda)s$,

A3'. $(p > q, r > s) \Rightarrow \alpha p + (1 - \alpha)s > \alpha q + (1 - \alpha)r$ *for some $0 < \alpha < 1$.*

Then there is a linear $u \colon P \to \mathbb{R}$ such that

$$\forall p, q \in P, p > q \Rightarrow u(p) > u(q).$$

Vincke (1980) assumes that $>$ on P is a semiorder and presents conditions necessary and sufficient for $p > q \Leftrightarrow u(p) > u(q) + \rho(q)$ with u linear, $\rho \geq 0$, and a few other things. Related results appear in Nakamura (1988). Weak order and full independence are retained by Hausner (1954), Chipman (1960) and Fishburn (1971a, 1982a), but A3 is dropped to obtain lexicographic linear representations for $>$ on P. Kannai (1963) discusses the axiomatization of partially ordered lexicographic linear utility, and Skala (1975) gives a general treatment of non-Archimedean utility. Other generalizations of the von Neumann–Morgenstern theory that substantially weaken the independence axiom A2 are described in the next section.

5. Nonlinear utility

Nonlinear utility theory involves representations for $>$ on P that abandon or substantially weaken the linearity property of expected utility and independence axioms like A2 and A2'. It was stimulated by systematic violations of independence uncovered by Allais (1953a,b) and confirmed by others [Morrison (1967), MacCrimmon (1968), MacCrimmon and Larsson (1979), Hagan (1979), Kahneman and Tversky (1979), Tversky and Kahneman (1986)]. For example, in an illustration of Allais's *certainty effect*, Kahneman and Tversky (1979) take $r(\$0) = 1$,

$\phantom{p' = \tfrac{1}{4}p + \tfrac{3}{4}r:}$ p: $\$3000$ with probability 1
$\phantom{p' = \tfrac{1}{4}p + \tfrac{3}{4}r:}$ q: $\$4000$ with probability 0.8, nothing otherwise;
$p' = \tfrac{1}{4}p + \tfrac{3}{4}r$: $\$3000$ with probability 0.25, $\$0$ otherwise
$q' = \tfrac{1}{4}q + \tfrac{3}{4}r$: $\$4000$ with probability 0.20, $\$0$ otherwise,

and observe that a majority of 94 respondents violate A2 with $p > q$ and $q' > p'$.

Other studies challenge the ordering axiom A1 either by generating plausible examples of preference cycles [Flood (1951, 1952), May (1954), Tversky (1969), MacCrimmon and Larsson (1979)] or by the preference reversal phenomenon [Lichtenstein and Slovic (1971, 1973), Lindman (1971), Grether and Plott (1979), Pommerehne et al. (1982), Reilly (1982), Slovic and Lichtenstein (1983), Goldstein and Einhorn (1987)]. A *preference reversal* occurs when p is preferred to q but an individual in possession of one or the other would sell p for less. Let $p(\$30) = 0.9$, $q(\$100) = 0.3$, with $\$0$ returns otherwise. If $p > q$, p's *certainty equivalent* (minimum selling price) is $\$25$, and q's certainty equivalent is $\$28$, we get $p > q \sim \$28 > \$25 \sim p$. Tversky et al. (1990) argue that preference reversals result more from overpricing low-probability high-payoff gambles, such as q, than from any inherent disposition toward intransitivities.

Still other research initiated by Preston and Baratta (1948) and Edwards (1953, 1954a,b,c), and pursued in Tversky and Kahneman (1973, 1986), Kahneman and Tversky (1972, 1979), Kahneman et al. (1982), and Gilboa (1985), suggests that people subjectively modify stated objective probabilities: they tend to overrate small chances and underrate large ones. This too can lead to violations of the expected utility model.

Nonlinear utility theories that have real valued representations fall into three main categories:

(1) Weak order with monetary outcomes;
(2) Weak order with arbitrary outcomes;
(3) Nontransitive with arbitrary outcomes.

The last of these obviously specialize to monetary outcomes.

Weak order theories designed specifically for the monetary context [Allais (1953a,b, 1979, 1988), Hagan (1972, 1979), Karmarkar (1978), Kahneman and Tversky (1979), Machina (1982a,b, 1983), Quiggin (1982), Yaari (1987), Becker and

Sarin (1987)] are constructed to satisfy first-degree stochastic dominance ($p >_1 q \Rightarrow p > q$) with the exception of Karmarkar (1978) and Kahneman and Tversky (1979). These two involve a quasi-expectational form with objective probabilities transformed by an increasing functional $\tau: [0, 1] \to [0, 1]$, and either violate first-degree stochastic dominance or have $\tau(\lambda) = \lambda$ for all λ. Quiggin (1982) also uses τ, but applies it to cumulative probabilities to accommodate stochastic dominance. His model entails $\tau(1/2) = 1/2$, and this was subsequently relaxed by Chew (1984) and Segal (1984). Yaari (1987) independently axiomatized the special case in which $u(x) = x$, thus obtaining an expectational form that is "linear in money" instead of being "linear in probability" as in the von Neumann–Morgenstern model.

The theories of Allais and Hagen are the only ones in category **1** that adopt the Bernoullian approach to riskless assessment of utility through difference comparisons. Along with weak order and satisfaction of stochastic dominance, they use a principle of uniformity that leads to

$$u(p) = \sum_X p(x)u(x) + \Theta(p^*),$$

where the sum is Bernoulli's expected utility, Θ is a functional, and p^* is the probability distribution induced by p on the differences of outcome utilities from their mean. A recent elaboration by Allais (1988) yields a specialization with transformed cumulative probabilities, and a different specialization with a "disappointment" interpretation is discussed by Loomes and Sugden (1986).

Machina's (1982a, 1987) alternative to von Neumann and Morgenstern's theory assumes that u on P is smooth in the sense of Fréchet differentiability and that u is linear locally in the limit. Allen (1987) and Chew et al. (1987) use other notions of smoothness to obtain economically interesting implications that are similar to those in Machina (1982a,b, 1983, 1984).

The main transitivity theory for arbitrary outcomes in category **2** is Chew's weighted linear utility theory. The weighted representation was first axiomatized by Chew and MacCrimmon (1979) with subsequent refinements by Chew (1982, 1983), Fishburn (1983a, 1988b) and Nakamura (1984, 1985). One representational form for weighted linear utility is

$$\forall p, q \in P, \ p > q \Leftrightarrow u(p)w(q) > u(q)w(p),$$

where u and w are linear functionals on P, $w \geq 0$, and $w > 0$ on $\{p \in P: q > p > q'$ for some $q, q' \in P\}$. If w is positive everywhere, we obtain $p > q \Leftrightarrow u(p)/w(p) > u(q)/w(q)$, thus separating p and q. If, in addition, $v = u/w$, then the representation can be expressed as $p > q \Leftrightarrow v(p) > v(q)$ along with

$$v(\lambda p + (1 - \lambda)q) = \frac{\lambda w(p)v(p) + (1 - \lambda)w(q)v(q)}{\lambda w(p) + (1 - \lambda)w(q)},$$

a form often seen in the literature. The ratio form $u(p)/w(p)$ is related to a ratio representation for preference between events developed by Bolker (1966, 1967).

Generalizations of the weighted linear representation are discussed by Fishburn (1982b, 1983a,b, 1988b), Chew (1985) and Dekel (1986). One of these weakens the ordering axiom to accommodate intransitivities, including preference cycles. Potential loss of transitivity can be accounted for by a two-argument functional ϕ on $P \times P$. We say that ϕ is an *SSB functional* on $P \times P$ if it is skew symmetric $[\phi(q,p) = -\phi(p,q)]$ and *bilinear*, i.e., linear separately in each argument. For example, $\phi(\lambda p + (1-\lambda)q, r) = \lambda\phi(p,r) + (1-\lambda)\phi(q,r)$ for the first argument. The so-called SSB model is

$$\forall p, q \in P, p > q \Leftrightarrow \phi(p,q) > 0.$$

When P is convex and this model holds, its ϕ is unique up to multiplication by a positive real number.

The SSB model was axiomatized by Fishburn (1982b) although its form was used earlier by Kreweras (1961) to prove two important theorems. First, the minimax theorem [von Neumann (1928), Nikaidô (1954)] is used to prove that if Q is a finitely generated convex subset of P, then $\phi(q^*, q) \geq 0$ for some q^* and all q in Q. Second, if all players in an n-person noncooperative game with finite pure strategy sets have SSB utilities, then the game has a Nash equilibrium (1951). These were independently discovered by Fishburn (1984a) and Fishburn and Rosenthal (1986).

When p and q are simple measures on X and $\phi(x, y)$ is defined as $\phi(p,q)$ when $p(x) = q(y) = 1$, bilinearity of ϕ implies the expectational form $\phi(p,q) = \sum_x \sum_y p(x)q(y)\phi(x,y)$. Additional conditions [Fishburn (1984a)] are needed to obtain the integral form $\phi(p,q) = \iint \phi(x,y) \mathrm{d}p(x) \mathrm{d}q(y)$ for more general measures. Fishburn (1984a,b) describes how SSB utilities preserve first- and second-degree stochastic dominance in the monetary context by simple assumptions on $\phi(x, y)$, and Loomes and Sugden (1983) and Fishburn (1985b) show how it accommodates preference reversals. These and other applications of SSB utility theory are included in Fishburn (1988b, ch. 6).

6. Subjective probability

Although there are aspects of subjective probability in older writings, including Bayes (1763) and Laplace (1812), it is largely a child of the present century. The two main strains of subjective probability are the *intuitive* [Koopman (1940), Good (1950), Kraft et al. (1959)], which takes $>$ (*is more probable than*) as a primitive binary relation on a Boolean algebra $\mathscr{A}$ of subsets of a state space S, and the *preference-based* [de Finetti (1931, 1937), Ramsey (1931), Savage (1954)] that ties *is more probable than* to preference between uncertain acts. In this section we suppress reference to preference in the latter strain and simply take $>$ as a comparative probability relation on $\mathscr{A}$. Surveys of axiomatizations for representations of $(\mathscr{A}, >)$ or $(\mathscr{A}, \gtrsim)$ are given by Fine (1973) and Fishburn (1986b). Kyburg and Smokler (1964) provide a collection of historically important articles on subjective probability, and Savage (1954) still merits careful reading as a leading proponent of the preference-based approach.

We begin with Savage's (1954) axiomatization of $(\mathscr{A}, >)$ and its nice uniqueness consequence. The following is more or less similar to other infinite-S axiomatizations by Bernstein (1917), Koopman (1940) and de Finetti (1931).

Theorem 6.1. *Suppose $\mathscr{A} = 2^S$ and $>$ on $\mathscr{A}$ satisfies the following for all $A, B, C \in \mathscr{A}$:*

S1. $>$ on $\mathscr{A}$ is a weak order,

S2. $S > \emptyset$,

S3. $\text{not}(\emptyset > A)$,

S4. $(A \cup B) \cap C = \emptyset \Rightarrow (A > B \Leftrightarrow A \cup C > B \cup C)$,

S5. $A > B \Rightarrow$ there is a finite partition $\{C_1, \ldots, C_m\}$ of S such that $A > (B \cup C_i)$ for $i = 1, \ldots, m$.

Then there is a unique probability measure $\pi: \mathscr{A} \to [0, 1]$ such that

$$\forall A, B \in \mathscr{A}, A > B \Leftrightarrow \pi(A) > \pi(B),$$

and for every $A \in \mathscr{A}$ with $\pi(A) > 0$ and every $0 < \lambda < 1$ there is a $B \subset A$ for which $\pi(B) = \lambda \pi(A)$.

The first four axioms, S1 (order), S2 (nontriviality), S3 (nonnegativity), and S4 (independence), were used by de Finetti (1931), but Savage's Archimedean axiom S5 was new and replaced the assumption that, for any m, S could be partitioned into m equally likely ($\sim$) events. S2 and S3 are uncontroversial: S2 says that something is more probable than nothing, and S3 says that the empty event is not more probable than something else. S1 may be challenged for its precision [Keynes (1921), Fishburn (1983c)], S4 has been shown to be vulnerable to problems of vagueness and ambiguity [Ellsberg (1961), Slovic and Tversky (1974), MacCrimmon and Larsson (1979)], and S5 forces S to be infinite, as seen by the final conclusion of theorem.

When π satisfies $A > B \Leftrightarrow \pi(A) > \pi(B)$, it is said to *agree* with $>$, and, when $\gtrsim$ is taken as the basic primitive relation, we say that π *almost agrees* with $\gtrsim$ if $A \gtrsim B \Rightarrow P(A) \geqslant P(B)$. Other axiomatizations for a uniquely agreeing π are due to Luce (1967), Fine (1971), Roberts (1973) and Wakker (1981), with Luce's version of special interest since it applies to finite as well as infinite S. Unique agreement for finite S is obtained by DeGroot (1970) and French (1982) by first adjoining S to an auxiliary experiment with a rich set of events, and a similar procedure is suggested by Allais (1953b, 1979). Necessary and sufficient conditions for agreement, which tend to be complex and do not of course imply uniqueness, are noted by Domotor (1969) and Chateauneuf (1985); see also Chateauneuf and Jaffray (1984). Sufficient conditions on $>$ for a unique almost agreeing measure are included in

Savage (1954), Niiniluoto (1972) and Wakker (1981), and nonunique almost agreement is axiomatized by Narens (1974) and Wakker (1981).

The preceding theories usually presume that π is only finitely additive and not necessarily *countably additive*, which holds if $\pi(\bigcup_{i=1}^{\infty} A_i) = \sum_{i=1}^{\infty} \pi(A_i)$ whenever the A_i are mutually disjoint events in $\mathscr{A}$ whose union is also in $\mathscr{A}$. The key condition needed for countable additivity is Villegas's (1964) monotone continuity axiom

S6. $\forall A, B, A_1, A_2, \ldots \in \mathscr{A}, (A_1 \subseteq A_2 \subseteq \ldots; A = \bigcup_i A_i; B \gtrsim A_i \text{ for all } i) \Rightarrow B \gtrsim A$.

The preceding theories give a countably additive π if and only if S6 holds [Villegas (1964), Chateauneuf and Jaffray (1984)] so long as S6 is compatible with the theory, and axiomatizations that explicitly adopt S6 with $\mathscr{A}$ a *σ-algebra* (a Boolean algebra closed under countable unions) are noted also by DeGroot (1970), French (1982) and Chuaqui and Malitz (1983).

As mentioned earlier, necessary and sufficient axioms for agreement with finite S were first obtained by Kraft et al. (1959), and equivalent sets of axioms are noted in Scott (1964), Fishburn (1970a) and Krantz et al. (1971). The following theorem illustrates the type of generalized independence axiom needed in this case.

Theorem 6.2. *Suppose S is nonempty and finite with $\mathscr{A} = 2^S$. Then there is a probability measure π on $\mathscr{A}$ for which*

$$\forall A, B \in \mathscr{A}, A > B \Leftrightarrow \sum_{s \in A} \pi(s) > \sum_{s \in B} \pi(s),$$

if and only if S2 and S3 hold along with the following for all $A_j, B_j \in \mathscr{A}$ and all $m \geq 2$:

S4'. *If $|\{j: s \in A_j\}| = |\{j: s \in B_j\}|$ for all $s \in S$ with $1 \leq j \leq m$, and if $A_j \gtrsim B_j$ for all $j < m$, then not$(A_m > B_m)$.*

Since the hypotheses of S4' imply $\sum_j \pi(A_j) = \sum_j \pi(B_j)$ for any measure π on $\mathscr{A}$, S4' is clearly necessary for finite agreement. Sufficiency, in conjunction with S2 and S3, is established by a standard result for the existence of a solution to a finite set of linear inequalities [Kuhn (1956), Aumann (1964), Scott (1964), Fishburn (1970a)]. Conditions for unique π in the finite case are discussed by Luce (1967), Van Lier (1989), Fishburn and Roberts (1989) and Fishburn and Odlyzko (1989).

Several people have generalized the preceding theories to accommodate partial orders, interval orders and semiorders, conditional probability, and lexicographic representations. Partial orders with $A > B \Rightarrow \pi(A) > \pi(B)$ are considered by Adams (1965) and Fishburn (1969) for finite S and by Fishburn (1975b) for infinite S. Axioms for representations by intervals are in Fishburn (1969, 1986c) and Domotor and Stelzer (1971), and related work on probability intervals and upper and lower probabilities includes Smith (1961), Dempster (1967, 1968), Shafer (1976), Williams (1976), Walley and Fine (1979), Kumar (1982) and Papamarcou and Fine (1986). Conditional probability $\gtrsim_0$ on $\mathscr{A} \times \mathscr{A}_0$, where $\mathscr{A}_0$ is a subset of "nonnull" events in $\mathscr{A}$ and $A|B \gtrsim_0 C|D$ means "A given B is at least as probable as C given D," has been axiomatized for representations like

$$A|B \gtrsim_0 C|D \Leftrightarrow \pi(A \cap B)/\pi(B) \geq \pi(C \cap D)/\pi(D)$$

by Koopman (1940), Aczel (1961), Luce (1968), Domotor (1969), Krantz et al. (1971) and Suppes and Zanotti (1982). Lexicographic representations for comparative probability are discussed by Hammond (1987) and Blume et al. (1989) among others.

A few of the representations of the preceding paragraph do not assume that probability is finitely additive, but virtually all presume *monotonicity*. To distinguish this case, I use σ instead of π, so monotonicity says that $\sigma(A) \leq \sigma(B)$ whenever $A \subseteq B$. Monotonic but not necessarily additive probability arises in theories mentioned in Section 8 [Davidson and Suppes (1956), Schmeidler (1989), Gilboa (1987)]. The probability part of Davidson–Suppes also requires $\sigma(A) + \sigma(A^c) = 1$, and Gilboa (1989) has arguments that favor this condition of *complementary additivity* in his and Schmeidler's utility models.

A generalization of additive subjective probability that accommodates intransitivities is discussed by Fishburn (1983c,d).

Finally, it is appropriate to note that subjective probability has received increasing attention in game theory following the contributions of Harsanyi (1967) and Aumann (1974). The recent article by Aumann (1987) provides further access to the topic.

7. Expected utility and subjective probability

In Savage's (1954) approach to decision under uncertainty, the individual is presumed to be uncertain about which state in S is the true state, or *obtains*, and his beliefs regarding this are assumed not to depend on the act chosen although the outcome $f(s) \in X$ depends on both the act f and the state s that obtains. Comments on the latter assumption and other interpretational aspects of Savage's theory are provided by Shafer (1986) and his discussants.

We describe Savage's merger of utility and subjective probability under the structural assumptions that $\mathscr{A} = 2^S$ and $F = X^S$, i.e., the set of acts $\{f, g, \ldots\}$ is the set of all mapping from S into X. Given $>$ on F, it is extended to outcomes in X through constant acts, so $x > y$ is tantamount to $f > g$ when $f(s) \equiv x$ and $g(s) \equiv y$. For convenience, write

$f =_A x$ if $f(s) = x$ for all $s \in A$,

$f =_A g$ if $f(s) = g(s)$ for all $s \in A$,

$f = xAy$ when $f =_A x$ and $f =_{A^c} y$.

An event $A \in \mathscr{A}$ is *null* if $f \sim g$ whenever $f =_{A^c} g$, and the set of null events is denoted by $\mathscr{N}$. For each $A \in \mathscr{A}$, a *conditional preference relation* $>_A$ is defined on

F by

$$f >_A g \text{ if, } \forall f', g' \in F, (f' =_A f, g' =_A g, f' =_{A^c} g') \Rightarrow f' > g'.$$

This reflects Savage's notion that preference between f and g should depend only on states for which $f(s) \neq g(s)$. Similar definitions hold for $\gtrsim_A$ and $\sim_A$.

Savage uses seven axioms. They apply to all $f, g, f', g' \in F$, all $x, y, x', y' \in X$ and all $A, B \subseteq S$.

P1. $>$ on F is a weak order.
P2. $(f =_A f', g =_A g', f =_{A^c} g, f' =_{A^c} g') \Rightarrow (f > g \Leftrightarrow f' > g')$.
P3. $(A \notin \mathcal{N}, f =_A x, g =_A y) \Rightarrow (f >_A g \Leftrightarrow x > y)$.
P4. $(x > y, x' > y') \Rightarrow (xAy > xBy \Leftrightarrow x'Ay' > x'By')$.
P5. $z > w$ for some $z, w \in X$.
P6. $f > g \Rightarrow$ [given x, there is a finite partition of S such that, for every member E of the partition, $(f' =_E x, f' =_{E^c} f) \Rightarrow f' > g$, and $(g' =_E x, g' =_{E^c} g) \Rightarrow f > g'$].
P7. $(\forall s \in A, f >_A g(s)) \Rightarrow f \gtrsim_A g; (\forall s \in A, f(s) >_A g) \Rightarrow f \gtrsim_A g$.

Axiom P2 says that preference should not depend on s for which $f(s) = g(s)$, and P3 (the other half of Savage's *sure thing principle*) ties conditional preference for nonnull events to outcome preference in a natural way. P4 provides consistency for subjective probability since $\pi(A) > \pi(B)$ will correspond to $(x > y, xAy > xBy)$. P5 ensures nontriviality, and P6 (cf. S6 in Section 6) is Savage's Archimedean axiom. P7 is a dominance axiom for extending expectations to arbitrary acts. It is not used in the basic derivation of π or u. Seidenfeld and Schervish (1983) and Toulet (1986) comment further on P7.

Theorem 7.1. *Suppose* P1 *through* P7 *hold for* $>$ *on* F. *Then there is a probability measure* π *on* $\mathcal{A} = 2^S$ *that satisfies the conclusions of Theorem 6.1 when* $A > B$ *corresponds to* $xAy > xBy$ *whenever* $x > y$, *and for which* $\pi(A) = 0 \Leftrightarrow A \in \mathcal{N}$, *and there is a bounded* $u: X \to \mathbb{R}$ *such that*

$$\forall f, g \in F, f > g \Leftrightarrow \int_S u(f(s)) \, d\pi(s) > \int_S u(g(s)) \, d\pi(s).$$

Moreover, such a u is unique up to a positive affine transformation.

Savage (1954) proves all of Theorem 7.1 except for the boundedness of u, which is included in Fishburn's (1970a) proof. In the proof, π is first obtained via Theorem 6.1, u is then established from Theorem 4.1 once it has been shown that A1–A3 hold under the reduction of *simple* (finite-image) acts to lotteries on X [$f \to p$ by means of $p(x) = \pi\{s: f(s) = x\}$], and the subjective expected utility representation for simple acts is then extended to all acts with the aid of P7.

Savage's theory motivated a few dozen other axiomatizations of subjective expected utility and closely related representations. These include simple modifications of the basic Ramsey/Savage theory [Suppes (1956), Davidson and Suppes (1956), Pfanzagl (1967, 1968), Toulet (1986)]; lottery-based theories [Anscombe and Aumann (1963), Pratt et al. (1964, 1965), Fishburn (1970a, 1975c, 1982a)] including one [Fishburn (1975c)] for partial orders; event-conditioned and state-dependent theories that do [Fishburn (1970a, 1973b), Karni et al. (1983), Karni (1985)] or do not [Luce and Krantz (1971), Krantz et al. (1971)] have a lottery feature; and theories that avoid Savage's distinction between events and outcomes [Jeffrey (1965, 1978), Bolker (1967), Domotor (1978)]. Most of these are reviewed extensively in Fishburn (1981), and other theories that depart more substantially from standard treatments of utility or subjective probability are noted in the next section.

The lottery-based approach of Anscombe and Aumann (1963) deserves further comment since it is used extensively in later developments. In its simplest form we let P_0 be the set of all *lotteries* (simple probability distributions) on X. The probabilities used for P_0 are presumed to be generated by a random device independent of S and are sometimes referred to as *extraneous scaling probabilities* in distinction to the subjective probabilities to be derived for S. Then X in Savage's theory is replaced by P_0 so that his act set F is replaced by the set $\mathbf{F} = P_0^S$ of *lottery acts*, each of which assigns a lottery $f(s)$ to each state is S. In those cases where outcomes possible under one state cannot occur under other states, as in state dependent theories [Fishburn (1970a, 1973b), Karni et al. (1983), Karni (1985), Fishburn and LaValle (1987b)], we take $X(s)$ as the outcome set for state s, $P(s)$ as the lotteries on $X(s)$, and $\mathbf{F} = \{\mathbf{f}: S \to \bigcup P(s): \mathbf{f}(s) \in P(s) \text{ for all } s \in S\}$.

We note one representation theorem for the simple case in which $\mathbf{F} = P_0^S$. For $\mathbf{f}, \mathbf{g} \in \mathbf{F}$ and $0 \le \lambda \le 1$, define $\lambda \mathbf{f} + (1 - \lambda)\mathbf{g}$ by $(\lambda \mathbf{f} + (1 - \lambda)\mathbf{g})(s) = \lambda \mathbf{f}(s) + (1 - \lambda)\mathbf{g}(s)$ for all $s \in S$. Then $\mathbf{F}$ is convex and we can apply A1–A3 of Theorem 4.1 to $>$ on $\mathbf{F}$ in place of $>$ on P. Also, in congruence with Savage, take $\mathbf{f} =_A p$ if $\mathbf{f}(s) = p$ for all $s \in A$, define $>$ on P_0 by $p > q$ if $\mathbf{f} > \mathbf{g}$ when $\mathbf{f} =_S p$ and $\mathbf{g} =_S q$, and define $\mathcal{N} \subseteq 2^S$ by $A \in \mathcal{N}$ if $\mathbf{f} \sim \mathbf{g}$ whenever $\mathbf{f} =_{A^c} \mathbf{g}$.

Theorem 7.2. *With $\mathcal{A} = 2^S$ and $\mathbf{F} = P_0^S$, suppose $>$ on $\mathbf{F}$ satisfies A1, A2 and A3 along with the following for all $A \subseteq S$, all $\mathbf{f}, \mathbf{g} \in \mathbf{F}$ and all $p, q \in P_0$:*

A4. $p' > q'$ *for some* $p', q' \in P_0$,

A5. $(A \notin \mathcal{N}, \mathbf{f} =_A p, \mathbf{g} =_A q, \mathbf{f} =_{A^c} \mathbf{g}) \Rightarrow (\mathbf{f} > \mathbf{g} \Leftrightarrow p > q)$.

Then there is a unique probability measure π on $\mathcal{A}$ with $A \in \mathcal{N} \Leftrightarrow \pi(A) = 0$, and a linear functional u on P, unique up to a positive affine transformation, such that, for all simple (finite-image) $\mathbf{f}, \mathbf{g} \in \mathbf{F}$,

$$\mathbf{f} > \mathbf{g} \Leftrightarrow \int_S u(\mathbf{f}(s)) \, d\pi(s) > \int_S u(\mathbf{g}(s)) \, d\pi(s).$$

Axioms A4 and A5 correspond respectively to Savage's P5 and a combination of P2 and P3. Proofs of the theorem are given in Fishburn (1970a, 1982a), which also discuss the extension of the integral representation to more general lottery acts. In particular, if we assume also that $(\mathbf{f}(s) > \mathbf{g}$ for all $s \in S) \Rightarrow \mathbf{f} \gtrsim \mathbf{g}$, and $(\mathbf{f} > \mathbf{g}(s)$ for all $s \in S) \Rightarrow \mathbf{f} \gtrsim \mathbf{g}$, then the representation holds for all $\mathbf{f}, \mathbf{g} \in F$, and u on P_0 is bounded if there is a denumerable partition of S such that $\pi(A) > 0$ for every member of the partition.

As noted in the state-dependent theories referenced above, if the $X(s)$ or $P(s)$ have minimal or no overlap, it is necessary to introduce a second preference relation on lotteries over $\cup X(s)$ without explicit regard to states in order to derive coherent subjective probabilities for states. The pioneering paper by Anscombe and Aumann (1963) in fact used two such preference relations although it also assumed the same outcome set for each state.

Various applications of the theories mentioned in this section, including the Bayesian approach to decision making with experimentation and new information, are included in Good (1950), Savage (1954), Schlaifer (1959), Raiffa and Schlaifer (1961), Raiffa (1968), Howard (1968), DeGroot (1970), LaValle (1978) and Hartigan (1983). Important recent discussions of the Bayesian paradigm are given by Shafer (1986) and Hammond (1988).

An interesting alternative to Savage's approach arises from a tree that describes all possible paths of an unfolding decision process from the present to a terminal node. Different paths through the tree are determined by chance moves, with or without known probabilities, at some branch-point nodes, along with choices by the decision maker at decision nodes. A *strategy* tells what the decision maker would do at each decision node if he were to arrive there. Savage's approach collapses the tree into a normal form in which acts are strategies, states are combinations of possibilities at chance nodes, and consequences, which embody everything of value to the decision maker along a path through the tree, are outcomes. Consequences can be identified as terminal nodes since for each of these there is exactly one path that leads to it.

Hammond (1988) investigates the implications of dynamically consistent choices in decision trees under the consequentialist position that associates all value with terminal nodes. Consistent choices are defined within a family of trees, some of which will be subtrees of larger trees that begin at an intermediate decision node of the parent tree and proceed to the possible terminal nodes from that point onward. A given tree may be a subtree of many parent trees that have very different structures outside the given tree. The central principle of *consistency* says that the decision maker's decision at a decision node in any tree should depend only on the part of the tree that originates at that node. In other words, behavior at the initial node of a given tree should be the same within all parent trees that have the given tree as a subtree. The principle leads naturally to backward recursion for the development of consistent strategies. Beginning at the terminal nodes, decide what to do at each closest decision node, use these choices to decide what to do at each preceding decision node, and continue the process back through the tree to the initial decision node.

Hammond's seminal paper (1988) shows that many interesting results follow from the consistency principle. Among other things, it implies the existence of a revealed preference weak order that satisfies several independence axioms that are similar to axiom A3 in Section 4 and to Savage's sure thing principle. If all chance nodes have known probabilities (no Savage type uncertainty) and a continuity axiom is adopted, consistency leads to the expected utility model. When Savage-type uncertainty is present, consistency does not imply the subjective expected utility model with well defined subjective probabilities, but it does give rise to precursors of that model that have additive or multiplicative utility decompositions over states. However, if additional structure is imposed on consequences that can arise under different states, revealed subjective probabilities do emerge from the additive form in much the same way as in Anscombe and Aumann (1963) or Fishburn (1970a, 1982a).

8. Generalizations and alternatives

In addition to the problems with expected utility that motivated its generalizations in Section 5, other difficulties involving states in the Savage/Ramsey theory have encouraged generalizations of and alternatives to subjective expected utility.

One such difficulty, recognized first by Allais (1953a,b) and Ellsberg (1961), concerns the part of Savage's sure thing principle which says that a preference between f and g that share a common outcome x on event A should not change when x is replaced by y. When this is not accepted, we must either abandon additivity for π or the expectation operation. Both routes have been taken.

Ellsberg's famous example makes the point. One ball is to be drawn randomly from an urn containing 90 balls, 30 of which are red (R) and 60 of which are black (B) and yellow (Y) in unknown proportion. Consider pairs of acts:

$$\begin{cases} f: & \text{win } \$1000 \text{ if } R \text{ drawn} \\ g: & \text{win } \$1000 \text{ if } B \text{ drawn} \end{cases}$$

$$\begin{cases} f': & \text{win } \$1000 \text{ if } R \text{ or } Y \text{ drawn} \\ g': & \text{win } \$1000 \text{ if } B \text{ or } Y \text{ drawn} \end{cases}$$

with \$0 otherwise in each case. Ellsberg claimed, and later experiments have verified [Slovic and Tversky (1974), MacCrimmon and Larsson (1979)], that many people prefer f to g and g' to f' in violation of Savage's principle. The specificity of R relative to B in the first pair (exactly 30 are R) and of B or Y to R or Y in the second pair (exactly 60 are B or Y) seems to motivate these preferences. In other words, many people are averse to ambiguity, or prefer specificity. By Savage's approach, $f > g \Rightarrow \pi(R) > \pi(B)$, and $g' > f' \Rightarrow \pi(B \cup Y) > \pi(R \cup Y)$, hence $\pi(B) > \pi(R)$, so either additivity must be dropped or a different approach to subjective probability (e.g., Allais) must be used and, if additivity is retained, expectation must be abandoned.

Raiffa (1961) critiques Ellsberg (1961) in a manner consistent with Savage. Later discussants of ambiguity include Sherman (1974), Franke (1978), Gärdenfors and Sahlin (1982), Einhorn and Hogarth (1985) and Segal (1987), and further remarks on nonadditive subjective probability are given in Edwards (1962).

Even if additive subjective probabilities are transparent, problems can arise from event dependencies. Consider two acts for one roll of a well-balanced die with probability $\frac{1}{6}$ for each face:

	1	2	3	4	5	6
f	$600	$700	$800	$900	$1000	$500
g	$500	$600	$700	$800	$900	$1000

For some people, $f > g$ because of f's greater payoff for five states; if a 6 obtains, it is just bad luck. Others have $g > f$ because they dread the thought of choosing f and losing out on the $500 difference should a 6 occur. Both preferences violate Savage's theory, which requires $f \sim g$ since both acts reduce to the same lottery on outcomes. The role of interlinked outcomes and notions of regret are highlighted in Tversky (1975) and developed further by Loomes and Sugden (1982, 1987) and Bell (1982).

Theories that accommodate one or both of the preceding effects can be classified under three dichotomies: additive/nonadditive subjective probability; transitive/nontransitive preference; and whether they use Savage acts or lottery acts. We consider additive cases first.

Allais (1953a, 1953b, 1979, 1988) rejects Savage's theory not only because of its independence axioms and non-Bernoullian assessment of utility but also because Savage ties subjective probability to preference. For Allais, additive subjective probabilities are assessed independently of preference by comparing uncertain events to other events that correspond to drawing balls from an urn. No payoffs or outcomes are involved. Once π has been assessed, it is used to reduce acts to lotteries on outcomes. Allais then follows his approach for P outlined in Section 5. A more complete description of Allais versus Savage is given in Fishburn (1987).

Another additive theory that uses Bernoullian riskless utility but departs from Allais by rejecting transitivity and the reduction procedure is developed by Loomes and Sugden (1982, 1987) and Bell (1982). Their general representation is

$$f > g \Leftrightarrow \int_S \phi(f(s), g(s)) \, d\pi(s) > 0,$$

where ϕ on $X \times X$ is skew-symmetric and includes a factor for regret/rejoicing due to interlinked outcomes. Depending on ϕ, this accommodates either $f > g$ or

$g > f$ for the preceding die example, and it allows cyclic preferences. It does not, however, account for Ellsberg's ambiguity phenomenon.

Fishburn (1989a) axiomatizes the preceding representation under Savage's approach thereby providing an interpretation without direct reference to regret or to riskless utility. The main change to Savage's axioms is the following weakening of P1:

> P1*. $>$ on F is asymmetric and, for all $x, y \in X$, the
> restriction of $>$ to $\{f \in F: f(S) \subseteq \{x, y\}\}$ is a weak order.

In addition, we also strengthen the Archimedean axiom P6 and add a dominance principle that is implied by Savage's axioms. This yields the preceding representation for all simple acts. Its extension to all acts is discussed in the appendix of Fishburn (1989a).

A related axiomatization for the lottery-acts version of the preceding model, with representation

$$\mathbf{f} > \mathbf{g} \Leftrightarrow \int_S \phi(\mathbf{f}(s), \mathbf{g}(s)) \, d\pi(s) > 0,$$

is given by Fishburn (1984c) and Fishburn and LaValle (1987a). In this version ϕ is an SSB functional on $P_0 \times P_0$. This model can be viewed as the SSB generalization of the one in Theorem 7.2. Fishburn and LaValle (1988) also note that if weak order is adjoined to the model then it reduces to the model of Theorem 7.2 provided that $0 < \pi(A) < 1$ for some A.

The first general alternative to Savage's theory that relaxes additivity to monotonicity for subjective probability $[A \subseteq B \Rightarrow \sigma(A) \leqslant \sigma(B)]$ was developed by Schmeidler (1989) in the lottery-acts mode, following the earlier finite-sets theory of Davidson and Suppes (1956). Schmeidler's representation is

$$\mathbf{f} > \mathbf{g} \Leftrightarrow \int_S u(\mathbf{f}(s)) \, d\sigma(s) > \int_S u(\mathbf{g}(s)) \, d\sigma(s),$$

where u is a linear functional on P_0, unique up to a positive affine transformation, and σ is a unique monotonic probability measure on 2^S. Integration in the representation is Choquet integration [Choquet (1953), Schmeidler (1986)] defined by

$$\int_S w(s) \, d\sigma(s) = \int_{c=0}^{\infty} \sigma(\{s: w(s) \geqslant c\}) \, dc - \int_{c=-\infty}^{0} [1 - \sigma(\{s: w(s) \geqslant c\})] \, dc.$$

Schmeidler's axioms involve a weakening of the independence axiom A2 (for $>$ on $\mathbf{F}$) to avoid additivity, but A1 and A3 are retained. The only other assumptions needed are nontriviality and the dominance axiom $(\forall s \in S, \mathbf{f}(s) \gtrsim \mathbf{g}(s)) \Rightarrow \mathbf{f} \gtrsim \mathbf{g}$. His representation accommodates Ellsberg's phenomenon, but its linearity for u does not account for Allais-type violations of independence.

The Savage-acts version of Schmeidler's model is axiomatized by Gilboa (1987).

In Gilboa's representation,

$$f > g \Leftrightarrow \int_S u(f(s)) \, \mathrm{d}\sigma(s) > \int_S u(g(s)) \, \mathrm{d}\sigma(s),$$

u is a bounded functional on X, unique up to a positive affine transformation, and σ is monotonic and unique. His axioms make substantial changes to Savage's, and because of this his proof of the representation requires new methods. As noted earlier, Gilboa (1989) argues for the complementary additivity property $\sigma(A) + \sigma(A^c) = 1$ although it is not implied by his axioms. In particular, if equivalence is required between maximization of $\int u \, \mathrm{d}\sigma$ and minimization of $\int (-u) \, \mathrm{d}\sigma$, complementary additivity follows from Choquet integration. He also suggests that a consistent theory for *conditional* probability is possible in the Schmeidler–Gilboa framework only when σ is fully additive.

Additional results for the Schmeidler and Gilboa representations are developed by Wakker (1989), and other representations with nonadditive subjective probability are discussed by Luce and Narens (1985), Luce (1986, 1988) and Fishburn (1988a).

References

Aczel, J. (1961) 'Über die Begründung der Additions- und Multiplikationsformeln von bedingten Wahrscheinlichkeiten', *A Magyar Tudományos Akadémia Matematikai Kutató Intézetének Közleményei*, **6**: 110–122.
Adams, E.W. (1965) 'Elements of a theory of inexact measurement', *Philosophy of Science*, **32**: 205–228.
Allais, M. (1953a) 'Le comportement de l'homme rationnel devant le risque: Critique des postulats et axiomes de l'école américaine', *Econometrica*, **21**: 503–546.
Allais, M. (1953b) 'Fondements d'une théorie positive des choix comportant un risque et critique des postulats et axiomes de l'école américaine', Colloques Internationaux du Centre National de la Recherche Scientifique. XL, *Econométrie*: pp. 257–332; Translated and augmented as 'The foundations of a positive theory of choice involving risk and a criticism of the postulates and axioms of the American school', in: Allais and Hagen (1979).
Allais, M. (1979) 'The so-called Allais paradox and rational decisions under uncertainty', in: Allais and Hagen (1979).
Allais, M. (1988) 'The general theory of random choices in relation to the invariant cardinal utility function and the specific probability function', in: B.R. Munier, ed., *Risk, decision and rationality*. Dordrecht: Reidel.
Allais, M. and O. Hagen, eds. (1979) *Expected utility hypotheses and the Allais paradox*. Dordrecht, Holland: Reidel.
Allen, B. (1987) 'Smooth preferencs and the approximate expected utility hypothesis', *Journal of Economic Theory*, **41**: 340–355.
Alt, F. (1936) 'Über die Messbarkeit des Nutzens', *Zeitschrift für Nationaloekonomie*, **7**: 161–169; English translation: 'On the measurement of utility', in: Chipman, Hurwicz, Richter and Sonnenschein (1971).
Anscombe, F.J. and R.J. Aumann (1963) 'A definition of subjective probability', *Annals of Mathematical Statistics*, **34**: 199–205.
Armstrong, W.E. (1939) 'The determinateness of the utility function', *Economic Journal*, **49**: 453–467.
Armstrong, W.E. (1948) 'Uncertainty and the utility function', *Economic Journal*, **58**: 1–10.
Armstrong, W. E. (1950) 'A note on the theory of consumer's behaviour', *Oxford Economic Papers*, **2**: 119–122.
Arrow, K.J. (1958) 'Bernoulli utility indicators for distributions over arbitrary spaces', *Technical Report 57*, Department of Economics, Stanford University.

Arrow, K.J. (1959) 'Rational choice functions and orderings', *Economica*, **26**: 121–127.
Arrow, K.J. (1974) *Essays in the theory of risk bearing*. Amsterdam: North-Holland.
Aumann, R.J. (1962) 'Utility theory without the completeness axiom', *Econometrica*, **30**: 445–462; **32** (1964): 210–212.
Aumann, R.J. (1964) 'Subjective programming', in: M.W. Shelly and G.L. Bryan, eds., *Human judgments and optimality*. New York: Wiley.
Aumann, R.J. (1974) 'Subjectivity and correlation in randomized strategies', *Journal of Mathematical Economics*, **1**: 67–96.
Aumann, R.J. (1987) 'Correlated equilibrium as an expression of Bayesian rationality', *Econometrica*, **55**: 1–18.
Barbera, S. and P.K. Pattanaik (1986) 'Falmagne and the rationalizability of stochastic choices in terms of random orderings', *Econometrica*, **54**: 707–715.
Baumol, W.J. (1958) 'The cardinal utiliy which is ordinal', *Economic Journal*, **68**: 665–672.
Bawa, V.S. (1982) 'Stochastic dominance: a research bibliography', *Management Science*, **28**: 698–712.
Bayes, T. (1763) 'An essay towards solving a problem in the doctrine of chances', *Philosophical Transactions of the Royal Society*, **53**: 370–418; Reprinted in: W.E. Deming, ed., *Facsimiles of two papers of Bàyes*. Washington, DC: Department of Agriculture, 1940.
Becker, G.M., M.H. DeGroot and J. Marschak (1963) 'Stochastic models of choice behavior', *Behavioral Science*, **8**: 41–55.
Becker, J.L. and R.K. Sarin (1987) 'Lottery-dependent utility', *Management Science*, **33**: 1367–1382.
Bell, D. (1982) 'Regret in decision making under uncertainty', *Operations Research*, **30**: 961–981.
Bell, D.E. (1986) 'Double-exponential utility functions', *Mathematics of Operations Research*, **11**: 351–361.
Bell, D.E. (1987) 'Multilinear representations for ordinal utility functions', *Journal of Mathematical Psychology*, **31**: 44–59.
Bernoulli, D. (1738) 'Specimen theoriae novae de mensura sortis', *Commentarii Academiae Scientiarum Imperialis Petropolitanae*, **5**: 175–192; Translated by L. Sommer as 'Exposition of a new theory on the measurement of risk', *Econometrica*, **22**(1954): 23–36.
Bernstein, S.N. (1917) 'On the axiomatic foundations of probability theory' [in Russian], *Soobshcheniya i Protokoly Khar'kovskago Matematicheskago Obshchestra*, **15**: 209–274.
Blackwell, D. and M.A. Girshick (1954) *Theory of games and statistical decisions*. New York: Wiley.
Blume, L., A. Brandenburger and E. Dekel (1989) 'An overview of lexicographic choice under uncertainty', *Annals of Operations Research*, **19**: 231–246.
Bolker, E.D. (1966) 'Functions resembling quotients of measures', *Transactions of the American Mathematical Society*, **124**: 292–312.
Bolker, E.D. (1967) 'A simultaneous axiomatization of utility and subjective probability', *Philosophy of Science*, **34**: 333–340.
Bridges, D.S. (1983) 'A numerical representation of preferences with intransitive indifference', *Journal of Mathematical Economics*, **11**: 25–42.
Bridges, D.S. (1986) 'Numerical representation of interval orders on a topological space', *Journal of Economic Theory*, **38**: 160–166.
Burness, H.S. (1973) 'Impatience and the preference for advancement in the timing of satisfactions', *Journal of Economic Theory*, **6**: 495–507.
Burness, H.S. (1976) 'On the rolc of separability assumptions in determining impatience implications', *Econometrica*, **44**: 67–78.
Campello de Souza, F.M. (1983) 'Mixed models, random utilities, and the triangle inequality', *Journal of Mathematical Psychology*, **27**: 183–200.
Cantor, G. (1895) 'Beiträge zur Begründung der transfiniten Mengenlehre', *Mathematische Annalen*, **46**: 481–512; **49**(1897): 207–246; Translated as *Contributions to the founding of the theory of transfinite numbers*. New York: Dover.
Chateauneuf, A. (1985) 'On the existence of a probability measure compatible with a total preorder on a Boolean algebra', *Journal of Mathematical Economics*, **14**: 43–52.
Chateauneuf, A. and J.-Y. Jaffray (1984) 'Archimedean qualitative probabilities', *Journal of Mathematical Psychology*, **28**: 191–204.
Chew, S.H. (1982) 'A mixture set axiomatization of weighted utility theory', *Discussion Paper 82–4*, College of Business and Public Administration, University of Arizona.
Chew, S.H. (1983) 'A generalization of the quasilinear mean with applications to the measurement of

income inequality and decision theory resolving the Allais paradox', *Econometrica*, **51**: 1065–1092.
Chew, S.H. (1984) 'An axiomatization of the rank dependent quasilinear mean generalizing the Gini mean and the quasilinear mean', *mimeographed*, Department of Political Economy, Johns Hopkins University.
Chew, S.H. (1985) 'From strong substitution to very weak substitution: mixture-monotone utility theory and semi-weighted utility theory', *mimeographed*, Department of Political Economy, Johns Hopkins University.
Chew, S.H., E. Karni and Z. Safra (1987) 'Risk aversion in the theory of expected utility with rank-dependent probabilities', *Journal of Economic Theory*, **42**: 370–381.
Chew, S.H. and K.R. MacCrimmon (1979) 'Alpha-nu choice theory: a generalization of expected utility theory', *Working Paper 669*, Faculty of Commerce and Business Administration, University of British Columbia.
Chipman, J.S. (1960) 'The foundations of utility', *Econometrica*, **28**: 193–224.
Chipman, J.S. (1971) 'Consumption theory without transitive indifference', in: Chipman, Hurwicz, Richter and Sonnenschein (1971).
Chipman, J.S., L. Hurwicz, M.K. Richter and H.F. Sonnenschein, eds. (1971) *Preferences, utility, and demand*. New York: Harcourt Brace Jovanovich.
Choquet, G. (1953) 'Theory of capacities', *Annales de l'Institut Fourier*, **5**: 131–295.
Chuaqui, R. and J. Malitz (1983) 'Preorderings compatible with probability measures', *Transactions of the American Mathematical Society*, **279**: 811–824.
Cohen, M. and J.C. Falmagne (1990) 'Random utility representation of binary choice probabilities: a new class of necessary conditions', *Journal of Mathematical Psychology*, **34**: 88–94.
Cohen, M., J.Y. Jaffray and T. Said (1985) 'Individual behavior under risk and under uncertainty: an experimental study', *Theory and Decision*, **18**: 203–228.
Corbin, R. and A.A.J. Marley (1974) 'Random utility models with equality: an apparent, but not actual, generalization of random utility models', *Journal of Mathematical Psychology*, **11**: 274–293.
Cozzens, M.B. and F.S. Roberts (1982) 'Double semiorders and double indifference graphs', *SIAM Journal on Algebraic and Discrete Methods*, **3**: 566–583.
Dagsvik, J.K. (1983) 'Discrete dynamic choice: an extension of the choice models of Thurstone and Luce', *Journal of Mathematical Psychology*, **27**: 1–43.
Davidson, D. and P. Suppes (1956) 'A finitistic axiomatization of subjective probability and utility', *Econometrica*, **24**: 264–275.
Debreu, G. (1958) 'Stochastic choice and cardinal utility', *Econometrica*, **26**: 440–444.
Debreu, G. (1959) *Theory of value*. New York: Wiley.
Debreu, G. (1960) 'Topological methods in cardinal utility theory', in: K.J. Arrow, S. Karlin and P. Suppes, eds., *Mathematical methods in the social sciences, 1959*. Stanford: Stanford University Press.
Debreu, G. (1964) 'Continuity properties of Paretian utility', *International Economic Review*, **5**: 285–293.
Debreu, G. and T.C. Koopmans (1982) 'Additively decomposed quasiconvex functions', *Mathematical Programming*, **24**: 1–38.
de Finetti, B. (1931) 'Sul significato soggettivo della probabilità', *Fundamenta Mathematicae*, **17**: 298–329.
de Finetti, B. (1937) 'La prévision: ses lois logiques, ses sources subjectives', *Annales de l'Institut Henri Poincaré*, **7**: 1–68; Translated by H.E. Kyburg as 'Foresight: its logical laws, its subjective sources', in: Kyburg and Smokler (1964).
DeGroot, M.H. (1970) *Optimal statistical decisions*. New York: McGraw-Hill.
Dekel, E. (1986) 'An axiomatic characterization of preferences under uncertainty: weakening the independence axiom', *Journal of Economic Theory*, **40**: 304–318.
Dempster, A.P. (1967) 'Upper and lower probabilities induced by a multivalued mapping', *Annals of Mathematical Statistics*, **38**: 325–339.
Dempster, A.P. (1968) 'A generalization of Bayesian inference', *Journal of the Royal Statistical Society, Series B*, **30**: 205–247.
Diamond, P.A. (1965) 'The evaluation of infinite utility streams', *Econometrica*, **33**: 170–177.
Doignon, J.-P. (1984) 'Generalizations of interval orders', in: E. Degreef and J. Van Buggenhaut, eds., *Trends in mathematical psychology*. Amsterdam: North-Holland.
Doignon, J.-P. (1987) 'Threshold representations of multiple semiorders', *SIAM Journal on Algebraic and Discrete Methods*, **8**: 77–84.

Doignon, J.-P., A. Ducamp and J.-C. Falmagne (1984) 'On realizable biorders and the biorder dimension of a relation', *Journal of Mathematical Psychology*, **28**: 73–109.

Doignon, J.-P., B . Monjardet, M. Roubens and Ph. Vincke (1986) 'Biorder families, valued relations, and preference modelling', *Journal of Mathematical Psychology*, **30**: 435–480.

Domotor, Z. (1969) 'Probabilistic relational structures and their applications', *Technical Report 144*, Institute for Mathematical Studies in the Social Sciences, Stanford University.

Domotor, Z. (1978) 'Axiomatization of Jeffrey utilities', *Synthese*, **39**: 165–210.

Domotor, Z . and J. Stelzer (1971) 'Representation of finitely additive semiordered qualitative probability structures', *Journal of Mathematical Psychology*, **8**: 145–158.

Drèze, J. and F. Modigliani (1972) 'Consumption decisions under uncertainty', *Journal of Economic Theory*, **5**: 308–335.

Edgeworth, F.Y. (1881) *Mathematical psychics*. London: Kegan Paul.

Edwards, W. (1953) 'Probability-preferences in gambling', *American Journal of Psychology*, **66**: 349–364.

Edwards, W. (1954a) 'The theory of decision making', *Psychological Bulletin*, **51**: 380–417.

Edwards, W. (1954b) 'The reliability of probability preferences', *American Journal of Psychology*, **67**: 68–95.

Edwards, W. (1954c) 'Probability preferences among bets with differing expected values', *American Journal of Psychology*, **67**: 56–67.

Edwards, W. (1962) 'Subjective probabilities inferred from decisions', *Psychological Review*, **69**: 109–135.

Eilenberg, S. (1941) 'Ordered topological spaces', *American Journal of Mathematics*, **63**: 39–45.

Einhorn, H.J. and R.M. Hogarth (1985) 'Ambiguity and uncertainty in probabilistic inference', *Psychological Review*, **92**: 433–461.

Ellsberg, D. (1961) 'Risk, ambiguity, and the Savage axioms', *Quarterly Journal of Economics*, **75**: 643–669.

Farquhar, P.H. (1975) 'A fractional hypercube decomposition theorem for multiattribute utility functions', *Operations Research*, **23**: 941–967.

Farquhar, P.H. (1977) 'A survey of multiattribute utility theory and applications', *TIMS Studies in the Management Sciences*, **6**: 59–89.

Farquhar, P.H. (1978) 'Interdependent criteria in utility analysis', in: S. Zionts, ed., *Multiple criteria problem solving*. Berlin: Springer-Verlag.

Fine, T. (1971) 'A note on the exisence of quantitative probability', *Annals of Mathematical Statistics*, **42**: 1182–1186.

Fine, T. (1973) *Theories of probability*. New York: Academic Press.

Fishburn, P.C. (1964) *Decision and value theory*. New York: Wiley.

Fishburn, P.C. (1965) 'Independence in utility theory with whole product sets', *Operations Research*, **13**: 28–45.

Fishburn, P.C. (1967) 'Bounded expected utility', *Annals of Mathematical Statistics*, **38**: 1054–1060.

Fishburn, P.C. (1969) 'Weak qualitative probability on finite sets', *Annals of Mathematical Statistics*, **40**: 2118–2126.

Fishburn, P.C. (1970a) *Utility theory for decision making*. New York: Wiley.

Fishburn, P.C. (1970b) 'Intransitive indifference in preference theory: a survey', *Operations Research*, **18**: 207–228.

Fishburn, P.C. (1971a) 'A study of lexicographic expected utility', *Management Science*, **17**: 672–678.

Fishburn, P.C. (1971b) 'One-way expected utility with finite consequence spaces', *Annals of Mathematical Statistics*, **42**: 572–577.

Fishburn, P.C. (1972a) 'Interdependent preferences on finite sets', *Journal of Mathematical Psychology*, **9**: 225–236.

Fishburn, P.C. (1972b) 'Alternative axiomatizations of one-way expected utility', *Annals of Mathematical Statistics*, **43**: 1648–1651.

Fishburn, P.C. (1973a) 'Binary choice probabilities: on the varieties of stochastic transitivity', *Journal of Mathematical Psychology*, **10**: 327–352.

Fishburn, P.C. (1973b) 'A mixture-set axiomatization of conditional subjective expected utility', *Econometrica*, **41**: 1–25.

Fishburn, P.C. (1974) 'Lexicographic orders, utilities and decision rules: a survey', *Management Science*, **20**: 1442–1471.

Fishburn, P.C. (1975a) 'Unbounded expected utility', *Annals of Statistics*, **3**: 884–896.

Fishburn, P.C. (1975b) 'Weak comparative probability on infinite sets', *Annals of Probability*, **3**: 889–893.
Fishburn, P.C. (1975c) 'A theory of subjective expected utility with vague preferences', *Theory and Decision*, **6**: 287–310.
Fishburn, P.C. (1976a) 'Cardinal utility: an interpretive essay', *International Review of Economics and Business*, **23**: 1102–1114.
Fishburn, P.C. (1976b) 'Axioms for expected utility in n-person games', *International Journal of Game Theory*, **5**: 137–149.
Fishburn, P.C. (1977) 'Multiattribute utilities in expected utility theory', in: D.E. Bell, R.L. Keeney and H. Raiffa, eds., *Conflicting objectives in decisions*. New York: Wiley.
Fishburn, P.C. (1978a) 'Ordinal preferences and uncertain lifetimes', *Econometrica*, **46**: 817–833.
Fishburn, P.C. (1978b) 'A survey of multiattribute/multicriterion evaluation theories', in: S. Zionts, ed., *Multiple criteria problem solving*. Berlin: Springer-Verlag.
Fishburn, P.C. (1980a) 'Lexicographic additive differences', *Journal of Mathematical Psychology*, **21**: 191–218.
Fishburn, P.C. (1980b) 'Continua of stochastic dominance relations for unbounded probability distributions', *Journal of Mathematical Economics*, **7**: 271–285.
Fishburn, P.C. (1981) 'Subjective expected utility: a review of normative theories', *Theory and Decision*, **13**: 139–199.
Fishburn, P.C. (1982a) *The foundations of expected utility*. Dordrecht, Holland: Reidel.
Fishburn, P.C. (1982b) 'Nontransitive measurable utility', *Journal of Mathematical Psychology*, **26**: 31–67.
Fishburn, P.C. (1983a) 'Transitive measurable utility', *Journal of Economic Theory*, **31**: 293–317.
Fishburn, P.C. (1983b) 'Utility functions on ordered convex sets', *Journal of Mathematical Economics*, **12**: 221–232.
Fishburn, P.C. (1983c) 'A generalization of comparative probability on finite sets', *Journal of Mathematical Psychology*, **27**: 298–310.
Fishburn, P.C. (1983d) 'Ellsberg revisited: a new look at comparative probability', *Annals of Statistics*, **11**: 1047–1059.
Fishburn, P.C. (1984a) 'Dominance in SSB utility theory', *Journal of Economic Theory*, **34**: 130–148.
Fishburn, P.C. (1984b) 'Elements of risk analysis in non-linear utility theory', *INFOR*, **22**: 81–97.
Fishburn, P.C. (1984c) 'SSB utility theory and decision-making under uncertainty', *Mathematical Social Sciences*, **8**: 253–285.
Fishburn, P.C. (1985a) *Interval orders and interval graphs*. New York: Wiley.
Fishburn, P.C. (1985b) 'Nontransitive preference theory and the preference reversal phenomenon', *International Review of Economics and Business*, **32**: 39–50.
Fishburn, P.C. (1986a) 'Ordered preference differences without ordered preferences', *Synthese*, **67**: 361–368.
Fishburn, P.C. (1986b) 'The axioms of subjective probability', *Statistical Science*, **1**: 345–355.
Fishburn, P.C. (1986c) 'Interval models for comparative probability on finite sets', *Journal of Mathematical Psychology*, **30**: 221–242.
Fishburn, P.C. (1987) 'Reconsiderations in the foundations of decision under uncertainty', *Economic Journal*, **97**: 825–841.
Fishburn, P.C. (1988a) 'Uncertainty aversion and separated effects in decision making under uncertainty', in: J. Kacprzyk and M. Fedrizzi, eds., *Combining fuzzy imprecision with probabilistic uncertainty in decision making*. Berlin: Springer-Verlag.
Fishburn, P.C. (1988b) *Nonlinear preference and utility theory*. Baltimore: Johns Hopkins University Press.
Fishburn, P.C. (1989a) 'Nontransitive measurable utility for decision under uncertainty', *Journal of Mathematical Economics*, **18**: 187–207.
Fishburn, P.C. (1989b) 'Retrospective on the utility theory of von Neumann and Morgenstern', *Journal of Risk and Uncertainty*, **2**: 127–158.
Fishburn, P.C. (1991) 'Nontransitive additive conjoint measurement', *Journal of Mathematical Psychology*, **35**: 1–40.
Fishburn, P.C. and J.-C. Falmagne (1989) 'Binary choice probabilities and rankings', *Economics Letters*, **31**: 113–117.
Fishburn, P.C. and P.H. Farquhar (1982) 'Finite-degree utility independence', *Mathematics of Operations Research*, **7**: 348–353.

Fishburn, P.C. and G.A. Kochenberger (1979) 'Two-piece von Neumann–Morgenstern utility functions', *Decision Sciences*, **10**: 503–518.
Fishburn, P.C. and I.H. LaValle (1987a) 'A nonlinear, nontransitive and additive-probability model for decisions under uncertainty', *Annals of Statistics*, **15**: 830–844.
Fishburn, P.C. and I.H . LaValle (1987b) 'State-dependent SSB utility', *Economics Letters*, **25**: 21–25.
Fishburn, P.C. and I.H. LaValle (1988) 'Transitivity is equivalent to independence for states-additive SSB utilities', *Journal of Economic Theory*, **44**: 202–208.
Fishburn, P.C., H. Marcus-Roberts and F.S. Roberts (1988) 'Unique finite difference measurement', *SIAM Journal on Discrete Mathematics*, **1**: 334–354.
Fishburn, P.C. and A.M. Odlyzko (1989) 'Unique subjective probability on finite sets', *Journal of the Ramanujan Mathematical Society*, **4**: 1–23.
Fishburn, P.C. and F.S. Roberts (1989) 'Axioms for unique subjective probability on finite sets', *Journal of Mathematical Psychology*, **33**: 117–130.
Fishburn, P.C. and R.W. Rosenthal (1986) 'Noncooperative games and nontransitive preferences', *Mathematical Social Sciences*, **12**: 1–7.
Fishburn, P.C. and A. Rubinstein (1982) 'Time preference', *International Economic Review*, **23**: 677–694.
Fishburn, P.C. and R.G. Vickson (1978) 'Theoretical foundations of stochastic dominance', in: Whitmore and Findlay (1978).
Fisher, I. (1892) 'Mathematical investigations in the theory of values and prices', *Transactions of Connecticut Academy of Arts and Sciences*, **9**: 1–124.
Flood, M.M. (1951, 1952) 'A preference experiment', *Rand Corporation Papers P-256, P-258 and P-263*.
Foldes, L. (1972) 'Expected utility and continuity', *Review of Economic Studies*, **39**: 407–421.
Franke, G. (1978) 'Expected utility with ambiguous probabilities and 'irrational' parameters', *Theory and Decision*, **9**: 267–283.
French, S. (1982) 'On the axiomatisation of subjective probabilities', *Theory and Decision*, **14**: 19–33.
Friedman, M. and L.J. Savage (1948) 'The utility analysis of choices involving risk', *Journal of Political Economy*, **56**: 279–304.
Friedman, M. and L.J. Savage (1952) 'The expected-utility hypothesis and the measurability of utility', *Journal of Political Economy*, **60**: 463–474.
Frisch, R. (1926) 'Sur un problème d'économie pure', *Norsk Matematisk Forenings Skrifter*, **16**: 1–40; English translation: 'On a problem in pure economics', in: Chipman, Hurwicz, Richter and Sonnenschein (1971).
Gärdenfors, P. and N.-E. Sahlin (1982) 'Unreliable probabilities, risk taking, and decision making', *Synthese*, **53**: 361–386.
Georgescu-Roegen, N. (1936) 'The pure theory of consumer's behavior', *Quarterly Journal of Economics*, **50**: 545–593; Reprinted in Georgescu-Roegen (1966).
Georgescu-Roegen, N. (1958) 'Threshold in choice and the theory of demand', *Econometrica*, **26**: 157–168; Reprinted in Georgescu-Roegen (1966).
Georgescu-Roegen, N. (1966) *Analytical economics: issues and problems*. Cambridge: Harvard University Press.
Gilboa, I. (1985) 'Subjective distortions of probabilities and non-additive probabilities', *Working Paper 18-85*, Foerder Institute for Economic Research, Tel-Aviv University.
Gilboa, I. (1987) 'Expected utility with purely subjective non-additive probabilities', *Journal of Mathematical Economics*, **16**: 65–88.
Gilboa, I. (1989) 'Duality in non-additive expected utility theory', *Annals of Operations Research*, **19**: 405–414.
Gilboa, I. (1990) 'A necessary but insufficient condition for the stochastic binary choice problem', *Journal of Mathematical Psychology*, **34**: 371–392.
Goldstein, W. and H.J. Einhorn (1987) 'Expression theory and the preference reversal phenomena', *Psychological Review*, **94**: 236–254.
Good, I.J. (1950) *Probability and the weighing of evidence*. London: Griffin.
Gorman, W.M. (1968) 'Conditions for additive separability', *Econometrica*, **36**: 605–609.
Gossen, H.H. (1854) Entwickelung der Gesetze des menschlichen Verkehrs, und der daraus fliessenden Regeln fur menschlichcs Handeln. Braunschweig: Vieweg and Sohn.
Grandmont, J.-M. (1972) 'Continuity properties of a von Neumann–Morgenstern utility', *Journal of Economic Theory*, **4**: 45–57.

Grether, D.M. and C.R. Plott (1979) 'Economic theory of choice and the preference reversal phenomenon', *American Economic Review*, **69**: 623–638.
Hadar, J. and W.R. Russell (1969) 'Rules for ordering uncertain prospects', *American Economic Review*, **59**: 25–34.
Hadar, J. and W.R. Russell (1971) Stochastic dominance and diversification', *Journal of Economic Theory*, **3**: 288–305.
Hagen, O. (1972) 'A new axiomatization of utility under risk', *Teorie A Metoda*, **4**: 55–80.
Hagen, O. (1979) 'Towards a positive theory of preferences under risk', in: Allais and Hagen (1979).
Hammond, P.J. (1976a) 'Changing tastes and coherent dynamic choice', *Review of Economic Studies*, **43**: 159–173.
Hammond, P.J. (1976b) 'Endogenous tastes and stable long-run choice', *Journal of Economic Theory*, **13**: 329–340.
Hammond, P.J. (1987) 'Extended probabilities for decision theory and games', mimeographed, Department of Economics, Stanford University.
Hammond, P.J. (1988) 'Consequentialist foundations for expected utility', *Theory and Decision*, **25**: 25–78.
Hanoch, G. and H. Levy (1969) 'The efficiency analysis of choices involving risk', *Review of Economic Studies*, **36**: 335–346.
Hardy, G.H., J.E. Littlewood and G. Polya (1934) *Inequalities*. Cambridge: Cambridge University Press.
Harsanyi, J.C. (1967) 'Games with incomplete information played by Bayesian players, Parts I, II, III', *Management Science*, **14**: 159–182, 320–334, 486–502.
Hartigan, J.A. (1983) *Bayes theory*. New York: Springer-Verlag.
Hausner, M. (1954) 'Multidimensional utilities', in: R.M. Thrall, C.H. Coombs and R.L. Davis, eds., *Decision processes*. New York: Wiley.
Herden, G. (1989) 'On the existence of utility functions', *Mathematical Social Sciences*, **17**: 297–313.
Hershey, J.C. and P.J.H. Schoemaker (1980) 'Prospect theory's reflection hypothesis: a critical examination', *Organizational Behavior and Human Performance*, **25**: 395–418.
Herstein, I.N. and J. Milnor (1953) 'An axiomatic approach to measurable utility', *Econometrica*, **21**: 291–297.
Hicks, J.R. and R.G.D. Allen (1934) 'A reconsideration of the theory of value: I; II', *Economica*, **1**: 52–75; 196–219.
Houthakker, H.S. (1950) 'Revealed preference and the utility function', *Economica*, **17**: 159–174.
Houthakker, H.S. (1961) 'The present state of consumption theory', *Econometrica*, **29**: 704–740.
Howard, R.A. (1968) 'The foundations of decision analysis', *IEEE Transactions on System Science and Cybernetics*, **SSC-4**: 211–219.
Hurwicz, L. and M.K. Richter (1971) 'Revealed preference without demand continuity assumptions', in: Chipman et al. (1971).
Jeffrey, R.C. (1965) *The logic of decision*. New York: McGraw-Hill.
Jeffrey, R.C. (1978) 'Axiomatizing the logic of decision', in: C.A. Hooker, J.J. Leach and E.F. McClennen, eds., *Foundations and Applications of Decision Theory*, Vol. I. *Theoretical foundations*. Dordrecht, Holland: Reldel.
Jensen, N.E. (1967) 'An introduction to Bernoullian utility theory. I. Utility functions', *Swedish Journal of Economics*, **69**: 163–183.
Jevons, W.S. (1871) *The theory of political economy*. London: Macmillan.
Jones, R.A. and J.M. Ostroy (1984) 'Flexibility and uncertainty', *Review of Economic Studies*, **51**: 13–32.
Kahneman, D., P. Slovic and A. Tversky, eds. (1982) *Judgement under uncertainty: heuristics and biases*. Cambridge: Cambridge University Press.
Kahneman, D. and A. Tversky (1972) 'Subjective probability: a judgment of representativeness', *Cognitive Psychology*, **3**: 430–454.
Kahneman, D. and A. Tversky (1979) 'Prospect theory: an analysis of decision under risk', *Econometrica*, **47**: 263–291.
Kannai, Y. (1963) 'Existence of a utility in infinite dimensional partially ordered spaces', *Israel Journal of Mathematics*, **1**: 229–234.
Karamata, J. (1932) 'Sur une inegalité relative aux fonctions convexes', *Publications Mathématiques de l'Université de Belgrade*, **1**: 145–148.
Karmarkar, U.S. (1978) 'Subjectively weighted utility: a descriptive extension of the expected utility model', *Organizational Behavior and Human Performance*, **21**: 61–72.

Karni, E. (1985) *Decision making under uncertainty: the case of state-dependent preferences*. Cambridge: Harvard University Press.
Karni, E., D. Schmeidler and K. Vind (1983) 'On state dependent preferences and subjective probabilities', *Econometrica*, **51**: 1021–1032.
Kauder, E. (1965) *A history of marginal utility theory*. Princeton: Princeton University Press.
Keeney, R.L. (1968) 'Quasi-separable utility functions', *Naval Research Logistics Quarterly*, **15**: 551–565.
Keeney, R.L. and H. Raiffa (1976) *Decisions with multiple objectives: preferences and value tradeoffs*. New York: Wiley.
Keynes, J.M. (1921) *A treatise on probability*. New York: Macmillan. Torchbook edn., 1962.
Koopman, B.O. (1940) 'The axioms and algebra of intuitive probability', *Annals of Mathematics*, **41**: 269–292.
Koopmans, T.C. (1960) 'Stationary ordinal utility and impatience', *Econometrica*, **28**: 287–309.
Koopmans, T.C., P.A. Diamond and R.E. Williamson (1964) 'Stationary utility and time perspective', *Econometrica*, **32**: 82–100.
Kraft, C.H., J.W. Pratt and A. Seidenberg (1959) 'Intuitive probability on finite sets', *Annals of Mathematical Statistics*, **30**: 408–419.
Krantz, D.H., R.D. Luce, P. Suppes and A. Tversky (1971) *Foundations of Measurement*, Vol. I. New York: Academic Press.
Kreps, D.M. and E.L. Porteus (1978) 'Temporal resolution of uncertainty and dynamic choice theory', *Econometrica*, **46**: 185–200.
Kreps, D.M. and E.L. Porteus (1979) 'Temporal von Neumann–Morgenstern and induced preferences', *Journal of Economic Theory*, **20**: 81–109.
Kreweras, G. (1961) 'Sur une possibilité de rationaliser les intransitivités', La Décision, Colloques Internationaux du Centre National de la Recherche Scientifique, pp. 27–32.
Kuhn, H.W. (1956) 'Solvability and consistency for linear equations and inequalities', *American Mathematical Monthly*, **63**: 217–232.
Kumar, A. (1982) 'Lower probabilities on infinite spaces and instability of stationary sequences', Ph.D. Dissertation, Cornell University.
Kyburg, H.E., Jr. and H.E. Smokler, eds. (1964) *Studies in subjective probability*. New York: Wiley.
Lange, O. (1934) 'The determinateness of the utility function', *Review of Economic Studies*, **1**: 218–224.
Laplace, P.S. (1812) *Théorie analytique des probabilités*. Paris; Reprinted in Oeuvres Complétes, **7**: 1947.
LaValle, I.H. (1978) *Fundamentals of decision analysis*. New York: Holt, Rinehart and Winston.
Ledyard, J.O. (1971) 'A pseudo-metric space of probability measures and the existence of measurable utility', *Annals of Mathematical Statistics*, **42**: 794–798.
Lichtenstein, S. and P. Slovic (1971) 'Reversals of preferences between bids and choices in gambling decisions', *Journal of Experimental Psychology*, **89**: 46–55.
Lichtenstein, S. and P. Slovic (1973) 'Response-induced reversals of preferences in gambling: an extended replication in Las Vegas', *Journal of Experimental Psychology*, **101**: 16–20.
Lindman, H.R. (1971) 'Inconsistent preferences among gambles', *Journal of Experimental Psychology*, **89**: 390–397.
Loomes, G. and R. Sugden (1982) 'Regret theory: an alternative theory of rational choice under uncertainty', *Economic Journal*, **92**: 805–824.
Loomes, G. and R. Sugden (1983) 'A rationale for preference reversal', *American Economic Review*, **73**: 428–432.
Loomes, G. and R. Sugden (1986) 'Disappointment and dynamic consistency in choice under uncertainty', *Review of Economic Studies*, **53**: 271–282.
Loomes, G. and R. Sugden (1987) 'Some implications of a more general form of regret theory', *Journal of Economic Theory*, **41**: 270–287.
Luce, R.D. (1956) 'Semiorders and a theory of utility discrimination', *Econometrica*, **24**: 178–191.
Luce, R.D. (1958) 'A probabilistic theory of utility', *Econometrica*, **26**: 193–224.
Luce, R.D. (1959) *Individual choice behavior: a theoretical analysis*. New York: Wiley.
Luce, R.D. (1967) 'Sufficient conditions for the existence of a finitely additive probability measure', *Annals of Mathematical Statistics*, **38**: 780–786.
Luce, R.D. (1968) 'On the numerical representation of qualitative conditional probability', *Annals of Mathematical Statistics*, **39**: 481–491.
Luce, R.D. (1978) 'Lexicographic tradeoff structures', *Theory and Decision*, **9**: 187–193.

Luce, R.D. (1986) 'Uniqueness and homogeneity of ordered relational structures', *Journal of Mathematical Psychology*, **30**: 391–415.
Luce, R.D. (1988) 'Rank-dependent, subjective expected-utility representations', *Journal of Risk and Uncertainty*, **1**: 305–332.
Luce, R.D. and D.H. Krantz (1971) 'Conditional expected utility', *Econometrica*, **39**: 253–271.
Luce, R.D., D.H. Krantz, P. Suppes and A. Tversky (1990) *Foundations of measurement*, Vol. III. New York: Academic Press.
Luce, R.D. and L. Narens (1985) 'Classification of concatenation measurement structures according to scale type', *Journal of Mathematical Psychology*, **29**: 1–72.
Luce, R.D. and H. Raiffa (1957) *Games and decisions*. New York: Wiley.
Luce, R.D. and P. Suppes (1965) 'Preference, utility, and subjective probability', in: R.D. Luce, R.R. Bush and E. Galanter, eds., *Handbook of mathematical psychology*, III. New York: Wiley.
Luce, R.D. and J.W. Tukey (1964) 'Simultaneous conjoint measurement: a new type of fundamental measurement', *Journal of Mathematical Psychology*, **1**: 1–27.
MacCrimmon, K.R. (1968) 'Descriptive and normative implications of the decision-theory postulates', in: K. Borch and J. Mossin, eds., *Risk and uncertainty*. New York: Macmillan.
MacCrimmon, K.R. and S. Larsson (1979) 'Utility theory: axioms versus 'paradoxes'', in: Allais and Hagen (1979).
Machina, M.J. (1982a) ''Expected utility' analysis without the independence axiom', *Econometrica*, **50**: 277–323.
Machina, M.J. (1982b) 'A stronger characterization of declining risk aversion', *Econometrica*, **50**: 1069–1079.
Machina, M.J. (1983) 'Generalized expected utility analysis and the nature of observed violations of the independence axiom', in: B. Stigum and F. Wenstøp, eds., *Foundations of utility and risk theory with applications*. Dordrecht, Holland: Reidel.
Machina, M.J. (1984) 'Temporal risk and the nature of induced preferences', *Journal of Economic Theory*, **33**: 199–231.
Machina, M.J. (1985) 'Stochastic choice functions generated from deterministic preferences over lotteries', *Economic Journal*, **95**: 575–594.
Machina, M.J. (1987) 'Decision-making in the presence of risk', *Science*, **236**: 537–543.
Machina, M.J. and W.S. Neilson (1987) 'The Ross characterization of risk aversion: strengthening and extension', *Econometrica*, **55**: 1139–1149.
Manders, K.L. (1981) 'On JND representations of semiorders', *Journal of Mathematical Psychology*, **24**: 224–248.
Manski, C.F. (1977) 'The structure of random utility models', *Theory and Decision*, **8**: 229–254.
Markowitz, H. (1952) 'The utility of wealth', *Journal of Political Economy*, **60**: 151–158.
Marley, A.A.J. (1965) 'The relation between the discard and regularity conditions for choice probabilities', *Journal of Mathematical Psychology*, **2**: 242–253.
Marley, A.A.J. (1968) 'Some probabilistic models of simple choice and ranking', *Journal of Mathematical Psychology*, **5**: 311–332.
Marschak, J. (1950) 'Rational behavior, uncertain prospects, and measurable utility', *Econometrica*, **18**: 111–141; Errata, 1950, p. 312.
Marschak, J. (1960) 'Binary-choice constraints and random utility indicators', in: K.J. Arrow, S. Karlin and P. Suppes, eds., *Mathematical methods in the social sciences, 1959*. Stanford: Stanford University Press.
Marshall, A. (1890) *Principles of economics*. London: Macmillan.
Mas-Colell, A. (1974) 'An equilibrium existence theorem without complete or transitive preferences', *Journal of Mathematical Economics*, **1**: 237–246.
May, K.O. (1954) 'Intransitivity, utility, and the aggregation of preference patterns', *Econometrica*, **22**: 1–13.
Menger, C. (1871) *Grundsätze der Volkswirthschaftslehre*. Vienna: W. Braumuller; English translation: *Principles of economics*. Glencoe, IL: Free Press, 1950.
Menger, K. (1967) 'The role of uncertainty in economics', in: M. Shubik, ed., *Essays in mathematical economics*. Princeton: Princeton University Press; Translated by W. Schoellkopf from 'Das Unsicherheitsmoment in der Wertlehre', *Zeitschrift fur Nationaloekonomie*, **5**(1934): 459–485.
Meyer, R.F. (1977) 'State-dependent time preferences', in: D.E. Bell, R.L. Keeney and H. Raiffa, eds., *Conflicting objectives in decisions*. New York: Wiley.

Morgan, B.J.T. (1974) 'On Luce's choice axiom', *Journal of Mathematical Psychology*, **11**: 107–123.
Morrison, D.G. (1967) 'On the consistency of preferences in Allais' paradox', *Behavioral Science*, **12**: 373–383.
Nachman, D.C. (1975) 'Risk aversion, impatience, and optimal timing decisions', *Journal of Economic Theory*, **11**: 196–246.
Nakamura, Y. (1984) 'Nonlinear measurable utility analysis', Ph.D. Dissertation, University of California, Davis.
Nakamura, Y. (1985) 'Weighted linear utility', *mimeographed*, Department of Precision Engineering, Osaka University.
Nakamura, Y. (1988) 'Expected ability with an interval ordered structure', *Journal of Mathematical Psychology*, **32**: 298–312.
Narens, L. (1974) 'Measurement without Archimedean axioms', *Philosophy of Science*, **41**: 374–393.
Narens, L. (1985) *Abstract measurement theory*. Cambridge: MIT Press.
Nash, J. (1951) 'Non-cooperative games', *Annals of Mathematics*, **54**: 286–295.
Newman, P. and R. Read (1961) 'Representation problems for preference orderings', *Journal of Economic Behavior*, **1**: 149–169.
Niiniluoto, I. (1972) 'A note on fine and tight qualitative probability', *Annals of Mathematical Statistics*, **43**: 1581–1591.
Nikaidô, H. (1954) 'On von Neumann's minimax theorem', *Pacific Journal of Mathematics*, **4**: 65–72.
Papamarcou, A. and T.L. Fine (1986) 'A note on undominated lower probabilities', *Annals of Probability*, **14**: 710–723.
Pareto, V. (1906) *Manuale di economia politica, con una intraduzione alla scienza sociale*. Milan: Società Editrice Libraria.
Peleg, B. and M.E. Yaari (1973) 'On the existence of a consistent course of action when tastes are changing', *Review of Economic Studies*, **40**: 391–401.
Pfanzagl, J. (1959) 'A general theory of measurement: applications to utility', *Naval Research Logistics Quarterly*, **6**: 283–294.
Pfanzagl, J. (1967) 'Subjective probability derived from the Morgenstern–von Neumann utility concept', in: M. Shubik, ed., *Essays in mathematical economics*. Princeton: Princeton University Press.
Pfanzagl, J. (1968) *Theory of measurement*. New York: Wiley.
Pollak, R.A. (1967) 'Additive von Neumann–Morgenstern utility functions', *Econometrica*, **35**: 485–494.
Pollak, R.A. (1968) 'Consistent planning', *Review of Economic Studies*, **35**: 201–208.
Pollak, R.A. (1976) 'Habit formation and long-run utility functions', *Journal of Economic Theory*, **13**: 272–297.
Pommerehne, W.W., F. Schneider and P. Zweifel (1982) 'Economic theory of choice and the preference reversal phenomenon: a reexamination', *American Economic Review*, **72**: 569–574.
Pratt, J.W. (1964) 'Risk aversion in the small and in the large', *Econometrica*, **32**: 122–136.
Pratt, J.W., H. Raiffa and R. Schlaifer (1964) 'The foundations of decision under uncertainty: an elementary exposition', *Journal of the American Statistical Association*, **59**: 353–375.
Pratt, J.W., H. Raiffa and R. Schlaifer (1965) *Introduction to statistical decision theory*. New York: McGraw-Hill.
Preston, M.G. and P. Baratta (1948) 'An experimental study of the auction value of an uncertain outcome', *American Journal of Psychology*, **61**: 183–193.
Quiggin, J. (1982) 'A theory of anticipated utility', *Journal of Economic Behavior and Organization*, **3**: 323–343.
Quirk, J.P. and R. Saposnik (1962) 'Admissibility and measurable utility functions', *Review of Economic Studies*, **29**: 140–146.
Rader, J.T. (1963) 'The existence of a utility function to represent preferences', *Review of Economic Studies*, **30**: 229–232.
Raiffa, H. (1961) 'Risk, ambiguity, and the Savage axioms: comment', *Quarterly Journal of Economics*, **75**: 690–694.
Raiffa, H. (1968) *Decision analysis: introductory lectures on choice under uncertainty*. Reading: Addison-Wesley.
Raiffa, H. and R. Schlaifer (1961) *Applied statistical decision theory*. Boston: Harvard Graduate School of Business Administration.

Ramsey, F.P. (1931) 'Truth and probability', in: *The foundations of mathematics and other logical essays*. London: Routledge and Kegan Paul; Reprinted in Kyburg and Smokler (1964).
Reilly, R.J. (1982) 'Preference reversal: further evidence and some suggested modifications in experimental design', *American Economic Review*, **72**: 576–584.
Restle, F. (1961) *Psychology of judgment and choice: a theoretical essay*. New York: Wiley.
Richter, M.K. (1966) 'Revealed preference theory', *Econometrica*, **34**: 635–645.
Roberts, F.S. (1971) 'Homogeneous families of semiorders and the theory of probabilistic consistency', *Journal of Mathematical Psychology*, **8**: 248–263.
Roberts, F.S. (1973) 'A note on Fine's axioms for qualitative probability', *Annals of Probability*, **1**: 484–487.
Roberts, F.S. (1979) *Measurement theory*. Reading: Addison–Wesley.
Ross, S. (1981) 'Some stronger measures of risk aversion in the small and the large with applications', *Econometrica*, **49**: 621–638.
Rothschild, M. and J.E. Stiglitz (1970) 'Increasing risk I: a definition', *Journal of Economic Theory*, **2**: 225–243.
Rothschild, M. and J.E. Stiglitz (1971) 'Increasing risk II: its economic consequences', *Journal of Economic Theory*, **3**: 66–84.
Samuelson, P.A. (1938) 'A note on the pure theory of consumer's behaviour', *Economica*, **5**: 61–71, 353–354.
Samuelson, P.A. (1947) *Foundations of economic analysis*. Cambridge: Harvard University Press.
Samuelson, P.A. (1977) 'St. Petersburg paradoxes: defanged, dissected, and historically described', *Journal of Economic Literature*, **15**: 24–55.
Sattath, S. and A. Tversky (1976) 'Unite and conquer: a multiplicative inequality for choice probabilities', *Econometrica*, **44**: 79–89.
Savage, L.J. (1954) *The foundations of statistics*. New York: Wiley.
Schlaifer, R. (1959) *Probability and statistics for business decisions*. New York: McGraw-Hill.
Schmeidler, D. (1986) 'Integral representation without additivity', *Proceedings of the American Mathematical Society*, **97**: 255–261.
Schmeidler, D. (1989) 'Subjective probability and expected utility without additivity', *Econometrica*, **57**: 571–587.
Schoemaker, P.J.H. (1980) *Experiments on decisions under risk*. Boston: Martinus Nijhoff.
Scott, D. (1964) 'Measurement structures and linear inequalities', *Journal of Mathematical Psychology*, **1**: 233–247.
Scott, D. and P. Suppes (1958) 'Foundational aspects of theories of measurement', *Journal of Symbolic Logic*, **23**: 113–128.
Segal, U. (1984) 'Nonlinear decision weights with the independence axiom', *Working Paper 353*, Department of Economics, University of California, Los Angeles.
Segal, U. (1987) 'The Ellsberg paradox and risk aversion: an anticipated utility approach', *International Economic Review*, **28**: 175–202.
Seidenfeld, T. and M.J. Schervish (1983) 'A conflict between finite additivity and avoiding Dutch Book', *Philosophy of Science*, **50**: 398–412.
Sen, A.K. (1977) 'Social choice theory: a re-examination', *Econometrica*, **45**: 53–89.
Shafer, G. (1976) *A mathematical theory of evidence*. Princeton: Princeton University Press.
Shafer, G. (1986) 'Savage revisited', *Statistical Science*, **1**: 463–485.
Shafer, W. and H. Sonnenschein (1975) 'Equilibrium in abstract economies without ordered preferences', *Journal of Mathematical Economics*, **2**: 345–348.
Sherman, R. (1974) 'The psychological difference between ambiguity and risk', *Quarterly Journal of Economics*, **88**: 166–169.
Skala, H.J. (1975) *Non-Archimedean utility theory*. Dordrecht, Holland: Reidel.
Slovic, P. and S. Lichtenstein (1983) 'Preference reversals: a broader perspective', *American Economic Review*, **73**: 596–605.
Slovic, P. and A. Tversky (1974) 'Who accepts Savage's axiom?', *Behavioral Science*, **19**: 368–373.
Slutsky, E. (1915) 'Sulla teoria del bilancio del consumatore', *Giornale degli Economisti e Rivista di Statistica*, **51**: 1–26.
Smith, C.A.B. (1961) 'Consistency in statistical inference and decision', *Journal of the Royal Statistical Society*, Series B, **23**: 1–37.

Sonnenschein, H.F. (1971) 'Demand theory without transitive preferences, with applications to the theory of competitive equilibrium', in: Chipman et al. (1971).
Spence, M. and R. Zeckhauser (1972) 'The effect of the timing of consumption decisions and the resolution of lotteries on the choice of lotteries', *Econometrica*, **40**: 401–403.
Stigler, G. J. (1950) 'The development of utility theory: I; II', *Journal of Political Economy*, **58**: 307–327; 373–396.
Strauss, D. (1979) 'Some results on random utility models', *Journal of Mathematical Psychology*, **20**: 35–52.
Strotz, R.H. (1953) 'Cardinal utility', *American Economic Review*, **43**: 384–397.
Strotz, R.H. (1956) 'Myopia and inconsistency in dynamic utility maximization', *Review of Economic Studies*, **23**: 165–180.
Suppes, P. (1956) 'The role of subjective probability and utility in decision making', *Proceedings of the Third Berkeley Symposium on Mathematical Statistics and Probability*, 1954–1955, **5**: 61–73.
Suppes, P. (1961) 'Behavioristic foundations of utility', *Econometrica*, **29**: 186–202.
Suppes, P., D.H. Krantz, R.D. Luce and A. Tversky (1989) *Foundations of measurement*, Vol. II. Academic Press: New York.
Suppes, P. and M. Winet (1955) 'An axiomatization of utility based on the notion of utility differences', *Management Science*, **1**: 259–270.
Suppes, P. and M. Zanotti (1982) 'Necessary and sufficient qualitative axioms for conditional probability', *Zeitschrift für Wahrscheinlichkeitstheorie und verwandte Gebiete*, **60**: 163–169.
Suppes, P. and J.L. Zinnes (1963) 'Basic measurement theory', in: R.D. Luce, R.R. Bush and E. Galanter, eds., *Handbook of mathematical psychology*, I. New York: Wiley.
Toulet, C. (1986) 'An axiomatic model of unbounded utility functions', *Mathematics of Operations Research*, **11**: 81–94.
Tutz, G. (1986) 'Bradley–Terry–Luce models with an ordered response', *Journal of Mathematical Psychology*, **30**: 306–316.
Tversky, A. (1969) 'Intransitivity of preferences', *Psychological Review*, **76**: 31–48.
Tversky, A. (1972a) 'Elimination by aspects: a theory of choice', *Psychological Review*, **79**: 281–299.
Tversky, A. (1972b) 'Choice by elimination', *Journal of Mathematical Psychology*, **9**: 341–367.
Tversky, A. (1975) 'A critique of expected utility theory: descriptive and normative considerations', *Erkenntnis*, **9**: 163–173.
Tversky, A. and D. Kahneman (1973) 'Availability: a heuristic for judging frequency probability', *Cognitive Psychology*, **5**: 207–232.
Tversky, A. and D. Kahneman (1986) 'Rational choice and the framing of decisions', in: R.M. Hogarth and M.W. Reder, eds., *Rational choice*. Chicago: University of Chicago Press.
Tversky, A., P. Slovic and D. Kahneman (1990) 'The causes of preference reversal', *American Economic Review*, **80**: 204–217.
Van Lier, L. (1989) 'A simple sufficient condition for the unique representability of a finite qualitative probability by a probability measure', *Journal of Mathematical Psychology*, **33**: 91–98.
Villegas, C. (1964) 'On qualitative probability σ-algebras', *Annals of Mathematical Statistics*, **35**: 1787–1796.
Vincke, P. (1980) 'Linear utility functions on semiordered mixture spaces', *Econometrica* **48**: 771–775.
Vind, K. (1991) 'Independent preferences', *Journal of Mathematical Economics*, **20**: 119–135.
von Neumann, J. (1928) 'Zur theorie der gesellschaftsspiele', *Mathematische Annalen*, **100**: 295–320.
von Neumann, J. and O. Morgenstern (1944) *Theory of games and economic behavior*. Princeton: Princeton University Press; second edn. 1947; third edn. 1953.
Wakker, P. (1981) 'Agreeing probability measures for comparative probability structures', *Annals of Statistics*, **9**: 658–662.
Wakker, P. P. (1989) *Additive representations of preferences*. Dordrecht, Holland: Kluwer.
Walley, P. and T.L. Fine (1979) 'Varieties of modal (classificatory) and comparative probability', *Synthese*, **41**: 321–374.
Walras, L. (1874) *Eléments d'économie politique pure*. Lausanne: Corbas and Cie.
Whitmore, G.A. (1970) 'Third-degree stochastic dominance', *American Economic Review*, **60**: 457–459.
Whitmore, G.A. and M. C. Findlay (eds.) (1978) *Stochastic dominance*. Lexington, MA: Heath.
Wiener, N. (1914) 'A contribution to the theory of relative position', *Proceedings of the Cambridge Philosophical Society*, **17**: 441–449.

Williams, P.M. (1976) 'Indeterminate probabilities', in: *Formal methods in the methodology of empirical sciences*. Dordrecht, Hollands: Ossolineum and Reidel.

Wold, H. (1943) 'A synthesis of pure demand analysis: I, II, III', *Skandinavisk Aktuarietidskrift*, **26**: 85–118, 220–263; **27**(1944): 69–120.

Wold, H. and L. Jureen (1953) *Demand analysis*. New York: Wiley.

Yaari, M.E. (1987) 'The dual theory of choice under risk', *Econometrica*, **55**: 95–115.

Yellott. J.I. Jr. (1977) 'The relationship between Luce's choice axiom, Thurstone's theory of comparative judgment, and the double exponential distribution', *Journal of Mathematical Psychology*, **15**: 109–144.

Yokoyama, T. (1956) 'Continuity conditions of preference ordering', *Osaka Economic Papers*, **4**: 39–45.

Chapter 40

COMMON KNOWLEDGE

JOHN GEANAKOPLOS*

Yale University

Contents

1.	Introduction	1438
2.	Puzzles about reasoning based on the reasoning of others	1439
3.	Interactive epistemology	1441
4.	The puzzles reconsidered	1444
5.	Characterizing common knowledge of events and actions	1450
6.	Common knowledge of actions negates asymmetric information about events	1453
7.	A dynamic state space	1455
8.	Generalizations of agreeing to disagree	1458
9.	Bayesian games	1461
10.	Speculation	1465
11.	Market trade and speculation	1467
12.	Dynamic Bayesian games	1469
13.	Infinite state spaces and knowledge about knowledge to level N	1476
14.	Approximate common knowledge	1480
15.	Hierarchies of belief: Is common knowledge of the partitions tautological?	1484
16.	Bounded rationality: Irrationality at some level	1488
17.	Bounded rationality: Mistakes in information processing	1490
References		1495

*About 60% of the material in this survey can be found in a less technical version "Common Knowledge" that appeared in the *Journal of Economic Perspectives*. I wish to acknowledge many inspiring conversations, over the course of many years, I have had with Bob Aumann on the subject of common knowledge. I also with to acknowledge funding from computer science grant IRI-9015570. Finally I wish to acknowledge helpful advice on early drafts of this paper from Barry Nalebuff, Tim Taylor, Carl Shapiro, Adam Brandenburger, and Yoram Moses.

Handbook of Game Theory, Volume 2, Edited by R.J. Aumann and S. Hart
© *Elsevier Science B.V., 1994. All rights reserved*

1. Introduction

People, no matter how rational they are, usually act on the basis of incomplete information. If they are rational they recognize their own ignorance and reflect carefully on what they know and what they do not know, before choosing how to act. Furthermore, when rational agents interact, they think about what the others know and do not know, and what the others know about what they know, before choosing how to act. Failing to do so can be disastrous. When the notorious evil genius Professor Moriarty confronts Sherlock Holmes for the first time he shows his ability to think interactively by remarking, "All I have to say has already crossed your mind." Holmes, even more adept at that kind of thinking, responds, "Then possibly my answer has crossed yours." Later, Moriarty's limited mastery of interactive epistemology allowed Holmes and Watson to escape from the train at Canterbury, a mistake which ultimately led to Moriarity's death, because he went on to Paris after calculating that Holmes would normally go on to Paris, failing to deduce that Holmes had deduced that he would deduce what Holmes would normally do and in this circumstance get off earlier.

Knowledge and interactive knowledge are central elements in economic theory. Any prospective stock buyer who has information suggesting the price will go up must consider that the seller might have information indicating that the price will go down. If the buyer further considers that the seller is willing to sell the stock, having also taken into account that the buyer is willing to purchase the stock, the prospective buyer must ask whether buying is still a good idea.

Can rational agents agree to disagree? In this question connected to whether rational agents will speculate in the stock market? How might the degree of rationality of the agents, or the length of time they talk, influence the answer to this question?

The notion of common knowledge plays a crucial role in the analysis of these questions. An event is common knowledge among a group of agents if each one knows it, each one knows that the others know it, each one knows that each one knows that the others know it, and so on. Thus, common knowledge is the limit of a potentially infinite chain of reasoning about knowledge. This definition of common knowledge was suggested by the philosopher D. Lewis in 1969. A formal definition of common knowledge was introduced into the economics literature by Robert Aumann in 1976.

Public events are the most obvious candidates for common knowledge. But events that the agents create themselves, like the rules of a game or contract, can also plausibly be seen as common knowledge. Certain beliefs about human nature might also be taken to be common knowledge. Economists are especially interested, for example, in the consequences of the hypothesis that it is common knowledge that all agents are optimizers. Finally, it often occurs that after lengthy conversations or observations, what people are going to do is common knowledge, though the reasons for their actions may be difficult to disentangle.

The purpose of this chapter is to survey some of the implications for economic behavior of the hypotheses that events are common knowledge, that actions are common knowledge, that optimization is common knowledge, and that rationality is common knowledge. The main conclusion is that an apparently innocuous assumption of common knowledge rules out speculation, betting, and agreeing to disagree. To try to restore the conventional understandings of these phenomena we allow for infinite state spaces, approximate common knowledge of various kinds including knowledge about knowledge only upto level n, and bounded rationality. We begin this survey with several puzzles that illustrate the strength of the common knowledge hypothesis.

2. Puzzles about reasoning based on the reasoning of others

The most famous example illustrating the ideas of reasoning about common knowledge can be told in many equivalent ways. The earliest version that I could find appears in Littlewood's Miscellania, (edited by Bollobás) published in 1953, although he noted that it was already well-known and had caused a sensation in Europe some years before. The colonial version of the story begins with many cannibals married to unfaithful wives, and of course a missionary. I shall be content to offer a more prosaic version, involving a group of logical children wearing hats.[1]

Imagine three girls sitting in a circle, each wearing either a red hat or a white hat. Suppose that all the hats are red. When the teacher asks if any student can identify the color of her own hat, the answer is always negative, since nobody can see her own hat. But if the teacher happens to remark that there is at least one red hat in the room, a fact which is well-known to every child (who can see two red hats in the room) then the answers change. The first student who is asked cannot tell, nor can the second. But the third will be able to answer with confidence that she is indeed wearing a red hat.

How? By following this chain of logic. If the hats on the heads of both children two and three were white, then the teacher's remark would allow the first child to answer with confidence that her hat was red. But she cannot tell, which reveals to children two and three that at least one of them is wearing a red hat. The third child watches the second also admit that she cannot tell her hat color, and then reasons as follows: "If my hat had been white, then the second girl would have

---

[1] These versions are so well-known that it is difficult to find out who told them first. The hats version appeared in Martin Gardner's collection (1984). It had already been presented by Gamow and Stern (1958) as the puzzle of the cheating wives. It was discussed in the economics literature by Geanakoplos–Polemarchakis (1982). It appeared in the computer science literature in Halpern–Moses (1984).

answered that she was wearing a red hat, since we both know that at least one of us is wearing a red hat. But the second girl could not answer. Therefore, I must be wearing a red hat." The story is surprising because aside from the apparently innocuous remark of the teacher, the students appear to learn from nothing except their own ignorance. Indeed this is precisely the case.

The story contains several crucial elements: it is common knowledge that everybody can see two hats; the pronouncements of ignorance are public; each child knows the reasoning used by the others. Each student knew the apparently innocuous fact related by the teacher – that there was at least one red hat in the room – but the fact was not common knowledge between them. When it became common knowledge, the second and third children could draw inferences from the answer of the first child, eventually enabling the third child to deduce her hat color.

Consider a second example, also described by Littlewood, involving betting. An honest but mischievous father tells his two sons that he has placed 10^n dollars in one envelope, and 10^{n+1} dollars in the other envelope, where n is chosen with equal probability among the integers between 1 and 6. The sons completely believe their father. He randomly hands each son an envelope. The first son looks inside his envelope and finds $10 000. Disappointed at the meager amount, he calculates that the odds are fifty–fifty that he has the smaller amount in his envelope. Since the other envelope contains either $1 000 or $100 000 with equal probability, the first son realizes that the expected amount in the other envelope is $50 500. The second son finds only $1 000 in his envelope. Based on his information, he expects to find either $100 or $10 000 in the first son's envelope, which at equal odds comes to an expectation of $5 050. The father privately asks each son whether he would be willing to pay $1 to switch envelopes, in effect betting that the other envelope has more money. Both sons say yes. The father then tells each son what his brother said and repeats the question. Again both sons say yes. The father relays the brothers' answers and asks each a third time whether he is willing to pay $1 to switch envelopes. Again both say yes. But if the father relays their answers and asks each a fourth time, the son with $1 000 will say yes, but the son with $10 000 will say no.

It is interesting to consider a slight variation of this story. Suppose now that the very first time the father tells each of his sons that he can pay $1 to switch envelopes it is understood that if the other son refuses, the deal is off and the father keeps the dollar. What would they do? Both would say no, as we shall explain in a later section.

A third puzzle is more recent.[2] Consider two detectives trained at the same police academy. Their instruction consists of a well-defined rule specifying who to

---

[2] This story is originally due to Bacharach, perhaps somewhat embellished by Aumann, from whom I learned it. It illustrates the analysis in Aumann (1976), Geanakoplos and Polemarchakis (1982), and Cave (1983).

arrest given the clues that have been discovered. Suppose now that a murder occurs, and the two detectives are ordered to conduct independent investigations. They promise not to share any data gathered from their research, and begin their sleuthing in different corners of the town. Suddenly the detectives are asked to appear and announce who they plan to arrest. Neither has had the time to complete a full investigation, so they each have gathered different clues. They meet on the way to the station. Recalling their pledges, they do not tell each other a single discovery, or even a single reason why they were led to their respective conclusions. But they do tell each other who they plan to arrest. Hearing the other's opinion, each detective may change his mind and give another opinion. This may cause a further change in opinion.

If they talk long enough, however, then we can be sure that both detectives will announce the same suspect at the station! This is so even though if asked to explain their choices, they may each produce entirely different motives, weapons, scenarios, and so on. And if they had shared their clues, they might well have agreed on an entirely different suspect!

It is commonplace in economics nowadays to say that many actions of optimizing, interacting agents can be naturally explained only on the basis of asymmetric information. But in the riddle of the detectives common knowledge of each agent's action (what suspect is chosen, given the decision rules) negates asymmetric information about events (what information was actually gathered). At the end, the detectives are necessarily led to a decision which can be explained by a common set of clues, although in fact their clues might have been different, even allowing for the deductions each made from hearing the opinions expressed in the conversation. The lesson we shall draw is that asymmetric information is important only if it leads to uncertainty about the action plans of the other agents.

3. Interactive epistemology

To examine the role of common knowledge, both in these three puzzles and in economics more generally, the fundamental conceptual tool we shall use is the state of the world. Leibnitz first introduced this idea; it has since been refined by Kripke, Savage, Harsanyi, and Aumann, among others. A "state of the world" is very detailed. It specifies the physical universe, past, present, and future; it describes what every agent knows, and what every agent knows about what every agent knows, and so on; it specifies what every agent does, and what every agent thinks about what every agent does, and what every agent thinks about what every agent thinks about what every agent does, and so on; it specifies the utility to every agent of every action, not only of those that are taken in that state of nature, but also those that hypothetically might have been taken, and it specifies what everybody thinks about the utility to everybody else of every possible action, and so on; it specifies not only what agents know, but what probability they assign to every event, and what probability they assign to every other agent assigning some probability to each event, and so on.

Let Ω be the set of all possible worlds, defined in this all-embracing sense. We model limited knowledge by analogy with a far-off observer who from his distance cannot quite distinguish some objects from others. For instance, the observer might be able to tell the sex of anyone he sees, but not who the person is. The agent's knowledge will be formally described throughout most of this survey by a collection of mutually disjoint and exhaustive classes of states of the world called cells that *partition* Ω. If two states of nature are in the same cell, then the agent cannot distinguish them. For each $\omega \in \Omega$, we define $P_i(\omega) \subset \Omega$ as all states that agent i cannot distinguish from ω.

Any subset E contained in Ω is called an *event*. If the true state of the world is ω, and if $\omega \in E$, then we say that E *occurs* or is *true*. If every state that i thinks is *possible* (given that ω is the true state) entails E, which we write as $P_i(\omega) \subset E$, then we say that agent i knows E. Note that at some ω, i may know E, while at other ω, i may not. If whenever E occurs i knows E, that is, if $P_i(\omega) \subset E$ for all states ω in E, then we say that E is *self-evident* to i. Such an event E cannot happen unless i knows it.

So far we have described the knowledge of agent i by what he would think is possible in each state of nature. There is an equivalent way of representing the knowledge of agent i at some state ω, simply by enumerating all the events which the information he has at ω guarantees must occur. The crispest notation to capture this idea is a *knowledge operator* K_i taking any event E into the set of all states at which i is sure that E has occurred: $K_i(E) = \{\omega \in \Omega : P_i(\omega) \subset E\}$. At ω, agent i has enough information to guarantee that event E has occurred iff $\omega \in K_i(E)$. A self-evident event can now be described as any subset E of Ω satisfying $K_i(E) = E$, i.e., the self-evident events are the fixed points of the K_i operator.

As long as the possibility correspondence P_i is a partition, the knowledge operator applied to any event E is the union of all the partition cells that are completely contained in E. It can easily be checked that the knowledge operator K_i derived from the partition possibility correspondence P_i satisfies the following five axioms: for all events A and B contained in Ω,

(1) $K_i(\Omega) = \Omega$. It is self evident to agent i that there are no states of the world outside of Ω.
(2) $K_i(A) \cap K_i(B) = K_i(A \cap B)$. Knowing A and knowing B is the same thing as knowing A and B.
(3) $K_i(A)$ contained in A. If i knows A, then A is true.
(4) $K_i K_i(A) = K_i(A)$. If i knows A, then he knows that he knows A.
(5) $-K_i(A) = K_i(-K_i(A))$. If i does not know A, then he knows that he does not know A.

Kripke (1963) called any system of knowledge satisfying the above five axioms S5. We shall later encounter descriptions of knowledge which permit less rationality.

In particular, the last axiom, which requires agents to be just as alert about things that do not happen as about things that do, is the most demanding. Dropping it has interesting consequences for economic theory, as we shall see later. Note that axiom (5) implies axiom (4); $K_i(K_iA) = K_i(-(-K_iA)) = K_i(-(K_i(-(K_iA))) = -K_i(-K_iA)) = -(-K_iA)) = K_iA$.

The most interesting events in the knowledge operator approach are the fixed point events E that satisfy $K_i(E) = E$. From axiom (4), these events make up the range of the $K_i : 2^\Omega \to 2^\Omega$ operator. Axioms (1)–(4) are analogous to the familiar properties of the "interior operator" defined on topological spaces, where Int E is the union of all open sets contained in E. To verify that $(\Omega, \text{Range } K_i)$ is a topological space, we must check that Ω itself is in Range K_i [which follows from axiom (1)], that the intersection of any two elements of Range K_i is in Range K_i [which follows from axiom (2)], and that the arbitrary union $E = \bigcup_{\alpha \in I} E_\alpha$ of sets E_α in Range K_i is itself in Range K_i. To see this, observe that by axiom (2), for all $\alpha \in I$,

$$E_\alpha = K_i(E_\alpha) = K_i(E_\alpha \cap E) = K_i(E_\alpha) \cap K_i(E) \subset K_i(E)$$

hence $E = \bigcup_{\alpha \in I} E_\alpha \subset K_i(E)$, and therefore by axiom (3), $E = K_i(E)$. Thus we have confirmed that $(\Omega, \text{Range } K_i)$ is a topological space, and that for any event $A \subset \Omega$, $K_i(A)$ is the union of all elements of Range K_i that are contained in A.

Axiom (5) gives us a very special topological space because it maintains that if E is a fixed point of K_i, then so is $-E$. The space Range K_i is a complete field, that is, closed under complements and arbitrary intersections. Thus the topological space $(\Omega, \text{Range } K_i)$ satisfies the property that every open set is also closed, and vice versa. In particular, this proves that an arbitrary intersection of fixed point events of K_i is itself a fixed point event of K_i. Hence the minimal fixed point events of K_i form a partition of Ω.

The partition approach to knowledge is completely equivalent to the knowledge operator approach satisfying S5. Given a set Ω of states of the world and a knowledge operator K_i satisfying S5, we can define a unique partition of Ω that would generate K_i. For all $\omega \in \Omega$, define $P_i(\omega)$ as the intersection of all fixed point events of the operator K_i that contain ω. By our analysis of the topology of fixed point events, $P_i(\omega)$ is the smallest fixed point event of the K_i operator that contains ω. It follows that the sets $P_i(\omega)$, $\omega \in \Omega$, form a partition of Ω. We must now check that P_i generates K_i, that is we must show that for any $A \subset \Omega$, $K_i(A) = \{\omega \in A : P_i(\omega) \subset A\}$. Since $K_i(A)$ is the union of all fixed point events contained in A, $\omega \in K_i(A)$ if and only if there is a fixed point event E with $\omega \in E \subset A$. Since $P_i(\omega)$ is the smallest fixed point event containing ω, we are done.

We can model an agent's learning by analogy to an observer getting closer to what he is looking at. Things which he could not previously distinguish, such as for example whether the people he is watching have brown hair or black hair, become discernible. In our framework, such an agent's partition becomes finer when he learns, perhaps containing four cells {{female/brown hair}, {female/black hair}, {male/brown hair}, {male/black hair}} instead of two, {{female}, {male}}.

Naturally, we can define the partitions of several agents, say i and j, simultaneously on the same state space. There is no reason that the two agents should have the same partitions. Indeed different people typically have different vantage points, and it is precisely this asymmetric information that makes the question of common knowledge interesting.

Suppose now that agent i knows the partition of j, i.e., suppose that i knows what j is able to know, and vice versa. (This does not mean that i knows what j knows; i may know that j knows her hair color without knowing it himself.) Since the possibility correspondences are functions of the state of nature, each state of nature ω specifies not only the physical universe, but also what each agent knows about the physical universe, and what each agent knows each agent knows about the physical universe and so on.

4. The puzzles reconsidered

With this framework, let us reconsider the puzzle of the three girls with red and white hats. A state of nature ω corresponds to the color of each child's hat. The table lists the eight possible states of nature.

STATES OF THE WORLD

		a	b	c	d	e	f	g	h
	1	R	R	R	R	W	W	W	W
PLAYER	2	R	R	W	W	R	R	W	W
	3	R	W	R	W	R	W	R	W

In the notation we have introduced, the set of all possible states of nature Ω can be summarized as $\{a, b, c, d, e, f, g, h\}$, with a letter designating each state. Then, the partitions of the three agents are given by: $P_1 = \{\{a, e\}, \{b, f\}, \{c, g\}, (d, h)\}$, $P_2 = \{\{a, c\}, \{b, d\}, (\{e, g\}, \{f, h\}), P_3 = \{\{a, b\}, \{c, d\}, \{e, f\}, \{g, h\}\}$.

These partitions give a faithful representation of what the agents could know at the outset. Each can observe four cells, based on the hats the others are wearing: both red, both white, or two combinations of one of each. None can observe her own hat, which is why the cells come in groups of two states. For example, if the true state of the world is all red hats – that is $\omega = a = RRR$ – then agent 1 is informed of $P_1(a) = \{a, e\}$, and thus knows that the true state is either $a = RRR$, or $e = WRR$. In the puzzle, agent i "knows" her hat color only if the color is the same in all states of nature ω which agent i regards as possible.

In using this model of knowledge to explain the puzzle of the hats, it helps to represent the state space as the vertices of a cube, as in Diagram 1a.[3] Think of R

---

[3] This has been pointed out by Fagin, Halpern, Moses, and Vardi (1988) in unpublished notes.

Ch. 40: Common Knowledge   1445

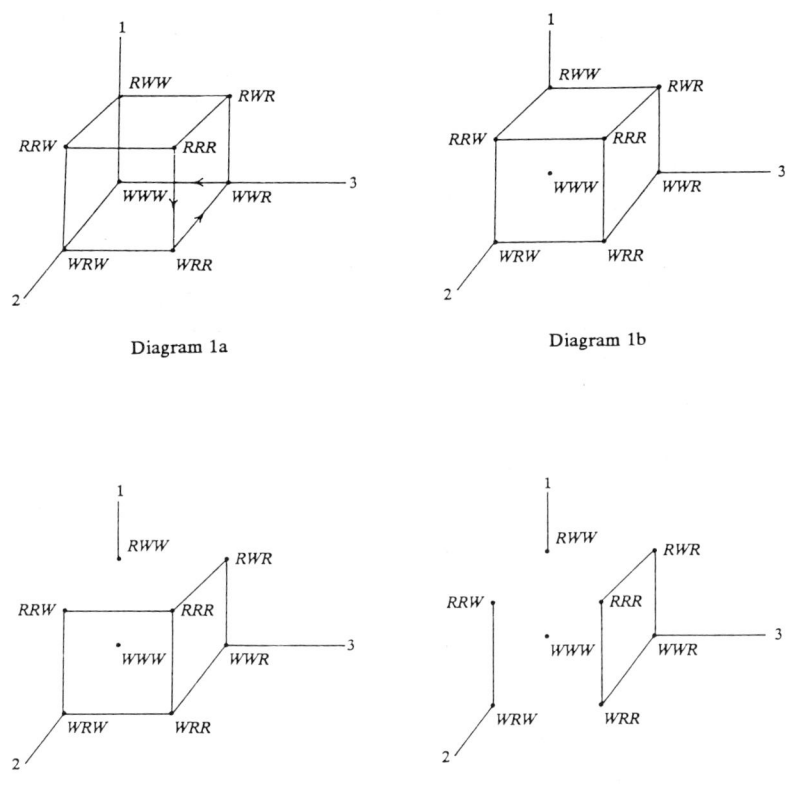

Diagram 1a

Diagram 1b

Diagram 1c

Diagram 1d

as 1 and W as 0. Then every corner of a cube has three coordinates which are either 1 or 0. Let the ith coordinate denote the hat color of the ith agent. For each agent i, connect two vertices with an edge if they lie in the same information cell in agent i's partition. These edges should be denoted by different colors to distinguish the agents, but no confusion should result even if all the edges are given by the same color. The edges corresponding to agent i are all parallel to the ith axis, so that if the vertical axis is designated as 1, the four vertical sides of the cube correspond to the four cells in agent 1's partition.

An agent i knows her hat color at a state if and only if the state is not connected by one of i's edges to another state in which i has a different hat color. In the original situation sketched above, no agent knows her hat color in any state.

Note that every two vertices are connected by at least one path. Consider for example the state RRR and the state WWW. At state RRR, agent 1 thinks WRR is possible. But at WRR, agent 2 thinks WWR is possible. And at WWR agent 3 thinks WWW is possible. In short, at RRR agent 1 thinks that agent 2 might think that agent 3 might think that WWW is possible. In other words, WWW is reachable from RRR. This chain of thinking is indicated in the diagram by the path marked by arrows.

We now describe the evolution of knowledge resulting from the teacher's announcement and the responses of the children. The analysis proceeds independent of the actual state, since it describes what the children would know at every time period for each state of the world. When the teacher announces that there is at least one red hat in the room, that is tantamount to declaring that the actual state is not WWW. This can be captured pictorially by dropping all the edges leading out of the state WWW, as seen in Diagram 1b. (Implicitly, we are assuming that had all the hats been white, the teacher would have said so.) Each of the girls now has a finer partition than before, that is, some states that were indistinguishable before have now become distinguishable. There are now two connected components to the graph: one consisting of the state WWW on its own, and the rest of the states.

If, after hearing the teacher's announcement, the first student announces she does not know her hat color, she reveals that the state could not be RWW, since if it were, she would also be able to deduce the state from her own information and the teacher's announcement and therefore would have known her hat color. We can capture the effect of the first student's announcement on every other agent's information by severing all the connections between the set $\{WWW, RWW\}$ and its complement. Diagram 1c now has three different components, and agents 2 and 3 have finer partitions.

The announcement by student 2 that she still does not know her hat color reveals that the state cannot be any of $\{WWW, RWW, RRW, WRW\}$, since these are the states in which the above diagram indicates student 2 would have the information (acquired in deductions from the teacher's announcement and the first student's announcement) to unambiguously know her hat color. Conversely, if 2 knows her hat color, then she reveals that the state must be among those in $\{WWW, RWW, RRW, WRW\}$. We represent the consequences of student 2's announcement on the other student's information partitions by severing all connections between the set $\{WWW, RWW, RRW, WRW\}$ and its complement, producing Diagram 1d. Notice now that the diagram has four separate components.

In this final situation, after hearing the teacher's announcement, and each of student 1 and student 2's announcements, student 3 knows her hat color at all the states. Thus no more information is revealed, even when student 3 says she knows her hat color is red.

If, after student 3 says yes, student 1 is asked the color of her hat again, she will still say no, she cannot tell. So will student 2. The answers will repeat indefinitely as the question for students 1 and 2 and 3 is repeated over and over. Eventually, their responses will be "common knowledge": every student will know what every other student is going to say, and each student will know that each other student knows what each student is going to say, and so on. By logic alone the students come to a common understanding of what must happen in the future. Note also that at the final stage of information, the three girls have different information.

The formal treatment of Littlewood's puzzle has confirmed his heuristic analysis. But it has also led to some further results which were not immediately obvious. For example, the analysis shows that for any initial hat colors (such as RWR) that involve a red hat for student 3, the same no, no, yes sequence will repeat indefinitely. For initial hat colors RRW or WRW, the responses will be no, yes, yes repeated indefinitely. Finally, if the state is either WWW or RWW, then after the teacher speaks every child will be able to identify the color of her hat. In fact, we will argue later that one student must eventually realize her hat color, no matter which state the teacher begins by confirming or denying, and no matter how many students there are, and no matter what order they answer in, including possibly answering simultaneously.

The second puzzle, about the envelopes, can be explored along similar lines, as a special case of the analysis in Sebenius and Geanakoplos (1983); it is closely related to Milgrom–Stokey (1982). For that story, take the set of all possible worlds Ω to be the set of ordered pairs (m, n) with m and n integers between 1 and 7; m and n differ by one, but either could be the larger. At state (m, n), agent 1 has 10^m dollars in his envelope, and agent 2 has 10^n dollars in his envelope.

We graph the state space and partitions for this example below. The dots correspond to states with coordinates giving the numbers of agent 1 and 2, respectively. Agent 1 cannot distinguish states lying in the same row, and agent 2 cannot distinguish states lying in the same column.

The partitions divide the state space into two components, namely those states reachable from $(2, 1)$ and those states reachable from $(1, 2)$. In one connected component of mutually reachable states, agent 1 has an even number and 2 has an odd number, and this is "common knowledge" – that is, 1 knows it and 2 knows it and 1 knows that 2 knows it, and so on. For example, the state $(4, 3)$ is reachable from the state $(2, 1)$, because at $(2, 1)$, agent 1 thinks the state $(2, 3)$ is possible, and at $(2, 3)$ agent 2 would think the state $(4, 3)$ is possible. This component of the state space is highlighted by the staircase where each step connects two states that agent 1 cannot distinguish, and each rising connects two states that agent 2 cannot distinguish. In the other component of mutually reachable states, the even/odd is reversed, and again that is common knowledge. At states $(1, 2)$ and $(7, 6)$ agent 1 knows the state, and in states $(2, 1)$ and $(6, 7)$ 2 knows the state. In every state in which an agent i does not know the state for sure, he can narrow down the possibilities to two states. Both players start by believing that all states are equally likely. Thus, at $\omega = (4, 3)$ each son quite rightly calculates that it is preferable to switch envelopes when first approached by his father. The sons began from a symmetric position, but they each have an incentive to take opposite sides of a bet because they have different information.

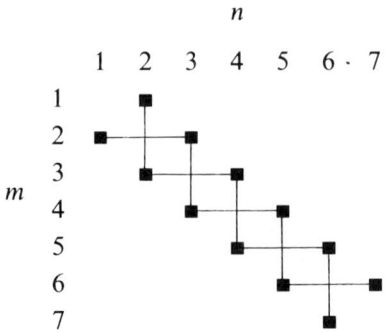

When their father tells each of them the other's previous answer, however, the situation changes. Neither son would bet if he had the maximum $10 million in his envelope, so when the sons learn that the other is willing to bet, it becomes "common knowledge" that neither number is 7. The state space is now divided into four pieces, with the end states (6, 7) and (7, 6) each on their own. But a moment later neither son would allow the bet to stand if he had $1 million in his envelope, since he would realize that he would be giving up $1 million for only $100 000. Hence if the bet still stands after the second instant, both sons conclude that the state does not involve a 6, and the state space is broken into two more pieces; now (5, 6) and (6, 5) stand on their own. If after one more instant the bet is still not rejected by one of the sons, they both conclude that neither has $100 000 in his envelope. But at this moment the son with $10 000 in his envelope recognizes that he must lose, and the next time his father asks him, he voids the bet.

If in choosing to bet the sons had to ante a dollar knowing that the bet would be cancelled and the dollar lost if the other son refused to bet in the same round, then both of them would say that they did not want the bet on the very first round. We explain this later.

Here is a third example, reminiscent of the detective story. Suppose, following Aumann (1976) and Geankoplos and Polemarchakis (1982), that two agents are discussing their opinions about the probability of some event, or more generally, of the expectation of a random variable. Suppose furthermore that the agents do not tell each other why they came to their conclusions, but only what their opinions are.

For example, let the set of all possible worlds be $\Omega = \{1, 2, \ldots, 9\}$, and let both agents have identical priors which put uniform weight $1/9$ on each state, and let $P_1 = \{\{1,2,3\}, \{4,5,6\}, \{7,8,9\}\}$ and $P_2 = \{\{1,2,3,4\}, \{5,6,7,8\}, \{9\}\}$. Suppose that a random variable x takes on the following values as a function of the state:

1	2	3	4	5	6	7	8	9
17	−7	−7	−7	17	−7	−7	−7	17

We can represent the information of both agents in the following graph, where heavy lines connect states that agent 1 cannot distinguish, and dotted lines connect states that agent 2 cannot distinguish.

$$1\overline{\cdots}2\overline{\cdots}3\cdots 4\overline{}5\overline{\cdots}6\cdots 7\overline{\cdots}8\overline{}9$$

Suppose that $\omega = 1$. Agent 1 calculates his opinion about the expectation of x by averaging the values of x over the three states, 1, 2, 3 that he thinks are possible, and equally likely. When agent 1 declares that his opinion of the expected value of x is 1, he reveals nothing, since no matter what the real state of the world, his partition would have led him to the same conclusion. But when agent 2 responds with his opinion, he is indeed revealing information. For if he thinks that $\{1, 2, 3, 4\}$ are possible, and equally likely, his opinion about the expected value of x is -1. Similarly, if he thought that $\{5, 6, 7, 8\}$ were possible and equally likely, he would say -1, while if he knew only $\{9\}$ was possible, then he would say 17. Hence when agent 2 answers, if he says -1, then he reveals that the state must be between 1 and 8, whereas if he says 17 then he is revealing that the state of the world is 9. After his announcement, the partitions take the following form:

$$1\overline{\cdots}2\overline{\cdots}3\cdots 4\overline{}5\overline{\cdots}6\cdots 7\overline{\cdots}8 \quad 9$$

If agent 1 now gives his opinion again, he will reveal new information, even if he repeats the same number he gave the last time. For 1 is the appropriate answer if the state is 1 through 6, but if the state were 7 or 8 he would say -7, and if the state were 9 he would say 17. Thus after 1's second announcement, the partitions take the following form:

$$1\overline{\cdots}2\overline{\cdots}3\cdots 4\overline{}5\overline{\cdots}6 \quad 7\overline{\cdots}8 \quad 9$$

If agent 2 now gives his opinion again he will also reveal more information, even if he repeats the same opinion of -1 that he gave the first time. Depending on whether he says -1, 5, or -7, agent 1 will learn something different, and so the partitions become:

$$1\overline{\cdots}2\overline{\cdots}3\cdots 4 \quad 5\overline{\cdots}6 \quad 7\overline{\cdots}8 \quad 9$$

Similarly if 1 responds a third time, he will yet again reveal more information, even if his opinion is the same as it was the first two times he spoke. The evolution of the partitions after 2 speaks a second time, and 1 speaks a third time are given below:

$$1\overline{\cdots}2\overline{\cdots}3 \quad 4 \quad 5\overline{\cdots}6 \quad 7\overline{\cdots}8 \quad 9$$

Finally there is no more information to be revealed. But notice that 2 must now have the same opinion as 1! If the actual state of nature is $\omega = 1$, then the responses of agents 1 and 2 would have been $(1, -1), (1, -1), (1, 1)$.

Although this example suggests that the partitions of the agents will converge, this is not necessarily true – all that must happen is that the opinions about expectations converge. Consider the state space below, and suppose that agents assign probability 1/4 to each state. As usual, 1 cannot distinguish states in the same row and 2 cannot distinguish states in the same column.

$$\begin{array}{cc} a & b \\ c & d \end{array}$$

Let $x(a) = x(d) = 1$, and $x(b) = x(c) = -1$. Then at $\omega = a$, both agents will say that their expectation of x is 0, and agreement is reached. But the information of the two agents is different. If asked *why* they think the expected value of x is 0, they would give different explanations, and if they shared their reasons, they would end up agreeing that the expectation should be 1, not 0.

As pointed out in Geanakoplos and Sebenius (1983), if instead of giving their opinions of the expectation of x, the agents in the last two examples were called upon to agree to bet, or more precisely, they were asked only if the expectation of x is positive or negative, exactly the same information would have been revealed, and at the same speed. In the end the agents would have agreed on whether the expectation of x is positive or negative, just as in the envelopes problem. This convergence is a general phenomenon. In general, however, the announcements of the precise value of the expectation of a random variable conveys much more information than the announcement of its sign, and so the two processes of betting and opining are quite different. When there are three agents, a bet can be represented by a vector $x(\omega) = (x_1(\omega), x_2(\omega), x_3(\omega))$, denoting the payoffs to each agent, such that $x_1(\omega) + x_2(\omega) + x_3(\omega) \leq 0$. If each agent i is asked in turn whether the expectation of x_i is positive, one agent will eventually say no. Thus eventually the agents will give different answers to different questions, as in the hats example. Nevertheless, in the next three sections we shall show how to understand all these examples in terms of a general process of convergence to "agreement."

5. Characterizing common knowledge of events and actions

To this point, the examples and discussion have used the term common knowledge rather loosely, as simply meaning a fact that everyone knows, that everyone knows that everyone knows, and so on. An example may help to give the reader a better grip on the idea.

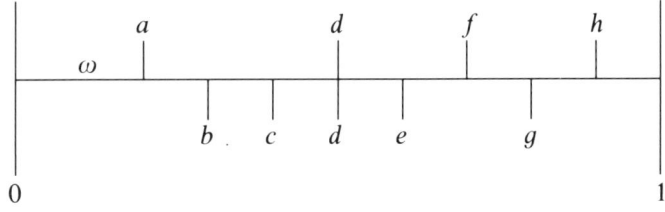

The whole interval $(0,1]$ represent Ω. The upper subintervals with endpoints $\{0, a, d, f, h, 1\}$ represent agent 1's partition. The lower subintervals with endpoints $\{0, b, c.d, e, g, 1\}$ represent agent 2's partition. At ω, 1 thinks $(0, a]$ is possible; 1 thinks 2 thinks $(0, b]$ is possible; 1 thinks 2 thinks 1 might think $(0, a]$ is possible or $(a, d]$ is possible. But nobody need think outside $(0, d]$. Note that $(0, d]$ is the smallest event containing ω that is both the union of partition cells of agent 1 (and hence self-evident to 1) and also the union of partition cells of player 2 (and hence self-evident to 2).

How can we formally capture the idea of i reasoning about the reasoning of j? For any event F, denote by $P_j(F)$ the set of all states that j might think are possible if the true state of the world were somewhere in F. That is, $P_j(F) = \bigcup_{\omega' \in F} P_j(\omega')$. Note that F is self-evident to j if and only if $P_j(F) = F$. Recall that for any ω, $P_i(\omega)$ is simply a set of states, that is it is itself an event. Hence we can write formally that at ω, i knows that j knows that the event G occurs iff $P_j(P_i(\omega)) \subset G$. The set $P_i(\omega)$ contains all worlds ω' that i believes are possible when the true world is ω, so i cannot be sure at ω that j knows that G occurs unless $P_j(P_j(\omega)) \subset G$.

The framework of Ω and the partitions (P_i) for the agents $i \in I$ also permits us to formalize the idea that at ω, i knows that j knows that k knows that some event G occurs by the formula $P_k(P_j(P_i(\omega))) \subset G$. (If $k = i$, then we say that i knows that j knows that i knows that G occurs). Clearly there is no limit to the number of levels of reasoning about each others' knowledge that our framework permits by iterating the P_i correspondences. In this framework we say that the state ω' is reachable from ω iff there is a sequence of agents $i, j, \ldots, k$ such that $\omega' \in P_k \cdots (P_j(P_i(\omega)))$, and we interpret that to mean that i thinks that j may think that $\ldots k$ may think that ω' is possible.

Definition. The event $E \subset \Omega$ is *common knowledge* among agents $i = 1, \ldots, I$ at ω if and only if for any n and any sequence $(i_1, \ldots, i_n)$, $P_{i_n}(P_{i_{n-1}} \cdots (P_{i_1}(\omega))) \subset E$, or equivalently, $\omega \in K_{i_1}(K_{i_2} \cdots (K_{i_n}(E)))$.

This formal definition of common knowledge was introduced by R. Aumann (1976). Note that an infinite number of conditions must be checked to verify that E is common knowledge. Yet when Ω is finite, Aumann (1976) showed that there is an equivalent definition of common knowledge that is easy to verify in a finite number of steps [see also Milgrom (1981)]. Recall that an event E is self-evident to i iff $P_i(E) = E$, and hence iff E is the union of some of i's partition cells. Since there are comparatively few such unions, the collection of self-evident events to a particular agent i is small. An event that is simultaneously self-evident to all agents i in I is called a public event. The collection of public events is much smaller still.

Characterizing common knowledge Theorem. *Let P_i, $i \in I$, be possibility correspondences representing the (partition) knowledge of individuals $i = 1, \ldots, I$ defined over a common state space Ω. Then the event E is common knowledge at ω if and* only if $M(\omega) \subset E$, where $M(\omega)$ is set of all states reachable from ω. Moreover, $M(\omega)$ can be described as the smallest set containing ω that is simultaneously self-evident to every agent $i \in I$. In short, E is common knowledge at ω if and only if there is a public event occurring at ω that entails E.

Proof. Let $M(\omega) = \bigcup_n \bigcup_{i_1,\ldots,i_n} P_{i_1} P_{i_2} \cdots P_{i_n}(\omega)$, where the union is taken over all strings $i_1,\ldots,i_n \in I$ of arbitrary length. Clearly E is common knowledge at ω if and only if $M(\omega) \subset E$. But notice that for all $i \in I$, $P_i(M(\omega)) = P_i(\bigcup_n \bigcup_{i_1,\ldots,i_n} P_{i_1} P_{i_2} \cdots P_{i_n}(\omega)) = \bigcup_{n+1} \bigcup_{i_1,\ldots,i_n} P_i P_{i_1} P_{i_2} \cdots P_{i_n}(\omega) \subset M(\omega)$, so $M(\omega)$ is self-evident for each i. □

Before leaving the characterization of common knowledge we define the meet M of the partitions $(P_i, i \in I)$ as the finest partition that is coarser than every P_i. (M is coarser than P_i if $P_i(\omega) \subset M(\omega)$ for all $\omega \in \Omega$; M is finer if the reverse inclusion holds.) To see that the meet exists and is unique, let us define the complete field $\mathscr{F}$ associated with any partition Q as the collection of all self-evident events, that is, the collection of all unions of the cells in Q. [A complete field is a collection of subsets of Ω that is closed under (arbitrary) intersections and complements.] Every complete field $\mathscr{F}$ defines a partition Q where $Q(\omega)$ is the intersection of all the sets in $\mathscr{F}$ that include ω. Given the partitions $(P_i, i \in I)$, let the associated complete fields be $(\mathscr{F}_i, i \in I)$, and define $\mathscr{F} = \bigcap_{i \in I} \mathscr{F}_i$ as the collection of public events. Since the intersection of complete fields is a complete field, $\mathscr{F}$ is a complete field and associated with $\mathscr{F}$ is the meet M of the partition $(P_i, i \in I)$. Clearly $M(\omega)$ is the smallest public event containing ω. Hence we have another way of saying that the event E is common knowledge at ω: at ω the agent whose knowledge is the meet M of the partitions $(P_i, i \in I)$ knows E.

Since self-evident sets are easy to find, it is easy to check whether the event E is common knowledge at ω. In our three puzzles, the public event $M(\omega)$ appears as the connected component of the graph that contains ω. An event E is common knowledge at ω iff it contains $M(\omega)$.

A state of nature so far has described the prevailing physical situation; it also describes what everybody knows, and what everybody knows about what everybody knows etc. We now allow each state to describe what everybody does. Indeed, in the three puzzles given so far, each state did specify at each time what each agent does. Consider the opinion puzzle. For all ω between 1 and 8, at first agent 2 was ready to announce the expectation of x was -1, while at $\omega = 9$, he was ready to announce the expectation of x was 17. By the last time period, he was ready to announce at ω between 1 and 3, the expectation of x was 1, at $\omega = 4$ it was -7 and so on. We now make the dependence of action on the state explicit. Let A_i be a set of possible actions for each agent i. Each ω thus specifies an action $a_i = f_i(\omega)$ in A_i for each agent i in I.

Having associated actions with states, it makes sense for us to rigorously describe whether at ω i knows what action j is taking. Let a_j be in A_j, and let E be the set of states at which agent j takes the action a_j. Then at ω, i knows that j is taking the action a_j iff at ω, i knows that E occurs. Similarly, we say that at ω it is common knowledge that j is taking the action a_j iff the event E is common knowledge at ω.

Let us close this section by noting that we can think of the actions an agent i takes as deriving from an *external action rule* $\psi_i : 2^\Omega / \phi \to A_i$ that prescribes what to do as a function of any information situation he might be in. The first girl could not identify her hat color because she thought both RRR and WRR were possible states. Had she thought that only the state RRR was possible, she would have known her hat color. The second detective expected x to be -1 because that was the average value x took on the states $\{1, 2, 3, 4\}$ that he thought were possible. Later, when he thought only $\{1, 2, 3\}$ were possible, his expectation of x became 1. Both the girl and the detective could have answered according to their action rule for any set of possible states.

6. Common knowledge of actions negates asymmetric information about events

The external action rules in our three puzzles all satisfy the sure-thing principle, which runs like this for the opinion game: If the expectation of a random variable is equal to "a" conditional on the state of nature lying in E, and similarly if the expectation of the same random variable is also "a" conditional on the state lying F, and if E and F are disjoint, then the expectation of the random variable conditional on $E \cup F$ is also "a". Similarly, if the expectation of a random variable is positive conditional on E, and it is also positive conditional on a disjoint set F, then it is positive conditional on $E \cup F$.[4] In the hat example, the sure-thing principle sounds like this: An agent who cannot tell his hat color if he is told only that the true state of nature is in E, and similarly if he is told it is in F, will still not know if he is told only that the true state is in $E \cup F$. Similarly if he could deduce from the fact that the state lies in E that his hat color is red, and if he could deduce the same thing from the knowledge that the states is in F, then he could also deduce this fact from the knowledge that the state is in $E \cup F$. (Note that we did not use the fact that E intersection F is empty).

An *Agreement Theorem* follows from this analysis, *that common knowledge of actions negates asymmetric information about events*. If agents follow action rules satisfying the sure-thing principle, and if with asymmetric information the agents

---

[4]Or in other terms, we say that an external action rule $\psi : 2^\Omega / \phi \to A$ satisfies the sure-thing principle iff $\psi(A) = \psi(B) = a$, $A \cap B = \phi$ implies $\psi(A \cup B) = a$. If Ω is infinite we require that $\psi(\bigcup_\alpha E_\alpha) = a$ whenever the E_α are disjoint, and $\psi(E_\alpha) = a$ for all α in an arbitrary index set. The sure-thing principle could have this interpretation in the detectives example: if a detective would have arrested the butler if the blood type turned out to be A, given his other clues, and if he also would have arrested the butler if the blood type turned out to be O, given those other clues, then he should arrest the butler as soon as he finds out the blood type must be A or O, given those other clues.

i are taking actions a_i, then if those actions are common knowledge, there is symmetric information that would lead to the same actions. Furthermore, if all the action rules are the same, then the agents must be taking the same actions, $a_i = a$ for all i.

Theorem. *Let $(\Omega, (P_i, A_i, f_i)_{i \in I})$ be given, where Ω is a set of states of the world, P_i is a partition on Ω, A_i is an action set, and $f_i: \Omega \to A_i$ specifies the action agent i takes at each $\omega \in \Omega$, for all $i \in I$. Suppose that f_i is generated by an action rule $\psi_i: 2^\Omega \to A_i$ satisfying the sure-thing-principle. [Thus $f_i(\omega) = \psi_i(P_i(\omega))$ for all $\omega \in \Omega$, $i \in I$.] If for each i it is common knowledge at ω that f_i takes on the value a_i, then there is some single event E such that $\psi_i(E) = a_i$ for every $i \in I$.*[5]

Corollary. *Under the conditions of the theorem, if $\psi_i = \psi$ for all i, then $a_i = a$ for all i.*

Proof. Let $E = M(\omega)$. Since it is common knowledge that f_i takes on the value a_i at ω, $\psi_i(P_i(\omega')) = f_i(\omega') = a_i$ for all $\omega' \in E$. Since E is self-evident to each i, it is the disjoint union of cells on which ψ_i takes the same action a_i. Hence by the sure-thing principle, $\psi_i(E) = a_i$ for all $i \in I$. □

To illustrate the theorem, consider the previous diagram in which at ω the information of agent 1, $(0, a]$, is different from the information of agent 2, $(0, b]$. This difference in information might be thought to explain why agent 1 is taking the action a_1 whereas agent 2 is taking action a_2. But if it is common knowledge that agent 1 is taking action a_1 at ω, then that agent must also be taking action a_1 at $(a, d]$. Hence by the sure-thing principle he would take action a_1 on $(0, d]$. Similarly, if it is common knowledge at ω that agent 2 is taking action a_2 at ω, then not only does that agent do a_2 on $(0, b]$, but also on $(b, c]$ and $(c, d]$. Hence by the sure-thing principle, he would have taken action a_2 had he been informed of $(0, d]$. So the symmetric information $(0, d]$ explains both actions. Furthermore, if the action rules of the two agents are the same, then with the same information $(0, d]$, they must take the same actions, hence $a_1 = a_2$.

The agreement theorem has the very surprising consequence that whenever logically sophisticated agents come to common knowledge (agreement) about what each shall do, the joint outcome does not use in any way the differential information about events they each possess. This theorem shows that it cannot be common knowledge that two or more players with common priors want to bet with each other, even though they have different information. Choosing to bet (which

---

[5] A special case of the theorem was proved by Aumann (1976), for the case where the decision rules $\psi_i = \psi =$ the posterior probability of a fixed event A. The logic of Aumann's proof was extended by Cave [1983] to all "union consistent" decision rules. Bacharach (1985) identified union consistency with the sure-thing principle. Both authors emphasized the agreement reached when $\psi_i = \psi$. However the aspect which I emphasize here is that even when the ψ_i are different, and the actions are different, they can all be explained by the same information E.

amounts to deciding that a random variable has positive expectation) satisfies the sure-thing principle, as we saw previously. Players with common priors and the same information would not bet against each other. The agreement theorem then assures us that even with asymmetric information it cannot be common knowledge that they want to bet [Milgrom and Stokey (1982)].

Similarly, agents who have the same priors will not agree to disagree about the expectation of a random variable. Conditional expectations satisfy the sure-thing principle. Agents with identical priors and the same information would have the same opinion. Hence the agreement theorem holds that they must have the same opinion, even with different information, if those opinions are common knowledge [Aumann (1986)].

7. A dynamic state space

We now come to the question of how agents reach common knowledge of actions. Recall that each of our three puzzle illustrated what could happen when agents learn over the course of time from the actions of the others. These examples are special cases of a *getting to common knowledge theorem*, which we state loosely as follows. Suppose that the state space Ω is finite, and that there are a finite number of agents whose knowledge is defined over Ω, but suppose that time goes on indefinitely. If all the agents see all the actions, then at some finite time period t^* it will be common knowledge at every ω what all the agents are going to do in the future.

The logic of the getting to common knowledge theorem is illustrated by our examples. Over time the partitions of the agents evolve, getting finer and finer as they learn more. But if Ω is finite, there is an upper bound on the cardinality of the partitions (they cannot have more cells than there are states of nature). Hence after a finite time the learning must stop.

Apply this argument to be betting scenario. Suppose that at every date t each agent declares, on the basis of the information that he has then, whether he would like to bet, assuming that if he says yes the bet will take place (no matter what the other agents say). Then eventually one agent will say no. From the convergence to common knowledge theorem, at some date t^* it becomes common knowledge what all the agents are going to say. From the theorem that common knowledge of actions negates asymmetric information, at that point the two agents would do the same thing with symmetric information, provided it were chosen properly. But no choice of symmetric information can get agents to bet against each other, if they have the same priors. Hence eventually someone must say no [Sebenius and Geanakoplos (1983)].

The same argument can be applied to the detectives' conversation, or to people expressing their opinions about the probability of some event [Geanakoplos and Polemarchakis (1982)]. Eventually it becomes common knowledge what everyone is going to say. At that point they must all say the same thing, since opining is an action rule which satisfies the sure-thing principle.

Let us show that the convergence to common knowledge theorem also clarifies the puzzle about the hats. Suppose RRR is the actual state and that it is common knowledge (after the teacher speaks) that the state is not WWW. Let the children speak in any order, perhaps several at a time, and suppose that each speaks at least every third period, and every girl is heard by everyone else. Then it must be the case that eventually one of the girls knows her hat color. For if not, then by the above theorem it would become common knowledge at RRR by some time t^* that no girl was ever going to know her hat color. This means that at every state ω reachable from RRR with the partitions that the agents have at t^*, no girl knows her hat color at ω. But since 1 does not know her hat color at RRR, she must think WRR is possible. Hence WRR is reachable from RRR. Since 2 does not know her hat color at any state reachable from RRR, in particular she does not know her hat color at WRR, and so she must think WWR is possible there. But then WWR is reachable from RRR. But then 3 must not know her hat color at WWR, hence she must think WWW is possible there. But this implies that WWW is reachable from RRR with the partitions the agents have at time t^*, which contradicts the fact that it is common knowledge at RRR that WWW is not the real state.

The hypothesis that the state space is finite, even though time is infinite, is very strong, and often not justified. But without that hypothesis, the theorem that convergence to common knowledge will eventually occur is clearly false. We shall discuss the implications of an infinite state space in the next section, and then again later.

It is useful to describe the dynamic state space formally. Let T be a discrete set of consecutive integers, possibly infinite, denoting calendar dates. We shall now consider an expanded state space $\bar{\Omega} = \Omega \times T$. A state of nature ω in Ω prescribes what has happened, what is happening, and what will happen at every date t in T. An event $\bar{E}$ contained in $\bar{\Omega}$ now specifies what happens at various dates. The simplest events are called dated events and they take the form $\bar{E} = E \times \{t\}$ for some calendar time y, where E is contained in Ω.

Knowledge of agent i can be represented in the dynamic state space precisely as it was in the static state space as a partition $\bar{P}_i$ of $\bar{\Omega}$. We shall always suppose that agent i is aware of the time, i.e., we suppose that if (ω', t') is in $\bar{P}_i(\omega, t)$, then $t' = t$. It follows that at each date t we can define a partition P_{it} of Ω corresponding to what agent i knows at date t about Ω, i.e., $P_{it}(\omega) = \{\omega' \in \Omega : (\omega', t) \in \bar{P}_i(\omega, t)\}$. The snapshot at time t is exactly analogous to the static model described earlier. Over time the agent's partition of Ω evolves.

In the dynamic state space we can formalize the idea that agent i knows at time t about what will happen later at time t', perhaps by applying the laws of physics to the rotation of the planets for example. We say at that at some (ω, t), agent i knows that a (dated) event $E = E \times \{t'\}$ will occur at time $t' > t$ if $P_{it}(\omega) \subset E$. We say that it is common knowledge among a group of agents i in I at time t that the event E occurs at time t' iff $E = \{\omega : (\omega, t') \in E\}$ is common knowledge with respect to the information partitions P_{it}, i in I.

We now describe how actions and knowledge co-evolve over time. Let A_i be the action space of agent i, for each $i \in I$. Let S_i be the signal space of agent i, for each i in I. Each player $i \in I$ receives a signal at each time $t \in T$, depending on all the actions taken at time t and the state of nature, given by the function $\sigma_{it} : A_1 \times \cdots \times A_I \times \Omega \to S_i$. At one extreme σ_{it} might be a constant, if i does not observe any action. At the other extreme, where $\sigma_{it}(a_1, \ldots, a_I, \omega) = (a_1, \ldots, a_I)$, i observes every action. If some action is observed by all the players at every state, then we say the action is public. If a_{it} depends on the last term ω, without depending at all on the actions, then agent i does not observe the actions, but he does learn something about the state of the world. If each agent whispers something to the person on his left, then $\sigma_{it}(a_1, \ldots, a_I, \omega) = a_{i+1}$ (take $I + 1 = 1$).

Agents take actions $f_{it} : \Omega \to A_i$ depending on the state of nature. The actions give rise to signals which the agents use to refine their information. On the other hand, an agent must take the same action in two different states that he cannot distinguish.

We say that $(\Omega, (A_i, S_i), (\sigma_{it}, P_{it}, f_{it}))_{i \in I}^{t \in T}$ is a dynamically consistent model of action and knowledge (DCMAK) iff for all $t \in T$, $i \in I$

(1) $[P_{it}(\omega) = P_{it}(\omega')] \to [f_{it}(\omega) = f_{it}(\omega')]$ and
(2) $[\omega' \in P_{it+1}(\omega)] \Leftrightarrow [\{\sigma_{it}(f_{1t}(\omega), \ldots, f_{It}(\omega), \omega)$
$= \sigma_{it}(f_{1t}(\omega'), \ldots, f_{It}(\omega'), \omega')\}$ and $\omega' \in P_{it}(\omega)]$.

Condition (1) says that an agent can take action only on the basis of his current stock of knowledge. Condition (2) says that an agent puts together what he knows at time t and the signal he observes at time t to generate his knowledge at time $t + 1$.

We can describe condition (2) somewhat differently. Let $g : \Omega \to S$, where g is any function and S any set. Then we say that g generates the partition G of Ω defined by $\omega' \in G(\omega)$ iff $g(\omega') = g(\omega)$. Furthermore, given partitions Q and G of Ω, we define their join $Q \vee G$ by

$$[Q \vee G](\omega) = Q(\omega) \cap G(\omega).$$

If we have a family of partitions $(Q_i, i \in I)$, we define their join $J = \vee_{i \in I} Q_i$ by

$$J(\omega) = [\vee_{i \in I} Q_i](\omega) = \bigcap_{i \in I} Q_i(\omega),$$

provided that Ω is finite. Note that J is a finer partition than each Q_i in the sense that $J(\omega) \subset Q_i(\omega)$ for all i and ω. But any partition R that is also finer than each Q_i must be finer than J; so J is the coarsest common refinement of the Q_i.

Let Σ_{it} be the partition generated by the function $\omega \to \sigma_{it}(f_{1t}(\omega), \ldots, f_{It}(\omega), \omega)$. Then condition (2) becomes

$$P_{it+1} = P_{it} \vee \Sigma_{it}.$$

Notice that over time the partitions grow finer, that is, each cell of $P_{i\tau}$ is the disjoint union of cells in P_{it} if $\tau < t$.

We now state a rigorous version of the getting to common knowledge theorem. Let $\#P_{i1}$ denote the number of cells in the first partition P_{i1}.

Theorem. *Let $(\Omega, (A_i, S_i), (\sigma_{it}, P_{it}, f_{it}))_{i\in I}^{t\in T}$ be a dynamically consistent model of action and knowledge. Let $T^* = \sum_{i\in I} (\#P_{i1} - 1)$. Suppose T^* is finite and suppose $T > T^*$. Suppose for all $i \in I$ and $t \in T$, σ_{it} does not depend on ω. Then there is some $t \leqslant T^*$ at which it is common knowledge at every $\omega \in \Omega$ that every agent i knows the signal he will observe in period t. If T is infinite, then there is some finite period t^* at which it is common knowledge at every $\omega \in \Omega$ that each agent already knows all the signals he will receive for all $t \geqslant t^*$. In particular, if some agent's actions are always public, and T is infinite, then at some time t^* it will already be common knowledge what action that agent will take for all $t \geqslant t^*$.*

8. Generalizations of agreeing to disagree

To conclusion that agents with common priors who talk long enough will eventually agree can be generalized to infinite state spaces in which the opinions may never become common knowledge. Moreover, the convergence does not depend on every agent hearing every opinion.

Let π be a probability measure on Ω, and let $x: \Omega \to \mathbb{R}$ be a random variable. For convenience, let us temporarily assume that Ω is finite and that $\pi(\omega) > 0$ for all $\omega \in \Omega$. Given any partition P on Ω, we define the random variable $f = E(x|P)$ by $f(\omega) = [1/\pi(P(\omega))] \sum_{\omega' \in P(\omega)} x(\omega') \pi(\omega')$. Notice that if F is the partition generated by f, then $E(f|Q) = f$ if and only if Q is finer than F. If so, then we say f is measurable wrt Q. If Q is finer than P, then $E(E(x|P)|Q) = E(x|P) = E(E(x|Q)|P)$.

A *martingale* $(f_t, P_t), t = 1, 2, \ldots$ is a sequence of random variables and partitions such that f_t is measurable wrt P_t for all t, and P_{t+1} is finer than P_t for all t, and such that for all t,

$$E(f_{t+1}|P_t) = f_t.$$

The martingale convergence theorem guarantees that the martingale functions must converge, $f_t(\omega) \to f(\omega)$ for all ω, for some function f. The classic case of a martingale occurs when x and the increasingly finer partitions P_t are given, and f_t is defined by $E(x|P_t)$. In that case $f_t \to f = E(x|P_\infty)$ where P_∞ is the join of the partitions $(P_t, t = 1, 2, \ldots)$. Furthermore, if (f_t, P_t) is a martingale and if F_t is the partition generated by $(f_1, \ldots, f_t)$, then (f_t, F_t) is also a martingale.

The foregoing definitions of conditional expectation, and the martingale convergece theorem, can be extended without change in notation to infinite state spaces Ω provided that we think of partitions as σ-fields, and convergence $f_t \to f$ as convergence π-almost everywhere, $f_t(\omega) \to f(\omega)$ for all $\omega \in A$ with $\pi(A) = 1$. (We must also assume that the f_t are all uniformly bounded, $|f_t(\omega)| \leq M$ for all t, and π-almost all ω.) The join of a family of σ-fields, $\mathscr{F}_i$, $i \in I$, is the smallest σ-field containing the union of all the $\mathscr{F}_i$. We presume the reader is familiar with σ-fields. Otherwise he can continue to assume Ω is finite.

We can reformulate the opinion dialogue described in Geanakoplos and Polemarchakis (1982) in terms of martingales, as Nielsen (1984) showed. Let the DCMAK $(\Omega, (A_i, S_i), (\sigma_{it}, P_{it}, f_{it}))_{i \in I}^{t \in T = N}$ be defined so that for some random variable $x: \Omega \to \mathbb{R}$ and probability π,

$$f_{it} = E(x|P_{it}),$$
$$A_i = S_i = \mathbb{R},$$
$$\sigma_{it}(a, \ldots, a_I, \omega) = (a_1, \ldots, a_I).$$

It is clear that (f_{it}, P_{it}) is a martingale for each $i \in I$. Hence we can be sure that each agent's opinion converges, $f_{it}(\omega) \to f_{i\infty}(\omega) = E[x|P_{i\infty}]$ where $P_{i\infty} = \vee_{t \in T} P_{it}$.

Let F_{it} be the σ-field generated by the functions $f_{i\tau}$, for $\tau \leq t$. Then (f_{it}, F_{it}) is also a martingale, and $f_{i\infty} = E(x|F_{i\infty})$ where $F_{i\infty} = \vee_{t \in T} F_{it}$. If agent j hears agent i's opinion at each time t, then for $\tau > t$, $P_{j\tau}$ is finer than F_{it}. Hence for $\tau > t$,

$$E(f_{j\tau}|F_{it}) = E(E(x|P_{j\tau})|F_{it}) = E(x|F_{it}) = f_{it}.$$

Letting $t \to \infty$, we get that $E(f_{j\infty}|F_{i\infty}) = f_{i\infty}$, from which it follows that the variance $\text{Var}(f_{j\infty}) > \text{Var}(f_{i\infty})$ unless $f_{j\infty} = f_{i\infty}$ (π-almost everywhere). But since i hears j's opinion at each period t, the same logic shows also that $\text{Var}(f_{i\infty}) > \text{Var}(f_{j\infty})$ unless $f_{i\infty} = f_{j\infty}$. We conclude that for all pairs, $f_{i\infty} = f_{j\infty}$. Thus we have an alternative proof of the convergence theorem in Geanakoplos–Polemarchakis, which also generalizes the result to infinite state space Ω.

The proof we have just given does not require that the announcements of opinions be public.

Following Parikh and Krasucki (1990), consider $I < \infty$ agents sitting in a circle. Let each agent whisper his opinion (i.e., his conditional expectation of x) in turn to the agent on his left. By our getting to common knowledge theorem if Ω is finite, then after going around the circle enough times, it will become common knowledge that each agent knows the opinion of the agent to his immediate right. (Even if Ω is infinite, the martingale property shows that each agent's own opinion converges.) It seems quite possible, however, that an agent might not know the opinion of somebody two places to his right, or indeed of the agent on his left to whom he does all his speaking but from whom he hears absolutely nothing. Yet all the opinions must eventually be the same, and hence eventually every agent does in fact know everybody else's opinion.

To see this, observe that if in the previous proof we supposed $\sigma_{it}(a_1, \ldots, a_I, \omega) = a_{i+1\,t}$, then we could still deduce that $E(f_{i\infty}|F_{i+1,\infty}) = f_{i+1,\infty}$, and hence $\text{Var}(f_{i\infty}) > \text{Var}(f_{i+1,\infty})$ unless $f_{i\infty} = f_{i+1,\infty}$. But working around the circle, taking $I + 1 = 1$, we get that $\text{Var}(f_{1\infty}) > \cdots > \text{Var}(f_{1\infty})$ unless all the $f_{i\infty}$ are the same.

The reader may wonder whether the convergence holds when the conversation proceeds privately around a circle if the actions f_{it} are not conditional expectations, but are derivable from external action rules $\psi_i: 2^\Omega \to A_i$ satisfying the sure-thing principle. Parikh and Krasucki show that the answer is no, even with a finite state space. When Ω is finite, then convergence obtains if the action rule satisfies $A_i = \mathbb{R}$ and if $E \cap F = \phi$, $\psi_i(E \cup F) = \lambda \psi_i(E) + (1 - \lambda) \psi_i(F)$ for some $0 < \lambda < 1$.

Following McKelvey and Page (1986), suppose that instead of whispering his opinion to the agent on his left, each agent whispers his opinion to a poll-taker who waits to hear from everybody and then publicly reveals the average opinion of the I agents. (Assume as before that all the agents have the same prior over Ω.) After hearing this pollster's announcement, the agents think some more and once again whisper their opinions to the pollster who again announces the average opinion, etc. From the convergence to common knowledge theorem, if the state space is finite, then eventually it will be common knowledge what the average opinion is even before the pollster announces it. But it is not obvious that any agent i will know what the opinion of any other agent j is, much less that they should be equal. But in fact it can be shown that everyone must eventually agree with the pollster, and so the opinions are eventually common knowledge and equal.

We can see why by reviewing the proof given in Nielsen et al. (1990). Continuing with our martingale framework, let $\sigma_{jt}(a_1, \ldots, a_I, \omega) = (1/I) \sum_{i \in I} a_i$. Let $f_t(\omega) = (1/I) \sum_{i \in I} f_{it}(\omega)$.

From the getting to common knowledge theorem for finite Ω, or the martingale convergence theorem for infinite Ω, we know that $E(x|P_{it}) \to_t f_{i\infty} \equiv E(x|P_{i\infty})$ for all $i \in I$, π-almost everywhere. Hence $f_t \to f_\infty = (1/I) \sum_{i \in I} f_{i\infty}$ π-almost everywhere. Note that f_t is measurable with P_{it+1}, hence $P_{i\infty} = \vee_{t=0}^\infty P_{it}$ is finer than the partition $\mathscr{F}$ generated by f_∞ for all $i \in I$. Then

$$E((x - f_\infty)(f_{i\infty} - f_\infty)|\mathscr{F})$$
$$= E(E((x - f_\infty)(f_{i\infty} - f_\infty)|P_{i\infty})|\mathscr{F})$$
$$= E((f_{i\infty} - f_\infty)^2|\mathscr{F}) \geq 0,$$

with equality holding only if $f_{i\infty} = f_\infty$ π-almost everywhere.

It follows that

$$0 \leq \frac{1}{I} \sum_{i \in I} E((x - f_\infty)(f_{i\infty} - f_\infty)|\mathscr{F})$$
$$= E\left((x - f_\infty)\left(\frac{1}{I} \sum_{i \in I} (f_{i\infty} - f_\infty)\right)|\mathscr{F}\right)$$
$$= 0,$$

where equality holds only if $f_{i\infty} = f_\infty$ π-almost everywhere for each $i \in I$. But that is exactly what we wanted to prove.

9. Bayesian games

The analysis so far has considered action rules which depend on what the agent knows, but not on what he expects the other agents to do. This framework was sufficient to analyze the puzzles about the hat color, and the expectation of the random variable x, and also for betting when each son assumed that the bet would be taken (perhaps by the father) as long as he himself gave the OK. But in the envelopes puzzle when the first son realizes that the bet will be taken only if the other son also accepts it, he must try to anticipate the other son's action before deciding on his own action, or else risk that the bet comes off only when he loses. To take this into account we now extend our model of interactive epistemology to include payoffs to each agent depending on any hypothetical action choices by all the agents.

So far the states of nature describe the physical universe, the knowledge of the agents, and the actions of the agents. Implicitly in our examples the states of nature also described the payoffs to the agents of their actions, for this is the motivation for why they took their actions. We now make this motivation more explicit. At each ω, let us associate with every vector of actions $(a_1, \ldots, a_I)$ of all the I players a payoff to each agent i. In short, each ω defines a game $\Gamma(\omega)$ among the I agents. Since the players do not know the state ω, we must say more before we can expect them to decide which action to take. We suppose, in accordance with the Bayesian tradition, that each agent has a prior probability on the states of nature in Ω, and that at ω the agent updates his prior to a posterior by conditioning on the information that ω is in $P_i(\omega)$. This defines a Bayesian game. The agents then choose at each ω the actions which maximizes their expected utility with respect to these posterior probabilities, taking the action rules of the others as given. If the mapping of states to actions satisfies this optimizing condition, then we refer to the entire framework of states, knowledge, actions, payoffs, and priors as a Bayesian Nash equilibrium.

Formally, a (Bayesian) game is a vector $\Gamma = (I, \Omega, (P_i, \pi_i, A_i, u_i)_{i \in I})$ where $I = \{1, \ldots I\}$ is the set of players, Ω is the set of states of the world, P_i is a partition of Ω, π_i is a prior probability on Ω, A_i is the set of possible actions for player i, and $u_i : A \times \Omega \to \mathbb{R}$, where $A = A_1 \times \cdots \times A_I$, is the payoff to player i. For any product $Y = Y_1 \times \cdots \times Y_I$, the notation Y_{-i} means $Y_1 \times \cdots \times Y_{i-1} \times Y_{i+1} \times \cdots \times Y_I$.

A (Bayesian) Nash equilibrium for the game Γ is a vector $f = (f_1, \ldots, f_I)$ where $\forall i, f_i : \Omega \to A_i$ and

(1) $[P_i(\omega) = P_i(\omega')] \to [f_i(\omega) = f_i(\omega')]$, $i = 1, \ldots, I$ and

(2) $\forall i,\ \forall a \in A_i,\ \forall \omega \in \Omega,$

$$\sum_{\omega' \in P_i(\omega)} u_i(f_i(\omega), f_{-i}(\omega'), \omega') \pi_i(\omega') \geq \sum_{\omega' \in P_i(\omega)} u_i(a, f_{-i}(\omega'), \omega') \pi_i(\omega').$$

Condition (1) means that if player i cannot distinguish ω' from ω, then he must choose the same action $f_i(\omega) = f_i(\omega')$ in both states. Condition (2) (called ex post optimality) means that at each ω to which agent i assigns positive probability agent i prefers the action $f_i(\omega)$ to any other action $a \in A_i$, given his information $P_i(\omega)$ and given the action rule f_{-i} of the other agents. [Condition (2) is deliberately vacuous when $\pi_i(P_i(\omega)) = 0$.] Implicit in the definition is the idea that each player i knows the decision functions f_{-i} of all the other players. Without f_{-i} it would be impossible to define the payoff to agent i, since u_i depends on the choices of the other players, as well as the state and agent i's choice. This is not the place to explain how BNE arises, and hence how player i comes to know f_{-i}.

For example, in the last version of the envelopes puzzle the payoffs to the sons depend on what they *both* do. Below we list the payoffs to each son at a state $\omega = (m, n)$, depending on whether each decides to bet (B) or to stick with his own envelope and not bet (N). Note also the dependence of $\Gamma(\omega)$ on ω.

	B	N
B	$10^n - 1, 10^m - 1$	$10^m - 1, 10^n$
N	$10^m, 10^n - 1$	$10^m, 10^n$

We consider two more examples of Bayesian games in which $\Gamma(\omega)$ does not depend on ω. The first is based on the payoff matrix G, called Matching Pennies, given below:

	Left	Right
Top	$1, -1$	$-1, 1$
Bottom	$-1, 1$	$1, -1$

We know that there is a unique mixed strategy Nash equilibrium to G in which each player randomizes with equal probability over both of his strategies. This Nash equilibrium, like all others, is a special kind of Bayesian Nash equilibrium. Consider a state space Ω with four elements arranged in a 2×2 matrix. The first player has a partition of the state space consisting of the two rows of Ω. Similarly the second player has a partition of Ω given by the two columns of Ω. Both players have prior $1/4$ on each state. Let $\Gamma(\omega) = G$ for all $\omega = \Omega$. This defines the Bayesian game of Matching Pennies. The Bayesian Nash equilibrium for Matching Pennies is for each player to play the move corresponding to what he sees: if player 1 sees Top, he plays Top, etc.

When the games $\Gamma(\omega) = G$ are independent of the state, and there is a common prior $\pi = \pi_i$ on Ω given by a product of individual priors, then a Bayesian Nash equilibrium for Γ gives a slightly different interpretation to behavior from the usual mixed strategy Nash equilibrium for G. In a mixed strategy Nash equilibrium each player is flipping a coin to decide how to play. In Bayesian Nash equilibrium, there is one actual state. Thus each player is making a unique choice of (pure) move, namely the one assigned by that state. But the other player does not know which move that is, so to him the choice seems random. This reinterpretation of mixed strategy Nash equilibrium in terms of Bayesian Nash equilibrium is due to Ambruster and Boge (1979).

When there is a common prior $\pi = \pi_i$, and the games $\Gamma(\omega) = G$ are independent of the state but the conditional distribution of opponent's actions is allowed to depend on the state, then Bayesian Nash equilibrium reduces to what has been called a correlated equilibrium of G. The notion of correlated equilibrium was invented by Aumann in 1974. An elementary but important example of a correlated equilibrium is a traffic light, which provides our third example of Bayesian Nash equilibrium.

Each of two agents sees the color of his own light. There are four states: (green, green), (green, red), (red, green), and (red, red). Both players assign prior probability 1/2 to (green, red) and to (red, green) and probability zero to the other two states. In every state the choices (stop and go) and the payoffs are the same:

	Stop	Go
Stop	(1, 1)	(1, 2)
Go	(2, 1)	(0, 0)

This describes the Bayesian Nash game. The Bayesian Nash equilibrium actions for each state are symmetric for each player: Stop if he sees red, Go if he sees green.

In a Bayesian Nash equilibrium it is tautological (and hence common knowledge at every state ω) that each agent's knowledge is described by a partition, and that each agent has a prior probability over the states of the world. I refer to the partition/individual prior representation of knowledge as Bayesian rationality. In a Bayesian Nash equilibrium agents are always optimizing, that is choosing their actions to maximize their conditional expected utility, hence this must be common knowledge. In short, we may describe the situation of Bayesian Nash equilibrium as common knowledge of Bayesian rationality, and of optimization. The Harsanyi doctrine asserts that all agents must have the same prior. (We briefly discuss the merits of this doctrine in a later section.) Accepting the Harsanyi doctrine, let us suppose that the game $\Gamma(\omega) = G$ is the same for all $\omega \in \Omega$. Then, as Aumann (1987) pointed out, common knowledge of rationality and optimization is tantamount to correlated equilibrium.

At this point it is worth emphasizing that the structure of Bayesian Nash equilibrium extends the framework of interactive epistemology that we developed earlier. For example, we can turn the hats puzzle into a Bayesian game by specifying that the payoff to player i if she correctly guesses her hat color is 1, and if she says she does not know her payoff is 0, and if she guesses the wrong hat color her payoff is -infinity. Similarly, in the opinion game (in which the random variable x that the players are guessing about is given) we can define the payoff at ω to any player i if he chooses the action a to be $-[a - x(\omega)]^2$. It is well-known from elementary statistical decision theory that a player minimizes the expected squared error by guessing the conditional expectation of the random variable. Hence these payoffs motivate the players in the opinion game to behave as we have described them in our previous analysis.

Nowadays it is conventional wisdom to assert that many phenomena can only be explained via asymmetric information. A buyer and seller of a house may make peculiar seeming bids and offers, it is suggested, because they have different private information: each knows what the house is worth to him, but not to the other. But our analysis shows that this wisdom depends on there being uncertainty about the actions of the players. If their actions were common knowledge (for example if the bid and offer were common knowledge) then asymmetric information would have no explanatory power. Bayesian optimal decisions (i.e., maximizing expected utility) satisfy the sure-thing principle. Hence an argument similar to that given in the section on common knowledge of actions proves the following agreement theorem for Bayesian games: Suppose that in Bayesian Nash equilibrium it is common knowledge at some ω what actions the players are each taking. Then we can replace the partitions of the agents so that at ω all the agents have the same information, without changing anything else including the payoffs and the actions of the agents at every state, and still be at a Bayesian Nash equilibrium. In particular, any vector of actions that can be common knowledge and played as part of a Bayesian Nash equilibrium with asymmetric information can also be played as part of a Bayesian Nash equilibrium with symmetric information.

Theorem. *Let $(f_1, \ldots, f_I)$ be a Bayesian Nash equilibrium for the Bayesian game $\Gamma = (I, \Omega, (P_i, \pi_i, A_i, u_i)_{i \in I})$. Suppose at ω it is common knowledge that $(f_1, \ldots, f_I) = (a_1, \ldots, a_I)$. Then there are partitions $\hat{P}_i$ of Ω such that $\hat{P}_i(\omega) = \hat{P}_j(\omega)$ for all $i, j \in I$ and such that $(f_1, \ldots, f_I)$ is a Bayesian Nash equilibrium for the Bayesian game $\Gamma = (I, \Omega, (\hat{P}_i, \pi_i, A_i, u_i)_{i \in I})$.*

This theorem is surprising and it explains the puzzles discussed earlier. Of course, its application to Bayesian games is limited by the fact that the actions need not be common knowledge in a Bayesian Nash equilibrium (and in these games asymmetric information does have explanatory power. We return to this question later when we discuss games in extensive form). Consider again the Bayesian Nash game with the envelopes. One common knowledge component of the state space Ω consists of all (m, n) with m even and n odd. (The other common knowledge component reverses the parity.) Hence the agreement theorem for Bayesian Nash equilibrium assures us that there cannot be a Bayesian Nash equilibrium in which both brothers *always* choose to bet when m is even and n odd, for if there were, then the brothers would bet against each other with the same information, which is impossible. (Looked at from the point of view of identical information, both would agree that one brother had an expected payment at least as high as the other, so that taking into account the one dollar betting fee, one brother would not want to bet.) On the other hand, this is a trivial result, since we know at a glance that if the second brother sees that he has the maximum number of dollars in his envelope, he will not bet. A much stronger result would be that there is only one Bayesian Nash equilibrium. Since there is one Bayesian Nash equilibrium in which each brother chooses not to bet at every state of the world, this would rule out any Bayesian Nash equilibrium of the envelopes game in which both brothers bet in even one state. Such a result indeed is true, and we shall prove it later when we discuss speculation. But it cannot be directly derived from the agreement theorem, which itself depends only on the sure-thing principle. It must be derived from another property of Bayesian optimal decisions, namely that more information cannot hurt.

10. Speculation

The cause of financial speculation and gambling has long been put down to differences of opinion. Since the simplest explanation for differences of opinion is differences in information, it was natural to conclude that such differences could explain gambling and speculation. Yet, we now see that such a conclusion was premature.

To understand why, begin by distinguishing speculation from investing. With an investment, there are gains to trade for all parties that can be perceived by all sides when they have the same information. An agent who buys a stock from another will win if the stock price rises dramatically, while the seller will lose. This appears to be a bet. But another reason for trading the stock could be that the seller's marginal utility for money at the moment of the transaction is relatively high (perhaps because children are starting college), whereas the buyer's marginal utility for money is relatively higher in the future when the stock is scheduled to pay dividends. Even with symmetric information, both parties might think they are benefiting from the trade. This is not speculation. It appears, however, that only a small proportion of the trades on the stock market can be explained by such savings/investment reasons. Similarly if one agent trades out of dollars into yen, while another agent is trading yen for dollars, it might be because the first agent plans to travel to Japan and the second agent needs dollars to buy American goods. But since the volume of trade on the currency markets is orders of magnitude greater than the money purchases of goods and services, it would seem that speculation and not transactions demand explains much of this activity.

In this discussion, speculation will mean actions taken purely on account of differences of information. To formalize this idea, suppose that each agent has a status quo action, which does not take any knowledge to implement, and which guarantees him a utility independent of what actions the others choose. Suppose also that if every agent pursued the status quo action in every state, the resulting utilities would be Pareto optimal. In other words, suppose that it is common knowledge that the status quo is Pareto optimal. At a Pareto optimum there can be no further trade, if agents have symmetric information. A typical Pareto optimal situation might arise as follows. Risk averse agents (possibly with different priors) trade a complete set of Arrow–Debreu state contingent claims for money, one agent promising to deliver in some states and receive money in others, and so on. At the moment the contracts are signed, the agents do not know which state is going to occur, although they will recognize the state once it occurs in order to carry out the payments. After the signing of all the contracts for delivery, but before the state has been revealed, the status quo action of refusing all other contracts is well known to be Pareto optimal.

But now suppose that each agent receives additional information revealing something about which state will occur. If different agents get different information, that would appear to create opportunities for betting, or speculative trade.

Here we must distinguish between two kinds of speculation. One involves two agents who agree on some contingent transfer of money, perhaps using a handshake or a contract to give some sign that the arrangement is common knowledge between them, and that the payoffs do not depend on their own future actions. The other kind of speculation occurs between many agents, say on the stock market or at a horse race or a gambling casino, where an agent may commit to risk money before knowing what the odds may be (as at a horse race) or whether anyone will take him up on the bet (as in submitting a buy order to a stockbroker). In the second kind of speculation, the payoffs depend partly on the actions of the speculators, and what the agents are doing is not common knowledge. We reserve the term betting for (the first kind of) common knowledge speculation.

If it is common knowledge that the agents want to trade, as occurs when agents bet against each other, then our theorem that common knowledge of actions negates asymmetric information about events implies that the trades must be zero. But even if the actions are not common knowledge, there will be no more trade. Since the actions are not common knowledge, what is? Only the facts that the agents are rational, i.e., their knowledge is given by partitions, and that they are optimizing, and that the status quo is Pareto optimal.

Nonspeculation theorem. *Common knowledge of rationality and of optimization eliminates speculation. Let $\Gamma = (I, \Omega, (P_i, \pi_i, A_i, u_i)_{i \in I})$ be a Bayesian game. Suppose each player i in I has an action $z_i \in A_i$ such that for all $(f_1, \ldots, f_I)$, $\sum_{\omega \in \Omega} u_i(z_i, f_{-i}(\omega), \omega) \pi_i(\omega) = \bar{u}_i$. Furthermore, suppose that $(z_1, \ldots, z_I)$ yields a Pareto optimal outcome in the sense that if any $(f_1, \ldots, f_I)$ satisfies $\sum_{\omega \in \Omega} u_i(f(\omega), \omega) \pi_i(\omega) \geq \bar{u}_i$ for all $i \in I$, then $f_j(\omega) = z_j$ for all $\omega \in \Omega$, $j \in I$. Then Γ has a unique Bayesian Nash equilibrium $(f_1^*, \ldots, f_I^*)$ and $f_i^*(\omega) = z_i$ for all $\omega \in \Omega$, $i \in I$.*

Proof. The following lemma needs no proof. We emphasize, however, that it relies on the properties of partitions. □

Lemma (Knowledge never hurts a Bayesian optimizer). *Consider two single-player Bayesian Nash games $\Gamma_A = (I = \{i\}, \Omega, P_i, \pi_i, A_i, u_i)$ and $\Gamma_B = (I = \{i\}, \Omega, Q_i, \pi_i, A_i, u_i)$ that differ only in that P_i is finer than Q_i. Let f_i be a Bayesian Nash equilibrium for Γ_A, and let g_i be a Bayesian Nash equilibrium for Γ_B. Then*

$$\sum_{\omega \in \Omega} u_i(f_i(\omega), \omega) \pi_i(\omega) \geq \sum_{\omega \in \Omega} u_i(g_i(\omega), \omega) \pi_i(\omega).$$

Indeed the above inequality holds for any g satisfying $[P_i(\omega) = P_i(\omega')] \to [g_i(\omega) = g_i(\omega')]$ for all $\omega, \omega' \in \Omega$.

Proof of nonspeculation theorem. Let $(f_1, \ldots, f_I)$ be a Bayesian Nash equilibrium for Γ. Fix f_j, $j \neq i$, and look at the one-person Bayesian game this induces for player i. Clearly f_i must be a Bayesian Nash equilibrium for this one-person game. From the fact that knowledge never hurts a Bayesian optimizer we conclude that i could not do better by ignoring his information and playing $f_i^*(\omega) = z_i$ for all $\omega \in \Omega$. Hence

$$\sum_{\omega \in \Omega} u_i(f(\omega), \omega) \pi_i(\omega) \geq \sum_{\omega \in \Omega} u_i(z_i, f_{-1}(\omega), \omega) \pi_i(\omega) = \bar{u}_i.$$

But this holds true for all $i \in I$. Hence by the Pareto optimality hypothesis, $f_i = f_i^*$ for all $i \in I$. □

In the envelopes example the action z_i corresponds to not betting N. (We are assuming for now that the agents are risk neutral.) The sum of the payoffs to the players in any state is uniquely maximized by the action choice (N, N) for both players. A bet wastes at least a dollar, and only transfers money from the loser to the winner. It follows that the sum of the two players' ex ante *expected* payoffs is uniquely maximized when the two players (N, N) at every state. Hence by the nonspeculation theorem, the unique Bayesian Nash equilibrium of the envelope game involves no betting (N, N) at every state.

11. Market trade and speculation

We define an economy $E = (I, \mathbb{R}_+^L, \Omega, (P_i, U_i, \pi_i, e_i)_{i \in I})$ by a set of agents I, a commodity space $\mathbb{R}_+^L$, a set Ω of states of nature, endowments $e_i \in \mathbb{R}_+^{L\Omega}$ and utilities $U_i: \mathbb{R}_+^L \times \Omega \to \mathbb{R}$ for $i = 1, \ldots, I$, and partitions P_i and measures π_i for each agent $i = 1, \ldots, I$. We suppose each U_i is strictly monotonic, and strictly concave.

Definition. A *rational expectations equilibrium* (REE) $(p, (x_i)_{i \in I})$ for $E = (I, \mathbb{R}_+^L, (P_i, U_i, \pi_i, e_i)_{i \in I})$ is a function $p: \Omega \to \mathbb{R}_{++}^L$, such that for each $i \in I$, $x_i \in \mathbb{R}_+^{L\Omega}$ and if $z_i = x_i - e_i$, then

(i) $\sum_{i=1}^I z_i = 0$.
(ii) $p(\omega) z_i(\omega) = 0$, for all $i = 1, \ldots, I$, and all $\omega \in \Omega$.
(iii) $[P_i(\omega) = P_i(\omega')$ and $p(\omega) = p(\omega')] \to [z_i(\omega) = z_i(\omega')]$ for all $i = 1, \ldots, I$, and all $\omega, \omega' \in \Omega$.
(iv) Let $Q(p) = \{\omega: p(\omega) = p\}$. Then $\forall \omega \in \Omega$, and all i, if $e_i(\omega') + y \in \mathbb{R}_+^L$, $\forall \omega' \in P_i(\omega) \cap Q(p(\omega))$, and $p(\omega) y = 0$, then

$$\sum_{\omega' \in P_i(\omega) \cap Q(p(\omega))} U_i(x_i(\omega'), \omega') \pi_i(\omega') \geq \sum_{\omega' \in P_i(\omega) \cap Q(p(\omega))} U_i(e_i(\omega') + y, \omega') \pi_i(\omega').$$

The reference to rational in REE comes from the fact that agents use the subtle information conveyed by prices in making their decision. That is, they not only use the prices to calculate their budgets, they also use their knowledge of the function p to learn more about the state of nature. If we modified (iv) above to

(iv') $\sum_{\omega' \in P_i(\omega)} U_i(x_i(\omega'), \omega') \pi_i(\omega') \geq \sum_{\omega' \in P_i(\omega)} U_i(e_i(\omega') + y, \omega') \pi_i(\omega')$ for all $i = 1, \ldots, I$, for all $\omega \in \Omega$ and all $y \in \mathbb{R}^L$ with $p(\omega)y = 0$ and $e_i(\omega') + y \geq 0 \; \forall \omega' \in P_i(\omega)$

then we would have the conventional definition of competitive equilibrium (CE). The following nonspeculation theorem holds for REE, but note for CE. For an example with partition information in which agents do not learn from prices, and so speculate, see Dubey, Geanakoplos and Shubik (1987). We say that there are only speculative reasons to trade in E if in the absence of asymmetric information there would be no perceived gains to trade. This occurs when the initial endowment allocation is ex ante Pareto optimal, that is if $\sum_{i=1}^{I} y_i(\omega) \leq \sum_{i=1}^{I} e_i(\omega)$ for all $\omega \in \Omega$, and if for each $i = 1, \ldots, I$, $\sum_{\omega \in \Omega} u_i(y_i(\omega), \omega) \pi_i(\omega) \geq \sum_{\omega \in \Omega} u_i(e_i(\omega), \omega) \pi_i(\omega)$, then $y_i = e_i$ for all $i = 1, \ldots, I$.

Theorem (Nonspeculation in REE). *Let $E = (I, \mathbb{R}^L_{++}, \Omega, (P_i, U_i, \pi_i, e_i)_{i \in I})$ be an economy, and suppose the initial endowment allocation is ex ante Pareto optimal. Let $(p, (x_i)_{i \in I})$ be a rational expectations equilibrium. Then, $x_i = e_i$ for all $i = 1, \ldots, I$.*

This theorem can be proved in two ways. A proof based on the sure-thing principle was given by Milgrom and Stokey (1982). Proofs based on the principle that more knowledge cannot hurt were given Kreps (1977), Tirole (1982), Dubey, Geanakoplos and Shubik (1987).

First proof. Let $A_i = \{B, N\}$, and define $u_i(B, B, \ldots, B, \omega) = U_i(x_i(\omega), \omega)$, and $u_i(a, \omega) = U_i(e_i(\omega), \omega)$ for $a \neq (B, B, \ldots, B)$. This gives a Bayesian Nash game $\Gamma = (I, \Omega, (P_i, \pi_i, A_i, u_i)_{i \in I})$ which must have a Nash equilibrium in which $f_i(\omega) = B \; \forall i \in I$, $\omega \in \Omega$. Since each $f_i = B$ is common knowledge, by the agreement theorem each player would be willing to play B even if they all had the same information, namely knowing only that $\omega \in \Omega$. But that means each agent (weakly) prefers x_i ex ante to e_i, which by the Pareto optimality hypothesis is impossible unless $x_i = e_i$.

A **second proof** based on the principle that knowledge cannot hurt is given by ignoring the fact that the actions are common knowledge, and noting that by playing N at each ω, without any information agent i could have guaranteed himself $e_i(\omega)$. Hence by the lemma that knowledge never hurts a Bayesian optimizer, x_i is ex ante at least as good as e_i to each agent i, and again the theorem follows from the Pareto optimality hypothesis on the e_i. □

It is interesting to consider what can be said if we drop the hypothesis that the endowments are ex ante Pareto optimal. The following theorem is easily derived from the theorem that common knowledge of actions negates asymmetric information about events.

Theorem. Let $E = (I, \mathbb{R}_+^L, \Omega, (P_i, U_i, \pi_i, e_i)_{i \in I})$ be an economy, and suppose $(p, (x_i)_{i \in I})$ is a rational expectations equilibrium. Suppose at some ω that the net trade vector $z_i(\omega) = x_i(\omega) - e_i(\omega)$ is common knowledge for each i. Then P_i can be replaced by $\hat{P}_i$ for each i such that $\hat{P}_i(\omega) = \hat{P}_j(\omega)$ for all $i, j \in I$, without disturbing the equilibrium.

When it is common knowledge that agents are rational and optimizing, differences of information not only fail to generate a reason for trade on their own, but even worse, they inhibit trade which would have taken place had there been symmetric information. For example, take the two sons with their envelopes. However, suppose now that the sons are risk averse, instead of risk neutral. Then before the sons open their envelopes each has an incentive to bet – not the whole amount of his envelope against the whole amount of the other envelope – but to bet half his envelope against half of the other envelope. In that way, each son guarantees himself the average of the two envelopes, which is a utility improvement for sufficiently risk averse bettors, despite the $1 transaction cost. Once each son opens his envelope, however, the incentive to trade disappears, precisely because of the difference in information! Each son must ask himself what the other son knows that he does not.

More generally, consider the envelopes problem where the sons may be risk neutral, but they have different priors on Ω. In the absence of information, many bets could be arranged between the two sons. But it can easily be argued that no matter what the priors, as long as each state got positive probability, after the sons look at their envelopes they will not be able to agree on a bet. The reason is that the sons act only on the basis of their conditional probabilities, and given any pair of priors with the given information structure it is possible to find a single prior, the same for both sons, that gives rise to the conditional probabilities each son has at each state of nature. The original (distinct) priors are then called consistent with respect to the information structure. Again, the message is that adding asymmetric information tends to suppress speculation, rather than encouraging it, when it is common knowledge that agents are rational. [See Morris (1991).]

12. Dynamic Bayesian games

We have seen that when actions are common knowledge in (one-shot) Bayesian Nash equilibrium, asymmetric information becomes irrelevant. Recall that a dynamically consistent model of action and knowledge $(\Omega, (A_i, S_i), (\sigma_{it}, P_{it}, f_{it})_{i \in I}^{t \in T})$ specifies what each agent will do and know at every time period, in every state of nature. Over time the players will learn. From our getting to common knowledge theorem for DCMAK we know that if Ω is finite and the time horizon is long enough, there will be some period t^* at which it is common knowledge what the players will do that period. If the time period is infinite, then there will be a finite time period t^* when it will become common knowledge what each player will do at every future time period t. One might therefore suppose that in a Bayesian

Nash equilibrium of a multiperiod (dynamic) game with a finite state space, asymmetric information would eventually become irrelevant. But unlike DCMAK, dynamic Bayesian Nash equilibrium must recognize the importance of contingent actions, or *action plans* as we shall call them. Even if the immediately occurring actions become common knowledge, or even if all the future actions become common knowledge, the action *plans* may not become common knowledge since an action plan must specify what a player will do if one of the other players deviates from the equilibrium path. Moreover, in dynamic Bayesian games it is common knowledge of action plans, not common knowledge of actions, that negates asymmetric information.[6] The reason is that a dynamic Bayesian game can always be converted into a Bayesian game whose action space consists of the action plans of the original dynamic Bayesian game.

We indicate the refinement to DCMAK needed to describe dynamic Bayesian games and equilibrium. An action plan is a sequence of functions $\alpha_i = (\alpha_{i1}, \alpha_{i2}, \ldots, \alpha_{it} \ldots)$ such that $\alpha_{i1} \in A_i$, and for all $t > 1$, $\alpha_{it}: \times_{t=1}^{t-1} S_i \to A_i$. At time t, agent i chooses his action on the basis of all the information he receives before period t. Denote by $\mathscr{A}_i$ the space of action plans for agent $i \in I$.

Action plans $(\alpha_i, i \in I)$ generate signals $s(\omega) \in (S_1 \times \cdots \times S_I)^T$ and actions $a(\omega) \in (A_1 \times \cdots \times A_I)^T$ for each $\omega \in \Omega$ that can be defined recursively as follows.

Let $a_{i1}(\omega) = \alpha_{i1}$, and let $s_{i1}(\omega) = \sigma_{i1}(a_{i1}(\omega), \ldots, a_{I1}(\omega), \omega)$ for $\omega \in \Omega$, $i \in I$.

For $t > 1$, let $a_{it}(\omega) = \alpha_{it}(s_{i1}(\omega), \ldots, s_{it-1}(\omega))$ and $s_{it}(\omega) = \sigma_{it}(a_{1t}(\omega), \ldots, a_{It}(\omega), \omega)$ for $\omega \in \Omega$, $i \in I$.

Define payoffs u_i that depend on any sequence of realized actions and the state of the world: $u_i: (A_1 \times \cdots \times A_I)^T \times \Omega \to \mathbb{R}$. We say that the payoffs are additively separable if there are functions $v_{it}: A_1 \times \cdots \times A_I \times \Omega \to \mathbb{R}$ such that for any $a \in (A_1 \times \cdots \times A_I)^T$,

$$u_i(a, \omega) = \sum_{t \in T} v_{it}(a_{1t}, \ldots, a_{It}, \omega).$$

A strategy is a function $\tilde{\alpha}_i: \Omega \to \mathscr{A}_i$ such that $[P_{i1}(\omega) = P_{i1}(\omega')]$ implies $[\tilde{\alpha}_i(\omega) = \tilde{\alpha}_i(\omega')]$. We may write $\tilde{\alpha}_i \in \tilde{\mathscr{A}}_i = \mathscr{A}_i^{\text{Range } P_{i1}}$.

Given a probability π_i on Ω, the strategies $(\tilde{\alpha}_1, \ldots, \tilde{\alpha}_I)$ give rise to payoffs $U_i(\tilde{\alpha}_1, \ldots, \tilde{\alpha}_I) = \sum_{\omega \in \Omega} u_i(a(\omega), \omega) \pi_i(\omega)$ where $a(\omega)$ is the outcome stemming from the action plans $(\alpha_1, \ldots, \alpha_I) = (\tilde{\alpha}_1(\omega), \ldots, \tilde{\alpha}_I(\omega))$.

A dynamic Bayesian game is given by a vector $\Gamma = (I, T, \Omega, (P_{i1}, \pi_i, A_i, u_i, \sigma_i)_{i \in I})$. A (dynamic) Bayesian Nash equilibrium is a tuple of strategies $(\tilde{\alpha}_1, \ldots, \tilde{\alpha}_I)$ such that for each $i \in I$, $\tilde{\alpha}_i \in \text{Arg Max}_{\beta \in \tilde{\mathscr{A}}_i} U_i(\tilde{\alpha}_1, \ldots, \beta, \ldots, \tilde{\alpha}_I)$. Clearly any (dynamic) Bayesian Nash equilibrium gives rise to a dynamically consistent model of action and knowledge. In particular, P_{it} for $t > 1$ can be derived from the agent's action plans and the signals σ_{it}, as explained in the section on dynamic states of nature.

---

[6] Yoram Moses, among others, has made this point.

Any dynamic Bayesian game Γ and any Bayesian Nash equilibrium $\tilde{\alpha} = (\tilde{\alpha}_1, \ldots, \tilde{\alpha}_I)$ for Γ defines for each t the truncated dynamic Bayesian game $\Gamma_t = (I, T_t, \Omega, (P_{it}, \pi_i, A_i, \bar{u}_i, \sigma_i)_{i \in I})$ where T_t begins at t and the P_{it} are derived from the Bayesian Nash equilibrium. The payoffs $\bar{u}_i : (A_1 \times \cdots \times A_I)^{T_t} \to \mathbb{R}$ are defined on any $b \in (A_1 \times \cdots \times A_I)^{T_t}$ by

$$\bar{u}_i(b, \omega) = u_i(a_1(\omega), \ldots, a_{t-1}(\omega), b, \omega),$$

where $a_1(\omega), \ldots, a_{t-1}(\omega)$ are the Bayesian Nash equilibrium actions played at ω that arise from the Bayesian Nash equilibrium $\tilde{\alpha}$.

We say that a dynamic Bayesian Nash equilibrium of a Bayesian game Γ does not depend on asymmetric information at ω if we can preserve the BNE and replace each P_{i1} with $\hat{P}_{i1}$ in such a way that $\hat{P}_{i1}(\omega)$ is the same for all $i \in I$. (We say the same thing about Γ_t if $\hat{P}_{it}(\omega)$ is the same for all $i \in I$.)

One can imagine an extensive form Bayesian game which has a Bayesian Nash equilibrium in which it is common knowledge at some date t what all the players are going to do in that period, and yet it is not common knowledge at t what the players will do at some subsequent date. In such a game one should not expect to be able to explain the behavior at date t on the basis of symmetric information. The classic example is the repeated Prisoner's Dilemma with a little bit of irrationality, first formulated by Kreps et al. (1982).

The two players have two possible moves at every state, and in each time period, called cooperate (C) and defect (D). The payoffs are additively separable, and the one-shot payoffs to these choices are given by

	C	D
C	5, 5	0, 6
D	6, 0	1, 1

Let us suppose that the game is repeated T times. An action plan for an agent consists of a designation at each t between 1 and T of which move to take, as a function of all the moves that were played in the past. One example of an action plan, called grim, is to defect at all times, no matter what. Tit for tat is to play C at $t = 1$ and for $t > 1$ to play what the other player did at $t - 1$. Trigger is the action plan in which a player plays C until the other player has defected, and then plays D for ever after. Other actions plans typically involve more complicated history dependence in the choices.

It is well-known that the only Nash equilibrium for the T-repeated Prisoner's Dilemma is defection in every period.

Consider again the Prisoner's Dilemma, but now let there be four states of exogenous uncertainty, SS, SN, NS, NN. S refers to an agent being sane, and N to him not being sane. Thus NS means agent 1 is not sane, but agent 2 is sane. Each agent knows whether he is sane or not, but he never finds out about the other agent. Each agent is sane with probability 4/5, and insane with probability

1/5, and these types are independent across agents, so for example the chance of NS is 4/25. The payoff to a sane agent is as before, but the payoff to an insane agent is 1 if his actions for $1 \leqslant t \leqslant T$ are consistent with the action plan trigger, and 0 otherwise. A strategy must associate an action plan to each partition cell. Let each agent play trigger when insane, and play trigger until time T, when he defects for sure, when sane. The reader can verify that this is a Bayesian Nash equilibrium. For example, let $\omega = SS$. In the second to last period agent 1 can defect, instead of playing C as his strategy indicates, gaining in payoff from 5 to 6. But with probability 1/5 he was facing N who would have dumbly played C in the last period, allowing 1 to get a payoff of 6 by playing D in the last period, whereas by playing D in the second to last period 1 gets only 1 in the last period even against N. Hence by defecting in the second to last period, agent 1 would gain 1 immediately, then lose 5 with probability 1/5 in the last period, which is a wash.

The getting to common knowledge theorem assures us that so long as $T > (\#P_1 - 1) + (\#P_2 - 1) = (2 - 1) + (2 - 1) = 2$, in any Bayesian Nash equilibrium there must be periods t at which it is common knowledge what the agents are going to do. Observe that in this Bayesian Nash equilibrium it is already common knowledge at $t = 1$ what the players are going to do for all $t \leqslant T - 1$, but not at date T. Yet as we have noted, we could not explain cooperative behavior at period 1 in state SS on the basis of symmetric information. If both players know the state is SS, then we are back in the standard repeated Prisoner's Dilemma which has a unique Nash equilibrium – defect in each period. If neither player knows the state, then in the last period by defecting a player can gain 1 with probability 4/5, and lose at most 1 with probability 1/5. Working backwards we see again there can be no cooperation in equilibrium. Thus we have a game where asymmetric information matters, because some *future* actions of the players do not become common knowledge before they occur.

By adding the chance of crazy behavior in the last period alone (the only period N's actions differ from S's actions), plus asymmetric information, we get the sane agents to cooperate all the way until the last period, and the common sense view that repetition encourages cooperation seems to be borne out. Note that in the above example we could not reduce the probability of N below 1/5, for if we did, it would no longer be optimal for S to cooperate in the second to last period. Kreps, Milgrom, Roberts, and Wilson (1982) showed that if the insane agent is given a strategy that differs from the sane agent's strategy for periods t less than T, then it is possible to support cooperation between the optimizing agents while letting the probability of N go to 0 as T goes to infinity. However, as the probability of irrationality goes to zero, the number of periods of nonoptimizing (when N and S differ) behavior must go to infinity.

In the Prisoner's Dilemma game a nontrivial threat is required to induce the optimizing agents not to defect, and this is what bounds the irrationality just described from below. A stronger result can be derived when the strategy spaces of the agents are continuous. In Chou and Geanakoplos (1988) it is shown that for generic continuous games, like the Cournot game where agents choose the quantity to produce, an arbitrarily small probability of nonoptimizing behavior in the last round alone suffices to enforce cooperation. The "altruistic" behavior in the last round can give the agents an incentive for a tiny bit of cooperation in the second to last round. The last two rounds together give agents the incentive for a little bit more cooperation in the third to last round, and so on. By the time one is removed sufficiently far from the end, there is a tremendous incentive to cooperate, otherwise all the gains from cooperation in all the succeeding periods will be lost. The nonoptimizing behavior in the last period may be interpreted as a promise or threat made by one of the players at the beginning of the game. Thus we see the tremendous power in the ability to commit oneself to an action in the distant future, even with a small probability. One man, like a Gandhi, who credibly committed himself to starvation, might change the behavior of an entire nation.

Even if it is common knowledge at $t=1$ what the agents will do at every time period $1 \leqslant t \leqslant T$, asymmetric information may still be indispensable to explaining the behavior, if $T > 1$. Suppose for example that player 1 chooses in the first period which game he and player 2 will play in the second period. Player 1 may avoid a choice because player 2 knows too much about that game, thereby selecting a sequence of forcing moves that renders the actions of both players common knowledge. The explanation for 1's choice, however, depends on asymmetric information. Consider the Match game. Let $\Omega = \{1,\ldots,100\}$ where each $\omega \in \Omega$ has equal probability. Suppose at $t=1$, agent i must chose L or R or D. If i chooses L, then in period 2 player j must pick a number $n \in \Omega$. If player j matches and $n = \omega$, then player j gets 1 and player i gets -1. Otherwise, if $n \neq \omega$, then player j gets -1 and player i gets 1. If i chooses R, then again player j must choose $n \in \Omega$, giving payoff $n - 100$ to j, and $100 - n$ to agent i, for all ω. If i chooses D, then in period 2 i must choose $n \in \Omega$; if $n = \omega$, then i gets 2 and j gets -1, while if $n \neq \omega$, then i gets -1 and j gets 1.

Suppose finally that $P_{i1} = \{\Omega\}$, while j knows the states, $P_{j1} = \{\{\omega\}, \omega \in \Omega\}$. A Bayesian Nash equilibrium is for i to play R, and for player j to choose $n = 100$ if R, and to choose $n = \omega$ if L. There can be no other outcome in Bayesian Nash equilibrium. Note that it is common knowledge at each state ω before the first move what actions all the agents will take at $t = 1$ and $t = 2$. But i does not know the action plan of agent j. Without knowing the state, i cannot predict what j *would* do if i played L.

Asymmetric information is crucial to this example. If agent j were similarly uninformed, $P_{j1} = \{\Omega\}$, then i would choose L and get an expected payoff of 98/100. If both parties were completely informed, i would choose D and get an expected payoff of 2. Symmetric information could not induce i to choose R.

Despite these examples to the contrary, there are at least two important classes of Bayesian games in extensive form where common knowledge of actions (rather than action plans) negates asymmetric information about events: nonatomic games and separable two-person zero-sum games.

Suppose the action plans of the agents are independent of the history of moves of any single agent. For example, the action plans may be entirely history independent. Or they may depend on a summary statistic that is insensitive to any single agent's action. This latter situation holds when there is a continuum of agents and the signal is an integral of their actions. In any BNE, once it becomes common knowledge at some date t^* what all the agents will do thereafter, the partitions P_{it^*} can be replaced by a common, coarser partition $\hat{P}_{it^*} = \hat{P}_{t^*}$ and each player will still have enough information to make the same responses to the signals he expects to see along the equilibrium path. However, he may no longer have the information to respond according to the BNE off the equilibrium path. But in the continuum of agents situations, no single agent can, by deviating, generate an off-equilibrium signal anyway. Hence if there was no incentive to deviate from the original equilibrium, by the sure-thing principle there can be no advantage in deviating once the information of all the agents is reduced to what is common knowledge. Without going into the details of defining nonatomic (i.e., continuum) games, these remarks can serve as an informal proof of the following informal theorem:

Theorem (Informal). *For nonatomic Bayesian games in extensive form where the state space is finite, if the time horizon is infinite, there will be a time period t^* such that the whole future of the equilibrium path can be explained on the basis of symmetric information. If the time horizon is finite but long enough, and if the payoffs are additively separable between time periods, then there will be a finite period t^* whose equilibrium actions can be explained on the basis of symmetric information in a one-period game.*

The three puzzles with which we began this paper can all be recast as nonatomic games with additively separable payoffs. We can simply replace each agent by a continuum of identical copies. The report that each agent gives will be taken and averaged with the report all his replicas give, and only this average will be transmitted to the others. Thus in the opinion game, each of the type 1 replicas will realize that what is transmitted is not his own opinion of the expectation of x, but the average opinion of all the replicas of type 1. Since a single replica can have no effect on this average, he will have no strategic reason not to maximize the one-shot payoff in each period separately. Similarly we can replace each girl in the hats puzzle with a continuum of identical copies. We put one copy of each of the original three girls in a separate room (so that copies of the same girl cannot see each other). Each girl realizes that when she says "yes, I know my hat color" or "no, I do not know may hat color," her message is not directly transmitted to the other girls. Instead the proportion of girls of her type who say yes is transmitted to the other girls. A similar story could be told about the boys who say yes or no about whether they will bet with their fathers (or with each other).

All three puzzles can be converted into nonatomic games in which the Bayesian

Nash equilibrium generates exactly the behavior we described. The reason this is possible for these three puzzles, but not for the repeated Prisoner's Dilemma or the Match game, is that the behavior of the agents in the puzzles was interpersonally myopic; no agent calculated how changes in his actions at period t might affect the behavior of others in future periods. This interpersonal myopia is precisely what is ensured by the nonatomic hypothesis. By contrast, the repeated Prisoner's Dilemma with a little bit of irrationality hinges entirely on the sane player's realization that his behavior in early periods influences the behavior of his opponent in later periods. In contrast to the puzzles, in the repeated Prisoner's Dilemma game and in the Match game, asymmetric information played an indispensable role even after the actions of the players became common knowledge.

Consider now a sequence of two-person zero-sum games in which the payoff to each of the players consists of the (separable, discounted) sum of the payoffs in the individual games. The game at each time t may depend on the state of nature, and possibly also t. The players may have different information about the state of nature. We call this the class of repeated zero-sum Bayesian games. In the literature on repeated games, the game played at time t is usually taken to be independent of t. We have the same basic theorem as in the nonatomic case:

Theorem. *Consider a (pure strategy) Bayesian Nash equilibrium of a repeated zero-sum Bayesian game with a finite set of states of the world. If the time horizon T is infinite, there will be a time period t^* such that the whole future of the equilibrium path can be explained on the basis of symmetric information. If the time horizon T is finite but $T > T^* = \#P_1 - 1 + \#P_2 - 1$, then there must be some period $t \leqslant T^*$ whose actions can be explained on the basis of symmetric information.*

Proof. In any Bayesian equilibrium the equilibrium strategies define a Bayesian Nash equilibrium for the truncated Bayesian game obtained by considering the time periods from $t+1$ onward beginning with the equilibrium partitions P_{it+1}. Since the games are zero-sum, and the payoffs are additively separable, the fact that player 2 cannot improve his payoff from period $t+1$ onward if 1 sticks to his equilibrium strategy means that player 1 can guarantee his payoff from period $t+1$ onward by sticking to his equilibrium strategy no matter what he does in period t, provided that he does not reveal additional information to player 2. Hence we deduce from the fact that we began with a Bayesian Nash equilibrium, that player 1 cannot find an action function $b(\omega)$ at any time t that improves his expected payoff at time t and that uses (and hence reveals) only information that both players already had at time t. (The reader should note that there could be actions that agent 1 can take on the basis of his own information at time t that would improve his time t payoff that he will not undertake, because those actions would reveal information to player 2 that could be used against player 1 in subsequent periods.)

From our getting to common knowledge theorem we know that there must be some time $t \leqslant T^*$ such that the actions of the players are common knowledge before they occur, at every state of the world. We can thus find a partition P of the state space that is coarser than the partition P_{it} of each of the agents at time t, such that the action functions $a_{it}(\omega)$ at time t of each of the players i is measurable with respect to P.

It follows from the last two paragraphs that there is some time $t \leqslant T^*$ and an information partition P such that if all the agents had the same information P, their actions a_{it} would form a Bayesian Nash equilibrium for the one-shot game defined at time t. This proves the second part of the theorem.

If the game is infinitely repeated, then there must be a t^* such that at t^* all the current and future equilibrium actions are common knowledge. Hence restricting both the players' actions to some common partition P for all periods t^* and onward will not disturb the equilibrium. □

Aumann and Maschler (1966) considered infinite repeated zero-sum games in which agent i has a finer partition P_{i1} than agent j's partition P_{j1}. They supposed that $\sigma_{kt}(a_i, a_j, \omega) = (a_i, a_j)$, for all $k \in \{1, 2\}$, $t \in T$, and $(a_i, a_j, \omega) \in A_i \times A_j \times \Omega$. They took as the payoffs the limit of the average of the one-shot payoffs, which has the consequence that payoffs in any finite set of time periods do not influence the final payoff.

Consider a (pure strategy) Bayesian Nash equilibrium of an Aumann–Maschler game. At each t, $P_{it} = P_{i1}$, while P_{jt} is intermediate between P_{j1} and P_{i1}. Once t^* is reached at which all subsequent moves are common knowledge, $P_{jt} = P_{jt^*}$ for all $t \geqslant t^*$. From the foregoing theorem, we know that if we replaced P_{it^*} with P_{jt^*}, we would not affect the equilibrium. In fact, since t^* is finite, this equilibrium gives the same payoffs as the game in which $\hat{P}_{i1} = \hat{P}_{j1} = P_{jt^*}$. In effect, player i chooses how much information P_{jt^*} to give player j, and then the two of them play the symmetric information game with partitions P_{jt^*}.

13. Infinite state spaces and knowledge about knowledge to level N

If we allow for random (exogenous) events at each date $t \in T$, such as the possibility that a message (or signal) might fail to be transmitted, and if the states of the world are meant to be complete descriptions of everything that might happen, then there must be at least as many states as there are time periods. If we allow for an arbitrarily large number of faulty messages, then we need an infinite state space.

The assumption that the state space Ω is finite played a crucial role in the theorem that common knowledge of actions must eventually be reached. With an infinite state space, common knowledge of actions may never be reached, and one wonders whether that calls into question our conclusions about agreement, betting, and speculation. The answer is that it does not. We have already seen via martingale theory that when agents are discussing the expectation of a random variable, their opinions must converge even with an infinite state space. We now turn to betting.

Consider the envelopes problem, but with no upper bound to the amount of money the father might put in an envelope. More precisely, suppose that the father chooses $m > 0$ with probability $1/2^m$, and puts $\$10^m$ in one envelope and $\$10^{m+1}$ in the other, and randomly hands them to his sons. Then no matter what amount he sees in his own envelope, each son calculates the odds are at least $1/3$ that he has the lowest envelope, and that therefore in expected terms he can gain from switching. This will remain the case no matter how long the father talks to him and his brother. At first glance this seems to reverse our previous findings. But in fact it has nothing to do with the state space being infinite. Rather it results because the expected number of dollars in each envelope [namely the infinite sum of $(1/2^m)(10^m)$] is infinite. On close examination, the same proof we gave before shows that with an infinite state space, even if the maximum amount of money in each envelope is unbounded, as long as the expected number of dollars is finite, betting cannot occur.

However, one consequence of a large state space is that it permits states of the world at which a fact is known by everybody, and it is known by all that the fact is known by all, and it is known by all that it is known by all that the fact is known by all, up to N times, without the fact being common knowledge. When the state space is infinite, there could be for each N a (different) state at which the fact was known to be known N times, without being common knowledge.

The remarkable thing is that iterated knowledge up to level N does not guarantee behavior that is anything like that guaranteed by common knowledge, no matter how large N is. The agreement theorem assures us that if actions are common knowledge, then they could have arisen from symmetric information. But this is far from true for actions that are N-times known, where N is finite. For example, in the opinion puzzle taken from Geanakoplos–Polemarchakis, at state $\omega = 1$, agent 1 thinks the expectation of x is 1, while agent 2 thinks it is -1. Both know that these are their opinions, and they know that they know these are their opinions, so there is iterated knowledge up to level 2, and yet these opinions could not be common knowledge because they are different. Indeed they are not common knowledge, since the agents do not know that they know that they know that these are their respective opinions.

Recall the infinite state space version of the envelopes example just described, where the maximum dollar amount is unbounded. At any state (m, n) with $m > 1$ and $n > 1$, agent 1 believes the probability is $1/3$ that he has the lower envelope, and agent 2 believes that the probability is $2/3$ that agent 1 has the lower envelope! (If $m = 1$, then agent 1 knows he has the lower envelope, and if $n = 1$, agent 2 knows that agent 1 does not have the lower envelope.) If $m > N + 1$, and $n > N + 1$, then it is iterated knowledge at least N times that the agents have these different opinions. Thus, for every N there is a state at which it is iterated knowledge N times that the agents disagree about the probability of the event that 1 has the lower dollar amount in his envelope. Moreover, not even the size of the disagreement depends on N. But of course for no state can this be common knowledge.

Similarly, in our original finite state envelopes puzzle, at the state (4, 3) each son wants to bet, and each son knows that the other wants to bet, and each knows that the other knows that they each want to bet, so their desires are iterated knowledge up to level 2. But since they would lead to betting, these desires cannot be common knowledge, and indeed they are not, since the state (6, 7) is reachable from (4, 3), and there the second son does not want to bet. It is easy to see that by expanding the state space and letting the maximum envelope contain $\$10^{5+N}$, instead of $\$10^7$, we could build a state space in which there is iterated knowledge to level N that both agents want to bet at the state (4, 3).

Another example illustrates the difficulty in coordinating logically sophisticated reasoners. Consider two airplane fighter pilots, and suppose that the first pilot radios a message to the second pilot telling him they should attack. If there is a probability $(1-p)$ that any message between pilots is lost, then even if the second pilot receives the message, he will know they should attack, but the first pilot will not know that the second pilot knows they should attack, since the first pilot cannot be sure that the message arrived. If the first pilot proceeds with the plan of attacking, then with probability p the attack is coordinated, but with probability $(1-p)$ he flies in with no protection. Alternatively, the first pilot could ask the second pilot for an acknowledgement of his message. If the acknowledgement comes back, then both pilots know they should attack, and both pilots know that the other knows they should attack, but the second pilot does not know that the first pilot knows that the second pilot knows they should attack. The potential level of iterated knowledge has increased, but has the degree of coordination improved? We must analyze the dynamic Bayesian game.

Suppose the pilots are self-interested, so each will attack if and only if he knows they should attack and the odds are at least even that the other pilot will be attacking. Suppose furthermore that the first pilot alone is able to observe whether they should attack. In these circumstances there is a trivial Bayesian Nash equilibrium where neither pilot ever attacks because each believes the other will not attack. If it were common knowledge whether they should attack, then there would be another BNE in which they would both attack when they should, and not when they should not. Unfortunately for the pilots, it can never be common knowledge that they should attack. The only other Bayesian Nash equilibrium is where each pilot attacks if and only if every possible message he might have gotten telling him to attack is received.

The second pilot clearly will not attack if he gets no message, for without the message he could not know that they should attack. At best, he will attack if he gets the message, and not otherwise. He will indeed be willing to do that if he expects the first pilot to attack if he gets the second pilot's acknowledgement (assuming that $p > 1/2$). Given the second pilot's strategy, the first pilot will indeed be willing to attack if he gets the acknowledgement, since he will then be sure the second pilot is attacking. Thus there is a BNE in which the pilots attack if every message is successfully transmitted. Notice that the first pilot will not attack if he does not get the acknowledgement, since, based on that fact (which is all he has to go on), the odds are more likely [namely $(1-p)$ versus $(1-p)p$] that it was his original message that got lost, rather than the acknowledgement. The chances are now p^2 that the attack is coordinated, and $(1-p)p$ that the second pilot attacks on his own, and there is probability $(1-p)$ that neither pilot attacks. (If a message is not received, then no acknowledgement is sent.)

Compared to the original plan of sending one message there is no improvement. In the original plan the first pilot could simply have flipped a coin and with probability $(1-p)$ sent no message at all, and not attacked, and with probability p sent the original message without demanding an acknowledgement. That would have produced precisely the same chances for coordination and one-pilot attack as the two-message plan. (Of course the vulnerable pilot in the two-message plan is the second pilot, whereas the vulnerable pilot in the one-message plan is the first pilot, but from the social point of view, that is immaterial. It may explain however why tourists who write to hotels for reservations demand acknowledgements about their reservations before going.)

Increasing the number of required acknowledgements does not help the situation. Aside from the trivial BNE, there is a unique Bayesian Nash equilibrium, in which each pilot attacks at the designated spot if and only if he has received every scheduled message. To see this, note that if to the contrary one pilot were required to attack with a threshold of messages received well below the other pilot's threshold, then there would be cases where he would know that he was supposed to attack and that the other pilot was not going to attack, and he would refuse to follow the plan. There is also a difficulty with a plan in which each pilot is supposed to attack once some number k less than the maximum number of scheduled messages (but equal for both pilots) is received. For if the second pilot gets k messages but not the $(k+1)$st, he would reason to himself that it was more likely that his acknowledgement that he received k messages got lost and that therefore the first pilot only got $(k-1)$ messages, rather than that the first pilot's reply to his acknowledgement got lost. Hence in case he got exactly k messages, the second pilot would calculate that the odds were better than even that the first pilot got only $k-1$ messages and would not be attacking, and he would therefore refuse to attack. This confirms that there is a unique non-trivial Bayesian Nash equilibrium. In that equilibrium, the attack is coordinated only if all the scheduled messages get through. One pilot flies in alone if all but the last scheduled message get through. If there is an interruption anywhere earlier, neither pilot attacks. The outcome is the same as the one message scenario where the first pilot sometimes withholds the message, except to change the identity of the vulnerable pilot. The chances for coordinated attack decline exponentially in the number of scheduled acknowledgements.

The most extreme plan is where the two pilots agree to send acknowledgements back and forth indefinitely. The unique non-trivial Bayesian Nash equilibrium is for each pilot to attack in the designated area only if he has gotten all the messages. But since with probability one, some message will eventually get lost, it follows that neither pilot will attack. This is exactly like the situation where only one message is ever expected, but the first pilot chooses with probability one not to send it.

Note that in the plan with infinite messages [studied in Rubinstein (1989)], for each N there is a state in which it is iterated knowledge up to level N that both pilots should attack, and yet they will not attack, whereas if it were common knowledge that they should attack, they would indeed attack. This example is reminiscent of the example in which the two brothers disagreed about the probability of the first brother having the lowest envelope. Indeed, the two examples are isomorphic. In the pilots example, the states of the world can be specified by ordered integer pairs (m, n), with $n = m$ or $n = m - 1$, and $n \geq 0$. The first entry m designates the number of messages the first pilot received from the second pilot, plus one if they should attack. The second entry n designates the number of messages the second pilot received from the first pilot. Thus if $(m, n) = (0, 0)$, there should be no attack, and the second pilot receives no message. If $m = n > 0$, then they should attack, and the nth acknowledgement from the second pilot was lost. If $m = n + 1 > 0$, then they should attack, and the nth message from the first pilot was lost. Let $\text{Prob}(0, 0) = 1/2$, and for $m \geq 1$, $\text{Prob}(m, n) = \frac{1}{2}p^{m+n-1}(1-p)$. Each pilot knows the number of messages he received, but cannot tell which of two numbers the other pilot received, giving the same staircase structure to the states of the world we saw in the earlier example.

The upshot is that when coordinating actions, there is no advantage in sending acknowledgements unless one side feels more vulnerable, or unless the acknowledgement has a higher probability of successful transmission than the previous message. Pilots acknowledge each other once, with the word "roger," presumably because a one word message has a much higher chance of successful transmission than a command, and because the acknowledgement puts the commanding officer in the less vulnerable position.

14. Approximate common knowledge

Since knowledge up to level N, no matter how large N is, does not guarantee behavior that even approximates behavior under common knowledge, we are left to wonder what is approximate common knowledge?

Consider a Bayesian game $\Gamma = (I, \Omega, (P_i, A_i, \pi_i, u_i)_{i \in I})$, and some event $E \subset \Omega$

and some $\omega \in \Omega$. If $\pi_i(\omega) > 0$, then we say that i p-believes E at ω iff the conditional probability $[\pi_i(P_i(\omega) \cap E)]/[\pi_i(P_i(\omega))] \geq p$, and we write $\omega \in B_i^p(E)$. Monderer and Samet (1989) called an event E p-self-evident to i iff for all $\omega \in E$, i p-believes E; an event E is p-public iff it is p-self-evident to every agent $i \in I$. Monderer and Samet called an event C p-common knowledge at ω iff there is some p-public event E with $\omega \in E \subset \bigcap_{i \in I} B_i^p(C)$.

We can illustrate this notion with a diagram.

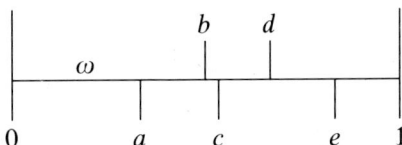

The only public events are ϕ and Ω. But $[0, b)$ is p-public where $p = \text{Prob}[a, b)/\text{Prob}[a, c)$. Any event C containing $[0, b)$ is p-common knowledge at ω.

In our first theorem we show that if in a Bayesian game with asymmetric information the players' actions are p-common knowledge at ω, then we can define alternative partitions for the players such that the information at ω is symmetric, and such that with respect to this alternative information, the same action function for each player is "nearly" optimal at "nearly" every state ω', including at $\omega' = \omega$, provided that p is nearly equal to 1.

Theorem. *Let $(f_1, \ldots, f_I)$ be a Bayesian Nash equilibrium for the Bayesian game $\Gamma = (I, \Omega, (P_i, A_i, \pi_i, u_i)_{i \in I})$. Suppose $\sup_{i \in I} \sup_{a, a' \in A} \sup_{\omega, \omega' \in \Omega} [u_i(a, \omega) - u_i(a', \omega')] \leq M$. Suppose $\pi_i(\omega) > 0$ for all $i \in I$, and suppose that at ω it is p-common knowledge that $(f_1, \ldots, f_I) = (a_1, \ldots, a_I)$. Then there is a Bayesian game $\hat{\Gamma} = (I, \Omega, (\hat{P}_i, A_i, \pi_i, u_i)_{i \in I})$ with symmetric information at ω, $\hat{P}_i(\omega) = E$ for all $i \in I$, and sets $\omega \in E \subset \Omega_i \subset \Omega$ with $\pi_i(\Omega_i) \geq p$ such that for all $\omega' \in \Omega_i$ with $\pi_i(\omega') > 0$, and all $b_i \in A_i$,*

$$\frac{1}{\pi_i(\hat{P}_i(\omega'))} \sum_{s \in \hat{P}_i(\omega')} [u_i(f(s), s) - u_i(b_i, f_{-i}(s), s)] \pi_i(s) \geq -M \frac{(1-p)}{p}.$$

Proof. Let E be a p-public event with $\omega \in E \subset \bigcap_{i \in I} B_i^p(F) \subset F \equiv \{\omega' \in \Omega : f(\omega') = a\}$. Define

$$\hat{P}_i(\omega') = \begin{cases} E, & \text{if } \omega' \in E, \\ -E \cap P_i(\omega'), & \text{if } \omega' \notin E. \end{cases}$$

Then $\hat{P}_i(\omega) = E \: \forall i \in I$. Note that since $f_i(s) = a_i$ for all $s \in E$, f_i is a feasible action function given the information $\hat{P}_i$.

Consider any ω' such that $P_i(\omega') \cap E = \phi$. Then $\hat{P}_i(\omega') = P_i(\omega')$, so $f_i(\omega')$ is optimal for i. Consider $\omega' \in E$. Then since $(f_1, \ldots, f_I)$ is a BNE,

$$\sum_{s \in P_i(\omega') \cap E} [u_i(f(s), s) - u_i(b_i, f_{-i}(s), s)] \pi_i(s)$$

$$\geq - \sum_{s \in P_i(\omega') \setminus E} [u_i(f(s), s) - u_i(b_i, f_{-i}(s), s)] \pi_i(s)$$

$$\geq - M \pi_i(P_i(\omega') \setminus E)$$

so

$$\frac{1}{\pi_i(E)} \sum_{s \in E} [u_i(f(s), s) - u_i(b_i, f_{-i}(s), s)] \pi_i(s)$$

$$\geq \frac{-M \pi_i(P_i(E) \setminus E)}{\pi_i(E)} \geq \frac{-M(1-p)}{p},$$

where $P_i(E) = \bigcup_{\omega' \in E} P_i(\omega')$.

Finally, the set $P_i(E) \setminus E$ on which i may not be optimizing even approximately has π_i probability at most $1 - p$. So let $\Omega_i = E \cup (\Omega \setminus P_i(E))$. □

As an immediate corollary we deduce a proposition in Monderer and Samet (1989) that if it is p-common knowledge that two agents with the same priors believe the probabilities of an event G are q_i, respectively, then $|q_i - q_j| \leq 2\sqrt{(1-p)/p}$. To see this, note that the optimal action for i at ω in $\hat{\Gamma}$ is to choose $r = \pi(G \cap E)/\pi(E)$. Since with the loss function $u_i(a_i, a_{-i}, \omega) = -[a_i - \chi_G(\omega)]^2$, $M = 1$, we know that q_i cannot do worse than r by any more than $(1-p)/p$. Hence $(q_i - r)^2 \leq (1-p)/p$, hence $|q_i - q_j| \leq 2\sqrt{(1-p)/p}$. Thus as $p \to 1$, the agents must nearly agree. This result stands in contrast to the example in the last section where the opinions 2/3 and 1/3 stayed bounded away from each other no matter how many levels of knowledge about knowledge were reached.

An alternative definition of approximate common knowledge which allows for more p-public events suggests itself. We could say that an event E with $\pi_i(E) > 0$ is weakly p-self-evident to agent i iff

$$\frac{1}{\pi_i(E)} \sum_{\omega' \in E} \frac{\pi_i(P_i(\omega') \cap E)}{\pi_i(P_i(\omega'))} \pi_i(\omega') \geq p.$$

Instead of requiring at every $\omega' \in E$ that agent i should think that the probability of E is at least p, this requires the same thing only on average. In the previous diagram the event $[0, c)$ is weakly p-self-evident to each player, but not p-self-evident to the first agent. Notice that under this more generous definition of weakly p-self-evident (and hence weakly p-public and weakly p-common knowledge) exactly the same proof can be used to prove the preceding theorem.

The preceding theorem can be generalized in a second way. Suppose that the action spaces A_i are compact metric spaces, and that the utilities u_i are continuous in A. Then we can replace the hypothesis that (it is weakly p-common knowledge that) the actions are $(a_1,\ldots,a_I)$ with the hypothesis that it is weakly p-common knowledge that the actions are within ε of $(a_1,\ldots,a_I)$. This explains why different agents' opinions must converge to each other in an infinite state space even though the opinions do not become common knowledge in finite time, but for brevity we omit the details.

The preceding theorem says that any BNE at which the actions are weakly p-common knowledge is an approximate BNE with symmetric information about events. The converse is also of interest. The following theorem [adapted from Monderer and Samet (1989)] shows that any conventional Nash equilibrium (which by definition can be achieved with symmetric information that the game is G) can be approximately achieved whenever it is p-common knowledge that the game is G.

Let M be as in the previous theorem, the maximum payoff difference for any player at any ω. Suppose now that the action spaces A_i are convex and compact, and that u_i is continuous in a, and concave in a_i for any fixed a_{-i} and ω.

Theorem. *Let $(\hat{f}_1,\ldots,\hat{f}_I)$ be a BNE for the Bayesian game $\hat{\Gamma}=(I,\Omega,\hat{P}_i,A_i,\pi_i,u_i)_{i\in I}$. Suppose that at some $\omega\in\Omega$, with $\pi_i(\omega)>0$ for all $i\in I$, $\hat{P}_i(\omega)=E$ for all $i\in I$, and $\hat{\Gamma}(\omega')=G$ for all $\omega'\in E$. Suppose that in the Bayesian game $\Gamma=(I,\Omega,(P_i,A_i,\pi_i,u_i)_{i\in I})$ E is p-common knowledge at ω. Then there exists $(f_1,\ldots,f_I)$ such that $f_i(\omega')=a_i=\hat{f}_i(\omega)$ for $\omega'\in E$ and all $i\in I$, and such that for all $\omega'\in\Omega$, $b_i\in A_i$,*

$$\frac{1}{\pi_i(P_i(\omega'))}\sum_{s\in P_i(\omega')}[u_i(f_i(s),f_{-i}(s),s)-u_i(b_i,f_{-i}(s),s)]\pi_i(s)\geq -M(1-p).$$

Proof. Define $f_i(\omega')=a_i$ if $P_i(\omega')\cap E\neq\phi$. Having fixed these actions, the Bayesian game Γ with these actions fixed defines a restricted Bayesian Game Γ^*. By our hypothesis on A_i and u_i, Γ^* must have a BNE $(f_1,\ldots,f_I)$. Observe that for ω' with $P_i(\omega')\cap E=\phi$, $f_i(\omega')$ is optimal in Γ. For ω' with $P_i(\omega')\cap E\neq\phi$,

$$\frac{1}{\pi_i(P_i(\omega'))}\sum_{s\in P_i(\omega')}[u_i(a_i,f_{-i}(s),s)-u_i(b_i,f_{-i}(s),s)]\pi_i(s)$$

$$\geq \frac{1}{\pi_i(P_i(\omega'))}\sum_{s\in P_i(\omega')\cap E}[u_i(a,s)-u_i(b_i,a_{-i},s)]\pi_i(s)+-M(1-p)$$

$$\geq 0+-M(1-p). \quad \square$$

The two theorems explain the coordinated attack problem. Suppose p is close to 1, so messages are quite reliable. Recalling our description from the last section, let $E=\{(m,n):m\geq 1\}$. For $(m,n)\geq(1,1)$, $\{\pi(E\cap P_i(m,n))\}/\{\pi(P_i(m,n))\}=1$. Only in the very unlikely state $(1,0)\in\bar{E}$ where the first message to the second pilot failed to arrive can it be true that it is appropriate to attack but pilot 2 does not know it. Hence E is weakly p-public, but not p-public. We conclude first that the BNE of never attacking, in which the actions are common knowledge but there is asymmetric information, can be (approximately) achieved when there is symmetric information and $P_i(\omega) = E$ for all $i \in I$ and $\omega \in E$. And indeed, not attacking is a (Pareto inferior) Nash equilibrium of the coordinated attack problem when $P_i(\omega) = E$ for all i. On the other hand, although attacking is a (Pareto superior) Nash equilibrium of the common information game where $P_i(\omega) = E$ for all i, because in the asymmetric information attack game E is only *weakly* p-common knowledge, attacking is not even an approximate BNE in the asymmetric informatio game.

15. Hierarchies of belief: Is common knowledge of the partitions tautological?

Our description of reasoning about the reasoning of others (and ultimately of common knowledge) is quite remarkable in one respect which has been emphasized by Harsanyi (1968), in a Bayesian context. We have been able to express a whole infinite hierarchy of beliefs (of the form i knows that j knows that m knows, etc.) with a finite number of primitive states $\omega \in \Omega$ and correspondences P_i. One might have been tempted to think that each higher level of knowledge is independent of the lower levels, and hence would require another primitive element.

The explanation of this riddle is that our definition of i's knowledge about j's knowledge presupposes that i knows how j thinks; more precisely, i knows P_j. Our definition that i knows that j knows that m knows that A is true at ω, presupposes that i knows P_j, j knows P_m, and i knows that j knows P_m. Thus the model does include an infinite number of additional primitive assumptions, if not an infinite number of states. We refer to these additional assumptions collectively as the hypothesis of mutual rationality.

In order to rigorously express the idea that an event is common knowledge we apparently must assume mutual rationality and take as primitive the idea that the information partitions are "common knowledge." This raises two related questions. Are there real (or actually important) situations for which mutual rationality is plausible? Is mutual rationality an inevitable consequence of universal individual rationality?

As for the first question, the puzzles we began with are clear situations where it is appropriate to assume common knowledge of knowledge operators. Each child can readily see that the others know his hat color, and that each of them knows that the rest of them know his hat color and so on. In a poker game it is also quite appropriate to suppose that players know their opponents' sources of information about the cards. But what about the even slightly more realistic settings, like horse races? Surely it is not sensible to suppose that every bettor knows what facts each other bettor has access to? This brings us to the second question.

One influential view, propounded first by Aumann (1976) along lines suggested by Harsanyi (1968), is that mutual rationality is a tautological consequence of individual rationality once one accepts the idea of a large enough state space. One could easily imagine that i does not know which of several partitions j has. This realistic feature could be incorporated into our framework by expanding the state space, so that each new state specifies the original state and also the kind of partition that j has over the original state space. By defining i's partition over this expanded state space, we allow i not only to be uncertain about what the original state is, but also about what j's partition over the original state space is. (The same device also can be used if i is uncertain about what prior j has over the original state space). Of course it may be the case that j is uncertain about which partition i has over this expanded state space, in which case we could expand the state space once more. We could easily be forced to do this an infinite number of times. One wonders whether the process would ever stop. The Harsanyi–Aumann doctrine asserts that it does. However, if it does, the states become descriptions of partition cells of the state space, which would seem to be an inevitable self-referential paradox requiring the identification of a set with all its subsets.

Armbruster and Boge (1979), Boge and Eisele (1979), and Mertens and Zamir (1985) were the first to squarely confront these issues. They focused on the analogous problem of probabilities. For each player i, each state is supposed to determine a conditional probability over all states, and over all conditional probabilities of player j, etc., again suggesting an infinite regress. Following Armbruster and Boge, Boge and Eisele and Mertens and Zamir, a large literature has developed attempting to show that these paradoxes can be dealt with. [See for example, Tan and Werlang (1985), Brandenburger and Dekel (1987), Gilboa (1988), Kaneko (1987), Shin (1993), Aumann (1989), and Fagin et al. (1992).]

The most straightforward analysis of the Harsanyi–Aumann doctrine (which owes much to Mertens and Zamir) is to return to the original problem of constructing the (infinite) hierarchy of partition knowledge to see whether at some level the information partitions are "common knowledge" at every ω, that is defined tautologically by the states themselves.

To be more precise, if $\Omega_0 = \{a, b\}$ is the set of payoff relevant states, we might be reluctant to suppose that any player $i \neq j$ knows j's partition of Ω_0, that is whether j can distinguish a from b. So let us set $\Omega^1 = \Omega_0 \times \{y_1, n_1\} \times \{y_2, n_2\}$. The first set $\{y_1, n_1\}$ refers to when player 1 can distinguish a from b (at y_1), and when he cannot (n_1). The second set $\{y_2, n_2\}$ refers to the second player. Thus the "extended state" $(a, (y_1, n_2))$ means that the payoff relevant state is a, that player 1 knows this, $y_1(a) = \{a\}$, but player 2 does not, $n_2(a) = \{a, b\}$. More generally, let Ω_0 be any finite set of primitive elements, which will define the payoff relevant universe. An element $\omega_0 \in \Omega_0$ might for example specify what the moves and payoffs to some game might be. For any set A, let $\mathbf{P}(A)$ be the set of partitions of A, that is $\mathbf{P}(A) = \{P: A \to 2^A \mid \omega \in P(\omega) \text{ for all } \omega \in A \text{ and } [P(\omega) = P(\omega')] \text{ or } [P(\omega) \cap P(\omega')] = \phi$ for all $\omega, \omega' \in A\}$. For each player $i = 1, \ldots, I$, let $\Omega_{1i} = \mathbf{P}(\Omega_0)$ and let $\Omega_1 = \times_{i=1}^I \Omega_{1i}$. Then we might regard $\Omega^1 = \Omega_0 \times \Omega_1$ as the new state space.

The trouble of course is that we must describe each player's partition of Ω^1. If for each player i there was a unique conceivable partition of Ω^1, then we would say that the state space Ω^1 tautologically defined the players' partitions. However, since Ω^1 has greater cardinality than Ω_0 it would seem that there are more conceivable partitions of Ω^1 than there were of Ω_0. But notice that each player's rationality restricts his possible partitions. In the example, if $\omega' = (a, (y_1, n_2))$ then player 1 should recognize that he can distinguish a from b. In particular, if P is player 1's partition of Ω^1, then $(c, (z_1, z_2)) \in P(a, (y_1, n_2))$ should imply $z_1 = y_1$ and $c = a$. (Since player 1 might not know 2's partition, z_2 could be either y_2 or n_2.) Letting Proj denote projection, we can write this more formally as

$$\text{Proj}_{\Omega_{1i}} P(a, (y_1, n_2)) = \{y_1\} \text{ and } \text{Proj}_{\Omega_0} P(a, (y_1, n_2)) = y_1(a).$$

In general, suppose we have defined Ω_0, and $\Omega_k = \Omega_{k1} \times \cdots \times \Omega_{kI}$ for all $0 < k < n$. This implicitly defines $\Omega^{n^*} = \times_{0 \leq k < n} \Omega_k$, and for each $k < n$, $\Omega^k = \times_{0 \leq r \leq k} \Omega_r$. Define $\Omega_{ni} = \{P_{ni} \in \mathbf{P}(\Omega^{n^*}) : \forall (\omega_0, \ldots, \omega_k, \ldots) \in \Omega^{n^*}, \forall k < n,$

(1) $\text{Proj}_{\Omega_{ki}} P_{ni}(\omega_0, \ldots, \omega_k, \ldots) = \{\omega_{ki}\}$,

(2) $\text{Proj}_{\Omega^k} P_{ni}(\omega_0, \ldots, \omega_k, \ldots) = \{\omega_{k+1,i}(\omega_0, \ldots, \omega_k)\}$.

Condition (1) says that i knows his partitions at lower levels, and condition (2) says that he uses his information at lower levels to refine his partition at higher levels.

Let $\Omega_n = \times_{i \in I} \Omega_{ni}$. By induction Ω_n is defined for all integers n. In fact, by transfinite induction, Ω_n is defined for all finite and transfinite ordinals.

The Harsanyi–Aumann question can now be put rigorously as follows. Is there any n, finite or infinite, such that the state space Ω^{n^*} defines the partitions of itself tautologically, i.e., such that Ω_{ni} contains a unique element P_{ni} for each $i \in I$?

The most likely candidate would seem to be $n = \alpha$, where α is the smallest infinite ordinal. In that case $\Omega^{\alpha^*} = \Omega_0 \times \Omega_1 \times \Omega_2 \times \cdots$. However, as shown in Fagin et al. (1992), following the previous work in Fagin et al. (1991), the cardinality of Ω_{ni} is not only greater than one, it is infinite for all infinite ordinals n, including $n = \alpha$. This shows that the Harsanyi–Aumann doctrine is false. Properly expanded, the state space does not tautologically define the partitions.

To see why, reconsider our simple example with two payoff relevant states. Since the cardinality of Ω_0 is 2, the number of partitions of Ω_0 is also 2, and so the cardinality of Ω^1 is $2 \times (2 \times 2) = 8$. Taking into account the restrictions imposed by player 1's own rationality, the number of possible partitions player 1 could have of Ω^1 is equal to the number of partitions of the four elements $\{a, b, y_2, n_2\}$, namely 15. Hence the cardinality of Ω^2 is $8 \times (15 \times 15) = 1800$.

As we go up the hierarchy, the restrictions from individual rationality become more biting, but the cardinality of the base of states grows larger. Indeed it is evident from the analysis just given that if the cardinality of Ω_{ki} is at least two, then the cardinality of $\Omega_{k+1,i}$ is at least two. It follows that the cardinality of Ω^{k+1} must be at least 2^I times the cardinality of Ω^k, for all finite $k \geq 0$, if $\#I \geq 2$. It would be astonishing if there were only one partition of Ω^{α^*} consistent with player i's rationality.

The fact that there may be at least two partitions $P_i \neq Q_i$ in $\Omega_{\alpha i}$, that is partitions of Ω^{α^*} that are consistent with the rationality of agent i, raises an important question: how different can P_i and Q_i be? To answer this question we introduce a topology on Ω^{α^*}. Note that for each finite k, Ω_k is a finite set, hence it is natural to think of using the discrete topology on Ω_k. Since $\Omega^{\alpha^*} = \times_{k=1}^{\infty} \Omega_k$, it is also natural to take the product topology on Ω^{α^*}. With this topology we can state the following theorem adapted from Fagin et al. (1992).

Theorem. *Let Ω_0 be finite. Then the Harsanyi–Aumann expanded state space Ω^{α^*} allows for each agent $i \in I$ one and only one partition $P_i \in \Omega_{\alpha i}$ of Ω^{α^*} such that every partition cell $P_i(\omega)$, $\omega \in \Omega^{\alpha^*}$, is a closed subset of Ω^{α^*}.*

If we are willing to restrict our attention to partitions with closed cells, then this theorem can be considered a vindication of the Harsanyi–Aumann doctrine. The proof of the theorem is not difficult. Let P_i and Q_i be in $\Omega_{\alpha i}$, and suppose $P_i(\omega)$ and $Q_i(\omega)$ are closed subsets of Ω^{α^*}. From conditions (1) and (2), we know that for each finite k,

$$\operatorname{Proj}_{\Omega^k} P_i(\omega) = \omega_{k+1,i}(\omega_1, \ldots, \omega_k) = \operatorname{Proj}_{\Omega^k} Q_i(\omega).$$

But since $P_i(\omega)$ and $Q_i(\omega)$ are closed in the product topology, this implies that $P_i(\omega) = Q_i(\omega)$, and the theorem follows.

We can state an analogous theorem [from Fagin et al. (1992)] that may also give the reader a sense of how close to true one might consider the Harsanyi–Aumann doctrine.

Theorem. *Let $(P_1, \ldots, P_I)$ and $(Q_1, \ldots, Q_I)$ be in Ω_α, that is let P_i and Q_i be partitions of the expanded state space Ω^{α^*} that are consistent with i's rationality, for each $i \in I$. Let $\omega \in \Omega^{\alpha^*}$, and let $E \subset \Omega^{\alpha^*}$ be closed. Then i knows E at ω with respect to P_i, $P_i(\omega) \subset E$, if and only if i knows E at ω with respect to Q_i, $Q_i(\omega) \subset E$. Furthermore, E is common knowledge at ω with respect to the partitions $(P_1, \ldots, P_I)$ if and only if E is common knowledge at ω with respect to the partitions $(Q_1, \ldots, Q_I)$.*

Note that any event E which depends only on a finite number of levels of the hierarchy is necessarily closed. These elementary events are probably of the most interest to noncooperative game theory. They include for example any description of the payoff relevant states, or what the players know about the payoff relevant states, and so on.

Here is an example of an event E that is not necessarily closed. Let A be a description of some payoff relevant states. Then E is defined as the set of ω at which i knows that it is not common knowledge between j and k that A happens.

In conclusion, we can say that the Harsanyi–Aumann doctrine can be partially vindicated by a rigorous construction of a knowledge hierarchy. If the partitions of the expanded state space Ω^{α^*} are restricted to have closed cells, then the state space Ω^{α^*} tautologically (uniquely) defines each agent's partition. A similar (positive) result was obtained by Mertens and Zamir (1985). If the state space is sufficiently enlarged, and if attention is restricted to *countably* additive *Borel* probabilities, then each state uniquely defines a conditional probability for each player. However, if more general (finitely additive) probabilities were allowed, then there would be many conditional probabilities consistent with a player's rationality.

Using our restrictions on potential partitions and probabilities, the knowledge of the players can always be described as in the first sections of this paper $(\Omega^*, P_1, \ldots, P_I)$, in which each player's knowledge pertains only to the state space Ω^* (and not to each other), and the partitions P_i are "common knowledge." As before, the universal state space Ω^* is the disjoint union of common knowledge components. In some of these there are a finite number of states, in others an infinite (uncountable) number. In some common knowledge components the players' conditional beliefs can all be explained as coming from a common prior; in others they cannot.

The restrictions to common priors, and finite Ω are nontrivial.[7] The "Harsanyi doctrine" asserts that it is reasonable that all agents should have the same prior, and many would agree. But the hierarchical argument we have just given does not provide any justification for this second doctrine.

16. Bounded rationality: Irrationality at some level

Common knowledge of rationality and optimization (interpreted as Bayesian Nash equilibrium) has surprisingly strong consequences. It implies that agents cannot agree to disagree; it implies that they cannot bet; and most surprising of all, it banishes speculation. (Here speculation is distinguished from betting because it may not be common knowledge that the deal is agreed, as for example, the moment at which a stock market investor places a buy order.) Yet casual empiricism suggests that all of these are frequently observed phenomena. This section explores the possibility that it is not really common knowledge that agents optimize, though in fact they do.

---

[7]Mertens and Zamir (1985) show that any common knowledge component of Ω^* that is infinite can be "approximated" by a finite common knowledge component.

We have already seen that if actions are iterated knowledge to some large but finite level N, then we may observe behavior which is very different from that which could be seen if the actions were common knowledge. In particular, agents could disagree. We hesitate to say that agents would bet, since in agreeing to the wager it might become common knowledge that they are betting. The significance of common knowledge, however, is lost on all but the most sophisticated reasoner, who would have to calculate that "he wants to bet, he wants to bet knowing that I am betting against him, he wants to bet knowing that I want to bet knowing that he wants to bet against me knowing that I want to bet etc." Most agents do not have the computing power, or logical powers, to make this calculation. To allow for this limitation, we suppose that somewhere in the calculation agents no longer can deduce any significance from the fact the other fellow is betting, which is to say that they no longer restrict attention to cases where the other fellow is optimizing. Suppose that at the actual state of the world ω, agents optimize, and know that they optimize, and know that they know that they optimize, but only a finite number of times rather than the infinity required by common knowledge. That is, suppose that it is common knowledge what the agents are doing, but only iterated knowledge to level N that they are optimizing. Then what the agents wish to do at ω might not be the same as when the actions and optimality are common knowledge.

As an example, reconsider the first version of the envelopes puzzle. Imagine that the two sons have $10 000 and $1 000 in their envelopes respectively. Suppose it were common knowledge that had son 2 seen $10 000 000 in his envelope, then he would bet (even though he could only lose in that case). Then the sons would be willing to bet against each other at state (4, 3), and in every other state (with m even and n odd). At state (4, 3), it is common knowledge that they are betting. Both sons are acting optimally given their information, both sons know that they are acting optimally, and they each know that the other knows that each is acting optimally. Of course it is not common knowledge that they are optimizing, since the state (6, 7) is reachable from (4, 3), and there the second son is not optimizing. The ex ante probability of nonoptimization here is only 1/12, and by extending the maximum amount in the envelopes we can make the probability of nonoptimal behavior arbitrarily small, and still guarantee that the sons bet against each other at (4, 3). (The astute reader will realize that although the prior probability of error can be made as small as possible in the envelopes example, the size of the blunder grows bigger and bigger. Indeed the expected error cannot vanish.) The same logic, of course, applies to the question of agreeing to disagree. [For more on this, see Aumann (1992).]

The possibility of nonoptimal behavior can also have dramatic consequences for dynamic Bayesian games. We have already seen in the N-repeated Prisoner's Dilemma that even when both players are optimizing, the possibility that the other is not can induce two optimizing agents to cooperate in the early periods, even though they never would if it were common knowledge that they were optimizing.

Simply by letting the time horizon N be uncertain, Neyman (in unpublished work) has shown that the two agents can each be optimizing, can each know that they are optimizing, and so on up to $m < N$ times, yet still cooperate in the first period. Of course this is analogous to the envelopes example just discussed.

Games in extensive form sometimes give rise to a backward induction paradox noted by Binmore (1987), Reny (1992), and Bicchieri (1988), among others. Consider the following extensive form game:

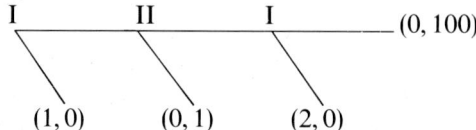

In the unique dynamic Bayesian equilibrium, I plays down immediately. We usually explain this by suggesting that I figures that if he played across instead, then II would play down in order to avoid putting I on the move again. But if I is "irrational" enough to play across on his first move, why should not II deduce that I is irrational enough to play across on his second move? It would appear, according to these authors, that to interpret fully a dynamic Bayesian game one needs a theory of "irrationality," or counterfactual reasoning. The beginning of such a theory is provided by Selten's (1975) notion of the trembling hand, which is discussed at length in other chapters of this volume.

17. Bounded rationality: Mistakes in information processing

When agents are shaking hands to bet, it seems implausible that the bet is not common knowledge. It might seem even less plausible that the agents do not fully realize that they are all trying to win, i.e., it seems plausible to suppose that it is also common knowledge they are optimizing. In the last part of this section we return to the assumption that it is common knowledge that agents optimize, but we continue to examine the implications of common knowledge by weakening the maintained hypothesis that agents process information perfectly, which has been subsumed so far in the assumption that knowledge has exclusively been described by a partition. We seek to answer the question: How much irrationality must be permitted before speculation, betting, and agreements to disagree emerge in equilibrium?[8]

There are a number of errors that are typically made by decision makers that suggest that we go beyond the orthodox Bayesian paradigm. Agents often forget, or ignore unpleasant information, or grasp only the superficial content of signals.

---

[8] Much of this section is taken from Geanakoplos (1989), which offers a fuller description of possible types of irrationality and derives a number of theorems about how they will affect behavior in a number of games.

Many of these mistakes turn on the idea that agents often do not know that they do not know. For example, in the story of Silver Blaze, Sherlock Holmes draws the attention of the inspector to "the curious incident of the dog in the night-time." "The dog did nothing in the night-time," protested the inspector, to which Holmes replied "That was the curious incident." Indeed, the puzzle of the hats was surprising because it relied on the girls being so rational that they learned from each other's ignorance, which we do not normally expect. As another example, it might be that there are only two states of nature: either the ozone layer is disintegrating or it is not. One can easily imagine a scenario in which a decaying ozone layer would emit gumma rays. Scientists, surprised by the new gamma rays would investigate their cause, and deduce that the ozone was disintegrating. If there were no gamma rays, scientists would not notice their absence, since they might never have thought to look for them, and so might incorrectly be in doubt as to the condition of the ozone.

We can model some aspects of non-Bayesian methods of information processing by generalizing the notion of information partition. We begin as usual with the set Ω of states of nature, and a possibility correspondence P mapping each element ω in Ω into a subset of Ω. As before, we interpret $P(\omega)$ to be the set of states the agent considers possible at ω. But now P may not be derived from a partition. For instance, following the ozone example we could imagine $\Omega = \{a, b\}$ and $P(a) = \{a\}$ while $P(b) = \{a, b\}$. A perfectly rational agent who noticed what he did not know would realize when he got the signal $\{a, b\}$ that he had not gotten the signal $\{a\}$ that comes whenever a is the actual state of the world, and hence he would deduce that the state must be b. But in some contexts it is more realistic to suppose that the agent is not so clever, and that he takes his signal at face value.

We can describe the knowledge operator generated by the possibility correspondece P just as we did for partitions: $K(E) = \{\omega \in \Omega : P(\omega) \subset E\}$ for all events E. In the ozone example, $K(\{a\}) = \{a\}$, $K(\{b\}) = \phi$, and $K(\{a,b\}) = \{a,b\}$. The reader can verify that K satisfies the first four axioms of S5 described earlier, but it fails the fifth: $\sim K(\{a\}) = \{b\} \neq \phi = K \sim K(\{a\})$. Non-partitional information has been discussed by Shin (1993), Samet (1990), Geanakoplos (1989), and Brandenburger et al. (1992).

The definitions of Bayesian game and Bayesian Nash equilibrium do not rely on partitions. If we substitute possibility correspondences for the partitions, we can retain the definitions word for word. At a Bayesian Nash equilibrium, for each state $\omega \in \Omega$ agents use their information $P_i(\omega)$ to update their priors, and then they take what appears to them to be the optimal action. Nothing in this definition formally requires that there be a relationship between $P_i(\omega)$ and $P_i(\omega')$ for different ω and ω'. Moreover, the definitions of self-evident and public also do not rely on any properties of the correspondences P_i. We can therefore investigate our agreement theorems and nonspeculation theorems when agents have nonpartitional information.

Imagine a doctor Q who initially assigns equal probability to all the four states

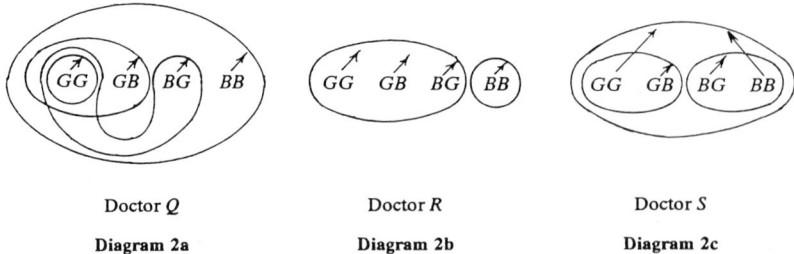

Doctor Q
Diagram 2a

Doctor R
Diagram 2b

Doctor S
Diagram 2c

describing whether each of two antibodies is in his patient's blood (which is good) or not in the blood (which is bad). If both antibodies are in the blood, i.e., if the state is *GG*, the operation he is contemplating will succeed and the payoff will be 3. But if either is missing, i.e., if the state is any of *GB*, *BG*, or *BB*, the operation will fail and he will lose 2. Suppose that in his laboratories his assistants are looking for the antibodies in blood samples. The doctor does not realize that if an antibody is present, the lab will find it, whereas if it is missing, the lab will never conclusively discover that it is not there. The possibility correspondence of doctor *Q* is therefore described by Diagram 2a. The laboratory never makes an error, always reporting correct information. However, though the doctor does not realize it, the way his laboratory and he process information is tantamount to recognizing good news, and ignoring unpleasant information.

A lab technician comes to the doctor and says that he has found the first antibody in the blood, and his proof is impeccable. Should the doctor proceed with the operation? Since doctor *Q* takes his information at face value, he will assign equal probability to *GG* and *GB*, and decide to go ahead. Indeed for each of the signals that the doctor could receive from the actual states *GG*, *GB*, *BG* the superficial content of the doctor's information would induce him to go ahead with the operation. (If the actual state of the world is *BB*, doctor *Q* will get no information from his lab and will decide not to do the operation). Yet a doctor R^0 who had no lab and knew nothing about which state was going to occur, would never choose to do the operation. Nor would doctor *R*, who can recognize whether or not both antibodies are missing. Doctor *R*'s information is given by the partition in Diagram 2b. From an ex ante perspective, both doctors R^0 and *R* do better than doctor *Q*. We see from this example that when agents do not process information completely rationally, more knowledge may be harmful. Furthermore, *Q* does not satisfy the sure-thing principle. At each ω in the self-evident set $M = \{GG, GB, BG\}$, doctor *Q* would choose to operate, given his information $Q(\omega)$. Yet if told only $\omega \in M$, he would choose not to operate.

Doctor *Q* does not know what he does not know; the knowledge operator K_Q derived from his possibility correspondence *Q* does not satisfy $\sim K_Q = K_Q \sim K_Q$. But like the ozone scientist *P*, doctor *Q* does satisfy the first four S5 axioms. In particular, doctor *Q* knows what he knows. He can recognize at *GB* that the first antibody is present, and whenever that condition obtains, he recognizes it. Formally, the possibility correspondence P_i gives rise to a knowledge operator K_i satisfying the "know that you know" (KTYK) axiom iff for all $\omega, \omega' \in \Omega$, $\omega' \in P_i(\omega)$ implies $P_i(\omega') \subset P_i(\omega)$. Doctor Q's error is that he sometimes overlooks the condition of the first antibody while recognizing the condition of the second antibody, and at other states he does the reverse. If he paid attention to the antibodies in the same order, his knowledge would satisfy the "nested" property. Formally, a possibility correspondence satisfies the memory property called nested if for all ω and ω' with $P_i(\omega) \cap P_i(\omega') \neq \phi$, either $P_i(\omega) \subset P_i(\omega')$ or $P_i(\omega') \subset P_i(\omega)$.

Theorem. *Consider the one-person Bayesian games $(\Omega, (P_i, A_i, \pi_i, u_i))$ and $(\Omega, (Q_i, A_i, \pi_i, u_i))$ where Q_i is a partition and P_i is finer than Q_i, but is not necessarily a partition. Let f be a BNE in the first game, and g in the second. Then if P_i is non-deluded ($\omega \in P_i(\omega)$ for all ω), and satisfies knowing that you know and nested, then $\sum_{\omega \in \Omega} u_i(f(\omega), \omega) \pi_i(\omega) \geq \sum_{\omega \in \Omega} u_i(g(\omega), \omega) \pi_i(\omega)$. Conversely, if P_i fails any one of these properties, then there exist A_i, u_i, π_i, and BNE f and g such that the above inequality fails.*

We could also give necessary and sufficient conditions (nondeluded plus a property called positively balanced, which is implied by nested) for a nonpartitional decisionmaker to satisfy the sure-thing principle. Doctor Q's information processing leads to decisions that violate both the principle that more information is better and the sure-thing principle. By contrast, ozone scientist P will satisfy both the sure-thing principle and the principle that more information is better, despite not being perfectly rational. It turns out that more rationality is required for the principle that information is good than for the sure-thing principle.

Imagine now another doctor S, contemplating the same operation, but with a different laboratory. Doctor S's lab reveals whether or not the first antibody is in the blood of his patient, except when both antibodies are present or when both are absent, in which cases the experiment fails and reveals nothing at all. If doctor S takes his laboratory results at face value, then his information is described by Diagram 2c. The superficial content of doctor S's information is also impeccable. But if taken at face value, it would lead him to undertake the operation if the state were GB (in which case the operation would actually fail), but not in any other state.

Doctor S is thus worse off ex ante with more information. Note that S fails KTYK. But S satisfies nested and therefore positively balanced, which implies that S's behavior will satisfy the sure-thing principle.

Since Bayesian games are formally well-defined with nonpartitional information, we can put the doctors together and see whether they would speculate, or bet, or agree to disagree. In general, as we saw in the last section, we would have to expand the state space to take into account each doctor's uncertainty about the other doctor's information. But for illustration, let us suppose that the state space

Ω of four states already takes this into account. Again, this formally makes sense because in BNE each player thinks about the states and actions, not directly about the other players. Once we assign actions for all players to each state, we are back in that framework.

Furthermore, it might appear that it would make no sense to ask when an event $E \subset \Omega$ is common knowledge when agents have nonpartitional information. Since the agents do not fully understand their own information precisely, how can they think about what the others think about their information processing? The answer is that common knowledge can, as we saw, be understood in a way that depends only on self-evident events. Each of the doctors Q, R, S can be perfectly aware of which events E are self-evident, i.e., satisfy $P_i(\omega) \subset E$ for all $\omega \in E$. This degree of mutual awareness would not induce them to refine their possibility correspondences, except at $\omega = BB$.

After getting their information at any $\omega \in M = \{GG, GB, GB\}$, doctors Q and R would be willing to make a wager in which doctor R pays doctor Q the net value the operation turns out to be worth if doctor Q performs it. At each of the states in M doctor Q will decide to perform the operation, and therefore the bet will come off. Moreover, the event M is public, so we can say that the bet is common knowledge at each $\omega \in M$. The uninformed but rational doctor R would in fact come out ahead, since 2 out of every 3 times the operation is performed he will receive 2, while 1 out of every 3 times he will lose 3.

Doctors S and R would also be willing to sign a bet in which R paid S the net value of the operation if doctor S decides to perform it. Again doctor R will come out ahead, despite having less information. In this BNE it is not known by doctor R that the bet is going to come off when doctors S and R set their wager. Doctor R is put in a position much like that of a speculator who places a buy order, but does not know whether it will be accepted. One can show that there is no doctor (with partition information, or even one who makes the same kinds of errors as doctor S) who doctor S would bet with and with whom it would be common knowledge that the bet was going to come off.

It can also be shown that ozone scientist P would not get lured into any unfavorable bets (provided that the ozone layer was the only issue on which he made information processing errors). Furthermore, it can be shown that none of the four agents P, Q, R, S would agree to disagree with any of the others about the probability of some event.

Geanakoplos (1989) establishes necessary and sufficient conditions for the degree of rationality of the agents (i.e., for the kinds of information processing errors captured by the nonpartitional possibility correspondences) to allow for speculation, betting, and agreeing to disagree. There is a hierarchy here. Agents can be a little irrational (satisfying nondeluded, KTYK, and nested), and still not speculate, bet, or agree to disagree. But if agents are a little more irrational (satisfying nondeluded and positively balanced), they will speculate, but not bet or agree to disagree. If they get still more irrational (satisfying nondeluded and balanced), they will speculate and bet, but not agree to disagree about the probability of an event.[9] Finally, with still more irrationality, they will speculate, bet, and agree to disagree.

References

Armbruster, W. and W. Boge (1979) 'Bayesian game theory', in: O. Moeschlin and D. Pallaschke, eds., *Game theory and related topics*. Amsterdam: North Holland.
Aumann, R. (1974) 'Subjectivity and correlation in randomized strategies', *Journal of Mathematical Economics*, 1: 67–96.
Aumann, R. (1976) 'Agreeing to disagree', *The Annals of Statistics*, 4: 1236–1239.
Aumann, R. (1987) 'Correlated equilibrium as an expression of Bayesian rationality', *Econometrica*, 55: 1–18.
Aumann, R. (1992) 'Irrationality in game theory', in P. Dasgupta, D. Gale, O. Hart and E. Maskin, eds., *Economic analysis of markets and games*, pp. 214–227.
Aumann, R. (1989) 'Notes on interactive epistemology', mimeo.
Aumann, R. and A. Brandenburger (1991) 'Epistemic conditions for Nash equilibrium', Working Paper 91-042, Harvard Business School.
Aumann, R. and M. Maschler (1966) 'Game theoretic aspects of gradual disarmament', Chapter V in Report to the US Arms Central and Disarmament Agency, Washington, DC.
Bacharach, M. (1985) 'Some extensions of a claim of Aumann in an axiomatic model of knowledge', *Journal of Economic Theory*, 37: 167–190.
Bicchieri, C. (1988) 'Common knowledge and backward induction: a solution to the paradox', in: M. Vardi, ed., *Proceedings of the Second Conference on Reasoning about Knowledge*. Los Altos: Morgan Kaufmann Publishers, pp. 381–394.
Binmore, K. (1987–88) 'Modeling rational players', *Economics and Philosophy*, 3: 179–214, 4: 9–55.
Boge, W. and Th. Eisele (1979) 'On solutions of Bayesian games', *International Journal of Game Theory*, 8(4): 193–215.
Bollobás, B., ed., *Littlewood's miscellany*. Cambridge: Cambridge University Press, 1953.
Brandenburger, A. and E. Dekel (1987) 'Common knowledge with probability 1', *Journal of Mathematical Economics*, 16(3): 237–245.
Brandenburger, A. and E. Dekel (1993) 'Hierarchies of belief and common knowledge', *Journal of Economic Theory*, 59(1): 189–198.
Brandenburger, A., E. Dekel and J. Geanakoplos (1992) 'Correlated equilibrium with generalized information structures', *Games and Economic Behavior*, 4(2): 182–201.
Cave, J. (1983) 'Learning to agree', *Economics Letters*, 12: 147–152.
Chou, C. and J. Geanakoplos (1988) 'The power of commitment', Cowles Foundation Discussion Paper No. 885.
Dubey, P., J. Geanakoplos and M. Shubik (1987) 'The revelation of information in strategic market games: a critique of rational expectations equilibrium', *Journal of Mathematical Economics*, 16(2): 105–138.
Fagin, R., J. Geanakoplos, J. Halpern and M. Vardi, 'The expressive power of the hierarchical approach to modeling knowledge and common knowledge', in M. Vardi, ed., *Fourth Symposium on Theoretical Aspects of Reasoning about Knowledge*. Los Altos: Morgan Kaufmann Publishers, 1992.
Fagin, R., J. Halpern and M. Vardi (1991) 'A model-theoretic analysis of knowledge', *Journal of the ACM*, 91(2): 382–428.
Gamow, G. and M. Stern (1958) 'Forty unfaithful wives', in: *Puzzle math*. New York: The Viking Press, pp. 20–23.
Gardner, M. (1984) *Puzzles from other worlds*. Vintage.

---

[9] Samet (1990) had previously shown that non-deduced and KTYK are sufficient conditions to rule out agreeing to disagree. Non-deluded and balanced are necessary as well as sufficient conditions to rule out agreeing to disagree.

Geanakoplos, J. (1989) 'Game theory without partitions and applications to speculation and consensus', Cowles Foundation Discussion Paper No. 914, Yale University, forthcoming *Journal of Economic Theory*.

Geanakoplos, J. (1992) 'Common knowledge', *Journal of Economic Perspectives*, **6**(4): 58–82.

Geanakoplos, J. and H. Polemarchakis (1982) 'We can't disagree forever', *Journal of Economic Theory*, **28**: 192–200.

Gilboa, I. (1988) 'Information and meta-information', in M. Vardi, ed., *Theoretical aspects of reasoning about knowledge*. Los Altos: Morgan Kaufmann Publishers, pp. 1–18.

Green, J. and J. Laffont (1987) 'Posterior implementability in a two-person decision problem', *Econometrica*, **55**: 69–94.

Halpern, J.Y. (1986) 'Reasoning about knowledge: an overview', IBM Research Report RJ-5001.

Halpern, J. and Y. Moses (1984) 'Knowledge and common knowledge in a distributed environment', in: *Proceedings of the 3rd ACM Conference on Principles of Distributed Computing*, pp. 50–61.

Harsanyi, J. (1967, 1968) 'Games with incomplete information played by "Bayesian" players', Parts I–III, *Management Science*, **14**(3): 159–183; **14**(5): 320–334; **14**(7): 486–502.

Kaneko, M. (1987) 'Structural common knowledge and factual common knowledge', RUEE Working Paper 87-27, Department of Economics, Hitosubashi University.

Kreps, D. (1977) 'A note on fulfilled expectations equilibrium', *Journal of Economic Theory*, 32–43.

Kreps, D., P. Milgrom, J. Roberts and R. Wilson (1982) 'Rational cooperation in the finitely repeated Prisoner's Dilemma', *Journal of Economic Theory*, **27**: 245–23.

Kripke, S. (1963) 'Semantical analysis of model logic', *Z. Math Logik Grundlag der Math*, **9**: 67–96.

Lewis, D. (1969) *Convention: a philosophical study*. Cambridge: Harvard University Press.

McKelvey, R. and T. Page (1986) 'Common knowledge, consensus and aggregate information', *Econometrica*, **54**: 109–127.

Mertens, J.F. and S. Zamir (1985) 'Formulation of Bayesian analysis for games with incomplete information', *International Journal of Game Theory*, **14**: 17–24.

Milgrom, P. (1981) 'An axiomatic characterization of common knowledge', *Econometrica*, **49**: 219–222.

Milgrom, P. and N. Stokey (1982) 'Information, trade and common knowledge', *Journal of Economic Theory*, **26**: 17–27.

Monderer, D. and D. Samet (1989) 'Approximating common knowledge with common beliefs', *Games and Economic Behavior*, **1**: 170–190.

Morris, S. (1991) 'The role of beliefs in economic theory', PhD. Dissertation, Yale University.

Nielsen, L. (1984) 'Common knowledge, communication and convergence of beliefs', *Mathematical Social Sciences*, **8**: 1–14.

Nielsen, L., A. Brandenburger, J. Geanakoplos, R. McKelvey and T. Page (1990) 'Common knowledge of an aggregate of expectations', *Econometrica*, 1235–1239.

Parikh, R. and P. Krasucki (1990) 'Communication, consensus and knowledge', *Journal of Economic Theory*, **52**(1): 178–189.

Rubinstein, A. (1989) 'The electronic mail game: strategic behavior under "almost common knowledge"', *American Economic Review*, **79**(3): 385–391.

Rubinstein, A. and A. Wolinsky (1990) 'Remarks on the logic of "agreeing to disagree" type results', *Journal of Economic Theory*, **51**: 184–193.

Reny, P.J. (1992) 'Rationality in extensive form games', *Journal of Economic Perspectives*, **6**(4): 103–118.

Samet, D. (1990) 'Ignoring ignorance and agreeing to disagree', *Journal of Economic Theory*, **52**(1): 190–207.

Savage, L. (1954) *The foundations of statistics*. New York: Wiley.

Sebenius, J. and J. Geanakoplos (1983) 'Don't bet on it: contingent agreements with asymmetric information', *Journal of the American Statistical Association*, **78**: 424–426.

Selten, R. (1975) 'Reexamination of the perfectness concept for equilibrium points in extensive games', *International Journal of Game Theory*, **4**(1): 25–55.

Shin, H. (1993) 'Logical structure of common knowledge', *Journal of Economic Theory*, **60**(1): 1–13.

Tan, T. and S. Werlang (1985) 'The Bayesian foundations of solution concepts of games', *Journal of Economic Theory*, **45**: 379–391.

Tirole, J. (1982) 'On the possibility of speculation under rational expectations', *Econometrica*, **50**: 1163–1181.

INDEX

a posteriori expectation 773
a priori distribution 770, 773
Abbey, R. 996
Abdou, J. 1092, 1121
Abreu, D. 892–896, 898–900, 1118, 1123
abstract core 1324–1326
abstract stable set 1324–1326, 1329
abstract system 1324, 1325
acceptable correlated equilibria 846
accuracy functions 762
Achen, C. 1002
Ackoff, R. 1015
Aczel, J. 1416
Adams, E.W. 1405, 1406, 1415
additional-member 1085
additional-member systems 1081
additive 1214, 1217
additive utility 1405, 1409
adjusted district voting (ADV) 1082, 1085
Admati, A. 850
admissibility 1061, 1063, 1064
admissible 1059, 1060, 1062, 1064, 1068, 1073
admissible (or dominant) 1065
admissible strategy 1066, 1067, 1071
adverse selection 840
advertising 912, 913, 921
Afek, Y. 1372
Affuso, P. 1132
agenda control 1148
aggregate activity 920
aggregated cost function 1216
aggregation invariant 1221, 1224
Agreement Theorem 1453
Ahn, B. 1002
Aho, A.V. 1371
AI (Artificial Intelligence) 1387
aiming and evasion, games of 1021, 1022
aircraft allocation, see tactical air war models
Ajtai, M. 1376, 1382, 1383, 1386
Akin, E. 986
Albers, W. 1327
Albon, S.D. 979
Alchian, A. 899
Alesina, A. 1330
Alexander, J. 998
Alfieri, M. 982
algebraic circuits 1359

algorithm 1350
Alker, H. 1006
Allais, M. 1411, 1412, 1414, 1420
Allais-type 1422
Allan, P. 1009
Allen, B. 1412
Allen, F. 885, 898
Allen, R.G.D. 1400
Allentuck, A. 1020
alliances, military 1014, 1016, 1017
 – financial contributions to 1016, 1017
Alon, N. 1376, 1381, 1383, 1384
α and β cores 1323, 1329
α and β theories 1308
α-effectiveness (coalitional) form 1300
α–β pruning 1387
Alt, F. 1400
Alt, J. 1000
alternate cost avoided 1201
ambiguity 1420
Ambruster, W. 1462, 1485
ambush games 1026
Amit, I. 1005
amortized cost 1390
Anant, T.C.A. 1251
Anbar, D. 1275
Anbarci, N. 1245, 1257, 1275, 1276
Anckar, D. 1084
Anderson, L. 1027
Andersson, M. 979
Anglin, P.M. 908
animal display behavior 979
animal fighting 972–975
anonymity 1246
anonymously equitable 1224
Anscombe, F.J. 1418, 1420
ant colonies 980
Anti-Ballistic Missile (ABM) Treaty 1023
antisubmarine warfare, see submarine warfare
approval 1066, 1084
approval voting 1062, 1064, 1065, 1067–1069, 1071, 1072
approximate common knowledge 1480
approximate core 1174–1176, 1180
approximate equilibrium 1178, 1180
Arak, A. 984
Archimedean axiom 1408, 1414, 1417, 1422

Aristotle 1057
arms building 1004–1006
arms control 1003–1005
arms control negotiations, *see* negotiations, international
arms control verification 1017, 1018
arms procurement 1006
arms race 818, 984, 1005, 1020
Armstrong, W.E. 1403, 1404
Arnott, R. 898
Aron, R. 1013
Arosalo, U. 1016
Arrington, T.S. 1084
Arrow, K.J. 774, 779, 870, 897, 1001, 1007, 1058, 1092, 1096, 1404, 1408, 1409
Arrow–Debreu 1465
Arthur Lewis, W. 1220
Asheim, G. 1331
Ashkenazi, V. 1019
aspiration 1327
assesment 978
asymmetric animal contest 974–978
asymmetric conflicts 962
asymmetric war of attrition 978
asynchronous 1350
asynchronous networks 1345
atom of a game 1301
auction 922
Aumann, R.J. 736, 776, 779, 828, 830, 831, 865, 938, 1001, 1003, 1005, 1009, 1168, 1171, 1183–1185, 1188, 1189, 1206, 1222, 1299–1302, 1306, 1308, 1309, 1311, 1322, 1323, 1326, 1327, 1329, 1405, 1415, 1416, 1418–1420, 1438, 1440, 1448, 1451, 1454, 1455, 1463, 1476, 1485
Aumann–Shapley prices 1222–1224
Austad, S.N. 979
Avenhaus, R. 999, 1005, 1014, 1017, 1018
Awerbuch, B. 1367
Axell, B. 912
Axelrod, R. 982, 1000, 1013, 1328, 1387

Babai, L. 1364
babbling equilibria 834
Bacharach, M. 1454
backward recursion 1419
Baire Category Theorem 754
balance of power, international 1016
balanced cover game 1327
balanced games 1293
balancedness 1316
Balas, E. 1009
Baldwin, R. 1001
Balinski, M.L. 1206
ballistic missile defence 997, 1023, 1024, 1027
ballistic missile defence, *see also* Strategic Defense Initiative
bankruptcy schemes 882

Banks, J.S. 857, 860
Bansal, T. 1028
Banzhaf, J. 1130, 1132, 1144
Banzhaf power index 1128, 1130, 1133, 1137
Banzhaf–Coleman 1373
Bar-Noy, I. 1346
Barany, I. 829, 1353
Baratta, P. 1411
Barbera, S. 1094, 1102, 1104, 1105, 1123, 1407
bargaining 913, 918, 920
– theory of 921
bargaining power 919
bargaining problem 1238
– n-person 1240
bargaining set 1311, 1315
bargaining solution 919
bargaining theory 907, 916, 921
bargaining under strike 819
Baron, D. 898
barrier-to-entry 816
Barron, E.N. 788, 789, 796
Bartels, L.M. 1084
Basar, T. 803, 804, 821, 823, 1028
Baton-passing game 1382
Bau, T. 998
Baumol, W.J. 1220, 1224, 1401
Bawa, V.S. 1409
Bayes, T. 1413
Bayes approach 770, 773
Bayes strategy 773
Bayesian games 1461
Becker, G.M. 1407, 1412
Beckner, W. 1378
behavior strategy 963
behavioral assumptions 1332
Bell, D.E. 1405, 1409, 1421
bell shaped kernels 757
Bellany, I. 1018
Bellman, R. 1021, 1028
Belzer, R. 1023
Belzer, T. 1022
Ben-David, S. 1390
Ben-Or, M. 1351, 1352, 1368, 1369, 1371, 1377, 1380–1382, 1389, 1390
Bénabou, R. 912, 915
Bendor, J. 1002
Bengtsson, B.O. 983
Benhabib, J. 823, 908
Benjamin, C. 998
Bennet, E. 1327
Bennett, J. 1006
Bennett, P. 996, 998
Benoit, J.P. 1331
Beresford, R. 1020
Berge's intersection theorem 747
Bergin, J. 1331
Berkovitz, L.D. 790, 793, 794, 1025

Berlekamp, E.R. 1387
Berman, P. 1348
Bernheim, B.D. 1329
Bernoulli, D. 1400, 1407, 1412
Bernoulli, Nicholas 1400
Bernoulli trials 777
Bernstein, S.N. 1401, 1414
Beryl, J. 1328
Besanko, D. 898
Best, M. 1014
best reply correspondence 806
Bester, H. 912, 916, 919, 920
β-effectiveness (coalitional) form 1300
Bewley, T. 1178, 1181
Bible 1057
Bicchieri, C. 1490
Bierlein, D. 1017
Bigelow, F. 1245
Bikhchandani, S. 909
bilinearity 1413
Billera, L.J. 1221, 1222, 1274, 1299
binary choice probabilities 1407
binary relation 1401
binding commitment 1307
Binmore, K.G. 916, 920, 1260, 1275, 1490
biological applications of game theory 971
biological opportunity cost 977
biological payoffs 975, 976
biological players 936
Bird, C. 1001
Birmingham airport 1213
Bishop, D.T. 953, 978
Bixby, R.E. 1274, 1299
Black, D. 1069, 1079, 1097
Blackett, P.M.S. 995, 996, 1030
Blackwell, D. 747, 765, 770, 774, 779, 1020–1022, 1028, 1408
Blair, D. 1093, 1098, 1103
Blau, J. 1098
Bloch, F. 1331
Blume, L. 1408, 1416
Boddy, M. 1387
Boehm, G.A.W. 1065
Boehm, V. 1223, 1308, 1316
Boes, D. 1223
Boge, W. 1462, 1485
Bohnenblust, H. 746, 755, 1022
Boiteux, M. 1220
boldness 1188
Bolger, E.M. 1078
Bolker, E.D. 1412, 1418
Bollobás, B. 1375
Bomze, I.M. 931, 942, 944
Boolean algebra 1398
Boorman, S.A. 934
Boppana, R.B. 1381, 1384, 1388
Borda, J.C. 1069, 1094

Borda count 1078, 1084, 1085
Border, K. 1105
Bordley, R.F. 1072
Borel, E. 738, 1022
Borowski, E. 1350
Bossert, W. 1242, 1248, 1262, 1276
Boulding, K. 997
bounded rationality 1488, 1490
bounded utility 1417, 1423
Bourgain, J. 1379
Bourqui, P. 1010
Boyd, R. 982
Boyer, M. 1015
Bracha, G. 1384
Bracken, J. 1014, 1023–1025, 1028
Bracken, P. 1010
Bradford, D. 1220
Brams, S.J. 998–1000, 1004–1006, 1009, 1015–1018, 1020, 1030, 1056, 1057, 1062, 1063, 1065, 1068, 1069, 1072, 1073, 1076–1078, 1081–1084, 1131, 1132, 1139
Brandenburger, A. 1485, 1491
Braverman, A. 898, 912, 914
break-even 1219
break-even constraint 1220
Brenner, S. 1084
Brennnan, D. 1011
Brickell, E. 1028
Bridges, D.S. 1404
Brito, D.L. 802, 1001, 1002, 1006, 1008, 1009, 1242
Brockmann, H.J. 977
Brodie, B. 1011
Brooks, P. 1022–1024
Brouwer's fixed point theorem 738
Brown, E.R. 982
Brown, G. 1025
Brown, G.W. 739, 744
Brown, J.L. 982
Brown, J.S. 972, 986
Brown, M. 1001
Brown, R. 1021
Brown, R.L.W. 984
Brown, S. 997
Brown, T. 1014, 1015
Bruce, N. 1016
Bryson, A.E. 817
Bucovetsky, S. 912
Bueno de Mesquita, B. 1001, 1002, 1009
Bull, C. 908
Bull, H. 1011
Bull, J.J. 985
Bunn, G. 1000, 1003
Burdett, K. 908, 909, 912–914, 921
Burness, H.S. 1405
Burr, S. 1024
Busch, M. 1003

Butrim, B.I. 1245
Butters, G.R. 913, 921

C^∞ 754
C-stable 1167, 1168, 1179, 1182
Call, G. 997
Callaham, M. 1014
callers and satellites 983
Calvert, R. 1000
Campello de Souza, F.M. 1407
Canadian Constitutional amendment scheme 1132
candidate strategies 1056
Cannings, C. 940, 978
Cantor, G. 1402
Canty, M. 1017
Caplow, T. 1325
capture time 794
Caraco, T. 982
Carathéodory theorem 747
cardinal utility 1400
Carlisle, T.R. 983
Carlson, J. 908, 912
cartel 1315, 1316
Caryl, P.G. 979
Case, J.H. 802, 817, 1005
Cave, J. 1440, 1454
Cederman, L. 1010
Cegielski, A. 767
central assignment games 1317
certainty equivalent 1411
Chakravorti, B. 1330
Chamberlin, J.R. 1068, 1072, 1079, 1084
Champsaur, P. 1161, 1163
Chan, C.I. 821
Chan, Y.-S. 912, 914
character displacement 986
Charnes, A. 1202
Charnov, E.L. 985
Chateauneuf, A. 1414, 1415
Chatterjee, P. 1016
Chattopadhyay, R. 1010
Chaum, D. 1369, 1371
cheap signals 979
cheap-talk 864
Chen, C.I. 821
Chen, S.F. 821
Chen, S.F.H. 819
Cheney, D.L. 981
Cheng, U. 1028
Cheon, W. 998
Chew, S.H. 1412, 1413
Chichilnisky, G. 1105
chicken models in international relations 1009, 1011, 1015
Chin, H. 754
China/Taiwan conflict, models of 997
Chipman, J.S. 1403–1406, 1408, 1410

Cho, I-K. 855, 857, 860, 861
choice function 1404
Choquet, G. 1422
Choquet integration 1422, 1423
Chor, B. 1343, 1345, 1367, 1370, 1371, 1389
Chou, C. 1472
Christensen, R. 909
chromosome 953, 954
Chuaqui, R. 1415
Chudnovsky, D. 1026
Chudnovsky, G. 1026
Chun, S. 1018
Chun, Y. 1242, 1245, 1251, 1253, 1254, 1257, 1260, 1262, 1263, 1273, 1275, 1276
Church–Turing thesis 1359
Chwe, M. 1332
Clark, C. 815
Clark, C.W. 815
Clark, R.A. 984
Clarke, R. 1001
classical search problem 906
Clemhout, S. 804, 807, 811–813, 816, 818, 819, 823
closed-loop equilibrium strategy 798
clubs 1177
Clutton-Brock, T.H. 979
coalition 1286
coalition production economies 1308
coalition proof Nash equilibrium 1321, 1329–1331
coalition structure 1174, 1177, 1307
coalition structure core 1316–1320, 1326
coalition structure game 1179, 1182
coalitional contingent threats 1330
coalitional function 1286
coalitional monotonicity 1210, 1211, 1218, 1219
coalitional Shapley value (CS-value) 1323
coalitions, international, *see* military alliances
Cochran, M.E. 986
codominated strategies 846
coevolutionary game 986
Cohen, D. 985
Cohen, M. 1407, 1409
Cohen, M.D. 1072
Colman, A. 1079
Colonel Blotto games 1022–1024
Comins, H.N. 985
command, control and communication games 1027, 1028
commitment 917, 1355
commodity redistribution 1183
commodity redistribution game 1185
commodity tax allocation 1185–1187
common knowledge 1344, 1386, 1437, 1451
common-property resource 812
communication complexity 1388
communication equilibrium 838
communication game 833

Index

comparable preference differences 1401
comparative probability relation 1413
competitive mechanism 907
competitive public equilibrium 1178
competitive solution 1326, 1327
competitive tax equilibrium 1186
complementary additivity 1416, 1423
completely mixed 745
complexity theory 1357
computability theory 1357
computer networks 1340
computer simulations 1072
concave cost function 1215
concavity of the distribution function 877
conditional preference relation 1416
conditional probability 1423
conditionally compact 752
Condon, A. 1386
Condorcet, Marquis de 1069, 1094
Condorcet candidate 1070, 1073–1075, 1084
Condorcet winner 1097, 1112, 1115
Condorcet's criterion 1070
conflict 1057
conflict analysis 998, 999
conflict over parental investment 983
Conley, J. 1248, 1251
connected 1059
consecutive 1317
consecutive coalitions 1319, 1328
consecutive game 1182, 1317, 1318
consensus 1345
consistency 1205, 1206, 1222, 1267
consistency principle 1420
constant-sum game 1290
constraint 1219
consumer surplus 1220
contestable market theory 1224
contingency 1061, 1068
continuity 1248
continuous utility 1403, 1409
contraction independence 1246
contraction of a game 1297
control function 783
conventional settlement of contests 972, 974–977
convex effectivity functions 1120
convex game 1290
convex payoffs 755
Conybeare, J. 1000
Cook, D. 997
cookie-cutter games 1026
Cooper, J. 1385
cooperation structure 1322, 1323
cooperative game 1307
core 1146, 1160, 1165, 1167, 1168, 1174, 1176, 1178–1180, 1182, 1195, 1196, 1199, 1200, 1214, 1311, 1325, 1329–1331
core of the cost function 1224

correlated equilibrium 831, 1330, 1463
cost allocation method 1197, 1203, 1222
cost complementarity 1225
cost contribution of project 1214
cost elements 1212
cost function 1197
cost objects 1215
cost of delay 919
cost-savings game 1198
cost-sharing game 1197
Courant, P.N. 1079
Cover, T.M. 1390
cover of a game 1295
covers total cost 1227
Cozzens, M.B. 1404
Crandall, M.G. 788
Crawford, V.P. 840, 863, 865, 944, 1012, 1275, 1276
credible threats 1307
Crespi, B.J. 196
Cressman, R. 950
Cressy, R.C. 912
crisis instability, *see* first-strike instability
Crott, H.W. 1245
Cruz, J. 1005
Cruz, J.B. 818, 821
CS-value, *see* coalitional Shapley value
Cuban Missile Crisis, models of 997–999
Cudd, A. 1000
cumulative voting 1080, 1081, 1085
curious players 1352
cyclical preferences 1058

Dacey, R. 1010, 1018
Dagsvik, J.K. 1407
Dahl, R. 1001
Dalkey, N. 1028
Damage Limitation Study 1023, 1024
Dando, M. 998
Danforth, J. 908
Danilov, V. 1118, 1121
Danskin, J. 1022, 1026
Dantzig, G.B. 739, 743
Dasgupta, P. 862
Dash, A.T. 950
D'Aspremont, C. 1316
Daughety, A. 916
David, P.A. 906
Davidson, D. 1416, 1418, 1422
Davies, N.B. 973, 975, 981
Davis, M. 1017, 1018
Davis, M.D. 1056
Davis, O. 1146
Dawkins, R. 934, 983
De Finetti, B. 1401, 1413, 1414
De Koster, R. 1265

De Maio, G. 1084
De Zeeuw, A. 1005
De Zeeuw, A.J. 818
Dean, T. 1387
Deb, R. 1098, 1321
Debreu, G. 1403, 1405–1407
decision problem 1358
decision procedures 771
decision theoretic and game theoretic analyses 1085
decision theory 1056, 1057
decision trees 1419
decision under uncertainty 1416
decomposability 1252
decomposition principle 1212
Deegan, J. 1144
Deegan–Packel power indices 1144
Deere, D.R. 916, 919, 920
DeGroot, M.H. 908, 1408, 1414, 1415, 1419
Deissenberg, C. 802
Dekel, E. 1245, 1413, 1485
Dekmajian, M. 1017, 1018
Demange, G. 1015, 1104, 1318, 1319
DeMarzo, P.M. 1331
democracy 1056–1058
democratic 1067, 1078, 1084
Dempster, A.P. 1415
Demsetz, H. 899
Deng, X. 1386
Denzau, A.T. 1314
DeSwaan, A. 1328, 1329
deterrence models 1006–1010
Deutsch, K. 996
devisive asymmetry 977
Dewdney, K. 1006
Deyer, S. 1005
Diamond, H. 1017
Diamond, P.A. 906, 909, 912, 915–917, 920, 1405
dichotomous preferences 1070
Dickins, D.W. 984
dictatorial solution 1243
Dierker, E. 1223
Diesing, P. 997
differential game 784
– many player 802
– n-person 796
– non-zero sum 817
– tractable 808, 815
– two-person, zero-sum 802
Diffie, W. 1361
Digby, J. 1012
digger wasps 977
dimension relations 746
dimension theorem 746
diminishing marginal utility 1400
Dion, D. 1387
diploid population 953

diplomacy 1387
Dirac measures 764
direct costs 1203
disagreement point concavity 1262
– weak 1263
disagreement point monotonicity 1261
Dishington, R. 1025
Diskin, A. 1016
distributed system 1340
divide-and-choose in arms bargaining 1004
Dobbie, J. 1026
Dockner, E. 815, 816
Dockner, E.G. 824
Doherty, A. 998
Doigon, J.-P. 1404
dollar auction 1015
dominance 1058–1063, 1408
dominance axiom 1422
dominance relation 1324, 1325, 1330
dominant player 1322, 1328, 1329
dominant strategy Nash equilibrium 1190
domination 1294
dominion 1324
Domotor, Z. 1414–1416, 1418
Don, B. 1001, 1016
Donsimoni, M. 1316
dormancy 986
Doron, G. 997, 1078
Downs, G. 1002, 1003, 1005
Doyle, J. 1387
Dresher, M. 736, 1017, 1019, 1021–1023, 1025
Drèze, J. 1308, 1309, 1311, 1313, 1315, 1322, 1323, 1409
drift resistance 961
dual linear programs in standard form 739
dual of a game 1291
duality theorem 739
Dubey, P. 1134, 1135, 1468
Dubin, G. 1019
Dubiner, M. 1380
Dudley, L. 1016
duels, theory of 1021
Dugatkin, L.A. 982
Dummett, M. 1056
dummy 1131, 1134, 1214
Dunz, K. 1314
Dupuit's bridge 1220
Dutta, B. 1118, 1121, 1315
Dutta, P.K. 882, 884–886, 897, 898
Dvoretzky, A. 775, 779
Dwork, C. 1343, 1345, 1349
Dym, H. 1377
dynamic Bayesian games 1469
dynamic consistency 1419
dynamic foundations of evolutionary game theory 948
dynamic solution 1325

Easwaran, M. 899
Eavy, C.L. 1328
Eckler, R. 1024
econometric policy evaluation 820
Edgeworth, F.Y. 1238
Edison, T. 1019
Edison, Thomas, and military gaming 1019
Edwards, W. 1411, 1421
efficacy 1059, 1069
efficiency 1196
efficient 1229
egalitarian solution 1243, 1251
egg trading 983
Ehrenfeucht, A. 1386
Eilenberg, S. 1403
Einhorn, H.J. 1411, 1421
Einy, E. 1330
Eisele, Th. 1485
Eisner, R. 1002
El-Yaniv, R. 1391
elections 1056
Elliott, R.J. 783, 784, 787, 789, 794
Ellner, S. 986
Ellsberg, D. 1011, 1414, 1420, 1421
Ellsberg example 1420
empirical tests of international relations models 1001, 1015
endogenous growth 802, 814, 817
Enelow, J. 1150
Enelow, J.M. 1056
Engelman, W. 1005
Engers, M. 861–863
Enquist, M. 978, 979, 983
envelope 1363
Epple, D. 1314
ε-core 1175, 1176, 1178, 1180, 1181
ε_0-core 1181
ε-equilibrium 1180, 1181
equal area solution 1244
equal treatment property 1176
equalizer theorem 740
equilibrium
 – correlated 938
 – credible message equilibria 864
 – E2 862
 – multiple 907, 920
 – Nash 850, 852
 – neologism proof 864
 – Pareto-dominated equilibria 920
 – perfect 918–920
 – perfect public 893
 – point 937–939, 942, 970
 – pooling 853
 – reactive 862
 – refinements in deterrence models 1008
 – separating 853
 – sequential 851, 856
 – stable 851
 – undefeated 861
 – value 797
equilibrium for differential games
 – a continuum 813
 – multiplicity of 814
 – open-loop Nash 811
 – Pareto-ranked 813
 – Stackelberg 819
 – subgame perfect Nash 804, 806
equiprobable 1069
equitable 1069, 1229
equity 1196
equivalence 1401
equivalence classes 1402
escalated contest 972, 974
escalation of conflict 1007, 1014–1016
Eshel, I. 947, 954–956, 986
ESS, see evolutionarily stable strategy 941
estimation 771
estraneous scaling probabilities 1418
ethics and nuclear weapons 1010
Euclidean preferences 1321
European Economic Community 1131
Evangelista, M. 1000, 1011
Evans, L.C. 789
evolution of cooperation 980–983
evolutionarily stable strategy (ESS) 931–933, 939–941, 943, 944, 946–954, 958, 963, 964, 968, 975, 982, 984
 – continuously stable 947
 – direct 968, 970
 – irregular 941
 – limit 969–971
 – of an asymmetric bimatrix game 965
 – regular 941, 942
 – strict 964
evolutionary game 985
evolutionary game theory 931, 932, 935, 936, 948, 971, 979, 986
evolutionary restriction of aggression 973
evolutionary stability 932, 936, 938, 940, 968
 – in extensive two-person games 965
Ewens, W.J. 954
ex-post strong correlated equilibrium 1330
exaustible common-property resource 804
 – ASAP exploitation 808
excludable public goods 1159
expected utility 1069, 1400, 1407
extended 774
extended Bayes 775
extended game 775
extensions of games 1295
extensive game with two symmetries 967
extensive two-person games 965
external stability 956, 958, 1325

externalities 907, 920, 1154, 1160, 1162, 1163, 1174, 1181, 1308, 1329
extreme optimal strategies 742
extreme points adjacent 744

face in international affairs 1008
Fagin, R. 1386, 1444, 1485
Fain, W. 1022, 1028
fair 1056, 1203
Falk, J. 1024, 1025, 1028
Falmagne, J.C. 1407
Fan, Ky. 749
Farquhar, P.H. 1409
Farquharson, R. 1056, 1092, 1093, 1105
Farrell, J. 841, 845, 857, 864, 1310, 1319, 1331
farsighted coalitional stability 1332
Faulhaber, G. 1199
fear of ruin 1188
Fearon, J. 1001, 1008
feasible 1061, 1063, 1064
feasible solution to the primal 739
Featherston, F. 1072
feedback control 791, 792, 797
Feichtinger, G. 824
Feiwel, G. 1002
Feld, S. 1146
Feldman, J.A. 1387
Feldman, M.W. 954–956
Feldman, P. 1351, 1380, 1385
Fellingham, J. 881
Felsenthal, D.S. 1065, 1069, 1072, 1084
Fenster, M.J. 1084
Fernandez, L. 862, 863
Fershtman, C. 816, 817
Fiat, A. 1390, 1391
Fichtner, J. 1014, 1017, 1018
fictitious play 744
Fiesenthal, D. 1010
Filar, J.A. 746, 1017
Findlay, M.C. 1409
Fine, T. 1413, 1414
Fine, T.L. 1415
Finn, M. 1019
Fiorina, M. 1328
first-strike instability 1013, 1014
Fischer, E.A. 931, 934, 982–984
Fischer, M.J. 1349, 1350
fish war 813, 817
Fishburn, P.C. 1056, 1062, 1063, 1068–1070, 1072, 1073, 1076, 1078, 1080, 1082–1084, 1399–1406, 1408–1410, 1412–1423
fitness 934, 935, 938, 954, 976, 977
fixed background approach 978
fixed point theorems for set valued maps 749
fixed-sample-size rule 913
Fleming, W. 789
Flood, M.M. 1411

focal voter 1061
Fogarty, T. 1010
Foldes, L. 1409
Foley, D.K. 1156–1160, 1162
food sharing 982
Forges, F. 829, 835, 842, 865
formula to compute the value 745
Forward, N. 996, 997
Foster, D. 948
Foster, D.P. 1390
Foster, J. 898
Fourier, J.B. 743, 1377
Fourier transform 1377
Fox, M. 765
Fraisse–Ehrenfeucht game 1386
franchise fee 874
Franck, R. 1014
Frank, S.A. 985
Franke, G. 1421
Fraser, K. 998
Fraser, N. 998, 999
Freedman, L. 1011
Freeman, J. 1000
Freimer, M. 1245
French, S. 1414, 1415
frequency-dependent selection 935, 954
Fretwell, S.D. 985
Frey, B. 1004
Frick, H. 1018
frictional unemployment 906
Friedman, A. 784–798
Friedman, J. 983
Friedman, M. 1408, 1409
Frisch, R. 1400, 1406
Fuchs, G. 1165
Fudenberg, D. 881, 885, 893, 895–898, 900, 916
Funk, S.G. 1328
funnel web spiders 975
Füredi, Z. 1353
Futia, C. 1386

Gafni, E. 1350
Gal, S. 908, 912, 1026
Gale, D. 746, 916, 920, 1026, 1317
game in characteristic function form 1307
game in coalitional form 1287, 1307
game in partition function form 1301
game of capitalism 819
game of survival 794
game theory 1056, 1057
game tree models in international relations 997, 998
game with communication 828
game with cooperation structure 1301
game with side payments 1307
games in normal form 736
games in partition form 1308

Gamow, G. 1439
Gamson, W.A. 1325
Gandhi 1473
Gantmacher, F. 764
Garay, J. 1348
Garay, M.R. 1348, 1357
garbling 865
Gärdenfors, P. 1421
Gardner, M. 1439
Gardner, R. 1016, 1183, 1189
Garey, M.R. 1359, 1365, 1387
Garfinkel, M. 1001
Gary Cox, W. 1056, 1084
Gately, D. 1202
Gates, S. 996, 1001, 1010, 1015
Gauthier, D. 1010
Geanakoplos, J. 1386, 1439, 1440, 1447, 1448, 1450, 1455, 1468, 1472, 1490, 1491, 1494
Gehrlein, W.V. 1070, 1072
Gekker, R. 999, 1010
gene locus 953
generalized correlated equilibrium 838
Genesereth, M.R. 1387
genotype 935, 954
genotype frequencies 954
George, A. 996, 1008, 1011
Georgescu-Roegen, N. 1403, 1405
Gerèb-Grauss, M. 1389
germination games 986
Giamboni, L. 1025
Gibbard, A. 1068, 1092, 1094, 1102, 1122, 1123
Gigandet, C. 997
Gilboa, I. 1407, 1411, 1416, 1422, 1485
Gilles, R.P. 1327
Gillespie, J.H. 935, 1001, 1005, 1016
Gillespie, J.V. 818
Girshick, A. 770, 774, 779
Girshick, G.A. 747, 765
Girshick, M.A. 1408
Glass, G.E. 984
Glassey, G. 1026
Glicksberg, I. 749, 754, 756
God 1057
Golberg, A. 1108
Gold, V. 997
Goldhamer, H. 1012, 1020
Goldreich, O. 1343, 1350, 1352, 1360, 1363, 1368, 1369, 1371, 1384, 1385, 1389
Goldstein, J. 1000
Goldstein, W. 1411
Goldwasser, S. 1361, 1364, 1367, 1384
Good, I.J. 1413, 1419
Gorman, W.M. 1405
Gossen, H.H. 1400
Gourvitch, V. 1108
Gowa, J. 1000, 1001
Grafen, A. 861, 934, 974, 977, 980, 983, 984

Graham, R.L. 1350
Grandmont, J.-M. 1409
graph coloring 1359
Greece 1057
Green, E. 893, 899, 900
Green, J. 863, 1004, 1227
Green, P. 1013
Greenberg, J. 1014, 1158, 1165, 1167, 1178, 1182, 1306, 1308, 1310, 1311, 1313–1318, 1320, 1321, 1324, 1327, 1329, 1330, 1332
Greenwald, B. 898
Grether, D.M. 1411
Gretlein, R. 1107
Grieco, J. 1000
GRIT procedure for conflict deescalation 1006
Grofman, B. 1146, 1148, 1150
Groneau, R. 909
Gross, O. 754, 1022, 1023
Grossman, S.J. 845, 850, 857, 875, 877, 878, 880, 892, 897–899
Grotte, J. 1022, 1028
Grotte, J.H. 1210
group selection 934
Groves, T. 900, 1227
Guckenheimer, J. 816
Guesnerie, R. 1165–1167, 1182, 1309, 1310
Gul, F. 1331
Guner, S. 1016
"guns versus butter" decision 1001, 1006
Gupta, S. 1242
Gurevich, Y. 1386
Gurk, H.M. 1292
Guth, W. 998, 1003–1005, 1016
Gutman, S. 802
Guyer, M. 997

Hadar, J. 1409
Hagen, O. 1411, 1412
haggling 921
Haigh, J. 941
Haigh's criterion 942
Hall, J. 908
Halliday, T.R. 973
Halperin, A. 1005
Halpern, J. 1439, 1444
Hamilton, W.D. 931, 942, 972, 980, 982, 985
Hamilton–Jacobi equations 787, 797
Hamiltonians 788
Hammerstein, P. 938, 965, 972–975, 977, 978, 982, 983, 985, 986
Hammond, P.J. 1405, 1416, 1419
Hammond, T. 1002
handicap principle 979, 980
Hanoch, G. 1409
hard-core bit 1363
Hardin, R. 996, 1010
Hardy, G.H. 1362, 1409

Hare system of single transferable vote (STV) 1075, 1084
harmonic analysis 1377
Harper, D.G.C. 985
Harrington, L. 1386
Harris, M. 897
Harrison, G.W. 908
Harsanyi, J.C. 830, 835, 837, 1001, 1003, 1005, 1013, 1206, 1241, 1247, 1267, 1275, 1289, 1332, 1416, 1484, 1485
Harsanyi coalitional form 1185
Harsanyi doctrine 1488
Harsanyi–Aumann doctrine 1485, 1486
Harsanyi–Aumann question 1486
Harsanyi–Shapley 1170, 1173
Harsanyi–Shapley NTU value 1184, 1185
Hart, O.D. 875, 877, 878, 880, 892, 897–899
Hart, S. 831, 865, 1206, 1297, 1310, 1313, 1323, 1331, 1375
Hartigan, J.C. 1419
Hartley, D. 998
Harvey, F. 1001, 1015
Hashim, H. 998
Hastad, J. 1389
Hastings, A. 985
Haurie, A. 816, 823
Hausch, D.B. 922
Hausfater, G. 973, 984
Hausner, M. 1408, 1410
Haviv, M. 1010
Hawk–Dove game 963, 974, 978
Hayek, O. 1026
Hayes, R. 1020
Haywood, O. 1020
Heal, G. 1105
Heaney, J.P. 1202
Heath, D.C. 1221, 1222
hedonic aspect 1308–1310, 1317
Heims, S. 1012, 1013
Helin, P. 1028
Hellman, M.E. 1361
Hellwig, M. 863
Helly's theorem 748
Helmer, O. 1001, 1023
Henriet, D. 1318
Herden, G. 1403
Herlihy, M. 1350
Herrero, C. 1242, 1245, 1248
Herrero, M.J. 916, 920
Hershey, J.C. 1409
Herstein, I.N. 1407, 1408
Hessel, M. 998
Hewitt, J. 1015
Hey, J.D. 912, 921
Hicks, J.R. 1400
hierarchies of belief 1484
Hildenbrand, W. 1174

Hillas, J. 865
Hillstead, R. 1025
Hines, W.G.S. 953, 972, 974
Hinich, M. 1150
Hinich, M.J. 1056
Hipel, K. 998, 999
Hipel, N. 998
history dependence in the second-best contract 880
Ho, Y. 1027
Ho, Y.C. 802, 805, 817, 823
Hodges, W. 1386
Hofbauer, J. 948–952
Hoffman, D.T. 1069
Hogarth, R.M. 1421
Hölldobler, B. 980
Hollis, M. 996
Holmes, P. 816
Holmes, R. 1023
Holmes, Sherlock 1491
Holmstrom, B. 878, 881, 888, 889, 897–899
Holt, C.F. 803, 817
Holt, R.D. 985
Holzman, R. 1078
homogeneity assumption 1136
Hone, T. 1020
honor in international relations 1008
Hopfinger, E. 1017
Hopmann, T. 1003
Horowitz, A.D. 1328
Horowitz, I. 1011, 1013
Horvitz, E.J. 1387
hotspot games 1022
Houston, A.I. 978, 981, 985
Houthakker, H.S. 1404
Hovi, J. 998, 1016
Howard, N. 998, 999
Howard, R.A. 1419
Howitt, P. 916, 920
Hrdy, S.B. 973
Hsieh, J. 997
Huber, R. 1005
Humes, B. 996, 1000
Hurwicz, L. 1093, 1189, 1227, 1403
Hurwitz, R. 996, 1006
Hutchinson, H. 1006
Hylland, A. 1122
hypergames 998, 999
hypothesis testing 771

i.s.e. 1315
Ichiishi, T. 1120, 1121, 1316
ideal free distribution 985
ideal point 1144
ideological space 1140, 1144
Iida, K. 1000, 1002
Imai, H. 1183, 1251, 1254
immunity 1383

Index

Impagliazzo, R. 1389
implementation 1331
implicit contract 915
impossibility theorem 1058
imputation 1294
Inada, K. 1070
incentive compatibility in international negotiations 1004
incentive compatible 1190, 1226–1228
incentive constraint 872
incentive to cooperate 1196
incentives 1196, 1229
inclusive fitness 974
income redistribution 1183, 1184, 1186
income tax allocation 1186, 1187
incomparability 1402
incomplete information 851
inconsistent beliefs 830
incremental cost test 1199
independence assumption 1136
independence axiom 1408
independence of non-individually rational alternatives 1248
independence violations 1411
indifference classes 1402
indifference relation 1402
individual monotonicity 1249
individual rationality 1228, 1248
– strong 1248
individual-rationality constraint 872
individual selection 933, 934
individually stable contractual equilibrium (i.s.c.e.) 1315
individually stable equilibrium 1313, 1314
infant killing 984
infanticide 972, 973
infinite spectrum 754
information acquisition 921
inheritance of resources 980
insincere 1084
inspection, arms, *see* arms control verification
integral equation with a compact operator 764
intensities of preference 1080
interactive proofs 1364
interdependent preferences 1405
intermational asymmetries 922
internal stability 1325
international political economy 1000, 1001
international relations game theory 997–1018, 1029, 1030
intersection theorems 747
interval order 1404
Intriligator, M. 1001, 1002, 1006, 1008, 1009
intuitive criterion 857
invariance 1254
invariant in direct costs 1203
IP 1365

irrelevant candidate 1080
Isaacs, R. 790, 793, 795, 802, 805, 1021, 1025–1027
Isard, W. 818, 1004
Isbell, J.R. 1022, 1292
Ishi, H. 788
isoperimetric inequality 1375
Israelites 1057
issue linkage in international negotiations 1000
Iwai, K. 916

Jackson, M. 1104, 1118, 1228
Jaffray, J.-Y. 1415
James, L.D. 1201
James, P. 1001, 1015
Jansen, M.J.M. 1249
Jansen, W. 952
Jaworsky, J. 998
Jeffrey, R.C. 1414, 1418
Jensen, N.E. 1408
Jervis, R. 996, 1000, 1004, 1007
Jevons, W.S. 1400
Jewitt, I. 897
Jian Jia-he 941
John, A. 1006
Johnson, D.S. 1357, 1359, 1365, 1387
Johnson, Michael 1178
Johnson, S. 1021
Johr, M. 997
Jones, R.A. 1409
Jonker, L.B. 948, 949
Jonsson, C. 1003
Jordan, J. 1105
Jorgenson, S. 824
Josko de Gueron, E. 1000
Joxe, A. 997
Judd, K.L. 912–914
Junne, G. 996
Jureen, L. 1403

K-equilibrium 1321
Kahan, J.P. 1328
Kahn, C. 1330
Kahn, H. 1011
Kahn, J. 1375, 1376, 1379
Kahneman, D. 1409, 1411, 1412
Kaiser, H. 984
Kaitala, V. 823
Kaitala, Y. 823
Kakutani, S. 738, 749
Kakutani's fixed point theorem 749
Kalai, E. 1098, 1104, 1216, 1217, 1239, 1245, 1248, 1249, 1251, 1252, 1274, 1275, 1297, 1325, 1342, 1355, 1388
Kalai, G. 1375, 1376, 1379
Kalai–Smorodinsky solution 1243, 1249
Kalman, P. 1009

Kalton, N. 783, 784, 789
Kamien, M.I. 805, 817
Kaneko, M. 1485
Kannai, Y. 1174, 1276, 1410
Kantorovich, L.V. 743
Kaplan, F. 1011, 1012, 1023
Kaplan, M. 1011–1013
Kaplansky, I. 745
Karamata, J. 1409
Karchmer, M. 1388
Karlin, S. 746, 755–757, 764, 768, 954, 985, 1029
Karmarkar, U.S. 1411, 1412
Karni, E. 1418
Karp, R.M. 1391
Karr, A. 1023, 1025
Katznelson, Y. 1379
Kauder, E. 1400
Kaufmann, W. 1011
Kawara, Y. 1025
Keeney, R.L. 1409
Keiding, H. 1092, 1120, 1121
Kelly, J.S. 1076
Kent, G. 1014, 1019, 1023
Kent, Glenn, and ballistic missile defence 1023, 1024
Keohane, R. 999, 1000
kernel 754, 1311
Kettelle, J. 1004
Keynes, J.M. 1402, 1414
Kidd, A. 1000
Kihlstrom, R.E. 1264, 1265
Kilgour, D.M. 1008, 1009, 1015, 1017, 1018, 1133, 1138
Kimball, G. 1020
Kimeldorf, G.S. 765, 768
Kimmeldorf, G. 1021
kin selection theory 980
Kirby, D. 997
Kirman, A. 1327
Kissinger, H. 1011
Kitchen, J. 1021
Kivikari, U. 1001
Klemisch-Ahlert, M. 1242, 1265
know 1442
Knuth, D. 1387–1389
Koc, E.W. 1084
Kochenberger, G.A. 1409
Koenig, W.D. 980
Kohlberg, E. 858, 1245
Kohlleppel, L. 861
Kohn, M. 908, 909
Koller, D. 1385
Konishi, H. 1314
Koopman, B.O. 1401, 1413, 1414, 1416
Koopmans, T.C. 743, 1405
Kotwal, A. 899
Kraft, C.H. 1401, 1413, 1415

Krantz, D.H. 1399, 1402, 1406, 1415, 1416, 1418
Krasner, S. 1000
Krasovskii, N.N. 787
Krasucki, P. 1459
Kratochwil, F. 1003
Kraus, S. 1387
Krein, M.G. 764
Krelle, W. 997
Kreps, D.M. 857, 860, 1409, 1468, 1471, 1472
Kreweras, G. 1413
Kripke, S. 1442
Krishna, V. 1331
Kroll, J. 1000
Kronick, R. 1078
Kugler, J. 1010
Kuhn, H.W. 736, 776, 779, 1004, 1017, 1415
Kuhn, J. 998
Kumar, A. 1415
Kummer, H. 975
Kupchan, C. 1016
Kuratowsky–Knaster–Mazurkievicz theorem 752
Kurisu, T. 767
Kurz, M. 1168, 1171, 1183–1185, 1188, 1189, 1310, 1313, 1323
Kushilevitz, E. 1363, 1370
Ky Fan's minimax theorem 750
Kydland, F.E. 819

La Valle, I.H. 1418, 1419, 1422
labor-managed economies 1310
Laffond, G. 1105
Laffont, J.-J. 1004, 1006, 1227
Lake, M. 1056
Lalley, S. 1026
Lalman, D. 1001, 1002, 1009
Lambert, R. 879, 880, 898
Lamport, L. 1344, 1346
Lancaster, K. 819
Lanchester models of combat 1025
Landau, H.J. 916, 919
Lang, J.P. 768
Lange, O. 1400
Langlois, J. 1015
language 1358
Lapan, H. 1009
Lapidot, E. 1292
Laplace, P.S. 1413
large public goods 1164
largest consistent set 1332
Larson, D. 996, 1000
Larsson, S. 1411, 1414, 1420
LaSalle, J. 1028
law of one price 912
Lawlor, L.R. 986
Lazarus, L. 982
Lazear, E. 899
Le Breton, M. 1105, 1321

Index

least core 1204
least favorable 774
Ledyard, J. 900
Lee, K. 1022
Lee, L. 1022
Lee, R.R. 1201
Lee, T. 922
legislative voting procedures 1056
Legros, P. 890, 892, 899
Lehrer, E. 1292
Leimar, O. 978, 979, 983, 986
Leininger, W. 1015
Leites, N. 1012, 1020
Leitmann, G. 802, 816, 819, 823
Leitzel, J. 996
Leland, H. 912, 914
Leng, R. 998, 1015
Lenin, V. 1002
Lensberg, T. 1206, 1239, 1248, 1265, 1267, 1268, 1271, 1272
Lenstra, A.K. 1362
Lenstra Jr, H.W. 1362
Leonard, R. 1019
Lessard, S. 985
lethal injury 972, 973
Levesque, T.J. 1133
Levhari, D. 804
Levin, L.A. 1363, 1389
Levin, S.A. 985
Levine, D. 896
Levins, R. 934
Levitt, P.R. 934
Levy, H. 1409
Levy, M. 997
Lewis, D. 1438
Lewis, H.R. 1342, 1357
lexicographic order 1402
lexicographic utility 1405
Li, L. 922
liberal 1058
Liberman, U. 955, 956
Lichbach, M. 1005
Lichtenstein, D. 1389
Lichtenstein, S. 1411
Licklider, R. 1011
likelihood function 776
Lima, S.L. 985
Lin, J. 789
Lindahl, E. 1158, 1189
Lindahl equilibrium 1156, 1158–1163, 1175, 1178
Lindahl prices 1160, 1162
Lindahl taxes 1162
Lindman, H.R. 1411
Lindskold, S. 1006
linear inequalities 1415
linear order 1401
linear-quadratic games 817

Lines, Marji 1084
Linial, N. 1375–1377, 1379–1385
Linster, B. 1016
Lions, P.L. 788, 790
Lippman, S.A. 908, 921
Lipson, C. 1000
Little, J.C.D. 1084
Littlechild, S.C. 1202, 1213
Littlewood 1439, 1440, 1447
Liu, P.T. 819
Livne, Z.A. 1242, 1249, 1258, 1260, 1262, 1263, 1275
$L^\infty = L_1^*$ 754
local public goods 1156, 1164, 1167, 1174, 1177, 1309
Loehman, E.T. 1216, 1217
Lohmann, S. 1002
Lomnicki, A. 985
Loomes, G. 1412, 1413, 1421
Loomis, I.H. 738
Loomis, L. 1379, 1382
Lorenz, K. 972
loss function 772
lotteries 1418
lottery acts 1418, 1422
Lovász, L. 1388
lower δ-game 785
lower δ-strategy 785
lower δ-value 785
lower semicontinuous 750
lower value 784, 786
Luby, M. 1389
Lucas, H.L. 985
Lucas, R. 906
Lucas, R.E. 820
Lucas, W.F. 1129, 1130, 1301, 1308
Luce, R.D. 1245, 1252, 1313, 1399, 1401, 1403–1405, 1407, 1408, 1414–1416, 1418, 1423
Ludwig, R. 996
Ludwin, W.G. 1079
Lumsden, M. 997
Lund, C. 1365
Luterbacher, U. 1016
Luther Martin 1130
Luxemburg, R. 1002
Lynch, N.A. 1349, 1350
Lynn-Jones, S. 1011

m-equilibrium 1320
Maccoby, M. 1013
MacCrimmon, K.R. 1411, 1412, 1414, 1420
MacDonald, S. 1132, 1143, 1144
Machina, M.J. 1407, 1409, 1411, 1412
Macleod, W.B. 1331
MacMinn, R. 912
macro-economics 819

macroeconomic models   920
MacWilliams, J.   1371
Mahmond, E.   1390
Mailath, G.   857, 861
Majeski, S.   1005
Makins, C.   1003
Malcomson, J.   885, 898
male armament   984
Maleug, D.A.   908
malicious players   1352
Malitz, J.   1415
Manasse, M.S.   1390
Manders, K.L.   1404
Mangasarin condition   809
manipulability   1068, 1077, 1078, 1085
manipulable   1072
manipulate   1084
manipulation   1068
manipulative   1075
manipulative strategy   1079
Mann, I.   1132
Manne, A.   1220
Manning, R.   908
Mansfield, E.   1000, 1001
Manski, C.F.   1407
Mansour, Y.   1363
Maoz, Z.   996, 997, 1010, 1016, 1072
Marco, M.C.   1245
Mares, D.   1001
marginal cost   1199
marginal savings   1201
market games   1178, 1287, 1293
Markowitz, H.   1409
Marley, A.A.J.   1407
Marlow, W.   1022
marriage game   1317
Marschak, J.   1407, 1408
Marshall, A.   1012, 1020, 1400
Martin, B.   996
Martin, L.   1000, 1001
Martinez Oliva, J.   1000
Mas-Colell, A.   1206, 1297, 1299, 1331, 1404
Maschler, M.   736, 1017, 1018, 1206, 1245, 1259, 1476
Maskin, E.   845, 862, 865, 896, 916, 917, 1116–1118, 1331
mate desertion   983
mate guarding   984
mate searching   983
Matheson, J.   1024
mating system   984
Matlin, S.   1024
Matsubara, N.   1001
Matsushima, H.   889, 890, 892, 895, 899, 900, 1123
Matthews, S.   1331
Matthews, S.A.   922

Mattli, W.   999
maximin criterion   1204
May, K.   1094
May, K.O.   1411
May, R.M.   942, 985
Mayberry, J.   1001, 1027
Mayer, F.   1002
Maynard Smith, J.   931–934, 937, 938, 940, 943, 958, 972–975, 978, 979, 983, 984, 986
McAfee, R.P.   898, 908, 912, 916, 920
McCall, J.J.   908, 921
McCardle, K.F.   831
McCarty, C.   1016
McDonald, J.   996
McGarvey, D.   1024
McGeoch, L.A.   1390
McGinnis, M.   1000, 1002, 1005, 1010
McGuire, C.   1028
McGuire, M.   1006
McKean, H.P.   1377
McKelvey, R.D.   1108, 1147, 1148, 1327, 1328, 1460
McKelvey's theorem   1147
McKenna, C.J.   916, 921
McKinsey, J.   1331
McLennan, A.   916, 920
McMillan, J.   898, 912, 915, 916
McNamara, J.   978, 985
Mearsheimer, J.   1001
Measor, N.   997
mechanism design   865
median line   1142
mediation mechanism   836
mediation plan   836
– incentive compatible   837, 841
– incentive constraints   837
– mechanism   841
– revelation principle   837
mediator   829
Medlin, S.M.   1310
Mefford, D.   1006
Megiddo, N.   1210, 1385, 1386, 1389
Mehlman, A.   803
Mehlmann, A.   803
Melese, F.   1005
Melumad, N.   900
Mengel, A.   1025
Menger, C.   1400
Menger, K.   1348, 1400
Mensch, A.   1019
Merrill, S.   1063, 1068–1070, 1072, 1075, 1084
Merrit, M.   1349
Mertens, J.-F.   858, 1485, 1488
Mertens, J.W.   751
metagames   998, 999
Metcalfe, N.B.   982
methods of aggregation   1058

Meyer, R.F. 1409
Micali, S. 1351, 1361, 1367, 1385
Michel, P. 1005
Michener, H.A. 1328
Middle East conflicts, models of 997–999
midpoint domination 1254
Miercort, F. 1028
Milgrom, P. 850, 855, 877, 881, 1447, 1451, 1455, 1468, 1472
Milgrom, P.R. 920, 922
Milinski, M. 982, 985
military doctrine 1029, 1030
military game theory 1018–1030
– in the Soviet Union 1019
Mill, John Stuart 1076
Miller, D. 1132
Miller, G.J. 1328
Miller, N. 1108, 1111, 1112
Miller, N.R. 1058
Milleron, J.C. 1155, 1159–1161, 1163, 1175, 1176
Milnor, J. 1134, 1407, 1408
minimal winning coalition 1128
minimax theorem 738, 1413
Mino, K. 818
Mirman, L.J. 804, 1221–1223, 1225
Mirowski, P. 1019
Mirrlees, J.A. 863, 875, 876, 897
Mishal, D. 998
Mishal, S. 1016
missile defence, *see* ballistice missile defence
missile launch timing 1021
mixed extension 737
mixed strategies 770, 912
Mo, J. 1002
mobilization, military 1014
Modigliani, F. 1409
Moglewer, S. 1017, 1025
Mohler, J.D. 984
Molander, P. 1029
Moldovanu, B. 1331
Monderer, D. 1224, 1481–1483
monomorphic 932
monotone likelihood ratio conditions 877
monotonic in the aggregate 1209
monotonicity 1197, 1209, 1290
Mookherjee, D. 892, 899, 1330
Moore, J. 998, 1093, 1118
Moore, R.N. 1387
Mor, B. 998, 1009, 1016
moral hazard 840, 870
morality 973
Moran, P.A.P. 954
Moreno, D. 1330
Morgan, P.B. 908, 909, 912, 915
Morgenstern, O. 738, 828, 1013, 1289, 1292, 1306, 1307, 1324–1327, 1329, 1401, 1407, 1409, 1410

Moriarty, G. 1005
Morris, C. 1013
Morris, S. 1469
Morrison, D.G. 1411
Morrow, J. 996, 1002, 1003, 1015, 1017
Morse, P. 1020
Mortenson, D.T. 906, 908, 916, 920, 921
Moses, Y. 1344, 1348, 1439, 1470
Mosschovakis, Y.N. 1386
Motro, U. 985
Moulin, H. 835, 1010, 1078, 1092, 1093, 1095, 1097, 1100, 1102, 1104, 1105, 1107, 1108, 1110, 1112–1114, 1116, 1118–1120, 1206, 1228, 1254, 1276
Mueller, D. 1108
Muench, T.J. 1161, 1163
Muller, E. 1093, 1098, 1103, 1104, 1117
multi-dimensional 861
multiparty equilibrium 1319
multiple attributes 1405
multiplicative utilities 1409
multiplier effect 977
Murdoch, J. 1016
mutation 954, 955
Muzzio, D. 1084
Myerson, R.B. 832, 834, 845–847, 865, 1241, 1252, 1257, 1258, 1260, 1301, 1306, 1322, 1323, 1326, 1327
myopic search 910

Nachman, D.C. 1409
Nagel, J. 1075, 1084
Nagel, J.H. 1084
Nakamura, K. 1098, 1099
Nakamura, Y. 1410, 1412
Nakamura number 1099
naked mole-rat 980
Nalebuff, B. 899, 1006, 1007, 1014
Naor, M. 1363, 1381, 1383, 1384
Narayanan, B.O. 1381, 1384
Narens, L. 1399, 1415, 1423
Nash, J.F. 738, 819, 1238, 1243, 1247, 1275, 1413
Nash bargaining model 917
Nash cooperative solution 918, 919
Nash equilibrium 797, 1190, 1313, 1314, 1329
Nash solution 1243, 1245
Nassau County Board 1130
natural monopoly 1224
natural selection 931, 933
Nau, R.F. 831
Nechyba, T. 1314
negative 1066
negative voting 1065–1067
negotiation 916, 919
negotiations, international 1002–1004

negotiations in crises 998, 1015
Neilson, W.S. 1409
neologism 864
neorealism, *see* realist theory in international relations
Newman, P. 1403
Neyman, A. 1168, 1222, 1224, 1490
Nicholson, M. 996–998, 1000, 1005, 1010, 1013, 1016
Nielsen, L. 1459, 1460
Nielsen, L.T. 1256, 1264
Niemi, R. 1108
Niemi, R.G. 1068, 1084
Nieto, J. 1275
Niiniluoto, I. 1415
Nikaido, H. 750, 1413
Niou, E. 998, 1003, 1016, 1029
no-benefit constraint 1083, 1085
no-show paradox 1078, 1084
noisy duels asymmetric strengths 764
noisy signals 866
non-Archimedean utility 1410
non-atomic Bayesian games 1474
non-atomic game 1301, 1302
non-convexity 1174
non-discrete duels 767
non-linear utility 1411
non-monotonicity 1076, 1078
non-ranking system 1072
non-ranking voting system 1063, 1084
non-rewarding orchids 986
non-sequential search 912
non-speculation in REE 1468
non-speculation theorem 1466
non-standard utility 1408
non-transferable utility game (NTU-game) 1298
non-transitive indifference 1403
non-transitive preference 1406, 1411, 1421
non-voting game 1169, 1172, 1173
(0,1)-normalization 1290
normalized nucleolus 1210
normalized power index 1130
North Atlantic Treaty Organization 997, 1009
Nowak, M.A. 982
NP 1357
NTU Harsanyi–Shapley value 1183
nuclear arms race, models of 997, 998
nuclear exchange models, *see* nuclear war models
nuclear war models 1028
nuclear weapons planning 1010–1012, 1020
nucleolus 1322
null event 1416
Nunnigham, J.K. 1328
nuptial gifts 984
Nurmi, H. 1001, 1056, 1068, 1070, 1072, 1079, 1084
Nye, J. 1011

objective functions 739
Oddou, C. 1165–1167, 1182, 1308–1310, 1316
Odlyzko, A.M. 1415
Oelrich, I. 1014
Okada, A. 1017
Olsder, G. 1027
Olsder, G.T. 803–805, 821, 823
O'Meara, N. 1024
on-line computation 1390
one shot noisy duel 762, 763
one shot silent duel 762, 763
one-way functions 1361
O'Neill, B. 996, 997, 1003–1011, 1013–1016, 1018–1020, 1027, 1030
open-loop equilibrium point 798
opportunity cost 1196
optimal decision rule 908
optimal income taxation 863
optimal mixed strategies 738
optimal search 918
optimal search rule 906, 908, 910
optimal search strategy 909
optimal strategies 1081
optimistic and conservative stable standards of behavior 1332
optimistic stable standard of behavior 1324, 1329, 1330
order dense 1402
ordered subfield 743
Ordeshook, P.C. 1003, 1016, 1029, 1138, 1328
ordinal invariance 1255
ordinal utility 1401
ordnance selection, games of 1022
Osborne, M.J. 921, 1183, 1188
Ostmann, A. 1292
Ostrom, C. 1001, 1010, 1015
Ostroy, J.M. 1409
out-of-equilibrium message 856
outcome 785, 786, 1061
overinvestment in signalling 859
Owen, G. 738, 1140, 1148, 1292, 1297, 1310, 1313, 1323, 1327, 1373, 1386
ownership 974, 975
Oye, K. 1000

P 1357
Packel, E. 1144
Packer, C. 981, 983
Page, T. 1460
Pahre, R. 1000
pairwise consistent 1206
Pak, S. 997
Palfrey, T. 1118
Panzar, J.C. 1224
Papadimitriou, C.H. 1342, 1357, 1386
Papamarcou, A. 1415
paradox 1075

Index

paradox of new members 1132
paradox of voting 1058, 1070
parameter 771
parental care 983
Pareto, V. 1400
Pareto optimality 1245
- weak 1250
Parikh, R. 1459
Parker, G.A. 938, 972–974, 978, 983–985
Parks, R.P. 1314
Parseval 1377
Parthasarathy, T. 741, 747, 752, 754
partial homogeneity assumptions 1137
partial order 1401
partially ordered linear utility 1410
partially ordered preferences 1403
partition 1442
partnership 870, 1216
partnership consistent 1217
partnership model 886
patent race 816
Pattanaik, P.K. 1407
Patterson, M. 1350
Pauly, M. 1178
Payne, C. 1025
Payne, R. 1003
payoff 736, 783, 786
payoff in biological games 934
payoff irrelevant asymmetry 974
Pearce, D. 894, 1331
Pease, M. 1346
Pecchenino, R. 1006
Peck, R. 1183
Peleg, B. 1056, 1092, 1102, 1111, 1114, 1116–1118, 1120, 1206, 1292, 1322, 1328, 1330, 1405
per capita prenucleolus 1210
perception theory of deterrence 1009
perfect competition 921
perfect equilibrium 917, 918
perfect voting precedure 1084
perfection 856
Perles, M.A. 1245, 1259
Perles–Maschler solution 1244
permanence 951, 952
Perry, M. 845, 850, 857, 1331
Peters, H.H. 1105
Peters, H.J.M. 1239, 1242, 1247, 1248, 1251, 1254, 1257, 1260, 1263, 1265
Peters, M. 916, 920
Petersen, W. 1001
Petrank, E. 1350
Petschke, C. 1017
Pfanzagl, J. 1399, 1406, 1418
phenotype 935
phenotypic external stability 956
Phillips, J. 1022

Phillips, W. 1020
Picard, L. 1003
Pilisuk, M. 1006
pin-down 1021
Pindyck, R.S. 819
Pissarides, C.A. 916, 920
plan, self-enforcing 829
plant pollinator interactions 986
Plato 1057
playing the field 931, 942
Plon, M. 996
Plott, C.R. 1146, 1328, 1411
Plous, S. 997, 1006
plurality voting 1062, 1064–1067, 1069, 1084
Poethke, H.J. 984
Pohjola, M. 819
Poisson process 911, 912
Polemarchakis, H. 1439, 1440, 1448, 1455
poll 1063
Pollak, R. 1098
Pollak, R.A. 1405, 1409
Pólya-type payoffs 755
Polydoros, A. 1028
polyhedral constraints and matrix games 746
polymorphic 932
Pommerehne, W.W. 1411
Ponssard, J. 1015
popular will 1058
population game 943–945, 964, 983
population genetics 952, 953
population monotonicity 1269
population state 932
Porter, R. 893, 900
Porteus, E.L. 1409
positive 1217
positive affine transformation 1406
positive linear operator 753
postmodernist theory in international relations 1000, 1003
Pötscher, B. 944
Poundstone, W. 1012
Pountney, I. 1079
Powell, C. 998
Powell, R. 1003, 1005–1007, 1013–1015
power and resources 1003
power in international relations 1001, 1003, 1016
power index 1128, 1134
practical voting procedures 1056
Pratt, J.W. 1401, 1409, 1418
predator inspection 982
preference cycles 1413
preference relation 1059
preference reversal 1411, 1413
preference truncation 1079, 1084
preferential system 1077
preferential voting 1075
prekernel 1206

prenucleolus   1205, 1206
preplay-communications   1331
Prescott, E.C.   819, 906
Preston, M.   1020
Preston, M.G.   1411
Price, G.R.   931–933, 937, 973
price ceiling   818
price dispersion   912–916
price distribution   907, 912, 914, 915
price telephone calls   1222
pricetaking   920
Prim–Read theory   1023, 1024, 1030
principal–agent problem   870, 872
principle of fair division   1206
prisoner's dilemma   1387, 1471, 1472, 1489
prisoner's dilemma game in biology   982
prisoner's dilemma models of international relations   1000, 1003, 1005, 1006, 1015
private computation   1368
private goods   1155–1157, 1159, 1160, 1165, 1171
probability, nonadditive   1416
probability constraints   832, 836
probability measure   1398
probability vectors   738
problem of pursuit and evasion   802
processors   1340
product topology   1405
production and consumption   1158
propensity to disrupt   1202
proper game   1128
proportional income tax   1314
proportional representation (PR)   1057, 1075, 1076, 1079–1082, 1085
proportional taxation   1318
proportional taxation game   1309
proportionally   1081
proto-game-theory   1029
Ψ-stability   1313
PSPACE   1365
public competitive equilibrium   1157, 1158, 1160, 1162, 1165
public goods   1154–1163, 1168, 1169, 1171, 1174, 1314, 1318
public goods economy   1169, 1172, 1175, 1177
public goods game   1161, 1170–1172, 1183
public goods with exclusion   1156
Pugh, G.   1027
Pulliam, H.R.   985
pure exchange economy   1287
pure public goods   1155, 1156, 1159, 1162, 1164, 1165, 1168, 1175
pure strategies   736
pursuit games   1026, 1027

Quandt, R.   996
quasi-concave   751
quasi-convex   751

quasi-cores   1174
Quester, G.   1009
question of individual effect   1136
Quiggin, J.   1411, 1412
Quinones, S.   1001, 1015
Quinzii, M.   861
Quirk, J.P.   1409
quota   1076, 1128, 1321

Rabani, Y.   1390
Rabin, M.   865
Rabin, M.O.   1351, 1371, 1386, 1388
Rabinowitz, G.   1132, 1143, 1144
Rader, J.T.   1403
Radford, K.   998
Radner, R.   823, 882, 884–886, 889, 890, 896–900
Radzik, T.   766, 768
Raghavan, T.E.S.   741, 745–747, 752, 754
Raiffa, H.   1005, 1245, 1252, 1313, 1401, 1408, 1409, 1419, 1421
Raiffa solution, discrete   1243
Ramey, G.   860, 861
Ramsey, F.   1220
Ramsey, F.P.   1401, 1413
Ramsey formula   1220
Ramsey prices   1219, 1220
RAND Corporation   1011, 1012, 1020, 1021
Randall, R.   1026
random number generation   1388
randomized   1350
rank candidates   1058
ranking procedures   1057
ranking systems   1084
Ransmeier, J.S.   1199
Rapoport, A.   996, 997, 1006, 1011, 1013, 1017, 1069, 1084, 1305, 1311, 1327–1329
Rasmussen, E.   899
ratio representation for preference   1412
rational   1057
rational expectations equilibrium   1467
rationality assumptions in international relations   1002, 1003
Ravid, I.   997, 1020
Ravid, Y.   1390
Raviv, A.   897
Rawls, John   1204
Ray, D.   1078, 1330
Read, J.   1023
Read, R.   1403
realist theory in international relations   999–1003
reciprocal altruism   931, 981
recombination rate   954
reconnaissance games   1022
recruitment function   812
redistribution   1183
reduced cost function   1205
reduction of a game   1297

Index

regime theory 1000–1003
regret 1421
Rehm, A. 1019
Reichelstein, S. 900
Reilly, R.J. 1411
Reinganum, J.F. 815, 816, 850, 855, 912, 914, 916, 918
Reinhardt, E. 1003
Reisinger, W. 1010
renegotiation-proofness 1331
Reny, P.J. 1331, 1490
repeated games in biology 981
repeated moral hazard 870, 879
repeated principal–agent model 873
repetitiveness of purchases 912, 915
replicator dynamics 948, 949, 951, 952
representational theory of measurement 1399
reproductive fitness 861
Repullo, R. 1118
reputation in nuclear deterrence 1007
rescaling axiom 1221
reservation price 912, 914, 915
reservation price strategy 910
reservation property 908, 909
reservation value 908, 909, 911, 912, 918
reservation value property 910
resilient computation 1368
resolve, showing 1007, 1008
resource-allocation strategies 1056
Restle, F. 1407
Restrepo, R. 767
restricted monotonicity 1250
restriction of a game 1297
revealed preference 1404, 1407
revelation principle 834, 865
Reyer, H.U. 980, 981
Ricart i Costa, J. 898
Richardson's arms race equations 1005
Richelson, J. 998
Richerson, P.J. 982
Richter, D. 1178
Richter, D.K. 1314
Richter, M.K. 1403
Ridley, M. 984
Riechert, S.E. 972, 975, 977, 978, 986
Riker, W.H. 996, 1016, 1056, 1058, 1070, 1072, 1084, 1130, 1138, 1139, 1148
Riley, J.G. 859, 862, 916
risk attitudes 1409
risk averse 1409
risk function 772
risk seeking 1409
risk-sensitivity 1264
riskless intensive utility 1401
Ritz, Z. 1245
Rivest, R. 1360, 1389
Rob, R. 912

Robbins, H. 1026
Robert, J. 1016
Roberts, F.S. 1399, 1404, 1414, 1415
Roberts, J. 850, 855, 920, 1472
Robinson, J. 745
Robinson, S. 1027
Robinson, T. 996
Rochet, J.-C. 861, 1107
Rock County, Wisconsin 1138
Rocke, D. 1002, 1003, 1005
Rodin, E. 1026
Roemer, J. 1275
Rogerson, W. 879, 897, 1006
Rome 1057
root competition in plants 986
root game 986
Rose, G. 1016
Rose-Ackerman, S. 1314
Rosen, S. 899
Rosenau, J. 996, 1002
Rosenfeld, D.B. 909, 910
Rosenschein, J.S. 1387
Rosenthal, H. 1330
Rosenthal, R.W. 912, 916, 919, 1161, 1245, 1248, 1251, 1413
Ross, S. 897, 1409
Roth, A.E. 1016, 1239, 1247, 1248, 1250–1252, 1255, 1256, 1264, 1265, 1328
Rothblum, U. 1265
Rothschild, M. 850, 859, 862, 909–912, 921, 1409
Roxin, E. 789
Rubinstein, A. 819, 850, 885, 898, 916, 920, 921, 1248, 1275, 1405, 1409, 1480
Ruckle, W. 1019, 1026
Rudnianski, M. 1010
running time 1358
runoff 1084
runoff approval 1075
runoff approval voting 1073, 1074
runoff plurality 1074, 1075
runoff system 1059, 1072
Russell, B. 1011
Russell, S. 1387
Russell, W.R. 1409
Rustichini, A. 899
Rutman, M.A. 764

S-games 747
Saari, D.G. 1068
Saaty, T. 1003, 1005
Sadanand, A. 912, 914
saddle point 737, 786
Safra, Z. 1265
Sahlin, N.-E. 1421
Said, A. 998
Sakawa, M. 1002
Saks, M. 1350, 1382, 1387

Sakurai, M.M. 1328
Salonen, H. 1245, 1249, 1251, 1254
Salop, S. 912, 914
Salter, S. 1004
Samet, D. 1216, 1217, 1222, 1297, 1481–1483, 1491, 1495
Sampson, M. 1001
Samuel 1057
Samuelson, P.A. 1400, 1403, 1404, 1407
sanctions, economic, and international influence 1001
Sandler, T. 1009, 1016
Sandretto, R. 1002
Saposnik, R. 1409
Sarin, R.K. 1412
Satterthwaite, M.A. 1068, 1092, 1098, 1102, 1117
Sauermann, H. 1328
Saul 1057
Savage, L.J. 773, 779, 1398, 1400, 1401, 1408, 1409, 1413–1416, 1419
Savage's model 1399
scale invariance 1246
Scalzo, R.C. 796, 798
Scarf, H. 1316, 1320, 1327
scenario bundles 999
Schaffer, M.E. 948
Schaster, K.G. 1328
Scheaffer, J. 1388
Schelling, T., and nuclear strategy 1012
Schenker, S. 1228
Schervish, M.J. 1417
Schlaifer, R. 1419
Schmeidler, D. 831, 1205, 1228, 1264, 1416, 1422
Schmitz, N. 1239
Schoemaker, P.J.H. 1409
Schuster, P. 951
Schwaghof Papers 999
Schwartz, A. 912, 914
Schwartz, N.L. 805
Schwartz, T. 1070
Schwarz, G. 778, 779
scoring methods 1079
Scotchmer, S. 1310, 1319
Scott, D. 1399, 1404–1406, 1415
screening 852, 862
search and bargaining 916
search games 1026
search problem 907
Sebenius, J. 1447, 1450, 1455
second-best solution 872
second-best tax game 1165
secrecy, military 1006, 1018, 1019
secret-sharing 1367
secure communications 1353
security dilemma 1004, 1005
seed dispersal 985
Segal, U. 1412, 1421

Seidenberg, A. 1401
Seidenfeld, T. 1417
Seidman, D. 864
self-evident 1442
self-generating 894
Selten, R. 846, 941, 947, 962, 964, 965, 967–971, 974, 982, 986, 1247, 1307, 1327, 1328
semi-infinite games 747
semicore 1202
semiorder 1404
Sen, A.K. 1092, 1094, 1099, 1105, 1118, 1121, 1123, 1205, 1404
sender–receiver game 840
sequential analysis 771
sequential assessment game 978
sequential bargaining 906, 917
sequential sampling 913
sex-ratio theory 984
Seyfart, R.M. 981
Shafer, G. 1404, 1415, 1416, 1419
Shaked, A. 1317
Shamir, A. 1351, 1365, 1367, 1370
Shank, C.C. 972
Shapiro, C. 899
Shapiro, R.D. 909, 910
Shapley, L.S. 742, 746, 755, 1128, 1129, 1131, 1132, 1134, 1135, 1140, 1144, 1148, 1163, 1174, 1176, 1180, 1214, 1216, 1217, 1222, 1255, 1256, 1292, 1294, 1302, 1317
Shapley value 1214, 1215, 1219, 1311, 1322, 1323, 1331
Shapley–Owen power index 1142–1144, 1148, 1149
Shapley–Shubik power index 1128, 1129, 1133, 1135, 1137, 1140
Shapley–Snow theorem 754
Sharkey, W. 1330
Sharkey, W.W. 1199, 1224
Sharma, S. 909
Shavell, S. 878, 897, 908, 909
Shavit, N. 1350
Shaw, R.F. 984
Shenoy, P.P. 1325, 1326
Shepsle, K. 1108, 1321
Sherlock Holmes 1438
Sherman, P.W. 980
Sherman, R. 1421
Sherman, S. 746
Shiffman, M. 764
Shilony, Y. 912
Shin, H. 1485, 1491
Shitovitz, B. 1305, 1314
Shmida, A. 947, 986
Shostak, R. 1346
Shubik, M. 1129, 1131, 1134, 1144, 1163, 1174, 1176, 1180, 1217, 1255, 1286, 1294, 1308, 1331, 1468

siblicide 984
Sibly, R.M. 983
side payments and transferable utility 1287
σ-algebra 1415
Sigmund, K. 948, 950–952, 982
signalling 849, 978
silent duels with asymmetric strengths 767
silent–noisy discrete duels 767
Simaan, M. 817, 818, 821
Simmons, L.W. 984
Simon, H. 897
Simon, L.K. 819
simple contracts 881
simple game 1128, 1291, 1321, 1327, 1373
simple quota games 1321
simplex algorithm 743
sincere 1059, 1067, 1068, 1084
sincere admissible 1073
sincere voting 1067
Singh, N. 898
single-ballot approval 1075
single-ballot plurality 1074, 1075
single-ballot voting systems 1059
single-crossing property 854
single-peaked 1096, 1319
Sion, M. 751
Sipser, M. 1388
size principle 1138
Sjöström, T. 1118, 1123
Skala, H.J. 1408, 1410
Skåne 1207
Skåne region 1206
skew-symmetric payoffs 740
skew-symmetry 1406
Sleator, D.D. 1390
Sloane, N. 1371
Slovic, P. 1411, 1414, 1420
Slutsky, E. 1400
Smith, A. 897
Smith, C.A.B. 1415
Smith, J.H. 1078
Smokler, H.E. 1414
Smorodinsky, M. 1239, 1245, 1249
Snow, R.N. 742
Snyder, J.M. 1056
Sobel, J. 840, 850, 855, 857, 860, 861, 863, 865, 1254, 1265, 1276
Sobolev, A.I. 1206
social choice 1056
social choice theory 1058
social situation 1324
– theory of 1306, 1307, 1331
social utility 1072
Socrates 1057
solution 1238
– axiomatic characterization 1238

Sondermann, D. 1308, 1316
Sonnenschein, H.F. 916, 920, 1404
Sorin, S. 1342, 1355, 1388
Souganidis, P.E. 789, 790
space 1358
Spafford, D. 1083
spatial models of voting 1056
spatial voting games 1326
Spear, S. 898
species welfare 986
species welfare paradigm 933, 973
speculation 1465
Spence, A.M. 850, 859, 897, 1409
Spencer, J. 1376
spider society 977
spillovers 1177
Spinnewyn, F. 885, 898
Spitzer, F. 884
Sproull, R.F. 1387
Sprumont, Y. 1305
Srivastava, S. 898, 1118
SSB functional 1413, 1422
St. Petersburg paradox 1400
stabilization of an uncertain system 802
stable structure 1167, 1168
Stacey, P.B. 980
Stachetti, E. 894
Stackelberg leader 906
Stag Hunt game 1004, 1005, 1014
Stahl, D.O. 912
stand-alone cost test 1199
star-shaped inverse 1263
Starr, A.W. 802
state dependent theories 1418
state of the world 1441
state substitutability 944
statistical decision theory 747
Stavely, E.S. 1057
Stelzer, J. 1415
Stenseth, N.C. 985
Stern, M. 1439
sticklebacks 982
sticky price duopoly 817
Stigler, G.J. 906, 1315, 1400, 1403
Stiglitz, J.E. 850, 859, 862, 863, 897–899, 911, 912, 914, 921, 1409
Stinchcombe, M.B. 819
stochastic dominance 1409, 1412, 1413
stochastic game 1386
stochastic utility 1407
stockpiling 912
Stokey, N.L. 815, 863, 899, 1447, 1455, 1468
stopping rules 907
stopping time 772
Straffin, P.D. 1070, 1132, 1134, 1136–1140, 1150, 1202
strategic calculations 1057

Strategic Defense Initiative (Star Wars)   1014, 1024
  see also ballistic missile defence
strategic equivalence   1288
strategic form   736
strategic incentive constraints   832
strategic stability   858
strategies for differential games
 – closed loop   803
 – dominant   802
 – feed back   804
 – Markovian   804
 – memory   821
 – mixed   805
 – Nash   802
 – open loop   803
strategy   783, 786, 1061
strategy-elimination property   846
strategy proofness   1067, 1068, 1083
strategyproof   1059, 1068, 1084
strategyproof game form   1101, 1103, 1108
Strauss, D.   1407
strictly stable sets   1332
strong game   1128
strong indifference   1403
strong monotonicity   1219, 1224, 1251
strong Nash equilibrium   1321, 1323, 1329
strong point   1148
strong separation theorem   741
Strotz, R.H.   1400, 1405
STV   1075, 1084
Styszyński, A.   767
subadditivity   1197, 1309
Subbotin, A.I.   787
subgame perfect equilibria
 – stationary   1331
subjective probability   1398, 1413
subjective probability, conditional   1416
submarine warfare   1019–1022
Subotnik, A.   1305
Sugden, R.   1412, 1413, 1421
superadditive cover game   1316
superadditivity   1290, 1308, 1310, 1313, 1314, 1316
supermodularity   920
superparasitism   986
Suppes, P.   1399, 1404, 1406, 1407, 1416, 1418, 1422
support of alternative   1320
sure thing principle   1417
surprise attack   1014
sustainable   1225
sustainable configuration   1318
Sutherland, W.J.   985
Sutton, J.   912
Suzumura, K.   1315
swing voter   1129
Sydvatten Company   1206

symmetric extensive game   966, 968
symmetry   1203, 1214, 1245
synchronous networks   1345
Szwajkowski, E.   1328

tactical airwar models   1024, 1025
Taft, Robert   230
Takayama, T.   817
Tan, G.   1276
Tan, T.   1485
Tarjan, R.E.   1390
Tarsi, M.   1387
Tauman, Y.   1221–1223, 1225, 1228
tax allocation   1186, 1187
tax game   1167, 1182
Taylor, P.D.   947–949, 985
Tchebyshef   1362
Telser, L.G.   1224
Tennessee Valley Authority   1198, 1201
terminal set   794
testban negotiations, nuclear   1016
Thompson, G.F.   1213
Thomson, W.   683, 1206, 1239, 1241, 1242, 1248, 1250, 1252, 1257, 1258, 1261–1263, 1265, 1268–1276
Thrall, R.M.   1301, 1308
threat credibility in deterrence   1007, 1008
Tidman, R.   1019
Tiebout, M.   1177, 1178, 1189
Tiebout equilibrium   1178, 1318
Tiebout-like equilibrium   1179, 1180
Tietz, R.   1009, 1328
Tijs, S.   747, 1242, 1249, 1251, 1264, 1265
Tillmann, G.   1223
Tirole, J.   845, 865, 898, 1006, 1468
tit for tat   982
Tolwinski, B.   821, 823
top cycle set   1147
topology   1403
totally balanced game   1294
Toulet, C.   1417, 1418
trajectory   783
transactional friction   920
transferable utility, see TU
transitive   1059
trapdoor   1362
triangle condition   1407
Tribes   1375
trilinear games   816
Trivers, R.L.   931, 981–983
truncation of preferences   1076
truthful revelation   1058
Tsebelis, G.   1001
Tsuji, N.   985
Tsutsui, S.   818
TU (transferable utility) game   1307
Tukey, J.W.   1405

Tullock, G.   1001
Tulowitzki, V.   1010
Turing Machine   1357
Tutz, G.   1407
Tversky, A.   1407, 1409, 1411, 1412, 1414, 1420, 1421
two-level games   1002
two-locus model for viability selection   953
two-person zero-sum game   797

unanimity game   1134
unbiased estimators   776
uncertain feasible set   1258
undecidable   1342
underrepresentation   1082
underrepresented parties   1085
undominated   1063
unemployment   920
uniform invasion barrier   944
uniform quota   1319
unitary actor hypothesis in international relations   1001, 1002
United Nations Security Council   1131
United States Electoral College   1132, 1143
universal CPNE   1330
universal divinity   860
universal Turing Machine   1358
unnormalized Banzhaf index   1135
unnormalized power index   1130
upper δ-game   785
upper δ-strategy   784
upper δ-value   785
upper semicontinuous   750
upper value   784, 786
Uscher, A.   1015
utilitarian solution   1244
utility   1398
utility in n-person games   1410
Uusi-Heikkilä, Y.   1072

Vachon, G.   1017
Vaidya, K.C.   1202
value   738
vampire bats   981
Van Alphen, J.J.M.   986
Van Damme, E.   864, 931, 941, 942, 949, 970, 1003, 1005, 1263, 1276, 1331
Van Der Ploeg, F.   818, 1005
Van Lier, L.   1415
Van Newenhizen, J.   1068
Van Stengel, B.   1014, 1017
Varaiya, P.   789, 798
Vardi, M.   1444
variable   1082
variable background approach   978
variable number of agents   1265
Varian, H.   1276

Varian, H.R.   912
Vazirani, U.V.   1389
Vazirani, V.V.   1389
verifiable secret-sharing   1367
verification, see arms control verification
veto core   1114
Vial, J.-P.   835
Vickers, G.T.   940
Vickrey, W.   1004
Vickson, R.G.   1409
Vietnam war, models of   998
vigilance game   985
Ville, J.   738, 749
Villegas, C.   1415
Vincent, T.L.   972, 986
Vincke, P.   1410
Vind, K.   1406
viscosity solution   787
Vishwanath, T.   908, 909
Visser, M.E.   986
Vohra, R.   1330
Von Neumann, J.   738, 739, 749, 828, 1012, 1289, 1292, 1306, 1307, 1324–1327, 1329, 1401, 1407, 1409, 1410, 1413
Von Neumann, J., and nuclear planning   1012
Von Neumann and Morgenstern solutions   1311
Von Neumann–Morgenstern axioms   1408
Von zur Muehlen, P.   912
Vorha, R.V.   1390
voting game   1128, 1169, 1170, 1172, 1173
voting in international institutions   1000
voting measure   1168–1172
voting procedures   1056
voting rules   1056
voting vs. non-voting game   1172

Waartz, O.   1348
Waddington, C.   1020
Wagner, H.   996, 998, 1001, 1009, 1014, 1016
Wagner, R.   1022
Wakker, P.   1248, 1414, 1415, 1423
Wald, A.   752, 770, 774–776, 779
Walker, M.   1227
Walley, P.   1415
Walras, L.   1400
Walrasian equilibrium   920
Waltz, E.C.   984
Waltz, K.   996
Wan, H.   898
Wan, H.Y.   802, 804, 807, 811–813, 816, 818, 819, 823
Wang, M.   998
war of attrition   978
Wärneryd, K.   865
Washburn, A.   999, 1022, 1026
weak consistency   1222
weak ε-core   1174

weak gross substitutability   1225
weak order   1401
weak topologies   750
weakly aggregation invariant   1221
weapons effectiveness measures   1027
Weber, R.J.   1022, 1072, 1299
Weber, S.   1000, 1011, 1016, 1167, 1182, 1305, 1310, 1317, 1318, 1320, 1321, 1327
Wefald, E.   1387
Weg, E.   1329
Weidner, M.   998
weighted linear utility   1412
weighted majority game   1321–1323
weighted Shapley value   1215, 1216
weighted voting game   1128, 1291
Weiss, A.   862
Weiss, M.   1028
Weissenberger, S.   1018
Weissing, F.J.   949, 950
Weitzman, M.L.   908
Wendroff, B.   1005
Weres, L.   997
Werlang, S.   1485
Westhoff, F.   1178, 1314
Weyl, H.   739
Weymark, J.   1105
Whinston, A.B.   1216, 1217
Whinston, M.   1329
Whitmore, G.A.   1409
Whitney, H.   1379, 1382
Wiener, N.   1404
Wiengast, B.   1108
Wiesmuth, H.   1000, 1016
Wiess, H.   1025
Wigderson, A.   1387, 1388
Wilde, L.   850, 855
Wilde, L.L.   912–914, 920
Wilkenfeld, J.   1015, 1387
Wilkie, S.   1248, 1251
Wilkinson, G.S.   972, 981
Williams, G.C.   934
Williams, J.   1002, 1012
Williams, P.M.   1415
Williams, S.   889, 890, 899
Willig, R.D.   1224
Willkie, Wendell   1139
Wilson, C.   850, 862
Wilson, E.O.   980
Wilson, F.   975
Wilson, R.   897, 921, 1007, 1009, 1096, 1472
Winet, M.   1406
winning coalition   1321, 1322, 1327
winning conditions   1128
Winter, E.   1321, 1323, 1324, 1327, 1331
Wittman, D.   1018
Wohlstetter, A.   1011–1013
Wold, H.   1403
Wolf, L.L.   984

Wolfowitz, J.   774–776, 779
Wolinsky, A.   916, 917, 919–921
Wooders, J.   1330
Wooders, M.H.   1163, 1174, 1316, 1327
Woodwaro, P.   1010
Wright, E.M.   1362
Wright, S.G.   1072
Wright, W.   998
Wu, Y.   1028
Wu Wen-tsün   941
Wynne-Edwards, V.C.   934

Xue, L.   1305, 1332

Yaari, M.E.   885, 898, 1405, 1411, 1412
Yamamura, N.   984, 985
Yamato, T.   1118
Yao, A.C.   1350, 1352, 1369, 1371, 1388, 1389
Yellot Jr, J.I.   1407
Yokoyama, T.   1403
York, H.   1012
Young, D.   997
Young, H.P.   948, 1079, 1206, 1208, 1210, 1211, 1219, 1221, 1224, 1228
Young, O.   1016
Young, P.   1268
Yu, P.   1016
Yu, P.L.   1245
Yu solution   1245

Z-matrices   745
Zagare, F.   998, 999, 1008–1010, 1015
Zagare, F.C.   1068
Zaharoglon, F.   1350
Zahavi, A.   850, 861, 979
Zajac, E.E.   1199
Zame, W.   1327
Zamir, S.   1014, 1017, 1182, 1310, 1485, 1488
Zang, I.   1225
Zanotti, M.   1416
Zauberman, A.   1022
Zeckhauser, J.   916
Zeckhauser, R.   897, 1409
Zeeman, E.C.   949
Zellner, A.   1001
zero-knowledge proofs   1365
zero-monotonic game   1290
zero sum   736
zero-sum Bayesian game   1475
Zeuthen, F.   1275
Zimmerman, C.   1021
Zinnes   818
Zinnes, J.L.   1406
Zinnes, P.   1001, 1005, 1016
ZK   1366
Zlotkin, G.   1387
Zuckerman, D.   1389
Zuckermann, S.   1016